FOURTH EDITION

ECKERT

ANIMAL PHYSIOLOGY

MECHANISMS AND ADAPTATIONS

FOURTH EDITION

ECKERT
ANIMAL
PHYSIOLOGY

MECHANISMS AND ADAPTATIONS

DAVID RANDALL
UNIVERSITY OF BRITISH COLUMBIA

WARREN BURGGREN
UNIVERSITY OF NEVADA, LAS VEGAS

KATHLEEN FRENCH
UNIVERSITY OF CALIFORNIA, SAN DIEGO

WITH CONTRIBUTIONS BY
RUSSELL FERNALD
STANFORD UNIVERSITY

W. H. Freeman and Company

New York

ACQUISITIONS EDITOR: Deborah Allen

DEVELOPMENT EDITOR: Kay Ueno

PROJECT EDITOR: Kate Ahr

COVER DESIGNER: Michael Mendelsohn, Design 2000, Inc.

BACK COVER ILLUSTRATION: Roberto Osti

TEXT DESIGNER: Michael Mendelsohn, Design 2000, Inc.; Victoria Tomaselli

ILLUSTRATION COORDINATOR: Bill Page

ILLUSTRATION: Fine Line Illustrations; Medical and Scientific Illustration,
William C. Ober, MD and Claire Garrison, RN, BA

PRODUCTION COORDINATOR: Maura Studley

COMPOSITION: Progressive Information Technologies

MANUFACTURING: RR Donnelley & Sons Company

Library of Congress Cataloging-in-Publication Data
Randall, David J., 1938–
    Eckert animal physiology: mechanisms and adaptations/David Randall, Warren
Burggren, Kathleen French.—4th ed.
        p.   cm.
    Includes bibliographical references and index.
    ISBN 0-7167-2414-6 (hardcover)
    1. Physiology.   I. Burggren, Warren.   II. French, Kathleen.   III. Title.
QP31.2.R36   1997
591.1—dc20   96-31713
                                                                     CIP

Printed in the United States of America

First printing 1997, RRD

For our families, and, of course,
for Roger

ABOUT THE AUTHORS

DAVID RANDALL

A prominent fish physiologist and a leading expert in respiratory and circulatory physiology, David Randall collaborated with the late Roger Eckert on the earlier editions of *Animal Physiology* and continues his contribution in the fourth. A faculty member at the University of British Columbia in Vancouver, Canada, since 1963, and full professor since 1973, Randall was appointed Associate Dean of Graduate Studies in 1990. Elected a fellow of the Royal Society of Canada in 1981, Randall has been both a Guggenheim and a Killam fellow, and was awarded the prestigious Fry Medal for research contributions to zoology by the Canadian Society of Zoology in 1993. In 1995, he received the Award of Excellence from the American Fisheries Society for contributions to the field of fish physiology. A frequent symposium lecturer on fish physiology and other subjects, most recently in Brazil, France, Germany, Italy, the People's Republic of China, Russia, and the United States. He has worked with both the World Health Organization and the United States Environmental Protection Agency in developing ammonia criteria. Widely published as author and co-author in leading journals, Randall is co-editor of the noted series *Fish Physiology* (Academic Press), of which 15 volumes are in print. Volume 16, subtitled "Deep-Sea Fish," will appear in 1997. Along with his other duties, Randall co-teaches third year courses in vertebrate physiology and environmental physiology. His research interests concern the interactions between gas and ion exchange across fish gills.

WARREN BURGGREN

Warren Burggren has taught in physiology for 23 years, and has been a professor of biological sciences at the University of Nevada at Las Vegas since 1992. Courses he has taught at UNLV and at the University of Massachusetts, where he was Professor of Zoology from 1987 through 1991, include Human Anatomy and Physiology, Bioenergetics, Introductory Zoology, and Comparative Physiology. Burggren's research interest include developmental physiology, comparative animal physiology, and environmental and ecological physiology. In particular, his research focuses on the ontogeny of respiratory and cardiovascular systems, and how the systems that regulate them change over the course of development. Burggren has been actively involved in symposia, seminars, and formal extramural research/training activities in many countries. A co-author of *The Evolution of Air Breathing in Vertebrates* (Cambridge University Press, 1981), Burggren has been a frequent contributor since 1980 to edited collections of physiology, including Prosser's *Comparative Animal Physiology, Fourth Edition* (Wiley-Liss, 1991). Burggren co-edited *Environmental Physiology of the Amphibia* (University of Chicago Press, 1992), and more recently co-edited *Development of Cardiovascular Systems; Molecules to Organisms* (Cambridge University Press, 1997).

KATHLEEN FRENCH

A neurobiologist at the University of California at San Diego since 1985, Kathleen French has for 10 years taught upper division courses in embryology, mammalian physiology for premedical students, and cellular neurobiology. In addition, at UCSD, French participates in a training program to instruct science teaching assistants in the techniques and philosophy of teaching. She also serves on the faculty of the Neuroscience and Behavior Course at the Marine Biological Laboratories in Woods Hole, Massachusetts, an intensive course designed primarily for graduate students and post-doctoral fellows. French brings her expertise in—and love of—teaching to her role as co-author of the current edition of *Animal Physiology*, along with a lifelong interest in the nervous systems of organisms from a broad range of phyla. As an Associate Project Scientist at UCSD, French's research focuses on the control of neuronal development, a topic that she has studied in various invertebrate species. Her current research concerns the cellular events that control differentiation of identified neurons in the medicinal leech, with an emphasis on the cellular physiology of embryonic neurons and the effects of cell-cell contacts. She has been the author and co-author of numerous published research and review articles in journals including the *Journal of Neuroscience* and *Journal of Neurophysiology*.

BRIEF CONTENTS

CONTENTS

PREFACE

It is nearly ten years since the third edition of *Animal Physiology* first appeared, written by Roger Eckert with the help of David Randall. Roger died in 1986 while revising the third edition, which was completed by George Augustine and David Randall. That book formed the basis for the fourth edition, which is fittingly referred to as *Eckert Animal Physiology*. Although this new edition has been extensively revised and redesigned, the approach that so successfully characterized earlier editions has been maintained: the use of comparative examples to illustrate general principles, often supported by experimental data. In addition, we have emphasized the principle of homeostasis, and we have updated the molecular and cellular coverage. Retained in this edition is the comprehensive coverage of tissues, organs, and organ systems. Cellular and molecular topics are integrated early in the book so that common threads are developed to explain and compare the interactions between regulated physiological systems that produce coordinated responses to environmental change in a wide variety of animal groups. The basic principles and mechanisms of animal physiology and the adaptations of animals that enable them to exist in so many different environments form the central theme of this book.

The diversity and adaptations of the several million species that make up the animal kingdom provide endless fascination and delight to those who love nature. Not the least of this pleasure derives from a consideration of how the bodies of animals function. At first it might appear that with so many kinds of animals adapted to such a variety of lifestyles and environments, the task of understanding and appreciating the physiology of animals would be overwhelming. Fortunately (for scientist and student alike), the concepts and principles that provide a basis for understanding animal function are relatively few, for evolution has been conservative as well as inventive.

A beginning course in physiology is a challenge for both teacher and student because of the interdisciplinary nature of the subject, which integrates chemistry, physics, and biology. Most students are eager to come to grips with the subject and get on with the more exciting levels of modern scientific insight. For this reason, *Eckert Animal Physiology* has been organized to present the essential background material in a way that allows students to review it on their own and go on quickly to consider animal function and to understand its experimental elucidation.

Eckert Animal Physiology develops the major concepts in a simple and direct manner, stressing principles and mechanisms over the compilation of information and illustrating the functional strategies of animals that have evolved within the bounds of chemical and physical possibility. Common principles and patterns, rather than exceptions, are emphasized. Examples are selected from the broad spectrum of animal life, consciously illustrating similarities between organisms; for example, similar compounds are associated with reproduction in both humans and yeast. Thus, the more esoteric and peripheral details receives only passing attention, or none at all, so as not to distract from central ideas. We use the device of a narrative, describing experiments, to provide a feeling for methods of investigation while presenting information.

ORGANIZATION OF THE BOOK

For the first time, the chapters are organized into three parts, which we feel will promote an understanding of animals as integrated systems at every level of organization. Each part is introduced by an opening statement that gives students an overview of the material to follow. Part I contains four chapters and is concerned with the central principles of physiology. Part II (Chapters 5–11) deals with physiological processes, while Part III (Chapters 12–16) discusses how these basic processes are integrated in animals living in a variety of environments. All 16 chapters have been extensively reworked and reorganized to stay abreast of new scientific developments.

NEW TO THIS EDITION

- A new chapter on methodology (Chapter 2) in Part I, in which some of the latest molecular techniques are discussed and illustrated, along with traditional methods.

- This emphasis on molecular coverage continues throughout the book; Chapters 5, 6, and 7, for example, are updated with recent insights into the cellular and molecular underpinnings of membrane excitation, synaptic transmission, and sensory transduction.

- Part II features a new chapter (Chapter 8, Glands: Mechanisms and Costs of Secretion), which brings together information on an important, but frequently neglected, effector system.

- In Part II, Chapter 11 (Behavior: Initiation, Patterns, and Control) preserves and expands the descriptions of vertebrate and invertebrate nervous systems found in previous editions, presenting an up-to-date view of systems neurobiology, one of the fastest-growing areas of neurobiology. Several concepts from neuroethology, which bridges the gap between the pure study of behavior and the study of cellular function in the nervous system, are introduced, along with examples of important recent neuroethological studies.

- The role of the nervous system in maintaining homeostasis through the modulation of all systems has been incorporated into Part III, which further advances the integrated approach of the book.

- There is an increased emphasis throughout the book on environmental adaptations, and specific examples of environmental adaptation (such as water balance in elephant seals in Chapter 14) illustrate the general principles of comparative physiology.

- Some of the new topics introduced in the fourth edition include a section on the immune response in Chapter 12 (Circulation), and a section on biorhythms in Chapter 13 (Using Energy: Meeting Environmental Challenges).

PEDAGOGY

- The ideas developed in the text are illuminated and augmented by liberal use of illustrations and figure legends. For the first time, full color drawings have been added, creating a high quality visual program to further motivate students.

- Spotlights provide in-depth information about the experiments and individuals associated with important advances in the subject matter, the derivation of some equations, or simply historical background on a topic under discussion.

- Thought questions within chapter text (look for the 📖) encourage problem-based learning and stimulate discussion on various aspects of the material presented.

The text narrative includes effective, integrated examples to support principles; while presenting information, it provides consistent thematic coverage and a feeling for methods of investigation. References to the literature within the body of the text and in figure legends are made unobtrusively, but with sufficient frequency that students can become aware of the role of scientists and their literature as a subject is developed. Further pedagogical aids include key terms that are explained and appear in boldface type at their first mention in the text, and that are formally defined in a useful, comprehensive glossary. End of chapter materials include a summary, which provides the student with a quick review of important points covered in the chapter, review questions, and an annotated list of suggested further readings. Students will find the following resources at the back of the book: appendixes that provide information on units, equations, and formulas; the glossary; and a bibliography that includes the full citations of all references cited in the chapters. Our goal has been to produce a balanced, up-to-date treatment of animal function that is characterized by its clarity of exposition. We hope that readers will find *Eckert Animal Physiology* valuable, and we welcome your constructive criticism and suggestions.

September 1996   DAVID RANDALL
                         WARREN BURGGREN
                         KATHLEEN FRENCH

ACKNOWLEDGMENTS

Eckert Animal Physiology has benefitted greatly from the contributions of several people whose efforts we gratefully acknowledge. Russell Fernald, Stanford University, participated in the planning and reorganization of the book and the initial revision of many of the chapters. We consulted with Lawrence C. Rome, University of Pennsylvania, on the update and expansion of Chapter 10 (Muscles and Animal Movement). Harold Atwood, University of Toronto, was involved in some early discussions of the revision. Further, we are grateful for the informal comments of associates around the world, and for the formal reviews of manuscript, which were provided by the following colleagues from across the country:

Joseph Bastian, *University of Oklahoma*
Robert B. Barlow, *Institute for Sensory Research, Syracuse University*
Francisco Bezanilla, *UCLA School of Medicine— Center for the Health Sciences*
Phillip Brownell, *Oregon State University*
Richard Bruch, *Louisiana State University*
Wayne W. Carley, *National Association of Biology Teachers*
Ingrith Deyrup-Olsen, *University of Washington*
Dale Erskine, *Lebanon Valley College*
A. Verdi Farmanfarmaian, *Rutgers University*
Robert Full, *University of California at Berkeley*
Carl Gans, *University of Michigan*
Edwin R. Griff, *University of Cincinnati*
Kimberly Hammond, *University of California at Riverside*
David F. Hanes, *Sonoma State University*
Michael Hedrick, *California State University—Hayward*
James W. Hicks, *University of California at Irvine*
Sara M. Hiebert, *Swarthmore College*
William H. Karasov, *University of Wisconsin at Madison*
Mark Konishi, *California Institute of Technology*
Bill Kristan, *University of California at San Diego*
Paul Lennard, *Emory University*

Jon E. Levine, *Northwestern University*
Harvey B. Lillywhite, *University of Florida, Gainesville*
Duane R. McPherson, *SUNY at Geneseo*
Duncan S. MacKenzie, *Texas A&M University*
Eric Mundall, late, *University of California at Los Angeles*
Kenneth Nagy, *University of California at Los Angeles*
Richard A. Nyhoff, *Calvin College*
Richard W. Olsen, *UCLA School of Medicine*
C. Leo Ortiz, *University of California at Santa Cruz*
Harry Peery, *University of Toronto*
J. Larry Renfro, *University of Connecticut*
Marc M. Roy, *Beloit College*
Roland Roy, *Brain Research Institute, UCLA School of Medicine*
Jonathon Scholey, *University of California at Davis*
C. Eugene Settle, *University of Arizona*
Michael P. Sheetz, *Washington University School of Medicine*
Gregory Snyder, *University of Colorado at Boulder*
Joe Henry Steinbach, *Washington University School of Medicine*
Curt Swanson, *Wayne State University*
Malcolm H. Taylor, *University of Delaware—School of Life and Health Sciences*
Ulrich A. Walker, *Columbia University—College of Physicians*
Eric P. Widmaier, *Boston University*
Andrea H. Worthington, *Siena College*
Ernest M. Wright, *UCLA School of Medicine—Center for the Health Sciences*

Finally, the good sense and kind words of Kay Ueno and Kate Ahr, our editors, have much improved the book and ensured its publication.

DAVID RANDALL
WARREN BURGGREN
KATHLEEN FRENCH

PRINCIPLES OF PHYSIOLOGY

Animal physiology is the study of how living animals function. Both the cheetah racing after a gazelle and the rattlesnake striking at a desert rat coordinate specialized anatomical features and physiological processes to capture their prey and, in turn, to evade predators and prolong life. The arctic fox on the cover of this book possesses a luxurious coat of fur, as well as finely tuned physiological mechanisms, to protect it from the bitter cold of its environment. Even animals living in an apparently ideal environment, with benign temperatures year round, ample food sources, and regular day/night cycles, face challenges, which include the pressures of sharing a habitat with members of their own, and with other, species. Meeting the demands of survival has resulted in numerous evolutionary variations on the basic theme of life, and the environments in which life expresses itself are equally varied. As a result, animal physiologists have a vast array of animals and environments available in which to investigate how animals work. (There is now some indication that life may have existed on Mars, so the environments open

to physiological study may not be limited to Earth.) Even so, the broad range of philosophical and technological approaches in the study of animal physiology rest on a relatively small number of fundamental concepts, which are presented in Part I of this book. These concepts are essential for understanding the physiological processes that underlie the behavior of predators like the cheetah and the rattlesnake and the evolution of physiological control mechanisms that allow animals—such as the desert rat and the arctic fox—to maintain internal body conditions that enable them to survive, even in very hostile environments.

Chapter 1 explores the central themes in animal physiology, including the close relationship between structure and function, the processes of adaptation and acclimation, and the concepts of homeostasis and its maintenance by feedback control systems. In science, all knowledge is based on experiments; consequently, in Chapter 2 we discuss the nature of experimentation and the various perspectives animal physiologists adopt in designing hypotheses and testing them. We briefly describe many of the major experimental methods that are currently used by physiologists, including new and rapidly evolving molecular techniques.

From the beginning, physiology has been grounded in physics and chemistry, and Chapter 3 reviews basic physical and chemical principles that underlie the physiological mechanisms discussed in the rest of the book. This chapter focuses particularly on the processes of metabolism, the biochemical reactions that are the basis for all physiological processes. The membranes that surround cells and their internal organelles provide an important example of how physical and chemical mechanisms combine in living cells to produce biological processes. In Chapter 4 we investigate the nature of cell membranes. We pay specific attention in this chapter to how the outer membrane of a cell helps to stabilize its internal environment. Active transport of materials across cell membranes is discussed in detail, because this process is crucial for numerous physiological processes as diverse as conduction of nerve impulses, regulation of body fluids, and uptake of nutrients, all discussed in later sections. How fundamental biochemical, molecular, and cellular processes are combined to produce the integrated regulation of physiological systems throughout an animal's body is discussed in Parts II and III of the book.

CHAPTER

1

STUDYING ANIMAL PHYSIOLOGY

Animal physiology focuses on the *function* of the tissues, organs, and organ systems of multicellular animals. The animal physiologist attempts to understand in physical and chemical terms the mechanisms that operate in living organisms at all levels, ranging from the subcellular to the integrated whole organism. Understanding how animals work requires detailed knowledge of the molecular interactions that set the stage for cellular processes. Armed with this knowledge, animal physiologists design experiments and test hypotheses to learn about the control and regulation of processes within groups of cells, and how the overall activities of these cell groups then affect the overall function of the animal. The coordinated activities of cells, collected into specialized organs, provides the basis for the behavioral capabilities and physiological processes that distinguish animals from plants. These distinguishing features include movement, relative independence from environmental conditions, sophisticated sensory information about the world, and complex social interactions, to name but a few.

Above all, animal physiology is an *integrative science*. Physiologists examine and try to understand how physiological systems (usually in the central nervous system) sort through and differentiate between the vast amounts of information about the external and internal environment typically received by an animal. For example, the seemingly straightforward process of a mammal maintaining a stable body temperature requires the temperature-control system of the brain to integrate information on the multitude of factors affecting body temperature: the heat load or heat sink presented by the external environment; the rate of heat production by metabolically active tissues; the transport of heat by blood flow between the body core and its periphery; the contribution of evaporative cooling; the insulative nature of its fur; and many other such physiological and anatomic variables. It is this theme of integration that sets physiology apart from other sciences—adding to its complexity but also to its fascination.

Physiology is firmly rooted in the laws and concepts of chemistry and physics. Your background in these disciplines will greatly help you learn about and understand animal physiology. Consider the following chemical and physical laws and concepts that relate to various physiological processes:

- Ohm's Law—blood flow and pressure; ionic current; capacitance of membranes

- Boyle's Law and the Ideal Gas Law—respiration

- Gravity—blood flow

- Kinetic and potential energy—muscle contraction; chest movements during exhalation

- Inertia, momentum, velocity, and drag—animal locomotion

These are just a few of the many chemical and physical laws and concepts that you will be using throughout this book.

The principles of evolution—natural selection and speciation—underlie animal physiology, just as they do any other study of life on earth. For instance, natural selection has led to the tolerance of mammalian and bird enzymes to high body temperature; to the use of modified gills of land crabs for air breathing, and to the tolerance of migrating salmon to both freshwater and seawater. The array of all possible physiological adaptations to all possible distinctive environments in the more than one million described animal species is staggering. The history of animal physiology is rich in studies of the adaptations by individual species to particular environmental constraints and demands; such studies have produced a great "knowledge matrix" of environments and adaptations. More recently, however, animal physiologists have sought to identify *patterns* in this vast array of physiological data by incorporating powerful new tools from evolutionary biology and molecular biology into their studies. One of the goals of this book is to describe the general patterns in animal physiology even as we use specific examples to illustrate physiological principles.

THE SUBDISCIPLINES OF ANIMAL PHYSIOLOGY

The overall discipline of animal physiology has several important subdisciplines or fields. *Comparative physiology* embraces that area of physiology that uses comparisons between species to discern physiological and evolutionary patterns; this "comparative approach" is described later. This also is commonly used in reference to physiological studies in animals other than those typically investigated by medical physiologists (e.g., rats, mice, cats, rabbits). Thus, a physiologist working on kidney function in rats would not think of herself as a comparative physiologist, while a physiologist working on kidney function in armadillos or trout might choose that label. *Environmental physiology* examines animals in the context of the environment they inhabit. Environmental physiology focuses on the evolutionary adaptations (e.g., the thick fur in Arctic animals; the high blood volume of diving seals; the waterproof cuticle of cockroaches) to environments that can range from benign to the supremely hostile. *Evolutionary physiology*, a relative newcomer on the physiological block, uses methods and techniques of evolutionary biology and systematics (e.g., the construction of taxonomic family trees, or cladograms) to understand the evolution of animals from a physiological viewpoint using physiological markers (e.g., maintenance of constant body temperature) rather than anatomic markers (e.g., feathers).

WHY STUDY ANIMAL PHYSIOLOGY?

The study of animal physiology can be traced back to the earliest writings of learned persons. Aristotle documented the rate of heart beat in a developing chick embryo. Renaissance chemists of Europe often looked to the metabolism of animals and plants to understand oxygen-consuming and oxygen-producing reactions. Several reasons account for the great fascination of animal physiology over the millennia.

Scientific Curiosity

Underlying all studies of animal physiology—even those designed for very practical, applied purposes—is curiosity about how animals work.

> *How can a hummingbird heart beat 20 times a second during hovering flight?*
>
> *How do insects see in the ultraviolet spectrum?*
>
> *How do kangaroo rats survive in the desert with no access to drinking water?*

Such questions fuel the curiosity of animal physiologists. There is no bound to this curiosity, for a common opinion among animal physiologists is that the more we learn, the more we realize how little we know about physiological systems of animals.

Commercial/Agricultural Applications

Animal physiology studies have yielded knowledge central to many commercial and agricultural advances during the last few decades. Veterinarians, for instance, now can offer veterinary care that in many instances rivals the medical treatment available to humans. Farmers have been able to improve the yield and quality of the milk, eggs, and meat they produce. And improved breeding techniques now include widespread use of artificial insemination.

Insights into Human Physiology

Finally, animal physiology can teach us much about physiological processes in humans. This is hardly surprising, for the human species shares with all other animal species (a) the same fundamental biological processes that, in total, are called "life"; (b) a common set of laws of physics and chemistry; (c) the same principles and mechanisms of Mendelian and molecular genetics; and (d) a linked evolutionary history. Thus, the beating heart in the human body results from physiological mechanisms fundamentally no different from those that underlie heart function in fish, frogs, snakes, birds, or apes. Likewise, the molecular events that produce an electrical nerve impulse in the human brain are fundamentally the same as those that produce an impulse in the nerve of a squid, crab, or rat. For these reasons, animal physiology has made innumerable contributions to understanding human physiology. In fact, most of what we have learned about the function of human cells, tissues, and organs was known first (or is still only known) through the study of various species of vertebrate and invertebrate animals.

Animal physiology, especially as it applies to the human body, is the cornerstone of scientific medical practice. Understanding of the functioning and malfunctioning of living tissue provides the foundation for developing effective, scientifically sound treatment for human disease. The contributions of animal physiology to medicine have been greatly expanded by new techniques for generating unique animal models for specific human diseases (e.g., diabetic mice, congenitally fat rats, zebrafish embryos with heart defects). These models allow a wide range of experiments that were previously only dreamed about. The insights provided by such animal models hinge upon a fundamental understanding of underlying physiological processes. A physician and medical researcher who understands animal physiology—both its potential contributions and limitations—is better equipped to make intelligent and perceptive use of information from such model systems.

CENTRAL THEMES IN ANIMAL PHYSIOLOGY

Our goals for this book are to explore physiological processes that are basic to all animal groups and to show how they have been shaped by selective forces during evolution. Comparing and contrasting how different organisms have adapted to survive similar environmental challenges provides useful insights about patterns of physiological evolution and the adaptive value of physiological processes. As you study animal physiology and physiological adaptation, you will notices several basic tenets, or themes, repeatedly emerging. We briefly discuss a few major themes here. Others will become obvious as you progress through the following chapters.

Structure-Function Relationships

Function is based on structure. We can illustrate this central principle of animal physiology with a familiar example. A frog leaps for a passing fly by contracting powerful skeletal muscles attached to the leg bones of its skeleton. Once the fly has been eaten, the smooth muscle in the frog's stomach slowly massages and mixes the stomach's contents. The nutrients derived from the fly are absorbed into the blood, where energy provided by the regular beating of the cardiac muscle of the heart propels blood throughout the body. Throughout this daily occurrence in a frog's life, three structurally distinct forms of muscle carry out three distinct functions. Such relationships between tissue structure and function are found not only in muscle but also in other tissues (e.g., bone, epithelium, glandular tissue), in fact, in every tissue in an animal's body.

That function depends on structure can be demonstrated at all levels of biological organization. As shown in Figure 1-1, structure-function relationships are clearly evident at the molecular level of muscle tissue. Indeed, the contractile machinery of skeletal muscle represents one of the most intensively studied examples of the dependence of function on structure at the molecular and biochemical levels. As you will learn in Chapter 10 (on muscles), the movement of a frog's leg is the culmination of a chain of biochemical events that depend critically on interactions between thousands of rodlike structures composed of the contractile proteins actin and myosin within each muscle cell. Each of these proteins has a molecular structure that permits them to temporarily interact in a way that moves one protein relative to the other. These protein movements lead to contraction (shortening) of activated individual muscle cells. Compounded over the thousands of muscle cells that form each leg muscle, muscle cell contraction causes overall shortening of the leg muscle. Because of the structural relationship between the powerful contracting muscle and the long bones of the frog's leg, muscle shortening moves the leg. This produces a jumping movement, which moves the frog's entire body.

The principle that function depends on structure holds true across the whole range of physiological processes. In fact, you will see that a consideration of structure-function relationships is virtually unavoidable for each of the physiological processes explored in this book.

Adaptation, Acclimatization, and Acclimation

The physiology of an animal is usually very well matched to the environment it occupies, thereby ensuring its survival. Evolution by natural selection is the accepted explanation for this condition, called **adaptation**. Adaptation occurs extremely slowly in a species over thousands of generations; it generally is not reversible. Adaptation is frequently confused with two other processes, acclimatization and acclimation. **Acclimatization** is a physiological, biochemical, or anatomic change within an individual animal that results from the animal's chronic exposure to new, naturally occurring environmental conditions in the animal's native environment. **Acclimation** refers to the same process as acclimatization, but the changes are induced experimentally in the laboratory or field by the investigator.

Generally, both acclimation and acclimatization are reversible. For example, if an animal voluntarily migrates from a mountain valley to the high slopes of a tall mountain (a voluntary change in a natural environment), its lung ventilation rate typically will increase initially to acquire adequate oxygen. Within a few days or weeks, however, lung ventilation begins to drop back towards sea-level rates as other physiological mechanisms that facilitate gas exchange at high altitude begin to operate. After several days, that individual animal is said to be *acclimatized* to the new high-altitude conditions. However, if an animal physiologist places that same animal in a hypobaric chamber, thus simulating high-altitude conditions, the animal becomes *acclimated* to the experimental conditions within a few days. Contrast these short-term responses with the bar-headed goose, which is able to fly above the peaks of Mt. Everest. This species of goose has become *adapted* to high altitude due to natural selection on the species.

Up until the last decade or so, animal physiologists operated under the assumption that animals were optimally adapted, and that every physiological process they observed was in some way maximized to ensure the survival of the animal. More recently, armed with theories and observations developed by evolutionary biologists, animal physiologists are now realizing that even though evolution by natural selection leads to change in physiological processes, many of these changes are good enough to ensure survival of the animal, but not necessarily perfectly suited to the task. For example, mammals typically control their body temperature within $1-2°C$. Given the precision of some known physiological controlling systems, it is conceivable that a more precise temperature-control system could exist, but it has not been fixed by selection. That is, a $1-2°C$ temperature range is tolerable, or good enough for survival.

Natural selection favors anatomic and physiological characteristics that ensure an animal's survival. What reasons can you think of to account for the fact that any given physiological trait may not be "optimally evolved" to carry out its function? (Hint: Many physiological systems perform multiple functions, and all physiological processes have metabolic costs.)

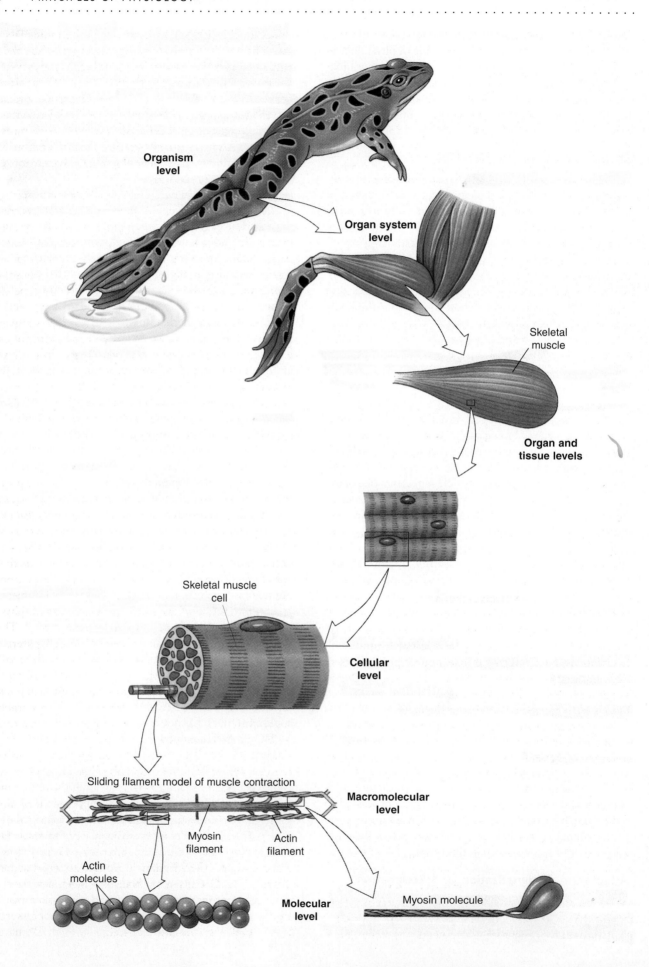

Organism level

Organ system level

Skeletal muscle

Organ and tissue levels

Skeletal muscle cell

Cellular level

Sliding filament model of muscle contraction

Macromolecular level

Myosin filament

Actin filament

Actin molecules

Molecular level

Myosin molecule

Figure 1-1 *(At left)* Function depends on structure at all levels of biological organization. Biological function at each level of organization depends on the structure of that level and that at more microscopic levels. Beginning with the whole animal, this principle can be traced from complexly structured physiological systems through cells down to macromolecular assemblages. In this example, the dozens of muscle systems present in the adult frog allow it to move its eyes, swallow, move and carry out all of the numerous activities of frogs. Groups of skeletal muscles form a system for moving the frog's leg. Skeletal muscles themselves are composed of skeletal muscles cells, which in turn are formed from thousands of macromolecular assemblages formed from a pair of contractile proteins—actin and myosin. These macromolecular assemblages form the basic unit of muscle contraction.

Clearly, adaptation is a central concept in animal physiology, but establishing whether some characteristic of an animal is actually of adaptive value in practice can be difficult. A physiological process is **adaptive** if it is present at high frequency in the population because it results in a higher probability of survival and reproduction than do alternative states. Definitive proof of the adaptive value of some physiological processes can be difficult to obtain, but the comparative approach may yield evidence to support the adaptive value of a process. In this approach, the researcher examines a physiological process in *distantly related species living in identical environments* and/or in *closely related species living in strikingly different environments.*

The presence of a similar physiological process supported by a similar anatomic structure in several distantly related animal species occupying a single environment would suggest that that process-structure combination is adaptive. Such comparative studies are more powerful if they are coupled with the examination of closely related species in different environments. A classic example of the power of this approach involves the llama and its close relative the camel. Originally, researchers were convinced that the unusually high affinity of llama blood for oxygen was an adaptation to the rarefied air at the high altitudes at which llamas live. To their surprise, animal physiologists discovered that camels, who live at low altitudes, also have high-affinity blood. Thus, the llama's high-oxygen affinity blood is *not* a specific adaptation to high altitude. That is, the blood characteristics of llamas and camels have little to do with the altitude at which they live, and much to do with their being in the camel family. Such indirect criteria for the adaptiveness of a particular physiological trait are typically accepted in judging adaptiveness, particularly when set in the framework of a carefully designed comparative study.

Physiological and anatomic adaptations to environments are genetically based, passed on from generation to generation and constantly shaped and maintained by natural selection. Animals inherit this genetic information from their parents in the form of **deoxyribonucleic acid (DNA)** molecules. Spontaneous alterations (**mutations**) can occur in the nucleotide sequence of DNA, potentially causing changes in the properties of the encoded proteins or ribonucleic acids (RNAs). Mutations in the germ-line DNA that enhance the survival of organisms and thus their chances to reproduce are retained by selection and increase in frequency of occurrence in the population of organisms over time. Conversely, those mutations that render organisms less well adapted to their environment will lessen their chances to reproduce; if deleterious enough, such mutations generally are eliminated over time. A small proportion of "neutral" mutations appear to neither enhance nor reduce the chances of survival.

Genetic material in the form of DNA is passed on from multicellular parents to their offspring. This DNA is contained in a line of germ cells that, in each generation, are derived directly from parent germ cells, creating an uninterrupted lineage. The blind, nondirected process of evolution is centered on the survival of the germ-line DNA, since the information it encodes defines a species. Failure by a species to reproduce that genetic information leads to the species' immediate, irreversible extinction. From the biological viewpoint, then, the major goal in an animal's life is to reproduce and propagate its DNA, and all behaviors, physiological processes, and anatomic structures are ultimately subservient to the survival of the germ line. Adaptation to the constraints and demands of the environment is best appreciated and understood in this context of an animal's struggle to maintain and reproduce its DNA.

Homeostasis

Despite the fact that many animals seem to live comfortably in their environment, most habitats are actually quite hostile to animal cells. For many aquatic animals, for example, the surrounding water is more dilute (freshwater) or more salty (seawater) than their own body fluids. Both terrestrial and aquatic animals may live in environments that are too hot or too cold. Moreover, with only a few exceptions (e.g., the deep abyss of the oceans) most environments are characterized by at least small fluctuations in their physical and chemical properties (especially temperature). The environmental changes raging about an animal's exterior would be a major disruptive force to internal cellular, tissue, and organ function were it not for the physiological control systems that are directed towards maintaining *relatively stable conditions within an animal's body tissues.* This tendency of organisms to maintain relative internal stability is called **homeostasis.**

Claude Bernard, the nineteenth-century French pioneer of modern physiology, first recognized the importance to animal function of maintaining stability in the *milieu intérieur,* or internal environment. Bernard noted the ability of mammals to regulate the condition of their internal environment within rather narrow limits. This ability is familiar to most of us from measurements of our body temperature, which in healthy individuals is maintained within a degree of 37°C. Cells in our body experience a relatively consistent environment with respect to not just temperature, but also glucose concentration, pH, osmotic pressure, oxygen level, ion concentrations, and so forth. Bernard

(1872) concluded, "Constancy of the internal environment is the condition of free life," arguing that the ability of animals to survive in often stressful and varying environments directly reflects their ability to maintain a stable internal environment. In the early 1900s Walter Cannon extended Bernard's notion of internal consistency to the organization and function within cells, tissues, and organs. It was Cannon (1929), in fact, who coined the term *homeostasis* to describe the tendency towards internal stability, and his research into how physiological systems maintain homeostasis earned him a Nobel Prize (see Chapter 9).

Homeostasis, one of the most influential concepts in the history of biology, provides a conceptual framework in which to interpret a wide range of physiological data. This phenomenon is nearly universal in living systems, allowing animals and plants to survive in stressful and varying environments (Figure 1-2). The evolution of homeostasis and the physiological systems that maintain it are thought to have been the essential factors that allowed animals to venture from physiologically friendly environments and invade environments more hostile to life processes. One fascination of physiology is discovering how different groups of animals have adapted through natural selection to maintain homeostasis in the face of particular environmental challenges.

Although complex, multi-organ physiological mechanisms are often involved in maintaining homeostasis, homeostasis is evident at the cellular level. In fact, a varying degree of homeostasis is found in the most simple unicellular organisms. Protozoa, for instance, have been able to invade freshwater and other osmotically stressful environments because the concentrations of salts, sugars, amino acids, and other solutes in their cytoplasm are regulated by selective membrane permeability, active transport, and other mechanisms. These processes maintain intracellular conditions, typically quite different from the extracellular environment, within limits favorable to the metabolic requirements of all cells, including the one-celled protozoa.

Feedback-Control Systems

The regulatory processes that maintain homeostasis in cells and multicellular organisms depend on **feedback**, which occurs when sensory information about a particular variable (e.g., temperature, salinity, pH) is used to control processes in cells, tissues, and organs that influence the internal level of this variable. Homeostatic regulation requires continuous sampling of controlled variables and corrective action, a process termed **negative feedback.** For example, suppose that an experienced driver is placed in a car on an absolutely straight, 10-mile-long stretch of traffic-free highway, is allowed to position the car, is blindfolded, and is then required to drive the 10 miles without deviating from his lane. The slightest asymmetry in both the neuromuscular or sensory systems of the driver and in the steering mechanism of the car—not to mention wind or unevenness of the road surface—makes this an impossible task. On the other hand, if the blindfold is removed, the driver will use visual information to stay in the lane. A gradual drift to one side of the lane or the other, due to whatever internal or external perturbations, will be corrected by a compensatory movement applied to the steering wheel. The visual system of the driver acts as the sensor in this case, and the neuromuscular system, by causing a correctional movement in the direction opposite to the perceived error, acts as an inverting amplifier that corrects for deviations from the set point (i.e., the center of the lane in this case).

Another example of regulation by negative feedback can be demonstrated with a thermostatic device that maintains the temperature of a hot-water bath at or near the set point (Figure 1-3). When the water temperature is below the set point, the sensor maintains the heater switch in the closed "on" position. As soon as the set-point temperature is achieved, the heater switch opens, and further heating

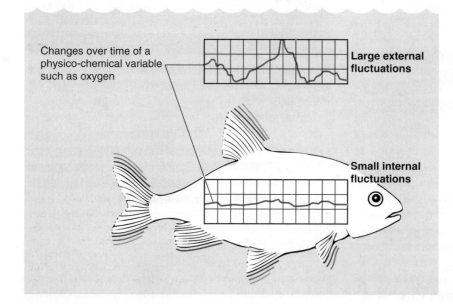

Changes over time of a physico-chemical variable such as oxygen

Large external fluctuations

Small internal fluctuations

Figure 1-2 Physiological regulatory systems maintain internal conditions within a relatively small range. Large variations in the external environment induce equally large responses of the control system to offset the disturbance. The net effect is that the internal fluctuations of a variable in an animal are usually far less than the environmental fluctuations in that variable. In other words, homeostasis is maintained.

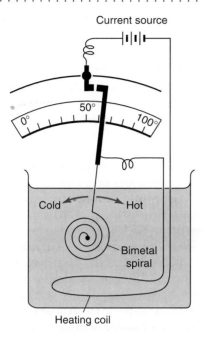

Figure 1-3 The temperature of a thermoregulated water bath is maintained by a negative-feedback control system. The bimetal coil, whose center is attached to the water bath wall, winds slightly as the temperature of the water bath drops below the desired (set-point) temperature. This winding action causes electrical contacts to touch, completing an electric circuit and allowing electric current to flow through a heating coil in the bath. As the water warms, the coil unwinds slightly and the contacts separate. The water temperature, at or slightly above the set point, now stops rising. When the bath begins to cool off again, the cycle is repeated. Many physiological control systems conceptually operate the same way as this thermoregulated water bath.

ceases until the temperature again drops below the set point. This example suggests that regulation of body temperature requires a "thermostat" whose information must be provided to a temperature-control system, which either heats or cools the body depending on the temperature signal. Physiological investigations have discovered a great deal about temperature regulation, including the location of the thermostat, as we'll discuss in Chapter 16.

The characteristics of negative and positive feedback are summarized in Spotlight 1-1, on page 12. You will find feedback-control systems appearing throughout this text, especially in our discussions of intermediary metabolism (Chapter 3), endocrine control (Chapter 9), neural control of muscle (Chapter 11), circulatory and respiratory control (Chapter 13), and regulation of ionic balance (Chapter 14). Indeed, the concept of feedback is pervasive in the study of physiological systems.

Conformity and Regulation

When an animal is confronted with changes in its environment (e.g., changes in oxygen availability or salinity), it shows one of two broad categories of responses: conformity or regulation. In some species, these challenges induce internal body changes that parallel the external conditions (Figure 1-4). Such animals, called *conformers*, are unable to maintain homeostasis for internal conditions like body fluid salinity or tissue oxygenation. For example, echinoderms like the starfish *Asterias* are **osmoconformers**, whose internal body fluids come to equilibrium with their environment, showing an increase in body fluid salinity when placed in high-salt water and a decrease in body fluid salinity when placed in low-salt water. Similarly, the oxygen consumption of **oxyconformers** like annelid worms rises

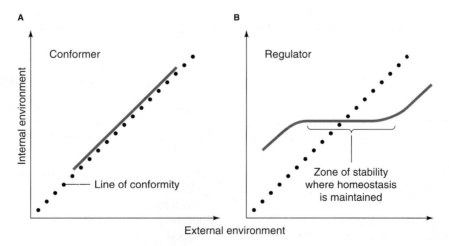

Figure 1-4 Conformers adjust their internal conditions to reflect external environmental conditions, whereas regulators maintain internal stability even as external conditions change. **(A)** Plot of the external environmental value of a variable (e.g., salinity, oxygen availability) versus its internal value for conformers (red line) typically is a straight line with a slope of 1. When an animal is unable to mount the physiological or other responses necessary to counteract external changes in a variable, so its internal value varies directly with its external value, then the response will mimic the "line of conformity" (black dotted line). **(B)** Plot of the external value versus internal value of a variable for regulators (red line) shows that they are able to maintain internal stability over a wide range of external change. The line of conformity (black dotted line) is shown for comparison. At environmental extremes, however, regulators are unable to regulate internal conditions and become conformers. The breadth of the zone of stability of a regulator depends on the species and the environmental variable.

and falls as oxygen availability waxes and wanes. The degree to which conformers survive these changing environments depends upon the tolerance of their body tissues to internal changes.

Regulators, as their name implies, use biochemical, physiological, behavioral, and other mechanisms to regulate their internal environment over a broad range of external environmental change—that is, to maintain homeostasis. Thus, an **osmoregulator** maintains the ion concentrations of body fluids above environmental levels when placed in dilute water and below environmental levels when placed in concentrated water. **Oxyregulators,** which include crayfish, most mollusks, and almost all vertebrates, maintain their oxygen consumption at near-steady levels as environmental oxygen availability falls. Eventually, however, oxygen may become so limited that oxygen consumption cannot be maintained, and the animal reverts to oxygen conformity.

It is tempting to make broad generalizations about taxonomy based on whether animals are conformers or regulators. Although most invertebrates are conformers and almost all vertebrates are regulators, there are many exceptions. For instance, decapod crustaceans (e.g., crabs, crayfish, shrimps, lobsters) tend to be accomplished regulators, as are many mollusks and most insects. Moreover, the zone of stability in which homeostasis is maintained in regulators may be very broad or very narrow depending on species.

LITERATURE OF PHYSIOLOGICAL SCIENCES

All the information described in this book is based on the experimental results of individual scientists as described in published reports and articles, commonly referred to as *scientific papers.* Such papers, which include descriptions of the experimental methods, a summary of the results, and discussion of the results, are published in scientific journals, many of which focus on particular disciplines or specialized research areas. Before publication, the journal's editor sends each submitted manuscript to two or more other scientists expert in the topic of the paper for their review and critical comments. The reviewers recommend acceptance or rejection on grounds of scientific quality and often make numerous suggestions for improvement of acceptable papers. This process, called *peer review,* helps assure that published papers are based on accepted research methods and their conclusions are valid. Once a paper is published, members of the scientific community are free to test its conclusions by repeating key experiments and to accept or reject individually the conclusions stated in the paper. Healthy skepticism and attempts to improve on the work of other scientists are of central importance to the self-correcting nature of an experimental science such as physiology.

Numerous scientific journals publish papers on research in animal physiology. Many of the most widely read journals are listed in Table 1-1. Some journals accept papers dealing with a wide range of topics, whereas numerous specialty journals provide in-depth coverage of more limited areas of interest. In addition, several review-type journals, which come out annually, publish articles that summarize and evaluate the findings related to particular topics that have been published previously in other journals. Another category of journals deals with organisms primarily from a taxonomic perspective. These publish physiological and other research papers dealing with specific animal groups. Finally, weekly scientific news journals publish preliminary reports on physiological research that the editors believe will capture the interest of the general scientific community.

As you become familiar with the physiological research literature, be aware that journals, like animals, have undergone evolution from their original forms. This is especially apparent when considering the names of journals, which in a few instances seem to only generally reflect their contents. For example, the *Journal of General Physiology* primarily publishes cellular physiology and biophysics, while the *Journal of Experimental Biology* publishes papers on animals only, excluding plant physiology. Likewise, the *Proceedings of the New York Academy of Sciences,* the *Midland Naturalist, Canadian Journal of Zoology, Australian Journal of Zoology,* and *Israel Journal of Zoology* publish papers from scientists around the world even while retaining their regional themes.

The sampling of periodicals listed in Table 1-1 represents only a tiny percentage of the literally thousands of journals that currently publish biological and biochemical research papers, with dozens of new journals being established every year. How can any one person—student or researcher—possibly hope to follow developments in a specific area of biology? Fortunately, along with the explosion of information over the last few decades, technology has been developed that allows us to sit in front of a computer terminal and, with a few keystrokes, search hundreds of thousands of documents, roaming freely through the libraries of thousands of universities and research institutes worldwide. Additionally, the rapidly expanding use of the Internet during the last few years is leading to a slow but inexorable shift in the way scientific information is disseminated. Traditional journals delivered to individuals through the mail gradually are being supplemented, or replaced, by electronic journals that post new articles as they are accepted. Couple this with World Wide Web pages that herald a laboratory's latest research projects and findings to all who can access such information, and we see a coming revolution in how information is accessed and processed.

What has not been replaced by technology is the need of the student to read and understand descriptions of original experiments in order to understand the process that generates physiological data. Consequently, at the end of each chapter of this text you will find a Suggested Readings section, listing a few key articles that offer greater detail on certain topics. In addition, the original sources for much of material presented in the text, figures, and tables are listed

TABLE 1-1
Sampling of scientific journals that publish physiological research papers

Name	Abbreviation*	Topics covered
General journals		
American Journal of Physiology	*Am. J. Physiol.*	Broad areas of physiology from the cell to organ systems
Pflügers Archive für Physiologie (now European Journal of Physiology)	*Pflugers Arch. Physiol. (Eur. J. Physiol.)*	
Journal of General Physiology	*J. Gen. Physiol.*	Physiological and biophysical studies at the cellular and subcellular level
Journal of Physiology	*J. Physiol.*	Many different areas with emphasis on lower vertebrates and invertebrates
Comparative Physiology and Biochemistry	*Comp. Physiol. Biochem.*	
Journal of Comparative Physiology	*J. Comp. Physiol.*	
Journal of Experimental Biology	*J. Exp. Biol.*	
Physiological Zoology	*Physiol. Zool.*	
Specialty journals		
Brain, Behavior and Evolution	*Brain Behav. Evol.*	Research related to specific areas or processes indicated by journal's name
Cell		
Circulation Research	*Circ. Res.*	
Endocrinology		
Gastroenterology		
Journal of Cell Physiology	*J. Cell Physiol.*	
Journal of Membrane Biology	*J. Membr. Res.*	
Journal of Neurophysiology	*J. Neurophysiol.*	
Journal of Neuroscience	*J. Neurosci.*	
Molecular Endocrinology	*Mol. Endocrinol.*	
Nephron		
Respiration Physiology	*Respir. Physiol.*	
Annual reviews		
Annual Review of Neuroscience	*Annu. Rev. Neurosci.*	Summaries and evaluations of original papers on particular topics published in other journals
Annual Review of Physiology	*Annu. Rev. Physiol.*	
Federation Proceedings	*Fed. Proc.*	
Physiological Reviews	*Physiol. Rev.*	
Taxonomy-oriented journals		
Auk		Physiology and other topics related to birds
Condor		
Emu		
Crustaceana		Physiology and other topics related to crustaceans
Copeia		Amphibian and reptile physiology
Herpetologica		
Journal of Herpetology	*J. Herpetol.*	
Journal of Mammology	*J. Mammol.*	Physiology and other topics dealing with mammals
Weekly journals		
Nature		Preliminary reports about topics of general interest to scientific community
Science		

*Single-word journal names are not abbreviated.

SPOTLIGHT 1-1

THE CONCEPT OF FEEDBACK

Any effective control system, whether the human brain, a computer, or a household thermostat, is vitally dependent on **feedback**, which is the return of sensory information to a controller that regulates a controlled variable. Feedback can be either positive or negative, each producing profoundly different effects. Feedback is widely employed by both biological and engineering control systems to maintain a preselected level of the controlled variable.

Negative Feedback

Consider the model system shown in part A of the accompanying figure. Assume for the moment that the *controlled system* experiences a new *disturbance* (e.g., a change in length, temperature, voltage, concentration). The output of this system is detected by a *sensor*, which sends a signal to an *amplifier*. Now, imagine an amplifier that inverts the signal it receives, so that the "sign" of its output is opposite to that of its input (e.g., plus changed to minus, or vice versa). Such *signal inversion* provides the basis for negative feedback, which can be used to regulate the *controlled variable* (e.g., length, temperature, voltage, concentration) within a limited range.

When the sensor detects a change in state (e.g., change in length, temperature, voltage, concentration) of the controlled system, it produces an *error signal* proportional to the difference between the *set point* to which the system is to be held and the actual state of the system. The error signal is then both amplified and inverted (i.e., changes in sign). The inverted output of the amplifier, fed back to the system, counteracts the disturbance. The inversion of sign is the most fundamental feature of negative-feedback control. The inverted output of the amplifier, by counteracting the disturbance, reduces the error signal, and the system tends to stabilize near the set point.

A hypothetical negative-feedback loop with infinite amplification would hold the system precisely at the set point, because the slightest error signal would result in a massive output from the amplifier to counteract the disturbance. Since no amplifier, electronic or biological, produces infinite amplification, negative feedback only approximates the set point during disturbance. The less amplification the system has, the less accurate is its control.

Positive Feedback

In the model system shown in part B of the figure, an applied disturbance acts on the controlled system just as in part A. However, now suppose that the signal input is amplified but that its sign (plus or minus) remains unchanged. In this case the output of the amplifier, when fed back to the controlled system, has the same effect as the original disturbance, reinforcing the disturbance to the controlled system. Such a positive-feedback system is highly unstable because the output becomes progressively stronger as it is fed back and reamplified. A familiar example involves public address systems: When the output of the loudspeaker is picked up and reamplified by the microphone, a loud squeal is generated. Thus, a tiny disturbance at the input can cause a much larger effect at the output. The output of the system is usually limited in some way; for example, in the public address system, the intensity of the output is limited by the power of the audio amplifier and speakers or by saturation of the microphone signal. In biological systems, the response may be limited by the amount of energy or substrate available.

In normal, healthy animals, positive feedback generally is used to produce a regenerative, explosive, or autocatalytic effect. This type of control often is used to generate the rising phase of a cyclic event, such as the upstroke of the nerve impulse or the explosive growth of a blood clot to prevent blood loss. The rapid emptying of a body cavity (e.g., expulsion of the fetus from the uterus, vomiting, swallowing) also often begins with positive-feedback processes.

Positive feedback, however, most often is encountered in pathological conditions that affect normal negative feedback control. Congestive heart failure provides a classic example. In this disorder, the inability of the heart to pump blood causes blood to accumulate in the ventricles, which further impairs their ability to pump blood, which causes more blood to accumulate in the ventricles, which further impairs their pumping ability, and so on. Unless such vicious cycles of positive feedback are interrupted, they will quickly lead to complete failure of the controlled system.

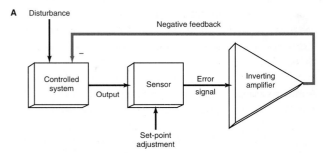

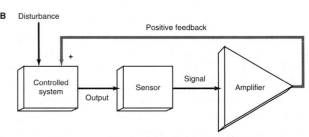

Biological control systems depend on negative or positive feedback. The unique features of negative-feedback systems **(A)** are generation of an error signal and sign inversion by the amplifier. Positive-feedback systems **(B)**, which amplify the signal without sign inversion, lead to a rapid increase in the controlled variable. The basic elements illustrated here occur in a number of variations in biological systems. In some cases, sensor and amplifier functions are performed by a single element (e.g., response of a single cell to environmental disturbances). Another variation occurs in the middle of the estrous cycle of female mammals, when sign inversion shifts from the amplifier to the sensor, which has the effect of converting a negative-feedback to a positive-feedback system.

in the References Cited in the back of the book. Although we recognize that as a student you have only limited time to spend on any one course, we nonetheless encourage you to visit the periodical section of your library and examine some of these articles.

ANIMAL EXPERIMENTATION IN PHYSIOLOGY

Much of the information presented in this book has come from experiments in which animals were used to answer specific questions about how physiological processes work. Because basic physiological properties are similar among different animal species, the results from animal experiments are broadly applicable to other species including humans. Indeed, almost all of the important medical treatments in use today are based directly on animal experimentation. Even though data from experiments on humans would be the most scientifically reliable, in most cases such research would be ethically unacceptable. Consequently, animal experimentation is of central importance in medical research as well as in its own right.

The first three editions of this book operated on the assumption that the benefits of animal experimentation were understood by all, but this assumption can no longer be made. Despite the general benefits animal research has produced for all humans, controversy has arisen about the ethics of using animals in experiments. In describing this controversy, it is important at the outset to distinguish between *animal welfare* and *animal rights*. The former refers to the humane treatment of animals with respect for their comfort and well-being. The latter refers to the idea that animals have intrinsic and unassailable "rights" to "life, liberty, and the pursuit of happiness," much like the inalienable human rights stated in our Declaration of Independence. The concept of animal rights, in its most extreme interpretation, would prohibit domestication of animals for food production and their being kept as pets.

The scientific community strongly supports the concept of animal welfare. Scientists recognize their professional obligations to safeguard and improve the welfare of laboratory animals. Indeed, scientific researchers concerned about the care and treatment of laboratory animals were the first to set voluntary care standards at the turn of the century, long before federal regulations were instituted in the United States. Now, governmental regulations strictly control animal research, requiring all animal care facilities to meet strict standards of cleanliness and animal care. Professional societies as well as scientific journals have stringent requirements regarding animal experimentation, and publication of research results requires evidence that these regulations have been met.

All research institutions that receive federal funding are required to have Laboratory Animal Care Committees, which evaluate proposed experiments to ensure that they are performed with the minimum pain and discomfort to animal subjects, and that they use the minimum number of animals necessary to achieve conclusive results. Such committees include scientists, veterinarians, and community representatives. These individuals are empowered to limit or prohibit experiments that do not adequately address the issue of animal pain, and such ill-conceived proposed experiments are quickly rejected. Scientists themselves are concerned about animal care and realize that experiments performed on animals that are suffering in any way may not yield meaningful or useful results. Proper animal care is essential in the search for accurate scientific data.

Despite effective internal and external safeguards, the issue of animal "rights" appeals to some who oppose animal research in any form. The presence of sanctioned animal care facilities and of strictly enforced regulations does not satisfy those advocating animal rights, nor does the overwhelming evidence of the benefits of animal research sway them. The ongoing debate over animal welfare versus animal rights is a healthy one, if the available data is evaluated objectively and the intentions of the participants are clear. Through such debate and evaluation we can ensure that the needs for animal experimentation are balanced by broadly shared concerns for the well-being of animals.

SUMMARY

Animal physiology deals with the functions of tissues, organs, and organ systems, particularly how these functions are controlled and regulated. Although this text concentrates on presenting the functional basis of animal physiology, a major theme is to understand the environmental constraints that have shaped the evolution of physiological processes through natural selection.

Biologists study animal physiology because they are curious about how animals work and also can learn much about human physiology by observing other animals. Animal physiology is a cornerstone of scientific medical practice as well as veterinary practice, and the study of animals with common physiological and evolutionary features has provided great insight into human physiology.

Several major themes characterize animal physiology. First, function depends on structure at all levels, from atoms to organisms. Specialized structures often produce specialized functions. Second, natural selection has led to physiological adaptation, that is, processes well suited for helping animals survive in often challenging environments. The adaptive cell, tissue, and organ functions that have arisen during evolution are genetically determined and encoded in DNA. Third, many animals exhibit homeostasis, the tendency toward a relative stability of the internal environment of an organism. Without homeostasis, fluctuations and nonoptimal levels of temperature, pH, oxygen, and other physicochemical characteristics may disrupt the basic chemical reactions underlying physiology, anatomy, and behavior. Fourth, feedback-control systems are critical to maintaining homeostasis. Finally, animals can respond to changes in external environmental conditions in two general ways. In conformers, the internal environment adjusts to reflect external conditions; that is, they cannot maintain homeostasis. In contrast, regulators can adjust their internal environment within narrow limits as environmental conditions change; that is, they can maintain homeostasis.

Understanding the experimental techniques used in physiological research is essential for appreciating how physiological knowledge advances. Research results are published in peer-reviewed journals, many of which can now be readily accessed by electronic searches carried out in libraries. Reading original research papers presenting specific results as well as review articles will help you grasp the essence of scientific research.

Almost all data about animal physiology—and most of what we know about human physiology—are derived from studies with experimental animals designed to answer specific questions about how physiological processes work. Meaningful results can only be acquired if animals are well cared for and their pain and discomfort are minimized. Numerous regulations, strictly enforced by local, state, and federal agencies, have been adopted to assure that researchers follow accepted standards for animal experimentation.

REVIEW QUESTIONS

1. Give an example of a simple structure-function relationship in physiology, and describe its conditions of operation.

2. What evolutionary advantage does successful maintenance of relative internal stability confer on an animal?

3. Compare and contrast negative and positive feedback, giving an example of each. Explain why negative rather than positive feedback is required for maintenance of homeostasis.

4. Go to your library and use an electronic data base to search for the word *homeostasis* among the catalogued books and articles.

5. Distinguish between the concepts of animal welfare and animal rights. Ask your professor about the makeup of the Laboratory Animal Care Committee at your college or university.

SUGGESTED READINGS

Benison, S. A., A. C. Barger, and E. L. Wolfe. 1987. *Walter B. Cannon: The Life and Times of a Young Scientist.* Cambridge: Harvard University Press. (An intriguing, insightful biography about a distinguished scientist who introduced the concept of homeostasis in 1929.)

Dworkin, B. R. 1993. *Learning and Physiological Regulation.* Chicago: University of Chicago Press. (A very thorough treatment of the theory and mechanisms underlying physiological regulation and behaviors.)

Futuyama, D. J. 1986. *Evolutionary Biology.* 2d ed. Sunderland, Mass.: Sinauer Associates. (One of several comprehensive undergraduate textbooks that introduce the basic concepts of evolutionary biology as they apply to physiological process.)

C H A P T E R

2

EXPERIMENTAL METHODS FOR EXPLORING PHYSIOLOGY

Our knowledge of animal physiology is based on information *(data)* derived from experimentation. Since the ultimate goal of animal physiology is to understand how a process operates within an organism, experiments must be designed to allow the measurement of key *variables* (e.g., metabolic rate, blood flow, urine production, muscle contraction) in the animal (or its cells or tissues) while it is in a known state such as resting, exercising, digesting, or sleeping. This kind of experimentation is particularly challenging and requires the use of a variety of techniques and methods. Many of the experimental techniques and measuring devices common in animal physiology are "time-honored." These include pressure transducers to measure pressure, catheter implantation to draw blood or inject samples, respirometers for determining metabolic rates, and numerous others. A description of each of these is beyond the scope of this chapter, especially since such fundamental techniques are well described in texts such as J. N. Cameron's *Principles of Physiological Measurement.* In this chapter, we will focus on a few of the many molecular and cellular techniques that have recently been added to the physiologist's tool box, briefly describing them and illustrating their use in physiological research. First, however, we consider the nature of hypotheses and the general principles that apply in testing them.

By knowing why and how experiments in animal physiology are performed—whether they employ traditional or emerging methods—you will be much better able to evaluate the strengths and limitations of the information you will learn in this book.

FORMULATING AND TESTING HYPOTHESES

Scientists use experimental data to create general laws of physiology—some literally centuries old, and some still emerging. These general laws, in turn, serve as the basis for formulating new **hypotheses,** which are specific predictions that can be tested by performing further experiments. An example of a general "law" supported by much existing

data is that water-breathing animals regulate acid-base balance by modifying the excretion of HCO_3^- in exhaled water, while air-breathing animals regulate acid-base balance by modifying the elimination of CO_2 gas in exhaled air. The following testable hypothesis could be derived from this general law: *A transition from HCO_3^- elimination to CO_2 elimination occurs when water-breathing tadpoles metamorphose into air-breathing frogs.* Although hypotheses are framed as statements rather than as questions, the goal of experimentation is to test the validity of hypotheses, and thus answer the implied questions.

Physiological experiments should begin with a well-formed, specific hypothesis that focuses on a particular level of analysis and is amenable to a verifiable experimental approach. Although a hypothesis such as *killer whales have a very high cardiac output while in pursuit of seals* may be interesting and in fact true, it is merely an intellectual exercise to suggest this hypothesis unless a feasible experimental approach exists for gathering data necessary to accept (prove) or reject (disprove) it. However, the search for means to test novel hypotheses has been an important stimulus for development of new experimental techniques and measuring instruments. For example, telemetry devices currently available for gathering data on blood flow in small to medium-sized animals like ducks, fishes, and seals are being modified for use on even larger animals.

The August Krogh Principle

August Krogh was a Danish animal physiologist with extremely broad interests in comparative physiology. Dozens of key research articles bearing his name have served as the basis for whole areas of further experimentation in the area of respiration and gas exchange. Indeed, Krogh's work in the late 1800s and early 1900s eventually led to his winning the Nobel Prize for physiology. One of the reasons for Krogh's extraordinary success as a physiologist was his uncanny ability to choose just the right experimental animal with which to test his hypotheses. His view was that for every defined physiological problem, there was an optimally suited animal that would most efficiently yield an answer.

The design of experiments based on the unusual characteristics of an animal has come to be known as the August Krogh principle (Krebs, 1975). Illustrations of this principle abound in this book and throughout modern animal physiology. For example, in the 1970s a group of animal physiologists, interested in the evolution of air breathing in crustaceans, were studying relatively tiny intertidal crabs, but they were frustrated because the small size of these animals kept them from "giving up" their physiological secrets. Evoking the August Krogh principle, which suggested that there was an ideal animal with which to carry out their studies, these physiologists organized an expedition to the Palau Islands in the South Pacific. These islands are home to the "coconut," or "robber," crab, a terrestrial hermit crab weighing up to 3 kg. The monstrous size of these animals (for a terrestrial crab) allowed numerous experiments yielding important new data during the one-month expedition.

As another example, animal physiologists interested in cardiac performance in fishes often have a difficult time measuring pressure and flow and sampling blood from the heart because of its typical location in bony fish (i.e., teleosts). Yet, the sea robin, a deep-water (benthic) marine teleost that is quite unremarkable in most respects (although it is down-right ugly!), has an unusually large heart, which is much easier to access than in other fishes. By following the August Krogh principle and using the sea robin as the basis for their experiments, comparative cardiovascular physiologists now know more about heart function in fishes than they would if they had continued to struggle with the relatively unforgiving anatomy of the trout, salmon, or catfish.

Experimental Design and Physiological Level

In designing an experiment, the first and most important decision a physiologist must make is about the level at which the physiological problem will be analyzed. This choice of level determines the methodology (and choice of animal) appropriate for measuring the experimental variables of interest.

Historically, techniques for exploring physiological problems at the level of the whole animal were developed first; subsequently, and with increasing rapidity in recent decades, have come new techniques for experimenting at the cellular and now at the molecular level. Conceptually, however, we generally operate in the reverse order: starting at the molecular level, then moving successively to the cellular, tissue, organ, and finally whole-animal levels, much as outlined in Figure 1-1. Consequently, in the following sections, we describe some representative experimental methods for studying physiological processes, beginning at the molecular level. Much of the information presented in other chapters of this book is based on experimental results obtained with these various techniques. Only by learning how and why these methods work, as well as some of their limitations, can you adequately assess the information presented.

Note that no level of analysis is intrinsically more valuable or important than any other. Indeed, the best understanding of animal physiology comes from integrating knowledge about the contributing components from the molecular through organ-system level. Having said this, we recognize the strong trend in animal physiology (as in all of biology) during the last decade towards "reductionism," the study of cellular and molecular mechanisms in an attempt to explain more complex processes at higher organizational levels. Ultimately, some of the most valuable experiments are those at a level of analysis that allows insights about processes at adjacent organizational levels.

Although researchers and students often are fascinated by new and frequently expensive methodologies, incisive results can be obtained with well-designed experiments using relatively simple instruments and techniques. In other words, a well-conceived experimental design often can compensate for the lack of the latest, cutting-edge equipment and techniques.

MOLECULAR TECHNIQUES

The past few decades have seen a veritable explosion in the number and sophistication of available techniques for probing molecular events, with new methods and refinements constantly emerging. The variety of molecular techniques available have had major implications for biological research in general, and animal physiology has certainly benefited from molecular approaches. In this section we describe just a few of the powerful molecular techniques that have been used to answer questions in animal physiology. More detailed discussion of these and related techniques are presented in textbooks such as *Molecular Cell Biology* by H. D. Lodish et al.

Tracing Molecules with Radioisotopes

Greater understanding of physiological processes can often be achieved by knowing the movements of molecules within and between cells. For example, we can more easily understand the role of a particular neurohormone in regulating physiological processes if its movements can be traced from its site of synthesis to its site of release and on to its site of action. Many types of experiments that follow the movement of physiologically important molecules employ **radioisotopes,** the relatively unstable, disintegrating radioactive isotopes of the chemical elements. The natural disintegration of radioisotopes is accompanied by release of high-energy particles, which can be detected by appropriate instruments. With the exception of ^{125}I, which emits γ particles, the isotopes commonly used in biological research emit β particles.

Although radioisotopes occur naturally, those normally used in experimental studies are produced in nuclear reactors. The most commonly used isotopes in biological research are ^{32}P, ^{125}I, ^{35}S, ^{14}C, ^{45}Ca, and ^{3}H. A radioisotope of an element normally present in the molecule of interest can be incorporated *in vitro* or *in vivo* either directly into the molecule or into a precursor molecule that will eventually be converted into the molecule of interest. The resulting *radiolabeled* molecule has the same chemical and biochemical properties as the unlabeled molecule. An amazing array of so-called radiolabeled biologically active molecules (e.g., amino acids, sugars, hormones, proteins) are now readily available (at a substantial price) from companies that specialize in their production. Once a molecule has been radiolabeled, the particles emitted from the radioisotope can be used to detect the presence of the molecule, even at very low concentrations.

In one type of tracing experiment, the radiolabeled molecule of interest or its precursor is administered to an animal, isolated organ, or cells growing *in vitro* culture, and then samples are removed periodically for measurement of particle emission. Two types of instruments are used to detect emitted particles. A **Geiger counter** detects ionization produced in a gas by emitted energy. A **scintillation counter** detects and counts tiny flashes of light that these particles create as they pass through a specialized "scintillation fluid." The amount of radiation detected by either instrument is related directly to the amount of the radiolabeled molecule present in the sample.

In another type of experiment, the location of radiolabeled molecules within a tissue section is pinpointed by **autoradiography**. In this technique, which literally "takes a picture" of the radioisotopes in tissues, a thin tissue slice containing a radioisotope is laid on a photographic emulsion. Over the course of days or weeks, particles emitted from the radioisotope expose the photographic emulsion, producing black grains that correspond to the location of the labeled molecules in the tissue (Figure 2-1). This qualitative record can be quantitated by measuring the amount of exposure of the emulsion in a **densitometer** and comparing it with exposures caused by standards of known concentration; in this way, the actual concentration of a radiolabeled molecule in the tissue or portions of it can be determined. Autoradiography has been particularly useful in neurobiology, endocrinology, immunology, and other areas of physiology involving cell-to-cell communication.

Tracing Molecules with Monoclonal Antibodies

Examination of a biological structure in a fixed tissue slice on a microscope slide can be daunting. Even when the tissue has been stained so that the cell nuclei are dark purple and the cell membranes a somewhat lighter shade, for example, it remains difficult to discern much about the details of the tissue. Much better visualization of the structural details of cells is possible with *antibody staining*. This remarkable technique permits localization of molecules present in such extremely low concentrations that they are difficult to study by other techniques.

Antibody staining generally involves covalently linking a flourescent dye to an antibody that recognizes a specific determinant on an antigen molecule. (Although we often think of antigens as disease-causing microbes or invading foreign materials like pollen, normal, biologically active molecules,

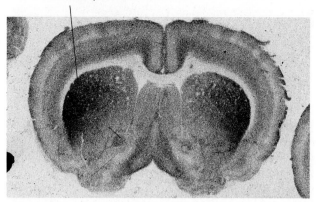

Caudate-putamen

Figure 2-1 Autoradiograms can reveal biochemical and structural details that cannot be seen with traditional techniques for tissue fixation and staining. This autoradiograph shows a frontal section through the rat brain after cannabinoid receptors have been bound by a radiolabelled synthetic cannabinoid (closely resembling the active ingredient of marijuana). The most radioactive areas (that is, the areas with the most cannabinoid receptors) have most heavily exposed the photograph film on which the brain slice was laid, and show up primarily as the dark areas in the striatum (caudate-putamen), which mediate motor functions. [Courtesy of Miles Herkenham, NIMH.]

such as neurotransmitters and cell growth regulators, can act as antigens and induce production of specific antibodies when injected into an appropriate animal.) Identical antibodies produced in response to an antigen are called **monoclonal antibodies;** however, most natural antigens have multiple, rather than single determinants, thus the production of several different antibodies is likely. A mixture of antibodies that recognize different determinants on the same antigen is called **polyclonal.** Once antibodies that recognize discrete sites on a molecule of interest have been produced and linked to a flourescent dye, thay can be injected into the cells or tissues under study. Over the past decade, researchers increasingly have used a combination of monoclonal and polyclonal antibodies for antibody staining, particularly in immunofluorescent microscopy (Figure 2-2).

Alternatively, radiolabeled monoclonal antibodies can be used and the location of any antigen-antibody complexes that form in a sample detected by autoradiography. This approach has been used to localize the hormones epinephrine and norepinephrine within certain cells of the adrenal medulla, as described in Chapter 8 on glands. Monoclonal antibodies can be used not only to track down specific molecules within cells but also to purify them, as described in a later section. Such purified molecules are suitable for detailed studies on their structure and function.

The crucial advance that made antibody staining feasible was development of a method for producing large amounts of monoclonal antibody. Isolation and purification of a single type of monoclonal antibody from antiserum taken from animals exposed to the corresponding antigen is not practical, because each type of antibody is present only in very small amounts. Moreover, the B

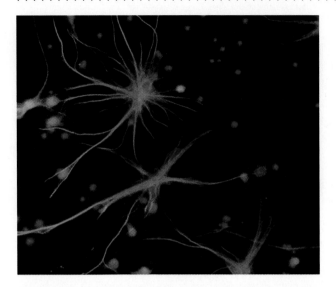

Figure 2-2 Both monoclonal and polyclonal antibodies are frequently used in antibody staining. In this immunofluorescent micrograph of rat spinal cord cultured 10 days, a mouse monoclonal antibody (green) and a rabbit polyclonal antibody (red) that is specific for a single protein, along with a blue fluorescent dye that binds DNA directly, are used. Here we see neurons (red), astrocytes (green), and DNA (blue). [Courtesy of Nancy L. Kedersha/Immuno Gen.]

lymphocytes (or B cells) that produce antibodies normally die within a few days, and thus cannot be grown for extended periods in culture. In the mid-1970s, G. Kohler and C. Milstein discovered that normal B cells could be fused with cancerous lymphocytes, called myeloma cells, which grow indefinitely in culture (i.e., they form an "immortal" cell line). The resulting hybrid cells, termed **hybridomas,** are spread out on a solid growth medium in a culture dish. Each cell grows into a **clone** of identical cells, with each clone secreting a single monoclonal antibody. Clones are then screened to identify those that secrete the desired antibody; these self-perpetuating cell lines can be maintained in culture and used to obtain large quantities of homogeneous monoclonal antibody (Figure 2-3). Although individual investigators can make and maintain their own hybridoma cell lines, many now choose to have specific monoclonal antibodies prepared by companies specializing in their production. (The next time you are in your university or college library, find the journal *Science* and take a look at the classified ads in the back.) The development of monoclonal antibody technology by Kohler and Milstein so revolutionized molecular studies that they received the Nobel Prize for their research.

Genetic Engineering

Genetic engineering encompasses various techniques for manipulating the genetic material of an organism. This approach is increasingly used in both agriculture and medicine, and it offers considerable promise for investigators in animal physiology. These techniques make it possible to produce large quantities of biologically important molecules (e.g., hormones) normally present at very low concentrations, animals with mutations that affect specific physiological processes, and animals that synthesize above- or below-normal amounts of specific gene products.

Genetic engineering begins with identification of the structural gene that codes for a specific protein within the DNA isolated from an organism of interest. For example, the gene that encodes human insulin can be identified in DNA isolated from human cells. The section of DNA containing the insulin gene of interest can be "clipped out" of the original very long human DNA strands and then inserted into a *cloning vector,* which is a DNA element that can replicate within appropriate host cells independently of the host cells' DNA. Insertion of a fragment of foreign DNA (e.g., the human insulin gene) into a cloning vector yields a **recombinant DNA,** which is any DNA molecule containing DNA from two or more different sources.

Bacterial *plasmids* are a common type of cloning vector. These are extrachromosomal circular DNA molecules that replicate themselves within bacterial cells. Under certain conditions a recombinant plasmid containing a gene of interest is taken up by the common bacterium *Escherichia coli,* a process called *transformation* (Figure 2-4). Normally, only a single plasmid molecule is taken up by any one bacterial cell. Within a transformed cell, the incorporated plasmid can replicate, and as the cell divides a group of identical cells, or clone, develops. Each cell in a clone contains at least one plasmid with the gene of interest. This general genetic engineering procedure, called *DNA* or *gene cloning,* can be used to obtain a DNA "library" consisting of multiple bacterial clones, each of which contains a specific gene from humans or other species. Several variations of DNA cloning are used depending on the size and number of the genes in the organism being studied.

Clonal populations for medicine and research
Under appropriate environmental conditions, the recombinant DNA in an "engineered" *E. coli* clone is transcribed into messenger RNA, which is used to direct synthesis of the encoded protein. Commercial companies, for example, grow *E. coli* cells carrying recombinant DNA containing the gene for human insulin or other hormones in huge vats; after the bacterial cells are harvested, large quantities of the human hormone can be isolated relatively easily. In the past, hormones needed for treating humans with endocrine disorders were extracted from the tissues of other mammals such as cows and pigs. Because hormones are present in quite low concentrations, this is a time-consuming and expensive process. Producing these hormones with genetically engineered bacteria has proven to be far less expensive and yields a purer product. Moreover, hormones isolated from other mammalian species often induce an immune response in humans, a complication not encountered with human hormones obtained from engineered bacteria.

Recombinant DNA technology is also a powerful tool in basic research on human genetic disorders. By isolating and studying genes associated with hereditary diseases, scientists can determine the molecular basis of these diseases. This will

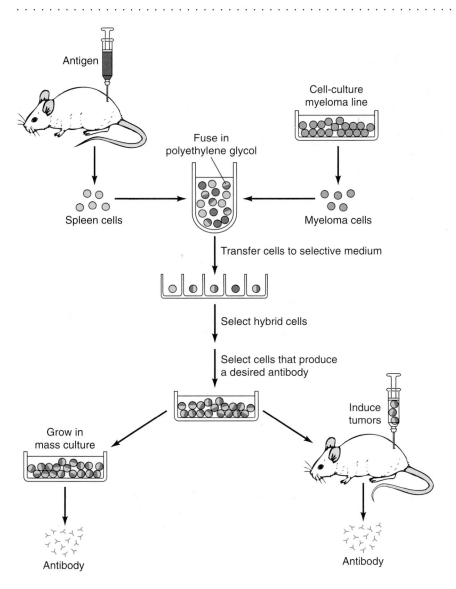

Figure 2-3 Hybridoma cell lines secrete "pure" (homogeneous) monoclonal antibodies. To prepare monoclonal antibodies, antibody-producing spleen cells first are fused with myeloma cells originally derived from B lymphocytes. The hybrid cells, or hybridomas, that secrete antibody specific for the protein of interest are separated out. They can be maintained in cell culture, where they secrete large quantities of the specific antibody, or injected into a host mouse, where they induce the production of the antibody.

certainly lead to better methods of controlling or even curing them. Over the last several years numerous laboratories worldwide have been engaged in a massive project to "map" the locations of all human genes on the long strands of DNA in human chromosomes and determine their nucleotide sequences. This Human Genome Project is providing invaluable data for researchers studying genetic diseases.

DNA cloning and recombinant DNA technology also form the basis for *gene therapy.* In this approach to treating those with genetic disorders, the normal form of the gene that is missing or defective is introduced into patients. For example, persons with cystic fibrosis have a defective *CFTR* gene and thus cannot produce the normal protein encoded by this gene. One result of this defect is production of a very thick mucus in the lungs' airways, which leads to potentially lethal breathing problems. Molecular biologists have engineered common cold viruses with the normal *CFTR* gene. When some cystic fibrosis patients were infected with an engineered cold virus, the viral particles carried the normal human gene into the patients' lung cells, where it became established. Subsequent synthesis of the normal gene product helped alleviate most of the symptoms of cystic fibrosis in the treated patients.

Considerable controversy has surrounded the use of genetic engineering in producing biochemical products, primarily because of fears that genetically modified microorganisms might escape into the environment and produce unexpected effects such as human or domestic crop diseases. However, many genetically engineered microorganisms are modified to make them unable to live outside of the chemical factory for which they were designed. Assuming you could modify anything about a bacterium's physiology or biochemistry (e.g., the temperature range it tolerates or the chemicals it uses for metabolic substrates), how would you go about ensuring that a genetically engineered bacterium could not flourish in the natural environment if it escaped?

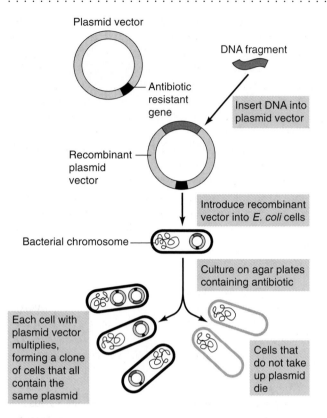

Plasmid vector

DNA fragment

Antibiotic resistant gene

Insert DNA into plasmid vector

Recombinant plasmid vector

Introduce recombinant vector into *E. coli* cells

Bacterial chromosome

Culture on agar plates containing antibiotic

Each cell with plasmid vector multiplies, forming a clone of cells that all contain the same plasmid

Cells that do not take up plasmid die

Figure 2-4 DNA cloning is a way to isolate and maintain individual genes. In the cloning procedure illustrated, a specific DNA fragment to be cloned is inserted into a plasmid vector, which also contains a gene conferring resistance to the antibiotic ampicillin. When the resulting recombinant plasmids are mixed with *E. coli* cells, a few cells take up a plasmid, which can replicate within the cells. If the cells are placed in media containing ampicillin, only those that have taken up the vector will grow. As each selected cell multiplies, it eventually forms a colony of cells (clone) all containing the same recombinant plasmid.

"Made-to-order" mutants

As mentioned in Chapter 1, **mutations** are permanent changes in the nucleotide sequence of DNA. Mutations, which can occur spontaneously or be induced experimentally, are duplicated and passed on to daughter cells at the time of cell division. Mutant genes can tell us a great deal about how physiological processes work. The specific disruption in a physiological process resulting from a single mutant gene can pinpoint the functions controlled by particular genes, information that may not be revealed by conventional physiological techniques.

For example, cardiovascular physiologists are producing and analyzing the effects of mutations in zebrafish to understand heart development. In research described by J-N. Chen and M. Fishman (1997), dozens of specific cardiovascular mutations have been produced in zebrafish. The process starts when adult zebrafish are exposed to powerful *mutagens*—compounds that produce permanent mutations in the germ cell line. Subsequent matings of the F_1 and F_2 generations lead to embryos with large numbers of mutations. Very rarely, an embryo will appear with just one specific mutation in a structure or process of interest. The Fishman group, for instance, has identified

mutations that cause a heart with abnormally thin ventricular walls and another with a constriction of the arterial outflow tract of the heart. Both of these conditions mimic human disease states.

Mutations often produce abnormal effects only in the homozygous state (i.e., when an individual receives a mutated form of a gene from each parent). Even when a mutation causes a lethal condition incompatible with long-term survival, it can be "preserved" in the parents, who are heterozygous for the mutation, carrying one normal and one mutated form of the gene. Each time these parents breed, some of the offspring will be homozygous and show the abnormal effects. Thus, the heterozygous parents are a "living gene library" of these mutations.

Transgenic animals

Transgenic animals are another type of genetically engineered organism with the potential for making great contributions to physiology. A **transgenic** animal is one whose genetic constitution has been experimentally altered by the addition or substitution of genes from other animals of the same or other species. Transgenic animals (especially, mice) are at the forefront of the menagerie of animal models that are helping researchers understand basic physiological processes and the disease states that results from their dysfunction.

Numerous techniques have been employed to produce transgenic animals. In one method, "foreign" DNA containing a gene of interest, called a *transgene*, is injected into a pronucleus of fertilized eggs (commonly from mice), which then are implanted into pseudopregnant females. At a relatively low frequency, the transgene is incorporated into the chromosomal DNA of the developing embryos, leading to offspring that carry the transgene in all their germ-line cells and somatic cells (Figure 2-5). Mice expressing the transgene then are mated to produce a transgenic line. This approach is used to add functional genes, either extra copies of a gene already present in the animal or a gene not normally present, leading to overexpression of the gene product. Subsequent analysis of the morphology and physiology of the transgenic animals can provide considerable insight into physiological processes that cannot easily be investigated in other ways.

Transgenic animals characterized by underexpression or complete lack of expression of a particular gene can be equally informative. M. R. Capecchi (1994) has reviewed a procedure for replacing a functional gene with a defective one, thereby producing so-called *knockout mice*. These mice cannot express the protein originally coded for by the replaced gene and thus lack the functions mediated by the missing protein. The molecular and genetic basis of physiological processes can be determined by examining the effect of such functional ablation of genes. Knockout mice are used extensively to unravel human physiological processes, because human and mouse genes are greater than 98% identical. To cite a couple of examples, researchers are investigating the normal genes that regulate

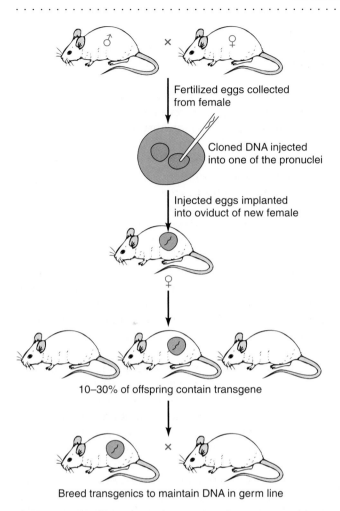

Fertilized eggs collected
from female

Cloned DNA injected
into one of the pronuclei

Injected eggs implanted
into oviduct of new female

10–30% of offspring contain transgene

Breed transgenics to maintain DNA in germ line

Figure 2-5 A transgenic animal is produced by adding or substituting genes from another animal of the same or different species. To introduce a transgene into mice, cloned "foreign" DNA is injected into fertilized eggs, which then are implanted into a female. A proportion of the viable offspring will retain the transgene, which can be maintained in the germ line by selective breeding.

early heart development in the embryo and the oncogenes responsible for some types of cancer in studies with knock-out mice.

CELLULAR TECHNIQUES

Understanding cells and cellular behavior is a goal of many experiments in physiology. With a knowledge of cellular behavior and communication, we can begin to understand how communities of cells function as tissues, and tissues as organs. Physiological analysis at the cellular level has been pursued most vigorously using several now-standard techniques. In this section we discuss three very common and productive cellular techniques: recording with microelectrodes, microscopy, and cell culture.

Uses of Microelectrodes and Micropipettes

Many experiments in cellular physiology make use of micropipettes or various types of microelectrodes. These tiny glass "needles," which can be inserted into tissues or even individual cells, are used to measure various properties of cells or inject materials into them. Although cellular physiologists employ these devices in a variety of ways, the technology used to make them is decades old. Essentially, a region in the middle of a glass capillary tube is heated to the point of melting. The ends of the tube are then pulled apart, which draws the soft spot in the glass down to an invisibly small diameter before it breaks and separates. Two micropipettes, each with a drawn-down tip as small as just a micron in diameter, are produced as a result. When a micropipette is filled with an appropriate solution, it can function as a microelectrode. Typically, a micropipette (or microelectrode) is mounted in a *micromanipulator,* a mechanical device that holds the pipette steady and allows its tip to be moved incrementally in three different planes.

Measuring electrical properties

Since neurons communicate via electrical signals, microelectrodes can be used to "eavesdrop" on their communication by measuring the electrical signals across the cell membrane and changes in these signals under different conditions. The microelectrodes used to measure the electric potential (voltage) across the cell membrane cause virtually no flow of current from the cell into the electrode. Thus little or no disruption of the nerve cell occurs even as its communication with neighboring cells is being detected.

A microelectrode for recording electrical signals from neurons or muscle cells is made by filling a micropipette with an ionic conducting solution (typically KCl) and connecting it to an appropriate amplifier. A second electrode connected to the amplifier is placed in the fluid or organism in the vicinity of the first electrode. When the tip of the first electrode is pushed through the cell membrane into the cytoplasm, it completes an electric circuit whose properties (voltage, current flow) can be measured. Since microelectrode recording techniques were introduced in the 1950s, our understanding about the electrical activities within a cell have increased dramatically.

One of the most revolutionary advances in microelectrode recording methodology is *patch clamping*. With this technique, the behavior of a single protein molecule constituting an ion channel can be recorded *in situ* (Latin for "in its normal place"), as illustrated in Figure 2-6. This method lies at the heart of the recent explosion of knowledge about membranes, including their channels and how they regulate the movement of materials across the membrane (see Chapters 4–6).

Measuring ion and gas concentrations

Specially constructed microelectrodes can be used to probe the intracellular concentration of common inorganic ions including H^+, Na^+, K^+, Cl^-, Ca^{2+}, and Mg^{2+}. Because cells use movements of ions across cell membranes to communicate and to do work, the magnitude, direction, and time course of ion movements provide important information about certain processes. Microelectrodes that measure the

A

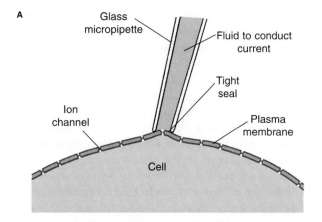

Glass micropipette

Fluid to conduct current

Tight seal

Ion channel

Plasma membrane

Cell

B

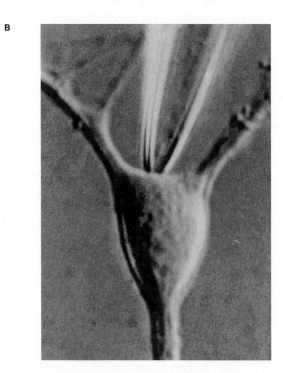

Figure 2-6 Patch-clamp recording permits determination of ion movement across a small patch of membrane containing transmembrane ion channels. **(A)** Diagram of patch clamp in place. When a fire-polished microelectrode is placed against the cell surface, a very high resistance seal forms between the electrode tip and the membrane. This tight seal allows direct measurement of the membrane features beneath the tip. Typically, only a few transmembrane ion channels lie beneath the tip, allowing the current flow through them to be measured directly. **(B)** Photomicrograph showing tip of a patch micropipette abutting the cell body of a nerve cell. The tip has a diameter of about 0.5 μm. [Part B from Sakmann, 1992.]

partial pressure of gases (e.g., O_2 and CO_2) dissolved in a fluid also are now available.

The tip of a microelectrode for measuring the concentration of a particular ion (e.g., Na^+) is plugged with an ion-exchange resin that is permeable only to that ion. The remainder of the electrode (the "barrel") is filled with a known concentration of the same ion. The electrical potential measured by the microelectrode when no current flows reflects the ratio of the ion concentrations on the two

sides of the ion-exchange barrier in the tip. Proton-selective microelectrodes are particularly useful for measuring the pH of blood and other body fluids.

Measuring intracellular and blood pressure

Microelectrodes are now being used to measure hydrostatic pressures within individual cells and microscopic blood vessels—indeed, in any fluid-filled space into which the tip of a microelectrode can be inserted. To understand the principle of such micropressure systems, let's consider a small blood vessel. A microelectrode, filled with at least a 0.5 M NaCl solution and mounted in a micromanipulator, is inserted into the vessel of interest. The higher pressure inside the vessel causes the interface between the plasma and the solution filling the electrode to move into the electrode. This results in increased resistance across the electrode tip, because the resistance of plasma is higher than that of the NaCl solution. The change in resistance is measured and is proportional to the change in blood pressure. A motor-driven pump associated with the micropressure system produces a pressure in the microelectrode that just offsets the pressure in the vessel. This opposing pressure keeps the interface at a stationary position; therefore, it is called a servo-null system. The required offsetting pressure generated in the micropressure system is then monitored with a conventional pressure transducer such as would be used for measuring blood pressure in much larger vessels.

Micropressure systems have greatly extended our knowledge of the development of cardiovascular function in developing embryos and larvae. These techniques have also allowed direct cardiovascular measurements in adults of very small animals like insects.

Microinjecting materials into cells

In addition to their use as microelectrodes, micropipettes also can be used to inject substances into individual cells. These substances may be active molecules that produce a measurable change in cell or tissue function. For instance, drugs that influence blood pressure and heart rate can be injected into very small blood vessels (e.g., those lining the shell of a bird egg) or into the microscopic heart of a frog embryo.

Alternatively, the injected substance may be a dye used to mark injected cells, helping to reveal cell processes or to trace cells as they divide. A classic variation of this technique involves horseradish peroxidase, an enzyme derived from the horseradish plant that forms a colored product from specific colorless substrates. When this enzyme is injected via a micropipette into the extensions (especially axons) of neurons, it is taken up and transported back to the neuron's cell body; subsequent injection of the substrate generates a colored "trail" between the injection site and cell body. By this technique, peripheral nerves can be traced back to their origin in the central nervous system, a task that would defy even the most skilled neuroanatomist using more traditional techniques.

Structural Analysis of Cells

Cellular function is dependent on cellular structure, reaffirming the central theme discussed in Chapter 1 that strong structure-function relationships govern animal physiology. Physiologists commonly use structural analyses at the cellular level to complement physiological measurements in order to discover how animals function. Such analyses depend on various types of *microscopy,* because animal cells are typically about 10–30 μm in diameter, which is well below the smallest particle visible to the human eye.

Light microscopy

Light microscopy, as its name implies, uses the photons of visible or near visible light to illuminate specially prepared cells. Under optimal conditions, the *resolution,* or resolving power, of light microscopes is a few microns; two objects that are located closer to each other than a microscope's resolution will appear as one. As the resolution of microscopes has been improved, our understanding of the structure of cells and their components has increased.

Since cells removed from a living animal rapidly die, tissue must be prepared quickly to prevent degradation of cellular constituents. *Fixation* is the addition of a specialized chemical (e.g., formalin) that kills the cells and immobilizes their constituents, typically by cross-linking amino groups of proteins with covalent bonds. The fixed cells then are treated with dyes or other reagents that *stain* particular cellular features, allowing visualization of the cells, which otherwise are colorless and translucent.

Fixation and staining of large blocks of tissue is impractical and does not allow visualization of individual cells. Typically, small blocks of tissue are cut into ultrathin sections or slices just 1–10 μm thick using a special knife called a *microtome.* Because most tissue is fragile even when fixed, it is embedded in some medium (e.g., wax, plastic, gelatin) to support it while it is sectioned. Such media surround and infiltrate the tissue and then harden to make sectioning possible. The tissue sections are then placed on glass slides for staining and subsequent viewing in a microscope (Figure 2-7A). In some instances, tissue embedding compromises the structure of the cell or its contents such that they can no longer be stained or labeled with special compounds prior to viewing. An alternative method is to freeze the tissue rather than embed it, allowing the ice to provide support for the tissue as it is subsequently sectioned. Once prepared, tissue is viewed with a compound optical microscope, the simplest type of light microscope (Figure 2-7B).

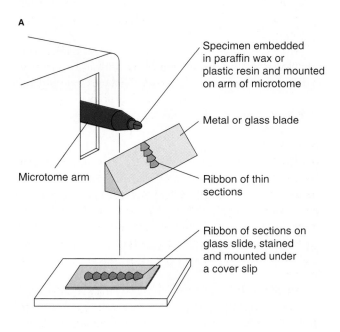

A

Specimen embedded in paraffin wax or plastic resin and mounted on arm of microtome

Metal or glass blade

Microtome arm

Ribbon of thin sections

Ribbon of sections on glass slide, stained and mounted under a cover slip

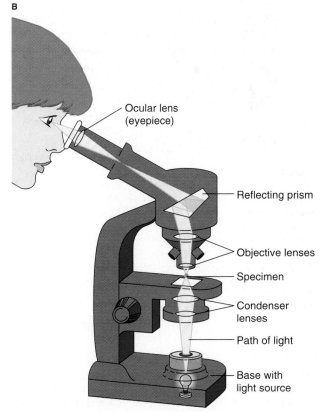

B

Ocular lens (eyepiece)

Reflecting prism

Objective lenses

Specimen

Condenser lenses

Path of light

Base with light source

Figure 2-7 Specimens are prepared for light microscopy by cutting them into thin sections and staining. **(A)** Cells and tissue removed from living organisms first are fixed to preserve their structure and then cut into thin sections using a metal or glass knife. These sections are mounted on a glass slide, where they can then be stained for subsequent viewing through a compound light microscope. **(B)** The compound optical microscope transmits light vertically up through a condenser lens, the specimen on the slide, an objective lens, and finally the ocular lens in the eyepiece, from which the specimen is viewed. [Adapted from Lodish et al., 1995.]

As improvements have been made in the available optics, staining techniques also have improved. Many organic dyes, originally developed for use in the textile industry, were discovered through trial and error to selectively stain particular cellular constituents. Some of these dyes stain according to the charge, such as hematoxylin, which marks negatively charged molecules like DNA, RNA, and acidic proteins. However, the basis of the specificity of many dyes is not known.

Staining with fluorescent-labeled reagents, rather than traditional dyes, increases the sensitivity of visualization. Fluorescent molecular labels absorb light of one wavelength and emit it at another, longer wavelength. When a specimen treated with a fluorescent reagent is viewed through a *fluorescence microscope,* only those cells or cellular constituents to which the label has bound are visualized (Figure 2-8). Probably the most common and useful type of fluorescence microscopy is *immunofluorescence microscopy* in which specimens are treated with fluorescent-labeled monoclonal and polyclonal antibodies. A good example of the images obtained with this technique is shown in Figure 2-2.

Because immunofluorescence microscopy gives poor results with fixed thin sections, this technique usually is applied to whole cells. However, the images obtained by standard fluorescence microscopy of whole cells represent a supposition of emitted light coming from labeled molecules located at many depths in the cell. For this reason, the images often are blurred. The *confocal scanning microscope* eliminates this problem, providing sharp images of fluores-

cent-labeled specimens without the need for thin sectioning. In this microscope, the specimen is illuminated with exciting light from a focused laser beam, which rapidly scans different areas of the specimen in a single plane. The light emitted from that plane is assembled by a computer into a composite image. Repeated scanning of a specimen in different planes provides data with which the computer can then create serial sections of the fluorescent images. Figure 2-9 compares the images obtained with conventional and confocal fluorescence microscopes.

Visualization by other types of microscopy depends on the specimen changing one or more properties of the light passing through the tissue on the slide, rather than on fixation and staining. Since these methods do not re-

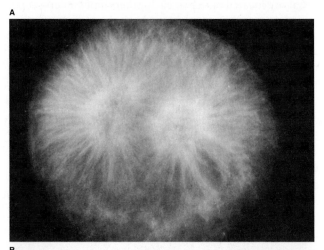

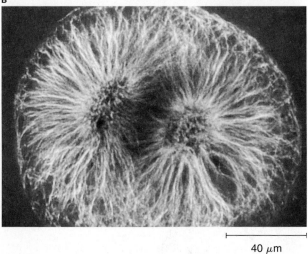

40 µm

Figure 2-8 A specimen stained with a fluorescent label is viewed through a fluorescence microscope, which produces an image only of structures that bind the label. The incident light source is passed through an exciter filter that passes blue light (450–490 nm) to provide optimal illumination for the specimen. The incident light is directed towards the specimen by a beam-splitting mirror that reflects light below 510 nm downwards but transmits light above 510 nm upwards. The fluorescent signals emitted from the labeled specimen pass upward through a barrier filter that removes unwanted fluorescent signals not corresponding to the wavelengths emitted by the label used to stain the specimen.

Figure 2-9 Conventional and confocal microscopy provide different images of biological material. These photomicrographs are of a lysed mitotic fertilized egg from a sea urchin. A fluorescein-tagged antibody was used to bind an antibody for tubulin, a major structural component of the mitotic spindle. **(A)** Conventional fluorescence microscopy shows a blurred image as a result of fluorescein molecule above and below the plane of focus. **(B)** Confocal microscopy, which detects fluorescence only from within the plane of focus, produces a much sharper image of the same sea urchin egg. [From White et al., 1987.]

Incident light source

Exciter filter

Eyepiece

Barrier filter

Dichromatic filter

Objective lens

Specimen

quire staining, they can be used on living tissue, provided it is thin enough to allow sufficient light to pass through. *Bright field microscopy* (Figure 2-10A) reveals few details compared to *phase-contrast microscopy*, in which the image has varying degrees of brightness and directness due to differential light refraction by different components of the specimen (Figure 2-10B). In *Nomarski microscopy*, also called *differential-interference contrast microscopy*, an illuminating beam of plane polarized light is split into closely parallel beams before it passes through the tissue specimen and the exiting beams are reassembled into a single image. Slight differences in the refractive index or thickness of adjacent parts of the specimen are converted into a bright image, if the beams are in phase when they recombine, or a dark image, if they are out of phase. The final image gives an illusion of depth to the specimen (Figure 2-10C). In *dark-field microscopy*, light is directed towards the specimen from the side so the observer sees only light scattered from cellular constituents. The image therefore appears as if the specimen has numerous sources of light within it.

In addition to direct viewing through the microscope, images can be stored electronically after collection by digital or video cameras. With a digital camera, a color image is collected in its entirety on a two-dimensional array of photosensitive elements. Although digital cameras provide a very high resolution, the required light intensity can be high. Alternatively, a video camera, which has lower light requirements, can be used to sample the image according to a preset scanning pattern. Because of the high light sensitivity of the video camera, it permits viewing of cells for long periods without associated light damage. Such image intensification is particularly important for viewing live cells that contain fluorescent labels, which can be toxic to cells at high light intensities.

Electron microscopy

For all imaging devices, the limit of resolution is directly related to the wavelength of the illuminating light. That is, the shorter the wavelength of the illumination, the shorter the minimal distance between two distinguishable objects (i.e., the greater the resolution). In electron microscopy, a high-velocity electron beam, rather than visible light, is used for illumination. Because the wavelength of electron beams is much shorter than that of visible light, electron microscopes have much better resolution. Indeed, modern transmission electron microscopes typically have a resolution of 0.5 nm (5 angstroms, Å), whereas light microscopes have a resolution of no less than about 1000 nm (1 μm). Because the effective wavelength of an electron beam decreases as its velocity increases, the limit of resolution of an electron microscope depends on the voltage available to accelerate the illuminating electrons.

The *transmission electron microscope* forms images by sending electrons through a specimen and focusing the resulting image on an electron-sensitive fluorescent screen or photographic film (Figure 2-11). The electron beam is modified by magnets, which bring the electrons into alignment and focuses them on the specimen, much like the condenser lens in a compound light microscope. Image formation depends on the differential scattering of electrons by different regions of the specimen; scattered electrons cannot be focused by the objective lenses and thus do not impinge on the viewing screen. Because the electron beam passes through an unstained sample nearly uniformly, little differentiation of its components is possible without staining. The most common stains for electron microscopy are salts of heavy metals (e.g., osmium, lead, or uranium), which increase electron scattering. In photographs of an electron microscope image, components stained with such electron-dense materials appear dark.

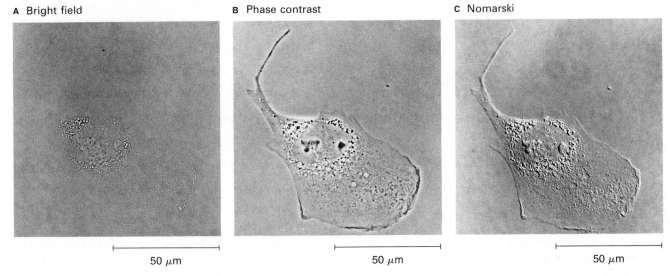

A Bright field   **B** Phase contrast   **C** Nomarski

50 μm   50 μm   50 μm

Figure 2-10 Different light microscopic techniques give strikingly different images. **(A)** Bright-field image of a cell, typical of that obtained with an unstained specimen viewed through a compound light microscope, exhibits little contrast and few details. **(B)** Phase-contrast image heightens the visual contrast between different regions of the specimen. **(C)** Nomarski (differential-interference contrast) microscopy provides the sensation of depth to the image. [Courtesy of Matthew J. Footer.]

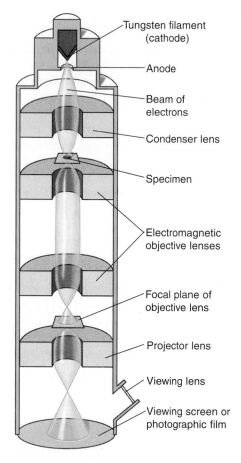

Tungsten filament (cathode)

Anode

Beam of electrons

Condenser lens

Specimen

Electromagnetic objective lenses

Focal plane of objective lens

Projector lens

Viewing lens

Viewing screen or photographic film

Figure 2-11 Electron microscopes share features such as lenses with compound optical microscopes, but use an electron beam rather than a light beam to illuminate the specimen. In a transmission electron microscope, shown here, an image is formed by passing electrons through an object and projecting them onto a fluorescent screen. In a scanning electron microscope, electrons reflected from the surface of a specimen coated with a reflective metal film are collected by lenses and viewed on a cathode ray tube.

Since air would deflect the focused electron beam aimed at the sample, the specimen must be held in a vacuum during imaging. Specimens must be very well fixed to preserve their biological structure during viewing in the electron microscope. Glutaraldehyde is used to covalently cross-link proteins and osmium tetroxide to stabilize lipid bilayers. After fixing, the specimens are infiltrated with a plastic resin. Thin sections cut from the resin block then are stained and finally placed on a metal grid in the transmission electron microscope. Specimens must be sectioned into extremely thin slices (50–100 nm thick) to allow penetration by the electron beam. Only diamond or glass knives are sharp enough to cut tissue sections into such thin slices. Glass knives are formed by breaking on the diagonal a 2.5-cm glass square that is about 5 mm thick. Because glass is actually a slow-moving liquid, the edge formed is only sharp enough to cut tissue for a few hours before molecular flow of glass dulls the edge. Although extremely expensive, diamond knives do not suffer from this problem and thus are the preferred tool for cutting thin sections.

The exquisite detail available from the transmission electron microscope can provide important insights into the structure of biological tissue (Figure 2-12A). Unfortunately, the size of the specimen that can be examined is very small, because it must be thin sectioned. Consequently, it is difficult to develop an understanding of the three-dimensional structure without the truly tedious procedure of reconstructing an image from a series of individual sections. Development of various techniques for preparing specimens for transmission electron microscopy have extended the range of objects that can be visualized and the information available from images.

The *scanning electron microscope*, like the transmission electron microscope, uses electrons rather than photons to form images of the specimen. However, the scanning electron microscope collects electrons scattered from the surface of a specially prepared specimen. This instrument provides excellent three-dimensional images of the surface of cells and tissues, but it cannot reveal features beneath the surface (Figure 2-12B). Before examination in a scanning electron microscope, the specimen is coated in an extraordinarily thin film of a heavy metal like platinum. The tissue is then dissolved away with acid, leaving a metal replica of the tissue's surface, which is viewed in the microscope. Scanning electron microscopes have a resolution of about 10 nm, considerably less than the resolution of transmission electron microscopes.

Cell Culture

The rearing of cells *in vitro* (Latin for "in glass") in glass or plastic containers is known as *cell culture*. This technique has revolutionized our ability to study cells and the physiological processes they support at the tissue and organ level. Historically, explants (small pieces of tissue removed from a donor animal) were kept alive and grown in a flask filled with an appropriate mixture of nutrients and other chemicals. Today, the most common procedure is to break up (dissociate) small pieces of tissues and then suspend the dissociated cells in a nourishing chemical broth in which they grow and divide as separate entities.

Successfully growing cells *in vitro* requires the right *culture medium*, the liquid in which the cells are suspended. Up until the early 1970s, cells from all animals were routinely grown in liquid medium consisting largely of either *serum* (a clear component of blood plasma) from horses or fetal calves or of an unrefined chemical extract made from ground-up chick embryos. However, these media were poorly defined chemically, containing numerous unidentified compounds. Moreover, it was difficult to predict whether cells from a particular source would grow in one of these media, or what components might be added if the first attempt was unsuccessful. Growing cells *in vitro* was largely a matter of trial and error (and luck). Today, defined culture media manufactured according to precise chemical formulas are available for research. However, the successful culture of many cell types requires addition of a small trace (less than 5%) of horse serum to such defined media. This observation

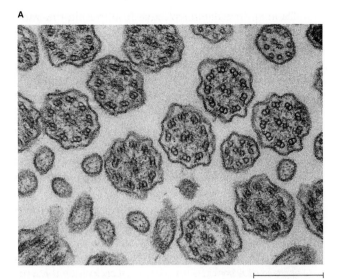

0.2 μm

Figure 2-12 Transmission electron microscopy provides a view of the interior of biological tissue, while scanning electron microscopy emphasizes surface features. **(A)** Transmission electron micrograph of cilia in the mouse oviduct. **(B)** Scanning electron micrograph of cilia in the mouse oviduct. [Courtesy of E. R. Dirksen.]

suggests that some growth factor in blood is necessary for the growth and division of animal cells *in vitro* (Figure 2-13).

Even with the availability of defined culture media, growing animal cells *in vitro* is a demanding technique. Normal animal cells generally can grow only for a few days *in vitro*, then stop multiplying and eventually die out. A relatively homogeneous population of such cells is referred to as a *cell strain*. Cultured cell strains are useful for many kinds of experiments but their limited life span makes them unsuitable for other studies. In addi-

tion, many types of animal cells have not yet been successfully cultured. However, the list of cells that can be cultured is constantly growing, the result of refinements in media and culture techniques. For example, cell strains derived from the following tissues and organs can be grown in culture:

- Bone and connective tissue
- Skeletal, cardiac, and smooth muscle
- Epithelial tissue from liver, lung, breast, skin, bladder, and kidney
- Some neural tissue
- Some endocrine glands (e.g., adrenal, pituitary, islets of Langerhans in pancreas)

In contrast to normal animal cells, cancer cells commonly exhibit rapid, uncontrolled growth in the body and are capable of indefinite growth in culture. Treatment of some normal cultured cells with certain agents may cause transformation, a process that makes them behave like cancer cells that have been isolated from tumors. Such transformed cells also can be cultured indefinitely. Homogeneous populations of such "immortal" cells are termed *cell lines*. Although normal cells differ from cancer cells and transformed cells in many ways, cell culture of the latter has permitted certain types of studies that are not feasible with primary cell cultures of normal cells.

Cell culture has numerous potential uses in animal physiology. New devices such as silicon wafer sensors for measuring acidity and other variables have been combined with cell culture techniques to provide important insights into cellular and organismal physiology. For example, the hormonal regulation of H^+ secretion from a variety of cells grown *in vitro* can be studied by stimulating the cultured cells with agonists and antagonists and measuring the changes in the rate of medium acidification. This approach has also been used to study tissues and organs with unusual rates or properties of H^+ secretion, such as the swim bladder tissues of fishes.

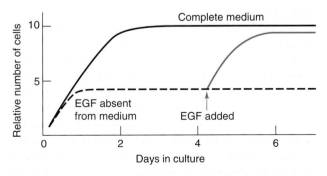

Figure 2-13 Cells grown in culture often require specific factors to stimulate maximal rates of division and growth. In the culture indicated, maximal cell numbers are reached only in the presence of epidermal growth factor (EGF). Addition of EGF (arrow) to a culture lacking this substance results in immediate further growth of the cell colony. [Adapted from Lodish et al., 1995.]

Smooth, skeletal, and cardiac muscle cells can all be cultured *in vitro*. How would you determine if an individual muscle cell that you were observing under a microscope as it lay in a culture dish was still actually capable of contracting, or had lost its ability to contract as a consequence of being grown in cell culture?

BIOCHEMICAL ANALYSIS

Most biochemical processes occur in aqueous solution and require the exchange of gases. For this reason, physiologists often need to measure the chemical composition of the fluid in various body compartments and/or the concentration of its constituents. For example, to assess whether a crab can regulate its internal salt concentration when swimming in dilute estuarine waters, a physiologist would need to know the salt concentration in the water surrounding the crab, in the crab's hemolymph (blood), and in the urine produced by the crab. With these data, the ability of the crab to maintain homeostasis can be evaluated. Biochemical analyses of biologically relevant fluids, gases, and structures typically are based on some physical or chemical attribute of the materials of interest (e.g., Na^+ in the crab's urine). The substantial increase in the sensitivity and accuracy of such measurements in the recent past has allowed physiologists to ask questions about subtle physiological functions that previously could not even be measured.

Both qualitative and quantitative analyses can be important for physiological studies. The objective of the first is to determine the *composition* of some fluid or structure — that is, the elements, ions, and compounds that compose it. The objective of the second analytical approach is to measure the *concentration* of particular substances in the fluid or structure of interest. Many analytical instruments and techniques provide both composition and concentration data.

Measuring Composition: What Is Present

Numerous time-honored and emerging methods are available for measuring chemical composition. Sometimes, animal physiologists are interested only in knowing whether a particular substance (e.g., ammonia or hemoglobin) is present in a sample. At other times, they may want to identify all the different proteins or carbohydrates or other molecular species in a sample. In other words, the nature of the problem being studied determines which compositional data are relevant. Rarely is a biological sample subjected to a full compositional analysis similar to that assigned in a beginning chemistry laboratory course.

A wide variety of *colorimetric assays* have been developed for determining the presence or absence of specific substances in a solution. These assays depend on the substance of interest undergoing a chemical reaction that changes its ability to absorb visible light or ultraviolet (UV)

radiation at different wavelengths. As a result, the transmission of light or UV radiation by the solution changes, which can be detected by a *spectrophotometer*. Many biochemical assays employ an enzyme that catalyzes a reaction involving the substance of interest. For example, a common assay for lactate (a product of the anaerobic metabolism of glucose) makes use of an enzyme that converts lactate into products with different UV absorption properties. To perform this assay, a solution suspected of containing lactate is placed in a small reaction vial along with the enzyme and other reaction components. After a short time, the vial is placed in a spectrophotometer, and the UV transmission of the solution is measured. The transmission of a control reaction vial lacking the enzyme also is measured. A difference in UV transmission through the control and assay vial indicates that lactate is present in the sample.

Chromatography is a widely used technique for separating proteins, nucleic acids, sugars, and other molecules present in a mixture. In its simplest form, *paper chromatography,* the components of the sample move at different rates in chromatography paper, depending on their relative solubility in the solvent, as illustrated in Figure 2-14A. In order to visualize the separated components, the chromatogram commonly is sprayed with a colorimetric reagent that produces a visible color with the components of interest. More complex mixtures can be separated by *column chromatography,* in which the sample solution is passed through a column packed with a porous matrix of beads (Figure 2-14B). The different components of the sample pass through the column at varying rates, and the resulting fractions are collected in series of tubes. Depending on the nature of the sample, different assays are used to determine the presence of the separate components in the fractions collected.

Many different kinds of matrix are employed in column chromatography depending on the composition of the solution being separated. For example, matrices are available that sort components according to their charge, size, insolubility in water (hydrophobicity), or binding affinity for the matrix. The last type of matrix is used in *affinity chromatography,* in which the matrix beads are coated with molecules (e.g., antibodies or receptors) that bind to the component of interest. When a sample mixture is applied to the column, all the components except the one recognized by the affinity matrix pass through. This is a very powerful technique for purifying proteins and other biological molecules present at very low concentrations.

Electrophoresis is a general technique for separating molecules in a mixture based on their rate of movement in an applied electric field. The net charge of a molecule, as well as its size and shape, determines its rate of migration during electrophoresis. Small molecules such as amino acids and nucleotides are well separated by this technique, but by far the most common use of electrophoresis is to separate mixtures of proteins or nucleic acids. In this case, the sample is placed at one end of an agarose or polyacrylamide gel, an inert matrix with fixed diameter pores that impedes or allows migration of molecules when an electric field is applied. Protein mixtures usually are exposed to SDS, a negatively charged detergent, before and during electrophoresis. The rate of migration of the resulting SDS-coated proteins in the gel is proportional to their molecular weight; the lower the molecular weight of a protein, the faster it moves through the gel (Figure 2-15). When a protein-binding stain is applied to the gel, the separated proteins are visualized as distinct bands.

Three slightly different, but basically similar, procedures employing gel electrophoresis are used to separate and detect specific DNA fragments, messenger RNAs (mRNAs), or proteins. Each of these procedures involves three steps (Figure 2-16):

1. Separation of the sample mixture by gel electrophoresis
2. Transfer of the separated bands to a nitrocellulose or other type of polymer sheet, a process called *blotting*
3. Treatment of the sheet (or blot) with a "probe" that reacts specifically with the component of interest

The first of these procedures to be developed, named *Southern blotting* after its inventor Edward Southern, is used to identify DNA fragments containing specific nucleotide sequences. *Northern blotting* is used to detect a particular mRNA within a mixture of mRNAs. Specific proteins

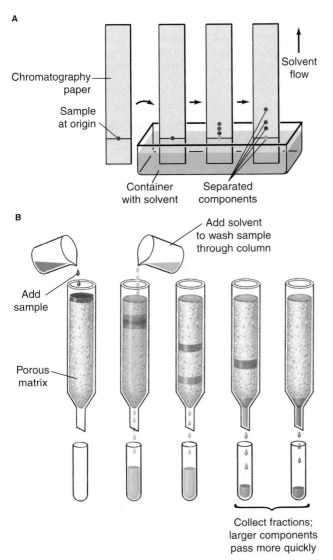

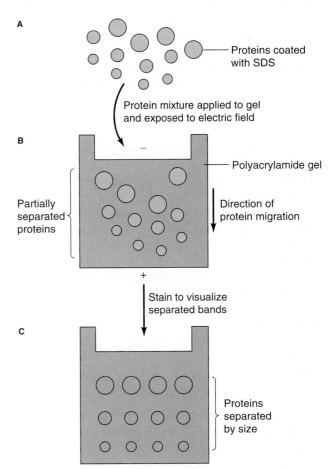

Figure 2-14 Chromatography is a powerful technique for separating the components of a mixture in solution. **(A)** In paper chromatography, the sample is applied to one end of a piece of chromatographic paper and dried. The paper is then placed into a solution containing two or more solvents, which flow upward through the paper via capillary action. Different components of the sample move at different rates in the paper because they have different relative solubilities in the solvent mixture. After several hours, the paper is dried and stained to determine the location and relative amounts of the separated components. **(B)** In column chromatography, the sample is applied to the top of a column that contains a permeable matrix of beads through which a solvent flows. Then solvent is pumped slowly through the column and is collected in separate tubes (called fractions) as it emerges from the bottom. Components of the sample travel at different rates through the column and are thus sorted into different fractions.

Figure 2-15 Gel electrophoresis separates the components of a mixture based on their charge and their mass. Proteins commonly are separated by SDS-polyacrylamide gel electrophoresis, as illustrated here. **(A)** SDS, a negatively charged detergent, is added to the sample to coat the proteins. **(B)** The sample then is placed in a well in the polyacrylamide gel and an electric field is applied. Small proteins move more easily and faster along the length of the gel than larger ones. **(C)** After a period of time, the proteins in the mixture are separated into bands composed of proteins of different sizes. These can be visualized by various protein-staining reagents. [Adapted from Lodish et al., 1995.]

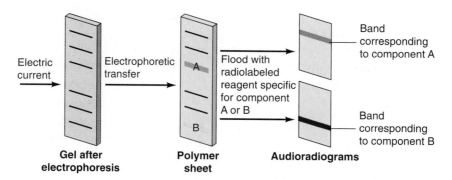

Figure 2-16 Southern, Northern, and Western blotting are similar procedures for separating and identifying specific DNA fragments, mRNAs, and proteins, respectively, within a mixture. In each method, the components of a sample first are separated by gel electrophoresis; the separate bands are transferred to a polymer sheet, which then is flooded with a radiolabeled reagent specific for the component of interest. The presence and in some cases the quantity of the labeled component is determined by autoradiography. See Table 2-1 for details of each procedure.

within a complex mixture can be detected by *Western blotting*, also known as immunoblotting. (As yet, there are no Eastern, South-Western, etc., blots, but it is probably just a matter of time.) Table 2-1 summarizes the unique features of the three blotting procedures.

Many of the common methods of determining composition are applicable to solutions but not gases. The *mass spectrometer*, however, can distinguish the different gases composing a gaseous mixture based on their mass and charge. Animal physiologists most often use this instrument to determine the composition of respiratory gases while an animal is resting or exercising in an experimental setting. Figure 2-17 illustrates the basic design of a mass spectrometer. The gas sample first is ionized by intense heating and passage through an electron beam. The charged ions then are focused and accelerated by an electric field into an analyzer where the beam of ions is deflected by either an applied magnetic field or by passage through tuned rods emitting specific radio frequencies that will deflect ions. The lighter the ion mass and smaller its charge, the smaller will

be the deflection of the ions from a standard path as they head towards the analyzer. The degree of deflection is detected by an array of detectors, which then allows determination of the presence and quantity of gases in the gas sample introduced into the mass spectrometer.

The various techniques for measuring chemical composition described in this section are widely used by physiologists, but many others also are employed in physiological research. To learn about additional methods and further details about those discussed here, you can refer to chemistry and biochemistry textbooks.

Measuring Concentration: How Much Is Present

Most instruments or analytical techniques used to determine the composition of a fluid or gas mixture also provide data about the concentration of the components present. For example, the degree of color change produced in a colorimetric assay depends on how much of the substance being measured is present in the sample. Likewise, the output signal from a mass spectrometer depends not only on the

TABLE 2-1
Electrophoretic blotting procedures

	Molecules detected	Separation and detection procedure*
Southern blotting	DNA fragments produced by cleavage of DNA with restriction enzymes	Electrophorese mixture of dsDNA fragments on agarose or polyacrylamide gel; denature separated fragments into ssDNA and transfer bands to polymer sheet; use radiolabeled ssDNA or RNA to label fragment of interest; detect labeled band with autoradiography
Northern blotting	Messenger RNAs	Denature sample mixture; electrophorese on polyacrylamide gel and transfer separated bands to polymer sheet; use radiolabeled ssDNA to label mRNA of interest; detect labeled band with autoradiography
Western blotting	Proteins	Electrophorese sample mixture on SDS-polyacrylamide gel and transfer separated bands to polymer sheet; use radiolabeled monoclonal antibody to label protein of interest; detect labeled band with autoradiography[†]

*dsDNA = double-stranded DNA; ssDNA = single-stranded DNA.

[†]If radiolabeled monoclonal antibody is not available, then the band containing the antibody-protein complexes can be detected by adding a secondary antibody that binds to any monoclonal antibody. This secondary antibody is covalently linked to an enzyme, such as alkaline phosphatase, that catalyzes a colorimetric reaction. When substrate is added, a colored product forms over the band with the protein of interest, generating a visible colored stain in this region of the blot.

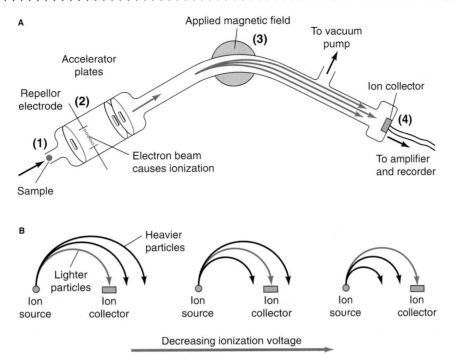

Figure 2-17 The identity of gases in a mixture and their concentrations can be determined by mass spectrometry. **(A)** The fixed-collector mass spectrometer, which detects how much an ionized sample is deflected by an imposed magnetic field, has four essential parts. First is a carefully constructed inlet device (1) through which the sample is delivered to the system with a constant viscous flow. Second is an ionization chamber (2), kept under vacuum and at high temperature (about 190°C), where the sample passes through an electron beam and is accelerated via application of an electric field. The gas molecules leave this chamber as negatively charged ions. Third is an analyzer tube in which the accelerated beam of ions are subjected to a magnetic field that causes the ions to flow in a curved path (3). Finally, the ion beam is detected using an ion collector situated at the end of the analyzer tube (4). The extent to which an ion is bent in the applied magnetic field depends on the strength of the field and the ion's mass, charge, and velocity. Only those ion species that are bent so their paths parallel the sides of analyzer tube will reach the ion collectors and be detected. **(B)** At a constant magnetic field, the specific type of particles detected in the mass spectrometer is determined by the strength of the ionization voltage, which can be varied. The lower the ionizing voltage, the heavier is the particle detected. [Adapted from Fessenden and Fessenden, 1982.]

types of gases present in a mixture, but how much of each is present. Thus the output signal produced by an analytical instrument—be it a transmission spectrophotometer, densitometer, or mass spectrometer—is directly related to the concentration of the substance responsible for the signal.

Typically, the analytical technique being used to determine the concentration of a particular substance is carried out on several samples of the substance at different known concentrations; the output signals then are plotted against the concentrations, yielding a *standard curve*. The actual concentration of an experimental sample corresponding to the output signal it produces is determined by comparison with this standard curve.

EXPERIMENTS WITH ISOLATED ORGANS AND ORGAN SYSTEMS

All animals have several different major organ systems that must be coordinated and controlled to help maintain homeostasis. As we'll examine in later chapters, the functions of these organ systems are regulated primarily by neural and/or hormonal inputs. To understand physiological control mechanisms, the key controlling inputs and their sources must be characterized. In many cases, this is difficult, if not impossible, to do by studying the intact organs *in situ;* instead, experiments are conducted on isolated organs removed from the animal by surgery and maintained in an *in vitro* artificial environment. Two examples will illustrate the power of this experimental approach.

When the heart of almost all vertebrates, including mammals, is isolated and placed in a bath of saline, it continues to beat and perform work by pumping saline or other fluid supplied to it. The isolated vertebrate heart will continue to beat in the absence of neural input if it is kept at an appropriate temperature and is perfused with an oxygenated solution that has the correct ionic composition and contains an energy source such as glucose. With the heart isolated, physiologists can measure the effect of chemical stimulation by drugs and hormones or electrical stimulation of nerves within the heart on the heart rate, amplitude, flow rate, and mechanical movements. Experiments performed on isolated hearts have been fundamental in advancing our knowledge of the cardiovascular system.

A second example is the vertebrate pineal gland, a small organ found at the top of the vertebrate brain. The pineal, which plays a key role in regulating daily (circadian) rhythms in physiological processes, is sensitive to light-related stimuli and releases various amounts of regulatory chemicals into the bloodstream as a function of the time of day. When the pineal gland is isolated and placed into an appropriate culture system, it continues to exhibit a circadian rhythm. Direct experimentation with this *in vitro* preparation has provided answers to specific questions concerning pineal gland regulation of physiological systems.

OBSERVING AND MEASURING ANIMAL BEHAVIOR

Scientists studying animal physiology frequently supplement their experiments with observations of animal behavior. Useful behavioral experiments are difficult to perform, however, because the animals must be in an appropriate physiological state (e.g., breeding, rearing young, digesting a meal, to name a few). Further, the experiment must exploit natural behavioral tendencies in the animal. Despite these difficulties, experimental methods to control and stimulate specific behavioral states can provide important insights into physiological processes that are not always amenable to direct physiological investigation. The prerequisite for such experimentation is a thorough knowledge of the natural behavior of the animal in its habitat.

The Power of Behavioral Experiments

Research in the 1950s and 1960s on the retrieval behavior of ground-nesting birds illustrates how behavioral studies can contribute to physiological knowledge. K. Z. Lorenz and N. Tinbergen discovered that geese not only recognized their eggs and recovered them if they lay outside the nest, but also would retrieve a wide variety of objects (e.g., grapefruits, light bulbs, baseballs) laying near their nest. Tinbergen and his students subsequently conducted ingenious experiments with gulls in which they offered pairs of objects to the birds and recorded which one was recovered first. By exploiting the process of pairwise comparison, they could define the properties that gulls use to choose what to retrieve. Although the birds would retrieve many different objects, these experiments showed that real eggs are always preferred over unnatural objects. The relative size, color, and speckling of an egg were found to contribute independently to the likelihood that an egg would be retrieved. Taken together, these experiments revealed that for gulls, eggs provide a powerful natural stimulus that induces specialized retrieval behavior. Armed with knowledge of the exact properties of the stimuli causing this behavior, physiologists have been better able to conduct physiological experiments about the nature of vision in birds.

Behavioral experiments often analyze the total time the animal under study spends performing each behavior and the temporal sequence of behaviors. These data, in conjunction with information about the behavior of other animals and key environmental variables, frequently reveal how closely behavior is related to the internal state of the animal. The great majority of information about animal behavior collected in this way pertains to reproduction and feeding, two of the most important behaviors engaged in by any animal; both reproductive and feeding behaviors are greatly affected by an animal's physiological state. Careful observations usually can reveal which behavioral patterns of one individual influence another and may suggest why this might be so. For example, in stickleback fish, the display of a red belly by a male signals to other male sticklebacks that he is defending a nest and to females that he is interested in spawning. Thus, the meaning of this signal depends on the sex of the receiver. The red belly arises from physiological processes triggered by the onset of the breeding season. The coordination between behavior and physiology in this species was investigated using behavioral analysis to guide physiological investigation.

Methods in Behavioral Research

A variety of instruments are used to record and analyze the physiological basis of specific behavioral acts. In some experiments, high-speed video cameras are used in conjunction with electrophysiological detectors of neural or muscular activity to capture simultaneously both the behavior and its physiological underpinnings. Since behavioral acts of interest are often rapid and fleeting, these events typically are recorded at high speed and the video tape played back at slow speed to aid analysis. The use of x-ray cameras allows examination of the interaction of skeletal components during specific behaviors (e.g., feeding, running on a treadmill). As in so many other aspects of physiology, the availability of inexpensive, fast computers with ever-increasing data storage capabilities has revolutionized the acquisition and analysis of data.

Figure 2-18 illustrates how many of the commonly used techniques to study animal behavior and its underlying physiological processes can be brought to bear on a single behavior—the feeding strike of a venomous snake. To discover how such a strike proceeds, the motion of the body and jaws must be related to the forces exerted by contraction of the jaw muscles. The rapid strike is recorded in two views, dorsal and lateral, using a video camera viewing the animal directly and via a mirror set at 45° above the snake. Quantification of the position of the animal is possible because of a grid image in the background that is included in the video images. The snake is placed on a platform that records the force exerted along three orthogonal axes. By measuring this set of external parameters, the investigator can record the forces associated with the snake's movement across the surface. The force exerted by the jaws of the snake are recorded by a strain gauge mounted on the head, and muscle activity is measured by electrodes into the four lateral jaw muscles. All of the data are recorded on both tape and in a computer using data-acquisition hardware and software.

The values of the measured variables are typically displayed as a function of time and related to the behavioral analysis recorded on videotape. Data from such an experiment reveals how contraction of the muscles results in positioning of the fang and closing of the jaws around the prey.

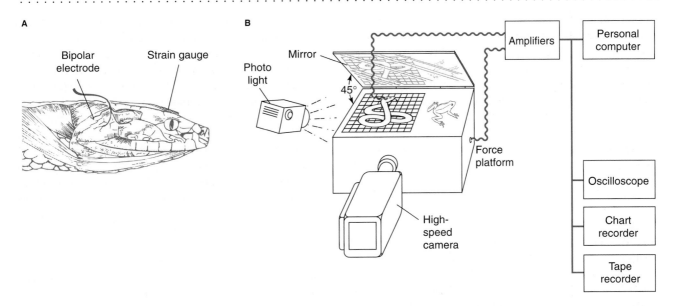

Figure 2-18 The feeding strike of a venomous snake can be analyzed to determine the muscles used and the pattern in which they contract. **(A)** To record electrical potentials from the jaw musculature, fine bipolar wire electrodes are placed surgically into the four lateral jaw muscles in a procedure that is performed under anesthesia. A strain gauge is also attached to the top of the snake's head to measure the motion of underly-ing skull bones. **(B)** The snake is placed on a force-recording table and videotaped as it strikes its prey. The leads from the electrodes and strain gauge are connected to electronic amplifiers, which boost their low-voltage signals. The amplified signals are displayed on an oscilloscope and chart recorder and stored on a tape recorder and computer.

These experimental measurements can be used to test hypotheses about which structures and muscles are involved in a strike and how their temporal relationships change during the behavior. The experiment also suggests how many physiological systems contribute to production of a complex behavioral act. A more complete functional analysis of this prey-capture behavior and a greater understanding of the performance of the animal is possible if other variables are measured in repeat experiments performed under identical conditions. This experimental setup could be used to measure differences in the strike behavior as a function of the size and type of prey species. Such measurements also can form the basis for formulation of hypotheses about the neural control of muscle activity, visual feedback guiding the behavior, and a host of other interesting topics.

 Humans can be instructed to behave in certain ways during physiological experiments (e.g., to breathe deeply, to run, or to flex muscles). Some animals can be trained to perform as needed for a particular experiment (e.g., to run on a treadmill), whereas others will only engage in the behavior of interest at irregular intervals while the investigator watches and waits to capture data at hopefully the right moment. What are the relative strengths and weaknesses of data from experiments in which the subject is instructed to perform, is trained to perform, or is simply allowed to behave spontaneously?

IMPORTANCE OF PHYSIOLOGICAL STATE IN RESEARCH

Research studies at all physiological levels—molecular to behavioral—must take into account the animal's *physiological state* at the time of experimentation (or tissue sampling). Some physiological states are quite obvious to the investigator, as when an animal is diving (breath holding), actively moving, or hibernating. Other physiological states may be far more subtle, but have just as great an influence on physiological processes. Of course, the obvious or subtle nature of a physiological state depends on the animal. For instance, a mouse that is curled up with eyes closed and is showing relatively regular breathing with no locomotor activity can probably be assumed to be asleep. But what about a relatively sluggish species of fish that is motionless? Is it asleep, or merely not showing locomotor activity? Physiological state may be greatly influenced by environmental variables such as the season and time of day. To illustrate, stimulation of the vagus nerve causes a much greater slowing of the heart in temperate frogs examined at night in the spring than in frogs examined in the afternoon in the autumn. Thus the outcome of an experiment can be greatly influenced by the time of day and year when it is performed experiments are carried out.

To characterize the physiological states of an animal, one or more variables can be measured while the animal is in different behavioral states and their values compared. For example, blood pressure, pulse rate, and skeletal muscle activity might all be measured simultaneously while the animal is observed in several different states such as sleeping, moving, digesting a meal, or hibernating. Such measurements usually do not allow identification of cause-and-effect relationships among the measured variables. However, inferences can be drawn and testable hypotheses about the relationships among the measured variables can be formulated based on such data. Because multiple physiological states can exist simultaneously (e.g., sleeping in winter, breath holding during hibernation), experiments to determine physiological states are often complex and time consuming. However, such experiments, if carefully planned to reveal the influences of physiological state on the basic physiological processes of an animal, can greatly elevate our knowledge of physiological systems.

A typical experiment, for example, might measure key physiological variables during intermittent bouts of hibernation of the ground squirrel. Comparison of body temperature and metabolic rate recorded over time with observed behavioral activity reveals that the increased activity in the awake state is correlated with increased body temperature and metabolic rate. These correlations suggest that as the animal becomes active, key physiological systems also become active at about the same time. Although it makes intuitive sense that the animal will need more blood circulating when it is physically active, it is not clear from these data how that increase in blood flow is achieved or how it is regulated. Does the physiological change precede the behavioral activity or result from it?

Distinguishing among these and other possible explanations requires experiments that focus on the causal relationships between specific behaviors and the physiological systems that underlie changes in physiological state. Observations of correlations between physiology and behavior such as those discussed here usually form the basis for such subsequent experiments. Perhaps as important, they can provide a way to characterize specific physiological states. In a study designed to probe the causal relationship between blood flow and heart rate during hibernation, for example, variables such as body temperature or metabolic rate could be used to assure that the animal was indeed hibernating during the experimental tests.

SUMMARY

Physiological research should begin with a well-formed, specific hypothesis related to a particular level of analysis and capable of being tested experimentally. Testing of hypotheses is greatly facilitated by employing the August Krogh principle, that is, choosing the optimally suited animal for carrying out those experiments needed to answer particular questions. A key issue in designing physiological experiments is the level at which each physiological problem studied will be analyzed. The choice of level determines the methodology and experimental animal appropriate for measuring the physiological variables of interest.

Techniques that detect or analyze events at the molecular level have greatly benefited animal physiology. Radioisotopes can be incorporated into physiologically important molecules or their precursors. After a radiolabeled molecule is injected into the animal, its movements can be determined by subsequently sampling tissue and measuring the particles emitted by the radioisotope using either a Geiger counter or scintillation counter. The presence and location of radiolabeled molecules in thin tissue slices can be detected by autoradiography. Monoclonal antibodies covalently labeled with a fluorescent dye or radioisotope are another powerful tool for tracing the movement of specific proteins within physiological systems. Because of their great specificity, monoclonal antibodies permit detection of a single protein (e.g., nerve growth factor or a neurotransmitter) even when it is present at a very low concentration in the cells or tissues under study.

Genetic engineering, which involves recombinant DNA technology and gene cloning, is also revolutionizing animal physiology. Genes cloned in easily grown bacterial cells can be used to produce large quantities of the gene products (e.g., human insulin and other hormones). Genetic engineering techniques also allow production of transgenic animals (commonly mice) that contain additional copies of a gene of interest. In knockout mice, a normal gene is replaced with a mutant form of the same gene, so the animals cannot produce a functional protein. Analysis of the effects of either the addition or deletion of specific genes can provide insights into the mechanism and regulation of a physiological process.

Microelectrodes and micropipettes have many uses in cellular physiology. The most common use of microelectrodes is to record electrical signals from neurons or muscle cells. The concentration of ions and some gases and the fluid pressure within cells or blood vessels can be determined with specially constructed microelectrodes. Micropipettes are used to inject materials (e.g., dyes, radiolabeled compounds) into individual cells or fluid-filled tissue spaces.

Structural analysis of cells, and the physiological processes that derive from these cells, depends heavily on microscopy. Light microscopy uses photons of visible or near visible light to illuminate specially prepared tissue samples. Specimens are first fixed (preserved), embedded in plastic or wax, and then cut into extremely thin slices (sections) with a microtome. Finally, the sections are treated with organic dyes or fluorescent-labeled antibodies that differentially bind to and stain various cell components. Once prepared, tissue is typically viewed with one of a variety of light microscopes. The advent of electron microscopes, which use electrons to form images, greatly increased the resolution of microscopic analysis, permitting visualization of intracellular structural details not apparent with light microscopes. In transmission electron microscopes, a beam of electrons is directed straight through ultrathin tissue slices stained with electron-dense heavy metals. In scanning electron microscopes, electrons are reflected from the surface of the specimen, producing a three-dimensional image of the surface features of cells and other structures.

Cell culture, the rearing of cells *in vitro,* allows the propagation of relatively short-lived cell strains and "immortal" cell lines, which can grow indefinitely. Cultured cells, which usually are quite homogeneous, are very useful in experiments designed to examine the functions, secretions, responses, and other properties of particular cell types. Such experiments depend on biochemical analysis to determine the composition of sample mixtures derived from cells as well as the concentration of the constituents present. Among the most commonly used techniques in biochemical analyses are colorimetric assays, transmission spectrophotometry, paper and column chromatography, electrophoresis, and mass spectrometry.

At an increasing level of organizational structure, maintenance of isolated organs or entire organ systems *in vitro* allows the function of intact tissues to be examined in an artificial, controlled environment. Important variables such as temperature, oxygen availability, and nutrient levels can be controlled, mimicking homeostasis, or can be varied to test particular hypotheses.

Animal physiologists frequently supplement their experiments with observations of animal behavior. Experimental methods to control and stimulate specific behaviors can provide important insights into physiological processes that are not always amenable to direct physiological investigation. Also, analysis of the total time spent performing each behavior and the temporal sequence of behaviors in conjunction with information about the behavior of other animals and key environmental variables may reveal how closely behavior is related to the internal physiological state of the animal.

Finally, in all experimental approaches, from those conducted at the simplest (molecular) level to those suitable at most complex (behavioral) level, the animal's physiological state at the time of experimentation (or tissue sampling) is an important consideration. Physiological state can depend upon internally regulated factors (sleep, hibernation, activity, etc.) or environmental influences. To characterize the physiological states of an animal, one or more variables can be measured and the values of these key variables correlated with different behavioral states.

REVIEW QUESTIONS

1. What is the difference between a scientific question, a hypothesis, a theory, and a law?
2. An investigator carries out experiments on crickets, bullfrogs, and rattlesnakes, but is testing a single hypothesis related to a single physiological process. Explain how this investigator could be embracing the August Krogh principle.
3. What are radioisotopes and monoclonal antibodies? What common feature makes them of use to physiologists?
4. What is a clone, and how is it produced?
5. If an interesting and useful mutation to a physiological system is ultimately lethal before an animal reaches the reproductive stage of its life cycle, how can it be perpetuated in the laboratory to allow repeated experiments for its long-term study?
6. Why would an air bubble within a microelectrode used in recording nerve action potentials disrupt the recording?
7. What are the major differences between light and electron microscopy? What are the advantages and disadvantages of each?
8. Describe the difference between an experiment done *in vivo, in vitro,* and *in situ.* What are the advantages and disadvantages of each experimental approach?
9. How would you determine if the resting heart rate of an animal was influenced by daily rhythms?

SUGGESTED READINGS

Burggren, W. W. 1987. Invasive and noninvasive methodologies in physiological ecology: a plea for integration. In M. E. Feder, A. F. Bennett, W. W. Burggren, and R. Huey, eds., *New Directions in Physiological Ecology.* New York: Cambridge University Press, pp. 251–272. (Description of two major approaches to animal experimentation.)

Burggren, W. W., and R. Fritsche. 1995. Cardiovascular measurements in animals in the milligram body mass range. *Brazil. J. Med. Biol. Res.* 28:1291–1305. (Description of methods for extending cardiovascular techniques to microscopic animals.)

Cameron, J. N. 1986. *Principles of Physiological Measurement.* New York: Academic Press. (A short but detailed introduction to several important methods of physiological measurement.)

Hall, Z. 1992. *An Introduction to Molecular Neurobiology.* Sunderland, Mass.: Sinauer Associates. (A comprehensive discussion of how a molecular approach can provide rich insight into a vitally important organ system.)

Lodish, H., et al. 1995. *Molecular Cell Biology.* 3d ed. New York: Scientific American Books. (A well-written, very comprehensive text that describes many techniques used for molecular analyses of the cell.)

Lorenz, K. Z. 1970. *Studies in Animal and Human Behavior.* Vol. 1. Cambridge, Mass.: Harvard University Press. (Collection of research papers, translated from the original German, describing the early research of Lorenz, who won the Noble Prize in physiology in 1973.)

MOLECULES, ENERGY, AND BIOSYNTHESIS

The living organisms found on our planet form a vast and varied array, ranging from viruses, bacteria, and protozoa to flowering plants, invertebrates, and the "higher" animals. In spite of this immense diversity, all forms of life as we know it consist of the same chemical elements and share similar types of organic molecules. Moreover, all life processes take place in a milieu of water and depend on the physicochemical properties of this extremely abundant and very special solvent. That all living organisms share a common biochemistry is one of the powerful evidences in support of their evolutionary kinship, the common thread that runs through all areas of biological study.

ORIGIN OF KEY BIOCHEMICAL MOLECULES

Biologists generally agree that life arose through processes of chance and natural selection under appropriate environmental conditions on the primitive Earth. Experiments first performed by Stanley Miller in 1953 show that certain molecules essential for primitive life (e.g., amino acids, peptides, nucleic acids) can be formed by the action of lightning-like electric discharges on an experimental atmosphere of methane, ammonia, and water. This simple atmosphere is believed to be similar in composition to that of the primitive atmosphere of Earth about 4 billion years ago. Earth's early atmosphere was modified during subsequent eons by photosynthetic plants, which added the currently immense quantities of oxygen and which took up nitrogen compounds for incorporation into nitrogenous biological compounds.

The experimental formation of simple organic molecules under conditions similar to those that may have prevailed in the primeval atmosphere suggests that such molecules may have accumulated in ancient shallow seas, forming an "organic soup" in which life may then have undergone its first evolutionary stages of organization. The combination and recombination of these molecules eventually led to the most simple life forms capable of producing and arranging more complex molecules into informa-

tional assemblages like nucleic acids and enzymes. Critical in the process of producing primitive cell-like organisms was the formation of small liquid droplets with membranes around them. Lipid (fat) molecules will spontaneously form a double layered "molecular skin" around microscopic fluid droplets. When these skins began to incorporate other materials (simple nucleotides, etc.), then the first steps were under way for the formation of a true cell membrane—thin structures that enclose the contents of a cell, control movement of molecules between the cell interior and the surrounding environment, and provide a potential structure for organizing its contents. Many, many such additional steps defined the path towards the current vast array of species in the more than 35 phyla now found on Earth.

This hypothetical scenario of the first stages towards the evolution of life raises many questions. To what degree did the origin of life depend on the "right" conditions? Would life of another sort have appeared on Earth if the chemical and physical environment had been quite different? What if there had been no carbon atom? As we will see shortly, the occurrence of life as we know it (and can imagine it) depends heavily on the chemical nature of the Earth's environment. Life would be either nonexistent or at least vastly different if some of the fundamental properties of matter of the early atmosphere had been different.

A controversy once raged between the vitalists, who believed life was based on special "vital" principles not found in the inanimate world, and the mechanists, who maintained that life can ultimately be explained in physical and chemical terms. Until the early part of the nineteenth century, students of the natural world supposed that the chemical composition of living matter differed fundamentally from that of inanimate minerals. The vitalist view held that "organic" substances can be produced only by living organisms, setting them apart in a mysterious way from the inorganic world. This concept met its end in 1828, when Friedrich Wöhler reacted lead cyanate and ammonia, both obtained from nonliving mineral sources, to synthesize the simple organic molecule **urea:**

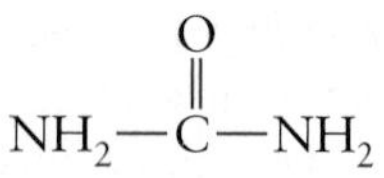

$$NH_2-\overset{\displaystyle\overset{O}{\|}}{C}-NH_2$$

His successful organic synthesis set the stage for modern chemical and physical studies aimed at elucidating the mechanisms of life processes. Modern biochemists can now duplicate *in vitro* in isolated cell-free systems nearly every synthetic and metabolic reaction normally performed by living cells.

The biochemical and physiological processes of the living organism ultimately depend on the physical and chemical properties of the elements and compounds it contains. At first glance, the properties of living systems seem far too marvelous and complex to be explained by a mere mixture of elements and compounds. Yet, living systems are not simple chemical "soups"; rather, they are highly organized structures composed of often very large and complex molecules called *macromolecules*. Macromolecules of many kinds participate in the regulation and direction of chemical activities within living cells. *Organelles* such as the plasma membrane, lysosomes, and mitochondria lend structural organization to the *cell,* the basic unit of living systems, by differentiating it from the surrounding environment and internally separating it into compartments and subcompartments. Organelles also hold molecules in functionally important spatial relations to one another. Cells are organized into tissues, tissues into organs, and those into interacting systems. Thus, the organism consists of an organizational hierarchy with each higher level imparting further functional complexity to the whole (see Figure 1-1). In this chapter, we begin with the most basic level—the chemical level—and learn how simple prin-

ciples of chemical reactions apply to the assembly of macromolecules and more complex cellular organelles that constitute the cell.

ATOMS, BONDS, AND MOLECULES

All matter is composed of chemical elements, which can be arranged into the familiar periodic table of the naturally occurring elements and dozens of artificially synthesized elements created fleetingly in the laboratory (Figure 3-1). Of all the chemical elements, only a very small subset naturally occurs in animal tissue. Table 3-1 compares the major components of the Earth's mineral crust and seawater with those in the human body. About 99% of the human body is made up of just four elements: hydrogen, oxygen, nitrogen, and carbon. This holds true for all organisms. Is the preponderance of these elements in living systems simply a matter of chance, or is there a mechanistic explanation for their uniform prevalence in the great diversity of organisms that have evolved over the past 3 billion years?

George Wald, a biologist who contributed much to our understanding of the chemical basis of vision, suggested that the biological predominance of hydrogen, oxygen, nitrogen, and carbon is not at all a matter of chance, but is the inevitable result of certain fundamental atomic properties of these elements—properties that render them especially suited for the chemistry of life. We will review briefly the factors that influence the chemical behavior of atoms, and then return to consider Wald's ideas.

Atomic structure is far more complex and subtle than can be fully described here; for our purposes, we need consider only some basic features that affect the formation of chemical bonds between atoms and molecules. The basic

First shell	1 H																	2 He
Second shell	3 Li	4 Be											5 B	6 C	7 N	8 O	9 F	10 Ne
Third shell	11 Na	12 Mg											13 Al	14 Si	15 P	16 S	17 Cl	18 Ar
Fourth shell	19 K	20 Ca	21 Sc	22 Ti	23 V	24 Cr	25 Mn	26 Fe	27 Co	28 Ni	29 Cu	30 Zn	31 Ga	32 Ge	33 As	34 Se	35 Br	36 Kr
Fifth shell	37 Rb	38 Sr	39 Y	40 Zr	41 Nb	42 Mo	43 Tc	44 Ru	45 Rh	46 Pd	47 Ag	48 Cd	49 In	50 Sn	51 Sb	52 Te	53 I	54 Xe
Sixth shell	55 Cs	56 Ba	57 La	72 Hf	73 Ta	74 W	75 Re	76 Os	77 Ir	78 Pt	79 Au	80 Hg	81 Tl	82 Pb	83 Bi	84 Po	85 At	86 Rn
Seventh shell	87 Fr	88 Ra	89 Ac	104	105	106												

58 Ce	59 Pr	60 Nd	61 Pm	62 Sm	63 Eu	64 Gd	65 Tb	66 Dy	67 Ho	68 Er	69 Tm	70 Yb	71 Lu
90 Th	91 Pa	92 U	93 Np	94 Pu	95 Am	96 Cm	97 Bk	98 Cf	99 E3	100 Fm	101 Md	102 No	103 Lw

Figure 3-1 In the periodic table of the elements, each row corresponds to a different electron orbital shell. The elements in colored lettering are physiologically important in their ionic forms.

TABLE 3-1
Comparison of the chemical composition of the human body with that of seawater and the Earth's crust*

Human body		Seawater		Earth's crust	
H	63	H	66	O	47
O	25.5	O	33	Si	28
C	9.5	Cl	0.33	Al	7.9
N	1.4	Na	0.28	Fe	4.5
Ca	0.31	Mg	0.033	Ca	3.5
P	0.22	S	0.017	Na	2.5
Cl	0.03	Ca	0.006	K	2.5
K	0.06	K	0.006	Mg	2.2
S	0.05	C	0.0014	Ti	0.46
Na	0.03	Br	0.0005	H	0.22
Mg	0.01			C	0.19
All others	<0.01	All others	<0.1	All others	<0.1

*Values are percentages of total numbers of atoms. Because figures have been rounded off, totals do not amount to 100.

Source: *Biology: An Appreciation of Life*, 1972.

chemical building blocks of all matter, the elements, are comprised of still smaller particles, each of which has distinctive properties. Understanding the behavior of three of these particles—*protons, neutrons,* and *electrons*—is im-

portant to animal physiology because they dominate interactions among elements central to organic life. Indeed, the interactions amongst these three particles dictate the attraction among elements necessary for life itself.

Each *atom* consists of a dense nucleus of protons and neutrons surrounded by a "cloud" of electrons equal in number to the protons in the nucleus. The atomic particles have the following charge and mass (in daltons, Da):

- Proton: +1; 1.672 Da
- Neutron: 0; 1.674 Da
- Electron: −1; 0.001 Da

Since the negatively charged electrons are equal in number to the positively charged protons, each atom in its elemental state carries no net electric charge. Although the mass of an atom is determined largely by the number of protons and neutrons in the nucleus, its chemical reactivity depends on the surrounding electrons. The electrons do not occupy fixed orbits, but their statistical distribution is such that they occupy some positions with greater probabilities than others. This distribution is quite systematic, so that in an atom with one or two electrons, the orbital paths are virtually confined to a single "shell" around the nucleus, as in the hydrogen and helium atoms (Figure 3-2). In atoms with

Number of electrons in atom

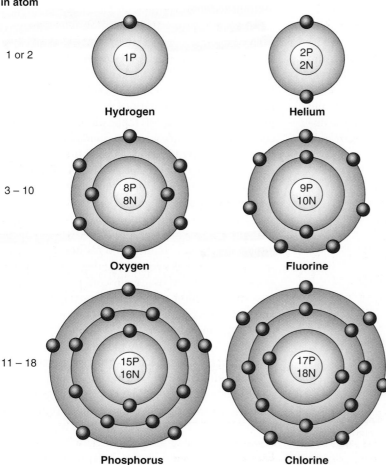

1 or 2

Hydrogen

Helium

3 – 10

Oxygen

Fluorine

11 – 18

Phosphorus

Chlorine

Figure 3-2 The electrons surrounding the nucleus of each atom are statistically distributed in orbital shells. If the outer shell of an atom does not contain the maximal number of electrons, the atom tends to share electrons with other atoms, thereby forming chemical bonds. In contrast, atoms in which the outer shell is filled (e.g., helium) are chemically inert. Chemical reactivity also is influenced by the size of an atom. Other things being equal, a smaller atom (e.g., fluorine) is more reactive and forms stronger, more stable chemical bonds than does a larger atom (e.g., chlorine).

3 to 10 electrons (e.g., carbon, nitrogen, and oxygen), this first shell is occupied by two electrons; the remaining electrons occupy a second shell, which can contain up to eight electrons, located farther from the nucleus. In atoms with 11 to 18 electrons (e.g., sodium, phosphorus, and chloride), a third shell is formed, which also can accommodate up to eight electrons. The fourth and fifth shells each can hold up to 18 electrons (Figure 3-3).

When the outermost shell of an atom contains the maximal number of electrons possible in that shell—that is, when it cannot accommodate additional electrons—the atom is highly stable and resists reactions with other atoms. This is true of all the noble gases, such as helium and neon, which appear at the far right of the periodic table. Most elements, however, have incomplete outer electron shells and are therefore reactive with certain other atoms. Hydrogen, for example, has one rather than two electrons in its only shell, and oxygen has only six, instead of eight, electrons in its outer shell. Thus, the hydrogen atom and the oxygen atom both have a tendency to share electrons so as to fill their respective outer shells and bring them into more stable configurations.

Although the number of electrons in the outer shell has an important influence on the physical characteristics and reactivity of an atom, other physical features are also important in determining its chemical properties. One of these is the size (or weight) of the atom. The heavier an atom (i.e., the more protons and neutrons in its nucleus), the more electrons surround the nucleus. As the number of electrons exceeds ten and a third shell of electrons appears, the *valence electrons* (i.e., those in the outermost shell) are correspondingly more distant from the compact nucleus, and hence less strongly attracted by it than the valence electrons of atoms with only two shells. This is because the electrostatic interactions between the single charged electrons and protons (**monopoles**) diminish with the square of the distance between them. Thus, chlorine, with seven electrons in

its third and outer shell, is less reactive than fluorine, which has seven electrons in its second and outer shell (see Figure 3-2). Both atoms have a tendency to gain one electron to complete the outermost shell, but this tendency is greater in fluorine, since its outermost shell feels a stronger electrostatic pull from its nucleus than the larger chlorine atom. As a result, with all other things being equal, a small atom forms stronger and hence more stable bonds with other atoms than does a large atom.

THE SPECIAL ROLES OF H, O, N, AND C IN LIFE PROCESSES

Let's return to Wald's view that hydrogen (H), oxygen (O), carbon (C), and nitrogen (N) dominate the composition of biological systems because they lend themselves especially well to the chemistry of living systems. Examination of the periodic table reveals that these four elements have one or two electron shells. Of the other elements with only one or two electron shells, helium and neon are virtually inert, rare gases; boron and fluorine form relatively rare salts; and the metals lithium and beryllium form easily dissociated ionic bonds. In contrast, H, O, N, and C will form strong covalent bonds by sharing one, two, three, and four electrons, respectively, to complete their outer electron shells.

Why are strong bonds important in living systems? Without strong bonding, subtle changes in temperature, pH, or other variables in the environment surrounding a molecule could cause it to break down or rearrange. Consider, for example, the biological chaos that would result if the chemical bonds in the hereditary material formed by DNA were easily dissociated and subject to alterations (i.e., mutations). In fact, mutations are quite rare (less than one per gene in every 10,000 replications) because the atoms composing DNA are strongly bonded to each other in a multitude of combinations. The short-term integrity of each organism and each species depends upon the stable bonds holding together the structures of DNA and other macromolecules.

Of the four major biologically important elements, three (O, N, C) are among the very few to form double or triple bonds. These multiple bonds not only increase the stability of molecules containing them, but also greatly increase the variety of molecular configurations that can be formed by reaction of these elements (Figure 3-4). For example, oxygen can react with carbon to form carbon dioxide, CO_2. Since the two double bonds connecting the O atoms to the C atom satisfy the tendencies of these atoms to react, the CO_2 molecule is relatively inert. Therefore, CO_2 can diffuse readily from its source of production to become available for recycling through the photosynthetic process of green plants. Because the carbon atom has a **valence** of four, it can form four single bonds, two double bonds, and combinations of single with double or triple bonds, endowing it with the ability to form a great diver-

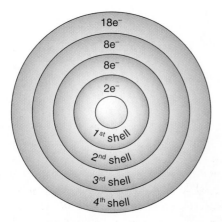

Figure 3-3 Each orbital shell in atomic structure can accommodate a set maximal number of negatively charged electrons, as shown here for the first four orbital shells (numbered 1, 2, 3, and 4). Atoms with more electrons than needed to fill, for example, shells 1 and 2, will form an additional shell or shells.

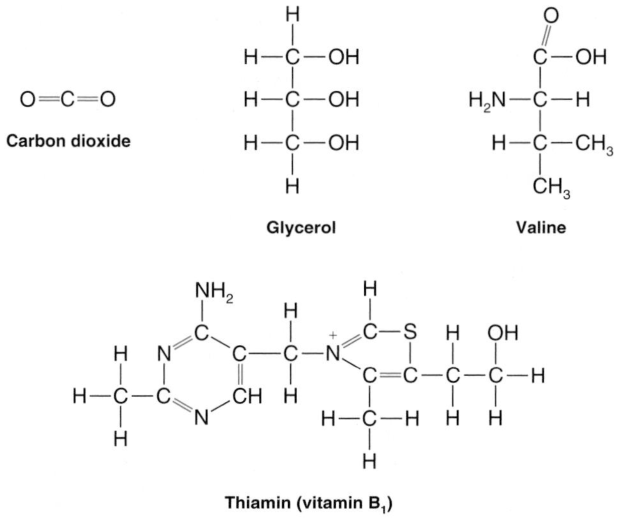

Carbon dioxide

$O = C = O$

Glycerol

Valine

Thiamin (vitamin B₁)

Figure 3-4 The ability of carbon, oxygen, and nitrogen to form double bonds (in red), in addition to single bonds, greatly increases the diversity of molecular structures including these elements. Glycerol is a constituent of fats, and valine is one of the amino acids present in natural proteins.

sity of atomic combinations with itself and other atoms, including straight and branched chains, as well as ring structures (see Figure 3-4).

Silicon (Si), which is in the same column and just below carbon in the periodic table, has some properties similar to those of carbon. Unlike carbon, however, it is larger and does not form double bonds. Therefore, it combines with two atoms of oxygen by two single bonds only:

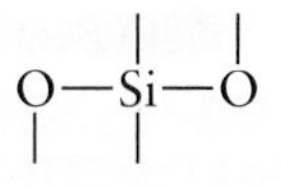

$$O - Si - O$$

Since the outer electron shells in all three atoms of silicon dioxide (SiO_2) are not filled, the SiO_2 molecule readily bonds with others of its kind, forming the huge polymeric molecules that make up silicate rocks and sand. Thus, it is evident that silicon, even though it has some properties similar to those of carbon, is far more suited for the formation of stone than it is for large-scale participation in the organization of biological molecules.

Besides its important role in combining with hydrogen to form water, oxygen acts as the final electron acceptor in the sequence of **oxidation** reactions through which chemical energy is released by cell metabolism. This important ability to oxidize (accept electrons from) other atoms and molecules is due to the oxygen atom's incomplete outer electron shell and relatively low atomic weight.

In addition to the four major biological elements, numerous other elements participate in cell chemistry, though in lesser numbers (see Table 3-1). These include phosphorus (P) and sulfur (S), the ions of four metallic elements (Na^+, K^+, Mg^{2+}, and Ca^{2+}), and the chloride ion (Cl^-). We will return to these later.

WATER: THE UNIQUE SOLVENT

We live on the "water planet." Because water is so common, it is often regarded with indifference, as some sort of inert filler occupying space in living systems. The truth is that water is directly and intimately involved in all details of animal physiology. Water is a highly reactive substance, quite different both physically and chemically from most other liquids. Water possesses a number of unusual and special properties of great importance to living systems. Indeed, life as we know it would be impossible if water did not have these properties. The first living systems presumably arose in the aqueous environment of shallow seas. It is therefore not surprising that the living organisms of the present are intimately adapted at the molecular level to the special properties of water. Today even terrestrial animals consist of 75% or more of water. Much of their energy expenditure and physiological effort is devoted to the conservation of body water and the regulation of the chemical composition of the internal aqueous environment.

The special properties of water so important to life stem directly from its molecular structure. Therefore, we should begin with a brief consideration of the water molecule.

The Water Molecule

Water molecules are held together by *polar covalent bonds* between one O and two H atoms. The polarity (i.e., uneven charge distribution) of the covalent bonds results from the strong tendency of the O atom to acquire electrons from other atoms, such as H. This high **electronegativity** causes the electrons of the two H atoms in the water molecule to occupy positions statistically closer to the O atom than to the parent H atoms. The O—H bond is therefore about 40% ionic in character, and the water molecule has the following partial charge distributions (δ represents the local partial charge of each atom):

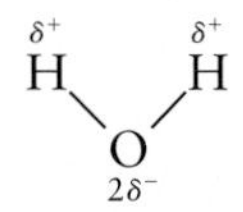

$$H^{\delta^+} \qquad H^{\delta^+}$$
$$O^{2\delta^-}$$

The angle between the two O—H bonds in the water molecule, rather than being 90° as predicted for purely covalent bonding, is found to be 104.5° (Figure 3-5). The increased angle can be ascribed to the mutual repulsion of the two positively charged H nuclei, which tends to force them apart. In contrast, the S—H bonds in the hydride of sulfur, H_2S, are purely covalent; so there is no asymmetrical charge distribution as in H_2O. Thus, the bond angle in H_2S is closer to 90°. Because of the semipolar nature of O—H bonds, H_2O differs greatly, both chemically and physically, from H_2S and other related hydrides. Why is this?

The uneven distribution of electrons in the water molecule causes it to act like a **dipole**. That is, it behaves somewhat like a bar magnet, but instead of having two opposite magnetic poles, it has two opposite electric poles, positive and negative (see Figure 3-5). As a result, the

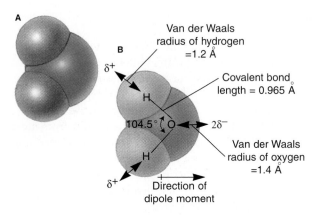

Figure 3-5 In the water molecule, the electron density is greater around the oxygen atom than around the hydrogen atoms, giving the O—H bond a semipolar character. The mutual repulsion between the resulting partial positive charges on the hydrogen atoms causes the angle between the two O—H bonds to be greater than that characteristic of purely covalent bonds. δ^+ and δ^- indicate a partial positive and negative charge, respectively.

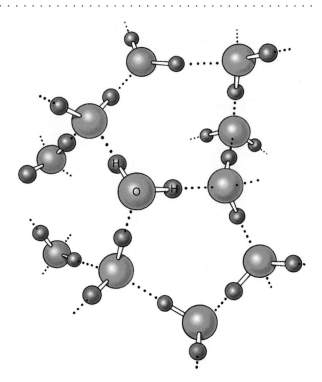

Figure 3-6 Because of the semipolar nature of the O—H bonds in water, adjacent water molecules form hydrogen bonds. These noncovalent bonds (indicated by black dots) represent the electrostatic interaction between the partially positively charged hydrogens on one molecule and electronegative oxygens in neighboring molecules.

water molecule tends to align with an electrostatic field. The **dipole moment** is the turning force exerted on the molecule by an external field. The high dipole moment of water (4.8 debyes) is the most important physical feature of the molecule and accounts for many of its special properties.

The most important chemical feature of water is its ability to form **hydrogen bonds** between the nearly electron-bare, positively charged protons (H atoms) of one water molecule and the negatively charged electron-rich oxygen atom of neighboring water molecules (Figure 3-6). In each water molecule, four of the eight electrons in the outer shell of the oxygen atom are covalently bonded with the two hydrogen atoms. This leaves two pairs of electrons free to interact electrostatically (i.e., to form hydrogen bonds) with the electron-poor H atoms of neighboring water molecules. Since the angle between the two covalent bonds of water is about 105°, groups of hydrogen-bonded water molecules form tetrahedral arrangements. This arrangement is the basis for the crystalline structure of the most common form of ice.

Properties of Water

The hydrogen-bonded structure of water is highly labile and transient, for the lifetime of a hydrogen bond in liquid water is only about 10^{-10} to 10^{-11} seconds. This transience is due to the relatively weak nature of the hydrogen bond. It takes only 4.5×10^3 **calories** (4.5 kcal) of energy to break a mole of hydrogen bonds, whereas $110 \text{ kcal} \cdot \text{mol}^{-1}$ are required to break the covalent O—H bonds within the water molecule. As a result of the weakness of hydrogen bonds, no specific groups of H_2O molecules remain hydrogen bonded for more than a brief instant. Statistically speaking, however, a constant fraction of the population is joined together by hydrogen bonding at all times at a given temperature.

In spite of the only modest strength of the hydrogen bond, it increases the total energy (i.e., heat) required to separate individual molecules from the rest of the population. For this reason, the **melting point, boiling point,** and **heat of vaporization** of water are much higher than those of other common hydrides of elements related to O (e.g., NH_3, HF, H_2S). Of the common hydrides, only water has a boiling point (100°C) far above temperatures common to the surface of the Earth.

The statistical loose bonding between water molecules also endows water with an unusually high **surface tension and cohesiveness,** which has major implications for both biochemical and biological events that occur, or depend upon, air/water interfaces. Ice has an open crystalline lattice, whereas liquid water has a much more random molecular organization, giving it a more closely packed, dense molecular organization. As a result, water is unusual in that its solid form, ice, is less dense than its liquid form. If ice were denser (heavier) than liquid water, it is widely agreed that the oceans and lakes would have turned to solid ice, except at the surface, as ice would have formed from the bottom up. Clearly, this property of water has had a major impact on life on Earth.

Water as a Solvent

The medieval alchemists, looking for the universal solvent, were never able to find a more effective and "universal" solvent than water. The solvent characteristics of water are due largely to its high **dielectric constant**, a manifestation of its electrostatic polarity. The dielectric constant is a measure of the effect that water or any polar dielectric substance has in diminishing the electrostatic force between two charges separated by water or another dielectric medium.* This is illustrated especially well by the behavior of ionic compounds, or **electrolytes**, which dissociate (ionize) when placed in water, thereby increasing the conductivity of the solution. Common electrolytes include salts, acids, and bases. In contrast, solutes that undergo no dissociation, and therefore do not increase the conductivity of a solution, are called nonelectrolytes. Common examples of nonelectrolytes are the sugars, alcohols, and oils.

Figure 3-7A illustrates the arrangement of the ions Na^+ and Cl^- in a sodium chloride crystal. The highly structured array is held together firmly by the electrostatic attraction between the positively charged sodium ions and the negatively charged chloride ions. A *nonpolar* liquid, such as hexane, cannot dissolve the crystal, because no source of energy exists in the nonpolar solvent to break an ion away from the rest of the crystal. Water, however, can dissolve the NaCl crystal, just as it can dissolve most other ionic compounds. The dissolving power of water arises because the dipolar water molecules can overcome the electrostatic interactions between the individual ions (Figure 3-7B). Weak electrostatic binding occurs between the partial negative charge of the oxygen atoms and the positively charged cations (Na^+ in this case). Such binding also occurs between the partial positive charge on the hydrogen atoms and the negatively charged ions (Cl^- in this case). The clustering of water molecules about individual ions and polar molecules is called **solvation**, or **hydration**.

As water molecules surround ions, they orient themselves so that their positive poles face **anions** (negatively charged ions) and their negative poles face **cations** (positively charged ions). This orientation further reduces the electrostatic attraction between the dissolved cations and anions of an ionic compound. In a sense, the H_2O molecules act as "insulators." The first shell of water molecules surrounding an ion attracts a second shell of less tightly bound, oppositely oriented water molecules. The second shell may even attract more water in a third shell. Thus, the ion may carry a considerable quantity of *water of hydration*. The effective diameter of hydrated ions of a given charge varies inversely with their diameter. For example, the ionic radii of Na^+ and K^+ are 0.095 and 0.133 nm, respectively, whereas their effective hydrated radii are 0.24 and 0.17 nm, respectively. The reason for this inverse relationship is that the electrostatic force between the nucleus

---

* The electrostatic force between two charges separated by water or another dielectric medium is given by Coulomb's law:

$$f = \frac{q_1 q_2}{\epsilon d^2}$$

where f is the force (in dynes) between the two electrostatic charges q_1 and q_2 (in electrostatic units), d is the distance (in centimeters) between the charges, and ϵ is the dielectric constant.

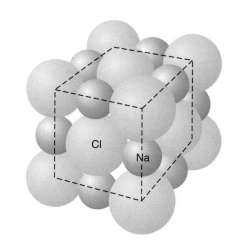

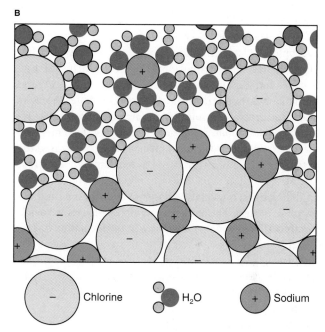

Chlorine H_2O Sodium

Figure 3-7 Water disrupts the crystalline structure of salts by interacting electrostatically with the ions composing the salt. **(A)** Representation of the highly organized crystalline structure of sodium chloride showing the relative ionic sizes of Na^+ and Cl^-. **(B)** Hydration of sodium chloride. The oxygen atoms of the water molecules are attracted to the cations, and the hydrogen atoms are attracted to the anions.

of the ion and the dipolar water molecule decreases markedly with distance between the water molecule and the nucleus of the ion (Figure 3-8). Thus, the smaller ion binds water more strongly and thereby carries a large number of water molecules with it.

Water also dissolves certain organic substances (e.g., alcohols and sugars) that do not dissociate into ions in solution but do have polar properties. In contrast, water does not dissolve or dissolve in compounds that are completely nonpolar (e.g., fats and oils), for it cannot form hydrogen bonds with such molecules. Water does, however, react partially with **amphipathic** compounds, which have a polar group and a nonpolar group. A good example is sodium oleate, a common constituent of soap, which has a

A

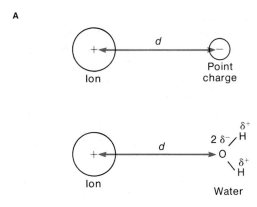

B

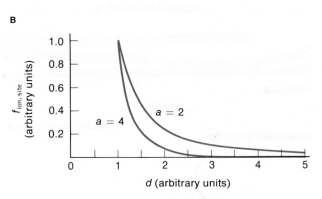

Figure 3-8 Interactions between ions and charged sites are influenced by the distance separating them. The electrostatic force, *f*, between an ion and a site of opposite charge varies inversely with the distance, *d*, raised to some power, *a*, between them: $f \propto 1/d^a$. **(A)** For a point charge, or monopole, the exponent *a* equals 2.0, so that the force drops inversely with the square of the distance. For a dipole such as the water molecule the value of *a* can be as high as 4.0. **(B)** The drop in electrostatic force as a function of distance is illustrated for these two values of *a*. In the case of water and a positive point charge, the actual value of *a* is closer to 3.0.

hydrophilic (water-attracting) polar head and a **hydrophobic** (water-repelling) nonpolar tail (Figure 3-9A). If a mixture of water and sodium oleate is shaken, the water will disperse the latter into minute droplets. The sodium oleate molecules in such a droplet, or **micelle,** are arranged with their hydrophobic, nonpolar tail groups huddled in the center and their hydrophilic, polar head groups arranged around the perimeter, facing outward, so as to interact with the water (Figure 3-9B). The same behavior is exhibited by **phospholipid** molecules, which also consist of hydrophobic and hydrophilic groups. The tendency of amphipathic molecules to form micelles in water is important in the formation of bio-

logical membranes in living cells. Micelle formation may have provided the basis for the first cell-like organization of a living system in the organic-rich shallow seas in which life is believed to have undergone its first evolutionary stages.

PROPERTIES OF SOLUTIONS

As noted already, water plays a critical role in living systems. Indeed, many of the physical and chemical processes of cells occur in water solution. The fluids within the cells and tissues of animals, as well as the aqueous environment in which aquatic animals live, are critically influenced by the solutes—particularly the electrolytes—that they contain.

A

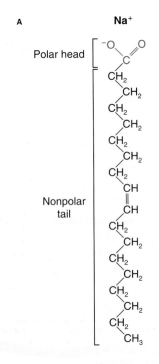

B

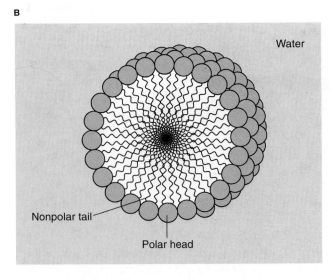

Figure 3-9 Sodium oleate is an amphipathic lipid that forms circular structures called micelles in a polar solvent, such as water. **(A)** Chemical structure of sodium oleate with the polar (hydrophilic) head shown in red and long nonpolar (hydrophobic) tail shown in black. **(B)** Diagram of a micelle with amphipathic lipid molecules represented by conventional symbols. The hydrophobic ends of the molecule tend to avoid contact with the polar solvent by grouping at the center of the micelle.

Concentration, Colligative Properties, and Activity

Conventionally, the quantity of a pure substance is expressed in **moles** (abbreviated mol). A mole is Avogadro's number of molecules (6.022×10^{23}) of an element or compound; it is equivalent to the molecular weight expressed in grams. Thus, 1 mol of ^{12}C consists of 12.00 g of the pure nuclide ^{12}C, or 6.022×10^{23} carbon atoms. Likewise, there are 6.022×10^{23} molecules in 2.00 g (1 mol) of H_2, in 28 g (1 mol) of N_2, and in 32 g (1 mol) of O_2.

For biological processes that involve molecules in solution, the amount of a solute in relation to the amount of solvent—that is, the concentration—is the most relevant measure of solute quantity. Sometimes physiologists express solute concentration in terms of **molality** *(m)*—the number of moles of solute in 1000 g of solvent (*not* total solution). For example, a one molal sucrose solution is produced by dissolving 1 mol (342.3 g) of sucrose in 1000 g of water. Although 1 liter (L) of water equals 1000 g, the total volume of 1000 g of water plus 1 mol of solute will be somewhat greater or less than 1 liter by some unpredictable amount. Molality, therefore, is generally an inconvenient way of stating concentration. A more useful measure of concentration in physiology is **molarity** (M). A one molar solution is one in which 1 mol of solute is dissolved in a total final volume of 1 liter; this is written as 1 mol/L, $1 \; mol \cdot L^{-1}$, or 1 M. In the laboratory, a 1 M solution is made by simply adding enough water to 1 mol of the solute to bring the volume of the final solution up to 1 liter. A millimolar (mM) solution contains $1/1000$ mole per liter, and a micromolar (μM) solution contains 10^{-6} mole per liter. If a solution contains equimolar concentrations of two solutes, then the number of molecules of one solute equals the number of molecules of the other solute per unit volume of solution.

The **colligative properties** of a solution depend on the number of solute particles in a given volume, irrespective of their chemical nature. These properties include osmotic pressure, depression of the freezing point, elevation of the boiling point, and depression of the water vapor pressure. All of these colligative properties are intimately related to one another, and are all quantitatively related to the number of solute particles dissolved in a given volume of solvent. Thus, 1 mol of an ideal solute—that is, one in which the particles neither dissociate nor associate—dissolved in 1000 g of water at standard pressure (760 mm Hg) depresses the freezing point by 1.86°C, elevates the boiling point by 0.54°C, and exhibits an osmotic pressure of 22.4 atm at standard temperature (0°C) when measured in an ideal apparatus. Measurement of any of these colligative properties can be used to determine the sum of the concentrations of solutes in a solution. Concentrations determined in this way are expressed in **osmoles** per liter, or the osmolarity (osM). In theory, osmolarity and molarity are equivalent for solutions of ideal nondissociating solutes exhibiting the same colligative properties.

The theoretical equivalence of osmolarity and molarity, however, does not hold for electrolyte solutions because of ionic dissociation. This is true because a dissociating electrolyte solution will contain more individual particles than a nonelectrolyte solution of the same molarity. As an example, a 10 mM NaCl solution contains nearly twice as many particles as the same volume of a 10 mMM glucose solution, because NaCl is a strongly dissociating electrolyte. Thus the colligative properties, and hence the osmolarity, of a 10 mM NaCl solution will be nearly equivalent to those of a 20 mM glucose solution.

Because of electrostatic interaction between the cations and anions of a dissolved electrolyte, there is a statistical probability that at any instant some cations will be associated with anions. For this reason, an electrolyte in solution behaves as if it were not 100% dissociated. The effective free concentration of an electrolyte, as indicated by its colligative properties, is referred to as **activity**. The **activity coefficient**, γ, of an electrolyte is defined as the ratio of its activity, *a*, to its molal (*not* molar) concentration, *m* ($\gamma = a/m$). As we saw earlier, however, the electrostatic force between ions decreases with the distance between them (see Figure 3-8A). Thus as an electrolyte solution becomes more dilute, the extent of dissociation increases. In other words, an electrolyte's activity and activity coefficient depend on both its tendency to dissociate in solution and on its total concentration. The lower the concentration, the higher the activity coefficient. Table 3-2 lists the activity coefficients of some common electrolytes. Those electrolytes that dissociate to a large extent (i.e., have a large activity coefficient) are called *strong electrolytes* (e.g., KCl, NaCl, HCl); those that dissociate only slightly are called *weak electrolytes* (e.g., $MgSO_4$). It should be noted that although the activity coefficient is useful as an index of a solute's tendency to dissociate and thus of its ability to impart colligative properties upon a solution, the activity coefficient is not directly related to the osmotic pressure or other colligative properties of that solute. This value is given by the osmotic coefficient, which must be determined empirically for each solution.

TABLE 3-2
Activity coefficients of representative electrolytes at various molal concentrations*

| Electrolyte | Molal concentration | | | | |
	0.01	0.05	0.10	1.00	2.00
KCl	0.899	0.815	0.764	0.597	0.569
NaCl	0.903	0.821	0.778	0.656	0.670
HCl	0.904	0.829	0.796	0.810	1.019
$CaCl_2$	0.732	0.582	0.528	0.725	1.555
H_2SO_4	0.617	0.397	0.313	1.150	0.147
$MgSO_4$	0.150	0.068	0.049	—	—

*Activity coefficients are given at various molal concentrations rather than molar concentrations. At low concentrations, however, molality and molarity are nearly identical.

Source: West, 1964.

Ionization of Water

The bonding between molecules of water is very dynamic, with covalent and hydrogen bonds alternating from one instant to the next. Because of the ever-changing nature of the bonding relations between adjacent water molecules, there is a finite probability that a hydrogen atom from one water molecule will become covalently bonded to the oxygen atom of another molecule, forming a **hydronium ion,** H_3O^+. The water molecule that loses a hydrogen atom is converted to a **hydroxyl ion,** OH^- (Figure 3-10A). The probability of H_3O^+ and OH^- ions forming is actually quite small. At any given time, a liter of pure water at 25°C contains only 1.0×10^{-7} mol of H_3O^+ and an equal number of OH^- ions. The positive charges on the hydrogen atoms of the hydronium ion form hydrogen bonds with the electronegative (oxygen) ends of surrounding nondissociated water molecules, yielding a stable, hydrated hydronium ion (Figure 3-10B).

The dissociation of water is conventionally written as

$$H_2O \rightleftharpoons H^+ + OH^-$$

Nevertheless, bear in mind that the proton (H^+) is not, in fact, free in solution but becomes part of the hydronium ion. A proton can, however, migrate to a surrounding H_2O molecule, converting it briefly to a H_3O^+ ion, which in turn loses one of its protons to another water molecule (Figure 3-11). A sequence of such migrations and displacements can, in the fashion of falling dominoes, conduct over relatively long distances, with any one proton traveling but a short distance. There is some evidence that such *proton conduction* may play an important role in some biochemical processes, such as photosynthesis and respiratory-chain phosphorylation.

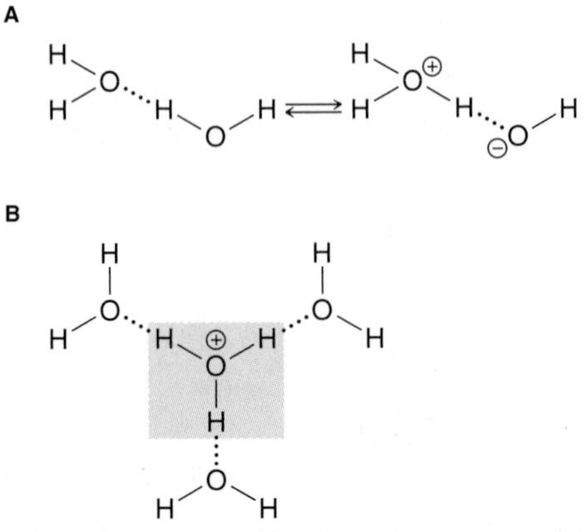

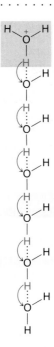

Figure 3-11 Protons migrate between water molecules. In the process of proton conduction, each water molecule exists briefly as a hydronium ion *(top)* but soon donates one of its protons to a neighboring water molecule, thereby converting it into a hydronium ion. [Adapted from Lehninger, 1975.]

Acids and Bases

Any substance that can donate a proton is called an **acid,** and any substance that combines with a proton is called a **base.** An acid-base reaction always involves such a *conjugate acid-base pair*—the proton donor and the proton acceptor (H_3O^+ and OH^- in the case of water). Water is said to be **amphoteric,** meaning it can act as either an acid or base. Amino acids also have amphoteric properties. Common acids include hydrochloric acid, carbonic acid, ammonium ion, and water:

hydrochloric acid	$HCl \rightleftharpoons H^+ + Cl^-$
carbonic acid	$H_2CO_3 \rightleftharpoons H^+ + HCO_3^-$
ammonium	$NH_4^+ \rightleftharpoons H^+ + NH_3$
water	$H_2O \rightleftharpoons H^+ + OH^-$

Common bases include ammonia, sodium hydroxide, phosphate ion, and water:

ammonia	$NH_3 + H^+ \rightleftharpoons NH_4^+$
sodium hydroxide	$NaOH + H^+ \rightleftharpoons Na^+ + H_2O$
phosphate	$HPO_4^{2-} + H^+ \rightleftharpoons H_2PO_4^-$
water	$H_2O + H^+ \rightleftharpoons H_3O^+$

The dissociation of water into H^+ and OH^- ions is an equilibrium process that can be described by the *law of mass action*. This law states that the rate of a chemical reaction is proportional to the active masses of the reacting

Figure 3-10 The bonding between adjacent water molecules is highly dynamic. **(A)** Resonance can cause separation of charges, producing hydronium ions, H_3O^+, and hydroxyl ions, OH^-. **(B)** In solution the hydronium ion (color) is associated by hydrogen bonds (dotted lines) to three water molecules.

substances. For example, the equilibrium constant for the reaction

$$H_2O \rightleftharpoons H^+ + OH^-$$

is given by

$$K_{eq} = \frac{[H^+][OH^-]}{[H_2O]} \qquad (3\text{-}1)$$

The concentration of water remains virtually unaltered by its partial dissociation into H^+ and OH^-, because the concentration of each of the dissociated products is only 10^{-7} M (10^{-7} mol·L^{-1}), whereas the molar concentration of water in a liter of pure water (equal to 1000 g) is 1000 g·L^{-1} divided by the gram molecular weight of water (18 g·L^{-1}), or 55.5 M (55.5 mol·L^{-1}). Equation 3-1 can thus be simplified to

$$55.5\, K_{eq} = [H^+][OH^-]$$

Recall that a consequence of the law of mass action is the reciprocal relation between the concentrations of two compounds in an equilibrium system. This reciprocity is apparent in the constant $[H^+][OH^-]$, which may be lumped with the molarity of water (55.5) into a constant that will be termed the *ion product* of water, K_w. At 25°C this has a value of 1×10^{-14}:

$$K_w = [H^+][OH^-] = 10^{-14}$$

This equation follows from the fact, noted above, that $[H^+]$ and $[OH^-]$ each equal 10^{-7} mol·L^{-1}. If $[H^+]$ for some reason increases, as when an acid substance is dissolved in water, $[OH^-]$ will decrease so as to keep $K_w = 10^{-14}$. This reaction is the basis for the **pH scale,** the standard for acidity and basicity, measured as the concentration of H^+ (actually H_3O^+) and defined as

$$pH = -\log_{10}[H^+]$$

Note that the pH scale is logarithmic and typically ranges from 1.0 M H^+ to 10^{-14} M H^+ (Table 3-3). Thus, a 10^{-3} M solution of a strong acid such as HCl, which dissociates completely in water, has a pH of 3.0. A solution in which $[H^+] = [OH^-] = 10^{-7}$ M has a pH of 7.0, and so forth. A solution with a pH of 7 is said to be neutral—that is, neither acidic nor basic. However, the ratio of $[H^+]$ to $[OH^-]$ depends on temperature, so the true "neutral pH" (called the **pN**) at which $[H^+] = [OH^-]$ actually rises above 7.0 at temperatures below 25°C and falls below 7.0 at temperatures above 25°C. The pH of a solution can be conveniently measured as the voltage produced by H^+ diffusing through the proton-selective glass envelope of an electrode immersed in the solution (Figure 3-12).

TABLE 3-3
The pH scale

	pH	$[H^+]$ (mol·L^{-1})	$[OH^-]$ (mol·L^{-1})	Examples
	0	10^0	10^{-14}	
	1	10^{-1}	10^{-13}	Human gastric fluids
⇑ Increasing acidity	2	10^{-2}	10^{-12}	
	3	10^{-3}	10^{-11}	Household vinegar
	4	10^{-4}	10^{-10}	
	5	10^{-5}	10^{-9}	Interior of lysosomes
	6	10^{-6}	10^{-8}	Cytoplasm of working muscle
Neutrality	**7**	**10^{-7}**	**10^{-7}**	Pure water at 25°C
	8	10^{-8}	10^{-6}	Seawater
	9	10^{-9}	10^{-5}	
⇓ Increasing alkalinity	10	10^{-10}	10^{-4}	Alkaline lakes
	11	10^{-11}	10^{-3}	Household ammonia
	12	10^{-12}	10^{-2}	Saturated lime solution
	13	10^{-13}	10^{-1}	
	14	10^{-14}	10^0	

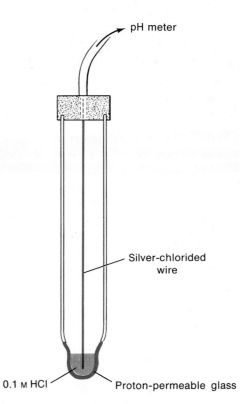

Figure 3-12 An electrode with a proton-selective tip is a convenient device for measuring the pH of solutions. The tip of a pH electrode contains a solution of pH 7 (i.e., $[H^+] = 10^{-7}$ M). When the tip is immersed in a solution of different $[H^+]$, the potential difference set up across the envelope of proton-selective glass is proportional to the log of the ratio of H^+ concentrations on the two sides of the glass.

pH meter

Silver-chlorided wire

0.1 M HCl

Proton-permeable glass

The pN (pH at neutrality) rises as temperature falls. Many animals that normally experience temperature fluctuations in their body fluids have homeostatic mechanisms to maintain pH at a constant fraction of a pH unit above neutral pH, rather than at a set pH *per se*. Some types of open heart surgery in humans require cooling the body temperature by several degrees, which changes the neutral pH of water-based fluids like blood. Should the anesthesiologist maintain the patient's blood pH at 7.4, the normal level for humans, or allow blood pH to rise as body temperature falls?

The Biological Importance of pH

The concentrations of H^+ and OH^- ions are important in biological systems because protons freely move from H_3O^+ to associate with and thereby neutralize negatively charged groups, and OH^- ions are available to neutralize positively charged groups. This ability to neutralize is especially important in amino acids and proteins, which are amphoteric molecules containing both carboxyl (i.e., –COOH) and amino (i.e. –NH_2) groups.

In solution, amino acids normally exist in a dipolar configuration called a *zwitterion*:

$$R-\underset{\underset{H}{|}}{\overset{\overset{NH_2}{|}}{C_\alpha}}-COOH \qquad\qquad R-\underset{\underset{H}{|}}{\overset{\overset{NH_3{}^+}{|}}{C_\alpha}}-COO^-$$

Undissociated   Zwitterion

Each amino acid and other amphoteric molecules has a characteristic **isoelectric point**, which is the pH at which the net charge of both the undissociated and zwitterion forms is zero. If the pH of an amino acid solution is decreased, the H^+ concentration of the solution increases. As a result, the probability of a proton neutralizing a carboxyl group will be greater than the probability of a hydroxyl ion removing the extra proton from the amino group. A large proportion of the amino acid molecules will then bear a net positive charge:
Raising the pH will, of course, have the opposite effect, with many of the amino acid molecules bearing a net negative charge.

$$R-\underset{\underset{H}{|}}{\overset{\overset{NH_3{}^+}{|}}{C_\alpha}}-COO^- + H^+ \Longleftrightarrow R-\underset{\underset{H}{|}}{\overset{\overset{NH_3{}^+}{|}}{C_\alpha}}-COOH$$

The only amphoteric groups in some amino acids are the –COOH and –NH_3 attached to the alpha-carbon atom (C_α); these groups enter into peptide bonds. Other amino acids, however, have additional carboxyl or amino side groups that can become acidic or basic. Dissociable side groups in a macromolecule will determine to a large extent the electrical properties of the molecule and will additionally render it sensitive to the pH of its environment. This sensitivity is most dramatically evident in the influence of pH on the properties of an enzyme's active site. Since the binding of a substrate to the active site of an enzyme generally includes electrostatic interactions, the formation of the enzyme-substrate complex is highly pH dependent. The highest probability of binding occurs at a particular pH, the optimal pH.

Henderson-Hasselbalch Equation

Some acids, such as HCl, dissociate completely, whereas others, such as acetic acid, dissociate only partially. The generalized chemical equation for the dissociation of an acid can be written as

$$HA \Longleftrightarrow H^+ + A^-$$

in which A^- is the anion of the acid HA. Accordingly, the dissociation constant derived from the law of mass action is given by

$$K' = \frac{[H^+][A^-]}{[HA]} \qquad (3\text{-}2)$$

It is convenient to use the logarithmic transformation of K', namely pK', which is analogous to pH:

$$pK' = -\log_{10} K'$$

Hence, if $pK' = 11$, then $K' = 10^{-11}$. A low pK' indicates a strong acid; a high pK' indicates a weak acid.

Acid-base problems can be simplified by rearranging equation 3-2. Taking the log of both sides, we obtain

$$\log K' = \log[H^+] + \log \frac{[A^-]}{[HA]} \qquad (3\text{-}3)$$

Rearranging gives us

$$-\log[H^+] = -\log K' + \log \frac{[A^-]}{[HA]} \qquad (3\text{-}4)$$

Substituting pH for $-\log[H^+]$ and pK' for $-\log K'$, we obtain

$$pH = pK' + \log \frac{[A^-]}{[HA]} \qquad (3\text{-}5)$$

In other words,

$$pH = pK' + \log \frac{[\text{proton acceptor}]}{[\text{proton donor}]}$$

Equation 3-5 is the **Henderson-Hasselbalch equation,** which permits the calculation of the pH of a conjugate acid-base pair, given the pK' and the molar ratio of the pair. Conversely, it permits the calculation of the pK', given the pH of a solution of known molar ratio.

Buffer Systems

Changes in pH affect the ionization of basic and acidic groups in enzymes and other biological molecules. Consequently, the pH of intra- and extracellular fluids must be held within the narrow limits in which enzyme systems have evolved if these enzymes are to carry out their normal functions. Deviations of one pH unit or more generally disrupt the biochemistry of organisms. This sensitivity to the pH of the aqueous intracellular environment exists in part because reaction rates of different enzyme systems become mismatched and uncoordinated. Maintaining the pH of blood is a major goal of the body's homeostatic mechanisms, because large changes in blood pH can be rapidly transmitted to other body fluids including intracellular fluids.

The pH of body fluids is maintained within normal ranges with the help of natural pH **buffers.** A buffered system is one that tolerates the addition of relatively large amounts of an acid or a base with little change in pH over a certain pH range. A buffer must contain an acid (HA) to neutralize added bases and a base (A^-) to neutralize added acids. (We have already seen that HA is an acid because it acts as an H^+ donor and that A^- is a base because it acts as an H^+ acceptor.) The properties of buffered systems are determined by adding small amounts of an acid or base, and recording the pH after each addition. A plot of pH versus the amount of added acid or base is a *titration curve.* The greatest buffering capacity of a conjugate acid-base pair occurs when [HA] and [A^-] are both large and equal. Referring to the Equation 3-5, we see that this situation exists when pH = pK' (since $\log_{10} 1 = 0$). This point corresponds to that portion of a titration curve along which there is the smallest change in pH (Figure 3-13).

The most effective buffer systems are combinations of weak acids and their salts. Weak acids dissociate only slightly, thus ensuring a large reservoir of HA. The salts of weak acids dissociate completely, providing a large reservoir of A^-. Added H^+ therefore combines with A^- to form HA, and added OH^- combines with H^+ to form H_2O. As H^+ is thereby removed, it is replaced by dissociation of HA. The most important inorganic buffer systems in the body fluids are the bicarbonates and phosphates. Amino acids, peptides, and proteins, because of their weak-acid side groups, form an important class of organic buffers in the cytoplasm and extracellular plasma.

Electric Current in Aqueous Solutions

Water conducts **electric current,** which is why we are frequently cautioned against using electrical appliances in wet conditions. Water's **conductivity,** the rate of charge transfer caused by the migration of ions under a given potential, is

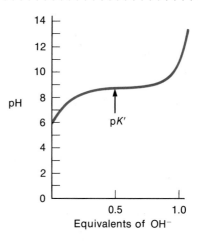

Figure 3-13 The greatest buffering capacity of a conjugate acid-base system is obtained when pH = pK'. On the graph, this point corresponds to the part of the curve with the shallowest slope (small pH changes with large amounts of OH^- added).

far greater than that of oils or other nonpolar liquids. The conductivity of water depends entirely on the presence of charged atoms or molecules (ions) in solution. Electrons, which carry electric current in metals and semiconductors, play no direct role in the flow of electric current in aqueous solutions. Because the concentrations of H^+ and OH^-, the ions present in pure water, are quite low (10^{-7} M at 25°C), the electrical conductivity of pure water is relatively low, though far higher than that of nonpolar liquids. The conductivity of water is greatly enhanced by the addition of electrolytes, which dissociate into cations (positive ions) and anions (negative ions) in water (see Figure 3-7B). Thus, seawater conducts electric current far more readily than freshwater. Spotlight 3-1 reviews some common terms, units, and conventions that apply to electrical properties.

The role of ions in conducting electric current in solution is illustrated in Figure 3-14. In this example, two electrodes are immersed in a solution of KCl and connected by wires to a source of electromotive force (emf), the two

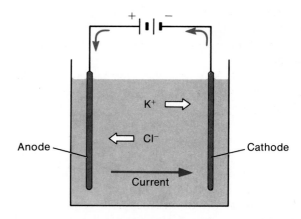

Figure 3-14 In aqueous solution, electric current is carried by the movement of the dissociated ions of electrolytes. Colored arrows indicate direction of current flow. Open arrows indicate direction of ion flow.

SPOTLIGHT 3-1

ELECTRICAL TERMINOLOGY AND CONVENTIONS

The main electrical properties and their units that you will encounter in this and later chapters are defined here. Common symbols used to diagram electrical circuits are shown in the accompanying figure.

- **Electric charge,** q, is measured in units of **coulombs** (C). To convert 1 g equivalent weight of a monovalent ion to its elemental form (or vice versa) requires a charge of 96,500 C (1 **faraday,** 1 F). Thus, in loose terms, a coulomb is equivalent to 1/96,500 g equivalent of electrons. The charge on one electron is -1.6×10^{-19} C. If this value is multiplied by Avogadro's number, the total charge is one faraday (i.e., $-96,487$ C·mol^{-1}).

- **Current,** I, is the flow of charge, which is measured in **amperes** (A). A current of 1 C·s^{-1} equals 1 ampere. By convention, the direction of current flow is the direction in which a positive charge moves (i.e., from the anode to the cathode).

- **Voltage,** V or E, is the electromotive force (emf) or electric potential expressed in **volts.** When the work required to move 1 C of charge from one point to a point of higher potential is 1 joule (J), or about 0.24 calories (cal), the potential difference between these points is said to be 1 volt (V).

- **Resistance,** R, is the property that hinders the flow of current measured in **ohms** (Ω). A resistance of 1 Ω allows exactly 1 A of current to flow when a potential drop of 1 V exists across the resistance. An ohm is equivalent to the resistance of a column of mercury 1 mm^2 in cross-sectional area and 106.3 cm long. R = resistivity × length/cross-sectional area.

- **Resistivity,** ρ, is the resistance of a conductor 1 cm in length and 1 cm^2 in cross-sectional area.

- **Conductance,** g, is the reciprocal of resistance, $g = 1/R$. The unit is the **siemens** (S) (formerly the mho).

- **Conductivity** is the reciprocal of resistivity.

Ohm's law states that current is proportional to voltage and inversely proportional to resistance:

$$I = \frac{V}{R} \quad \text{or} \quad = I \times R$$

Thus, a potential of 1 V across a resistance of 1 Ω will result in a current of 1 A. Conversely, a current of 1 A flowing through a resistance of 1 Ω produces a potential difference across that resistance of 1 V.

Capacitance, C, is the property of a nonconductor to store electric charge. A capacitor (or conductor) consists of two plates separated by an insulator. If a battery is connected in parallel with the two plates, charges will move up to one plate and away from the other until the potential difference between the plates is equal to the emf of the battery, or until the insulation breaks down. No charges move "bodily" across the insulation between the plates in an ideal capacitor, but charges of one sign accumulating on one plate electrostatically repel similar charges on the opposite plate. The **capacity,** or charge-storing ability, of a capacitor is given in **farads** (F). If a potential of 1 V is applied across a capacitor and 1 C of positive charge is thereby accumulated by one plate and lost by the other plate, the capacitor is said to have a capacity of 1 F:

$$C = \frac{q}{V} = \frac{1 \text{ coulomb (C)}}{1 \text{ volt (V)}} = 1 \text{ farad (F)}$$

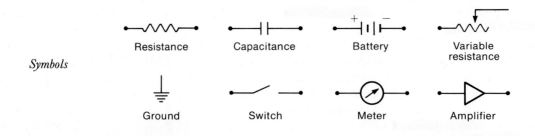

Symbols

Resistance   Capacitance   Battery   Variable resistance

Ground   Switch   Meter   Amplifier

---

terminals marked + and −. The emf causes a current (i.e., a unidirectional displacement of positive electric charge) to flow through the electrolyte solution from one electrode to the other. What does this electric current consist of? In the wire, it consists of the displacement of electrons from the outer shell of one metal atom to another, then to another, and so on. In the KCl solution, electric charge is carried pri-

marily by K$^+$ and Cl$^-$; because the concentrations OH$^-$, H$_3$O$^+$, and H$^+$ are so low, their contribution to the current will be ignored. When a potential difference (voltage) is applied to an electrolyte solution, the cations migrate toward the **cathode** (electrode with the negative potential) and the anions migrate toward the **anode** (electrode with positive potential).

The rate at which each species of ion migrates in solution is termed its **electrical mobility**. This mobility is determined by the ion's hydrated mass and the amount of charge (monovalent, divalent, or trivalent) that it bears. The mobility of H^+ is considerably higher than the mobilities of other common ions. The movement of ions that constitutes an ionic current is crudely analogous to a wave of falling dominoes, in which each domino (ion) is displaced just enough to cause a displacement of the next domino. Instead of interacting mechanically, like falling dominoes, ions influence each other through electrostatic interactions, with like charges repelling each other.

The current in a solution is said by convention to flow in the direction of cation migration. Anions flow in the opposite direction. The rate at which positive charges are displaced past a given point in the solution, plus the rate at which negative charges are displaced in the opposite direction, determines the *intensity* of the electric current, that is, the number of unit charges flowing past a point in 1 second. Thus electric current is analogous to the volume of water that flows in a second past a point in a pipe (Figure 3-15).

An electric current always meets some electrical **resistance** to its flow, just as water meets a mechanical resistance owing to such factors as friction during its flow through a pipe. In order for the charges to flow through an electrical resistance, there must be an electrostatic force acting on the charges. This force (analogous to hydrostatic pressure in a water-filled pipe) is the difference in electric pressure, or potential, V, between the two ends of the resistive pathway (see Figure 3-15A). A difference in potential, or **voltage,** exists between separated negative $(-)$ and positive $(+)$ charges. This potential difference, or emf, is related to the current, I, and resistance, R, as described by **Ohm's law** (see Spotlight 3-1). To force a given current through a pathway of twice the resistance requires twice the voltage (Figure 3-16A). Similarly, the current will be reduced to half its value if the resistance it encounters is doubled while the voltage is kept constant (Figure 3-16B).

Three major factors determine the resistance to current flow in a solution:

1. The availability of charge carriers in the solution (i.e., the ion concentration): The more dilute an electrolyte solution, the higher its resistance, and thus the lower its conductivity (Spotlight 3-1). This makes sense, since fewer ions are available to carry current.

2. The cross-sectional area of the solution in a plane perpendicular to the direction of current flow: The smaller this cross-sectional area, the higher the resistance encountered by the current. This, again, is analogous to the effect of the cross-sectional area of a pipe carrying water.

3. The distance traversed in solution by the current: The total resistance encountered by a current passing through an electrolyte solution is directly proportional to the distance the current traverses.

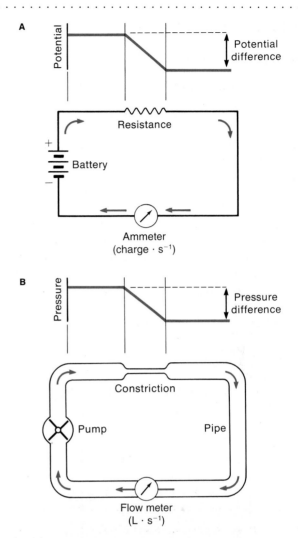

Figure 3-15 The flow of electrons in a wire **(A)** can be compared to the flow of water in a pipe **(B)**. An electric current always meets some resistance, analogous to a constriction in a water pipe.

The ions carrying current are distributed evenly throughout a solution. Yet current flowing between two electrodes does arche out in curved paths rather than flowing in a direct pathway (Figure 3-17). This behavior brings far more ions into play than are present in a direct path between the electrodes, thus providing a lower effective resistance to the flow of electric current (point 1, above), even though the curved pathway is longer.

The importance of electrical phenomena in animal physiology will become abundantly apparent in later chapters, especially those dealing with the nervous system. Familiarity with basic concepts of electricity is also useful for an appreciation of laboratory instruments.

Binding of Ions to Macromolecules

Ions free in solution inside or outside living cells interact electrostatically with one another and with a variety of ionized or partially ionized portions of molecules, especially proteins. The **ion-binding sites** of macromolecules carry electric charges, and their interactions with free inorganic

A

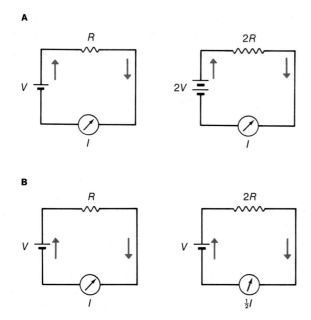

B

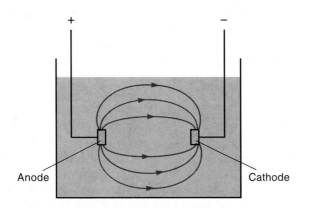

Figure 3-16 Ohm's law describes the relationship between electric current, *I* (number of charges moving past a point per unit time), potential difference, *V*, and resistance, *R*. **(A)** The current intensity, indicated by readings on the ammeter, remains unchanged if both voltage and resistance are doubled. **(B)** Current drops by half if resistance alone is doubled.

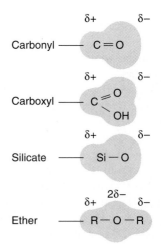

Figure 3-17 Current flow through a volume of electrolyte solution spreads so as to decrease current density.

ions are based on the same principles that determine ion exchange at sites on such nonbiological materials as soil particles, glass, and certain plastics. Interactions between fixed ion-binding sites and various ions are highly important in certain physiological mechanisms, such as enzyme activation and the selectivity of membrane channels and carriers for particular ions.

The energetic basis for interaction between an ion and an ion-binding site is the electrostatic attraction between the two and is identical in principle to the interactions that occur between anions and cations in free solution. Thus, a site with a negative charge or a partial negative charge (recall the partial charge on the oxygen atom of the water molecule) attracts cations; a site with a positive charge attracts anions. Two or more species of cations in solution will compete with each other to bind electrostatically to an

anionic (i.e., electronegative) site. The negatively charged site will show an order of binding preference among cation species, ranging from those that bind most strongly to those that bind least strongly. This order of preference is called the **affinity sequence,** or **selectivity sequence,** of the site.

Cation-binding sites on organic molecules are generally oxygen atoms in such groups as silicates ($-SiO^-$), carbonyls ($R-C=O$), carboxylates ($R-COO^-$), and ethers (R_1-O-R_2). As was noted earlier, the oxygen atom is strongly electron hungry, and draws electrons from surrounding atoms in the molecule. The oxygen atoms in such neutral groups as the carbonyls or ethers can be treated as having a partial negative charge due to the statistically higher number of electrons around them (Figure 3-18). Since the group itself is neutral, there must also, of course, be partial positive charges on the other atoms. When silicate and carboxylate groups are ionized, their oxygen atoms carry a full negative charge.

The energetics of electrostatic interaction of a site with an ion are expressed in terms of potential energy—namely, the energy, U, of bringing together two charges, q^+ and q^- in a vacuum from a separation of infinity to the new distance of separation d:*

$$d = \frac{(q^+q^-)}{d^a} \qquad (3\text{-}6)$$

The exponent a equals 1 in the case of two monopoles each carrying a full charge (i.e., a monovalent anion and a monovalent cation). For a dipolar molecule such as water, in which there are centers of both negative and positive charge (but no net charge), the energy of interaction falls off

Figure 3-18 Many biological molecules contain groups that exhibit a partial charge separation. Most common are oxygen-containing groups in which the highly electronegative oxygen atom draws electrons from neighboring atoms. The electron-cloud distributions of several molecular side groups are indicated by shading. Although not present in animals, silicate is a major component of the skeleton of diatoms.

* Equation 3-6 should not be confused with Coulomb's law, which is given in the footnote on page 43.

more rapidly with distance (i.e., a in Equation 3-6 is greater than 1). This plays an important role in the electrostatic tug-of-war experienced by an ion dissolved in water attracted to a site of opposite charge.

In an aqueous environment (i.e., in a solution as opposed to a vacuum), the coulombic relation (Equation 3-6) between the atomic radius of a cation and its affinity for a given fixed electronegative site is modified by the electrostatic interaction of the cation with dipolar water molecules. The cation is attracted to both the electron-rich oxygen atom of the fixed monopolar site and the electron-rich oxygen atom of the dipolar water molecule. Thus water and the site engage in competition for binding of the cation. The more successfully the site competes with water for a given ionic species, the greater the "selectivity" of the site for that ionic species (Figure 3-19). The selectivity sequence of a site for a group of different ions will be determined by the field strength and the polar/multipolar distribution of electrons near the site. In addition, the nucleus of a small atom can more closely approach another atom than can the nucleus of a large atom. Thus, *small* monovalent cations will interact more strongly with a particular electronegative site than will *large* monovalent cations because they carry the same unit charge but have a smaller distance of closest approach.

In addition to the principles of electrostatic interaction briefly described here, there are steric constraints on the binding of ions with some sites. If, for example, a site is situated so that an interacting ion must squeeze into a narrow depression or hollow in or between molecules, the hydrated size of the ion will then also have an effect on the total energy required to reach and interact with the site.

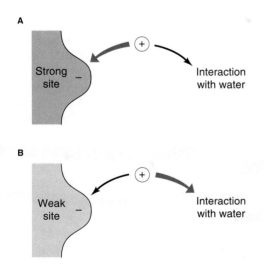

Figure 3-19 The ability of a fixed anionic site to compete with water molecules for a cation depends on the ion-attracting strength of the site as well as the size of the cation, because smaller cations allow a smaller distance of closest approach. **(A)** The force of attraction of a small monovalent cation to a strong anionic point site is greater than its attraction to water (and vice versa for a large monovalent cation). **(B)** The force of attraction of a small monovalent cation to a weak point site is less than its attraction to water (and vice versa for a large monovalent cation).

Having discussed some of the basics of the interactions of atoms, elements, and molecules, let us now turn our attention to those molecules of specific importance to organisms.

BIOLOGICAL MOLECULES

Even a "simple" unicellular organism has an almost indescribably complex molecular makeup. This complexity is further compounded by the fact that no two animal species have the same molecular composition. In fact, the molecular composition of no individual of a species is identical to that of any other in the same species except in those reproduced by cellular fission (e.g., the two daughter cells of an amoeba or monozygotous mammalian twins). Such biochemical diversity is a major factor in evolution, for it provides an enormous number of variables in a population of organisms and acts as the raw material, so to speak, on which natural selection operates. This diversity is in part made possible by the great potential for structural variability exhibited by the carbon atom, with its ability to form four highly stable bonds. In fact, carbon is the "backbone" molecule for the four major classes of organic compounds found in living organisms: **lipids, carbohydrates, proteins,** and **nucleic acids.** We will review the chemical structures of these four classes of substances and consider some properties important to their roles in physiology. More specialized texts on biochemistry should be consulted for further details (see Suggested Readings).

Lipids

Lipids comprise a diverse group of water-insoluble biological molecules with relatively simple chemical structures. The various lipids have a variety of functions. For example, fats serve as energy stores, while phospholipids and sterols are major components of membranes (see Chapter 4).

Fats are composed of **triglyceride** molecules, each of which consists of a glycerol molecule connected through ester bonds with three fatty acid chains. When triglycerides are hydrolyzed (i.e., digested) by the insertion of H^+ and OH^- into the ester bonds, they break down into glycerol and three fatty acid molecules (Figure 3-20). The three fatty acids in a triglyceride may or may not be the same, but they all contain an even number of carbon atoms. If all the carbon atoms in a fatty acid chain are linked by single bonds (i.e., each carbon except the carboxyl carbon bears two hydrogens), the fatty acid is said to be **saturated.** If the fatty acid chain contains one or more double bonds between carbon atoms, the fatty acid is said to be **unsaturated.** The degree of saturation and the length of the fatty acids (i.e., number of carbon atoms) composing a fat determine its physical properties.

Fats containing unsaturated fatty acids generally have low melting points and form oils or soft fats at room temperature, whereas fats containing saturated fatty acids form solids at room temperatures (Table 3-4). That is why the process of hydrogenation (saturating the fatty acid chains

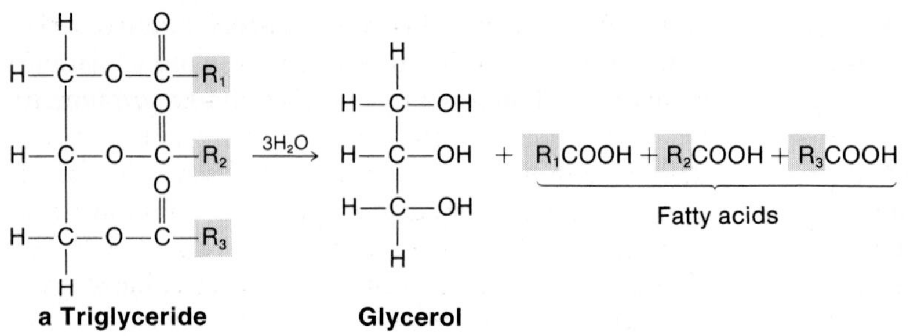

Figure 3-20 Fats are composed of triglyceride molecules, which are hydrolyzed to glycerol and fatty acids. This reaction is catalyzed by the enzyme lipase. R represents a fatty acid radical. The fatty acid groups in a particular triglyceride may be the same or different.

with hydrogens and thereby breaking the double bonds) converts vegetable oil into Crisco, for example. In addition, if the number of double bonds is constant, the shorter the chain length of a fatty acid, the lower its melting point, as illustrated by the saturated fatty acids in Table 3-4. Saturated fatty acids are more readily converted by metabolic processes into sterols such as cholesterol. Because excess cholesterol appears to be a risk factor for cardiovascular disease in humans, many dietary guidelines recommend limiting consumption of saturated fats. Cholesterol, however, is a component of biological membranes and also is the precursor for synthesis of the steroid hormones (see Figure 9-23).

Triglycerides typically accumulate in the fat vacuoles of specialized adipose cells in vertebrates. Because of their low solubility in water, these energy-rich molecules can be stored in large concentrations in the body without requiring large quantities of water as a solvent. Triglyceride energy stores are also rendered highly compact by the relatively high proportions of hydrogen and carbon and low proportions of oxygen in the molecule. Thus, 1 g of triglyceride will yield about two times as much energy upon oxidation as 1 g of carbohydrate (Table 3-5).

In phospholipids one of the outer fatty acid chains of a triglyceride is replaced with a phosphate-containing group (see Figure 4-3). Thus the phospholipids are amphipathic molecules, with a hydrophilic portion (the phosphate-containing group) and a hydrophobic portion (the fatty acid chains), which is soluble in lipids (or lipophilic).

This property allows phospholipid molecules in biological membranes to form a transition layer between an aqueous phase and a lipid phase. As discussed in the next chapter, biological membranes consist largely of two phospholipid layers, with the nonpolar "tails" of each layer oriented inward toward each other and all the polar "heads" oriented toward the aqueous phases (see Figure 4-6).

Other types of lipids found in membranes are glycolipids, which contain one or more sugar groups, and sphingolipids, which contain a long-chain amino alcohol called sphingosine. Sphingolipids are present in particularly high concentrations in brain and nerve tissue. Waxes constitute another group of lipid; they form an important waterproofing layer in certain insects (see Chapter 14).

Carbohydrates

Carbohydrates are polyhydroxyl aldehydes and ketones with the general chemical formula of $(CH_2O)_n$. The simplest carbohydrates are the monosaccharide sugars, the most common of which contain six carbons (hexoses) or five carbons (pentoses). Monosaccharides typically exist as ring structures containing four or five carbon atoms and one oxygen, with the remaining carbon(s) outside the ring (Figure 3-21A). Green plants manufacture the hexose glucose from H_2O and CO_2 by the process of photosynthesis. All the energy trapped by photosynthesis and transmitted as chemical energy to the living world (i.e., all plant and animal tissues) is channeled through such six-carbon sugars as glucose. As noted later in this next chapter, the complete or partial degradation of glucose to H_2O and CO_2 during cellular respiration releases the chemical energy that was stored in its molecular structure during photosynthesis. The two most important pentose sugars are ribose and 2-deoxyribose (see Figure 3-21A). These pentoses, which occur in the backbones of all nucleic acid molecules, are essential for replication of DNA and synthesis of proteins.

TABLE 3-4
Melting points of various fatty acids

Fatty acids	No. of carbon atoms	No. of double bonds	Melting point (°C)
Saturated			
Lauric acid	11	0	44
Palmitic acid	16	0	63
Arachidic acid	20	0	75
Lignoceric acid	24	0	84
Unsaturated			
Oleic acid	18	1	13
Linoleic acid	18	2	−5
Arachidonic acid	20	4	−50

TABLE 3-5
The energy content of the three major categories of foodstuffs

Substrate	Energy content (kcal · g⁻¹)
Carbohydrates	4.0
Proteins	4.5
Fats	9.5

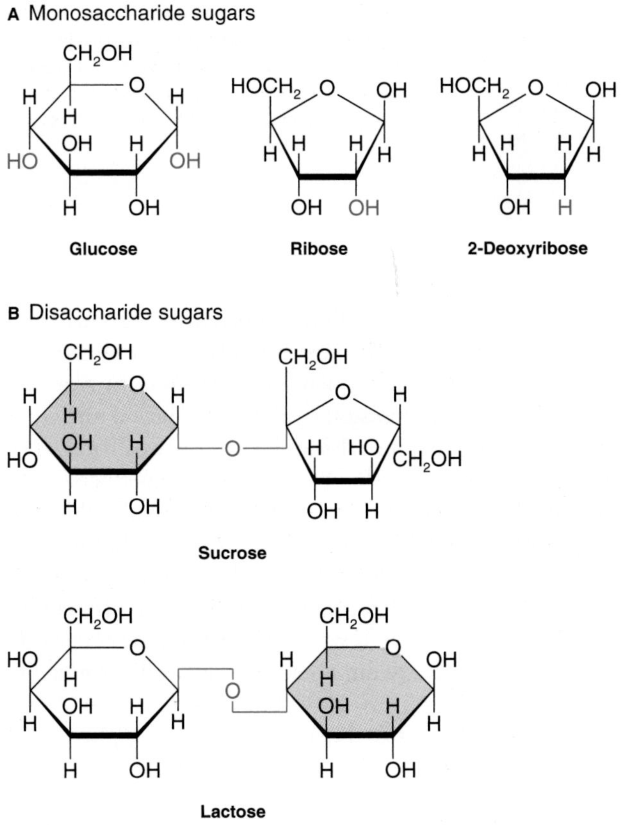

A Monosaccharide sugars

Glucose   Ribose   2-Deoxyribose

B Disaccharide sugars

Sucrose

Lactose

Figure 3-21 The simple sugars are monosaccharides and disaccharides. **(A)** Glucose, the most prevalent hexose in cells, is degraded to provide energy. The hydroxyl groups in red can form a covalent bond with another sugar molecule, forming a disaccharide. Two pentoses, ribose and 2-deoxyribose, are constituents of nucleic acids. **(B)** Disaccharides are formed by condensation of two monosaccharide units. Sucrose and lactose both contain one glucose unit (shaded) plus a second monosaccharide. The glycosidic bond (red) linking two monosaccharide units can have two different orientations, designated α and β.

Cells contain enzymes that can convert glucose to other monosaccharides or link two monosaccharide molecules to form a **disaccharide sugar** such as sucrose or lactose (Figure 3-21B). Cells also can synthesize various carbohy-

drate **polymers** containing large numbers of monosaccharide units. Two branched polymers of D-glucose—**starch** in plant cells and **glycogen** in animal cells—are the primary forms for storing carbohydrate (Figure 3-22). Like fats, these high-molecular-weight carbohydrate polymers require a minimum of water as a solvent and constitute a concentrated form of food reserve in the cell. In vertebrates, glycogen is found in the form of minute intracellular granules, primarily in liver and muscle cells.

Carbohydrate polymers also form structural substances. The main structural substance in plants, for example, is cellulose—an unbranched polymer of D-glucose. **Chitin**, which is a major constituent of the exoskeletons of insects and crustaceans, is a cellulose-like polymer of N-acetyl glucosamine, an amino derivative of D-glucose (Figure 3-23). Both the plant polymer cellulose and chitin are flexible, elastic, and insoluble in water.

Proteins

Proteins are the most complex and the most abundant organic molecules in the living cell, making up more than half the mass of a cell as measured by dry weight. Although the basic structure of all proteins is similar, a vast array of different proteins with diverse functions is found in biological systems. Table 3-6 lists the major functional types of proteins with several examples of each type. Enzymes constitute the largest functional group of proteins, with more than 1000 already identified and many unknown ones presumably to be discovered.

Primary structure

Proteins are composed of linear chains of amino acids, which are amphoteric molecules containing at least one carboxyl group and one amino group. The 20 common amino acids that make up proteins are all alpha-amino acids, in which the amino group is bonded to the alpha-carbon (C_α) atom, that is, the carbon atom adjacent to the carboxyl group. Amino acids differ from one another in the structure of their side groups, generically referred to as R

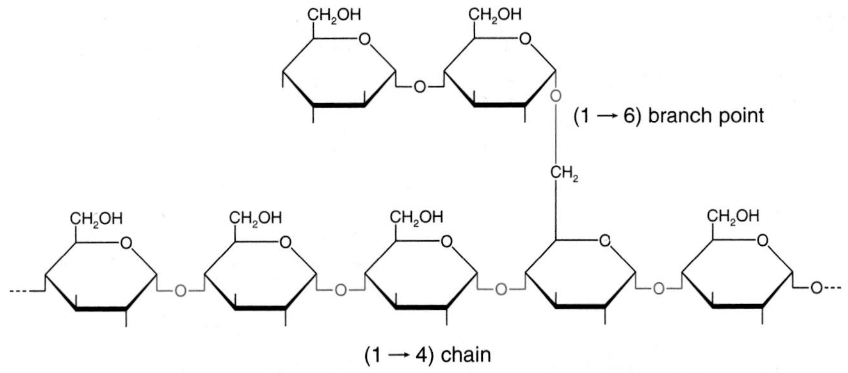

$(1 \rightarrow 6)$ branch point

$(1 \rightarrow 4)$ chain

Figure 3-22 Glycogen, a large glucose polymer, is the primary carbohydrate storage form in animal cells. A glycogen molecule is a long chain of glucose residues, in which carbons 1 and 4 in adjacent molecules are linked, with branches extending from carbon 6 every eight to ten glucose residues. Only a small portion of a glycogen molecule is depicted.

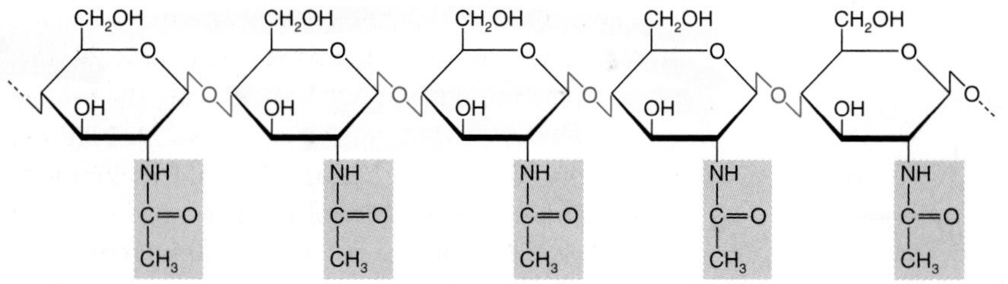

Figure 3-23 Chitin, a structural carbohydrate polymer, consists of *N*-acetyl glucosamine units joined by 1 ⟶ 4 glycosidic bonds. In *N*-acetyl glucosamine, an acetamide group (color shading) replaces the hydroxyl group on carbon 2 of glucose.

groups (Figure 3-24A). The protein-synthesizing machinery of cells joins amino acid molecules via covalent **peptide bonds,** forming long **polypeptide chains.** Adjacent C_α atoms in a polypeptide chain are separated by a planar **amide group** (Figure 3-24B). The specific linear sequence of amino acid residues of a polypeptide is termed its **primary structure.** Since the amino acid residues of a polypeptide chain differ only in their side groups, these groups are like letters in the protein alphabet, defining the primary structure of a protein (Table 3-7). A protein molecule may consist of one, two, or several polypeptide chains, either covalently linked or held together by weaker bonding.

The amino acid sequence of a polypeptide (i.e., its primary structure) is encoded in an organism's genetic mate-rial. Indeed, all the hereditary information carried in the genetic material is translated initially into protein molecules: the amino acid sequence laid down during protein synthesis is the expression of this information and is the primary determinant of the properties of any protein molecule. Since there are about 20 different amino acid building blocks, an impressive variety of different amino acid sequences is possible. Suppose, for example, that we were to construct a polypeptide molecule consisting of one of each of those 20 building blocks. How many different linear arrangements could we make without ever repeating the same sequence of amino acids? This is determined by multiplying $20 \times 19 \times 18 \times 17 \times 16 \times \ldots \times 2 \times 1$ (i.e., 20!), or 10^{18}. But this startlingly large figure, which

TABLE 3-6

Classification of proteins and peptides according to biological function

Type/examples	Occurrence or function	Type/examples	Occurrence or function
Enzymes		**Protective proteins in vertebrate blood**	
Cytochrome *c*	Transfers electrons	Antibodies	Form complexes with foreign proteins
Ribonuclease	Hydrolyzes RNA		
Trypsin	Hydrolyzes some peptides	Fibrinogen	Precursor of fibrin in blood clotting
Regulatory proteins		Thrombin	Component of clotting mechanism
Calmodulin	Intracellular calcium-binding modulator		
Tropomyosin	Contraction regulator in muscle	**Toxins**	
Troponin C	Calcium-binding contraction regulator in muscle	Bungarotoxin	Agent in cobra venom that blocks neurotransmitter receptors
Storage proteins		*Clostridium botulinum* toxin	Blocks neurotransmitter release
Casein	Milk protein	**Hormones**	
Ferritin	Iron storage in spleen	Adrenocorticotropic hormone	Regulates corticosteroid synthesis
Myoglobin	O_2 storage in muscle		
Ovalbumin	Egg-white protein	Growth hormone	Induces growth of bones
Transport proteins		Insulin	Regulates glucose metabolism
Hemocyanin	Transports O_2 in hemolymph of some invertebrates	**Structural proteins**	
Hemoglobin	Transports O_2 in blood of vertebrates	Alpha-keratin	Skin, feathers, nails, hoofs
		Collagen	Fibrous connective tissue (tendons, bone, cartilage)
Serum albumin	Transports fatty acids in blood	Elastin	Elastic connective tissue (ligaments)
Contractile proteins		Fibroin (beta-keratin)	Silk of cocoons, spider webs
Actin	Moving filaments in myofibril	Glycoproteins	Cell coats and walls
Dynein	Cilia and flagella	Sclerotin	Exoskeletons of insects
Myosin	Stationary filaments in myofibril		

A General structure of alpha-amino acids

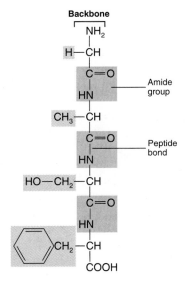

B Structure of a tetrapeptide

Figure 3-24 The primary structure of proteins is a linear sequence of alpha-amino acids linked by peptide bonds. **(A)** All the amino acids found in proteins have a common structure. Each has a characteristic side group commonly indicated by R (see Table 3-7). **(B)** The peptide bonds (red lines) linking the amino acid residues in polypeptides have a partial double-bond character. As a result, the amide group (gray shading) is planar. Although the polypeptide backbone is the same in all proteins, they differ in the sequence of side groups. This sequence, the primary structure, is the defining property of each protein.

applies to a relatively small protein with a molecular weight of about 2400, pales in comparison to the possibilities for a more typical protein with a molecular weight of 35,000. For a protein of this size, containing just 12 kinds of amino acids, the number of possible sequences exceeds 10^{300}.

Higher levels of structure

The primary structure of a polypeptide chain determines the three-dimensional conformation, or shape, that it assumes in a given environment. This conformation depends on the nature and position of the side groups that project from the peptide backbone. In addition to the primary structure (i.e., the amino acid sequence), proteins exhibit additional levels of structure, designated secondary, tertiary, and quaternary. **Secondary structure** refers to the local organization of parts of the polypeptide chain, which can assume several different arrangements; **tertiary structure** refers to the foldings of the chain to produce globular or rodlike molecules; and **quaternary structure** refers to the joining of two or more polypeptide chains to form dimers, trimers, and occasionally even larger aggregates.

Because the C_α—N peptide bond has a partial double-bond character, it is not free to rotate; hence, the atoms of the amide group are confined to a single plane (see Figure 3-24B). However, the remaining bonds of the peptide backbone are free to rotate. Linus Pauling and Robert Corey, using precisely constructed atomic models, found that the simplest stable secondary structure of a polypeptide chain is a helical arrangement called the **alpha (α) helix** (Figure 3-25). In this structure, the plane of each amide group is parallel to the major axis of the helix and there are

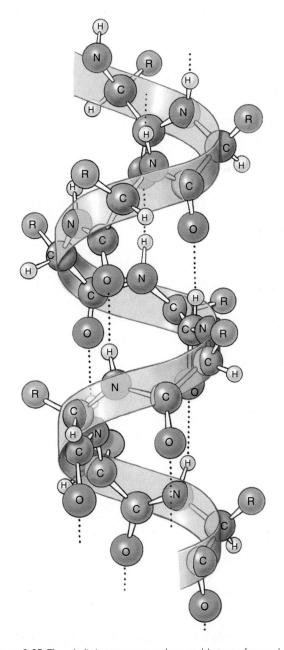

Figure 3-25 The α helix is a common and very stable type of secondary structure in proteins. This helical arrangement, containing 3.6 amino acids per turn, is stabilized by hydrogen bonds (black dots) between the oxygen atom of a carbonyl group and the hydrogen atom of the amide group four residues away in the backbone. The side groups (R) extend outward from the axis of the backbone.

TABLE 3-7
Side groups or radicals of the 20 common alpha-amino acids

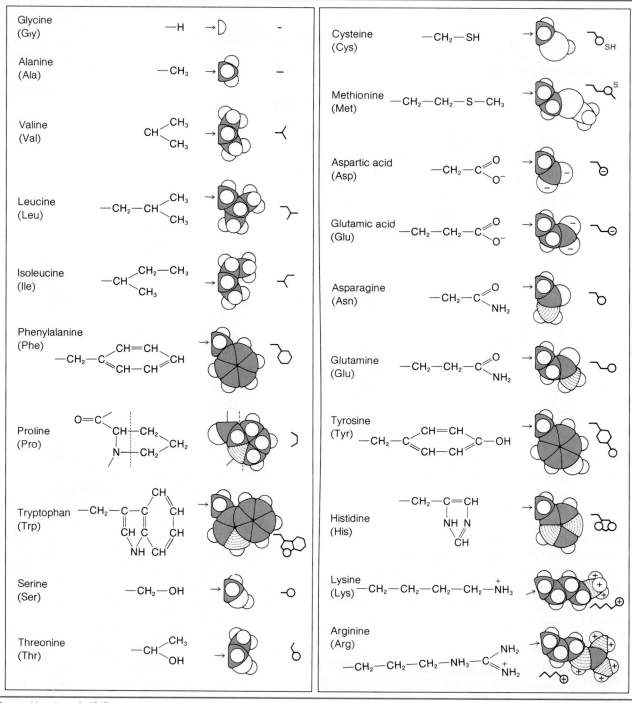

Source: Haggis et al., 1965.

3.6 amino acid residues per turn. The side group of each amino acid residue extends outward from the helical backbone, free for interaction with other side groups or other molecules. The stability of the α helix is enhanced substantially by hydrogen bonding between the oxygen atom of a carbonyl group and the hydrogen atom of the amide group four residues ahead. Because of the stability of the α helix, a polypeptide chain spontaneously assumes this conformation, provided that the side groups do not interfere. In the

amino acid proline, for example, the side group is a rigid ring that includes the alpha-nitrogen atom (see Table 3-7). Thus, the C_α—N bond in proline (and hydroxyproline) cannot rotate; as a result, whenever proline (or hydroxyproline) occurs in a peptide chain, it interrupts the α helix, causing the peptide backbone to bend.

Another major type of protein secondary structure is the **beta (β) pleated sheet** (Figure 3-26). This consists of laterally associated β strands, which are fairly short, nearly

Face view

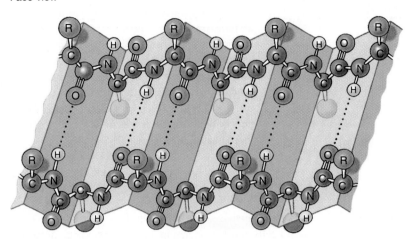

Side view

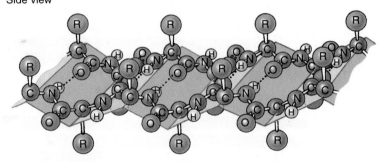

Figure 3-26 The β pleated sheet is an element of secondary structure in silk fibers and some other fibrous proteins. Pleated sheets are formed by the lateral association of two β strands stabilized by hydrogen bonds (black dots). The side groups (R) extend above and below the plane of the sheet. The broad arrows represent β strands. [Adapted from Lodish et al., 1995.]

fully extended stretches of the polypeptide chain. Hydrogen bonding between carbonyl oxygen atoms and amide hydrogen atoms in adjacent β strands forms a pleated sheet with the side groups of the amino acid residues projecting above or below the plane of the sheet. Association of β strands within the same polypeptide chain contributes to secondary structure, whereas association of β strands in different polypeptide chains contributes to quaternary structure.

Long polypeptide chains with an uninterrupted α-helix conformation are characteristic of fibrous proteins, such as the **alpha-keratins** that form hair, fingernails and claws, wool, horn, and feathers. **Beta-keratins** are an exception, having a secondary structure consisting of β pleated sheets rather than α helices. Beta-keratins are the major constituent of spider webs and silk, which is produced by caterpillars. Nonstructural intracellular proteins typically have a random-coil secondary structure, although these proteins may contain short segments in the α-helix or β-pleated-sheet conformation.

Regions of a polypeptide chain with long α helices often assume a rodlike tertiary structure, whereas those lacking this feature have a globular tertiary structure. Two types of relatively weak, noncovalent interactions help stabilize tertiary structure: coulombic (electrostatic) interactions between the charged side groups and **van der Waals forces** between hydrophobic side groups. Another major contribu-

tor to the conformation of proteins is the sulfhydryl side group (–SH) of the amino acid cysteine. The reaction of two cysteine residues forms a **disulfide linkage** (S–S), which covalently joins the residues (Figure 3-27). A disulfide linkage can covalently cross-link different portions of a polypeptide chain, thereby stabilizing its folded tertiary structure, or connect two separate chains. Since the sulfhydryl group is highly reactive, it is not surprising that one or more cysteine residues frequently occupy the active sites of enzymes. The toxicity of mercury and other heavy metals is due, in part, to their reaction with the sulfur atom of cysteine, displacing the hydrogen atoms. This reaction can poison (i.e., render catalytically inoperative) the active site of an enzyme.

Some, but not all, proteins undergo *self-assembly*, forming a quaternary structure. The amino acid sequence of a polypeptide chain—and hence the positions of the different amino acid side groups—not only determines the secondary and tertiary structure of the molecule, but also may allow interaction with other polypeptide chains, thereby forming protein molecules with two or more subunits. The association of subunits can involve covalent disulfide linkages between them as well as noncovalent interactions between complementary regions on their surfaces. For example, negatively charged groups of one subunit fit against positively charged groups of another subunit; hydrophobic, nonpolar side groups on the

$$2 \; SH—CH_2—\overset{\overset{\displaystyle H}{|}}{\underset{\underset{\displaystyle NH_2}{|}}{C}}—COOH$$ **Cysteine**

$$COOH—\overset{\overset{\displaystyle H}{|}}{\underset{\underset{\displaystyle NH_2}{|}}{C}}—CH_2—S—S—CH_2—\overset{\overset{\displaystyle H}{|}}{\underset{\underset{\displaystyle NH_2}{|}}{C}}—COOH$$ **Cystine**

Disulfide
linkage

Figure 3-27 A disulfide bond can contribute to the tertiary structure of proteins by linking cyteine residues present in different portions of the same polypeptide chain. Disulfide bonds also can form between cysteine residues in different polypeptide chains, thereby contributing to quaternary structure.

subunits meet to the mutual exclusion of water molecules; or residues in each subunit are oriented so they can form hydrogen bonds. Some enzymes, the respiratory pigment hemoglobin, and many other proteins consist of more than one polypeptide chain held together by noncovalent bonds. In some multi-subunit proteins, β pleated sheets connect the subunits. The three subunits of collagen, the major protein in connective tissue, are twisted into a characteristic superhelix (Figure 3-28). The subunits of all these proteins will assemble themselves spontaneously if added separately to an aqueous solution and mixed. The associated and dissociated subunits of hemoglobin are illustrated in Figure 13-2A.

Except for covalent disulfide linkages between cysteine residues, the secondary, tertiary, and quaternary structure of proteins depends on coulombic interactions, hydrogen bonding, and van der Waals forces. All of these noncovalent interactions are relatively weak and heat labile. Heating a protein disrupts these interactions leading to alterations in its conformation, called **denaturation**. Hair curling irons work in this way, temporarily heating the proteins in the hair shaft and then letting them cool in slightly new configurations that alter the shaft's orientation. In this same way, high temperatures can change the shape of enzymes, rendering them inactive and killing the cells in which they reside.

The proteins of most animals begin to denature at temperatures above 43–45°C. Yet, some species of fishes, insects, algae, and bacteria inhabit hot-water springs in the range of 48°C. A few species of bacteria live at temperatures of up to 54°C! What structural specializations do you think could account for the continuing function in high temperatures of the proteins of these heat-tolerant species?

Nucleic Acids

Deoxyribonucleic acid (DNA) was first isolated from white blood cells and fish sperm in 1869 by Friedrich Miescher. During the next decades the chemical composition of DNA was gradually worked out, and evidence slowly accumulated that implicated it in the mechanisms of heredity. We now know that DNA, which is associated with the chromosomes, carries coded information, arranged into **genes**, that is passed from each cell to its daughter cells and from one generation of organisms to the next. A second group of nucleic acids, **ribonucleic acid** (RNA), was subsequently discovered. RNA is now known to be instrumental in translating the coded message of DNA into sequences of amino acids during synthesis of protein molecules.

The nucleic acids are polymers of **nucleotides**, which consist of a **pyrimidine** or **purine** base, a pentose sugar, and a phosphoric acid residue (Figure 3-29). The nucleotides composing DNA contain deoxyribose, whereas those composing RNA contain ribose (see Figure 3-21A). The major nucleotides found in nucleic acids contain the following bases: **adenine, thymine, guanine, cytosine, and uracil.** Thymine occurs only in DNA, and uracil only in RNA; the other three bases are found in both nucleic acids. Stable *base pairs*, linked by hydrogen bonds, can form between adenine and thymine (A–T), guanine and cytosine (G–C), and adenine and uracil (A–U), as depicted in Figure 3-30. In a polynucleotide chain, **phosphodiester bonds** link the 3′ carbon of one pentose ring and the 5′ carbon of the next pentose (Figure 3-31). The purine and pyrimidine bases extend outward from the polynucleotide backbone and are not involved in the repetitive, nonvarying backbone.

Native DNA consists of two chains (or strands) in which the sequence of bases is *complementary* (e.g., an ade-

α-helix 2

α-helix 1   α-helix 3

Figure 3-28 The quaternary structure of collagen is a "superhelix" composed of three polypeptide chains, each in the α-helical conformation. Hydrogen bonds (not shown) hold the three chains together.

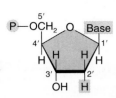

Figure 3-29 The four nucleotides composing the nucleic acids have a common structure consisting of a purine or pyrimidine base, a pentose sugar, and a phosphoric acid residue P_i. In DNA, the pentose is 2-deoxyribose, which has two hydrogen atoms attached to the $C_{2'}$ atom; in RNA, one of these hydrogens, indicated by shading, is replaced by a hydroxyl group.

nine in one strand is matched by a thymine in the other). Each complementary strand is coiled into a helical staircase, and the two strands are intertwined, forming the familiar DNA *double helix*, with the hydrogen-bonded base pairs on the inside of the molecule (Figure 3-32). During replication of DNA, the strands separate from each other, and each of the strands acts as a template for the formation of its complementary strand, thereby yielding two molecules of double-stranded DNA.

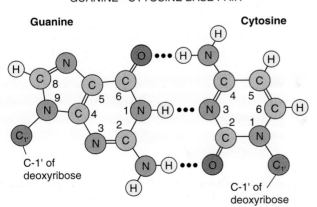

ADENINE- THYMINE BASE PAIR

GUANINE - CYTOSINE BASE PAIR

Figure 3-30 In nucleic acids, hydrogen bonding (black dots) between purine and pyrimidine bases forms the stable base pairs G-C, A-T (in DNA), and A-U (in RNA). The structure of uracil (U) is the same as thymine except that the methyl (–CH_3) group on C_5 is replaced with a hydrogen.

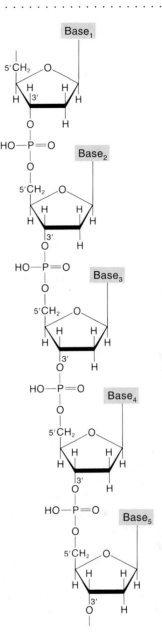

Figure 3-31 The backbone structure of polynucleotide chains consists of pentose residues linked by phosphodiester bonds. The bases extend away from the backbone. This diagram shows a small portion of a single strand of DNA. [Lehninger, 1975.]

The genetic information of an organism is encoded in the sequence of bases in its DNA. In a process called **transcription**, a DNA strand acts as a template for synthesis of **messenger RNA** (mRNA) in the nucleus (see Figure 3-32, *bottom*). The mRNA strand, which contains the informational sequence present in its DNA template, leaves the nucleus and enters the cytoplasm to be decoded by a ribosome into the amino acid sequence of a polypeptide chain. In this process of **translation**, certain sequences of three bases in the DNA code for certain amino acids. For example, GGU, GGC, CGA, and GGC all code for the amino acid glycine; and GCU, GCC, GCA, and GCT code for the amino acid alanine. Thus, the genetic code consists

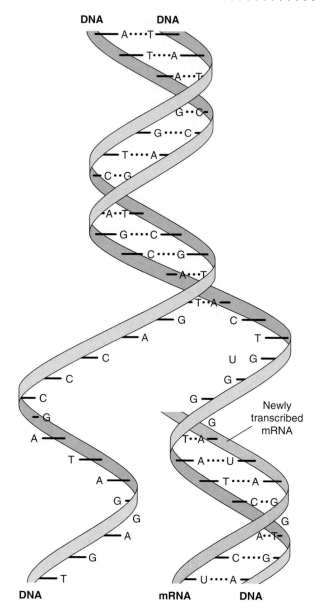

Figure 3-32 Native DNA contains two strands wound around each other in a double helix. Hydrogen bonding (black dots) between complementary bases stabilizes the structure. The molecule unwinds during transcription, and one of the strands acts as a template for synthesis of mRNA complementary to the DNA.

of a four-letter alphabet (A, G, C, T) combined into three-letter words.

As discussed in the previous section, the primary structure (amino acid sequence) of a polypeptide determines its final three-dimensional conformation. Thus once a polypeptide chain is synthesized, it curls and folds, assuming the characteristic secondary and tertiary structure of a protein molecule, and in some cases associates with other chains to form a multi-subunit protein.

More detailed accounts of the major steps relating protein synthesis to nucleic acids may be found in the references listed under Suggested Readings at the end of the chapter.

ENERGETICS OF LIVING CELLS

The biochemical reactions of animal cells are in many senses those of a chemical machine. As in all machines, each event is accompanied by an energy transaction. For a machine or a cell to perform work, energy must be transferred from one part of the system to another, usually with the conversion of at least part of the energy from one form to another. This holds true even when the parts of the system are as minute as reacting molecules.

Animals are fueled by the intake of organic food molecules and their subsequent degradation by digestive and metabolic processes. During these processes the chemical energy inherent in the molecular structures of foods is released and made available for the energetic needs of the organism. Energy is required for such obvious activity as muscle contraction, ciliary movement, and the active transport of molecules by membranes. However, chemical energy is also required for the synthesis of complex biological molecules from simple chemical building blocks and for the subsequent organization of these molecules into organelles, cells, tissues, organ systems, and complete organisms. A living organism must frequently take in fuel and continuously expend energy to maintain its function and structure at all levels of organization. If energy intake drops below the amount required for maintenance, the organism will consume its own energy stores. When these are exhausted, it no longer has any source of energy. The organism dies because it cannot stave off the tendency to become disorganized, nor can it continue to perform the necessary energy-requiring functions.

The material and energy transactions that take place in an organism constitute its **metabolism.** At the intracellular level, these transactions take place via intricate reaction sequences called **metabolic pathways,** which in a single cell can involve thousands of different kinds of reactions. These reactions do not occur randomly, but in orderly sequences, regulated by a variety of genetic and chemical control mechanisms. The organization of atoms and molecules into highly specific structures plus the ability to carry out cellular metabolism distinguishes living systems from the nonliving.

The processes of cell metabolism in animals are of two kinds:

- *Extraction* of chemical energy from foodstuff molecules and the channeling of that energy into useful functions.

- Chemical *alteration* and *rearrangement* of nutrient molecules into small precursors of other kinds of biological molecules.

An example of extraction is the acquisition of amino acids during the digestion of foodstuff proteins and their subsequent oxidation within cells, which releases their chemical energy. An example of alteration and arrangement is the incorporation of amino acids into newly synthesized protein molecules in accord with the specifications the genetic information of the cell. We are concerned here less with the biochemical details of cell metabolism than with the thermodynamic and chemical principles that underlie the transfer and utilization of chemical energy within the cell. Thus, we will consider the mechanisms by which chemical energy is extracted from foodstuff molecules and the manner in which it is made available for the energy-requiring processes discussed in subsequent chapters.

Energy: Concepts and Definitions

Energy may be defined as the capacity to do **work**. Work, in turn may be defined as the product of force times distance ($W = F \times d$). As an example, when a force lifts a 1-kg mass a height of 1 m, the force is 1 kg, and the mechanical work done is 1 m·kg. The energy expended to do this work (i.e., the useful energy, not including that expended in overcoming friction or expended as heat) is also 1 m·kg. Once the kilogram mass is raised to the height of 1 m, it possesses, by virtue of its position, a **potential energy** of 1 m·kg. This potential energy can be converted to **kinetic energy** (energy of movement) if the mass is allowed to drop. Thus, we see that energy exists in different forms, including the following:

- Mechanical potential energy (e.g., a stretched spring or a lifted weight)
- Chemical potential energy (e.g., gasoline, glucose)
- Mechanical kinetic energy (e.g., a falling weight)
- Thermal energy (actually kinetic energy at the molecular level)
- Electrical energy
- Radiant energy

The various forms of energy can power different types of work, as summarized in Table 3-8. We will be concerned in this chapter primarily with **chemical energy,** the potential energy stored in the structure of molecules. Before delving into the energy relationships involved in the biochemical reactions of cellular metabolism, it will be useful to review the first and second laws of thermodynamics and the concept of free energy.

Thermodynamic laws

The **first law of thermodynamics** states that energy is neither created nor lost in the Universe. Thus, if we burn wood or coal to fuel a steam engine, this does not create new energy, but merely converts one form to another—in this example, chemical energy to thermal energy, thermal energy to mechanical energy, and mechanical energy to work.

The **second law of thermodynamics** states that all the energy of the Universe will inevitably be degraded to heat and that the organization of matter will become totally randomized. In more formal terms, the second law states that the **entropy,** a measure of the randomness, of a closed system will progressively increase and that the amount of energy within the system capable of performing useful work will diminish. A system that is ordered (nonrandom) contains energy in the form of its orderliness, because in becoming disordered (i.e., as a result of an increase in entropy), it can perform work. This is illustrated in Figure 3-33A, which shows gas molecules in thermal motion in a hypothetical system consisting of two compartments open to each other. Initially the gas is confined almost entirely to compartment I, in which case the system possesses a certain degree of order. Clearly, this situation has a very small probability of occurring spontaneously if in the starting condition the gas molecules are evenly distributed between the two compartments. The gas molecules can all be forced into one compartment only by the expenditure of energy (e.g., a piston pushing the gas from one compartment to the other). As the gas is permitted to escape from compartment I into compartment II, the entropy of the system increases (i.e., the system becomes more random). The movement of molecules from compartment I to compartment II is a form of useful energy that can be made to do work on an appropriate apparatus placed near the opening between the two compartments. Once the system is fully randomized (i.e., entropy is maximal), no further work can be extracted from the system, even though the gas molecules remain in constant thermal motion (Figure 3-33B).

Orderliness increases as the organism develops from a fertilized egg to the adult. In this sense, living systems temporarily defy the second law. It should be recalled, however, that the second law refers to a *closed* system (e.g., the Universe), and animals are not closed systems. Living organisms maintain a relatively low entropy at the expense of energy obtained from their environment. For example, a rhinoceros eating, digesting, and metabolizing grass in quantities just sufficient to maintain constant weight ultimately increases the entropy of the matter it ingests. The highly ordered carbohydrate, protein, and fat molecules in the grass are converted in the animal to CO_2, H_2O, and low-molecular-weight nitrogen compounds, releasing energy trapped in the organization of the larger molecules (Figure 3-34). The carbon, hydrogen, and oxygen atoms in cellulose, for instance, are in a much more highly ordered state than they are in CO_2 and H_2O; thus the metabolic breakdown of cellulose in the grass represents an increase in entropy. At the same time, the cells of the rhinoceros utilize for their own energy requirements a portion of the chemical energy originally stored in the molecular

TABLE 3-8
Various kinds of work

Type of work	Driving force	Displacement variable
Expansion work	P (pressure)	Volume
Mechanical work	F (force)	Length
Electrical work	E (electric potential)	Electric charge
Surface work	Γ (surface tension)	Surface area
Chemical work	μ (chemical potential)	Mole numbers

A

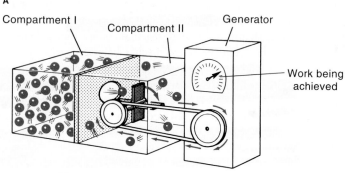

Compartment I   Compartment II   Generator

Work being achieved

B

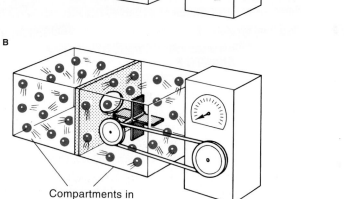

Compartments in equilibrium

organization of foodstuff molecules. This is not in conflict with the second law because the decrease in entropy that results from the animal's synthesis of complex molecules occurs at the expense of increasing the entropy of ingested foodstuff molecules produced by plants with the energy of the sun. Ultimately, of course, the rhinoceros dies, and the entropy of its body greatly increases as it decays or is consumed by other animals.

Free energy

Living systems must function at relatively uniform temperatures and pressures, for there can be only minor temperature or pressure gradients between the various parts of an organism. For this reason, biological systems can utilize only that component of the total available chemical energy capable of doing work under isothermal conditions. This component is called the **free energy**, symbolized by the letter *G*. Changes in free energy are related to changes in heat and entropy by the equation

$$\Delta G = \Delta H - T \Delta S \qquad (3\text{-}7)$$

in which ΔH is the heat (or *enthalpy*) produced or taken up by the reaction, T is absolute temperature, and ΔS is the change in entropy (in units of $cal \cdot mol \cdot K^{-1}$). From this equation it is evident that in a chemical reaction that pro-

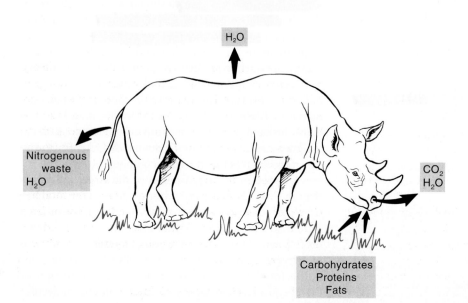

H_2O

Nitrogenous waste H_2O

CO_2 H_2O

Carbohydrates Proteins Fats

duces no change in temperature ($\Delta H = 0$), there will be a decline in free energy (i.e., ΔG is negative) if there is a rise in entropy (i.e., ΔS is positive), and vice versa. Since the direction of energy flow is toward increased entropy (second law), chemical reactions proceed spontaneously if they produce an increase in entropy (and thus a decrease in free energy). In other words, the reduction of free energy is the driving force in chemical reactions.

The inevitable trend toward increased entropy, with the inevitable degradation of useful chemical energy into useless thermal energy, requires that living systems must trap or capture new energy from time to time in order to maintain their structural and functional status quo. In fact, the ability to extract useful energy from their environment is one of the remarkable features that distinguish living systems from inanimate matter.

With the exceptions of chemoautotrophic bacteria and algae, which obtain energy by the oxidation of inorganic compounds, and those animals that obtain their nourishment from these organisms, all life on Earth ultimately depends on radiant energy from the sun. This electromagnetic energy (including visible light) has its origin in nuclear fusion, a process in which the energy of atomic structure is converted to radiant energy. In this process, four hydrogen nuclei are fused to form one helium nucleus, with the release of an enormous amount of radiant energy. A very small fraction of this radiant energy reaches the planet Earth, and a small portion of that is absorbed by chlorophyll molecules in green plants and algae. The energy trapped by photically activated chlorophyll molecules eventually is used in the energy-requiring synthesis of glucose from H_2O and CO_2. The chemical energy stored in the structure of glucose is available to the plant for controlled release during the processes of cellular respiration.

All animals directly or indirectly obtain the energy they need from the carbohydrates, lipids, and proteins manufactured by green plants. Herbivores (e.g., grasshoppers, cattle) obtain these energy-rich compounds by feeding directly on plant materials, whereas predators (e.g., spiders, cats) and scavengers (e.g., lobsters, vultures) obtain them second, third, or fourth hand. The transfer of chemical energy between various *trophic levels* of the living world is diagrammed in Figure 3-35.

Later in this chapter we will consider the metabolic pathways by which animal cells release energy through the oxidation of food molecules. First, however, it will be useful to examine some general principles of energy transfer in biochemical reactions, and also some features of enzymes, the cellular proteins that allow biochemical reactions to proceed rapidly at biological temperatures.

Transfer of Chemical Energy by Coupled Reactions

There are several categories of biochemical reactions, but the features of reaction rates and kinetics can be illustrated by a simple combination reaction in which two reactant molecules, A and B, react to form two new product molecules, C and D:

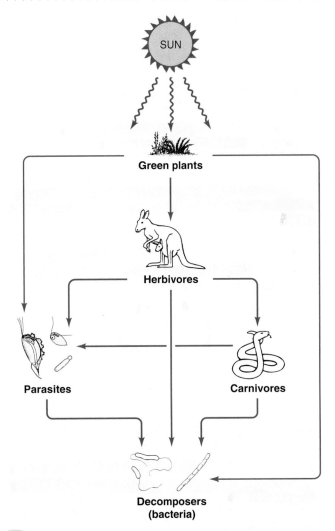

Figure 3-35 Trophic energy levels are highly interlinked by energy flow (arrows). Note the central position of green plants and herbivores. Bacterial decomposers are important in the recycling of organic matter.

$$\mathrm{A + B \rightleftharpoons C + D} \qquad (3\text{-}8)$$
$$\text{(reactants)} \qquad \text{(products)}$$

As the arrows indicate, this reaction is *reversible*. In theory, any chemical reaction can proceed in either direction provided that the products are not removed from the solution. Sometimes, however, the tendency for a reaction to go forward (reactants $\longrightarrow$ products) is so much greater than the tendency to go in reverse that for practical purposes the reaction may be considered irreversible.

A reaction tends to go forward if it shows a free-energy change, ΔG, that is negative. In other words, the total free energy of the reactants exceeds that of the products. Such reactions are said to be **exergonic** (or **exothermic**) and typically liberate heat. The oxidation of hydrogen to water is a simple exergonic reaction:

$$2\,H_2 + O_2 \longrightarrow 2\,H_2O + \text{heat}$$

The energy-requiring reverse reaction occurs during photosynthesis, the energy being supplied by chlorophyll-trapped light quanta:

$$2\,H_2O \xrightarrow{\text{light quantum}} 2\,H_2 + O_2$$

This reaction, which requires the input of energy, is an example of an **endergonic** (or endothermic) reaction. Exergonic and endergonic reactions sometimes are referred to as "downhill" and "uphill" reactions, respectively.

The amount of energy liberated or taken up by a reaction is related to the equilibrium constant, K'_{eq}, of the reaction. This is a constant of proportionality relating the concentrations of the products to the concentrations of the reactants when the reaction has reached equilibrium—that is, when the forward rate is equal to the reverse rate, and the concentration of reactants and products has stabilized:

$$K'_{eq} = \frac{[C][D]}{[A][B]} \qquad (3\text{-}8a)$$

Here [A], [B], [C], and [D] are the equilibrium molar concentrations of the reactants and products in Equation 3-7. It is evident that the greater the tendency for the reaction in Equation 3-7 to go to the right, the higher the value of its K'_{eq}. As noted already, this tendency depends on the difference in free energy, ΔG, between the products C and D and the reactants A and B. The greater the drop in free energy, the more completely the reaction proceeds to the right and the higher its K'_{eq}. The equilibrium constant is related to the change in standard free energy, $\Delta G°$, of the system by the equation

$$\Delta G° = -RT \ln K'_{eq} \qquad (3\text{-}9)$$

It is evident from this equation that if K'_{eq} is greater than 1.0, $\Delta G°$ will be negative; and if K'_{eq} is less than 1.0, $\Delta G°$ will be positive. Exergonic reactions have a negative $\Delta G°$ and therefore occur spontaneously without the need of external energy to "drive" them. Endergonic reactions have a positive $\Delta G°$; that is, they require the input of energy from a source other than the reactants.

Some biochemical processes in living cells are exergonic and others are endergonic. Exergonic processes, since they proceed on their own under the appropriate conditions, present relatively few problems in cellular energetics. Endergonic processes, however, must be "driven." This is generally done in the cell by means of *coupled reactions,* in which *common intermediates* transfer chemical energy from a molecule of relatively high energy content to a reactant of lower energy content. As a result, the reactant is converted into a molecule of higher energy content and can undergo the required reaction by releasing some of this energy.

A mechanical analogy of a coupled reaction is seen in Figure 3-36. The 10-kg weight on the left can lose its po-

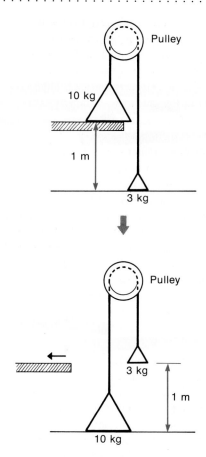

Figure 3-36 In this mechanical analogy of a coupled reaction, the fall of the 10-kg weight provides the energy required to lift the 3-kg weight. The pulley and rope connecting the two weights is the mechanism for coupling the energy of the falling 10-kg weight to the other weight.

tential energy ($10\,m \cdot kg$) by dropping a distance of 1 m, in which case it will lift the 3-kg weight on the right the same distance. Because the two weights are connected with a rope over a pulley, the fall of the 10-kg weight is coupled to the rise of the 3-kg weight, which initially had no potential energy of its own. It is evident that the falling weight can raise the other one only if it weighs more. Likewise, an exergonic reaction can "drive" an endergonic reaction only if the former liberates more free energy than the latter requires. As a consequence, some energy is lost, and the efficiency is, of necessity, less than 100%.

ATP: Energy Carrier of the Cell

The most ubiquitous energy-rich common intermediate in cellular metabolism is the nucleotide **adenosine triphosphate** (ATP), which can donate its terminal energy-rich phosphate group to any of a large number of organic acceptor molecules (e.g., sugars, amino acids, nucleotides). This **phosphorylation** reaction raises the free-energy level of the acceptor molecule, allowing it to react exergonically in enzyme-catalyzed biochemical reactions.

The ATP molecule consists of an adenosine group, made up of the pyrimidine base adenine, the five-carbon sugar residue ribose, and three linked phosphate groups

(Figure 3-37A). Much of the free energy of the molecule resides in the mutual electrostatic repulsion of the three phosphate units, with their positively charged phosphorus atoms and negatively charged oxygen atoms. The mutual repulsion of these phosphate units is analogous to the repulsion of bar magnets, with their north and south poles aligned, held together by a sticky wax (Figure 3-38). If the wax, which is analogous to the $O \sim P$ bonds in ATP, is softened by warming, the energy stored by virtue of the proximity of the mutually repelling magnets is released as the magnets spring apart. Likewise, the breaking of the bonds between the phosphate units of ATP results in the release of free energy (Figure 3-37B). Once the terminal phosphate group of ATP is removed by hydrolysis, the mutual repulsion of the two products, adenosine diphosphate (ADP) and inorganic phosphate (P_i), is such that the probability of their recombining is very low. That is, their recombination

is highly endergonic. The standard free-energy change, $\Delta G°$, for the hydrolysis of ATP under standard conditions is -7.3 kcal$\cdot$mol^{-1}.

The role of ATP in driving otherwise endergonic reactions by means of coupled reactions is illustrated by the condensation of the two compounds X and Y to yield Z:

$$X + ATP \rightleftharpoons X\text{—phosphate} + ADP$$
$$\Delta G° = -3.0 \text{ kcal} \cdot \text{mol}^{-1}$$
$$X\text{—phosphate} + Y \rightleftharpoons Z + P_i$$
$$\Delta G° = -2.3 \text{ kcal} \cdot \text{mol}^{-1}$$

The total free energy liberated in these two reactions (-5.3 kcal$\cdot$mol^{-1}) will be equal to the sum of the free-energy changes of the two parent reactions:

$$ATP + HOH \rightleftharpoons ADP + P_i$$
$$\Delta G° = -7.3 \text{ kcal} \cdot \text{mol}^{-1}$$
$$X + Y \rightleftharpoons Z$$
$$\Delta G° = \frac{+2.0 \text{ kcal} \cdot \text{mol}^{-1}}{-5.3 \text{ kcal} \cdot \text{mol}^{-1}}$$

Note that the $\Delta G°$ for the condensation of X and Y has a positive value ($+2.0$ kcal$\cdot$mol^{-1}); thus normally this reaction would not proceed. However, because $\Delta G°$ for the

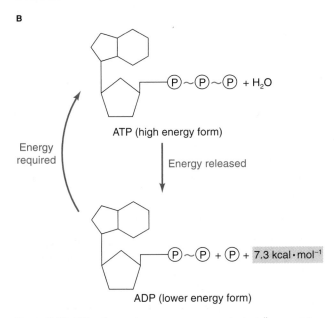

A

Adenosine group

Triphosphate group

B

Energy required

ATP (high energy form)

Energy released

ADP (lower energy form)

7.3 kcal$\cdot$mol^{-1}

Figure 3-37 ATP—the most common energy carrier in cells—contains two high-energy phosphate bonds (red $\sim$). **(A)** Structural formula of ATP with adenosine and triphosphate groups highlighted. **(B)** Schematic depiction of interconversion of charged and uncharged forms of ATP. Hydrolysis of ATP to ADP and inorganic phosphate, P_i, releases about 7.3 kcal of free energy per mole of ATP. This reaction is conveniently monitored by measuring the concentration of inorganic phosphate.

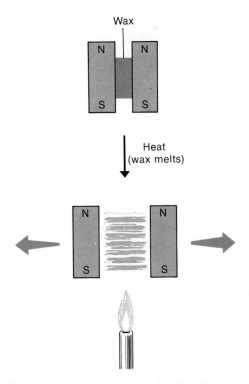

Wax

Heat (wax melts)

Figure 3-38 The high-energy phosphate bond can be portrayed in a magnetic analogy. Energy is stored in pushing the magnets together against bonding wax. When the wax melts (or ATP is hydrolyzed), the magnets fly apart, releasing the energy. In this analogy, the flame supplies the activation energy for melting the wax.

hydrolysis of ATP is larger and negative ($-7.3 \, \text{kcal} \cdot \text{mol}^{-1}$), the net $\Delta G°$ of the coupled reaction is negative, allowing it to proceed.

Although ATP and other nucleotide triphosphates (e.g., **guanosine triphosphate,** GTP) are responsible for the transfer of energy in many coupled reactions, it should be stressed that the mechanism of a common intermediate is widely employed in biochemical reaction sequences. Thus, portions of molecules—and even atoms, such as hydrogen—are transferred, along with chemical energy, from one molecule to another by common intermediates in consecutive reactions. The high-energy nucleotides are special only in that they act as a *general energy currency* in a large number of energy-requiring reactions. In this role, ADP is the "discharged" form, and ATP is the "charged" form (see Figure 3-37B). Numerous other high-energy phosphorylated compounds occur in the cell, some with higher free energies of hydrolysis than ATP (Figure 3-39). The cell can use these compounds in the formation of ATP. As we will see later on, the cell also has other biochemical mechanisms for channeling chemical energy into the formation of ATP.

Phosphoarginine and **phosphocreatine** are special reservoirs of chemical energy for the rapid phosphorylation of ADP to reconstitute ATP during vigorous muscle contraction. These compounds are called **phosphagens.** In vertebrate muscle, which contains only creatine phosphate, the following transphosphorylation reaction occurs:

$$\text{creatine phosphate} + \text{ADP} \underset{\text{enzyme}}{\overset{\text{transphosphorylase}}{\rightleftharpoons}} \text{creatine} + \text{ATP}$$

$$\Delta G° = -3.0 \, \text{kcal} \cdot \text{mol}^{-1}$$

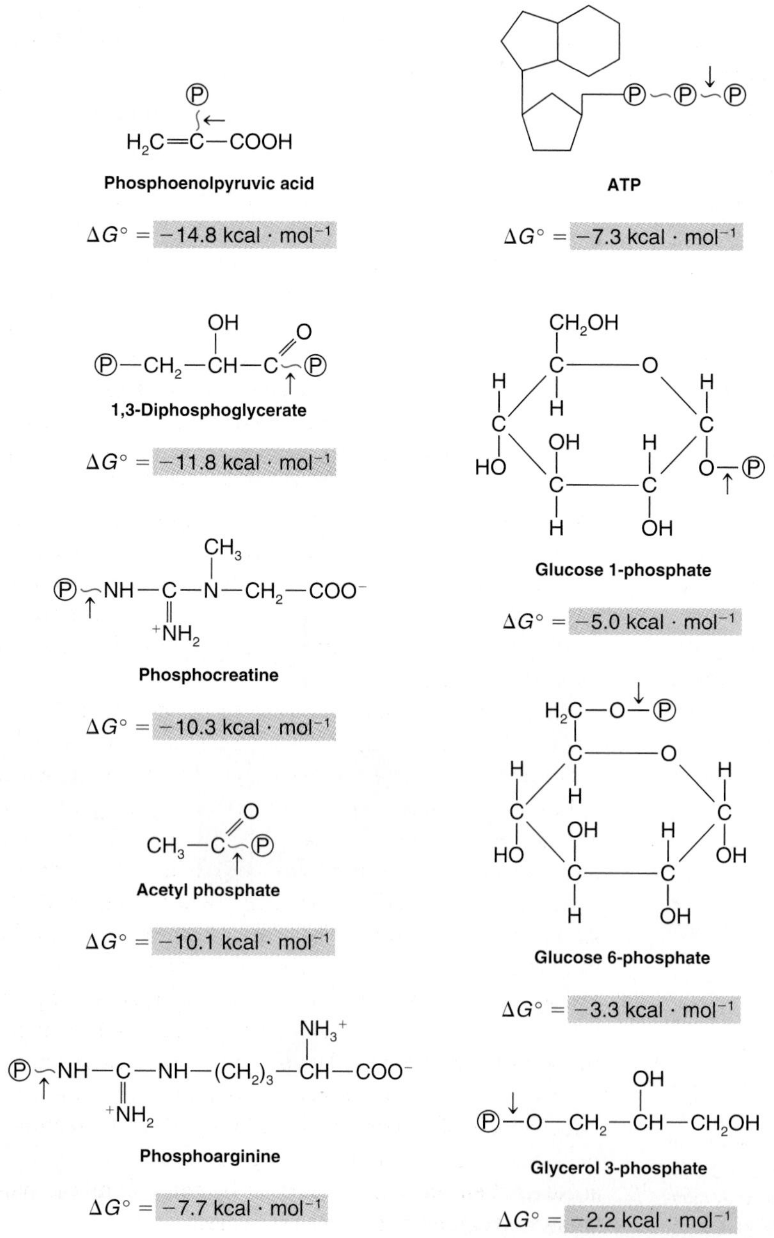

Figure 3-39 Hydrolysis of compounds containing high-energy phosphate bonds (red ~) provides cells with energy for energy-requiring reactions and processes. Although ATP is the most common energy currency in biological systems, several other phosphorylated compounds have higher free energies of hydrolysis. These compounds, shown on the left, can be used by cells to synthesize ATP from ADP and inorganic phosphate. The $\Delta G°$ values are the standard free energies at pH 7 for hydrolysis of the bonds indicated by the small arrows.

Phosphoenolpyruvic acid

$\Delta G° = -14.8 \, \text{kcal} \cdot \text{mol}^{-1}$

ATP

$\Delta G° = -7.3 \, \text{kcal} \cdot \text{mol}^{-1}$

1,3-Diphosphoglycerate

$\Delta G° = -11.8 \, \text{kcal} \cdot \text{mol}^{-1}$

Glucose 1-phosphate

$\Delta G° = -5.0 \, \text{kcal} \cdot \text{mol}^{-1}$

Phosphocreatine

$\Delta G° = -10.3 \, \text{kcal} \cdot \text{mol}^{-1}$

Acetyl phosphate

$\Delta G° = -10.1 \, \text{kcal} \cdot \text{mol}^{-1}$

Glucose 6-phosphate

$\Delta G° = -3.3 \, \text{kcal} \cdot \text{mol}^{-1}$

Phosphoarginine

$\Delta G° = -7.7 \, \text{kcal} \cdot \text{mol}^{-1}$

Glycerol 3-phosphate

$\Delta G° = -2.2 \, \text{kcal} \cdot \text{mol}^{-1}$

Both creatine phosphate and arginine phosphate are found singly or together in invertebrate muscle.

Temperature and Reaction Rates

The rate at which a chemical reaction proceeds depends on the temperature. This is not surprising, because temperature is an expression of molecular motion. As temperature increases, so does the average molecular velocity. This greater velocity increases the number of collisions per unit time and thereby increases the probability of successful interaction of the reactant molecules. Furthermore, as their velocities increase, the molecules possess higher kinetic energies and thus are more likely to react on collision. The kinetic energy required to cause two colliding molecules to react is called the free energy of activation, or **activation energy**. It is measured as the number of calories required to bring all the molecules in a mole of reactant at a given temperature to a reactive (or *activated*) state.

The requirement for activation applies to exothermic as well as endothermic reactions. Although a reaction may have the potential for liberating free energy, it will not proceed unless the reactant molecules possess the necessary kinetic energy. This situation can be compared to one in which it is necessary to push an object over a low ridge before it is free to roll downhill (Figure 3-40).

The dashed curve in Figure 3-41 shows the relationship between free energy and the progress of a reaction in which a reactant, or substrate (S), is converted to a product (P). The substrate must first be raised to an energy state sufficient to activate it, allowing it to react. Since the reaction yields free energy, the energy state of the product is lower than that of the substrate. Note that the overall free-energy change of the reaction is independent of the activation energy required to produce the reaction.

In many industrial processes both the reaction rate and activation energy (i.e., the temperature) are significantly reduced by the use of **catalysts**—substances that are neither consumed nor altered by a reaction, but facilitate the interaction of the reactant particles. Reactions in the living cell are similarly aided by biological catalysts called **enzymes**. The solid curve in Figure 3-41 shows how an enzyme affects the progress of the reaction S $\longrightarrow$ P. Note that the presence of the enzyme has no effect on the overall free-

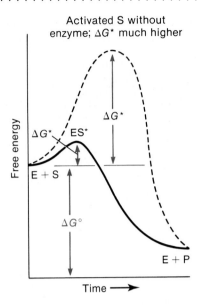

Figure 3-41 The activation energy, ΔG^*, of a reaction is lowered by the catalytic action of an enzyme. Note that the overall free-energy change, $\Delta G°$, is the same in the nonenzymatic reaction (dashed curve) and enzymatic reaction (solid curve). E, enzyme; S, substrate; ES*, activated enzyme-substrate complex; P, product.

energy change (and hence the equilibrium constant) of the reaction; it merely reduces the activation energy of the reaction and hence increases the rate of reaction.

The increase in reaction rates produced by enzymes is extremely useful biologically because it allows reactions that would otherwise proceed at imperceptibly slow rates to proceed at far higher rates at biologically tolerable temperatures. Within any population of reactant molecules at a given temperature, only those possessing sufficient kinetic energy to be activated will react. If an enzyme that reduces the energy required for activation is added, a far larger number of molecules can react in a given time at the same temperature. The rates of various enzyme-catalyzed reactions range from 10^8 to 10^{20} times the rate of the corresponding uncatalyzed reactions, an enormous acceleration of reaction rates.

An extremely important advantage of catalyzed reactions is the possibility for regulating the rate of reaction by varying the concentration of catalyst. For example, when H_2 and O_2 are burned noncatalytically, they explode in an uncontrolled manner, because the heat released by the rapid combustion of the H_2 produces a rapid ignition of the remaining unburned H_2. In contrast, when H_2 is oxidized slowly at a low temperature with small quantities of the catalytic agent platinum, the release of heat is slowed enough so that no explosion occurs. The quantity of platinum relative to the fuel (H_2) and oxidant (O_2) regulates the rate of combustion. Likewise, most biological reactions are regulated by modulation of the quantity or the catalytic effectiveness of certain enzymes. In the next two sections, we'll first discuss how enzymes operate and then how cells regulate their metabolic reactions by controlling the synthesis and catalytic activity of enzymes.

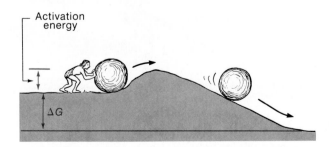

Figure 3-40 The activation energy is the energy required to bring the reactants into position to interact. In this analogy, the potential energy of the rock cannot be liberated until some energy, referred to as the activation energy, is expended to bring it into position at the crest of the hill.

ENZYMES: GENERAL PROPERTIES

Substances that increased the rate of alcoholic fermentation were first isolated from living cells by water extraction of yeast about a century ago. These substances, now called enzymes, were found to be inactivated by heating, whereas the reactants in fermentation reactions were unaffected by heating. This finding was the first indication that enzymes are protein molecules. It was subsequently discovered that, without exception, each species of enzyme molecule is a protein of very specific amino acid composition and sequence. All of these proteins, or at least their enzymatically active portions, have a globular conformation. Each cell in an organism contains literally thousands of species of enzyme molecules catalyzing all the synthetic and metabolic reactions of the cell. The work of molecular geneticists has demonstrated that enzymes are the primary gene products of major significance. By specifying the structure of each enzyme molecule that is produced, the genetic apparatus is indirectly responsible for all enzymatic reactions in a cell.

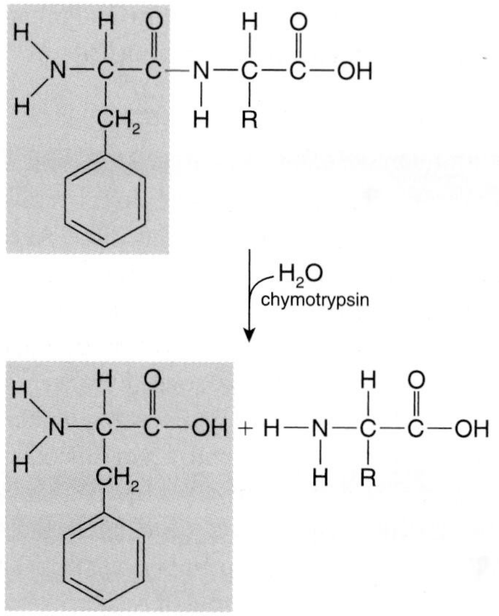

Enzyme Specificity and Active Sites

Each enzyme is, to some degree, specific for a certain **substrate** (reactant molecule). Some enzymes act at certain types of bonds and may therefore act on many different substrates having such bonds. For example, trypsin, a **proteolytic enzyme** found in the digestive tract, catalyzes the hydrolysis of any peptide bond in which the carbonyl group is part of an arginine or lysine residue, regardless of the position of the bonds in the polypeptide chain of a protein. Another intestinal proteolytic enzyme, chymotrypsin, specifically catalyzes hydrolysis of the peptide bond in which the carbonyl group belongs to a phenylalanine, tyrosine, or tryptophan residue (Figure 3-42).

Most enzymes, however, exhibit far more substrate specificity than do proteolytic enzymes. For instance, the enzyme sucrase catalyzes hydrolysis of the disaccharide sucrose into glucose and fructose, but it cannot attack other disaccharides, such as lactose and maltose. These substrates are hydrolyzed by enzymes specific for them (lactase and maltase, respectively). Many enzymes differentiate between optical isomers, that is, molecules that are chemically and structurally identical except that one is the mirror image of the other. For example, the enzyme L-amino oxidase catalyzes the oxidation of the L-isomer of an α-keto acid but is totally ineffective for the D-isomer of these molecules.

The highly specific nature of most enzymes just noted is consistent with the concept that a substrate molecule "fits" a special portion of the enzyme surface called the **active site.** The enzyme molecule is made up of one or more peptide chains folded about so as to form the tertiary structure of a more or less globular protein of a specific conformation. The active site is thought to consist of the side groups of certain amino acid residues that are brought into proximity in the tertiary structure, even though they may be widely separated in the amino acid sequence of the enzyme

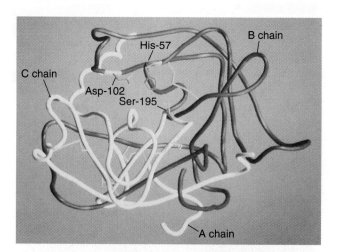

Figure 3-42 Chymotrypsin hydrolyzes any peptide bond in which the carbonyl carbon belongs to a phenylalanine, tyrosine, or tryptophan residue. Shown here is its action on a dipeptide containing phenylalanine (shading).

(Figure 3-43). Because interaction of the active site and substrate involves relatively weak attractive forces (i.e., electrostatic bonds, van der Waals forces, and hydrogen bonds), the substrate molecule must have a conformation that closely fits into the active site.

The steric specificity of enzyme active sites has been well established by experiments with substrate analogs (i.e., molecules similar to but slightly different from the substrate molecule). The ability of an enzyme's active site to interact with analogs decreases as the interatomic distances, num-

Figure 3-43 This computer-generated model of the enzyme chymotrypsin illustrates how amino acid residues that are widely separated in the primary structure are brought together by folding of the protein to form the active site. In chymotrypsin the three residues shown in red are needed for catalytic activity. This globular protein contains three polypeptide chains (A, B, and C) and five disulfide bonds shown in yellow. [Adapted from Tsukada and Blow, 1995; courtesy of Gareth White.]

ber and position of charged groups, and bond angles of the analog molecules depart from those of the normal substrate.

Mechanism of Catalysis by Enzymes

Enzyme activity, the catalytic potency of an enzyme, can be expressed as the *turnover number,* which is the number of molecules of substrate per second with which one molecule of the enzyme reacts to produce product molecules. In an enzymatic reaction, the substrate(s) first interacts with the active site of the enzyme, forming an *enzyme-substrate complex* (ES). As noted earlier, this interaction reduces the activation energy of the reaction, thereby increasing the probability and rate of the reaction (see Figure 3-41).

Several catalytic mechanisms are employed to accelerate the reaction rates of enzymatic reactions:

- An enzyme may hold the substrate molecules in a particular orientation in which the reacting groups are sufficiently close to one another to enhance the probability of reaction.

- An enzyme may react with the substrate molecule to form an unstable intermediate that then readily undergoes a second reaction, forming the final products.

- Side groups within the active site may act as proton donors or acceptors to bring about general-acid or general-base reactions.

- Binding of the enzyme to the substrate may cause internal strain in the susceptible bond of the substrate, increasing its probability of breaking.

Whatever the precise catalytic mechanism for a particular reaction, once the substrate molecules have reacted, the enzyme separates from the products, freeing the enzyme molecule to form an ES complex with a new substrate molecule. Since the ES persists for a finite time, all the enzyme can become tied up as ES if the substrate concentration is high enough relative to the enzyme concentration.

Effect of Temperature and pH on Enzymatic Reactions

Any factor that influences the conformation of an enzyme, and hence the arrangement of amino acid side groups in the active site, will alter the activity of the enzyme. Temperature and pH are two common factors that influence the rates of enzymatic reactions in this way.

As we saw previously, an increase in temperature increases the probability of protein denaturation, which disrupts the conformation of polypeptide chains. In the case of enzymes, denaturation destroys catalytic activity. For this reason, enzyme-catalyzed reactions exhibit a characteristic curve of reaction rate versus temperature (Figure 3-44A). As temperature increases, the reaction rate initially increases due to the increased kinetic energy of the substrate molecules. As temperature increases further, how-

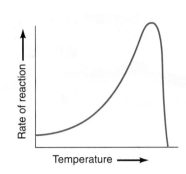

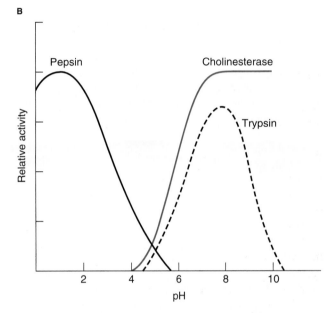

Figure 3-44 Both temperature and pH influence the activity of enzymes. **(A)** The effect of temperature on the reaction rate is generally similar for most enzymes. **(B)** The effect of pH on catalytic activity varies among enzymes, but nearly all have a distinct optimal pH.

ever, the rate of enzyme inactivation due to denaturation also increases. At the optimal temperature, the rate of enzyme destruction by heat is balanced by the increase in enzyme-substrate reactivity, and the two effects of elevated temperature cancel. At that temperature the reaction rate is maximal. At higher temperatures enzyme destruction becomes dominant, and the rate of reaction rapidly decreases. The temperature sensitivity of enzymes and other protein molecules contributes to the lethal effects of excessive temperatures.

Electrostatic bonds often participate in the formation of an ES. Since H^+ and OH^- can act as counterions for electrostatic sites, a drop in pH exposes more positive sites on an enzyme for interaction with negative groups on a substrate molecule. Conversely, a rise in pH facilitates the binding of positive groups on a substrate to negative sites on the enzyme. Thus, it is not surprising that the activity of enzymes typically varies with the pH of the medium and that each enzyme has an optimal pH range (Figure 3-44B).

Cofactors

Some enzymes require the participation of small molecules called **cofactors** in order to perform their catalytic function. In that case, the protein moiety is called the **apoenzyme**. One class of cofactors consists of small organic molecules called **coenzymes,** which activate their apoenzymes by accepting hydrogen atoms (protons) from the substrate. For example, the enzyme glutamate dehydrogenase requires the coenzyme **nicotinamide adenine dinucleotide** (NAD) to catalyze the oxidative deamination of the amino acid glutamic acid:

$$\text{glutamate} + NAD^+ \rightleftharpoons$$
(oxidized)

$$\alpha\text{-ketoglutarate} + NADH + NH_4^+$$
(reduced)

A number of coenzymes contain vitamins as part of the molecule. Since an apoenzyme cannot function without its coenzyme, it is not surprising that vitamin deficiencies can have profound pathological effects because of their action at the enzyme level.

Other enzymes require monovalent or divalent metal ions as cofactors, generally in a highly selective manner. The main ions that function as cofactors, along with examples of enzymes that require them, are listed in Table 3-9. Especially interesting is Ca^{2+}, whose intracellular concentration ($<10^{-6}$ M) is much lower than that of most other common physiologically important ions. Unlike Mg^{2+}, Na^+, K^+, and other cofactor ions, which generally are present in nonlimiting concentrations, Ca^{2+} is present in limiting concentrations for certain enzymes. As discussed in Chapter 9, the Ca^{2+} concentration of the cytosol is regulated by the surface membrane and by internal organelles, such as the endoplasmic reticulum. In this way, the activity of calcium-activated enzymes can be regulated by the cell. Cellular processes regulated by the Ca^{2+} concentration include muscle contraction, secretion of neurotransmitters and hormones, ciliary activity, assembly of microtubules, and ameboid movement.

Enzyme Kinetics

The rate at which an enzymatic reaction proceeds depends on the concentrations of substrate, product, and active enzyme. For purposes of simplicity, we will assume that the product is removed as fast as it is formed. In that case, the rate of reaction will be limited by the concentration of either the enzyme or the substrate. If we further assume that the enzyme is present in excess, then the rate at which a single substrate, A, is converted to the product, P, is determined by the concentration of A:

$$A \xrightarrow{k} P$$

where k is the **rate constant** of the reaction. The rate of conversion of A to P can be expressed mathematically as

$$\frac{-d[A]}{dt} = k[A] \tag{3-10}$$

in which [A] is the instantaneous concentration of the substrate, k is the rate constant of the reaction, and $d[A]/dt$ is the rate at which A is converted to P with respect to time. The disappearance of A and the appearance of P are plotted as functions of time in Figure 3-45. Note that the concentration of A decreases exponentially as the concentration of P increases exponentially. An exponential time function is always generated when the rate of change of a quantity ($d[A]/dt$ in this example) is proportional to the instantaneous value of that quantity ([A] in this example).

The relationship expressed by Equation 3-10 is more usefully presented as

$$\log \frac{a}{a - x} = \frac{k_1 t}{2.303} \tag{3-10a}$$

TABLE 3-9
Metal ions that function as cofactors

Metal ion	Some enzymes requiring metal cofactor
Ca^{2+}	Phosphodiesterase
	Protein kinase C
	Troponin
Cu^{2+} (Cu^+)	Cytochrome oxidase
	Tyrosinase
Fe^{2+} or Fe^{3+}	Catalase
	Cytochromes
	Ferredoxin
	Peroxidase
K^+	Pyruvate phosphokinase (also requires Mg^{+2})
Mg^{2+}	Phosphohydrolases
	Phosphotransferases
Mn^{2+}	Arginase
	Phosphotransferases
Na^+	Plasma membrane ATPase (also requires K^+ and Mg^{2+})
Zn^{2+}	Alcohol dehydrogenase
	Carbonic anhydrase
	Carboxypeptidase

Source: Lehninger, 1975.

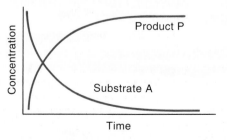

Figure 3-45 The concentration of substrate A and product P change in a nonlinear fashion during the reaction A $\longrightarrow$ P.

where a is the initial concentration of substrate and x is the amount of substrate that has reacted within time t. A plot of the left side of Equation 3-10a versus time yields a straight line whose slope is proportional to the rate constant, k_1 (Figure 3-46A). A reaction that exhibits this behavior is said to have **first-order kinetics.** The rate constant of a first-order reaction has the dimension of reciprocal time—that is, "per second," or s^{-1}. The rate constant can be inverted to yield the **time constant,** which has the dimension of time. Thus, a first-order reaction with a rate constant of $10 \cdot s^{-1}$ has a time constant of 0.1 seconds.

In a reaction with two substrates, A and B, in which excess enzyme is present and the product, P, does not accumulate, the rate of disappearance of A will be proportional to the product [A][B].

$$A + B \xrightarrow{k} P$$

This reaction will proceed with **second-order kinetics.** It is noteworthy that the order of the reaction is not determined by the number of substrate species participating as reactants, but instead by the number of species present in *rate-limiting* concentrations. Thus, if B were present in great excess over A, the reaction $A + B \longrightarrow P$ would become first order, since its rate would be limited by only one substrate concentration.

The rate of an enzymatic reaction is independent of substrate concentrations when the enzyme is present in limiting concentrations and all the enzyme molecules are complexed with substrate (i.e., the enzyme is *saturated*). Such reactions proceed with **zero-order kinetics** (Figure 3-46B).

Figure 3-47 shows plots of the initial rate, v_0, of an enzymatic reaction ($S \longrightarrow P$) as a function of substrate concentration, [S], at two different enzyme concentrations. At both enzyme levels, the reaction is first order (i.e., v_0 is proportional to [S]) at low substrate concentrations. At higher substrate concentrations, however, the reaction becomes zero order, because all the enzyme molecules are complexed with substrate; in this situation, the concentration of enzyme, not substrate, limits v_0. In the living cell all orders of reaction, as well as mixed-order reactions, occur.

The maximum rate of any enzymatic reaction, V_{max}, occurs when all the enzyme molecules catalyzing that reaction are tied up, or saturated, with substrate molecules, that is, when the substrate is present in excess and the enzyme concentration is rate limiting (see Figure 3-47). For each enzymatic reaction, there is a characteristic relationship between V_{max} and enzyme concentration. Although all enzymes can become saturated, they show great variation in the concentration of a given substrate that will produce saturation. The reason for this is that enzymes differ in affinity for their substrates. The greater the tendency for an enzyme and its substrate to form a complex, ES, the higher the percentage of total enzyme, E_t, tied up as ES at any given concentration of substrate. Thus, the higher this affinity, the lower the substrate concentration required to saturate the enzyme.

The relationship between the kinetics of an enzyme-catalyzed reaction and the affinity of the enzyme for substrate was worked out early in the 20th century. The general theory of enzyme action and kinetics was proposed by Leonor Michaelis and Maud L. Menten in 1913, and later extended by George E. Briggs and John B. S. Haldane. The **Michaelis-Menten equation** is the rate equation for a reaction catalyzed by a single enzyme:

$$v_0 = \frac{V_{max}[S]}{K_M + [S]} \tag{3-11}$$

A First order

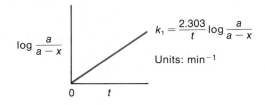

$$k_1 = \frac{2.303}{t} \log \frac{a}{a-x}$$

Units: min^{-1}

B Zero order

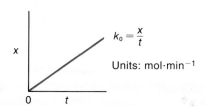

$$k_0 = \frac{x}{t}$$

Units: $mol \cdot min^{-1}$

Figure 3-46 The kinetic order of enzymatic reactions is revealed in graphical plots. In these plots, the x represents the amount of substrate A reacting within time t, and a represents the initial amount of A at time zero. The slope is proportional to the rate constant. **(A)** Because a first-order reaction has an exponential time course (see Figure 3-45), a plot of $\log(a/a - x)$ versus t gives a straight line. **(B)** When the enzyme concentration is limiting, the reaction exhibits zero-order kinetics, which gives a straight line in plots of x versus t.

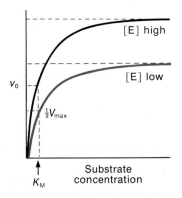

Figure 3-47 At a given enzyme concentration, the initial rate, v_0, of the reaction $S \longrightarrow P$ rises linearly with increasing substrate concentration. Eventually all of the enzyme becomes saturated, at which time the enzyme (E) becomes rate limiting. The Michaelis-Menten constant, K_M, is equal to the substrate concentration at which the reaction rate is one-half maximum. The black and red curves are for different enzyme concentrations. Note that K_M is independent of the enzyme concentration, [E], whereas V_{max} depends directly on [E].

where ν_0 is the initial reaction rate at substrate concentration [S], V_{max} is the reaction rate with excess substrate, and K_M is the **Michaelis-Menten constant**. Consider the special case when $\nu_0 = \frac{1}{2}V_{max}$. Substituting for ν_0, we get

$$\frac{V_{max}}{2} = \frac{V_{max}[S]}{K_M + [S]} \qquad (3\text{-}12)$$

Dividing by V_{max} gives

$$\frac{1}{2} = \frac{[S]}{K_M + [S]} \qquad (3\text{-}13)$$

On rearranging, we obtain

$$K_M + [S] = 2[S] \qquad (3\text{-}14)$$

or

$$K_M = [S] \qquad (3\text{-}15)$$

Therefore, K_M equals the substrate concentration at which the initial reaction rate is half what it would be if the substrate were present to saturation.

Thus, the Michaelis-Menten constant, K_M (in units of moles per liter), depends on the affinity of the enzyme for a substrate. For a given enzyme and substrate, K_M is equal to the substrate concentration at which the initial reaction is $\frac{1}{2}V_{max}$. By inference, then, K_M represents the concentration of substrate at which half the total enzyme present is combined with substrate in the ES; that is, $[E_t]/[ES] = 2$. The value of K_M can be determined from a plot of ν_0 versus [S], as illustrated in Figure 3-47. The greater the affinity between an enzyme and its substrate, the lower the K_M of the enzyme-substrate interaction. Stated inversely, $1/K_M$ is a measure of the affinity of the enzyme for its substrate. As illustrated by the plots for two enzyme concentrations in Figure 3-47, K_M is independent of the enzyme concentration, whereas V_{max} is dependent on enzyme concentration.

The relationship between ν_0 and substrate concentration described by the Michaelis-Menten equation (Equation 3-11) is a hyperbolic function. For this reason, numerous data points are required to accurately plot ν_0 against [S] as in Figure 3-47. The Michaelis-Menten equation can be algebraically transformed, however, into a linear form called the **Lineweaver-Burk equation**:

$$\left(\frac{1}{\nu_0} = \frac{K_M}{V_{max}[S]} + \frac{1}{V_{max}} \right) \qquad (3\text{-}16)$$

It's clear from this equation that a plot of $1/\nu_0$ versus $1/[S]$ will give a straight line with a slope of K_M/V_{max} and with intercepts of $1/V_{max}$ on the vertical axis and $-1/K_M$ on the horizontal axis (Figure 3-48). Because of the linear nature of this curve, only two experimental data points (i.e., ν_0 at

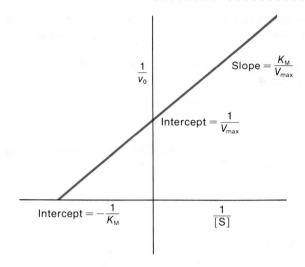

Figure 3-48 In a Lineweaver-Burk plot, the reciprocal of the reaction rate, $1/\nu_0$, is plotted against the reciprocal of the substrate concentration, $1/[S]$. For an enzymatic reaction with first-order kinetics, this plot intercepts the horizontal axis at $-1/K_M$, and the vertical axis at $1/V_{max}$.

two values of [S]) are needed to construct a Lineweaver-Burk plot. V_{max} and K_M can be determined from the two intercepts.

Note that the Michaelis-Menten analysis is not limited to enzyme-substrate interactions, but can be (and often is) applied to any system that displays hyperbolic saturation kinetics as illustrated in Figure 3-47.

Enzyme Inhibition

The activity of most enzymes can be inhibited by certain molecules, and as we'll see in the next section, this property of enzymes is used in the living cell as a means of controlling enzymatic reactions. By studying the molecular mechanisms of inhibition, using both physiological and nonphysiological inhibitors, enzymologists have discovered important features of the active site of enzymes and of the mechanism of enzyme action. The therapeutic effect of many drugs depends on their ability to inhibit specific enzymes, thereby blocking metabolic or physiologic processes involved in disease. For example, the drug saralasin blocks the action of the enzyme angiotensin II and helps lower blood pressure in humans with severe hypertension.

Enzymes can be poisoned by agents that form highly stable covalent bonds with groups inside the active site and thereby block formation of the enzyme-substrate complex, ES. Such agents can produce *irreversible inhibition*, which cannot be alleviated by removal of the inhibitor; that is, the enzyme is, in effect, rendered permanently inactive. More relevant to normal cell function, however, are two types of *reversible inhibition*:

- **Competitive inhibition** is caused by substances that appear to react directly with the active site of the enzyme; it can be reversed by an increase in substrate concentration.

- **Noncompetitive inhibition** appears to be caused by substances that bind to a region(s) of the enzyme outside of the active site; it is not reversed by an increase in substrate concentration, but can be reversed by dilution or removal of the competitor.

Most competitive inhibitors are substrate analogs and compete with substrate molecules for the active site (Figure 3-49). Thus, increasing the concentration of the substrate reduces the probability of the inhibitor binding; this is why competitive inhibition is reversible by an increase in substrate concentration. Because noncompetitive inhibitors do not bind to the active site, their chemical structure typically differs from that of the substrate. In addition, as a noncompetitive inhibitor and the substrate interact with different sites on the enzyme, they do not compete directly; this is why noncompetitive inhibition is not reversed by an increase in substrate concentration. As illustrated in Figure 3-50, competitive and noncompetitive inhibitors produce readily distinguishable alterations in Lineweaver-Burk plots. Although both types of inhibitors increase the slope of a Lineweaver-Burk plot, reflecting the decrease in reaction rate, they have opposite effects on the intercepts.

A competitive inhibitor does not change the V_{max} of an enzymatic reaction; that is, when the substrate concentration is extrapolated so that 1/[S] approaches 0, the substrate will displace all of the inhibitor molecules from the enzyme. Thus the intercept on the $1/v_0$ axis of a Lineweaver-Burk plot, which equals $1/V_{max}$, is unaffected by a competitive inhibitor (see Figure 3-50A). On the other hand, the intercept on the 1/[S] axis is shifted toward a

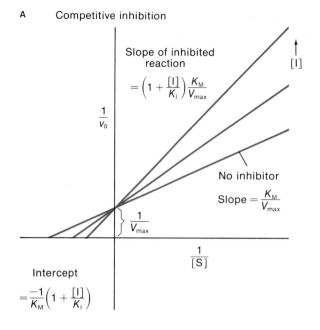

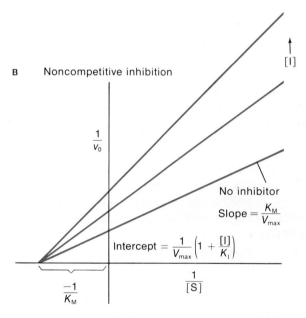

Figure 3-50 Competitive and noncompetitive inhibitors produce different effects on Lineweaver-Burk plots. **(A)** A competitive inhibitor increases the K_M but does not affect the V_{max} of an enzyme. **(B)** Conversely, a noncompetitive inhibitor produces no change in K_M but decreases the V_{max}. It is kinetically similar to a reduction in enzyme concentration. [I], inhibitor concentration; [S], substrate concentration; K_I, dissociation constant of inhibitor-enzyme complex.

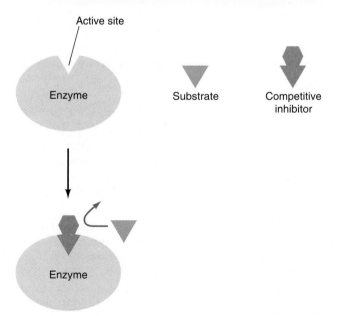

Figure 3-49 Binding of a competitive inhibitor to the active site of an enzyme interferes with the binding of the substrate. If the substrate concentration is increased, however, substrate molecules can displace bound inhibitor molecules. Thus competitive inhibition is reversible. Small red arrows indicate relative concentrations of substrate and inhibitor.

higher substrate concentration in the presence of a competitive inhibitor. That is, it takes a higher concentration of substrate in the presence of the competitive inhibitor to keep half of the enzyme molecules at any instant complexed with the substrate. In effect, a competitive inhibitor increases the "apparent" K_M by an amount related to the concentration of inhibitor, [I], and the dissociation constant, K_I, of the inhibitor-enzyme complex. As [I] increases and/or K_I decreases (i.e., the tighter the binding of inhibitor to enzyme), the greater the displacement of the intercept on the 1/[S] toward a higher [S] value.

Because a noncompetitive inhibitor does not directly interfere with binding of substrate to an enzyme, it has no effect on the intercept on the 1/[S] axis in a Lineweaver-Burk plot (Figure 3-50B). In other words, the K_M of an enzyme is not affected by a noncompetitive inhibitor. On the other hand, the value of V_{max}, indicated by the intercept on the $1/\nu_0$ axis, increases by an amount related to [I] and the K_I. This effect results from the failure of the substrate at high concentrations to displace a noncompetitive inhibitor from its binding site on the enzyme. Thus the kinetic effect of a noncompetitive inhibitor is to reduce the catalytic potency, or turnover number, of an enzyme; that is, it reduces the effective concentration of the enzyme. It is not surprising, then, that the K_M of an enzyme is unaltered by addition of a noncompetitive inhibitor, for, as noted above, K_M is independent of the enzyme concentration.

REGULATION OF METABOLIC REACTIONS

Without any regulation of reaction rates, cellular metabolism would be uncoordinated and undirected. Growth, differentiation, and maintenance would be impossible, to say nothing of subtle homeostatic compensatory responses of the biological machine to externally imposed stresses. Most metabolic control is exerted via mechanisms that regulate the quantity or activity of the various enzymes that catalyze nearly all biochemical reactions. We will now consider the major types of metabolic control.

Control of Enzyme Synthesis

The quantity of an enzyme present in a cell is a function of the rate of synthesis and the rate of destruction of enzyme molecules. As discussed earlier, enzymes are denatured at elevated temperature and are broken down by the action of proteolytic enzymes. The rate of synthesis of an enzyme can be limited under certain conditions (e.g., inadequate diet or the unavailability of amino acid precursors) that reduce protein synthesis generally. Normally, however, the rate of synthesis of a particular enzyme is regulated at the molecular level by modulation of the rate of transcription of the gene encoding it.

Figure 3-51 depicts the model of the control of enzyme synthesis proposed by Francois Jacob and Jacques Monod in 1961 based on studies in bacteria. They found that the **structural genes** encoding several enzymes involved in some metabolic pathways are located adjacent to each other in the cell's DNA. Next to the first of such linked genes is a short stretch of DNA called an **operator.** An operator and its associated contiguous structural genes constitute an **operon.** Transcription of the structural genes into mRNA, which is necessary for enzyme synthesis, can be "turned off" or "turned on" by the action of a **repressor protein,** encoded by a **regulator gene.** Binding of the repressor protein to the operator controls transcription of all the associated structural genes. Thus synthesis of all the enzymes encoded by the operon is collectively controlled by interaction of the repressor protein with the operator. In the case of some operons, combination of the repressor protein with a particular small organic molecule, called **an inducer,** renders it incapable of binding to the operator (as illustrated in Figure 3-51). In other operons, the repressor protein can bind to the operator only when it is associated with a small molecule called a **corepressor.**

Some cells synthesize certain enzymes (e.g., those involved in metabolizing lactose) only after they are exposed to the initial substrate (or related molecules) in the reaction pathway, a phenomenon called **enzyme induction.** This phenomenon can be explained in terms of the Jacob-Monod model. In this case, the substrate acts as an inducer, and binding of the substrate to the repressor protein relieves the repression of the structural genes. As a result, cells

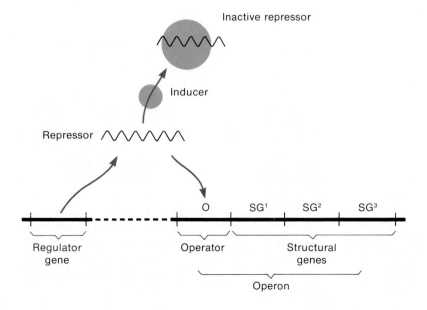

Figure 3-51 The Jacob-Monod operon model can explain the induction and repression of enzyme synthesis. As illustrated here, binding of the repressor protein to the operator prevents transcription of the adjacent structural genes. In the presence of an inducer, this repression is relieved and the enzymes encoded by the structural genes are produced. In some operons, the repressor is inactive until it combines with a small corepressor molecule. In this case, enzyme synthesis proceeds as long as the corepressor concentration is low. See text for further discussion. [Adapted from Goldsby, 1967.]

begin to synthesize the enzymes needed to metabolize the substrate, which previously had been repressed. This process is an example of metabolic economy, for inducible enzymes are synthesized only when needed (i.e., when the substrate is present).

Synthesis of the enzymes involved in a biosynthetic reaction sequence may be regulated by the end product of the pathway. In this situation, the repressor protein, called an **aporepressor,** is inactive until it combines with a small organic molecule—the corepressor—produced at the end of a biosynthetic reaction sequence. Binding of the active repressor (i.e., aporepressor-corepressor complex) to the operator prevents transcription of the structural genes in the operon, and synthesis of all the encoded enzymes drops. Sometimes, the genes for all the enzymes in a biosynthetic reaction are not located next to each other in an operon. But if synthesis of an enzyme that acts early in the biosynthetic pathway is regulated, then operation of the entire synthetic pathway and the rate of production of its end product are kept in check. Again this is an example of metabolic economy. If the end product begins to accumulate for any reason, such as a reduction in its rate of incorporation into cell structures, the entire synthetic pathway is slowed by a drop in the rate of synthesis of the regulated enzyme (Figure 3-52).

In addition to these mechanisms, cells possess other mechanisms for regulating the transcription of genes encoding enzymes, and hence the quantity of various enzymes present in cells. All of these mechanisms are of great importance in the development of an organism. Each somatic cell in an organism contains the same information coded in its DNA. Different cell types in different tissues, however, contain widely divergent amounts of the different enzymes coded by the genetic material. It is evident that in any given tissue some genes are turned on, while others are turned off. This situation may occur in part through mechanisms of enzyme induction and repression in response to differences in the local chemical environments of different cells and tissues in the developing organism.

Control of Enzyme Activity

The activity of some enzymes can be regulated by *modulator* (regulator) molecules, which interact with a part of the enzyme molecule distinct from the active site. This part of the enzyme, called the **allosteric site,** can bind a modulator molecule, causing a change in the tertiary structure of the enzyme that changes the conformation of the active site (Figure 3-53). As a result, the affinity of the enzyme for its substrate decreases or increases. Allosterically regulated enzymes operate at key points in metabolic pathways, and modulation of their activities plays an important role in the regulation of these pathways. Let's take a closer look at several mechanisms for controlling enzyme activity.

End-product (feedback) inhibition

Some metabolic pathways have built-in mechanisms for regulating the rate of the reaction sequence, independent of

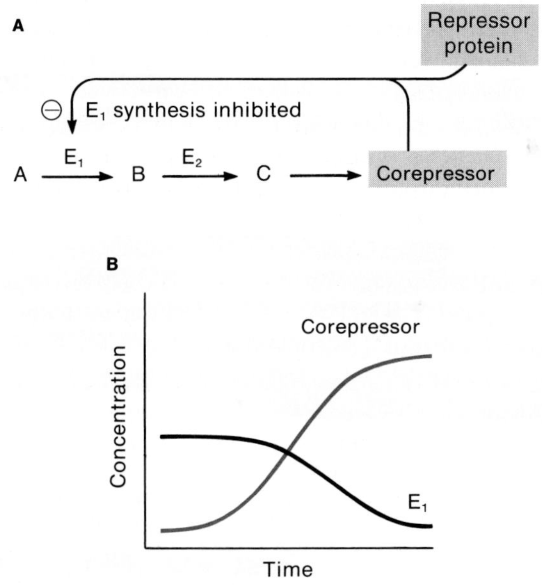

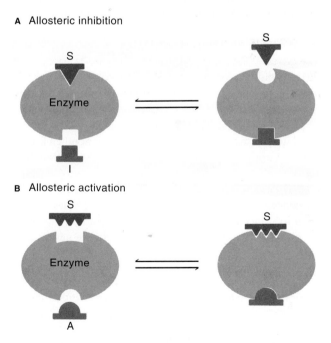

Figure 3-52 The end product in a biosynthetic pathway may have a repressive effect on synthesis of an early enzyme in the pathway. **(A)** Example of a negative-feedback loop in which synthesis of E_1 is repressed by accumulation of a product (C) several steps farther along the pathway. The end product acts as a corepressor, which binds to the repressor protein (or aporepressor). Binding of this complex to the gene for E_1 inhibits transcription of the gene. **(B)** Time course of E_1 and corepressor concentration. The level of E_1 drops as the level of the end-product corepressor rises.

Figure 3-53 Allosteric interactions can result in either activation or inhibition of enzyme activity. **(A)** Binding of an allosteric inhibitor molecule (I) to an allosteric site can indirectly alter the configuration of the active site of an enzyme (E), thereby rendering the enzyme inactive. Noncompetitive inhibitors act by this mechanism. **(B)** Conversely, binding of an allosteric activator (A) can alter the active site so that the enzyme becomes catalytically active.

the quantity of enzymes present. In these pathways, it is usually the first enzyme of the sequence that acts as a regulatory enzyme. Most commonly, interaction of the end product of the pathway with this enzyme inhibits the activity of that enzyme (Figure 3-54). Such **end-product inhibition** limits the rate of accumulation of the end product by slowing the entire sequence. In most cases, a regulatory enzyme catalyzes a reaction that is virtually irreversible under cellular conditions; for this reason, accumulation of the product does not slow the rate of reaction.

The interaction of the end product in a biosynthetic pathway with a regulatory enzyme has been shown to occur at an allosteric site. Thus the end product acts as an allosteric inhibitor (see Figure 3-53A). An example of this control mechanism occurs in the biosynthesis of the catecholamine norepinephrine, which functions as both a neurotransmitter and a hormone. High concentrations of catecholamine inhibit the enzyme tyrosine hydroxylase, an essential enzyme in the sequence of reactions leading to production of norepinephrine.

Enzyme activation

The requirement for cofactors exhibited by some enzymes provides the cell with another means of regulating enzyme activity and hence the rate of biochemical reactions. As noted earlier, Ca^{2+} and several other cations act as cofactors for various enzymes (see Table 3-9). In the case of some enzymes, cation cofactors appear to act as allosteric activators (see Figure 3-53B). However, no single mechanism appears to explain the effect of cofactor ions on enzyme activity.

The intracellular free concentration of certain ions depends on diffusion and active transport across membranes separating the cell exterior from intracellular ion storage sites. By regulating the levels of cofactor ions, the cell can modulate the concentration of cofactor ions and, in turn, the activity of certain enzymes. An important and common regulatory cofactor is Ca^{2+}, which is present at much lower concentrations than other cofactor cations within the **cytosol,** the unstructured fluid phase of the cytoplasm in which many metabolic reactions occur. Because extracellular concentrations of Ca^{2+} are typically 1000 times higher than its concentration in the cytosol (commonly far less than 10^{-6} M), extremely small changes in the net flux of

Ca^{2+} across the cell membrane (or the membranes of cytoplasmic organelles) can produce substantial *percentage* changes in the intracellular free Ca^{2+} concentration (see Figure 9-16). The special role of Ca^{2+} as an intracellular regulatory molecule is discussed in Chapter 9.

Now that we've discussed the principles underlying cellular energetics and the characteristics of enzyme-catalyzed reactions, we will consider in detail how cells produce ATP, the key molecule in the energy transactions of cells.

METABOLIC PRODUCTION OF ATP

If we compare the energy utilization of an animal with that of an automobile (continuing our analogy between animal and machine), we note that both types of machines require the intermittent intake of chemical fuel to energize their activities. Their use of fuel differs, however, in at least one very important aspect. In the automobile engine the organic fuel molecules in the gasoline are oxidized (ideally) to CO_2 and H_2O in one explosive step. The heat generated by the rapid oxidation produces a great increase in the pressure of gases in the engine's cylinders. In this way the chemical energy of the fuel is converted to mechanical movement (kinetic energy). This conversion depends on the high temperatures produced by the burning gasoline, for the chemical energy of the gasoline is converted directly into heat, and heat can be used to do work only if there is a temperature and pressure difference between two parts of the machine.

Since living systems are capable of sustaining only small temperature and pressure gradients, the heat provided by the simple one-step combustion of fuel would be essentially useless for powering their activities. For this reason, cells have evolved metabolic mechanisms with a series of discrete reactions for the stepwise conversion of chemical energy. The energy of foodstuff molecules is recovered for useful work through the formation of intermediate compounds of progressively lower energy content. At each exergonic step some of the chemical energy is liberated as heat, while the rest is transferred as free energy to the reaction products. Chemical energy conserved and stored in the structure of intermediate compounds is then transferred to the general-purpose, high-energy intermediate ATP and to other high-energy intermediates. The chemical energy of these molecules is readily available for a wide variety of cellular processes (Figure 3-55).

As mentioned earlier, carbohydrates, lipids, and proteins ingested in foodstuffs are the primary fuels for animals. After digestion, these molecules generally enter the circulatory system as five- or six-carbon sugars, fatty acids, and amino acids, respectively (Figure 3-56). These small molecules then enter the tissues and cells of the animal, where they may (1) be immediately broken down into smaller molecules for the extraction of chemical energy or for rearrangement and recombination into other types of molecules, or (2) be built up into larger molecules, such as polysaccharides (e.g., glycogen), fats, or proteins. With few

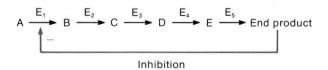

Inhibition

Figure 3-54 Products of a reaction sequence can allosterically inhibit a rate-limiting enzyme in the pathway. The metabolic end product in this case directly affects the activity of the initial enzyme, as illustrated in Figure 3-53A. In contrast, in the mechanism depicted in Figure 3-52 the end product affects the amount of enzyme by controlling transcription of its gene.

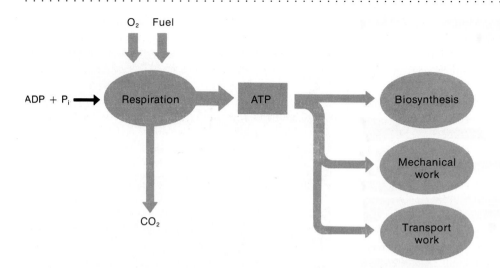

Figure 3-55 The hydrolysis of ATP powers numerous energy-requiring processes in biological systems. The ADP produced by hydrolysis is recycled to ATP by rephosphorylation energized by the oxidation of foodstuff molecules to CO_2 and H_2O.

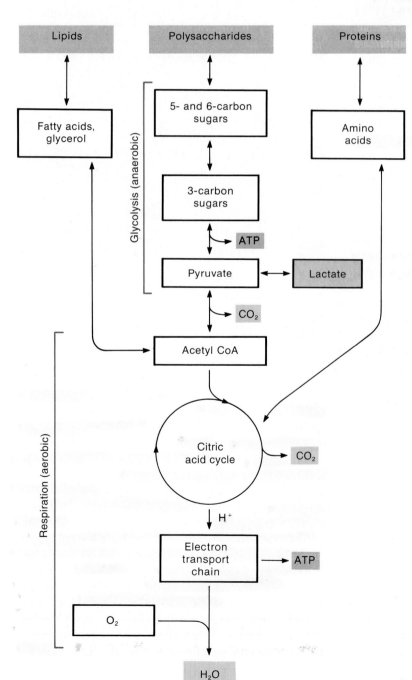

Figure 3-56 Carbohydrates, lipids, and proteins all can be degraded by cells to yield usable energy in the form of ATP. All three classes of foodstuffs are interlinked in intermediary metabolism and feed intermediates into the citric acid cycle (see Figure 3-68), which is linked to the electron-transport chain. During aerobic metabolism, complete oxidation to CO_2 and H_2O occurs. During anaerobic metabolism, however, cellular respiration is impossible and lactate accumulates.

exceptions, these too will eventually be broken down and eliminated as CO_2, H_2O, and urea. Nearly all molecular constituents of a cell are in dynamic equilibrium, constantly being replaced by components newly synthesized from simpler organic molecules.

Some simple organisms, including certain bacteria and yeasts, as well as a few invertebrate species, can live indefinitely under totally **anaerobic** (i.e., essentially oxygen-free) conditions. The anaerobes fall into two groups: obligatory anaerobes, which cannot grow in the presence of oxygen (e.g., botulism bacterium, *Clostridium botulinum*), and facultative anaerobes, which survive and reproduce well either in the absence or in the presence of oxygen (e.g., yeasts). All vertebrates and most invertebrates, however, require molecular oxygen for cellular respiration and are therefore termed **aerobic.** Even these aerobic animals generally possess tissues that can metabolize anaerobically for periods of time, building up an **oxygen debt** that is repaid when sufficient oxygen becomes available.

As these observations suggest, there are two kinds of energy-yielding metabolic pathways in animal tissues (Figure 3-57):

- **Aerobic metabolism,** in which foodstuff molecules are finally oxidized completely to carbon dioxide and water by molecular oxygen

- **Anaerobic metabolism,** in which foodstuff molecules are oxidized incompletely to lactic acid

The energy yield per molecule of glucose in anaerobic metabolism is only a fraction of the energy yield in aerobic metabolism. For this reason, cells with high metabolic rates and low energy stores survive only briefly when deprived of oxygen. The nerve cells of the mammalian brain are a familiar example. An oxygen deficiency lasting only a few minutes will lead to massive cell death and permanently impaired brain function.

Aerobic metabolism in animal cells is intimately associated with the **mitochondria.** Detailed description of these organelles, which are just visible in the light microscope, had to await development of the electron microscope

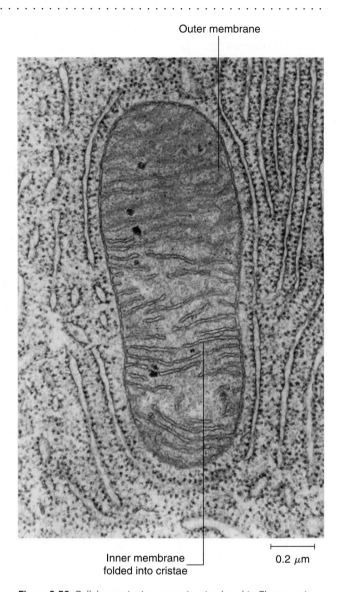

Figure 3-58 Cellular respiration occurs in mitochondria. Electron micrograph of a mitochondrion in a bat pancreas cell revealing the outer membrane and the inner membrane, which is folded to form cristae. [Courtesy of K. R. Porter.]

(Figure 3-58). Mitochondria consist of an outer membrane and an inner membrane, which are not connected with each other. These two membranes carry out completely different functions. The inner membrane is thrown into folds called **cristae,** which increase the area of the inner membrane relative to the outer membrane. The interior space bounded by the inner membrane is called the matrix compartment, and the space between the two membranes is the intermembrane space. As we shall see later on, the inner membrane and matrix compartment contain enzymes that catalyze final oxidation of foodstuffs and production of ATP during cellular respiration. The matrix compartment contains ribosomes, dense granules (consisting primarily of salts of calcium), and mitochondrial DNA, which is involved in replication of the mitochondria. Mitochondria are quite numerous in most cells, with estimates ranging from 800 to 2500 for a liver cell. They also tend to congre-

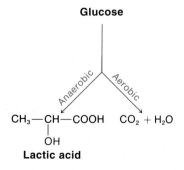

Figure 3-57 Glucose catabolism can take either an anaerobic or aerobic pathway. The energy yield from aerobic metabolism is much greater than that from anaerobic metabolism.

gate most densely in those portions of a cell that are most active in utilizing ATP.

Oxidation, Phosphorylation, and Energy Transfer

Before going on to consider the biochemical pathways in cellular energy metabolism, we will examine how chemical energy, liberated during metabolism, is conserved and channeled into high-energy intermediates. You will remember that when a complex organic molecule is taken apart, free energy is liberated, thus increasing the entropy (degree of randomness) of the constituent matter. This situation occurs when glucose is oxidized to carbon dioxide and water by combustion in the overall reaction

$$C_6H_{12}O_6 + 6\,O_2 \longrightarrow 6\,CO_2 + 6\,H_2O$$

$$\Delta G° = -686\ \text{kcal} \cdot \text{mol}^{-1}$$

The 686 kcal liberated by the oxidation of 1 mol of glucose is the difference between the free energy incorporated into the structure of the glucose molecule during photosynthesis and the total free energy contained in the carbon dioxide and water produced. If 1 mol of glucose is oxidized to carbon dioxide and water in a one-step combustion (i.e., if it is burned), the free-energy change will appear simply as 686 kcal of heat. During cellular respiration, however, a portion of this energy, instead of appearing as heat, is conserved as useful chemical energy and is channeled into ATP through the phosphorylation of ADP. The overall reaction for the metabolic oxidation of glucose by the cell, including the coupled conversion of ADP to ATP, can be written as

$$C_6H_{12}O_6 + 38\,P_i + 38\,\text{ADP} + 6\,O_2 \longrightarrow$$
$$6\,CO_2 + 6\,H_2O + 38\,\text{ATP}$$

$$\Delta G° = -420\ \text{kcal (as heat)}$$

Thus, 266 kcal (686 − 420) is incorporated into 38 mol of ATP (7 kcal·mol^{-1} ATP).

How is the free energy of the glucose molecule transferred to ATP? To understand this, we must first recall that oxidation of a molecule is most broadly defined as the transfer of electrons from that molecule to another molecule; conversely, **reduction** of a molecule is the acceptance of electrons from another molecule. In an oxidation-reduction reaction the **reductant** (electron donor) is oxidized by the **oxidant** (electron acceptor). Together they form a **redox pair:**

$$\text{electron donor} \rightleftharpoons e^- + \text{electron acceptor}$$

or

$$\text{reductant} \rightleftharpoons n\,e^- + \text{oxidant}$$

where *n* is the number of electrons transferred. Whenever electrons are accepted from a reductant by an oxidant, en-

ergy is liberated, for the electrons move into a more stable (higher-entropy) situation when transferred to the oxidant. This is akin to water dropping from one level to a lower level. It is the difference between the two levels that determines the amount of energy liberated.

Thus, chemical energy is liberated when electrons are transferred from a compound of a given *electron pressure* (tendency to donate electrons) to one of a lower electron pressure. If a molecule has a higher electron pressure than the molecule with which it undergoes a redox reaction, it is said to have a greater **reduction potential,** and will act as a reducing agent; if it has less electron pressure, it will act as an oxidant. The free-energy change in each reaction is proportional to the difference between the electron pressures of the two molecules of the redox pair.

In aerobic cell metabolism, electrons move sequentially from compounds of higher electron pressure to compounds of lower electron pressure. The *final electron acceptor* in aerobic metabolism is molecular oxygen. Since oxygen acts merely as an electron acceptor, it is possible in theory to support aerobic metabolism without oxygen, provided that a suitable electron acceptor is supplied in place of oxygen.

In being transferred from glucose to oxygen, electrons undergo an enormous drop in both reduction potential and free energy. One of the functions of cell metabolism is to transport electrons from glucose to oxygen in a series of small steps instead of one large drop. This transport is implemented by two mechanisms found in all cells. First, as we noted earlier, the chemical conversion of foodstuff molecules such as glucose to the fully oxidized end products (e.g., CO_2 and H_2O) occurs in steps and involves many intermediate compounds. Second, electrons removed from substrate molecules are passed to oxygen via a series of electron acceptors and donors of progressively lower electron pressure. As we will see shortly, these mechanisms allow energy to be channeled into the synthesis of ATP in "packets" of appropriate size.

 What reasons can you think of to explain why animal life evolved with O_2 as the final electron acceptor?

Electron-transferring coenzymes

During certain biochemical reactions, electrons, together with protons (i.e., hydrogens), are removed from substrate molecules by enzymes collectively called **dehydrogenases.** These enzymes all function in conjunction with pyridine or flavin coenzymes. The most common are nicotinamide adenine dinucleotide (NAD$^+$), noted earlier, and **flavin adenine dinucleotide (FAD).** Their structural formulas are shown in Figure 3-59. These coenzymes act as electron acceptors in their oxidized form and as electron donors in their reduced form:

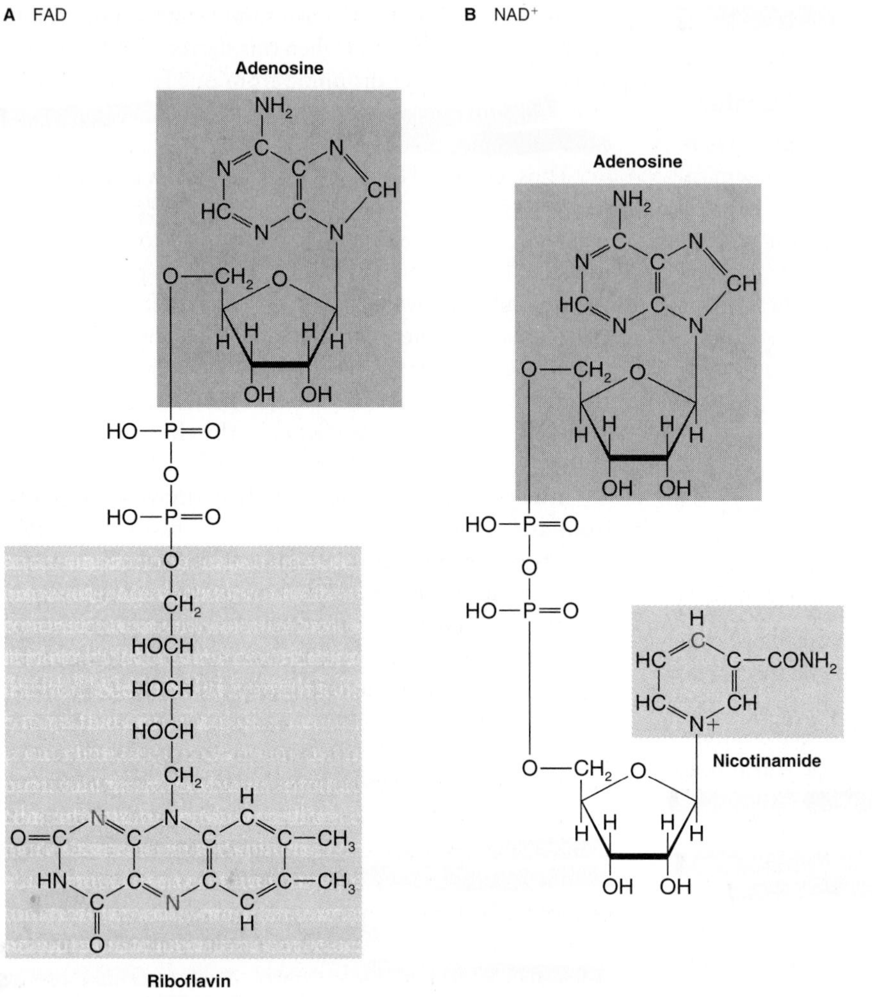

A FAD

B NAD$^+$

Figure 3-59 The two most common electron-carrying coenzymes—flavin adenine dinucleotide (FAD) and nicotinamide adenine dinucleotide (NAD$^+$)—contain an adenosine group and a vitamin-derived group (color shading). Riboflavin, one of the B vitamins, is part of FAD, and nicotinamide, one form of the vitamin niacin, is part of NAD$^+$. In the oxidized coenzyme molecules, shown here, the atoms in red type can accept protons and electrons.

reduced substrate + NAD$^+$ $\rightleftharpoons$
NADH + H$^+$ + oxidized substrate

reduced substrate + FAD $\rightleftharpoons$
FADH$_2$ + oxidized substrate

A very convenient property of these coenzymes for experimental purposes is that their reduced and oxidized forms have different ultraviolet absorption spectra (Figure 3-60). They also undergo a change in ultraviolet-excited fluorescence upon oxidation and reduction. These two properties have permitted physiologists and biochemists to use photometric methods to monitor changes in the amount of reduced coenzyme under experimental conditions in living cells.

The free energy of the reduced forms of both coenzymes, NADH and FADH$_2$, is very high relative to oxygen. As a result, the transfer of electrons from the reduced coenzymes to oxygen is accompanied by a large decrease in free energy. For example, the $\Delta G°$ for the reaction NADH + $\frac{1}{2}$ O$_2$ $\longrightarrow$ NAD$^+$ + H$_2$O, in which two electrons are transferred from NADH to O$_2$, is about -52 kcal $\cdot$ mol^{-1}. The $\Delta G°$ for transfer of electrons from FADH$_2$ to oxygen is about the same. During the oxidation of 1 mol of glucose in cells, 10 mol of reduced NAD and

2 mol of reduced FAD are produced. Thus, of the 686 kcal of free energy available from the oxidation of 1 mol of glucose, about 624 kcal (12 × 52), or 91%, is transferred to the electron-transferring coenzymes to be released in subsequent stages of electron transfer. As noted earlier, 266 kcal of this free energy is eventually retained by the synthesis of ATP.

Electron-transport chain
In spite of the large difference in electron pressure between NADH or FADH$_2$ and oxygen, there is no enzymatic mechanism by which these reduced coenzymes can be directly oxidized by oxygen. Instead, an elaborate **electron-transport chain**, or **respiratory chain**, has evolved in which the electrons move through about seven discrete steps from the high reducing potential of NADH and FADH$_2$ to the final electron acceptor, molecular oxygen. This sequence of electron transfers is the final common pathway for all electrons during aerobic metabolism. Its function, as we will see, is to utilize the energy of electron transfer efficiently for the phosphorylation of ADP to ATP.

The electron-transport chain consists of a series of proteins and coenzymes that can exist in oxidized and reduced forms. Several of the electron carriers in the chain are iron-containing proteins called **cytochromes**, each of which con-

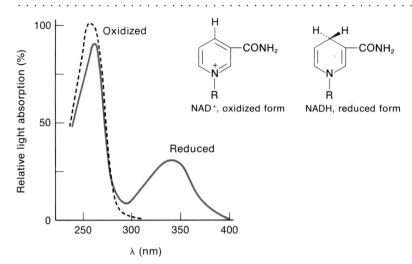

tains a substituted **heme** group. The deeply colored heme group consists, essentially, of a **porphyrin** ring with an iron atom at its center; the heme group is also present in the hemoglobin molecule of vertebrate red blood cells (see Figure 13-2B). The substituted heme groups in the various cytochromes differ in the side chains attached to the porphyrin ring (Figure 3-61). The cytochromes show characteristic absorption spectra in their oxidized and reduced forms, absorbing more strongly in the red when reduced. This behavior led to the first discovery of their function by David Keilin in 1925. Using a spectroscope, he discovered that the flight muscles of insects contain compounds that are oxidized and reduced during respiration. He named these compounds *cytochromes* and hypothesized that they transfer electrons from energy-rich substrates to oxygen.

The functional order of the components of the electron-transport chain is diagrammed in Figure 3-62. From left to right in this figure, each successive electron carrier has a lower electron pressure than its predecessor. As a result, elec-

trons are transferred from NADH down the electron-transport chain in a series of coupled reactions, ending with the reduction of molecular oxygen. Only the last enzyme in the chain, cytochrome aa_3, is able to transfer its electrons directly to oxygen. The order in which the various electron carriers function was worked out in studies with several poisons that block the flow of electrons at specific points in the chain. In these studies, the presence of the oxidized or reduced forms of the electron carriers was monitored by spectrophotometric methods. For example, when the final step—electron transfer by *cytochrome oxidase* (composed of subunits of a and a_3) to O_2—is blocked by cyanide, the effect on electron transport is identical to the removal of molecular oxygen (see Figure 3-62). Electrons pile up, so to speak, because transport is interrupted along the chain, reducing all the cytochromes and other electron carriers above the point of the block. Another poison, antimycin, blocks the flow of electrons from cytochrome b to c, causing the electron carriers above the block to become fully reduced and those below the block to become completely oxidized.

The cascading of electrons through a series of small discrete steps, contrasted with the direct reduction of oxygen by NADH or $FADH_2$, confers a great energetic advantage on the cell. The "logic" of the electron-transport system becomes apparent when it is recalled that the energy required to convert ADP to ATP, the standard currency in biological energy exchange, is small compared with the total change in free energy produced by the transfer of electrons from NADH to oxygen. As we saw earlier, it requires a minimum of 7.3 kcal to synthesize ATP from ADP and inorganic phosphate, whereas 52 kcal are released in the transfer of two electrons from NADH to oxygen. Oxidation of NADH in a one-step reaction could be coupled to the formation of only one molecule of ATP. Such a coupled reaction would be quite inefficient, conserving only 14% (7.3/52) of the available free energy in NADH in the form of ATP, the rest being lost as heat. In contrast, the electron-transport system is a much more efficient mechanism that releases the total energy available in NADH (and $FADH_2$) in small doses just large enough for synthesis of ATP.

Figure 3-61 Heme A acts as the electron acceptor-donor group of cytochrome aa_3. At the center of the porphyrin ring is the iron atom that is oxidized or reduced during transport. The side groups highlighted in color differ in the other cytochromes.

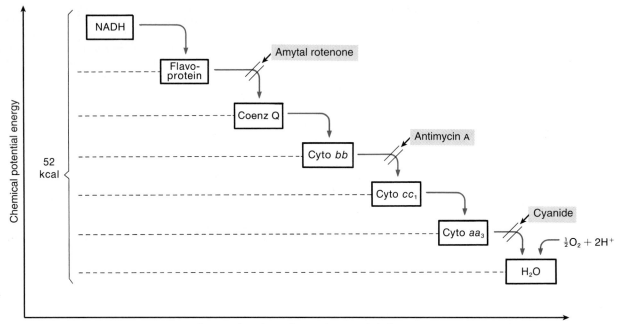

Figure 3-62 In the electron-transport chain, electrons cascade from one carrier to the next in order of their decreasing electron pressures. Respiratory poisons (red shading), which blockx specific steps in this sequence, were useful in determining the order in which the electron carriers function.

The actual synthesis of ATP from ADP and inorganic phosphate (P_i) during electron transport is called **oxidative phosphorylation,** or respiratory-chain phosphorylation. The phosphorylation of ADP to ATP occurs as a consequence of the transfer of electrons between three pairs of electron carriers:

- Flavoprotein to coenzyme Q
- Cytochrome b to cytochromes c and c_1
- Cytochromes aa_3 (cytochrome oxidase) to molecular oxygen

At each of these three steps in the electron-transport chain, the drop in free energy is adequate to drive the phosphorylation of ADP (Figure 3-63). Thus, for each pair of electrons that passes along the entire chain, three molecules of ATP are generated from three molecules of ADP and three molecules of P_i. Each pair of electrons finally reduces one-half molecule of O_2 to form one molecule of water:

$$2e^- + 2H^+ + \tfrac{1}{2}O_2 \rightleftharpoons H_2O$$

By comparing the amount of O_2 consumed (i.e., converted to water) and the amount of P_i consumed (i.e., incorporated into ATP), we can establish the *P/O ratio* (ratio of P_i to atomic oxygen). For example, if one oxidative phosphorylation occurs at each of the three steps noted above, 3 mol of P_i will be incorporated into ATP for each mol of

oxygen atoms ($\tfrac{1}{2}O_2$) consumed in the formation of H_2O. Thus, P/O = 3. Reduced FAD, however, transfers electrons directly to coenzyme Q, bypassing the first phosphorylation reaction; thus, for each pair of electrons transferred from $FADH_2$ to atomic oxygen, only two ATP molecules are formed, for a P/O = 2.

Of the several theories proposed to explain how the synthesis of ATP is *coupled* at the molecular level to the free energy liberated during electron transfer, the *chemiosmotic theory* of energy transduction is the most widely accepted. This mechanism is discussed in Chapter 4. It is interesting to note here that oxidative phosphorylation becomes *uncoupled* from electron transport whenever anything happens that makes the inner mitochondrial membrane "leaky," that is, more permeable than normal to H^+ or other cations. In this case, the production of ATP drops or ceases while both electron transport and the reduction of O_2 to H_2O continue. All the released energy is given off as heat. Oxidative phosphorylation is also uncoupled from electron transport by certain drugs, such as dinitrophenol (DNP). Because this drug reduces the efficiency of energy metabolism, it was once prescribed by physicians to help patients lose weight. Its use as a weight-reducing drug was discontinued when it was found to produce pathological side effects.

Glycolysis

The term **glycolysis,** which means "breakdown of sugar," refers to the sequence of reactions by which the six-carbon

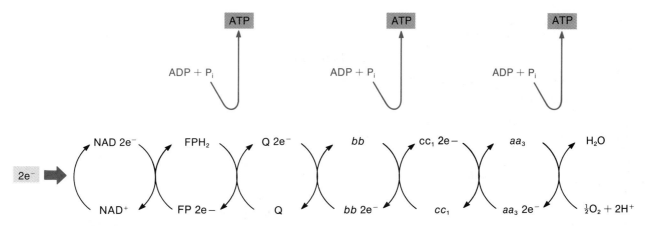

Figure 3-63 As two electrons from a molecule of NADH flow through the entire electron-transport chain, three molecules of ADP are phosphorylated to ATP. Formation of ATP occurs at the indicated steps. FP, flavo- protein; Q, coenzyme Q. The letters b, c, c_1, a, and a_3 refer to the respective cytochromes, shown working in pairs transporting pairs of electrons.

sugar glucose is converted to pyruvic acid, a three-carbon molecule (see Figure 3-56). This sequence of reactions, the most fundamental in the energy metabolism of animal cells, is required for both anaerobic and aerobic release of energy from foodstuffs. The glycolytic pathway is also called the **Embden-Meyerhof pathway** after the two German biochemists who worked out the details of glycolysis in the 1930s.

The first step in the glycolytic pathway is the phosphorylation of glucose by ATP to give *glucose 6-phosphate* (Figure 3-64). The breakdown of glycogen also yields phosphorylated glucose, which then can enter the glycolytic pathway (see Figure 9-13). Glucose 6-phosphate then is converted to fructose 6-phosphate, which is phosphorylated to *fructose 1,6-diphosphate* at the expense of a second ATP (steps 2 and 3). At first glance, it would seem uneconomical for the cell to expend 2 mol of ATP to phosphorylate 1 mol of hexose, since the object of glycolysis is to produce ATP. On closer examination, however, the phosphorylation of glucose does make sense. At physiological pH, phosphorylated hexose and **triose** (three-carbon sugar) molecules become ionized and thus have very low membrane permeabilities. Although unphosphorylated glucose is free to enter (or leave) the cell by diffusion through the surface membrane, the phosphorylated form is conveniently trapped along with its phosphorylated derivatives within the cell. The 2 mol of ATP expended in these so-called *priming phosphorylations* is, in fact, not really lost, for later in the glycolytic pathway these phosphate groups—and their intramolecular free energies—are used to regenerate ATP, thereby conserving the energy of the phosphate groups utilized in the priming phosphorylations.

Fructose 1,6-diphosphate is cleaved in step 4 into two triose sugars, *glyceraldehyde 3-phosphate* and *dihydroxy-*

acetone phosphate. The latter molecule is enzymatically rearranged into the former, so that each mole of glucose yields 2 mol of glyceraldehyde 3-phosphate. This completes the first stage of glycolysis, in which 1 mol of the six-carbon glucose is converted into 2 mol of the three-carbon glyceraldehyde 3-phosphate (steps 1–5 in Figure 3-64).

The second stage of glycolysis begins with the oxidation of glyceraldehyde 3-phosphate and addition of a phosphate group to yield *1,3-diphosphoglycerate* (step 6). In this crucial reaction the energy that would otherwise be released by oxidation of the aldehyde group is captured in a high-energy acyl-phosphate bond linking the second phosphate group to the carbonyl carbon atom (see Figure 3-39). The elucidation of the mechanism of this reaction and of the following one (step 7), in which ADP is directly phosphorylated to ATP by the substrate, is considered among the most important contributions to modern biology. Through these discoveries, Otto Warburg and his colleagues provided, in the late 1930s, the first insight into a mechanism by which chemical energy of oxidation is conserved in the form of ATP. This process is called **substrate-level phosphorylation** to distinguish it from the process of oxidative phosphorylation coupled to electron flow in the respiratory chain.

In steps 8–10 of glycolysis, 3-phosphoglyceric acid is converted to *2-phosphoglyceric acid*, water is removed to form *phosphoenolpyruvate*, and finally the latter yields its phosphate group to ADP, forming ATP and pyruvic acid, another substrate-level phosphorylation. Thus, the glycolytic pathway ends with 2 mol of pyruvic acid produced from each mole of glucose. The phosphorylation of each mole of hexose consumes 2 mol of ATP, and each mole of triose generates 2 mol of ATP (steps 7 and 10). Since each mole of glucose yields 2 mol of triose, the net gain per mole of glucose in glycolysis is 2 mol of ATP.

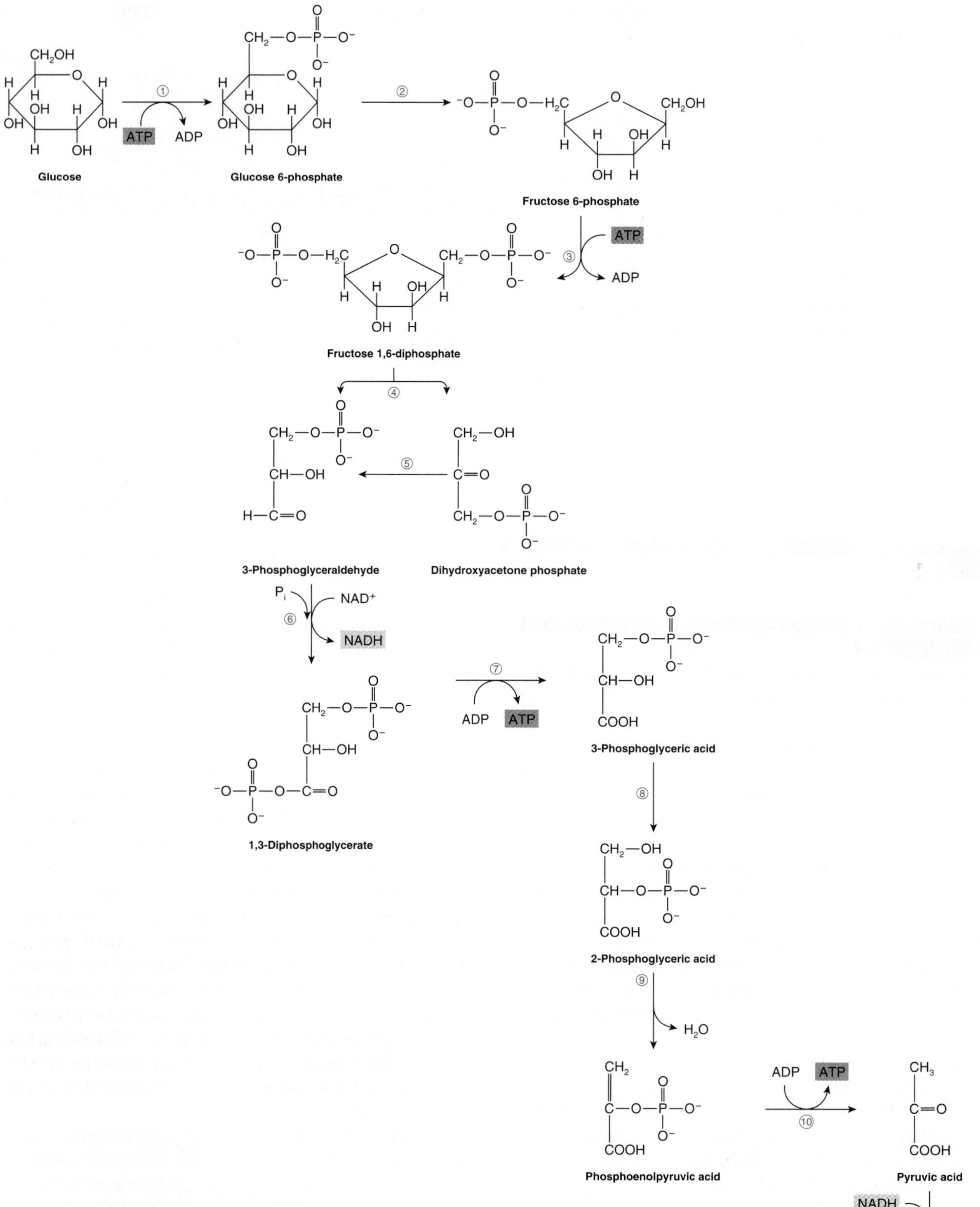

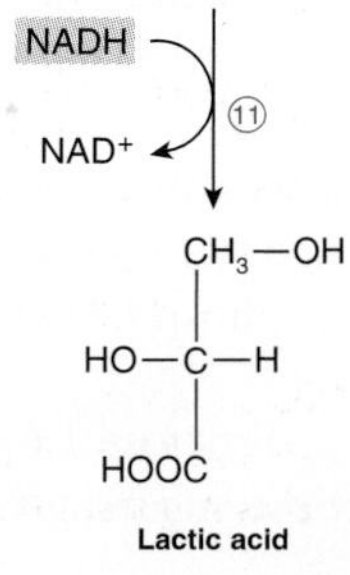

Figure 3-64 In the Embden-Meyerhof glycolytic pathway, glucose is degraded to pyruvic acid. Under aerobic conditions, pyruvic acid is further oxidized; in the absence of oxygen, however, it is reduced to lactate. Note that at step 4 each hexose molecule is split into two triose molecules. This doubles the molarities of reactants in the remainder of the pathway. The common energy intermediates, ATP and NADH, are shaded.

As indicated in Figure 3-65, glycolysis of 1 mol of glucose also yields 2 mol of NADH. In the presence of oxygen—that is, during aerobic metabolism—each mole of NADH eventually is oxidized by molecular oxygen, via the electron-transport chain, with concomitant production of 3 mol of ATP (see Figure 3-63). In this way, NAD^+ is regenerated for use in the glycolytic pathway. In the absence of oxygen—that is, during anaerobic metabolism—pyruvic acid resulting from glycolysis is reduced to lactate (Figure 3-64, step 11) or, in certain microorganisms such as yeast, to ethanol. This substrate reduction is coupled to oxidation of NADH, thereby replenishing the NAD^+ reduced to NADH in step 6 of glycolysis. In this case the electrons of NADH are accepted by pyruvate instead of oxygen. Without this anaerobic oxidation of the reduced coenzyme, there would be a depletion of the oxidized form of the coenzyme (NAD^+), and glycolysis would be blocked for lack of an electron acceptor at step 6 (the oxidation of 3-phosphoglyceraldehyde to 1,3-diphosphoglycerate) in the absence of molecular oxygen. The anaerobic $NAD^+ \rightleftharpoons$ NADH cycle that operates between steps 6 and 11 of the glycolytic pathway is shown in Figure 3-66.

Citric Acid Cycle

Under aerobic conditions, pyruvic acid is decarboxylated (i.e., 1 mol of CO_2 is removed), leaving a two-carbon *acetate* residue, which then reacts with **coenzyme A** (CoA) to yield acetyl coenzyme A (acetyl CoA). In this coupled reaction, NAD^+ accepts one hydrogen atom from pyruvic acid and one from coenzyme A (Figure 3-67). Coenzyme A acts as a carrier for the acetate residue, transferring it to *oxaloacetic acid* to form *citric acid* in the first reaction of the **citric acid cycle** (Figure 3-68). This reaction releases free CoA, which is recycled to repeatedly transfer acetate residues from pyruvic acid to oxaloacetate.

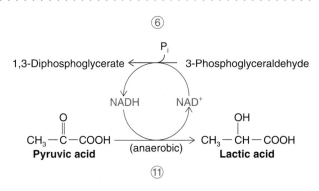

Figure 3-66 Under anaerobic conditions, the NAD^+ consumed in the breakdown of glucose to pyruvic acid is replenished by reduction of pyruvic acid to lactate. This $NAD^+ \longleftrightarrow$ NADH cycle permits glycolysis to proceed in the absence of oxygen. Circled numbers refer to steps in glycolytic pathway shown in Figure 3-64.

All the reactions of the glycolytic pathway leading to formation of pyruvic acid occur in free solution in the cytosol. The formation of acetyl CoA and CO_2 from pyruvic acid and the eight major reactions composing the citric acid cycle are all catalyzed by enzymes confined to the matrix compartment or inner membrane of mitochondria. For each acetate residue that enters the cycle, two additional molecules of CO_2 and two molecules of H_2O are produced. The overall reaction for the complete oxidation of pyruvate via the citric acid cycle and electron-transport chain is written as follows:

$$2 \ CH_3COCOOH + 5 \ O_2 \longrightarrow 6 \ CO_2 + 4 \ H_2O$$

The citric acid cycle is also known as the **Krebs cycle** after Hans Krebs, who in the early 1940s elucidated the major features of the reaction sequence and its cyclic nature. (It's also known as the tricarboxylic acid cycle because several intermediates have three carboxyl groups.) The two-carbon acetate residue of acetyl CoA first condenses with the four-carbon oxaloacetic acid to form the six-carbon citric acid (see Figure 3-68, step 1). In steps 4 and 5, two carboxyl groups of isocitric acid are removed to form two molecules of CO_2. Moreover, four hydrogen atoms are transferred to NAD^+ to form two molecules of NADH. Step 6 is catalyzed by succinic dehydrogenase, which is bound to the inner mitochondrial membrane. In this reaction, two hydrogen atoms are transferred from succinic

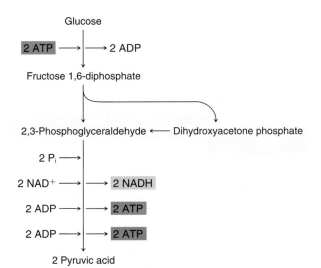

Figure 3-65 ATP is both consumed and produced during glycolysis. Oxidation of 1 mol of glucose to pyruvic acid yields a net of 2 mol of ATP as well as 2 mol of NADH.

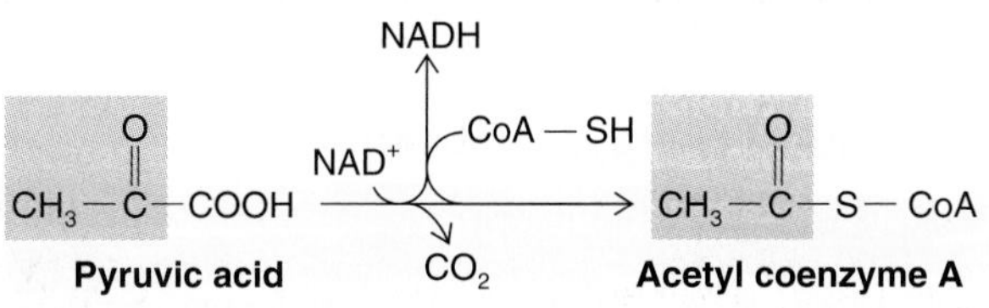

Figure 3-67 Pyruvic acid is decarboxylated and its two-carbon acetyl group (shaded) is transferred to coenzyme A (CoA) in a coupled reaction that yields acetyl CoA. This reaction occurs in the matrix compartment of mitochondria; it links glycolysis, which occurs in the cytosol, with cellular respiration within mitochondria.

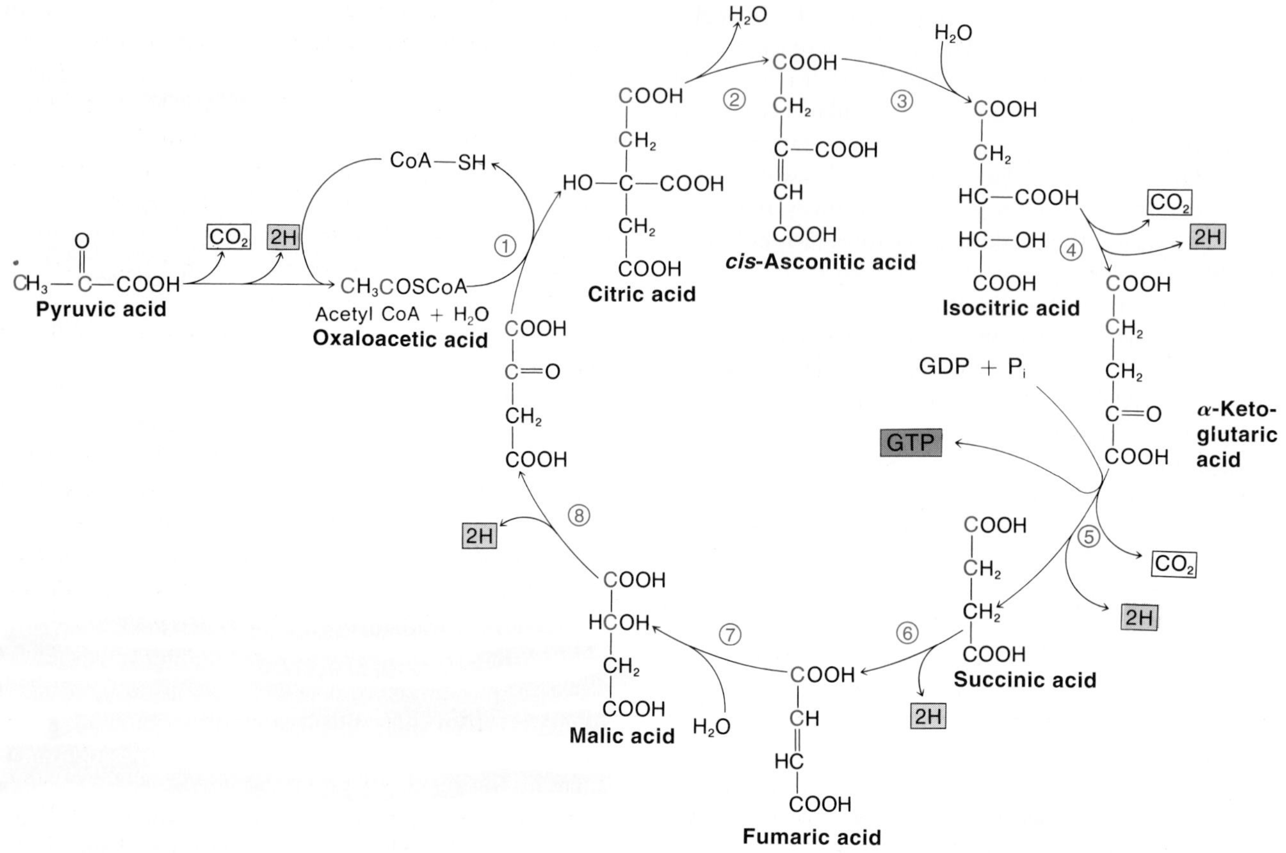

Figure 3-68 In the course of each circuit of the citric acid (Krebs) cycle, an acetate group transferred from acetyl CoA moves through several intermediates with the production of two CO_2 molecules and transfer of four pairs of protons (shaded) to electron-carrier coenzymes. The carbons of each entering acetyl group (red type) remain intact during its initial circuit around the cycle. Note that one molecule of GTP is produced.

acid to FAD, forming fumaric acid and $FADH_2$. Another oxidation occurs when malic acid is converted to oxaloacetic acid by the transfer of two hydrogen atoms to NAD^+ (step 8). A new acetate residue then condenses with the oxaloacetate to reconstitute the citric acid molecule, thus beginning another repetition of the cycle.

Each time one circuit of the citric acid cycle is completed, two carbon atoms and four oxygen atoms are removed as two molecules of CO_2 and eight hydrogen atoms are removed, two at a time (Figure 3-69). These hydrogens (as electrons accompanied by protons), carried by NADH and $FADH_2$, are fed into the respiratory chain and eventually oxidized to H_2O by molecular oxygen (see Figure 3-62). The CO_2 leaves the mitochondrion and then the cell by simple diffusion; it is finally eliminated as a gas via the circulatory and respiratory systems (see Chapter 13).

Efficiency of Energy Metabolism

The direct oxidation (burning) of glucose and its metabolic oxidation both liberate the same amount of free energy — namely, 686 kcal·mol^{-1}. If water is boiled by the heat of burning glucose to produce steam pressure for a steam engine, the mechanical output of the engine divided by the free-energy drop of 686 kcal·mol^{-1} represents the efficiency of the conversion of chemical to mechanical energy.

Steam engines have attained efficiencies of approximately 30%. Now let us see how efficiently the living cell transfers chemical energy from glucose to ATP.

Under standard conditions, it takes about 7 kcal to phosphorylate 1 mol of ADP to form ATP. If the free energy of glucose were conserved with an efficiency of 100%, each mole of glucose could energize the synthesis of 98 mol of ATP (686/7 = 98) from ADP and inorganic phosphate. In fact, during the metabolic oxidation of 1 mol of glucose only 38 mol of ATP are synthesized, giving an overall efficiency of about 42% or more.* The remaining free energy is liberated as metabolic heat, which accounts for a part of the heat that warms and thereby increases the metabolic rate of the tissue. Essentially all the energy incorporated into ATP and transferred to other molecules is eventually degraded to heat. The oxidation of fossil fuels represents a long-delayed return of stored energy to the original low-energy, high-entropy state of CO_2 and water.

---

* The 42% calculated here is for standard conditions. The efficiency of energy conservation by the cell may in fact be as high as 60%, because the free energy of hydrolysis of ATP under intracellular conditions has been estimated to be greater than that under standard conditions. The energetic efficiency of ATP production is therefore substantially better than that of a steam engine, in fact better than that of any other method yet devised by humans for converting chemical energy to mechanical energy.

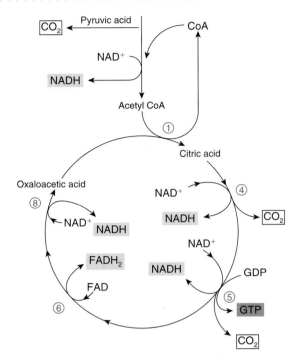

Figure 3-69 The oxidation of pyruvic acid produces CO_2, NADH, $FADH_2$, and GTP. The CO_2, a waste product, is eliminated. Both NADH and $FADH_2$ enter the electron-transport chain where their electron pressure is converted into ATP during oxidative phosphorylation (see Figure 3-63). GTP is formed in a substrate-level phosphorylation and like ATP can drive energy-requiring reactions. The circled numbers refer to the corresponding steps in Figure 3-68. [Adapted from Vander et al., 1975.]

It is interesting to compare the efficiencies of anaerobic and aerobic glucose metabolism, keeping in mind that since each mole of glucose yields 2 mol of the three-carbon derivatives, it is necessary to double all the molarities beyond step 5 of glycolysis. In anaerobic glycolysis there is a net production of 2 mol of ATP per mole of glucose (see Figure 3-65), because 2 of the 4 mol of ATP produced by substrate-level phosphorylation of ADP are consumed in the priming phosphorylations. As noted previously, the 2 mol of NADH produced by the oxidation of 3-phosphoglyceraldehyde is reoxidized to NAD^+ when the two pairs of hydrogen atoms are transferred to 2 mol of pyruvic acid to form 2 mol of lactic acid under anaerobic conditions (see Figure 3-66).

Under aerobic conditions, each of the 2 mol of NADH produced in glycolysis by the oxidation of 3-phosphoglyceraldehyde yields 3 mol of ATP during oxidative phosphorylation (Figure 3-63). Pyruvic acid goes on to fuel the citric acid cycle, yielding a total of 10 pairs of hydrogen atoms for every 2 mol of pyruvic acid (Figure 3-69). Eight pairs are carried by NAD^+, yielding 24 mol of ATP, while two pairs are carried by FAD, yielding 4 mol of ATP. Finally, 2 mol of GTP is produced by substrate-level phosphorylation of guanosine diphosphate (GDP) during the oxidation of α-ketoglutaric acid to succinic acid in step 5 of the citric acid cycle (see

Figure 3-68). This adds up to 38 mol of nucleotide triphosphate per mole of glucose during aerobic respiration. As noted earlier, only 2 mol is produced during anaerobic respiration. Thus, although aerobic respiration conserves a minimum of about 42% of the free energy of the glucose molecule, anaerobic respiration conserves only about 2%. Stated differently, the energy conservation of glucose metabolism via aerobic glycolysis and the citric acid cycle is about 20 times as efficient as that via anaerobic glycolysis. It is not surprising, then, that most animals carry on aerobic metabolism and require molecular oxygen for survival.

Oxygen Debt

When animal tissue, such as active muscle, receives less oxygen than is required to produce adequate amounts of ATP by respiratory-chain phosphorylation, some of the pyruvic acid, instead of going on to fuel the citric acid cycle, is reduced to lactic acid. For every 2 mol of pyruvic acid reduced, 2 mol of NADH is oxidized, costing 6 mol of ATP that might have been synthesized by respiratory-chain phosphorylation. If the oxygen deficiency is maintained, lactic acid concentrations rise, and some may enter the extracellular space and circulatory system. When the muscle stops its strenuous activity and oxygen is again available, the enzyme lactate dehydrogenase oxidizes the accumulated lactic acid back to pyruvic acid coupled to the reduction of NAD^+ (Figure 3-70). The NADH produced in this reaction then is oxidized in the respiratory chain, recouping some of the ATP forfeited by the anaerobic formation of lactic acid. Moreover, some of the pyruvic acid regenerated from lactic acid goes on to fuel the citric acid cycle, eventually recouping more of the forfeited ATP. (Some of the recovered pyruvic acid may be used for the synthesis of alanine and glucose.)

In other words, when muscle is starved for oxygen, it switches to anaerobic metabolism in which ATP is formed with low efficiency. But the unused chemical energy is stored in the tissue as lactic acid, and later becomes available for aerobic metabolism when sufficient oxygen is available. With cessation of heavy exercise, the respiratory and circulatory systems continue for some time to supply large amounts of oxygen in order to "repay" the oxygen debt that was built up as an accumulation of lactic acid.

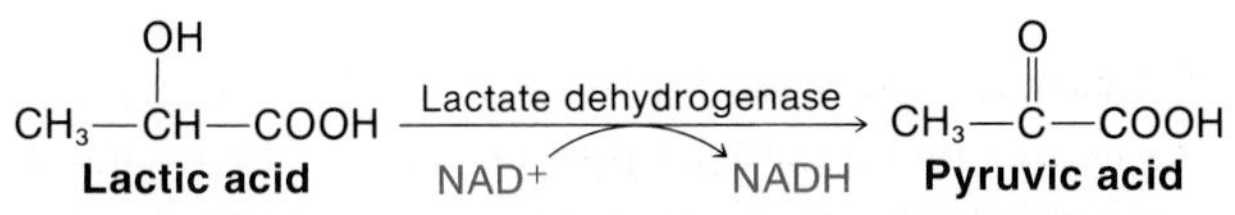

$$\underset{\textbf{Lactic acid}}{CH_3-\overset{\overset{\displaystyle OH}{|}}{C}H-COOH} \xrightarrow[NAD^+ \quad NADH]{\text{Lactate dehydrogenase}} \underset{\textbf{Pyruvic acid}}{CH_3-\overset{\overset{\displaystyle O}{\|}}{C}-COOH}$$

Figure 3-70 The lactic acid that accumulates in muscle during strenuous activity is oxidized to pyruvate following cessation of activity when the oxygen supply increases. Oxidation of the NADH, as well as some of the pyruvate produced in this reaction, eventually fuels respiratory-chain phosphorylation, thereby recovering some of the ATP forfeited during anaerobic production of lactic acid.

SUMMARY

Biologists generally accept the hypothesis that life on Earth arose spontaneously in shallow seas under special conditions that no longer exist. It is believed that organic molecules, synthesized in the primitive atmosphere by reactions energized by lightning discharges or radiation, accumulated in the water over long periods, providing the raw material for primordial living cells.

Living matter is composed primarily of carbon, nitrogen, oxygen, and hydrogen in stable (covalently bonded) associations. Carbon, nitrogen, and oxygen are capable of forming double and triple bonds, which greatly increase the structural variety of biological molecules.

The polarity of the water molecule is responsible for hydrogen bonding. Besides linking hydrogen and oxygen atoms of adjacent water molecules, hydrogen bonding confers on water many special properties that have profoundly shaped the evolution and survival of animal organisms. Water dissociates spontaneously into H^+ and OH^-; in 1 liter of pure water there is 10^{-7} mol of each ion. Many substances in solution contribute to an imbalance in the concentrations of H^+ and OH^-, giving rise to acid-base behavior (i.e., donation and acceptance of protons). Their concentrations are measured by the pH system. The pH of biological fluids influences the charges carried by amino acid side groups and hence the conformation and activity of proteins. Physiological buffering systems maintain the intra- and extracellular pH within a narrow range.

The electrostatic force attracting an ion to a site of opposite charge is determined by the distance of closest approach of the ion to the site. The ion selectivity of a site depends on the relative ability of the site to compete with the dipolar water molecules in binding different ion species.

Four major groups of organic molecules compose animal cells. The lipids, which include triglycerides (fats), fatty acids, waxes, sterols, and phospholipids, are important as energy stores and as constituents of biological membranes. The carbohydrates include sugars, storage carbohydrates (glycogen and starch), and structural polymers such as chitin and cellulose. The sugars, glycogen, and starch are major sources of substrate for energy metabolism by cells. Proteins, made up of linearly arranged amino acid residues, form many structural materials such as collagen, keratin, and subcellular fibrils and tubules. Enzymes are specialized proteins bearing catalytically active sites and are important in nearly all biological reactions. The nucleic acids DNA and RNA encode the genetic information necessary for the orderly synthesis of all the protein molecules in the cell.

A major characteristic of biological systems is that they maintain a low state of entropy—that is, they are highly and improbably organized. A steady expenditure of energy must therefore be derived from foodstuff molecules by the processes of energy metabolism. In the living cell, metabolism occurs as orderly, regulated sequences of chemical reactions catalyzed by enzymes. Chemical reactions spontaneously tend to go down an energy gradient, decreasing free energy and increasing entropy. Living systems appear to defy entropy, but they do not; they merely exist at the expense of chemical energy obtained from their environment.

Energy-requiring biological reactions utilize ATP, a triply phosphorylated nucleotide that serves as a common intermediate capable of contributing chemical energy stored in the form of its terminal phosphate bond. This energy transfer is accomplished by means of coupled reactions in which an endergonic (energy-requiring) reaction is driven by an exergonic (energy-releasing) reaction. ATP is reconstituted from ADP by the oxidation of foodstuff molecules, which largely have their origin in the radiant solar energy trapped during the process of photosynthesis in green plants. Thus, animals depend on energy ultimately derived from the sun.

Enzymes act as biological catalysts; that is, they lower the energy required to activate reactants sufficiently for reaction and thereby increase the rate of reaction at a given temperature. With the aid of enzymes, cell chemistry can proceed at reasonable body temperatures. The catalytic action of an enzyme arises from its ability to bind specific reactant molecules at the active site; the close steric fit required for this interaction is largely responsible for enzyme specificity. This binding produces favorable spatial relations between the reacting molecules. Regulation of enzyme concentrations appropriate to the needs, function, and environment of the cell is performed at the level of DNA through enzyme induction and repression. The activity of some enzymes can also be controlled by the binding of regulatory molecules or ions to the enzyme molecule at an allosteric site that is distinct from the active site of the enzyme. This binding results in a conformational change that affects the properties of the active site.

Liberation of the free energy stored in foodstuffs during metabolism occurs by transfer of electrons from an electron donor (reducer) to an oxidizer. The release of free energy in the cell is budgeted in small steps compatible with the amounts of free energy required to phosphorylate ADP to ATP. For example, electrons from the reduced coenzymes NADH and $FADH_2$ are transported in increments down a chain of electron acceptors and electron donors, yielding enough energy for synthesis of ATP at three points in the chain. An electron-pressure gradient exists along the electron-transport chain of cytochromes, so that electrons flow to the ultimate electron acceptor, molecular oxygen. It is the electron-hungry nature of the oxygen atom, and its abundance on the surface of the Earth, that makes it the ideal terminal electron acceptor in living systems.

During glycolysis, each glucose molecule is broken down into two molecules of the three-carbon pyruvic acid with net formation of 2 ATP molecules and 2 NADH molecules. During anaerobic metabolism, the pyruvic acid is reduced to lactate, regenerating the NAD^+ consumed during glycolysis. During aerobic metabolism, pyruvic acid produced by glycolysis is oxidized completely to CO_2 and H_2O, via the citric acid cycle and respiratory chain. Oxidation of 2 molecules of pyruvic acid is accompanied by the formation of 34 more ATP molecules and 2 GTP molecules. Biological systems therefore attain efficiencies of at least 42%, considerably better than those of any manufactured engine energized by the oxidation of organic fuels.

REVIEW QUESTIONS

1. What evidence is there that the molecular building blocks of life might have arisen spontaneously on the primordial Earth?

2. What determines the reactivity of a given atom? Why?

3. What properties of carbon, hydrogen, oxygen, and nitrogen make them especially well adapted for the construction of biological molecules?

4. Why is oxygen of such biological importance?

5. What important physical and chemical characteristics of H_2O can be directly related to the dipole nature of the water molecule?

6. What is the pH of a 1 M solution of an acid that is 10% dissociated?

7. Why is a weak acid rather than a strong one required for a pH buffer system?

8. What is the difference between molality and molarity?

9. How many grams does a mole of CO_2 weigh?

10. Approximately how many particles are in a 1 M solution of NaCl?

11. What is the approximate boiling point of a 1 molal solution of NaCl?

12. Why do some liquids conduct electricity whereas others do not?

13. How many ions flow past a point (equivalents per second) at a current of 1 mA?

14. What are the primary factors that govern the binding of two cations, a and b, to an electronegative binding site? Write the expression that integrates these factors into a meaningful quantity.

15. Does the force of attraction fall off most rapidly with distance between a monovalent cation and (a) a monopolar binding site or (b) a multipolar site? Give the expression relating force and distance for each site.

16. What factors determine each level of protein structure—primary, secondary, tertiary, and quaternary protein structure?

17. What special characteristic does cysteine have that makes it a likely participant in the active sites of enzyme molecules?

18. Why do proteins become denatured (structurally disorganized) at elevated temperatures?

19. Living systems would appear to defy the second law of thermodynamics because of their high degree of maintained order. How do you reconcile the low entropy of an organism with this fundamental physical law?

20. At a particular temperature, will a reaction with $\Delta S > \Delta H$ be endergonic or exergonic?

21. Under what conditions will an endergonic reaction proceed?

22. What is ΔG for a system at equilibrium?

23. How does ATP "donate" stored chemical energy to an endergonic reaction?

24. What is meant by the term *coupled reaction*?

25. How does increased temperature increase the rate of a chemical reaction?

26. What factors can influence the temperature optimal for an enzymatic reaction?

27. How does a catalyst increase the rate of a reaction?

28. Why is catalysis necessary in living organisms?

29. How do enzymes exhibit substrate or bond specificity?

30. How does pH affect the activity of an enzyme?

31. How was the "steric-fit" theory of active-site specificity shown to be correct?

32. What factors can influence the rate of enzyme-catalyzed reactions?

33. The Michaelis-Menten constant, K_M, is equal to the substrate concentration at which a particular reaction proceeds at half its maximum velocity, V_{max}. Does a high K_M indicate a greater or a lesser enzyme-substrate affinity?

34. Why does a high substrate concentration reverse the effects of a competitive inhibitor and yet have no effect on a noncompetitive inhibitor?

35. How does each type of inhibition affect the Michaelis-Menten constant, K_M? Explain why.

36. Why does aerobic metabolism yield much more energy per glucose molecule than anaerobic metabolism?

37. What is the advantage of incremental drops in electron pressure compared with a single large drop in electron pressure in the electron-transport chain?

38. How is energy liberated in discrete amounts in the electron-transport chain?

39. How does the mechanism of energy release by the citric acid cycle differ from that during glycolysis?

SUGGESTED READINGS

Atkins, P. W. 1994. *Physical Chemistry*. New York : W. H. Freeman and Company. (A complete treatment at the undergraduate level of many of the basic concepts introduced in this chapter.)

Lehninger, A. L., et al. 1993. *Principles of Biochemistry*. 2d ed. New York: Worth. (A short, straightforward book about biochemical principles.)

Lodish, H. D., et al. 1995. *Molecular Cell Biology*. 3d ed. New York: Scientific American Books. (Comprehensive textbook describing many of the basic biochemical processes that occur in the cell.)

Stryer, L. 1995. *Biochemistry*. New York: W. H. Freeman and Co. (Highly readable reference for information about biochemical structures and mechanism.)

MEMBRANES, CHANNELS, AND TRANSPORT

The complex chemical reactions that ultimately are responsible for animal life will proceed only under stable, restricted conditions. Such constancy is maintained within cells largely through the action of biological membranes, which form a protective barrier that allows only certain materials to pass into or out of the cell. Animal tissue contains an astounding amount of biological membrane. The chimpanzee brain, for example, is estimated to have about 100,000 m² of cell membrane, an area equal to three full-sized soccer fields. Though cell membranes are a major constituent of all living matter and essential to all life

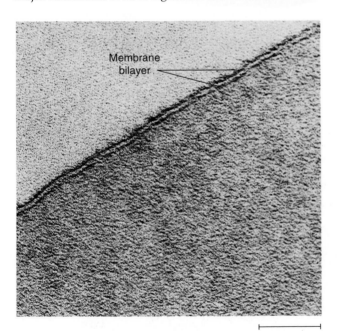

Membrane bilayer

| 10 nm

Figure 4-1 The plasma membrane creates a barrier between the interior and exterior of the cell, as revealed in this electron micrograph. The cell interior (lower right) is separated from the cell exterior by the surface membrane bilayer, which is seen in cross-section as a dark-light profile about 10 nm thick. The dark-light-dark sandwich-like appearance is due to the differential staining of the "unit membrane" by an electron-opaque substance during preparation of the tissue. [Courtesy of J. D. Robertson.]

process, their existence was questioned up until the 1930s. There was little or no direct anatomical evidence for biological membranes at the time, so their existence could only be inferred from physiological studies. The first important observations on the diffusion-limiting properties of the cell surface were made in the mid-nineteenth century by Karl Wilhelm von Nägeli, who noticed that the cell surface acted as a barrier to free diffusion of dyes into the cell from the extracellular fluid. From these experiments he deduced the presence of a "plasma membrane." He also discovered the osmotic behavior of cells, noting that they swell when placed in dilute solutions and shrink in concentrated solutions. Structural evidence for the existence of a distinct cell membrane was first available after the development of electron microscopy (see Chapter 2). At the surface of every cell type is a continuous double-layered membrane ranging in thickness from 6 to 23 nm (Figure 4-1).

Understanding membrane structure and function is critical to the study of animal physiology. In this chapter we discuss membrane structural features and their critical role in maintaining cell integrity and controlling cell activities. In the next chapter, we discuss the electrical behavior of cell membranes that is responsible for cell-to-cell signaling, which, in turn, coordinates action in animals.

MEMBRANE STRUCTURE AND ORGANIZATION

At their surfaces, cells are surrounded by a **plasma membrane,** an extraordinarily thin, complex, lipid-based structure that encloses the **cytoplasm** (including the **cytosol** and all cell organelles) and the cell nucleus. (The internal organelles, such as the ATP-producing mitochondria, which we discussed in Chapter 3, have their own surface membranes.) This enclosing feature of the plasma membrane is its most obvious function, and also its most critical. With the help of various metabolic mechanisms, described later, the membrane regulates molecular traffic between the orderly interior of the cell and the more disorderly, potentially disruptive external environment.

Membrane Composition

The cell membrane is spanned by **integral proteins**. These proteins act as both selective filters and active transport devices responsible for getting nutrients into and cellular products and waste out of the cell. Other proteins contained within the membrane sense external signals that direct the cell responses to environmental changes.

Cell membranes sustain different concentrations of certain ions on their two sides, leading to **concentration gradients** of several ionic species across membranes. The channel proteins contained in cell membranes actively participate in the translocation of substances between compartments and ultimately regulate the cytoplasmic concentration of dissolved ions and other molecules rather precisely. This allows maintenance of an **intracellular milieu** required for the finely balanced metabolic and synthetic chemical reactions of the cell.

All biological membranes, including the internal membranes of organelles of eukaryotic cells, have essentially the same structure: lipid and protein molecules kept together by noncovalent interactions. The lipid molecules are arranged in a continuous double layer, called the **lipid bilayer**, which is relatively impermeable to passage of most water-soluble molecules. In 1925, using a simple but elegant experiment, Gorter and Grendel provided the first evidence that cell membranes are lipid bilayers. First, they dissolved the lipids from red blood cell ghosts, the empty membrane sacs left when red blood cells have been induced to burst open. The extracted membrane lipids were then allowed to spread out on the surface of water in a trough. Because of their asymmetry, the lipid molecules became oriented so that their po-

lar head groups formed hydrogen bonds with the water and their hydrophobic hydrocarbon chains stuck up into the air. When the dispersed film of lipid molecules was gently compressed into a continuous monomolecular film, it occupied an area about twice the surface area of the original red blood cells. Since the only membrane in mammalian red blood cells is the plasma membrane, it was concluded that the lipid molecules in the membrane must be a continuous bilayer. As we saw in Figure 4-1, the bilayer has since been visualized in cross-section by the electron microscope, as well as with freeze fracture methods, in which the membrane is split through the center of the bilayer (see Chapter 2). The chemical properties of lipid molecules, which cause them to assemble spontaneously into bilayers even under artificial conditions (see Chapter 3), are responsible for the structure of the membrane.

Membranes are remarkably fluid structures, in which most of the lipid and protein molecules "float" around in the plane of the bilayer (Figure 4-2). The relative proportion of lipids and proteins present in a membrane depends on the kind of cell or organelle the membrane encloses. Lipids, which are far smaller and simpler molecules than proteins, provide the primary structure of the membrane. Integral proteins embedded in the membrane play more specialized roles such as transporting molecules through the membrane, catalyzing reactions, and transducing chemical signals. Other proteins connect the membrane to the cytoskeleton or to adjacent cells. Some proteins are intimately associated with lipid molecules because of lipophilic groups exposed on the surface of the protein molecule. The protein-lipid complex is called a **lipoprotein.**

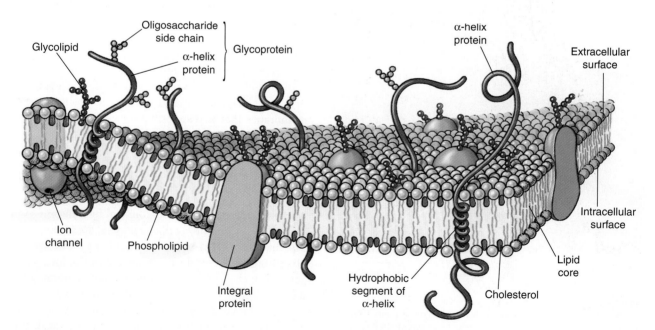

Figure 4-2 The Singer-Nicolson fluid mosaic model of the membrane is widely accepted. The globular integral proteins embedded in the lipid bilayer provide a mechanism for transmembrane transport. The inner mitochondrial membrane would have an even higher protein content and thus less lipid bilayer than this figure represents. The glycoproteins bear oligosaccharide side chains and are vital for cell recognition and communication. Cholesterol molecules lie close to the heads of the phospholipid molecules, where they reduce membrane flexibility. The inner ends of the phospholipid tails are highly mobile, giving the membrane fluidity.

Lipid molecules are insoluble in water but can be dissolved in organic solvents. They comprise about half the mass of plasma membranes in animal cells, the rest being essentially protein. Each square micron of membrane has about 10^6 lipid molecules, meaning typical small cells have about 10^9 lipid molecules. The three primary types of lipids in cell membranes are

- **Phosphoglycerides,** characterized by a glycerol backbone

- **Sphingolipids,** which have backbones made of sphingosine bases

- **Sterols,** such as cholesterol, which are nonpolar and only slightly soluble in water

The first two lipid types are **amphipathic,** meaning that they have a hydrophilic (water-soluble) and a hydrophobic (water-insoluble) end (Figure 4-3). The dual nature of these amphipathic membrane lipids, with their hydrophilic heads and hydrophobic tails, is crucial to the organization of biological membranes. Their polar heads seek water and their nonpolar tails seek one another (see Figure 3-14), being mutually attracted by van der Waals forces. Thus, these molecules are ideally suited to form an interface between a nonaqueous lipid environment (phase) within the membrane itself and the aqueous intra- and extracellular phases in contact with the inner and outer membrane surfaces. These same forces cause lipid bilayers to reseal themselves when they are torn, which gives cells a **self-repair capability.** Differences in the length of the two fatty acid chains and in their composition (see Figure 3-20) influence lipid packing and hence fluidity, causing subtle differences in lipid bilayer characteristics. The hydrophobic properties of the phospholipid hydrocarbon tails are responsible for the low permeability of membranes to polar substances (e.g., inorganic ions and polar nonelectrolytes such as sucrose and inulin) and for their correspondingly greater permeability to nonpolar substances (e.g., steroid hormones).

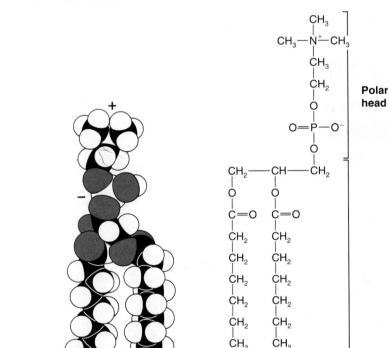

Polar head

Nonpolar tails

Figure 4-3 Phosphatidyl choline, a phosphoglyceride, has charges that give the head group its polar character. Note that the left hydrocarbon chain in this figure is unsaturated. In order to distinguish the unsaturated fatty acid chain from the saturated one, in this figure (and in those that follow) the unsaturated fatty acid chain is drawn with a distinct bend in it rather than with a small kink. Actually, only the double bond is rigid in the unsaturated fatty acid. Because the single carbon-carbon bonds in the rest of the chain are free to rotate, both the saturated and unsaturated fatty acid chains tend to pack in parallel arrays in each phospholipid monolayer. [Stryer, 1988.]

The third class of membrane lipids, the sterols, are largely nonpolar and only slightly soluble in water (Figure 4-4). In aqueous solution they form complexes with proteins that are far more water-soluble than the sterols are alone. Once in the membrane, the sterol molecule fits snugly between the hydrocarbon tails of the phospholipids and glycolipids (Figure 4-5) and increases the viscosity of the hydrocarbon core of the membrane.

Fluid Mosaic Membranes

The concept of a lipid bilayer membrane enclosing most cells gained wide acceptance by the early 1950s because of compelling evidence from a variety of measurement techniques (Spotlight 4-1). Chemical fractionation of membranes and immunochemical studies confirmed that proteins are also an important component of membranes. Moreover, the enzymatic properties of membranes, such as active transport and other metabolic functions, require the participation of proteins. An example is the protein complexes responsible for electron transport and oxidative phosphorylation described in Chapter 3.

Despite this early progress in characterizing the membrane, it wasn't until 20 years later that researchers recognized how fluid and heterogeneous membranes really are. It was discovered that some of the protein molecules are free to diffuse laterally along the membrane, presumably because of the fluidity of the lipid matrix. In addition, labeling studies demonstrated that protein molecules or parts of molecules facing one side of the membrane differ from those facing the other side, and that they normally do not "flip-flop" across the membrane as previously suspected.

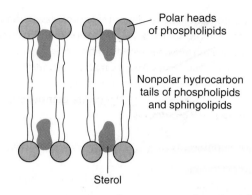

Figure 4-5 Nonpolar sterols insert themselves between the hydrocarbon tails and the polar head groups of the phospholipids in the membrane.

Additionally, in many membranes, the distribution of lipid species differs in the two lipid layers.

The fluid mosaic model

On the basis of evidence that had emerged during the 1950s and especially the 1960s, Singer and Nicolson (1972) proposed the **fluid mosaic model** of the membrane, in which globular proteins are integrated with the lipid bilayer, with some protein molecules penetrating the bilayer completely and others penetrating only partially (see Figure 4-2). These integral proteins are thought to be amphipathic, their nonpolar portions buried in the hydrocarbon core of the bilayer and their polar portions protruding from the core to form a hydrophilic surface with charged amino acid side groups in the aqueous phase. Uncharged hydrophobic side groups, on the other hand, are associated with the hydrocarbon bilayer (Figure 4-6). The hydrophobic nature of these side groups is important in keeping the integral proteins from leaving the lipid bilayer. Evidence continues to emerge in support of this model, which is now widely accepted a quarter of a century after first being described.

Membrane fluidity

A variety of techniques have been used to demonstrate that lipid molecules in the membrane very rarely move from one

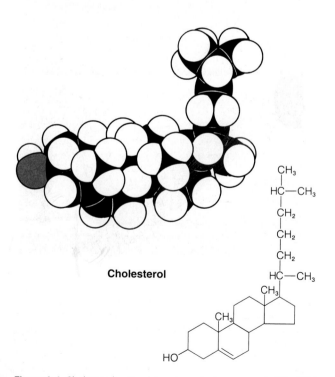

Cholesterol

Figure 4-4 Cholesterol, a sterol, is an important component of the lipid membrane. [From Lehninger, 1975.]

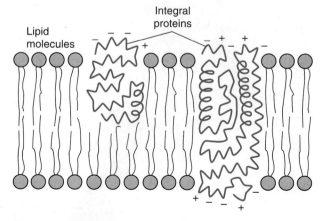

Figure 4-6 A cross-sectional view shows the complexity of the mosaic bilayer model. The charged hydrophilic amino acid side groups of the proteins project into the aqueous phase, and the uncharged hydrophobic groups are buried in the lipid phase of the bilayer.

side of the membrane to the other (about once a month), but exchange places with adjacent molecules in a monolayer about 10^7 times per second. This brisk lipid exchange within a membrane results in rapid migration along the plane of the membrane but not across it.

Membrane fluidity depends on its composition, and **cholesterol** plays an important role in governing this membrane characteristic. Plasma membranes in eukaryotic organisms contain lots of cholesterol, up to one molecule for every phospholipid. Cholesterol, when present, binds weakly to adjacent phospholipids, making lipid bilayers significantly less fluid, but stronger (Figure 4-7). The incorporation of too much cholesterol into cell membranes, however, causes the membranes to lose flexibility. This is the mechanism underlying "hardening of the arteries," a major cause of cardiovascular disease, in which the cell membranes of the endothelial cells lining the arteries become abnormally rigid (with additional cholesterol plaques stored in the intimal layers, as well).

Lipid composition of biological membranes varies among tissue types. While most membranes contain a significant fraction of cholesterol (>18%), other types of lipids may be present in much higher or lower fractions. Lipids also differ in their head groups (see Figure 3-13), which in turn influences their interactions with proteins. Indeed, some integral proteins function only in the presence of a specific ratio of lipid types. Hence, cells must regulate the distribution of lipid species in their membranes during cell development and rearrange lipid concentrations according to specific functional needs.

Heterogeneity of the integral membrane proteins

The integral proteins found in the plasma membrane (see Figure 4-2) take many functional forms, including the ion channels, various carriers and membrane pumps, receptor molecules, and recognition molecules. The number of integral proteins varies, but in some membranes the protein content is so high that only about three lipid molecules separate the proteins at the point of closest approach.

Morphological evidence for the heterogeneous **mosaic** arrangement of globular proteins in a lipid bilayer is seen in freeze-etch electron micrographs of the surface of a membrane (Figure 4-8). When subjected to digestion by

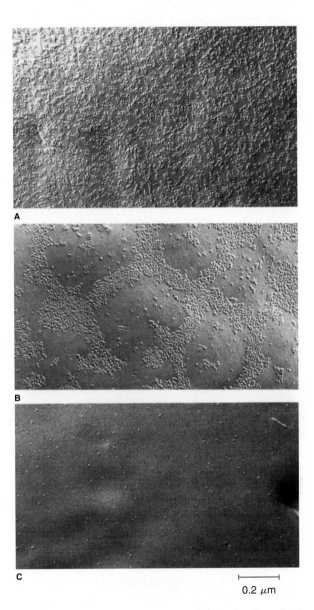

Figure 4-8 Freeze-etch methods yield morphological evidence for the mosaic membrane model. In these freeze-etch electron micrographs, the plasma membrane has been split along the middle of the bilayer, exposing membrane-embedded particles with diameters of 5 to 8 nm. Digestion with a proteolytic enzyme produced progressive loss of these particles, indicating that they are globular proteins inserted into the lipid phase of the membrane. **(A)** Control. **(B)** 45% of the particles digested. **(C)** 70% digested. [Courtesy of L. H. Engstrom and D. Branton.]

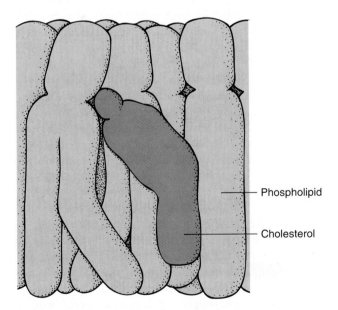

Phospholipid

Cholesterol

Figure 4-7 Cholesterol interacts weakly with adjacent phospholipids in the membrane, partially immobilizing their fatty acyl chains. As a result, the membrane is less fluid but mechanically stronger. The amount of cholesterol present in the lipid bilayer varies widely with cell type. In some cells, the membranes have nearly as many cholesterol molecules as phospholipids, while the membranes of other cells are almost devoid of cholesterol. The structural formula of cholesterol is shown in Figure 4-4.

SPOTLIGHT 4-1

THE CASE FOR A LIPID BILAYER MEMBRANE

There is a large body of accumulated evidence that points to the existence of the lipid bilayer membrane:

1. The **lipid content of membranes** is consistent with a bilayer of oriented lipid molecules, as first shown by Gorter and Grendel in 1925.
2. The **ease of passage** of nonelectrolytes through the membrane is consistent with the presence of a lipid membrane barrier, given the tendency of such molecules to leave an aqueous phase for a lipid phase, as when oil and water separate. The greater this tendency, the more permeant the molecule. Moreover, certain lipid-insoluble substances must first be converted to a lipid-soluble form (by attachment of a lipid-soluble molecule) before they can cross the membrane.
3. The **capacitance** of biological membranes, typically 10^{-6} F·cm^{-2}, is the same as that of a layer of lipid the thickness of two phospholipid molecules placed end to end (i.e., 6.0–7.5 nm).

4. When fixed with permanganate, membranes appear as triple-layered profiles: a lightly staining central zone sandwiched between two electron-dense outer layers (see Figure 4-1), with a total thickness of about 7.5 nm. In 1955, J. David Robertson (1960) named this three-layered structure the **unit membrane.** The unit-membrane concept is consistent with a bimolecular layer of lipid between two layers of protein.
5. The **thickness** of a lipid bilayer, calculated as twice the length of a single membrane lipid molecule, agrees roughly with the dimensions of the unit membrane seen in electron micrographs.
6. Freeze-etch electron microscopy shows that membranes have a **preferential plane of splitting** down the middle, which is consistent with separation of a bilayer into two monolayers.
7. **Artificial lipid bilayers** (see Spotlight 4-2), reconstituted lipid bilayers of similar thickness and presumed structure to the bimolecular lipid core of the fluid mosaic membrane model, have permeabilities and electrical properties fundamentally similar to those of cell membranes. Those differences that exist can be attributed to special channels and carriers present in natural membranes.

---

proteolytic (protein-digesting) enzymes, the globular units seen in the membrane are progressively removed, demonstrating that they are indeed proteins.

Variation in Membrane Form

Membrane composition varies greatly between cell types. At one extreme is the metabolically inert myelin sheath surrounding the axons of some nerve cells, in which the lipid bilayer is largely uninterrupted. At the other extreme are cells with membranes that have structures of repeating nonlipid macromolecular units that nearly obliterate the lipid bilayers. Such membranes evolved for highly specialized purposes such as signaling or enzymatic activity. In visual receptor cells, for example, the repeating macromolecular units are molecules of the visual pigment opsin. Mitochondrial membranes, specialized for enzymatic activity, are composed almost entirely of repeating subunits of ordered enzymatic aggregates. Between these extremes are the plasma membrane and most intracellular membranes, in which the bilayer is interrupted frequently by integral protein molecules. Thus the basic structure of the lipid bilayer with integral proteins is highly modified as required for functional specialization.

CROSSING THE MEMBRANE: AN OVERVIEW

The structure of membranes makes them quite selective about which molecules can pass through them. The hy-

drophobic interior of the lipid bilayer makes membranes highly impermeable to most polar molecules. This prevents water-soluble components of the cell from easily entering or escaping. However, such movement may at times be necessary or desirable, so mechanisms for transferring these molecules across membranes have evolved in all cells. Macromolecules like proteins and large particles must also be transported across plasma membranes using specialized mechanisms.

To understand these special means of membrane transport in living cells, we will first review the physical principles of solute and solvent displacement in solution and across semipermeable membranes. Such membranes closely resemble those found in living cells, and the principles explained here apply in many physiological situations.

Diffusion

Random thermal motion of suspended or dissolved molecules causes their dispersion from regions of higher concentration to regions of lower concentration, a process called **diffusion.** Diffusion is extremely slow when viewed on a tissue, rather than a cellular, scale. For example, a crystal of copper sulfate dissolves in unstirred water so slowly that it may take a whole day to color a liter of water completely. When viewed in the microscopic dimensions of the cell, however, diffusion times can be as short as a fraction of a millisecond.

The rate of diffusion of a solute s can be defined by the **Fick diffusion equation:**

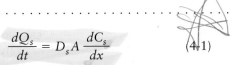

$$\frac{dQ_s}{dt} = D_s A \frac{dC_s}{dx} \qquad (4\text{-}1)$$

in which dQ_s/dt is the rate of diffusion (i.e., quantity of s diffusing per unit time), D_s is the **diffusion coefficient** of s, A is the cross-sectional area through which s is diffusing, and dC_s/dx is the concentration gradient of s (i.e., the change in concentration with distance). The gradient factor dC_s/dx is clearly very important, because it determines the rate at which s will diffuse down the gradient. D_s varies with the nature and molecular weight of s and of the solvent, which is water in most physiological situations.

Membrane Flux

If a solute occurs on both sides of a membrane through which it can diffuse, it will exhibit a unidirectional flux in each direction (Figure 4-9A). The flux, or rate of diffusion, J is the amount of the solute that passes through a unit area of membrane every second in one direction, so that

$$J = \frac{dQ_s}{dt} \qquad (4\text{-}2)$$

where J would typically have units of moles per square centimeter per second ($M \cdot cm^{-2} \cdot s^{-1}$). The flux in one direction (say, from cell exterior to cell interior) is considered *independent* of the flux in the opposite direction. Thus, if the **influx** and **efflux** are equal, the **net flux** is zero. If the unidirectional flux is greater in one direction, there is a net flux, which is the difference between the two unidirectional fluxes (Figure 4-9B).

The **permeability** of the membrane to a substance is the rate at which that substance passively penetrates the membrane under a specified set of conditions. A greater permeability will be accompanied by a greater flux if other factors remain equal. If we assume that the membrane is a homogeneous barrier and that a continuous concentration gradient exists for a nonelectrolyte substance between the side of high concentration (I) and the side of low concentration (II), then

$$\frac{dQ_s}{dt} = P(C_I - C_{II}) \qquad (4\text{-}3)$$

in which dQ_s/dt is, again, the amount of substance s crossing a unit area of membrane per unit time (e.g., moles per square centimeter per second), C_I and C_{II} are the respective concentrations (e.g., $M \cdot cm^{-2}$) of the substance on the two sides of the membrane, and P is the **permeability constant** of the substance, with the dimension of velocity ($cm \cdot s^{-1}$).

Note that equation 4-3 applies *only* to molecules that are not being actively transported or influenced by any forces other than simple diffusion. This excludes electrolytes, since they are electrically charged when dissociated, and consequently their flux depends not only on the

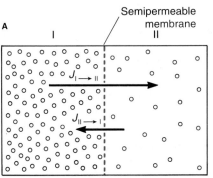

Figure 4-9 Solutes can move through a membrane in either direction, depending on prevailing physical and chemical conditions. **(A)** The arrows represent the actual fluxes of a substance between compartments I and II. **(B)** The single arrow indicates the resulting net flux, from compartment I to II.

concentration gradient, but also on the electrical gradient (i.e., the electric potential difference across the membrane). As is evident from equation 4-3, the flux of a nonelectrolyte should be a linear function of the concentration gradient ($C_I - C_{II}$). This linear relationship is characteristic of simple diffusion, and can be used in experiments to distinguish between passive diffusion of a substance and any other mechanism. The permeability constant incorporates all the factors inherent in the membrane and the substance in question. These factors will determine the probability that a molecule of a particular substance will cross the membrane. This relationship can be expressed formally as

$$P = \frac{D_m K}{x} \qquad (4\text{-}4)$$

where D_m is the **diffusion coefficient** of the substance within the membrane, K is the partition coefficient of the substance, and x is the thickness of the membrane. The more viscous the membrane or the larger the molecule, the lower the value of D_m.

Permeability constants for different substances vary greatly. For example, the permeability of red blood cells to different solutes ranges from 10^{-12} cm·s^{-1} to 10^{-2} cm·s^{-1}. Importantly, the permeability of some membranes to certain substances can be altered greatly by hormones and other molecules that react with receptor sites on the membrane and thereby influence channel size or carrier mechanisms. Antidiuretic hormone, for example, can increase the water permeability of the renal collecting duct in mammals by as much as 10 times. Similarly, neurotransmitters, acting on specialized integral membrane proteins in nerve and muscle cells, induce large increases in permeability to ions such as Na$^+$, K$^+$, Ca^{2+}, or Cl$^-$.

Osmosis

In 1748 Abbé Jean Antoine Nollet noted that if pure water is placed on one side of an animal membrane (e.g., a bladder wall) and a solution of water containing electrolytes or other molecules is placed on the other side, the water passes through the membrane into the solution. This movement of water down its concentration gradient was called **osmosis** (from the Greek *osmos*, "to push"). We don't usually think of "water concentration," but in fact water acts just like any other substance by diffusing down its concentration gradient. It was later found that osmosis produces a hydrostatic pressure gradient. Osmosis is the colligative property of greatest importance to living systems. As can be seen in Figure 4-10, the pressure difference causes a rise in the level of the solution as water diffuses through the **semipermeable membrane** into the solution. The rise in the level of the solution continues until the net rate of water movement (net flux) across the membrane becomes zero. This occurs when the hydrostatic pressure of the solution in compartment II is sufficient to force water molecules back through the membrane to compartment I at the same rate that osmosis causes water molecules to diffuse from I to II. The hydrostatic back pressure required to cancel the osmotic diffusion of water from compartment I to compartment II is called the **osmotic pressure** of the solution in compartment II.

In 1877, Wilhelm Pfeller made the first quantitative studies of osmotic pressure. He deposited a "membrane" of copper ferrocyanide on the surface of porous clay cups,

producing membranes that would allow water molecules to diffuse through them far more freely than sucrose molecules could. These artificial membranes were also strong enough to withstand relatively high pressures without rupturing because of the clay substratum. Using these membranes, Pfeller was able to make the first direct measurements of osmotic pressure. Some of his results are shown in Table 4-1. Note in the table that the osmotic pressure is proportional to the solute concentration.

Osmosis is responsible for the net movement of water across cell membranes and epithelia. To understand this, consider a 1.0 M aqueous solution of sucrose carefully layered under a 0.01 M aqueous solution of sucrose. There would be net diffusion of water molecules from the solution of lower sucrose concentration (the 0.01 M solution) into the 1.0 M sucrose solution, and sucrose would show net diffusion in the opposite direction until equilibrium was achieved. If these two solutions were separated by a membrane permeable to water but not to sucrose, the water molecules would still show a net diffusion from the solution in which H$_2$O is more concentrated (the 0.01 M sucrose solution) into the 1.0 M sucrose solution, in which the H$_2$O concentration is lower. Since the sucrose could not cross the membrane, there would be a net diffusion of water (**osmotic flow**) through the membrane from the solution of lower solute concentration to the solution of higher solute concentration.

Osmotic pressure π is proportional not only to the concentration of the solute, C (moles of solute particles per liter

TABLE 4-1
Osmotic pressure of sucrose solutions of various concentrations*

Sucrose (%)	Osmotic pressure (atm)	Ratio of osmotic pressure to percentage of sucrose
1	0.70	0.70
2	1.34	0.67
4	2.74	0.68
6	4.10	0.68

*Results were obtained by Pfeffer (1877) in experimental measurements.

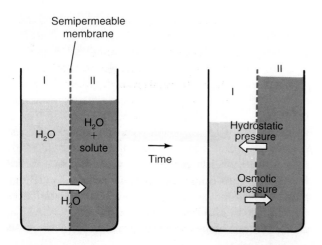

Figure 4-10 Water flow produced by osmosis through a semipermeable membrane generates hydrostatic pressures. Compartment I contains pure water; compartment II, water with impermeant solute. Osmotic pressure forces water to enter compartment II from compartment I until the hydrostatic pressure difference equals the opposing osmotic pressure difference. When the pressures are equal, the flux is zero.

of solvent = osmolarity), but also to its absolute temperature T:

$$\pi = K_1 C \qquad (4\text{-}5)$$

and

$$\pi = K_2 T \qquad (4\text{-}6)$$

where K_1 and K_2 are constants of proportionality. Jacobus van't Hoff related these observations to the gas laws and showed that solute molecules in solution behave thermodynamically like gas molecules. Thus,

$$\pi = RTC$$

or

$$\pi = \frac{nRT}{V} \qquad (4\text{-}7)$$

where n is the number of mole equivalents of solute, R is the molar gas constant ($0.082 \ \text{L} \cdot \text{atm} \cdot \text{K}^{-1} \cdot \text{mol}^{-1}$),* and V is the volume in liters. Like the gas laws, however, this expression for osmotic pressure holds true only for dilute solutions and for completely dissociated electrolytes.

Large concentration gradients across cell membranes can generate surprisingly high osmotic pressures—on the order of several atmospheres. Such pressures, if allowed to develop, would be large enough literally to explode a cell. Consequently, mechanisms for regulating osmotic balance have evolved that minimize osmotic pressure gradients across cell membranes and through tissues (see Chapter 14).

Osmolarity and Tonicity

Two solutions that exert the same osmotic pressure through a membrane permeable only to water are said to be **isosmotic** to each other. If one solution exerts less osmotic pressure than the other, it is **hypoosmotic** with respect to the other solution; if it exerts greater osmotic pressure, it is **hyperosmotic**. **Osmolarity** is thus defined on the basis of an ideal osmometer in which the osmotic membrane allows water to pass but completely prevents the solute from passing. All solutions with the same number of dissolved particles per unit volume have the same osmolarity and are thus defined as isosmotic.

The **tonicity** of a solution, in contrast to its osmolarity, is defined by the response of cells or tissues immersed in the solution. A solution is considered to be **isotonic** with a given cell or tissue if the cell or tissue immersed in it neither shrinks nor swells. If the tissue swells, the solution is said to be **hypotonic** to the tissue; if it shrinks, the solution is said

to be **hypertonic** to it. These effects result from movement of water across the cell membrane in response to osmotic pressure differences between the cell interior and the extracellular solution.

If cells actually behaved as ideal osmometers, tonicity and osmolarity would be equivalent, but this is not generally true. For example, sea urchin eggs maintain a constant volume in a solution of NaCl that is isosmotic relative to seawater, but they swell if immersed in a solution of $CaCl_2$ that is isosmotic relative to seawater. The NaCl solution therefore behaves isotonically relative to the sea urchin egg, whereas the $CaCl_2$ solution behaves hypotonically. The tonicity of a solution depends on the rate of intracellular accumulation of the solute in the tissues in question, as well as on the concentration of the solution. The more readily the solute accumulates, the lower the tonicity of a solution of a given concentration or osmolarity. This is because as the cell gradually loads up with the solute, water follows according to osmotic principles, causing the cell to swell. Thus, the terms isotonic, hypertonic, and hypotonic are meaningful only in reference to actual experimental determinations on living cells or tissues.

Electrical Influences on Ion Distribution

Membrane permeability to charged particles depends both on the membrane permeability constant and on the electrical potential across the membrane. Understanding the interaction of charged particles with membranes is extremely important for understanding how electrically excitable cells function. Neurons are the most highly specialized of this class of cells. Since neurons will be discussed in the next couple of chapters, only a few important observations will be summarized here.

Two forces can act on charged atoms and molecules (such as Na^+, K^+, Cl^-, Ca^{2+}, amino acids) to produce a net passive diffusion of each species across a membrane:

1. The chemical gradient arising from differences in the concentration of the substance on the two sides of the membrane
2. The electric field, or difference in electric potential across the membrane

An ion will move away from regions of high concentration, and if that ion is positively charged it will also move toward increasing negative potential. The sum of the combined forces of concentration gradient and electrical gradient determine the net **electrochemical gradient** acting on that ion.

When an ion is at equilibrium with respect to a membrane (that is, when there is no net transmembrane flux of that ion species), there will exist a potential difference just sufficient to balance and counteract the chemical gradient acting on the ion. The potential at which an ion is in electrochemical equilibrium is called the **equilibrium potential**, measured in volts (or millivolts). Several factors influence the value of the equilibrium potential, but the most prominent is the ratio of the ion concentrations on opposite sides

---

* R is the constant of proportionality in the gas equation $PV/T = R$ when referring to 1 mol of a perfect gas, and it has the value of $1.985 \ \text{cal} \cdot \text{mol}^{-1} \cdot \text{K}^{-1}$; P is in atmospheres and V is in liters.

of the membrane. For a monovalent ion such as Na^+ or K^+ at 18°C, the equilibrium potential (in volts) is equal to $0.058 \times \log_{10}$ of the ratio of the extracellular to intracellular concentrations of the ion. Thus, a 58 mV potential difference across the membrane has the same effect on the net diffusion of that ion as a transmembrane concentration ratio of $10:1$.

An apparently paradoxical situation therefore arises in which an ion species can passively diffuse *against* its chemical concentration gradient (that is, move "uphill" to an area of higher concentration) *if* the electrical gradient (i.e., potential difference) across the membrane is in the opposite direction to and exceeds the concentration gradient. For example, if the interior of a cell has a greater negative charge than the equilibrium potential for K^+, potassium ions will diffuse into the cell even though the intracellular concentration of K^+ is much higher than the extracellular concentration. The distribution of ions across membranes and the attendant equilibrium potential is described by the Nernst relationship, which is discussed in detail in the following chapter.

Electrical forces cannot act directly on uncharged molecules such as sugars. These substances will be influenced primarily by the concentration gradient they experience.

Donnan Equilibrium

If diffusible solutes are separated by a membrane that is freely permeable to water and electrolytes but totally impermeable to one species of ion, the diffusible solutes become unequally distributed between the two compartments. This phenomenon was discovered in 1911 by Frederick Donnan, who first described how the solutes would be distributed and hence has been commemorated by having this equilibrium state named for him.

To understand a Donnan equilibrium, imagine starting with pure water in two compartments and adding some KCl to one of them (Figure 4-11). The dissolved salt (K^+ and Cl^-) will diffuse through the membrane until the system is in equilibrium—that is, until the concentrations of K^+ and Cl^- become equal on both sides of the membrane (Figure 4-11A). Now imagine adding the potassium salt of a nondiffusible anion (a macromolecule A^-, having multiple negative charges) to the solution in compartment I. The K^+ and Cl^- quickly become redistributed until a new equilibrium is established by movement of some K^+ and some Cl^- from compartment I to compartment II (Figure 4-11B). **Donnan equilibrium** is characterized by a reciprocal distribution of the anions and cations so that

$$\frac{[K^+]_I}{[K^+]_{II}} = \frac{[Cl^-]_{II}}{[Cl^-]_I}$$

At equilibrium, the diffusible cation, K^+, is more concentrated in the compartment in which the nondiffusible anion, A^-, is confined than in the other, whereas the diffusible anion, Cl^-, becomes less concentrated in that compartment than in the other.

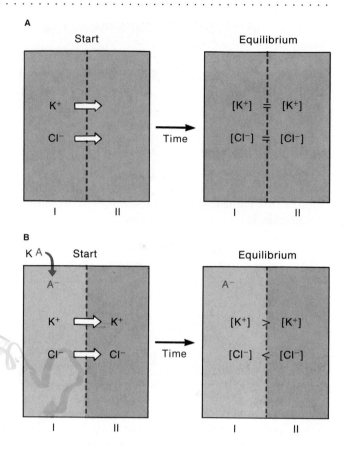

Figure 4-11 The Donnan equilibrium describes ion distribution across a semipermeable membrane. **(A)** When KCl is added to compartment I of a container divided by a permeable membrane, K^+ and Cl^- diffuse across the membrane until the concentrations are equal on either side. **(B)** If the potassium salt of an impermeant anion is added to compartment I, some K^+ and Cl^- diffuse into compartment II until electrochemical equilibrium is reestablished. It should be noted that these chambers (unlike the living cell) are not distensible.

We can understand this situation by considering the consequences of the following physical principles:

1. There must be electroneutrality within both compartments; that is, in each compartment the total number of positive charges must equal the total number of negative charges. Thus, in this example, $[K^+] = [Cl^-]$ in compartment II.

2. Considered statistically, the diffusible ions K^+ and Cl^- cross the membrane in pairs to maintain electrical neutrality. The probability that they will cross together is proportional to the product $[K^+] \times [Cl^-]$.

3. At equilibrium the rate of diffusion of KCl in one direction through the membrane must equal the rate of KCl diffusion in the opposite direction. It follows, then, that at equilibrium the product $[K^+] \times [Cl^-]$ in one compartment must equal the product in the other compartment. Letting x, y, and z represent the concentrations of the ions in compartments I and II, as shown in Figure 4-12, we can express the equilibrium

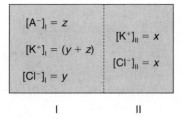

Figure 4-12 The Donnan equilibrium can be described algebraically. The equilibrium condition established in Figure 4-11B after the salt of an impermeant anion is added to compartment I is shown.

condition (i.e., equality of the product $[K^+] \times [Cl^-]$ in the two compartments) algebraically:

$$x^2 = y(y + z) \qquad (4\text{-}8)$$

This equation also holds, of course, if A^- is not present. In that case, K^+ and Cl^- are equally distributed, and $z = 0$ and $x = y$. By rearranging equation 4-8, we can see that, at equilibrium, the distributions of the diffusible ions in the two compartments are reciprocal:

$$\frac{y + z}{x} = \frac{x}{y} \quad \text{or} \quad \frac{[K^+]_I}{[K^+]_{II}} = \frac{[Cl^-]_{II}}{[Cl^-]_I} \qquad (4\text{-}9)$$

From this relation, it is clear that as the concentration of the nondiffusible anion, z, is increased, the concentrations of the diffusible ions (x and y) will become increasingly divergent. This unequal distribution of diffusible ions is the hallmark of Donnan equilibrium.

At Donnan equilibrium, the osmotically unequal distribution of solute particles makes water move in the direction of the compartment of higher osmolarity (compartment I in Figure 4-11). This osmotic pressure difference plus any resultant increase in hydrostatic pressure of that compartment is called the **oncotic pressure**. This concept is important in understanding the balance of hydrostatic and osmotic pressures across certain biological barriers such as capillary walls.

The explanation of a Donnan equilibrium depends on an ideal set of conditions for the sake of simplicity. The living cell and its surface membrane are, of course, far more complex. For example, the cell membrane is somewhat permeable to a variety of ions and molecules, and there will almost never be a single "nondiffusible anion," which here represents various anionic side groups of proteins and other large molecules. Although the physical and mathematical principles recognized by Donnan play a role in regulating the distribution of electrolytes in living cells, clearly nonequilibrium mechanisms must modify the distribution of many substances across the cell membrane. In particular, the permeability of the cell membrane to particular ions can change over time, changing the conditions dramatically. Thus, cells cannot be considered passive "osmometers," and the distribution of substances across biological membranes cannot be predicted entirely by Donnan equilibrium principles except in certain cases.

OSMOTIC PROPERTIES OF CELLS

We can now use the physical principles outlined above to analyze properties of the cell membrane that maintain different concentrations of ions inside and outside the cell (Figure 4-13). Cell membranes ultimately must closely regulate cell volume and thus intracellular osmotic pressure.

Ionic Steady State

Every cell maintains concentrations of inorganic solutes inside the cell that are different from those outside the cell (Table 4-2). The most concentrated inorganic ion in the cytosol is K^+, which is typically 10–30 times as concentrated there as in the extracellular fluid. Conversely, the internal concentrations of free Na^+ and Cl^- are typically less (approximately one-tenth or less) than the external concentrations. Another important generalization is that the intracellular concentration of Ca^{2+} is maintained several orders of magnitude below the extracellular concentration. This difference is due in part to active transport of Ca^{2+} out across the cell membrane and in part to the sequestering of this ion within such organelles as the mitochondria and endoplasmic reticulum. As a result, the concentration of Ca^{2+} in the cytosol is generally well below 10^{-6} M.

Cell membranes typically are about 30 times more permeable to K^+ than to Na^+. Membrane permeability to chloride ions varies. In some cells it is similar to that of K^+ while in others it is lower. The permeability of the cell membrane to Na^+ is low, but it is not low enough to prevent Na^+ from leaking steadily into the cell.

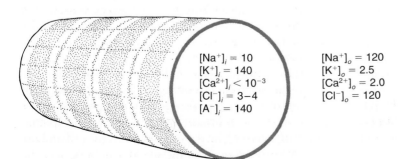

$[Na^+]_i = 10$
$[K^+]_i = 140$
$[Ca^{2+}]_i < 10^{-3}$
$[Cl^-]_i = 3\text{--}4$
$[A^-]_i = 140$

$[Na^+]_o = 120$
$[K^+]_o = 2.5$
$[Ca^{2+}]_o = 2.0$
$[Cl^-]_o = 120$

Figure 4-13 Concentrations of common ions are very different inside and outside a vertebrate skeletal muscle cell. The concentrations shown are in millimoles per liter. The concentration given for intracellular Ca^{2+} is for the free, unbound, and unsequestered ion in the myoplasm. Because the list of ions is incomplete, the totals do not balance out perfectly. $[A^-]_i$ represents the molar equivalent negative charges carried by various impermeant anions.

TABLE 4-2
Internal and external concentrations of some electrolytes in specific nerve and muscle tissues

Tissue	Internal concentrations (mM)			External concentrations (mM)			Ratios, inside/outside		
	Na$^+$	K$^+$	Cl$^-$	Na$^+$	K$^+$	Cl$^-$	Na$^+$	K$^+$	Cl$^-$
Squid nerve	49	410	40–100	440	22	560	1/9	19/1	1/14–1/6
Crab leg nerve	52	410	26	510	12	540	1/10	34/1	1/21
Frog sartorius muscle	10	140	4	120	2.5	120	1/12	56/1	1/30

Certain features of the cell membrane, particularly the differential permeability of the membrane to different ion species, suggest that under some conditions the Donnan equilibrium might apply. To understand when the Donnan equilibrium is useful in determining membrane characteristics of living cells, three related factors are important:

1. Inside the cell, carboxyls and other anionic sites found on nonpermeant peptide and protein molecules contribute most of the net negative charge. These charges must be balanced by positively charged counterions such as Na$^+$, K$^+$, Mg^{2+}, and Ca^{2+}.

2. These anionic sites trapped inside the cell make it similar to the artificial case presented above (see Figure 4-11) in which Donnan equilibrium applies. If K$^+$ and Cl$^-$ were the only diffusible ions, an equilibrium situation similar to that shown in Figure 4-11B would indeed develop in the cell. However, the cell membrane is leaky to Na$^+$ and other inorganic ions, and with time the cell would load up with these ions if they were simply allowed to accumulate. This, in turn, would cause osmotic movement of water into the cell, causing it to swell.

3. Such osmotic disasters are avoided because the cell pumps out Na$^+$, Ca^{2+}, and some other ions at the same rate as they leak in, keeping the intracellular Na$^+$ concentration about an order of magnitude lower than the extracellular concentration. This active pumping, which will be discussed later, is equivalent to an effective impermeability to Na$^+$ and Ca^{2+}. As a result, the concentrations of these ions are not allowed to come into equilibrium, and the cell in fact behaves very much as if it were in a state of Donnan equilibrium. Actually, the unequal distribution of ions represents a steady state requiring the continual expenditure of energy (to pump ions) rather than a true equilibrium.

Since K$^+$ and Cl$^-$ are by far the most concentrated and most permeant ions in the tissue, they distribute themselves in a way similar to that in an ideal Donnan equilibrium. That is, the KCl product [K$^+$] × [Cl$^-$] of the cell interior will approximately equal the KCl product of the extracellular solution (Figure 4-14), provided the membrane permeabilities of K$^+$ and Cl$^-$ are both high relative to those of other ions present.

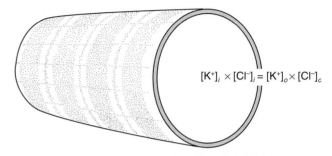

$$[K^+]_i \times [Cl^-]_i = [K^+]_o \times [Cl^-]_o$$

Figure 4-14 The KCl product is governed by the Donnan equilibrium. The distribution of K$^+$ and Cl$^-$ will follow Donnan equilibrium principles, provided the membrane is permeable to both K$^+$ and Cl$^-$.

Cell Volume

Plant and bacterial cells have rigid walls secreted by the cell membrane. These walls place an upper limit on the size of the cell, allowing the osmotic buildup of turgor pressure in these cells. In contrast, animal cells do not have rigid walls and therefore cannot resist any buildup of large intracellular pressure. As a result, cells will change size when placed in different concentrations of impermeable substances dissolved in water. This shrinkage or swelling is due to osmotic movement of water (Figure 4-15). There are two ways in which the surface membrane might prevent osmotic swelling in the cell. One is to pump water out as fast as it leaks in. There is no evidence that this occurs, although a similar effect is achieved by the contractile vacuole of certain protozoans. The other, which appears to be the major mechanism for regulation of cell volume, is to pump out solutes that leak into the cell (Figure 4-16). Thus, at steady state, Na$^+$, the major osmotic constituent outside the cell, is expelled from the cell by active transport as rapidly as it leaks in. In effect, there is no net entry. The situation is osmotically equivalent to complete sodium impermeability, with a relatively fixed concentration of Na$^+$ trapped in the cell. Because Na$^+$ is not allowed to further accumulate in the cell, there is no compensatory osmotic influx of water.

The low intracellular (relative to extracellular) sodium concentration is important in balancing the other osmotically active solutes in the cytoplasm. The importance of active transport in maintaining the sodium gradient, and thereby the osmolarity of the cell and the cell volume, is seen when the energy metabolism of the

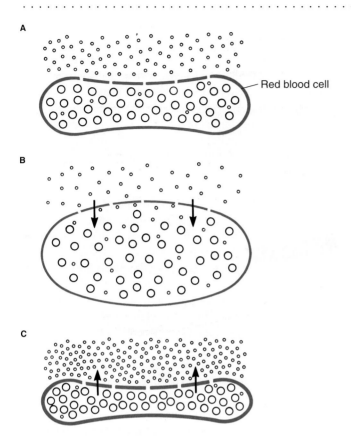

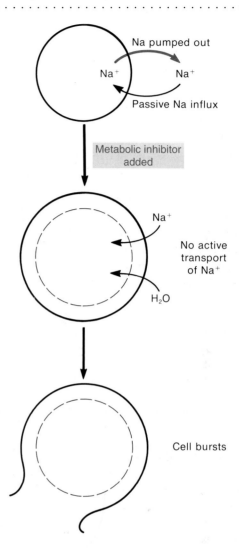

Figure 4-15 Osmotic changes alter the volume of a red blood cell. **(A)** Isotonic solution: the cell volume remains unchanged. **(B)** Hypotonic solution: water (arrows) enters the cell because of the higher osmoticity of the cytoplasm with respect to the solution, producing swelling. **(C)** Hypertonic solution: in a more concentrated medium, water leaves the cell, causing shrinkage.

cell is interrupted by metabolic poisons (Figure 4-17). Without ATP to energize uphill extrusion of Na⁺, the sodium ion, together with its chloride counterion, leaks into the cell, and water follows osmotically, causing the cell to swell.

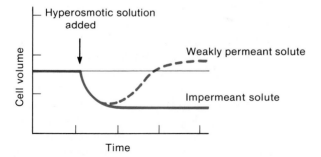

Figure 4-16 Hyperosmotic solutions with impermeant and weakly impermeant solute both cause initial cell shrinkage. If the solute is completely impermeant, it causes maintained cell shrinkage because the solution is basically hypertonic in this situation. If the solute is only weakly impermeant, however, the solution is hypotonic and enters the cell slowly followed by the osmotic flow of water. This process eventually produces swelling, in spite of the fact that the solution is hyperosmotic.

Figure 4-17 A metabolic inhibitor interferes with Na⁺ pumping and, therefore, with maintenance of cell volume. Under normal circumstances, levels of Na⁺ are maintained at equilibrium inside and outside a cell: the ion passively enters the cell and then is pumped out of the cell. With the addition of a metabolic inhibitor, however, the cell is rendered unable to pump out the Na⁺ that steadily leaks into the cell. As a result, $[Na^+]_i$ rises inside the cell and water follows osmotically, increasing cell volume above its initial volume (dashed line). Eventually the cell bursts because of massive swelling.

PASSIVE TRANSMEMBRANE MOVEMENTS

Molecules can cross membranes without the direct input of energy, that is, passively, in several different ways. Note that while these processes do not directly require metabolically energized processes, they ultimately depend on a concentration or electrical gradient across a cell membrane that has at some point required energy for its creation and maintenance. The energy stored in such gradients is ultimately responsible for the translocation of molecules across the membrane. Nonetheless, it is useful and appropriate to think of these processes as passive.

There are three basic routes for passive transmembrane movements of molecules or ions (Figure 4-18). In the first, a

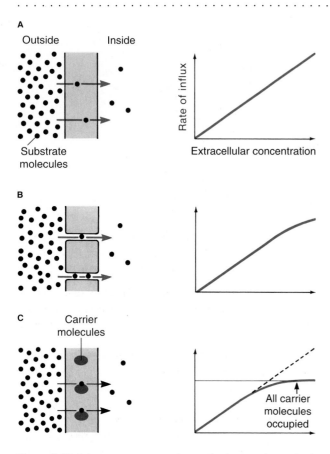

Figure 4-18 Substances cross membranes by three major methods. **(A)** Dissolving in lipid phase. **(B)** Diffusion through labile or fixed aqueous channels. **(C)** Carrier-mediated transport (either facilitated transport or active transport).

molecule simply diffuses through the membrane. It leaves the aqueous phase on one side of the membrane, dissolves directly in the lipid layer of the membrane, diffuses across the thickness of the lipid or protein layer, and finally enters the aqueous phase on the opposite side of the membrane. In the second, the solute molecule remains in the aqueous phase and diffuses through *aqueous channels,* water-filled pores in the membrane. In the third route, the solute molecule combines with a **carrier molecule** dissolved in the membrane. This carrier "mediates" or "facilitates" the movement of the solute molecule across the membrane. Carriers can "mask" even a polar solute and because of their lipid solubility, allow the solute to diffuse more readily across the membrane, down its concentration or electrochemical gradient. This is called **carrier-mediated** (or facilitated) **transport** and may take any of several different forms. Let's now consider each of these three major pathways in turn.

Simple Diffusion through the Lipid Bilayer

If a solute molecule comes into contact with the lipid layer of the membrane and its thermal energy is high enough, it may enter and cross the lipid phase and finally emerge into the aqueous phase on the other side of the membrane. To

leave the aqueous phase and enter the lipid phase, a solute must first break all its hydrogen bonds with water. This requires about 5 kcal of kinetic energy per hydrogen bond. Moreover, the solute molecule crossing the lipid phase of the membrane must dissolve in the lipid bilayer. So, its lipid solubility will also play a major role in determining whether or not it will cross the membrane. Consequently, those molecules having a minimum of hydrogen bonding with water will most readily enter the lipid bilayer, whereas polar molecules such as inorganic ions will almost never dissolve in the bilayer.

A number of factors, such as molecular weight and molecular shape, influence the mobility of nonelectrolytes within the membrane, but the empirically measured **partition coefficient** is the primary predictor of the diffusion of a nonelectrolyte across the lipid bilayer. To measure this property, a test substance is shaken in a closed tube containing equal amounts of water and olive oil, and the coefficient K is determined from the relative solubilities in water and oil at equilibrium, using the equation

$$K = \frac{\text{solute concentration in lipid}}{\text{solute concentration in water}} \quad (4\text{-}10)$$

Is nonelectrolyte membrane permeability related to the lipid-water partition coefficient of the solute? Collander (1937) systematically tested this idea in the giant algal cell *Chara* by plotting the permeability coefficient (equation 4-4) against the partition coefficient (equation 4-10). Lipid solubility is almost linearly related to permeability of a substance, independent of molecule size (Figure 4-19).

Nonelectrolytes exhibit a wide range of partition coefficients. For example, the value for urethane is 1000 times that for glycerol (see Figure 4-19). These differences depend on particular features of the molecular structure, as il-

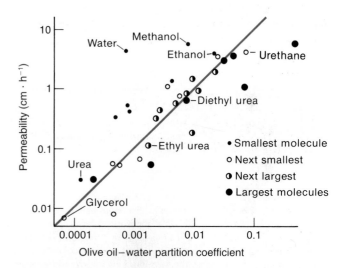

Figure 4-19 Membrane permeability of nonelectrolytes is linearly related to their respective oil-water partition coefficients. Note that the permeability of nonelectrolytes is independent of molecular size.

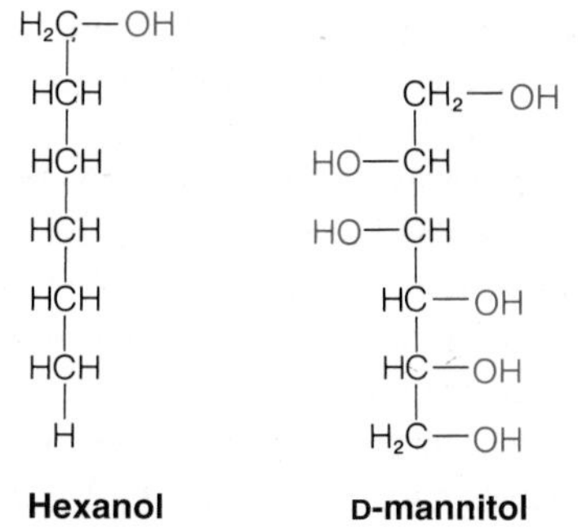

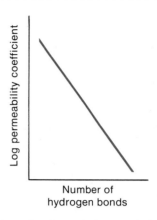

Figure 4-20 The structure of six-carbon molecules determines their water and lipid solubility. Note the difference between hexanol and mannitol in the number of hydroxyl groups. Hexanol, with its weak hydrogen-bonding capacity, is poorly soluble in water and highly soluble in lipids; mannitol, with its strong hydrogen-bonding capacity, is highly soluble in water and poorly soluble in lipids.

Figure 4-21 Hydrogen bonds greatly decrease the lipid solubility and, hence, the permeability of a membrane.

lustrated in Figure 4-20, which compares two molecules with different solubilities. Hexanol and mannitol have similar structures except that hexanol contains only one —OH group while mannitol contains six. These —OH groups facilitate hydrogen bonding to water and therefore decrease lipid solubility. In fact, each additional hydrogen bond results in a fortyfold decrease in the partition coefficient, which is reflected in a decrease in permeability (Figure 4-21). Consequently, hexanol diffuses across membranes much more readily than mannitol does.

Water exhibits a much higher permeability across cellular membranes than predicted from its partition coefficient (see Figure 4-19). This is partly because water can pass through selective permanent channels that penetrate the lipid bilayer. Structural evidence of this is seen in certain epithelial cells, where water permeability depends on water channels in the plasma membrane. However, even in channel-free, artificial lipid bilayers, water permeability is still several times higher than that predicted from the solubility of water in long-chain hydrocarbons. A possible explanation is that the small, uncharged water molecules may pass through temporary channels between lipid molecules. Other small, uncharged polar molecules, such as CO_2, NO,

and CO, also have relatively high permeabilities across artificial and natural membranes, though it is not known whether this is explained by specialized channels or ones lacking selectivity.

Simple diffusion through the lipid bilayer exhibits **non-saturation kinetics** (see Figure 4-18A), meaning that the rate of influx increases in proportion to the concentration of the solute in the extracellular fluid. This is because the net rate of influx is determined only by the difference in the number of solute molecules on the two sides of the cell membrane. This proportionality between external concentration and rate of influx over a large range of concentrations distinguishes simple diffusion from channel permeation or carrier-mediated transport mechanisms (Figure 4-18B and C).

Diffusion through Membrane Channels

Charged molecules can cross membranes by diffusing through specific water-filled channels. Since inorganic ions such as Na^+, K^+, Ca^{2+}, and Cl^- cannot diffuse through lipid bilayers, special protein molecules have evolved that extend across the cell membranes and act as pores. When these pores are open, they allow specific solutes to pass through them (Figure 4-22A).

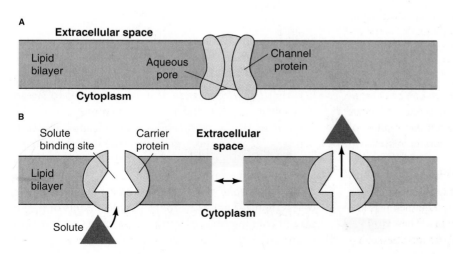

Figure 4-22 Membrane transport proteins act as carriers or form channels in the membrane. **(A)** A channel protein forms a water-filled pore across the bilayer, through which specific ions can diffuse. **(B)** In contrast, a carrier protein alternates between two conformations, so that the solute binding site is sequentially accessible on one side of the bilayer and then on the other.

SPOTLIGHT 4-2

ARTIFICIAL BILAYERS

Many of our ideas of how molecules and ions pass across membranes have grown out of experiments and observations on artificial bilayers that are similar to the bimolecular leaflet that forms the basis of the cell membrane. Artificial bilayers are extremely useful in studies of permeation mechanisms because they can be made from chemically defined mixtures of lipids. Selected substances can be added to test their effects on permeability. Channel-forming substances, such as the antibiotic ionophores (molecules that facilitate the diffusion of ions across membranes) and membrane channel components of excitable tissues, have been incorporated into artificial bilayers, allowing their properties to be studied in isolation under the highly controlled conditions shown in the accompanying figure.

The principle of bilayer formation is shown in the figure (part B). The most stable configuration attained consists of two layers of lipid molecules whose hydrophobic, lipophilic hydrocarbon tails are loosely associated to form a liquid-lipid phase sandwiched between the hydrophilic polar ends of the molecules, which are directed outward toward the aqueous medium. The thickness of the lipid film is easily determined from the interference color of light reflected from the two surfaces of the film. Membranes with thicknesses of approximately 7 nm (black interference color) are most commonly used. These membranes have electrical conductances (ion permeabilities) and capacitances consistent with their thickness and lipid composition. Although their permeability to ions is much lower than that of cell membranes, the addition of certain ionophores increases it to values that are characteristic of cell membranes.

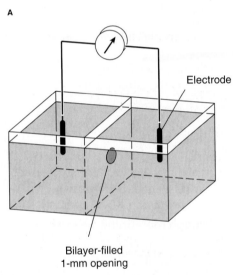

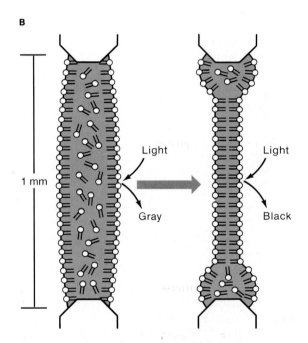

Lipid bilayers can be induced to form across a 1-mm opening between two fluid-filled chambers. **(A)** The permeability of the bilayer to electrolytes in the chamber can be measured electrically by placing test solutions with different electrolyte concentrations in each of the chambers. **(B)** The bilayer is formed by filling the opening with a small amount of the

lipid in a solvent such as hexane. Initially, while the bilayer is forming, its interference color is gray (left). As the membrane assumes the more stable bilayer configuration (right), the interference color changes to black. [From Kotyk and Janácek, 1970.]

The functioning of membrane channels can be demonstrated directly in artificial lipid bilayer membranes that are by themselves highly impermeable to even the smallest of charged molecules (Spotlight 4-2). A dramatic increase in ion permeability occurs upon addition of small amounts of channel proteins extracted from cellular membranes. This increase is measured as discrete pulses of current carried by ions from one side of the membrane to the other, just like those measured in biological membranes. These unitary currents are due to the sudden opening of individual chan-

nels that allow thousands of ions per second to stream down their gradients and across the membrane.

Studies of the permeabilities of cell membranes to other polar substances give an estimated 0.7 nm for the **equivalent pore size**—the pore diameter that would account for the rate of diffusion across the membrane. Thus, membrane channels presumably have diameters of less than 1.0 nm, close to the practical limits of resolution of contemporary electron microscopes and fixation methods.

As an example, rod-shaped molecules of the antibiotic nystatin applied to both sides of an artificial or a natural membrane aggregate to form channels. These pores permit the passage of water, urea, and chloride, all of which are less than 0.4 nm in diameter. Larger molecules cannot penetrate the channels. Cations also are excluded, presumably because there are fixed positive sites along the channel walls. Incorporation of nystatin into artificial membranes produces a negligible increase in membrane area occupied by fixed channels (0.001%–0.01%), but it produces a 100,000-fold increase in membrane permeability to chloride ions. This means that very little membrane area need be devoted to channels to account for the ion permeabilities of natural membranes. This conclusion is supported by the fact that the electrical capacitance of the cell membrane remains relatively unchanged during large changes in the permeability exhibited during the excitation of some membranes. (This phenomenon is discussed further in Chapter 5.)

Facilitated Transport across Membranes

Membranes are permeable to various polar molecules such as sugars, amino acids, nucleotides, and certain cell metabolites that would cross lipid bilayers by diffusion only very slowly. This is because of facilitated transport, the movement of molecules through membranes by the action of **membrane transport proteins** (see Figure 4-22B). Facilitated transport, unlike active transport discussed later, does not require energy in the form of ATP. Membrane transport proteins, which exist in many forms in all types of membranes, are exquisitely selective about which species of molecules they transport. Carrier proteins that transport a single solute from one side of the membrane to the other are called **uniporters,** while those that transfer one solute and simultaneously or sequentially transfer a second solute are called **coupled transporters.** Coupled transporters that transfer two solutes in the same direction are called **symporters,** while those that transfer solutes in opposite directions are called **antiporters** (Figure 4-23). These terms can also be applied to active transport systems.

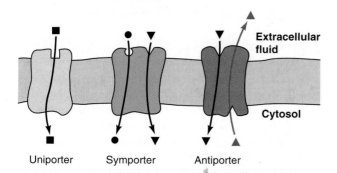

Figure 4-23 Membrane carrier proteins can be configured as uniporters, symporters, or antiporters. Uniporters transport a single type of ion in one direction across the membrane, while symporters simultaneously transport two different ions in the same direction. Antiporters also transport two ions, but create an exchange of ions by moving the two in opposite directions across the membrane.

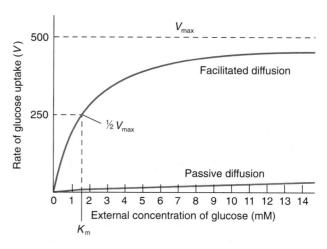

Figure 4-24 The kinetics of simple diffusion differ from those of carrier-mediated (facilitated) diffusion. In this example of glucose movements, the rate of simple diffusion is always proportional to the glucose concentration. However, the rate of carrier-mediated glucose diffusion reaches a maximum (V_{max}) when the glucose carrier protein is saturated. The binding constant (K_m) of the carrier for glucose, which is analogous to the K_m of an enzyme for its substrate solute, is measured when transport is at half its maximal value. [Adapted from Lodish et al., 1995.]

The existence of such transporters was initially inferred from kinetics studies of molecule transfer across membranes (Figure 4-24). For some solutes, the measured rate of influx reaches a plateau beyond which an increase in solute concentration produces no further increase. This reveals that a rate-limiting step must occur in permeation. Experiments elucidating the kinetics of such permeation led to the conclusion that transport occurs through the formation of a carrier-substrate complex similar in concept to an enzyme-substrate complex. Each carrier protein has a characteristic binding constant for its solute equal to the concentration of solute when the transport rate is half its maximum value (see Figure 4-24). As in enzyme reactions, the solute binding can be blocked by specific competitive inhibitors as well as by noncompetitive inhibitors. The carrier and solute molecule temporarily form a complex based on bonding, stearic specificity, or both.

The specificity of these transporters was first established in studies where single gene mutations abolished the ability of bacteria to transport specific sugars across their cell membranes. Similar mutations have now been found in many cases, including human inherited diseases that affect the transport of specific solutes across kidney, intestine, or lungs. For example, in cystic fibrosis, a defect in the chloride transport channel protein (CFTR) appears to be responsible for fluid imbalance in the lungs.

ACTIVE TRANSPORT

All channel proteins and most carrier proteins allow solutes to cross the membrane passively at no energetic cost (other than the original cost of generating the potential energy in the form of different solute concentrations on opposite sides of the membrane, as mentioned earlier). The concentration gradient determines the direction of passive transport. As diffusion proceeds, the solute concentrations in the two compartments approach equilibrium, at which point no further net diffusion will occur.

For charged molecules, transport is influenced by both the concentration gradient and the electrical gradient (i.e., the electrochemical gradient) across the membrane. All plasma membranes have an electrical potential difference across them, where the inside is negative relative to the outside of the cell. This favors the entry of positively charged ions and opposes entry of negatively charged ions. In this case, as above, passive processes will continue until the membrane is in equilibrium.

The distribution of ions across cell membranes is at true equilibrium only in dead cells. All living cells continually expend chemical energy to maintain the transmembrane concentrations of solutes far away from equilibrium. This energy is typically supplied in the form of ATP. Mechanisms that actively transport substances against a gradient are collectively called **membrane pumps**. When the source of energy for such pumps is cut off, active uphill transport ceases and passive diffusion governs the distribution of substances. The concentrations of these substances gradually redistribute toward equilibrium.

The Na$^+$/K$^+$ Pump as a Model of Active Transport

Many of the features of active transport are demonstrated in the system that maintains steep concentration gradients for Na$^+$ and K$^+$ in the cell. The concentration of K$^+$ is about 10–20 times higher inside cells than outside, while the opposite is true for Na$^+$ (see Figure 4-13). These concentration differences are sustained by a **Na$^+$/K$^+$ pump** found in the plasma membrane of virtually all animal cells. This pump is an ATPase with binding sites for Na$^+$ and ATP on its cytoplasmic surface and binding sites for K$^+$ on its external surface. In the steady state, the number of Na$^+$ ions pumped, or transported, out of the cell is equal to the number of Na$^+$ ions that leak in. Thus, even though there is a continual turnover of Na$^+$ (and other ion species) across the membrane, the net Na$^+$ flux over any period of time is zero. There are two factors that determine the size of a Na$^+$ concentration gradient that will be built up between the cell interior and cell exterior: the rate of active transport of Na$^+$ and the rate at which Na$^+$ can leak (i.e., diffuse passively) back into the cell. The rate at which the membrane allows Na$^+$ to leak back into the cell determines, of course, the rate at which the Na$^+$ pump has to work in order to maintain a given ratio of extracellular-to-intracellular Na$^+$. There is evidence that an increase in the intracellular concentration of Na$^+$ leads to an increase in the rate of Na$^+$ expulsion by the pump (which may merely be a mass action effect due to the increased availability of intracellular Na$^+$ to the carrier molecules in the membrane).

Several important features of active transport should be noted:

1. *Transport can take place against substantial concentration gradients.* The most commonly studied membrane pump is the one that transports Na$^+$ from the cell interior to the external fluid against a 10:1 Na$^+$ concentration gradient.

2. *The active transport system generally exhibits a high degree of selectivity.* The Na$^+$ pump, for example, fails to transport lithium ions, which have ionic properties very similar to those of sodium ions.

3. *ATP or other sources of chemical energy are required.* Metabolic poisons that stop the production of ATP bring active transport to a halt.

4. *Certain membrane pumps exchange one kind of molecule or ion from one side of the membrane for another kind of molecular or ion from the other side.* The Na$^+$/K$^+$ antiport features active outward transport of Na$^+$ concomitant with the inward transport of K$^+$ by the sodium-potassium pump. This process involves the obligatory exchange of two potassium ions from outside the cell for three sodium ions from inside the cell (Figure 4-25). When external K$^+$ is absent, the Na$^+$ ions that normally would have been exchanged for K$^+$ ions are no longer pumped out.

5. *Some pumps perform electrical work by producing a net flux of charge.* For example, the Na$^+$/K$^+$ exchange pump just mentioned produces a net outward movement of one positive charge per cycle in the form of three Na$^+$ exchanged for only two K$^+$. Ionic pumps that produce net charge movement are said to be **rheogenic** because they produce a transmembrane electric current. If the current produces a measurable effect on the voltage across the membrane, the pump is also said to be **electrogenic**.

6. *Active transport can be selectively inhibited by specific blocking agents.* The cardiac glycoside ouabain, applied to the extracellular surface of the membrane, blocks the potassium-dependent active extrusion of Na$^+$ from the cell. It does this by competing for the K$^+$ binding sites of the Na$^+$/K$^+$ pump at the outside surface of the membrane.

7. *Energy for active transport is released by the hydrolysis of ATP by enzymes (ATPases) present in the membrane.* Active transport exhibits Michaelis-Menten kinetics and competitive inhibition by analog molecules. Both behaviors are characteristic of enzymatic reactions. Calcium-activated ATPases have been associated with calcium-pumping membranes. Associated with the Na$^+$/K$^+$ pump are Na$^+$ and K$^+$–activated ATPases isolated from red blood cell membranes and other tissues. These enzymes catalyze the hydrolysis of ATP into ADP and inorganic phosphate only in the presence of Na$^+$ and K$^+$, and they bind the specific Na$^+$ pump inhibitor ouabain. The fact that ouabain binds to the mem-

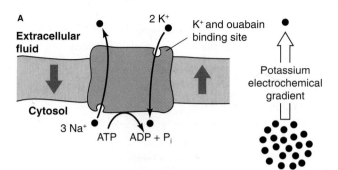

A

Extracellular fluid

Cytosol

2 K⁺

K⁺ and ouabain binding site

3 Na⁺

ATP   ADP + P$_i$

Potassium electrochemical gradient

Figure 4-25 The Na⁺/K⁺ ATPase actively pumps Na⁺ out of the K⁺ into a cell against their respective electrochemical gradients. **(A)** For every molecule of ATP hydrolyzed directly to power transmembrane transport, three Na⁺ ions are pumped out and two K⁺ ions are pumped in. The specific pump inhibitor ouabain and K⁺ compete for the same sites on the external side of the ATPase. **(B)** This schematic model of Na⁺/K⁺ ATPase shows the movement of Na⁺ and K⁺ by a single protein. The binding of Na⁺ (step 1) and the subsequent phosphorylation of the cytoplasmic face of the ATPase by ATP (step 2) cause a conformational change in the protein resulting in a transfer of the Na⁺ across the membrane (step 3). The Na⁺ is released to the cell's exterior and K⁺ is bound (step 4). Subsequent dephosphorylation of the protein (step 5) induces a return to the protein's original conformation and consequent transfer of the K⁺ across the membrane (step 6), where it is released into the cytosol (step 7).

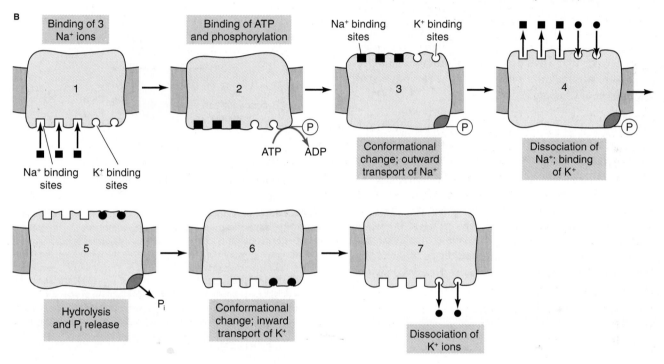

B

Binding of 3 Na⁺ ions

1

Na⁺ binding sites   K⁺ binding sites

Binding of ATP and phosphorylation

2

ATP   ADP   (P)

Na⁺ binding sites   K⁺ binding sites

3

Conformational change; outward transport of Na⁺   (P)

4

Dissociation of Na⁺; binding of K⁺   (P)

5

Hydrolysis and P$_i$ release   P$_i$

6

Conformational change; inward transport of K⁺

7

Dissociation of K⁺ ions

brane and blocks the Na⁺/K⁺ pump is evidence that these ATPases are involved in active transport of Na⁺ and K⁺.

The operation of the Na⁺/K⁺ ATPase is thought to depend on a series of conformational changes in the transport protein that allow the cotransport of K⁺ and Na⁺ across the cell membrane (see Figure 4-25).

The actual process of metabolically energized transport takes place across the cell membrane, pumping molecules either into or out of the cell. However, the organization of cells into an epithelial sheet makes possible the active transport of substances from one side of the epithelial sheet to the other because the cell surfaces at each side are asymmetrical in their transport properties. One side of the cell may tend to import a substance, while the other side tends toward export, thus effecting transfer of the substance to opposite sides of the cell. This characteristic enables the epithelia of amphibian skin and bladder, fish gills, the vertebrate cornea, kidney tubules, the intestine, and many other tissues to move salts and other substances across the tissues.

Active transport and facilitated transport both show saturation kinetics. What does this tell you about the mechanisms underlying these two kinds of transport?

Ion Gradients as a Source of Cell Energy

Electrochemical gradients across biological membranes provide an important energy source immediately available to cells. This energy can be used to drive passive or secondary active transport and is also used to store or conduct information along the surface of cell membranes (see Chapter 5). The amount of free energy stored in an electrochemical gradient depends on the ratio of ion concentrations—or, more accurately, the ratio of the chemical activities of an ion species—on the two sides of the membrane. Energy release occurs when the ions are allowed to flow down their gradient across the membrane. Three important cellular processes utilize the free energy of biological gradients: production of electrical signals, chemiosmotic energy transduction, and uphill transport of other molecules.

Production of electrical signals

Electrochemical energy is stored across the membrane primarily as Na^+ and Ca^{2+} gradients. The release of this electrical energy is under the control of "gated" channels. These channels are normally closed, but in response to certain chemical or electrical signals, they switch to an open state in which they exhibit selective permeability to specific ions. These ions then flow passively across the membrane down their electrochemical gradients. Because of the charges it carries, an ion species, as it moves across a membrane, produces an electric current and changes the potential difference that exists across the membrane. This electrical activity is the functional basis of the nervous system (the subject of Chapter 5).

Chemiosmotic energy transduction

The energy released by the metabolism of foodstuffs culminates in the passage of electrons along the respiratory chain in mitochondria. This, in turn, releases their energy, which is stored as an electrochemical proton gradient across the mitochondrial inner membrane (see Chapter 3). This novel energy storage mechanism, which does not use the conventional high-energy chemical intermediates, puzzled cell biologists for many years until Peter Mitchell proposed the chemiosmotic coupling hypothesis. *Chemiosmotic* refers to the direct link between chemical ('chemi') and transport ('osmotic') processes.

Two ideas are central to the chemiosmotic theory:

- The redox enzymes are oriented within the inner membrane of the mitochondrion so that the electron-transport system of the respiratory chain pumps hydrogen ions from inside the mitochondrial matrix across the inner membrane into the intermembrane space (Figure 4-26). The inner mitochondrial membrane has a low intrinsic permeability to H^+, so that this active pumping produces an excess of OH^- (and, therefore, a high pH) within the mitochondrial matrix and an excess of H^+ (and low pH) within the intermembrane space.

- The energy-rich H^+ gradient set up in this way across the inner membrane provides the free energy that removes HOH from ADP + P_i, as required for the production of ATP:

$$ADP + P_i \longrightarrow ATP + H_2O$$

$$\Delta G^{\circ\prime} = +7.3 \ kcal \cdot mol^{-1}$$

This reaction also requires that an ATPase complex be oriented on the inner mitochondrial membrane so as to take advantage of the separation of H^+ and OH^- across the membrane. The H^+ that is enzymatically removed from

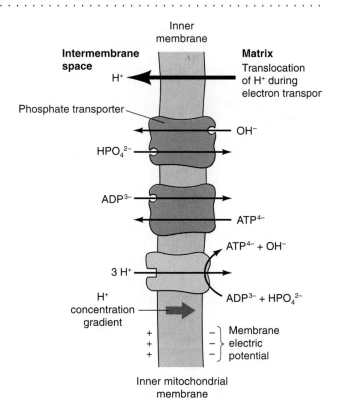

Figure 4-26 The phosphate and ATP-ADP transport system that generates ATP is located in the inner mitochondrial membrane. The phosphate transporter couples the uptake of one HPO_4^{2-} (inorganic phosphate) to the outward movement of one OH^- anion. At the same time, the ATP-ADP antiporter exchanges one incoming ADP^{3-} for one ATP^{4-} exported from the matrix. The exported OH^- combines with an H^+ translocated outward by respiration. As a result, there is a net uptake of one ADP^{3-} and one HPO_4^{2-} in exchange for one ATP^{4-}. This process is powered by the outward translocation of one H^+ during electron transport. For every four H^+ translocated outward, three are used to synthesize one ATP molecule and one is used to export ATP in exchange for ADP and P_i. [Adapted from Lodish et al., 1995.]

ADP is thought to be "siphoned off" into the OH^--rich mitochondrial interior to form HOH (Figure 4-27). The OH^- removed from the inorganic phosphate molecule is shunted outside the mitochondrion to react with the excess H^+ to form HOH. Thus, the H^+/OH^- gradient provides the energy needed to remove the water during the phosphorylation. Following the dehydration, phosphate bond formation goes forward on the active site of the ATPase without further need for energy input.

$$ADP + P_i \longrightarrow ATP$$

Chemiosmotic energy transduction similar to that proposed for oxidative phosphorylation in mitochondria has been implicated as the mechanism for energy transduction during photosynthesis in chloroplasts and photosynthetic bacteria. In addition, there is evidence that the Na^+/K^+ pump, which normally utilizes ATP to produce the Na^+ gradient, can in special circumstances run in reverse, so that the movement of Na^+ down its gradient will cause the pump to synthesize ATP from ADP and P_i.

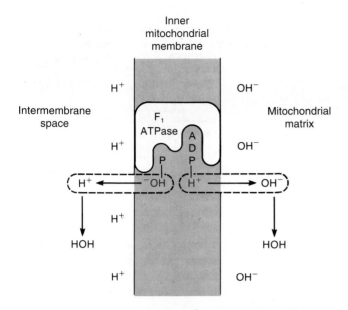

Figure 4-27 The second phase of Mitchell's chemiosmotic theory explains energy transduction in the mitochondrion. With the catalytic aid of F_1 ATPase located in the inner mitochondrial membrane, ADP and P_i have H^+ and OH^-, respectively, stripped away by high OH^- levels in the mitochondrial matrix and by the relatively high concentration of H^+ in the intermembrane space. This process allows P_i to condense with ADP to form ATP.

Coupled Transport

Movement of some molecules up a concentration gradient is driven by movement of another substance *down* its concentration gradient. Thus, the ubiquitous Na^+ gradient is used to carry certain sugars and amino acids along through the membrane by a *symport* mechanism and to drive Ca^{2+} out of the cell by an *antiport* mechanism. The next section will consider such coupled transport in detail.

Symporters

Symporters are one form of coupled active transport systems that run on energy stored in ion gradients. An example is the transport of the amino acid alanine, which is coupled to Na^+ (Figure 4-28). In the presence of Na^+, the amino acid is taken up by the cell until the internal concentration is 7–10 times that of the external concentration. In the absence of Na^+, the intracellular concentration of alanine merely approaches the extracellular concentration. In both cases the rate of influx shows saturation kinetics, indicating a carrier mechanism. The effect of extracellular Na^+ is to enhance the activity of the alanine carrier. Increasing the intracellular Na^+ concentration by blocking the Na^+ pump with ouabain has the same effect as decreasing the extracellular Na^+ concentration. Thus, it appears to be the Na^+ gradient that is important for inward alanine transport, and not merely presence of Na^+ in the extracellular fluid.

The transport of amino acids and sugars is coupled to inward Na^+ leakage by means of a **common carrier.** The carrier molecule must bind both Na^+ and the organic sub-

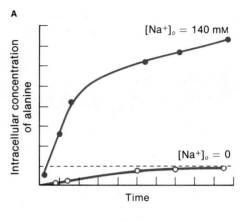

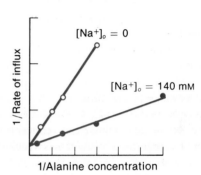

Figure 4-28 The cellular uptake of an amino acid such as alanine depends on Na^+ concentration. **(A)** Intracellular concentration of alanine, an amino acid, as a function of time with and without extracellular Na^+ present. The dashed line represents the extracellular concentration of alanine. **(B)** Lineweaver-Burk plots of alanine influx with and without extracellular Na^+. The abscissa is the reciprocal of the extracellular alanine concentration. The common intercept indicates that at infinite concentration of alanine, the rate of transport is independent of $[Na^+]$. [From Schultz and Curran, 1969.]

strate molecule before it can transport either (Figure 4-29). The tendency for Na^+ to diffuse down its concentration gradient drives this carrier system. Anything that reduces the concentration gradient of Na^+ (low extracellular Na^+ or increased intracellular Na^+) reduces the inwardly directed driving force and thereby reduces the coupled transport of amino acids and sugars into the cell. If the direction of the Na^+ gradient is experimentally reversed, the direction of transport of these molecules is also reversed. The carrier-mediated transport of Na^+ in this case also depends on the presence of amino acids and sugars. In the absence of amino acids and sugars, the common carrier will transport Na^+ only very weakly, and as a result the inward leakage of Na^+ is reduced.

The common carrier appears to shuttle between the two sides of the membrane passively, without direct utilization of metabolic energy. The coupled uphill transport of organic molecules derives its energy from the downhill diffusion of Na^+. However, the potential energy stored in the Na^+ gradient is ultimately derived from metabolic energy that drives the Na^+ pump. The Na^+ concentration gradient is thus an intermediate form of energy that can be

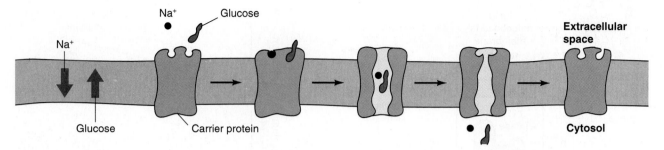

Figure 4-29 Sugar and amino acid transport may be achieved by sodium-mediated cotransport. The carrier protein must bind both the Na$^+$ and the organic substrate before it will transport either. Net transport is inward because of the Na$^+$ gradient, indicated by the arrow. Note that glucose is moving against its gradient.

used to drive several energy-requiring processes in the membrane.

Antiporters

The Na$^+$ concentration gradient also plays a role in the maintenance of a very low intracellular Ca^{2+} concentration in certain cells via the Na$^+$/Ca^{2+} antiport system. In most, if not all cells, the intracellular Ca^{2+} concentration is several orders of magnitude below the extracellular concentrations (less than 10^{-6} M), and many cell functions are regulated by changes in the intracellular Ca^{2+} concentration. Efflux of Ca^{2+} from cells is reduced when extracellular Na$^+$ is removed, because Ca^{2+} is expelled from the cell in exchange for Na$^+$ leaking in. The opposing movements of these two ions are coupled to each other by an **antiporter**. One view is that Ca^{2+} and Na$^+$ both compete for the carrier, but that Ca^{2+} competes more successfully inside the cell than on the outer surface, so that there is a net efflux of Ca^{2+}. Here, again, the immediate source of energy is the Na$^+$ gradient, which ultimately depends on the ATP-energized active transport of Na$^+$. Ca^{2+} is also transported independently of the Na$^+$ gradient by an ATP-energized Ca^{2+} pump, which is the major source of Ca^{2+} extrusion under normal conditions.

The Na$^+$/H$^+$ antiporter in the proximal tubule of the mammalian kidney is another example of cotransport in opposite directions (see Chapter 14). Here the extrusion of H$^+$ from inside the cells lining the renal tubule into the urine contained within the tubule is coupled to Na$^+$ uptake into the cell in a 1:1 stoichiometry. That is, for each H$^+$ expelled, one Na$^+$ is taken up into the cell. This has the advantages of (1) avoiding the expenditure of energy to perform electrical work, since two equivalent positive charges are exchanged and (2) enabling the kidney to reclaim Na$^+$ from the urine and excrete excess protons. The Na$^+$/H$^+$ exchanger, unlike the Na$^+$/K$^+$ pump, is oriented so as to move Na$^+$ out of the lumen and into the cell. Also unlike the Na$^+$/K$^+$ pump, this mechanism is not an example of primary active transport, in which ATP is the immediate source of energy. Instead, the Na$^+$/H$^+$ exchanger is an example of secondary active transport, in which the source of energy is the electrochemical gradient of one or both exchanged ions. In this case the energy driving the exchange arises from the Na$^+$ concentration gradient, directed from the lumen into the cell. This gradient is maintained by the removal of Na$^+$ from the cell by the Na$^+$/K$^+$ pump located in the membrane on the other side of the cell, which faces the plasma and blood.

MEMBRANE SELECTIVITY

The utility of cell membranes lies in their selectivity—their ability to allow the passage of only specific types of molecules. This selectivity is important because a nonselective membrane will not protect the contents of the cell from intrusion by unwanted chemicals. Each kind of membrane transport system also displays selectivity, which differs in a given membrane for different transport systems. For example, when the Na$^+$ in a physiological saline solution used to bathe a nerve cell is replaced with lithium ions, the Li$^+$ readily passes through the Na$^+$ channels, which open during electrical excitation of the nerve cell membrane. The other alkali metal cations, K$^+$, Rb$^+$, and Cs$^+$, are essentially impermeant through these channels. On the other hand, the ATPase of the Na$^+$ pump in the same membrane is highly specific for intracellular Na$^+$ and is not activated by Li$^+$. Lithium ions, passing through the Na$^+$ channels, will therefore gradually accumulate in the cell until it comes into electrochemical equilibrium. This is an example of electrolyte selectivity by the transport system, but not by the membrane channels. We will now consider how this selectivity for both electrolytes and nonelectrolytes is achieved.

Selectivity for Electrolytes

How do channels discriminate between different ions? Although enzymes recognize substrates via distinct shapes or chemical structures, membranes can distinguish ions of essentially identical shape and size. For example, Na$^+$ and K$^+$ have almost the same shape and size (K$^+$ is a little larger), yet the resting nerve cell membrane is about 30 times more permeable to K$^+$ than to Na$^+$. At first glance, we might conclude that these ions are distinguished on the basis of their hydrated size, with K$^+$ passing freely through channels that are too small for Na$^+$. Size can explain how the K$^+$ channel excludes Cs$^+$ or Rb$^+$ (Table 4-3) but not Na$^+$,

TABLE 4-3
Ionic radii and hydration energies of the alkali metal cations.

Cation	Ionic radius (Å)	Free energy of hydration (kcal · mol^{-1})
Li$^+$	0.60	−122
Na$^+$	0.95	−98
K$^+$	1.33	−80
Rb$^+$	1.48	−75
Cs$^+$	1.69	−67

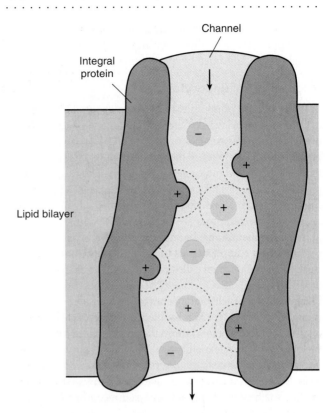

Figure 4-30 Positive charges lining the membrane channel allow anions to pass but retard the diffusion of cations through the channel, as shown in this hypothetical cross-section through a highly simplified membrane channel.

particularly in light of the fact that permeability to Na$^+$ can change dramatically. For example, during the excitation of nerve or muscle membrane, the Na$^+$ permeability of the membrane increases about 300-fold to a value about 10 times greater than the K$^+$ permeability at rest. If, during excitation, the membrane were suddenly to develop channels that pass the Na$^+$ ion on the basis of size alone, there should be a simultaneous increase in permeability to K$^+$ through the same channels, given their comparable sizes. Since this increase does not occur, the membrane's selectivity must rest on properties other than size. Indeed, estimated pore sizes for different membrane channels illustrate that size alone cannot be the agent of membrane selectivity.

Two interesting features other than size appear to be important in governing membrane pore selectivity: ease of dehydration and interaction with charges within the pore. For an ion to enter a pore, it must dissociate from water molecules. Ease of dehydration appears to be an important factor in governing selectivity, particularly if the charges within the pore are weak. Since large ions dehydrate more easily than small ones (see Table 4-3), a pore with weak polar sites along it will admit large ions preferentially over small ones.

In channels with strongly charged sites, the interaction of the dehydrated ion with these sites may be more important for conferring specificity than ease of dehydration. Thus, a channel lined with predominantly positively charged residues will selectively repel positively charged ions, but permit negatively charged ions to pass through (Figure 4-30). In such cases, smaller ions can approach the polar sites more closely and hence interact more strongly than can large ions, exaggerating the effect.

Selectivity for Nonelectrolytes

Virtually all nonelectrolytes cross the membrane by dissolving in the lipid bilayer and simply diffusing across it. Since the relationship between permeability and partition coefficient K is essentially linear (see Figure 4-19), selectivity is completely determined by molecular properties responsible for the partition coefficient. Those few nonelectrolytes that deviate from the linear relation between partition coefficient and permeability all have greater than predicted permeability. Some of these substances cross the membrane by carrier-mediated transport. Alternatively,

small molecules such as ethyl alcohol, methyl alcohol, and urea can cross via both the lipid layer and water-filled channels. All these deviant molecules are small and water-soluble regardless of their relative solubilities in water versus lipid (i.e., their partition coefficients). It is important to note that mechanisms for precise control of nonelectrolyte access through membranes have not evolved, making cells vulnerable to penetration by these molecules. Drugs applied to the human skin, such as anti-nauseants delivered by skin patches placed behind the ear, can use this route to enter the body.

ENDOCYTOSIS AND EXOCYTOSIS

The transport processes described above for small polar molecules across membranes cannot transport macromolecules such as proteins, polynucleotides, or polysaccharides. Yet, cells do manage to ingest and secrete macromolecules, using mechanisms very different from those used for small solutes and ions. Transmembrane movement of macromolecules is accomplished through the sequential formation and fusion of membrane-bounded vesicles. The intake of material into the cell is given the general term **endocytosis**. The process is more specifically called **pinocytosis** if fluid is ingested and **phagocytosis** if solids are ingested. The secretion from the cell of macromolecules is called **exocytosis**. In both exocytosis and endocytosis, the fusion of separate regions of the lipid bilayer occurs in at least two steps: the bilayers come into close apposition and then they fuse. Both processes are thought to be controlled by specialized proteins.

Mechanisms of Endocytosis

Transfer of macromolecules across membranes by endocytosis requires specialized control mechanisms. **Receptor-mediated endocytosis** depends on the presence of receptor molecules embedded in the cell membrane (Figure 4-31A). These bind certain ligand molecules or particles, including plasma proteins, hormones, viruses, toxins, immunoglobulins, and certain other substances that cannot pass through membrane channels. The receptors are free to diffuse laterally in the plane of the membrane, but upon binding of ligand, the receptor-ligand complex tends to accumulate within depressions in the membranes called **coated pits**. The coated pit internalizes the ligand. One theory of how it does this is by the formation of a vesicle that pinches off into the cytoplasm, as shown (Figure 4-31B). This is called a **coated vesicle** because of a layer of the protein **clathrin** that covers the cytoplasmic surface of the vesicle membrane. The clathrin is organized into pentagonal or hexagonal lattice-like arrays on the membrane surface, and appears to have several functions. These include the binding of ligand-occupied receptor molecules

and the subsequent budding off of the vesicle from the surface membrane. Once the coated vesicle buds off into the cytoplasm, it is believed to fuse with and deliver its contents to other organelles, such as lysosomes. The clathrin and receptors are recycled into the surface membrane.

Mechanisms of Exocytosis

The release of chemicals from cell membranes through exocytosis plays a crucial role in the endocrine and nervous systems. For example, the presynaptic terminals of nerve cells contain many membrane-delimited internal vesicles about 50 nm in diameter, which contain the neural transmitter substance. These vesicles coalesce with the surface membrane of the nerve terminal and release their contents to the cell exterior, the typical method of exocytosis. This activity occurs with greatly enhanced probability when the terminal is invaded by a nerve impulse, and serves to release the synaptic transmitter that interacts with the postsynaptic membrane. Similar mechanisms are involved in the secretion of hormones.

An important feature of exocytosis (as well as endocytosis) is that the secreted or ingested macromolecules are sequestered in vesicles and hence do not mix with macromolecules or organelles in the cell. Since the vesicles can fuse with

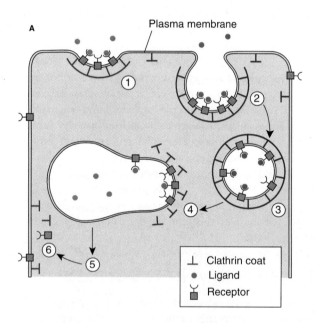

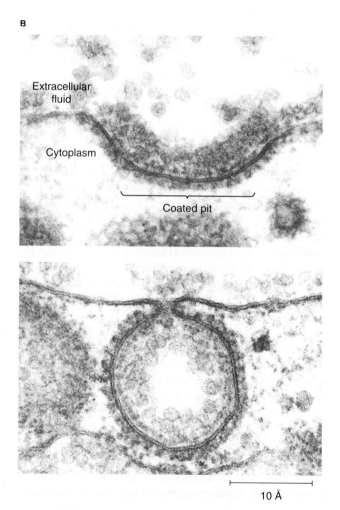

Figure 4-31 Coated vesicles form during receptor-mediated endocytosis. **(A)** There are six major steps involved in the process: (1) ligand molecules bind to surface receptor molecules located in coated pits formed by clathrin molecules that are bound to surface membrane; (2) the coated pit is invaginated; (3) the coated vesicle is formed; (4) the coated vesicle fuses with an existing vacuole, shedding the clathrin molecules; (5) the fused complex undergoes further processing, depending on its contents, while (6) clathrin and receptor molecules are recycled for reuse in the plasma membrane. **(B)** Electron micrographs of coated pit (top) and coated vesicle (bottom). These two stages, taken from a chicken oocyte, show the dense clathrin coat on the cytoplasmic surface of the membrane. The surface membrane can be seen pinching off the vesicle. [Part A from Pearse, 1980; part B from Bretscher, 1985.]

only specific membranes, they assure the directed transfer of their contents in the cell. In exocytosis, once the membrane of the vesicle is incorporated into the surface membrane, the freed contents—hormones, neurotransmitters, and accessory molecules—diffuse away into the interstitial space.

Exocytosis requires a method for recovering the relatively large amounts of secretory vesicle membrane that initially surrounds the macromolecules being expelled. In the absence of retrieval of this newly incorporated membrane, the surface area of the plasma membrane would continually grow. However, endocytosis is thought to be responsible for the eventual recovery of this excess membrane through its reformation into new secretory vesicles. Evidence for such **membrane recycling** through endocytosis comes from experiments in which electron-opaque protein molecules, such as **horseradish peroxidase**, are introduced into the extracellular fluid and their movement into the cell determined with electron-microscopic methods. In these experiments, horseradish peroxidase shows up inside the cell, but only within vesicles. Since the large size of the horseradish peroxidase molecule prevents its penetration by direct passage across biological membranes, it must have been taken up in bulk during the endocytotic formation of vesicles budding off from the plasma membrane into the cytoplasm.

The calcium ion is responsible for the exocytotic secretion of neurotransmitter substances from nerve cells and of hormones from endocrine cells. Although the precise role of Ca^{2+} in initiating secretion is unknown, it appears that an elevation of intracellular Ca^{2+} somehow enhances the probability of exocytotic activity, perhaps by permitting the coalescence of vesicles with the inner surface of the membrane. The membrane regulates exocytotic activity by regulating the intracellular accumulation of Ca^{2+}. As enhanced calcium influx allows Ca^{2+} levels to rise, the rate of exocytotic secretion increases.

The vesicle membrane itself may participate actively in the initial steps leading to exocytosis. The secretory granules (or vesicles) of the adrenal medulla have been found to be rich in an unusual phospholipid, **lysolecithin,** that facilitates the fusion of membranes and thus may help the vesicle membrane fuse with the surface membrane. Before fusion of the two membranes can take place, the secretory granule must come into contact with the plasmalemma. Release of secretory products from glandular secretory cells can be blocked by **colchicine**, an anti-mitotic agent that leads to the disassembly of microtubules, or by **cytochalasin,** an agent that disrupts microfilaments. This pharmacological evidence has led to the suggestion that microtubules or microfilaments participate in the movement of secretory granules toward sites of exocytotic release on the inner side of the surface membrane.

JUNCTIONS BETWEEN CELLS

Cells in animals are organized into cooperative assemblies called **tissues.** In certain tissues, including epithelium, smooth muscle, cardiac muscle, central nervous tissues, and many embryonic tissues, neighboring cells are connected by special adaptations of their abutting surfaces. These specialized surfaces are divided into two major groups: gap junctions and tight junctions. Gap junctions enhance cell-cell communication through minute, water-filled channels that connect adjacent cells, while tight junctions "sew" cells involved in transepithelial transport into sheets.

Gap Junctions

Gap junctions provide communication between cells by allowing inorganic ions and small water-soluble molecules to pass directly from the cytoplasm of one cell to the cytoplasm of the other. These junctions couple cells electrically and metabolically, with important functional consequences for the tissue. The distance between two membranes of a gap junction is only 2 nm (Figure 4-32). The two adjoining membranes each contain clusters of hexagonal arrays of six subunits that span the narrow space between the two membranes (Figure 4-33A). The subunit arrays are about 5 nm in diameter and resemble miniature doughnuts whose hollow centers form passageways between the interiors of the neighboring cells (Figure 4-33B). The continuity of the cell-cell passageways through the gap junction has been demonstrated by injecting fluorescent dyes, such as fluorescein (molecular weight 332) and procion yellow (molecular weight 500), into one cell and following their diffusion into neighboring cells (Figure 4-34). This continuity has been corroborated for direct exchange of ions by the finding that electric current readily passes directly from one cell into another if gap junctions are present. The intercellular channels in these junctions pass molecules with a molecular weight of at least 500, so that small molecules, such as ions,

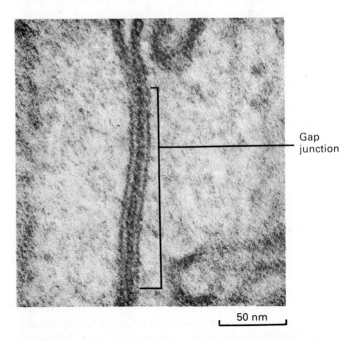

Gap junction

50 nm

Figure 4-32 The gap junction interval (2 nm) between neighboring cells is at the lower limit for electron microscopic resolution. The electron micrograph reveals a gap junction between membranes of two neighboring mouse liver cells. [Courtesy of D. Goodenough.]

A

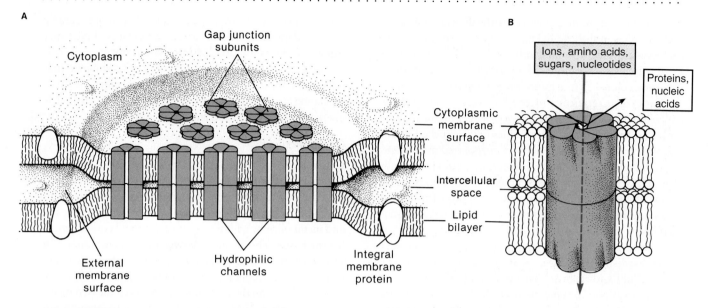

B

Figure 4-33 Gap junctions permit passage of molecules between neighboring cells. **(A)** The two membranes belonging to neighboring coupled cells both contain an array of hexagonal subunits, each of which connects with a matching subunit in the apposed membrane. A central channel penetrates both subunits, providing a path of communication between the connected cells. **(B)** Detail of a channel complex. Molecules smaller than about 2 nm can pass between coupled cells through the channel. Molecules larger than 2 nm, such as proteins and nucleic acids, are too large to penetrate the channel. [Part A adapted from Staehelin, 1974; part B adapted from Bretscher, 1985.]

amino acids, sugars, and nucleotides, are easily exchanged between cells (see Figure 4-33B). This exchange of small molecules is responsible for gap junction-mediated cell-cell communication.

Gap junctions are labile and close rapidly (within seconds) in response to any treatment that increases intracellular Ca^{2+} or H^+ concentration. Uncoupling of cells from their neighbors can be produced by injecting Ca^{2+} or H^+

into a coupled cell, by lowering temperature, or by using poisons that inhibit energy metabolism. The subsequent loss of electrical transmission between cells confirms the uncoupling. Thus, gap junctions are maintained intact only if the metabolic activity of the surface membrane maintains sufficiently low concentrations of intracellular free Ca^{2+} and H^+. The mechanism of closing of the gap junction channel is not clearly understood, but the channel appears to be open or shut depending on the relative positions of the six subunits of the channel.

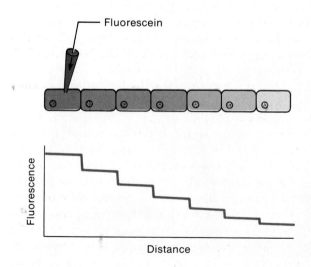

Figure 4-34 Gap junctions between coupled cells can be demonstrated by following the flow of fluorescent dye injected into one of a group of coupled epithelial cells. Subsequent diffusion of the dye into neighboring cells without loss into extracellular space indicates that there are direct pathways from the cytoplasm of one cell to the cytoplasm of the adjacent cell.

 Gap junctions, which are common in invertebrate nervous systems, allow the exchange of many types of cytoplasmic materials between adjacent cells. How do assemblages of cells linked by gap junctions—freely exchanging ions, amino acids, sugars, and nucleotides—challenge the concept of cells? What is the functional difference between a single cell, a series of cells linked by gap junctions, and a tissue?

Tight Junctions

Tight junctions seal cells together in an epithelial cell sheet so that even small molecules cannot get from one side of the sheet to the other. The two apposing cell membranes make intimate contact, fully occluding the extracellular space in between. Tight junctions are found most commonly in ep-

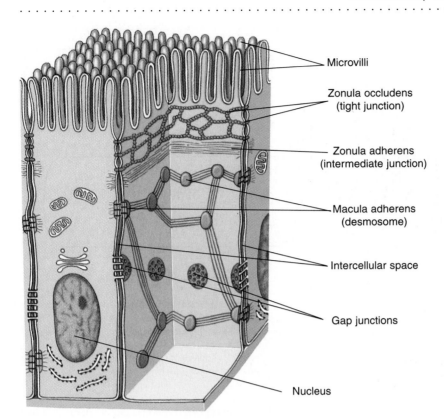

Figure 4-35 Adjacent epithelial cells like those that line the mammalian small intestine are connected by intercellular junctions. The membranes and associated structures are drawn disproportionately large in this reconstruction of the cell-cell junctions.

Microvilli

Zonula occludens (tight junction)

Zonula adherens (intermediate junction)

Macula adherens (desmosome)

Intercellular space

Gap junctions

Nucleus

ithelial tissues as a **zonula occludens,** a thin band of protein molecules that encircles a cell like a gasket. The zonula occludens is in tight contact with the zonulae of the surrounding cells, forming an impermeable seal that prevents passage of substances from one side of the epithelium to the other via leakage down along the sides of the cells (Figure 4-35). Considered en masse, the zonulae are conceptually like a continuous rubber sheet, penetrated only by the ends of the epithelial cells. Substances can pass through the ends of the cells (the **transcellular pathway**), but not around them (the **paracellular pathway**). In tissues such as the mammalian small intestine, the gallbladder, and the proximal tubule of the nephron, these zonulae are not fully continuous and thus not really very "tight." These tissues are so leaky, they do not produce a transepithelial potential difference, even though their cells contain ion pumps capable of generating transepithelial ion fluxes. Unlike gap junctions, tight junctions appear to have no special channels for cell-cell communication.

Two other types of cell junctions are shown in Figure 4-35: the **zonula adherens** and the **desmosome** serve primarily to aid the structural bonding of neighboring cells.

EPITHELIAL TRANSPORT

Epithelial cell sheets line the cavities and free surfaces of animal bodies and form barriers to the movement of water, solutes, and cells from one body compartment to another. Each organ or compartment within an animal has such a lining of surface cells. Some of these sheets serve only as passive barriers between compartments and do

not preferentially transport solutes and water. In other cases they are involved in active transport, performing regulatory functions. For example, osmoregulatory activities of animals are carried out by actively transporting epithelia in a variety of specialized tissues and organs (see Chapter 15).

Epithelia have several features in common. First, they occur at surfaces that separate the internal space of the organism from the environment. Included are the surfaces lining deep invaginations such as in the lumen of the intestine, which nevertheless comprise external space. Second, the cells forming the outermost layer of the epithelium are generally sealed together by tight junctions, which to varying degrees in different epithelia obliterate paracellular pathways between the **serosal** (inner) and **mucosal** (external) sides of the epithelium (Figure 4-36). In epithelia such as that of capillary walls, leaky junctions permit water and solute molecules to cross the epithelial layer by diffusion within the passages that exist between the cells. Such diffusion through paracellular pathways is not coupled to any metabolically energized transport mechanism, so such passages allow only passive movement of water and ions. Substances that are *actively* transported across an epithelium must follow transcellular pathways, in which the cell membrane participates. Such substances must cross the cell membrane first on one side of the cell and then on the other. As discussed in the next section, the functional properties of the surface membrane of an epithelial cell are dissimilar in some respects on its serosal and mucosal surfaces. This asymmetry is important to epithelial active transport.

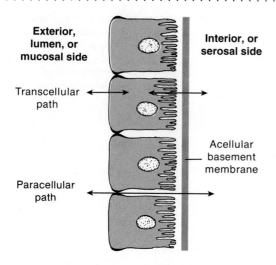

Figure 4-36 Substances cross epithelial layers by two pathways: paracellular and transcellular. Active transport takes place only across cell membranes, suggesting that all actively transported molecules follow the transcellular pathway.

Active Salt Transport across an Epithelium

Energy-requiring transport of ions from one side of an epithelium to the other has been demonstrated in a number of epithelial tissues, including amphibian skin and urinary bladder, the gills of fishes and aquatic invertebrates, insect and vertebrate intestine, and vertebrate kidney tubule and gallbladder. Much of the initial work on epithelial active transport was done on the skin of the frog. In amphibians, the skin acts as a major osmoregulatory organ. Salt is actively transported from the mucosal side (i.e., the side facing the pond water) to the serosal side of the skin to compensate for the salt that leaks out of the skin into the freshwater surrounding the frog. Similar uptake occurs in the gut. Water that enters the skin because of the osmotic gradient between the hypotonic pond water and the more concentrated internal fluid is eliminated in the form of a copious dilute urine that is hypotonic relative to the body fluids (see Chapter 14).

Frog skin was first used in the study of epithelial transport in the 1930s and 1940s by the German physiologist Ernst Huf and the Danish physiologist Hans Ussing. In their procedure, a piece of abdominal skin several square centimeters in area is removed from an anesthetized and decapitated frog and placed between two halves of an **Ussing chamber** (Figure 4-37). The dissection is very simple, since the skin of the frog lies largely unattached over an extensive lymph space. Once the skin is gently clamped between the two half-chambers, a test solution—for example, frog Ringer's solution (a solution that mimics the ionic composition of frog plasma)—is introduced, with the frog skin acting as a partition between the two compartments. The compartment facing the mucosal side of the skin can be designated as the outside compartment and the one facing the serosal side as the inside compartment. Air is bubbled through the two solutions to keep them well oxygenated.

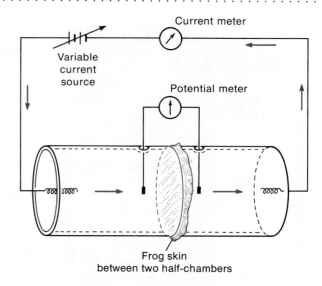

Figure 4-37 A frog skin separates the two halves of this Ussing chamber. Each half is filled with a physiological saline or other test solution. The current source is adjusted until the potential difference across the skin is zero. Under those conditions, the current flowing through the circuit (and thus through the skin) is equivalent to the rate of charge transferred by the active movement of sodium ions across the skin.

In 1947 Ussing reported the first experiments in which two isotopes of the same ion were used to measure bidirectional fluxes (i.e., the simultaneous movement of that ionic species in opposing directions across the epithelium). The Ringer's solution in the outside compartment was prepared using isotope $^{22}Na^+$, and the Ringer's in the inside compartment was prepared using $^{24}Na^+$. The appearance of each of the two isotopes on the opposite side of the skin was tracked over time. The two isotopes were switched around in other experiments of the same type to rule out any effects due to possible (but unlikely) differences in transport rates inherent in the isotopes themselves. In all experiments it was found that Na^+ shows a net movement across the skin from the outside compartment to the inside one. That the Na^+ flux is the result of active transport is seen in the fact that it

- Occurs without any concentration gradient, and even against an electrochemical gradient.

- Is inhibited by general metabolic inhibitors, such as cyanide and iodoacetic acid, and by specific transport inhibitors, such as ouabain.

- Displays a strong temperature dependence.

- Exhibits saturation kinetics.

- Shows chemical specificity. For example, Na^+ is transported while the very similarly structured lithium ion is not.

How can an active movement of ions be produced across a layer of cells contained in an epithelium? Adjacent cells of transport epithelium are intimately tied together with tight junctions. Assume for the sake of simplicity that this closeness eliminates all extracellular passageways for the diffusion of ions between the two sides of the epithelium. This would force all substances that cross the epithelium to traverse the epithelial cell membrane twice, first crossing the membrane on one side of the cell and then leaving through the membrane on the other side. Active transport by this route requires that the surface membrane of each epithelial cell be differentiated, so that the portion of the cell membrane facing the serosal side of the epithelium differs in functional properties from the portion facing the mucosal side. Experiments on frog skin in Ussing chambers have provided several lines of evidence to support a hypothesis of a differentiated membrane. For example,

- Ouabain, which blocks the Na^+/K^+ pump, inhibits transepithelial sodium transport only when applied to the inner (serosal) side of the epithelium. It is ineffective on the outer (mucosal) side. Conversely, the drug amiloride, a powerful inhibitor of passive carrier-facilitated transport, blocks Na^+ movements across the skin only when applied to the outer side of the skin.

- For active Na^+ transport to take place, K^+ must be present in the solution on the inner side of the chamber, but is not required on the outer side.

- Transport of Na^+ exhibits saturation kinetics as a function of Na^+ concentration in the outer solution; it is unaffected by Na^+ concentration in the inner solution.

Such evidence led to the model of epithelial Na^+ transport shown in Figure 4-38. According to this model, a Na^+/K^+ exchange pump is located in the membrane of the serosal side of the epithelial cell (together with Na^+/H^+ and Na^+/NH_4^+ exchange pumps in the intact animal). This membrane behaves in the manner typical of many cell membranes, pumping Na^+ out in exchange for K^+, thus maintaining a high intracellular K^+ concentration and a low intracellular Na^+ concentration. The outward diffusion of K^+ across the membrane on this side of the cell produces an inside negative resting potential.

The situation on the mucosal side must be different. The cell membrane on this side of the cell is relatively impermeable to K^+. Moreover, a net inward diffusion of Na^+ across this membrane (apparently facilitated by carriers or through channels in the membrane) replaces the Na^+ pumped out of the cell on the serosal side. This model explains why Na^+ pump inhibitors exert an effect only from the serosal side of the epithelium, and why only changes in the concentration of K^+ on that side influence the rate of Na^+ transport.

Thus, there is a net flow of Na^+ across the frog skin from the mucosal side to the serosal side as a result of the functional asymmetries of the membranes on the two sides. The driving force is none other than the active transport of Na^+ that is common to cell membranes of all tissues.

The frog skin has served as a model system for the general problem of epithelial salt transport. Although details

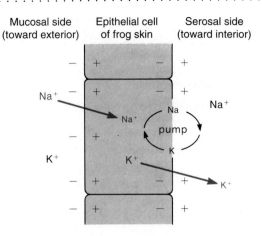

Figure 4-38 Transepithelial sodium transport depends on a combination of both diffusion and active transport. In this model of an isolated frog skin that has been bathed in Ringer's solution, sodium diffuses passively down its concentration gradient into the cell from the mucosal solution. K^+ diffuses out of the cell into the serosal space as it is displaced by Na^+ influx. In the face of these leaks, a Na^+/K^+ exchange pump in the serosal membrane of the cell maintains the high internal K^+ and low internal Na^+ concentrations. [From Koefoed-Johnsen and Ussing, 1958.]

may differ from one type of epithelial tissue to another, the major features, listed below, are probably common to all transport epithelia.

1. To varying degrees, tight junctions obliterate paracellular pathways. As a result, transport through transcellular pathways assumes major importance in epithelial transport.
2. Mucosal and serosal portions of the cell membranes exhibit functional differences, being asymmetrical in both pumping activity and membrane permeabilities.
3. The active transport of cations across an epithelium is typically accompanied by transport (passive or active) of anions in the same direction or by exchange for another species of cation, minimizing the buildup of electric potentials. The converse applies to actively transported anions.
4. Epithelial transport is not limited to the pumping of Na^+ and Cl^- ions. Various epithelia are known to transport H^+, HCO_3^-, K^+, and other ions.

Transport of Water

To function properly, animal tissues require the right amount of water in the right place at all times. This is achieved by regulation of water via epithelial sheets. A number of epithelia absorb or secrete fluids. For example, the stomach secretes gastric juice, the choroid plexus secretes cerebrospinal fluid, the gallbladder and intestine transport water, and the kidney tubules of birds and mammals absorb water from the glomerular filtrate. In some of these tissues, water moves across an epithelium in the absence of or against an osmotic gradient existing between the bulk solutions on either side of the epithelium. A number of possible explanations for the uphill movement of water have been given, but all these hypotheses can be placed in one of two major categories:

- Water is transported by a specific water-carrier mechanism driven by metabolic energy.
- Water is transported secondarily as the consequence of solute transport.

The latter includes classic osmosis, in which water undergoes a net diffusion in one direction owing to concentration gradients built up by solute transport. So far, there has been no convincing evidence to indicate that water is actively transported by a primary water-carrier pump.

The osmotic hypothesis of water transport received a boost by Curran, who showed that an osmotic gradient produced by active salt transport from one subcompartment of the epithelium into the other could, in theory, result in a net flow of water across the epithelium (Figure 4-39). Biological correlates of Curran's model were subsequently found in the epithelium of mammalian gallbladder. This finding led to the **standing-gradient hypothesis** of solute-coupled water transport, presented by Diamond and Bossert (1967). A simplified schematic version is shown in Figure 4-40. Two anatomical features are of major importance. First, the tight junctions near the luminal (mucosal) surface block extracellular pathways through the epithelium. Second, the lateral intercellular spaces, or intercellular clefts, between adjacent cells are restricted at the luminal ends by the tight junctions and are freely open at the basal ends.

The basis for the standing-gradient hypothesis is the active transport of salt across the portions of the epithelial cell membranes facing the intercellular clefts. The membranes bordering the lateral clefts have been shown to be especially active in pumping Na^+ out of the cell. It is suggested that as salt is transported out of the cell into these long, narrow clefts, the salt concentration will set up an osmotic gradient

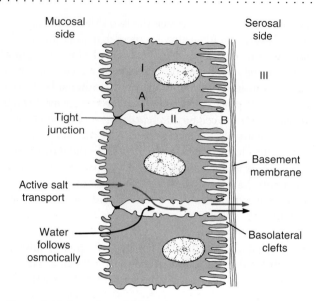

Figure 4-40 Curran's model for solute-linked water transport has a biological counterpart. The compartments corresponding to those in Figure 4-39 are numbered I, II, and III. Salt transported actively into the intercellular clefts produces a high osmolarity within the clefts. Water flows osmotically into the clefts across the cell, and the bulk solution flows through the freely permeable basement membrane and into the bulk fluid of the interstitium. The barriers A and B are analogous to A and B in Figure 4-39. [From Diamond and Tormey, 1966.]

between the extracellular spaces on either side of the tight junctions that join the epithelial cells. There may also be an osmotic gradient within the cleft, with the salt concentration highest near the closed ends of the clefts, diminishing toward the open ends of the clefts, where it comes into equilibrium with the bulk phase. As a consequence of the high extracellular osmolarity in the clefts, water is osmotically drawn into the cleft across the "not so tight" tight junction, or possibly from within the cell across the cell membrane into the intercellular space. The water leaving the cell would have to be replaced by water drawn osmotically into the cell at the mucosal surface. The water that enters the clefts gradually moves, together with solute, out into the bulk phase. In this way, the steady, active extrusion of salt by one surface membrane of the cell produces an elevated concentration in the narrow intercellular spaces. This, in turn, results in a steady osmotic flow of water from one side of the epithelium to the other.

The general applicability of the standing-gradient mechanism of solute-coupled water transport is supported by ultrastructural studies showing that the necessary cellular geometry—namely, narrow intercellular spaces closed off at the luminal end by tight junctions—is present in all the water-transporting epithelia that have been examined. Also important in this regard are deep basolateral clefts and infoldings typical of transporting epithelial cells (see Figure 4-40). These spaces are dilated in epithelia fixed during conditions that produce water transport. In epithelia fixed in the absence of water transport, the intercellular clefts are largely obliterated.

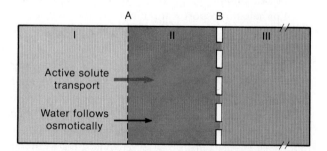

Figure 4-39 Curran's model for solute-linked water transport depends on active transport of a solute across a water-permeable membrane. A solute (e.g., Na^+) is pumped through barrier A from compartment I to compartment II. Semibarrier B slows diffusion of solute into compartment III and thereby keeps the osmolarity high in II. The rise in osmolarity in compartment II causes water to be drawn from I to II. In the steady state, both water and solutes diffuse into compartment III at the same rate at which they appear in II. Compartment III is much larger than II, as indicated by the breaks in the walls of the compartment. [From Curran, 1965.]

SUMMARY

Lipid bilayer membranes are fundamental structures in the formation of various cellular organelles, as well as the surface membrane. Their roles include (1) cellular and subcellular compartmentalization, (2) maintenance of the intracellular milieu using selective permeability and transport mechanisms, (3) regulation of intracellular metabolism by maintaining concentrations of intracellular enzyme cofactors and substrates, (4) metabolic activities carried out by enzyme molecules present in ordered arrays in or on the membrane, (5) sensing and transduction of extracellular chemical signals by means of surface receptor molecules and regulatory molecules located in the membrane, (6) propagation of electrical signals that conduct messages, regulate the transport of substances across the membrane, or both, (7) endo- and exocytosis of bulk material. The foundation structure of membranes is a lipid bilayer in which the hydrophilic heads of the phospholipid molecules face outward and the lipophilic tails face inward, toward the center of the bilayer. The most widely accepted model of membrane structure proposes that a mosaic of globular proteins, including enzymes, penetrates the bilayer.

Because of an unequal distribution of solutes between cell interior and exterior, water enters the cell, following its tendency to flow from a region of lower to a region of higher osmotic pressure. Osmotic pressure is equal to the hydrostatic pressure necessary to balance osmotic flow (water movement across a semipermeable membrane) down a concentration gradient at equilibrium. The concept of tonicity describes the osmotic effects that a solution has on a given tissue, whereas osmolarity describes the number of dissolved particles per volume of solvent, as well as the behavior of a solution in an ideal osmometer.

Permeability is a measure of the ease with which a substance traverses a membrane. There are several ways in which substances cross the membrane. Nonpolar molecules can diffuse readily through the lipid phase of the membrane. Water and some small polar molecules diffuse through transient aqueous channels created by thermal motion. There is a great deal of evidence for the existence of fixed channels that are more or less specific for certain ions and molecules. Diffusion across the membrane of some substances can occur via carrier molecules that complex with the substance, facilitating its transport across the membrane by shuttling it within the lipid phase of the membrane. In addition to these passive mechanisms, several active transport systems also move substances across the membrane.

Active transport of a substance occurs by means of carriers and requires metabolic energy, usually provided by ATP. It is responsible for the movement of a substance across a membrane against a concentration gradient. The most familiar active transport system is the sodium-potassium pump, which maintains the intracellular Na^+ concentration below that of the cell exterior. The energy stored in the form of an extracellular-intracellular Na^+ concentration gradient is utilized to drive the uphill movement of a number of other substances, such as calcium ions, amino acids, and sugars, by means of exchange diffusion and coupled transport. The Na^+ and K^+ gradients are also important for the production of electrical signals, such as nerve impulses.

Another important function of active transport is to compensate for the tendency of certain substances, such as Na^+, to leak into cells and thereby cause uncontrolled increases in osmotic pressure and subsequent swelling of the cell. Continual removal of Na^+ by the Na^+/K^+ pump is therefore a major factor in controlling cell volume. Transepithelial transport depends on an asymmetry in the permeability and pumping activities of the mucosal and serosal portions of epithelial cell membranes. On the serosal side of the cell, ions are actively transported across the membrane against an electrochemical gradient; on the mucosal side, ions cross the membrane by diffusion or facilitated transport. Diffusion of ions back through the epithelial layer is slow because the spaces between cells are restricted by tight junctions. Water is transported across some epithelia by being drawn osmotically down a standing salt concentration gradient built up by active salt transport between the epithelial cell interior and the intercellular clefts. There is no evidence for true active transport of water.

REVIEW QUESTIONS

1. What are some of the physiological functions of membranes?
2. What is the evidence for the existence of membranes as real physical barriers?
3. What is the evidence for the lipid bilayer model of the membrane?
4. What is the evidence for a mosaic of globular proteins set into the lipid bilayer of the membrane?
5. Explain the meanings of isotonic and isosmotic. How can a solution be isosmotic but not isotonic to another solution?
6. What factors determine the permeability of the membrane to a given electrolyte? Nonelectrolyte?
7. Describe the probable mechanisms by which water and other small (less than 1 nm in diameter) polar molecules pass through the membrane?
8. Why do nonpolar substances diffuse more easily than polar substances through the membrane?
9. There is no convincing evidence for direct active transport of water. Explain one way in which water is moved by epithelia against a concentration gradient, that is, from a concentrated salt solution to a more dilute salt solution.
10. How does facilitated transport differ from simple diffusion?
11. What factors influence the rate of facilitated transport of ions across a membrane?

12. How does active transport differ from facilitated transport?

13. Why can the sodium ion concentration gradient be considered a common cellular energy currency?

14. What are some parameters by which the membrane discriminates between ions of the same charge?

15. Explain the osmotic consequences of poisoning the metabolism of a cell.

16. How does the cell maintain a higher concentration of K^+ inside the cell than in the extracellular fluid?

17. What are the morphological and functional distinctions between gap and tight junctions?

18. A given cell is 40 times as permeable to K^+ and Cl^- as to any other ions present. If the inside-to-outside ratio of K^+ is 25, what would the approximate inside-to-outside ratio of Cl^- be?

19. Given that cell membranes can transport substances only into or out of a cell, explain how substances are transported *through* cells.

20. Describe the experiments that first demonstrated active transport of Na^+ across an epithelium.

21. What is some of the evidence that active transport of Na^+ and K^+ occurs only across the serosal membranes of epithelial cells?

SUGGESTED READINGS

Bretscher, M. S. 1985. The molecules of the cell membrane. *Sci. Am.* 253:100–108. (The complex, heterogeneous nature of the cell membrane is explored in this well-illustrated article.)

Goodsell, D. S. 1991. Inside a living cell. *Trends in Biochem. Sci.* 16:203–206. (This article takes the reader on a tour through the amazing structures and processes found in the living cell.)

Lodish, H., et al. 1995. *Molecular Cell Biology*. 3d ed. New York: W. H. Freeman. (This comprehensive textbook describes many of the basic biochemical processes that occur in the cell.)

Singer, S. J., and G. L. Nicolson. 1972. The fluid mosaic model of the structure of cell membranes. *Science* 175:720–731. (This is the original paper proposing the fluid mosaic model of the structure of cell membranes.)

Verkman, A. S. 1992. Water channels in cell membranes. Ann. Rev. Physiol. 54:97–108. (This review focuses specifically on the channels involved in transmembrane movements.)

Yeagle, P. L. 1993. *The Membranes of Cells*. (Both the morphology and the molecular physiology of plasma and organelle membranes are highlighted in this book.)

PHYSIOLOGICAL PROCESSES

If an animal is to survive and prosper it must respond appropriately and effectively to its environment and to its own internal states. Effective responses often require that different parts of the body, which may be quite far apart, act in a coordinated fashion. The nervous and endocrine systems act together to initiate coordinated responses, and the muscles and glands generate an animal's behavioral responses. In Part II, we focus on both the signaling systems (nervous and endocrine) and the effector systems (muscles and glands). These different types of tissues are composed of highly specialized cells that work together in groups to integrate information and to generate responses that are suitable to the perceived situation.

The task of collecting information from outside and inside the body, and the task of integrating that information, belong largely to the cells of the nervous system. In Chapter 5 we discuss the properties of nerve cells that allow them to gather, transform, and transmit information. Neurons in all species that have been studied share a remarkable number of features, from the nature of the molecules underlying their function to the physical principles that determine how they work. Thus it has been possible to study nerve cells that are particularly convenient for experimental manipulations with the conviction that the knowledge gained will be broadly applicable.

All nervous systems consist of large numbers of cells, which must share information in order to function effectively. Chapter 6 considers the processes that allow signals to travel along the cell membrane of a single neuron and the processes that allow signals to be passed between neurons. Within a single neuron, a signal is encoded electrically, and in some cases transmission between neurons also is accomplished electrically. In most cases, however, the electrical signal in one neuron must be transformed into a chemical signal if it is to be passed on to another cell. Understanding the mechanisms that allow neurons to communicate with one another and with other cells provides a basis for understanding the power and the limitations of the nervous system.

At the interface between an animal and its environment are many nerve cells that are specially tuned to receive information; other specialized nerve cells monitor conditions within the body. These cells, whose properties make them particularly suited for gathering information, are called sensory neurons, and they are discussed in Chapter 7.

The second major system that contributes to coordinating processes within an animal's body is the endocrine system. The cells of this system are collected into organs called endocrine glands, and their signals are molecules that are released into the blood stream of the animal, a process called secretion. Other glands, the exocrine glands, produce chemicals that are secreted into particular locations. The properties of endocrine and exocrine glands are discussed in Chapter 8.

The signaling molecules of endocrine glands, called hormones, can influence widely separate parts of the body simultaneously because they are carried throughout the body in the circulatory system. Hormones act on their target cells through specific receptor molecules, and the effect of a hormone on its targets depends upon both the nature of the receptor molecules and their effect on the internal processes of the target cell. The mechanisms that control the release of hormones and the mechanisms by which hormones act on their targets is discussed in Chapter 9.

The outwardly visible behavior of an animal, as well as much of the activity that goes on inside the body, depends upon contractions of muscle cells. In Chapter 10 we discuss the cellular properties of muscles that allow them to move the body or to change the shape of internal organs. We then turn to how muscular movements can be coordinated to produce effective behavior.

Finally, in Chapter 11 we consider some examples of how specific behaviors are actually produced. Intensive experimental investigation has elucidated details of how particular behaviors are initiated and shaped, following information from sensory input, through processing in the nervous system, to the production of movements that allow an animal to find food or a mate or to flee a potential predator.

One emphasis in Part II is on the properties of single cells that allow them to perform their particular tasks and to work together effectively and in harmony. Another emphasis is on the mechanisms that coordinate cellular function into larger levels of organization that enhance an animal's overall fitness.

CHAPTER
5

THE PHYSICAL BASIS
OF NEURONAL FUNCTION

Animal activity depends on the precisely coordinated performance of many individual cells. Perhaps the most important cells for producing this coordination are nerve cells, called **neurons,** which communicate information using a combination of electrical and chemical signals. The membranes of most neurons are *electrically excitable;* that is, signals are generated in and transmitted along them without decrement as the result of the movement of charged particles (**ions**). The properties of electrical signals allow neurons to carry information rapidly and accurately to coordinate actions involving many parts, or even all, of an animal's body. All of the neurons in an organism's body, along with supporting cells called **glial cells** (or neuroglia) make up the **nervous system,** which collects and processes information, analyzes it, and generates coordinated output to control complex behaviors.

Although the complete patterns of neuronal activity that underlie behavior are understood only for some very small and relatively simple circuits (see Chapter 11), neurons themselves are among the most thoroughly studied of all cell types for several reasons. First, neurons transmit information electrically, which allows scientists to monitor the activity of individual neurons by using various instruments originally developed for the physical sciences. These measurement techniques, some of which are described in Chapter 2, have facilitated research on neurons and enormously increased the information available about them. Second, recordings of electrical activity in neurons have revealed that the properties of individual neurons from nearly all animals are very similar. In other words, the mechanisms involved in transmitting information along and between neurons is essentially the same whether the neurons come from an ant or from an anteater. Finally, neurons process information in a highly sophisticated manner, but in doing so they rely on a surprisingly small number of physical and chemical processes, making it possible to formulate general principles about their function. In this chapter, we introduce the physical and molecular mechanisms that allow neurons to function so effectively in acquiring and transmitting information.

OVERVIEW OF NEURONAL STRUCTURE, FUNCTION, AND ORGANIZATION

Neurons have evolved specialized properties that allow them to receive information, process it, and transmit it to other cells. These functions, which are reflected in the overall shape and size of neurons, are performed by identifiable and anatomically distinct regions of the cell, characterized by specializations within the membrane and in the subcellular architecture. Although neurons vary greatly in shape and size, every neuron typically has a **soma,** or cell body, which is responsible for metabolic maintenance of the cell and from which emanate several thin processes (Figure 5-1). There are two main types of processes: dendrites and axons. Most neurons possess multiple dendrites and a single axon.

Dendrites generally are branched, extend from the cell body, and serve as the receptive surface that brings signals from other neurons *toward* the cell body. Information from other neurons is usually transmitted to the dendrites and soma of a neuron, so neurons with an extensive and complex dendritic tree typically receive many inputs. The location and branching pattern of dendrites can reveal from where each neuron gets its information.

Axons (also called *nerve fibers*) are specialized processes that conduct signals *away* from the cell body. Although many neurons have relatively short axons, the axons of some nerve cells extend surprisingly long distances. Axons have evolved mechanisms that allow them to carry information for long distances with high fidelity and without loss. In a whale, for example, the axon of a single **motor neuron** (or motoneuron), which carries information from the nervous system out to muscle fibers, may extend many meters from the base of the spine to the muscles that it controls in the tail fin. (Notice that bundles of axons running through the tissues of an animal's body are called **nerves.**) At its termination, each axon may divide into numerous branches, allowing its signals to be sent simultaneously to many other neurons, to glands, or to muscle fibers (see Figure 5-1).

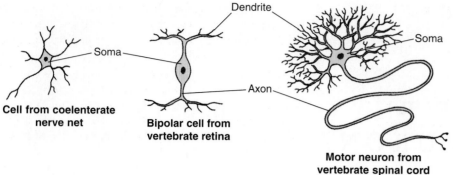

Dendrite

Soma

Soma

Axon

**Cell from coelenterate
nerve net**

**Bipolar cell from
vertebrate retina**

**Motor neuron from
vertebrate spinal cord**

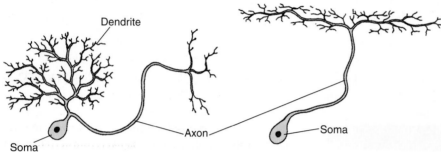

Dendrite

Soma

Axon

Soma

Two different insect neurons

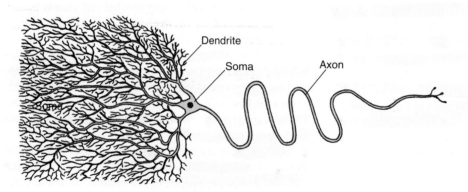

Dendrite

Soma

Axon

Soma

Purkinje cell from mammalian cerebellum

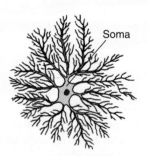

Soma

**Association cell from
mammalian thalamus**

Figure 5-1 The general morphology of neurons varies from simple to very complex, but most neurons have certain identifiable regions, including dendrites, a soma, and an axon. Notice that there is little correlation between phylogeny and the complexity of neuronal structure. Although simple animals have simple neurons (e.g., the coelenterate neuron), some neurons in higher animals also have a simple structure (e.g., the vertebrate retinal bipolar cell). Higher animals also have neurons with a very complex structure (e.g., the Purkinje cell from the mammalian cerebellum), but so do insects and other simpler animals. In some neurons (e.g., cerebellar Purkinje cells and vertebrate motor neurons), the dendrites and the axon are easily distinguished. In other neurons (e.g., retinal bipolar cells and mammalian association cells), no morphologic features readily distinguish the axon from the dendrites.

During the embryonic development of each neuron, dendrites and the axon grow out from the soma. Throughout the life of the cell, the maintenance of processes depends on the slow, but steady, flow down the processes of proteins and other constituents that are synthesized in the soma. If an axon in an adult organism is badly damaged or severed, it typically degenerates within a few days or weeks. In mammals, regeneration or regrowth of axons is limited to nerves in the periphery of the body, whereas in cold-blooded vertebrates, some regeneration also may occur within the central nervous system (i.e., the brain and spinal cord). Damaged neurons in some invertebrates readily regenerate and reinnervate their original targets.

Transmission of Signals in a Single Neuron

A typical vertebrate spinal motor neuron, which has its soma in the spinal cord and carries signals to skeletal mus-

cle fibers, displays the structural and functional features that characterize many neurons (Figure 5-2). The surface membrane of motor-neuron dendrites and the membrane of the soma receive signals from, or are *innervated by,* the terminals of other nerve cells. Typically, but not always, the soma integrates (sums) messages from the many input neurons to determine whether the neuron will initiate its own response, an **action potential** (AP), also called a *nerve impulse.* The axon carries an AP from its point of origin, the **spike-initiating zone,** to the **axon terminals,** which contact skeletal muscle cells in the case of motor neurons. Typically the spike-initiating zone is located near the **axon hillock,** the junction of the axon and cell body, although in many neurons the spike-initiating zone lies elsewhere. Many axons are surrounded by supporting cells, which provide an insulating layer called the **myelin sheath** (see Chapter 6).

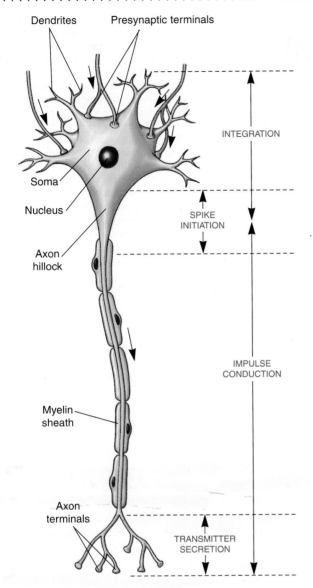

Figure 5-2 A vertebrate spinal motor neuron exemplifies the functionally specialized regions of a typical neuron. The flow of information is indicated by small black arrows. Information is received and integrated by the membrane of the dendrites. In some neurons the soma also receives information and contributes to integration. In spinal motor neurons, action potentials are initiated at the spike-initiating zone, in or near the axon hillock, and then travel along the axon to the axon terminals, where a chemical neurotransmitter is released to carry the signal on to another cell. In other types of neurons, the spike-initiating zone may have a different location. The axon and surrounding myelin sheath cells are shown in longitudinal section.

The physiological behavior of a neuron depends on the properties of its surface membrane. All entities that can carry electrical signals, including wires and cell membranes, possess *passive electrical properties* such as capacitance and resistance. Unlike wires, the membranes of neurons also possess *active electrical properties* that allow these cells to conduct electrical signals without decrement. The active electrical responses of neurons and other excitable cells depend on the presence of specialized proteins, known as **voltage-gated ion channels,** in the cell membrane.

Voltage-gated ion channels allow ions to move across the cell membrane in response to changes in the electrical field across the membrane. Neurons possess various types of ion channels, each permitting passage of a particular ionic species. The different types of ion channels found in neurons are not distributed uniformly over the surface of the cell. Instead, they are localized in different regions that have specialized signaling functions. For example, the axonal membrane is specialized for the conduction of APs by virtue of its fast-acting, voltage-gated ion channels that selectively allow Na$^+$ and K$^+$ to cross the membrane. In addition, the membrane of axon terminals contains voltage-gated Ca^{2+} channels and other specializations that allow neurons to transmit signals to other cells when APs invade the terminals.

Transmission of Signals between Neurons

Information processing by any nervous system begins when *sensory neurons* collect information and send it to other neurons. The axon of an information-gathering neuron is called an **afferent fiber.** Sensory neurons pass information on to other neurons, and the signal is transferred from neuron to neuron in the animal's nervous system. **Interneurons** lie entirely within the central nervous system and carry information between other neurons. Information is passed between neurons, or between neurons and other target cells, at specialized locations called **synapses.** Eventually, if the animal is to respond in any way to the sensory information, signals must activate neurons that control effector organs, such as muscles or glands. Neurons that carry information out to effectors are called **efferent neurons.** Together the afferent neurons and efferent neurons, along with any interneurons that participate in processing the information, make up a **neuronal circuit** (Figure 5-3).

A cell that carries information *toward* a particular neuron is said to be **presynaptic** to that neuron, while a cell that receives information transmitted across a synapse *from* a particular neuron is said to be **postsynaptic** to that neuron. Synaptic transmission affects the postsynaptic neuron via mechanisms discussed in Chapter 6. Briefly, most synaptic transmission is carried by **neurotransmitters,** which are specific molecules released by the axon terminals of the presynaptic neuron in response to APs in the axon. The membrane of the postsynaptic neuron's dendrites and soma, the part of the postsynaptic cell that typically receives synaptic signals, contains **ligand-gated ion channels,** which bind neurotransmitters and react to them. The postsynaptic effects of the numerous synaptic inputs are integrated to produce a net postsynaptic potential in the dendrites, soma, and axon hillock. As indicated in Figure 5-4, information is carried in neuronal circuits via alternating graded and all-or-none electrical signals.

Organization of the Nervous System

The nervous system is composed of two basic cells types: neurons and supporting glial cells. As we have seen, neurons are classified functionally into three types:

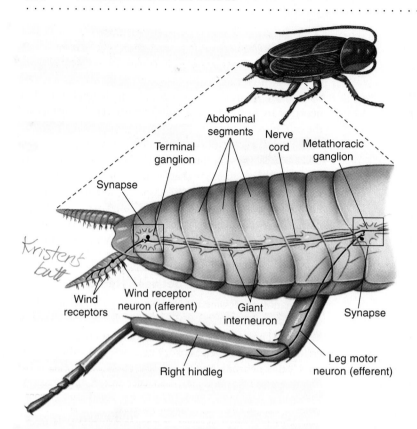

Figure 5-3 In a simple neuronal circuit, an afferent neuron carries sensory information to interneurons in the central nervous system, and an efferent neuron carries the processed information to effector organs. This figure of the posterior of a cockroach illustrates the neuronal circuit consisting of wind-receptor (afferent) neurons in the tail, giant interneurons in the central nervous system, and motor (efferent) neurons controlling the muscles of the legs. The wind-receptor neurons contact the giant interneurons at synapses in the terminal ganglion of the nervous system, and the giant interneurons contact the leg motor neurons at synapses in the thoracic ganglia. Stimulation of the wind-receptor neurons causes the cockroach to run away from the stimulus.

- Sensory neurons, which transmit information collected from external stimuli (e.g., sound, light, pressure, and chemical signals) or respond to stimuli inside the body (e.g., the blood oxygen level, the position of a joint, or the orientation of the head)

- Interneurons, which connect other neurons within the central nervous system

- Motor neurons, which carry signals to effector organs, causing contraction of muscles or secretion by gland cells

The essential function of sensory neurons is to transform the physical energy available in a stimulus into the electrical signals used by the nervous system. Networks of interneurons, which are the most numerous type of neuron, exchange information and perform most of the complex computations that produce behavior. Motor neurons constitute the output portion of a neuronal circuit, carrying specific instructions to muscles or to other effector organs that they innervate. (Neurons that innervate gland cells and other effector targets are also called "motor" cells, even though they do not control bodily movement.)

Neurons are grouped into clusters in almost all phyla. Typically, the cell bodies of many—even most—neurons are contained within the central nervous system (CNS), although the cell bodies of some neurons are located in the periphery. In most animals, the CNS consists of a brain, located in the head, and a nerve cord, which extends posteriorly along the midline of the animal. Many invertebrates have a brain located in the head and collections of neuronal cell bodies, called **ganglia,** distributed along the nerve cord; these control local regions of the animal (Figure 5-5). Vertebrates also have ganglia, consisting of peripheral neuronal cell bodies, outside the CNS. In vertebrates the nerve cord, called the **spinal cord,** is located along the dorsal midline, whereas in many invertebrates (e.g., insects, crustaceans, and annelids), the major nerve cord lies along the ventral midline. Many neurons that have cell bodies in the CNS send processes into the rest of the body (the periphery) to collect sensory information or to deliver motor information to control the activity of muscles or glands.

The other main cell type in the nervous system, glial cells, fills all of the space between neurons, except for a very thin extracellular space (about 20 nm wide) that separates the glial and neuronal membranes. In general, more complex animals have a larger number of glial cells relative to neurons than do less complex animals. The vertebrate central nervous system, for instance, contains 10–50 times more glial cells than neurons, and these cells occupy about half the volume of the nervous system. The proportion of glial cells to neurons is considerably smaller in most invertebrates.

Glial cells provide intimate structural, and perhaps metabolic, support for neurons. Several different cell types function as glial cells. In vertebrates, for example, **oligodendrocytes** in the CNS and **Schwann cells** in the periphery wrap each axon in an insulating myelin sheath, which contributes to assuring reliable and rapid transmission of APs (see Figure 5-2).

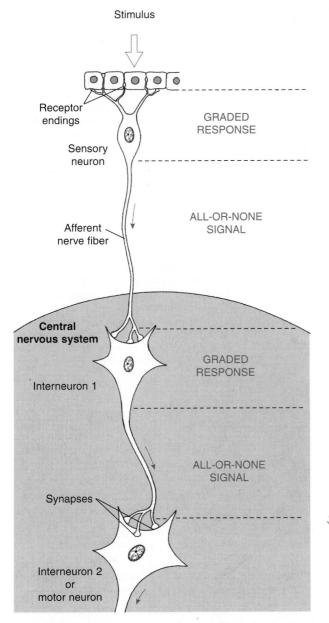

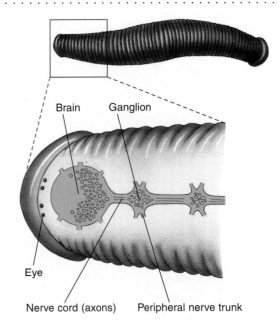

Dorsal view — Anterior end

Figure 5-5 The central nervous system, typically consisting of the brain and nerve cord, is the site of most information processing and usually contains most of an animal's neuronal cell bodies as well as many axons. The brain, usually in the head of the animal, contains a large number of neurons and their interconnections. In many animals, such as the medicinal leech *Hirudo medicinalis* shown in this figure, the somata of other neurons are grouped within the nerve cord, in structures called ganglia. In a segmented animal like the leech, each segment commonly contains a ganglion. Nerves containing axons connect structures within the CNS and connect the CNS to peripheral structures.

Figure 5-4 Information is typically carried through a neuronal circuit via electrical action potentials and chemical synaptic potentials. Colored arrows indicate the direction of information flow. In response to a stimulus, the receptor endings of the sensory neuron produce a graded response that is proportional to the stimulus intensity, although it is neither linearly proportional to the stimulus nor identical in shape. This graded potential spreads to a spike-initiating zone, where it may elicit one or more all-or-none signals (action potentials, or APs), which are propagated in the axon. When these signals arrive at the terminals, they cause the release of a chemical neurotransmitter substance from the presynaptic cell. The neurotransmitter produces a graded potential in the next (postsynaptic) neuron. If the change in membrane potential in the postsynaptic neuron is large enough, one or more all-or-none APs will occur in the postsynaptic neuron. Thus, graded and all-or-none electrical signals alternate in the pathway. A single pathway through the central nervous system may include only a few or a large number of neurons connected by synapses. The sensory neuron is presynaptic to interneuron 1; interneuron 2 (or motor neuron) is postsynaptic to interneuron 1. Thus interneuron 1 is both postsynaptic and presynaptic to other neurons.

Although some glial cells have voltage-gated ion channels in their membranes, glial cells generally do not produce APs, and their role in the nervous system has long been a puzzle. One suggestion has been that glial cells help to regulate the concentration of K^+ and the pH in the extracellular fluid of the nervous system. Glial cell membranes are highly permeable to K^+, and adjacent glial cells are often electrically coupled by junctions that allow K^+ to flow between them. This flux permits glial cells to take up and redistribute extracellular K^+, which otherwise could build up to high concentrations in narrow extracellular spaces following activity in neurons. Glial cells also may take up neurotransmitter molecules from the extracellular space, thereby limiting the amount of time a neurotransmitter could be active at synapses. Research continues into the function of glial cells, and it seems likely that even more roles will be discovered for these important components of the nervous system.

MEMBRANE EXCITATION

Although a stable **voltage** (or potential difference) exists across the plasma membrane of all animal cells, only the membranes in electrically excitable cells (e.g., neurons, muscles, and sensory cells) can respond to changes in the potential difference by generating APs. The study of electrically

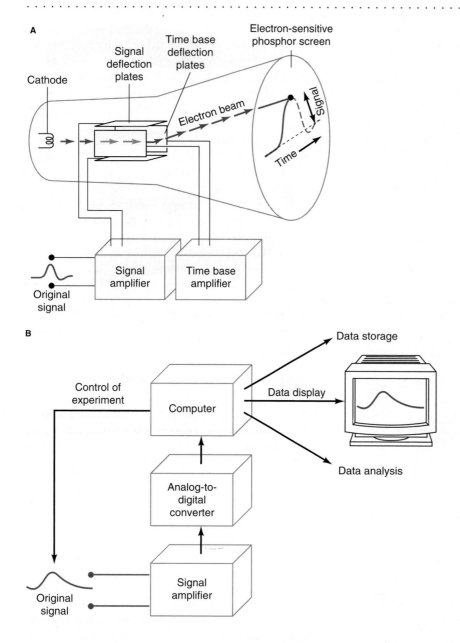

Figure 5-6 Potential differences measured across cell membranes are typically amplified and displayed using one of two types of electrical instrumentation. **(A)** An oscilloscope amplifies and plots the amplitude of an electrical signal in the vertical displacement of a beam of electrons emitted by a cathode ray tube; the passage of time is indicated along the horizontal axis. The beam of electrons is driven from left to right by a time base generator, and its position is visualized as it "writes" on the phosphor screen. Thus, the size of a signal fed into the oscilloscope is plotted on the screen as a function of time. **(B)** A digital computer can be used to simulate the essential functions of an oscilloscope with the advantage that data can be digitized and stored directly as they are recorded. An analog signal sensed by electrodes placed across a cell membrane is converted to digital information by sampling the signal at discrete time intervals that are sufficiently short to capture all the necessary information. Once the data are stored in the computer, they can be readily displayed or processed as required. Computers also can be used to *control* many aspects of an experiment, including the pattern of stimulation to cells.

excitable tissues has a long history, which is reviewed in Spotlight 5-1. To understand both the basis for this excitability and its consequences for how a neuron functions, we need to be able to measure the details of the electrical changes across the membrane as a function of time.

Measuring Membrane Potentials

Electric currents are generated in living tissues whenever there is a *net flux of charged particles* across the membrane. Such currents can be detected directly by using two electrodes to measure the change in electrical potential that is caused by current flow across the cell membrane. One sensing electrode is placed in electrical contact with the fluid inside the cell, and the other is placed in contact with the extracellular medium, so the two electrodes indicate the voltage, or potential difference, between the cytosol and the extracellular fluid. The potential difference measured

across the cell membrane (the **membrane potential**, V_m) is electronically amplified and displayed on a recording instrument. Until recently, recordings of membrane potentials were typically visualized using an oscilloscope, but researchers now rely extensively on digital computers to control equipment and display data during experiments, as well as to analyze and store data later (Figure 5-6).

Much of what we know about how APs are generated rests on experiments carried out by A. L. Hodgkin and A. F. Huxley in the 1940s and 1950s. They recorded membrane potentials from squid axons, which are large enough that a silver wire can be inserted and run longitudinally down the inside of this cylindrical process (Hodgkin and Huxley, 1952). The electrical activity of neurons that are smaller than the squid giant axon has been studied using glass capillary microelectrodes, which are described in Chapter 2. A cell is not damaged when it is impaled by a microelectrode because the lipid bilayer of the membrane seals itself around the pipette tip after penetration. Insertion of the electrode tip through the plasma membrane into the cell brings the cell interior into electrical continuity with the voltage-recording amplifier. By convention, the membrane potential is always taken as the intracellular potential relative to the extracellular potential. In other words, the extracellular potential is arbitrarily defined as zero. In practice the amplifier subtracts the extracellular potential from the intracellular potential to give the potential difference.

A simple arrangement for measuring membrane potential is shown in Figure 5-7. In this experiment, the neuron is immersed in a physiological saline solution that is in contact with a *reference electrode*. Before the tip of the *recording microelectrode* enters the cell, the microelectrode and the reference electrode are both in the saline solution and are therefore at the same electrical potential; the potential difference recorded between the two electrodes is zero (see Figure 5-7A). As the tip of the microelectrode penetrates the cell membrane and enters the cytosol, a downward shift of the voltage trace suddenly appears, indicating that the electrode has entered the cell and is now measuring the potential difference across the cell membrane (see Figure 5-7B). By electrophysiological convention, inside-negative potentials are shown as *downward* displacements of the oscilloscope trace. The steady inside-negative potential recorded by the electrode tip after it enters the cytosol

is the **resting potential**, V_{rest}, of the cell membrane and is most conveniently expressed in millivolts (mV, or thousandths of a **volt**). This resting potential is the membrane potential measured when no active events or postsynaptic events are occurring. Virtually all cells that have been investigated have a negative resting potential with a value between −20 mV and −100 mV.

The potential sensed by the intracellular electrode does not change as the tip is advanced further into the cell, because in the steady state the interior of the cell has the same potential everywhere. Thus, the entire potential difference between the cell's interior and exterior is localized across the surface membrane and in the regions immediately adjacent to the inner and outer surfaces of the membrane. This potential difference constitutes an electrical gradient that is available as an energy source to move ions across the membrane. The electric field, E, equals the voltage in volts divided by the distance, d, in meters ($E = V/d$). Since d, the thickness of the membrane, is approximately 5 nm (5×10^{-9} m), the actual electric field across cell membranes is very large.

Distinguishing Passive and Active Membrane Electrical Properties

As noted earlier, the membranes of neurons and other excitable cells display both passive and active electrical properties. To understand the basis of electrical integration and

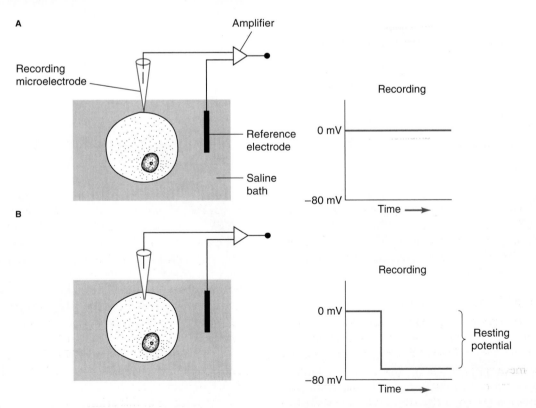

Figure 5-7 When a microelectrode penetrates a neuron's cell membrane, there is a shift in the recorded potential difference. **(A)** No potential difference is recorded between the reference electrode and the recording microelectrode when both are in the saline bathing the neuron. **(B)** As soon as the microelectrode tip penetrates the neuron's membrane, the electrode records an abrupt shift in the negative direction, which is shown as a downward deflection on the oscilloscope screen or the computer monitor. This deflection corresponds to the resting potential across the membrane.

SPOTLIGHT 5-1

THE DISCOVERY OF "ANIMAL ELECTRICITY"

Electrical excitability is a fundamental property of neurons and muscles. Today we understand this phenomenon in great detail, but electrically excitable animal tissues have been studied for centuries. Both the study of "animal electricity" and the origin of electrochemical theory can be traced to observations made late in the 18th century by Luigi Galvani, a professor of anatomy at Bologna, Italy. Working with a nerve-muscle preparation dissected out of a frog leg, Galvani noticed that the muscles contracted if the nerve and muscle were touched by metal rods in a particular way. The two rods had to be made of different metals (e.g., one of copper and the other of zinc). In Galvani's experimental setup, one rod contacted the muscle and the other contacted the nerve to that muscle; when the two rods were brought together, the muscle contracted, as illustrated in diagram A. Galvani and his nephew Giovanni Aldini, a physicist, ascribed this response to a discharge of "animal electricity" that was stored in the muscle and delivered by the nerves. They hypothesized that an "electrical fluid" passed from the muscle through the metal and back into the nerve, and that the discharge of electricity from the muscle triggered the contraction. We now understand that this creative interpretation, which was published in 1791, is largely incorrect. Nevertheless, this work stimulated many inquisitive amateur and professional scientists of the time to investigate two new and important areas of science: the physiology of excitation in nerve and muscle and the chemical origin of electricity.

Alessandro Volta, a physicist at Pavia, Italy, took up Galvani's experiments. In 1792 he proposed an alternative explanation for Galvani's results. He suggested that the electrical stimulus causing the muscle to contract in Galvani's experiments was generated outside the tissue by the contact between dissimilar metals and the saline fluids of the tissue. It took several years for Volta to demonstrate unequivocally the electrolytic origin of electrical current from dissimilar metals, because at that time no physical instrument was available that was sufficiently sensitive to detect these weak currents. Indeed, the nerve-muscle preparation from the frog leg was probably the most sensitive indicator of electric current available at that time and for many years afterward.

In his search for a method to produce stronger electrical currents, Volta found that he could increase the amount of electricity produced electrolytically by placing metal-and-saline cells in series. The fruit of his labor was the so-called voltaic pile, a stack of alternating silver and zinc plates separated by saline-soaked papers. This first "wet-cell" battery produced higher voltages than can be produced by a single silver-zinc cell, and the design is still commonly used in today's batteries.

Although Galvani's original experiments did not really prove the existence of "animal electricity," they did demonstrate that some living tissues can respond to minute electric currents. In 1840, Carlo Matteucci advanced the study of electricity in tissue by using the electrical activity of a contracting muscle to stimulate a second nerve-muscle preparation (diagram B). His experiment was the first recorded demonstration that excitable tissue actually produces electric current. Since the 19th century, it has become evident that the production of signals in the nervous system and in other excitable tissues depends on the electrical properties of cell membranes.

A

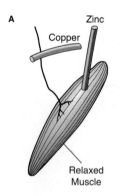

Copper / Zinc

Relaxed Muscle

Contracted Muscle

B

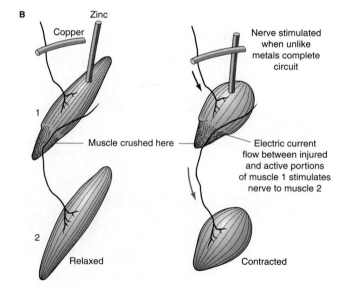

Copper / Zinc

Nerve stimulated when unlike metals complete circuit

Muscle crushed here

Electric current flow between injured and active portions of muscle 1 stimulates nerve to muscle 2

1

2

Relaxed

Contracted

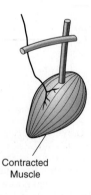

In Galvani's experiments **(A),** a muscle and its nerve were contacted by rods of two different metals, such as copper and zinc. When the two rods were touched together, the muscle contracted. **(B)** Matteucci elaborated on Galvani's experiment by connecting one muscle-nerve preparation (1) to a second one (2). When muscle 1 was stimulated by making the dissimilar metal rods contact each other, the electrical activity of muscle 1 stimulated the nerve to muscle 2, causing it to contract. In this experiment, muscle 1 had to be injured at the point of contact with the nerve from muscle 2 in order for the ionic currents flowing in the fibers of muscle 1 to stimulate the nerve.

signaling in neurons, both their passive and active electrical properties must be measured.

The passive electrical properties of the cell membrane can be measured by passing a pulse of current across the membrane to produce a slight perturbation of the membrane potential. To do this, two microelectrodes are inserted into one cell, as illustrated in Figure 5-8. One, the *current electrode,* delivers a current that can be made to flow across the membrane in either the inward (bath-to-cytosol) or the outward (cytosol-to-bath) direction, depending on the polarity (direction) of the electric current delivered from the electrode. The other electrode, the recording electrode, records the effect of this current upon V_m. Note that all the current carried in solution and across the membrane is in the form of migrating ions. By convention, the flow of ionic current is from a region of relative positivity to one of relative negativity and corresponds to the direction of cation migration. Thus, if the electrode is made positive, this current will, by definition, flow directly from the current electrode *into* the cell and out of the cell through its membrane. Conversely, if the electrode is made negative, it will draw positive charge out of the cell and cause current flow into the cell across the membrane; this situation is depicted in Figure 5-8A.

When a current pulse removes positive charge from inside the cell through the current electrode (i.e., when the current electrode is made more negative), the negative interior of the cell becomes still more negative; this increase in the potential difference across the cell membrane is called **hyperpolarization.** For example, the magnitude of the intracellular potential may change from a resting potential of -60 mV to a new hyperpolarized potential of -70 mV. Neuronal membranes generally respond passively to hyperpolarization, producing no response other than the change in potential caused by the applied current (Figure 5-9, traces 1 and 2). As discussed in the next section, the change in membrane potential that accompanies the passage of an applied current is given by Ohm's law (equation 5-1), which relates the potential in volts to the magnitude of the current in amperes flowing across the membrane and the membrane's resistance in ohms. Thus the passive electrical properties of the membrane can be represented by conventional electrical circuit elements.

When a current pulse adds positive charge to the interior of the cell through the current electrode, the additional positive charges will diminish the potential difference across the membrane, causing **depolarization** of the membrane (Figure 5-9, trace 3). That is, the intracellular

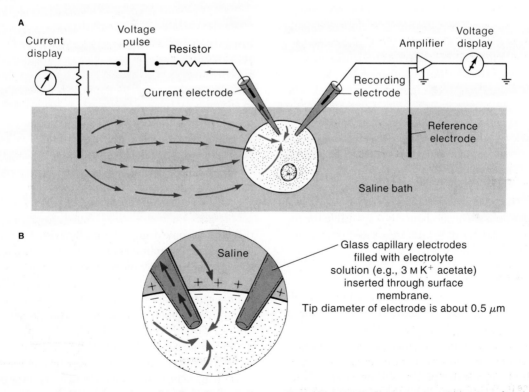

Figure 5-8 Impaling a cell with two microelectrodes allows the passive electrical properties of the membrane to be measured. **(A)** Diagram of an experimental setup to measure passive membrane properties. Current flows in a circuit through the wires, bathing saline, resistor, current electrode, and cell membrane. To maintain constancy of the stimulating current, the resistor is selected to have a far greater resistance than the other elements of the stimulating circuit. The recording amplifier has a very high input resistance, preventing any appreciable current from leaving the cell through the recording electrode. **(B)** Magnified view of the glass capillary microelectrodes inserted through the membrane of the cell. The electrode at the left is used to pass current into, or out of, the cell. The current changes the membrane potential, V_m, as it crosses the plasma membrane.

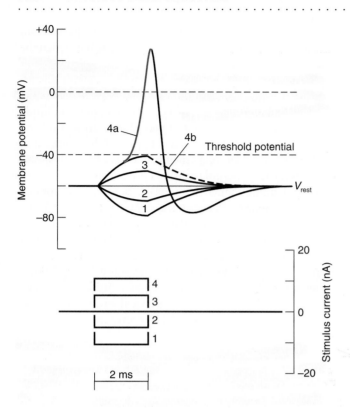

Figure 5-9 Some stimuli evoke only passive membrane responses; other stimuli may evoke active responses as well. Shown here are typical records from an experiment in which different amounts of current are passed into an excitable cell using the experimental setup depicted in Figure 5-8. The traces of the stimulus currents *(bottom)* and the corresponding changes in the membrane potential *(top)* are indicated by small numerals. If the introduced current causes the membrane potential to become more negative with respect to the outside of the cell, it is called hyperpolarizing (traces 1 and 2). If the introduced current causes the membrane potential to become more positive with respect to the outside of the cell, it is called depolarizing (traces 3 and 4). Hyperpolarizing currents and small depolarizing currents produce passive shifts in V_m (traces 1–3). With sufficient depolarization, an additional current enters the cell, producing a sudden active electrical response, the AP (trace 4a). Note that the size of the passive responses is more or less linearly proportional to the stimulus current, whereas in the active response the change in membrane potential is much larger than would be expected for a linear response. At threshold, some stimuli will produce APs, whereas other stimuli of the same magnitude will produce only a passive response (trace 4b).

potential becomes less negative (e.g., it may shift from −60 mV to −50 mV). As the amount of applied current is increased, the degree of depolarization will also increase. Depolarizing a neuron causes membrane channels that are selectively permeable to sodium ions (Na^+) to begin opening. When electrically excitable cells become sufficiently depolarized to open some critical number of Na^+-selective channels, an AP is triggered (Figure 5-9, trace 4a). The value of the membrane potential that triggers production of an AP is called the **threshold potential.** The opening of voltage-gated Na^+ channels in response to depolarization, and the resulting flow of Na^+ ions into the cell, is one example of membrane excitation. The opening of Na^+-selective channels reduces the electrical resistance across the membrane (or increases the conductance), which can allow more current to flow. The mechanisms responsible for the AP and other instances of membrane excitation are discussed in more detail later in this chapter.

Role of Ion Channels

Table 5-1 summarizes the properties of some ion channels that function in the passive and active electrical responses of neurons. The passive change in V_m that occurs in response to hyperpolarization or depolarization does *not* depend on opening or closing of ion-selective channels. Rather, the ionic current that produces passive electrical responses flows primarily through K^+-selective channels that are always open. These *resting potassium channels* are largely responsible for maintaining the resting potential, V_{rest}, across the cell membrane. These channels typically are uniformly distributed over the entire membrane of excitable cells.

The active change in V_m that occurs in excitable cells depends on the opening or closing *(gating)* of numerous ion-selective channels typically in response to depolarization of the membrane. The opening (or closing) of a population of ion channels controls the flow of an ionic current that is driven from one side of the membrane to the other by the electrochemical gradient for the ionic species permeating the channels. This gating of ion channels is the immediate cause for nearly all active electrical signals in living tissue. The simultaneous opening of many ion channels generates the currents that are measured across cell membranes. The voltage-gated ion channels that mediate active electrical responses may be localized to particular areas (e.g., the axon hillock or axonal membrane in neurons).

Most membrane channels allow one or a few species of ions to pass much more readily than all other ions; that is, they exhibit some degree of ion selectivity. The ion channels that make cell membranes excitable are named for the ionic species that normally moves through them. For example, Na^+ is the ion that normally moves through the fast-acting, *voltage-gated sodium channels,* which open in response to membrane depolarization, but certain other ions (e.g., lithium ions) also can pass through these channels. In addition to voltage-gated channels, which respond to changes in V_m, other ion channels are gated when messenger molecules (e.g., neurotransmitters) bind to receptor proteins on the cell surface. These ligand-gated channels are discussed in Chapter 6. Still other ion channels, found in sensory receptor cells, are gated by specific stimulus energies such as light (photoreceptors), chemicals (taste buds and olfactory neurons), or mechanical strain (mechanoreceptors). These channels are discussed in Chapter 7.

TABLE 5-1
Examples of ion channels found in axons

Channel	Current through channel	Characteristics	Selected blockers	Function
Resting K+ channel (open in resting axon)	$I_{K\,(leak)}$	Produces relatively high P_K of resting cell	Partially blocked by Tetraethylammonium (TEA)	Largely responsible for V_m
Voltage-gated Na+ channel	I_{Na}	Rapidly activated by depolarization; becomes inactivated even if V_m remains depolarized	Tetrodotoxin (TTX)	Produces rising phase of AP
Voltage-gated Ca2+ channel	I_{Ca}	Activated by depolarization but more slowly than Na+ channel; is inactivated as function of cytoplasmic [Ca2+] or V_m	Verapamil, D600, Co2+, Cd2+, Mn2+, Ni2+, La3+	Produces slow depolarization; allows Ca2+ to enter cell, where it can act as second messenger
Voltage-gated K+ channel ("delayed rectifier")	$I_{K\,(V)}$	Activated by depolarization but more slowly than Na+ channel; is inactivated slowly and not completely if V_m remains depolarized	Intra- and extracellular TEA, amino pyridines	Carries current that rapidly repolarizes the membrane to terminate an AP
Ca2+-activated K+ channel	$I_{K\,(Ca)}$	Activated by depolarization and elevated cytoplasmic [Ca2+]; remains open as long as cytoplasmic [Ca2+] is higher than normal	Extracellular TEA	Carries current that repolarizes the cell following APs based on either Na+ or Ca2+ and that balances I_{Ca}, thus limiting depolarization by I_{Ca}

PASSIVE ELECTRICAL PROPERTIES OF MEMBRANES

The capacitance and conductance of cell membranes, which account for their passive electrical responses, correspond to particular membrane structural elements. The lipid bilayer, which is impermeable to ions, acts as an insulator that separates charged ions, conferring on the membrane an electrical **capacitance.** Ion channels, through which electric charges cross the membrane, give the membrane its electrical **conductance.** These two membrane electrical properties can be represented by an equivalent circuit in which an electrical capacitor is connected in parallel with a resistor (Figure 5-10). The resistor represents the conductance of the ion channels, and the capacitor represents the capacitance of the lipid bilayer. Such an equivalent circuit is useful for making an explicit model of measured current flow through a membrane to determine likely values for membrane conductance and capacitance.

Membrane Resistance and Conductance

The **resistance** of a membrane is a measure of its impermeability to ions, whereas the conductance is a measure of its permeability to ions. For a given transmembrane voltage, the lower the resistance of the membrane (i.e., the greater its conductance), the more ionic charges will cross the membrane through ion channels per unit time. The relationship between current, resistance, and steady-state voltage across a membrane is described by **Ohm's law,** which states that the voltage drop produced across a membrane by a current passed across the membrane is directly pro-

portional to the current multiplied by the resistance of the membrane:

$$\Delta V_m = \Delta I \times R \qquad (5\text{-}1)$$

where ΔV_m is the voltage drop across the membrane, ΔI is the current (in **amperes**) across the membrane, and R (in **ohms, Ω**) is the electrical resistance of the membrane. The *input resistance* of a cell (i.e., the total resistance encountered by current flowing into or out of the cell) is a function of its membrane area, A, because the membrane of a larger cell will typically contain more ion channels than does the membrane of a smaller cell. Therefore, when membranes of different cells are compared, it is necessary to take into account the effect of membrane area. To do this, we define the *specific resistance*, R_m, of the membrane as

$$R_m = R \times A \qquad (5\text{-}2a)$$

where A is the membrane area and R_m is the resistance of a unit area of membrane. Rearranging Ohm's law gives

$$R = \frac{\Delta V_m}{\Delta I} \qquad (5\text{-}2b)$$

then substituting yields

$$R_m = \frac{\Delta V_m}{\Delta I} \times A \qquad (5\text{-}2c)$$

where $\Delta V_m/\Delta I$ is in ohms and area is in square centimeters; thus R_m is in units of ohms · cm^2. Note that membrane area and input resistance, R, are reciprocally related. The specific resistance, R_m, of the membrane is a property of the population of ion channels carrying ionic current across the membrane. Specific resistances of various cell membranes range from hundreds to tens of thousands of ohms · cm^2.

The reciprocal of resistance, R, is conductance, g (in units of **siemens**, S):

$$g = \frac{1}{R} \qquad (5\text{-}3a)$$

Substituting this expression into Ohm's law gives

$$\Delta V_m = \frac{\Delta I}{g_{input}} \qquad (5\text{-}3b)$$

The reciprocal of the specific resistance of a membrane is the *specific conductance*, g_m (in units of siemens · cm^{-2}). Conductance is closely related to the ionic permeability of the membrane, but is not strictly synonymous with it. Conductance for a given species of ion is defined by Ohm's law as the current carried by that species of ion divided by the electrical force acting on that species:

$$g_x = \frac{I_x}{emf_x} \qquad (5\text{-}3c)$$

where g_x is the membrane conductance for ion species X, I_x is the current carried by that species, and emf_x is the **electromotive force** (in volts) acting on that species. Although emf_x varies with membrane potential, it is not identical with membrane potential, as will be discussed later.

Even though a membrane may be permeable to ion X, the conductance, g_x, depends on the presence and concentration of this species in the solution, because the ion species cannot carry current unless it is actually present. In addition, the permeability of the membrane to nonelectrolytes does not contribute to conductance, because nonelectrolytes are uncharged and hence cannot carry current. Thus, the terms conductance and permeability are not strictly synonymous.

Membrane Capacitance

Although ions do not cross the lipid bilayer except through ion channels, they can interact across the membrane, which is a thin insulating layer, to produce a **capacitative current** (or charge displacement) even when no charges physically cross the membrane in the process. When a voltage is applied across a membrane, positive ions will tend to move towards the negative *(cathodal)* side from the positive *(anodal)* side in response to the force of the applied electric field. Because the ions cannot cross the lipid bilayer, they pile up on the two surfaces of the membrane, cations on the anodal side and anions on the cathodal side, producing a net excess of oppositely charged ions on opposite sides of the membrane. The membrane thus stores charges in the same way that charges are stored by a capacitor in an electrical circuit (Figure 5-11). Oppositely charged ions that

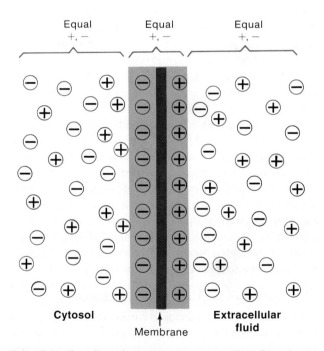

Figure 5-11 The cell membrane acts as a capacitor. The cell membrane can separate charges, with cations and anions forming a diffuse layer on opposite sides of the membrane and interacting electrostatically across the thin lipid barrier. This electrostatic interaction holds the charges in a narrow region immediately adjacent to the two surfaces of the cell membrane. Except for these few excess cations or anions on each side, the bulk solutions on either side of the membrane conform to the principle of electroneutrality, as does the shaded region in the neighborhood near the membrane.

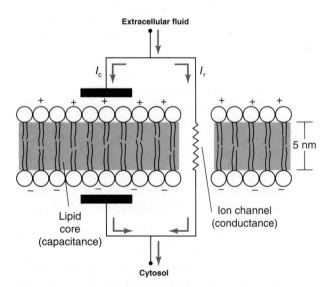

Figure 5-10 The passive electrical properties of a cell membrane can be represented by a simple electrical circuit. The membrane has a capacitance because the lipid core is an insulator. The membrane's ionic conductance depends on the presence of open ion channels. Arrows indicate flow of capacitative current, I_c, and resistive current, I_r.

have accumulated on the two sides of the membrane can interact electrostatically with one another because the membrane is so thin.

The ability of the bilayer to separate or store charges in this way is its capacitance, which is measured in units of coulombs per volt, or **farads** (F). The amount of charge that can be separated by a layer of insulating material depends on its thickness and on its **dielectric constant,** a property that reflects the ability of an insulator to store charge. It is possible to calculate an expected value for the capacitance of nerve cell membranes if the membrane thickness and dielectric constant of membrane lipids are known. Based on a lipid-layer thickness of about 5 nm and a dielectric constant of 3, which is about that of an 18-carbon fatty acid, the membrane capacitance has been calculated to be about 1 microfarad (μF) per square centimeter (1 μF = 10^{-6} F). Indeed, measured values of the capacitance of biological membranes have proven generally to be close to 1 μF$\cdot$cm^{-2}.

The capacitance of a membrane affects how it responds to a change in applied voltage or current. Positive ions piling up on one side of the membrane repel positive ions from the other side, generating a transient flow of capacitative current. Since the movement of charged particles near the membrane takes time, capacitance effectively limits how fast the voltage across a membrane can change under specified conditions. This effect can be illustrated by an equivalent circuit representing the neuronal membrane, as shown in Figure 5-12A. In this example, a steady current, I_m, of 1 ampere is applied suddenly to the membrane. According to equation 5-1a, such an applied current will produce a voltage drop across the membrane as shown in Figure 5-12B (trace V_m). This passive potential change, called an **electrotonic potential,** is produced by the current that passes across the cell membrane.

As shown in Figure 5-12A, the current passing across the membrane must be distributed between the resistive and the capacitative pathways, which are arranged in parallel across the membrane. When the steplike pulse of current, I_m, is forced across the membrane, the distribution of current between the membrane capacitance and resistance changes with time. At first, charges move relatively easily onto the capacitor component of the circuit, so most of the current is carried by the capacitance as it charges, and little current flows through the membrane resistance. The buildup of charge on the capacitor is one form of electrical current, although no charges physically move across the capacitor. Instead, the capacitative current, I_c, is carried by the displacement of charges onto, or off of, the opposite sides of the lipid bilayer. As time passes and the capacitance charges, a voltage appears across the capacitance, which slows the rate at which further charging occurs. This reduction in the rate at which charges can move onto the capacitor displaces an increasingly large fraction of current to flow through the membrane resistance. Hence, the capacitative current, I_c, *falls* along an exponential time course, while the membrane potential, V_m, *rises* with the same time

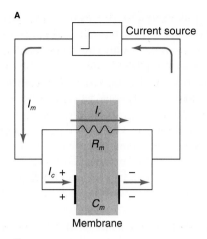

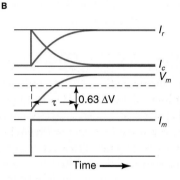

Figure 5-12 The capacitance and resistance of the cell membrane are responsible for the passive change in potential (electrotonic potential) in response to applied current. **(A)** Equivalent circuit for a cell membrane showing the current flow (colored arrows) when an abrupt, sustained pulse of constant current is passed across the membrane. **(B)** Time courses for the currents and voltage produced by the sustained stimulus shown in panel A. The top record shows how the total current is divided between the resistive current, I_r, and the capacitative current, I_c. The middle record shows the membrane potential, V_m, (i.e., the potential across both the membrane resistance and the membrane capacitance). The bottom record shows the total current across the membrane, $I_m = I_c + I_r$. The time required for I_r and V_m to reach 63% of their asymptotic values is proportional to the product of the resistance and the capacitance of the membrane. This product is the time constant, τ, of the membrane.

course (see Figure 5-12B). At the same time, the resistive current, I_r, which passes through the membrane conductance (i.e., through ion channels) increases exponentially, because the sum of I_r and I_c necessarily equals the total current applied.

The relationship between potential and time during the charging of the capacitance is given by the equation

$$V_t = V_\infty(1 - e^{-t/RC}) \qquad (5\text{-}4)$$

where V_∞ is the potential across the capacitor at time $t = \infty$ produced by a constant current applied to the network, t is the time in seconds after the onset of the current pulse, R is the resistance of the circuit in ohms, C is the capacitance of the circuit in farads, and V_t is the potential across the capacitor at any time, t.

When t is equal to the product RC, then $V_t = V_\infty(1 - 1/e) = 0.63\ V_\infty$. The value of t (in seconds) that equals RC is termed the **time constant** (τ) of the process. Note that τ is independent of both V_∞ and current strength. It is the time required for the voltage across a charging capacitor to reach 63% of the asymptotic value, V_∞ (see Figure 5-12B, V_m trace).

In summary, the passive electrical properties of the cell membrane are its equivalent resistance and capacitance, which are connected in parallel. Together they give the membrane a time-dependent response to changes in voltage; the size and time course of the response depends on the values of the resistance and the capacitance. To understand how these parameters influence changes in membrane potential, we need to examine the origin of potential differences across the membrane.

ELECTROCHEMICAL POTENTIALS

All electrical phenomena in neurons and other cells depend on the transmembrane potential difference, V_m. This voltage difference across the cell membrane is an **electrochemical potential** that arises from two features found in all eukaryotic cells. First, the concentrations of several ions inside the cell are different from the concentrations of those same ions in the fluids outside the cell, and these concentration gradients are maintained at the expense of metabolic energy. Second, the ion channels that span the membrane are selectively permeable to different ionic species. These two properties result in the transmembrane potential, V_m, which together with the electrical properties of the membrane that we have just discussed produces the signals neurons use in communication. In this section, we will examine how the electrochemical membrane potential arises and what determines its magnitude.

To begin, consider the situation depicted in Figure 5-13. A chamber containing 0.01 M KCl is divided into two compartments by a membrane permeable only to K^+ ions. Under these conditions, the voltage recorded across the membrane is zero. This hypothetical membrane is permeable to K^+ but *not* to Cl^-, so K^+ diffuses across the membrane without its *counterion* (the species of ion that counterbalances its positive charge, in this case Cl^-). However, the net flux of K^+ is zero, because the concentrations in the two compartments are equal, so on the average, for each K^+ ion that passes in one direction across the membrane, another will pass in the opposite direction. As a result, there will be no net flux of K^+ ions, and the transmembrane potential will be zero (see Figure 5-13A). If we now increase the concentration of K^+ in compartment I tenfold to 0.1 M, K^+ will diffuse from compartment I to compartment II, producing an increase in positive charge in the latter (see Figure 5-13B). A positive potential will thus quickly develop in compartment II, and the voltmeter will indicate a potential difference between the two compartments (see Figure 5-13C). This potential difference will reach an equilibrium value that will be maintained

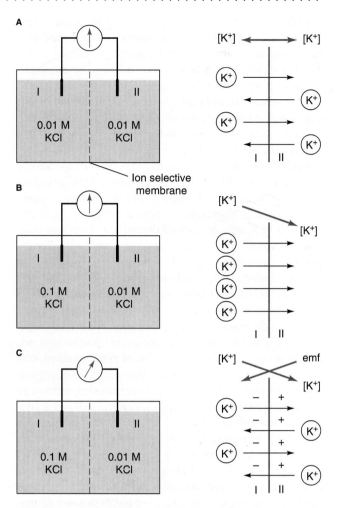

Figure 5-13 At electrochemical equilibrium, an electrical potential difference exactly balances the chemical concentration gradient. **(A)** A membrane that is permeable only to K^+ separates compartments I and II, each of which contains 0.01 M KCl. **(B)** When the KCl concentration in compartment I is increased to 0.1 M, there is a small net movement of K^+ into compartment II. **(C)** The electromotive force (emf) corresponding to the resulting charge separation acts to impede further movement of K^+. When the emf just balances the concentration gradient, the net movement of K^+ again becomes zero.

indefinitely, provided there is no leakage of Cl^- across the membrane.

How do we explain the maintenance of this potential difference? After we increase the KCl concentration in compartment I, for every K^+ that statistically is available for diffusion through K^+ channels from compartment II to compartment I, 10 K^+ ions in compartment I are available to pass across the membrane into compartment II. The difference in K^+ concentrations—a chemical gradient, or *chemical potential difference*—causes an initial net diffusion across the membrane from I to II. However, each additional K^+ that diffuses from compartment I to compartment II adds its positive charge to compartment II, because Cl^- cannot accompany K^+ across the membrane. As K^+ ions accumulate in compartment II, the potential difference across the membrane quickly rises, because the membrane then separates an excess of positive charges on one side from an excess of negative charges on the other (see Figure 5-11). As K^+ enters compartment II and produces this buildup of positive charge in that compartment, further movement of K^+ becomes more difficult, because the extra positive charges in compartment II repel additional positively charged ions, while the electrostatic attraction of the excess negative charges in compartment I holds K^+ back.

In this situation, every K^+ entering the membrane via a K^+ channel has two forces acting on it: (1) a *chemical gradient* favoring a net flux of K^+ from I to II, and (2) an *electrical potential difference* favoring a net flux of K^+ from II to I. After some time, the opposing forces come into equilibrium and remain balanced, with the electromotive force due to the electrical potential difference across the membrane precisely offsetting the tendency for K^+ to diffuse down its concentration gradient (see Figure 5-13C). At this point, K^+ is in **electrochemical equilibrium.** A potential difference across the membrane that is established in this way is called the **equilibrium potential** for the ion in question—in this case, the potassium equilibrium potential, E_K.

Once an ion is in electrochemical equilibrium, it shows no further *net* flux across the membrane, even if the membrane is freely permeable to that ion. Conversely, if an ion present in the system cannot diffuse across the membrane, its presence does not influence the equilibrium state. Thus, in our hypothetical system, Cl^- would contribute nothing to the membrane potential, even though it is far out of electrochemical equilibrium (its concentration gradient would favor movement from compartment I into compartment II), because it is unable to cross the membrane.

It is important to note that the process of establishing the equilibrium state involves the diffusion of only a very small number of ions across the membrane, from one compartment to the other, compared to the total number of ions present in solution. Virtually no change takes place in the *concentrations* of KCl in the two compartments during this process, because the number of K^+ ions that crossed into compartment II is insignificant compared to the number originally present in the solution. This situation is discussed further in Spotlight 5-2.

The Nernst Equation: Calculating the Equilibrium Potential for Single Ions

As the concentration gradient of an ion across a membrane increases, its equilibrium potential will also increase: a larger chemical gradient across the membrane requires a greater electrical potential difference across the membrane to offset the increased tendency for the ions to diffuse down their concentration gradient. In fact, the equilibrium potential is proportional to the *logarithm* of the ratio of the concentrations in the two compartments (for a brief review of logarithms, see Appendix 2). The relation between the chemical gradient and the electrical potential difference across a membrane at equilibrium was derived from the gas laws in the latter part of the nineteenth century by Walther Nernst. The **Nernst equation** states that the equilibrium potential depends on the absolute temperature, the valence of

the diffusible ion, and importantly, on the ratio of concentrations on the two sides of the membrane:

$$E_x = \frac{RT}{zF} \ln \frac{[X]_I}{[X]_{II}} \tag{5-5}$$

Here R is the gas constant, T is the absolute (Kelvin) temperature; F is the Faraday constant (96,500 coulombs/gram-equivalent charge); z is the valence of ion X; $[X]_I$ and $[X]_{II}$ are the concentrations (more accurately, the chemical activities) of ion X on sides I and II of the membrane; and E_x is the equilibrium potential for ion X (potential of side II minus side I). At a temperature of 18°C, for a monovalent ion, and converting from ln to log, the Nernst equation reduces to

$$E_x = \frac{0.058}{z} \log \frac{[X]_I}{[X]_{II}} \tag{5-6}$$

where E_x is expressed in volts. (At 38°C, which is approximately the body temperature of many mammals, the multiplication factor is 0.061/z.) Note that E_x will be positive if X is a cation and the ratio of $[X]_I$ to $[X]_{II}$ is greater than unity. The sign will become negative if the ratio is less than 1. Likewise, the sign will be reversed if X is an anion, rather than a cation, because z will be negative.

By convention, the electrical potential inside a living cell, V_i, is expressed relative to the potential outside the cell, V_o. That is, the membrane potential, V_m, is given as $V_i - V_o$, so the potential of the cell exterior is arbitrarily defined as zero. For this reason, when determining the equilibrium potential across the cell membrane, we place the extracellular concentration of the ion in the numerator and the intracellular concentration in the denominator of the concentration ratio. Applying the Nernst relation (equation 5-6), we can calculate the potassium equilibrium potential, E_K, in a hypothetical cell in which $[K]_o = 0.01$ M and $[K]_i = 0.1$ M:

$$\begin{aligned}
E_K &= \frac{0.058}{z} \log \frac{[K^+]_o}{[K^+]_i} \\
&= \frac{0.058}{1} \log \frac{0.01}{0.1} \\
&= 0.058 \times (-1) = -0.058 \text{ V} = -58 \text{ mV}
\end{aligned}$$

Note that E_K has a negative sign. The inside of the cell will become negative when a minute amount of K^+ leaks out of the cell, driven by the concentration gradient of K^+. The Nernst equation predicts a rise in the equilibrium potential difference of 58 mV when the concentration ratio of the permeating ion is increased by a factor of 10. When E_K is plotted as a function of log $[K^+]_o/[K^+]_i$, the relation has a slope of 58 mV per tenfold increase in the concentration ratio (Figure 5-14A). Equation 5-6 also implies that if the ion in question is a divalent cation (i.e., $z = +2$), the slope of the relation becomes 29 mV per tenfold increase in the concentration ratio.

SPOTLIGHT 5-2

A QUANTITATIVE CONSIDERATION OF CHARGE SEPARATION ACROSS MEMBRANES

It takes only a small number of ions diffusing across 1 cm^2 of the membrane in Figure 5-13B to allow the membrane potential to reach E_K. In fact, the actual number of excess ions that cross the membrane can be calculated for a system that contains only one diffusible ion. The number of excess K$^+$ ions that accumulate in compartment II (and hence the number of excess Cl$^-$ ions left behind in compartment I) depends on two factors: E_K and the capacitance, C, of the membrane. The charge, Q, that accumulates across a capacitor equals the capacitance times the voltage, V:

$$Q = C \times V$$

where Q is in **coulombs** (C), C is in farads (F), and V is in volts (V).

Biological membranes typically have a capacitance of about 1 μF·cm^{-2}. According to the Nernst equation (equation 5-6), the voltage at equilibrium, E_K, when the membrane separates a tenfold difference in the concentrations of a monovalent cation such as K$^+$ is 58 mV. By substituting these values in the above equation, we can calculate the coulombs of charge that diffuse across 1 cm^2 of a biological membrane when the membrane separates a tenfold difference in the concentrations of K$^+$:

$$Q = (10^{-6}\,F \cdot cm^{-2})(5.8 \times 10^{-2}\,V)$$
$$= 5.8 \times 10^{-8}\,C \cdot cm^{-2}$$

There is one Faraday of charge (= 96,500 coulombs) in 1 gram-equivalent weight (1 mole) of a monovalent ion. Thus the number of moles of K$^+$ required to transfer 5.8×10^{-8} C of charge across 1 cm^2 of membrane is calculated as follows:

$$\frac{5.8 \times 10^{-8}\,C \cdot cm^{-2}}{9.65 \times 10^{4}\,C \cdot (mol\,K^+)^{-1}} = 6 \times 10^{-13}\,mol\,K^+\,per\,cm^2$$

The number of excess K$^+$ ions that have accumulated in compartment II at equilibrium in Figure 5-13C is found by multiplying the number of moles of K$^+$ by Avogadro's number (6×10^{23} molecules/mol):

$$(6 \times 10^{-13}\,mol\,K^+)(6 \times 10^{23}) = 3.6 \times 10^{11}\,K^+\,per\,cm^2$$

An equal number of Cl$^-$ ions remains in excess in compartment I. This number is more than 10,000,000 times smaller than the number of K$^+$ ions in a cubic centimeter of solution II (which you can calculate is 6×10^{18} K$^+$ ions per cm^3). So the concentrations in compartments I and II are essentially unchanged as a result of the charge separation across the membrane. Even though there is a slight separation of anions from cations across the membrane, the segregation exists only on a microscopic scale, separated by about the thickness of the membrane. Electroneutrality (i.e., an equal number of + charges and of − charges) is maintained on the macroscopic scale.

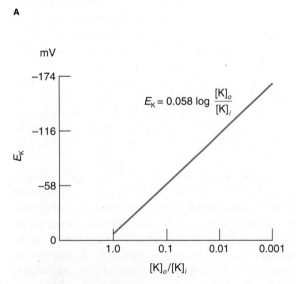

A

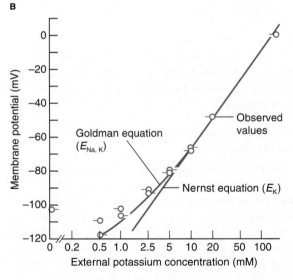

B

Figure 5-14 The concentration ratio of K$^+$ across the cell membrane affects both the calculated membrane potential and V_{rest} measured experimentally. **(A)** Semilog plot of the relationship between the equilibrium potential for K$^+$, E_K, and the ratio of K$^+$ concentrations on the two sides of a membrane, $[K]_o/[K]_i$, calculated from the Nernst equation. At 20°C, the slope of the line is 58 mV for every tenfold increase in the concentration ratio. **(B)** Semilog plots comparing dependence of the calculated $E_{Na,k}$ and the measured V_{rest} in frog muscle (observed values) on the ex-

ternal K$^+$ concentration. The values of $E_{Na,K}$ were calculated from the Goldman equation assuming P_{Na} is 1% of P_K, $[K^+]_i$ is 140 mm, and ignoring any contribution made by Cl$^-$. The straight line represents the predicted change in E_K as $[K^+]_o$ changes calculated from the Nernst equation. Note that both the relationship predicted by the Goldman equation and the experimental V_{rest} values deviate from this straight line at low $[K^+]_o$. At higher $[K^+]_o$, the calculated E_K and $E_{Na,K}$ are nearly identical. [Part B adapted from Hodgkin and Horowicz, 1960.]

The Goldman Equation: Calculating the Equilibrium Potential for Multiple Ions

If there is a concentration gradient across the membrane for a particular ion, the Nernst equation gives the equilibrium potential for that ion. However, the Nernst equation can be used for only one ionic species at a time. In contrast, essentially all cell membranes are permeable to several ionic species, all of which are distributed asymmetrically across their plasma membranes. If more than one ionic species can permeate the membrane, all of the permeating ionic species may contribute to setting the potential difference across the membrane. In these more complicated situations, the Nernst equation does not apply, and a different equation must be used to calculate a predicted equilibrium potential.

Based on the notion that the transmembrane potential is directly related to ion permeability, D. E. Goldman (1943) derived a quantitative representation of the membrane potential when more than one ionic species can cross the membrane. The **Goldman equation** can be considered as an approximate generalization of the Nernst equation that has been extended to include the relative permeability of each species of ion:

$$E_{\text{ions}} = \frac{RT}{F} \ln \frac{P_K[K^+]_o + P_{Na}[Na^+]_o + P_{Cl}[Cl^-]_i}{P_K[K^+]_i + P_{Na}[Na^+]_i + P_{Cl}[Cl^-]_o} \tag{5-7}$$

in which P_K, P_{Na}, and P_{Cl} are the permeability constants for the major ion species in the intracellular and extracellular compartments, and $[K^+]_o$ and $[K^+]_i$ indicate the concentrations outside and inside the cell, respectively.

In this equation, the probability that one ionic species will cross the membrane is taken to be proportional to the product of its concentration (more accurately, its thermodynamic activity) on that side and the permeability of the membrane to that ionic species. In frog muscle cells, the permeability constant for sodium is about $1/100$ that of potassium, and the membrane is nearly impermeable to chloride. In this situation, the Goldman equation can be simplified as follows:

$$E_{\text{Na,K}} = \frac{RT}{F} \ln \frac{1[K^+]_o + 0.01[Na^+]_o}{1[K^+]_i + 0.01[Na^+]_i}$$

If we then substitute into this equation the millimolar concentrations of K^+ and Na^+ inside and outside frog muscle cells, we get

$$E_{\text{Na,K}} = 0.058 \log \frac{2.5 + (0.01 \times 120)}{140 + (0.01 \times 10)}$$
$$= -0.092 \text{ V} = -92 \text{ mV}$$

If the membrane potential of frog muscle cells depends most heavily upon the diffusion of Na^+ and K^+, then the Goldman equation predicts that the value of V_m should be near -92 mV. Indeed, measured values for V_m in frog muscles do lie close to this value, but it is possible to test further the accuracy of the hypothesis that V_m depends on the diffusion of specific ionic species.

THE RESTING POTENTIAL

At equilibrium, every cell that is in a nonexcited or "resting" state has a potential difference, V_{rest}, across its membrane. Typically, V_{rest} lies between -30 and -100 mV, depending on the kind of cell and on the ionic environment. Two factors govern this potential: first, ion channels in the membrane that are permeable to some — but not necessarily all — of the ionic species present, and second, the unequal distribution of inorganic ions between the cell interior and cell exterior, maintained by active transport across the membrane and by the Donnan distribution (see Chapter 4). The unequal distribution of ions provides a chemical driving force allowing an equilibrium potential to be established.

Role of Ion Gradients and Channels

As predicted by the Goldman equation, ions influence the potential across a membrane roughly in proportion to the permeability of the membrane to each of the ionic species present. In the limiting case, if the membrane is impermeable to a particular ion species (as it is to the many large organic ions inside of cells), then the electrochemical gradient of that species has no effect on the membrane potential, because nonpermeating ions cannot carry charge from one side of the membrane to the other. Conversely, if a membrane is permeable to only one species of ion, the distribution of that ion will dictate the transmembrane potential, which can be predicted by using the Nernst equation for that ionic species. If a membrane is permeable to more than one ion — and most biological membranes are — and the permeabilities are known, it is possible to predict the value of V_{rest} using the Goldman equation. In the previous section, we calculated a predicted $E_{\text{Na,K}}$ of -92 mV using measured values for the permeabilities and ion concentrations in frog skeletal muscle fibers. In fact, measurements of V_{rest} in frog skeletal muscle fibers range from -90 to -100 mV, supporting the idea that the resting potential depends in large part on the diffusion of Na^+ and K^+ ions.

The contribution that a particular ionic species makes to the membrane potential diminishes as its concentration gradient is reduced. This point is illustrated in Figure 5-14B, in which values of the membrane potential calculated from the Goldman equation are plotted against the external K^+ concentration, assuming $P_{Na} = 0.01 P_K$. At high external K^+ concentrations, the slope of the plot is about 58 mV per tenfold increase in $[K^+]_o$, a value that is predicted by the Nernst equation for K^+. However, at low external K^+ concentrations the curve deviates from this slope because the product $P_{Na}[Na^+]_o$ approaches the value of the product $P_K[K^+]_o$, allowing Na^+ to make a more important contribution to the potential in spite of the low permeability of the membrane to Na^+. The measured values for V_{rest} of living frog muscle cells closely parallel the values calculated from the Goldman equation. It is interesting to note that these relationships apply equally well to neurons and muscle cells, suggesting conservation of important functional elements among excitable cells during the course of evolution.

> **?** Some aquatic animals do not control their internal osmolarity; instead the osmotic concentration of their internal fluids changes if they move into water with a different concentration. Would you expect V_{rest} of their cells to change when their internal fluids change? What would be required to keep V_{rest} constant?

Resting potentials of muscle, nerve, and most other cells have been found to be far more sensitive to changes in the $[K^+]_o$ than they are to changes in the concentrations of other cations. This experimental result is consistent with the relatively high permeability of cell membranes to K^+ as compared to other cations. The high permeability is thought to depend upon a set of K^+-selective channels that remain open in the resting membrane. Large changes in $[Na^+]_o$ have little effect on the resting potential, because the resting membrane is relatively impermeable to Na^+.

Role of Active Transport

The second factor contributing to V_{rest} is the asymmetric distribution of ions across cell membranes, which depends upon active transport of particular ions across the membrane. Because biological membranes are more or less leaky to solutes, cells must spend energy to maintain an asymmetric distribution of inorganic ions across their plasma membranes. Key ions are actively transported in the direction opposite to their movement down their concentration gradients.

Consider Na^+ in frog muscle. The concentrations of extracellular and intracellular Na^+ are about 120 mM and 10 mM, respectively. From these values we can calculate the sodium equilibrium potential, E_{Na}, from the Nernst equation (5-6) as follows:

$$E_{Na} = \frac{0.058}{1} \log \frac{120}{10}$$

$$= 0.063 \text{ V} = +63 \text{ mV}$$

Since V_m in frog muscle ranges from -90 to -100 mV, the sodium ions are more than -150 mV ($V_m - E_{Na}$) out of equilibrium. That is, there is a strong electrical force driving Na^+ into the cell. Even with only a small permeability to Na^+, there will be a steady influx of Na^+, driven by the large electric potential acting on that ion. If Na^+ were not removed from the cell's interior at the same rate at which it

leaks in, it would gradually accumulate in the cell. Such a rise in the intracellular Na^+ concentration would depolarize the cell; the resulting reduction in internal negativity would be less able to hold K^+ inside, and internal K^+ would leak out, moving down its concentration gradient. In fact, the high intracellular concentration of K^+ and low intracellular concentration of Na^+ are maintained by the action of a specific membrane protein, often called the **sodium pump**. This protein is a *Na$^+$/K$^+$ ATPase*, which transports Na^+ out of, and K^+ into, the cell at the cost of ATP hydrolysis. This active transport is not stoichiometrically balanced, however, because for each molecule of ATP hydrolyzed, three Na^+ ions are transported out of, and two K^+ ions are transported into, the cell.

The unequal stoichiometry of the sodium pump has some important consequences for V_{rest} (Figure 5-15). Because the pump produces a net transport of charge across the membrane, it is called **electrogenic** and could contribute to the membrane potential. The size of the pump's actual contribution to the value of V_{rest} depends on the rate at which charge, generally in the form of K^+ or Cl^- ions, can leak back across the cell membrane and partially compensate for the unequal numbers of charges transported into and out of the cell by the pump. The net effect of the pump would be to cause V_{rest} to be more negative than the equilibrium potential calculated, using the Goldman equation, for the highly permeant K^+ ions and less permeant Cl^-. However, it has been found that the sodium pump rarely makes

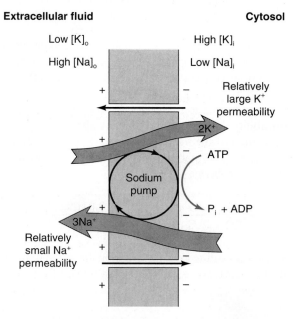

Figure 5-15 The Na$^+$/K$^+$ ATPase contributes to V_{rest} in two ways. The pump *indirectly* contributes to the resting potential by maintaining a high internal K^+ concentration. The major source of V_{rest} is the high permeability of the membrane to K^+ (relative to other ion permeabilities at rest), so that K^+ diffuses out of the cell until its exit is retarded by the excess negative charge it leaves behind in the cell. In addition, because the Na$^+$/K$^+$ ATPase transports Na$^+$ and K$^+$ in a ratio of 3Na:2K, it may also contribute *directly* to the resting potential by removing a small net amount of positive charge from the cell's interior.

a direct contribution to the value of V_{rest} that is more than several millivolts. This observation implies that some positive charge leaks back into the cell (or negative charge leaks out of the cell), partially offsetting the effect of the pump.

A second consequence of the unequal stoichiometry of the pump is that the passive fluxes of Na^+ and K^+ must be unequal. The influx of Na^+ through ion channels must equal the outward flux of Na^+ generated by the pump, and correspondingly, the efflux of K^+ through channels must match the influx of K^+ generated by the pump. The stoichiometry of the pump thus requires that the passive inward Na^+ current must be about 1.5 times the outward K^+ current, even though the passive permeability of the membrane to K^+ is larger than the passive permeability to Na^+. This discrepancy between Na^+ and K^+ ion fluxes can occur because the Na^+ equilibrium potential is so far from V_m that the large net electrical force acting on Na^+ ($V_m - E_{Na}$) drives more current through the smaller resting Na^+ conductance.

When metabolically driven sodium transport is eliminated by an inhibitor of oxidative metabolism (e.g., cyanide or azide) or by a specific inhibitor of sodium transport (e.g., ouabain), a net influx of Na^+ occurs and internal K^+ is gradually displaced. As a consequence, the resting potential decays as the ratio of $[K^+]_i$ to $[K^+]_o$ gradually decreases. Thus, over the long term, it is the metabolically energized transport of Na^+ and of K^+ that keeps the Na^+ and K^+ concentration gradients from running downhill to a final equilibrium where $V_m = 0$. By continuous maintenance of the potassium concentration gradient, the sodium pump plays an important *indirect* role in determining the resting potential.

To summarize, the major portion of the negative V_{rest} across a cell's membrane arises directly from the high internal K^+ concentration relative to the extracellular K^+ concentration, combined with a high P_K. As a result, K^+ tends to leak out of the cell through numerous K^+-selective channels that are open at rest, and a net negative charge is left behind. Because the resting membrane has relatively few open Na^+-selective channels, Na^+ makes only a very small contribution to the resting potential. In some cells, both the P_{Cl} and the electrochemical gradient for Cl^- are small, so Cl^- makes little contribution to V_{rest}. In other cells, the membrane is quite permeable to Cl^-, and the flux of Cl^- across the membrane contributes to stabilizing V_{rest}. The indirect but ultimate basis for the resting potential is the metabolically energized active transport of Na^+ out of the cell in exchange for K^+. By maintaining a low intracellular Na^+ concentration, the Na^+/K^+ exchange pump allows K^+ to be the dominant intracellular cation. In addition, a small fraction of the resting potential arises directly as the result of the pumping of a net amount of positive charge (Na^+) out of the cell.

ACTION POTENTIALS

All neurons use one type of signal, the action potential (AP), to send information to and along the output segments of the cell, often over long distances. Actions potentials are large, brief changes in V_m that are propagated along axons without decrement. That is, once an AP is initiated in a neuron, the signal travels along the cell membrane, producing the same amount of change in V_m at every point. In addition, the time course of the voltage change is constant as the AP travels along the axonal membrane. In a single axon, every time an AP is initiated, it will produce the same amount of change in V_m, with the same time course; there are no intermediate-sized APs. As a result, APs are said to be "all-or-none" events. These potential changes carry information over long distances in nerve and muscle tissue and can control effector responses, including the activation of electrically gated ion channels, of muscle contraction, and of exocytosis.

The production of an AP depends on three key elements:

- The active transport of ions, by specific proteins in the membrane, generates unequal concentrations of ionic species across the membrane.

- This unequal distribution of ions generates an electrochemical gradient across the membrane that provides a source of potential energy.

- The gating of ion channels that are selective for particular ionic species allows ionic currents to flow through the channels across the membrane, driven by electrochemical gradients.

Two types of voltage-gated ion channels, Na^+ channels and K^+ channels, are most important in the production of APs. These two channel types are quite different from the passive channels discussed in connection with V_{rest}. The Na^+ and K^+ channels responsible for the AP have different properties from one another, and their interdependent activity is responsible for essentially all features of the AP. Since APs in the nervous system are responsible for every sensation, every memory, every thought—indeed, every impulse to act in the environment—it is important to understand how they are formed and regulated within living cells.

General Properties of Action Potentials

Action potentials (also called *spikes* and *nerve impulses*) are generated by the membranes of neurons and muscle cells, as well as by some receptor cells, secretory cells, and protozoa. The shape, magnitude, and time course of all APs produced by a particular cell type are essentially identical. Depolarization of an excitable membrane past a threshold value triggers a rapid and continuously increasing depolarization until the cell briefly becomes inside-positive and then rapidly repolarizes to a potential near V_{rest}. In many types of cells, the repolarization continues until the cell is transiently hyperpolarized *(after-hyperpolarization)*, and then V_m slowly returns to its original resting value.

To illustrate the general features of an AP, assume that short pulses of depolarizing current are passed across the membrane of a nerve cell (Figure 5-16). These pulses will

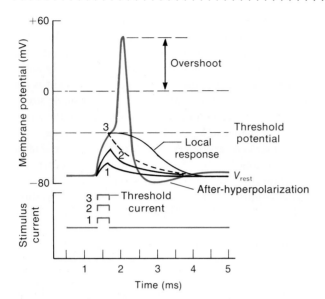

Figure 5-16 A neuron produces an action potential when a stimulus changes V_m enough to exceed the threshold potential. The magnitudes of three stimuli are shown at the bottom of the figure, and the corresponding responses in a neuron are shown above. The small numbers indicate which stimulus produced which response. Stimulus 3 depolarized the membrane sufficiently to trigger an AP. Smaller changes in V_m evoked much smaller responses in the neuron (traces 1 and 2). The dashed curve illustrates the change in V_m that would occur if the neuron produced only a passive response to stimulus 3, rather than an AP. Sometimes a stimulus that is just at the threshold value evokes an abortive, nonpropagated excitation, called a *local response*, rather than an all-or-none AP.

produce passive depolarization until the current delivered is strong enough to depolarize the membrane to its threshold potential, whereupon an AP is triggered. If the depolarization is just too small to reach threshold, there may be an abortive, nonpropagated excitation called a *local response,* which is simply the beginning of an AP that died out before it was irreversibly under way.

The *threshold current* is the intensity of stimulating current that is just sufficient to bring the membrane to the threshold potential and elicit an AP. Although most neurons have threshold potentials between -30 mV and -50 mV, no absolutely consistent value can be assigned either to the threshold current intensity or to the threshold potential, because the threshold depends on immediate past electrical events that can modify the state of the membrane.

Once the threshold potential is reached, the AP becomes *regenerative;* that is, the event becomes self-perpetuating, and V_m continues to change with no further stimulus required. As the cell interior rapidly gains positive ions, V_m becomes less negative until the intracellular potential actually exceeds zero (i.e., reverses polarity). The interior continues to become increasingly positive until it reaches a peak of $+10$ mV to $+50$ mV. The very brief period when V_m is inside-positive is called the *overshoot* (see Figure 5-16). In mammalian neurons, APs typically last only a millisecond or so, although in many invertebrate species, APs can last as long as 10—or even 100—milliseconds. In

other types of excitable cells in vertebrate animals (e.g., heart muscle cells), each AP can last as long as half a second (see Figure 12-7). When the interval between two APs is reduced, the second AP becomes progressively smaller, and it fails completely if a stimulus is delivered too soon after the first AP. During this **absolute refractory period** the neuron does not respond to a stimulus of any size.

The flushing of a toilet illustrates some of the features of an AP. Once it is initiated, the flush (AP) continues to completion, independent of pressure (stimulus) applied to the triggering lever. The flush is thus an all-or-none phenomenon, like an AP. On the other hand, if after one flush, the lever is pressed before the tank has filled completely, the second flush that is produced will be smaller than normal. If the lever is pressed too soon after one flush, a second flush may not occur at all.

In a neuron the situation is somewhat more complicated than in the flushing toilet analogy. No stimulus, however large, is sufficient to evoke a second AP during the *absolute refractory period,* which exists during and for a short time after an AP (Figure 5-17A). Following this period, the excitable cell membrane enters the *relative refractory period,* during which the threshold potential is elevated above normal, but a strong stimulus may evoke an AP. An AP initiated during this period may have a reduced amplitude (the overshoot is smaller). Excitability progressively increases (and the threshold potential decreases) during the relative refractory period until it returns to the level characteristic of the resting membrane before stimulation (Figure 5-17B). *Refractoriness,* or diminished excitability, during and immediately after the AP prevents fusion of impulses, but permits the propagation of discrete impulses.

If a neuron is stimulated by a series of subthreshold depolarizations, a time-dependent decrease in excitability occurs (i.e., the threshold potential increases). For example, if the membrane is depolarized gradually with a current of steadily increasing intensity, a greater depolarization is required to elicit an AP than when the stimulus has an abrupt onset (Figure 5-18). The slower the rate of increase in the intensity of the stimulating current, the greater the increase in threshold potential. This characteristic of excitable membranes, called **accommodation,** results from time-dependent changes in the sensitivity of membrane channels to depolarization.

When they are stimulated continuously by a current of constant intensity, some neurons accommodate rapidly and generate only one or two APs at the beginning of the stimulus period (Figure 5-19A). These neurons are said to have a *phasic response.* Other neurons accommodate more slowly and therefore fire repetitively, although with gradually decreasing frequency, in response to a prolonged constant-current stimulus (Figure 5-19B). These neurons are said to have a *tonic response.* This difference among neurons plays a key role in how sensory neurons transmit information (see Chapter 7). The reduction in the frequency of APs that is typically seen in a neuron that responds tonically during a sustained stimulus is termed **adaptation.**

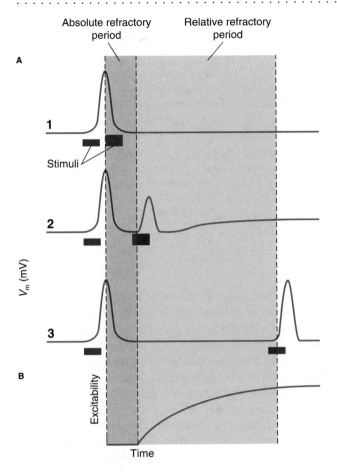

Figure 5-17 During and after one AP, the neuron is refractory to producing another AP. **(A)** Recorded changes in V_m in response to three pairs of stimuli that were delivered to a neuron. The timing of each stimulus is represented by gray bars under the V_m traces; the thickness of these bars indicate the strength of the stimulus required to produce an AP. In trace 1, the second stimulus produced no AP, indicating that it was delivered during the absolute refractory period. In trace 2, the second AP was smaller and a stronger-than-normal stimulus was required to reach threshold, indicating that the second stimulus was delivered during the relative refractory period. When the two stimuli were separated by a sufficiently long interval, both stimuli produced normally sized APs (trace 3). **(B)** Time course of the change in membrane excitability during the refractory period. During the absolute refractory period, the neuron cannot be excited to produce another AP, regardless of the size of the stimulus. During the relative refractory period, excitability is reduced (i.e., the threshold is elevated), so stronger stimuli are required to reach threshold. Over time, membrane excitability returns to normal.

All of the electrical features of the AP that we have described so far can be explained at the level of the molecular structure of the membrane. Indeed, they all depend on the selective regulation of ion channels in the nerve cell membrane.

Ionic Basis of the Action Potential

The electrical signature of an AP is the rapid depolarization of the membrane. We now know that this change in V_m depends on an inward Na^+ current caused by a sudden large increase in the Na^+ conductance (g_{Na}) of the membrane (Figure 5-20). During the rising phase of the AP, g_{Na} exceeds g_K (K^+ conductance), and the net inward current

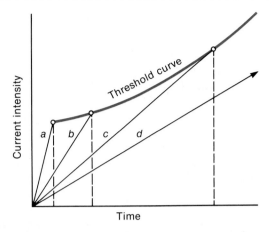

Figure 5-18 When a neuron is subjected to subthreshold stimuli of slowly increasing intensity, the threshold potential increases. The phenomenon is called accommodation. In this experiment, a ramplike stimulating current that was gradually increased in intensity was passed into a nerve cell, with different rates of rise on different trials (a–d). For the most rapid rate of rise (line a), the threshold was closest to the normal threshold potential. When the intensity of the stimulating current increased more slowly, the intensity of current necessary to reach threshold became larger (lines b and c). If the intensity of the stimulating current rose sufficiently slowly, threshold was never reached (line d).

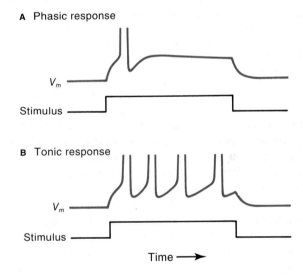

Figure 5-19 Many, but not all, neurons adapt to sustained stimulation. **(A)** Some neurons exhibit strong adaptation to a prolonged stimulus, producing only one or two impulses at the beginning of the stimulus, a phasic response. **(B)** Other neurons adapt very little, except for a progressive lengthening of the interspike interval, a tonic response. Note that in this figure the overshoot of the APs was cut off by the recording device.

moves the membrane potential towards E_{Na}. As the AP reaches its peak, g_{Na} declines and g_K increases. Very quickly, the outward K^+ current exceeds the inward Na^+ current, causing the membrane potential to repolarize back toward E_K. These dramatic changes in the ionic conductances result from molecular changes in the membrane. To summarize, at rest the membrane is most permeable to K^+, but in the early phase of an AP, it becomes very much more permeable to Na^+. When the permeability to Na^+ once again becomes small, the membrane is at first very permeable to

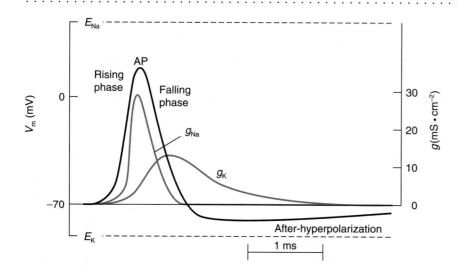

Figure 5-20 An action potential is caused by transient changes in ionic conductances across the membrane. Initially g_{Na} increases; then with some delay g_K increases, too. The AP in this figure, recorded from a squid giant axon, consists of three phases: a rising phase that depends upon the increase in g_{Na}; a falling phase that depends upon both a sudden drop in g_{Na} and the rise in g_K; and an after-hyperpolarization that occurs when g_K remains elevated for some time. The decrease in g_{Na} is due to inactivation of Na+ channels, whereas the decrease in g_K is caused by repolarization. When g_{Na} is high, V_m approaches E_{Na}; when g_K is high, V_m approaches E_K.

K+ because voltage-gated K+ channels are open. Later, the permeability to K+ drops, because only the passive resting-membrane K+ channels remain open. The permeability of the membrane to Cl− does not change during an AP. This brief overview of the mechanisms responsible for APs summarizes decades of painstaking work by many scientists who, through clever manipulation of the electrical and chemical environment of axons, discovered the physical basis of signaling in the nervous system.

To understand how changing membrane characteristics might account for the AP, it is useful to return to the model of a membrane as composed of a capacitance arranged in parallel with a conductance (see Figure 5-12A), only we will modify the model to include separate, parallel conductances for each species of ion. These ionic conductances represent open channels through membrane protein molecules, channels that carry the Na+ and K+ currents across the membrane. Like any current carried by charged entities, ionic currents through membrane protein channels obey Ohm's law (equation 5-1). Thus, the current carried by ion species X, I_x, is given as

$$I_x = g_x \times emf_x \qquad (5-8)$$

where g_x is the membrane conductance for X, which is proportional to the number of open X-selective channels, and emf_x is the electromotive force acting on X. The electromotive force acting on X is the difference between the membrane potential, V_m, and the equilibrium potential of X, E_x:

$$emf_x = V_m - E_x \qquad (5-9)$$

If E_x differs from V_m, there will be some driving force acting on X to force it across the membrane in one direction or the other; the direction the ion will move can be inferred from the sign of emf_x. Substituting this expression for emf_x into equation 5-8, we have a new form of Ohm's law:

$$I_x = g_x(V_m - E_x) \qquad (5-10)$$

This equation says that there is an ionic current across the membrane *only* if there is both a driving force and a conductance for ion X. If either g_x or the emf for X is zero, there will be no I_x. For example, if many X-selective channels were open, then g_x would be high. However, if $V_m = E_x$, there would be no emf driving X across the membrane, and I_x would equal zero. Likewise, if g_x is zero (i.e., no ion channels for X are open), I_x will be zero regardless of the emf.

What can be responsible for changes in ionic current across the membrane? To answer this, we can consider the elements of equation 5-10. E_x, the equilibrium potential for X, cannot change rapidly because it depends on the concentration gradient of ion X, and the concentrations of X inside and outside the cell remain essentially unchanged during an action potential. The variable in equation 5-10 that is most likely to change is the ionic conductance, g_x, which allows ion X to cross the membrane. Hence, changes in ionic conductance (i.e., changes in the number of ion channels that are open in the membrane) play a crucial role in controlling the electrical currents carried across biological membranes. Notice that changes in V_m will also affect the ionic current due to X, because the driving force on X is equal to $V_m - E_x$.

Our knowledge about how changes in the open state of ion channels generate APs is due in large part to the efforts of several pioneering physiologists. In 1936, a distinguished English zoologist, J. Z. Young, working at the marine station in Naples, Italy, first reported that longitudinal structures that had been described in squids and cuttlefish were not blood vessels, as had previously been thought, but were instead very large axons (Figure 5-21A). These axons, which control the speedy escape response of the animal, are thought to have evolved their large size to facilitate rapid conduction of action potentials. They came to be known as *giant axons* and proved to be a boon to biophysicists because their large diameter—up to 1 mm—allows electrode wires to be inserted through them longitudinally for stimulation and recording (Figure 5-21B).

Working with giant axons, two groups of experimenters—K. S. Cole and H. J. Curtis in Woods Hole,

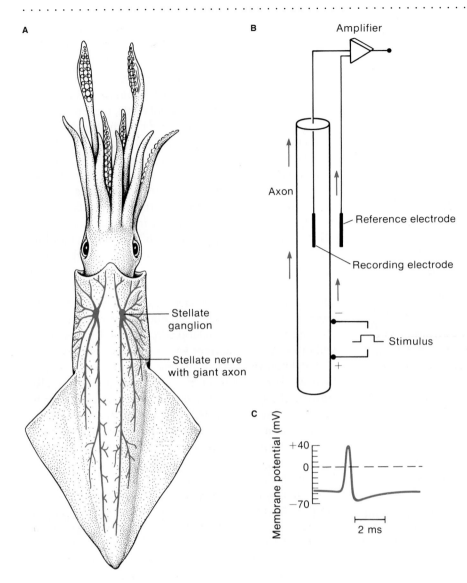

A

- Stellate ganglion
- Stellate nerve with giant axon

B

Amplifier

Axon

Reference electrode

Recording electrode

Stimulus

C

+40

0

−70

Membrane potential (mV)

2 ms

Figure 5-21 Giant axons of the squid, *Loligo*, played an important role in early biophysical studies of the AP. **(A)** Illustration of the squid showing giant axons. Each stellate nerve contains a giant axon that is several inches long and has a diameter of up to 1 mm. Because of their large size, giant axons conduct APs rapidly and ensure the relatively synchronous activation of all muscles in the mantle. When the squid is startled and the giant axons are activated, mantle muscles contract suddenly, producing a jet of water and rapidly propelling the squid backward and away from potential predators. **(B)** Schematic diagram of the experimental setup used by Hodgkin and Huxley (1939) to discover that V_m reverses its sign during an AP. The arrows indicate the direction of propagation of an AP past the recording electrodes. **(C)** Trace of V_m, recorded over time showing the AP as it passed the location of the two electrodes illustrated in part B. [Part A adapted from Keynes, 1958.]

Massachusetts, and Alan Hodgkin and Andrew Huxley in Plymouth, England—made major discoveries in 1939 about the mechanisms responsible for APs. Cole and Curtis demonstrated that during an AP membrane conductance increases, but membrane capacitance remains constant. This discovery implied that a change in conductance must be entirely responsible for the change in ionic currents. Hodgkin and Huxley found that V_m does not simply go to zero during an AP, but instead actually reverses sign during the impulse (Figure 5-21C). Although this observation might seem at first to be a mere detail, it was at odds with the prevailing belief that the increased ionic current measured during excitation was nonspecific, allowing all ions present to move according to their emf. Thus, before Hodgkin and Huxley observed that V_m overshoots zero during an AP, it was believed that a nerve impulse consisted of a simple collapse of V_m to zero. In fact, the overshoot of the AP approaches the E_{Na} calculated from the Nernst equation (equation 5-6) assuming a 10:1 ratio of external-to-internal Na$^+$ concentrations:

$$E_{Na} = \frac{0.058}{1} \log 10 = 0.058 \text{ V} = +58 \text{ mV}$$

Further confirmation of the role of Na$^+$ in the AP was obtained through experiments by Hodgkin and Bernard Katz (1949). In their experiments, a squid giant axon was bathed in artificial seawater in which choline chloride replaced NaCl. Choline, a large organic cation, cannot cross the membrane, and its presence decreased the magnitude of the AP, exactly as predicted if Na$^+$ were the major cation responsible for carrying ionic current across the membrane (Figure 5-22).

The results from these early biophysical experiments with squid giant axons provided four pieces of evidence that Na$^+$ is the major ionic species responsible for the AP:

- Because [Na$^+$]$_{out}$ exceeds [Na$^+$]$_{in}$ by a factor of about 10, the E_{Na} calculated by the Nernst equation is about +55 mV to +60 mV. Thus the emf acting on Na$^+$

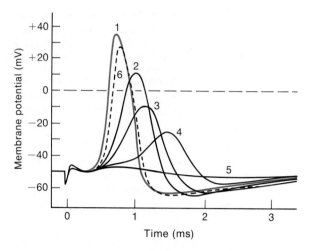

Figure 5-22 The amplitude of the AP rising phase depends upon the presence of Na⁺ in the solution bathing the outside of an axon. Trace 1 shows a control AP recorded in a squid giant axon bathed by normal sea water. Traces 2–5 show the progressive change over time in the amplitude and shape of APs in this axon after normal seawater (which contains approximately 470 mM NaCl) was replaced by artificial seawater containing choline chloride, instead of NaCl. The giant axon is enclosed by a layer of cells, and as time progressed, the Na⁺ concentration inside the coating and near the membrane gradually decreased. Trace 6 was made after the axon was held in normal seawater again at the end of the experiment. [Adapted from Hodgkin and Katz, 1949.]

$(V_m - E_{Na})$ will be large (about 100 mV) and will tend to drive Na⁺ into the cell across the membrane.

- Entry of positively charged Na⁺ into the cell would produce the positive shift in V_m that Hodgkin and Huxley reported in 1939.

- The observed overshoot of the AP approaches the calculated E_{Na}.

- The magnitude of the overshoot changes as a function of extracellular Na⁺ concentration.

After World War II, Hodgkin and Huxley (1952a, 1952b) continued their electrical measurements of action potentials in the squid giant axon and were able to measure directly the currents carried by individual ionic species. They did this using a newly invented electronic technique called **voltage clamping** (Spotlight 5-3). This method, first applied to the squid giant axon, employs a feedback circuit that allows the experimenter to change V_m abruptly and then to keep it at any preselected value. While V_m is held constant, the ionic transmembrane current that flows driven by the imposed voltage can be measured. This method overcame a central problem that made it difficult to study the rapidly occurring AP, namely that ionic currents could not be measured accurately. Moreover, because V_m could be held constant and current could be measured, Ohm's law could be used to calculate the changes in membrane conductances that occur during the AP. For these reasons, voltage clamping has proven invaluable for studies of the behavior of the voltage-gated channels through which

ions such as Na⁺ and K⁺ cross the membrane to produce electric signals.

In their experiments, Hodgkin and Huxley measured the currents that flow across the axonal membrane when V_m is abruptly changed from its resting level (Figure 5-23A). When V_m is depolarized, there is an initial transient inward current, followed by a sustained outward current (Figure 5-23B, trace *a*). This record is the characteristic total current associated with an AP. In a key experiment, Hodgkin and Huxley demonstrated that the transient initial inward current is due to the passage of Na⁺ ions across the membrane. In this experiment, they clamped the voltage at E_{Na}, a value at which there is no driving force on Na⁺ because $V_m - E_{Na}$ equals zero. In addition, they lowered the external Na⁺ concentration by substituting choline for Na⁺ in the seawater outside of the axon, so that less Na⁺ would be available. When the membrane then was clamped at E_{Na}, there was no measured inward current, but a delayed outward current remained (Figure 5-23B, trace *b*). Hodgkin and Huxley then found that this outward current was influenced, but not eliminated, by stepping the membrane potential to E_{Cl}, so they proposed that the outward current was carried by K⁺. Returning the axon to normal Na⁺-containing seawater restored the inward current. The reappearance of the inward current when Na⁺ was again present indicated that the inward current is produced by a transient influx of Na⁺

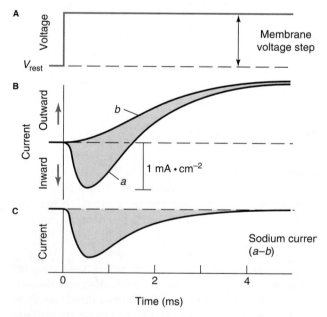

Figure 5-23 Voltage-clamping experiments allow the time course of ionic currents to be determined during an action potential. **(A)** In this experiment, the membrane of a squid giant axon was clamped at +60 mV for at least 5 ms. **(B)** Trace *a* shows the total transmembrane current during the voltage-clamp pulse shown in A; this current is carried by both Na⁺ and K⁺. Trace *b* shows the current carried by K⁺ alone, recorded in low-Na⁺ seawater with V_m held at E_{Na} by the voltage clamp. With this protocol, the sodium current, I_{Na}, equals zero, because there is no emf acting on Na⁺. **(C)** Subtracting trace *b* from trace *a* revealed the time course of I_{Na}. [Adapted from Hodgkin and Huxley, 1952a.]

THE VOLTAGE-CLAMP METHOD

The discovery that it was possible to hold a voltage difference across a membrane at a constant level by electronic feedback contributed greatly to our understanding of signal transmission along axons. This method, called voltage clamping, was first described by Kenneth Cole in 1949. With very finely tuned electrical circuitry, precisely controlled steplike changes in voltage are applied to the membrane; this procedure allows the experimental measurement of ionic currents that flow across the membrane when ion channels are activated. The results are interpreted based on Ohm's law, $I = V/R$. If voltage is held constant, any changes in the current, I, *must* reflect changes in conductance. We now know that the changes in conductance that occur during an AP depend upon the opening and closing of particular ion channels. Stepping the membrane to a series of different voltages and holding it at each voltage for some time produces a description of the *voltage dependence* of channel conductance. By performing voltage-clamp experiments on neurons that have been exposed to solutions of different ionic composition, or to agents that specifically block particular ion channels, researchers have directly determined the ionic currents that underlie neuronal activity.

In a voltage-clamp experiment, an electrode is inserted into the neuron, and the potential recorded across the membrane is compared by a control amplifier with an electronically generated "command" potential (diagram A). The experimenter can choose the command potential from a broad range of values. If V_m is different from the command signal, the amplifier produces a current that passes across the membrane in the direction that will make V_m equal the command signal. The adjustment of V_m occurs rapidly—within a fraction of a millisecond after the command pulse is initiated. Typical experiments include command potentials that produce both hyperpolarized and depolarized values of

V_m. When Na^+ channels (or other channels) open in response to a depolarizing step, ions move across the membrane driven by their electrochemical gradient. If positively charged ions enter the depolarized neuron, under normal circumstances they would make V_m more positive. However, in a voltage-clamped neuron, the clamping circuitry will pass a current that exactly counteracts the ionic current, holding V_m constant. The current supplied, or removed, by the control amplifier in order to maintain the selected membrane potential is recorded, and because it is exactly equal (and opposite) to the ionic current, it reflects the ionic currents over time. Examples are shown in diagram B.

A

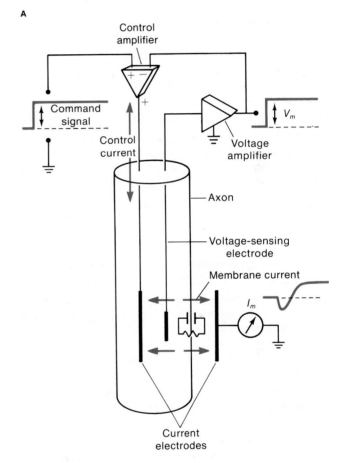

B

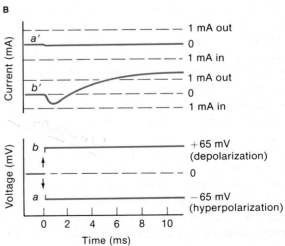

In voltage clamping, an electronic feedback circuit allows the potential across the membrane, V_m, to be held constant. **(A)** The control amplifier compares V_m with the command signal. If V_m is different from the preset command potential, an electrical current is quickly passed across the membrane to make V_m once again equal to the command potential. If the permeability of the membrane to ions changes, more or less current will be required to maintain V_m at a constant level. **(B)** Results of a voltage-clamp experiment illustrate the different effects of hyperpolarizing and depolarizing voltage steps. When V_m was hyperpolarized (*a*), a small, constant membrane current was induced (*a'*). When V_m was depolarized (*b*), the induced ionic current initially was inward; later the ionic current was outward (*b'*). (By convention, inward fluxes of positive ions are plotted downward from 0.) We now know that the early inward current is carried by Na^+ ions, and the later outward current is carried by K^+ ions. [Part B adapted from Hodgkin, Huxley, and Katz, 1952].

across the membrane into the cell. Later experiments showed that the delayed outward current was carried by K^+ (see below). When the delayed current was subtracted from the complex total current obtained in normal seawater, the difference between the two currents (shaded area, Figure 5-23B) shows the time course of the inward current carried by Na^+, which is plotted in Figure 5-23C.

These experiments led to the hypothesis that a sudden depolarization causes a significant number of Na^+-selective channels to open briefly, producing an increase in the Na^+ conductance through the membrane and allowing Na^+ to flow into the axon. In the normal extracellular environment, the electrochemical gradient acting on Na^+ ($V_m - E_{Na}$) should drive Na^+ into the cell. Thus, according to Ohm's law, when g_{Na} rises, so does I_{Na}:

$$I_{Na} = g_{Na}(V_m - E_{Na}) \qquad (5\text{-}11)$$

What does the time course of the Na^+ current tell us about the behavior of the membrane? Based on equation 5-11, the time course of I_{Na} depends both on changes in the conductance of the membrane to Na^+, g_{Na}, and on changes in the emf acting on Na^+, which equals $V_m - E_{Na}$. Clamping V_m at a constant value holds the emf on Na^+ constant, because Na^+ concentrations inside and outside the cell do not change. As a result, the time course of I_{Na} must directly reflect how g_{Na} changes over time in response to depolarization. An important feature of I_{Na}, illustrated in Figure 5-23C, is that even when V_m was held constant at a depolarized potential, I_{Na} reached a maximum within 1 ms and then rapidly returned to its low prestimulus value. Thus I_{Na}, and hence g_{Na}, must consist of two separate processes: *activation*, the time-dependent increase in g_{Na} caused by a depolarization, and *inactivation*, the time-dependent return of g_{Na} to its baseline level.

Hodgkin and Huxley used the voltage-clamp technique to make time-dependent measurements of the distinct ionic currents that contribute to the total current during an AP, concluding that each ionic current reflected a separate conductance through the membrane that was selective for the particular ion carrying that current. Using these data, they formulated equations to express each conductance as a function of V_m and time. These equations predicted the electrical behavior of a nerve membrane under many different conditions and demonstrated that properties of many excitable membranes could be entirely accounted for by time-dependent changes in Na^+ channels and K^+ channels in the membrane.

Voltage-gated sodium channels

Hodgkin and Huxley's detailed electrical measurements in squid giant axons revealed that g_{Na} and g_K change during an AP. They hypothesized that these changes in conductance permitted ions to move across the membrane and that the resulting ionic currents caused the AP. Their elegant experiments clearly defined the aggregate, macroscopic properties that must characterize membrane channels in

neurons. However, these experiments could not reveal the precise nature of these voltage-gated conductance channels, the basis for their selectivity, or how they are activated. In the time since Hodgkin and Huxley performed their experiments, two important advances have contributed significantly to our understanding of membrane channels. First, techniques were developed for measuring ionic currents across small regions of the cell membrane, even from one single ion channel. In addition, the techniques of protein chemistry and molecular biology have made it possible to identify the membrane proteins that constitute the channels. As a result, we now have a clear and consistent view of the molecular nature of ion channels, and we are developing a description of how modification in molecular structure can allow the channels to change the conductance of a membrane to particular ionic species.

To relate the molecular properties of channels to their role in the generation of APs, we need to understand four key features of ion channels: (1) the distribution of ion channels in neuronal membrane; (2) the nature of current flow through a single channel; (3) the mechanism by which depolarization of the membrane can open an electrically gated channel; and (4) the physical basis for how channels can select among ions. In this section, we will discuss each of these features, particularly in relation to voltage-gated Na^+ channels.

Localization and characterization of voltage-dependent channels has been facilitated by several naturally occurring neurotoxins that bind to specific channels. One particularly potent and useful toxin is **tetrodotoxin** (TTX) from the viscera of the Japanese puffer fish, *Sphoeroides rubripes*, and related species. TTX selectively blocks fast-acting, voltage-gated Na^+ channels. When radioactively labeled TTX molecules are added to the extracellular fluid, they bind to Na^+ channels. Examination of neurons labeled by this technique has allowed the density of bound molecules, and hence of Na^+ channels, to be estimated. More recently, antibodies to the channel proteins have been developed, allowing the molecules to be labeled and viewed directly (see Chapter 2). In nonmyelinated axons from a variety of different types of neurons, the density of Na^+ channels has been measured at about 500 Na^+ channels per μm^2; in these axons, the Na^+ channels occupy about $1/100$ of the total surface area. Although this value might seem to be a surprisingly low density of channels, other calculations have indicated that each channel can pass up to 10^7 Na^+ ions per second, providing enough I_{Na} to account for the macroscopic currents that have been measured in various neurons.

Observing the properties of current flow through a single membrane channel proved difficult, because ordinary voltage-clamp techniques cannot record currents through single channels. A conventional voltage-clamp device necessarily collects current from thousands of channels contained in a large area of membrane, and the precision of recording is limited by substantial background electrical noise arising from the passive flow of current through other membrane channels. However, in the late 1970s pioneering work by E. Neher and B. Sakmann led to the development of **patch clamping**, a technique for recording from individual ion channels using a modification of glass recording micropipettes. In this method, a micropipette with a tip diameter of $1–2 \ \mu m$ is placed in intimate contact with the membrane and gentle suction is applied, producing a very tight, high-resistance seal between the pipette and the cell surface (Figure 5-24A; see also Figure 2-6). This remarkable technical advance has profoundly influenced how we think about membrane channels. The patch-clamp technique has made it possible to record the activity of ion channels in the membranes of many types of neurons and from many species. The results of these experiments has indicated a remarkable amount of conservation throughout phylogeny. Neurons in animals as widely separated as jellyfish and mice share ionic mechanisms that underlie their APs.

Using the patch-clamp technique, researchers can record ionic currents through single membrane channels while V_m in that small region is clamped to a chosen value. When a sufficiently large depolarizing step voltage is applied, the recorded events occur as all-or-none currents having a square shape that indicates abrupt opening and closing of the ion channel whose activity is being recorded (Figure 5-24B). The currents are of similar amplitude for individual channels that share a particular ion selectivity and kinetics. The observed similarity in the records from channels with different ionic specificities suggests that all voltage-gated ion channels function in the same general way. The time that individual channels remain open varies randomly over a wide range. The conductance of a single sodium channel does not depend significantly on V_m; its value ranges from 5 to 25 picosiemens, pS ($10 \ \text{pS} = 10 \times 10^{-12} \ \text{S}$, or $10^{11} \ \Omega$ of resistance) for different channels.

From Ohm's law, Faraday's constant, and Avogadro's number, it can be calculated that one activated Na^+ channel carries Na^+ ions at a rate of about *6000 ions per millisecond* at an emf ($V_m - E_{Na}$) of -100 mV (approximately the driving force as the AP gets under way). The summed activity (i.e., openings and closings) of thousands of Na^+ channels, each contributing a minute pulsatile *unitary current*, gives rise to the macroscopic I_{Na} that produces the rising phase of the AP (Figure 5-24C). The number of Na^+ channels open at any instant depends on time (because of the time course of processes that lead to channel activation and inactivation), as well as on V_m. Thus macroscopic changes in the g_{Na} of the membrane, which occur as functions of V_m and time, reflect the behavior of thousands of Na^+ channels, each one opening and closing during depolarization in accord with certain probabilistic principles.

How could depolarization of the membrane influence the opening of voltage-gated ion channels? Hodgkin and Huxley originally suggested that changes in V_m might regulate g_{Na} and g_K by causing a conformational change in a gating molecule. According to their proposal, a gating molecule would bear a net charge at physiological pH values and a change in membrane potential would produce an emf on the charge, causing it to move in space, thus producing a conformational change in the molecule. Consider a typical resting neuron with a potential difference of about -75 mV across the membrane. A depolarization of 50 mV (to -25 mV) generally activates a large fraction of the Na^+ channels present in such a membrane. These channels consist of protein molecules inserted through the lipid bilayer of the membrane, which is about 5 nm thick. It can be calculated that the portions of the channel proteins within the 5-nm layer of the membrane sense a voltage change of 10^{-3} V per 10^{-8} cm, or $100,000 \ \text{V} \cdot \text{cm}^{-1}$, during the 50-mV depolarization. Charged groups on channel proteins certainly might move, driven by this huge membrane electric field and producing conformational changes in the proteins.

When Hodgkin and Huxley proposed that membrane depolarization would lead to a movement of the gating charge from inside to outside of the membrane, they also suggested that this movement of charge should correspond to a small *gating current* (I_g) that would be associated in time with the opening and closing of Na^+ channels. For technical reasons, this hypothetical I_g was not detected until the early 1970s, when the development of very sensitive techniques allowed it to be measured. I_g can be observed when the much larger ionic current through the Na^+ channels, I_{Na}, is pharmacologically blocked by tetrodotoxin or a similar agent. Gating currents also have been detected for K^+ channels and for Ca^{2+} channels in several tissues.

Figure 5-25 presents a model of the opening and closing of a voltage-gated Na^+ channel when a depolarizing step voltage is applied to the membrane. Detailed analysis of the Na^+ channel gating current has revealed that (1) gating of the channel takes place in several distinct steps, each of which is associated with charge movement, and (2) activation and inactivation of the channel are coupled processes. It appears that even though inactivation is not associated with a gating current, the voltage dependence of inactivation (see Figure 5-25, step e) occurs because the activation and inactivation processes are coupled.

Once a Na^+ channel is open, only certain ions can pass through it. A channel's selectivity is indicated by its relative permeability for various ion species. For instance, if the permeability of the Na^+ channel for Na^+ is set at 1.00, than the permeability for Li^+ is 0.93 and for K^+ is only 0.09. The channel acts as if it contains a filter that selects partly on the basis of size, but estimates of channel size relative to the size of permeating ions suggest that relative size cannot be the whole story (Figure 5-26A). Current hypotheses explaining how a channel selects among ions are based partly on ionic size and partly on other properties of the permeating species. Negative charges located at the outer mouth of a cation-selective channel, such as Na^+ and K^+ channels, attract cations and repel anions. Cations larger than 4 Å in diameter are too large to pass through the pore of either Na^+ or K^+ channels. Cations smaller than 4 Å pass through the pore, but only once they have

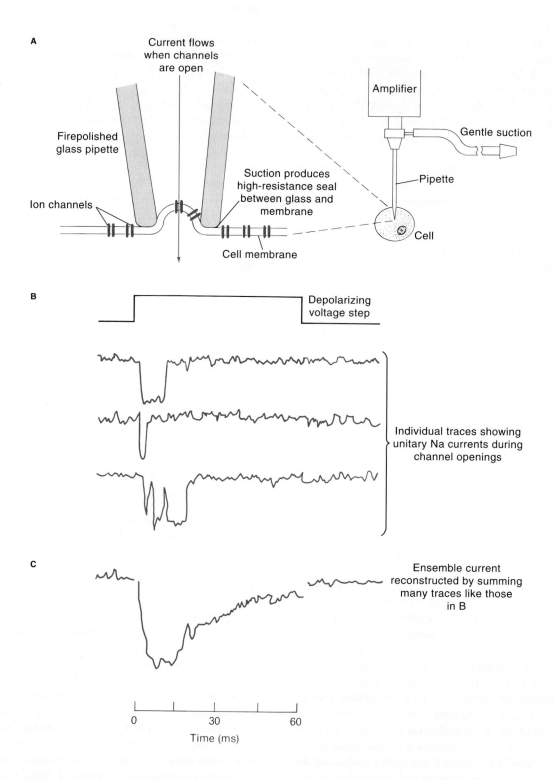

Figure 5-24 The patch-clamp technique allows current through a single membrane channel to be recorded. **(A)** A fire-polished patch pipette, with a tip diameter of about 2 μm and containing the same solution as the one bathing the rest of the neuron, is sealed against the clean surface membrane of a neuron until a very high-resistance contact is achieved, which prevents the loss of current from the pipette to the outside saline. Current flow through an open channel is detected by a sensitive electronic circuit. The voltage across the patch of membrane that is surrounded by the tip of the pipette is clamped using an electronic feedback circuit similar in principle to that used in the original voltage-clamp ex-

periments described in Spotlight 5-3. **(B)** A depolarization (black trace) of a patch of rat muscle fiber membrane caused a single sodium channel to open transiently several times, producing unitary sodium currents that varied in duration and latency (three colored traces). **(C)** Summation of 144 such records from one patch produced an "ensemble current," the time course of which reflects the temporal distribution of individual channel openings of that *one* channel following depolarization. This time course resembles that of a macroscopic sodium current, which depends on the activity of *many* channels in response to a single depolarizing step (see Figure 5-23C). [Parts B and C adapted from Patlack and Horn, 1982.]

Extracellular side

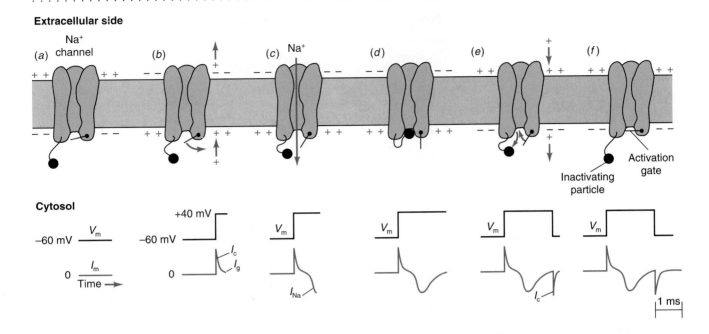

Figure 5-25 The conformation of the voltage-sensitive Na$^+$ channel changes during membrane excitation. A schematic model representing the opening and closing of the channel is shown, with cumulative current records corresponding to each step below. In the resting state (a), the activation gate is closed and the inactivating particle is located away from the pore. When V_m is voltage-clamped to a depolarized level (b), the gating charge (vertical arrows) moves in response to the new electric field across the membrane, producing a small gating current I_g, and the activation gate for the channel moves to the open configuration. I_g is su-

perimposed on an initial capacitative current, I_c, through the membrane capacitance (c). When the majority of the Na$^+$ channels have opened, the inward Na$^+$ current, I_{Na}, is maximal. As depolarization continues (d), open channels begin to close, because the inactivation particle moves to block each open channel. After the membrane is repolarized (e), the gating charges of the Na$^+$ channel again reorient, giving rise to another capacitative gating current. The inactivation particle moves out of the channel, and the activation particle moves away from the pore, returning the channel to its resting state (f).

lost the shell of water molecules (the *water of hydration*) that normally surround charged species in free aqueous solution (Figure 5-26B). How readily polar oxygen or charged functional groups that line the pore of the channel can substitute for the water of hydration determines how easily ions of the right size pass through the channel. According to this hypothesis, some mechanism must exist in the Na$^+$ channel for selectively replacing the water of hydration surrounding Na$^+$ ions, making the channel most permeable to Na$^+$.

In summary, in an AP Na$^+$ channels respond to an initial depolarization by opening, allowing Na$^+$ to enter the cell, which further depolarizes the membrane. This depolarization causes more channels to open, allowing still more Na$^+$ to enter the cell and triggering an explosive, regenerative event. This relationship between membrane potential and sodium conductance, termed the **Hodgkin cycle**, represents a type of positive-feedback system (Figure 5-27). Such systems are rarely found in biological tissue because they are inherently unstable. As noted previously, once an AP is started, it needs no additional stimulus to continue. The effects of this positive-feedback system on the membrane are limited, however, in two ways. First, as the membrane potential approaches E_{Na}, the driving force on Na$^+$ is reduced. Second, open Na$^+$ channels are inactivated, *independently* of V_m, after a short

time and no longer respond to depolarization. The spontaneous termination of the Na$^+$ current by the intrinsic inactivation of Na$^+$ channels would be sufficient to finish an AP. However, K$^+$ channels in the membrane accelerate the recovery of the membrane potential following depolarization.

Voltage-gated potassium channels

The neuronal membrane also contains voltage-gated potassium channels whose probability of opening is increased by depolarization. In comparison with the voltage-gated Na$^+$ channels, however, these K$^+$ channels respond more slowly to voltage changes. The membrane g_K does not begin to increase until the AP is near its peak, and g_K remains high in the falling phase. The net outward current through the K$^+$ channels brings V_m back toward its resting value, and in neurons that have an after-hyperpolarization, the membrane potential moves closer to E_K than it is at rest (see Figure 5-20).

Many types of voltage-gated K$^+$ channels do not intrinsically inactivate, as do voltage-gated Na$^+$ channels. Instead, their conductance to ions depends only on V_m, so as I_K returns V_m back toward its resting value, g_K decreases. As a neuron becomes depolarized, its K$^+$ channels open, thereby speeding repolarization. As the cell repolarizes, the K$^+$ channels close again. As noted above, K$^+$ channels are

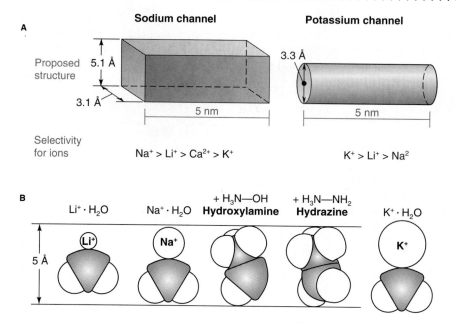

A

| Sodium channel | Potassium channel |

Proposed structure

5.1 Å

3.1 Å

3.3 Å

5 nm

5 nm

Selectivity for ions

$Na^+ > Li^+ > Ca^{2+} > K^+$

$K^+ > Li^+ > Na^2$

B

$Li^+ \cdot H_2O$ $Na^+ \cdot H_2O$ $+ H_3N—OH$ **Hydroxylamine** $+ H_3N—NH_2$ **Hydrazine** $K^+ \cdot H_2O$

5 Å

Figure 5-26 Ionic selectivity of channels is relatively specific and probably depends upon the relative size of some critical portion of the channel, along with the ease of dehydrating the permeating ions. **(A)** Schematic diagrams of the Na^+ and K^+ channels based on the relative permeabilities of various ions. The K^+ channel is thought to have a round selectivity filter that is smaller in at least one dimension than the selectivity filter of the Na^+ channel. The sizes and shapes of the organic ions that can pass through the selectivity filter of the Na^+ channel indicate that the channel may be rectangular, rather than round. The selectivity filter is thought to contact the dehydrated ions directly, so the channel must strip the water of hydration from the ions. **(B)** Schematic diagrams of several partially hydrated inorganic and organic ions. All of these ions can pass through the Na^+ channel, but only K^+ can pass through the K^+ channel. The channel sizes and ions are drawn to the same scale. [Part B adapted from "Ion channels in the nerve cell membrane" by Richard D. Keynes. Copyright © 1979 by Scientific American, Inc. All rights reserved.]

not needed by the membrane to generate APs, and indeed some myelinated mammalian neurons appear to lack them entirely. However, the acceleration of membrane repolarization mediated by K^+ channels does shorten the AP, al-

lowing neurons to generate APs at a higher frequency than they otherwise could.

Absolute and relative refractory periods

As we saw earlier, immediately following an AP the neuronal membrane first is nonresponsive and then less responsive to stimuli than it is at rest (see Figure 5-17). These absolute and relative refractory periods result from the same mechanisms responsible for the after-hyperpolarization: inactivation of Na^+ channels and continued activation of K^+ channels. The majority of Na^+ channels that participate in an AP remain inactivated, and thus cannot be opened by depolarization, during the falling phase and for a brief period thereafter. At a slightly later time—many, but not all—of the Na^+ channels are no longer inactivated, so a depolarization can open a few channels. However, because this smaller number of channels produces much less inward current, an above-normal depolarization is required to generate enough inward current to initiate an AP during this period. Moreover, outward current through the K^+ channels that are still open from the previous AP also opposes the inward Na^+ current. Together, these two ionic mechanisms produce the macroscopic effects measured as refractory periods.

Molecular structure of voltage-gated ion channels

A combination of protein chemistry and molecular biology has yielded detailed information about the channels that conduct action potentials. Voltage-gated Na^+ and K^+ chan-

Applied outward current

↓

Membrane depolarization

2

Na^+ enters through channels

Na channels open

1

Figure 5-27 The Hodgkin cycle is the positive-feedback loop between membrane depolarization and sodium conductance responsible for the rising phase of an AP. The cycle is normally initiated by a depolarization of the membrane that comes from outside the neuron and occurs independently of the voltage-gated Na^+ channels. The positive-feedback loop usually is interrupted by the intrinsic inactivation of Na^+ channels, which terminates the rising phase (black line 1). Alternatively, it is interrupted during a voltage-clamp experiment with the Na^+ channels open (black line 2).

nels have now been identified, cloned, and sequenced in a variety of species from *Drosophila* to mice and humans, and the homology among channels from different species is striking. Furthermore, there is considerable homology among the various types of voltage-gated cation channels.

A Na$^+$ channel has been found to consist of one large α protein (about 260 kilodaltons, kD) that contains four homologous transmembrane domains (Figure 5-28A, *left*).

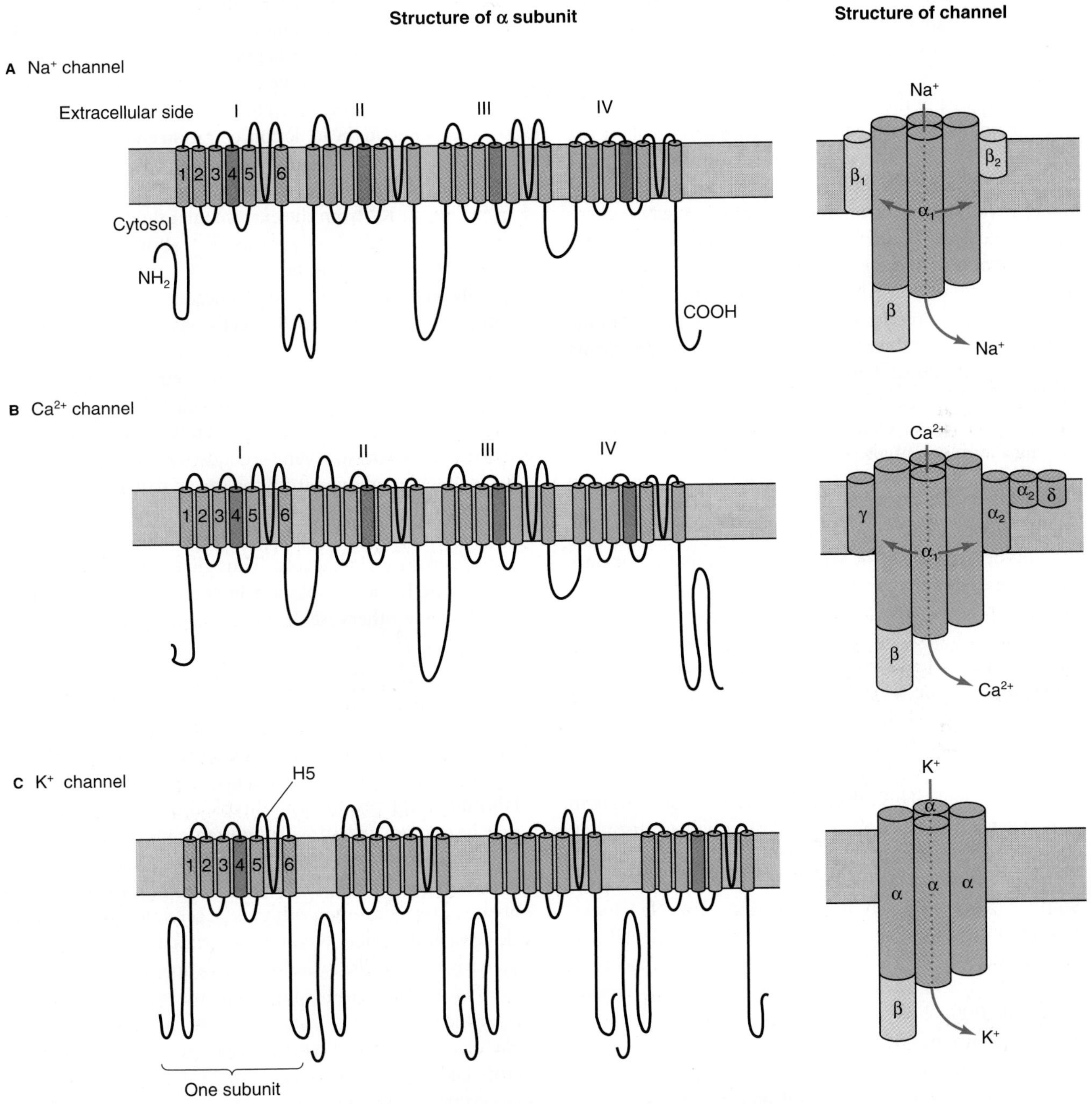

Structure of α subunit   **Structure of channel**

A Na$^+$ channel

B Ca^{2+} channel

C K$^+$ channel

One subunit

Figure 5-28 The molecular structures of the voltage-gated Na$^+$, K$^+$, and Ca^{2+} channels are similar. **(A,B)** Each Na$^+$ channel and Ca^{2+} channel contains a large protein subunit (α or α_1) that can form functional channels when expressed by itself in *Xenopus* oocytes (*left*). The α subunit in each channel contains four homologous repeats (I–IV) of α-helical regions (1–6), each of which is thought to span the lipid bilayer of the membrane. The voltage sensor is thought to be located on the α helix numbered 4. The four repeats of α-helical regions are thought to associate with one another to form the ion-conducting pore of the channel (*right*). Each long cylinder in the channel diagrams corresponds to one repeat in the α subunit. In addition, one or more smaller proteins (designated by various Greek letters) normally associate with the α subunit to form the complete channel complex. These small subunits, which differ between Na$^+$ and Ca^{2+} channels, contribute to the physiological properties of the channels. **(C)** The α subunit of K$^+$ channel is about one-quarter the size of the α subunits in Na$^+$ and Ca^{2+} channel. It contains six α-helical regions that are homologous to one repeat of the transmembrane helical regions in the Na$^+$ and Ca^{2+} channel α subunits. Several variations of the α-helical sequences are known for K$^+$ channels. A functional K$^+$ channel is formed by association of four α subunits, which may have identical or different sequences (*right*). Smaller β subunits that accompany the K$^+$ channel α subunits in native neuronal channels help determine the physiological properties of these channels. [Adapted from Lodish et al., 1995, and from Hall, 1992.]

Although the α protein normally is associated with several smaller β proteins, it can, by itself, exhibit voltage-gated channel function when it is expressed in *Xenopus* oocytes. The transmembrane domains of the α protein are thought to be arranged around a central pore, thus forming the actual path through which Na^+ ions can flow (Figure 5-28A, *right*). Further analysis has identified particular amino acids that are likely to confer the property of inactivation on these channels.

This model of the Na^+ channel is strengthened by the structure of the K^+ channel protein. With a molecular mass of about 70 kD, the K^+ channel protein is much smaller than the Na^+ channel α protein, but it exhibits strong homology with the transmembrane domains of the Na^+ channel α protein. A K^+ channel is thought to be formed by association of four of these smaller proteins in a manner very similar to the way in which the four transmembrane domains of the α protein associate to form a Na^+ channel (Figure 5-28C). Further molecular biological experiments employing site-directed mutagenesis are currently investigating the details of ion selectivity, voltage gating, and other properties of these voltage-gated channels involved in generating action potentials.

Changes in Ion Concentration during Excitation

As mentioned earlier, only a few ions move across a cell membrane to produce the voltage changes that occur during a single AP, and they *do not appreciably change the intracellular ionic concentrations* (see Spotlight 5-2). For example, 10^{-12} mol of Na^+ crossing 1 cm^2 of membrane with a capacitance of 1 $\mu F \cdot cm^2$ is sufficient to produce an AP of 100 mV amplitude. That is, only 160 Na^+ ions per μm^2 of membrane are needed. Since a single Na^+ channel can pass 10^7 Na^+ ions per second, the I_{Na} required for an AP could be supplied by a remarkably small number of channel openings. This calculated number of Na^+ ions that cross the membrane during one AP is probably somewhat of an underestimate, because some K^+ ions move out of the cell, partially canceling the influx of Na^+. The actual number of ions is likely to be closer to 500 Na^+ ions/ μm^2/impulse. If this value is correct, during a single AP in a squid giant axon with a diameter of 1 mm, the intracellular Na^+ concentration would change by only 1 part in more than 100,000. For this reason, a squid axon can generate thousands of impulses after its Na^+ pumps have been incapacitated by a metabolic poison. Eventually, of course, in a poisoned axon the concentrations, and hence the equilibrium potentials, of Na^+ and K^+ will change, and the axon will no longer be able to sustain normal action potentials.

In smaller axons, which have a greater surface-to-volume ratio than the squid giant axon, a more significant change in ionic concentrations occurs even following a single AP. For example, mammalian C fibers have a diameter of only about 1 μm; in these axons, ionic fluxes from a single impulse change the intracellular Na^+ and K^+ concentrations by about 1 percent. The result is a drop in the rest-

ing potential, V_{rest}, of about 0.3 mV; following 10 action potentials in close succession, there is a cumulative depolarization of about 2 mV. Thus, the ability of small-diameter axons to continue generating APs depends on the rapid restoration of the intracellular and extracellular resting concentrations of Na^+ and K^+ by active transport or other mechanisms before cumulative ion fluxes produce significant changes in ion gradients. Note that the metabolic pumping of ions across the cell membrane does not enter directly into the production or recovery of an AP, but does serve to maintain the ionic concentration gradients required for the production of all membrane currents.

Glial cells contribute to maintaining the V_{rest} of neurons by taking up K^+ from the extracellular fluids surrounding axons. They then slowly release K^+, allowing it to be reclaimed by the neurons. Because the V_{rest} of all cells depends heavily on the extracellular K^+ concentration, recording the membrane potential of glial cells has provided a useful measure of the changes in extracellular K^+ concentration that are produced when surrounding neurons conduct APs. Changes in the V_m of glial cells, which reflect the accumulation of K^+ in extracellular spaces following neuronal activity, subside within several seconds, indicating that excess K^+ is quickly removed from the very restricted extracellular space. This removal is thought to depend, at least in part, upon uptake of K^+ across the surface membranes of both neurons and glial cells. Thus glial cells probably help to prevent K^+ accumulation in the extracellular space, which would otherwise depolarize neurons.

OTHER ELECTRICALLY EXCITED CHANNELS

Transmembrane ion channels occur in virtually all cell types. Although voltage-gated Na^+ and K^+ channels collaborate in the production of typical APs, Ca^{2+} channels may be of more widespread importance in cell function (see Table 5-1). Ca^{2+}-selective channels carry at least part of the regenerative depolarizing current in crustacean muscle fibers; in vertebrate smooth muscle cells and cardiac muscle cells; in the cell bodies, the dendrites, and the terminals of many nerve cells; in embryonic neurons; and in ciliates like *Paramecium*, to list just a few examples. In many of these membranes, Ca^{2+} carries inward current along with Na^+, but in a few cell types it carries all of the inward current. The I_{Ca} typically is not strong enough to produce an all-or-none AP without help from an I_{Na}, and in most membranes that contain voltage-gated Ca^{2+} channels, the rising phase of an all-or-none AP is generated largely by a strong I_{Na} that rapidly depolarizes the membrane. The Ca^{2+} channels, which open more slowly and conduct less current, are activated by this depolarization. The Ca^{2+} ions that enter the cell through the Ca^{2+} channels, often have two functions: propagating an electrical signal and acting as an intracellular messenger that triggers subsequent intracellular events. For example, Ca^{2+} is responsible for the release of neurotransmitter substances from the presynaptic terminals and can also contribute to causing contraction of muscles (see Chapters 6 and 10).

The molecular structure of voltage-gated Ca^{2+} channels is strikingly similar to voltage-gated Na^+ channels. Like Na^+ channels, Ca^{2+} channels consist of a large protein (α_1) that includes four homologous transmembrane domains, which are thought to associate and form the ion-conducting pore similar to the Na^+ channel α protein. The Ca^{2+} channel α_1 protein also typically associates with several smaller proteins (Figure 5-28B). Embryonic neurons often express both voltage-gated Na^+ and voltage-gated Ca^{2+} channels. Usually the Ca^{2+} channels appear first, and the Na^+ channels become functional at later stages. The remarkably homologous molecular structure of Na^+ and Ca^{2+} channels, the frequency with which the two kinds of channels appear in the same cells, and the greater prevalence of Ca^{2+} channels in simpler organisms such as the protozoa suggest that Na^+ channels may be a more recent evolutionary specialization for impulse conduction. In addition, Ca^{2+} acts as an intracellular messenger in *most* types of cells, providing further evidence that Ca^{2+} channels may have had a more primitive evolutionary origin than the channels that pass monovalent cations.

Unlike voltage-gated Na^+ channels, many types of Ca^{2+} channels fail to inactivate fully under maintained depolarization. Instead, in at least one type of Ca^{2+} channel, the probability of inactivation increases as the intracellular concentration of free Ca^{2+} rises. These Ca^{2+} channels become inactivated during maintained depolarization because I_{Ca} through the channels increases the intracellular Ca^{2+} concentration near the membrane.

The intracellular concentration of free Ca^{2+} also regulates the function of a class of voltage-gated K^+ channels that differs from the delayed voltage-gated K^+ channels that we discussed previously. These K^+ channels, which are found in many different tissues, are activated by membrane depolarization, but only if the concentration of intracellular free Ca^{2+} is higher than normal (see Table 5-1). In these cells, Ca^{2+} enters through Ca^{2+} channels and accumulates near the inner surface of the membrane; it then causes these *Ca^{2+}-dependent voltage-gated K^+ channels* to open if V_m is depolarized. Where these channels are present, the entry of Ca^{2+} fosters repolarization as the result of the enhanced I_K, which carries positive charge (K^+) out of the cell. This I_K also contributes to the production of an after-hyperpolarization and to the refractoriness of neurons, thus setting limits on the maximum frequency of APs that can be produced by the neurons.

Each of these four kinds of voltage-gated ion channels is highly selective for a particular ion. The Na^+ and Ca^{2+} channels normally carry current into the cell, because there is a strong emf driving both ions into the cell. The delayed voltage-dependent K^+ channel and Ca^{2+}-dependent K^+ channel generally carry current out of the cell, because the emf on K^+ drives it out of the cell. The distribution of these and other types of voltage-gated channels determine the electrical behavior of excitable tissues.

 Some neurons have APs based on voltage-gated Ca^{2+} channels, but most do not. What kinds of effects might occur in neurons with Ca^{2+}-based APs? Could these effects produce selective pressure to change the ionic basis of APs to currents carried by monovalent cations?

SUMMARY

Specific electrical properties of the cell membrane affect the ability of neurons to carry information; these electrical properties depend on the molecular makeup of the membrane. The lipid bilayer membrane acts as an electrical capacitance: although it does not readily allow charge carriers (i.e., ions) to pass, it is very thin (about 5 nm) and thus can accumulate and store charge by means of electrostatic interaction between cations and anions on opposite sides of the membrane. Channels composed of protein molecules embedded in the lipid bilayer provide selective electrical conductances. These channels permit the physical passage of certain inorganic ions across the membrane; the flow of ions through such ion-selective channels constitutes an electric current. These two properties, capacitance and conductance, determine the time course of voltage changes that are produced by the flow of current across electrically active cell membranes.

An asymmetrical distribution of ions in solution on the two sides of a membrane can produce an electrical potential across the membrane, depending on how permeable the membrane is to the ions present. The size of the electrical potential can be predicted from the Nernst equation or the Goldman equation. Because resting cell membranes are most permeable to K^+ and Cl^-, the resting potential is typically close to the equilibrium potentials of these two ions, typically between -40 and -100 mV (inside negative with respect to outside).

Active transport of Na^+ and Ca^{2+} causes these ions to be less concentrated within the cytoplasm than outside the cell. For each of these ions, there is a large driving force into the cell and a low permeability, so to maintain the low intracellular concentration, they must be continually pumped out. Stimuli that increase the normally low permeability to either Na^+ or Ca^{2+} lead to an influx of one or the other, which makes the interior of the cell less negative. For example, the transient opening of Na^+ channels is responsible for the rising phase (depolarization) of the action potential (nerve impulse). Since this increase in Na^+ permeability is evoked by membrane depolarization, the upstroke of the nerve impulse is regenerative and causes the membrane potential to approach briefly the sodium equilibrium potential of $+50$ to $+60$ mV at the peak of the nerve impulse. A delayed rise in K^+ permeability is brought on by the voltage change across the membrane, and together with rapid inactivation of the Na^+ channels, it brings the membrane back to the resting potential, thus terminating the AP.

Thus the electrical behavior of excitable membranes depends on the passive properties of membrane capacitance and resting conductance, on metabolically sustained ion gradients across the membrane, and on the presence of ion-selective membrane channel proteins, some of which are activated by depolarization of the membrane. It is now possible to explain the details of changes in the membrane potential that occur during an AP in terms of the molecular properties of these ion channels.

REVIEW QUESTIONS

1. What are the major anatomic regions of a neuron, and what is the principle physiological function of each?

2. Does a signal that reaches the brain accurately encode all features of the original stimulus? Why or why not?

3. The cell membrane separates electrical charge and therefore has a potential difference across it. Does this violate the general physical requirement for electroneutrality? Why or why not?

4. What is the structural basis for membrane capacitance? For membrane conductance?

5. How is the time course of potential changes across the cell membrane related to the resistance and capacitance of the membrane?

6. Cells are typically inside-negative. Explain this observation in terms of diffusion potentials.

7. You have on your lab bench an artificial system of aqueous solutions separated by a semipermeable membrane. The solutions separated by the membrane contain the same ions, but in different concentrations, and the membrane is permeable to only one of the ions in solution. Will there be a stable electrical potential difference across the membrane? Why or why not?

8. Living cells are typically quite permeable to K^+ and at least slightly permeable to other cations as well. What maintains the high K^+ concentration inside the cell by preventing the K^+ inside from gradually being displaced by other cations?

9. What are the equilibrium potentials for each of the following ions of the given concentrations? (a) $[K^+]_o = 3$ mm, $[K^+]_i = 150$ mm; (b) $[Na^+]_o = 100$ mm, $[Na^+]_i = 10$ mm; and (c) $[Ca^{2+}]_o = 10$ mm, $[Ca^{2+}]_i = 10^{-3}$ mm.

10. For a typical cell that is 100 times more permeable to K^+ than to any other ion, use the Goldman equation to determine the potential change that would be produced by a doubling of the extracellular K^+ concentration.

11. In 1939 Cole and Curtis reported that membrane conductance increases, but capacitance remains essentially unchanged, during an AP. Relate these findings to membrane structure and to changes in structure thought to occur during excitation.

12. What are two observations suggesting that Na^+ carries the inward current responsible for the upstroke of the action potential?

13. Does the sodium pump play a *direct* role in any part of the AP? Explain. How is the sodium pump *indirectly* important in the production of an AP?

14. What limits the flux of Na^+ across the membrane during an AP? What limits the flux of K^+ across the membrane during an AP?

15. Calculate the approximate number of sodium ions entering through each square centimeter of axon surface during an AP having an amplitude of 100 mV. (Recall that 96,500 C is equivalent to 1 mol-equiv of charge; that membranes have a typical capacitance of 10^{-6} F·cm^{-2}; and that Avogadro's number is 6.022×10^{23} atoms·mol^{-1}.)

16. The following properties of action potentials were discovered decades before physiologists knew about ion channels and their role in action potentials. Explain each of the properties in terms of the behavior of ion channels: (a) threshold potential; (b) all-or-none overshoot; (c) refractoriness; and (d) accommodation.

17. Why is it that an axon of large diameter undergoes essentially no change in ionic concentration during several APs, whereas the very thinnest axons may undergo significant changes in concentration during several impulses? How would these changes affect the function of a small axon, and what mechanisms might help to prevent the changes in ionic concentration?

18. The rising phase of an AP is an example of positive feedback in a biological system. How does positive feedback occur in this situation? Positive feedback is inherently unstable, so how can you account for the limited amplitude of the upstroke?

19. How do the properties of ionic currents through a single channel compare with the properties of macroscopic ionic currents measured across the membranes of real cells?

20. Name four kinds of voltage-gated ion channels that contribute to neuronal function. How do channels select for particular ionic species?

SUGGESTED READINGS

Aidley, D. J. 1989. *The Physiology of Excitable Cells*. 3d ed. New York: Cambridge University Press. (A thorough examination of the physiology of nerves, muscles, and unusual features such as the electric organs of some fishes.)

Catterall, W. A. 1993 Structure and function of voltage-gated ion channels. *Trends Neurosci.* 16:500–506. (A readable review of recent work on the molecular features of ion channels written by a major contributor to the field.)

Hall, Z. 1992 *An Introduction to Molecular Neurobiology.* Sunderland, Mass.: Sinauer. (A description of neurophysiology that emphasizes the roles and properties of proteins and the regulation of their expression.)

Hille, B. 1992. *Ionic Channels of Excitable Membranes.* 2d ed. Sunderland, Mass.: Sinauer. (An excellent compendium of information about the biophysics of ion channels.)

Hodgkin, A. L. 1964. *The Conduction of the Nervous Impulse.* Springfield, Ill.: Thomas. (A classic description of the electrophysiology of signal conduction in neurons.)

Hodgkin, A. L. 1976. Chance and design in electrophysiology: An informal account of certain experiments on nerve carried out between 1934 and 1952. *J. Physiol.* (London) 263:1–21. (An insider's view of biophysical history.)

Kandel, E. R., J. H. Schwartz, and T. M. Jessell. 1991. *Principles of Neural Science.* 3d ed. New York: Elsevier. (An enormous and authoritative compendium of information about the function of the nervous system, from the biophysics of membrane channels to the physiological basis of memory and learning.)

C H A P T E R

6

COMMUNICATION ALONG AND BETWEEN NEURONS

The survival of all animals depends on their ability to respond to challenges from other animals and from the environment. Often, the response must be both rapid and well coordinated in order to be effective, and animals can produce such responses only when information is gathered, organized, and transmitted quickly throughout the body. Nervous systems evolved to allow such rapid and adaptive responses to occur. They are found in all animals, from the simplest coelenterates to the most complex mammals.

The complexity of nervous systems is amply illustrated by the human nervous system, which contains more than 10^{11} neurons plus an even larger number of supportive cells (*glial cells,* or *neuroglia*). The functional units that allow animals to respond effectively to their environments are sets of neurons that are connected with one another in a way that allows information to be passed among the cells. These arrays are called **neuronal circuits,** their interconnections being in many ways analogous to electrical circuits. All the complex capacities of the nervous system—movement, perception, learning, memory, and consciousness—arise from the physical and chemical processes that were considered in Chapter 5 and are examined further in this chapter. Understanding how neuronal activity results in behavior is undoubtedly one of the greatest of all challenges to biologists, and whether we will ever completely understand the physical and chemical basis of consciousness and creative thought remains an open question.

In spite of the enormous complexity of most nervous systems, a great deal has already been learned about the physiology and biophysics of single neurons, as described in Chapter 5. We know that all neurons carry information by means of electrical signals that are based on the movement of particular ions across the cell membrane. Ionic currents (see Chapter 5) encode signals that travel along axons by mechanisms that constitute the first topic of this chapter. Although we consider events in single neurons when we examine the transmission of information by action potentials (APs), behavior is never produced by the activity of a single neuron. Even in very simple animals, behavior depends on the activity of many neurons working together;

so, if the signal carried by any one neuron is to generate effective and adaptive behavior, it must be transmitted to other neurons. Information is passed along to other neurons at structures called synapses, and **synaptic transmission** (i.e., the mechanisms that allow information to be passed from one neuron to the next at synapses) constitutes the second major topic of this chapter.

TRANSMISSION OF SIGNALS IN THE NERVOUS SYSTEM: AN OVERVIEW

Signals move from point to point along the plasma membrane of a single neuron in either of two ways: *graded, electrotonically conducted potentials* and *action potentials* (all-or-none impulses). These two basic methods of transmission alternate as information passes along one neuron and is then transferred to another neuron, as illustrated schematically in Figure 6-1, which shows the path traveled by information received at the interface between an animal and its surroundings (e.g., at the skin). Energy from a physical stimulus is received and changes the membrane potential, V_m, at a specialized type of membrane in a sensory neuron (see Chapter 7 for further treatment of this process), producing a *receptor potential* that is graded (i.e., varies in a continuous fashion) in proportion to the strength of the stimulus. A low-intensity stimulus produces a small change in V_m; a more intense stimulus produces a larger change in V_m. This change in V_m at the receptor membrane generally continues, often with some attenuation, for the duration of the stimulus. Because the time course and amplitude of a receptor potential are closely related to the time course and intensity of the stimulus, the receptor potential is an electrical neuronal analog of the stimulus. For example, a pressure stimulus that persists for a long time typically produces a long, slightly attenuated depolarization in the receptor endings. This signal spreads along the cell membrane away from the location of the stimulus very much as an electrical signal spreads along a wire. The membrane that is specialized to receive sensory stimuli lacks the voltage-gated ion channels that produce all-or-none APs, so signals cannot be

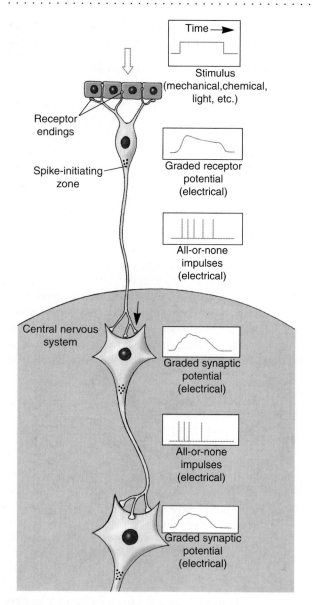

Figure 6-1 Graded and all-or-none electrical signals alternate with one another as a signal is carried along a neuronal circuit. A stimulus applied to the receptor endings of a sensory neuron produces a graded potential that reflects the amplitude and duration of the stimulus, although it is neither linearly proportional to the stimulus nor identical in form. This potential spreads passively through the first part of the sensory neuron and, if the change in V_m is sufficiently large at the spike-initiating zone, it elicits all-or-none propagated APs in the axon. When they arrive at the terminals of the sensory neuron, the APs cause the release of a chemical neurotransmitter that induces a graded change in V_m in the next neuron. If the potential in the second neuron reaches threshold, an AP or a train of APs will be produced. Thus, graded and all-or-none potentials alternate along the pathway, and the information is carried alternately by electrical and chemical signals. Within each neuron the signal is carried electrically, whereas between neurons the signal is carried by chemicals. Arrows indicate the sites of chemical signal transmission.

propagated regeneratively in this part of a sensory neuron. Just as electrical signals decay when they are carried through a lifeless electrical conductor away from a source, these receptor potentials progressively decay with distance

from the site of origin. This kind of signal transmission is called **passive electrotonic transmission,** to distinguish it from the regenerative transmission of APs. Passively transmitted signals from receptor endings decay over relatively short distances, so passive transmission is not effective for carrying signals to distant parts of the body. For long-distance transmission, the sensory signals must be transformed into APs, which can conduct signals without decrement for long distances. The membrane at the spike-initiating zone of a sensory neuron contains the voltage-gated ion channels that permit APs to start, so if the passively propagated depolarization of a receptor potential is large enough at the spike-initiating zone, the signal will be transformed from passive into active propagation and will be carried without decrement along the sensory neuron's axon, which may span many meters.

The next transformation of the signal takes place at the axon terminals of the sensory neuron, where it must be passed across synapses to other neurons. This transfer of information between neurons is typically (but not always) accomplished through chemical signals carried by molecules called *neurotransmitters*. If the transfer of information across a synapse is to be effective, the neurotransmitter must cause the membrane potential, V_m, of the postsynaptic neuron to change. The amount of transmitter released, and thus the amplitude of the response in the next neuron along the pathway, depends on the number and frequency of APs arriving in the terminals of the sensory neuron (called the *presynaptic neuron*). Within limits, higher frequencies and more APs in the presynaptic neuron cause more neurotransmitter molecules to be released, producing a greater change in V_m of the receiver neuron (called the *postsynaptic neuron*). The change in V_m of the postsynaptic neuron, called the *postsynaptic potential* (psp), is a graded signal, reflecting at least some properties of the original stimulus, although it may be quite distorted. If it is sufficiently large, this graded postsynaptic potential can bring the spike-initiating zone of the postsynaptic neuron to threshold, triggering one or more all-or-none APs in the postsynaptic neuron.

Thus, as a signal is received from the environment and transmitted by neurons, it is coded alternately in graded potentials and in all-or-none APs. Graded potentials are produced at sensory and postsynaptic membranes, and all-or-none nerve impulses are largely confined to structures that are specialized for long-distance conduction, such as axons. With minor exceptions, all of the signals falling into these two major categories of signal transmission—all-or-none impulses and graded potential changes—are generated by the gating of specific types of membrane channels (see also Chapters 5 and 7).

As it travels through the nervous system, information is transformed over and over again. Sometimes it is carried by graded changes in V_m, which may be translated into all-or-none, actively conducted APs. At synapses, the electrically encoded signals are changed into chemical signals in the form of neurotransmitter molecules, which carry the information from one cell to the next. The chemical signal is then reconverted into an electrical signal in the postsynaptic neuron. The interconversions between electrical and chemical methods of conducting signals along a chain of neurons change the character of a signal as it is transferred through the nervous system. Modifications in synaptic connections are thought to underlie such complicated behavioral changes as learning. Our knowledge about synapses and how they can influence neuronal communication within the body is growing very rapidly as more neurotransmitters are identified and the broad range of their activities is better understood.

TRANSMISSION OF INFORMATION WITHIN A SINGLE NEURON

Information spreads through a neuron, away from its point of origin, through the interaction of two basic mechanisms: passive electrotonic conduction and active regenerative APs. Electrotonic conduction takes place in all neurons, whereas only those neurons with functional voltage-gated ion channels (see Chapter 5) can carry APs. Electrotonic conduction depends on the physical properties of the neuron. In contrast,

a neuron's APs are shaped by a combination of its physical properties and the nature of its voltage-gated ion channels.

Passive Spread of Electrical Signals

The passive spread of changes in V_m occurs in all neurons. The resistance and capacitance of the plasma membrane govern how potentials and currents spread within a cell. In a hypothetical spherical cell, potentials would spread uniformly with minimal decay, because the electrical resistance of the saline cytosol would be much lower than the electrical resistance of the cell membrane (which depends on the number of open ion channels—see Chapter 5). Consequently, a current injected into a spherical cell would spread out and pass through the membrane with relatively uniform density over the entire cell surface. Neurons, however, have more complicated shapes, so the spread of a change in V_m is more complex. Many neurons have long processes that conduct signals over large distances. If current is injected at a point on the membrane of a long, thin, cylindrical region (e.g., in an axon, a dendrite, or a muscle fiber), an electrical signal can spread away from that point because cells possess **cable properties** (Figure 6-2). The cable

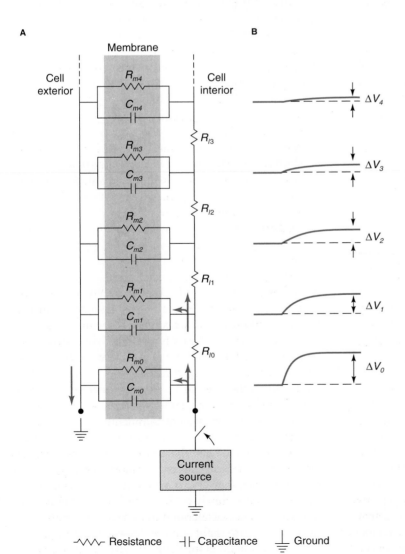

Figure 6-2 The cable properties of a long, thin cylinder determine how current spreads along the cylinder. **(A)** Equivalent circuit of a simple axon. Resistance and capacitance elements, representing properties of the membrane, are connected by the internal and external longitudinal resistances. The membrane resistance, R_m, the longitudinal resistance, R_l, and the capacitance, C_m, are arbitrarily divided into discrete circuit elements, 0, 1, 2, 3, 4. The red arrows show the direction of the current flow. **(B)** The electrical responses that would be recorded across each of the resistance and capacitance elements in the membrane in response to a step pulse of current that was produced by the current source when the switch (shown in Part A) was closed. The amplitudes of the potential changes, ΔV_0 to ΔV_4, decrease exponentially with distance from the source; in addition, the signals rise more slowly farther from the source.

properties depend on the physical parameters of the cells, and they imply that any current flowing longitudinally along an axon (or other long, narrow cylinder) decays with distance because (1) there is some resistance to the flow of electrical signals through the cytoplasm, and (2) the resistance of the cell membrane to the passage of electrical signals is high but finite. Unlike the current in an electrical wire, where the insulation surrounding the wire assures that longitudinal current is uniform, longitudinal current in a nerve cell decreases as it travels because part of the current can leak out of the cell across the plasma membrane at every point along the cylinder. The fraction that leaks out no longer flows in the cytoplasm; instead, it returns along extracellular pathways to complete the electrical circuit. Understanding the implications of the cable properties of neurons is important for understanding how current spreads through cells and how impulses are conducted along axons.

We can better understand how current is distributed along an axon by constructing an electrical circuit having properties equivalent to those found in the axon. For reasons that will become clear shortly, current flowing along axons travels significantly more slowly than does electrical current in wires. Nonetheless, the current in axons can be modeled with the use of circuit elements that have properties matching those of ionic currents. Modeling the electrical features of axons has contributed significantly to the design of experiments that have helped to elucidate how neurons work.

Current entering an axon distributes itself along the axon according to the passive electrical properties depicted in the equivalent circuit shown in Figure 6-2. The components R_m and C_m are the same as those illustrated in Figure 5-10 and represent the uniformly distributed passive resistance and capacitance of the inactive membrane. (In Figure 6-2, the elements are depicted as discrete entities for convenience.)

In an electrical circuit, as in any physical system, energy must be conserved. Conservation of energy in an electrical circuit requires that the sum of all the currents leaving a point within a circuit must equal the sum of all the currents entering that point (Kirchhoff's first law). In addition to meeting the conservation of energy requirement, the current flow must satisfy Ohm's law, which states that voltage equals current times resistance in a circuit (see Equation 5-1). Ohm's law implies that current distributes itself in inverse proportion to the resistances of the various routes open to it at each branch point. Thus, when the switch in the equivalent circuit of Figure 6-2 is closed, the pulse, ΔI, of constant-intensity current will flow across the "equivalent membrane," dividing at each branch point (0, 1, 2, 3, 4). At each branch point, a proportion of the current will pass through the membrane resistance (R_m), and the remainder will travel through the longitudinal resistance (R_l). The current along the axon will be diminished by each increment in R_l encountered, because longitudinal resistance is cumulative. The change in V_m that results from the current flow is not instantaneous; it takes a short, but finite, time to build up (see Figure 5-12). The time necessary for V_m to stabilize depends on the membrane capacitance, because charges must accumulate on either side of the membrane to produce a given V_m. Because of membrane capacitance, the square pulse delivered at $x = 0$ appears a few millimeters away as a slow, gradual rise and fall of potential. Thus, the membrane capacitance both slows the passive transmission of signals along the axon and distorts them, and the transmembrane current through the R_m's decreases exponentially with distance from the point of current injection. Because all the R_m's in this model circuit have the same value, Ohm's law requires the potential developed across them also to decrease exponentially with distance. Hence, the transmembrane steady-state potential (ΔV_m) will diminish exponentially with distance along the axon (Figure 6-3).

This decay with distance can be described mathematically as

$$V_x = V_0 e^{-x/\lambda} \tag{6-1}$$

Here, V_x is the potential change measured at a distance x from the point at which the current is injected, and V_0 is the potential change at the point $x = 0$. The symbol λ denotes the **length constant**, or *space constant*, which is related to the resistances of the axonal membrane, of the cytoplasm, and of the external solution by the expression

$$\lambda = \sqrt{\frac{R_m}{R_i + R_o}} = \sqrt{\frac{R_m}{R_l}} \tag{6-2}$$

in which R_m is the resistance of a unit length of axon membrane and R_l is the summed longitudinal internal and external resistances ($R_i + R_o$) over a unit length. Equation 6-1 reveals that when $x = \lambda$,

$$V_x = V_0 e^{-1} = V_0 \frac{1}{e} = 0.37 \, V_0 \tag{6-3}$$

As a result, λ is defined as the distance over which a steady-state potential shows a 63% drop in amplitude (see Figure 6-3). Length constants for real axons depend critically on R_m and range from 0.1 mm for a small axon with a low-resistance membrane to 5 mm for a large axon with a high-resistance (nonleaky) membrane.

Note that the value of λ is directly proportional to the square roots of both R_m and $1/R_l$ (equation 6-2), so the spread of electric current along the interior of an axon is enhanced by a high membrane resistance or by a low longitudinal resistance. The cable properties of neurons affect many aspects of neuronal function. For example, we will see shortly that the velocity at which APs are conducted along an axon is closely related to how effectively current can spread along the interior of the axon. Cable properties of nerve cells also shape the ways in which sensory information is processed in the nervous system (see Chapter 7).

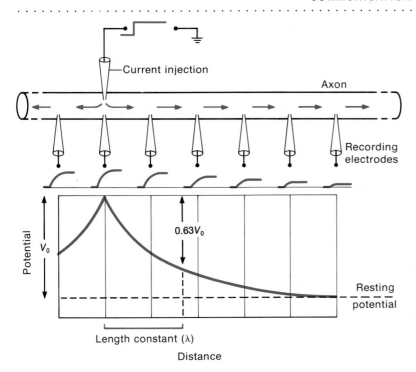

Figure 6-3 When current is transmitted passively, the change in the transmembrane steady-state potential decays exponentially with distance from the source. The steady-state ΔV_m that is caused by a square pulse of current injected at one point diminishes along a nerve or muscle fiber with increased distance along the fiber from the point of current injection. The length constant, λ, is defined as the distance over which the potential falls by $1 - (1/e)$, a reduction of 63% from its initial value at the point of injection (V_0).

Propagation of Action Potentials

Nerve cells typically have long axons (see, e.g., Figure 6-1), so they can carry information acquired in one part of the body to other, often distant, parts of the body by the propagation of APs. However, some neurons are sufficiently small that they accomplish many—even all—of their normal electrical functions without the aid of APs. In fact, many such cells are incapable of producing APs and are therefore referred to as **nonspiking neurons.** Their graded signals are conducted electrotonically to the axon terminals without the aid of all-or-none impulses. In these nonspiking, *local-circuit* neurons, the amplitude of the signals is attenuated as they spread through the cell, but the signals are still large enough at the terminals to modulate the release of a neurotransmitter. Local-circuit neurons are broadly distributed in the animal kingdom. For example, they can be found in such widely different locations as the vertebrate retina and other parts of the vertebrate central nervous system, in the barnacle eye, in the insect central nervous system, and in the crustacean stomatogastric ganglion. Local-circuit neurons are seldom more than a very few millimeters in overall length, and they are generally characterized by a high specific membrane resistance, which contributes to a large length constant and, hence, to the efficient and relatively undecremented electrotonic spread of signals.

More typically, however, communication between different parts of the nervous system depends critically on the propagation of APs along the axons of neurons, because the distances are too great for the electrotonic spread of signals to be effective. Similar propagation also takes place in many muscle cells. As discussed in Chapter 5, APs are large voltage changes across the neuronal membrane that have an invariant shape when plotted as a function of time and

that are propagated along axons without any decrement. In Chapter 5, we considered the events that produce an AP at a single location in an excitable cell. In order for APs to carry information, these events must occur over and over again along an axon. Basically, the signal is propagated when each activated patch of membrane excites neighboring patches. Typically, the change in V_m that occurs in an AP is about five times as great as the threshold depolarization, and a patch of axonal membrane that is producing an AP is able to excite the quiescent patch ahead of it, thereby causing the AP to propagate along the axon. The difference between the size of the depolarization during an AP and that of the threshold is known as the **safety factor** of the neuron.

The action potential is produced by two classes of ion channels, one selective for Na^+ and the other selective for K^+; these channels open and close with temporal precision to produce a transient reversal of the membrane potential (see Chapter 5). This reversal in the polarity of V_m can travel along the axon at high speed (e.g., as high as $120 \text{ m} \cdot \text{s}^{-1}$ in some large mammalian axons). At the initiation of an AP, voltage-gated Na^+-selective channels open, increasing the permeability of the plasma membrane to Na^+. When the Na^+ channels are open, Na^+ ions carry a large, but transient, current into the underlying region of the axon. This inward current spreads longitudinally along the axon and then leaks out across the membrane to complete the circuit of current flow. The longitudinal electrotonic spread of current depends on the cable properties of the axon described earlier. Thus, the influx of Na^+ that produces the upstroke of the AP (Figure 6-4A) supplies a current that spreads longitudinally, both forward and potentially backward as well, from the point of origin within the axon (Figure 6-4B). The electrotonic spread of current

A

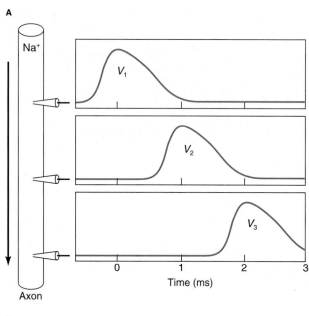

Figure 6-4 Electrotonic spread of current carries an action potential from one region of an axon to the next. **(A)** As an AP travels along an axon, electrodes placed along the axon record the same-sized change in V_m at successively later times. **(B)** An AP propagates along an axon as current from (1) an active patch of membrane (dark shading) spreads electrotonically to (2) a neighboring inactive patch of membrane, depolarizing it (light shading) and bringing it to threshold. Then current across (3) the newly active patch of membrane spreads to (4) a new inactive patch of membrane just ahead, depolarizing it and bringing it to threshold. This process is repeated over and over again along the axon. In this figure, the AP is traveling from left to right. A patch of membrane that has just returned to its polarized state is transiently incapable of becoming activated again, so the AP is propagated in only one direction. (See text for further discussion.)

B

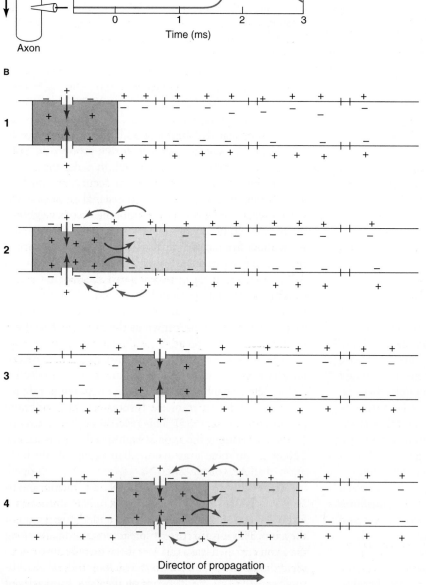

Director of propagation

away from active patches of membrane is responsible for the propagation of the AP.

When positively charged ions enter the patch of axonal membrane that is immediately ahead of the AP, that patch of membrane partially depolarizes (compare Figures 5-16 and 6-4A). In fact, as an undergraduate in 1937, Alan Hodgkin confirmed the hypothesis that the inactive membrane ahead of an AP becomes depolarized by electrotonically conducted current. His experiment is illustrated in Figure 6-5. Electrical current can flow in a circuit only if the circuit is complete, and this physical law must be obeyed in axons as well. The current that flows longitudinally within an axon—that is, in the direction of impulse propagation (to the right in Figure 6-4B)—must flow out of the axon—across unexcited parts of the membrane that lie ahead of the region of Na⁺ influx—and then back into the active region of the axon to complete the circuit. (Because the conductance of the resting membrane depends

primarily on open K^+ channels, the outward current is carried primarily by K^+.)

As the membrane ahead of the impulse becomes depolarized by local circuit current, voltage-gated Na^+ channels in that membrane open (i.e., g_{Na} increases), initiating an AP in this new patch of membrane. The newly excited region then generates a local-circuit current that depolarizes, and thereby excites, parts of the axon ahead of it. Thus, local-circuit current from each excited region depolarizes and excites the region immediately ahead of it. In this manner, the signal is continuously boosted and maintained at full strength as it travels along the axon. Note how different this method of signal transmission is from the decremented electrotonic conduction in passive transmission. The amount of depolarization required to bring an inactive membrane to threshold is about 20 mV, whereas the total depolarization during an AP is typically about 100 mV. Thus, an AP produces an approximately fivefold "boost" of the electrotonic signal.

Some of the current that enters an axon at the excited region spreads backward within the axon—that is, in the direction from which the impulse originated. However, under normal conditions, this backward-moving current cannot excite the membrane and produce a backward-traveling AP, because the membrane just behind a region of advancing excitation is in a refractory state (see Chapter 5). The Na^+ channels in that region are inactivated, and the K^+ channels are still open; so current is carried out of the cell as an efflux of K^+, preventing depolarization in that region. The delayed activation of K^+ channels is also important because it speeds up the repolarization of the membrane, priming it for future APs.

To summarize, propagation of a nerve impulse depends primarily on two factors:

1. The passive cable properties of an axon, which permit the electrotonic spread of local-circuit current from a region of Na^+ influx to neighboring regions of inactive membrane.
2. The electrical excitability of Na^+ channels in the axon's membrane. These Na^+ channels carry the current that produces regenerative amplification of the passive depolarization produced by local-circuit current.

You might ask why extracellular currents from an axon that is conducting APs do not excite other, nearby axons, creating "cross talk" between the axons. The answer, in short, is that the resistance of inactive membranes is so much greater than the resistance of the extracellular current path that only a tiny fraction of the total current produced by an active membrane flows into a neighboring inactive axon; this tiny current is not sufficient to bring the neighboring axon to threshold. However, extracellular currents generated by an AP can be detected by extracellular electrodes (Spotlight 6-1), which provides physiologists with a convenient way to monitor activity in the nervous system.

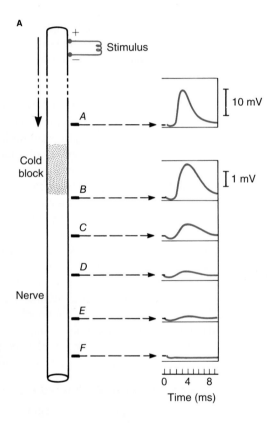

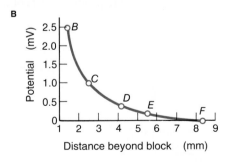

Figure 6-5 As an undergraduate, Alan Hodgkin demonstrated that the membrane ahead of an AP becomes depolarized electrotonically. **(A)** In this experiment, Hodgkin blocked the conduction of APs by cooling a small part of a nerve (stippled area). He then recorded the potentials at points (*B–F*) successively farther from the blocked region. Because no APs could travel through the cold block, any potential changes recorded beyond the blocked region must have been conducted electrotonically. **(B)** Under these conditions, the depolarization of the membrane decreased exponentially with distance from the region of the block. [Adapted from Hodgkin, 1937.]

Speed of Propagation

Johannes Müller, a leading nineteenth-century physiologist, declared in the 1830s that the velocity of the AP would never be measured. He reasoned that the AP, being an electrical impulse, must travel at a speed approaching that of light (3×10^{10} cm·s^{-1}), too fast to resolve over biological distances, even with the best instruments available at that time. His reasoning is understandable because he thought that all electrical signals are the same. However, as mentioned earlier, the AP is an electrical current carried by ions flowing through pores and charging a membrane capacitance, and this kind of signal travels much more slowly than electricity, which is carried by electrons flowing longitudinally through a conducting wire.

Indeed, within 15 years, one of Müller's own students, Hermann von Helmholtz, had measured the velocity of impulse propagation in frog nerves, using an elegantly simple method (Figure 6-6) that can easily be duplicated in a student laboratory with the use of a frog nerve-muscle preparation. The nerve is stimulated at each of two locations that are 3 cm apart, and the latency to the peak of the muscle twitch is determined. Suppose that the latency increases by 1 ms when the stimulating electrode is moved from location 1 to location 2 in the figure. The velocity of propagation, v_p, can then be calculated as

$$v_p = \frac{\Delta d}{\Delta t} = \frac{3 \text{ cm}}{1 \text{ ms}} = 3 \times 10^3 \text{ cm·s}^{-1} = 30 \text{ m·s}^{-1}$$

This value is seven orders of magnitude slower than the speed at which current flows through a copper wire or in an electrolyte solution. From such experiments, Helmholtz correctly concluded that the nerve impulse is more complex than a simple longitudinal flow of current along the nerve fiber.

The velocity of impulse propagation varies as a function of axon diameter and the presence of a myelin sheath. For example, in the large axons of vertebrates, APs can travel as fast as 120 m·s^{-1}; whereas, in very thin axons, they travel at only several centimeters per second (Table 6-1 and Figure 6-7).

The conduction velocity of an AP depends primarily on how fast the membrane ahead of the active region is brought to threshold by the local-circuit currents. The greater the length constant, the farther the local-circuit current flows before it becomes too weak to bring neighboring membrane regions to threshold and the more rapidly the membrane ahead of the excited region depolarizes. The effect of the length constant on conduction velocity is demonstrated when the length constant is decreased by placing an axon in a fluid that has a higher resistance than saline — for example, oil or air. This procedure leaves only a thin film of saline on the surface of the axon, and the length constant is decreased because the external longitudinal resistance (R_o in equation 6-2) is increased. These conditions decrease the rate of conduction from its value when the same axon is immersed in saline.

Among animals, the conduction velocity of APs has increased with evolutionary increases in the length constant of axons. One of the ways in which the length constant has been increased — typified by the giant axons of squid, arthropods, annelids, and teleosts — is by an increase in axonal diameter, which reduces the internal longitudinal resistance (R_i in equation 6-2). See Spotlight 6-2 for a more detailed explanation of this effect. Giant axons produce rapid and synchronous activation of locomotor reflexes with important roles in the escape or withdrawal responses of many species, among them some molluscs (e.g., squid), some arthropods (e.g., crayfish, the cockroach), and some annelids (e.g., the earthworm). The presence of a large number of axons in a bundle, however, severely limits the possible increases in axonal diameter. In the vertebrates, a single nerve can consist of tens of thousands of axons, and another mechanism, **myelination**, evolved to increase the length constant.

Rapid, Saltatory Conduction in Myelinated Axons

Some types of glial cells are wrapped around segments of axons many times (Figure 6-8A) to produce layers of fatty

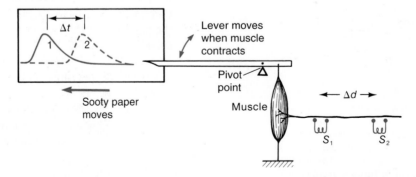

Figure 6-6 Hermann von Helmholtz measured the time from stimulus to contraction in a frog nerve-muscle preparation to determine conduction velocity along a nerve. Electrodes to stimulate the nerve were first placed at position S_1, and contraction of the muscle moved a lever that scratched a record on a rapidly moving sheet of sooty paper. (Notice that time is plotted as the paper moves.) Then the electrodes were moved to position S_2, the paper having been aligned so that the time of the second stimulus exactly coincided with the time of the first stimulus. The tracings made by the lever showed the change in latency, Δt, when the location of the electrodes was changed. The conduction velocity can be calculated from the difference in latency of the muscle twitch evoked by stimulating the nerve at the two different locations.

TABLE 6-1
The diameter of frog axons and the presence or absence of myelination control the conduction velocity

Fiber type	Axon diameter (μm)	Conduction velocity (m·s^{-1})
Myelinated fibers		
Aα	18.5	42
Aβ	14.0	25
Aγ	11.0	17
B	Approx. 3.0	4.2
Unmyelinated fibers		
C	2.5	0.4–0.5

Source: Fiber classification by Erlanger and Gasser, 1937. Adapted from Davson, 1964.

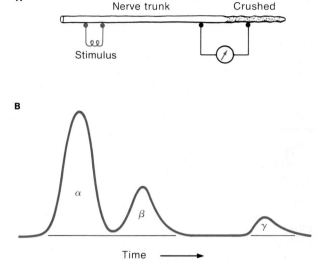

Figure 6-7 A single frog nerve contains axons with many different conduction velocities. **(A)** Experimental setup for stimulating and recording from a nerve, which contains many axons. **(B)** Externally recorded "compound APs" (i.e., the summed signals from all active fibers in the group). The α-fibers have the largest diameter and the highest conduction velocity. The γ-fibers have the smallest diameter and the lowest conduction velocity of those shown here (but see Table 6-1). The nerve was stimulated before the beginning of the record shown here. In practice, one end of the nerve is crushed to act as a neutral reference site (see Spotlight 6-1).

membranes (collectively known as *myelin*). These fatty layers have two effects on the electrical properties of neurons: they increase the transmembrane resistance, and they decrease the effective neuronal membrane capacitance. Myelin produces these effects because the layers of membrane act electrically as additional resistive and capacitative elements in the equivalent circuit of the membrane (see Figure 6-2). The resistance between the cytoplasm and the extracellular fluids increases as a function of the number of membrane layers wrapped around the axon, which may be as many as 200. The capacitance decreases because the

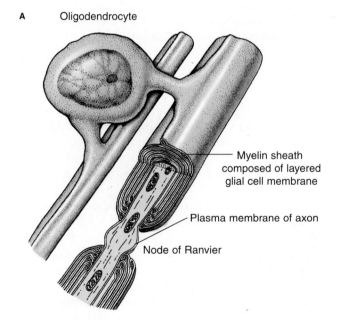

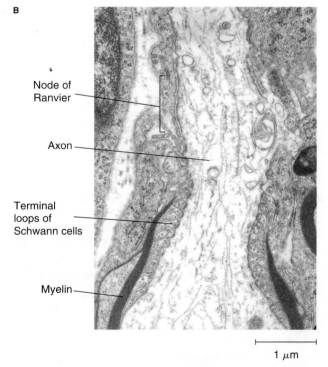

1 μm

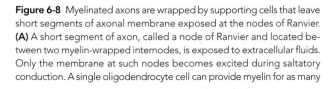

Figure 6-8 Myelinated axons are wrapped by supporting cells that leave short segments of axonal membrane exposed at the nodes of Ranvier. **(A)** A short segment of axon, called a node of Ranvier and located between two myelin-wrapped internodes, is exposed to extracellular fluids. Only the membrane at such nodes becomes excited during saltatory conduction. A single oligodendrocyte cell can provide myelin for as many as 50 internodes on several adjacent axons. This schematic illustration does not show how tightly the layers of myelin are wrapped around the axon. **(B)** Electron micrograph of a node of Ranvier in a spinal root of a young rat. In these nerves, a segment of axonal membrane about 2 μm long is exposed to extracellular fluid. [Part B courtesy of Mark Ellisman.]

SPOTLIGHT 6-1

EXTRACELLULAR SIGNS OF IMPULSE CONDUCTION

Nerve impulses can be recorded with a pair of extracellular electrodes (diagrams A–C). The recording electronics are arranged so that a negative potential in electrode 1 will cause an oscilloscope beam to go up, whereas a similar potential recorded by electrode 2 will make the beam go down; positive potentials would do the opposite. An AP passing along an axon is seen from the outside as a wave of negative potential, because the cell exterior becomes more negative than its surroundings when Na^+ ions flow into the cell during the rising phase of the AP. As a result, when an AP passes by the two electrodes, it produces a biphasic waveform on the oscilloscope (diagram A).

The recording is simplified if the AP can be prevented from invading the part of the axon that is in contact with electrode 2, which is accomplished by anesthetizing, cooling, or crushing that part of the axon (diagram B). A similar effect can be obtained by placing electrode 2 in the bath, at some distance from the axon (diagram C).

Extracellular recordings are frequently made from nerve bundles or tracts that contain many axons (diagram D). In this situation, the summed activity of many axons gives a compound recording, the characteristics of which depend on the number of axons conducting and on their relative timing and current strength. Larger axons generate larger extracellular currents, because the amount of current flowing across a membrane in-

Facing page The activity of axons in a nerve bundle can be recorded with extracellular electrodes. Stippled areas in each diagram show the location of the depolarized membrane at five different times, from t_1 to t_5. The records show what you would see—for example, on an oscilloscope—for each method of recording. Numbers on the records correspond to the five different times. In all cases, the oscilloscope signal is produced by electronic subtraction of the potential measured by electrode 2 from the potential measured by electrode 1. **(A)** A biphasic recording compares the electric potential at two locations along the nerve bundle. As an AP passes along the axons, first one and then the other electrode records the AP. **(B)** Recording with one electrode on a crushed part of the nerve. **(C)** Recording the electrical potential near the axon with reference to the potential at a distant point in the bath. **(D)** Extracellular recordings pick up signals from multiple axons in a nerve bundle. Sometimes the sizes of the extracellular signals enable an experimenter to distinguish between fibers of different diameters. For example, the signal from axon c is always large in the record shown here, whereas the signal from axon a is small. A larger current flowing along a larger axon can produce a larger voltage between the recording electrodes. (Notice, though, that the distance between the electrode and the axon also can affect the amplitude of the signal.)

creases in direct proportion to the area of the membrane. The size of an AP recorded with extracellular electrodes is proportional to the amount of current flowing through the extracellular fluid, so APs from axons with a large diameter appear bigger when recorded extracellularly, even though the value of ΔV_m measured inside the axons would be no larger than that of ΔV_m in smaller nerve fibers. Because of these amplitude differences in extracellularly recorded APs, signals carried by particular axons can often be distinguished by their size.

SPOTLIGHT 6-2

AXON DIAMETER AND CONDUCTION VELOCITY

The velocity at which an AP is propagated depends in part on the forward distance over which the current arising from the Na^+ influx can spread at any instant. This distance depends on the relation between the longitudinal resistance (within the axon) and the transverse resistance (across the axon membrane) encountered by currents flowing in a unit length of axon (equation 6-2). The transverse resistance, R_m, of a unit length, l, of axon membrane is *inversely proportional* to the radius, r, of the axon, because the area, A_s, of a cylinder of unit length is equal to $2\pi rl$. The longitudinal resistance, R_i, of a unit length of axoplasm is inversely proportional to the cross-sectional area, A_x, of the axon. Because $A = \pi r^2$, resistance R_i is *inversely proportional to the square of the radius*. It follows then that, for any increase in radius, the drop in R_i will be greater than the drop in R_m. The length constant $\lambda = \sqrt{R_m/(R_i + R_o)}$ (equation 6-2), so

the larger drop in longitudinal resistance, R_i, that accompanies an increase in axon diameter produces an increase in λ. Typically, $R_i \gg R_o$, so λ is proportional to k times the square root of r, where k is simply a constant; in other words, the length constant increases in proportion to the radius of the axon. As the radius increases, λ increases.

Because the velocity of propagation depends on the *rate* of depolarization at each point ahead of the AP, membrane capacitance cannot be ignored. Note that the time constant ($R_m \times C_m$) of a unit length of axon membrane remains constant as axon diameter changes, because capacitance (C_m) *increases* in direct proportion to surface area, whereas resistance (R_m) *decreases* in proportion to an increase in membrane area. The increased λ that accompanies increased axon diameter therefore occurs without changing the time constant of the membrane. Thus an increase in diameter produces a greater outward membrane current at distance x without an increase in membrane time constant, and the increased rate of depolarization brings the membrane to threshold sooner at every distance and increases the conduction velocity.

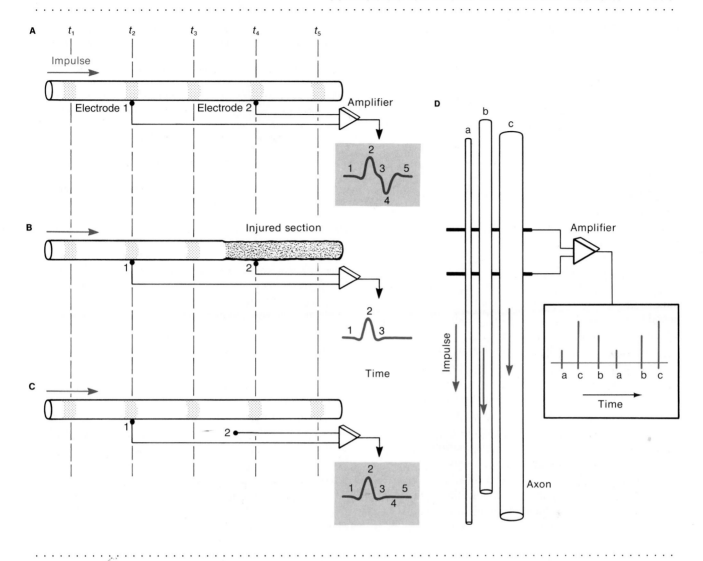

myelin layer is very thick. This reduction in C_m means that less capacitive current is required to change V_m, so more charge can flow down the axon to depolarize the next segment. The changes in resistance and capacitance greatly increase the length constant, λ, of the axonal membrane that is covered by myelin, thus enhancing the efficiency with which longitudinal current spreads. However, this insulation would not have this effect if it completely covered the axon, because the electrotonically conducted current would eventually decrease to zero as a function of distance. Instead, the length of the myelinated segments are typically about 100 times the external diameter of the axon, ranging from 200 μm to 2 mm long, and the segments are interrupted by short, unmyelinated gaps called **nodes of Ranvier**, at which about 10 μm of the excitable axon is exposed to the extracellular fluid (Figure 6-8B). The segments of axon that lie under the myelin wrapping are called **internodes**.

In the course of development, myelin is laid down around the axons of peripheral and central tracts in vertebrates by two kinds of glial cells: *Schwann cells* in periph-eral nerves and *oligodendrocytes* in the central nervous system. Between nodes of Ranvier, the sheath is so close to the axon membrane that it nearly eliminates the extracellular space surrounding the axon membrane. Moreover, the internodal axon membrane has been found to lack voltage-gated Na^+ channels. Thus, when a local-circuit current flows in advance of the AP, it exits the axon almost exclusively through the nodes of Ranvier. As noted earlier, very little current is expended in discharging membrane capacitance along the internodes, because of the low capacitance of the thick myelin sheath. An AP that is initiated at one node electrotonically depolarizes the membrane at the next node; thus, in myelinated axons, APs do not propagate continuously along the axonal membrane, as they do in nonmyelinated nerve fibers. Instead, APs are produced only in the small areas of the membrane exposed at the nodes of Ranvier. The result is **saltatory conduction,** a series of discontinuous and regenerative depolarizations that take place only at the nodes of Ranvier, as illustrated in Figure 6-9. The velocity of signal transmission is greatly enhanced because the electrotonic spread of local circuit current occurs

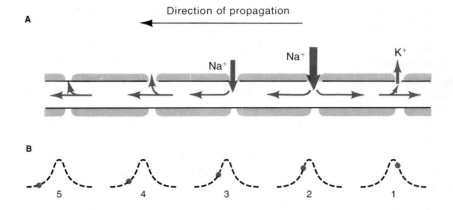

Figure 6-9 In saltatory conduction, the action potential jumps from node to node. **(A)** Current spreads longitudinally between nodes. The large red arrows indicate Na⁺ influx through activated Na⁺ channels that are located at the nodes. The smaller red arrow represents the later outflux of K⁺ through activated K⁺ channels. **(B)** Dots indicate the value of V_m at a single instant at each node shown in part A. At site 1, the membrane is in the falling phase of an AP; at site 2, the membrane is in the rising phase. At sites 3, 4, and 5, the membrane is in successively earlier phases of an AP.

rapidly over internodal segments. The conduction velocity of myelinated fibers varies from a few millimeters per second to more than 100 m·s⁻¹, in contrast with unmyelinated fibers of similar diameter, which conduct at a fraction of a meter per second (see Table 6-1).

The evolution of saltatory conduction and the resulting higher speed of AP propagation was probably crucial for the successful coordination of activity in the large muscles of vertebrates. Myelination allows APs to travel rapidly in many axons within a compact nerve trunk. The importance of the specialized insulation provided by myelin for coordinating neuronal information is particularly evident in such demyelinating diseases as multiple sclerosis. In this disease, the myelin sheath is reduced or eliminated in some nerves, making the velocity of neuronal transmission highly variable among neurons, which severely compromises the control of coordinated movement.

TRANSMISSION OF INFORMATION BETWEEN NEURONS: SYNAPSES

All information processing done by neurons depends on the transmission of signals from one neuron to another, which is accomplished at structures called *synapses*. At **electrical synapses,** the presynaptic neuron is electrically coupled to the postsynaptic neuron by particular proteins within the membranes. Transmission across electrical synapses proceeds very much like signal transmission along a single axon. However, electrical synapses are relatively rare. Most signaling between neurons takes place at **chemical synapses.** At a chemical synapse, APs in the presynaptic neuron cause the release of neurotransmitter molecules that diffuse across a narrow space (about 20 nm wide), called the **synaptic cleft,** that separates the membranes of pre- and postsynaptic neurons. As recently as the 1970s, only a handful of chemicals were known to be synaptic transmitters. All of them were thought to act in a similar fashion, in accord with the results obtained in studies of transmission at the synapses, called **neuromuscular junctions** (NMJs), between motor neurons and the skeletal muscles that they control. Today, more than 50

neurotransmitters have been identified in the wide range of animals studied, more are being discovered all the time, and we now know that their modes of action vary greatly. Initially, neurotransmitters were thought to act only by causing the voltage in a postsynaptic cell to change, either by hyperpolarizing the cell or by depolarizing it. However, neurotransmitters can also increase or decrease the number of ion channels inserted into the membrane of the postsynaptic cell, alter the excitability of the postsynaptic cell by changing the rate at which ion channels open and close, or modify their sensitivity to activating signals. The discovery of these varied modes of action has dramatically broadened our understanding of the role that synapses play in neuronal communication.

Synaptic transmission was a subject of controversy for a very long time. Early in the twentieth century, the great histologist Santiago Ramon y Cajal used the light microscope and a silver-based staining technique developed by neuroanatomist Camillo Golgi to show that neurons stain as discrete units. In spite of this observation, many anatomists continued to believe that the nervous system was a continuous reticulum, rather than a set of morphologically separate nerve cells. It was not until electron microscopy was developed in the 1940s that unequivocal evidence was obtained supporting the notions that neurons are indeed separate from one another and that particular regions of neurons are specialized for communication between cells.

However, in 1897, long before the ultrastructural basis of neuron–neuron interactions was determined, the functional junction between two neurons was given the name *synapse* (from the Greek, meaning "to clasp") by Sir Charles Sherrington, who is widely regarded as the founder of modern neurophysiology. It was his conclusion that ". . . the neurone itself is visibly a continuum from end to end, but continuity fails to be demonstrable where neurone meets neurone—at the synapse. There a different kind of transmission may occur" (Sherrington, 1906). Although Sherrington had no direct information about the microstructure or microphysiology of these specialized regions of interaction between excitable cells, he had extraordinary insight, the sources of which were his cleverly designed experiments on the spinal reflexes of animals, most of them mammals. Among other things, he deduced that some synapses are *excitatory,* increasing the probability that APs will arise in the postsynaptic cell, and that others are *inhibitory,* reducing the probability of APs in the postsynaptic cell.

In this section, we start by considering synaptic transmission across electrical synapses, which is similar to signal conduction along axons. We then turn to the topic of chemical synapses, dealing first with transmission at the neuromuscular junction and then with other, more recently discovered types of chemical synapses.

Synaptic Structure and Function: Electrical Synapses

Electrical synapses transfer information between cells by direct ionic coupling. At an electrical synapse, the plasma membranes of the pre- and postsynaptic cells are in close apposition and are coupled by protein structures called **gap junctions** (Figure 6-10A) through which electrical current can flow directly from one cell into the other (Figure 6-10B; see also Chapter 4). Because current travels across gap junctions, an electrical signal in the presynaptic cell produces a similar, although somewhat attenuated, signal in the postsynaptic cell by simple electrical conduction through the junction (Figure 6-10C). Thus, at an electrical synapse, the transfer of information occurs by purely electrical means, without the intervention of a chemical transmitter, and a key feature of electrical synaptic transmission is its rapidity. As we will soon see, signal transmission across chemical synapses is always slower than purely electrical signal transmission.

Electrical transmission can be illustrated experimentally by injecting current into one cell and measuring the effect in a connected cell (see Figure 6-10C). A subthreshold current pulse injected into cell *A* elicits a transient change in the membrane potential of that cell. If a significant fraction of the current injected into cell *A* spreads through gap junctions into cell *B*, it will cause a detectable change in the V_m of cell *B* as well. Because there is a potential drop as the current crosses the gap junctions, the potential change recorded across the membrane of cell *B* will always be less than that recorded in cell *A*. The gap junctions through which current flows from one cell to another are generally, but not always, symmetrical in resistance to the passage of current—that is, current generally meets the same resistance as it passes in either direction between the two cells. However, at some specialized synapses, the transfer of current between the two coupled cells occurs readily in one direction, but not in the other (Figure 6-10D). Such junctions are said to be *rectifying.*

The transmission of an AP through an electrical synapse is basically no different from propagation within one cell, because both phenomena depend on the passive spread of local circuit current beyond the AP to depolarize and excite the region ahead. Because the safety factor

of an AP (the ratio of the change in V_m during an AP to the change in V_m required to bring a cell to threshold) is typically about 5, the attenuation of the change in V_m from one cell to the next must be no greater than the safety factor if the depolarization of the postsynaptic cell is to reach threshold and initiate an impulse. Therefore, a single presynaptic action potential might be unable to provide enough local circuit current across an electrical synapse to elicit an action potential in the postsynaptic cell, which may be one evolutionary reason why electrical synapses are less common than chemical synapses. However, the fact that electrical synapses conduct signals much more rapidly than do chemical synapses gives them definite advantages where rapid signal transmission is important.

Electrical transmission between excitable cells was first discovered in 1959 by Edwin J. Furshpan and David D. Potter, who were studying the nervous system of crayfish at the time. They found that an electrical synapse between the crayfish lateral giant nerve fiber and a large motor axon has the unusual property of passing current preferentially in one direction (see Figure 6-10D). Since their early work, electrical transmission has been discovered between cells in the vertebrate central nervous system and in the vertebrate retina, between smooth muscle fibers, between cardiac muscle fibers, between receptor cells, and between axons. The rapidity with which current crosses electrical synapses makes this means of information transfer particularly effective in the synchronization of electrical activity within a group of cells. It is also effective for rapidly transmitting information across a series of cell–cell junctions—for example, in the giant nerve fibers of the earthworm, which are composed of many segmental axons connected in series along the worm's body, and in the myocardium of the vertebrate heart, in which signals are passed between muscle cells.

At some synapses, transmission is both electrical and chemical. Such combined synapses were first identified in cells of the avian ciliary ganglion, and they have also been found in a circuit controlling the fish escape response, in synapses made by some neurons onto spinal interneurons of the lamprey, and in synapses onto frog spinal motor neurons. However, as interesting as they are, combined synapses are unusual phenomena.

Synaptic Structure and Function: Chemical Synapses

A common mode of synaptic transmission is known as *fast chemical synaptic transmission,* which is found at many synapses in the central nervous system and at the neuromuscular junction. (Although this transmission is called "fast," it is in fact considerably slower than transmission across electrical synapses.) At neuromuscular junctions, the neurotransmitter **acetylcholine** (ACh) is stored in membrane-enclosed vesicles and is secreted by exocytosis into the extracellular fluid separating the neuron and the muscle. The sequence of events at these nerve terminals is

A

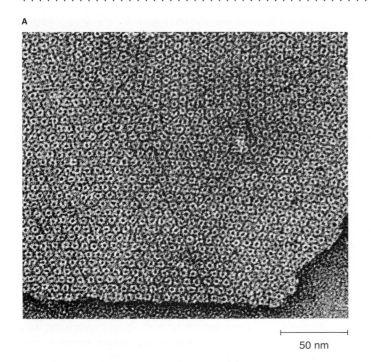

50 nm

B

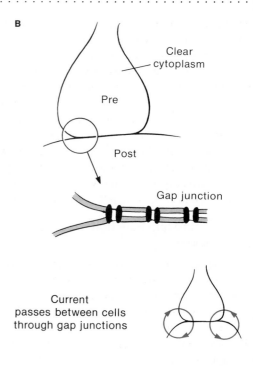

Clear cytoplasm

Pre

Post

Gap junction

Current passes between cells through gap junctions

C

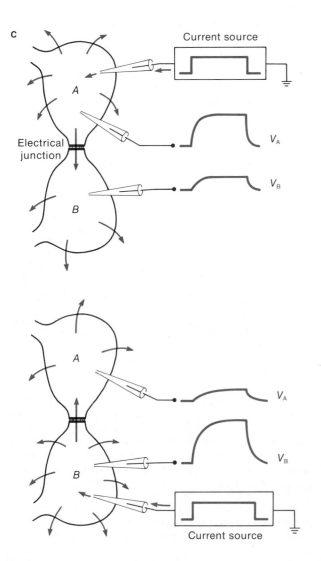

Current source

V_A

V_B

Electrical junction

A

B

A

B

V_A

V_B

Current source

D

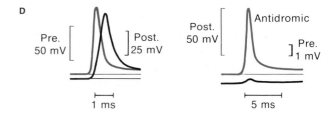

Pre. 50 mV

Post. 25 mV

1 ms

Post. 50 mV

Antidromic

Pre. 1 mV

5 ms

Figure 6-10 Pre- and postsynaptic cells are in electrical continuity at electrical synapses, permitting rapid signal transmission between the cells. **(A)** Electron micrograph of densely packed gap junctions in a membrane. Each "doughnut" in this micrograph is a protein complex that forms a pore, allowing ions and small molecules to move between the coupled cells. When cells are coupled by gap junctions, the membrane of each cell contains these protein complexes, and the complexes line up with one another, forming continuous channels between the cytosolic compartments of the two cells. **(B)** Gap junctions connecting pre- and postsynaptic membranes permit ionic currents to flow between cells. **(C)** In electrically coupled cells, the injection of current into one cell elicits a potential change in both cells. Usually, the electrical coupling at electrical synapses is symmetric, so the injection of current into either cell changes V_m in both cells, although V_m changes more in the cell into which current was injected than it does in the cell coupled to it. There are, however, exceptions, such as the one shown in part D. **(D)** The giant electrical synapse in the crayfish illustrates the relation between pre- and postsynaptic signals in an electrical synapse that has asymmetric electrical coupling. *(Left)* An AP in the presynaptic axon is transmitted across the electrical junction, bringing the postsynaptic cell to threshold and eliciting an AP with only a small delay. This recording is a typical example of signal transmission across an electrical synapse. *(Right)* At this asymmetric electrical synapse, however, an AP in the axon that was postsynaptic in the record at the left fails to produce a significant potential change in the neuron that was presynaptic in the record at the left. Injection of current pulses into one cell and then into the other showed that there is preferential flow of current in one direction between these two neurons, an arrangement that is unusual at an electrical synapse. [Part A courtesy of N. Gilula; part D adapted from Furshpan and Potter, 1959.]

summarized in Figure 6-11. Briefly, when an AP travels down an axon and spreads to the axon terminals, neurotransmitter molecules that are stored in membrane-bounded spheres, called **synaptic vesicles,** within the terminals are released into the synaptic cleft, the fluid-filled space separating the presynaptic and postsynaptic cells. The liberated neurotransmitter molecules bind to specific protein receptor molecules in the postsynaptic membrane, which in Figure 6-11 include ligand-gated ion channels. When neurotransmitter molecules bind to the receptor proteins, the result is a brief ionic current through the membrane of the postsynaptic cell. This mechanism is the basis for synaptic transmission in all animals.

The existence of chemical transmission and transmitter substances was the subject of intense scientific debate in the first six decades of the twentieth century. The first direct evidence for a chemical transmitter substance was obtained by Otto Loewi in 1921. In his experiments, he isolated a frog heart with the vagus nerve attached. When he electrically stimulated the vagus nerve, the heart rate slowed down, but he also found that when the stimulated vagus nerve caused the heart to beat more slowly, a substance was released into the surrounding saline solution that could cause a second frog heart to beat more slowly, too. Loewi's finding led to the subsequent discovery that acetylcholine is the transmitter substance released by postganglionic neurons of the parasympathetic nervous system in response to stimulation of the vagus nerve (see Chapter 11) and by motor neurons innervating skeletal muscle in vertebrates.

For decades, all synaptic transmission was thought to operate by mechanisms that were very similar to transmission at the neuromuscular junction. However, that view has changed. It is now known that, in addition to fast chemical synapses, most species also have synapses that produce *slow chemical synaptic transmission,* in which communication between pre- and postsynaptic cells is slower than at the neuromuscular junction and takes place by a different postsynaptic mechanism. In addition, although physiologists believed for decades that each synaptic terminal can contain only a single neurotransmitter, it has recently been discovered that many neurons synthesize and release more than one transmitter substance; in such neurons, one of the substances may produce fast transmission while the other produces slow transmission. In many respects, slow synaptic transmission is similar to rapid chemical transmission (Figure 6-12). The neurotransmitter molecules are packed into vesicles in the presynaptic terminal and are released by exocytosis that is triggered by APs. However, there are significant differences between these two synaptic mechanisms. In slow synaptic transmission, the neurotransmitters are typically synthesized from one or more amino acids and are called **biogenic amines,** if they contain a single amino acid, or **neuropeptides,** if they consist of several amino acid residues. As the name implies, the onset of the postsynaptic response is slower (hundreds of milliseconds), and it can last much longer (from seconds to hours). Vesicles used in the fast system are synthesized and packaged within the nerve terminals, whereas vesicles in the slow system are larger and are usually synthesized in the cell body, after

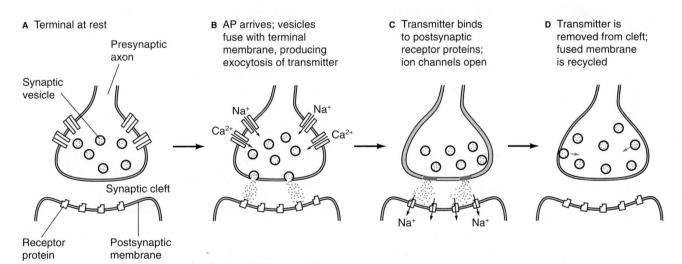

Figure 6-11 In fast chemical synaptic transmission, signals in the pre- and postsynaptic cells are linked by chemical neurotransmitters. The pre- and postsynaptic cells are not electrically coupled, and there is no direct flow of current between them. Ionic current flows across the postsynaptic membrane only when ligand-gated ion channels are open in the postsynaptic membrane. **(A)** At rest, transmitter molecules are packaged into membrane-bounded vesicles contained in the axon terminals. **(B)** When an AP enters the presynaptic terminal, it causes voltage-gated Ca^{2+} channels in the membrane to open, allowing Ca^{2+} ions to flow into the terminal. The increase in intracellular free Ca^{2+} causes synaptic vesicles to fuse with the presynaptic membrane, releasing neurotransmitter into the

synaptic cleft by exocytosis. **(C)** Neurotransmitter molecules diffuse across the synaptic cleft, driven by their concentration gradient, and bind to receptor proteins in the postsynaptic membrane, opening ligand-gated ion channels. In this case, Na^+ flows through the open channels into the presynaptic cell. The vesicle membrane remains fused with the membrane of the terminal, but it moves to the sides of the terminal. **(D)** Transmitter molecules are removed from the cleft, the postsynaptic ion channels close, and the membrane that was added to the presynaptic terminal when the synaptic vesicles fused is eventually recycled into the terminal (small arrows) and may be reused for more vesicles.

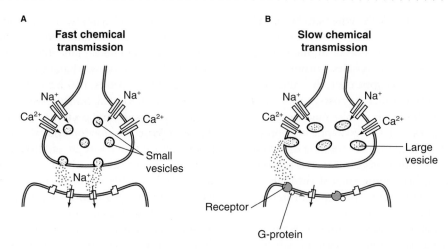

Figure 6-12 Fast chemical synaptic transmission and slow chemical synaptic transmission act through different postsynaptic mechanisms. **(A)** In fast chemical transmission, neurotransmitters are synthesized in the terminals and stored in small, clear vesicles. These transmitters are typically small molecules. The vesicles are located near the plasma membrane, and transmitters are released by exocytosis into the synaptic cleft through specialized sites on the membrane. After they are released, these neurotransmitters act on ligand-gated channels in the postsynaptic membrane. **(B)** In slow synaptic transmission, the transmitters are typically larger molecules—for example, peptides containing many amino acids. These transmitters are stored in large, distinctive vesicles and are released from sites that lack morphological specialization and that are located away from the sites from which the fast neurotransmitters are released. In the postsynaptic cell, these neurotransmitters typically act indirectly through G protein–linked receptors to modify channels and other intracellular processes. Single neurons may produce both kinds of transmission, and a single neurotransmitter may affect postsynaptic neurons both by means of ligand-gated channels and by means of G-protein–coupled receptors.

which they are transported to the nerve terminal. Vesicles that mediate slow synaptic transmission may release their transmitter molecules at many sites in the presynaptic terminal and usually affect the postsynaptic cell, not through ligand-gated channels but by altering the levels of intracellular second messengers through intermediate molecules called G proteins. Physiological and anatomical evidence indicates that single presynaptic neurons may participate in both kinds of neurotransmission.

The release of neurotransmitter into the synaptic cleft is controlled by mechanisms that are common to both fast and slow synaptic transmission. When an AP arrives at the axon terminals, it activates voltage-gated Ca^{2+} channels in the membrane of the terminals, allowing Ca^{2+} to enter the terminal (see Figure 6-11B). The increased concentration of Ca^{2+} inside the terminal initiates the exocytosis of vesicles containing the transmitter substance, dumping neurotransmitter molecules into the synaptic cleft where they diffuse away from the presynaptic terminal. In fast synaptic transmission, neurotransmitter-containing synaptic vesicles fuse with the plasma membrane at specialized sites called **active zones**. After crossing the synaptic cleft, some neurotransmitter molecules bind to receptor molecules in the postsynaptic membrane. When transmitter molecules bind to these receptor molecules, they modify ionic current through channels that are associated with the receptor molecules, allowing permeant ions to carry a postsynaptic current driven by electrochemical gradients. In slow transmission, the neurotransmitter affects the postsynaptic cell through G-protein intermediates to modify activities of intracellular second messengers that then influence ion channels or other intracellular processes (see Figure 6-12). The postsynaptic current produced by the neurotransmitter causes a change in the membrane potential of the postsynaptic cell. If the sum of the potential changes caused by many such synaptic events is sufficient to exceed the threshold potential in the postsynaptic cell, an AP will be initiated in the postsynaptic cell.

In fact, the currents that are generated in the postsynaptic cell may either increase or decrease the probability that APs will occur in that cell; that is, synaptic effects can be either excitatory or inhibitory. What makes a synaptic signal one or the other is examined later in this chapter.

 What are some relative advantages and disadvantages of chemical synaptic transmission compared with electrical synaptic transmission? Of fast compared with slow chemical transmission? Why might all of these different forms of synaptic transmission have evolved?

Fast Chemical Synapses

The most extensive studies of synaptic transmission have been done on fast chemical transmission at the neuromuscular junctions (also called *motor terminals* or *motor endplates*) of vertebrate skeletal muscle, where acetylcholine has been shown to be the neurotransmitter. We will use the neuromuscular junction as our primary example, because it is so well studied. It is a good example because fast chemical synaptic transmission between neurons within the central nervous system closely resembles transmission at the neuromuscular junction, although in many cases the transmitters are different.

Structural features

The frog motor endplate (Figure 6-13) includes structural specializations of the presynaptic terminal, of the postsynaptic membrane, and of associated Schwann cells. The axon of the presynaptic motor neuron terminal bifurcates, and the branches, each of which is approximately 2 μm in diameter, lie in a longitudinal depression along the surface of the muscle fiber. The muscle membrane lining the depression is thrown into transverse **junctional folds** at intervals of 1 to 2 μm. Directly above these folds within the nerve terminal are the active zones—transverse regions of slight thickening in the presynaptic membrane above which are clustered many synaptic vesicles. The vesicles are released along the active zones by the process of exocytosis (Figure 6-14). There

are thousands of vesicles, each about 50 nm in diameter, in one presynaptic terminal. For example, the branches of the nerve terminal innervating a single frog muscle fiber typically contain a total of about 10^5 synaptic vesicles. When the vesicles fuse with the plasma membrane and release transmitter molecules into the synaptic cleft, the transmitter molecules reach the postsynaptic membrane by diffusing down their concentration gradient. The cleft itself is filled with a mucopolysaccharide that "glues" together the pre- and postsynaptic membranes, both of which usually show some degree of thickening at the synapse. The vesicular membrane that fused with the plasma membrane of the terminal is taken up into the terminal and may be recycled (see Figure 6-11D).

When acetylcholine (ACh) is released into the synaptic cleft, it can bind to ACh-specific receptor molecules in the postsynaptic membrane of the endplate, causing ion channels that are selective for Na$^+$ and K$^+$ to open briefly.

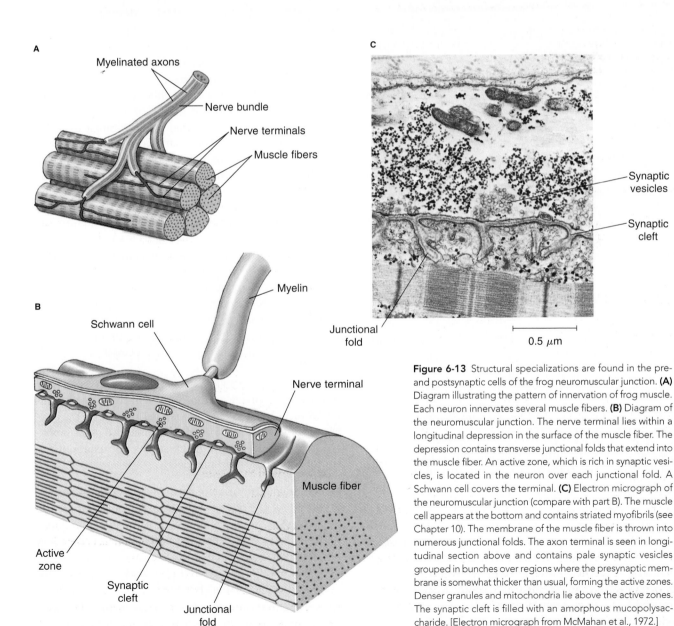

Figure 6-13 Structural specializations are found in the pre- and postsynaptic cells of the frog neuromuscular junction. **(A)** Diagram illustrating the pattern of innervation of frog muscle. Each neuron innervates several muscle fibers. **(B)** Diagram of the neuromuscular junction. The nerve terminal lies within a longitudinal depression in the surface of the muscle fiber. The depression contains transverse junctional folds that extend into the muscle fiber. An active zone, which is rich in synaptic vesicles, is located in the neuron over each junctional fold. A Schwann cell covers the terminal. **(C)** Electron micrograph of the neuromuscular junction (compare with part B). The muscle cell appears at the bottom and contains striated myofibrils (see Chapter 10). The membrane of the muscle fiber is thrown into numerous junctional folds. The axon terminal is seen in longitudinal section above and contains pale synaptic vesicles grouped in bunches over regions where the presynaptic membrane is somewhat thicker than usual, forming the active zones. Denser granules and mitochondria lie above the active zones. The synaptic cleft is filled with an amorphous mucopolysaccharide. [Electron micrograph from McMahan et al., 1972.]

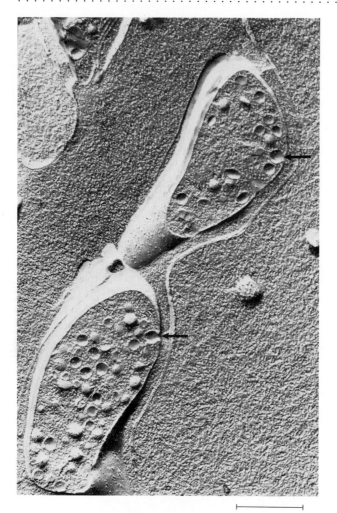

0.2 μm

Figure 6-14 The presynaptic terminal at a neuromuscular junction contains thousands of vesicles. A transverse section in a freeze-etched specimen from the electric organ of the ray *Torpedo*. Synaptic vesicles can be seen in the terminal. Two vesicles (black arrows) had fused with the presynaptic membrane when the tissue was fixed, illustrating the process of exocytosis. Calibration bar = 0.2 μm. [From Nickel and Potter, 1970.]

However, in the synaptic cleft, ACh is subject to hydrolysis by the enzyme *acetylcholinesterase* (AChE). This enzyme can be detected by histochemical methods and is located in the junctional folds. The removal of neurotransmitter molecules from the synaptic cleft is essential because its effect is to limit the time during which the transmitter can be active. In a cholinergic synapse, hydrolysis of ACh inactivates the transmitter and turns off the synaptic transmission. While many neurotransmitters are inactivated by enzymatic action, others are taken up by the presynaptic terminals, carried by specialized transporter molecules.

Synaptic potentials

In 1942, Stephen W. Kuffler recorded electrical potentials from single fibers of frog muscle and found depolarizations that were intimately associated with the motor endplate. These depolarizations occurred in response to motor neuron APs and preceded the AP generated in the muscle cell. The potential changes, recorded with extracellular elec-

trodes, were greatest in amplitude at the endplate and gradually became smaller with distance, so they were named **endplate potentials** (epps), or, more generally, *postsynaptic potentials* (psps). Kuffler correctly concluded that the arrival of an AP in the presynaptic terminal could cause local depolarization of postsynaptic membrane and thus initiate the propagation of an AP through the muscle.

The development of the glass capillary microelectrode in the late 1940s made it possible to record potentials produced within a much smaller tissue volume and, hence, to identify more exactly the source of endplate potentials. Numerous intracellular studies of synaptic transmission at the frog neuromuscular junction, performed largely at the laboratory of Bernard Katz in England, have provided a remarkably complete picture of electrical events at this synapse.

Like neurons, muscle fibers have a resting potential across their membranes (see Chapter 10). When a muscle fiber is impaled by a microelectrode at a point several millimeters from the motor endplate, the microelectrode records not only this resting potential, but also all-or-none muscle APs that arise with a delay of several milliseconds after APs arrive in the terminals of the innervating motor axon. Every time the motor axon is stimulated, a muscle AP will be recorded, and the muscle fiber will respond with a twitch. To understand the nature of the nerve-muscle synapse, Katz and others used pharmacological agents to interfere with its biochemical reactions. For example, if the South American poison *curare* (D-tubocurarine, Spotlight 6-3) is applied to a frog nerve-muscle preparation and the concentration of the curare is increased incrementally, at some particular concentration there is a sudden, all-or-none failure of the muscle AP and concomitantly the muscle fails to contract. The APs in the motor axon, however, remain unaffected, as does the ability of the muscle fiber to generate an AP and contract if an electrical stimulus is applied directly to the fiber. Because the presynaptic and postsynaptic APs remain unaffected by the poison, curare must interfere directly with synaptic transmission at the neuromuscular junction. Determining how curare works has been a source of insight into the processes of synaptic transmission.

For example, in an experiment designed to reveal how curare affects synaptic transmission, a microelectrode is inserted very close to (i.e., less than 0.1 mm from) the endplate region (Figure 6-15), and curare is added to the preparation. What do the following results tell us about the nature of synaptic transmission?

- As the concentration of curare is gradually increased, the muscle AP is seen to rise from a depolarization that is distinctly *slower* in time course and *lower* in amplitude than normal, and the initial slope of the rising phase is not as abrupt as in a normal muscle AP (see Figure 6-15B). This initial slow increase in V_m is an endplate, or postsynaptic, potential.

- Raising the concentration of curare further decreases the amplitude of the endplate potentials.

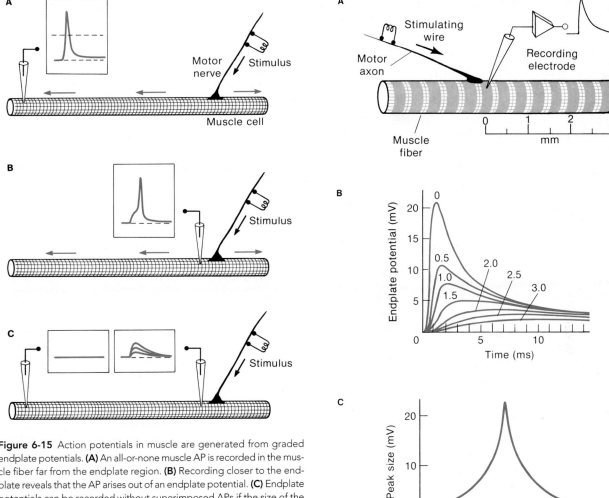

Figure 6-15 Action potentials in muscle are generated from graded endplate potentials. **(A)** An all-or-none muscle AP is recorded in the muscle fiber far from the endplate region. **(B)** Recording closer to the endplate reveals that the AP arises out of an endplate potential. **(C)** Endplate potentials can be recorded without superimposed APs if the size of the endplate potentials can be decreased to the point that they cannot bring the muscle fiber to threshold. Curare, a drug that blocks receptor channels in the postsynaptic membrane, provides one way to reduce the amplitude of endplate potentials. When a preparation is bathed in saline containing curare, the membrane far away from the endplate (left record) remains at its resting potential when the motor neuron fires, while, at the same time, graded endplate potentials are recorded near the endplate.

• The synaptic potential must reach a minimum level (the threshold potential) to trigger the muscle AP; so, when an increase in the concentration of curare causes the amplitude of the endplate potentials to drop below threshold, there is an abrupt failure of the AP.

These results suggest that curare interferes with synaptic transmission by blocking endplate potentials in proportion to its concentration. If the concentration of curare is sufficient to reduce the size of the synaptic potential in the muscle to just below threshold, the AP is eliminated, and the synaptic potential is revealed without a superimposed AP (see Figure 6-15C). If the recording electrode is now reinserted into the muscle fiber a number of times at progressively greater distances from the motor endplate, the amplitude of the measured postsynaptic potential drops approximately exponentially with distance from the endplate (Figure 6-16). In contrast with the AP, which propagates in

Figure 6-16 The amplitude of an endplate potential decays exponentially with distance from the motor endplate. **(A)** The endplate potential was recorded with a microelectrode that was sequentially inserted at 0, 0.5, 1.0, 1.5, 2.0, 2.5, and 3.0 mm from the endplate in a partly curarized frog muscle fiber. **(B)** Recordings of endplate potentials at each location. The distance away from the endplate (in millimeters) is given for each recording. **(C)** A plot of the peak potential of each recording shows that the amplitude of the endplate potential decreases approximately exponentially with distance from the endplate. [Adapted from Fatt and Katz, 1951.]

an unattenuated manner because it is regenerative, the synaptic potential spreads passively and thus decays with distance. In experiments like this one, curare enabled physiologists to distinguish among the elements of the synaptic response in vertebrate muscle fibers.

Synaptic currents

As described in Chapter 5, a change in membrane permeability for one or more species of ion (i.e., opening or closing a population of membrane channels that selectively passes those ions) typically shifts the membrane potential toward a new level. This change in V_m occurs because when

PHARMACOLOGICAL AGENTS USEFUL IN SYNAPTIC STUDIES

Studies of axonal and synaptic transmission have been greatly aided by the discovery and application of natural toxins—from animals, plants, or fungi—that selectively interfere with, or partially mimic, certain steps in the process of transmission. Toxins have been found that interact with ion channels, with receptors, and with enzymes important for nervous system function. Some of the commonly used agents that have been useful in studies of synaptic transmission are described here.

Channel Toxins

Several toxins are specific for particular types of ion channels. Tetrodotoxin (TTX) from the puffer fish (*Sphoeroides* sp.) binds to a site on voltage-gated Na^+ channels and blocks Na^+ current across the membrane. Similarly, saxitoxin (STX), derived from dinoflagellates, blocks voltage-gated Na^+ channels, although by a slightly different mechanism. Potassium ion channels can be blocked by several agents. For example, tetraethylammonium (TEA), a synthetic organic compound, blocks most types of K^+ channels from either inside or outside the membrane, and 4-amino pyridine blocks several types. Calcium ion channels can be blocked by any of several ω-conotoxins derived from the piscivorous (fish-eating) cone snail (*Conus geographus*). The various subtypes of this toxin block different classes of Ca^{2+} channels.

Glutamate-gated channel toxins have proved invaluable in distinguishing among the variety of channel types. Kainic acid, from a red alga (*Digenea simplex*), is an effective agonist for one subtype of glutamate receptors. Quisqualic acid, derived from seeds of the plant *Quisqualis indica,* is a second potent agonist that is selective for another subtype. One important antagonist is conatokins, from cone snails, which is a noncompetitive antagonist of a third class of glutamate receptors, called NMDA receptors for *N*-methyl-D-aspartate, which activates them.

Presynaptic Toxins

Several toxins act on presynaptic terminals to inhibit transmitter release. β-Bungarotoxin, derived from cobra venom, inhibits transmitter release by permeabilizing the nerve terminal. No-

texin from the tiger snake also inhibits transmitter release, causing lethal paralysis. The evolution of these toxins has made them highly effective for incapacitating a victim (i.e., only very small amounts are needed), and they must be handled with great caution in the laboratory.

Postsynaptic Receptor Toxins

Agonists and antagonists for receptor subtypes have contributed importantly to defining the role of these receptors in neuronal processing. γ-Aminobutyric acid (GABA), a largely inhibitory neurotransmitter, has been studied extensively with the use of a pair of chemicals, one an agonist of GABA and the other an antagonist. The agonist, muscimol, is derived from the mushroom *Amanita muscaria*. It specifically acivates $GABA_A$ type Cl^- channels. Bicuculline, produced from the plant *Dicentra cucullaria*, is a competitive antagonist of the same channel.

A huge collection of reagents exists for ACh receptors. Muscarine and other agents, including pilocarpine, activate muscarinic ACh receptors. In vertebrates, muscarinic ACh receptors are most prevalent in the visceral tissues that are innervated by the cholinergic axons of the parasympathetic system. Atropine (belladonna) is a plant-derived alkaloid that blocks muscarinic synaptic transmission.

Nicotine, another plant alkaloid, and certain other agents, such as carbachol, act as agonists of nicotinic ACh receptors. D-Tubocurarine is the active principle of curare, the South American blow-dart poison, made from the plant *Chondodendron tomentosum*. This molecule blocks transmission postsynaptically by competing with ACh for the ACh-binding site of the nicotinic receptors. It binds competitively to these sites without opening the channels, and it thereby interferes with the generation of a postsynaptic current. Similarly, α-Bungarotoxin (α-BuTX) is isolated from the venom of the krait, a member of the cobra family. This protein molecule binds irreversibly and with very high specificity to nicotinic ACh receptors. With the use of radioactively labeled α-BuTX, it has been possible to determine the number of ACh receptors present in a membrane as well as to isolate and purify the receptor protein.

Eserine (physostigmine) is an anticholinesterase; that is, it blocks the action of acetylcholinesterase. Use of this alkaloid has enabled physiologists to measure the amount of ACh released at a synapse, by preventing the rapid enzymatic destruction of the transmitter molecules. Partial doses accentuate the postsynaptic potential at cholinergic synapses.

---

channels open they allow a flow of ions that transfers charge from one side of the membrane to the other, which in turn causes the measured transmembrane voltage to change. In chemical synaptic transmission, postsynaptic channels in the membrane open when neurotransmitter binds to receptor proteins, and a **synaptic current** can then flow through these newly opened postsynaptic channels. (In some cases, postsynaptic channels close, reducing the flow

of ions through the membrane.) The direction and intensity of the synaptic current, which are controlled by the size of the conductance through the open channels and by the electrochemical driving force and charge on the permeant ions, determine the polarity and the amplitude of the postsynaptic potential. Because neurotransmitters activate channels with *selective* ionic permeabilities, they confer specificity on synaptic signal transmission by allowing only certain ionic species to cross the postsynaptic membrane in response to particular neurotransmitters.

The ionic currents that produce postsynaptic potentials can be recorded by voltage-clamping the postsynaptic membrane, thus holding the postsynaptic potential constant (see Spotlight 5-3). In a nerve-muscle preparation, this procedure must be carried out close to the motor endplate (Figure 6-17A). The motor nerve (the presynaptic element) is stimulated while V_m of the postsynaptic membrane is voltage-clamped at some predetermined value. The release of transmitter by the presynaptic nerve ending is quickly followed by a synaptic current (Figure 6-17B) that is produced when ions flow down their electrochemical gradients through open channels in the postsynaptic membrane.

The ions responsible for carrying the synaptic current at particular synapses have been identified through experiments in which the extracellular concentrations of specific ions were changed and the resulting effect on the synaptic current was measured. Such measurements demonstrated that the depolarizing synaptic current at the vertebrate neu-

romuscular junction consists of an influx of Na^+ that is partly canceled by a simultaneous, and somewhat smaller, efflux of K^+. At this synapse, both Na^+ and K^+ ions pass through the very same postsynaptic ACh-activated channels, indicating that these channels have a broader ion selectivity than do the highly selective, voltage-gated Na^+ and K^+ channels that underlie APs (see Figure 5-26).

Synaptic currents last a considerably shorter time than do synaptic potentials (see Figure 6-17B). Acetylcholine-activated channels open only briefly, because the transmitter at the neuromuscular junction is rapidly removed from the cleft by enzymatic destruction, after which the channels close and the synaptic current ceases to flow. A postsynaptic potential lasts longer than the synaptic current because its time course depends on the time constant of the membrane, as well as on the duration of the synaptic current.

Reversal potential At every fast chemical synapse, one (or more) species of ions carries current across the postsynaptic membrane, and the change in V_m caused by this current determines whether the synapse is excitatory or inhibitory. Measuring the properties of the synaptic current provides an experimenter with clues to the identity of the ions that carry the synaptic current. These measurements are made by injecting current into the postsynaptic cell to set the membrane potential at different values and then observing the sign and amplitude of the postsynaptic potential produced by synaptic inputs (Figure 6-18A and B). The amplitude and sign of the postsynaptic potential depend on the transmembrane voltage and on the species of ion or ions carrying the current. Remember that activation of membrane channels that select for a given ionic species, X, causes V_m to move closer to the equilibrium potential, E_x, for that ion (see Chapter 5).

Consider the experiment that is illustrated in Figure 6-18 for a synapse at which only one ionic species, X, carries the synaptic current. As the membrane potential, V_m, is shifted toward the equilibrium potential, E_X, the driving force on X ($V_m - E_X$), will decrease. When $V_m = E_X$, no current will flow across the membrane, even though the channels are open, because there is no driving force on the ions. If in the experiment V_m is set on the other side of E_X, current will once again flow, because $V_m - E_X$ will again be nonzero, but the sign will have changed, indicating that the driving force is in the opposite direction. As a result, X will flow through the open channels in the direction opposite to that of its previous flow, and the sign of the postsynaptic potential will be opposite to that of its previous value (Figure 6-18B and C). Because the direction of the ionic current and the sign of the postsynaptic potential reverse as V_m passes through E_X, E_X is called the **reversal potential**, E_{rev}. When synaptic channels open, the synaptic current causes V_m to shift toward the E_{rev} of the current, no matter where V_m was set experimentally before the synapse was activated. The reversal potential has proved to be a useful property of synaptic currents because it provides a hint about which ions carry the current. In fact, before

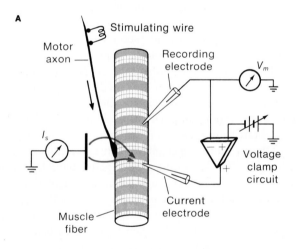

A

Stimulating wire

Motor axon

Recording electrode

V_m

I_s

Voltage clamp circuit

Current electrode

Muscle fiber

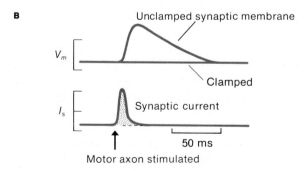

B

Unclamped synaptic membrane

V_m

Clamped

I_s

Synaptic current

50 ms

Motor axon stimulated

Figure 6-17 Voltage-clamping the postsynaptic membrane allows the synaptic current to be measured. **(A)** Setup for voltage-clamping the muscle membrane, which holds the postsynaptic potential constant while ionic current flowing across the postsynaptic membrane through channels opened by a neurotransmitter is recorded (see Spotlight 5-3). **(B)** The upper record shows an endplate potential when the neuron is stimulated and the muscle is not voltage-clamped. The lower record shows a synaptic current when the muscle fiber is voltage-clamped under the same conditions. The synaptic current decays much faster than does the endplate potential.

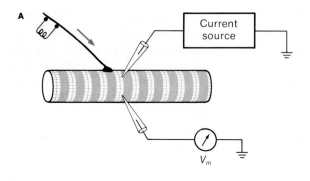

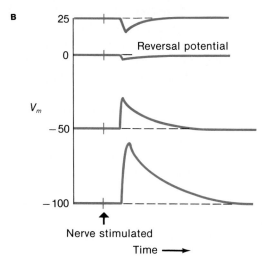

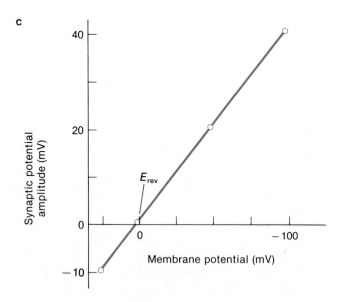

Figure 6-18 The synaptic reversal potential is measured by changing the membrane potential and recording the postsynaptic potential. **(A)** Method for determining the reversal potential (E_{rev}) at a synapse. Steady current is injected into the postsynaptic cell with an electrode to set V_m at different values. At each value of V_m, an endplate potential is produced by stimulating the presynaptic nerve. The endplate potentials are recorded with a second electrode in the postsynaptic muscle fiber. **(B)** At first (bottom), V_m is set at values that are more negative than the equilibrium potential, E_X, for the ions carrying the synaptic current. When V_m is set equal to E_X, no synaptic current flows, and the amplitude of the postsynaptic potential is zero, even though the postsynaptic ion channels are open. When V_m is set at values more positive than E_X, the driving force on the ions carrying the synaptic current is opposite to the direction of the driving force when V_m is more negative than E_X. As a result, when V_m is more positive than E_X, ions flow through the synaptic channels in the opposite direction from their direction when V_m is more negative than E_X, and the sign of the endplate potential reverses. **(C)** The results of this type of experiment can be plotted to show the amplitude of the endplate potentials as a function of the values of V_m. The line fitting the experimental points crosses the abscissa at E_{rev}. In this case, $E_{rev} = 0$ mV.

ter 5). However, if synaptic channels are permeable to several ionic species, as is the acetylcholine channel, E_{rev} depends on the concentrations and relative permeabilities of all of the participating ions. If the concentrations and permeabilities of the various ionic species are known, E_{rev} can be predicted by using the Goldman equation (see *Goldman Equation* in Chapter 5), rather than the Nernst equation. Alternatively, if the current is carried by only two ionic species, E_{rev} can be calculated from Ohm's law for the two ionic species (Spotlight 6-4). The ACh-activated channels at vertebrate neuromuscular junctions provide an example. When those channels open, they become permeable to both Na^+ and K^+. In such a case, the reversal potential, E_{rev}, of the current will lie between the equilibrium potentials of the two permeant ions (Figure 6-19). In Figure 6-19, V_m was electronically clamped at several different values and then the synapse was activated. When V_m was clamped at E_{Na} (trace *a*), the driving force on Na^+ was zero ($V_m - E_{Na} = 0$), but there was a large driving force on K^+ ($V_m - E_K$). Thus, the synaptic current at E_{Na} is carried entirely by an outward flux of K^+, which makes V_m more negative. In contrast, when V_m is set at E_K (trace *e*), there is no driving force on K^+, but there will be a large driving force on Na^+. In this case, all of the current through the ACh-activated channel will be carried by an influx of Na^+, and V_m will become more positive. Somewhere between E_{Na} and E_K, there must, then, be a value of V_m at which the Na^+ and K^+ currents through this channel will be equal and opposite to one another, so that, although both ions flow through the channel, there will be no net current (trace *c*). This value of V_m is the reversal potential for the ACh-activated current. In the frog endplate channel, the conductances for the two permeant ions, Na^+ and K^+, are approximately equal. Notice that the synaptic current cannot drive V_m past E_{rev}, regardless of how many channels become activated. When V_m reaches E_{rev}, the net driving force on the permeating ions drops to zero, and V_m cannot change further. As a result, E_{rev} sets the maximum change

membrane patch recording was introduced, measuring the reversal potential of a current was the primary method of distinguishing the ionic species that produced a particular postsynaptic potential, although it was—by itself—not conclusive.

If a single ion carries the synaptic current, the reversal potential, E_{rev}, can be calculated by using the Nernst equation for that ionic species (see *Nernst Equation* in Chap-

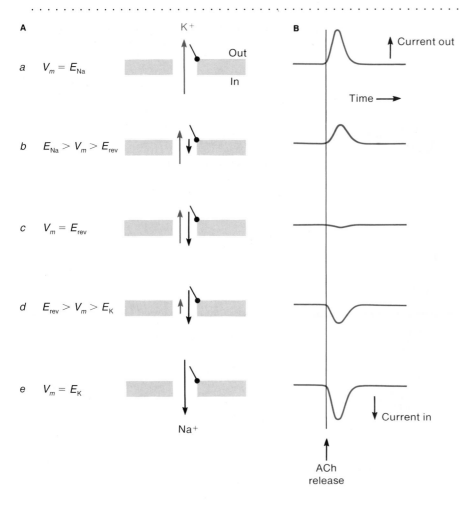

Figure 6-19 The synaptic current at the vertebrate neuromuscular junction is carried by both sodium and potassium ions. **(A)** Sodium and potassium currents through activated acetylcholine (ACh) channels at different membrane potentials, beginning at E_{Na}. The ACh-activated channels are approximately equally permeable to Na^+ and K^+, so the magnitudes of I_{Na} and I_K depend on the driving force on each ion. The relative magnitudes of the Na^+ and K^+ currents are represented by the lengths of the arrows. **(B)** The amplitude and time course of the net current through the ACh-activated channels are shown against time. At E_{rev} for the combined currents, the net current through the channels is zero.

in V_m that can be produced by the activation of synaptic channels (or indeed by the activation of any ion channels). The reversal potential also has a special functional significance at synapses, because the relation between E_{rev} and the threshold for excitation in the postsynaptic cell determines how synaptic events affect the postsynaptic cell.

Postsynaptic excitation and inhibition Any synaptic event that increases the probability that an AP will be initiated in the postsynaptic cell is called an **excitatory postsynaptic potential** (epsp); conversely, any synaptic event that reduces the probability of an AP in the postsynaptic cell is an **inhibitory postsynaptic potential** (ipsp). If the reversal potential (E_{rev}) of a synaptic current is more positive than the threshold of the postsynaptic cell, that synapse is excitatory (Figures 6-20A and 6-21A). If E_{rev} is more negative than threshold, the synapse is inhibitory. At fast chemical synapses, excitatory currents are typically carried through channels that conduct Na^+ or Ca^{2+}. These channels may be permeable to K^+ as well, as is the ACh channel of the vertebrate neuromuscular junction, but the K^+ current itself does not contribute to the excitatory nature of the synapse (see Figure 6-19). Inhibitory synaptic currents are typically carried by channels that are permeable either to K^+ or to Cl^-. The reversal potential, E_{rev}, for K^+ or Cl^- typically lies near V_{rest}, so it is more negative than the thresh-

old. If E_{rev} for inhibitory channels is more negative than V_{rest} in the postsynaptic cell, the synaptic current will cause V_m to become more negative than V_{rest}, hyperpolarizing the cell toward E_{rev} (see Figure 6-20A). Hyperpolarizing synaptic currents can add to depolarizing synaptic currents, reducing the net amount of depolarization in the postsynaptic cell.

Although all excitatory synapses generate depolarizing postsynaptic currents, there are special cases among inhibitory synapses. For example, if E_{rev} for a synaptic current happens to be identical with V_{rest} ($V_m - E_{rev} = 0$), no net synaptic current will flow even if postsynaptic channels open. The net current will be zero because the driving force on the ion, or ions, that can pass through the channels will be zero. In this case, when the synaptic channels open, V_m will not change. In some cases, E_{rev} is more positive than V_{rest} but more negative than threshold (Figure 6-21B). In this situation, the postsynaptic potential is depolarizing, but it is, nonetheless, *inhibitory* because it increases the difficulty of bringing V_m up to threshold. In each of these two special cases, the synapses have an inhibitory action, because activation of these channels can counteract a simultaneous activation of excitatory channels (Figure 6-21C). In effect, the opening of inhibitory postsynaptic channels "short-circuits" excitatory currents, because the positive charge carried into the cell by excitatory currents can leave

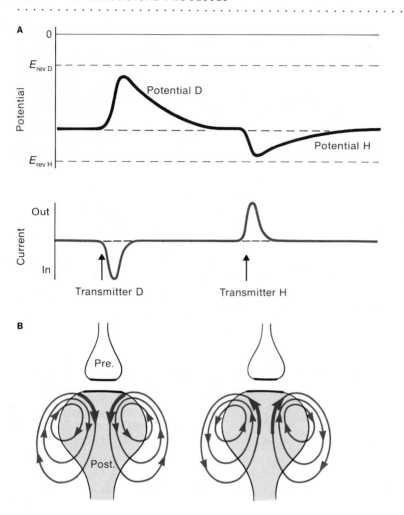

Figure 6-20 Synaptic currents can be excitatory or inhibitory. **(A)** Transmitter D evokes an excitatory depolarizing postsynaptic potential, because it increases an ionic conductance generating a net inward current that adds positive charge to the cell's interior. For example, transmitter D might increase the permeability to Na^+. Transmitter H produces an inhibitory hyperpolarizing synaptic potential, because it increases the conductance to ions that cause a net loss of positive charge from the cell. For example, transmitter H might increase the permeability to K^+ or to Cl^-. **(B)** The direction in which positive current flows through channels opened by transmitter D is opposite to the direction in which positive current flows through channels opened by transmitter H.

the cell through the inhibitory channels, preventing the positive charges from bringing V_m up to threshold.

Note that there is nothing *inherently* excitatory or inhibitory about any particular transmitter substance. Rather, the properties of the channels that are opened by the transmitter and the identities of the ions that flow through those channels, determine how a transmitter affects the postsynaptic cell. For example, ACh is an excitatory transmitter at the vertebrate neuromuscular junction, where it opens channels that allow Na^+ and K^+ to cross the postsynaptic membrane. In contrast, ACh is inhibitory at the terminals of parasympathetic neurons innervating the vertebrate heart, where it affects K^+-selective channels.

From this description, it follows that an inhibitory transmitter could be made excitatory if the ionic gradients across the postsynaptic membrane were changed. This experimental manipulation has been accomplished for neurons of the mammalian spinal cord and for neurons of a snail (Figure 6-22). In certain snail neurons, ACh increases g_{Cl} of the postsynaptic membrane. In one group of these cells (called H cells, or hyperpolarizing cells), the intracellular Cl^- concentration is relatively low, making E_{Cl}

more negative than V_{rest}. When ACh is applied to H cells, it opens Cl^- channels, allowing Cl^- to flow into the cell down its electrochemical gradient. The result is to shift V_m toward E_{Cl}, hyperpolarizing the cell (see Figure 6-22A). If all extracellular Cl^- is replaced by SO_4^{2-}, which cannot pass through the chloride channels, application of ACh leads to an efflux of Cl^-, because it now has an outwardly directed electrochemical gradient. This efflux of negative charge produces both a depolarization and an increase in the frequency of action potentials (see Figure 6-22B). Thus, ACh is normally inhibitory for these cells, but it can produce excitation if the electrochemical gradient for Cl^- is reversed. In fact, in this species of snail, there are other brain cells (called D cells, or depolarizing cells) that naturally maintain a high intracellular Cl^- concentration by actively accumulating Cl^-. Acetylcholine causes an increase in g_{Cl} for these cells, as it does in the H cells. However, in the D cells, the net effect is a depolarization, because the electrochemical gradient for Cl^- is normally outward. Hence, in this example, excitation and inhibition depend critically on the nature of ionic gradients, and not on the identity of the signaling molecule.

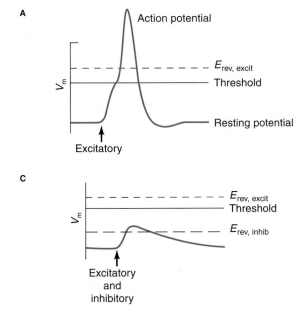

A

Action potential

$E_{rev, excit}$

Threshold

V_m

Resting potential

Excitatory

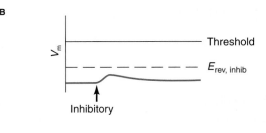

B

V_m

Threshold

$E_{rev, inhib}$

Inhibitory

C

$E_{rev, excit}$
Threshold

V_m

$E_{rev, inhib}$

Excitatory
and
inhibitory

Figure 6-21 Excitatory and inhibitory synaptic signals interact in the postsynaptic cell. **(A)** An action potential arises out of an excitatory postsynaptic potential if that postsynaptic potential brings the membrane potential V_m, above threshold. **(B)** A postsynaptic potential is inhibitory, even if it depolarizes V_m, if its E_{rev} is more negative than the threshold for impulse generation. **(C)** An inhibitory transmitter (as in part B) may reduce the depolarization produced by an excitatory transmitter (as in part A) sufficiently to keep the postsynaptic potential from reaching threshold.

Presynaptic inhibition Experiments performed in the 1960s on neurons of the mammalian spinal cord and on the crustacean neuromuscular junction revealed an additional inhibitory mechanism at some synapses. In this mechanism, called **presynaptic inhibition,** an inhibitory transmitter is released from a terminal that ends *on the presynaptic terminal* of an excitatory axon (Figure 6-23). In this case, the presynaptic terminal of the excitatory axon is itself a postsynaptic element. During presynaptic inhibition, the amount of transmitter released from the excitatory terminal is reduced, which reduces synaptic excitation in the cell that is postsynaptic to the excitatory neuron (see Figure 6-23B). In some cases, the presynaptic inhibitory transmitter increases g_K or g_{Cl} in the presynaptic terminals of the excitatory axon, which reduces the amplitude of any AP invading the excitatory terminal and hence diminishes the amount of

transmitter released from the terminal. In other examples of presynaptic inhibition, the inhibitory transmitter modifies some property of Ca^{2+} channels in the presynaptic membrane, rendering them less responsive to depolarization. Because the release of transmitter molecules depends on Ca^{2+} entry into the terminal (see the next section of this chapter), reducing Ca^{2+} entry reduces transmitter release. Regardless of the mechanism, the net effect of presynaptic inhibition is that the postsynaptic cell receives less transmitter and thus a smaller postsynaptic potential is generated.

Postsynaptic and presynaptic inhibition produce quite different consequences for the postsynaptic cell. Postsynaptic inhibition globally reduces the excitability of the postsynaptic cell, making it is less able to respond to all excitatory inputs. In contrast, presynaptic inhibition acts only on specific inputs to the cell, allowing the cell to remain

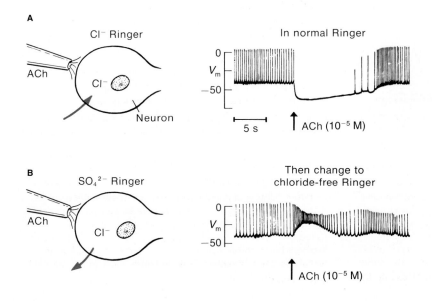

A

Cl⁻ Ringer

ACh

Cl⁻

Neuron

In normal Ringer

0

V_m

−50

5 s   ↑ ACh (10^{-5} M)

B

SO₄²⁻ Ringer

ACh

Cl⁻

Then change to
chloride-free Ringer

0

V_m

−50

↑ ACh (10^{-5} M)

Figure 6-22 Experimentally changing ionic gradients across the membrane of a postsynaptic cell can change the sign of a synapse. **(A)** Acetylcholine (ACh) applied to H-type cells in the snail brain activates Cl⁻ channels, producing a hyperpolarization because Cl⁻ brings negative charges into the cell as it moves down its electrochemical gradient. **(B)** When the extracellular Cl⁻ ions are entirely replaced by SO_4^{2-}, leaving Cl⁻ inside the cell, the electrochemical gradient for Cl⁻ is reversed. Reversing the electrochemical gradient causes the direction of the synaptic current to reverse. As a result, the postsynaptic potential becomes depolarizing, and the synapse becomes excitatory. Electrical activity of the cell before, during, and after synaptic activation is shown at the right. [Adapted from Kerkut and Thomas, 1964.]

SPOTLIGHT 6-4

CALCULATION OF REVERSAL POTENTIAL

The value of the reversal potential of an ionic current elicited by a stimulus or a neurotransmitter depends on the relative conductances of the ions carrying the current as well as on their equilibrium potentials. If we assume that only Na^+ and K^+ carry current in response to the stimulus, the reversal potential can be related to the conductances of these ions by using equation 5-10, with the values g_K and g_{Na} representing the respective transient changes in the two conductances.

$$I_K = g_K \times (V_m - E_K) \qquad (1)$$

$$I_{Na} = g_{Na} \times (V_m - E_{Na}) \qquad (2)$$

At the reversal potential, I_K and I_{Na} must be equal and opposite regardless of the relative conductances, because the net current must be zero. Thus, when V_m is at the reversal potential, E_{rev},

$$-I_K = I_{Na} \qquad (3)$$

Substituting from equations 1 and 2, at the reversal potential we have

$$-g_K(V_m - E_K) = g_{Na}(V_m - E_{Na}) \qquad (4)$$

From this equation, if g_K is greater than g_{Na}, then V_m must be closer to E_K than to E_{Na}, and vice versa. Solving equation 4 for $V_m = E_{rev}$ gives

$$E_{rev} = \frac{g_K}{g_{Na} + g_K} E_K + \frac{g_K}{g_{Na} + g_K} E_{Na} \qquad (5)$$

From equation 5, it is apparent that E_{rev} will not be simply the algebraic sum of E_{Na} and E_K, but will lie somewhere between the two, depending on the ratio g_{Na}/g_K. Thus, if g_{Na} and g_K become equal to each other (e.g., as they may when endplate channels are activated by ACh in frog muscle), the membrane potential will shift toward a reversal potential that lies exactly halfway between g_{Na}, and g_K:

$$E_{rev} = \left(\frac{1}{2}\right)E_K + \left(\frac{1}{2}\right)E_{Na} = \left(\frac{1}{2}\right)(E_K + E_{Na})$$

For frog muscle, E_K is about -100 mV, and E_{Na} is about $+60$ mV. Hence, we would predict that during synaptic activation of a frog muscle, $E_{rev} = \frac{1}{2}(-100 + 60) = -20$ mV. The measured reversal potential of the current at the frog neuromuscular synapse, -10 mV, is somewhat more positive than this value, possibly because g_{Na} is actually somewhat greater than g_K.

To summarize, the reversal potentials of membrane currents differ according to the species of ions that participate, the equilibrium potentials of those ions, and the relative conductances to each of the ionic species that participates in the current.

---

normally responsive to other inputs. Thus presynaptic inhibition provides a mechanism for narrowly targeted and subtle control of **synaptic efficacy** (the effectiveness of a presynaptic impulse in producing a postsynaptic potential change) among the many synaptic connections onto a particular neuron.

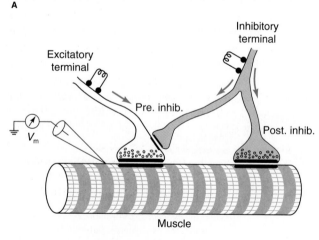

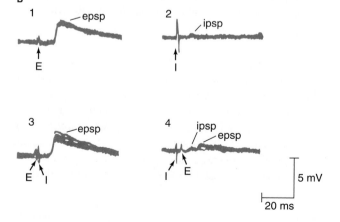

Figure 6-23 Neurons that produce inhibition at the crustacean neuromuscular junction also inhibit excitatory motor neurons presynaptically. **(A)** The morphological arrangement of excitatory and inhibitory terminals, showing the location of an inhibitory synapse that produces presynaptic inhibition and the arrangement for the experiment illustrated in part B. **(B)** Intracellular recording from the muscle fiber innervated by excitatory and inhibitory motor neurons. (1) Stimulation of the excitatory axon (labeled E on record) produced a 2 mV excitatory postsynaptic potential (epsp). (2) Stimulation of the inhibitory axon (labeled I on record) produced a depolarizing inhibitory postsynaptic potential (ipsp) of about 0.2 mV. (3) If the inhibitory neuron was stimulated a few milliseconds *after* the excitatory neuron, the excitatory postsynaptic potential was unaffected. (4) However, if the inhibitory neuron was stimulated a few milliseconds *before* the excitatory neuron, the excitatory postsynaptic potential was almost abolished. [Adapted from Dudel and Kuffler, 1961.]

PRESYNAPTIC RELEASE OF NEUROTRANSMITTERS

The properties of neurotransmitter release from presynaptic terminals determine the effectiveness of synaptic transmission, because the number of transmitter molecules released affects the size of the postsynaptic response. Understanding transmitter release is thus of central importance for understanding synaptic transmission and its normal role in neuronal communication. Besides its importance for physiology, the history of experimentation on transmitter release provides classic examples of the scientific method and experimental strategies. A particularly striking example is the demonstration, by Sir Bernard Katz and his co-workers, that neurotransmitters are generally released in tiny packets called *quanta*. More recent experiments have shown that synaptic release is closely related to other forms of exocytosis used by cells, such as glandular cells, to release chemicals (see Chapter 9). The conservation of this mechanism has facilitated experiments designed to understand the details of exocytosis in all cells.

Quantal Release of Neurotransmitters

In their investigation of neuromuscular transmission, Paul Fatt and Bernard Katz (1952) discovered that spontaneous "miniature" depolarizations (<1 mV in amplitude) could be recorded from the vicinity of the postsynaptic membrane of the motor endplate in the frog muscle (Figure 6-24). These spontaneous signals became progressively smaller when the intracellular recording electrode was inserted at greater distances from the endplate, and they eventually disappeared if the electrode was moved far enough away from the endplate. Because these potentials have a shape, time course, and drug sensitivity similar to those of endplate potentials, they were called **miniature endplate potentials** (mepps). The measurement of miniature endplate potentials has played a key role in analyzing and understanding the mechanisms of neurotransmitter release.

A key insight of Katz and his collaborators was to ask how these spontaneous miniature potentials might be related to the endplate potentials elicited by motor nerve stimulation. They asked whether miniature endplate potentials might represent a "unit" of transmitter release. If so, endplate potentials evoked by stimulating the motor nerve could simply be the result of many such units being released in unison subsequent to presynaptic APs. It had been found earlier that progressively increasing the concentration of extracellular Mg^{2+} or decreasing that of ex-

tracellular Ca^{2+}, or both, caused the evoked endplate potential to become smaller in amplitude. Katz and his co-workers used this observation to find concentrations of these cations at which the evoked endplate potential became as small as, or a small and simple multiple of, a single spontaneously occurring miniature endplate potential. They then recorded postsynaptic responses to presynaptic motor nerve impulses in this high Mg^{2+}–low Ca^{2+} solution and obtained the following statistical evidence:

- Some motor nerve impulses produced no response at all. They called these trials "failures."

- Some impulses produced endplate potentials that had approximately the same amplitude as those of single spontaneous miniature endplate potentials.

- Some impulses produced endplate potentials with amplitudes that were integral multiples (e.g., two, three, four, etc.) of the mean amplitude of single spontaneous miniature endplate potentials (Figure 6-25).

These findings lent further support to the hypothesis that a normal endplate potential is produced by a large number of transmitter units (equivalent to the units producing the spontaneous miniature endplate potentials) released in unison during a presynaptic action potential. Calculations showed that in frog muscle approximately 100–300 such units could account for the amplitude of normal evoked excitatory postsynaptic potentials.

Because synaptic release appeared to occur in the form of discrete units—packets, or quanta, of transmitter molecules—Katz and his associates called the process **quantal release.** They then asked another simple, direct, and important question: What was the makeup of the unit, or quantum, of transmitter release? Was it a single molecule of ACh? If not, how many molecules were in a quantum? If the cause of a spontaneous miniature endplate potential was a single ACh molecule that leaked out of the presynaptic terminal, then adding a very low concentration of ACh to the saline bathing the muscle should greatly increase the number of miniature endplate potentials. Beginning with very low concentrations of ACh and working their way up to higher concentrations, they never saw an increase in miniature endplate potentials, but they did observe a depolarization in the postsynaptic muscle fiber that smoothly increased in size with increased ACh concentration. They concluded that the miniature endplate potentials were not produced in response to single ACh molecules.

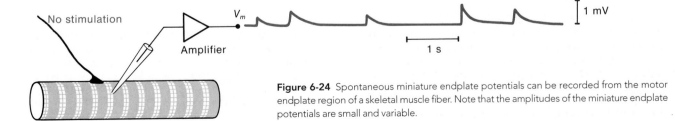

Figure 6-24 Spontaneous miniature endplate potentials can be recorded from the motor endplate region of a skeletal muscle fiber. Note that the amplitudes of the miniature endplate potentials are small and variable.

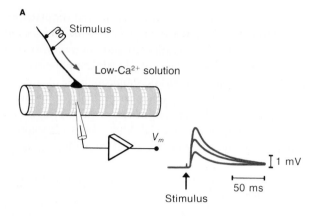

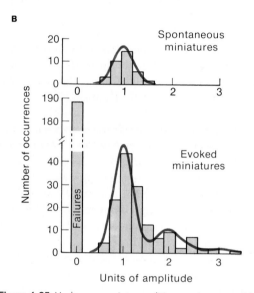

Figure 6-25 Under appropriate conditions, action potentials in a motor nerve produce small endplate potentials, similar to miniature endplate potentials, in the postsynaptic muscle fiber. **(A)** A nerve-muscle preparation is placed in a low-Ca²⁺ solution, which reduces the amount of transmitter that is released at the neuromuscular junction when the nerve is stimulated. The evoked endplate potentials then have small and variable amplitudes. **(B)** Amplitude distribution of spontaneous miniature endplate potentials *(top)* and of endplate potentials *(bottom)* that were evoked by motor nerve stimulation in low-Ca²⁺ saline. Note the many failures in the lower distribution. Most of the evoked endplate potentials had an amplitude distribution similar to that of the single spontaneous miniature endplate potentials. The continuous curves in the upper and lower histograms were calculated from the theoretical Poisson distribution, assuming that the evoked endplate potentials are made up of units corresponding to the spontaneous miniature endplate potentials. [Part B adapted from Del Castillo and Katz, 1954.]

Indeed, they calculated that each miniature endplate potential was produced by the release of a packet of transmitter molecules consisting of about 10,000 molecules of ACh, and that these molecules activated about 2000 postsynaptic channels. At about this time, electron-microscopic studies revealed the presence of membrane-bounded packets, or vesicles, in presynaptic endings (see Figure 6-13C). The vesicles constituted the anatomical basis for the packets of transmitter that had been inferred physiologically by Katz and his group. The release of transmitter from presy-

naptic vesicles (see Figure 6-14) could be the basis both for the miniature endplate potentials (each caused by the release of a single vesicle) and for evoked endplate potentials (each caused when many vesicles are released simultaneously). Various lines of evidence support this view, including electrical measurements of membrane capacitance at the presynaptic terminal, which increases during exocytosis because vesicles fuse with the plasma membrane and increase its surface area.

It is impossible to study individual quanta, so the quantal release of transmitter at the frog motor endplate has been studied extensively by statistical analysis, and we now have a coherent description of events that take place during transmitter release. At any given time, only a fraction of the population of vesicles inside a nerve terminal are available for immediate release. For any particular set of physiological conditions (e.g., Ca²⁺ and Mg²⁺ concentrations, temperature), there is some particular probability that any one of the available vesicles will be released. (In fast transmission, the available vesicles appear to be those located in the active zones.) If the concentration of Ca²⁺ in the extracellular fluid is reduced, entry of Ca²⁺ into the presynaptic terminal in response to an AP is reduced. An influx of Ca²⁺ into the terminal is essential for transmitter release; therefore, when the influx of Ca²⁺ drops, so does the probability that any presynaptic vesicle will be released. If the probability is sufficiently low, it produces the condition illustrated in Figure 6-25B, in which presynaptic stimulation leads to many failures (i.e., no vesicles are released) or to the release of only one, two, or a few vesicles to produce endplate potentials with amplitudes that correspond to multiples of one, two, three, and so forth, miniature endplate potentials. When the normal number of quanta in an endplate potential, called its **quantal content**, is reduced by a low concentration of extracellular Ca²⁺, it is possible to determine the number of vesicles that were released in response to each stimulus in a large series. Statistical analysis of those numbers has shown that the probability of vesicle release follows a **Poisson distribution** (see Figure 6-25B), which implies that release is a random process.

Depolarization-Release Coupling

According to the quantal theory of transmitter release, the probability that a given vesicle will undergo exocytosis and release its contents at any given instant is quite low when the presynaptic membrane potential, V_m, is at the resting level. When the presynaptic neuron is at rest, miniature endplate potentials are relatively rare and occur randomly; however, when the presynaptic membrane is depolarized, the probability of quantal release dramatically increases. The increased probability of release can be seen in the increased frequency of miniature endplate potentials that accompanies a steady depolarization in the presynaptic neuron (Figure 6-26).

The relation between presynaptic membrane potential and transmitter release was examined by Bernard Katz and Ricardo Miledi in an unusually large synapse onto the

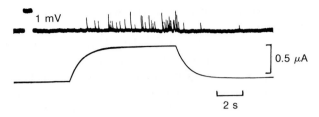

Figure 6-26 When the presynaptic terminal is depolarized, miniature endplate potentials in the postsynaptic muscle fiber become more frequent. Depolarization of the presynaptic terminal by application of current (lower trace) increases the probability of transmitter release, shown by an increase in the frequency of miniature endplate potentials recorded in the muscle fiber (upper trace). [Adapted from Katz and Miledi, 1967.]

squid giant axon (Figure 6-27A). Because of the presynaptic terminal's large size, microelectrodes for passing current and for recording membrane potential can be inserted into the terminal close to the synaptic region. This arrangement would be technically impossible in most other synapses, because most presynaptic terminals are tiny. In this experiment, Na$^+$ channels were blocked by *tetrodotoxin* (TTX), and K$^+$ channels were blocked by *tetraethylammonium* (TEA), so V_m of the presynaptic terminal could be controlled without interference from all-or-none APs. The postsynaptic potential, recorded with a third microelectrode inserted near the synapse, provided a highly sensitive bioassay of how much transmitter was released from the presynaptic cell. The following results are shown in Figure 6-27B and C:

- When the presynaptic membrane was depolarized, transmitter was released (detected as depolarization in the postsynaptic cell) even though the normal mechanism of the AP had been eliminated.

- As the presynaptic membrane was made increasingly depolarized, the amplitude of the postsynaptic potential increased, implying that the amount of transmitter released varied directly with depolarization of the presynaptic terminal: more depolarization caused more transmitter to be released.

- A given amount of presynaptic depolarization produced a smaller postsynaptic response when the concentration of Ca^{2+} was reduced in the extracellular fluid.

These three results supported the hypotheses that events in the presynaptic terminals depend on a depolarization of the membrane, but not on the specific ions that cause the depolarization, and that Ca^{2+} ions play a role in neurotransmitter release. Further support for the hypothesis that Ca^{2+} ions were responsible for transmitter release came from an experiment in which the presynaptic terminal was depolarized all the way to the Ca^{2+} equilibrium potential, E_{Ca} (see *Nernst Equation* in Chapter 5). In this condition, there could be no net flux through open Ca^{2+} channels in the presynaptic terminal, because there was no net driving force on Ca^{2+} ions. Correspondingly, no transmitter release was detected until the membrane potential, V_m, was allowed to return to its resting level after the

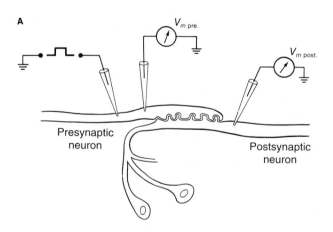

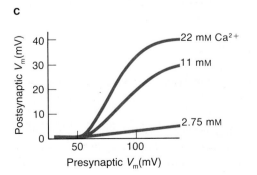

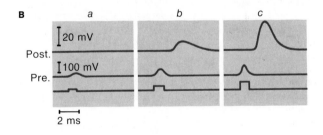

Figure 6-27 The relation between presynaptic depolarization and transmitter release can be studied directly at the squid giant synapse. **(A)** The experimental setup allowing the presynaptic membrane to be depolarized with current from an intracellular electrode while V_m of both the presynaptic and the postsynaptic regions are recorded with two more microelectrodes. **(B)** Depolarizing current into the presynaptic terminal increased from *a* to *b* to *c*, producing larger postsynaptic potentials in the postsynaptic neuron. **(C)** Depolarizing the presynaptic element or increasing extracellular Ca^{2+} or doing both increased the amplitude of the postsynaptic potentials recorded. For a constant value of [Ca^{2+}]$_{outside}$, a larger depolarization produced more transmitter release and hence larger postsynaptic potentials. As the concentration of Ca^{2+} was decreased in the extracellular solution, the amplitude of postsynaptic potentials dropped. [Adapted from Katz and Miledi, 1966; 1970.]

depolarization. This result supported the idea that Ca^{2+} plays a crucial role in causing neurotransmitter release.

A direct relation between Ca^{2+} entry and transmitter release has been demonstrated in the giant synapse of the squid axon. Calcium ion concentration was measured using *aequorin*, which is a Ca^{2+}-sensitive protein, extracted from a bioluminescent jellyfish, that emits light in the presence of free Ca^{2+}. Aequorin was injected into the presynaptic terminal, and V_m was recorded in the pre- and postsynaptic cells while light emission was monitored by a phototube (Figure 6-28). When depolarizing current was injected into the presynaptic terminal, a postsynaptic potential was elicited only when light was produced by aequorin, indicating that Ca^{2+} ions had entered the presynaptic terminal.

Other experiments have confirmed this result. The concentration of Ca^{2+} inside the presynaptic terminal of the neuromuscular junction must rise after an AP arrives in order for transmitter to be released. If conditions interfere with the entry of Ca^{2+} into the terminal (e.g., low concentration of extracellular Ca^{2+} or the presence of competing ions, such as Mg^{2+} and La^{3+}), no transmitter is released. Finally, the injection of Ca^{2+} into the presynap-

tic terminal of the squid giant synapse evokes the release of transmitter.

Taken together, these data indicate that Ca^{2+} ions enter the presynaptic terminal after an AP arrives, and their entry is necessary for triggering the release of transmitter. Studies of individual synaptic terminals have demonstrated that synaptic vesicles fuse with the inner surface of the cell membrane only if the intracellular concentration of free Ca^{2+} rises. Exocytosis of synaptic vesicles can occasionally be seen in electron micrographs (Figure 6-29). As essential as the rapid increase in Ca^{2+} concentration is within the presynaptic terminal, its rapid decrease is at least as important. Although ultimately the Ca^{2+} ions that entered must be pumped back into the extracellular space by an exchanger in the cell membrane, this process is too slow to account for the necessarily rapid disappearance of Ca^{2+} ions from the cytosol of the terminal. It seems likely that the extra Ca^{2+} ions are stored in internal compartments until they can be extruded.

Recently, there have been several significant discoveries concerning the mechanism of synaptic release. Several proteins that participate in some part of the process have been identified, cloned, and sequenced. Release proceeds

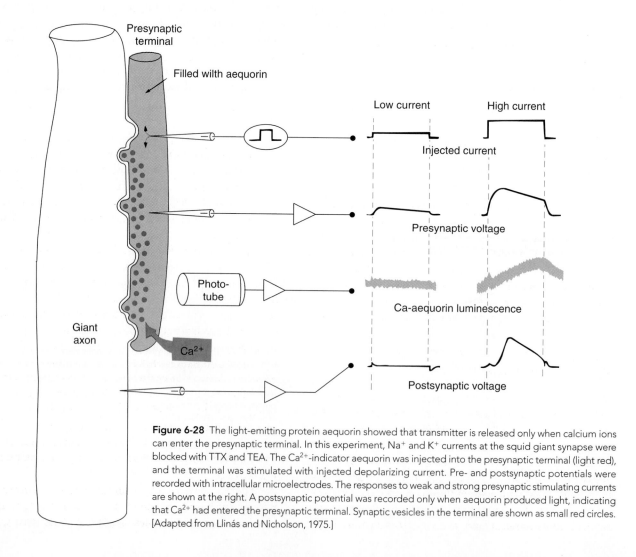

Figure 6-28 The light-emitting protein aequorin showed that transmitter is released only when calcium ions can enter the presynaptic terminal. In this experiment, Na^+ and K^+ currents at the squid giant synapse were blocked with TTX and TEA. The Ca^{2+}-indicator aequorin was injected into the presynaptic terminal (light red), and the terminal was stimulated with injected depolarizing current. Pre- and postsynaptic potentials were recorded with intracellular microelectrodes. The responses to weak and strong presynaptic stimulating currents are shown at the right. A postsynaptic potential was recorded only when aequorin produced light, indicating that Ca^{2+} had entered the presynaptic terminal. Synaptic vesicles in the terminal are shown as small red circles. [Adapted from Llinás and Nicholson, 1975.]

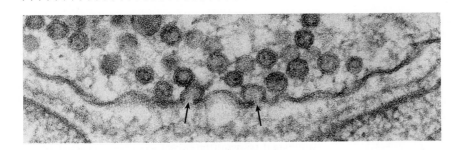

Figure 6-29 Exocytosis at the synaptic terminal can be visualized with an electron microscope. Exocytosis at a nerve terminal of a frog neuromuscular junction. The arrows point to vesicles emptying into the synaptic cleft. [Courtesy of J. Heuser.]

through identifiable steps. Filled synaptic vesicles dock preferentially at active zones after which a maturation process makes synaptic vesicles competent to undergo fast Ca^{2+}-dependent membrane fusion. Exocytosis takes less than 0.3 ms, but only one of many docked vesicles at an active zone fuses and not all APs lead to exocytosis from an active zone. Recently, Dieter Bruns and Reinhard Jahn (1995) reported the first real-time measurement of transmitter release from single synaptic vesicles in cultured leech neurons that contain two kinds of vesicles. Small, clear vesicles discharge about 4700 transmitter molecules with a time constant of about 260 μs, and large, dense-core vesicles release about 80×10^3 molecules with a time constant of about 1.3 ms. This technical advance suggests that molecular details of release may soon be discovered for many kinds of synapses, which will expand our understanding of how synaptic transmission regulates neuronal communication.

Nonspiking Release

Some neurons, in both vertebrates and invertebrates, have been shown to release neurotransmitter from their terminals even in the absence of APs, which is called **nonspiking release**. At least some of these neurons are unusual in that they may never carry APs; all information transfer is accomplished by electrotonically conducted voltage signals. The amount of transmitter that is released into the synaptic cleft by these cells depends on the membrane potential, V_m. When the cells are more strongly depolarized, they release more transmitter; when they are less strongly depolarized, less transmitter is released. As a result, the amount of transmitter released is a direct function of the membrane potential, V_m, of the neurons, which can indicate how strongly the neurons themselves were stimulated.

THE CHEMICAL NATURE OF NEUROTRANSMITTERS

Once it became clear that most synaptic transmission is carried by chemical signals, it became important to identify the molecular identities of these transmitter substances. By the mid-1960s, only three compounds had been unequivocally identified as neurotransmitters: *acetylcholine, norepinephrine,* and *γ-aminobutyric acid* (GABA). In the process of identifying and characterizing these compounds, three criteria were established that had to be met by any candidate molecule to determine that it was a neurotransmitter:

1. If the candidate substance is applied to the membrane of a postsynaptic cell, it must elicit precisely the same physiological effects in the postsynaptic cell as does presynaptic stimulation.
2. The substance must be released during activity of the presynaptic neuron.
3. The action of the substance must be blocked by the same agents that block natural transmission at that synapse.

Using these criteria, physiologists have identified many other neurotransmitters, but enormous effort has been required. The identification of the transmitters at most synapses in the vertebrate central nervous system has been very difficult, because very little transmitter is released at most synapses (only about 10^4 molecules per synapse per AP—compare this with Avogadro's number). Moreover, neuronal tissue is a nonhomogeneous collection of tightly packed and diverse cell types, which complicates collecting these molecules. However, a sufficient number of transmitter substances have been identified that patterns have begun to emerge.

The effect that a neurotransmitter exerts on a postsynaptic cell can depend on either of two very different mechanisms, and this difference forms the basis of one classification scheme for neurotransmitters. All transmitters ultimately modify the conductance of ion channels, but the change in conductance is produced in a variety of ways. Some transmitters act directly on ion channel proteins to change conductances through the postsynaptic membrane and thereby to change V_m; this type of transmission is *fast*, or *direct, synaptic transmission*. Other transmitters work through a biochemical pathway within the postsynaptic cell, changing the state of membrane-associated or cytosolic second messengers and thus producing changes in ion channel proteins. The shifts in V_m generated by this second type of transmitter occur more slowly, so this type of transmission is *slow*, or *indirect, synaptic transmission*. We have now identified more slow transmitters than fast ones. Neurotransmitters that work indirectly may also act as **neuromodulators**, affecting many neighboring neurons, if they are released more generally into extracellular fluids and are thus able to modify the behavior of many postsynaptic neurons at once.

Alternatively, neurotransmitters can be sorted into two groups based on their chemical structure. One group consists of small molecules (Table 6-2). The other group,

TABLE 6-2
Small neurotransmitter and neuromodulatory molecules are distributed broadly across many phyla

Transmitter	Site of action	Action
Acetylcholine (ACh)	Vertebrate NMJ	E
	Vertebrate autonomic nervous system: pre- to postganglionic neurons	E
	Parasympathetic neurons	E or I
	Vertebrate CNS	E
	Many invertebrates	E or I
Norepinephrine	Vertebrate postganglionic sympathetic neurons	E or I
	Vertebrate CNS	E or I
Glutamic acid	Vertebrate CNS	E
	Crustacean CNS and PNS	E
γ-Aminobutyric acid (GABA)	Vertebrate CNS	I
	Crustacean CNS and PNS	I
	Annelid CNS and PNS	I
Serotonin (5-hydroxy-tryptamine)	Vertebrate and invertebrate CNS	I, M
Dopamine	Vertebrate, annelid, arthropod CNS, PNS, or both	E or I (?)
Glycine	Vertebrate spinal cord	I

Abbreviations: NMJ, neuromuscular junction; CNS, central nervous system; PNS, peripheral nervous system; E, excitatory; I, inhibitory; M, modulatory.

the neuropeptides, consists of larger molecules that are constructed of amino acids. More than 40 neuropeptide transmitters have been identified in the mammalian central nervous system.

If we are to understand chemical synaptic transmission, we must understand how the different neurotransmitters exert effects on postsynaptic cells. In addition, we must understand how these signals are terminated. Postsynaptic neurons can pass along information about the firing frequency of their presynaptic partners only if the duration of each postsynaptic potential is limited. The mechanism that limits the length of each postsynaptic potential is the removal of transmitter from the synaptic cleft, which is achieved differently for different neurotransmitters. For some, such as ACh, a specific enzyme hydrolyzes the transmitter molecule, rendering it inactive. Other neurotransmitters, such as serotonin, are taken from the synaptic cleft back into the presynaptic neuron and at least potentially reused.

The same neurotransmitters have been found throughout the animal kingdom from nematodes to gnus, revealing that these molecules have been very highly conserved in evolution. In this section, we consider several identified neurotransmitters and the types of synaptic transmission to which they contribute.

Fast, Direct Neurotransmission

Among the low molecular weight neurotransmitters, only a few are known to mediate fast neurotransmission. Acetyl-

choline, glutamate, aspartate, and adenosine triphosphate (ATP) usually, but not always, participate in fast excitatory synaptic transmission. γ-Aminobutyric acid and glycine mediate fast inhibition. All of these transmitters, except aspartate and ATP, have been shown to open ion channels in the membrane of the postsynaptic cell, and it is expected that aspartate and ATP will be found to do so as well.

Acetylcholine (Figure 6-30) is the most familiar of the established transmitter substances. Neurons that release ACh, which are said to be *cholinergic*, are widely distributed throughout the animal kingdom. To list just a few of the known examples, ACh is the neurotransmitter used by vertebrate motor neurons, the preganglionic neurons of the vertebrate autonomic nervous system, the postganglionic neurons of the parasympathetic division of the autonomic nervous system, and many neurons of the vertebrate central nervous system. Acetylcholine is also the transmitter in a number of invertebrate neurons, including cells of the molluscan central nervous system, motor neurons of annelid worms, and sensory neurons of arthropods. Molecules that have crucial structural features in common with ACh can also act at cholinergic synapses (see Figure 6-30). For example, many molecules, such as carbachol, can activate cholinergic synapses; molecules that mimic the action of a neurotransmitter are said to be **agonists** at the synapse. Alternatively, molecules that have structural features in common with a transmitter can block transmission. For example, D-tubocurarine, the active agent in the South Amer-

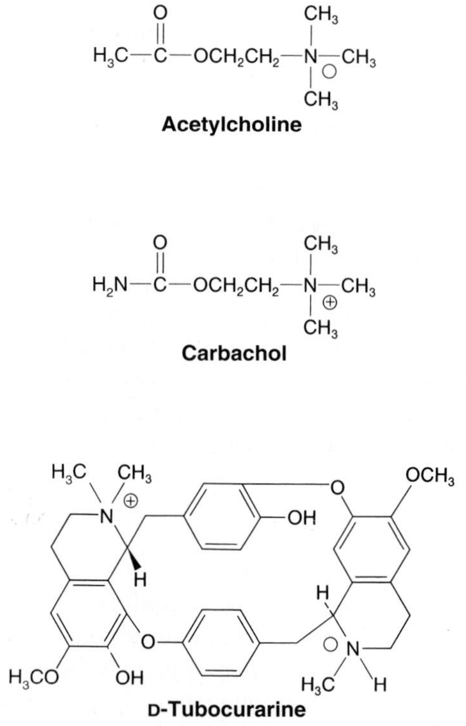

Acetylcholine

Carbachol

D-Tubocurarine

Figure 6-30 Several different molecular species can interact with postsynaptic receptors at fast acetylcholine synapses. Acetylcholine (ACh) is the natural ligand. The ACh analog carbachol is an agonist; it mimics the action of ACh at fast ACh synapses. D-Tubocurarine blocks the activation of receptors at these synapses.

ican blow-dart poison *curare*, blocks transmission at many cholinergic synapses by competing with ACh for receptor binding sites. Molecules that block the action of a neurotransmitter are called **antagonists.**

Transmission is terminated at cholinergic synapses when ACh is hydrolyzed to choline and acetate by the enzyme acetylcholinesterase (AChE), which is abundantly present in the synaptic cleft near the surface of the postsynaptic membrane (Figure 6-31). The choline that remains in the cleft is actively reabsorbed by the presynaptic membrane and recycled by condensation with acetyl coenzyme A (acetyl CoA) to form new molecules of ACh. Blocking the activity of AChE produces dramatic and dangerous effects, as illustrated by the action of certain nerve gases (such as Sarin, which was released into a Tokyo subway in the spring of 1995) and many insecticides. These agents block AChE activity; as a result, ACh lingers in the synaptic cleft, and its concentration can build up. In some cases, the postsynaptic cell cannot repolarize; but, at many synapses, the ACh receptor molecules become inactivated. In either case, function of the nervous and neuromuscular systems is disrupted, and death can follow. (In the vertebrates, death is typically caused by paralysis of the respiratory muscles.)

Several amino acids have been found to act as fast neurotransmitters (Figure 6-32; see also Table 6-2). Glutamate (glutamic acid) is released at excitatory synapses in the vertebrate central nervous system and at fast excitatory neuromuscular junctions in insects and crustaceans. γ-Aminobutyric acid is the transmitter at the inhibitory motor synapses onto crustacean and annelid muscle, and it plays a very important role as an inhibitory transmitter in the vertebrate central nervous system.

It has been found that each neuron synthesizes only one fast neurotransmitter, and each neuron releases that transmitter at all of its synapses. However, many neurons appear to synthesize other synaptically active agents as well, such as the molecules considered next.

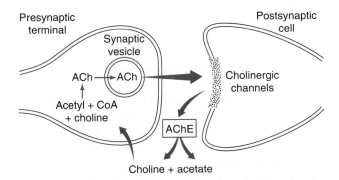

Figure 6-31 Choline is recycled at fast cholinergic synapses. Acetylcholine (ACh) that is released from the terminal is hydrolyzed to acetate and choline by the enzyme acetylcholinesterase (AChE), which is plentiful in the synaptic cleft. The liberated choline is taken up by the presynaptic terminal and reacetylated to form new ACh molecules. [Adapted from Mountcastle and Baldessarini, 1968.]

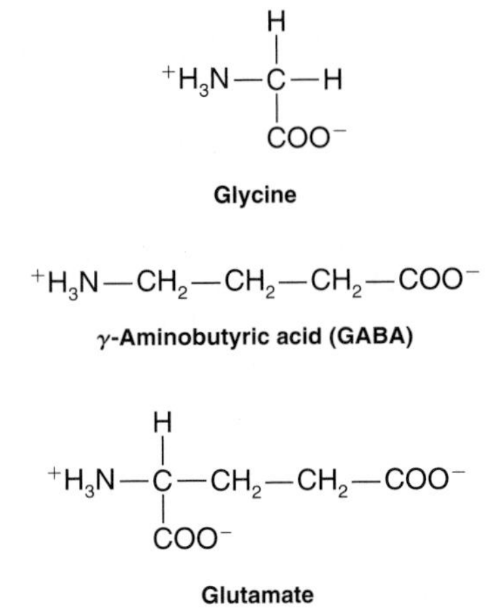

$$^{+}H_3N-\overset{\displaystyle H}{\underset{\displaystyle COO^-}{C}}-H$$

Glycine

$$^{+}H_3N-CH_2-CH_2-CH_2-COO^-$$

γ-Aminobutyric acid (GABA)

$$^{+}H_3N-\overset{\displaystyle H}{\underset{\displaystyle COO^-}{C}}-CH_2-CH_2-COO^-$$

Glutamate

Figure 6-32 Several amino acids act as fast neurotransmitters. Chemical structures of glycine, γ-aminobutyric acid (GABA), and glutamate. Glycine and GABA typically act as inhibitory neurotransmitters, whereas glutamate is often excitatory.

Slow, Indirect Neurotransmission

The biogenic amines constitute an important class of neurotransmitters (Figure 6-33) that act through second messengers to produce slow synaptic transmission. This class of neurotransmitters includes:

- *Epinephrine, norepinephrine,* and *dopamine,* classified as catecholamines on the basis of their chemical structure
- *Serotonin* (5-hydroxytryptamine, or 5-HT), an indolamine
- *Histamine,* an imidazole

These substances can be detected visually in individual neurons because they fluoresce in ultraviolet light after the tissue has been fixed with formaldehyde. They act as neurotransmitters in some invertebrate neurons and in the central and autonomic nervous systems of vertebrates (see Table 6-2).

Norepinephrine (also known as noradrenaline) is the primary excitatory transmitter released by postganglionic cells of the vertebrate sympathetic system (see Chapter 11). It is also released by the chromaffin cells of the vertebrate adrenal medulla (see Chapter 8). The chromaffin cells are derived embryologically from postganglionic neurons, and they secrete epinephrine (adrenaline) as well as norepinephrine. Epinephrine and norepinephrine are structurally very similar (see Figure 6-33), and they have similar pharmacological actions. Neurons that use epinephrine or norepinephrine as transmitters are *adrenergic neurons.* At some synapses, epinephrine is excitatory; at others, it is inhibitory. Its effect depends on the properties of the postsynaptic membrane.

Norepinephrine is synthesized from the amino acid phenylalanine (Figure 6-34A), and it is inactivated in several ways. It is taken up into the cytoplasm of the presynaptic neuron, where some of it is repackaged into synaptic vesicles for rerelease and some of it is inactivated by

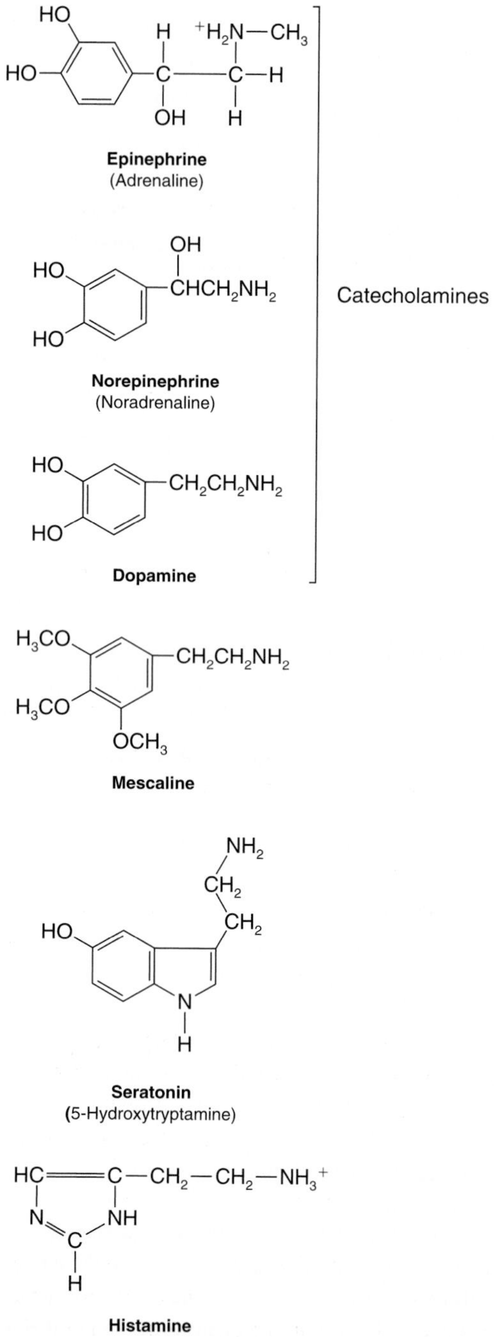

Figure 6-33 Several neurotransmitters are monoamines. These transmitters are each synthesized from single amino acid molecules and are classified on the basis of their molecular structure. Epinephrine, norepinephrine, and dopamine constitute one group, the catecholamines. Mescaline, a halucinogenic drug, has structural features in common with the catecholamines and appears to produce its effects by interacting with catecholamine receptors in the central nervous system. Serotonin (5-hydroxytryptamine) is an indolamine, and histamine is an imidazole. These transmitters are found in the nervous systems of vertebrates, as well as many invertebrates.

monoamine oxidase. In addition, it is deactivated by methylation within the synaptic cleft (Figure 6-34B). Several psychoactive drugs have molecular structures that are similar to the biogenic amines, which allows them to act at synapses that use these transmitters. For example, mesca-

line (see Figure 6-33), a psychoactive drug that is extracted from the peyote cactus, induces hallucinations, apparently by interfering with its analog norepinephrine at synapses in the central nervous system. Both amphetamines and cocaine exert their effects by interacting with adrenergic neurotransmission—amphetamines by mimicking norepinephrine and cocaine by interfering with the inactivation of norepinephrine.

In addition to the relatively small, "classic" transmitter molecules, there is a growing list (now more than 40) of peptide molecules that are produced and released in the vertebrate central nervous system. Many of these molecules, or very similar analogs, have also been found in the nervous systems of invertebrates. Some of these peptides act as transmitters; others act as modulators that influence synaptic transmission. Interestingly, a number of these neuropeptides are produced in many tissues, not just in neurons. Thus, a single molecular species may be released from intestinal endocrine cells, from autonomic neurons, from various sensory neurons, and in various parts of the central nervous system. In fact, some neuropeptides were initially discovered in visceral tissues and were only later identified in neurons. The gastrointestinal hormones glucagon, gastrin, and cholecystokinin (see Chapter 15) are prime examples.

It is not yet clear how many peptide neurotransmitters there are. We know that some neuropeptides act in a *neurosecretory* fashion; that is, they are liberated into the circulation and are carried by the blood to their targets, rather than being released into the confined space of a synaptic cleft. The pituitary hormone–releasing factors of the hypothalamus operate by neurosecretion (see Chapter 9). There is evidence that a single neuropeptide species may be liberated as a transmitter from some neurons, as a neurosecretory substance from other neurons, and as a hormone from nonneuronal tissue. This multiplicity of function is not really all that novel. It has long been known that norepinephrine (as well as its close relative, epinephrine) acts as a hormone when liberated by the adrenal medulla and as a transmitter when released at synapses. Recently, however, it has become clear—much to the surprise of neurophysiologists—that a neuropeptide can be released as a **cotransmitter** from nerve terminals that also release a more familiar transmitter such as ACh, serotonin, or norepinephrine. Several combinations of a classic transmitter and a paired cotransmitter have been identified in the mammalian brain (Table 6-3).

The first neuropeptide was discovered in 1931 by U. S. von Euler and John H. Gaddum while they were assaying for ACh in extracts of rabbit brain and intestine. The extracts stimulated contraction of the isolated intestine, much as ACh does, but the resulting contractions were not blocked by ACh antagonists. This observation led von Euler and Gaddum to discover that the contraction was produced in response to a polypeptide, which the researchers named *substance P*. Since then, substance P and a growing list of other neuropeptides have been found in various parts of the central, peripheral, and autonomic ner-

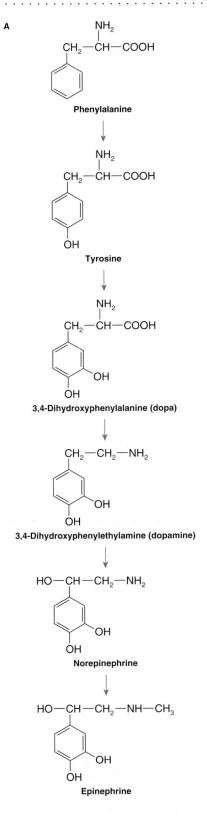

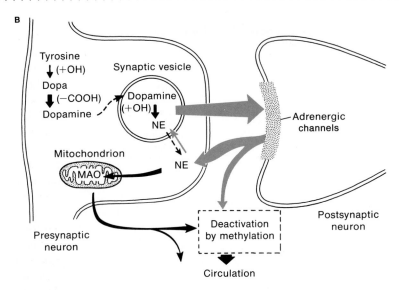

Figure 6-34 Epinephrine is synthesized from phenylalanine, with dopamine and norepinephrine as intermediates, and is inactivated by reuptake or by methylation. **(A)** Biosynthetic pathway leading to epinephrine. Each of the lower three molecules is used as a neurotransmitter by some neurons. **(B)** Norepinephrine is synthesized from the amino acid phenylalanine, through conversion into tyrosine, and stored in synaptic vesicles. After it is released into the synaptic cleft, some norepinephrine is taken back up into the presynaptic terminal, and some is deactivated by methylation and carried away in the blood. Cytoplasmic norepinephrine is either repackaged into synaptic vesicles or degraded by monoamine oxidase (MAO). [Part A adapted from Eiduson, 1974; part B adapted from Mountcastle and Baldessarini, 1968.]

vous systems of vertebrates and in many invertebrate nervous systems. To explore the localization of these molecules, investigators have typically used immunological labeling with fluorescent antibodies that recognize specific neuropeptides. In histological sections, this labeling can be detected with a fluorescence microscope, and it reveals the distribution of specific peptides in the nervous sys-

tem. Some well known neuropeptides are antidiuretic hormone (see Chapter 14), the hypothalamic-releasing hormones (see Chapter 9), and various gastric hormones (see Chapter 15).

Unlike small neurotransmitters, which may be synthesized in the synaptic terminals, neuropeptides are made in the cell body and are transported along the axons to the terminals. Neuropeptides are typically synthesized as a part of larger proteins, called **propeptides,** which may contain the sequences for many biologically active molecules. Specific enzymes cleave the propeptide into individual peptide molecules. This method of production can limit the amount of peptide neurotransmitter available at a synapse compared with a neurotransmitter synthesized on site. Peptides are, however, more potent than small neurotransmitters for three reasons. First, they bind to receptors at much lower concentrations than do other neurotransmitters (about 10^{-9} M versus 10^{-5} M for typical neurotransmitters), so very small amounts of neuropeptide can be effective. Second, they act through intracellular pathways that can provide significant amplification. Thus, even a small amount can produce a large effect. Third, the mechanisms that terminate their actions are slower than are those for other neurotransmitters, so they remain available to their receptors for a longer time.

Recent research has focused on two groups of naturally occurring neuropeptides, known as **endorphins** and **enkephalins,** that reduce the perception of pain and induce euphoria, much as exogenous opiates such as opium and heroin do. The levels of endorphin and enkephalin

TABLE 6-3
Examples of small and large neurotransmitter molecules that have been found together in neurons

Small neurotransmitter	Peptide in same neuron
Acetylcholine	CGRP
	Enkephalin
	Galanin
	GnRH
	Neurotensin
	Somatostatin
	Substance P
	VIP
Dopamine	CCK
	Enkephalin
	Neurotensin
Epinephrine	Enkephalin
	Neuropeptide Y
	Neurotensin
	Substance P
Norepinephrine	Enkephalin
	Neuropeptide Y
	Neurotensin
	Somatostatin
	Vasopressin
γ-Aminobutyric acid	CCK
	Enkephalin
	Somatostatin
	Neuropeptide Y
	Substance P
	VIP

Most of these data are based on immunocytochemistry, and the precise chemical nature of the immunoreactive peptide has not been determined. Abbreviations: CCK, cholecystokinin; CGRP, calcitonin gene–related peptide; GnRH, gonadotropin-releasing hormone; VIP, vasoactive intestinal peptide.

Source: Adapted from Hall, 1992.

molecules have been found to rise in the brain in response to eating, to listening to pleasant music, and to other activities generally perceived as pleasurable. Because of their properties and because these neuropeptides bind to the same receptors in the nervous system to which **opiates** such as opium and its derivatives bind, they are called **endogenous opioids.** Until these endogenous neuropeptides were discovered, it was very difficult to understand how alkaloids derived from plants—such as opium, morphine, and heroin—could so powerfully affect the nervous systems of animals. We now know that the surface membranes of many central neurons contain *opioid receptors,* and this class of receptors normally binds the enkephalins and endorphins that are produced within the central nervous system. Only secondarily, and perhaps coincidentally, do they bind exogenous opioids. However, when opioid molecules bind to the receptors, they elicit such intense feelings of pleasure that people learned to use opiate narcotics to stimulate the receptors. There is, however, a physiological problem associated with this intense pleasure: repeated doses of the exogenous opiates provoke compensatory changes in neuronal metabolism, such that removal of the opiate shifts the nervous system into a state

that is perceived as extreme *dis*comfort until the opiate is readministered. This metabolically induced dependence is termed **addiction.**

The drug *naloxone,* which acts as a competitive blocker of the opioid receptor, has proved to be a useful tool in studies of opioid receptors. Because naloxone interferes with the ability of either opiates or the opioid peptides to act on their target cells, it has allowed investigators to determine whether a response is mediated by opioid receptors. For example, naloxone has been found to block the analgesic effect that can be produced by a **placebo** (inert substance given to patients with the suggestion that it will relieve pain). Apparently the very fact that a subject believes a medication or other treatment will relieve pain can induce the release of endogenous opioid peptides, and this observation may have revealed the physiological basis for the well known "placebo effect" (i.e., almost anything that you do to research subjects will produce whatever effect that you promise, at least in some of the subjects.) Similarly, naloxone renders acupuncture ineffective in relieving pain, which has led to the suggestion that the stimulation of acupuncture causes the release of natural opioid peptides within the central nervous system.

There is some indication that the analgesic properties of the endogenous opioids may depend on the ability of these neuropeptides to block the release of transmitter from certain nerve endings. For example, the sensation of pain may be diminished if neuropeptides interfere with synaptic transmission along afferent pathways that carry information about noxious stimuli. Indeed, enkephalins and endorphins have been found in the dorsal horn of the vertebrate spinal cord, part of the pathway carrying sensory input within the spinal cord.

 Most vertebrate and invertebrate species have both fast and slow neurotransmission. What kinds of information processing would be served best by fast neurotransmission? What kinds by slow neurotransmission?

POSTSYNAPTIC MECHANISMS

Neurotransmitter molecules act through specific protein receptors in the membranes of postsynaptic cells. The properties of the postsynaptic molecules thus form a crucial link in the chain of events that begin when an action potential arrives at the terminal of a presynaptic neuron and end when the response of the postsynaptic neuron is complete. In this section, we will consider in detail the receptor molecules that mediate the two major classes—fast and slow—of chemical synaptic transmission and the events that take place after a neurotransmitter molecule has bound to these receptors.

Receptors and Channels in Fast, Direct Neurotransmission

As we have seen, chemical transmitters act by directly changing the permeability of the postsynaptic membrane to certain ions. (Typically the permeability increases.) This interaction requires two major events:

1. Transmitter molecules must bind to receptor molecules in the postsynaptic membrane.
2. When the transmitter molecules bind to the receptors, closed ion channels must open (or, more rarely, open channels must close) transiently. The receptor site may be located in the same molecular complex that forms the channel or it may be on a molecule that is separate from those that make up the channel.

When a synaptic channel opens, a minute ionic current passes through the open channel. Many such single-channel currents normally sum to form the macroscopic synaptic current that produces postsynaptic potentials in response to the release of tens—or even hundreds—of thousands of transmitter molecules from the presynaptic terminal. Most of what we know about these events has been revealed in studies of the ACh-activated channels at the vertebrate neuromuscular junction.

The acetylcholine receptor channel

The number of postsynaptic channel protein molecules is very small relative to other proteins in a membrane; as a result, the isolation, identification, and characterization of these important proteins was difficult. In early studies, physiologists used a variety of pharmacological agents to distinguish among receptor types, creating a pharmacological taxonomy of receptor types. As a result, various ion channels were named for substances that could modify the activity of the channels. For example, there are two types of acetylcholine receptors. *Nicotine,* an alkaloid produced by some plants, mimics the action of ACh on the channels found at the vertebrate neuromuscular junction, so these ACh receptors (AChRs) are called **nicotinic AChRs.** *Muscarine,* a toxin isolated from certain mushrooms, activates the other type of AChR, which is found in the target cells of parasympathetic neurons in the vertebrate autonomic nervous system. These AChRs are called **muscarinic AChRs.**

Our understanding of nicotinic AChRs was given a huge boost when it was discovered that specialized organs of certain elasmobranch and teleost fishes contain extremely high densities of these receptors. The receptors are found on one side of the **electroplax organ,** which consists of many flattened cells that originate from embryonic muscle tissue and produce the very high intensity electrical discharges used by these species to stun prey and to send navigational signals. The unusually high density of nicotinic AChRs in electroplax tissues allowed the nicotinic AChR to be the first ligand-gated channel to be purified chemically and studied electrically. More recently, its molecular structure has been resolved; we even have images of the form of the receptor channel as it opens.

A second important aid to the analysis of the AChR is its sensitivity to *α-Bungarotoxin* (αBuTX; see Spotlight 6-3), a component of cobra venom that binds irreversibly and with high specificity to nicotinic AChRs. α-Bungarotoxin can be isotopically labeled and used to tag AChR molecules, facilitating chemical isolation and purification. Physiological and biochemical studies have shown that the AChR and the postsynaptic channel that is activated by ACh are identical: the receptor site to which ACh molecules bind is an integral part of the channel protein complex.

Each nicotinic AChR consists of five homologous subunits that associate and form a channel at the center of the complex (Figure 6-35). There are two identical α subunits plus one each of three different subunits termed β, γ, and δ. Each subunit is a glycoprotein with a molecular mass of approximately 55 kD, giving the entire complex a total molecular mass of about 275 kD. This molecular weight agrees well with the size of channel structures that have been seen, using the electron microscope, to penetrate the surface membrane. The channels protrude from both sides of the membrane, with the funnel-shaped opening bulging outward from the cell surface.

Acetylcholine binds to the AChR where the receptor molecule extends into the extracellular space. This location was first deduced because ACh injected *into* a muscle cell near the endplate produced no electrical effect. Since then, experiments have shown that there are receptor sites on each of the two α-subunits. When both sites are occupied by *ligand* molecules (i.e., ACh or other agonists, such as carbachol or nicotine, that activate the channel), the channel shows a high probability of shifting from a closed to an open state. The nature of this gating process has been studied most extensively at the neuromuscular junction of frog skeletal muscle.

As described earlier, the postsynaptic ion channels of the frog neuromuscular junction become permeable to both K^+ and Na^+ when they are activated by ACh. The increased permeability permits the flow of an inward current with a reversal potential of about -10 mV. Normally, these channels and the associated AChRs are confined to the postsynaptic membrane in the endplate region. The density of ACh-activated channels in the postsynaptic membrane of the frog endplate is about 10^4 per square micrometer. Although this high density of channels proved useful for analyzing the summed activity of many ACh channels, for a long time little was understood about the activity of individual channels. Analysis of single channels was made possible by the invention of patch-clamp recording by Erwin Neher and Bert Sakmann (1976; see Figure 5-24), for which they were awarded a Nobel Prize in 1992. Their work on single AChR channels depended both on the development of the patch-clamp technique (see Chapter 2 and Figure 5-24) and on finding a region of muscle that had a sufficiently sparse distribution of AChR channels that they could isolate and record from a single channel. They

A

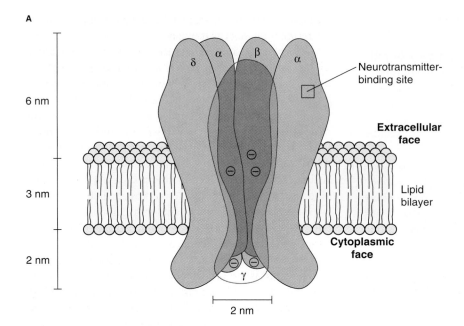

6 nm

3 nm

2 nm

2 nm

Neurotransmitter-binding site

Extracellular face

Lipid bilayer

Cytoplasmic face

B

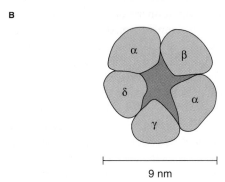

9 nm

Figure 6-35 Nicotinic acetylcholine receptors at the neuromuscular junction consist of five protein subunits associated to form a transmembrane channel. **(A)** The channel is inserted through the lipid bilayer, protruding into the extracellular space and into the cytoplasm. The α subunits contain the sites to which acetylcholine molecules bind to activate the channel. The entry to the channel from outside the cell is a broad funnel that becomes narrower and bears a net negative charge toward the cytoplasm, forming the selectivity filter—the region of the pore that controls which ions will permeate readily. In this diagram, the inside of the channel is darker in color than the surrounding region. The subunit facing you is a γ subunit. **(B)** Top view showing the five subunits associated to form the channel. These structural features have been inferred on the basis of electron microscopy and X-ray diffraction analysis. [Adapted from Unwin, 1993.]

generated this sparse distribution by taking advantage of the changes that occur in frog skeletal muscle after the motor nerve controlling the muscle has been cut.

When a muscle is *denervated* (i.e., it loses its neuronal input—experimentally the axons are crushed), the region of the membrane that responds to ACh gradually spreads across the surface. Initially, only the membrane at the endplate region can respond, but eventually most or all of the membrane contains AChRs and can respond to ACh. (The normal suppression of these *extrajunctional AChRs* is thought to be dependent on two factors: first, trophic action from the motor neuron that innervates each muscle fiber and, second, electrical and contractile activity that takes place in an innervated muscle fiber. If the motor axon is allowed to reinnervate the muscle, the extrajunctional receptors disappear, and sensitivity to ACh is again confined to the endplate.)

The broad, but sparse, distribution of extrajunctional ACh-activated channels that develops in denervated frog muscle was exploited by Neher and Sakmann to explore the gating of the channel, using their newly developed patch-clamp method. The muscle membrane was voltage-clamped (see Spotlight 5-3) at a hyperpolarized potential to increase the driving force for inward current. They used a

micropipette that had a smoothly polished tip with a tip diameter of 10 μm, which they filled with Ringer solution containing a low concentration of ACh or one of its agonists. They then brought the pipette up to the surface of the muscle fiber, exposing any AChRs under the pipette tip to the ACh. The pipette was connected to a highly sensitive, low-noise amplifier (Figure 6-36A) with which they could record currents flowing in the extracellular pipette. When it was applied snugly to the surface of the denervated muscle fiber, the pipette detected minute (less than 5×10^{-12} A) and transient inward currents (Figure 6-36B) produced by the transient opening of the ACh-activated channels. With this experiment, Neher and Sakmann produced the first recordings ever made of currents through single ion channels in a biological membrane. Indeed, this effort produced the first direct evidence that ionic currents cross the membrane through discrete, gated channels rather than by some other means, such as carrier molecules.

Single-channel currents such as those first recorded by Neher and Sakmann in 1976 are more or less rectangular in shape; they turn on abruptly and then turn off abruptly, and they are all-or-none. This observation strongly suggests that the channels can exist only in one of two states, completely shut or completely open. Moreover, the unitary cur-

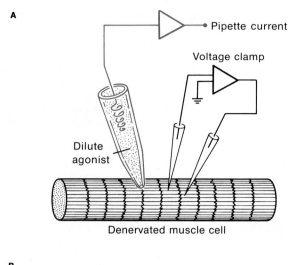

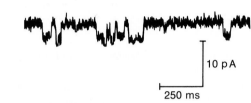

Figure 6-36 The patch recording technique reveals ionic currents through single acetylcholine receptor channels. **(A)** The muscle membrane is held at a hyperpolarized potential (−120 mV) by a voltage-clamp circuit, greatly increasing the driving force on ions through acetylcholine receptor (AChR) channels, while the surface of the muscle is explored with a patch pipette filled with Ringer solution containing 2×10^{-7} M suberylcholine (an ACh agonist). **(B)** When the pipette tip is sealed tightly against the membrane, brief, minute, inward currents are recorded. In this experiment, the pipette records current flow through the ion channel of a single AChR protein complex that opens transiently when agonist molecules are bound to receptor sites. [Adapted from Neher and Sakmann, 1976.]

rents recorded from each nicotinic ACh-activated channel are about the same size as the currents recorded from all other nicotinic ACh-activated channels, provided that the electrochemical driving force is kept constant. Ohm's law indicates that this result must mean that all individual nicotinic ACh channels have similar conductances. When two or more channels in the patch being recorded open with overlapping times, the individual, unitary single-channel currents sum linearly, producing a current two (or three, etc.) times as large as the individual unitary current. These currents are not present unless the pipette contains ACh or an agonist, and the frequency of their occurrence depends on the concentration of the transmitter or agonist in the pipette. From Ohm's law, the conductance of a single open nicotinic AChR channel was calculated to be about 2×10^{-11} S, which is usually expressed as 20 picosiemens (20×10^{-12} S; i.e., the channels have a resistance of 5×10^{10} Ω).

Since the pioneering patch-clamp experiments of Neher and Sakmann, many ligand-gated postsynaptic ion channels have been studied intensively with this method of recording single-channel currents. Statistical analyses of these unitary currents indicate that the channels can fluctuate between several closed states and at least one open state. Binding of an agonist molecule to the receptor sites of the closed channel greatly increases the probability that the channel will change to an open state and briefly allow ions to flow through the channel. The channel remains open for only about 1 ms and then closes, even though ACh is still bound to the receptor sites. After a short time, the agonist molecules leave the binding sites, and the channel remains closed until more molecules of ACh bind (Figure 6-37). The macroscopic currents and postsynaptic potentials recorded at a synapse represent the sum of many such single-channel events in the postsynaptic membrane.

Other ligand-gated channels

Since the ACh channel proteins were purified from electroplax organs, several types of ligand-gated channels have been isolated from neurons and characterized, including the glycine, $GABA_A$, and neuronal ACh receptors, all of which mediate rapid postsynaptic responses. These receptors have a common pentameric protein structure, and each is composed of two to four different kinds of subunits. As in the muscle ACh channel, only one type of subunit binds the ligand. The remarkable homologies among these different channel proteins have allowed the diversity of subunit types and their distribution in nervous tissue to be characterized at the molecular level. Somewhat surprisingly, for each type of receptor—ACh, glycine, and $GABA_A$—a number of different subunits are found to be assembled in different combinations to produce receptors with slightly different properties. Moreover, each type of receptor is expressed in a unique and characteristic pattern within the mammalian brain, indicating that expression of receptor subtypes is regulated differently in different regions of the nervous system. Recognizing the large number of permutations that are possible, even within receptors that respond to a single neurotransmitter, has helped us to understand how subtle the mechanisms that allow the brain to achieve its highly differentiated functional states can be. Furthermore, a comparison of the DNA sequences of the receptors for ACh, GABA, and glycine reveals them to be closely related, suggesting that all ligand-gated ion channels may have a common ancestral origin.

DNA sequence analysis has revealed that glutamate receptors belong to a separate family having only a slight resemblance to the nicotinic receptors. Currently, there is intense interest in this receptor family, both because glutamate is the most common excitatory neurotransmitter in the mammalian central nervous system and because glutamate receptors participate in modifications of synaptic strength, which may underlie learning and memory. At present, three types of fast-acting glutamate receptors have been identified and are named for their sensitivity to specific agonists. The agonists that typify the three receptor classes are *kainate*, *quisqualate* (α-amino-3-hydroxy-5-methylisoxazole-4-propionic acid), and *NMDA* (N-methyl-D-aspartate). These receptor types are considered further later

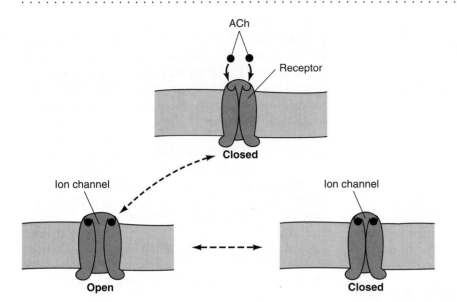

ACh

Receptor

Closed

Ion channel

Open

Ion channel

Closed

Figure 6-37 The nicotinic acetylcholine receptor channel exists in three functional states. The ion channel through the receptor opens when acetylcholine (ACh) or agonist molecules bind to the protein complex. After about 1 ms, the ion channel closes, even though the ACh molecules are still bound. The channel can "flicker" between closed and open states while ACh molecules remain bound. Then the ACh molecules unbind, the channel closes, and it remains in the closed state until two more ACh molecules bind.

in this chapter in the section concerning mechanisms of synaptic modification (see *Long-term potentiation*).

Receptors in Slow, Indirect Neurotransmission

A large family of receptors responds to the family of slow neurotransmitters. Interestingly, these receptors have many features in common with receptors that respond to light, to odor, to hormones, and to other extracellular messengers. Most such receptors act by activating members of a group of proteins, known as *G proteins,* that are associated with the cell membrane and that bind guanosine triphosphate (GTP). A G protein consists of three subunits, called α, β, and γ. The G protein–transmembrane signaling pathway was discovered and described by Alfred Gilman and Martin Rodbell, who studied its role in transduction of nonsteroid hormone signaling (for a more complete treatment of G proteins, see Chapter 9); for this work, they received the Nobel Prize in 1994. When GTP binds to a G-protein molecule, the protein is activated, and it catalyzes the hydrolysis of the bound GTP to GDP, which terminates its activation (Figure 6-38). When a membrane receptor molecule binds to its ligand, this cycle of binding and hydrolyzing GTP is facilitated, because the receptor-ligand complex catalyzes the release of GDP from the G protein, making the binding site more rapidly available to a new GTP molecule. Three separate proteins contribute to G protein–mediated synaptic transmission. The neurotransmitter receptor molecules span the membrane, binding the neurotransmitter on the extracellular face and catalyzing G-protein activation on the cytoplasmic face. The activated G protein can regulate the activity of effector proteins, which can be ion channels or enzymes that control the concentrations of intracellular second messengers or both. We now know of more than 100 receptors that act through G proteins, and these signaling molecules respond to a wide variety of external stimuli ranging from peptides to light and odors. The G proteins themselves constitute a family of at least 20 different proteins. The combinatorial richness of

this system provides yet another mechanism for producing subtle control within the nervous system.

A well studied example of indirect neurotransmission that regulates ion channels is found in heart atrial cells, the system Otto Loewi used more than 75 years ago in the first demonstration that neurons can transfer information by way of chemical signals. Acetylcholine acts on muscarinic receptors in the heart to hold K^+-selective channels open, prolonging a hyperpolarization. Several different kinds of experiments were needed to establish that this action of ACh depends on a G protein.

Some of the experimental results are described here. Acetylcholine has been found to act on heart atrial cells only if GTP is inside the cells, and the muscarinic activation of K^+ channels is known to be blocked by pertussis toxin,

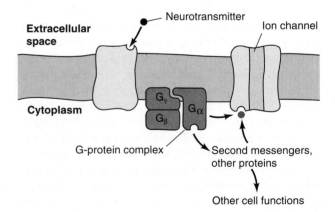

Extracellular space

Neurotransmitter

Ion channel

Cytoplasm

G_γ

G_β

G_α

G-protein complex

Second messengers, other proteins

Other cell functions

Figure 6-38 Intracellular second messengers modify channel conductances at slow chemical synapses. G proteins participate in signal transmission at many slow chemical synapses. At this kind of synapse, the receptor protein spans the plasma membrane. Neurotransmitter molecules bind to the extracellular domain of the receptor, which activates a G protein that resides on the cytoplasmic side of the membrane. The activated G protein regulates the activity of other intracellular proteins, which directly or indirectly changes the conductance through ion channels in the membrane. Activated G proteins can also modify other cellular functions, changing metabolic pathways or the structure of the cytoskeleton.

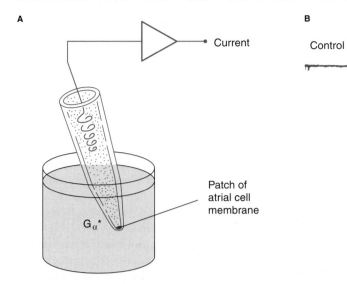

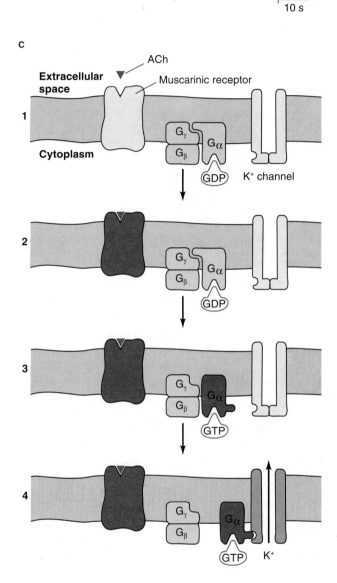

Figure 6-39 Muscarinic acetylcholine receptors in cardiac cells indirectly cause potassium ion channels in the membrane to open. **(A)** Experimental setup for measuring the effect of slow synaptic activation on guinea pig heart atrial cells. A nonhydrolyzable analog of GTP, GTPγS, was bound to α subunits of the G protein to activate them, and the activated α subunits (the activated state is indicated by an asterisk) were applied to the intracellular surface of an isolated patch of membrane from atrial cells. The net effect mimics the result of receptor-mediated activation of the endogenous G protein. **(B)** Typical recordings from an experiment like the one shown in part A. When the concentration of activated α subunits increased, the K$^+$ channels opened more frequently, producing more frequent current steps in the single-channel records. **(C)** Schematic representation of events at a muscarinic synapse in an intact cell. When ACh binds to muscarinic receptors, G proteins in the membrane are activated, and α subunits of the G proteins bind to K$^+$ channels, causing them to open. [Data adapted from Covina et al., 1987.]

which inactivates numerous G proteins. In a direct test of the hypothesis that ACh acts on these cells by means of a G protein, Codina and colleagues (1987) applied G-protein α subunits that had been activated by GTPγS, a nonhydrolyzable analog of GTP, to the inside of membrane patches from cardiac muscle cells (Figure 6-39A). The result mimicked the stable activation of G proteins in the membrane. As the amount of activated α subunit was increased in the bathing solution, the number of open channels increased, as shown by the increased number of single-channel currents (Figure 6-39B). Similar experiments have identified a large variety of K$^+$, Na$^+$, and Ca^{2+} channels whose activity is regulated similarly by receptor-activated α subunits of G proteins.

Neuromodulation

The postsynaptic response to fast synaptic transmitters is immediate, brief, and localized to specialized sites on the postsynaptic cell. In contrast, slow synaptic transmission not only is slow and long lasting, but can also be spatially widespread. In some cases, slow, or indirect, synaptic transmission can interact with and modulate the effects of fast synaptic transmission. The interaction can affect just one postsynaptic neuron, or it may affect many more post-

synaptic neurons, a phenomenon called **neuromodulation.** Neuromodulation (or, more precisely, modulation of synaptic transmission) refers to transient changes in how effectively a presynaptic neuron can control events in the postsynaptic neuron (i.e., its synaptic efficacy.) Neuromodulatory changes in synaptic efficacy last from seconds to minutes, and this time course distinguishes neuromodulation from *synaptic plasticity,* described later in this chapter, in which the effects are much longer lasting or even permanent.

One of the best understood examples of neuromodulation and its role in normal synaptic excitation is found in cells of sympathetic ganglia of frogs. The system is complex because these cells receive three distinct classes of synaptic inputs that are mediated by two different neurotransmitters acting on three distinct types of receptors. Three distinct excitatory postsynaptic responses are produced: a fast epsp, a slow epsp, and a late slow epsp. A typical experimental setup is shown in Figure 6-40A. Both the fast and the slow excitatory postsynaptic potentials are produced by ACh

from presynaptic nerve terminals. The postsynaptic cells have both nicotinic receptors (the fast response) and muscarinic receptors (the slow response) in their membranes. In contrast, the late, slow excitatory postsynaptic potential is produced by a neuropeptide that is very similar to the gonadotrophin-releasing hormone (GnRH—see Chapter 9) in mammals and is also released from presynaptic neurons, but not directly onto the postsynaptic neurons. The three postsynaptic potentials depolarize the postsynaptic cell by different amounts and at different times after stimulation, and they act through different, but not entirely independent, mechanisms.

When ACh binds to a nicotinic receptor, the ion channel in the receptor complex opens and Na^+ and K^+ can pass through, producing the fast response (Figure 6-40B). This type of excitatory postsynaptic potential can be elicited by a single stimulus that lasts only a few tens of milliseconds. The slow excitatory postsynaptic potential is produced when ACh binds to muscarinic receptors, and it can be elicited only after several trains of APs have arrived at the presynaptic site and released ACh. The muscarinic receptors act through a G protein to cause a type of K^+ channel, called an M channel, to close (Figure 6-41A). When these K^+ channels close, the steady state influx of Na^+ is no longer balanced by an efflux of K^+ and, as a result, the cell depolarizes. The depolarization is small (only about 10 mV; see Figure 6-40B), because it depends on the small steady-state Na^+ current. By itself, it cannot produce an AP in the postsynaptic cell, but it can significantly change the response of the cell to fast synaptic signals, particularly when it acts in concert with a late, slow excitatory postsynaptic potential. The late, slow epsp results from the release of a different kind of neurotransmitter, the GnRH-like peptide, that acts through a transmembrane receptor to close the same M channels that are affected by the muscarinic receptors. Adding exogenous GnRH to the postsynaptic neurons produces the same kind of response (see Figure 6-40C). The time course of the response to GnRH is even slower than the muscarinic response; it begins 100 ms after the stimulus and can last for 40 min (see Figure 6-40B).

The similarities and differences between these two slower responses are important for understanding how neuromodulation might act in animals. To explore the role of the slow excitatory postsynaptic potential in these sympathetic ganglion cells, the efficacy of an injection of current into the presynaptic cell was evaluated before and during a slow epsp (Figure 6-41B). Before the slow epsp, a presynaptic stimulus caused a single postsynaptic AP; whereas, during the slow epsp, the same stimulus elicited a burst of APs. Clearly, the slow epsp modified signal transmission across this synapse. Normally the K^+ current through M channels is activated by membrane depolarization and tends to repolarize the cell by shunting depolarizing currents that enter through synaptic channels, thus reducing the effectiveness of any excitatory postsynaptic potentials. When the M channels are kept closed by ACh

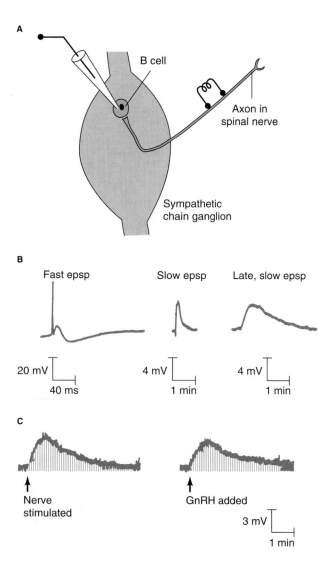

Figure 6-40 Postsynaptic potentials with very different time courses can be recorded in bullfrog sympathetic ganglion cells. **(A)** The sympathetic chain ganglia are located on either side of the spinal cord (see Chapter 11), and the responses of the large B cells (one class of neurons in the ganglia) can be recorded while nerves that innervate the ganglia are stimulated. Anterior is *up* in this diagram. **(B)** Three different kinds of synaptic responses can be recorded in B cells: (1) a fast excitatory postsynaptic potential, (with a latency of 30–50 ms) when ACh activates nicotinic receptors in the postsynaptic membrane; (2) a slow epsp (with a latency of 30–60 ms) when ACh binds to muscarinic receptors in the postsynaptic membrane; and (3) a late, slow epsp (with a latency of more than 100 ms) caused by a decapeptide messenger—found in the brains of cold-blooded vertebrates—that is closely related to the hypothalamic releasing factor GnRH. When GnRH binds to postsynaptic receptors, it produces a depolarization in the B cells that lasts many minutes. (Notice the calibration bars below the records.) **(C)** When exogenous GnRH is applied to B cells, the effect is identical in onset, magnitude, and duration with the late, slow epsp in part B. [Adapted from Jan and Jan, 1982.]

A

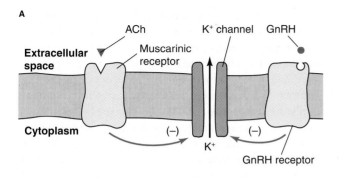

B

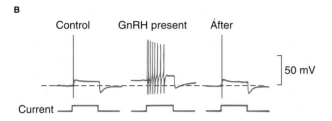

C

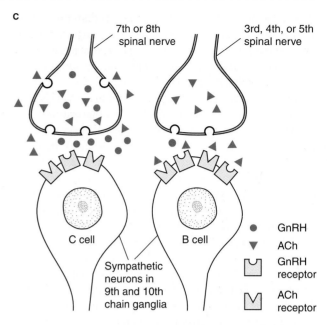

Figure 6-41 Both muscarinic acetylcholine receptors and GnRH receptors depolarize a postsynaptic cell by closing M-type potassium ion channels. **(A)** When acetylcholine (ACh) binds to muscarinic receptors or when the GnRH-like neuropeptide binds to its receptor, M-type channels close, reducing the K+ current across the membrane and depolarizing the neuron. **(B)** The effect of a fast excitatory postsynaptic potential (epsp) in a postsynaptic B cell before, during, and after a slow epsp. During the slow epsp, the decrease in the K+ current through M channels increased the excitability of the B cell, producing a train of action potentials in response to the fast epsp. **(C)** Cholinergic neurons from the seventh and eighth spinal nerves innervate C cells of the ninth and tenth sympathetic ganglia, whereas neurons from the third, fourth, and fifth nerves innervate only B cells in those ganglia. Only C cells receive terminals that are immunoreactive for GnRH, but stimulating the seventh and eighth spinal nerves produces a late, slow epsp in both B and C cells, suggesting that GnRH diffuses from its site of release at the surface of C cells and activates receptors on B cells. [Part B adapted from James and Adams, 1987; part C adapted from Jan and Jan, 1982.]

muscarinic receptors, repolarization of the membrane by the K+ current is prevented and further excitation is potentiated. The late, slow excitatory postsynaptic potential acts similarly, but with a longer latency and for a longer time, and it shares the M channel as a final common pathway. However, there is an additional twist because the peptide neurotransmitter diffuses to nearby neurons, which it can influence identically if the appropriate receptors are present (Figure 6-41C). Only some of the presynaptic neurons can release GnRH, but most postsynaptic cells seem to have GnRH receptors, which strongly suggests that neuromodulation is a normal part of this neuronal circuitry. Taken together, these mechanisms can produce a variety of postsynaptic effects following presynaptic transmitter release. A brief burst of activity in the presynaptic cells would typically generate only the fast excitatory postsynaptic response. More prolonged stimulation might activate the slow pathway in addition, which would effectively amplify the response of the postsynaptic cell to its fast excitatory postsynaptic potentials. With still greater stimulation, the late, slow pathway would additionally increase the effectiveness of fast excitatory postsynaptic potentials and could also potentiate responses in neighboring neurons (see Figure 6-41C), increasing the efficacy of neurotransmission in cells that are not directly postsynaptic to the GnRH-releasing neurons. Moreover, this modulation could be relatively long lived, given the long time constant of the late slow, response.

Within the past few years, studies in the crustacean stomatogastric ganglion have demonstrated the extreme power of neuromodulatory mechanisms. This ganglion contains only 30–40 identified neurons, whose interconnections have been characterized in detail and whose output patterns are well known. When certain neuromodulatory substances, such as proctolin or cholesytokinin, are added to the saline bathing the stomatogastric ganglion, the properties of at least some of the membrane channels change dramatically, effectively rewiring the entire ganglion and generating circuits and outputs that are never seen in the absence of the modulator. Thus, neuromodulators afford a means of remodeling neuronal circuitry, allowing a set of neurons to interact in several distinctly different ways, even though their physical synaptic relations remain unchanged.

INTEGRATION AT SYNAPSES

Only rarely are single neurons responsible for producing behavior. Even the simplest behavior requires that several hundred to many thousands of neurons act in a coordinated fashion. This coordination among neurons is called **neuronal integration.** Used in this sense, "to integrate" means "to combine into a whole." At the level of a single neuron, integration consists of responding to incoming synaptic inputs either by producing an AP or by not producing one, and every neuron integrates the various excitatory and inhibitory synaptic signals that impinge on it. The process of integration depends heavily on the passive electrical properties of the membrane that lies between the synapses and the spike-initiating zone. In addition, the density and voltage sensitivity of the Na^+ and K^+ channels determine the threshold and the rate of firing that is produced in response to a given synaptic depolarization.

Much of what we know about neuronal integration has been obtained from studies of the large *α-motor neurons* (Figure 6-42) in the vertebrate spinal cord. These neurons innervate groups of skeletal muscle fibers at neuromuscular junctions. In vertebrates, these are the only neurons that synapse directly onto skeletal muscle fibers, so they play an exceedingly important role in generating overt behavior (see Chapter 10). Thousands of inhibitory and excitatory synaptic terminals contact the dendrites and cell body of each α-motor neuron. The net effect of all synaptic activity is to control the frequency with which APs are generated in the cell. This frequency of firing (typically measured in impulses per second) determines the strength of contraction in the set of muscle fibers innervated by the motor neuron.

All of the integrative activity in a neuron is centered on producing APs (i.e., excitation) or suppressing them (i.e., inhibition). Because APs are the only events that can carry information over distances greater than a few millimeters, only synaptic inputs that cause APs in α-motor neurons can generate behavior. Any excitatory input that fails to bring a motor neuron to threshold, either by itself or by summation with other inputs, is lost because no AP is produced in the postsynaptic cell and the signal dies out.

In an α-motor neuron, APs are generated in the initial segment of the axon just beyond the axon hillock (see Figure 5-2). This region is more sensitive to depolarization than are the soma and dendrites (perhaps the membrane has a higher density of Na^+ channels in this location), so it has a lower threshold for producing APs. If it is to generate APs in the cell, a synaptic current must be able to bring the membrane of the spike-initiating zone up to threshold.

How do the many thousands of individual synaptic inputs onto a motor neuron influence its activity? Synaptic currents spread electrotonically from synapses on dendrites and the soma. How much the current decays over distance is determined by the cable properties of the neuron (Figure 6-43), but in all cases, synaptic potentials become smaller as they spread away from their sites of origin and toward the spike-initiating zone (see *Passive Spread of Electrical Signals* earlier in this chapter and Figure 6-16). Because the decrement depends on distance, a synaptic current set up at the end of a long, slender dendrite will decay more than will currents closer to the spike-initiating zone, so distant synapses exert a relatively smaller influence on the activity of the postsynaptic neuron. As a result, the location of synapses, as well as the initial size of synaptic currents, can influence how much control particular synapses have. (Interestingly, recent evidence suggests that in at least some neurons of the mammalian brain, there may be some Na^+ channels in dendritic membranes, and these channels can boost synaptic currents, preventing them from decaying as rapidly as they would if they were conducted only electrotonically.) In many cases, the density of inhibitory synapses is highest near the axon hillock, where these synapses can be most effective in preventing excitatory synaptic current from depolarizing the spike-initiating zone to threshold.

We have learned many of these concepts from experiments in frogs of the genus *Rana*. For example, in such an experiment, several segments of the spinal cord of an anesthetized frog are exposed by opening up the vertebral column. Then a microelectrode is lowered into the ventral horn of the gray matter and inserted into the soma of a single α-motor neuron. Small bundles of afferent axons dissected from the dorsal root are placed on silver-wire, stimulating electrodes, providing stimulation to some axons that cause the α-motor neurons to be excited and to others that cause the motor neurons to be inhibited.

Initially, the intracellular recording electrode will pick up randomly occurring postsynaptic potentials. These sig-

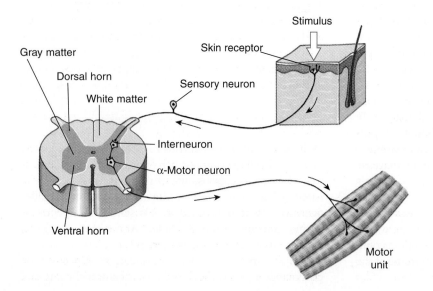

Figure 6-42 Neurons connected by synapses work together to process information. In this diagram, a spinal α-motor neuron, whose soma is located in the ventral spinal cord, is part of a disynaptic reflex arc (called the flexion reflex) in which a noxious stimulus applied to the skin causes excitation of a motor neuron that controls a flexor muscle. The pathway includes one interneuron between the sensory and motor neurons. Activation of the motor neuron causes the muscle fibers that it innervates to contract.

Gray matter

Dorsal horn

White matter

Stimulus

Skin receptor

Sensory neuron

Interneuron

α-Motor neuron

Ventral horn

Motor unit

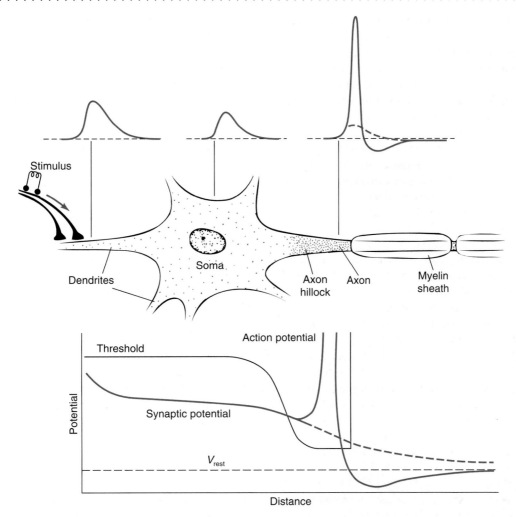

Figure 6-43 Each synaptic input decays with distance as it travels to the spike-initiating zone. An excitatory postsynaptic potential originating in a dendrite spreads electrotonically and gets smaller with distance *(top)*. The density of Na⁺ channels (red dots) in the membrane determines the threshold (black trace at *bottom*) for generating an AP. The synaptic potential gets smaller as it spreads toward the axon, and no AP is generated until the current reaches the dense Na⁺-channel distribution in the spike-initiating zone of the axon hillock (or at the first node of Ranvier), where the firing threshold is lowest. The graph shows the relative values of the threshold potential and the synaptic potential along the membrane between the synapse and the spike-initiating zone. The dashed line shows what the amplitude of the excitatory postsynaptic potential would be if the AP were blocked.

nals are caused by synaptic input onto the motor neuron that is not under experimental control. Typically, the activity consists of synaptic potentials with amplitudes of about 1 mV, which are similar to those of miniature endplate potentials recorded at the muscle endplate (see Figure 6-24). It has been shown that stimulation of a single neuron that is presynaptic to these motor neurons releases only from one to several quanta of transmitter in response to a presynaptic AP. In this respect, excitatory synapses ending on a motor neuron are quantitatively different from those on a neuromuscular junction, where a single motorneuron terminal releases approximately 100 to 300 quanta in response to a single presynaptic impulse and produces an excitatory postsynaptic potential of 60 mV or more. The transmitter released from a single synaptic ending onto an α-motor neuron depolarizes the neuron by only about 1 mV, far less than the amount required to shift the membrane potential to the firing level. Whereas the vertebrate neuromuscular

junction acts as a single relay synapse, transmitting in a one-to-one manner (i.e., one postsynaptic impulse for each presynaptic impulse), a motor neuron requires the more-or-less concurrent activation of *numerous* excitatory synaptic inputs impinging on it in order for the synaptic potential to reach the firing threshold for the initiation of a postsynaptic AP. Thus, the decision to fire is a response to a collection of presynaptic inputs, and although each small synaptic current is ineffective by itself, the activity at a single ending can contribute significantly to the integrative behavior of the neuron. This rather democratic behavior prevents activation of motor neurons by trivial input or spontaneous activity in their input neurons. More importantly, it provides a means of integrating inputs from various sources, both excitatory and inhibitory, to determine when the neuron will produce APs and how many there will be.

As the strength of the stimulus current applied to the presynaptic axons in the dorsal root is increased, more and more excitatory axons become active; that is, they are *recruited* by the increased stimulus. When these neurons fire in unison, the total amount of transmitter released onto the motor neurons rises, producing more individual synaptic currents that add together to cause a larger excitatory postsynaptic potential. When inputs from several individual synapses add together simultaneously to change V_m in the postsynaptic neuron, the process is called **spatial summation.** The summed synaptic inputs can lead to greater depolarization if they are all excitatory (Figure 6-44). If inhibitory transmitter is released simultaneously with excitatory transmitter, it also produces synaptic currents that sum with the excitatory currents (Figure 6-45). Open inhibitory synaptic channels can short-circuit the depolarizing current carried by Na^+ ions moving through excitatory channels; that is, as depolarizing positive charge is carried into the cell by Na^+ ions, some of that charge is promptly removed from the cell when K^+ ions move out or Cl^- ions move in through inhibitory synaptic channels. The activation of inhibitory synapses reduces the depolarization at the spike-initiating zone and decreases the probability that an AP will be produced.

When a second postsynaptic potential is elicited within a very short time after the first, it can add to, or "ride piggyback" on the first, even if the two synaptic events are caused by the same presynaptic neuron. This effect is called

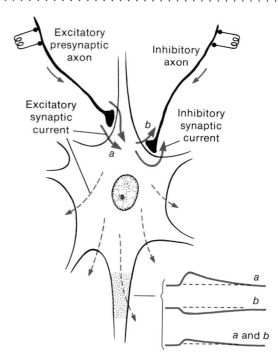

Figure 6-45 Excitatory and inhibitory synaptic currents sum. Stimulation of separate presynaptic pathways gives rise to excitatory (*a*) and inhibitory (*b*) synaptic currents. Traces at lower right show synaptic potentials recorded from the spike-initiating zone when either *a* or *b* was stimulated individually and then when they were stimulated together, illustrating the effects of summation. The dashed arrows indicate that some of the excitatory synaptic current is diminished by the open inhibitory channels.

temporal summation (Figure 6-46). The shorter the interval between two successive synaptic potentials, the higher the second response rides upon the first and so the bigger the postsynaptic potential can become. Further summation can be achieved if additional stimuli arrive in rapid succession, with the third synaptic potential riding on the second, and so forth. Under natural conditions, spatial and temporal summations often occur together. For example, if different excitatory synapses on one motor neuron are active at slightly different times, the effects will sum both spatially and temporally.

Both spatial and temporal summations of synaptic potentials depend on the passive electrical properties of the neurons. Spatial summation occurs because synaptic currents that originate at the same time, but at different synapses, each spread electrotonically away from the synapse (see Figure 6-43), so their effects on V_m can add together at the spike-initiating zone. Temporal summation, on the other hand, does not require the summation of synaptic *currents* and can occur even though the individual currents do not overlap (see Figure 6-46C), because the electrical time constant of the membrane is long relative to the time course of synaptic currents. The first synaptic current brings positive charge into the cell, partially discharging the negative resting potential of the cell membrane. The positive charge carried into the neuron by the synaptic current then slowly leaks out (through the resistance—K^+

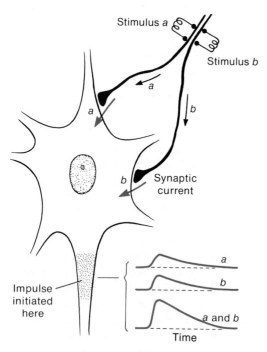

Figure 6-44 Synaptic inputs from several presynaptic neurons produce spatial summation onto a motor neuron. Two excitatory synaptic currents, from two separate neurons *a* and *b*, arise at spatially separated synapses. Traces at lower right show synaptic potentials recorded at the spike-initiating zone when each input acted alone and when the two inputs were active simultaneously, producing spatial summation. Spatial summation of currents from many synapses is required to produce a synaptic potential that exceeds the threshold of a motor neuron. If too few excitatory inputs are active simultaneously, V_m at the spike-initiating zone fails to reach threshold, and no APs are produced.

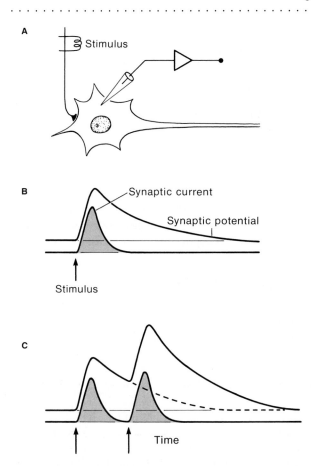

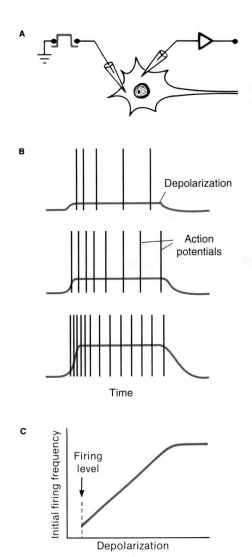

but randomly, changing V_m. Every now and then, excitatory inputs will sum to trigger an AP in the neuron, which in turn leads to an AP and twitch in each of the muscle fibers innervated by the neuron. The result of this activity is a constant background of low-level tension in skeletal muscles as first one and then another motor neuron fires and causes the contraction of the muscle fibers that it innervates. (See Chapter 10 for further discussion of muscle fibers and their control.)

The membrane at the spike-initiating zone in motor neurons typically never accommodates completely to maintained depolarization. Therefore, if synaptic input is both strong and maintained, it can cause the motor neuron to fire a sustained train of APs. The frequency of impulses in a train depends on how depolarized the spike-initiating zone becomes (Figure 6-47), which in turn depends on the

Figure 6-46 In temporal summation, presynaptic signals arrive at the synapse in rapid succession. **(A)** Setup for recording postsynaptic events. **(B)** A single stimulus evokes a synaptic current (shaded signal) and a more slowly decaying synaptic potential. **(C)** Summation of synaptic currents is not required for summation of synaptic potentials, because the time constant of the synaptic *potential* is longer than the time course of the synaptic *current*. Arrows indicate the time points at which presynaptic impulses arrived at the synapse.

channels—and capacitance of the membrane), and V_m gradually returns to rest after the synaptic current has ceased. Thus synaptic *potentials* outlast synaptic *currents* by milliseconds and, if a second synaptic *current* flows before the first synaptic *potential* has subsided, it will cause a second depolarization that adds to the falling phase of the first, even though the two synaptic currents do not overlap in time. Thus, the membrane's charge-storing capacity allows the voltage effect of synaptic currents to sum over time. The longer the time constant of the membrane is, the slower the decay of postsynaptic potentials will be, and the more effective the temporal summation of asynchronous synaptic inputs can be. The membrane time constant, (τ), of vertebrate motor neurons is about 10 ms, and it can range from 1 ms to 100 ms in other neurons.

Microelectrode recordings reveal that, under normal conditions, motor neurons are almost never electrically silent, but instead always exhibit some **synaptic noise** (irregular fluctuations in membrane potential) caused by ongoing activity in presynaptic neurons. The result is a constantly,

Figure 6-47 The initial frequency of impulses generated in a motor neuron is approximately proportional to the amplitude of the membrane depolarization. **(A)** Two electrodes, one for passing depolarizing current and one for recording membrane potential, are inserted into a spinal α-motor neuron. **(B)** Three idealized traces show that increased depolarization *(top to bottom)* causes an increased rate of firing. **(C)** Initial firing frequency plotted against amount of depolarization. As the depolarization is increased, the frequency of APs increases, up to some maximum value.

amplitude of summed synaptic inputs. Thus the number and frequency of APs produced in the motor neuron carry information about the input to the neuron. In fact, most information transfer in the nervous system depends on this frequency code.

To summarize, APs are generated in a neuron when the low-threshold initial segment or axon hillock is depolarized to threshold or beyond. The frequency of APs in the neuron rises as depolarization increases to some maximum firing frequency. The amount of depolarization at the spike-initiating zone depends on the relative timing of excitatory and inhibitory synaptic currents and on where those currents originate.

SYNAPTIC PLASTICITY

The nervous system would be much less useful to an animal if it could not be changed by experience. **Neuronal plasticity,** the modification of neuronal function as a result of experience, is of premier importance for the survival of any organism. Common examples of neuronal plasticity in our lives are learning and the development of motor skills and habits. This plasticity lies behind human intelligence, as well as the ability of all higher animals to respond adaptively to stimuli in ways that allow them to go beyond fixed reflexes programmed into their developing nervous systems by genetic mechanisms. Virtually all animals demonstrate a degree of behavioral plasticity, and the mechanisms that underlie synaptic plasticity are currently the subject of many experiments. Synaptic plasticity also takes place as the result of developmental events over the course of a lifetime. Synaptic connections that are established in embryos are later refined into adult patterns, and even later changes in synaptic strengths are thought to be important mechanisms for learning and memory at mature synapses. Interestingly, shaping the development of mature synapses and modifying them in learning and memory both appear to depend on a retrograde signal that is sent from the postsynaptic neuron to the presynaptic neuron.

In fully developed adult organisms, neuronal plasticity requires changes in *synaptic efficacy.* A change in synaptic efficacy is not the only way in which neuronal function might be modified, but, at present, it is the one for which there is the most experimental support. D. O. Hebb suggested in 1949 that the effectiveness of an excitatory synapse will increase if activity at this synapse is consistently and positively correlated with activity in the postsynaptic neuron. Since then, one challenge has been to identify the mechanisms that could underlie this kind of change.

Two large categories of mechanisms that could fill this role are (1) changes in the presynaptic terminals and (2) changes in the postsynaptic neuron. One example of a presynaptic mechanism would be a change in the amount of transmitter released from presynaptic terminals in response to a presynaptic AP. An example of a postsynaptic mechanism would be a change in the postsynaptic apparatus that altered the amplitude of depolarization produced

when a given amount of transmitter was released from the presynaptic endings. Relatively little is known about the mechanisms of postsynaptic plasticity, although it has been demonstrated in several tissues. We will consider presynaptic mechanisms of neuronal plasticity.

There are two major classes of presynaptic mechanisms that change synaptic efficacy. In one class, activity in the terminal itself causes a use–dependent change in the release of transmitter, so these mechanisms are called **homosynaptic modulation.** In the other class, changes in presynaptic function are induced by the action of a modulator substance released from another, closely apposed nerve terminal, so these mechanisms are called **heterosynaptic modulation.** Typically, heterosynaptic modulation lasts longer than homosynaptic modulation.

Homosynaptic Modulation: Facilitation

A use-dependent change in synaptic efficacy can be seen in a partly curarized endplate region of a frog skeletal muscle fiber if two stimuli are applied to the motor axon in rapid succession. If the second synaptic potential begins before the first has subsided, they will sum, but the amplitude of the second response will be greater than can be accounted for by summation alone. If the second synaptic potential begins soon after the first has completely subsided, precluding temporal summation, the second postsynaptic potential may still reach a higher amplitude than the first one. This effect, termed **synaptic facilitation,** lasts from 100 to 200 ms at the frog neuromuscular junction (Figure 6-48).

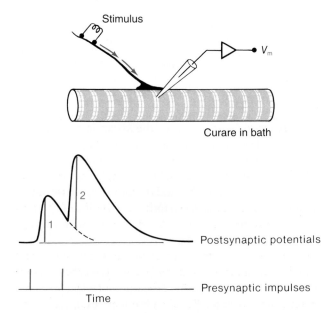

Figure 6-48 Synaptic facilitation occurs at the frog neuromuscular junction. In this experiment, curare in the bathing saline blocked some ACh receptors, reducing the amplitude of excitatory postsynaptic potentials to below the firing threshold. Two stimuli were delivered to the nerve in rapid succession. The second synaptic potential summed with the falling phase of the first, producing a larger postsynaptic potential; but, in addition, the amplitude of the second response (indicated by the line labeled 2) was greater than could be accounted for by summation alone.

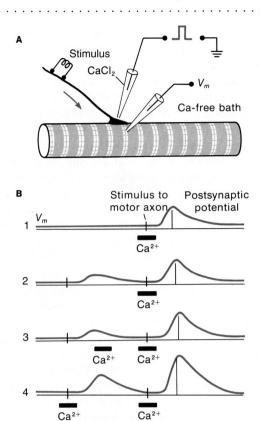

Figure 6-49 Synaptic facilitation depends on the presence of calcium ions in the extracellular fluid. **(A)** The motor neuron innervating the muscle fiber was stimulated, and the resulting postsynaptic potential was recorded. The bathing solution was calcium free, but small pulses of $CaCl_2$ were provided by a $CaCl_2$-containing pipette positioned just at the endplate region. In this experiment, the relative timing between stimuli to the motor neuron and the delivery of $CaCl_2$ was varied. **(B)** Recordings of postsynaptic potentials in the muscle fiber. Horizontal black bars show the timing of Ca^{2+} pulses. Thin vertical lines indicate stimuli to the presynaptic neuron. Trace 1 shows the amplitude of a postsynaptic potential in response to a single AP in the motor neuron. In the three other traces, the temporal relation between the first AP and the pulse of $CaCl_2$ was varied. In all cases, Ca^{2+} ions were available at the time of the second AP. Facilitation occurred only when Ca^{2+} ions were present at the endplate when both of the APs reached the endplate. [Adapted from Katz and Miledi, 1968.]

The best evidence indicates that synaptic facilitation depends on the amount of free Ca^{2+} within the presynaptic terminal. The concentration of intracellular free Ca^{2+} ions rises in the terminal when the first AP opens voltage-dependent Ca^{2+} channels, and this increase in Ca^{2+} ion concentration persists for a short time. When the second impulse arrives at the terminal, the Ca^{2+} concentration is still somewhat elevated, and the Ca^{2+} ions that enter as a result of the second AP add to the remaining Ca^{2+} ions, generating an even higher Ca^{2+} concentration in the terminal. Because the release of transmitter is a power function of the intracellular Ca^{2+} concentration near the presynaptic release sites, this small increase in Ca^{2+} concentration inside the terminal produces a large increase in the amount of transmitter released subsequent to the second impulse. Experimental evidence for this hypothesis was ob-

tained by Katz and Miledi (1968). They used a carefully positioned micropipette to supply pulses of Ca^{2+} ions to the external solution near a motor endplate of a frog muscle that was immersed in Ca^{2+}-free Ringer solution (Figure 6-49A). They found that facilitation of the postsynaptic potential evoked by the second stimulus was greatest when a pulse of extracellular Ca^{2+} ions was supplied to coincide with the arrival of the first AP (Figure 6-49B). The first pulse of Ca^{2+} did not significantly enhance facilitation if it was given after the first AP arrived at the terminals (see Figure 6-49B). Thus, if synaptic facilitation is to occur, Ca^{2+} must be available to enter the presynaptic terminal when an AP invades the terminal. If Ca^{2+} ions can enter the terminal from the external fluid, Ca^{2+} ions from the second AP add to any Ca^{2+} ions remaining from the first, leading to the release of more transmitter.

Homosynaptic Modulation: Posttetanic Potentiation

When a frog motor axon is stimulated *tetanically* (i.e., at a high frequency for a relatively long time), synaptic transmission at the neuromuscular junction is initially depressed after the stimulation. However, responses to test pulses applied at later times after the stimulation are found to be larger than normal. This increase in the amplitude of the response lasts for as long as several minutes and, during this time, the responses are said to be *potentiated*. This **posttetanic potentiation** is another example of a use-dependent change in presynaptic efficacy, and it is found in one form or another at many types of synapses. Figure 6-50 illustrates the results of one such experiment. Initially,

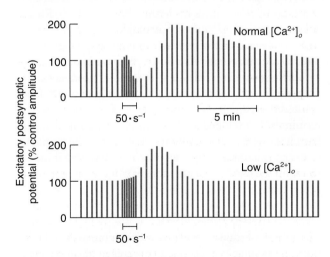

Figure 6-50 Tetanic stimulation of a frog motor nerve elicits depression and potentiation of excitatory postsynaptic potentials in muscle fibers. Curare was used to reduce the amplitude of synaptic potentials, blocking the production of APs and revealing the amplitude of the synaptic potentials. When the nerve and muscle were bathed in normal frog Ringer's solution, which has a Ca^{2+} concentration of about 2 mM *(top)*, stimulating the motor nerve at 50 stimuli per second for about a minute produced first a depression in subsequent excitatory postsynaptic potentials and then potentiation. When the concentration of extracellular Ca^{2+} was reduced to 0.225 mM, only potentiation was seen after high-frequency stimulation. [Adapted from Rosenthal, 1969.]

excitatory postsynaptic potentials (epsps) were evoked at a frog neuromuscular junction by stimulating the motor nerve at a low control rate (one stimulus every 30 seconds). The rate of stimulation was then increased to 50 per second for a period of 20 seconds, after which a series of test stimuli were administered at the initial rate of one every 30 seconds. In Ringer solution that contained a normal concentration of Ca^{2+} (see Figure 6-50, top), posttetanic depression of the evoked epsps occurred immediately after the tetanic stimulation. However, within 1 minute, the amplitude of the epsps increased; in other words, posttetanic potentiation had occurred. The amplitude of the epsps returned to the control level after about 10 minutes. In Ringer solution that contained a lower-than-normal concentration of Ca^{2+} (see Figure 6-50, bottom), there was no depression, and the posttetanic potentiation subsided more rapidly.

These results are thought to depend on events within the terminals. During high-frequency stimulation in a normal concentration of extracellular Ca^{2+} (1.8 mM), the available synaptic vesicles are released faster than they can be replaced, so the amount of transmitter available for release is depleted and remains low for a time immediately after the high-frequency stimulation. Later in the posttetanic period, quanta of transmitter available for release are restored, and the depression subsides. During tetanic stimulation, Ca^{2+} ions that have entered the terminals accumulate, load up the available Ca^{2+}-binding sites that ordinarily buffer the intracellular concentration of Ca^{2+}, and linger within the terminals until they are gradually pumped out by active transport across the cell membrane. It is believed that posttetanic potentiation and its slow decay reflect this increase and subsequent decrease in the concentration of Ca^{2+} inside the terminals. In low-Ca^{2+} Ringer solution, fewer Ca^{2+} ions are available to enter the terminals, so fewer synaptic vesicles can bind to the membrane and release transmitter. As a result, there is less depletion of available synaptic vesicles, and there is no posttetanic depression. Posttetanic potentiation is just as pronounced, because repeated stimulation does bring Ca^{2+} ions into the terminals, but the potentiation decays more rapidly, perhaps because the concentration of Ca^{2+} inside the terminals is less elevated or because the presynaptic terminal is able to pump the extra Ca^{2+} out more rapidly because less has accumulated.

Heterosynaptic Modulation

The release of transmitter from nerve terminals can be influenced at some synapses by the presence of certain neuromodulators. These modulatory agents include *serotonin* in mollusks and in vertebrates, *octopamine* in insects, and *norepinephrine* and *GABA* in vertebrates. All of these agents are also neurotransmitters (see Table 6-2). In addition, endogenous opioids have been shown to act as modulatory agents in vertebrate neurons. Such agents, released into the circulation or liberated by nerve endings near a synapse, are believed to modify the release of transmitter from presynaptic terminals. When they are liberated near,

but not at, a presynaptic ending, they are said to act heterosynaptically, because transmission through the synapse is altered by an additional, third neuron, which released the modulator. One class of heterosynaptic action that has already been discussed in this chapter is presynaptic inhibition; another, in which the amount of transmitter released is *increased* by the presence of the modulator, is called **heterosynaptic facilitation.**

In heterosynaptic modulation, the modulator is thought to alter the number of Ca^{2+} ions that enter the terminals subsequent to a presynaptic AP. Synaptic modulators usually do not directly open (or close) ion channels. Instead, they change how ion channels respond to another stimulus; by doing so, they increase or decrease ionic currents carried through channels that are activated by a presynaptic AP. This action by modulators is typically mediated by one or more intracellular messengers that act on the ion channels. In contrast, fast neurotransmitters bind to membrane receptors and open (or close) channels.

The most extensively studied example of heterosynaptic modulation at a synapse is found in the sea hare *Aplysia californica,* a sluglike gastropod mollusk that has been widely used in studies of neuronal plasticity. Eric Kandel and his associates have found that excitatory transmission between specific identified neurons in the central nervous system of *Aplysia* is enhanced during behavioral sensitization. They have found that this enhancement occurs through heterosynaptic facilitation of transmitter release triggered by the release of serotonin near the synapse (Figure 6-51). In this case, serotonin is thought to elevate the concentration of the intracellular messenger 3′,5′-cyclic adenosine monophosphate (cAMP), which has been found to influence the opening of a specific type of K^+ channel, known as the S channel. Specifically, when cAMP is elevated in the presynaptic neuron, S channels are more likely to be shut at any given V_m. The efflux of K^+ through S channels contributes to repolarization after an AP, so the closing of S channels will prolong the presynaptic AP and allow more Ca^{2+} ions to enter the terminal through voltage-gated Ca^{2+} channels. An increase in the influx of Ca^{2+} ions allows more transmitter to be released and increases the amplitude and duration of the postsynaptic potential.

Long Term Potentiation

In the past few years, intense interest has been focused on long-term changes in synaptic efficacy that have been identified in the mammalian hippocampus, the site of certain memories. High-frequency stimulation of inputs to the hippocampus produces an increase in the amplitude of excitatory postsynaptic potentials recorded in the postsynaptic hippocampal neurons. In an intact animal, the increased amplitude can last for hours—even days or weeks—after the potentiating stimulation. This prolonged facilitation of synaptic transmission, called **long-term potentiation,** has been shown to occur in many synaptic pathways. At different sites, long-term potentiation may require different

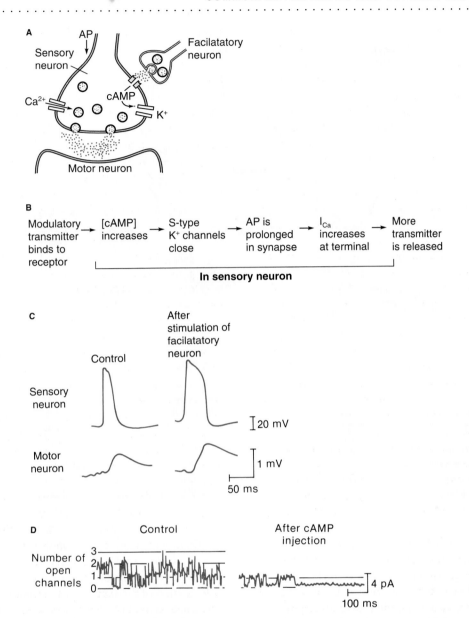

Figure 6-51 Heterosynaptic facilitation at a synapse in *Aplysia* occurs when potassium ion channels are closed in the synaptic terminals, allowing more calcium ions to enter and causing more transmitter to be released. **(A)** If the facilitatory neuron is active at the same time that APs arrive at the terminals of the sensory neuron, the net result is an increase in the amount of transmitter released by the sensory neuron. The transmitter of the facilitatory neuron binds to a receptor that causes an increase in the level of cAMP in the terminal. Elevated cAMP causes S-type K⁺ channels in the terminal to close, prolonging the depolarization of the AP and holding voltage-gated Ca²⁺ channels open. **(B)** Summary of events at the sensory neuron terminals. **(C)** An AP produced in the sensory neuron *(upper records)* produces an excitatory postsynaptic potential in the motor neuron *(lower records)*. Stimulating the facilitatory neuron prolongs the AP in the sensory neuron and produces a concomitant facilitation of the synaptic response in the motor neuron. **(D)** Currents through single channels in the presynaptic membrane of sensory neurons, recorded by the patch-clamp technique. Records were made before and after cAMP was injected into the neurons. The activity of S-type K⁺ channels was reduced after cAMP was injected. [Part C adapted from Kandel et al., 1983; part D adapted from Siegelbaum et al., 1982.]

patterns of stimulation, may decay at different rates, and may depend on different underlying mechanisms. In the cases studied, glutamate or a similar substance, has been found to be the principal excitatory transmitter. Of the three pharmacologically distinct glutamate receptors (see *Other ligand-gated channels* earlier in this chapter), only the type that responds to *N*-methyl-D-aspartate (NMDA) is thought to take part in long-term potentiation. At many synapses in the hippocampus, activation of NMDA recep-

tors is required for the induction of long-term potentiation, although it is not required for normal neurotransmission. To date, only a handful of experiments have been performed in intact animals, but the results have been consistent with the hypothesis that changes in the properties of NMDA receptors must occur if the strength of the synaptic connection is to be modified. Furthermore, the results of several experiments suggest that, in long-term potentiation, a retrograde signal from the postsynaptic cell travels back

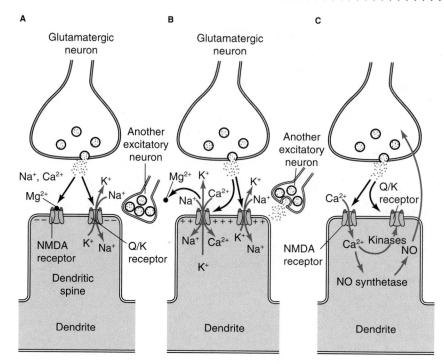

Figure 6-52 Long-term potentiation in the hippocampus depends on NMDA receptors, a class of glutamate receptors, in the postsynaptic membrane. The proposed role of NMDA receptors in long-term potentiation at a hippocampal dendritic spine. Glutamate is released from the presynaptic terminal and binds to both NMDA and quisqualate/kainate (Q/K) type glutamate receptors in the postsynaptic membrane. **(A)** If the postsynaptic dendrite is not depolarized, Na^+ and K^+ flow through the Q/K receptor channel, but not through the NMDA receptor channel, because Mg^{2+} ions block NMDA channels when V_m is near V_{rest}. **(B)** When the dendrite is depolarized, as it is when the postsynaptic neuron is stim-

ulated tetanically by another source, the Mg^{2+} block of the NMDA channels is removed, allowing Na^+, K^+, and Ca^{2+} ions to flow through the NMDA channel. **(C)** The resulting increase in cytoplasmic Ca^{2+} concentration within the dendritic spine activates intracellular second messengers, including kinases, which can produce long-term modification of the quisqualate/kainate channels, and nitric oxide synthetase, which catalyzes the increased production of nitric oxide (NO). Nitric oxide is thought to diffuse back to the presynaptic terminal and potentiate the release of transmitter in response to subsequent APs.

to the presynaptic terminal, producing a change in the behavior of the presynaptic cell. Recent experiments suggest that one such retrograde messenger could be the gas *nitric oxide* (NO, Figure 6-52). Nitric oxide may act in the presynaptic terminal by modifying the activation of enzymes, such as guanylyl cyclase or ADP ribosyltransferase. Whether this mechanism can account for the modifications in neuronal function that underlie learning remains to be seen, but recent progress in identifying the molecular substrates of memory has been encouraging, and the topic continues to be the target of an enormous research effort.

SUMMARY

Within a single neuron, information can be carried in two ways: (1) by graded, passively conducted potential changes and (2) by all-or-none, regenerative action potentials (APs). Between neurons, information is typically carried by chemical messenger molecules, although there are some electrical synapses as well. Within a neuron, graded potentials occur at specialized regions of the cell—for example, sensory receptor and postsynaptic membranes. Action potentials are confined largely to axons and axon terminals. Information about signal intensity is coded by the amplitude of graded potentials and by the frequency of APs.

Propagation along an axon depends on two phenomena: (1) the longitudinal spread of current, which depends on the cable properties of the axon; and (2) the continual regeneration of the signal by excitation of new Na^+ channels when resting membrane is depolarized by local-circuit current as it flows along the axon ahead of the impulse. Because APs are propagated without decrement along axons, they can carry information between even widely separated parts of the nervous system. The velocity at which APs are propagated depends on axon diameter and on the presence (in some vertebrate axons) of insulating segments of myelin sheath that are separated by nodes of Ranvier— short, bare sections of axonal membrane. In myelinated axons, saltatory conduction passes from node to node, skipping the intimately sheathed parts of the axon that lie between the nodes and thereby increasing the velocity of conduction.

There are two major types of synapses: electrical and chemical. The principle of electrical synaptic transmission is essentially identical with that of impulse propagation; the current flows from one cell into another through low-resistance channels in specialized regions called gap junctions, depolarizing the second cell. There are two kinds of chemical transmission: fast and slow. In fast synaptic transmission, presynaptic endings liberate a neurotransmitter that interacts with ligand-gated ion channels on the postsynaptic membrane, causing the channels to open and allowing ionic current to flow, which produces a synaptic potential across the postsynaptic membrane. In slow synaptic transmission and in a variant called neuromodulation, neurotransmitters bind to receptors, changing the state of G proteins associated with the membrane and indirectly modulating the function of ion channels. Ion channels can be directly regulated by G-protein subunits. Alternatively, neurotransmitters can modify the function of ion channels indirectly through cyclic nucleotides or through cyclic nucleotide–activated protein kinases that phosphorylate intracellular domains of the channels.

Chemical synapses have three advantages over electrical transmission: (1) the postsynaptic current can produce either an excitatory or an inhibitory action; (2) the postsynaptic membrane is the source of the synaptic current, and hence a tiny presynaptic axon can produce large postsynaptic currents through the influence of its transmitter on postsynaptic channels; and (3) there is a greater possibility of synaptic integration.

In excitatory synapses, the transmitter changes the membrane's ion permeability, tending to make V_m more positive than the threshold for the initiation of an AP. Inhibitory neurotransmitters change the conductance so as to prevent V_m from reaching threshold. The properties of excitation and inhibition are not inherent in the transmitter substances; rather, excitation and inhibition depend on the ion selectivity of postsynaptic channels activated by the transmitters and on the reversal potentials of the currents carried through those channels. Fast neurotransmitters can act within milliseconds, and their effect is transient. Slow neurotransmission or neuromodulation changes V_m of the postsynaptic cell for seconds or minutes, rather than for milliseconds.

Both excitatory and inhibitory transmitters are stored in, and released from, vesicles in the nerve terminal. The arrival of an AP depolarizes the presynaptic membrane, allowing Ca^{2+} to enter the terminal. The Ca^{2+} ions, in a manner that is not yet completely understood, increase the probability that synaptic vesicles will fuse with the terminal membrane and release their contents into the synaptic cleft. Vesicular membranes subsequently undergo endocytosis and are recycled as new vesicles. Vesicles containing neurotransmitters that are responsible for fast neurotransmission are released into the narrow synaptic cleft between pre- and postsynaptic neurons; neuromodulatory transmitters are typically released from the side of the synaptic terminals.

Temporal and spatial summation of synaptic potentials depends on the passive electrical properties of the postsynaptic cell, and the net effect of all synaptic currents is to determine whether the membrane at the spike-initiating zone will be sufficiently depolarized to reach threshold. The time constant of the postsynaptic cell allows temporal summation to occur, even when the synaptic currents do not overlap in time.

Some changes in synaptic efficacy are due to prior activity in the same synapse, which is called homosynaptic modulation. In some cases, it has been shown that a change in synaptic efficacy occurs when the amount of transmitter that is released in response to a presynaptic AP changes. Modulatory agents released from a third neuron or from endocrine glands also can alter the effectiveness of synaptic transmission by influencing the amount of Ca^{2+} entering the terminal during the presynaptic impulse and thereby increasing the amount of transmitter that is released from the terminal.

REVIEW QUESTIONS

1. Compare and contrast the two basic kinds of signal transmission found in the nervous system.
2. Action potentials are carried along neurons by electrical currents. Why are they so much slower than electricity traveling along a wire?
3. How can an AP travel over long distances without decrement, whereas synaptic potentials cannot?
4. Explain why, if all else is equal, an axon of large diameter will conduct impulses at higher velocity than will an axon of small diameter.
5. Calculate the relative conduction velocities for unmyelinated axons that are 10 μm and 25 μm in diameter, with all other parameters being equal in the two kinds of axons.
6. Explain why, if all else is equal, a myelinated axon will conduct impulses at a higher velocity than will an unmyelinated axon.
7. Explain how loss of myelination, which happens in the demyelinating disease multiple sclerosis, disrupts signal transmission in the nervous system.
8. Design an experiment to test whether a synapse between two neurons is electrical or chemical.
9. What determines if a neurotransmitter is excitatory or inhibitory?
10. What factors determine whether a transmitter depolarizes or hyperpolarizes the postsynaptic membrane?
11. Marine invertebrates typically have much higher concentrations of inorganic ions in their body fluids than do freshwater invertebrates. For example, the table below gives the intracellular and extracellular K^+ concentrations for two molluscs: *Limnaea*, a freshwater snail, and *Sepia*, the marine cuttlefish.

	Limnaea	Sepia
Intracellular	14.8 mM	188.3 mM
Extracellular	1.8 mM	21.9 mM

If these two species had in common a neurotransmitter that opened K^+ channels in the postsynaptic membrane, how would the values for E_{rev} at those synapses compare in the two species? If the postsynaptic neurons rested at −70 mV and had a threshold voltage of −55 mV, would the transmitter be excitatory or inhibitory in each of the two species?

12. How can a synapse produce a depolarizing post-synaptic potential and still be inhibitory?

13. What is the evidence that an endplate potential is composed of smaller units called miniature endplate potentials?

14. What places an absolute limit on the amplitude of a postsynaptic potential?

15. What prevents ACh released from the presynaptic terminal from persisting and interfering with subsequent synaptic transmission? What happens if ACh remains in the synaptic cleft?

16. The amplitude of postsynaptic potentials decays with distance, so where on a neuron is the most effective site for a synapse to be located?

17. Compare and contrast fast and slow chemical neurotransmission.

18. What is meant by neuromodulation?

19. Discuss the role of Ca^{2+} in each of the following events: depolarization-release coupling, facilitation, post-tetanic potentiation, heterosynaptic modulation of transmitter release, long-term potentiation.

20. How is signal intensity encoded in graded signals, such as receptor potentials and synaptic potentials? How is intensity encoded in APs?

21. Compare the synaptic effects of ACh at the neuromuscular junction and on heart atrial cells.

SUGGESTED READINGS

Cooper J. R., F. E. Bloom, and R. H. Roth. 1996. *The Biochemical Basis of Neuropharmacology*. 7th ed. New York: Oxford University Press. (A classic book describing the chemistry of neurotransmission and neuromodulation.)

Hall, Z. H. 1992. *Molecular Neurobiology*. Sunderland, Mass.: Sinauer. (A thorough description of the molecular basis of signal transmission along axons and across synapses.)

Hille, B. 1992. *Ionic Channels in Excitable Membranes*. 2d ed. Sunderland, Mass.: Sinauer. (An authoritative survey of the original work that revealed the mechanisms of axonal conduction.)

Hille, B. 1994. Modulation of ion-channel function by G-protein–coupled receptors. *Trends Neurosci.* 17:531–536.

Hodgkin, A. L. 1964. *The Conduction of the Nervous Impulse*. Springfield, Ill.: Thomas. (A summary of the original work on how APs are initiated and conducted, written by one of the primary contributors to the field.)

Kaczmarek, L. D., and I. B. Levitan. 1987. *Neuromodulation: The Biochemical Control of Neuronal Excitability*. New York: Oxford University Press. (A summary of this field written by two experts.)

Levitan, I. B., and L. K. Kaczmarek. 1997. *The Neuron: Cell and Molecular Biology*. 2d ed. New York: Oxford University Press. (This has particularly good chapters on synaptic transmission and neuromodulation, written by researchers who have contributed greatly to this field.)

Nicholls, J. A., R. Martin, and B. G. Wallace. 1992. *From Neuron to Brain*. 3d ed. Sunderland, Mass.: Sinauer. (A comprehensive treatment of function in the nervous system.)

Snyder, S. H. 1985. The molecular basis of communication between cells. *Scientific American* 253:114–123. (An introduction to the process of chemical synaptic transmission.)

Sudhof, T. C. 1995. The synaptic vesicle cycle: A cascade of protein–protein interactions. *Nature* 375:645–653. (A recent review of vesicular neurotransmitter release.)

Unwin, N. 1989. The structure of ion channels in membranes of excitable cells. *Neuron* 3:665–676. (An overview of the structure of ion channels in neurons and muscles.)

Unwin, N. 1995. Acetylcholine receptor channel imaged in the open state. *Nature* 373:37–43. (High resolution images of the nicotinic acetylcholine receptor channel as it opens.)

SENSING THE ENVIRONMENT

Everything an animal does depends on receiving and correctly interpreting information from its external and internal environments. A bird listening for calls from its competitors, a gazelle sniffing the air as a lion passes upwind, and a hawk hovering over a meadow and peering with one eye and then the other at the brush below—all need accurate information about their surroundings in order to decide what to do next. Their decisions can be appropriate only if data gathered from the environment are faithfully coded into signals that can be received and processed by neurons in the brain.

In fact, sensory organs provide the *only* channels of communication from the external world into the nervous system. Sensory input is gathered constantly from the environment, and it interacts with the organization and properties of the nervous system, which are inherited genetically and organized during embryogenesis, to provide each animal with its entire store of "knowledge." This concept was recognized two millennia ago by Aristotle when he said, "Nothing is in the mind that does not pass through the senses." An understanding of how environmental information is converted into neuronal signals and of how these signals are then processed is therefore of deep philosophical, as well as scientific, interest.

Sensory reception begins in organs containing cells, called **receptor cells**, that are specialized to respond to particular kinds of stimuli. Sensory organs are positioned at many locations both on the surface and inside of the body, and they constitute the first step in gathering sensory information. (Neurons that carry information from the periphery toward and into the central nervous system are called **afferent neurons.** Neurons that carry information away from the central nervous system are called **efferent neurons.**) In contrast with this initial coding step, **sensations** are part of our subjective experience, and they arise when signals that are initiated in sensory receptor cells are transmitted through the nervous system to particular parts of the brain, producing signals in the brain that we experience as subjective phenomena closely associated with the stimulus.

Stimulus types possess features that distinguish them from one another. For example, the mechanical stimulation that produces the sensation of touch is different from the light that produces a visual response. In addition, stimuli of a particular type may differ in some features. Light can be red or blue; sounds can be high or low. The features that characterize stimuli are called **qualities.**

Human subjects are able to describe the perceived sensation resulting from a particular kind of stimulus, and different subjects generally agree on the kind of sensation that is produced by that stimulus, even though such subjective sensations are not really inherent in the stimuli themselves. For example, when sugar is placed on the tongues of many subjects, all are likely to report that it is "sweet." Similarly, light with a wavelength of 650–700 nm is described by most subjects as being "red." In both cases, these perceptions are not inherent in the stimuli themselves. Instead, the perception depends entirely on the subjects' neuronal processing of the stimulus. Thus, a description of sensory physiology must include the properties of receptor cells that allow them to receive information from the environment and a consideration of how the nervous system processes information from the sensory cells to produce recognizable sensations. Notice that any distortions that are produced by the sensory cells or the subsequent processing will shape our perceptions of the stimuli and will seem to be intrinsic to the stimuli themselves.

A listing of **sensory modalities** (i.e., types of sensory information that we can distinguish) typically includes vision, hearing, touch, taste, and smell, but this list leaves out important internal sensory systems, as well as sensory modalities that nonhuman animals possess. For example, many **interoceptive** (internal) **receptors** respond to signals from within the body and communicate the information to the brain by pathways that often are not brought into consciousness. For example, **proprioceptors** monitor the position of muscles and joints, and other receptors monitor the orientation and chemical and thermal state of the body. These internal receptor systems play crucial roles in providing information to the brain about the state of the body and its position in space, but we normally are not consciously aware of these signals. Imagine how complicated walking would be if we had to pay conscious attention to the position of every muscle and joint taking part in the process.

Many species of animals use sensory modalities that are unavailable to human beings. For example, some species of snakes, the pit vipers, can detect emitted heat energy (infrared radiation), which they can use to locate their mammalian prey because these warm-bodied animals stand out against a cold background. The fish species that are called "weakly electric fish" (to distinguish them from the electric fish that can stun or kill prey by using electric shocks) use very low frequency electric signals to communicate in murky water, allowing them to find one another and to negotiate with one another regarding reproduction and territory. Some animals appear to sense Earth's magnetic field and use it as a navigational guide. (These examples are considered later in this chapter.) Obviously, we can have very little idea about the *subjective* quality of such sensory information, because we do not possess these receptors, but important principles of organization that apply to these systems also apply to other systems discussed in this chapter. In this chapter, we consider **chemoreceptors, mechanoreceptors, electroreceptors, thermoreceptors,** and **photoreceptors.** The names of receptors are based on the forms of energy to which they are most sensitive: chemical, mechanical, electrical, thermal, and light.

In the course of evolution, sensory systems have developed from single, independent receptors into specialized sense organs in which the receptor cells are arranged in well-organized spatial arrays and are associated with accessory structures. The cellular organization of sensory organs allows stimuli to be sampled more accurately than can be accomplished by isolated receptor cells. The vertebrate eye includes several structural adaptations (considered later in this chapter) that improve both our visual sensitivity and our ability to perceive images. The vertebrate eye can be contrasted with the simpler eyes of many invertebrates—for example, barnacles. Without a lens to form an image, a barnacle eye can detect changes in light, but it cannot form images. Barnacle photoreceptors can sense information only about changes in light intensity, so a barnacle's response to visual input can be based only on this simple form of information. In contrast, the vertebrate eye provides a remarkably high-quality optical image to the receptor cells. These cells, in turn, encode features of the scene and pass it along to the brain to be interpreted, which results in our subjective experience of "vision." Seeing well seems to have made an important contribution to evolutionary success, because about 85% of all living animal species have image-forming eyes.

Until very recently, the extraordinary diversity of stimuli and the corresponding receptor types were considered a tribute to the wide variety of solutions that could be generated by natural selection, because no unifying principles were apparent among these receptors. However, recent evidence has revealed surprising similarities among cellular mechanisms in sensory receptors. This chapter presents the general principles of how sensory receptors encode and transmit information and compares the events in the receptors of several major sensory systems. The ways in which sensory information is used to generate and to shape behavior are considered in Chapter 11.

GENERAL PROPERTIES OF SENSORY RECEPTION

Until recently, physiologists were struck by the wide variety of sensory receptors and the large functional differences observed among the receptor cells in different sensory modalities. However, we now know that several features are shared by many—even most—sensory receptors, regardless of modality. We begin this chapter with a discussion of several of these common features to provide background for the consideration of some specific sensory modalities that follows.

Properties of Receptor Cells

Sensation begins in receptor cells, or more exactly at the specialized membranes of these cells (Figure 7-1). Two general features are common to sensory receptor cells. First, each kind of receptor cell is highly *selective* for a specific kind of energy. Second, the receptors are exquisitely *sensitive* to their selected stimuli because they can amplify the signal that is being received. The form of energy (e.g., light, sound, mechanical pressure, etc.) to which a sensory receptor is most sensitive is called its *sensory modality,* and receptor cells are said to *transduce* the sensory input, because they change the stimulus energy into the energy of a nerve impulse.

Sensory receptor cells are selective because their membranes—or structures that are associated with their membranes—respond differentially to different types of energy. External energy, such as light, may strike any part of the external anatomy; but, in the mammals, only the eyes and the pineal gland (a small gland located in the brain) contain sensory cells that can transduce photons into neuronal energy. Often, transduction depends on a conformational change in particular receptor molecules, which typically are proteins.

For example, the cell membrane of a photoreceptor cell contains a visual pigment based on protein molecules called **opsins.** A functional pigment molecule, called **rhodopsin,** consists of an opsin plus an organic, light-absorbing molecule. Molecules of rhodopsin absorb photons, capturing their energy. When a photon is captured, it produces a transient structural change that activates a cascade of associated molecules, ultimately changing the functional state of ion channels in the receptor cell membrane. (For details of this process, see *Vision* later in this chapter.) Similarly, the membranes of mechanoreceptors contain molecules that respond to slight distortions in the cell membrane. Evidence from the molecular biological study of these receptor molecules has indicated that many of them have related structures, and they may have evolved from common ancestors.

Stimuli

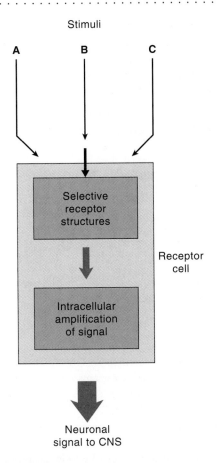

Figure 7-1 Sensory receptors are specialized to respond to only certain stimuli. Although many forms of energy may impinge on a receptor (represented by arrows A, B, and C), only one form—in this case, stimulus B—effectively activates the receptor at weak to moderate levels of stimulus energy. Other kinds of stimuli fail to activate the receptor at such low energy levels. Often, the signal is chemically amplified within the receptor cell and, in order for it to be effective, the intracellular chemical signal must cause membrane channels to open (or, in some cases, to close), producing a neuronal signal that can travel to the central nervous system (CNS).

Receptor cells can receive very weak signals of their selected form of energy and transduce these signals into nerve signals that contain much larger amounts of energy, because receptor cells contain intracellular machinery that amplifies weak stimuli. The initial activation of receptor molecules leads to different types of intracellular events, depending on the receptor type. In some receptors, activation of receptor molecules in the membrane initiates a cascade of chemical reactions in the cell that effectively amplifies the signal by many orders of magnitude. The final step in all receptor cells is the opening (or closing) of ion channels, which changes the amount of ionic current crossing the cell membrane and potentially modifies the number of APs produced in the receptor cells. (As we will soon see, some receptor cells produce only graded potential changes in response to sensory stimuli.) In summary, each receptor cell transduces a particular form of stimulus energy into a membrane current that produces a change in membrane potential, V_m, of the receptor cell. In this way, receptors are

analogous to common electrical devices—for example, a microphone or a photocell. A microphone transduces the mechanical energy of sound into modulated electrical signals, which can then be amplified. Similarly, a photocell converts light into an electrical signal.

Once again, vertebrate photoreceptor cells provide a convenient example. One photon of red light contains about 3×10^{-19} joules (J) of radiant energy, but capture of a single photon by a receptor cell has been found to produce a receptor current equivalent to about 5×10^{-14} J of electrical energy. The cell amplifies the signal by a factor of 1.7×10^5. The exquisite sensitivity of human photoreceptor cells allows a dark-adapted human subject to detect a flash containing as few as 10 photons delivered simultaneously over a small region of the retina, a feat that is equivalent to being able to see the light from a candle flame that is 19 miles away.

Common Mechanisms and Molecules of Sensory Transduction

All sensory transduction systems perform the same basic operations of detection, amplification, and transmission; it is now clear that many types of sensory receptors operate through similar cellular mechanisms and contain related molecules. Table 7-1 summarizes typical events in sensory transduction as it is carried out by many different kinds of receptors. Some of the processes occur within single receptor cells, whereas others depend on interactions among many cells. The basic events in a receptor cell are detection, amplification, and encoding of the sensory stimulus.

The initial event in all sensory transduction is *detection*, and the smallest amount of stimulus energy that will produce a response in a receptor 50% of the time is called the *threshold of detection*. Significant technical advances have enabled physiologists to measure transduction events at extremely low stimulus intensities, providing accurate estimates of the absolute threshold of detection and of the time constant for the response. Many sensory receptors are capable of detecting inputs that are very near the theoretical limits of the stimulus energy: photoreceptors can be activated by single photons, mechanoreceptor hair cells by displacements equal to the diameter of a hydrogen atom, and odor receptors by binding only a few molecules of the correct sort. The time constant of sensory reception is important because, in order for a sensory system to convey accurate information about rapidly changing stimuli, the receptors must be able to respond quickly and repeatedly. Alternatively, the receptors must be interconnected in a way that allows the population of receptors to extract information about very rapid events on the basis of their collective activity. Interestingly, the response latencies of the various known receptor cells vary over five log units. Hair cells in the auditory system respond within several microseconds; olfactory receptors respond only after several hundred milliseconds. It is intriguing to speculate about how such large differences in time constants might reflect

TABLE 7-1
General features and processes common to many types of sensory receptors

Transduction operations*	Found within single cells	Found in cell populations
Detection	Mechanisms that select stimulus modality: filters, carriers, tuning, inactivation	Mechanisms that select stimulus modality: filters, carriers, tuning, inactivation
Amplification	Positive feedback among chemical reactions or membrane channels	Positive feedback among cells
	Signal/noise enhancement	Signal/noise enhancement
	Active processes in membranes	
Encoding and discrimination	Intensity coding	Different dynamic ranges among cells
	Temporal differentiation	Independent coding of quality and intensity
	Quality coding	Center-surround antagonisms
		Opponent mechanisms
Adaptation and termination	Desensitization	Temporal discrimination
	Negative feedback	
	Temporal discrimination	
	Repetitive responses	
Gating of ion channels	Channels open or close	
Electrical response of membrane	Depolarization or hyperpolarization	
Transmission to brain	Electronic spread	Spatial patterns: maps and image formation
	Action potentials; number and frequency	Temporal patterns; directional selectivity, etc.
	Synaptic transmission	

*Arrows indicate that these operations are a series of steps.

fundamental differences in the roles played in the life of an animal by different sensory modalities.

Recent evidence indicates that in vertebrates the receptors for three of the senses—vision, olfaction, and, probably, sweet and bitter tastes—have in their cell membranes receptor proteins with a common structural motif. The secondary structure of these membrane proteins includes seven transmembrane α helix domains, and transduction in all three senses requires G proteins as intermediaries (Figure 7-2). This pattern is also found in several neurotransmitter receptor molecules, including the muscarinic acetylcholine receptor (see Chapter 6).

By far, the most detailed knowledge available is about the molecules responsible for detecting photons: the protein opsin and its associated molecules (Figure 7-3), which will be discussed in more detail later in this chapter. However, the close relation among sensory receptors was recently underscored when the DNA sequence that codes for opsin was used to identify putative olfactory receptor molecules (Chess et al., 1992). The sequences in this new family of olfactory receptor molecules are expressed only in the cells of the olfactory epithelium, and the family appears to be very large (containing several hundred different gene products—perhaps even as many as a thousand). The sequences

A   Muscarinic acetylcholine receptor

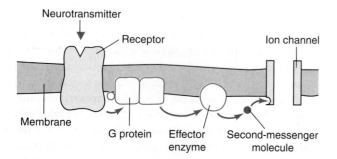

B   Photoreceptor

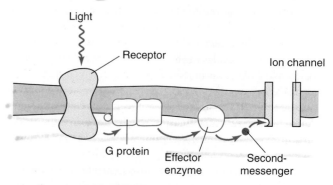

Figure 7-2 The molecular mechanism of sensory reception in visual receptors resembles the molecular mechanism of transmission at many synapses. For the examples shown, both processes begin with a structural change in a transmembrane protein (the receptor molecule), which interacts with a GTP-binding protein (G protein) to alter intracellular second-messenger pathways. The second-messengers modify conductance through ion channels, either directly or indirectly, and can thus modify the pattern of APs in afferent neurons. [Adapted from Bear et al., 1996.]

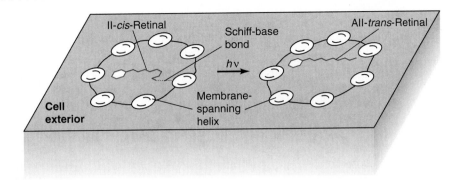

Figure 7-3 The visual pigment rhodopsin consists of a protein, opsin, and an associated light-capturing molecule, 11-*cis*-retinal. This diagram shows rhodopsin as it would appear from outside the cell. The seven-helix motif of this transmembrane receptor is found in other sensory receptor proteins and in receptor molecules that respond to hormones or to neurotransmitters (see Figure 7-2). The 11-*cis*-retinal molecule is lo-cated inside the transmembrane domain of the opsin protein, near the center of the bilayer. Light is captured by 11-*cis*-retinal, which causes the molecule to change into the all-*trans* configuration, initiating a cascade of intracellular events that finally produce a change in conductance through an ion channel (see Figure 7-53A).

of these molecules differ significantly from one another only in a particular region that is thought to be the site at which stimulus molecules bind.

Detection of salt and sour tastes occurs by way of much simpler mechanisms than those underlying vision and olfaction. Detection of these tastes depends on ion channels that are found on cells throughout the body. Sour sensation is mediated by a common pH-sensitive K^+ channel, and the salt response is caused by passive movement of Na^+ across the cell membrane, depolarizing the taste cell directly. In both of these cases, because the stimuli are themselves ions and are in great abundance, no intermediate amplification step is needed.

In some sensory systems, *amplification* of sensory signals occurs within the receptor cells, mediated by a number of different intracellular mechanisms that will be discussed in more detail later in this chapter. Interestingly, this amplification often occurs at the same time that noise is suppressed, so the signal-to-noise ratio improves in the process. Amplification is best understood for vertebrate photoreceptors (largely from the eyes of cattle). When a photon is captured by a visual pigment molecule (discussed later in this chapter), the net effect is to activate *transducin,* which is a GTP-binding protein, or G protein (see Figure 7-2 and Chapter 9). Transducin, in turn, activates a phosphodiesterase to hydrolyze cyclic guanosine monophosphate (cGMP), which in turn leads to modification of conductance through ion channels. Each photon captured leads to the hydrolysis of many cGMP molecules, producing a huge amplification in the signal. Although these steps have not yet been positively identified in the detection of odor and taste, several facets of the transduction cascade are probably similar. In each case, the energy available from a unitary stimulus at the receptor site is so low that amplification within the receptor cell is required to generate neuronal impulses that can carry the signal into the central nervous system.

Encoding sensory information into a neuronal signal to be transmitted to the brain depends on changes in the con-ductance through membrane ion channels. When channel conductance changes, it can shift the probability that the neuron will produce an AP, although it must be remembered that not all receptors transmit information through the use of APs. In photoreceptors, cGMP may act directly on a class of membrane channels to increase their conductance. The corresponding mechanisms for olfaction and taste (also called **gustation**) are still unknown, although channels that respond to cyclic nucleotides have been recently found in the olfactory system.

Responses within a single receptor neuron encode information about the intensity of a stimulus, but they cannot directly report the quality of the stimulus. For example, a single photoreceptor cannot report whether a stimulating light is red or blue. Additional information, such as the wavelength of light or the frequency of a sound, is conveyed by activity patterns within *combinations* of receptor cells that are activated by the stimulus. Typically, sensory organs contain a variety of receptor cells that respond differentially to stimuli with different qualities. For example, certain photoreceptors respond maximally to red light, whereas others respond maximally to blue light. Thus, when receptor cells are grouped into organs, significantly more information about the stimulus can be conveyed, including its absolute intensity, its spatial distribution, and other characteristics such as quality.

Each sensory system must be able to detect stimuli that persist in time, while at the same time retaining the ability to respond to further changes. The process of adaptation, described in Chapter 5 for single neurons, also occurs in the response of many receptor cells. From the perspective of an organism, adaptation allows detection of new sensory stimuli in the presence of ongoing stimulation, and it thus makes the sensory system much more useful. For example, wearing clothing stimulates touch receptors at all points where our garments touch the skin, and we typically adapt to the touch input from our garments. Yet, we can easily detect any *new* touch stimuli that impinge on our skin, even at locations covered by our clothing. Many mechanisms

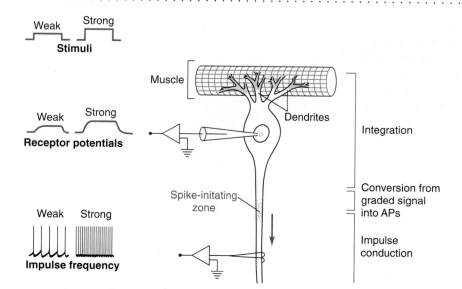

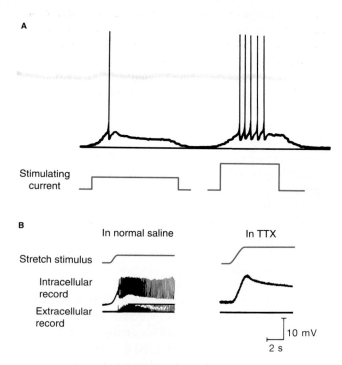

Figure 7-4 Stretch receptors in the tail of a crayfish transmit information about the amount that the tail muscles are stretched. The receptor consists of a sensory neuron that has its stretch-sensitive dendrites embedded in a special muscle bundle located on the dorsal surface of the tail muscles. When the tail of the crayfish bends, the muscle is stretched, and the receptor is activated. On the left side of the diagram, intracellular records from the soma and extracellular records from the axon show how the receptor responds to small and large stretches of the muscle. The parts of the neuron are functionally differentiated, as labeled at the right. Graded receptor potentials from the stretch-sensitive membrane of the dendrites are converted into all-or-none APs at the spike-initiating zone. The arrow indicates the direction of AP propagation.

underlying adaptation take place within individual receptor cells, and several of them appear to depend on Ca^{2+} (e.g., in vision, olfaction, and mechanoreception). In addition, some adaptation depends on negative feedback from higher brain centers.

From Transduction to Neuronal Output

Electrical measurements are important sources of insight into the steps that lie between sensory transduction and the generation of neuronal responses. One of the first such experiments was done on receptor cells, called **stretch receptors,** that sense muscle length in the abdomen of crayfish and lobsters (Figure 7-4). Because each stretch receptor is a relatively large cell, its soma can be impaled with microelectrodes. It is also possible to record extracellularly from the axon of the cell. The dendrites of each stretch receptor are attached across the surface of muscle fibers, and if the muscle is stretched, a steady train of impulses can be recorded from the axon. The frequency of the APs varies directly with the amount of stretch applied. To understand the source of the APs, the intracellular potential can be recorded by inserting a microelectrode into the cell body. A small stretch applied to the relaxed muscle leads to a small depolarization called the **receptor potential** (see Figure 7-4). A stronger stretch produces a larger depolarizing receptor potential. This change in V_m indicates that an underlying **receptor current** must flow across the membrane and that this receptor current must carry positive charge into the cell to produce the depolarization. If receptor potentials are sufficiently large, they trigger one or more APs in the cell (Figure 7-5).

What is the relation between stimulus, receptor current, receptor potential, and APs? Action potentials can be eliminated (see Figure 7-5B) by blocking the electrically excited sodium channels with tetrodotoxin. When APs are blocked, the receptor potential remains, indicating that it must be produced by a different mechanism from the one that generates the all-or-none upstroke of the AP.

Moreover, the size of the receptor potential varies with the strength of the stimulus, in contrast with APs that are all-or-none. In these respects, receptor potentials resemble excitatory postsynaptic potentials at the postsynaptic membrane of muscle and nerve cells and are quite distinct from APs.

Figure 7-5 The response of some crayfish stretch receptors is phasic; the response of others is tonic. **(A)** Responses of a *phasic* stretch receptor to a weak *(left)* and to a strong *(right)* stimulus. A stronger stimulus generates more APs than a weak stimulus. Even when the stimulus is maintained, the cell produces only one or a few APs. **(B)** Responses of a *tonic* stretch receptor in normal saline *(left)* and after the addition of tetrodotoxin (TTX, *right*). Tetrodotoxin blocks APs, revealing the underlying receptor potential. [Part A adapted from Eyzaguirre and Kuffler, 1955; part B adapted from Loewenstein, 1971.]

Sensory receptors differ in how faithfully they reproduce the timing of a stimulus. A **phasic** receptor produces APs during only part of the stimulation—often only at the onset or at the offset of the stimulus—and thus cannot by itself convey information about the duration of the stimulus. In contrast, **tonic** receptors continue to fire APs throughout the stimulation and can thus directly convey information about the duration of the stimulus. (See *Receptor Adaptation* later in this chapter for a more complete discussion.) Local stimulation of the stretch receptor cell was used to test the ability of various parts of a single cell to produce sustained trains of APs. In these experiments (Figure 7-6), a steady stimulus current produced a sustained discharge only when the current depolarized the low-threshold, spike-initiating zone of the receptor. When other regions of the cell were stimulated, APs were produced, but there was no sustained train of impulses. This difference implies that the ion channels at the spike-initiating zone have different properties from ion channels along the rest of the axon.

In summary, a general sequence of steps leading from a stimulus to a train of impulses in a sensory neuron can be formulated from the results obtained with the crayfish stretch receptor (Figure 7-7). Stimulus energy produces an alteration in a receptor protein, generally located in a membrane. The receptor protein may be part of an ion channel, or it may modulate the activity of membrane channels indirectly through an enzyme cascade, which amplifies the signal. In either case, the absorption of stimulus energy by a receptor molecule eventually causes a population of ion channels to open or to close. This change in membrane permeability produces a shift in V_m in accord with principles presented in Chapter 5. As the intensity of the stimulus in-

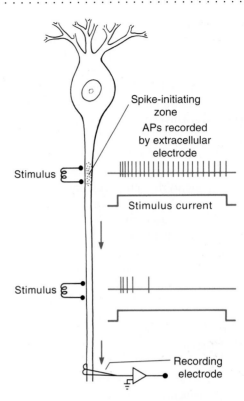

Figure 7-6 Sustained stimulation of a crayfish stretch receptor produces a long train of APs only when the stimulus depolarizes the spike-initiating zone. Other areas of the cell adapt rapidly to steady stimulation. [Adapted from Nakajima and Onodera, 1969.]

creases, more channels respond, producing an increased (or decreased) receptor current, and hence a larger receptor potential. Thus, all the steps leading to, and including, the receptor potential are graded in amplitude. Unlike the

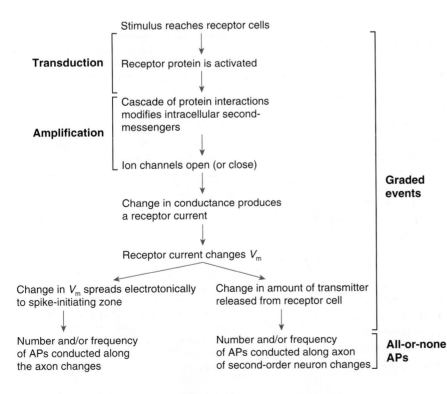

Figure 7-7 Several steps link the onset of a stimulus to the production of APs in a sensory pathway. In some sensory systems, the receptor cells themselves generate and carry APs into the central nervous system *(lower left)*. In other systems, receptor cells synaptically modulate APs in other neurons that carry the signal into the central nervous system *(lower right)*.

current carried by sodium during an AP, the receptor current is not regenerative, even if it is carried by Na⁺, and it therefore must spread through the cell electrotonically (see Chapter 6). If sensory information is to be propagated over long distances into the central nervous system, the information contained in the receptor potential must be converted into APs. This conversion occurs in one of two ways:

1.   In some receptors, a depolarizing receptor potential spreads electrotonically from its site of origin in the receptor zone to the spike-initiating zone in the axon membrane, which then generates APs. The receptor zone may be a part of the same neuron that carries APs to the central nervous system (Figure 7-8A and B). When the receptor potential spreads to the electrically excitable membrane without intervention of a synapse and directly modulates the generation of APs at the spike-initiating zone, it is sometimes also termed a **generator potential**. In a variation on this theme, the receptor zone in some systems is located in an inexcitable receptor cell that is electrically coupled to the afferent neuron (not shown in Figure 7-8).

2.   In other sensory systems, the receptor and the conducting elements are separated by a chemical synapse. In this case, a depolarizing or hyperpolarizing receptor potential spreads electrotonically from the sensory region of the receptor cell to the presynaptic part of that same cell, modulating the release of a neurotransmitter (Figure 7-8C). The transmitter produces a post-

synaptic potential in this second neuron of the chain, called a **second-order neuron**, modulating the frequency of APs in the postsynaptic neuron. In all cases, the cell that contains the receptor membrane is called the **primary sensory neuron**. The axon that carries APs to the central nervous system can be referred to as a *sensory fiber*, an *afferent fiber*, or a *sensory neuron*.

Encoding Stimulus Intensities

Individual APs originating in different sense organs are indistinguishable from one another, as first noted in the 1830s by Johannes Müller, who called this the *law of specific nerve energies*. Müller hypothesized that the modality of a stimulus is encoded not by any characteristics inherent in the individual APs but depends instead on the anatomical region in the brain to which the information is sent. Thus, stimulation of the photoreceptors in the eye produces the sensation of light, whether the photoreceptors are stimulated in the normal way by incident light or abnormally by an intense blow to the eye.

Because APs are all-or-none phenomena, the only way in which information carried along a single nerve fiber can be encoded, other than by the specificity of anatomical connectivity, is in the number and the timing of impulses. Thus, a high frequency of impulses normally represents a strong stimulus, and a reduced frequency of ongoing impulses signals a reduction in the strength of the stimulus. There is no simple rule for sensory coding, because the relations between a stimulus and the sensory response differ in differ-

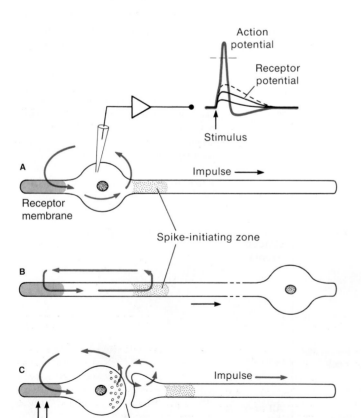

Figure 7-8 Sensory signals are carried into the central nervous system by action potentials. In two types of primary sensory neurons (**A** and **B**), the receptor current arises in the receptor membrane and spreads electrotonically to depolarize the spike-initiating zone. In both cell types, the axon of the receptor cell extends into the central nervous system. The difference between cells A and B is that the cell body of cell A is located relatively far away from the central nervous system, whereas that of cell B is located near the central nervous system. Many invertebrate sensory neurons are like cell A. Vertebrate touch receptors are like cell B. (**C**) In this arrangement, the receptor cell does not produce impulses but instead releases transmitter at a synapse that modulates AP production by an afferent neuron. This arrangement is found in the mammalian auditory and visual systems. Red arrows show the pattern of current flow, and the records at the upper right were made by an intracellular electrode in the soma of neuron A.

ent kinds of receptors. For example, some receptors that receive a particular kind of stimulus information are tonic, whereas others are phasic. Nevertheless, some generalizations can be made about how the intensity of stimuli can be encoded. As the intensity of a stimulus is increased, the receptor current increases, and a greater depolarization (or, in some cases, hyperpolarization) is produced. In many receptors, the spike-initiating zone (see Figure 7-4) will continue to produce a steady train of impulses as long as it is held depolarized.

Input-Output Relations

An ideal sensory system would be able to translate stimuli of all intensities into useful signals. However, biological sensory systems can actually encode stimulus intensity only over a limited range. The range of stimulus intensities over which a receptor can encode higher intensity by producing more APs at a higher frequency is called the **dynamic range** of the receptor (or sense organ). Three major factors serve to set the maximum response that a receptor cell can produce to strong stimuli:

1. There is an upper limit on the receptor current that can flow in response to a strong stimulus because there are a finite number of receptor current channels.
2. There is an upper limit on the amplitude of the receptor potential because it cannot exceed the reversal potential of the receptor current (see *Reversal Potential* in Chapter 6).
3. There is an upper limit on the frequency of APs carried along each axon, because refractoriness (see *General Properties of Action Potentials,* Chapter 5) determines the minimum time between APs propagating along the axon. Typically, the maximum frequency of APs is several hundred APs per second or less.

Biophysical properties cause most sensory responses to be linear with the logarithm of the stimulus intensity. The amplitude of the receptor potential in most receptor cells is approximately proportional to the logarithm of the stimulus intensity (Figure 7-9A), and the frequency of sensory impulses varies approximately linearly with the amplitude of the receptor potential (Figure 7-9B), up to the limit set by the length of the refractory period. As a consequence of these two relations, the frequency of APs in a slowly adapting receptor is typically a function of the logarithm of the stimulus intensity (Figure 7-9C). When the APs reach the central terminals of sensory neurons, they generate postsynaptic potentials that sum and undergo synaptic facilitation as a function of the frequency of impulses. Thus, the postsynaptic potential produced in the central sensory neuron is graded as a function of the stimulus intensity, and it remains an analog of the stimulus, although with somewhat altered characteristics.

The logarithmic relation between stimulus energy and the frequency of sensory impulses that is found in many sensory systems has important implications for how sensory information is processed. Most sensory systems encounter an enormous range of stimulus intensities. For example, sunlight is 10^9 times more intense than moonlight, and the human auditory system can perceive without significant distortion sounds that range over twelve orders of magnitude. This ability of sense organs to function over such enormous ranges of stimulus intensity is quite remarkable and is based on several physiological mechanisms. First, the transduction process itself has a broad dynamic range. In addition, prolonged exposure to a stimulus causes a change in the amplification of the receptor events, shifting the intensity-coding characteristics of the receptor, a process called **adaptation.** Furthermore, neuronal networks that process sensory signals have features that extend the dynamic range of the system beyond the capabilities of individual receptor neurons.

At low stimulus intensities, the receptor potential in a nonadapted receptor neuron represents a very large amplification of energy. The amplification factor is, however, progressively reduced as the intensity of the stimulus increases. The logarithmic relation between intensity of a stimulus

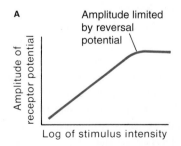
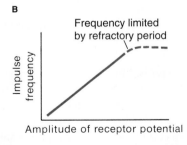
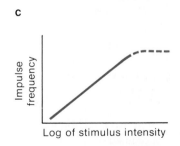

Figure 7-9 The response in most sensory receptors is proportional to the logarithm of stimulus intensity. **(A)** In many receptors, the amplitude of the receptor potential is linearly related to the logarithm of stimulus intensity over a large—but finite—range. The amplitude of the receptor potential cannot increase infinitely, because it is limited by the reversal potential of the receptor current and by other biophysical properties of the receptor cell. **(B)** Within a certain range, the frequency of APs in a re-

ceptor neuron depends linearly on the amplitude of the receptor potential. The refractory period of the neuron sets an upper limit on the frequency. **(C)** As a result of the relations illustrated in parts A and B, the frequency of APs in many sensory axons varies linearly with the log of the stimulus intensity. The broken parts of the curves in parts B and C indicate that refractoriness limits the maximum frequency of APs.

and the amplitude of the receptor potential is explained, at least in part, by the Goldman equation (see Chapter 5), which predicts that V_m should vary with the log of the membrane's permeability to the ion (or ions), P_{ion}, involved in the receptor potential. After stimulation, the change in V_m should be proportional to the log of the change in the permeability to sodium, (P_{Na}), that was produced by the stimulus. Normal stimulus intensities lie within the logarithmic part of the input-output curve (see Figure 7-9A). Not all receptors follow this general rule. In some receptors, there is, instead, a power-function relation: the log of the response amplitude is proportional to the log of the stimulus intensity. For practical purposes, either a logarithmic function or a power function describes the relation between stimulus intensity and receptor response very well in the range of stimulus intensities commonly encountered by animals. The differences between the two functions become apparent only at extreme values of stimulus intensity.

As a consequence of the logarithmic relation between the intensity of a stimulus and the amplitude of the receptor potential, any given *percentage* of change in stimulus intensity evokes the same *increment* of change (i.e., same number of millivolts) in the receptor potential over a large range of intensities. In other words, a doubling of the stimulus intensity at the low end of the intensity range will evoke the same increment in the amplitude of the receptor potential as will a doubling of the intensity of the stimulus near the high end of the range up to the limit where the receptor potential cannot increase. So we have

$$\frac{\Delta I}{I} = K$$

where I is the stimulus intensity and K is a constant. The logarithmic relation between the intensity of the stimulus and the intensity of the response (Figure 7-9C) thus "compresses" the high-intensity end of the scale, which greatly extends the range of discrimination. This relation is similar to that which governs subjectively perceived changes in stimulus intensity, a relation known in psychology as the Weber-Fechner law.

This feature of sensory systems confers great advantage. For example, this property allows us to recognize the objects in a particular scene, even when we see the scene under very different lighting conditions. If we observe the scene in very bright sunlight, each object is distinguished by its relative brightness. If we observe the same scene by moonlight, the absolute brightness of each object is very different from its brightness in sunlight—in fact, the difference in an object's brightness under the two conditions of lighting may be far greater than the difference in brightness of various objects in the scene when illuminated by bright sunlight. However, we are able to recognize the objects in the scene on the basis of their relative intensities, independent of the absolute level of illumination. Detecting *relative intensities* and *changes in intensity* within a given

scene are far more informative to a viewer than is the absolute energy content of each stimulus. Thus, a deer in a field or in a forest is acutely attuned to movements (changes in the distribution of visual stimuli) independent of changes in the lighting conditions.

Range Fractionation

The dynamic range of a multineuronal sensory system is typically much broader than is the range of any single receptor or afferent sensory fiber. The extended dynamic range of the entire system is possible because individual afferent fibers of a sensory system cover different parts of the full spectrum of sensitivity. The most sensitive receptors produce a maximal response at stimulus intensities that are below threshold or only slightly above threshold for other, less sensitive receptors in the population. Above that intensity, the most sensitive receptors become saturated, but the less sensitive receptors can take over at the higher intensities. Thus, at the lowest stimulus energies, a few especially sensitive sensory fibers will respond weakly. If the stimulus energy is increased a little, their discharge frequencies will increase, whereas new, less sensitive fibers will join in weakly. With still greater stimulus intensities, another, formerly quiescent lower-sensitivity population of afferents will join in. As the stimulus intensity is increased, receptors that are less and less sensitive will become active, a phenomenon called **recruitment,** until the least sensitive sensory fibers will finally be recruited, and all receptors will respond maximally. At that point, the system will be saturated and therefore unable to detect further increases in intensity. This **range fractionation,** in which individual receptors or sensory afferents cover only a fraction of the total dynamic range of the sensory system (Figure 7-10), enables the sensory processing centers of the central nervous system to discriminate stimulus intensities over a range much greater than that of any single sensory receptor. One example of range fractionation are the photoreceptors of the vertebrate eye. **Rod** photoreceptors are more sensitive to light and respond to dimmer stimuli; **cones** respond to bright light that saturates the rods (see *Visual receptor cells of vertebrates* later in this chapter).

Control of Sensory Sensitivity

How accurate are our sensations? That is, how do sense organs compare with physical transducers such as thermometers, light meters, and strain gauges? From our own experience, we know that biological sensory systems may not be very trustworthy indicators of absolute energy levels. Moreover, many sensations change over time. For example, when a person dives into an unheated pool for a swim, the water initially feels colder than it does a minute or two later. A pleasantly sunny day may seem painfully bright for a few minutes after a person emerges from a dimly lit house; for this reason, even an experienced photographer requires a light meter to make accurate judgments of camera exposure settings. These changes of *perceived* intensity, when the intensity of the stimulus

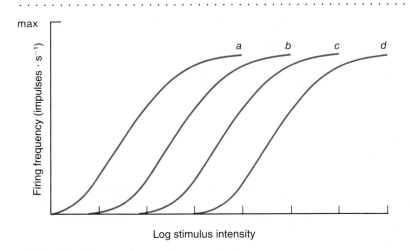

Figure 7-10 Range fractionation extends the dynamic range of sets of sensory receptors. Each curve in this graph represents the discharge frequency of an individual sensory afferent, plotted as a function of stimulus intensity. In this hypothetical example, each of the four sensory fibers, labeled *a* through *d*, has a dynamic range of about three to four log units of stimulus intensity, whereas the overall dynamic range of the four neurons taken together covers seven log units of intensity.

has not itself changed, are lumped under the general term **sensory adaptation.** Where does the adaptation take place? There is no simple answer. Some adaptation takes place in the receptor cells, some as a result of time-dependent changes in accessory tissues, and some in the central nervous system.

Mechanisms of adaptation
Different classes of receptors exhibit different degrees of adaptation. Tonic receptors continue to fire steadily in response to a constant stimulus, as illustrated in Figure 7-11A, which shows a receptor that responds to the displacement of a hair; this receptor produces APs at an almost constant frequency when the hair is displaced and

held in the new position. In contrast, phasic receptors adapt quickly. In one class of phasic receptors, for example, APs occur only during *changes* in the strength of the stimulus, as illustrated in Figure 7-11B by a mechanoreceptor that fires only when the displacement of the hair changes, and the frequency of APs depends on the rate of change.

Adaptation can take place at any of a number of different stages in the processes that link a stimulus to the output of a sensory neuron (Figure 7-12):

1.   The mechanical properties of the receptor cell may act as a filter that preferentially passes transient, rather than sustained, stimuli. This mechanism is common among mechanoreceptors.

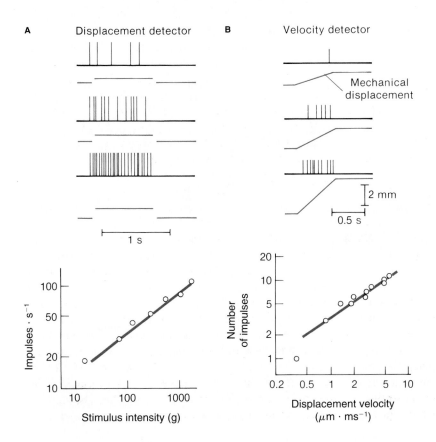

Figure 7-11 A displacement detector fires tonically, whereas a velocity detector fires phasically. **(A)** Behavior of a tonic displacement detector. This mechanoreceptor responded to a steady displacement (shown by red trace) with APs at a relatively constant frequency throughout the stimulation. *(Top)* Extracellularly recorded APs elicited by three different amounts of displacement, which increased from bottom to top. The amount of displacement is indicated below each record. *(Bottom)* Steady-state frequency of APs plotted against the number of grams of tension applied. **(B)** Behavior of a phasic velocity detector. This rapidly adapting mechanoreceptor responded to the *rate* at which the position changed. *(Top)* Action potentials elicited by three different rates of change (red traces). Higher velocities produced more APs. *(Bottom)* The number of impulses produced during a 0.5 second stimulus was proportional to the log of the velocity of displacement. [Adapted from Schmidt, 1971.]

Stimulus

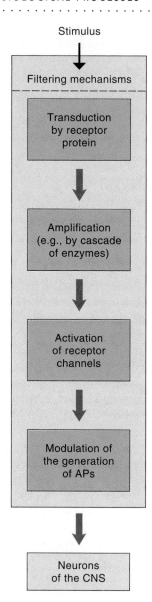

Figure 7-12 Sensory adaptation can take place at any of several stages in information processing. The dashed lines indicate that, in some systems, filtering or the modulation of AP frequency, or both, take place in the receptor cell itself; whereas, in other systems, these functions take place outside the receptor cell.

2. The transducer molecules themselves may "run down" during a constant stimulus. For example, a significant percentage of visual pigment molecules can become bleached when exposed to continuous light and must be regenerated metabolically before they can again respond to illumination.

3. The enzyme cascade activated by a transducer molecule may be inhibited by the accumulation of a product or an intermediate substance.

4. The electrical properties of the receptor cell may change in the course of sustained stimulation. In some receptors, activation of receptor channels diminishes because intracellular free Ca^{2+} increases during sustained stimulation. Accumulation of intracellular free

Ca^{2+} can also activate Ca-dependent K^+ channels, producing a shift in membrane potential, V_m, back toward the resting potential.

5. The membrane of the spike-initiating zone (see Figure 7-4) may become less excitable during sustained stimuli.

6. Sensory adaptation can also take place in higher-order cells in the central nervous system (which includes the vertebrate retina).

Both the first and fifth of these mechanisms of adaptation are illustrated by the muscle stretch receptors of crayfish and lobsters. These receptors are present in pairs in the abdominal musculature, each pair consisting of one phasic receptor and one tonic receptor. A stretch of the muscle fiber produces a transient response in the phasic receptor (Figure 7-13A) and a sustained response in the tonic receptor (Figure 7-13B). When these receptors are stimulated by direct injection of a depolarizing current through a microelectrode, rather than by stretching the muscle fiber, each cell retains some of its characteristic properties. That is, when the stimulating current is prolonged, the tonic receptor produces a longer train of APs than does the phasic receptor.

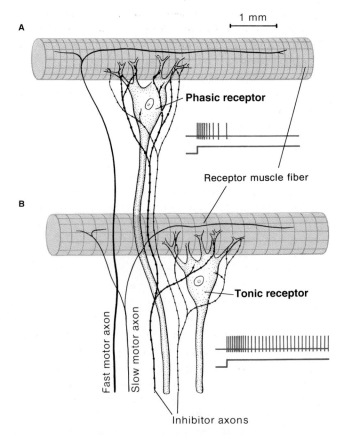

Figure 7-13 The phasic and tonic stretch receptors of the crayfish adapt differently to sustained stimuli. The phasic receptor **(A)** adapts quickly to a constant stretch, producing only a short train of impulses. The tonic receptor **(B)** fires steadily during a maintained stretch, although the frequency of APs is highest at the beginning of the stimulation and drops off during a sustained stretch. [Adapted from Horridge, 1968.]

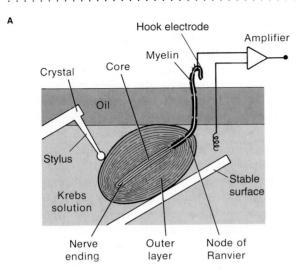

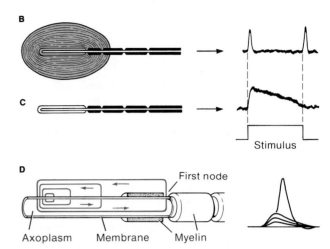

Figure 7-14 Adaptation in the Pacinian corpuscle depends on the mechanical properties of accessory structures. **(A)** Experimental arrangement for tapping a Pacinian corpuscle receptor with a piezoelectric crystal-driven stylus. The electrical recording was made between the hook electrode on the axon and the oil-water interface. **(B)** Electrical response of the intact corpuscle. The neuron depolarized transiently at the onset and the offset of the stimulus (dashed red lines). **(C)** In contrast, af- ter the lamellae were removed, the neuron remained depolarized during most of the stimulus. **(D)** Receptor current flow in response to deformation at the sensory zone of the axon. The generator potential is conducted electrotonically to the spike-initiating zone at the first node of Ranvier. If the generator potential is sufficiently large, it will bring the spike-initiating zone to threshold, producing APs in the axon. [Adapted from Loewenstein, 1960.]

A change in the filtering that is produced by accessory structures (mechanism 1) contributes importantly to the rapid adaptation of the **Pacinian corpuscle**, a pressure and vibration receptor found in the skin, muscles, mesentery, tendons, and joints of mammals (Figure 7-14A). Each Pacinian corpuscle contains a region of receptor membrane that is sensitive to mechanical stimuli and that is surrounded by concentric lamellae of connective tissue resembling the layers of an onion. When something presses on the corpuscle, deforming it, the disturbance is transmitted mechanically through the layers to the sensitive membrane of the receptor neuron. The receptor membrane normally responds with a brief, transient depolarization at both the onset and the offset of the deformation (Figure 7-14B). However, when the layers of the corpuscle are peeled away, permitting a mechanical stimulus to be applied directly to the naked axon, the receptor potential obtained is sustained much longer, producing a more accurate representation of the stimulus (Figure 7-14C). Although the receptor potential still shows some degree of adaptation (there is sag in the record shown in Figure 7-14C), there is no distinctive response at the offset of the stimulus. The mechanical properties of the intact corpuscle, which preferentially pass rapid *changes* in pressure, confer on the receptor neuron its normally phasic response. This behavior explains, in part, why we quickly lose awareness of moderate, sustained pressures, such as the stimuli that wearing clothing produces on our skin.

Regardless of its site or mechanism of origin, adaptation plays a major role in extending the dynamic range of sensory reception. Together with the logarithmic nature of the primary transduction process, sensory adaptation allows an animal to detect changes in stimulus energy against background intensities that range over many orders of magnitude.

Mechanisms that enhance sensitivity

Many receptor cells produce APs—or release neurotransmitter independent of APs—spontaneously in the absence of stimuli. (The amount of transmitter released from non-spiking receptors varies with the membrane potential, V_m.) When these spontaneously active receptors are stimulated, the frequency of their APs—or their non-spiking release of transmitter—is increased or decreased above its baseline level. Several mechanisms enhance the sensitivity of receptors to sustained stimuli, and one important mechanism modifies the properties of the receptor's on-going spontaneous activity. The spontaneous release of transmitter from receptor cells—whether it is mediated by APs or by graded changes in V_m—has two important consequences. First, any small increase in stimulus energy will produce an increase in the rate of firing above the spontaneous level. Small receptor currents in response to weak stimuli modulate the impulse frequency by shortening the intervals between impulses (Figure 7-15). This modulation of impulse frequency allows the receptors to be much more sensitive to changes in stimuli than would be feasible if the receptor current had to bring a completely quiescent spike-initiating zone to threshold. The input-output relations of such a sensory fiber are described by the sigmoid curve in Figure 7-16. In the unstimulated condition, the firing frequency is already on the steep part of the curve, so even a small input will produce a significant increase in firing frequency.

Second, in some spontaneously active sensory neurons, stimuli can either increase or decrease impulse frequency, permitting the receptor to convey information about the polarity or direction of a stimulus. For example, in some mechanoreceptors, such as hair cells, movement of the hair in one direction increases the rate of firing in the sensory

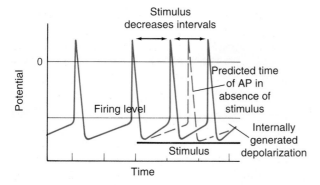

Figure 7-15 In a spontaneously firing receptor cell, the interval between APs depends on stimulus conditions. The interval can be decreased by extremely small stimuli, because a stimulus increases the slope of the internally generated depolarization. (In cases where APs are generated in second-order sensory fibers, synaptic potentials, rather than internally generated depolarizations, are enhanced.)

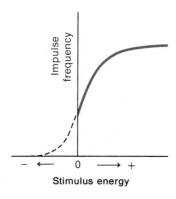

Figure 7-16 The input-output relation of a spontaneously active receptor cell (or second-order sensory fiber) is sigmoidal. In the absence of any input (0 on the abscissa), the receptor cell, or second-order sensory fiber, fires spontaneously. This spontaneous output lies on the steep part of the curve relating the stimulus intensity to the frequency of APs, so even a very small stimulus will increase the rate of firing. In some receptors, stimuli can also decrease the rate of firing, indicated by the black dashed part of the curve.

fiber, whereas movement in the other direction decreases the rate of firing. If these receptors were silent when they were not stimulated, it would be impossible to encode information about movement in the second direction.

The existence of numerous parallel sensory pathways provides another mechanism for enhancing the distinction between a signal and ongoing background noise. In this situation, signals from many receptor cells can be summed by the central nervous system. All of the signals produced by the stimulus will arrive at central neurons nearly simultaneously, whereas noise will be random and will tend to be canceled out at central synapses. By reducing noise, this arrangement allows small changes in input to be detected. For example, a human observer cannot reliably perceive a single photon absorbed by a single receptor cell; but, if each of several receptors simultaneously absorbs a single photon, the observer experiences the sensation of light.

Efferent control of receptor sensitivity

The responsiveness of some sense organs is influenced by the central nervous system through efferent axons that innervate the sense organ itself. For example, the muscles to which muscle stretch receptors in skeletal muscles of vertebrates and crustaceans are attached are innervated by efferent fibers. By controlling the length of the receptor muscles, this efferent innervation sets the sensitivity of the stretch receptor to changes in overall muscle length.

In crayfish and lobsters, when the tail extensor muscle shortens, the receptor muscles (which run parallel to the extensor muscle) shorten too, driven by efferent neurons. If there were no such mechanism, when the tail extensor muscle shortened, the stretch receptors would go slack and would be unable to detect any further change in the length of the extensor muscle. Instead, contraction of the receptor muscles, in response to the efferent input, maintains a fairly constant tension on the sensory parts of the receptor, and the receptor retains its sensitivity to flexion of the tail, regardless of the tail's position in space. In addition to this mechanism by which the receptors maintain high sensitivity, the abdominal stretch receptors are innervated by ef-

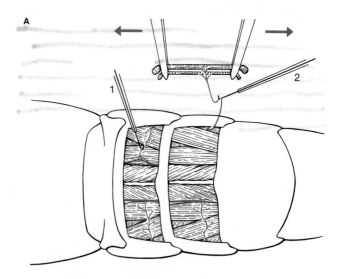

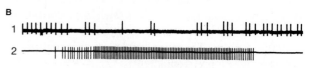

Figure 7-17 Steady stretching of crayfish muscles produces reflex inhibition of crayfish stretch receptors. **(A)** The muscle stretch receptors of one segment are removed with their innervation intact, and recordings are made as shown. Electrode 1 records activity in an intact receptor; electrode 2 records activity in a receptor that has been removed from the tail. **(B)** At the beginning of the record, electrode 1 records a steady train of APs from the intact tonic receptor in response to a steady stretch of the tail. When the isolated tonic receptor is stretched, electrode 2 records a train of sensory impulses, and there is a simultaneous drop in the frequency of APs recorded by electrode 1 in the intact receptor. Activity in the stretch receptor monitored by electrode 2 increases activity in an inhibitory neuron, which reduces the steady output of the neuron that is monitored by electrode 1. [Adapted from Eckert, 1961.]

ferent neurons that form inhibitory synapses directly on the stretch receptor cells (see Figure 7-13). When the inhibitory efferent neuron is active, the size of the receptor potential in the stretch receptor is diminished, reducing the frequency of APs in the axon or even abolishing them altogether. The interplay of these two mechanisms—one that enhances the responsiveness and the other that inhibits it—allows activity in the central nervous system either to increase or to decrease the sensitivity of these stretch receptors.

Feedback inhibition of receptors

The sensitivity of sensory receptors is also controlled through feedback inhibition. In this mechanism, activity in the receptors produces signals that are sent more or less directly back to the receptors, inhibiting them. The crustacean abdominal stretch receptors provide one example of this (Figure 7-17). Activation of the sensory neuron by stretch produces a reflex output—initiated in the central nervous system—that travels in the efferent inhibitory nerves leading to the stimulated sensory neuron *(autoinhibition)* and to its anterior and posterior neighbors *(lateral inhibition)*. At low stimulus intensities, the feedback plays little or no role, because it takes a relatively strong sensory signal to evoke the reflex activation of the inhibitory neurons. However, stronger stimuli produce stronger inhibitory feedback; as a result, strong stimuli are preferentially inhibited. This mechanism acts to keep the receptor within its operating range (i.e., it keeps the frequency of APs less than the maximum frequency possible in the cell), and the net effect of the inhibitory feedback is to extend the dynamic range of the receptor.

When receptors produce signals that inhibit their neighbors, as do the crayfish stretch receptors, this mutual inhibition between neighboring receptors can strongly influence sensory reception. For example, this **lateral inhibition** can enhance the contrast between activity in neighboring receptors (Figure 7-18). Although this phenomenon was first discovered in visual systems (see Chapter 11 for further discussion), it occurs in a number of sensory systems. The net effect of the interaction between neighboring cells is that the differences in activity levels found in weakly and strongly stimulated receptors are exaggerated, producing an increase in the perceived contrast between regions of weak and strong stimulation.

THE CHEMICAL SENSES: TASTE AND SMELL

Although unicellular organisms were on Earth 3.6 billion years ago, the first multicellular organisms did not arise until 2.5 billion years later. This enormous time lag may indicate, at least in part, how long it took to evolve mechanisms for cell–cell signaling, which are required to coordinate the development and activity of many cells acting in concert. Signaling between organisms may have appeared earlier than did signaling among cells within a single organism. The sensitivity of cells to specific molecules is widespread and includes metabolic responses of tissues to chemical messengers as well as the ability of lower organisms—such as bacteria—to detect and respond to certain substances in the environment. Although many cell types respond to molecules in their environment, *chemoreceptors* are receptor cells that are specialized for acquiring

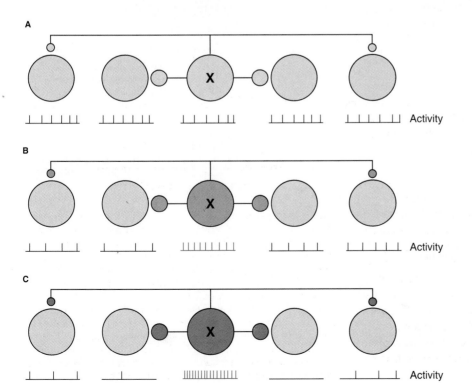

Figure 7-18 Lateral inhibition enhances the edges of stimuli by exaggerating the differences in activity of receptors near an edge. **(A)** Five adjacent receptors are spontaneously active in the absence of stimulation. **(B)** When the center receptor, marked X, is stimulated weakly, it inhibits its neighboring receptors. The strength of the inhibition drops with distance from the activated receptor (symbolized by large and small black circles representing inhibitory synapses). **(C)** Stronger stimulation of the center receptor increases its inhibition of its neighbors.

information about the chemical environment and transmitting it to other neurons. According to one widely accepted classification scheme, chemoreceptors can be divided into two categories: *gustatory* (taste) receptors, which respond to dissolved molecules, and *olfactory* (smell) receptors, which respond to airborne molecules. This dichotomy, however, rapidly breaks down. By these definitions, aquatic organisms, such as fishes, could have no smell receptors; all of their chemical sensation would be taste. Furthermore, even in terrestrial organisms, airborne molecules must pass through a layer of aqueous solution before they reach the olfactory receptors.

If taste and smell are legitimately different senses, there must be a more useful distinction between them. Indeed, as we will see, the receptors of taste and the receptors of smell operate quite differently from one another, making it possible to distinguish between the two senses at the cellular level. In addition, there are alternative schemes to distinguish between taste and smell at a more global level. For example, smell can be considered as the chemoreception of signals from distant sources and "taste" as the chemoreception of signals from material that is in direct contact with the receptive structure (e.g., masticated food in the mouth or material at the bottom of a pond in which a catfish is living).

Chemosensory systems can be extraordinarily sensitive. The antennal chemoreceptors of the male silkworm moth (*Bombyx mori*) for *bombykol*, the female's sex-attractant pheromone, provide a spectacular example. In the laboratory, a male moth responds behaviorally to concentrations of the pheromone that are as low as one molecule per 10^{17} molecules of air. These receptors are highly specific, responding only to bombykol and a few of its chemical analogs. This highly evolved stimulant-receptor system allows the male *Bombyx* moth to locate a single female at night from several miles downwind, an ability that confers obvious reproductive advantage in a widely dispersed species.

To investigate the sensitivity of the receptors to bombykol, the electrical responses of antennal olfactory receptors of *Bombyx* have been recorded . When only about 90 bombykol molecules per second impinge on a single receptor cell, the rate at which the cell fires APs increases significantly. However, a male moth reacts behaviorally (e.g., flaps his wings excitedly) when *only about 40 receptor cells (out of a total of 20,000 per antenna) each intercept one molecule per second*. No change can be detected in the frequency of APs fired by a single receptor cell in response to a single odorant molecule. As a result, it is inferred that the moth's central nervous system is capable of sensing very slight average increases in impulse frequency arriving along numerous chemosensory channels, as described earlier in this chapter (see *Mechanisms that enhance sensitivity*).

Mechanisms of Taste Reception

Electrophysiological studies of the *contact chemoreceptors* (taste hairs) of insects have revealed generally useful information about how chemoreception works. These receptor cells send fine dendrites to the tips of hollow hairlike projections of the cuticle, called **sensilla** (plural; singular, sensillum). Each sensillum has a minute pore that allows stimulant molecules to reach the sensory cells (Figure 7-19). In the proboscis or on the feet of an ordinary housefly, every sensillum contains several cells, each of which is sensitive to a different chemical stimulus (e.g., water, cations, anions, or carbohydrates).

The electrical activity of these chemoreceptors can be recorded through a crack made in the wall of the sensillum, and such records have revealed both a receptor potential and APs. The receptor potential is produced at the ends of

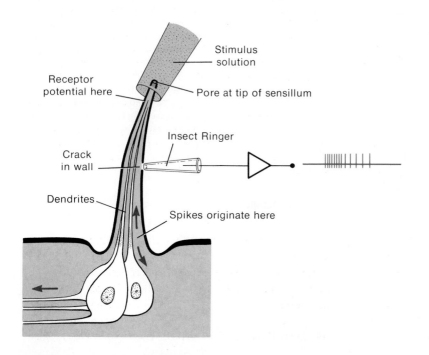

Figure 7-19 The response of a housefly's contact chemoreceptors can be recorded extracellularly. The dendrites of several neurons share a single sensillum. The dendrite of each individual neuron within the sensillum is sensitive to a particular class of substances (e.g., sugars, cations, anions, or water). Stimuli are presented through a cannula slipped over the tip of the sensillum, and electrical responses (in red at right) are recorded through a crack made in the cuticle covering the sensillum.

Stimulus solution

Receptor potential here

Pore at tip of sensillum

Insect Ringer

Crack in wall

Dendrites

Spikes originate here

the dendrites that extend to the tip of the sensillum, whereas the APs originate near the cell body.

Appropriate chemical stimulation of even a single sensillum will evoke a behavioral response in a fly. For example, a small drop of sugar solution applied to a single sensillum on the foot will cause the fly to lower its proboscis to feed, and the effectiveness of various compounds to evoke this stereotyped behavior has been tested. All compounds that release the feeding reflex also evoke electrical activity in the sugar receptor. This receptor cell is known to respond to only certain carbohydrates, and those compounds that do not trigger feeding behavior, such as D-ribose, also fail to stimulate the sugar receptor. Interestingly, the sugar receptor of the fly shows the same sequence of sensitivity (fructose > sucrose > glucose) as do the sweetness receptors of the human tongue.

Like insects, many vertebrates have taste receptors on the body. The bottom-dwelling sea robin fish, for example has modified pectoral (anterior) fins with taste receptors at the tips of the fin rays, which it uses to probe the muddy bottom for food. In terrestrial vertebrates, taste receptors are found on the tongue and epiglottis, in the back of the mouth, and in the pharynx and upper esophagus.

In vertebrates, taste receptor cells are located in taste buds, which have some organizational features in common with olfactory organs (Figure 7-20). The taste receptors are surrounded by support cells and by *basal cells*, which are progenitor cells that give rise to new taste receptors. Basal cells are derived from epithelial cells, and they regularly generate new sensory receptor cells; taste receptor cells live for only about 10 days. This remarkable turnover of primary sensory cells also occurs in vertebrate olfactory organs and in specialized parts, called *outer segments*, of photoreceptor cells. All of these cells, or cell parts, that are regularly renewed directly interact with physical stimuli from outside the organism: taste and smell molecules in gustatory and olfactory cells, respectively, and photons in photoreceptor outer segments. The turnover of all sensory cells poses a problem for the maintenance of sensory specificity in an organism because, unless the new cells were precisely integrated into the existing network, specificity would be lost. Just how the integrity of taste and smell sensations is maintained remains an unsolved, but actively studied, mystery.

Although our subjective experience would suggest that there is a wonderfully large spectrum of possible tastes, these sensations can be grouped into four distinct qualities: sweet, salt, sour, and bitter. In evolutionary terms, these categories may be related to some basic properties of food. Sweet foods are likely to be rich in calories and thus useful; salt is essential for maintaining water balance (see Chapter 14); a sour taste can signal danger if it is in excess; and many bitter substances are toxic. The discovery that vertebrates respond to only four fundamental categories of tastes suggests that all perceived tastes must depend on various combinations of these fundamental characteristics. In addition, it generated the hypothesis that there is a separate, identifiable sensory pathway associated with each of the four tastes.

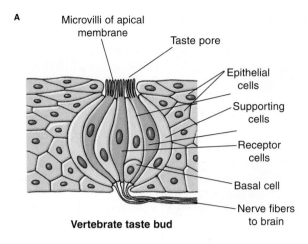

Vertebrate taste bud

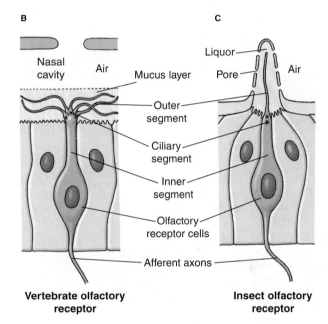

Vertebrate olfactory receptor   **Insect olfactory receptor**

Figure 7-20 Chemosensory organs typically consist of receptor cells surrounded by supporting structures. **(A)** In vertebrate taste buds, the receptor cells are surrounded by basal cells, which generate new receptor cells, and by supporting cells. Transduction takes place across the apical membrane. The receptor cells do not themselves send axons to the central nervous system, although they can produce APs. Instead, they synaptically excite afferent neurons that carry the information to the CNS. In contrast, vertebrate **(B)** and insect **(C)** olfactory receptors themselves send primary afferent axons to the CNS. Structures that are analogous between vertebrates and insects are drawn similarly in parts B and C. All three types of receptors extend fine processes into a mucous layer that covers the epithelium. In insects, these fine processes are true dendrites. [Part A adapted from Murray and Murray, 1970; part C adapted from Steinbrecht, 1969.]

How do molecules interact with membranes to produce distinct tastes? In the past few years, the use of patch-clamp recording has allowed the mechanisms responsible for each taste modality to be identified (Figure 7-21). Each individual taste receptor cell reacts to particular stimuli, and each class of taste stimuli activates a distinctive cellular pathway in the receptors that respond to it. Salty

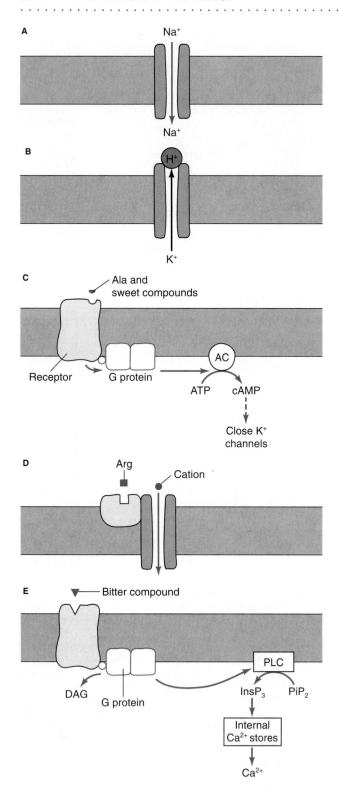

Figure 7-21 Each kind of taste is transduced by a distinctive mechanism. **(A)** In the transduction of salty and some sour tastes, Na$^+$ (or H$^+$) ions pass through ion channels in the apical membrane, directly depolarizing the receptor cell. **(B)** In the transduction of other sour tastes and some bitter tastes, protons (sour) or certain bitter compounds block K$^+$ channels, and residual leakage of cations into the cell depolarizes the receptor. **(C)** L-Alanine (Ala) and some other sweet compounds bind to receptors (R) and activate a G protein (G). The activated G protein activates adenylate cyclase (AC), and the resulting increase in cAMP closes K$^+$ channels in the basolateral membrane, depolarizing the cell. **(D)** L-Arginine (Arg) binds to, and opens, a ligand-gated, nonselective cation channel. **(E)** Some bitter compounds bind to a receptor and activate a G protein that is thought to be coupled to phospholipase C (PLC), producing an increase in intracellular inositol triphosphate (InsP$_3$), which could then release Ca^{2+} from intracellular stores. The net result is depolarization of the receptor cell, but the mechanism is not yet completely understood. PIP$_2$, phosphoinositol 4,5-biphosphate. [Adapted from Avenet et al., 1993.]

channel (observed in the hamster) or by blocking a K$^+$ channel (observed in the salamander *Necturus*). In either case, the membrane is depolarized. Sweet compounds and the amino acid alanine (Ala) bind to receptors coupled to an intracellular cascade that closes K$^+$ channels in the basolateral membrane, depolarizing the receptor. Other sweet substances, including the amino acid arginine (Arg) and monosodium glutamate, activate nonspecific cation-selective channels in taste cells. Some bitter substances, such as Ca^{2+} and quinine, close K$^+$ channels in the apical membrane, allowing the cell to depolarize. Transduction of other bitter substances is less well understood but appears to rely on intracellular second-messenger systems (either the InsP$_3$ or cAMP pathways) to excite the cell. It is hypothesized that the sweet and bitter pathways that act through second-messengers are mediated by G proteins, and recent reports have suggested candidate molecules. In all cases, the initial event in the receptor cell eventually causes an increase in the concentration of intracellular Ca^{2+} and thus increases the release of neurotransmitter onto the second-order cells in the pathway.

Taste receptors generate APs, but they have no axons, so they cannot themselves carry information to the central nervous system. Instead, they synapse onto, and modulate activity in, neurons whose axons run in the facial, glossopharyngeal, and vagus nerves (seventh, ninth, and tenth cranial nerves). The existence of four kinds of taste sensations and the specificity of membrane transduction mechanisms for each kind of taste suggest that each receptor subtype might be connected to a particular set of axons. In that arrangement, for example, information about "sweetness" would be carried by some specific subset of axons. Such a pattern is called **labeled line coding,** but recordings have revealed that taste information is not nearly this neatly organized. Recordings from single neurons show that a receptor will often respond optimally to a particular type of stimulus (Figure 7-22), but many receptors also respond suboptimally to stimuli in other classes. The data thus suggest that a single fiber innervating a taste bud receives information from receptors belonging to different subtypes.

cellular pathway in the receptors that respond to it. Salty stimuli, such as NaCl, readily dissociate in water, and the Na$^+$ ions enter receptors through Na$^+$ channels in the membrane to depolarize the membrane potential. These Na$^+$ channels are distinctive because they can be blocked by the drug amiloride, unlike the voltage-gated Na$^+$ channels that mediate most APs. Sour stimuli, which are characterized by excess H$^+$ ions, act either through this same

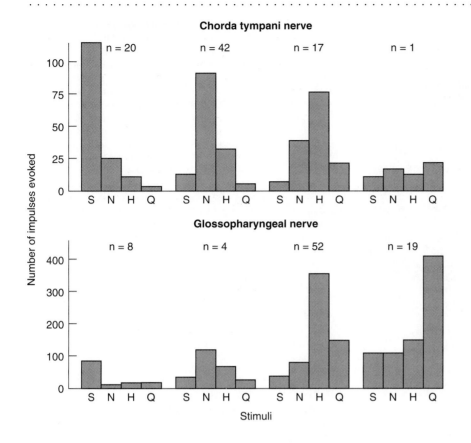

Figure 7-22 Each afferent taste neuron is most effectively stimulated by one type of stimulus, but it also responds to other stimuli. The responses to four different taste stimuli were recorded from single taste afferent axons in two different nerves in hamsters. Each neuron responded maximally to one of the four taste stimuli; different neurons responded maximally to different stimuli. However, all of the axons responded at least weakly to all four stimuli, indicating that each taste afferent is not restricted to carrying information about only one kind of taste. Abbreviations: S, sucrose (sweet); N, NaCl (salty); H, HCl (sour); Q, quinine HCl (bitter). The number of axons is indicated for each group. [Adapted from Hanamori et al., 1988.]

Rather than simple labeled line coding, sensory information about taste must depend on the analysis of many gustatory axons in parallel.

Mechanisms of Olfactory Reception

In mammals, olfactory receptors are located inside the nasal cavity, arranged so that a stream of air or water flows over them during respiration (Figure 7-23). Animals that are particularly dependent on olfactory cues have complex cavities that are lined by sheets of receptors. These cavities are called **turbinates,** and the mechanisms guaranteeing air flow through them remains unknown. Each receptor neuron has a long thin dendrite that terminates in a small knob at the surface (Figure 7-24A).

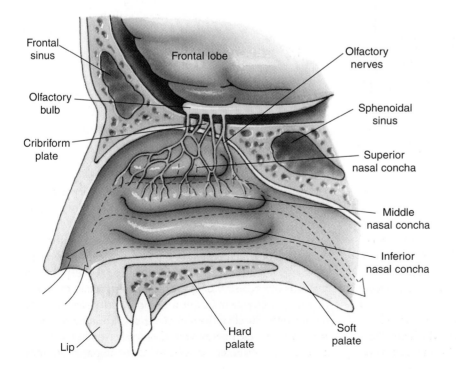

Figure 7-23 In vertebrate olfactory organs, air (or water) carrying odorants is moved past olfactory receptors. The olfactory epithelium of human beings covers part of the surface of air passages in the nose. The arrows indicate the route followed by air as it is breathed in through the nose. The dashed portion of each line shows the movement of air in the turbinates (shaded in red) where the olfactory receptors are located. Dashed lines also indicate eddy currents in the air that are created over the olfactory epithelium that lines the dorsal recesses of the nasal cavity.

A

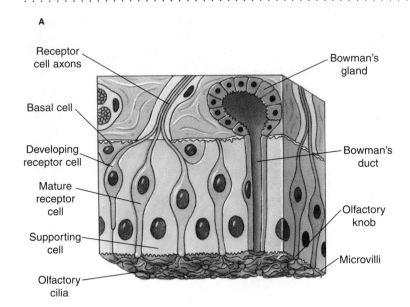

Receptor cell axons

Basal cell

Developing receptor cell

Mature receptor cell

Supporting cell

Olfactory cilia

Bowman's gland

Bowman's duct

Olfactory knob

Microvilli

B

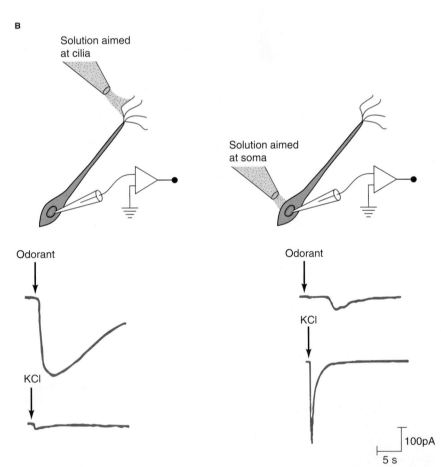

Solution aimed at cilia

Solution aimed at soma

Odorant

KCl

Odorant

KCl

100pA

5 s

Figure 7-24 Receptors of the vertebrate olfactory epithelium depolarize in response to oderant molecules. **(A)** The organization of cells within the mammalian olfactory epithelium. **(B)** Response of a cultured salamander olfactory receptor neuron to focal pulses of an odorant. *(Left)* When the stimulating chemical pulse was directed at the receptive membrane on the cilia, it produced a large current (top record). When a solution containing a high concentration of K^+ was focused at the same location, the response was small (bottom record). *(Right)* When the stimulating chemical pulse was directed at the soma, rather than at the cilia, the response was small (top record). However, when the solution containing a high concentration of K^+ was directed at the soma, it produced a large response (bottom record). [Part A adapted from Shepherd, 1994; part B adapted from Firestein et al., 1990.]

Emanating from the knob are several thin cilia (about 0.1 μm in diameter and about 200 μm long) that are covered by a protein solution called **mucus**. Molecules delivered to the nasal cavity are absorbed into the mucous layer and delivered to the cilia.

Two lines of evidence suggest that the cilia are the location of olfactory transduction. First, only ciliated neurons respond to odors, implying that the cilia must be the site of transduction. The second piece of evidence comes from experiments in which olfactory neurons were grown in culture and were exposed to odorants while the receptor current was recorded by an intracellular electrode in the soma (Figure 7-24B). If a solution of odorant molecules was ejected onto the cilia, the cell responded strongly; whereas, if the same solution was ejected onto the soma, there was only a small response. In contrast, ejecting a solution of

KCl (which would depolarize the membrane of the receptor) onto the cilia produced a small response, whereas ejecting KCl onto the soma produced a large response. These data imply that only the cilia were able to respond to the odorant, causing V_m to change significantly.

The olfactory transduction cascade includes an adenylate cyclase that is linked to a G protein. (See the discussion of transduction earlier in the chapter.) A very large family of proteins has recently been found to be expressed only in olfactory epithelial cells. The structure of each protein includes seven transmembrane domains, and other features also indicate that these molecules are homologous to proteins that mediate other transduction processes. The large size of this family of proteins suggests that there could be many individual receptor subtypes for distinct odors, in contrast with the small number of receptors that code for taste.

Olfactory coding in vertebrates has been studied electrically in the olfactory epithelium of the frog (Figure 7-25A) In these experiments, activity of single receptor axons was recorded by one electrode while the summed potential of large numbers of olfactory receptors in the epithelium (the **electro-olfactogram,** or EOG) were recorded simultaneously by another electrode (Figure 7-25B). Impulses from individual receptors were then superimposed electronically on the electro-olfactogram. This technique permits the activity in a single receptor to be compared with the total response of many receptors when a single odorant, or a combination of odorants, is presented.

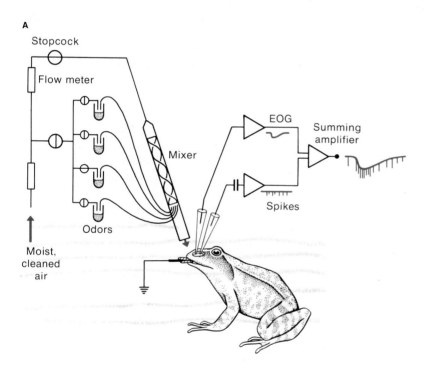

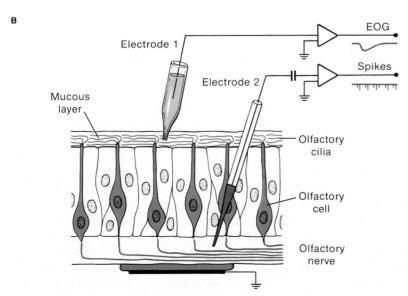

Figure 7-25 Olfactory reception can be studied at the cell and organ level simultaneously in the olfactory epithelium of the frog. **(A)** Various odorants can be applied to the nasal epithelium while the summed electro-olfactogram (EOG) and spikes from single receptor cells are recorded. The two kinds of records can then be electronically summed to give a composite recording *(right)*. **(B)** Detail of tissue and electrodes. Electrode 1 records the overall EOG potential, because it is far from any one axon, while electrode 2 records the activity of the single axon to which it is closest. [Adapted from Gesteland, 1966.]

The results indicate that stimulus coding in the vertebrate nose is far more complex than is coding in the contact chemoreceptors of houseflies. Different receptors respond differently to the same odorant. In some olfactory axons, a particular odorant increased the impulse frequency (Figure 7-26A). Some odorants that smell alike to humans have similar effects on some frog olfactory cells, suggesting that they would smell alike to frogs, too. However, these same odorants have different effects on other cells (see Figure 7-26A, cell *a* vs. cell *b*), suggesting that they would *not* smell alike to the frog. In the olfactory bulb, farther along the chain of olfactory neurons, neurons may respond to an odorant with decreased activity or with increased activity (Figure 7-26B). In fact, it has proved to be impossible to establish a one-to-one relation

between classes of odorants and types of olfactory cells in the frog. Instead, each olfactory receptor cell appears to express a mosaic of odorant receptor molecules with differing specificities. The response characteristics of a particular olfactory receptor must, then, depend on the proportions of its many types of receptor molecules. This situation implies that the ability of mammals to distinguish among a wide variety of odors must reside in the ability of the higher olfactory centers in the brain to decode a combinatorial signal from a large number of olfactory receptors.

MECHANORECEPTION

All animals can sense physical contact on the surface of their bodies. Such stimuli are detected by *mechanoreceptors,* the simplest of which consist of morphologically undifferentiated nerve endings found in the connective tissue of skin. More complex mechanoreceptors have *accessory structures* that transfer mechanical energy to the receptive membrane. These accessory structures often also filter the mechanical energy in some way, as described earlier in regard to the mammalian Pacinian corpuscle in which the sensitive ending is covered by a capsule (see Figure 7-14). Other mechanoreceptors include the muscle stretch receptors of various kinds found in arthropods and vertebrates, in which mechanically sensitive sensory endings are associated with specialized muscle fibers (see Figure 7-13), and the hairlike sensilla that extend from the exoskeletons of arthropods (Figure 7-27). The most elaborate accessory structures associated with mechanoreceptive cells are found in the vertebrate middle and inner ear and in the vestibular system, both of which are considered later in this chapter.

The stimulus that activates mechanoreceptive membrane is a stretch or distortion of the surface membrane. Indeed, stretch-sensitive channels are found in all types of organisms from the simplest to the most complex. Patch-clamp data indicate that these channels respond to changes in tension in the plane of the membrane, and they can be either activated or inactivated by stretch. Stretch-sensitive channels defy simple classification with regard to selectivity, because they show a wide range of conductances and fidelity. Possible transducers of mechanical stress include the cytoskeleton, enzymes, or the ion channels themselves. Mechanically sensitive channels are the only primary mechanotransducers that do not depend on enzymatic activity but instead directly use the free energy stored in the transmembrane electrochemical gradient. Mechanoreceptors can be exquisitely sensitive, responding to mechanical displacements of as little as 0.1 nm. It is a continuing challenge to understand how such small displacements can produce changes in ion permeability through the membrane.

Hair Cells

The **hair cells** of vertebrates are extraordinarily sensitive mechanoreceptors that are responsible for transducing me-

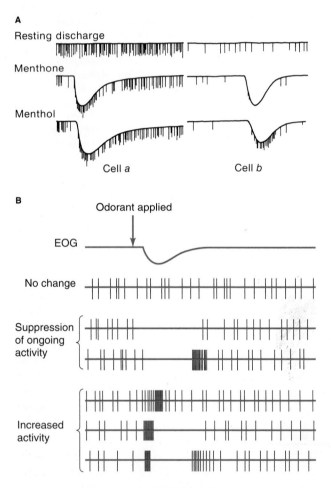

Figure 7-26 Olfactory receptors have complicated responses to individual odorants. **(A)** Recordings from two frog olfactory receptors. Menthone and menthol both slightly suppresssed ongoing activity in cell *a*, indicating that cell *a* could not distinguish between the two substances. In contrast, cell *b* responded differently to the two substances, producing more APs in response to menthol than it did at rest but fewer in response to menthone than it did at rest. Thus, cell *b* could potentially distinguish between the two substances, whereas cell *a* could not. Notice that the electro-olfactogram (EOG) has been summed with the individual record for each cell. **(B)** Recordings from second-order olfactory cells of the tiger salamander. Odorants may reduce or increase ongoing activity in these cells. [Part A adapted from Gesteland, 1966; part B adapted from Kauer, 1987.]

A

B

C

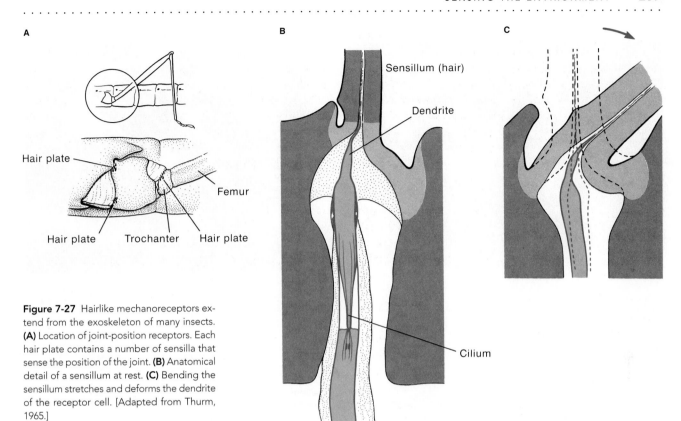

Sensillum (hair)

Dendrite

Hair plate

Femur

Hair plate   Trochanter   Hair plate

Cilium

Figure 7-27 Hairlike mechanoreceptors extend from the exoskeleton of many insects. **(A)** Location of joint-position receptors. Each hair plate contains a number of sensilla that sense the position of the joint. **(B)** Anatomical detail of a sensillum at rest. **(C)** Bending the sensillum stretches and deforms the dendrite of the receptor cell. [Adapted from Thurm, 1965.]

chanical stimuli into electrical signals (Figure 7-28). They are found in several locations. Fish and amphibians have a set of external receptors, called the **lateral-line system,** that are based on hair cells and that detect motion in the surrounding water (Figure 7-29). The vertebrate organs of hearing and the organs that report the position of the body with respect to gravity (called the **organs of equilibrium**) also are based on hair cells. Typically, the organs of equilibrium include the **semicircular canals** and the **vestibular apparatus.**

Hair cells are named for the many cilia that project from the apical end of each cell. These cilia fall into two classes: each hair cell typically has a single **kinocilium** and 20–300 nonmotile **stereocilia.** The kinocilium has a "9 + 2" arrangement of internal microtubules (see Figure 7-28A) that are similar to that of other motile cilia. The stereocilia contain many fine longitudinal actin filaments, and they are thought to be both structurally and developmentally distinct from the kinocilium.

Although the hair cells of the lateral line and of the organs of equilibrium have both a kinocilium and several stereocilia, some hair cells in the adult mammalian ear lack kinocilia. In addition, the technically remarkable feat of microsurgically removing the kinocilia from hair cells that normally have them does not block transduction. From these two observations, it appears that kinocilia must not be necessary for mechanotransduction. The stereocilia of a hair cell are arranged in order of increasing length from one side of the cell to the other (see Figure 7-28B and C). A plane of symmetry through the kinocilium bisects the

stereocilia, making a hair cell bilaterally symmetric, with the top beveled like a hypodermic needle. In most organs, the hair bundles are coupled to some kind of accessory structure through their kinocilia. Stimuli that affect the accessory structure are transmitted to bundles of stereocilia through bonds that connect the accessory structures and kinocilium to the stereocilia. In addition, if the tip of the bundle of stereocilia is touched with a fine probe, the bundle moves as a unit, regardless of the direction of stimulation.

The exact process by which pressure or force from the outside world moves bundles of stereocilia depends on the specific arrangement of hair cells and accessory structures within each sense organ, but ultimately it is the movement of the stereocilia that produces an electrical signal. When the cilia bend toward the tallest cilium, a hair cell depolarizes; whereas, when they bend in the opposite direction, the cell hyperpolarizes (see Figure 7-28D). (If the stereocilia bend to either side, rather than toward or away from the kinocilium, V_m remains the same.)

At rest, about 15% of the channels in a hair cell are open, producing a resting potential of about −60 mV. Hair cells do not produce APs. Instead, they form chemical synapses onto afferent neurons and release neurotransmitter in a graded fashion, depending on V_m in the receptor neuron; the afferent neurons then carry information into the central nervous system. The amount of transmitter released onto the afferent neurons determines their frequency of discharge. Notice that the input-output relation for hair cells is markedly asymmetrical (see Figure 7-28D); that is,

A

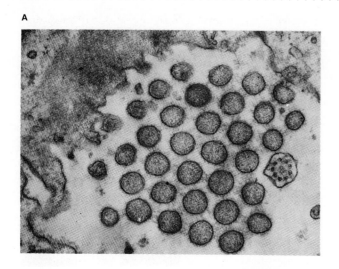

B

2 μm

C

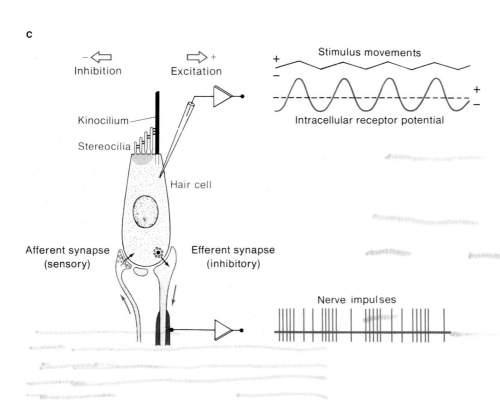

Inhibition

Excitation

Stimulus movements

Kinocilium

Stereocilia

Hair cell

Intracellular receptor potential

Afferent synapse (sensory)

Efferent synapse (inhibitory)

Nerve impulses

D

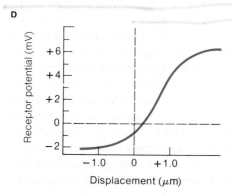

Figure 7-28 *(At left)* The membrane potential of hair cell receptors changes when the cilia are moved away from a rest position. **(A)** Electron micrograph of a cross section through the cilia of a hair cell. The large cilium, containing a typical 9 + 2 structure of microtubules, is the kinocilium; the others are stereocilia. **(B)** Scanning electron micrograph showing the structure of a hair cell from the neuromast of a giant Danio fish. **(C)** Diagram of a typical hair cell showing the anatomical relations of the stereocilia and the kinocilium The hair cell releases transmitter onto an afferent neuron, which carries the sensory signal to the central nervous system. It also receives synapses from efferent neurons. Depending on the direction in which the cilia are bent, the hair cell can either increase or decrease the frequency of APs in the afferent fiber. A linear back-and-forth motion applied to the cilia produces intracellular potential changes that can be recorded with a microelectrode. Extracellular recording from the afferent axon shows APs associated with changes in V_m in the receptor cell. **(D)** Input-output relation for a hair cell. Note that the depolarization produced by movement toward the kinocilium is larger than the hyperpolarization in response to movement away from the kinocilium. [Part A from Flock, 1967; part B courtesy of Christopher Braun; part C adapted from Harris and Flock, 1967; part D adapted from Russell, 1980.]

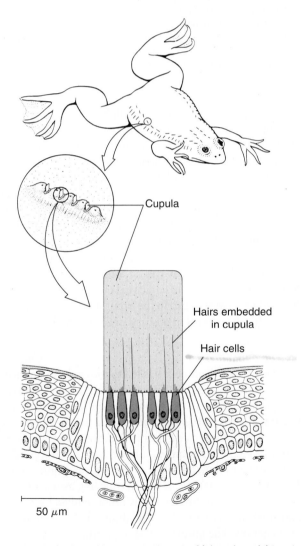

Cupula

Hairs embedded
in cupula

Hair cells

50 μm

Figure 7-29 The lateral-line sensory system of fish and amphibians is based on hair cells. The drawing shows the location of these receptive organs along the body of an African clawed frog, *Xenopus*. The lower diagram shows a cross-section through part of the lateral line, illustrating the cupula, an accessory structure that bends cilia when it is displaced by motion of the surrounding water. Compare the structure of this organ with the hair cells shown in Figure 7-28.

the depolarization produced by a given amount of displacement toward the tallest cilium is larger than the hyperpolarization produced by a similar displacement of the cilia in the opposite direction. This asymmetry is important because, when hair cells are subjected to symmetrical vibrations such as sound waves, changes in the membrane potential can faithfully follow the alternating phases of the stimulus only up to frequencies of several hundred hertz (Hz) but sound frequencies are often much higher than this value. At higher frequencies, the response to the vibrations fuses into a steady depolarization; even if the stimulus displaces the cilia in both directions by equal amounts away from zero displacement, the hair cell will depolarize. This steady depolarization in response to high frequency stimuli produces steady, rather than modulated, transmitter release by the hair cell and, hence, high frequency firing of the afferent neurons. The details of transduction by hair cells are presented later (see *Excitation of cochlear hair cells*).

Organs of Equilibrium

The simplest organ that has evolved to detect an animal's position with respect to gravity or its acceleration is the **statocyst**. Forms of this type of organ are found in a number of animal groups, ranging from jellyfish to vertebrates. (Interestingly, insects lack these sense organs and apparently depend entirely on other senses, such as vision or joint proprioceptors, for orientational information.) A statocyst consists of a hollow cavity lined with ciliated mechanoreceptor cells that make contact with a **statolith**, which can be sand grains, calcareous concretions, or some other relatively dense material (Figure 7-30A). The statolith is either taken up from the animal's surroundings or secreted by the epithelium of the statocyst. For example, a lobster loses its statoliths at every molt and replaces them with new grains of sand. In either case, the statolith must have a higher specific gravity than the surrounding fluid.

As the position of the animal changes, the statolith rests on different regions of the statocyst. When a lobster is tilted to the right about its longitudinal axis, the statolith rests on the receptor cells on the right side of the statocyst, stimulating them and causing a tonic discharge in the sensory fibers from the stimulated receptor cells (Figure 7-30B). Recordings from many different fibers of a statocyst reveal that each cell fires maximally in response to a certain orientation of the lobster (Figure 7-30C). Information from these receptors travels to the central nervous system and sets up reflex movements of the appendages. This pattern of information processing was confirmed in a clever experiment in which molting lobsters were presented with iron filings, rather than with sand. They replaced their statoliths with iron filings, allowing the position of the iron statolith to be manipulated by a magnet. As the magnet was moved through space, pulling on the iron statolith, the lobster—whose position with respect to gravity had not changed—produced a series of compensatory postural responses.

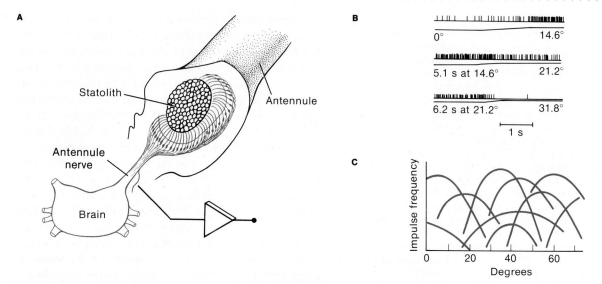

Figure 7-30 Statocysts sense acceleration and the position of an animal with respect to gravity. **(A)** Structure of a statocyst in a lobster. A statolith rests on the receptor part of an array of hair cells. **(B)** Action potentials recorded from individually dissected nerve fibers while the lobster was tilted. Each recording shown here was made from a different fiber. The trace below each recording indicates the time course of the tilt and the angle to which the animal was tilted. **(C)** Frequencies of APs recorded from different fibers plotted as a function of the position of the animal. Each cell responded with a maximum rate of discharge at a different position. [Adapted from Horridge, 1968.]

The Vertebrate Ear

The ears of vertebrates perform two sensory functions, each of which is based on the activity of hair cells. Some structures of the ear, the *organs of equilibrium*, perform like the statocysts of invertebrates, reporting on the animal's position with respect to gravity and acceleration through space. Other structures, the *organs of hearing*, provide information about vibrational stimuli in the environment—stimuli that are called *sound* if they fall within a particular frequency range.

Vertebrate organs of equilibrium

In vertebrates, the organs of equilibrium reside in a membranous labyrinth that develops from the anterior end of the lateral-line system. It consists of two chambers, the **sacculus** and the **utriculus**, that are surrounded by bone and filled with **endolymph**, a specialized fluid. Endolymph differs from most extracellular fluids because it is high in K^+ (about 150 mM in human beings) and low in Na^+ (about 1 mM in human beings); the significance of this unusual composition is considered in the section titled *The mammalian ear*. The utriculus gives rise to the three *semicircular canals* of the inner ear, which lie in three mutually perpendicular planes (Figure 7-31). Hair cells in the three orthogonal semicircular canals detect acceleration of the head. As the head is accelerated in one of the planes of the canals, the inertia of the endolymphatic fluid in the corresponding canal produces a relative motion of the endolymph past a gelatinous projection, the **cupula**, which moves the cupula. When the cupula moves with respect to the cilia of the hair cells at its base, V_m of the hair cells changes. All hair cells in the canal are oriented with the kinocilium on the same side, so all hair cells attached to the cupula are excited when the fluid moves in one direction and inhibited when it moves in the opposite direction. The orthogonal arrangement of the three canals allows them to detect any movement of the head in three-dimensional space.

Below the semicircular canals, larger bony chambers contain three more patches of hair cells called **maculae**. Mineralized concretions termed **otoliths** are associated with the maculae, similar to the statoliths associated with statocysts. The otoliths signal position relative to the direction of gravity; in the lower vertebrates, they can also detect vibrations, such as sound waves, in the surrounding medium. Sensory signals from the semicircular canals are integrated with other sensory input in the brain stem and in the cerebellum for the control of postural and other motor reflexes.

The mammalian ear

Sound in the environment has led to the evolution of hearing in many phyla. Hearing allows an animal to detect predators or prey and to estimate their location and distance while they are still relatively far away. Sound also plays an important role in intraspecific acoustic communication, which often requires subtle tuning of both production and reception. Sound is a mechanical vibration that propagates through air or water, traveling as waves of alternating high and low pressure, accompanied by a back-and-forth movement of the medium in the direction of propagation. The nature of sound, particularly the differences in how it is conducted through air and through water, has set distinct constraints on its detection. The evolution of hearing illustrates many different mechanisms that have evolved to solve the various problems presented by the physical nature of sound. A well-studied example, which we examine here, is the mammalian ear.

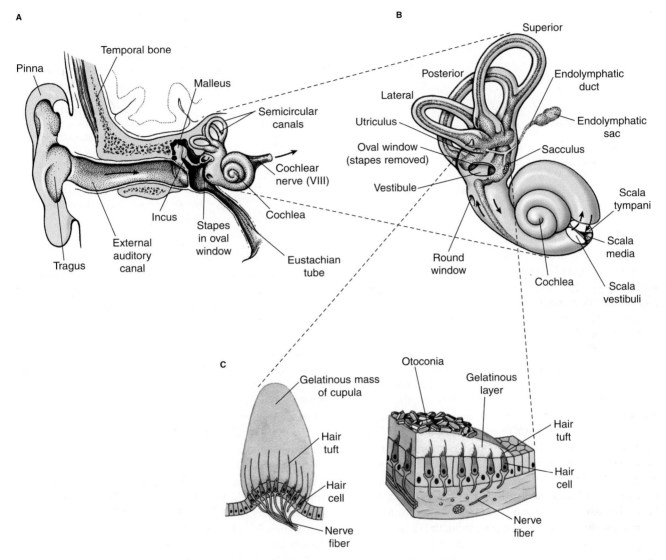

A

Pinna

Temporal bone

Malleus

Semicircular canals

Cochlear nerve (VIII)

Incus

Stapes in oval window

External auditory canal

Tragus

Cochlea

Eustachian tube

B

Superior

Posterior

Lateral

Utriculus

Oval window (stapes removed)

Vestibule

Round window

Endolymphatic duct

Endolymphatic sac

Sacculus

Scala tympani

Scala media

Scala vestibuli

Cochlea

C

Gelatinous mass of cupula

Hair tuft

Hair cell

Nerve fiber

Otoconia

Gelatinous layer

Hair tuft

Hair cell

Nerve fiber

Figure 7-31 The human auditory organs and organs of equilibrium are located in the ear. **(A)** The major parts of the ear. **(B)** The semicircular canals and cochlea. The stapes has been removed to reveal the oval window. The pathway taken by auditory signals is shown by red arrows. At the far right, a section has been removed from the cochlea to reveal the inner structure. (Figure 7-33 shows this structure in more detail.) **(C)** Detailed structure of two parts of the organs of equilibrium. The cilia of receptors in a semicircular canal are embedded in the gelatinous cupula. When fluid moves in the canal, the cupula bends the cilia *(left)*. Particles called otoconia rest on the cilia of receptors in the sacculus (one of the maculae). Changes in the position of the head cause the otoconia to shift position, changing how much the cilia are bent *(right)*. [Parts A and B adapted from Beck, 1971; part C adapted from Williams et al., 1995.]

External ear, auditory canal, and middle ear The structure of the external ear acts as a funnel that collects sound waves in the air from a large area and concentrates the oscillating air pressure onto a specialized surface, the eardrum or **tympanic membrane.** The external structures of the ear—the **pinna** and **tragus**—facilitate the collection of sound waves. The shell-like outer structures, and in some species the mobility of the shell, can modify the directional sensitivity of the auditory system. In some species, including human beings, the acoustical properties of the external ear amplify sound in particular frequency ranges. In addition, the human ear emphasizes the spatial distribution of stimuli by amplifying sounds coming from some directions more than it does sounds from other directions (Figure 7-32).

To be detected, airborne vibrations must be transmitted to the fluid-filled inner ear, where the receptor hair cells reside. The difficulty of communicating across a air-liquid interface can be appreciated by trying to talk with someone who is under water. Most of the sound energy generated in air is reflected back from the water's surface, so it is difficult to generate sufficient energy with airborne sounds to move the water at the required frequency and displacement. This kind of situation is called *acoustical impedance mismatch.* In the ear, this mismatch is partially overcome by a series of three small bones connected in series that are attached to the tympanic membrane at one end and to the oval window of the cochlea at the other. These bones, the **auditory ossicles** (labeled **incus, malleus,** and **stapes** in Figure 7-31A),

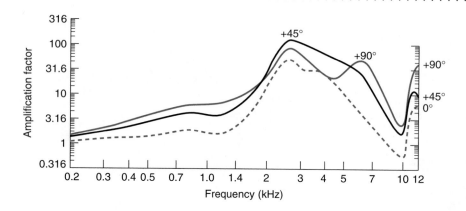

Figure 7-32 The structure of the human pinna and tragus selectively amplifies specific frequencies of sound. This graph shows the gain in pressure at the eardrum over what the pressure would be if sound impinged on the ear canal with all external ear structures removed. If there were no amplification, the graph would be a horizontal line intersecting with the ordinate at an amplification factor of 1. Values above 1 indicate amplification; values below 1 indicate suppression. The gain varies as a function of frequency, and sounds emanating from different directions are amplified differentially. Zero degrees is straight in front of the face. [Adapted from Shaw, 1974.]

evolved from the articulation points of the posterior jaw and are now located in the middle ear. Changes in air pressure produced by sound waves in the external auditory canal cause the eardrum to move, which transfers the energy first to the ossicles, and then to the structures of the inner ear. In the inner ear, the first structure to receive the mechanical input is the **oval window,** which forms the outermost surface of a fluid-filled chamber (the **cochlea**) that contains the receptor hair cells. At the other end of the fluid-filled compartment is another membrane, the **round window.**

There are two important consequences of this arrangement. First, the properties of the mechanical coupling between the eardrum, ossicles, and oval window amplify the signal by about 1.3 times. Second, the pressure of the signal is greatly amplified between the eardrum and the oval window because the eardrum has an area of about 0.6 cm^2, whereas the oval window is smaller, about 0.032 cm^2. This ratio of about 17:1 between the areas of the two membranes means that the sound pressure onto the typanum is concentrated onto the smaller area of the oval window, producing a much greater pressure, which is important because the inertia of the cochlear fluid on the other side of the oval window is greater than that of air. The increase in pressure helps to efficiently transfer airborne vibrations to the cochlear fluid. As a consequence of these two mechanical features, signals arriving at the eardrum are amplified by at least a factor of 22 by the time they reach the cochlea.

Structure and function of the cochlea This mechanically amplified sound input is transduced into neuronal signals by the hair cells of the inner ear. The hair cells of the mammalian ear are located in the **organ of Corti** in the cochlea (Figure 7-33). The movement of fluid in the cochlea causes a vibration of the hair cells, which displaces their stereocilia; the hair cells, in turn, excite the sensory axons of the auditory nerve. The hair cells in the organ of Corti resemble the hair cells of the lateral-line system in lower vertebrates, except that the kinocilium is absent from some cochlear hair cells in adults.

The cochlea, a tapered tube encased in the mastoid bone, is coiled somewhat like the shell of a snail (see

Figure 7-31A and B). It is divided internally into three longitudinal compartments (see Figure 7-33A). The two outer compartments (**scala tympani** and **scala vestibuli**) are connected by the **helicotrema,** an opening located at the apical end of the cochlea (see Figure 7-35B). The scala tympani and scala vestibuli are filled with an aqueous fluid called the **perilymph,** which resembles other extracellular fluids in having a relatively high concentration of Na$^+$ (about 140 mM) and a low concentration of K$^+$ (about 7 mM). Between these compartments—and bounded by the **basilar membrane** and **Reissner's membrane**—is another compartment, the **scala media,** which is filled with *endolymph* (high in K$^+$ and low in Na$^+$), the same type of fluid that surrounds the cilia of hair cells in the organs of equilibrium. The unusual ionic composition of endolymph contributes importantly to the process of auditory transduction. The organ of Corti, which bears the hair cells that transduce auditory stimuli into sensory signals, lies within the scala media and sits on the basilar membrane, and signal transduction by the cochlear hair cells depends in part on this anatomical arrangement.

Among the vertebrates, only mammals possess a true cochlea, although birds and crocodilians have a nearly straight cochlear duct that contains some of the same features, including the basilar membrane and the organ of Corti. The other vertebrates have no cochlea. Some of the lower vertebrates are able to detect sound waves through the activity of hair cells associated with the otoliths of the utriculus and sacculus and with the **lagena,** one of the three maculae in the organs of equilibrium.

The hair cells of the mammalian cochlea encode both the frequency (i.e., the pitch) and the intensity of sound. The adult human cochlea contains four rows of hair cells, one inner and three outer rows, with about 4000 hair cells in each row (see Figure 7-33B). The stereocilia of the hair cells contact the overlying **tectorial membrane.** The cilia are bent by shearing forces (i.e., a force perpendicular to the axis of the cilia) that arise when the hairs move through the gelatinous mucus that coats the tectorial membrane.

Sound vibrations are transferred by the ossicles to the oval window and then pass through the cochlear fluids and the membranes that separate the cochlear compartments

A

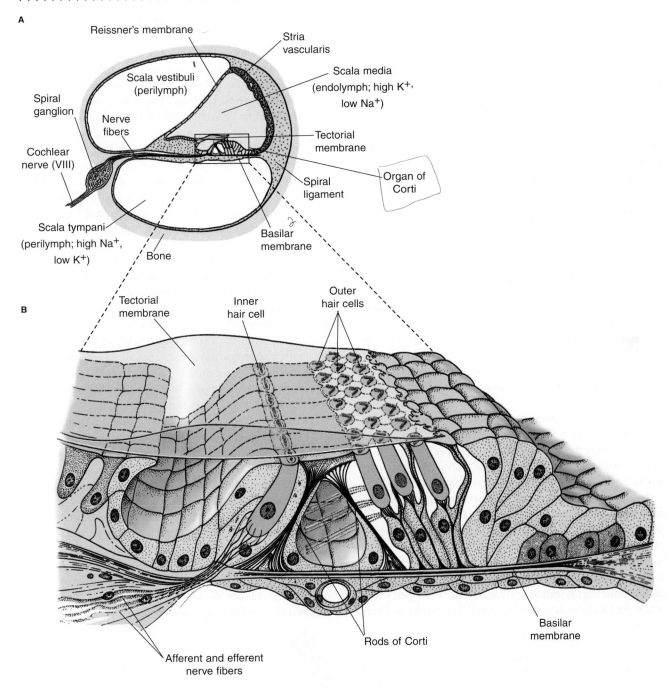

Reissner's membrane

Stria vascularis

Scala vestibuli (perilymph)

Scala media (endolymph; high K⁺, low Na⁺)

Spiral ganglion

Nerve fibers

Cochlear nerve (VIII)

Tectorial membrane

Organ of Corti

Spiral ligament

Scala tympani (perilymph; high Na⁺, low K⁺)

Bone

Basilar membrane

B

Tectorial membrane

Inner hair cell

Outer hair cells

Rods of Corti

Basilar membrane

Afferent and efferent nerve fibers

Figure 7-33 Sound stimuli are transduced by hair cells in the cochlea. **(A)** Cross section through cochlear canal, made at about the location illustrated in Figure 7-31B, showing the two outer chambers (the scala vestibuli and the scala tympani) and the organ of Corti attached to the basilar membrane in the central canal. **(B)** Enlargement of the organ of Corti. The cilia of the hair cells are embedded in the gelatinous layer of the tectorial membrane, whereas their cell bodies are fixed with respect to the basilar membrane.

(both Reissner's membrane and the basilar membrane) before their energy is dissipated through the membrane-covered round window. The distensibility of the round and oval windows is an important adaptation because, if the fluid-filled cochlea were encased entirely by solid bone, the displacements of the oval window, the fluid, and the internal tissues would be very small. The distribution of the perturbations within the cochlea depends on the frequencies of vibrations entering the oval window. To visualize this, imagine a displacement of the eardrum transferred through the ossicles of the middle ear to the oval window. Vibrations displace the incompressible perilymph along the scala vestibuli, through the helicotrema, and back through the scala tympani toward the round window.

Excitation of cochlear hair cells Electrical recordings made at various locations in the cochlea show fluctuations in electrical potential that are similar in frequency, phase, and amplitude to the sound waves that produced them. These *cochlear microphonics* result from the summation of receptor currents from the numerous hair cells that were stimulated by movements of the basilar membrane. The actual transduction event occurs when a perturbation of the basilar membrane forces the tips of the stereocilia to bend laterally, because the basilar membrane has moved with respect to the tectorial membrane (Figure 7-34). This mechanical deflection directly causes ion channels in the tips of the stereocilia to open. In the past few years, our understanding of these events has grown dramatically, although questions still remain about the details of transduction.

The perceptual threshold of cochlear hair cells corresponds to a deflection of 0.1–1.0 nm, which is equivalent to a change in membrane current of only about 1 pA through ion channels in the hair cell membrane. These channels have been experimentally shown to be permeable to many small, monovalent cations (e.g., Li^+, Na^+, K^+, Rb^+, and Cs^+). When they open *in vivo*, K^+ ions and some Ca^{2+} ions enter the cell from the endolymph. (The high concentration of K^+ in the endolymph produces a net inward driving force on K^+, unlike the usual situation in which $V_m - E_K$ is an outward force. This inward K^+ current depolarizes hair cells, because it adds positive charge to the inside of the cells.)

On the basis of measurements of current flow, it has been estimated that there are about 30–300 channels per bundle of stereocilia, which implies that as few as one to five channels per stereocilium may be responsible for transduction. The channels are thought to be opened directly by a mechanical stimulus because, when isolated bundles of stereocilia are abruptly deflected in experiments, the transduction current increases with an extremely short latency (about 40 μs). The brevity of this latency period makes it unlikely that an enzymatic or biochemical step is included in this process. This interpretation is reinforced by patch-clamp experiments indicating that the channels open faster when the deflection is larger, again suggesting a direct

mechanical influence on the conformational states of the channel.

Several factors affect the sensitivity of hair cells. Each hair cell in the cochlea appears to be tuned to a particular band of sound frequency as a result of both mechanical and channel properties. Each cell has a resonance frequency that is determined by the length of the stereocilia in the hair bundle. Cells with long hairs are most sensitive to low-frequency sounds, whereas cells with short hairs are tuned to high-frequency sounds. In addition, each cell responds maximally to a particular frequency of electrical stimulation. This electrical resonance frequency is determined by the balance of currents through voltage-gated Ca^{2+} channels and through Ca^{2+}-sensitive K^+ channels in the basal membrane (which is exposed to perilymph).

The outer hair cells of the cochlea may contribute to tuning in the cochlea by modifying the *mechanical* properties of the organ of Corti. The outer hair cells make few afferent connections, but they receive a large number of efferent synapses. When these cells are stimulated electrically during experiments, they shorten when depolarized and elongate when hyperpolarized. Thus, it is possible that the outer hair cells *could* modify the mechanical coupling between the inner hair cells and the tectorial membrane, which would change transduction. It remains to be demonstrated that this mechanism actually affects audition.

Hair cells adapt to changes in the position of their stereocilia, a process that has been particularly well studied in the bullfrog sacculus. When the cilia of frog hair cells are deflected by a probe and held at the new position, the operating range of the cell adapts within milliseconds to this new tonic position, causing the hair cell to then respond to small changes in position away from this new set point. Calcium ions have been shown to play a pivotal role in the process, evidently by modifying the tension in the spring that opens the transduction channels. Finally, efferent input onto a hair cell can decrease the cell's response to sound and broaden its frequency selectivity by opening inhibitory

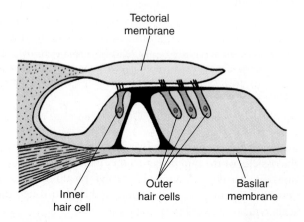

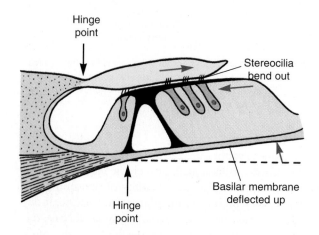

Figure 7-34 Movement of the basilar membrane with respect to the tectorial membrane produces shear on the stereocilia of cochlear hair cells. The tectorial membrane slides over the organ of Corti, because the tectorial membrane and the basilar membrane pivot about different points when they are displaced by waves traveling along the cochlea. The movements are greatly exaggerated in this diagram. [Adapted from Davis, 1968.]

K⁺ channels, which short-circuit the cell's electrical resonance. Taken together, the attributes of the hair cells reveal their exquisite tuning. However, all of the adaptations that make hair cells so extremely sensitive also make them highly vulnerable to overstimulation, which can cause rupture at the base of the stereocilia. Acoustic trauma can produce permanent hearing loss that is worst at the frequencies of sound that actually damaged the hair cells. Although some cold-blooded vertebrates can recover from such trauma, the loss is permanent in mammals.

 Can there be signal amplification in sensory receptor cells in which the stimulus directly opens ion channels without any intermediary intracellular messengers? What could provide a source of energy to generate the amplification?

The receptor currents of the hair cells faithfully transduce the movements of the basilar membrane over the whole spectrum of audible sound frequencies. The cells transmit their excitation through chemical synapses onto sensory axons of auditory neurons that have cell bodies located in the **spiral ganglion**. Release of neurotransmitter by the hair cells modulates the firing rate of these axons, which travel in the *vestibulocochlear* (eighth cranial) nerve and synapse onto neurons in the cochlear nucleus. In fact, the hair cells in the inner row receive about 90% of the contacts made by neurons of the spiral ganglion, suggesting that the inner row of cells is largely responsible for detecting sound. In contrast, hair cells of the outer three rows receive many efferent synapses and may participate in modulating the sensitivity of the cochlea by changing the mechanical relation between the basilar and tectorial membranes.

Frequency analysis by the cochlea Pioneering studies of the exposed cochlea, carried out by Georg von Békésy, greatly enhanced our understanding of how the auditory system can encode information about the frequency of stimuli. His studies showed that:

1. In response to a pure sine wave tone, the perturbations of the basilar membrane have the same frequency as the tone.
2. Low frequency perturbations move as a traveling wave along the whole length of the basilar membrane.
3. The location where the basilar membrane is displaced maximally by a tone is a function of the frequency of the tone. High frequencies displace only the initial parts of the membrane, whereas low frequencies displace more distant parts.

Thus, each point along the basilar membrane is displaced *most* effectively by some unique frequency, and that point

varies in an orderly fashion with higher frequencies displacing the basilar membrane close to the oval window and lower frequencies displacing the basilar membrane farthest from the oval window. For sounds up to about 1 kHz, APs in the auditory sensory axons appear to follow the fundamental frequency. Above this level, the time constant of the hair cells and the electrical properties of the axons in the auditory nerve prevent a one-to-one correspondence between sound waves and electrical signals. In this higher frequency range, some other mechanism must inform the central nervous system of the sound frequency.

Hermann von Helmholtz noted in 1867 that the basilar membrane consists of many transverse bands that increase gradually in length from the proximal end to the apical end of the basilar membrane (from about 100 μm long at the base to about 500 μm long at the apex), which reminded him of the strings of a piano and led him to propose his resonance theory. He proposed that various locations along the basilar membrane vibrate in resonance with a specific tonal frequency while the other locations remain stationary, just as the appropriate string of a piano resonates in response to a tone from a tuning fork. This theory was later challenged by von Békésy (1960), who found that the movements of the basilar membrane are *not* standing waves, as Helmholtz suggested, but consist instead of **traveling waves** that move from the narrow base of the basilar membrane toward the wider apical end (Figure 7-35). These waves have the same frequency as the sound entering the ear, but they move much more slowly than sound moves through air.

A familiar example of a traveling wave can be seen by moving the free end of a rope that is secured at the other end. Unlike a rope, however, the basilar membrane has mechanical properties that change along its length. The **compliance** of the membrane (the amount that the membrane will stretch in response to a given amount of force) increases from its narrow end to its broad end, which causes the amplitude of a traveling wave to change along the length of the membrane (see Figure 7-35). The position along the cochlea at which the displacement of the basilar membrane is maximal—causing the hair cells at that location to be stimulated maximally—depends on the frequency of the traveling waves and consequently on the frequency of the stimulating sound. The traveling waves produce maximum displacements near the basal end of the cochlea when the stimulating sound has a high frequency. The region of maximal displacement moves along the basilar membrane toward the apex as the frequency of the sound drops. The extent of membrane displacement at any point along the basilar membrane determines how strongly the hair cells are stimulated, so it also determines the rate of discharge in sensory fibers arising from different parts of the basilar membrane. Even at maximum amplitude, all of the movements are very tiny: the loudest sounds produce displacements of the basilar membrane of only about 1 μm. The movement of the hair cell cilia is much smaller, and the threshold of stimulus detection is at the limit of the thermal noise.

A

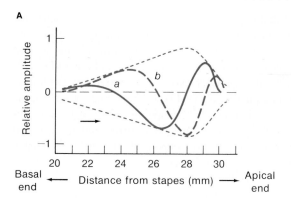

B

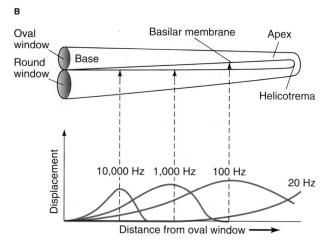

Figure 7-35 Sound sets up traveling waves along the basilar membrane. **(A)** Waves move in the direction shown by the arrow. Lines *a* and *b* indicate the shape of the membrane at two different times. The light dashed lines indicate the envelope generated by the movements, which in this case has the largest amplitude near the apical end. (The amplitudes of the waves are greatly exaggerated in this figure). **(B)** The cochlea drawn as if it had been straightened out. The locations that respond maximally to sounds of different frequency are indicated below. [Part A adapted from Von Bekesy, 1960; part B adapted from Moffett et al., 1993.]

An Insect Ear

Many organisms have ears that operate differently from the mammalian ear, and it is instructive to consider at least one of them to know the variation possible. Crickets find their mates through auditory communication: male crickets produce a song whose pattern is specific to their species, and female crickets are attracted to the song of their species. Cricket ears are located on the first thoracic legs, and they are associated with the respiratory passages, called **tracheae** (Figure 7-36). Each ear contains a **tympanum,** analogous in function to the tympanic membrane of the mammalian ear, and the changes in air pressure that constitute sound are carried through the tracheae to the tympanum. The tympanum is exposed to changes in air pressure from both outside the animal and from inside through the tracheae. If a sound arises on the right side of the cricket, it will directly cause the tympanum on the right side to vibrate. In addition, it will be carried through the tracheal system to the left tympanum, causing the left tympanum to vibrate as well.

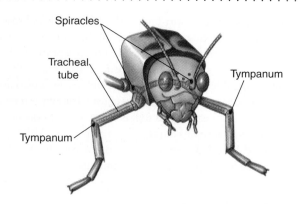

Figure 7-36 A cricket's ears are located on its anterior thoracic legs. The tympanum receives sound stimuli through the tracheal ducts and vibrates in response either to sounds from the outside or to sounds carried through the inside of the animal through the tracheae. Nerve cells associated with the tympanum transduce the sound stimuli.

The differences in the time at which the stimulus reached the right and the left tympani can be used to help localize sounds, a principle that is used by vertebrates as well (see Chapter 11). In some species, hair cells are associated with the tympanum, suggesting that excitation within the insect ear may be similar to excitation in the mammalian ear. The insect ear incorporates some of the features of the mammalian ear: a channel conducts sound waves to a movable surface, which vibrates in response to the sound waves. When the tympanum vibrates, it excites receptors either directly or indirectly, sending signals to the central nervous system. However, the tracheal system allows sounds to travel through the body of the animal and to move the tympanum from either the inside or the outside of the animal's body.

ELECTRORECEPTION

Hair cells located in the skin of certain species of bony and cartilaginous fishes have lost their cilia and become modified for the detection of electric currents in the water. The sources of these currents are either the fishes themselves or currents that originate in the active tissues of other animals in the vicinity. The weakly electric fishes (such as the *Mormyrids*) possess specialized electric organs that generate the fields sensed by these receptors; they can use the fields for communicating with one another and for navigating in turbid water. In fact, all electrically active tissue can generate electrical fields, and some sharks are especially adept at locating their prey by sensing the electric currents emanating from the active muscles of the animal. The electroreceptors of fishes are distributed throughout the head and body in the lateral-line system (Figure 7-37A).

In weakly electric fishes (as opposed to strongly electric fishes, such as the electric eel), electrical pulses produced by modified muscle or nerve tissue at one end of the body reenter the fish through epithelial pores in the lateral-line system. At the base of each pore, the current encounters an electroreceptor cell (Figure 7-37B), which makes synaptic

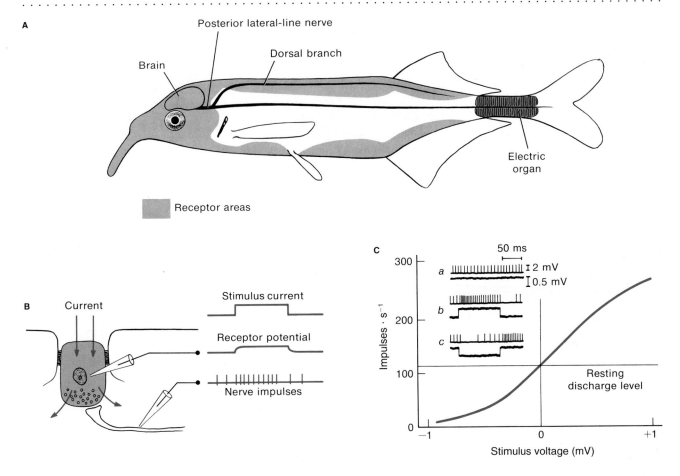

Figure 7-37 Electroreceptor cells are specialized hair cells located along the lateral line of many species of fishes. **(A)** Positions of the electric organ and of the lateral-line nerve trunk and distribution of electroreceptor pores in the weakly electric fish *Gnathonemus petersii*. **(B)** At the base of each electroreceptor pore there is an electroreceptor cell whose apical membrane has a low electrical resistance compared with that of the basal membrane. **(C)** Receptor cells release transmitter molecules spontaneously *(a)*. Current entering the cell *(b)* depolarizes it, increasing the rate of release and, hence, the frequency of APs in the sensory fiber innervating the cell. Current leaving the cell *(c)* decreases the rate of release. The amount of transmitter released by the receptor cells changes when V_m is altered by only a few microvolts. [Adapted from Bennett, 1968.]

contact with axons of the eighth cranial nerve that innervate the lateral-line system. The cell membrane facing the exterior has a lower electrical resistance than does the basal membrane, so most of the potential drop caused by the current moving across the cell occurs across the basal membrane, depolarizing it. Depolarization of the basal membrane activates Ca^{2+} channels in the membrane, and the resulting influx of Ca^{2+} at the base of the cell increases the release of synaptic transmitter by the receptor cell. This transmitter increases the frequency of APs in the sensory fiber that innervates the receptor. Conversely, a current flowing out of the body of the fish hyperpolarizes the basal membrane of the receptor cell and decreases the release of transmitter below the spontaneous rate. Thus, the firing frequency in the sensory fiber goes up or down, depending on the direction of the current flowing through the electroreceptor cell (see Figure 7-37B and C). The sensitivity of these receptors and their sensory fibers, like that of the hair cells of the vertebrate ear, is truly remarkable. As seen in Figure 7-37C, changes in sensory nerve discharge occur in response to changes in V_m of the receptor cell of as little as several microvolts.

The train of current pulses flows through the water from the posterior to the anterior end of the fish (Figure 7-38). Any object whose conductivity differs from that of water will distort the lines of current flow. The lateral-line electroreceptors detect the distribution of current flowing back into the fish through the lateral-line pores on the head and anterior end of the body and can detect the changes in field produced by objects in the water. This sensory information is then processed in the greatly enlarged cerebellum of the fish, enabling it to detect and locate objects in its immediate environment.

Electrical signals are produced by other species of fishes and used for quite a different task. In contrast with the weakly electric fishes that use electrical fields for navigation and signaling, some eels, torpedoes, and other fishes produce a powerful discharge of current to stun enemies and prey. These strongly electric fishes produce a continuous series of synchronous, relatively high frequency

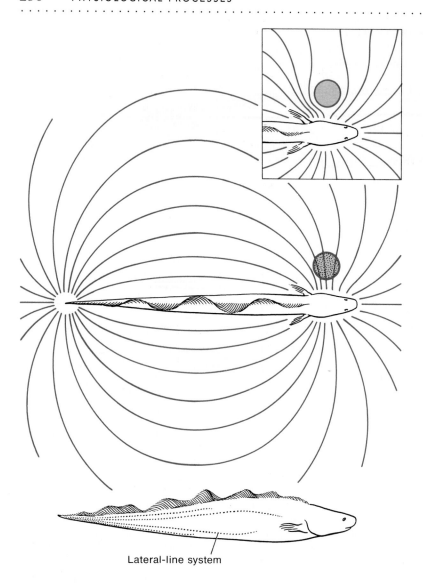

Lateral-line system

depolarizations with their electric organs, and the way in which the electrical discharges of these fishes are generated and used resembles the way in which muscles are controlled to produce movement (see Chapter 10).

THERMORECEPTION

Temperature is an important environmental variable, and many organisms acquire sensory information about temperature from the action of specialized nerve endings, or *thermoreceptors*, in the skin. Higher-order neurons receive input from thermoreceptors and contribute to the mechanisms that regulate the temperature of the body (see Chapter 16). In addition, some of the neurons in the hypothalamus of vertebrates are able to detect changes in body temperature.

Temperature receptors can be remarkably sensitive. The infrared (radiant heat) detectors in the facial pits of rattlesnakes provide an example (Figure 7-39A). The receptor membrane consists of the branched endings of sensory nerve fibers, with no readily apparent structural specializa-

tions, and the endings appear to detect changes in tissue temperature, rather than the radiant energy itself. The mechanisms by which temperature changes can alter receptor output are not known. The sensory axons from the pit organs of the rattlesnake increase their firing rate transiently when the temperature inside the pit increases as little as 0.002°C, and this change in receptor firing rate can modify behavior. For example, a rattlesnake can detect the radiant heat emitted by a mouse that is 40 cm away if the body temperature of the mouse is at least 10°C above the ambient temperature. Furthermore, the temperature receptors lie deep within the facial pits, and this arrangement allows the snake to detect the direction of a source of radiant heat (Figure 7-39B).

Both the external skin and the upper surface of the tongue of mammals contain two kinds of thermoreceptors: those that increase their firing when the skin is warmed ("warmth" receptors) and those that increase their firing when the skin is cooled ("cold" receptors). These receptors, too, are quite sensitive. Human beings can detect a change in skin temperature of as little as 0.01°C. The two cate-

A

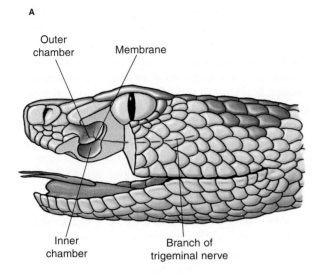

Outer chamber

Membrane

Inner chamber

Branch of trigeminal nerve

B

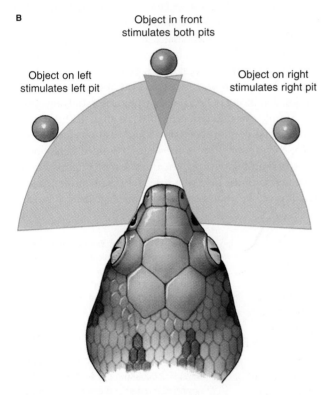

Object in front stimulates both pits

Object on left stimulates left pit

Object on right stimulates right pit

Figure 7-39 The facial pits of rattlesnakes contain extremely sensitive thermoreceptors. **(A)** Structure of a facial pit in the rattlesnake *Crotalus viridis*. **(B)** The position of the facial pits makes thermoreception by the pit organs directionally sensitive. [Adapted from Bullock and Diecke, 1956.]

gories of thermoreceptors are distinguished from one another in accord with how they respond to temperature changes near the normal temperature of the human body (about 37°C). Both warmth and cold receptors increase their firing rate as the temperature becomes increasingly different from 30°–35°C (Figure 7-40A): warmth receptors fire faster as the temperature gets warmer; cold receptors fire faster as the temperature gets colder. However, when the temperature becomes sufficiently different from

A

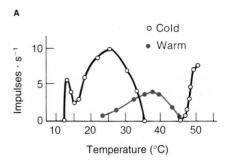

○ Cold
● Warm

B

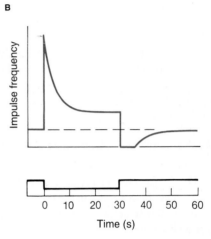

Figure 7-40 The frequency of APs in mammalian thermoreceptors varies with temperature at the surface of the body. **(A)** The steady-state firing rate of cold and warmth receptors that arborize in the surface of a mammal's tongue. **(B)** Time course of a cold receptor's response when the tongue was first cooled and then rewarmed as shown in black. [After Zotterman, 1959.]

30°–35°C, this pattern changes for both kinds of receptors, and the frequency of APs drops (see Figure 7-40A). The response of the thermoreceptors consists of a large transient change in firing rate, followed by a longer-lasting, steady-state phase. The transient phase is an accurate response to any change in temperature (Figure 7-40B), even though the steady-state phase behaves as shown in Figure 7-40A.

VISION

Since the Earth formed more than 5 billion years ago, sunlight has been an extremely potent selective force in the evolution of living organisms, and most organisms are able to respond to light in some way. Photoreception consists of transducing photons of light into electrical signals that can be interpreted by the nervous system, and photoreceptive organs—typically called eyes—have evolved into many shapes and sizes and with many distinct designs. Interestingly, although the physical structure of eyes varies greatly among species, visual transduction is based upon a very highly conserved set of protein molecules that provide an optical pathway leading light to the photoreceptive surface and that capture photons within the photoreceptors.

This conservation of visual molecules suggests that, when suitable biochemical means had evolved to solve the problem of capturing light's energy, the sequences were conserved even though they were packaged into organs with highly diverse structural properties. For example, the *opsins* are protein visual pigment molecules. Each molecule includes seven transmembrane domains. Opsins are coupled to **photopigment** molecules, which are structurally altered by the absorption of photons and which in turn modify the properties of the opsin protein (see Figure 7-3). Opsins are found widely in the animal kingdom, even in photoreceptive structures that are extremely simple and that lack the features that would constitute an eye. In many organisms, the structure of the eye has evolved to collect and focus incident light rays before they arrive at the site of transduction. Eyes refract light through highly concentrated soluble proteins that are formed into lenses, and these refractive structures also have an interesting evolutionary history. We first consider how eyes collect and focus light.

Optic Mechanisms: Evolution and Function

The physics of light tightly constrains the structure of an eye that will produce a usable image. Most of the possible designs have been "discovered" in the course of evolution, giving rise to similar structures in unrelated animals. One of the most well known examples of convergent evolution is the similarity of eyes in the phylogenetically unrelated squid and fishes. These eyes are similar in many details because optical laws have dictated convergent solutions to the problem of seeing under water. In contrast, the eyes of human beings and fishes are similar because they share common evolutionary descent, although they differ to some extent because the two species live in different optical media.

The evolution of eyes has proceeded in two stages. Virtually all major animal groups have evolved simple eyespots consisting of a few receptors in an open cup of screening pigment cells (Figure 7-41A). Some biologists estimate that such photon detectors have evolved independently between 40 and 65 times. Eyespots provide information about the surrounding distribution of light and dark, but they do not provide enough information to allow detection of either predators or prey. For pattern recognition or for controlling locomotion, animals need an eye with an optical system that can restrict the acceptance angle of individual receptors and form some kind of image. This stage of optical evolution happened less frequently, occurring in only 6 of the 33 metazoan phyla (Cnidaria, Mollusca, Annelida, Onychophora, Arthopoda, and Chordata). Because these phyla contain about 96% of all extant species, it is tempting to speculate that having eyes confers significant selective benefits.

Ten optically distinct designs for image-forming eyes have been discovered to date. They include nearly all the possibilities known from physical optics except Fresnel and zoom lenses. In addition, there are some variations, such as array optics, that have not been used by physicists studying optics.

Simple eyespots are typically less than 100 μm in diameter and contain between 1 and 100 receptors. Even simple eyespots allow some visually guided behavior. In protozoa and flatworms, the direction of a light source is detected with the help of a screening pigment that casts a shadow on the photoreceptors. Some flagellates, for exam-

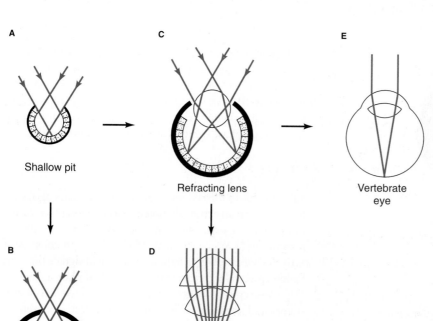

A

Shallow pit

B

Pinhole eye

C

Refracting lens

D

Many lenses in series

E

Vertebrate eye

Figure 7-41 The structures of eyes incorporate many different optical principles. **(A)** The simplest eye consists of a shallow open pit lined with photoreceptor cells. **(B)** In slightly more complicated eyes, the aperture of the eye is small in proportion to the size of the eye, and the eye operates similarly to a pinhole camera. **(C)** An alternative improvement allowing image formation is the addition of a refracting element between the aperture and the layer of photoreceptors. **(D)** Three lenses arranged in series improve the optical properties of the eye in *Pontella*, a copepod. **(E)** The vertebrate eye is an elaboration of a simple eye to which a small aperture and a lens have been added. Arrows indicate possible evolutionary relations among types of eyes. [Adapted from Land and Fernald, 1992.]

ple, have a light-sensitive organelle, near the base of the flagellum, that is shielded on one side by a pigmented eyespot. This shielded organelle provides a crude, but effective, indication of directionality. As the flagellate swims along, it rotates about its longitudinal axis roughly once per second. If it enters a beam of light shining from one side and perpendicular to its path of locomotion, the eyespot is shaded each time the shielding pigment passes between the source of the light and the photosensitive part of the base of the flagellum. Each time this happens, the flagellum moves just enough to turn the flagellate slightly toward the side bearing the shielding pigment. The net effect is to turn the flagellate toward the source of the light.

The simplest eyes are improvements on eyespots that have been achieved by reducing the size of the aperture to produce a pinhole eye (Figure 7-41B) or by adding a refracting structure (Figure 7-41C). The evolutionarily ancient cephalopod mollusc *Nautilis* has a pinhole eye that, except for the absence of a lens, is quite advanced. It is nearly 1 cm in diameter and the aperture is variable, expanding from 0.4 to 2.8 mm. In addition, extraocular muscles compensate for the rocking motion produced when the animal swims, stabilizing the eye.

Most aquatic animals have a single-chambered eye with a spherical lens (see Figure 7-41C). This type of lens provides the high refractive power needed to focus images under water, but it poses the problem of spherical aberration. The lenses found in fishes and cephalopods avoid this difficulty because the material of the lens is not homogeneous. Instead, it is dense with a high refractive index in the center and has a gradient of decreasing density and refractive index toward the periphery. This pattern was first noted in 1877 by Matthiessen, who showed that a consequence of the density gradient is a short focal length, about 2.5 times the radius (known as Matthiessen's ratio). This remarkable gradient in density has evolved eight times among aquatic animals, suggesting that it is a very good solution and perhaps the simplest. Other aquatic species have eyes with multiple lenses. For example, the eye of the copepod *Pontella* (Figure 7-41D) contains three lenses in series that together correct for spherical aberration.

The vertebrate eye (Figure 7-41E) combines a relatively small aperture with a refractile lens. These two features together provide a very high quality image that is focused on the layer of photoreceptors in the **retina,** located at the back of the eye.

Compound Eyes

The compound eyes of arthropods are image-forming eyes composed of many units, each of which has the features of the eye shown in Figure 7-41C. Each optic unit, called an ommatidium, is aimed at a different part of the visual field (Figure 7-42A), and each samples an angular cone that takes in about 2–3 degrees of the visual field. In contrast, in the vertebrate eye, each receptor may sample as little as 0.02 degree of the visual field. Because the receptive field of

Compound eye

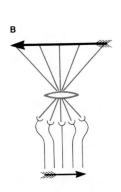

Vertebrate eye

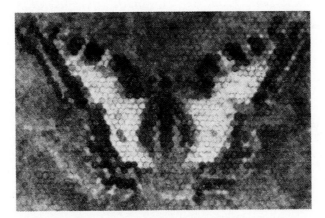

Figure 7-42 Compound eyes produce mosaic images. **(A)** In a compound eye, each ommatidium samples a different part of the visual field through a separate lens. The image at the right illustrates a mosaic image of a butterfly as it would be perceived by a dragonfly at a distance of 10 cm. **(B)** In a simple eye, each receptor cell samples part of the field through a lens that is shared by all receptor cells. For comparison, the image at the right illustrates the same butterfly as it would be perceived by a simple vertebrate eye. Arrows indicate that the optics of the vertebrate eye invert the image on the retina, whereas the optics of the compound eye do not. [Adapted from Kirschfeld, 1971, and Mazokhin-Porshnyakov, 1969.]

each unit in a compound eye is relatively large, compound eyes have lower visual acuity than do vertebrate eyes. However, although the mosaic image formed by this type of eye is coarser than the image produced by a vertebrate eye (Figure 7-42B), it is certainly recognizable.

The eyes of Limulus

Each ommatidium of a compound eye contains several photoreceptors. The most intensively studied invertebrate photoreceptors are those in the *lateral eyes* and the *ventral eye* of the horseshoe crab *Limulus polyphemus* (Figure 7-43). The two lateral eyes of *Limulus* are typical compound eyes, similar to those in Figure 7-42A, whereas the unpaired ventral eye is simpler in structure and more like the eyespot shown in Figure 7-41A. Most of the early electrical recordings made from single visual units were done with this lateral eye, because the eye was experimen-

tally accessible and its activity could be monitored with simple electrical recording techniques.

The visual receptor cells of the *Limulus* compound eye are located at the base of each ommatidium (Figure 7-43B and C). Each ommatidium lies beneath a hexagonal section of an outer transparent layer, the *corneal lens.* The primary photoreceptors are 12 **retinular cells,** which surround the dendrite of another neuron, the *eccentric cell.* Each retinular cell has a **rhabdomere,** in which the surface membrane of the cell is thrown into a dense profusion of **microvilli,** which are miniature tubular evaginations of the surface membrane (see Figure 7-43D). The microvilli greatly increase the surface area of the cell membrane in the rhabdomere. Light enters through the lens and is absorbed by molecules of the photopigment rhodopsin that are located in the receptor membrane within the rhabdomere. Transient, random depolarizations of the membrane po-

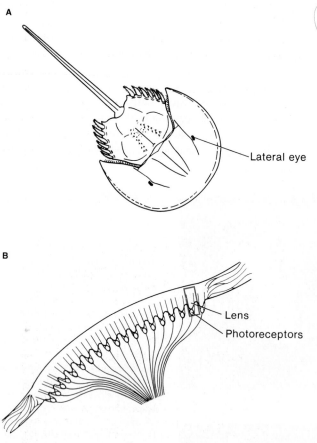

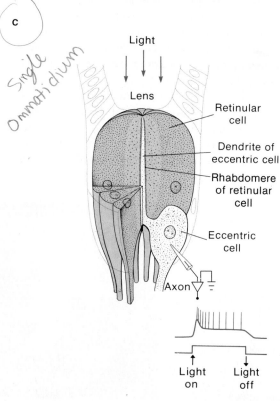

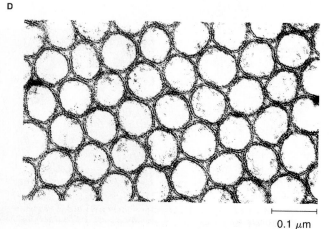

Figure 7-43 Early studies of the compound eyes of *Limulus polyphemus* provided insights into visual transduction. **(A)** The lateral eyes of the horseshoe crab *Limulus* are located on the dorsal carapace. **(B)** A cross section through a lateral eye, which is made up of ommatidia. **(C)** The structure of a single ommatidium (outlined in red in part B). Light enters through the lens and is intercepted by visual pigment in the rhabdomeres of the retinular cells. The cells are arranged like the segments of an orange around the dendrite of the eccentric cell. The eccentric cell depolarizes and generates APs when light shines on the rhabdomeres. **(D)** An electron micrograph of a cross section through the microvilli of a rhabdomere. [Part C from "How Cells Receive Stimuli," by W. H. Miller, F. Ratliff, and H. K. Hartline. Copyright © 1961 by Scientific American, Inc. All rights reserved. Part D courtesy of A. Lasansky.]

tential occur in the retinular cells when the eye is exposed to very dim, steady illumination. These "quantum bumps" in the recording increase in frequency when the light intensity is gradually increased, which causes more photons to impinge on the receptors. The transient depolarizations are electrical signals generated by the absorption of individual quanta of light by single photopigment molecules. A single photon captured by a single visual pigment molecule in *Limulus* produces a receptor current of 10^{-9} A. This transduction event amplifies the energy of the absorbed photon between 10^5 and 10^6 times.

How can capture of a single photon lead to the rapid release of so much energy? In this case, the amplification occurs through a cascade of chemical reactions inside the cell that include G-protein activation (see *From Transduction to Neuronal Output* earlier in this chapter). The net effect is to open ion channels, allowing cations to enter the cell. In *Limulus,* the receptor current through the light-activated channels is carried by Na^+ and K^+. This current causes a *depolarizing* receptor potential, by a mechanism similar to the depolarizing postsynaptic potential that is generated when acetylcholine activates the motor endplate channels in muscle (see Chapter 6). When the light goes off, these channels close again, and the membrane repolarizes. The sensitivity of individual photoreceptors drops with exposure to light, and this adaptation is thought to be mediated by Ca^{2+} ions, which enter the cells when light opens ion channels and which then reduce the current through light-activated channels.

Although retinular cells have axons, they apparently do not produce APs. Instead, the receptor current arising in retinular cells spreads through low-resistance gap junctions into the dendrite of the eccentric cell, and from there the depolarization spreads to the eccentric cell axon where it generates APs. The APs are conducted in the optic nerve to the central nervous system. Although the organization of the *Limulus* eye is simple in comparison with that of vertebrate eyes, the *Limulus* visual system is capable of generating electrical activity that parallels some of the more sophisticated features of human visual perception (Spotlight 7-1).

Perceiving the plane of polarized light

The arrangement of cells within the ommatidia of compound eyes confers special abilities on some arthropods. For example, some insects and crustaceans can orient behaviorally with respect to the sun, even when the sun itself is blocked from their view. This ability depends on the polarization of sunlight, which is different in different parts of the sky. It has been found that many arthropods can detect the plane of the electric vector of polarized light entering the eye, and some use this information for orientation and navigation. Measurements of the birefringence (the ability of a substance to absorb light polarized in various planes) of the retinular cells in the crayfish show that the absorption of polarized light is maximal when the plane of the electric vector of light is parallel to the longitudinal axis of the microvilli of the rhabdomeres. Each crayfish ommatidium consists of seven cells, and the rhabdomeres of the seven retinular cells interdigitate, forming the rhabdome. Within the rhabdome, the microvilli of some receptors are oriented at 90 degrees to the microvilli of a second group of receptors (Figure 7-44). If the photopigment molecules

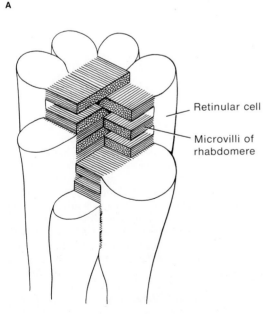

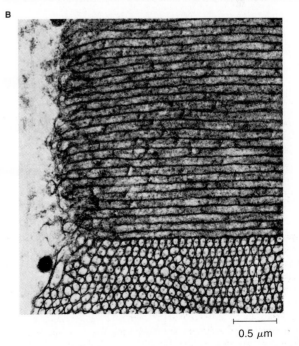

0.5 μm

Figure 7-44 The structure of ommatidia allows some arthropods to perceive the plane of polarized light. **(A)** The interdigitating rhabdomeres of separate retinular cells produce two sets of mutually perpendicular microvilli. **(B)** Electron micrograph of a section through the rhabdome formed by two sets of microvilli. The upper microvilli were sectioned parallel to their longitudinal axis, and those in the lower set were sectioned perpendicular to their longitudinal axis. [Part A adapted from Horridge, 1968; part B courtesy of Waterman et al., 1969.]

Retinular cell

Microvilli of rhabdomere

SPOTLIGHT 7-1

SUBJECTIVE CORRELATES OF PRIMARY PHOTORESPONSES

Studies on the *Limulus* eye carried out by H. Keffer Hartline and his associates in the 1930s revealed correlations between the activity in the photoreceptors and the parameters of the stimuli. Although these receptors differ in some features from human photoreceptors, they are similar in fundamental ways, such as the chemical identity of their visual pigment and the electrical properties of the cells. One of the most interesting results of Hartline's work is that many features of human visual perception, measured in psychophysical experiments, parallel the electrical behavior of single *Limulus* visual cells. This relation suggests that some properties of visual perception originate in the behavior of the photoreceptor cells themselves and remain relatively unmodified as information undergoes further processing by the nervous system. For example:

1. The frequency of APs recorded from the axons of single ommatidia is proportional to the logarithm of the intensity of the stimulating light (shown at the right in part A of the adjoining figure). This logarithmic relation is also typical of the judgments made by a human subject who is asked to compare the intensities of different light stimuli.

2. A receptor's response to flashes of light that are less than 1 second in duration is proportional to the total number of photons in the flash, regardless of the actual duration. That is, the number of impulses generated remains constant as long as the product of intensity and duration is kept constant. This result might be expected, because the response should be determined, within some limits, by the number of photopigment molecules isomerized by photons impinging on the receptor. For short flashes, a human observer cannot tell the difference when intensity and duration of the flash are changed reciprocally.

3. If a photoreceptor is stimulated with a flickering light, V_m will follow the frequency of the flashes up to nearly 10 Hz (part B of the adjoining figure). Beyond this frequency, the receptor potential can no longer follow the flashes; instead, the ripples in V_m fuse into a steady level of depolarization (see Figure 7-55 also). Action potentials in sensory fibers no longer follow the patterning of the flashes but, instead, are generated at a steady rate. When the patterning of APs no longer conforms to the frequency of flicker, the message sent to the central nervous system indicates that the light is constant, even though the actual stimulus is not. In fact, human beings cannot tell the difference between a steady light and one that flickers at a rate higher than the frequency that can no longer be encoded by receptors. The lowest frequency at which flickering lights produce constant stimulation of visual sensory fibers is called the *critical fusion frequency*. For example, a standard incandescent light bulb flickers at 60 Hz, but to us it appears as a constant light source. This characteristic of photoreceptors is very important to the film and television industries.

A

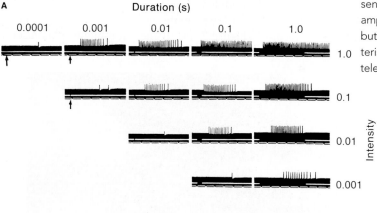

B

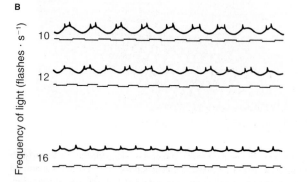

(A) When light flashes are shorter than 1 second, the product of intensity and duration determines the number of APs produced by a *Limulus* photoreceptor. Short, bright flashes can produce a response that is indistinguishable from the response to a dimmer, but longer, stimulus. **(B)** Flickering lights above a certain frequency cannot be distinguished from constant illumination. The on-off pattern of the stimulus is shown under the response to the stimulus recorded from a *Limulus* photoreceptor. At 10 Hz, the photoreceptor follows the flicker faithfully; at 12 Hz, the photoreceptor becomes less accurate in reporting the flicker; and, at 16 Hz, the response in the photoreceptor is continuous. [Part A adapted from Hartline, 1934; part B from "How Cells Receive Stimuli," by W. H. Miller, F. Ratliff, and H. K. Hartline. Copyright © 1961 by Scientific American, Inc. All rights reserved.]

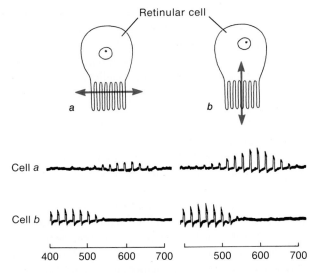

Figure 7-45 The response of crayfish photoreceptors to polarized light varies with the plane of polarization. Two cells, *a* and *b*, were presented with a series of equal-energy flashes of polarized light at various wavelengths. The color of light in each flash (indicated by its wavelength in nanometers) is indicated along the lower axis. Cell *a* responded maximally to light with a wavelength of about 600 nm; cell *b* responded maximally to light of 450 nm. When the plane of polarization (red arrow) was perpendicular to the microvilli, the responses in both cells were small (*left*). Responses of both cells were enhanced when the plane of polarization (red arrow) was rotated so as to lie parallel to the microvilli (*right*). [Adapted from Waterman and Fernandez, 1970.]

were oriented systematically in the microvilli and each preferentially absorbed light with its electric vector parallel to the microvilli, the anatomical arrangement within the rhabdome could provide a physical basis for the detection of

planes of polarized light by arthropods. In electrical recordings from single retinular cells in crayfish, the response to a given intensity of light did indeed vary with the plane of polarization in the stimulating light, consistent with this hypothesis (Figure 7-45).

The Vertebrate Eye

Vertebrate eyes (see Figure 7-41E) have certain structural features similar to those of a camera. In a camera, the image is focused on the film by moving the lens forward or backward along the optic axis. For example, to bring objects that are close to the camera into focus, the lens must be positioned relatively far away from the film. To focus on distant objects, the lens is moved forward. In the vertebrate eye, incident light is focused in two stages. In the initial stage, incident light rays are bent as they pass through the clear outer surface of the eye, called the **cornea** (Figure 7-46). They are further bent, or *refracted*, as they pass through a second structure, the *lens*, and finally form an inverted image on the rear internal surface of the eye, the *retina*. In fact, most of the refraction that occurs in the eye (about 85% of the total) takes place at the air-cornea interface, and the rest depends on the effect of the lens. Like a camera, certain bony fishes focus images on the retina by moving the lens of the eye with respect to the retina. (This principle of changing the distance between the lens and the light-receptive surface has also been adapted by some invertebrates. For example, in the eyes of jumping spiders, the position of the lens is fixed, and focusing depends on moving the retina.)

In contrast, neither the lens nor the retina can be moved in the eyes of higher vertebrates. Rather, the image is

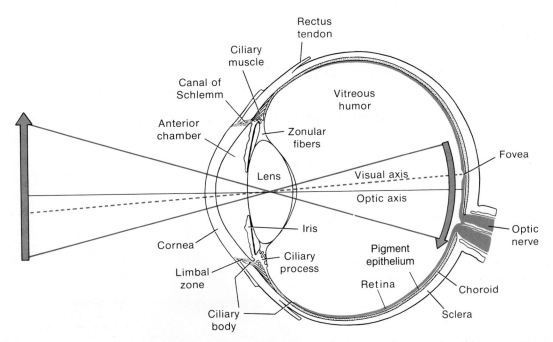

Figure 7-46 In the mammalian eye, incident light is refracted by the cornea and the lens and is focused on the photosensitive retina. In this diagram, the refraction of light has been simplified; refraction at the air-cornea interface is omitted. The image focused on the retina is inverted by the lens. The lens is held in place by the zonular fibers. When ciliary muscle fibers contract, tension on the zonular fibers is reduced, and the elastic properties of the lens cause it to become more rounded, shortening the focal length.

focused by changing the curvature and thickness of the lens. Changing the curvature of the lens surfaces changes the distance at which an image passed through the lens comes into focus, called the *focal length* of the lens. The shape of the lens is changed by modification of the tension exerted on the perimeter of the lens. The lens is held in place within the eye by the radially oriented fibers of the *zonula* (see Figure 7-46). The fibers of the zonula exert outwardly directed tension around the perimeter of the lens. Radially arranged *ciliary muscles* adjust the amount of tension exerted on the lens. When the ciliary muscles relax, the lens is flattened by elastic tension exerted by the fibers of the zonula, which pull the perimeter of the lens outward. In this state, objects far from the eye are focused on the retina, but objects close by would be fuzzy. Objects close to the eye are brought into focus on the retina when the ciliary muscles contract, relieving some of the tension on the lens and allowing the lens to become more rounded. This process is called *accommodation* to close objects. The ability to accommodate decreases with age in human beings as the lens becomes less elastic, producing a type of "far-sightedness," called *presbyopia*.

Perhaps the most remarkable thing about accommodation is not the mechanical mechanisms for altering the focal length of the lens, but the neuronal mechanisms by which a "selected" image—out of all the complexity in the visual environment—is correctly focused on the retina as a result of nerve impulses to the ciliary muscles. A related neuronal mechanism produces **binocular convergence,** in which the left and right eyes are positioned by the ocular muscles in such a way that the images received by the two eyes fall on analogous parts of the two retinas, regardless of the distance between the object and the two eyes. When an object is close, each of the two eyes must rotate toward the middle of the nose; when an object is far away, the two eyes rotate outward from the midline.

Responses to changes in light intensity

In a camera, the intensity of light admitted to the film is controlled by adjusting the aperture of a mechanical diaphragm through which light is admitted when the shutter opens. The vertebrate eye has an opaque iris with a variable aperture called the **pupil,** which is analogous to the mechanical diaphragm of a camera. When circular smooth muscle fibers in the iris contract, the pupillary diameter decreases, and the proportion of incident light that is allowed to enter the eye is reduced. Contraction of radially oriented muscle fibers enlarges the pupil. The contraction of these muscles—and, hence, the diameter of the pupil—is controlled by a central neuronal reflex that originates in the retina. This *pupillary reflex* can be demonstrated in a dimly lit room by suddenly illuminating a subject's eye with a flashlight.

Changes in pupillary diameter are transient. After a response to a sudden change in illumination, the pupil gradually returns after several minutes to its average size. Moreover, the area of the pupil can change only about fivefold, making it no match for the changes in the intensity of illumination normally encountered by the eye, which equal six or more orders of magnitude. Thus, although the pupil can produce rapid adjustments to moderate changes in light intensity, other mechanisms must be available. The eye adapts to extremes of illumination by changes in the state of visual pigments and by the processes of neuronal adaptation (see *Mechanisms of Adaptation* earlier in this chapter). Pupillary constriction provides an additional advantage: the quality of the image on the retina improves. The edges of the lens are optically less perfect than is the center; so, when the pupil is constricted, light is prevented from passing through the perimeter of the lens and optical aberrations are reduced. The depth of focus (the distances through which objects are in focus when the lens is in one fixed shape) increases with decreased pupillary diameter, just as it does in a camera when the aperture is reduced.

Visual receptor cells of vertebrates

The stimulus to all visual receptor cells is electromagnetic radiation that falls within a particular range of energy, called *visible light* (Figure 7-47). The energy of electromagnetic radiation varies inversely with its wavelength, and we perceive this variation in energy as variation in color. Violet light, the highest energy of light to which the human

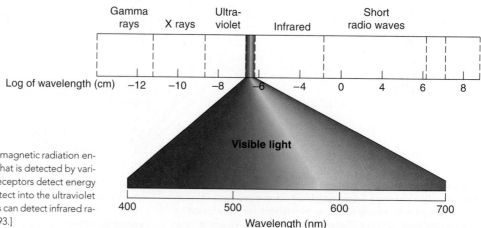

Figure 7-47 The spectrum of electromagnetic radiation encompasses a broad range of energy that is detected by various sensory modalities. Most visual receptors detect energy in the visible range, but some can detect into the ultraviolet as well. The pit organs of some snakes can detect infrared radiation. [Adapted from Lehninger, 1993.]

eye responds, has a wavelength of approximately 400 nm. Red light, at the low-energy end of the visible spectrum, has wavelengths between 650 and 700 nm. Bright light delivers more energy per unit of time than does dim light. The photoreceptor cells that capture the energy of light and transduce it into neuronal signals are located in the retina of the vertebrate eye.

In mammals, birds, and other vertebrates, the retina contains several types of cells that are interconnected into a network. The visual receptor cells themselves fall into two classes, **rods** and **cones,** which were named for the shapes of the cells as observed under a microscope (Figure 7-48). All neurons within the retina, as well as the epithelial cells, contribute to the light response of the vertebrate eye, but rods and cones have different physiological characteristics. For example, cones function best in bright light and pro-vide high resolution, whereas rods function best in dim light. These different capabilities are used by different animals to provide particular visual capabilities. For example, animals that live in flat, open environments (e.g., cheetahs and rabbits) usually have horizontal **visual streaks,** regions within the retina that contain unusually high densities of cone receptors. Such a region corresponds to the horizon in the visual world and is thought to confer maximal resolution in this part of the scene. The visual streak also contains a high-density population of **ganglion cells**—the cells that transmit visual information to the brain. In contrast, arboreal species (and human beings) typically have a radially symmetric density gradient of photoreceptors. An important feature of this kind of retina is the **fovea,** or *area centralis.* It is a small (about 1 mm^2) central part of many mammalian retinas, and it provides very detailed

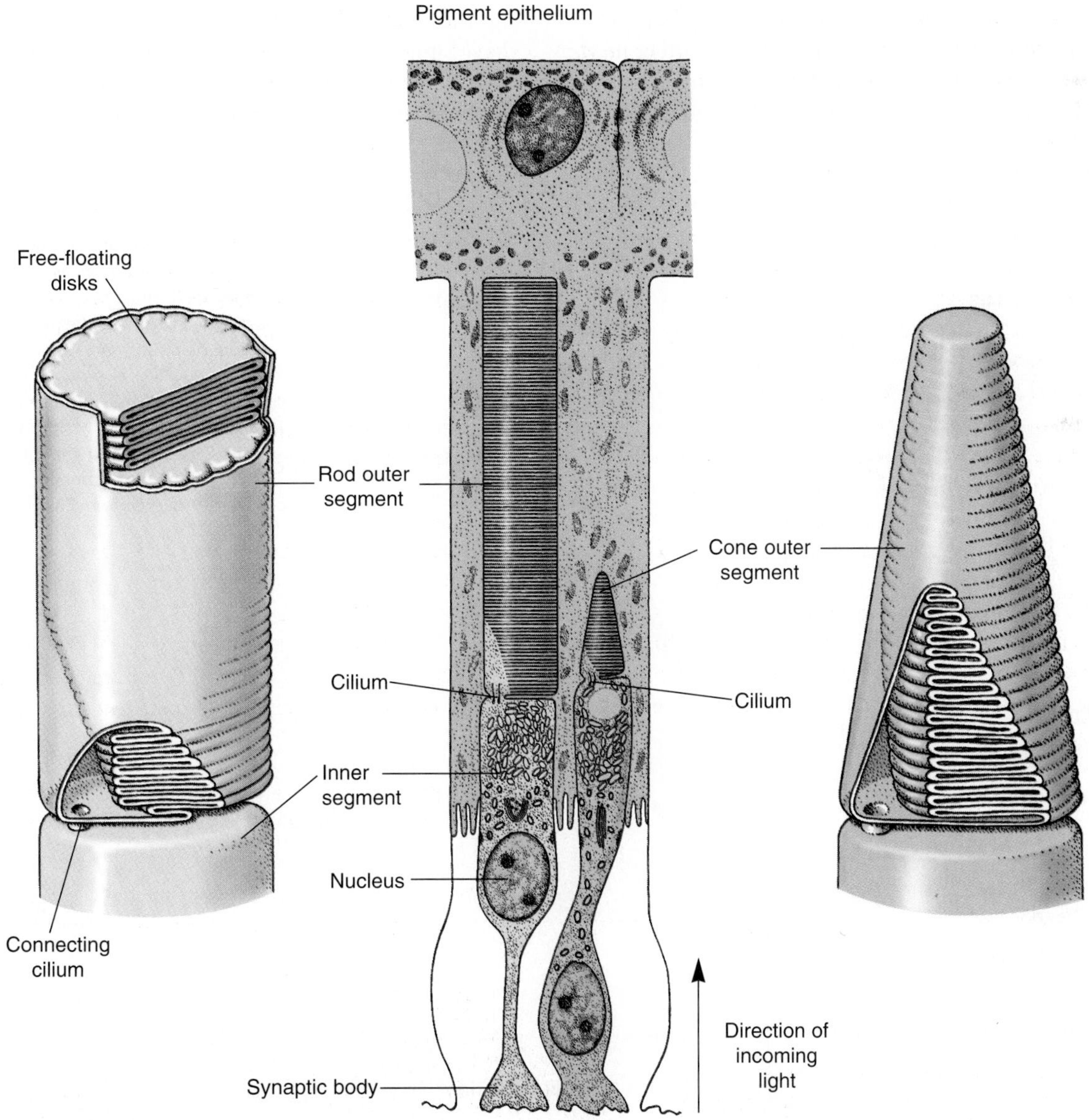

Figure 7-48 Vertebrate photoreceptors are classified as rods or cones on the basis of their morphological and physiological properties. The outer segments of rods and cones, where light is captured, face away from the source of light. The pigment that absorbs the light energy is contained in membrane lamellae, and the ends of the outer segments lie against the pigment epithelium.

⊃n about the visual world, a characteristic called
.. *visual acuity.* In human beings and in certain other
mammals, the fovea contains only cones, whereas the
remainder of the retina contains a mixture of rods and
cones, with the rods significantly outnumbering the cones.
In mammals, cones mediate color vision, and rods, which
are more light-sensitive, mediate only achromatic vision.
This distinction between rods and cones does not, however,
pertain to all vertebrates. In fact, the retina in some species
contains only rods, based on morphology, but may
nonetheless be capable of color vision.

Rods and cones are structurally and functionally more
similar to one another than are the wide variety of photoreceptors found in the invertebrates. Each vertebrate
receptor cell contains a segment with an internal structure similar to that of a cilium. This rudimentary cilium
connects the **outer segment,** which contains the photoreceptive membranes, to the **inner segment,** which contains
the nucleus, mitochondria, synaptic contacts, and so forth
(see Figure 7-48). The receptor membranes of vertebrate visual cells consist of flattened lamellae derived from the surface membrane near the origin of the outer segment. In the
cones of mammals and some other vertebrates, the lumen
of each lamella is open to the cell exterior. In rods, the
lamellae pinch off completely from the surface membrane
of the outer segment so as to form flattened sacs, or *disks,*
that are stacked like pita bread on top of one another
within the rod outer segment. The stack of disks is completely contained within the surface membrane of the visual
cell. The *photopigment* molecules are embedded in the
membranes of the disks. Because the photopigment lies in
the disk membranes of the rod outer segment but not in the
surface membrane, the primary step in photochemical
transduction must take place in the disk membranes, rather
than in the surface membrane.

The eyes of many invertebrates lack the ciliary structure
that connects the inner and outer segments of vertebrate
rods and cones (Figure 7-49). In these invertebrate eyes, the
photopigment is located in microvilli formed by the cell
membrane, and these pigment-containing microvilli make
up the rhabdomeres. Because many invertebrate species
have simple eyes in which the photoreceptors are of the
rhabdomeric type, it might be tempting to conclude that
rhabdomeric photoreceptors are found only in simple eyes.

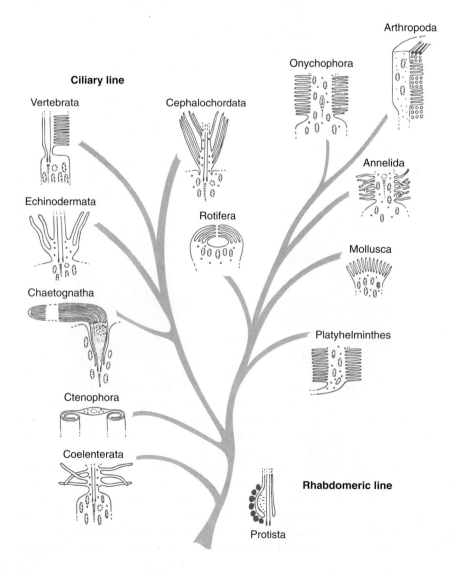

Figure 7-49 Vertebrate photoreceptors contain a typical 9 + 2 ciliary structure that connects the inner and outer segments, but many
invertebrate photoreceptors lack this ciliary
structure and instead contain many microvilli.
This diagram illustrates the phylogenetic distribution of ciliary and rhabdomeric eyes.
There are, however, exceptions. Both the scallop, *Pecten,* and the surf clam, *Lima,* have
complicated eyes with two layers of photoreceptors. One layer contains ciliary photoreceptors and the other contains rhabdomeric
receptors. [Adapted from Eakin, 1965.]

However, the eyes of the octopus are optically very complex and the photoreceptors are rhabdomeric. In addition, some bivalve mollusks (e.g., the scallop, *Pecten*, and the clam, *Lima*) have eyes with two separate layers of photoreceptors. One layer contains ciliary receptors, and the other contains rhabdomeric receptors.

How many different types of sensory receptor cells include cilia in their structures? Why might this cellular arrangement be so ubiquitous among sensory receptors?

In all photoreceptor cells, the transduction of light energy produces a change in the membrane potential; however, the effect of the transduction is different in vertebrate and invertebrate photoreceptors. Invertebrate photoreceptors depolarize in response to light (Figure 7-50A; see also Figure 7-45), but vertebrate rods and cones hyperpolarize in response to a light stimulus (Figure 7-50B). Membrane conductance measurements before and during illumination have shown that the effect of light on vertebrate photoreceptors is to *decrease* the conductance for sodium, g_{Na}, of the outer segment membrane. In the dark, the surface membrane of the vertebrate rod outer segment is nearly equally permeable to Na^+ and K^+, and V_{rest} lies about halfway between E_K and E_{Na}. In this state, Na^+ ions leak into the outer segment through channels that are steadily open in the dark (Figure 7-51A). The Na^+ ions that carry this inward current, which is called the **dark current** because

it is maximal in the dark, are kept from accumulating in the cell by the steady action of the metabolically driven Na^+,K^+ ATPase. A dark current is found only in vertebrate photoreceptor cells and not in invertebrate photoreceptors.

After light absorption by the photopigment, the conductance to sodium, g_{Na}, of the outer segment decreases, causing the dark current to decrease and V_m to hyperpolarize toward E_K (see Figures 7-50B and 7-51B). When the light stimulus stops, g_{Na} of the membrane returns to its high resting level, and V_m becomes more positive, returning to its resting level between E_{Na} and E_K. The change in V_m at the onset of light is carried electrotonically (see *Passive Spread of Electrical Signals* in Chapter 6) into the inner segment of the photoreceptor. In the inner segment, changes in V_m modulate the steady release of neurotransmitter from the presynaptic sites located in the basal part of the inner segment. Like vertebrate auditory receptors, vertebrate photoreceptors lack axons. They synapse onto other neurons, which carry the visual signal toward the central nervous system. The neuronal signal is passed along by other neuronal cells of the retina, ultimately influencing the activity of axons that project to the brain within the optic nerve. It is interesting that a *hyper*polarization, rather than a *de*polarization, is produced when a vertebrate photoreceptor absorbs light, because in most sensory systems reception of a stimulus depolarizes the receptor cell. In vertebrate photoreceptors, the inner segment steadily secretes a transmitter while it is being partially depolarized by the dark current. The hyperpolarization that occurs in response to illumination decreases the amount of transmitter released onto the next neuron in line, modifying the activity of that second-order neuron.

The change in membrane potential that is produced in a group of photoreceptors when they are illuminated can be recorded by extracellular electrodes, as can action potentials traveling down the axons of a nerve. Many photoreceptors are tiny cells making intracellular recording difficult, so this method of recording—called an *electroretinogram*—has been extremely useful in the study of vision (Spotlight 7-2).

Photoreception: Converting Photons into Neuronal Signals

When photons strike the photosensitive pigment molecules of a photoreceptor, the cell must generate APs, either by itself (in invertebrate photoreceptors) or in higher-order neurons (in vertebrate photoreceptors) if the signal is to be carried to the central nervous system. The process of visual transduction has received an enormous amount of research attention, and the features of the visual process have provided clues to physiologists studying sensory transduction in other sensory modalities. Studies of photoreception have been carried out in many different species spanning several phyla. Many similarities between vertebrate and invertebrate photoreception have been found, although it is now thought that invertebrate photoreception may be more complex because it relies on two light-activated pathways,

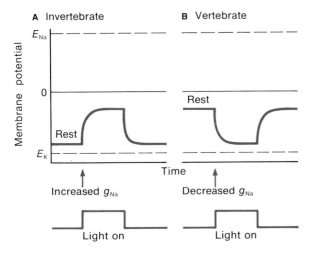

Figure 7-50 Most invertebrate photoreceptors depolarize in response to a stimulus, whereas vertebrate photoreceptors hyperpolarize. **(A)** Transduction of light energy into chemical energy within most invertebrate photoreceptors causes an *increase* in the permeability of the surface membrane to Na^+ and K^+, depolarizing the cell. **(B)** Vertebrate photoreceptors respond to light with a *decrease* in g_{Na} of the surface membrane, leaving a residual low g_K and shifting V_m toward E_K. As a result, the cell hyperpolarizes.

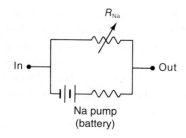

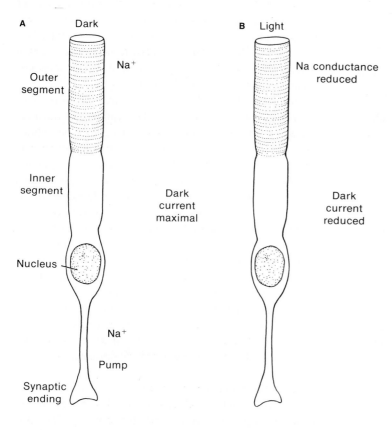

A   Dark

Outer segment   Na⁺

Inner segment   Dark current maximal

Nucleus

Na⁺

Pump

Synaptic ending

B   Light

Na conductance reduced

Dark current reduced

Figure 7-51 Illumination reduces the dark current in vertebrate rods. The g_{Na} of the rod outer segment is high in the dark **(A)** and becomes reduced in the light **(B)**. For this reason, the dark current, which is carried by Na⁺ ions that leak into the outer segment, drops during illumination. In the equivalent circuit *(top left)*, the battery is the Na⁺, K⁺ ATPase, and the light-activated variable resistor (R_{Na}) represents the g_{Na} of the outer segment. [Adapted from Hagins, 1972.]

rather than the single pathway found in vertebrates. There are other, perhaps related, differences as well. For example, capture of a single photon by a photoreceptor in *Limulus* produces a peak current of ~1 nA, whereas capture of a single photon by a vertebrate rod photoreceptor changes the current by ~1 pA, three orders of magnitude smaller. Moreover, invertebrate photoreceptors may respond to light intensities spanning seven orders of magnitude, whereas vertebrate rods respond to intensities only within four orders of magnitude. Despite these differences in detail, all types of photoreceptors have been shaped by evolution to convert the energy of photons into neuronal energy, and studies of all types of eyes have contributed to our understanding of the process.

Visual pigments

The spectrum of electromagnetic radiation extends from gamma rays, with wavelengths as short as 10^{-12} cm, to radio waves, with wavelengths greater than 10^6 cm (see Figure 7-47). The segment of the electromagnetic spectrum

with wavelengths between 10^{-8} cm and 10^{-2} cm is termed light. Only a small part of this segment of the spectrum—ranging from about 400 nm to about 740 nm—is visible to human beings. Below this range is the **ultraviolet** (UV) part of the spectrum, and above it the **infrared** (IR), neither of which is visible to human beings and other mammals.

There is nothing qualitatively special about those parts of the spectrum that renders them invisible to us. Rather, what we see depends on which wavelengths are absorbed by our visual pigments. For example, there is a condition called *cataract*, in which the lens becomes opaque. Treatment of the condition consists of surgically removing the lens; after this surgery, patients can see light into the UV range because it is absorption of UV light by the lens that prevents people from seeing those wavelengths. The compound eyes of many insects can detect light into the UV range, causing some flowers containing UV-reflecting pigments to look much less plain to insects than they do to mammals, but all animals are sensitive to only a part of the spectrum of electromagnetic radiation that is available in

THE ELECTRORETINOGRAM

In a teaching laboratory, it is sometimes useful to record the summed electrical activity of the eye, which is technically much less complicated than recording from single cells with micro-electrodes. The recording electrode (which can be a thread or a wick that is saturated with saline) is placed on the cornea, and the ground electrode is attached to another part of the body. When a light is flashed on the eye, a complex waveform is recorded by the electrode (as shown in the adjoining figure). This recording is called an **electroretinogram** (ERG), and it records the summed activity of the photoreceptor cells and other neurons in the retina. It took several years to sort out the source of each of the components of the ERG, but we now believe that the *a* wave is due to the receptor current produced by the visual receptor cells. The *b* wave follows the *a* wave, and it is produced by electrical activity of the second-order retinal neurons that receive input from the receptor cells. The *c* wave is found only in vertebrates and appears to be produced by the pigment epithelial cells against which the outer segments of the visual cells abut. In the developing eyes of tadpoles, the ERG consists of only an *a* wave before synaptic contacts are established. Similarly, in the eye of an adult frog, if synaptic transmission between the photo-receptors and the second-order neurons is blocked pharmaco-logically, the ERG consists of only an *a* wave.

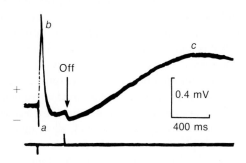

A vertebrate electroretinogram consists of several components, each from a different source. The timing of the stimulus is shown under the recording. [Adapted from Brown, 1974.]

sunlight. The visual pigments of vertebrates may absorb only a limited range of the electromagnetic spectrum of sunlight because vertebrate life evolved in water, which heavily filters electromagnetic radiation. The range of the spectrum to which vertebrate photopigments—including those of terrestrial mammals such as human beings—are sensitive matches rather closely the spectrum of light that is admitted through water.

All known organic pigments owe their ability to absorb light selectively to the presence of a carbon chain, or ring, that contains alternating single and double bonds. When a photon is captured by one of these molecules, the energy state of the molecule is changed. The energy contained in a quantum of radiation is equal to Planck's constant divided by the wavelength, λ (in centimeters):

$$E = \frac{2.854}{\lambda} \, \text{g} \cdot \text{cal} \cdot \text{mol}^{-1} \qquad (7\text{-}1)$$

Thus, the energy in a photon increases as the wavelength of radiation decreases. Quanta with wavelengths less than 1 nm contain so much energy that they break chemical bonds or even atomic nuclei; quanta with wavelengths greater than 1000 nm lack sufficient energy to affect molecular structure. The visual pigments absorb maximally between these two limits. When a quantum of radiation is absorbed by a photopigment molecule, it raises the energy state of the molecule by increasing the orbital diameter of the electrons associated with a conjugated double bond, the same process that is used in the photosynthetic conversion of radiant energy into chemical energy in plants.

Photochemistry of visual pigments

The energy content of visible light is just low enough to be absorbed by molecules without breaking them up. The concept that a pigment is essential for the process of absorbing light and transducing its electromagnetic energy into chemical energy originated with John W. Draper, who concluded in 1872 that, to be detected, light must be absorbed by molecules in the visual system. R. Boll found soon thereafter that the characteristic reddish purple color of the frog's retina fades (bleaches) when the retina is exposed to light. The light-sensitive substance, *rhodopsin*, that is responsible for the purple color, was extracted in 1878 by W. Kühne, who also found that, after the pigment has been bleached by light, its reddish purple color can be restored by keeping the retina in the dark, provided that the receptor cells remain in contact with the pigment epithelium at the back of the eye.

Since then, much has been learned about the chemical nature and physiological effects of rhodopsin. It absorbs light maximally at wavelengths of about 500 nm. It is found in the outer segments of rods in many vertebrate species and in the photoreceptors of many invertebrates. Rhodopsin molecules are packed at high density into receptor membranes; there may be as many as 5×10^{12} molecules per square centimeter, which is equivalent to an intermolecular spacing of about 5 nm.

All known visual pigments consist of two major components: a protein (opsin) and a light-absorbing molecule. In all instances, the light-absorbing molecule is either **retinal** or **3-dehydroretinal** (Figure 7-52). Retinal is the aldehyde of vitamin A_1, a carotenoid. Vitamin A_1 is an alcohol and is also called retinol; 3-dehydroretinal is the aldehyde of vitamin A_2, which is also called 3-dehydroretinol. In addition to its major components, rhodopsin includes a six-sugar polysaccharide chain and a variable number (as many as 30 or more) of phospholipid molecules. The lipoprotein opsin, which binds the phospholipids and the polysaccharide chain, appears to be an integral part of the photoreceptor membrane. Carotenoid molecules move back and forth between the photoreceptor membrane and the pigment epithelium at the back of the retina during bleaching and regeneration of the visual pigment. (Incidentally, the pigment that confers a dark color on the pigment epithelium is photochemically inactive and is unrelated to the visual pigment. Instead, it keeps light from scattering and reflecting diffusely back toward the retina.)

The retinal molecule assumes two sterically distinct states in the retina. In the absence of light, the opsin and the retinal are linked covalently by a Schiff's base bond, and retinal is in the 11-*cis* configuration (see Figure 7-3). When the 11-*cis* retinal captures a photon, it isomerizes into the all-*trans* configuration (see Figure 7-52). This *cis*–*trans* isomerization is light's only direct effect on the visual pigment. The conversion from 11-*cis* to all-*trans* retinal initiates a series of changes in the relation between the retinal and the opsin protein, including changes in the conformation of the opsin itself.

When light hits the photopigment, an intermediate, *metarhodopsin II*, is formed. Metarhodopsin II activates another protein that is associated with the membrane and that binds GTP in exchange for GDP. This protein, which we now know belongs to the family of G proteins, is called **transducin** in recognition of its key role in the transduction of light. The activated subunit of transducin diffuses in the plane of the membrane, encountering many phosphodiesterase molecules, which hydrolyze cGMP to 5'-GMP. In vertebrate photoreceptors, the dark current Na^+ channels are open only in the presence of cGMP; so, when cGMP is hydrolyzed, these channels close (Figure 7-53). The membrane of the rod outer segment contains a class of channels that are permeable to three cations: Na^+, Mg^{2+} and Ca^{2+}. When the level of cGMP drops, the conductance through these channels drops. Most importantly, the inward I_{Na} drops, and the residual K^+ current through other channels causes the cell to hyperpolarize. When the light stimulus ends, cGMP is regenerated by the action of another enzyme, guanylate cyclase. As the level of cGMP rises, the dark current channels open, and the receptor current returns to its full value in the dark. Activated transducin collides with, and activates, phosphodiesterase molecules at a rate of about 10^6 molecules per second, allowing the capture of a single photon to affect the conductance through an enormous number of ion channels. This numerical relation generates an impressive amplification between the capture of a single photon and the effect that the event produces on V_m.

After the *cis*–*trans* isomerization of retinal has occurred, further changes in the molecule appear to be irrel-

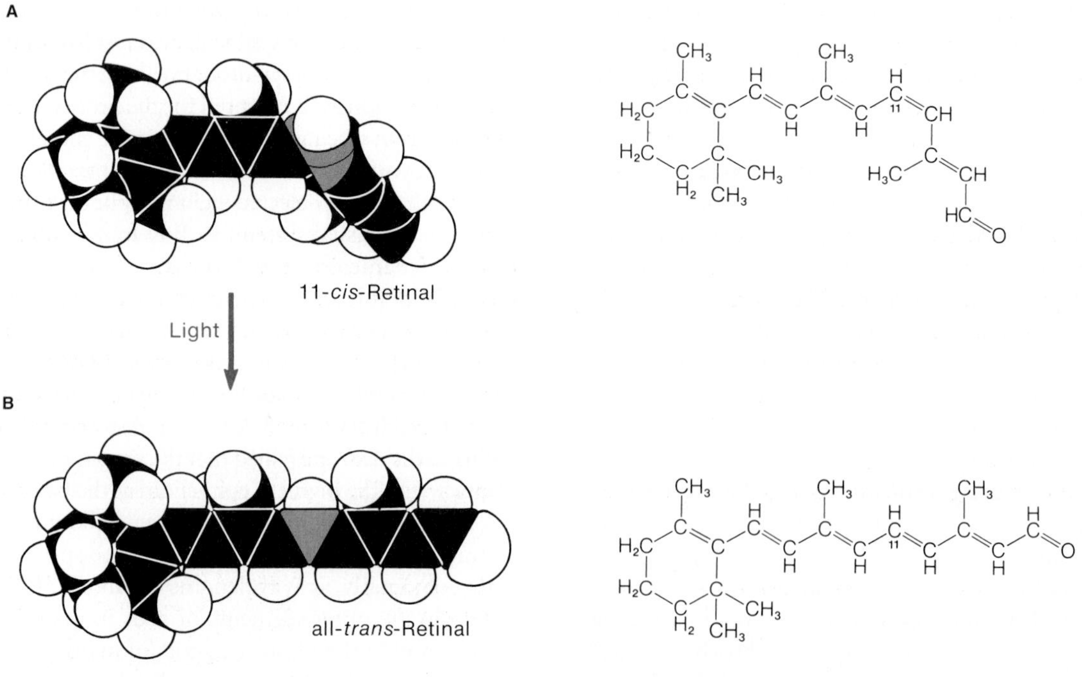

Figure 7-52 The carotenoid pigment retinal changes its steric conformation when it absorbs a photon. **(A)** In the dark, the bonds of carbon 11 are arranged in the *cis* configuration. **(B)** When a photon is captured, these bonds are converted into the straight, all-*trans* configuration. Both space-filling and line diagrams of the molecular structure are shown. [Part B from "Molecular Isomers in Vision," by R. Hubbard and A. Kropf. Copyright © 1967 by Scientific American, Inc. All rights reserved.]

A

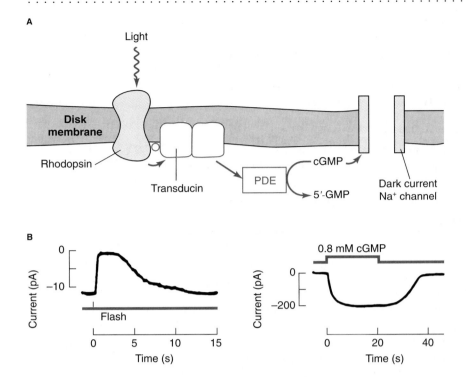

B

Figure 7-53 When light is absorbed by retinal, a series of reactions causes the Na+ channels that carry the dark current to close. **(A)** Activated rhodopsin increases the activity of a G protein, transducin. The activated G protein then activates many phosphodiesterase (PDE) molecules, reducing the intracellular concentration of cyclic guanosine monophosphate (cGMP), which leads to the closing of Na+ channels that carry the dark current. The receptor cell then hyperpolarizes. **(B)** Recordings of currents in single rod photoreceptors isolated from toad retina. *(Left)* A flash of light causes the inward dark current of 10 pA to drop to zero. *(Right)* The photoreceptor has been broken open, and the external saline has been changed to mimic intracellular ionic concentrations. When cGMP is added to the external saline (exposing the inside of the outer segment to a high concentration of cGMP), a very large inward current develops. [Part B adapted from Yau and Nakatani, 1985.]

evant to the excitation of visual receptor cells, but the subsequent reactions (Figure 7-54) are necessary for regenerating active rhodopsin. Activated rhodopsin is hydrolyzed spontaneously to retinal and opsin, which are both reused repeatedly. Free retinal is reisomerized back into the 11-*cis* form and reassembled with an opsin to form rhodopsin. Any retinal that is lost or chemically degraded in the process is replenished from vitamin A_1 (retinol) stored in the cells of the pigment epithelium, which actively take up the vitamin from the blood. A nutritional deficiency of vitamin A_1 decreases the amount of retinal that can be synthesized and, hence, decreases the amount of available rhodopsin. The result is reduced photosensitivity of the eyes, a condition that is commonly known as *night blindness*.

Rod photoreceptors can respond to the absorption of a single photon, partly because rhodopsin is so densely packed into their disc membranes. There are about 20,000 rhodopsin molecules per square micrometer in the rod outer segment, which is much closer packing than, for example, the density of acetylcholine receptors at the neuromuscular junction. By recording from a single rod, Denis Baylor, of Stanford University, has measured the response to the capture of a single photon (Figure 7-55). In these ex-

periments, rods are teased apart and one of them is drawn into a recording pipette where it is stimulated by a small beam of light. When the stimulating light is very dim, it is possible to record small current fluctuations, each of which occurs when a single rhodopsin molecule has been photoisomerized by a single photon. The properties of the current recorded under these conditions are similar to the properties of the current measured through a single acetylcholine receptor channel at the neuromuscular junction. (The change in current that is associated with the capture of one photon is about 1 pA.) Because photoreceptors can respond to a single photon, or quantum of energy, the sensitivity of photoreception is limited by the quantal nature of light; there is no smaller amount of light than one photon.

Elucidating the process of visual transduction has demonstrated the power of a comparative approach. Although vertebrate and invertebrate photoreceptors seem quite different from each other at the electrophysiological level, there are many similarities between them at the molecular level. The molecular genetics available in *Drosophila* and the physiological accessibility of the vertebrate retina have been combined to provide an array of enormously powerful experimental approaches to the question of how visual information is acquired and processed by

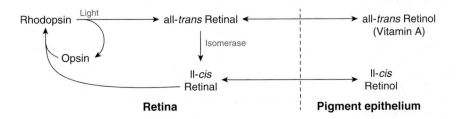

Figure 7-54 When rhodopsin is activated, the all-*trans* retinal separates from the opsin. Rhodopsin is reconstituted after an isomerase returns the retinal to the 11-*cis* configuration. Retinol (vitamin A) is stored in the pigment epithelium and can be delivered to the photoreceptors for generating new rhodopsin molecules.

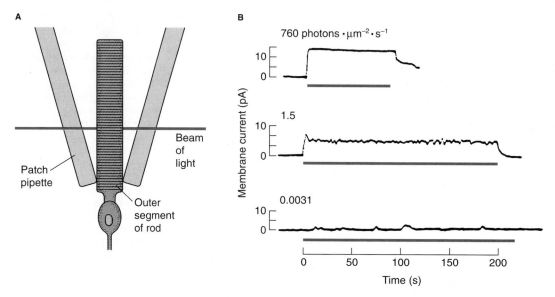

Figure 7-55 Rods can respond electrically to the capture of a single photon. **(A)** A single rod outer segment is sucked into a smooth glass pipette electrode and is illuminated by a narrow bar of light while the ionic current across the membrane is recorded by the pipette. **(B)** The recorded membrane current changes in response to illumination. In very dim light *(bottom record)*, small individual changes in the current ac-company the capture of single photons. As the light intensity is increased, (intensity is indicated above each recording), the responses become larger and smoother. The duration of the illumination is indicated by the bar under each recording. Membrane currents are in pA. [Adapted from Baylor et al., 1979.]

photoreceptors. If the molecular identities of the players had not been so strongly conserved through phylogeny, teasing out the details of visual transduction would probably have taken much longer.

Cones and rods

The ability to distinguish color, rather than just perceiving the visual world in shades of gray, is correlated with the possession of multiple visual pigments, each of which absorbs maximally at a different wavelength (Spotlight 7-3). In vertebrate species that have color vision, it has been found that different groups of photoreceptors contain spectrally identifiable visual pigments, and each class of photoreceptor has a distinctive **action spectrum.** That is, the electrical response of each photoreceptor, when it is illuminated, is maximal at a particular wavelength and falls off when the wavelength of incident light is either raised or lowered. In many species for which action spectra have been recorded, three classes of photoreceptors have been found. The action spectra for some species have then been compared with the absorption spectra of individual photoreceptors. Absorption spectra for single photoreceptors have been measured by a process, called *microspectrophotometry,* in which a tiny beam of light is focused on one photoreceptor at a time and the absorption properties of that cell are determined. Photoreceptors studied in this way fall into distinct classes for each species; there are no intermediate spectra, which implies that each photoreceptor synthesizes a single visual pigment (Figure 7-56). Both the action spectra and absorption spectra of photoreceptors have been determined in many species, and the two kinds of spectra match one another closely, confirming that the action spectrum of a photoreceptor depends upon the absorption properties of its visual pigment. In addition, results of such experiments confirm that each photoreceptor synthesizes only one of the visual pigments. Light that contains different wavelengths generates photochemical reactions in a particular photoreceptor cell in proportion to the amount of each wavelength absorbed; thus, a photoreceptor cell is excited by different wavelengths in proportion to the efficiency with which its pigment absorbs each wavelength. Any photon that is not absorbed can have no effect on the pigment molecule; any photon that is absorbed transfers part of its energy to the molecule as described in *Photochemistry of Visual Pigments* in this chapter. Thus, it is possible to restate Young's trichromacy theory (see Spotlight 7-3) in relation to cone photoreceptors and their photopigments: there are in the human retina three classes of cones, each of which contains one visual pigment that is maximally sensitive to blue, to green, or to orange light. The electrical output of each class depends on the number of quanta that are absorbed by the pigment and can thus contribute to the events of transduction. The sensation of color arises when higher-order neurons integrate signals received from the three classes of cones.

Knowledge about the molecular basis of color reception has grown enormously since 1984, when Jeremy Nathans described the molecular structure of human opsins and thus provided an explanation for hereditary color blindness. For example, point mutations in individual pigment genes cause defects in sensitivity to particular wavelengths. Indeed, the molecular basis for differential spectral sensitivity among the opsins has been characterized by using naturally occurring variants in visual pigments.

A

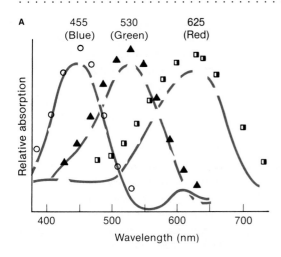

B

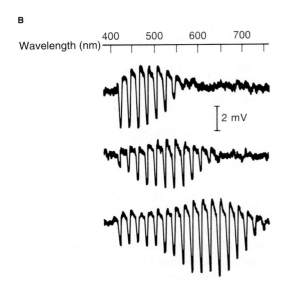

C

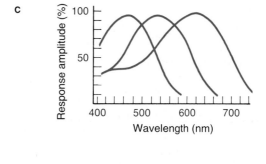

Figure 7-56 Each class of cones in the carp retina has a distinctive action spectrum. **(A)** Absorption spectra of individual cones in the retina of a goldfish indicate that there are three separate visual pigments, each with a distinctive absorbance peak. These measurements were made by microspectrophotometry, which allows the absorption spectrum of a single photoreceptor to be measured. In human beings, the class of cones that is equivalent to the red-absorbing cone in the goldfish absorbs maximally closer to 560 nm, which is in the yellow part of the spectrum. **(B)** Electrical responses of three single cones in the retina of a carp to flashes of different wavelengths, as shown by the scale at the top. The wavelength that produces a maximal response is different for each of the three cones. **(C)** When the amplitude of activity in each cell shown in part B was plotted as a function of wavelength, three classes of cones were revealed, each with an action spectrum approximating one of the absorption spectra in part A. [Part A adapted from Marks, 1965; parts B and C adapted from Tomita et al., 1967.]

It appears that 11-*cis*-retinal (or 11-*cis*-3-dehydroretinal) is the light-absorbing molecule in all visual pigments, and this prosthetic group is combined with different opsin molecules to produce visual pigments with different absorption maxima. Differences in the amino acid composition of opsins—rather than variation in the light-absorbing prosthetic group—produce rhodopsins with different absorption maxima. Nathans and his co-workers discovered three genes that encode for the opsins in human cones. The gene encoding the protein part of the blue-absorbing pigment is located on an autosomal chromosome, whereas the two genes for the "red"-absorbing and green-absorbing proteins are closely linked on the X chromosome. The "red" and green opsins differ at only 15 of 348 amino acids, and each shares about half its amino acids with rhodopsin in rods (Figure 7-57). On the basis of sequence similarity, we can surmise that the genes for these pigments probably arose from a common ancestral gene that underwent duplication and divergence. A comparison of the amino acid sequences suggests that, of the cone pigments, the blue-sensitive pigment arose first, followed by the red and the green. *Color blindness* is caused by an absence of,

or a defect in, one of the cone opsin genes. With the use of these molecular markers in conjunction with visual tests, it is now possible to define the molecular basis of this perceptual problem. For example, the high incidence of

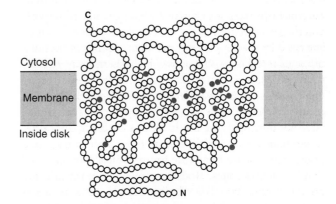

Figure 7-57 The two opsin proteins in the human rhodopsins that absorb maximally in the red and the green parts of the visible spectrum differ by only 15 amino acids, most of which are thought to be in membrane-spanning helices. In this diagram, these variable amino acids are red. [Adapted from Nathans et al., 1986]

SPOTLIGHT 7-3

LIGHT, PAINT, AND COLOR VISION

In 1666, Sir Isaac Newton demonstrated that white light is separated into a number of colors when it is passed through a prism. Each *spectral color* is monochromatic; that is, it cannot be separated into yet more colors. At that time, however, it was already known that a painter could match any spectral color (e.g., orange) by mixing two pure pigments (e.g., red and yellow), each of which reflects a wavelength different from that of the color produced. Thus, there seemed to be a paradox between Newton's demonstration that there is an infinite number of colors in light and the growing awareness of Renaissance painters that all colors could be produced by combinations of three primary pigments—red, yellow, and blue. This paradox appeared to be resolved by Thomas Young's suggestion, in 1802, that the receptors in the eye are selective for the three *primary colors:* red, yellow, and blue. Young reconciled the infinite variety of spectral colors that can be duplicated with the small number of painter's pigments required to produce all colors by proposing that each class of color receptor is excited to a greater or lesser degree by any wavelength of light: The "red" and "yellow" re-ceptors would be stimulated maximally by separate monochro-matic "red" and "yellow" wavelengths, respectively, and both would be stimulated to a lesser degree by monochromatic or-ange light. Young proposed, in other words, that the sensation for "orange" results from the simultaneous excitation of "red" receptors and "yellow" receptors. Young had no notion about the physiology of photoreceptors, making his insight truly remarkable.

Extensive psychophysical investigations carried out in the nineteenth century by James Maxwell and Hermann von Helmholtz supported Young's **trichromacy theory,** and ad-ditional support came from later investigations by William Rushton. However, direct evidence for the existence of three classes of color-receiving photoreceptors was missing. Then, in 1965, W. B. Marks and E. MacNichol measured the color ab-sorption of single cone photoreceptors in the goldfish retina (see Figure 7-56A). They found three classes of cones, each of which absorbed maximally at a unique wavelength. Subsequent mea-surements of the absorption spectra of cones in retinas from hu-man beings, monkeys, and other species of fishes reproduced these results. It thus appears that the retinas of species that can perceive and respond to color contain photoreceptor cells with different absorption spectra and that, in many of these species, there are three distinct classes of receptors.

---

red-green color blindness arises from the recombination of these closely linked red and green opsin genes.

Color vision has been demonstrated in some members of all classes of vertebrates. In general, retinas that include cone photoreceptors are associated with color vision, but examples of different color classes among rods have been found. For example, frogs have two kinds of rods—red (containing rhodopsin, which absorbs in the blue green) and green (containing a pigment that absorbs in the blue)—in addition to cones.

If color vision is mediated primarily by the cones, what do rods contribute? Rods are more sensitive to light than are cones (the records in Figure 7-55 were made from rods), and their connections to the next neurons in line are char-acterized by greater convergence than are the connections of cones (see *Visual Processing in the Vertebrate Retina,* Chapter 11), producing greater summation of weak stim-uli. Thus, rods are most effective in dim light. Because the cones produce color vision, when only our rods are stimu-lated by dim light, we see in shades of black and gray, rather than in color. In the human retina, most images are prefentially focused onto the fovea, which contains only tightly packed cones. Rods are found only outside the fovea. The differential distribution of rods and cones in the retina causes us to be most sensitive to dim light when an image is focused outside the fovea, onto parts of the retina where the rod population is higher. For example, a dim star will appear brighter if you adjust your gaze to make its im-age fall outside the fovea. If you shift your gaze to make the image fall on the fovea, the star will fade or even disappear. This increased sensitivity comes at a price: the broader con-nections among rods reduces the acuity of rod-based vision. Our visual *sensitivity* is greatest when an image is focused onto the rods outside the fovea; our visual *acuity* is great-est when we focus an image onto the cones of the fovea.

When visual pigments are explored throughout phy-logeny, some interesting patterns emerge. For example, all visual pigments for which retinal is the prosthetic group are called rhodopsins. All human visual pigments—the rod pigment and the three cone pigments—are rhodopsins. Vi-sual pigments in which 3-dehydroretinal is the prosthetic group are called **porphyropsins,** and the distribution of the rhodopsins and porphyropsins among species shows an interesting correlation with environment. All visual pig-ments of terrestrial vertebrates are rhodopsins. In addition, rhodopsins are found among invertebrates, including *Limulus,* insects, and crustaceans. In contrast, porphy-ropsins are found in the retinas of freshwater fishes, eury-haline fishes (see Chapter 14), and some amphibians. This distribution suggests that some feature of the porphy-ropsins makes them particularly well adapted to conditions found in freshwater. In fact, anadromous fishes, which mi-grate from freshwater to saltwater—or vice versa—during their life cycle, change their visual pigment between por-phyropsin and rhodopsin during the migration. They syn-thesize porphyropsin during their stay in freshwater and rhodopsin while they are in saltwater. The absorption max-ima of the porphyropsins is shifted toward the longer-wavelength, red end of the visual spectrum, whereas rhodopsins absorb maximally at shorter wavelengths. Perhaps the freshwater environment makes sensitivity to the red end of the spectrum important.

Tracing the transduction of information from the absorption of a photon to the production of neuronal signals leaves the question of how all this information about incident radiation is molded into a coherent view of the world unanswered. The collected information is passed on to higher neuronal centers, where it is integrated and can be used to shape behavior—a topic that is explored in Chapter 11.

LIMITATIONS ON SENSORY RECEPTION

To be most useful, a sensory receptor should be very sensitive to stimulation from the environment and should encode the information with perfect accuracy. In fact, no receptor meets these requirements, because of the physical properties of stimuli and of receptors; all receptors represent compromises in how they receive and encode sensory information. Some physical principles that apply to receptors of many sensory modalities necessarily limit the fidelity with which sensory information is received and transmitted by cells. In some cases, the accuracy of sensory reception is limited by the relative magnitudes of the signals and the background noise. This *signal-to-noise ratio* limits the performance of all systems that receive and transmit information, whether or not they are living. In other cases, the performance and sensitivity of a sensory system are limited by the form of energy to which the receptors are tuned. For example, light is by its nature quantized into photons. No receptor can receive less than one quantum of light, because light does not exist in fractions of quanta.

A major source of background noise arises from a corollary of the Third Law of Thermodynamics. That is, at all temperatures above $0°K$, molecules have kinetic energy and are in motion. Thermal energy is given by

$$E_{\text{therm}} = kT \qquad (7-2)$$

where k is Boltzmann's constant (1.3805×10^{-16} erg K^{-1}) and T is the absolute temperature. This equation gives the energy that is associated with the movement of molecules (i.e., Brownian motion) at an animal's body temperature. It sets a lower limit on the sensitivity of receptors in detecting signals because thermal energy provides a constant noise level against which stimulation occurs. To detect an external signal, a receptor must be able to distinguish the signal from this baseline thermal noise. How easily can receptor cells accomplish this task?

Photoreceptors provide an example. At a body temperature of 25°C, the thermal energy is about 0.58 kcal $\cdot$ mol^{-1}, or 4×10^{-14} erg. We must compare this energy to the energy of a typical sensory stimulus. The stimulus for a vertebrate photoreceptor is light in the visible

part of the electromagnetic spectrum (see Figure 7-47). The energy of a single photon of light is given by the Einstein relation:

$$E = h\nu = \frac{hc}{\lambda} \qquad (7-3)$$

where h is Planck's constant and ν, c, and λ are the frequency, speed, and wavelength of light, respectively. Substituting the values for a photon of blue light ($\lambda = 500$ nm), the energy is calculated to be about 57 kcal $\cdot$ mol^{-1}— almost 100-fold greater than the thermal energy. In vision, detection is definitely *not* limited by thermal energy within the detector. Instead, it has been found to be limited by the quantal nature of light itself.

In audition, the energy is given by the Einstein relation for single *phonons*, which are quantum units of sound energy analogous to photons of light. Animals hear sounds across a remarkably broad range of frequencies, from 10 to 10^5 Hz. The energy of phonons at these frequencies ranges from 7×10^{-26} to 7×10^{-23} erg. In the middle of this range, the energy of a single phonon is 10 orders of magnitude (10^{10}) *below* the limit of detection set by thermal energy. This result indicates that the detection of *acoustic* stimuli is fundamentally limited by thermal noise, and there must be special mechanisms that permit auditory sensory reception. Indeed, some advantage is gained by tuning the detectors to limit their range, a common feature found in the hair cells of auditory systems. Numerous mechanisms have evolved to combat the limitations of thermal noise, but direct measurements have also shown that sensory cells in auditory systems faithfully reproduce the thermal noise at their inputs.

As discussed earlier in this chapter, most chemical stimuli (olfaction, taste, chemotaxis) bind to specific receptors, rather than directly change ionic currents through membrane channels. In this case, the relation between binding energy and thermal energy determines the limits of detection. The binding energies that have been measured in chemical sensory systems are typically about 1 kcal $\cdot$ mol^{-1}. This energy is sufficiently greater than thermal energy that chemoreceptors could theoretically count single molecules. There is, however, an important constraint dictated by the physics of receptor binding. The greater the binding energy, the longer the molecule remains associated with its receptor. For a binding constant of 10^{-6} M, the association time is about 3×10^{-3} seconds; whereas, for a binding constant of 10^{-11} M (giving very high specificity), the association would last for more than 5 minutes. Because the performance of a receptor system depends at least in part on comparisons across many receptors, long binding times would require that comparisons be made over very long time periods, and evolution seems to have shunned such a mechanism. Instead, the binding constants between chemical stimuli and their receptor molecules are moderately high, reducing the binding energy but also the time required to transduce and interpret chemical signals.

Signal-to-noise properties can be predicted for stimuli that activate electroreceptors and thermoreceptors. Electroreception is relatively widespread in aquatic organisms, in which it is used for navigation, communication, and predation. The energy in the electric fields, carried through water at the frequencies used, is about 10 orders of magnitude below the thermal energy. Thus, the process of electroreception, like that of auditory reception, must be dominated by thermal noise in the detector. Thermoreception depends on the detection of photons in the infrared region of the electromagnetic spectrum (which has lower frequencies and longer wavelengths than those in the visible spectrum), and, by definition, it is limited by the temperature difference between the measured object and the measuring organ. Some animals, particularly beetles, have been found to perform at or near the theoretical limits; others have apparent adaptations that keep their detectors cooler than the rest of the body, decreasing the background thermal noise.

As scientists have explored the limits of detection achieved by animal senses, it has become clear that many modalities operate at or near the theoretical limits imposed by physical laws. To accomplish this prodigious task, many types of receptors have evolved similar molecular mechanisms, and the mechanisms that are available to be used in each sensory modality depend, at least to some extent, on whether sensory reception is limited by thermal energy or by the quantal nature of the stimulus.

SUMMARY

Receptor cells are highly sensitive to specific kinds of stimulus energy and relatively insensitive to other kinds of stimulation, and they transduce the stimulus into an electrical signal, usually (but not always) a depolarization. The lower limit to sensation often depends on how much energy is carried in the signal compared with the energy in the thermal noise within the organism. The transduction process is most sensitive to weak stimuli, producing receptor signals that contain several orders of magnitude more energy than the stimulus itself. This sensitivity drops off with increasing stimulus strength. In most receptor cells, the primary sites of reception and transduction are receptor molecules located in the cell membrane or in intracellular membranes.

Activation of receptor molecules causes the conductance of particular classes of ion channels in the membrane to change; typically, the change in conductance permits the flow of a receptor current, producing a receptor potential. In many sensory modalities, the receptor cells do not themselves produce APs. Instead, receptor potentials modulate the amount of neurotransmitter that the receptor cells release onto second-order neurons, which in turn initiates or modulates the number of APs in second-order neurons. Stimulus intensity is typically encoded in the frequency of impulses, which in many sensory fibers is roughly proportional to the logarithm of the intensity, up to a maximum frequency. The logarithmic relation between stimulus and response magnitudes permits reception over a large dynamic range while retaining high sensitivity to weak stimuli.

Parallel inputs from receptors that cover different parts of the intensity range increase the range of stimulus intensities that can be perceived. The time-dependent loss of sensitivity to a maintained stimulus, termed sensory adaptation, is a common property of receptor cells; some receptors adapt rapidly and others adapt slowly. The mechanisms responsible for sensory adaptation vary. Some take place in the receptor cell, others in the network of neurons that carry the sensory information. In at least one case (the *Limulus* photoreceptor), adaptation results in part from intracellular elevation of Ca^{2+}, which blocks the light-dependent activation of Na^+- K^+-selective ion channels.

Some receptor cells exist individually, but others are organized into sensory tissues and organs, such as the vertebrate nasal epithelium or the retina of the eye. Anatomical organization affects how a sensory organ functions. For example, the quality of the image formed by the vertebrate visual system depends on the presence of a lens and a huge population of photoreceptor cells in the retina.

Several sensory systems have features in common. In particular, many receptor molecules contain seven transmembrane domains, a feature that is also found in some neurotransmitter and hormone receptors. Many sensory systems also have common elements in the cascade of biochemical events that immediately follow signal detection and that amplify the signal.

Mechanoreception is a result of distortion or stretching of the receptor membrane, directly producing changes in ion conductances. The deflection of the stereocilia of hair cells provides directional information by modulating, upward or downward, the frequency of spontaneously occurring impulses of axons in the eighth cranial nerve. This function is the basis of reception in several sensory organs—the lateral-line system of fishes and amphibians, vertebrate audition, and the organs of equilibrium in both vertebrates and invertebrates. The mammalian cochlea analyzes sound frequencies according to their effectiveness in displacing different parts of the basilar membrane, which bears hair cells. Mechanical waves travel along the basilar membrane, set up by sound-driven movements of the eardrum and auditory ossicles; they stimulate the hair cells, which in turn modulate their synaptic activity onto auditory nerve fibers. Certain sound frequencies stimulate each location along the basilar membrane more strongly than do other frequencies, which is the basis for frequency discrimination in mammals.

Electroreceptors of fishes are modified hair cells that have lost their cilia. Exogenous currents flowing through electroreceptor cells produce changes in the transmembrane potential that modulate the release of transmitter at the base of the receptor cell, thus determining the rate of APs in sensory fibers.

Visual receptors employ pigment molecules, in specialized membranes, that undergo a conformational change after absorbing a photon. The change in the conformation of the photoreceptive molecules initiates a cascade of reactions that leads to a·change in the conductance of the receptor cell membrane. All visual pigments consist of a protein molecule (an opsin) combined with a carotenoid chromophore, either retinal (in rhodopsins) or 3-dehydro-retinal (in porphyropsins). The amino acid sequence of the opsin determines the absorption spectrum of each visual pigment. A *cis–trans* isomerization of the carotenoid initiates all visual responses. Absorption of photons is coupled to the opening (in invertebrates) or closing (in vertebrates) of ion channels by intracellular second-messengers. In vertebrate rods, photon capture by rhodopsin molecules leads to the activation of associated G-protein molecules located in the receptor membrane. Each G protein then activates many phosphodiesterase molecules, each of which hydrolyzes many molecules of the internal messenger cGMP. In the dark, cGMP continually activates Na^+ channels that carry the dark current. The light-dependent hydrolysis of cGMP reduces the dark current, and a residual K^+ current hyperpolarizes the photoreceptor cell, reducing the steady release of neurotransmitter at the inner segment. The reduced rate of transmitter release causes a change in activity in the next higher-order neuron.

Some vertebrates have three types of cones in the fovea, each containing a visual pigment that is maximally sensitive to a different part of the spectrum. The integration of activity from all of these cones produces color vision. Rods, which in human beings all contain only one type of photopigment, are present in great densities in the periphery of the retina outside the fovea, are more sensitive than cones, and show much greater synaptic convergence. As a result, they exhibit low acuity and high sensitivity.

REVIEW QUESTIONS

1. Visual receptor cells can be stimulated by pressure, heat, and electricity, as well as by light, as long as the intensity of these other stimuli is sufficiently great. How can this fact be reconciled with the concept of receptor specificity?

2. Choose one sensory modality and outline the steps from energy absorption by a receptor cell to the initiation of action potentials (APs) that will travel to the central nervous system.

3. Why must receptor potentials be converted into APs to be effective?

4. All sensory information enters the central nervous system in the form of APs having similar properties. How can we differentiate among various stimulus modalities?

5. What is the difference between sensory transduction and sensory amplification? Choose one sensory modality and describe how these two processes are related in that modality.

6. Discuss the relation between the intensity of a stimulus and the magnitude of the signal sent to the central nervous system by receptor cells. How is stimulus intensity encoded? How can a sensory system respond to stimuli whose intensity varies over many orders of magnitude?

7. Discuss three mechanisms that contribute to sensory adaptation.

8. Discuss one example in which efferent activity can regulate the sensitivity of receptor cells.

9. How are movements of the basilar membrane converted into auditory nerve impulses?

10. Discuss the function of inner and outer hair cells in the cochlea.

11. How does spontaneous firing enhance the sensitivity of certain receptor systems—for example, lateral-line electroreceptors?

12. How is the presence of an object perceived by electroreceptors of the weakly electric fishes?

13. What is the major difference between vertebrate and invertebrate photoreceptor cells in their electrical responses to illumination?

14. Compare the mechanisms that allow the auditory system to distinguish the frequency of incident sounds and the visual system to distinguish the frequency of incident light.

15. Outline the steps, as currently understood, in the transduction of light energy in vertebrate visual receptors.

16. How does our current understanding of the physiology of color vision corroborate Young's trichromacy theory?

17. Compare and contrast the morphological and functional properties of vertebrate rods and cones.

18. What allows some arthropods to respond to the orientation of polarized light? Human beings cannot do it. Why?

19. Compare the ways in which mammalian and teleost lenses focus images.

SUGGESTED READING

Corey, D. P., and S. D. Roper. 1992. *Sensory Transduction.* 45th Annual Symposium of the Society of General Physiologists. New York: Rockefeller University Press. A series of papers that discuss recent data from studies of transduction in many different sensory modalities.

Dowling, J. 1987. *The Retina.* Cambridge, Mass.: Belknap Press of Harvard University Press. A very readable compendium of information on the vertebrate retina written by a major contributor to our understanding of this sensory organ.

Finger, T. E., and W. L. Silver. 1987. *Neurobiology of Taste and Smell.* New York: Wiley. A collection of papers considering the chemical senses in a broad range of animals.

Hudspeth, A. J. 1989. How the ear's works work. *Nature* 341:397–404. A beautifully written account of auditory transduction by hair cells, written by a man who has played a major role in exploring the subject.

Kandel, E. R., J. H. Schwartz, and T. M. Jessell. 1991. *Principles of Neural Science*. 3d ed. New York: Elsevier. An enormous and authoritative compendium of information about the function of the nervous system, from the biophysics of membrane channels to the physiological basis of memory and learning. Several chapters consider sensory mechanisms, with some emphasis on vertebrates.

Land, M., and R. Fernald. 1992. The evolution of eyes. *Ann. Rev. Neurosci.*, 15:1–29. A consideration of the physical and optical properties of visual organs across all of animal phylogeny.

Shepherd, G. M. 1994. *Neurobiology*. 3d ed. New York: Oxford University Press. A concise text that considers several sensory modalities in both vertebrates and invertebrates.

GLANDS: MECHANISMS AND COSTS OF SECRETION

All cells secrete material into their surrounding environment, either to form a protective barrier around the cell, or to communicate with other cells. Cells secreting similar substances (e.g., hormones) are often collected together to form **glands**. The specialized cells composing a gland act as a unit, secreting and sometimes excreting material within the body or onto the body surface. Every animal has a large number of different glands, which vary in both structure and function. The particular types of glands possessed by an individual vary not only with the species, but also during various developmental stages. The venom glands of snakes, human sweat glands, wax glands of insects, the thyroid gland, and the pituitary are just a few of the wide variety of glands found in the animal world. Glandular secretions are synthesized by cells that form the secretory part of the gland and are released from the gland in response to an appropriate stimulus. The nature and extent of the secretion and the form of the stimulus vary greatly among different glands.

The secretions from glands constitute an important response of animals to a variety of situations. Feeding, for example, results in the massive activation of a whole range of digestive glands (see Chapter 15). In vertebrates the secretion of hydrochloric acid (HCl) by cells lining the stomach into the stomach lumen after eating can be so large as to cause a marked increase in blood pH, the so-called postprandial (after-dinner) alkaline tide. (Remember, an increase in pH corresponds to more alkaline conditions.) This increase in blood pH can be very large, especially in carnivorous animals like crocodiles, which are capable of eating a whole gazelle at a single sitting! The webs spun by spiders, which are used to catch prey, are another example of glandular secretions. The nature of the secretion and the pattern of the web vary with the species and the environmental conditions. Mucus nets secreted by deep-sea fish may play a role similar to that of spiders' webs in the interactions between predator and prey in the unique environment of the deep sea. Secretions from glands often play important roles in mating behavior as well as reproduction in general (see Chapter 9).

In this chapter, we first describe the nature and mechanism of cellular secretions. We then move to the organ level, discussing the adrenal medulla, the mammalian salivary gland, and the spinnerets glands of spiders that produce the silk for webs. Because glandular secretions are involved in all aspects of physiology, various glands are discussed throughout this book; only a few examples are described in detail in this chapter. Finally, a discussion of the energy expended in glandular activity rounds out this chapter.

CELLULAR SECRETIONS

A surface coat is secreted by most cells. Mucus is also secreted onto the external surface of epithelia, giving rise to the term *mucosal* referring to the external surface of the epithelium. The surface coat allows cells to recognize each other; the coat and mucus also form a protective barrier around epithelial cells, creating a somewhat controlled microenvironment between these cells and the surrounding extracellular space. In addition, cells secrete various signaling substances that are used for communication between cells. Such cellular communication varies from intimate contact to distant signaling.

Types and Functions of Secretions

Secretions involved in communication between cells can be categorized by the distance at which they have an effect (Figure 8-1):

- **Autocrine** secretion refers to a secreted substance that affects the secreting cell itself. An example is norepinephrine released from adrenergic nerve endings, which inhibits further release of norepinephrine from that nerve.

- **Paracrine** secretion refers to a substance that has an effect on neighboring cells. For example, during the inflammatory response localized vasodilation is induced mainly by histamine released from mast cells in the area of tissue damage.

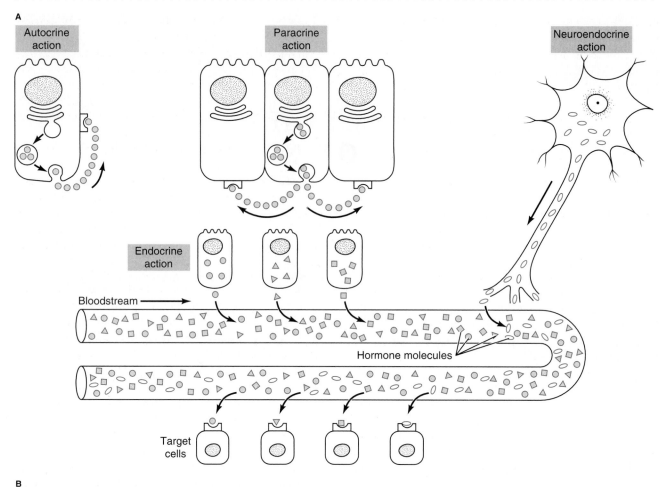

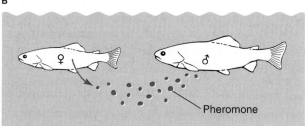

Figure 8-1 Cells communicate through a variety of pathways. **(A)** Autocrine and paracrine actions involve the same or neighboring cells. Simple endocrine action involves transport of hormones over distances via the bloodstream; neuroendocrine action, in which hormones are released from nerve endings into the circulation, is similar. **(B)** The release of pheromones into the environment by one animal to communicate with other animals. The female fish has liberated a pheromone into the water that is detected by, and alters the behavior of, the male fish.

- **Endocrine** secretion refers to a substance that is released into the bloodstream and acts on a distant target tissue.

- **Exocrine** secretion refers to a substance that is released onto the surface of the animal, including the surface of the gut and other internalized structures.

Some exocrine secretions, called **pheromones**, are produced by one animal to communicate with another and are involved in initiating a range of physiological responses. Individual members of many animal species utilize pheromones to communicate with each other. Among insects, for instance, pheromones function as a means of identifying the members of a colony. Pheromones also play an important role in reproduction for many species. For example, **bombykol** is a powerful sex attractant released by female silkworm moths. In certain marine invertebrates, such as clams and starfish, spawning of eggs and sperm is triggered by pheromones liberated along with the gametes. Thus, the spawning of one individual triggers spawning in others of both sexes. The adaptive value of such epidemic spawning is that it enhances the probability that sperm and egg will meet and that fertilization will occur. A steroid that induces molting in crabs also serves as the female sex attractant, producing behavioral responses in males at concentrations in the seawater as low as 10^{-13} M. These signals are an essential part of communicating about mating.

Pheromones also are used to repel predators. A common example is the foul-smelling musk that makes skunks unpalatable to their enemies. A pair of scent glands near the anus produce this yellowish, oily, malodorous secretion, which has been detected by humans at sea, as far as 30 kilometers from the nearest land. The repulsive odor is due to *trans*-2-butene-1-thiol, 3-methyl-1-butanethiol and *trans*-2-butenyl methyl disulfide in the volatile portion of the scent. The encapsulating muscles close to the anal opening are capable of propelling the musk over a meter.

It is not uncommon for hormones, or their breakdown products, to act as pheromones. Water in which ovulated female goldfish have lived contains reproductive hormones and/or their metabolites, which can induce male goldfish sexual behavior; that is, these substances cause the male goldfish to become much more active and investigate tankmates. Likewise, steroid glucuronides (e.g., estradiol glucuronide) released by zebra fish shortly after ovulation attract males. Mixtures of steroid glucuronides, however, are much more potent than single chemicals in eliciting male sexual behavior in fish.

Some animals produce secretions that act both locally and at a distance and, therefore, have autocrine, paracrine, and endocrine effects. For instance, calcitonin produced in the gills of Pacific salmon modulates calcium flux across the gills through calcitonin receptors in the gills; this calcitonin produced by, and acting on, the gills acts as both an autocrine and paracrine secretion. Pacific salmon also produce calcitonin in the ultimonbranchial gland; in this case, the calcitonin is released into the blood and subsequently acts on the gills as an endocrine secretion.

In addition to communication, secretions have numerous other functions. Saliva produced in the mouth helps food slide down the esophagus, and pancreatic secretions aid digestion. Snails secrete mucus with special elastic properties that enable them to slide and stick as they move over the ground. The path of a snail is marked by its secreted mucus, which is essential to this mode of locomotion.

Thus cells secrete material that can be detected, in some cases, by an adjacent cell or, in other instances, by other animals at a distance of 30 kilometers. Cells producing autocrine and paracrine secretions may be, but are not always, collected together to form glands. Cells secreting hormones or pheromones almost always occur together in glandular structures. As discussed below, many different cells produce a glycocalyx and mucus. Secretion of a wide variety of other substances (e.g., hormones, pheromones, and digestive juices) varies from cell type to cell type. Many of the substances secreted by different species have the same or similar functions, and these substances frequently have identical or related structures. In other words, there is a conservatism in the primary structure of secreted chemicals (Spotlight 8-1).

Surface Secretions: The Cell Coat and Mucus

The external surface of the plasma membrane of all cells is protected by a cell coat, or **glycocalyx,** which is secreted by the cell and is continuously renewed. The cell coat is composed of glycoproteins and polysaccharides, with negatively charged sialic acid termini. The glycocalyx can be visualized using appropriate dyes such as Alcian blue for light microscopy and ruthenium red for electron microscopy.

The oligosaccharides within the glycocalyx can be seen using lectins (e.g., concanavalin A) that have been labeled with fluorescent dyes or electron-dense material. Visualization of the glycocalyx clearly shows that it protects the cell membrane and creates a microenvironment around the cell, thereby modulating filtration and diffusion processes. Collagen in the cell coat serves as a mechanical support for tissues and offers surfaces on which cells may slide.

The chemical composition of the glycocalyx also permits certain cells to recognize each other and adhere, forming organized structures. For example, red blood cells have specific cell-surface antigens that are distinguished by their terminal carbohydrates and form the basis of the ABO blood groups. Red blood cells in the same blood group do not aggregate, whereas those in different groups will aggregate when mixed. In most tissues, similar cells aggregate to form organs, and cells in culture will aggregate with other like cells. Interestingly, reaggregation of chick embryonic tissue is organ rather than species specific.

The glycocalyx around some cells contains mucopolysaccharides that can associate with proteins to form *mucoproteins*. Mucoproteins have a much longer polysaccharide component than glycoproteins, are amorphous, and form gels able to hold large amounts of water. The jelly coats of frog eggs are a common example of such gels. Although distinct from the glycocalyx, *mucus* also contains mucoproteins with a large number of sialic acid termini and creates a protective environment around the cell.

Mucus is produced by specialized cells, called *goblet cells*, found in most epithelia. These cells secrete large amounts of mucus, which can cover the surface of many cells. For example, when a hagfish is disturbed, glands in its skin produce enormous volumes of mucus; once released from the body, the mucus expands very rapidly in water and can completely fill a bucket containing the hagfish within minutes. This slimy coat presumably protects the hagfish from attack. The mammalian lung provides a clear example of the protective role of mucus. The airways to the lungs continually secrete a sticky mucous layer that is propelled towards the mouth by the action of millions of tiny cilia. The mucus traps dirt particles and bacteria; this mixture of mucus, dirt, and bacteria is then carried away from the lung surfaces by ciliary action and, subsequently, is either swallowed or expectorated. This production and movement of mucus, along with antibodies secreted into the mucus and alveolar macrophages, keeps the lungs clean and free from infection. Tobacco smoke inhibits both ciliary activity in the airways and the action of the alveolar macrophages, but it enhances the production of mucus. The "smoker's cough" is an attempt to remove the accumulated mucus.

Packaging and Transport of Secreted Material

Cells that secrete specific substances are generally morphologically *polarized*; that is, the synthesis and packaging of the secreted substance take place in one part of the secretory cell and its secretion takes place in another part

SPOTLIGHT 8-1

SUBSTANCES WITH SIMILAR STRUCTURES AND FUNCTIONS SECRETED BY DIFFERENT ORGANISMS

Just as organisms have many biochemical pathways in common, there are many similarities in the chemical structure of substances secreted by different animals. Substances with similar functions that are secreted by quite different organisms often have similar structures. For example, α-factor secreted by yeast and gonadotropin-releasing hormone (GnRH) produced in the pituitary gland of mammals are small peptide hormones with similar amino acid sequences. Not only are their structures homologous, but both secretions function in reproductive processes. The α-factor acts as a mating pheromone in yeast,

whereas GnRH induces the release of luteinizing hormone (LH), which causes ovulation in humans (see Chapter 9). When injected into mice, yeast α-factor causes release of LH, but it has a lower affinity for GnRH receptors than native GnRH (i.e., GnRH from mice). In most but not all cases, a native hormone has a higher affinity for it's own receptors and is more effective than a similar but foreign substance.

Occasionally, a foreign substance has a greater effect than a native hormone. This is true of mammalian calcitonin, which is produced by the C (clear) cells of the thyroid gland. Calcitonin acts on bones, the major body store of calcium, to lower extracellular calcium levels under conditions when the calcium levels are raised (see Chapter 9). It accomplishes this by inhibiting osteoclastic (bone-resorbing) activity without affecting osteoblastic (bone-building) activity. Salmon and eels also produce a structurally similar calcitonin hormone in their ultimonbranchial glands. This foreign hormone is many times more effective than native human calcitonin in preventing bone resorption and thus lowering blood calcium in humans.

A

Yeast α-factor   Trp-*His*-*Trp*-Leu-Gln-*Leu*-*Lys*-*Pro*-*Gly*-Gln-Pro-Met-Tyr

Mammalian GnRH   pGlu-*His*-*Trp*-Ser-Tyr-Gly-*Leu*-*Arg*-*Pro*-*Gly*-NH₂

B

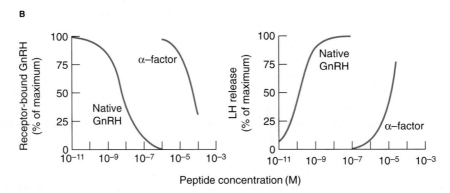

Many species produce secretions with similar structures and analogous functions. **(A)** The amino acid sequence of yeast α-factor and mammalian gonadotropin-releasing hormone (GnRH). These short peptide hormones contain several identical residues shown in italics. **(B)** Binding and activity curves for yeast α-factor and GnRH. The α-factor can bind to mammalian receptors for GnRH (*left graph*), and when injected into mice it induces release of luteinizing hormone (LH), the normal effect of GnRH (*right graph*). Compared with GnRH, however, a much higher concentration of α-factor is required for binding and LH release. [Adapted from Loumaye, Thorner, and Catt, 1982.]

(Figure 8-2). The nature of the synthesis and storage varies with the nature of the substance being secreted. For example, the steroid hormones appear to be secreted in diffuse molecular (i.e., unpackaged) form (see Chapter 9). Most substances, however, are packaged in membrane-bound vesicles within the secretory cell, later to be liberated into the extracellular space. Thus, electron microscopy of most secretory tissues reveals membrane-limited **secretory granules (secretory vesicles)**, 100 to 400 nm in diameter, that contain the substance to be secreted. The terms secretory granule and secretory vesicle are used interchangeably, depending on whether the emphasis is on the contents (granule) or the limiting membrane (vesicle). Secretory vesicles

are similar in many respects to synaptic vesicles, which are somewhat smaller (~50 nm in diameter).

Polymer gels

Mucus, a polymer gel, is packaged and stored along with various other chemicals as compact (condensed) granules in secretory vesicles within goblet cells. Mucus consists of extremely long mucoprotein strands with a large number of sulfate and sialic acid termini, which are negatively charged at neutral pH (Figure 8-3). Individual mucin chains are linked end-to-end by disulfide bonds between cysteine residues, creating extremely long mucin strands up to 4 to

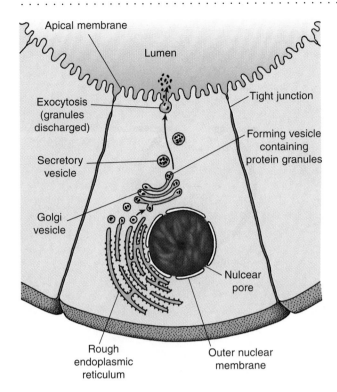

Figure 8-2 Secretory proteins are synthesized in the rough endoplasmic reticulum (ER), transferred in vesicles to the Golgi complex, and released from the apical surface. After the proteins are concentrated in secretory vesicles, the vesicles move to and fuse with the apical surface membrane, discharging their contents into the lumen of the gland by exocytosis.

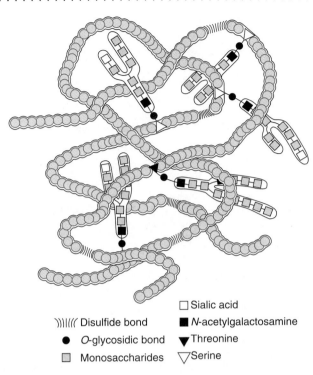

Figure 8-3 Mucus is a polymer gel consisting of mucoprotein strands joined end-to-end by disulfide bonds. The disulfide bonds do not form cross-bridges, which would restrict movement of the polymer chains. Notice the oligosaccharide side chains, many of which have a negatively charged sialic acid terminus. In the condensed phase, the mucin strands form a highly tangled network. [Adapted from Verdugo, 1990.]

6 μm in length. These mucin strands diffuse along their axes, rather like randomly moving snakes clustered in a ball. Within secretory granules, the mucin strands form a highly condensed polymer network arranged as a tangled web. When released into water, however, the mucin network can expand rapidly as much as several hundredfold and be diluted infinitely.

Polymer gels can exist in two phases: a condensed or an expanded hydrated phase. They can undergo phase transition from the condensed to the expanded phase and vice versa very rapidly, similar to the change from a liquid to a gas phase as water boils. High concentrations of calcium ions or hydrogen ions (i.e., low pH) can neutralize the negative charges on the sialic acid termini of mucin strands, thus favoring the condensed phase. Mucin vesicles appear to contain sufficient calcium to keep the mucin in the condensed phase. The low pH of mucin vesicles may also play a role in the condensation process, and lipids within the vesicles may help to ensure that the mucin network remains in the condensed phase while stored in the vesicle. Calcium is not the only shielding cation that keeps polymers in the condensed phase in vesicles. In chromaffin cell granules, for example, chromogranin is the polymer and catecholamines act as the cation, and in mast cell granules, heparin is the polymer and histamine is the cation.

Mucus is released by **exocytosis,** which entails fusion of the membranes surrounding secretory vesicles with the cell membrane so that the contents are expelled from the cell. In most cases, this process is regulated by the level of intracellular free calcium ions. Exocytosis appears to be the mechanism for releasing from cells all exocrine and endocrine secretions that are stored in vesicles. In the exocytosis of mucus, a vesicle moves to the cell surface and a pore forms as the vesicle fuses with the surface membrane. Increase in conductance of the pore has been shown to be independent of and, therefore, not caused by expansion of the polymer gel. Ion exchange occurs between the contents of the vesicle and the extracellular space, causing the calcium level to fall and perhaps the pH to rise within the vesicle. As a result, the release of mucus by exocytosis is explosive, the mucus expanding so rapidly that it pops out of the vesicle like the release of a jack-in-the-box (Figure 8-4A). The mucin network in the giant granule of the slug can expand as much as 600-fold in 20 to 30 ms. This extremely rapid rate of swelling is driven by repulsive forces between the negative charges uncovered by the loss of calcium and perhaps the rise in pH, rather than by the diffusion of water into the mucus, a process too slow to explain the rapid expansion that occurs (Figure 8-4B). It appears that mucus is present in many different types of secretory vesicles and probably assists in the release of stored material under most conditions. Chromogranin probably plays this role in chromaffin granules, aiding in the release of catecholamines into the blood.

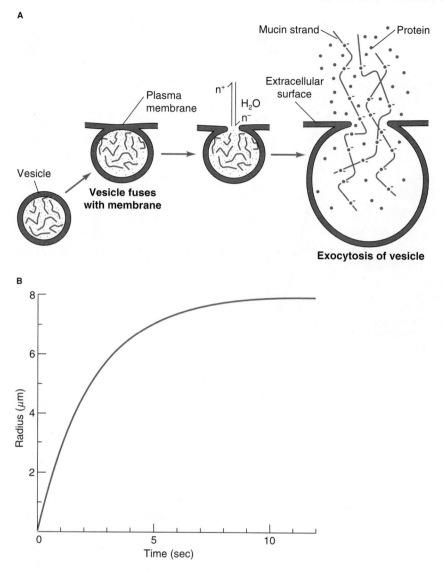

Figure 8-4 Mucus explodes from vesicles like a jack-in-the-box. **(A)** Model for product release by exocytosis. Following fusion of the vesicle to the plasma membrane, the shielding cations (n^+) are released from the mucus inside the vesicle and/or extracellular anions (n^-) flow inward. The net result is that the negative charges in the condensed polyionic mucin become unshielded, driving a fast volumetric expansion and release of the vesicle contents to the extracellular space. Water enters and enlarges the vesicle as the mucus swells. **(B)** Time course of the swelling of mucus in the extracellular space immediately after its release from a goblet cell *in vitro*. The change in radius as a function of time follows first order kinetics. [Adapted from Verdugo et al., 1987.]

The length of the polymer strands in the tangled web of a condensed polymer gel affects its rate of expansion: the shorter the strands, the more rapid the rate of expansion. Some polymer networks (but not mucin networks) have covalent disulfide bonds linking the strands, which limit expansion and may influence exocytotic release. The most important factor determining the rate of expansion of the mucin gel, however, is the nature of the environment into which the mucin network is released. Hyperosmotic solutions can inhibit swelling of the mucin granule, and the ionic composition, pH, and quantity of fluid into which mucus is released has a marked effect on the final state of hydration and , therefore, the flow properties of released mucus. For example, the properties of mucus secreted onto the surfaces of the lung airways are strongly influenced by the nature of the fluid lining the airways. The abnormally thick mucus found in humans suffering from cystic fibrosis has its origins in defective ion transport processes across the epithelium of the lung airways, which changes the ionic composition of the extracellular fluid and, therefore, the rheological properties of the mucus.

Secretory and membrane proteins

The intracellular movement of secretory proteins has been studied by pulse-chase radiography, a tracer technique in which radioactively labeled amino acids are incorporated for a brief period into newly synthesized proteins. Such studies reveal that secretory proteins are synthesized on messenger RNA templates on the polyribosomes (polysomes) of the *rough endoplasmic reticulum (ER)* and ac-

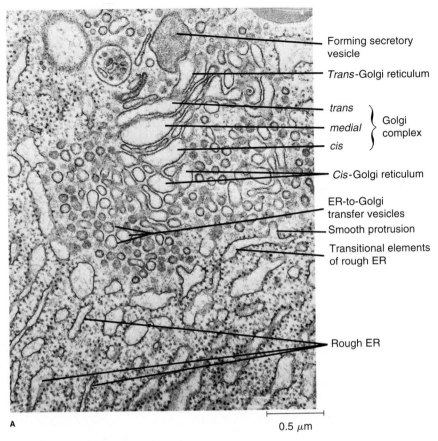

A

0.5 μm

Forming secretory vesicle

Trans-Golgi reticulum

trans
medial } Golgi complex
cis

Cis-Golgi reticulum

ER-to-Golgi transfer vesicles

Smooth protrusion

Transitional elements of rough ER

Rough ER

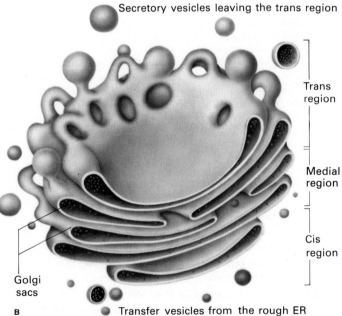

Secretory vesicles leaving the trans region

Trans region

Medial region

Cis region

Golgi sacs

B   Transfer vesicles from the rough ER

Figure 8-5 Intracellular vesicles transport secretory and membrane proteins. **(A)** Electron micrograph of the Golgi complex and rough ER in an exocrine pancreatic cell. Notice the stacked layers of the Golgi complex and a forming secretory vesicle, as well as the transfer vesicles that shuttle secretory and membrane proteins from the rough ER to the Golgi complex. **(B)** Three-dimensional model of the Golgi complex and intracellular vesicles. Transfer vesicles that have budded off from the rough ER fuse with the *cis* membranes of the Golgi complex. The secretory vesicles that bud off from sacs on the *trans* membranes store secretory and membrane proteins in concentrated form. [Part A courtesy of G. Palade; part B from Lodish et al., 1995, after a model by J. Kephardt.]

cumulate within the reticulum. The proteins then pass into the smooth, polysome-free portions of the ER, termed *transitional elements;* the membrane of these elements buds off, encapsulating the secretory products in transfer vesicles (Figure 8-5A). These transfer vesicles then migrate to the Golgi complex, which consists of a stacked set of slightly concave, nearly flat membrane saccules with closely associated free vesicles and vacuoles (Figure 8-5B). Microscopic studies indicate that the membranes of the transfer vesicles fuse with Golgi saccules. Within the Golgi complex, which contains enzymes bound to the luminal membrane surfaces, some proteins undergo alterations such as the addition of sugar residues or the excision of fragments and joining of two polypeptide chains.

The Golgi complex consists of at least three sets of cisternae—the *cis, medial,* and *trans* cisternae. The cis face of the Golgi complex receives transfer vesicles from the ER, whereas the trans face or the *trans-Golgi network (TGN)* produces secretory vesicles, which subsequently move to the surface of the cell (see Figure 8-5B). It is believed that in a process that begins in the Golgi cisternae, but which takes place mainly in the *condensing vacuoles,* water is drawn out of the future secretory vesicle, with the result that the effective concentration of the enclosed protein increases 20- to 25-fold. Mature secretory vesicles eventually reach the plasma membrane to await the appropriate signal to release their contents to the cell exterior.

Intracellular vesicles not only transport secretions to the cell surface for exocytosis but also deliver proteins to be incorporated into the plasma membrane. After synthesis in the rough ER and transport to the Golgi complex, such membrane proteins are incorporated into the vesicular membrane and then inserted into the plasma membrane when the vesicular membrane fuses with the plasma membrane of the cell. This vesicular system is capable of directing specific membrane proteins to different regions of the secretory cell. For example, Na^+-K^+ ATPase is delivered to the basolateral membrane, and proton ATPase to the apical membrane of the same cell. Thus, differences between the apical and basal regions of secretory cells are maintained by the specific proteins delivered by vesicular transport and incorporated into the surface membranes.

The TGN is responsible for directed delivery of material to the apical or basolateral surface. Some vesicles produced by the TGN migrate to the apical surface, whereas other vesicles migrate to the basolateral surface. Newly synthesized membrane proteins transported from the rough ER to the Golgi complex are not yet sorted by ultimate destination, but within the TGN they are sorted into vesicles destined for either the apical or basolateral membrane (Figure 8-6). Some proteins delivered to the basolateral membrane are ultimately delivered to the apical membrane; such transport is referred to as *transcytotic delivery*. The microtubular system appears to play a central role in the movement of vesicles to the cell surface, but the mechanism of sorting is not understood.

 Can you propose possible mechanisms within the cell for sorting vesicles to either the basal or apical membranes? How might such a system be influenced by hormonal action?

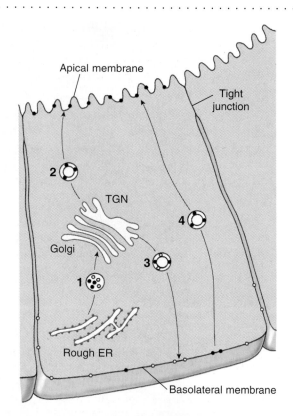

Figure 8-6 The *trans*-Golgi network (TGN) sorts newly synthesized membrane proteins into vesicles destined for the apical or basolateral membrane. After their synthesis in the rough ER, apical and basolateral membrane proteins move in common transfer vesicles to the Golgi complex (1), where they occupy the same compartments. In the TGN, the apical and basolateral proteins are sorted into vesicles that move, respectively, to the apical membrane (2) and the basolateral membrane (3). Some apically destined proteins first are delivered to the basolateral membrane but then are retrieved and transported to the apical surface, a process termed transcytotic delivery (4). Unlike secretory proteins released from the cell, membrane proteins are incorporated into the vesicular membrane; they are inserted into the plasma membrane by fusion of the vesicle with the plasma membrane.

Storage of Secreted Substances

Retention of a substance (e.g., a hormone) within secretory vesicles is accomplished by a variety of means. Large protein hormones are retained simply because of their size, which renders them incapable of crossing the vesicular membrane. Some small hormone molecules are bound to larger accessory molecules, usually proteins. There is evidence that the catecholamines (norepinephrine and epinephrine) are kept in their secretory vesicles at least in part by continual active uptake into the vesicles from the cytosol. The tranquilizing drug **reserpine** interferes with this uptake, thereby causing the catecholamines to leak out of their secretory vesicles and out of the secretory cells.

The duration of storage of a hormone within a secretory tissue varies widely. The steroid hormones, which are not packaged in vesicles and are lipid soluble, appear to diffuse out of secretory cells across the surface membrane in a matter of minutes after their synthesis. Most endocrine hormones are stored in vesicles until their release is stimulated by mechanisms discussed in the next subsection. The thyroid hormones are secreted into the extracellular spaces of spherical clusters of cells, termed *follicles*, and are stored there for up to several months (see Chapter 9). Even after being secreted into the circulation, a hormone is in effect stored in the bloodstream until it is taken up by cells or is degraded. The steroid and thyroid hormones, which are hydrophobic, are carried in the blood attached to *binding proteins*; these hormones remain inactive until they dissociate from the protein.

Secretory Mechanisms

There are several conceivable mechanisms by which substances stored within the cell might find their way to the cell exterior. For most substances stored in secretory vesicles, the most widely accepted theory is that the entire contents of a vesicle are delivered to the cell exterior by exocytosis. However, the details of the release mechanism depends on the animal group and the tissue:

- In *apocrine secretion,* the apical portion of the cell, which contains the secretory material, is sloughed off, and the cell then reseals at its apex. This occurs in some molluscan exocrine glands and certain sweat glands in hairy regions of the human body.

- In *merocrine secretion,* the apical portion of the cell pinches off, and this portion, containing the secretory products, breaks open in the lumen of the gland. This is characteristic of many digestive glands in mammals. Arthropod and annelid exocrine glands also utilize this mechanism.

- In *holocrine secretion,* the entire cell is cast off and breaks up to release its contents. This occurs in some insect and molluscan exocrine tissues and is characteristic of mammalian sebaceous glands in the skin.

Secretion occurs in response to stimulation of the cell. The stimulus may be a hormone or a neurotransmitter at the membrane of the secreting cell; for example, acetylcholine released from sympathetic neurons stimulates the chromaffin tissue of the adrenal medulla to secrete catecholamines. Secretion also may result from a nonhumoral stimulus; for example, an increase in plasma osmolarity stimulates certain hormone-secreting neurons. In neurosecretory nerve cells, the stimulus sets up action potentials (APs) that travel to the axon terminals and there elicit the release of hormone from the endings. This effect can be demonstrated experimentally by stimulating such cells electrically at a distance from their endings, so as to set up impulses, while monitoring the release of hormone from the endings. The rate of hormone secretion increases with increased frequency of impulses (Figure 8-7A). Membrane depolarization in the *absence* of action potentials can be achieved by experimentally increasing the extracellular K^+ concentration, which also increases the rate of hormone secretion. Secretion rises to a maximum with increasing K^+ concentration and hence with increasing depolarization (Figure 8-7B). The stimulation of secretion by depolarization suggests that the action potential also evokes secretion by virtue of its depolarization.

At still higher K^+ concentrations, membrane depolarization exceeds the value for maximal Ca^{2+} entry and thus secretion decreases (see Figure 8-7B). In view of the well-known role of Ca^{2+} in regulation of neurotransmitter release (described in Chapter 6), it should come as no surprise that Ca^{2+} has also been implicated in the coupling of hormone secretion to membrane stimulation. The evidence that Ca^{2+} is the **secretagogue** that couples stimulation to hormone secretion comes from experiments on several kinds of endocrine tissue. Any stimulus that leads to an increase in the internal Ca^{2+} concentration in the output portion of the cell is followed by an increase in secretory activity.

In neurosecretory cells and in ordinary nerve cells, the stimulus is sensed by specific receptors in the input region, which is separated from the output region by the interven-

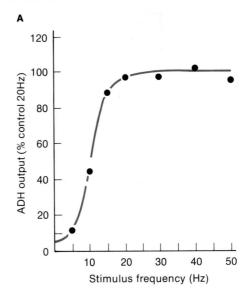

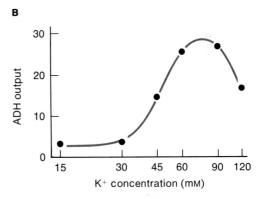

Figure 8-7 Both electrical stimulation (action potentials) and increased extracellular K^+ induce release of antidiuretic hormone (ADH) from neurosecretory cells. **(A)** Release of ADH from the rat neurohypophysis as a function of the frequency of electrical stimulation. The stimulus pulses at each frequency were continued for 5 minutes. **(B)** Release of ADH (arbitrary units) as a function of extracellular K^+ concentration. Freshly dissected neurohypophyses were placed in incubation media of different K^+ content (to produce varying degrees of depolarization) for 10 minutes after which the amount of release ADH in the medium was assayed. [Part A adapted from Mikiten, 1967; part B adapted from Douglas, 1974.]

ing conducting region (Figure 8-8A,B). Incoming stimuli (synaptic input, physical or chemical changes in the plasma) elicit an increased frequency of impulse firing in the axon. By invading and depolarizing the terminal membranes, the action potentials cause calcium-permeable channels in the surface membrane to open. The resulting influx of Ca^{2+} then triggers exocytosis.

Stimulation of some simple endocrine and exocrine cells leads to release of intracellular sequestered Ca^{2+} from the ER and to entry of Ca^{2+} from the extracellular medium. The resulting rise in free cytosolic Ca^{2+} induces hormone secretion (Figure 8-8C). For example, in pancreatic acinar cells, which secrete digestive enzymes, the stimulus induces production of inositol trisphosphate ($InsP_3$), a second messenger that then elicits the release of Ca^{2+} stored in the endoplasmic reticulum.

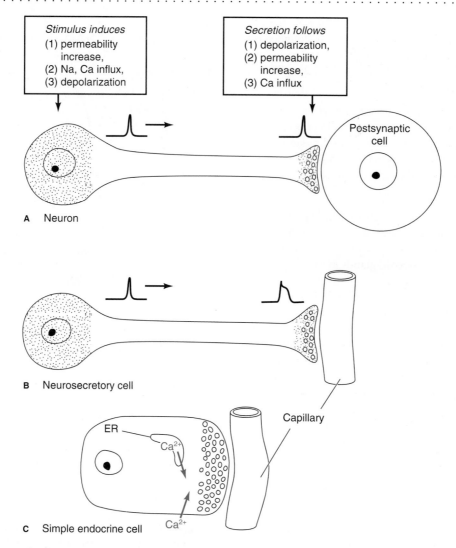

Figure 8-8 The elevation of Ca^{2+} in the output region triggers exocytosis in secretory cells. Depolarization is initiated in the input region and spreads to the output region by action potentials in ordinary neurons **(A)** and neurosecretory cells **(B)** or electronically in simple endocrine cells. Note the prolonged action potential characteristic of some neurosecre- tory terminals. Although some simple endocrine cells produce action potentials, many are activated to secrete without membrane depolarization. In these cells, the stimulus causes a release of Ca^{2+} stored in the ER, thereby increasing the cytosolic Ca^{2+} level **(C).**

GLANDULAR SECRETIONS

Glandular secretions result from the combined activity of a number of secretory cells. Secretion often occurs at a low "resting" level, which can be modulated up or down by signals acting on the gland. Some glands, however, exhibit no secretory activity until they are stimulated into action. For instance, the nasal gland of birds may be inactive while a bird drinks freshwater, but is activated to excrete salt following a drink of seawater. Various types of signals regulate glandular activity: neurotransmitters released from neurons innervating the glandular tissue or hormones released from other tissues. In addition, some glandular tissues respond directly or indirectly to conditions of the extracellular environment. For example, osmoregulatory neurons in the vertebrate hypothalamus respond to the osmotic pressure of the extracellular fluid bathing them, which, of course, reflects the osmotic pressure of the blood.

The salivary glands are under direct neural control, influenced by both conditioned and unconditioned reflexes. The sight and smell of food can cause a marked increase in salivation, especially if the animal is hungry. In fact even the thought of food can increase saliva production. By associating the ringing of bells with the appearance of food, Pavlov was able to train dogs to salivate when bells were rung (see Spotlight 15-1).

Types and General Properties of Glands

Glands are classified as either endocrine or exocrine glands (Figure 8-9). **Endocrine glands** are organs that secrete hormones directly into the circulatory system and modulate body processes; for example, the thyroid gland produces thyroid hormone, which modulates growth. Endocrine glands are sometimes referred to as the *ductless* glands. **Exocrine glands,** on the other hand, secrete fluids through a duct onto the epithelial surface of the body; for example,

A Exocrine glands

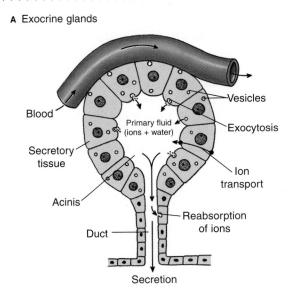

B Endocrine glands

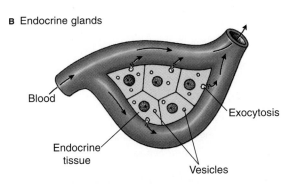

Figure 8-9 Glands can be divided into two broad structural types. **(A)** An exocrine gland, which releases secretions via a duct onto an epithelial surface. The primary fluid is formed by ion transport with water following osmotically. Mucin plus a variety of other compounds may be added to the primary fluid by exocytosis. The resulting primary secretion may be modified by reabsorption of material as the fluid passes down the duct. **(B)** An endocrine gland, which is ductless and releases secretions directly into the bloodstream. Water-soluble secretions are released by exocytosis of secretory vesicles, whereas lipid-soluble secretions may leave secretory cells by diffusion.

sweat glands produce sweat for evaporative cooling, and the gallbladder stores bile salts produced in the liver and excretes them into the gut via the bile duct.

Glands have been studied for centuries. Many symptoms of endocrine dysfunction were well known long before endocrine tissues had been identified and the function of their secretions determined. The study of endocrinology probably began in 1849 when A. A. Berthold reported his classical experiments in which he showed that castrated cockerels had small combs and wattles, showed little interest in hens or fighting, and had a weak crow (Figure 8-10). If he replaced the testes into the abdominal cavity, then the comb and wattle were large and the cock crowed and showed normal male behavior. Berthold speculated that the testes secreted some-

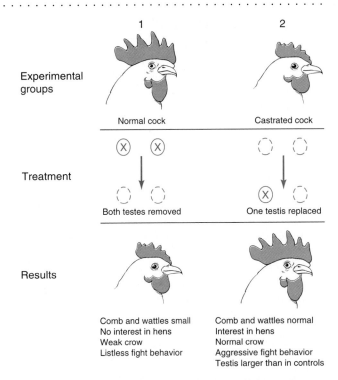

Figure 8-10 A. A. Berthold carried out some of the earliest experiments demonstrating endocrine action. When he removed the testes from chickens, they lost many of the characteristics of cocks (group 1). When one testis was replaced into the abdominal cavity, then the male characteristics were retained (group 2). Since Berthold's experiments, the functions of many endocrine glands have been identified by similar removal-and-replacement experiments. [Adapted from Hadley, 1992.]

thing that conditioned the blood and that the blood then acted on the cockerel to produce male characteristics.

The experiments carried out by Berthold paved the way for many other similar experiments in which the effects of removing and then replacing an organ were observed in order to demonstrate an endocrine function for that organ. Since Berthold's experiments were published, a large number of endocrine and exocrine glands, along with the chemical structure of their secretions and mode of action, have been described in detail. Although endocrine and exocrine glands usually can be distinguished by the presence (exocrine) or absence (endocrine) of ducts, there is no such thing as a typical gland.

 The effects of castration on cattle and humans were known long before Berthold's experiments. What was the important contribution made by Berthold?

Exocrine glands are more easily identified than endocrine glands because of their duct leading to the body surface. On the other hand, the various endocrine tissues are structurally and chemically diverse, and some contain more than one kind of secretory cell, each elaborating a different hormone. Endocrine glands exhibit no common morphologic plan or distinctive gross morphologic feature (other than a rich vascularization). For this reason, establishing unequivocally that a particular tissue suspected of having an endocrine function actually does and locating the site of secretion of a hormone have proved difficult in some cases.

The asymmetrical distribution of ion pumps on the apical and basolateral surfaces of secretory cells makes it possible for cells to pump ions from one side to the other. The movement of ions is followed by water. In many exocrine glands (e.g., the rectal gland of sharks, the nasal gland of birds, and the sweat glands of mammals), ions are transported into the lumen with water following osmotically, yielding a fluid secretion, termed the *primary fluid*. Sodium chloride often is the secreted salt, but occasionally potassium chloride is used. In some glands (e.g., sweat glands) some of the salt is reabsorbed in the duct leading from the secretory portion of the gland (see Figure 8-9A).

In many exocrine glands proteins, hormones, or other substances are added to the primary fluid by exocytosis of secretory vesicles. For example, mammary glands of mammals produce a primary fluid to which various substances, including hormones, are added before the milk is consumed by the young. In the salivary glands, amylase and glycoproteins are added to the primary fluid by exocytosis, as discussed in detail later. In some other exocrine glands, such as the sweat glands, the primary fluid does not contain much in the way of additives. Unlike exocrine glands, most endocrine glands release hormones directly into the bloodstream without the formation of a primary fluid.

In the following sections, one endocrine gland—the adrenal medulla—and two exocrine glands—the salivary gland and silk gland—are described in detail. Although presented as examples, each of these glands, like all glands, has its own special properties; thus they should not be considered as typical or characteristic of glands in general.

TABLE 8-1
Vertebrate endocrine glands and tissues

Gland/source	Hormone	Major physiological role*
Adrenal gland:		
Steroidogenic tissue (cortex)	Aldosterone	↑ Sodium retention
	Cortisol and corticosterone	↑ Carbohydrate metabolism and sympathetic function
Chromaffin tissue (medulla)	Epinephrine and norepinephrine	Multiple ↑ and ↓ effects on nerves, muscles, cellular secretions, and metabolism
Gastrointestinal tract	Cholecystokinin	↑ Secretion of enzymes by pancreatic acinar cell; ↑ gall bladder contraction
	Chymodenin	↑ Secretion of chymotrypsinogen from the exocrine pancreas
	Gastric inhibitory peptide	↓ Gastric acid (HCl) secretion
	Gastrin	↑ Gastric acid (HCl) secretion
	Gastrin-releasing peptide	↑ Gastrin secretion; ↓ gastric acid (HCl) secretion
	Motilin	↑ Gastric acid secretion and motility of intestinal villi
	Neurotensin	Enteric neurotransmitter
	Secretin	↑ Bicarbonate secretion by pancreatic acinar cells
	Substance P	Enteric neurotransmitter
	Vasoactive intestinal peptide	↑ Intestinal secretion of electrolytes
Heart (atrium)	Atrial natriuretic factor	↑ Salt and water excretion by kidney
Kidney	Calcitriol†	↑ Blood Ca^{2+}, bone formation, and intestinal absorption of Ca^{2+} and PO_4^{-3}
	Erythropoietin (erythrocyte-stimulating factor)	↑ Production of red blood cells (erythropoiesis)
Ovary:		
Preluteal follicle	Estradiol	↑ Female sexual development and behavior
Corpus luteum	Progesterone	↑ Growth of uterine lining and mammary glands, and maternal behavior
	Relaxin	↑ Relaxation of pubic symphysis and dilation of uterine cervix
Pancreas (islets of Langerhans)	Glucagon	↑ Blood glucose, gluconeogenesis, and glycogenolysis
	Insulin	↓ Blood glucose; ↑ protein, glycogen, and fat synthesis
	Pancreatic polypeptide	↑ ↓ Secretion of other pancreatic islet hormones

* ↑ means hormone stimulates or increases indicated effect; ↓ means hormone inhibits or decreases indicated effect.

†The final steps in synthesis of calcitriol from vitamin D_3 occur in the kidney, but the skin and liver also play a role in its synthesis.

Endocrine Glands

Table 8-1 lists the major endocrine glands and tissues in vertebrates, the hormones produced by each, and their physiological role. (The pituitary hormones and details of hormonal regulation and action are discussed in Chapter 9.) Endocrine tissues often are embedded in organs with nonendocrine functions. For example, cells within the atria of the heart produce atrial natriuretic peptide; this hormone is released into the bloodstream in response to factors such as a rise in venous pressure and helps regulate blood volume. Although the role of the atria of the heart in blood circulation has been known for centuries, its role in the production of atrial natriuretic peptide was realized only recently. Some hormone-like substances, including the prostaglandins and leukotrienes, are produced by all or nearly all tissues. Others, including some growth factors and the endorphins, are produced by various selected tissues.

Identifying and studying endocrine tissue

As noted already, the absence of discrete morphologic markers complicates the identification of endocrine glands. The following criteria have been used to establish whether a tissue has an endocrine function:

- Ablation (removal) of the suspected tissue should produce deficiency symptoms in the subject. Experimentally this may be difficult to demonstrate if the tissue is part of an organ that has more than one function (e.g., atrial tissue of the heart).

- Replacement (reimplantation) of the ablated tissue elsewhere in the body should prevent or reverse the deficiency symptoms. If the effects produced by removal of the tissue are due to the absence of a blood-borne substance produced by that tissue, replacement of the ablated tissue, which restores the missing hormone, should restore normal function. Misleading results,

Gland/source	Hormone	Major physiological role*
	Somatostatin	$\downarrow$ Secretion of other pancreatic islets hormones
Parathyroid glands	Parathormone	$\uparrow$ Blood Ca^{2+}; $\downarrow$ blood PO_4^{-3}
Pineal (epiphysis)	Melatonin	$\downarrow$ Gonadal development (antigonadotropic action)
Pituitary gland	See Chapter 9	
Placenta	Chorionic gonadotropin (choriogonadotropin)	$\uparrow$ Progesterone synthesis by corpus luteum
	Placental lactogen	$\uparrow$ Fetal growth and development (possibly); $\uparrow$ mammary gland development in the mother
Plasma angiotensinogen‡	Angiotensin II	$\uparrow$ Vasoconstriction and aldosterone secretion; $\uparrow$ thirst and fluid ingestion (dipsogenic behavior)
Testes:		
Leydig cells	Testosterone	$\uparrow$ Male sexual development and behavior
Sertoli cells	Inhibin	$\downarrow$ Pituitary FSH secretion
	Müllerian regression factor	$\uparrow$ Müllerian duct regression (atrophy)
Thymus gland	Thymic hormones	$\uparrow$ Proliferation and differentiation of lymphocytes
Thyroid gland:		
Follicular cells	Thyroxine and triiodothronine	$\uparrow$ Growth and differentiation; $\uparrow$ metabolic rate and oxygen consumption (calorigenesis)
Parafollicular cells (or ultimonbranchial glands)	Calcitonin	$\downarrow$ Blood Ca^{2+}
Most or all tissues	Leukotrienes	$\uparrow\downarrow$ Cyclic nucleotide formation
	Prostacyclins	$\uparrow$ Cyclic nucleotide (cAMP) formation
	Prostaglandins	$\uparrow$ Cyclic nucleotide (cAMP) formation
	Thromboxanes	$\uparrow$ Cyclic nucleotide (cGMP?) formation
Selected tissues	Endorphins	Opiate-like activity
	Epidermal growth factor	$\uparrow$ Epidermal cell proliferation
	Fibroblast growth factor	$\uparrow$ Fibroblast proliferation
	Nerve growth factor	$\uparrow$ Neurite development
	Somatomedins	$\uparrow$ Cellular growth and proliferation

‡Angiotensinogen is produced in the liver and circulates in the bloodstream, where it is cleaved by renin to form the active hormone angiotensin II.

Source: Adapted from Hadley, 1992.

however, may be obtained when ablation-and-reimplantation experiments are done with tissues closely associated with the nervous system, because of the interruptions of neural connections.

- The deficiency symptoms should be relieved by replacing the suspected hormone by injection. Successful replacement is the most important criterion for identification of a suspected endocrine tissue and its hormone. It is also the basis of replacement therapy for patients with a dysfunctional endocrine gland.

- Following purification of the suspected hormone, the chemical structure of the active substance is determined. The molecule is then synthesized and tested for biological potency.

- Immunohistochemistry can be used to determine the cellular location of the hormone in different tissues once it has been isolated.

A variety of techniques have resulted in the rapid development of endocrinology over the last two decades. For example, **radioimmunoassays** (RIAs) permit detection of specific hormones in minute concentrations with a high degree of accuracy. Antibodies are raised against the hormone in question, usually in a rabbit. A standard curve is then constructed to describe the binding of the hormone to the antibody, using a radiolabeled hormone and a known amount of antibody. Unlabeled hormone will compete with and therefore reduce the extent of binding of the labeled hormone. Thus the quantity of hormone in a sample can be determined by the extent to which it reduces the binding of labeled hormone to the antibody. The development of RIA has resulted in many new insights into the synthesis, secretion and function of many hormones and many other substances. The use of **monoclonal antibodies,** which recognize only one antigen, has improved the accuracy of detecting and quantifying hormones and their receptors by RIA.

Endocrinologists also use recombinant DNA techniques in various ways. For instance, genetic material can be inserted into bacteria to produce strains capable of synthesizing human hormones. Foreign genes have also been introduced into mammalian embryos; for example, when the structural gene for rat growth hormone is introduced into mouse embryos, the resulting mice grow much larger than normal.

Mammalian adrenal medulla

The paired mammalian adrenal glands, as the name implies, are situated near the kidneys, one attached to the superior end of each kidney (Figure 8-11). Each adrenal gland is in fact two glands in one: an outer layer, the adrenal cortex, surrounds an inner portion, the adrenal medulla (Figure 8-12). The two portions of the mammalian adrenals are of quite different origin. The cells of the cortex are

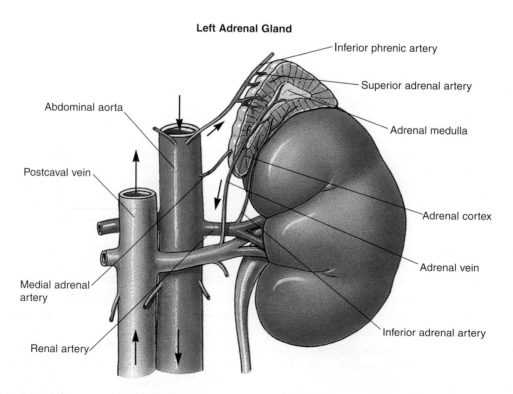

Left Adrenal Gland

Inferior phrenic artery

Superior adrenal artery

Adrenal medulla

Abdominal aorta

Postcaval vein

Adrenal cortex

Adrenal vein

Medial adrenal artery

Inferior adrenal artery

Renal artery

Figure 8-11 The adrenal glands in mammals are attached at the rostral ends of the kidneys. Two arteries enter the glands through the capsule and branch into smaller vessels, which pass into the centrally located medulla. Thus hormones produced in the cortex and released into the blood are carried into the medulla, which is drained by the inferior phrenic vein.

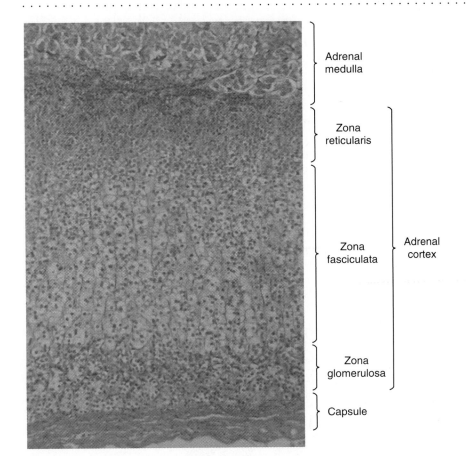

Adrenal
medulla

Zona
reticularis

Zona
fasciculata

Adrenal
cortex

Zona
glomerulosa

Capsule

Figure 8-12 Mammalian adrenal glands have a recognizable cortex and medulla, which produce different hormones. This light micrograph reveals the outer capsule, the three concentric layers of the cortex, and the underlying medulla. The zona glomerulosa, the outermost cortical layer, secretes mineralocorticoids; the zona fasciculata and the zona reticularis secrete glucocorticoids. The adrenal medulla secretes two catecholamines, epinephrine and norepinephrine. [Courtesy of Frederic H. Martini.]

derived from mesodermal tissue, whereas those of the medulla are derived from epidermal tissue. The adrenal cortex produces steroid hormones involved in blood ion and glucose regulation and anti-inflammatory reactions (see Chapter 9).

The cells of the adrenal medulla, on the other hand, produce **catecholamines**, namely, **epinephrine** and **norepinephrine**. Epinephrine and norepinephrine released from sympathetic nerves and the adrenal medulla have numerous cardiovascular and metabolic effects, which in total constitute the *fight-or-flight reaction*. For example, plasma epinephrine levels can be elevated in a cat when it hears a dog bark. This fight-or-flight reaction, or syndrome, is a response to stress in which various tissues are activated and the body is mobilized to either attack or flee from the object of stress. Catecholamines are not simply released during fight-or-flight situations, but are released under a wide variety of physiological conditions, for example, during heavy exercise or even when humans move from a sitting to a standing position.

Adrenal medulla cells are referred to as **chromaffin cells** because they stain easily with chromium salts. The chromaffin cells that produce norepinephrine have dark-staining irregular granules, whereas those that produce epinephrine have light-staining, spherical granules. Chromaffin cells are modified postganglionic sympathetic neurons. A small number of cells within the medulla, which are somewhat intermediate between chromaffin cells and neurons, are referred to as small-granule chromaffin cells. Under some conditions the chromaffin cells will grow into

typical postganglionic sympathetic neurons. They are prevented from doing so by the presence of high concentrations of glucocorticoid hormones released from the surrounding cortex into the blood flowing from the cortex to the medulla (see Figure 8-11).

Synthesis of catecholamines The production and release of catecholamines including epinephrine and norepinephrine is outlined in Figure 8-13. The secretory granules within a single chromaffin cell contain either norepinephrine or epinephrine, and each cell secretes one or the other catecholamine. The granules also contain enkephalin, ATP, and several acidic proteins called *chromogranins*. The catecholamines within the granule are probably bound to these chromogranins, which are polymers maintained in the condensed state by the shielding action of catecholamines within the granule. Once a pore is opened in the vesicle, the catecholamines begin to diffuse out and the chromogranin polymer rapidly expands, propelling the contents of the vesicle into the extracellular space.

Norepinephrine is synthesized from tyrosine, with dopa and dopamine as intermediate compounds (Figure 8-14). Conversion of tyrosine to dopamine occurs in the cytosol, catalyzed by tyrosine hydroxylase and dopa decarboxylase, which are cytosolic enzymes. Dopamine is then incorporated into the granules and converted to norepinephrine; this reaction is catalyzed by dopamine β-hydroxylase (DBH) contained in the secretory granules. Norepinephrine is methylated to epinephrine, a reaction catalyzed by

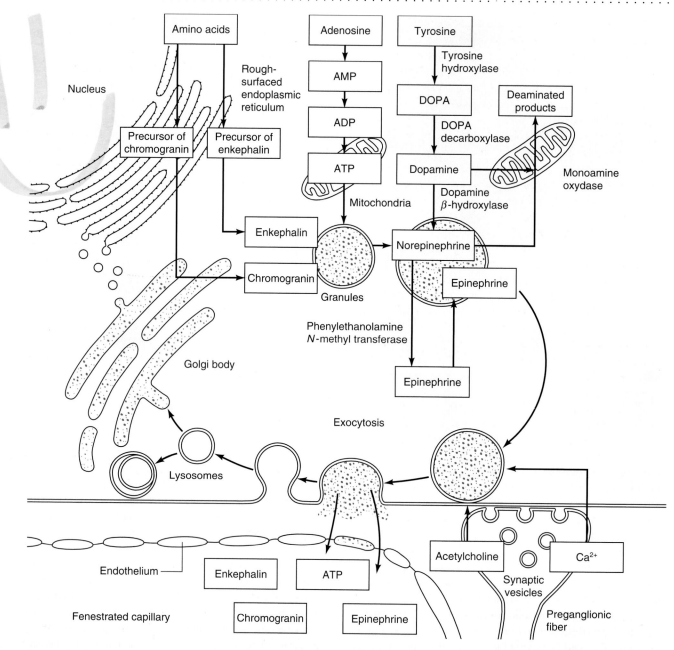

Figure 8-13 Secretory vesicles in chromaffin cells of the adrenal medulla contain catecholamines, enkephalin, ATP, and chromogranin, which are synthesized in different cellular compartments. In epinephrine-producing cells (shown here), norepinephrine leaves the secretory vesicles to be converted to epinephrine and then is reincorporated into the vesicles. Stimulation of chromaffin cells by acetylcholine, which is liberated from the endings of the preganglionic fibers, triggers release of the granule contents by exocytosis. The neural stimulus increases the membrane permeability for Ca^{2+}, leading to the increased intracellular Ca^{2+} required for exocytosis. [Adapted from Matsumoto and Ischii, 1992.]

phenylethanolamine N-methyl transferase, which is found in the cytosol. Thus, norepinephrine must leave the secretory granules to be converted to epinephrine, which then reenters the granules (see Figure 8-13).

Although chromaffin tissue and steroidogenic tissue are associated in the adrenal glands in mammals, this is not the case in all vertebrates. In fish, for instance, the chromaffin tissue is separate from the steroidogenic cells, but both are still in the general region of the kidney; the chromaffin tissue is associated with blood vessels, whereas the steroid-producing cells are embedded in the kidney. The close as-

sociation of the adrenal cortex and medulla in mammals is of functional significance. In those tissues where there is close association of the chromaffin cells (adrenal medulla) with steroidogenic tissue (adrenal cortex), as in the mammalian adrenals, most of the chromaffin cells produce epinephrine. As noted already, blood entering the medulla has passed through the cortex and thus carries high levels of glucocorticoid hormones (see Figure 8-11). In the medulla these glucocorticoids promote the synthesis of phenylethanolamine N-methyl transferase, the enzyme that catalyzes conversion of norepinephrine to epinephrine.

On the other hand, when chromaffin tissue isolated from the influence of steroidogenic tissue, as in the dogfish, it produces more norepinephrine than epinephrine. The human fetus has some isolated chromaffin tissue, which contains norepinephrine rather than epinephrine, presumably because of the absence of the action of steroidogenic tissue. Postganglionic sympathetic nerves also produce norepinephrine for the same reason, that is, the absence of a marked influence of steroid hormones.

Release of catecholamines The release of epinephrine and norepinephrine from the adrenal medulla is controlled by the action of preganglionic sympathetic nerves (Figure 8-15). These preganglionic fibers are cholinergic; that is, they release acetylcholine as a neurotransmitter. When the chromaffin cells are stimulated by acetylcholine, their membrane conductance for Ca^{2+} increases, resulting in an influx of Ca^{2+} and elevation of intracellular Ca^{2+} levels; this rise in intracellular Ca^{2+} in turn causes the further release of both epinephrine and norepinephrine by exocytosis (see Figure 8-13). Catecholamines cause an increase in blood flow to the adrenals, and this effect also augments catecholamine release from the adrenal medulla. Thus the release of catecholamines has a positive feedback on further catecholamine release. (The release of norepinephrine from postganglionic sympathetic nerves, however, inhibits further norepinephrine release from these nerve endings. In this case, negative feedback operates.) ATP is stored in the granules of chromaffin cells and released along with catecholamines. ATP and its breakdown product adenosine, which inhibit release of catecholamines by reducing calcium influx, provide negative-feedback control on catecholamine release from the medulla. Hypoxia also stimulates catecholamine release from chromaffin cells. When chromaffin cells are not innervated (e.g., those located in the hagfish heart), hypoxia is an important stimulus for catecholamine release.

Catecholamines released into the extracellular fluid are rapidly taken up and either stored in secretory vesicles or destroyed by monoamine oxidase located on the outer membrane of mitochondria (see Figure 8-13). Catecholamines in the extracellular space are catabolized by catecholamine-O-methyltransferase, especially in the liver and kidney, and the breakdown products are excreted. The actual level of catecholamines circulating in the blood thus depends on the balance between their release, uptake, and catabolism. Although the level of catecholamines in the blood is dominated by release from the adrenal medulla, release from postganglionic sympathetic nerves contributes significant amounts to blood levels. Adrenergic nerves release norepinephrine, whereas the medulla releases mainly epinephrine, so the relative activity of nerves and the medulla will also influence the relative levels of epinephrine and norepinephrine in the blood. Catecholamine levels in the blood can remain elevated for only a few minutes in humans, but they can remain high for several hours following exhaustive exercise in fish.

Effects and regulation of catecholamines Epinephrine and norepinephrine bind with **adrenergic receptors,** also termed **adrenoreceptors,** in cell membranes. This binding then activates one of a number of intracellular second messengers, leading to a particular tissue response. (These pathways are described in detail in Chapter 9.) In a paper published in 1948, R. P. Ahlquist concluded that there are two types of adrenoreceptors—α and β—that differ in their sensitivity to sympathetic amines. More recent studies have demonstrated several subtypes of both α- and β-adrenoreceptors based on the ability of various drugs to either activate or block receptor activity (Figure 8-16).

The α_1-adrenoreceptors mediate smooth muscle contraction in many tissues. Stimulation of these receptors results in activation of the inositol trisphosphate (InsP$_3$) pathway, leading to elevation of intracellular InsP$_3$ (Figure 8-17). Elevated InsP$_3$ causes release of calcium from stores within the cell; the resulting rise in cytosolic calcium causes muscle contraction (see Chapter 9). There is evidence that there are subtypes of α_1-adrenoreceptors in different tissues. The α_2-adrenoreceptors located in presynaptic cells at noradrenergic synapses cause inhibition of norepinephrine release, an action mediated by an inhibitory effect on adenylate cyclase. Thus, these receptors are part of a short negative-feedback loop in which the release of norepinephrine inhibits further release of norepinephrine. This is sometimes referred to as autoinhibition. There are also α_2-adrenoreceptors located on some postsynaptic sites in liver, brain, and some smooth muscle.

The β-adrenoreceptors also are divided into two subtypes, β_1- and β_2-adrenoreceptors, both of which activate adenylate cyclase, leading to an increase in cAMP (see Figure 8-17). Stimulation of β_1-adrenoreceptors, largely due to neuronal release of norepinephrine, results in increased contraction of cardiac muscle and the release of fatty acids from adipose tissue, whereas stimulation of β_2-adrenoreceptors, largely due to elevated levels of circulating catecholamines, mediates bronchodilation and vasodilation. The elevation of cAMP resulting from stimulation of β_1-adrenoreceptors increases calcium conductance, thereby raising the intracellular calcium level, which in turn augments muscle contraction. In contrast, elevation of cAMP following stimulation of β_2-adrenoreceptors causes activation of the calcium pump rather than an increase in calcium conductance. Thus, following β_2-adrenoreceptor stimulation, calcium is both sequestered within and extruded from the cell, so intracellular calcium levels fall, promoting muscle relaxation.

The following generalizations can be made concerning the roles of adrenoreceptors:

- α-Adrenoreceptors mediate smooth muscle contraction (except in intestinal smooth muscle) and (with a few exceptions) inhibit cellular secretion.

- β-Adrenoreceptors mediate smooth muscle relaxation (except in cardiac muscle) and stimulate cellular secretion.

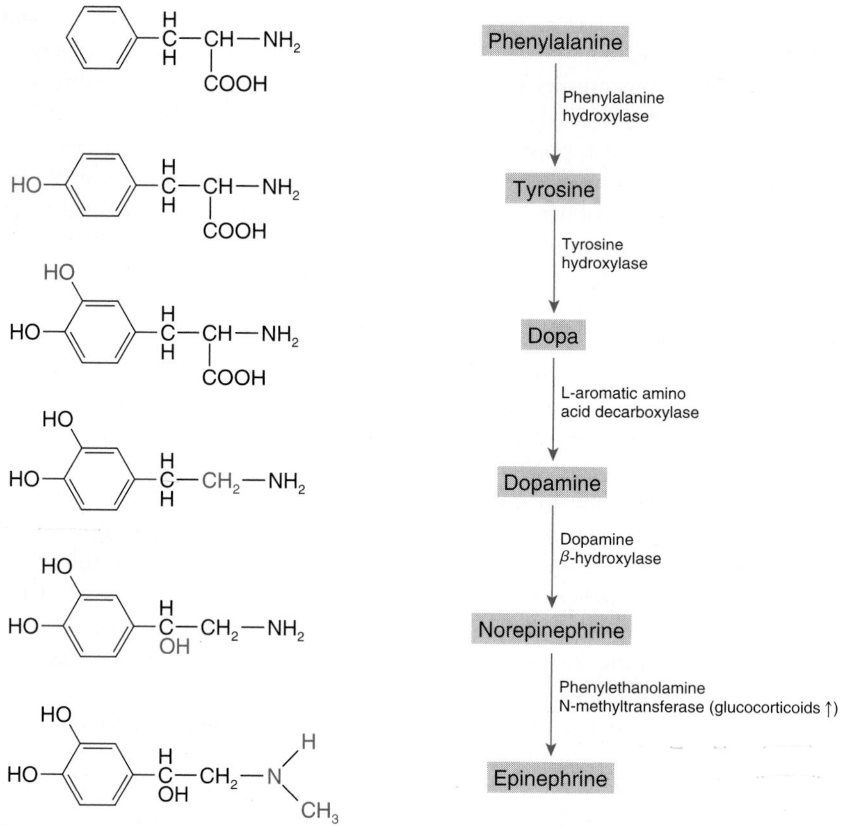

Phenylalanine

Phenylalanine
hydroxylase

Tyrosine

Tyrosine
hydroxylase

Dopa

L-aromatic amino
acid decarboxylase

Dopamine

Dopamine
β-hydroxylase

Norepinephrine

Phenylethanolamine
N-methyltransferase (glucocorticoids ↑)

Epinephrine

Figure 8-14 The catecholamines—dopamine, norepinephrine, and epinephrine—are synthesized from phenylalanine and tyrosine. Glucocorticoids produced by the adrenal cortex increase the activity of phenyl- ethanolamine *N*-methyltransferase and, therefore, promote conversion of norepinephrine to epinephrine.

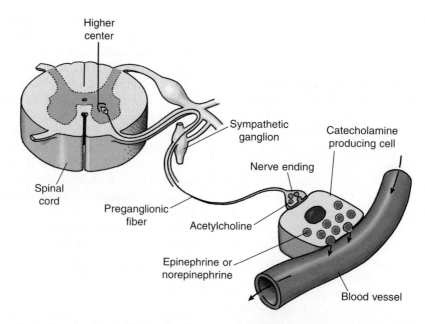

Figure 8-15 Hormone secretion by the adrenal medulla is regulated by neural stimuli. Sympathetic nerve axons originating in the spinal cord pass through the sympathetic ganglia without synapse formation, but then synapse on the catecholamine-producing cells. Acetylcholine liberated from these preganglionic nerve endings stimulate the secretion of medullary hormones.

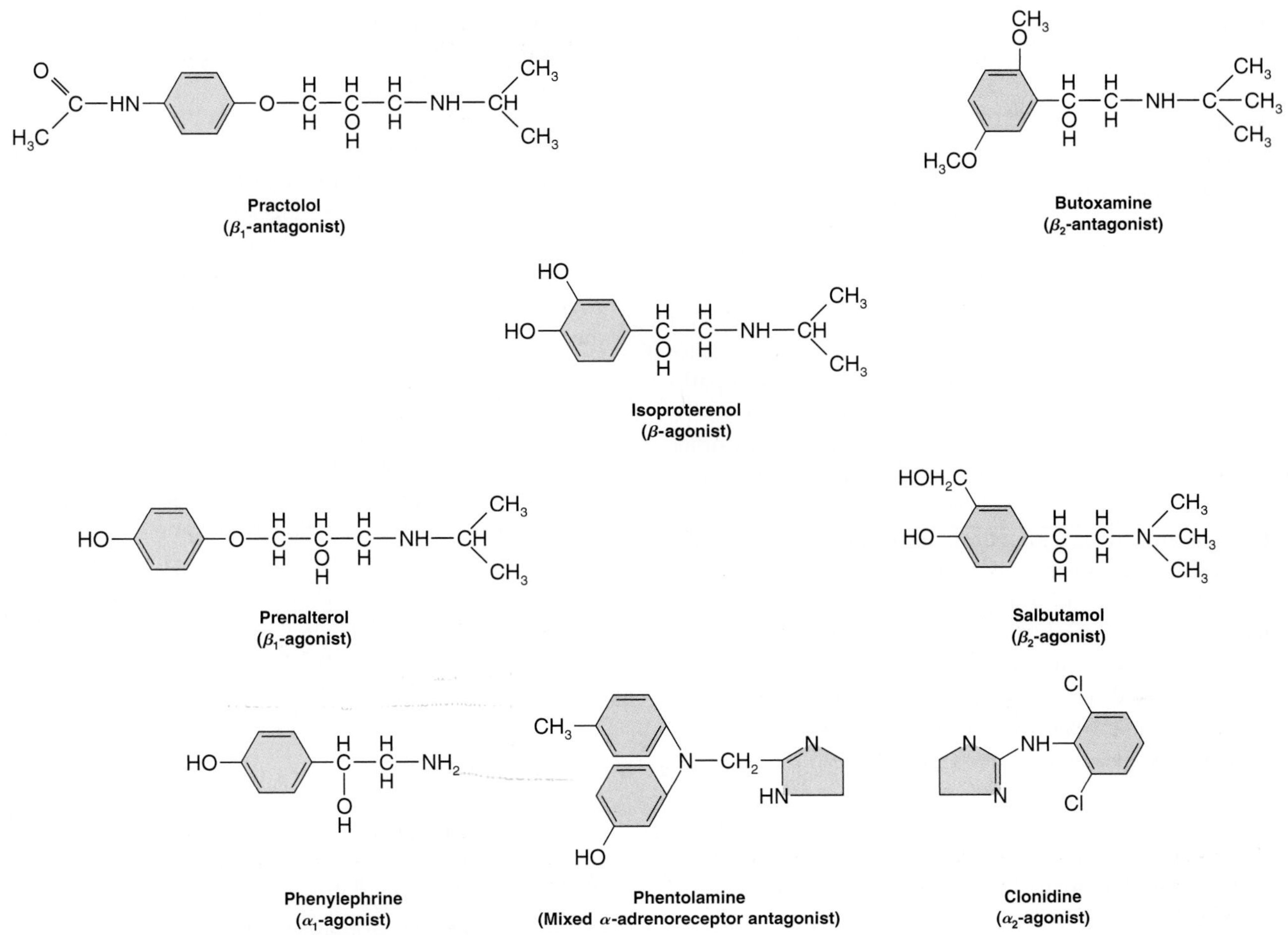

Practolol
(β_1-antagonist)

Butoxamine
(β_2-antagonist)

Isoproterenol
(β-agonist)

Prenalterol
(β_1-agonist)

Salbutamol
(β_2-agonist)

Phenylephrine
(α_1-agonist)

Phentolamine
(Mixed α-adrenoreceptor antagonist)

Clonidine
(α_2-agonist)

Figure 8-16 A variety of drugs can activate (agonists) or block (antagonists) adrenoreceptors. These drugs have been used to identify adrenoreceptor subtypes and to determine the effects of catecholamines on different tissues.

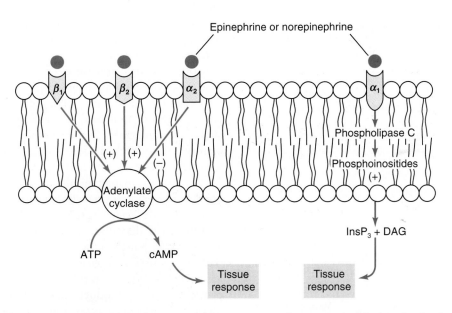

Figure 8-17 Binding of catecholamines to α_1-, α_2-, β_1-, or β_2-adrenoreceptors either activates (+) or inhibits (−) a second-messenger pathway. Adrenoreceptor signal transduction occurs via the adenylate cyclase or the membrane phospholipid pathway. The former involves cAMP as a second messenger and the latter involves inositol trisphosphate (InsP$_3$) and diacylglycerol (DAG) as second messengers. See Figures 9-11 and 9-13 for more details. [Adapted from Hadley, 1992.]

- If β-adrenoreceptors cause relaxation in a tissue, then cholinergic eceptors promote contraction in that same tissue.

- If α-adrenoreceptors cause contraction in a tissue, then cholinergic receptors normally regulate relaxation in that same tissue.

The physiological action of catecholamines is quite variable and is influenced by other factors. For example, neuropeptide Y (NPY), which sometimes is co-released with norepinephrine from adrenergic nerves, modulates the action of catecholamines on the InsP₃ second-messenger pathway, augmenting the action of catecholamines in some tissues and reducing their action in other tissues. Many other factors can modulate both the release and action of catecholamines. Adenosine, for instance, has been shown to inhibit catecholamine release from the bovine adrenal medulla by reducing calcium flux. Adenosine is released from tissues during hypoxia, but it is rapidly destroyed in the blood, so its effects are limited to the region of production.

Alterations in adrenoreceptor density in the membrane of target cells also can modify catecholamine action. An increase in receptor density is referred to as *up-regulation;* a decrease, as *down-regulation*. Continual exposure to catecholamines can lead to down-regulation of receptor concentration and, therefore, a decreased response. Sympathetic denervation can lead to up-regulation of receptors and, therefore, to increased sensitivity of a tissue to circulating catecholamines. Adrenoreceptor density also can be affected by other substances, in particular steroid hormones. Glucocorticoids not only modulate adrenoreceptor density but also have other effects on the action of catecholamines. In the estrogen-dominated uterus, stimulation of β-adrenoreceptors causes contraction, whereas it causes relaxation during pregnancy.

In summary, then, the action of catecholamines depends on the rate and site of their release (i.e., from the adrenal medulla or nerves) and the rate of uptake and/or breakdown of the catecholamines once released. In addition, the type and distribution of receptors on target tissues, and the up- or down-regulation of these receptors due to past experience, have a marked effect on the nature and magnitude of the response. The presence or absence of steroids can influence both adrenoreceptor density and the levels of enzymes involved in the conversion of norepinephrine to epinephrine; the latter effect modulates the ratio of these two compounds released into the blood. The level and nature of gonadal steroids, for example, can alter the response of the uterus to catecholamines from contraction to relaxation. Finally other substances, such as ATP, adenosine, and neuropeptide Y, can modulate the release and action of catecholamines. As a result of these various modulating mechanisms, physiological responses to epinephrine and norepinephrine vary widely depending on both the tissue in question and the physiological state of the animal (Table 8-2).

What are the advantages of having so many sites for modulating catecholamine activity? List the possible sites of modulation. Are all sites of modulation equally important?

Exocrine Glands

Unlike endocrine secretions, the output of an exocrine gland does not ooze into the circulation but generally flows through a duct into a body cavity (e.g., the mouth, gut, nasal passage, or urinary tract) that is in continuity with the exterior. As noted earlier, exocrine secretions usually are aqueous mixtures, consisting of a water-based primary fluid and added components, rather than a single substance. In the alimentary canal, these mixtures typically consist of water, ions, enzymes, and mucus. Exocrine tissues of the alimentary canal include the salivary glands, secretory cells in the stomach and intestinal epithelium, and secretory cells of the liver and pancreas.

An exocrine gland typically consists of an invaginated epithelium of closely packed secretory cells lining a blind cavity called the **acinus** (see Figure 8-9A). Several acini connect to a small duct that, in turn, connects to a larger duct leading to the lumen of the digestive canal or to some part of the body surface. The basal surfaces of the epithelial cells are usually in close contact with the circulation. Once the primary secretory products are free in the acinar lumen, they generally become secondarily modified in the secretory duct. This modification can involve further transport of water and electrolytes into or out of the duct to produce the final secretory juice.

TABLE 8-2
Physiological responses to epinephrine and norepinephrine

Variable	Response to	
	Epinephrine	Norepinephrine
Heart rate	Increase	Decrease*
Cardiac output	Increase	Variable
Total peripheral resistance	Decrease	Increase
Blood pressure	Rise	Greater rise
Respiration	Stimulation	Stimulation
Skin vessels	Constriction	Constriction
Muscles vessels	Dilatation	Constriction
Bronchus	Dilatation	Less dilatation
Metabolism	Increase	Slight increase
Oxygen consumption	Increase	Slight effect
Blood sugar	Increase	Slight increase
Kidney	Vasoconstriction	Vasoconstriction

*This effect is secondary to peripheral vasoconstriction that raises blood pressure. In the isolated heart, norepinephrine increases the rate.

Source: Bell et al., 1972.

(margin handwritten note: *Adenosine*)

Exocrine glands (e.g., sweat glands) are classified as apocrine or eccrine glands based on their structure. An *eccrine* gland has a coiled, unbranched duct that leads from the secretory region; the duct opening lies perpendicular to the body surface. Eccrine glands respond to elevated temperatures by secreting a clear fluid that evaporates and cools the body. An *apocrine* gland has a branched duct leading from the secretory region to the surface. Apocrine glands often produce a turbid or white secretion, which can be released by an apocrine, merocrine, or holocrine secretory mechanism, not simply by apocrine secretion, as the name of these glands might suggest. This extensive, and often confusing, terminology is a reflection of the vast array of glands with a multiplicity of functions found in animals.

Vertebrate salivary gland

The saliva present in the human mouth is a complex mixture consisting of secretions from a number of salivary glands, bacteria normally resident in the mouth, epithelial cells, and the remains of food and drink and whatever else has been in the mouth. This complex liquid is referred to as *whole saliva* to distinguish it from *duct saliva* released from an individual gland. Whole saliva is around 99.5% water, has a pH of 5.0–8.0, and contains a variety of ions (Table 8-3).

Functions and flow of saliva Saliva has many functions. First, it *lubricates the mouth and surrounding regions*, thereby facilitating speech, eating, and swallowing. By dissolving and diluting food and drink placed in the mouth, saliva assists in swallowing and allows food to be tasted. Second, saliva *controls the bacterial flora* in the mouth by inhibiting the growth of some bacteria and promoting the growth of others. The antibacterial action of saliva depends on three components: (a) lysozyme, which causes bacterial lysis; (b) lactoferrin, which removes from saliva the free iron needed for growth of some bacteria; and (c) sialoperoxidase, which oxidizes thiocyanate to hypothiocyanate, a potent antibacterial agent. Third, the enzyme salivary amylase present in saliva *initiates the digestion of starch*. The high pH of saliva promotes this action and, in addition, is an effective buffer that protects the tissues of the oral cavity. Although saliva is not essential for digestion of food, impairment of saliva production makes chewing and swallowing difficult, and leads to rotten teeth. In addition, as most of us know, it is difficult to talk or sing with a dry mouth. Finally, saliva permits some soccer and baseball players to express themselves by spitting, and it allows babies to blow bubbles and gurgle.

A slow resting flow of saliva keeps the mouth moist. A circadian rhythm maintains minimum flow during the night especially during sleep. Dehydration and stress activate the sympathetic system and reduce flow, causing the dry mouth that typically accompanies fear or anxiety. Flow increases on the anticipation or the sight and smell of food, especially if hungry. The activation of muscle and tendon stretch receptors associated with jaw movements during

TABLE 8-3
Inorganic constituents of whole saliva (mg · 100 ml⁻¹)

Constituent	Range	Mean
Sodium	0–80	15 resting 60 stimulated
Potassium	60–100	80
Calcium	2–11	6
Phosphorus (inorganic)	6–71	17 resting 12 stimulated
Chloride	50–100	—
Thiocyanate	—	9 (smokers) 2 (nonsmokers)
Fluoride (parts/10⁶)	0.01–0.04	0.03 resting 0.01 stimulated
Bicarbonate	0–40	6 resting 36 stimulated
pH	5.0–8.0	—

Source: Edgar, 1992.

eating also causes an increase in saliva production. The increase in flow during eating depends partly on taste sensations. In general, sour tastes cause the greatest increase in salivation, followed by sweet, salt, and bitter in decreasing order of effectiveness in promoting flow. Salivation also is common before vomiting; this increased flow presumably protects the oral membranes by diluting and buffering the vomit.

Formation of saliva Saliva is formed as a *primary secretion* in the acini and is then modified during passage through the ducts. NaCl is secreted into the acinus with water following down an osmotic gradient. Amylase, mucous glycoproteins, and proline-rich glycoproteins are added to this fluid by exocytosis. The production of saliva, unlike other digestive exocrine secretions, is solely under neural control, and the salivary glands are innervated by both sympathetic and parasympathetic nerves. The sympathetic nerves innervating the glands release norepinephrine, which increases the production of amylase and other proteins but causes vasoconstriction and a decrease in saliva production. Parasympathetic stimulation, which is mediated by acetylcholine, substance P, and vasointestinal polypeptide (VIP), causes vasodilation and increases salivation. Chewing tobacco mimics the effects of parasympathetic stimulation causing a large increase in the flow of saliva.

The binding of acetylcholine, substance P, and norepinephrine to the appropriate receptors on the basal membrane of an acinar cell leads to activation of phospholipase C, which catalyzes formation of diacylglycerol (DAG) and inositol trisphosphate (InsP$_3$) from phosphatidyl inositol biphosphate (Figure 8-18). The inositol trisphosphate so formed stimulates the release of Ca^{2+} from the endoplasmic reticulum, which in turn causes opening of potassium channels in the plasma membrane, leading to increased K^+ conductance and K^+ efflux from the cell. A rise in the external K^+ level activates a Na-KCl cotransporter, and K^+, Na^+,

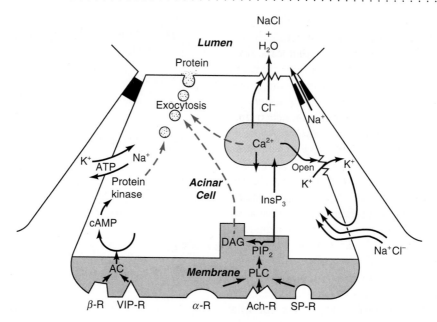

Figure 8-18 Production and release of the primary secretion by salivary gland acinar cells is under neural control. Stimulation of α-adrenoreceptors (α-R), acetylcholine receptors (Ach-R), and substance P receptors (SP-R) activates phospholipase C (PLC). This enzyme splits phosphatidyl inositol biphosphate (PIP_2) into diacylglycerol (DAG) and inositol trisphosphate ($InsP_3$), leading to release of stored Ca^{2+} and opening of potassium channels. As a result of various ion movements, NaCl and water enter the lumen. Exocytosis of amylase and glycoproteins stored in secretory granules is promoted by activation of the adenylate cyclase (AC) pathway due to stimulation of receptors for vasoactive intestinal peptide (VIP-R) and β-adrenoreceptors (β-R). DAG and increased cytosolic Ca^{2+} also promote exocytosis. The primary secretion released into the acinus lumen is modified as it passes through the salivary gland duct. See text for further discussion. [Adapted from Edgar, 1992.]

and Cl^- enter the cell. These movements of Na^+ and K^+ are counteracted by a Na^+-K^+ ATPase, which maintains Na^+ and K^+ levels in the cell. Thus, Na^+ and K^+ are cycled through the membrane, and the only net transfer is the inward movement of Cl^- ions, which move across the cell and leave via the apical (luminal) membrane. That is, there is a net movement of Cl^- across the acinar cell from the blood to the lumen of the gland. This generates a transepithelial potential that is positive on the blood side and creates the driving force for diffusion of Na^+ through paracellular channels from the blood to the lumen. This movement of NaCl into the lumen establishes the osmotic gradient generating the flow of water into the lumen.

Binding of norepinephrine to β-adrenoreceptors or of vasoactive intestinal peptide to peptidergic receptors activates the adenyl cyclase pathway (see Figure 8-18). This results in the formation of cAMP, which in turn activates a protein kinase that stimulates exocytosis. Diacylglycerol, formed in the phospholipase C pathway, also promotes exocytosis of amylase, mucous glycoproteins, and proline-rich glycoproteins into the lumen.

The primary secretion, which consists of water, sodium chloride, amino acids, proteins, and glycoproteins, is forced into the salivary duct by the formation of more fluid. As it passes down the duct, potassium bicarbonate is added to the fluid and some sodium is reabsorbed. Because less sodium is reabsorbed at high flow rates, the final product leaving the duct approaches the composition of the primary secretion during heavy salivation. In contrast, the bicarbonate level in the final secretion does not fall, but in fact rises, with increased flow. The addition of bicarbonate to the fluid must somehow be coupled to the flow rate, such that an increase in flow promotes bicarbonate addition to the duct fluid.

Invertebrate silk gland

The number and variety of glands in invertebrates is probably larger than that in vertebrates. The silk gland is described here not so much because it is representative of a large number of invertebrate exocrine glands, but because it is reasonably well understood. Many insects and spiders produce silk threads from silk glands to make webs and spin cocoons. The silkworm, *Bombyx mori*, is raised commercially for its larvae, which spin a protective cocoon. Each cocoon produced by a pupating larva consists of about 275 m of silk thread. Commercial silk thread is made by weaving threads from several cocoons together.

The production of silk cloth began in China some 4000 years ago. Silk brought to Europe over the Silk Road from China was used to make robes for Roman emperors. Silk cloth was worn by the horsemen of Genghis Khan for protection, because arrows do not penetrate silk easily and arrows can be removed by pulling on the silk thrust into the wound. This strong light material was one of several reasons for the great military success of the Mongols. Silk from the silk moth is still used to produce fabric for today's consumers. Although spider silk is stronger than that of the silk moth, there has been little or no commercial use of the silk made by spiders; certainly European knights never went into battle covered by spider silk for protection!

Spider silk and webs Spiders are a very prolific group. It has been estimated that an acre of meadow in England may contain over two million spiders. One reason for the success of this group is that the over 30,000 species of spiders spin silk. Spider silk is made by *spinneret glands* on the underside of the abdomen, which exude a liquid that hardens into silk threads once it leaves the gland. These silk threads are used as a dragline and to make a variety of webs, silken egg cases, and silk-lined tunnels. The main, but not the only, function of webs is to capture prey, such as insects and other small animals, which become entangled in, or stuck to, the web. The webs vibrate easily, and the spider can detect the position and the nature of the animal in the web by the pattern of vibration. Different behavioral responses are evoked by different patterns of vibration of the web. A male, wishing to be recognized as a potential mate rather than food, vibrates the web in a species-specific pattern to evoke the appropriate response from the female.

Not all spiders make webs. The large, nonpoisonous tarantula spider, *Lycosa tarantula,* relies instead on its speed to capture prey. Most spiders bite their captured prey, us.ng their fangs to inject poisons to subdue and digest their victims, after which the spider sucks out the digested juices. Only a few species of spider are dangerous to humans. The male black widow spider, *Lactrodectus mactans,* is harmless, but the much larger female has a venomous bite; although a bite causes pain and fever, human victims usually survive. Female black widow spiders are about 1.3 cm long and have a large red patch below the abdomen.

Spiders continually produce silk thread, which can trail behind them as a dragline and is fastened at intervals to the substratum. Spiders can become aerial by swinging from their dragline attached to a bush or tree; bungie-jumping is a common event in the daily life of many spiders. Spider webs undoubtedly evolved from this dragline, with the simplest webs being sticky threads hanging in the breeze to catch insects. More complex webs are constructed in either two or three dimensions and have intricate designs for snaring prey. Webs are set up in the flight paths of insects and other small animals. Some webs are vertical, whereas others are horizontal to catch insects flying up from the ground. In many webs, some of the threads break as the insect hurtles into the web; the more the insect struggles, the more tangled it becomes in the web. In others, portions of the web are sticky and the prey becomes attached to the web through these sticky strands. Webs are designed so that an insect flying into a web does not bounce out due to elastic recoil, as in a trampoline.

Obviously the tensile properties of the threads and the pattern of the web determines the characteristics of the web. There are many web designs, and spider species are often named for the type of web they construct: ladder-web spiders, funnel spiders, filmy dome spiders, and mesh-web spiders, to name just a few (Figure 8-19). A familiar web in temperate climates is that of the garden cross spider, *Araneus diadematus,* which spins a two-dimensional web with spokes radiating from the center that are joined together by

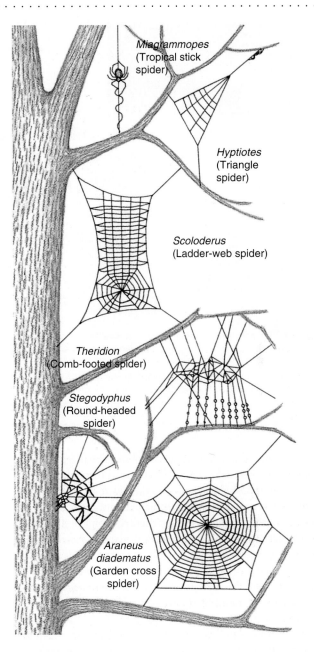

Figure 8-19 Different species of spider build webs with a characteristic design. The common names of spiders often are based on the appearance of their webs. [Adapted from "Spider Webs and Silks," by Fritz Vollrath. Copyright © 1992 by Scientific American, Inc. All rights reserved.]

a single spiral thread starting at the center and gradually spiraling outwards.

 What is the size of the largest object caught in a spider's web? What are the limitations to building larger webs?

Silk production by spiders The abdominal silk glands are large and open via modified appendages, the spinnerets, each of which has several spigots. The spinnerets are like

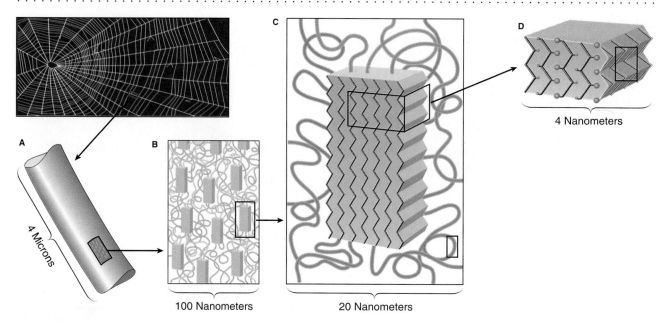

Figure 8-20 Spider silk thread **(A)** is a composite material composed of alpha-keratin crystals embedded in a disordered matrix of amino acid chains **(B** and **C)**. Each alpha-keratin crystal is composed of several amino acid chains that are pressed into an accordion-like structure called a β-pleated sheet **(D)**. The contracted disarray of the matrix provides silk with its elasticity. Most of what is known about the molecular structure of silk comes from studies of silkworm silk. In this illustration, it is assumed that spider silk resembles that of the silkworm. [Adapted from "Spider Webs and Silks," by Fritz Vollrath. Copyright © 1992 by Scientific American, Inc. All rights reserved.]

conical gun towers and are very mobile. The threads, produced in abdominal glands and extruded through spigots, consist of alpha-keratin crystals embedded in a rubber-like matrix of amino acid chains, which are not cross-linked to the crystalline alpha-keratin structures (Figure 8-20). In many cases the extruded thread is dry and may remain so because of an oily lipid covering. These dry threads are elastic and have great strength, but they can only be extended to about 25% of their length before breaking. The silk threads are stiff and brittle when dry, but become pliable when wet.

Dry threads in three-dimensional webs break when an insect collides with the web, thereby snaring the prey by tangling it in a network of threads. In two-dimensional webs, like that of the garden cross spider, dry threads form the spokes of the web, but wet threads form the sticky continuous spiral that forms the shape of the web. The spiral thread has glue droplets surrounding glycoprotein doughnuts at intervals along its length; these make the thread adhesive so that insects stick to the web. The cribellate spiders, on the other hand, produce an adhesive dry thread by covering it with a loose network of entangling amino acid chains, rather like Velcro.

Each spider has several different silk glands, each of which produces a unique silk characterized by the composition of its amino acid matrix. Spiders probably can adjust the valving mechanism at the exit of the gland to create thicker threads. But if threads with a different amino acid matrix are required, they use a different gland. Thus, the quality of the silk can be changed either by altering the valving mechanism of the spigot or by changing to another gland, which enables spiders to produce silk for a wide variety of purposes (Figure 8-21). Spiders also coat the threads with fungicides and bactericides, which prevent microorganisms from consuming webs. The presence of these substances probably accounts for the use of spider webs in folk medicine to heal cuts and abrasions to the skin.

Spiders expend a great deal of energy constructing webs, which often are damaged rather quickly. Spiders will eat their own damaged webs, an important source of amino acids, and construct new webs daily, often overnight. The garden cross spider, for example, can construct a web in less than an hour using about 20 m of thread.

ENERGY COST OF GLANDULAR ACTIVITY

Glands can have very high rates of secretory activity. For example, the extra energy expended by a nursing mother for milk production can be equivalent to the energy expenditure of a long-distance runner. In litter-bearing animals the energy cost of lactation is even greater. Breast milk is the sole source of nourishment for newborn mice until they are almost half the size of the mother mouse. For a litter of eight, the total weight of the babies at weaning is four times that of the mother. Thus the mother must eat enough food to supply all the nutrients for four times her body weight, 75% of which is diverted to lactation to supply the litter. Food intake by lactating mice increases with litter size as illustrated in Figure 8-22.

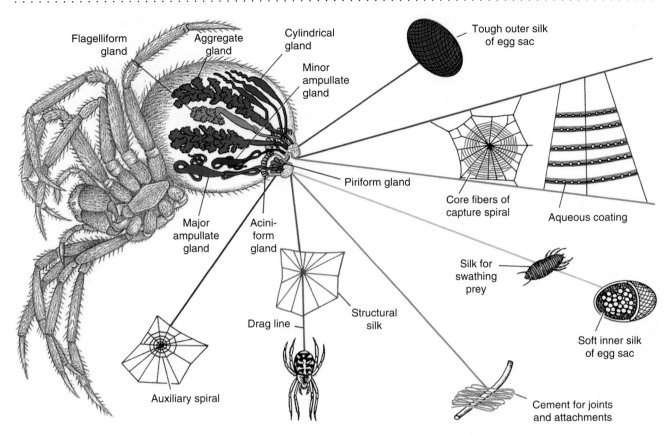

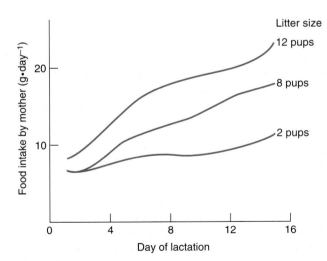

Figure 8-21 Different silks having different functions can be produced by the same spider. The garden cross spider, *A. diadematus,* has seven different abdominal glands each of which produces silk with a characteristic amino acid matrix composition. The various glands open into common spigots, but silk from only one gland is extruded at one time. By switching from one gland to another, a spider can produce the silk appropriate to the task at hand. [Adapted from "Spider Webs and Silks," by Fritz Vollrath. Copyright © 1992 by Scientific American, Inc. All rights reserved.]

Figure 8-22 Food intake by lactating mice increases with litter size and the number of days of lactation after giving birth. Maternal food intake parallels the demands that growing pups make on maternal milk production. Milk output reaches its peak on day 15, after which the pups begin to meet their food requirements by nibbling solid food. [Adapted from Diamond and Hammond, 1992.]

The graphs in Figure 8-23 show that on average food intake doubles in lactating ground squirrels during the nursing period. The mother does not gain weight because most of the ingested energy is transferred as potential energy to the young in the breast milk. Only a small fraction of the increased energy intake is used by the mother to sustain the increased metabolism associated with high rates of milk production. When these laboratory experiments were repeated at a lower temperature, the mothers increased their food intake and metabolism to maintain body temperature and still keep milk production at the same rate, indicating that food intake was not the limiting factor. Because food availability varies in the normal environment of ground squirrels, breeding is timed to occur during periods of high food availability and in warm weather so that much of the ingested energy can be used for milk production.

Why do squirrels produce young only in the spring, whereas humans produce young throughout the year? Is it more expensive in terms of energy turnover to be a male or a female of a species?

A

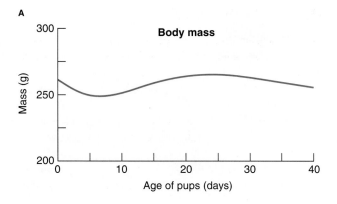

B

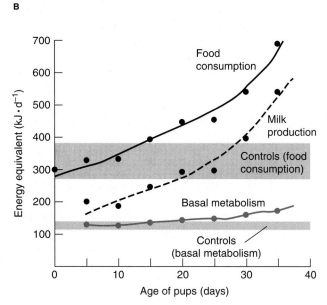

Figure 8-23 Most of the increased energy intake by lactating ground squirrels is stored as potential energy in the milk produced and thus is transferred to the pups. Data are mean values for females producing an average litter of four pups. **(A)** Maternal body mass as function of age of pups. Note that mothers did not gain weight despite their increased food consumption during the nursing period. **(B)** Comparison of maternal food consumption, milk production, and basal metabolism, expressed as daily energy equivalents in kilojoules (kJ · d⁻¹), during the nursing period. Food consumption and basal metabolism of nonlactating control squirrels are indicated by shaded bars. Note that milk production accounts for about 75% of the energy intake by mothers. [Adapted from Kenagy, Stevenson, and Masman, 1989.]

Glandular secretions often are critical for survival of animals. For example, the sweat secreted onto the body surface of mammals and some other mammals is important in temperature regulation. Evaporation of sweat from the body surface dissipates heat and is one mechanism of heat loss in mammals. The American statesman, politician, diplomat, newspaper publisher, and scientist, Benjamin Franklin (1706–1790), who was elected a Fellow of the Royal Society for his work on electricity, was also interested in temperature regulation of the body. He realized that skin temperature was lower than deep body temperature due to evaporative cooling of sweat. About 20% of the heat loss

from humans results from evaporative cooling, two thirds of which is associated with sweating; the remaining one third is associated with respiration. Evaporative heat loss is increased greatly by sweating. The prodigious volume of sweat produced during vigorous exercise or when it is hot can be sufficient to prevent excessive heating of the body. Humans can survive in a sauna at temperatures high enough to cook meat because of the activity of sweat glands, which produce enough sweat and, therefore, sufficient evaporative cooling to maintain body temperature below 40°C. If the humidity of the sauna is increased by pouring water on the heated rocks, evaporative cooling is decreased, leading to an immediate rise in body surface temperature.

 What is the ratio of your energy expenditure for glandular secretion and locomotor activity over the last hour? How might you determine this ratio?

SUMMARY

Glands are organs composed of specialized cells that act as a unit. Secretions are synthesized by cells that form the secretory part of the gland. In response to an appropriate stimulus, glandular secretions are released into the bloodstream or onto the surface of an internal cavity or the body. The nature and extent of secretions and the form of the stimulus varies greatly among glands. There are a multitude of glands that vary not only among species but also during development.

Most glandular secretions contain mucus, which is packaged in vesicles along with other secretory products and subsequently released from the gland by exocytosis. The contents of these vesicles often are released into a primary fluid formed by the active transport of ions (e.g., NaCl) followed by water into the lumen of the gland. Secretory vesicles are formed in the Golgi complex and are directed to either the apical or basal membranes of the secretory cell by the *trans*-Golgi network. Exocytosis of vesicles usually is triggered by an increase in intracellular calcium levels resulting from neural or hormonal stimulation of the secreting cell. In some cases secreting cells are stimulated by environmental changes.

Many secretions function in communication between cells. Such secretions are classified into four types based on the distance at which they exert an effect. An autocrine secretion affects the secreting cell itself. A paracrine secretion has an effect on neighboring cells. An endocrine secretion is released into the bloodstream and acts on a distant target tissue. Exocrine secretions used for communication, called pheromones, are released through a duct onto the epithelial surface of the body; these secretions permit one animal to communicate with another. Some secretions can act both locally and at a distance and, therefore, have autocrine, paracrine, and endocrine effects. Cells that produce autocrine and paracrine secretions may be, but are not always, collected together to form glands. Cells that produce endocrine secretions or pheromones are almost always collected together into glandular structures.

Glands can be characterized as either endocrine or exocrine glands. Because endocrine glands lack any characteristic morphologic markers, a variety of techniques have been used to identify such glands. The development of radioimmunoassays (RIAs) and recombinant DNA techniques has led to rapid developments in endocrinology over the past several decades. Exocrine glands are easier to recognize than endocrine glands because they all possess a duct and secrete material onto the body surface.

The adrenal medulla, an endocrine gland, secretes the catecholamines epinephrine and norepinephrine into the bloodstream. In the mammalian adrenal medulla the catecholamine-producing chromaffin cells are associated with steroidogenic tissue and most cells produce epinephrine. However, in some species (e.g., dogfish), chromaffin cells are not associated with steroidogenic tissue as in sharks and as a result norepinephrine production dominates. Catecholamines have a large number of effects, both on the circulation and metabolism. They act through both α- and β-adrenoreceptors, which are linked to the inositol trisphosphate and adenylate cyclase second-messenger systems, respectively. ATP, neuropeptide Y and adenosine can modulate the release and activity of catecholamines. In addition, the density of adrenoreceptors on target tissues can be up- or down-regulated, thereby modulating catecholamine action.

The products of exocrine glands flow through ducts onto the body surface. The surface may be enclosed as in the case of the mouth or gut. Salivary glands are exocrine glands that secrete saliva into the mouth. Saliva is about 99.5% water and contains a number of ions. Except for some hydrolysis of polysaccharides into disaccharides by salivary amylase, digestion in the mouth is minimal. Saliva serves as a lubricant, assisting in eating, swallowing, and talking. It also has an antibacterial action, which helps reduce tooth decay.

Spiders have abdominal exocrine glands that produce silk threads, which often are used to form webs. These webs are made from both dry and wet silk threads and can have very complex designs. The tensile properties of the threads and the pattern of the web determines the characteristics of the web. Spider species are often named for the type of web they produce. Silk threads are extruded through spigots and consist of alpha-keratin crystals embedded in a rubber-like matrix of amino acid chains, which are not cross-linked to the crystalline alpha-keratin structure. Different silk glands in a spider produce silk with a different amino acid composition. A spider can change the characteristics of the silk produced either by altering the valving mechanism of the spigot or by changing to another gland.

Glands are critical for survival, but they can be expensive to maintain and operate. A lactating mouse must eat enough food to supply nutrients for herself and her offspring, which may have a collective mass of four times that of her own body before weaning.

REVIEW QUESTIONS

1. What role does mucus play in exocytosis? Explain the rapid expulsion of mucus from vesicles associated with exocytosis.

2. What is the role of the *trans*-Golgi network in determining cell polarity?

3. Discuss the differences between autocrine, paracrine, neurocrine, and endocrine, secretion. What are pheromones?

4. What criteria must be met before a tissue can be unequivocally identified as having an endocrine function?

5. What is the significance of having the adrenal medulla and cortex collected together in a single organ? How does the circulatory pattern in the adrenal glands affect the relative secretion of epinephrine and norepinephrine?

6. Explain how catecholamines can have so many different actions.

7. Explain how differential activation of sympathetic and parasympathetic nerves can influence the composition of saliva.

8. How does a spider alter the composition of silk produced for making a web and other structures?

9. Describe the differences between dry and wet silk threads.

10. How much of the total energy budget of an animal is spent on secretion? How could this be estimated?

11. Discuss from an energetics point of view why many mammals give birth to young in the spring.

SUGGESTED READINGS

Edgar, W. M. 1992. Saliva: Its secretions, compositions and functions. *Br. Entomol. J.* 172:305–312. (A concise description of the functional organization of the mammalian salivary gland.)

Hadley, M. E. 1992. *Endocrinology.* 3d ed. Chapters 1, 2, and 14. Englewood Cliffs, N.J.: Prentice Hall. (A useful general text on endocrinology.)

Matsumoto, A., and S. Ischii, eds. 1992. *Atlas of Endocrine Organs.* Heidelberg: Springer-Verlag. (Beautiful diagrams of the structure of vertebrate endocrine organs.)

Pimplakar, S. W., and K. Simons. 1993. Role of heterotrimeric G proteins in polarized membrane transport. *J. Cell Sci.* 17(Suppl):27–32. (Plenty of information, but not for the beginner.)

Verdugo, P. 1994. Molecular biophysics of mucin secretion. In T. Takishima and S. Shimura, eds., *Airway Secretion: Physiological Basis for the Control of Mucous Hypersecretion.* New York: Marcel Dekker, pp. 101–121. (A good review of the subject.)

Vollrath, F. 1992. Spider webs and silks. *Sci. Am.* 266(3):70–76. (All you wanted to know about spider webs.)

HORMONES: REGULATION AND ACTION

One of the great advances in evolutionary history was the appearance of **metazoa**—multicellular organisms in which different tissues specialized in different functions. This division of labor required that each type of tissue be able to communicate with other types so as to coordinate their activities to promote the survival of the organism. (Table 9-1 summarizes the various types of messenger and regulatory molecules found in metazoa.)

The French physiologist Claude Bernard (1813–1878) emphasized the differences between the external environment that surrounds an animal from the internal environment, the *milieu interieur,* that bathes the cells of the body. He concluded that animals became more independent of the surrounding environment as they became more able to control the composition of the internal environment bathing the cells. Walter Cannon (1871–1945), who taught at Harvard University, coined the term **homeostasis** to describe the tendency of the normal body to maintain steady states, especially the constancy of the *milieu interieur* (see Chapter 1). Homeostasis is achieved by coordination of a complex set of physiological processes via chemical and/or electrical communication between tissues that

TABLE 9-1
Classification of chemical messengers and regulators

Type	Origin	Mode of action	Examples
Intracellular messengers	Intracellular	Regulation of intracellular reactions; phosphorylation of enzymes, etc.	Ca^{2+} cGMP cAMP Inositol triphosphate ($InsP_3$) Diacylglycerol (DAG)
Neurotransmitters	Nerve cells	Synaptic transmission: transported short distances; brief duration of activity	Acetylcholine Serotonin Norepinephrine
Neuromodulators	Nerve cells	Alteration of responses of ion channels to stimuli	Norepinephrine Neuropeptides
Neurohormones	Nerve cells	Endocrine function: transported by circulation; tropic effect common	Vertebrate neurohypophyseal hormones Arthropod developmental hormones
Glandular hormones	Nonneural endocrine tissues	Endocrine function: transported throughout body to distant target organs	Epinephrine Ecdysone Juvenile hormone Insulin
Local hormones	Various tissues	Endocrine function: paracrine actions on nearby targets	Prostaglandins Histamine
Pheromones	Glands opening to the environment	Intraspecific communication between individuals	Bombykol

elicits appropriate responses. Hormones play a central role in this communication and thus are critical to homeostasis.

As discussed in Chapter 8, chemical signaling can involve autocrine, paracrine, endocrine, or exocrine secretions (see Figure 8-1). In each type of signaling, the target cells bind the signaling molecules via special proteins called **receptors,** which are specific for a particular molecule. This binding initiates the response of the target cell. We saw in Chapter 6 that neurotransmitters are released from nerve cells and act over short distances to activate receptors on postsynaptic cells, an example of paracrine action. In contrast, **hormones** released from various glands travel through the bloodstream to act on distance target cells, the prototype of simple endocrine action.

It should be noted that chemical regulation of cellular processes can be found in even the most primitive plants and animal species, and undoubtedly preceded the origin of the metazoa. For example, individual amoebas of the cellular slime molds exhibit aggregating behavior in response to cAMP, which is liberated by individual amoebas. (In higher organisms, cAMP is a ubiquitous regulatory molecule, generally involved in intracellular signaling.) An even more primitive kind of chemical regulation occurs in the freshwater coelenterate *Hydra*. The water from a crowded culture of *Hydra* induces the differentiation of reproductive tissues in that animal. This effect is mediated by the elevated concentration of CO_2, which accumulates as a normal by-product of metabolism. Thus, chemical regulatory agents include relatively nonspecific molecules (e.g., NO, CO_2, H^+, O_2, and Ca^{2+}) and the more complex **messenger molecules,** produced specifically for mediating cellular communication and regulation.

This chapter focuses on the actions of glandular hormones and neurohormones. Hormones coordinate the functions of animal tissues and organs on a time scale of minutes to days. The functions under hormonal control include growth, maintenance, osmoregulation, reproduction, and behavior, among others.

How sharp are the divisions between exocrine, endocrine, and neurocrine secretions? Is this a useful classification?

ENDOCRINE SYSTEMS: OVERVIEW

William Bayliss and Ernest H. Starling described the first hormone to be discovered, **secretin,** a substance liberated from the mucosa of the small intestine that causes increased flow from the pancreas (see Chapter 15). Starling (1908) introduced the term *hormone,* derived from the Greek for "I arouse." He proposed that three characteristic properties define hormones:

- Hormones are synthesized by specific tissues or glands.

- Hormones are secreted into the bloodstream, which carries them to their site(s) of action.

- Hormones change the activities of target tissues or organs.

Although hormone molecules come into contact with all the tissues in the body, only cells that contain receptors specific for a particular hormone are affected by it. Binding of many hormones to their receptors triggers a cascade of two or more intracellular signaling molecules, called **second messengers,** that lead to a specific response in the target tissue.

The amount of hormone produced by an endocrine gland generally is small, and it is diluted in the blood and interstitial fluid. Thus hormones must be effective at very low concentrations (typically between 10^{-8} and 10^{-12} M). By way of comparison, if human taste buds could detect sugar at 10^{-12} M, we would be able to taste a pinch of sugar dissolved in a large swimming pool full of coffee or tea. [In contrast, the local concentrations of synaptic neurotransmitters are much higher ($\sim 5 \times 10^{-4}$ M), and these substances are effective only at such concentrations.] The high sensitivity of hormonal signaling is due to the high affinity of target cell receptors for hormones. As discussed later, binding of a hormone molecule to its receptor leads to a cascade of enzymatic steps that amplify the effect; thus just a few hormone molecules can influence thousands or millions of molecular reactions within a cell.

Chemical Types and General Functions of Hormones

Although hormones exhibit diverse chemical structures, most of the known hormones found in metozoans belong to one of four structural categories (Figure 9-1):

- **Amines,** including the catecholamines epinephrine and norepinephrine, as well as the thyroid hormones are small molecules derived from amino acids.

- **Prostaglandins** (or eicosanoids) are cyclic unsaturated hydroxy fatty acids synthesized in membranes from 20-carbon fatty acid chains.

- **Steroid hormones** (e.g., testosterone and estrogen) are cyclic hydrocarbon derivatives synthesized in all instances from the precursor steroid cholesterol.

- **Peptide** and **protein** hormones (e.g., insulin) are the largest and most complex hormones.

In contrast to neurotransmitters, which signal rapidly over short distances, hormones signal more slowly over longer distances. Thus endocrine systems are well suited for regulatory functions that are sustained for minutes, hours, or days. These include the maintenance of blood osmolarity (antidiuretic hormone) and of blood sugar (insulin), regulation of metabolic rates (growth hormone and thyrox-

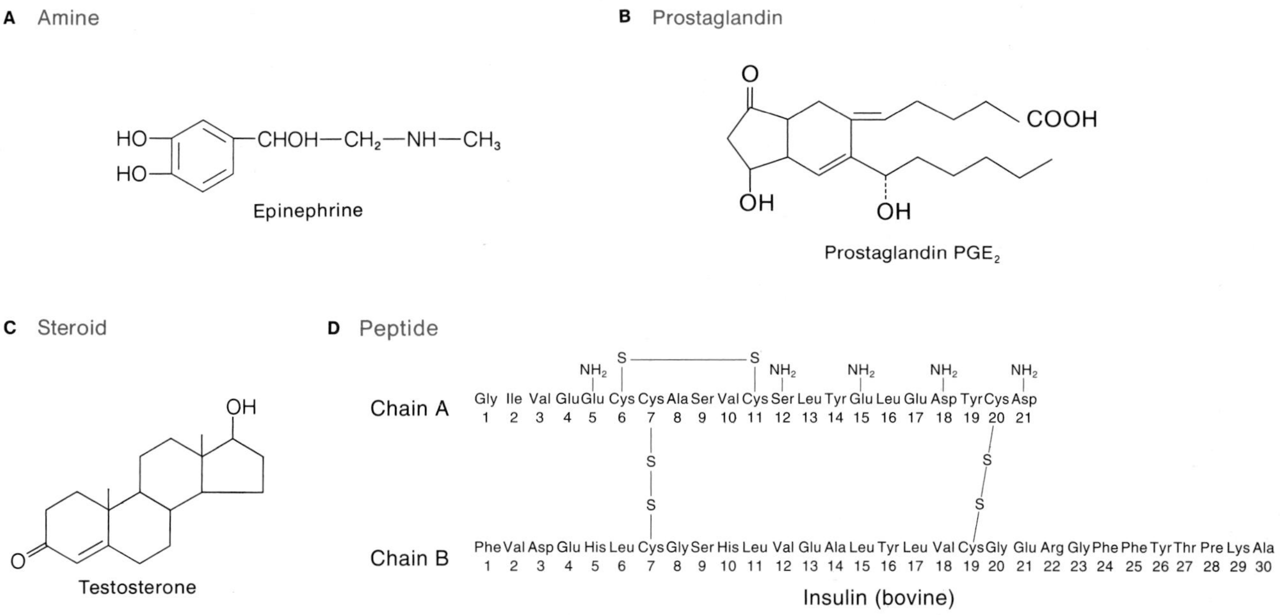

A Amine

HO—⟨ring⟩—CHOH—CH₂—NH—CH₃
HO—

Epinephrine

B Prostaglandin

COOH

OH   OH

Prostaglandin PGE₂

C Steroid

OH

O

Testosterone

D Peptide

Chain A

	S————S		NH₂		NH₂		NH₂		NH₂

NH₂   S————S   NH₂   NH₂   NH₂   NH₂

Gly Ile Val Glu Glu Cys Cys Ala Ser Val Cys Ser Leu Tyr Glu Leu Glu Asp Tyr Cys Asp
 1   2   3   4   5   6   7   8   9 10 11 12 13 14 15 16 17 18 19 20 21

Chain B

Phe Val Asp Glu His Leu Cys Gly Ser His Leu Val Glu Ala Leu Tyr Leu Val Cys Gly Glu Arg Gly Phe Phe Tyr Thr Pre Lys Ala
 1   2   3   4   5   6   7   8   9 10 11 12 13 14 15 16 17 18 19 20 21 22 23 24 25 26 27 28 29 30

Insulin (bovine)

Figure 9-1 Most hormones belong to one of four structural categories. Amine hormones (with the exception of thyroid hormones) and peptide hormones are lipid insoluble, whereas steroid hormones and prostaglandins are lipid soluble.

ine), control of sexual activity and the reproductive cycles (sex hormones), and modification of behavior (various hormones). In fact, the rapid-acting activity of the nervous system and the slower, more-sustained activity of the endocrine system complement one another in the overall integration of physiological and metabolic functions in a body. A given molecule may serve as a neurotransmitter in some instances, and the same or a closely related molecule may act as a hormone in other instances. In fact, there is an exceedingly close and overlapping relationship between the nervous and the endocrine systems. Indeed, in many respects the nervous system can be viewed as perhaps the most important endocrine organ, for it produces certain hormones that regulate the activity of many endocrine tissues.

Regulation of Hormone Secretion

The secretion of hormones occurs generally at a resting level that is modulated up or down by signals acting on the endocrine tissue. These signals often are **neurohormones,** which are released from specialized neurons and act directly on the endocrine tissue, as discussed in the next section. In some cases, the endocrine tissue responds directly to conditions of the extracellular environment (e.g., changes in osmolarity). Endocrine tissues are part of either feedforward or feedback circuits. In a feedforward circuit, secretion is *not* modulated by any consequences of the secreted hormone, whereas in a **feedback** circuit, secretion is modulated by one or more consequences of the secreted hormone.

The secretory activities of endocrine tissues generally are modulated by *negative feedback* (Figure 9-2 on page 304). That is, the increasing concentration of the hormone itself, or a response to the hormone by the target tissue (e.g., reduced blood glucose levels in the insulin loop), has an inhibitory effect on either the synthesis or release of the same

hormone. Such feedback can involve either a short loop or long loop. In *short-loop feedback,* the concentration of the hormone itself, or an effect produced by it, acts directly back on the endocrine tissue to reduce secretion, thereby keeping hormone secretion in check. *Long-loop feedback* operates on similar principles, but it includes more elements in series.

When an extremely rapid response is required, the endocrine tissue may be subject to *positive feedback;* that is, the secretion of a hormone leads directly or indirectly to its increased secretion. Positive feedback is most common in the early phases of the response. For instance, positive feedback occurs early in the reproductive cycle of some vertebrates (and perhaps also invertebrates) when responses (e.g., increase in level of luteinizing hormone) must peak relatively quickly. Ultimately, of course, this must be countered by a process that ends the rapid increase.

 Although feedback control is discussed in this chapter on hormones, it is a general regulatory mechanism in physiological systems. What are some examples of feedforward and feedback control in other physiological systems?

NEUROENDOCRINE SYSTEMS

As discussed earlier, secretion of hormones from some endocrine tissues is regulated by neurohormones, which are produced by specialized nerve cells called **neurosecretory cells.** Some neurohormones secreted from neurosecretory

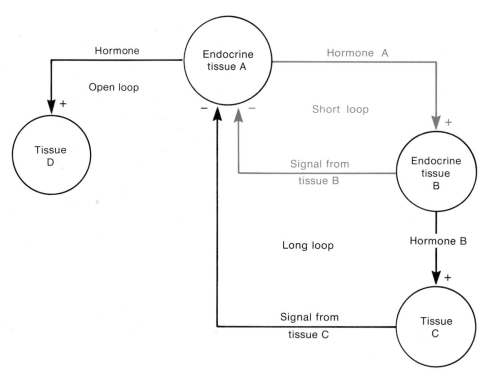

Figure 9-2 Most endocrine tissues are subject to negative-feedback control. In short-loop feedback, the response of the primary target tissue (B) feeds back onto the endocrine gland. In long-loop feedback, a sig- nal from secondary target tissues (C) control secretory activity. In an open loop there is no feedback.

cells in the **hypothalamus** regulate secretion of various glandular hormones from the nonneural anterior **pituitary gland.** In contrast, the neurohormones released from the posterior pituitary gland act directly on various target tissues; these hormones are produced in neurosecretory cell bodies located in the anterior hypothalamus. This close relation between neural and endocrine systems is the basis of the neuroendocrine reflex (Figure 9-3). The neurosecre- tory cells in the hypothalamus respond to sensory input from various parts of the body. The pituitary gland, also termed the **hypophysis,** is a small appendage lying below the hypothalamus. Because it secretes at least nine hor- mones, the pituitary gland has been called the "master gland."

Although ordinary nerve cells and most neurosecretory cells are generally similar, they exhibit several differences. First, the secretory vesicles containing neurosecretory hor- mones typically are 100–400 nm in diameter, whereas the presynaptic vesicles containing neurotransmitters in ordi- nary nerve cells are much smaller, 30–60 nm in diameter (Figure 9-4). Second, although ordinary nerve cells use both slow and fast axonal transport systems, neurosecretory cells appear to use only fast axonal transport, moving neu- rohormones at rates of up to 2800 mm a day. Third, ordi- nary nerve cells form synapses with other cells at their terminals, whereas neurosecretory axons generally termi- nate in clusters in a bed of capillaries, forming a discrete **neurohemal organ** (see Figure 9-3). The neurohormones re- leased into the interstitial space diffuse into the capillaries

and are carried in the bloodstream to target endocrine tis- sues or other target tissues.

Hypothalamic Control of the Anterior Pituitary Gland

The axons of some hypothalamic neurosecretory cells have their endings in the **median eminence** at the floor of the hypothalamus (Figure 9-5). These cells secrete at least seven hormones that control secretion of various hor- mones by the anterior pituitary gland (Table 9-2). All but one of these hypothalamic **releasing hormones** (RHs) and **release-inhibiting hormones** (RIHs) are peptides (Spotlight 9-1). The discovery of these hypothalamic hormones has proved to be one of the most important developments in vertebrate endocrinology, opening investigations into the orchestration of virtually the entire vertebrate endocrine system.

As early as the 1930s, studies revealed that capillaries within the median eminence converge to form a series of portal vessels that carry blood directly from the neurose- cretory tissue of the median eminence to the glandular se- cretory tissue of the anterior pituitary gland. There they break up again into a capillary bed before finally recon- verging to join the venous system. This portal system en- hances chemical communication from the hypothalamus to the anterior pituitary gland by carrying the hypothalamic RHs and RIHs directly to the interstitium of the anterior pituitary gland, also known as the **adenohypophysis** (see Figure 9-5). Here these hypothalamic hormones come into

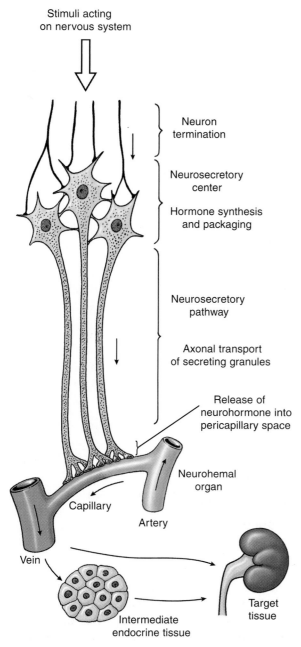

Figure 9-3 Neurohormones are released from the terminals of neurosecretory cells into a bed of capillaries, forming a neurohemal organ. After entering the bloodstream, some neurohormones (e.g., oxytocin) act directly on a somatic target tissue, but most activate an intermediate endocrine gland, stimulating secretion of another hormone that acts on the target tissue.

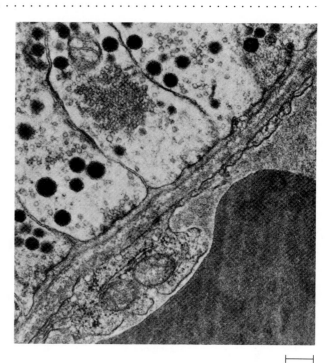

Figure 9-4 The secretory vesicles in neurosecretory axons are much larger than neurotransmitter vesicles in ordinary presynaptic neurons. In this electron micrograph of the hamster posterior pituitary gland, the large dark bodies are the secretory vesicles (or granules). The terminals end on an endothelial basement membrane that separates the terminals from a fenestrated capillary wall. The large dark object at the lower right corner is a red blood cell in the capillary. [Douglas et al., 1971.]

The first physiological evidence for the neurohumoral control of the anterior pituitary gland came in the late 1950s with the discovery of a substance that stimulates the release of **adrenocorticotropic hormone** (ACTH) from the anterior pituitary gland. This substance, obtained by the extraction of the hypothalamus from thousands of pigs, was given the name corticotropin-releasing hormone (CRH). Minute amounts of CRH are liberated from neurosecretory cells in the hypothalamus when they are activated by neural input in response to a variety of stressful stimuli to the organism (e.g., cold, fright, sustained pain). The ACTH released from the anterior pituitary gland in response to CRH stimulation circulates in the bloodstream to its target tissue, the adrenal cortex, where it stimulates release of adrenocortical hormones.

Glandular Hormones Released from the Anterior Pituitary Gland

The anterior lobe of the pituitary gland consists of the pars distalis, pars tuberalis, and pars intermedia. Six hormones are released from the pars distalis and one from the pars intermedia in mammals (see Figure 9-5). Although all the glandular secretory cells of the adenohypophysis are generally similar in appearance, they can be classified into two histochemically distinct types:

contact with the glandular endocrine cells that secrete *adenohypophyseal hormones*, either stimulating or inhibiting their secretory activity. Because of the direct portal connection from the hypothalamus to the anterior pituitary gland, very low concentrations of the RHs and RIHs can produce effects on the anterior pituitary gland. Once these hormones enter the general circulation, they are diluted to ineffective concentrations and are enzymatically degraded within several minutes.

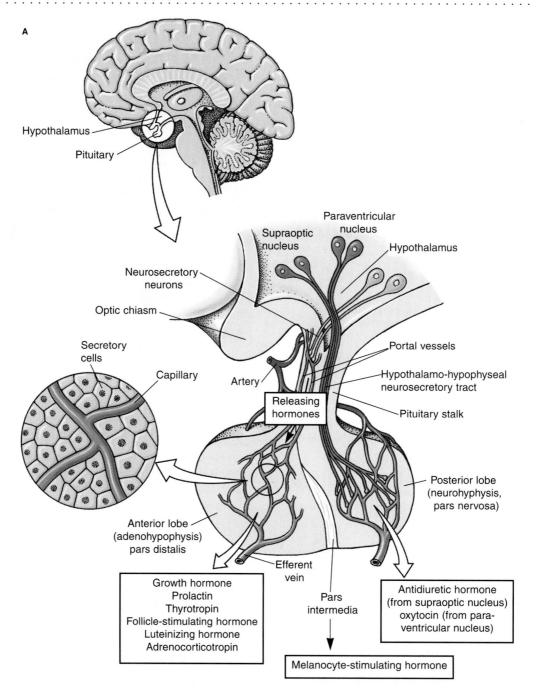

A

Hypothalamus

Pituitary

Paraventricular nucleus

Supraoptic nucleus

Hypothalamus

Neurosecretory neurons

Optic chiasm

Secretory cells

Capillary

Artery

Portal vessels

Hypothalamo-hypophyseal neurosecretory tract

Releasing hormones

Pituitary stalk

Posterior lobe (neurohyphysis, pars nervosa)

Anterior lobe (adenohypophysis) pars distalis

Efferent vein

Pars intermedia

Growth hormone
Prolactin
Thyrotropin
Follicle-stimulating hormone
Luteinizing hormone
Adrenocorticotropin

Antidiuretic hormone (from supraoptic nucleus) oxytocin (from para-ventricular nucleus)

Melanocyte-stimulating hormone

Figure 9-5 Hormonal secretion from the primate pituitary gland (hypophysis) is controlled by the hypothalamus. The anterior lobe of the pituitary gland (adenohypophysis) consists of the pars distalis, pars intermedia, and pars tuberalis. (The pars tuberalis, not shown, consists of a thin layer of cells surrounding the pituitary stalk.) The posterior lobe (neurohypophysis), an extension of the brain, consists of neural tissue, whereas the anterior lobe consists of nonneural glandular tissue. Releasing or release-inhibiting hormones secreted by hypothalamic neurosecretory endings in the median eminence are carried via the portal vessels (hypothalamo-hypophyseal portal system) to the anterior pituitary gland, where they stimulate (or inhibit) secretion of several glandular hormones. Two neurohormones produced in hypothalamic cell bodies are released from terminals of the neurosecretory cells in the posterior pituitary glands.

- Acidophils, which stain orange or red with acid dyes, secrete **growth hormone** (GH; also termed somatotropin) and **prolactin** (PRL).

- Basophils, which stain blue with basic dyes, secrete ACTH, **thyroid-stimulating hormone** (TSH), **melano-cyte-stimulating hormone** (MSH), **luteinizing hormone** (LH), and **follicle-stimulating hormone** (FSH).

    Like ACTH, TSH, LH, and FSH are primarily tropic in their actions (Table 9-3). That is, they act on other endocrine tissues (e.g., thyroid, gonads, and adrenal cortex),

TABLE 9-2
Hypothalamic hormones that stimulate or inhibit release of adenohypophyseal hormones*

Hormone	Structure	Primary action in mammals	Regulation
Stimulatory			
Corticotropin-releasing hormone (CRH)	Peptide	Stimulates ACTH release	Stressful neural input increases secretion; ACTH inhibits secretion
GH-releasing hormone (GRH)	Peptide	Stimulates GH release	Hypoglycemia stimulates secretion
Gonadotropin-releasing hormone (GnRH)	Peptide	Stimulates release of FSH and LH	In the male, low blood testosterone levels stimulate secretion; in the female, neural input and decreased estrogen levels stimulate secretion; high blood FSH or LH inhibits secretion
TSH-releasing hormone (TRH)	Peptide	Stimulates TSH release and prolactin secretion	Low body temperatures induce secretion; thyroid hormone inhibits secretion
Inhibitory			
MSH-inhibiting hormone (MIH)	Peptide	Inhibits MSH release	Melatonin stimulates secretion
Prolactin-inhibiting hormone (PIH)	Amine	Inhibits prolactin release	High levels of prolactin increase secretion; estrogen, testosterone, and neural stimuli (suckling) inhibit secretion
Somatostatin (GH-inhibiting hormone, GIH)	Peptide	Inhibits release of GH and many other hormones (e.g., TSH, insulin, glucagon)	Exercise induces secretion; hormone is rapidly inactivated in body tissues

*ACTH = adrenocorticotropin hormone; FSH = follicle-stimulating hormone; GH = growth hormone; LH = luteinizing hormone; MSH = melanocyte-stimulating hormone; TSH = thyroid-stimulating hormone.

regulating the secretory activity of these target glands. LH and FSH, which act on the gonads, often are referred to as **gonadotropins.** Thus the effect of these tropic hormones on nonendocrine somatic tissues is indirect, operating through the hormones released from their target glands. The remaining adenohypophyseal hormones—growth hormone, prolactin, and MSH—are direct-acting hormones; that is, they act directly on somatic target tissues without the intervention of other hormones. The actions of growth hormone and prolactin are discussed in later sections. MSH, whose release is regulated by MIH from the hypothalamus, acts on pigment cells in the skin to increase the synthesis and dispersal of *melanin*, leading to darkening of the skin.

The relations between the hypothalamus and the anterior pituitary gland are summarized in Figure 9-6. The three release-inhibiting hormones from the hypothalamus suppress the release from the anterior pituitary gland of MSH, prolactin, and growth hormone. Growth hormone is also under the control of a releasing hormone. Note the short and long feedback loops, involving ACTH, TSH, FSH, and LH, which control the hypothalamo-anterior pituitary system and the long feedback loop, involving growth hormone, prolactin, and MSH, which controls the hypothalamus.

TABLE 9-3
Tropic hormones of the anterior pituitary gland

Hormone	Structure	Target tissue	Primary action in mammals	Regulation*
Adrenocorticotropin (ACTH)	Peptide	Adrenal cortex secretion of steroid hormones by adrenal cortex	Increases synthesis and slows release of CRH	CRH stimulates release; acth
Follicle-stimulating hormone (FSH)	Glycoprotein	Ovarina follicles (female) Seminiferous tublues male In male, increases sperm production	In female, stimulates maturation of ovarian follicles release	GnRH stimulates release; inhibin and steroid sex hormones inhibit
Luteinizing hormone (LH)	Glycoprotein	Ovarian interstitial cells (female) Testicular interstitial cells (male)	In female, induces final maturation of ovarian follicles, estrogen secretion, ovulation, corpus luteum formation, and progesterone secretion In male, increases synthesis and secretion of androgens	GnRH stimulates release; inhibin and steroid sex hormones inhibit release
Thyroid-stimulating hormone (TSH)	Glycoprotein	Thyroid gland	Increases synthesis and secretion of thyroid hormones	TRH induces secretion; thyroid hormones and somatostatin slow release

*See Table 9-2 for abbreviations.

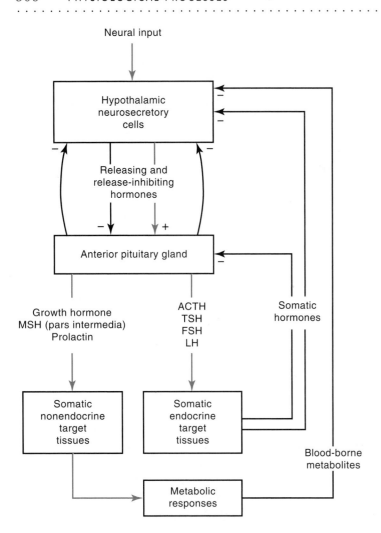

Neural input

Hypothalamic
neurosecretory
cells

Releasing and
release-inhibiting
hormones

Anterior pituitary gland

Growth hormone
MSH (pars intermedia)
Prolactin

ACTH
TSH
FSH
LH

Somatic
hormones

Somatic
nonendocrine
target
tissues

Somatic
endocrine
target
tissues

Blood-borne
metabolites

Metabolic
responses

Figure 9-6 Secretion of adenohypophyseal hormones is regulated by hypothalamic release and/or release-inhibiting hormones and further modulated by feedback. Growth hormone, melanocyte-stimulating hormone (MSH), and prolactin act directly on nonendocrine somatic (nonneural) tissues. The tropic hormones—adrenocorticotropin (ACTH), thyroid-stimulating hormone (TSH), follicle-stimulating hormone (FSH), and luteinizing hormone (LH)—all stimulate the secretory activity of somatic endocrine tissues. Once released, the corresponding somatic hormones exert negative feedback on the hypothalamic neurosecretory cells and in some cases on the corresponding adenohypophyseal cells themselves. The circulating products of some somatic metabolic responses (e.g., blood glucose) also act on the hypothalamic centers, providing additional negative feedback.

Neurohormones Released from the Posterior Pituitary Gland

The posterior lobe of the pituitary gland, also called the **neurohypophysis** and pars nervosa, stores and releases two neurohormones, **antidiuretic hormone** and **oxytocin.** These *neurohypophyseal hormones* are synthesized and packaged in the cell bodies of two groups of neurosecretory cells that comprise the supraoptic and paraventricular nuclei in the anterior portion of the hypothalamus (see Figure 9-5). After their synthesis, the hormones are transported within the axons of the hypothalamo-hypophyseal tract to nerve terminals in the neurohypophysis, where they are released into a capillary bed. This was the first neurosecretory system discovered in the vertebrates.

Both antidiuretic hormone (ADH), also known as vasopressin, and oxytocin are peptides containing nine amino acid residues. Both are mildly effective in fostering contractions of the smooth-muscle tissue in arterioles and the uterus (Figure 9-7). In mammals, however, oxytocin is best known for stimulating uterine contractions during parturition and for stimulating release of milk from the mammary gland; in birds, it stimulates motility of the oviduct. The foremost function of ADH is to promote water retention in the kidney, as discussed in a later section.

The amino acid sequences of mammalian oxytocin and arginine vasopressin differ at only positions 3 and 8 in the peptide chain. Likewise, the sequences of the neurohypophyseal hormones from different vertebrate groups exhibit variations at positions 3, 4 and 8 (Table 9-4). The sequence of amino acid residues in each pituitary nonapeptide is, of course, genetically determined. Substitution of amino acid residues at positions 3, 4, and/or 8 during evolution has resulted in several forms of these peptide hormones. The residues that are highly conserved (never undergo substitution) in these hormones are presumably necessary for function; those that are not conserved (positions 3, 4, and 8) seem to be functionally neutral and probably serve only to place the essential residues in the positions appropriate for the biological activity of these neuropeptides.

Within their respective neurosecretory cells, the neurohypophyseal hormones are covalently linked in a 1 : 1 ratio to cysteine-rich protein molecules termed **neurophysins,** which exist in two major types, neurophysin I and neurophysin II. Oxytocin is associated with type I and vasopressin with type II. Neurophysins have no hormone activity, even though they are secreted along with the neurohypophyseal hormones. It is conjectured that

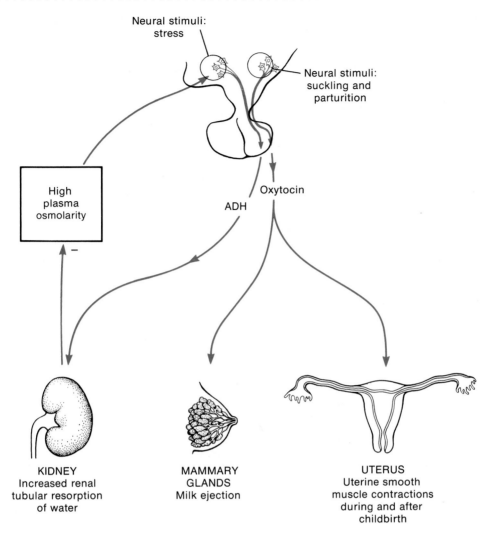

Figure 9-7 The two neurohormones released from the mammalian posterior pituitary gland function primarily in reproduction (oxytocin) and regulation of water balance (ADH). Osmoreceptors in the hypothalamus, baroreceptors in the aorta, and exteroceptive sensory input all influence the neurosecretion of antidiuretic hormone (ADH). High plasma-solute concentration and low blood pressure resulting from low plasma volume stimulate ADH output. Oxytocin is released during labor and nursing.

TABLE 9-4
Variant forms of neurohypophyseal nonapeptide hormones

Peptide	Positions of amino acid residues*										Animal group
	1	2	3	4	5	6	7	8	9		
Lysine vasopressin	Cys—	Tyr—	Phe—	Gln—	Asn—	Cys—	Pro—	Lys	—Gly—	(NH$_2$)	Pigs and relatives
Arginine vasopressin	Cys—	Tyr—	Phe—	Gln—	Asn—	Cys—	Pro—	Arg	—Gly—	(NH$_2$)	Mammals
Oxytocin	Cys—	Tyr—	Ile	—Gln—	Asn—	Cys—	Pro—	Leu	—Gly—	(NH$_2$)	Mammals
Arginine vasotocin	Cys—	Tyr—	Ile	—Gln—	Asn—	Cys—	Pro—	Arg	—Gly—	(NH$_2$)	Reptiles, fish, and birds
Isotocin	Cys—	Tyr—	Ile	—Ser—	Asn—	Cys—	Pro—	Ile	—Gly—	(NH$_2$)	Some teleosts
Mesotocin	Cys—	Tyr—	Ile	—Gln—	Asn—	Cys—	Pro—	Ile	—Gly—	(NH$_2$)	Reptiles, amphibians, and lungfish
Glumitocin	Cys—	Tyr—	Ile	—Ser—	Asn—	Cys—	Pro—	Gln	—Gly—	(NH$_2$)	Some elasmobranchs

*The cysteine residues in positions 1 and 6 of each peptide are bridged by a disulfide bond.

Source: Frieden and Lipner, 1971.

SPOTLIGHT 9-1

PEPTIDE HORMONES

An interesting example of the opportunism in biochemical evolution is evident in the distribution and structure of a group of hormones and neurotransmitters consisting of small polypeptide chains. These can range from as few as three or four amino acid residues to as many as two or three dozen residues (see figure). Collectively called peptide hormones, most of these substances are widely distributed in the human body and throughout the animal kingdom. Thus, we find some peptide hormones in visceral tissues, such as the digestive tract (see Chapter 15), and the central nervous system (see Chapter 6). For example, insulin and somatostatin, both of which were originally discovered in the pancreas, are now known to be present in hypothalamic neurons. TSH-releasing hormone (TRH), the hypothalamic hormone originally found to cause the release of thyroid-stimulating hormone (TSH) from the anterior pituitary gland, has recently also been found in lampreys (which produce no TSH) and snails (which have no thyroid or pituitary glands), as well as in many other invertebrates.

The discovery in the 1970s that peptide hormones originally thought to be confined to tissues of the mammalian gut also occurred in various parts of the CNS initially was surprising. Now the concept of "brain-gut" hormones is no longer unusual, and we have grown used to the idea that the gene coding for a regulatory molecule performing one task in one type of body tissue is also utilized by another tissue to make the same hormone, but for a different function. Recall that the action of a hormone depends on the nature of the enzyme cascade linked to the hormone's receptor, as well as the effector molecules expressed in a particular tissue.

An interesting feature of the peptide hormones is that some of them are produced in variant forms both within an individual and among different taxonomic groups. This is well illustrated by the vasopressin-oxytocin family (see Table 9-4). Another example is cholecystokinin: variants of this hormone with 33, 39, or 58 amino acid residues, are present in the mammalian digestive tract, but small 4- or 8-residue fragments, cleaved from the carboxyl end of the larger cholecystokinin variants are found in the brain.

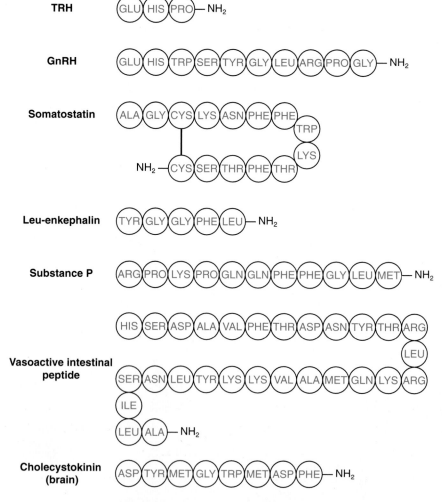

Peptide hormones range in length from as few as three amino acid residues to several dozen residues. Of the representative peptide hormones shown here, the upper three are releasing or release-inhibiting hormones produced by hypothalamic neurons and the lower four are brain-gut hormones. The circles represent individual residues identified with the three-letter amino acid code (see Table 3-7). TRH = TSH-releasing hormone; GnRH = gonadotropin-releasing hormone.

the hormone-neurophysin molecules are enzymatically cleaved upon release into the blood, yielding the neurohypophyseal hormone and the neurophysin moiety. Thus neurophysins appear to act as storage proteins, serving to retain the hormones in the secretory granules until release.

CELLULAR MECHANISMS OF HORMONE ACTION

As noted already, hormones produce their specific effects on their target tissues via specialized receptor proteins located either inside the cell or on the surface of the cell. Most lipid-soluble (hydrophobic) hormones, such as the steroid and thyroid hormones, readily penetrate the plasma membrane and bind to receptors in the cytoplasm of target cells. In contrast, lipid-insoluble (hydrophilic) hormones cannot penetrate the plasma membrane and therefore bind to receptors on the cell surface.

The intracellular mechanism of action of a hormone depends on whether it binds to cytoplasmic or cell-surface receptors (Figure 9-8):

- Lipid-soluble hormones bind to cytoplasmic receptors, forming hormone-receptor complexes that translocate to the nucleus and act directly on the DNA of the cell to effect long-term changes lasting hours or days.

- Lipid-insoluble hormones bind to cell-surface receptors, often leading to production of one or more second mes-

sengers, which amplify the signal and mediate rapid, short-lived responses via various effector proteins.

The prostaglandins are the exception proving the rule that the nature of the receptor, not the hormone, determines the mode of action. Although lipid-soluble, prostaglandins bind to cell-surface receptors and have a rapid, short-lasting effect, similar to that of lipid-insoluble hormones. Table 9-5 summarizes some properties characteristic of the major types of lipid-soluble and lipid-insoluble hormones.

Lipid-Soluble Hormones and Cytoplasmic Receptors

The lipid-soluble steroid and thyroid hormones are carried in the bloodstream complexed with carrier proteins. Without these carriers, only small amounts of these hormones could dissolve in the blood, which is an aqueous solution, and they would be taken up completely by the first encountered lipids in the circulation. The association of lipid-soluble hormones with carriers thus markedly increases the amounts of these hormones that can be carried in the blood. The binding constants of different carriers varies, ensuring adequate rates of hormone delivery to all target tissues.

Once steroid and thyroid hormones dissociate from their carrier proteins, they can readily enter and leave neighboring cells by diffusing across the plasma membrane. Initially, nearly all the hormone appears in the cytoplasm, but in target cells hormone-receptor complexes form and move

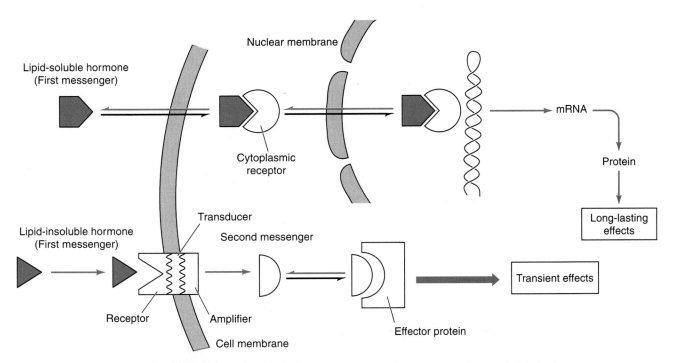

Figure 9-8 Lipid-soluble and lipid-insoluble hormones differ in their primary intracellular mode of action. **(A)** Most lipid-soluble hormones move through the plasma membrane and combine with intracellular receptor proteins, forming active complexes that act on the genetic machinery to modulate gene expression. **(B)** Lipid-insoluble hormones bind to cell-surface receptors, triggering an intracellular signaling pathway that may involve a second messenger, which in turn combines with another molecule to produce a metabolically active complex. Although they are lipid soluble, the prostaglandins bind to cell-surface receptors. [Concept by M. J. Berridge.]

TABLE 9-5
Comparison of lipid-soluble and lipid-insoluble hormones

Property	Lipid soluble		Lipid insoluble	
	Steroids	Thyroxine	Peptides and proteins	Catecholamines
Feedback regulation of synthesis	Yes	Yes	Yes	Yes
Binding to plasma proteins	Yes	Yes	Rarely	No
Lifetime in blood plasma	Hours	Days	Minutes	Seconds
Time course of action	Hours to days	Days	Minutes to hours	Seconds or less
Receptor location	Cytosolic or nuclear	Nuclear	Plasma membrane	Plasma membrane
Mechanism of action	Receptor-hormone complex stimulates or inhibits gene expression		Hormone binding triggers second-messenger or activates intrinsic catalytic activity	Hormone binding causes change in membrane potential of triggers second-messenger pathway

Source: Adapted from Smith et al., 1983, p. 358. Used with permission of McGraw-Hill.

into the nucleus, so that in time more and more hormone appears in the nucleus (Figure 9-9A). Although diffusion of lipid-soluble hormones in and out of cells is a random process, these hormones exert effects only on specific target cells. Understanding how target and nontarget cells differed came from several types of evidence. First, autoradiographic studies in the 1960s showed that steroid hormones accumulate in the nuclei of their target cells, but not in the nuclei of other cells. This specific accumulation occurs very rapidly and persists for some time after the labeled steroid is removed from the circulation. These findings indicated that target cells must contain steroid hormone-specific receptors, which are absent from nontarget cells.

Such receptor molecules were found by fractionating a target tissue incubated with a radiolabeled hormone and separating the components with different molecular weights by sucrose density-gradient centrifugation. The receptor-hormone complex then could be identified by the radioactivity of the labeled hormone. Roger Gorski (1979) and associates identified the estradiol receptor in this way, using labeled estradiol and rat uterus as the target tissue. They found that the receptor, a protein with a molecular weight of about 200,000, binds estradiol very strongly, and is present in uterine tissues but not in other tissues. Most significant was the observation that substances that mimic the hormonal action of estradiol in the uterus are all bound by this same receptor protein. Other receptor proteins have since been identified in target tissues of other lipid-soluble hormones.

All the cytoplasmic receptors that bind lipid-soluble hormones share a highly conserved *DNA-binding domain* (Figure 9-9B). In the absence of hormone, these receptors are bound to an inhibitor protein that blocks the DNA-binding domain of the receptor, making it inactive. Binding of hormone to the receptor causes the inhibitor protein to dissociate, thereby activating the receptor by exposing its DNA-binding site. Once the receptor-hormone complex moves into the nucleus, the DNA-binding domain of the receptor binds specific *regulatory sequences* within the DNA,

thereby regulating the transcription of specific genes and ultimately production of their encoded proteins. Since the lipid-soluble hormones act on the cell's DNA to stimulate or inhibit production of particular proteins, their effects persist for hours to days, whereas the effects of lipid-insoluble hormones usually last only minutes to hours.

What are the differences in the regulation and action of lipid-soluble and lipid-insoluble hormones?

Lipid-Insoluble Hormones and Intracellular Signaling

As noted already, binding of hormone to many cell-surface receptors triggers production of second messengers, which transduce the extracellular hormonal signal into the cell's response. Despite the large number of known hormones that stimulate second-messenger formation, the most important second messengers fall into only three distinct groups (Figure 9-10):

- Cyclic nucleotide monophosphates (cNMPs), such as adenosine 3′,5′-cyclic monophosphate (cAMP) and the closely related guanosine 3′,5′-cyclic monophosphate (cGMP)

- Inositol phospholipids, including inositol trisphosphate ($InsP_3$) and diacylglycerol (DAG)

- Ca^{2+} ions

We'll first examine intracellular signaling systems employing each of these second messengers as well as membrane-bound enzyme signaling systems, which don't involve second messengers. Then we'll see how multiple systems can interact to produce complex tissue responses.

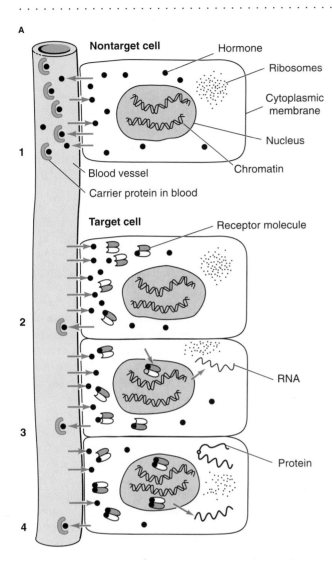

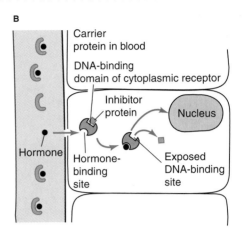

Figure 9-9 Steroid and thyroid hormones move from the bloodstream into target cells, where they can bind to specific receptors that have a common domain structure. **(A)** Mechanism of action of lipid-soluble hormones with cytoplasmic receptors. These hormones diffuse randomly into and out of nontarget cells without interaction or binding (1). Target cells for a particular hormone contain cytoplasmic receptors, consisting of two subunits, specific for that hormone. Formation of hormone-receptor complexes preferentially retains hormone molecules within target cells (2). These complexes then accumulate in the nucleus, where they bind to regulatory elements in the DNA, in most cases stimulating transcription of specific genes into the corresponding messenger RNA (3). The resulting messenger RNA is then translated into protein by the ribosomes (4). **(B)** Model of common domain structure of cytoplasmic receptors for lipid-soluble hormones. In its inactive state the receptor is bound to an inhibitor protein that blocks the DNA-binding domain of the receptor. The binding of hormone to the receptor causes the inhibitor protein to dissociate, thereby activating the receptor by exposing the DNA-binding site. [Part A adapted from O'Malley and Schrader, 1976.]

Cyclic nucleotide signaling systems

The advance of science generally depends on two forms of progress. One is the everyday growth of scientific knowledge by the slow but steady accumulation of data in thousands of laboratories. Such small-scale incremental progress represents by far the major effort expended by the entire community of scientists. This type of progress generally builds upon the infrequent and often unanticipated breakthroughs that provide revolutionary new insights or points of departure. Such breakthroughs open new paths of inquiry, which are then explored in detail by the slow step-by-step mode of progress until at some unexpected time another major breakthrough provides new insight and again alters the course of daily investigation.

An example of such a giant leap occurred in the mid-1950s when the late Earl W. Sutherland and associates discovered the role of **cyclic AMP** (cAMP) as an intracellular regulatory agent. In his initial studies on cAMP, Sutherland noted that the activity of **adenylate cyclase,** which catalyzes conversion of ATP to cAMP, is enhanced when certain hormones are added to cell-free liver homogenates or preparations of intact cells. He then separated the cell-free ho-

mogenate into fractions and found that the adenylate cyclase activity disappeared if the cell-membrane fragments of the homogenate were removed. It was subsequently discovered that adenylate cyclase is intimately associated with a hormone receptor in the membrane. Note that the hormones that stimulate adenylate cyclase activity do so without entering the cell; moreover, neither ATP nor cAMP readily penetrate the plasma membrane when placed in the extracellular fluid.

The discovery of the hormonal stimulation of adenylate cyclase provided the first evidence for a link between extracellular hormones and intracellular regulatory agents and led to the *second-messenger hypothesis.* The hormone acts on the outer surface of the cell membrane, whereas cAMP is produced enzymatically from ATP at the inner surface of the membrane. The hormone conveys its signal through the surface membrane without having to penetrate the membrane. Sutherland's findings opened the way for a totally new understanding of regulatory processes in many areas of biochemistry and cell biology. Other researchers subsequently accumulated vast amounts of data that confirmed the role of cAMP as an intracellular regulatory agent

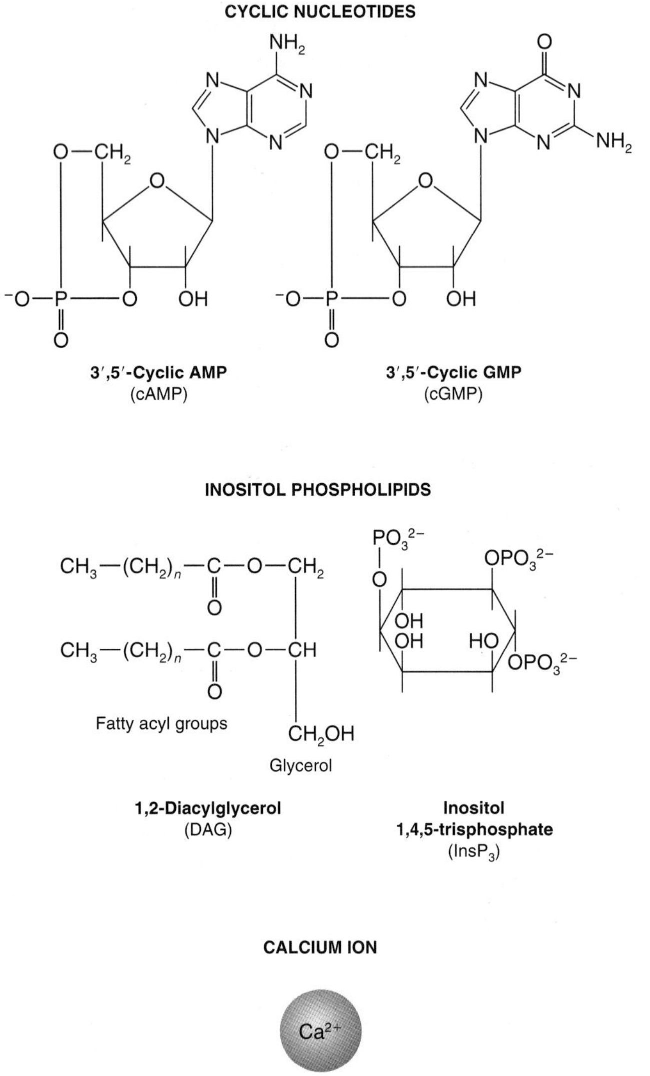

CYCLIC NUCLEOTIDES

3′,5′-Cyclic AMP
(cAMP)

3′,5′-Cyclic GMP
(cGMP)

INOSITOL PHOSPHOLIPIDS

$CH_3-(CH_2)_n-C-O-CH_2$

$CH_3-(CH_2)_n-C-O-CH$

Fatty acyl groups

Glycerol

CH_2OH

1,2-Diacylglycerol
(DAG)

Inositol
1,4,5-trisphosphate
(InsP$_3$)

CALCIUM ION

Ca^{2+}

Figure 9-10 The three classes of second messengers have very different structures. The cyclic nucleotides are synthesized from ATP and GTP. DAG and InsP$_3$ are produced by hydrolysis of a common precursor.

mediating the actions of many hormones and other extracellular messengers in a wide variety of cellular responses.

The general model of the events in the cAMP signaling system is shown in Figure 9-11. The left panel outlines the series of coupling steps in this system, which are similar to those in the inositol phospholipid system. Binding of an external signal (i.e., the first messenger) with a specific receptor molecule projecting from the outer surface of the target cell membrane activates a *transducer protein* that carries signals through the membrane. The transducer protein then activates an *amplifier* that catalyzes the formation of a second messenger. The second messenger binds to an internal *regulator* that controls various *effectors*, leading to a cellular response(s).

As shown in the right side of Figure 9-11, the pathway employing cAMP as the second messenger has a stimula-

tory receptor (R_s) and an inhibitory one (R_i), which both communicate with the amplifier adenylate cyclase by way of **transducer G proteins;** stimulatory G protein (G_s) and inhibitory G protein (G_i). Thus, the message is carried through the membrane by interactions of three membrane-bound proteins: the receptors, G proteins, and adenylate cyclase. Hormone binding stimulates guanosine triphosphate (GTP; a close relative of ATP) to bind to the G proteins (hence their name). Figure 9-12 shows that the G proteins remain activated as long as they bind GTP; they are inactivated when the GTP is hydrolyzed to guanosine diphosphate (GDP). The hydrolysis of ATP into cAMP by adenylate cyclase requires Mg^{2+} and a trace amount of Ca^{2+}. As cAMP is produced, it binds to an inhibitory regulatory subunit of protein kinase A, causing the subunit to dissociate. This leaves the catalytic subunit of protein kinase A free to phosphorylate effectors proteins using ATP as the source of high-energy phosphate groups. Phosphorylation of these effector proteins may either increase or inhibit their activity, thereby inducing a cellular response(s). Some effector proteins are enzymes, which catalyze further chemical reactions; others are nonenzymatic proteins such as membrane channels, structural proteins, or regulatory proteins (see bottom of Figure 9-11).

Signal amplification in the cAMP pathway One major problem in intracellular signaling pathways is how to amplify the signal produced by hormone binding so that a few hormone molecules can influence the function of many molecules within the cell. Because binding of the hormone by the receptor occurs in a one-to-one manner, no amplification occurs at this step. However, amplification occurs at several later stages in the cAMP pathway. First, a single activated receptor protein can activate many G-protein molecules, which then activate many molecules of adenylate cyclase, thereby amplifying the extracellular signal. In some cases, the hormone may remain bound to its receptor for less than 1 second, not enough time for this amplification mechanism to operate. But as noted above, G protein remains active as long as it binds GTP (e.g., 10–15 seconds), providing sufficient time for amplification to occur. Second, each molecule of activated adenylate cyclase catalyzes the conversion of many high-energy ATP molecules into energy-poor cAMP. Because this reaction involves a large drop in free energy, it favors production of cAMP (see Chapter 3); adenylate cyclase, like all enzymes, accelerates the rate of the reaction. Thus, one hormone molecule bound to a receptor for a short period (i.e., 1 second or less) can elicit the generation of hundreds of cAMP molecules. Each molecule of cAMP binds to the regulatory subunit of a protein kinase A molecule, liberating a catalytic subunit that in turn catalyzes activation of many effector molecules, further amplifying the effect. Finally, many effectors are themselves enzymes and thus a fourth amplification occurs as they act on many substrate molecules. The role of enzyme cascades in amplifying extracellular signals is discussed in more detail in Spotlight 9-2.

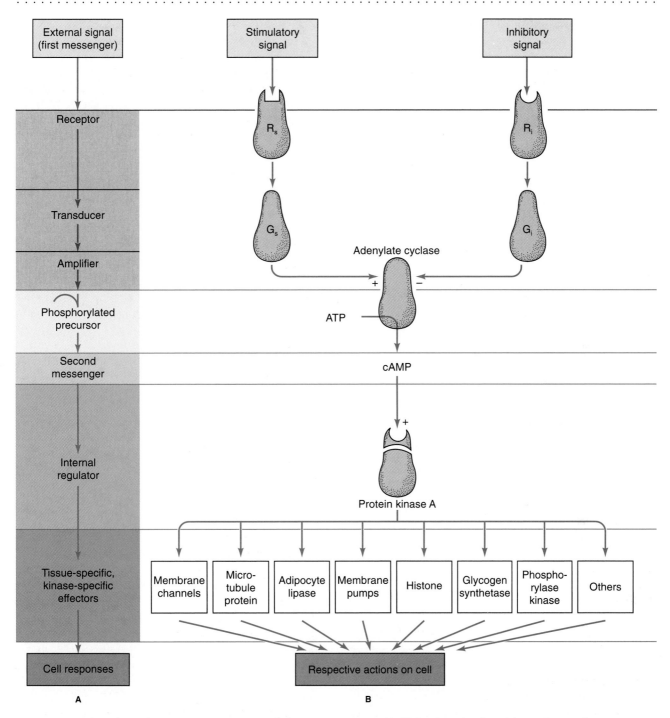

Figure 9-11 Binding of many hormones to G protein–coupled receptors stimulates or decreases production of the second-messenger cAMP, which transduces the signal into cellular responses. **(A)** Generic steps leading from hormone binding by the surface receptor to the cell response(s). **(B)** Condensed outline of the cAMP second-messenger system. Stimulating and inhibiting receptors are denoted by R_s and R_i, respectively; transducer proteins by G_s and G_i.

Control of cellular responses Another problem with second-messenger systems is how to reduce or terminate the signal induced by binding of the hormone so that only an appropriate response is elicited for an appropriate amount of time. Three control mechanisms operate in the cAMP pathway. As we've seen, there are two kinds of receptors, R_s and R_i, that bind stimulatory and inhibitory hormones, respectively. Of course, the two types of trans-

ducer proteins, G_s and G_i, are linked to R_s and R_i, respectively. Thus, the activity of adenylate cyclase can be increased by a stimulatory signal (R_s through G_s) or reduced by an inhibitory signal (R_i through G_i). Both stimulation and inhibition of adenylate cyclase can occur in the same cell, the end result depending on the intensity of each signal. For example, lipid breakdown in fat cells is accelerated in response to binding of epinephrine to the stimulatory

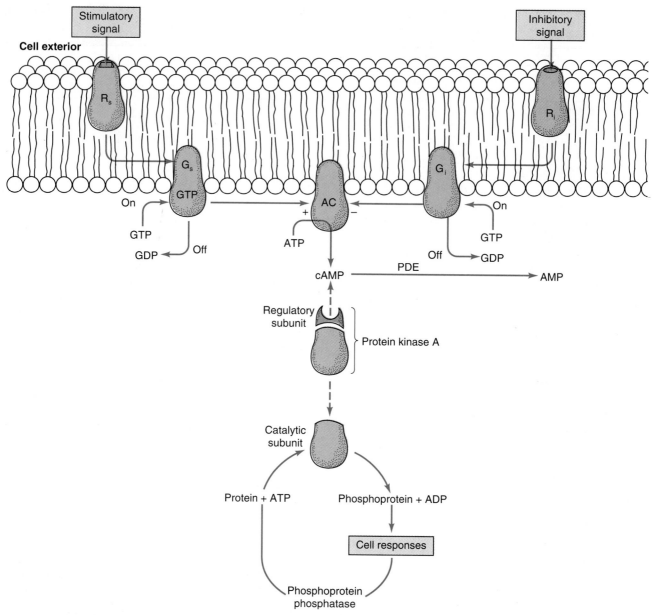

Figure 9-12 Hormone-stimulated regulation of adenylate cyclase (AC) within the membrane leads to an increase or decrease in cytosolic cAMP level. Binding of hormones or other ligands to their stimulatory or inhibitory receptors (R_s and R_i, respectively) induce binding of GTP to the respective transducer proteins, G_s and G_i. The GTP-activated G proteins are then able to either activate or inhibit the catalytic activity of adenylate cyclase until the GTP is dephosphorylated enzymatically to guanosine diphosphate (GDP) and the effect on the cyclase ceases. Activated adenylate cyclase catalyzes the conversion of ATP to cAMP, which binds to and removes the regulatory subunit of protein kinase A. The catalytic subunit, once free of the inhibitory regulatory subunit, can phosphorylate various intracellular effector proteins, yielding activated phosphoproteins that mediate cellular responses. In time, cAMP is degraded to AMP by a phosphodiesterase (PDE), and the phosphorylated effector proteins often are dephosphorylated to their inactive forms. Both of these mechanisms reduce or terminate the effects of the external signal. [Adapted from Berridge, 1985.]

β-adrenoreceptors, whereas it is decreased in response to binding of either epinephrine or adenosine to the inhibitory α-adrenoreceptors or adenosine receptors.

> **?** What factors, other than hormones, can activate the second-messenger systems described in this chapter?

A second control mechanism involves the intracellular level of cAMP, which depends not only on its rate of synthesis from ATP, but also on its rate of degradation to adenosine 5′-phosphate (AMP):

$$ATP \xrightarrow{\;1\;} cAMP \xrightarrow{\;2\;} AMP$$

The balance between the rate of synthesis of cAMP (step 1) and its rate of hydrolysis (step 2) determines the cAMP level in the cell. Step 1, in many tissues, is under the control of extracellular signals that modulate the activity of adenylate cyclase. Step 2 is catalyzed by phosphodiesterase (PDE), which is activated by Ca^{2+} (see Figure 9-12). The activity of PDE is slowed by methyl xanthines, caffeine, or theophylline; thus these agents increase the intracellular concentration of cAMP. The basal concentration of cAMP within cells ranges from 10^{-12} M to more than 10^{-7} M. The regeneration of ATP from AMP is energized by intermediary metabolism, as described in Chapter 3.

Finally, the cellular response to extracellular signals transduced by cAMP can be controlled by dephosphorylation of the phosphorylated effector proteins. These effector proteins, which directly mediate the cellular response, are phosphorylated by the catalytic subunit of protein kinase A following hormone binding. They are dephosphorylated by phosphoprotein phosphatase, whose activity can affect the magnitude and duration of the cellular response to hormone stimulation. Interestingly, the activity of phosphoprotein phosphatase is indirectly dependent on the cAMP level, decreasing as the cAMP level increases.

Diversity of cAMP-mediated responses Since the discovery by Sutherland that cAMP acts as the second messenger linking glucose mobilization to hormone action in liver cells, cAMP has been shown to function as a second messenger for numerous other hormones. To confirm that cAMP is the intracellular second messenger of a hormone, researchers often have used dibutyryl cAMP, a lipid-soluble analog that, unlike cAMP, can penetrate the plasma membrane. For example, observation that application of dibutyryl cAMP to tissues mimics the effects normally induced by a particular hormone indicates that the hormone is linked to cAMP. A second approach is to treat tissues with methyl xanthines, which block phosphodiesterase, thereby elevating cAMP levels. The finding that such treatment increases the response to a particular hormone also provides evidence that it operates via the cAMP pathway.

The various hormones linked to cAMP induce multiple physiological effects (Table 9-6). How, you might ask, can the same second messenger mediate such diverse biochemical and physiological responses? The key to the specificity of hormonal effects lies in the tissue distribution of effector proteins that that can be phosphorylated by cAMP-dependent protein kinase A. Even though cAMP can mediate activation of a wide variety of effectors (see bottom of Figure 9-11), not all tissues contain all effectors. For example, various hormone-stimulated effector proteins involved in the process of secretion are present in secretory tissues but not in nonsecretory tissues.

At one time it was postulated that cAMP activates a number of different protein kinases, each specific for a different phosphoprotein. However, more recent studies show that the catalytic subunit isolated from one kind of tissue in one animal species can replace the native catalytic subunit in the tissues of completely unrelated animal

species. These findings suggest that there is essentially only one kind of cAMP-dependent protein kinase, protein kinase A, the structure of which has been remarkably well preserved through the course of evolution.

Hormone-stimulated mobilization of glucose Let's take a closer look at the cAMP pathway involved in the hormone-stimulated mobilization of glucose from glycogen. The sequence of reactions in this system, originally studied by Sutherland and his associates, has been worked out in great detail. The hormone glucagon stimulates the breakdown of glycogen to glucose 6-phosphate (**glycogenolysis**) in the liver, and epinephrine does the same in skeletal and cardiac muscle; these hormones also inhibit synthesis of glycogen from glucose (glycogenesis) and stimulate formation of glucose from lactate and amino acids (**gluconeogenesis**). Thus the net effect of hormone stimulation is a rise in blood glucose.

Figure 9-13 outlines the steps between binding of glucagon (in liver) and epinephrine (in skeletal and cardiac muscle) and the resulting increase in blood glucose. Binding of each hormone to the membrane-bound β-adrenoreceptor activates adenylate cyclase, resulting in an increased rate of cAMP synthesis from ATP (steps 1 and 2). The immediate action of cAMP is activation of protein kinase A (step 3). These three steps appear to be the common to all cAMP-regulated systems. Once activated, protein kinase A can catalyze phosphorylation of another

TABLE 9-6
Some hormone-induced responses mediated by the cAMP pathway

Signal	Tissue	Cell response
Stimulatory		
Epinephrine (β receptors)	Skeletal muscle	Breakdown of glycogen
	Fat cells	Increased breakdown of lipids
	Heart	Increased heart rate and force of contraction
	Intestine	Fluid secretion
	Smooth muscle	Relaxation
Thyroid-stimulating hormone (TSH)	Thyroid gland	Thyroxine secretion
Vasopressin	Kidney	Resorption of water
Glucagon	Liver	Breakdown of glycogen
Serotonin	Salivary gland (blowfly)	Fluid secretion
Prostaglandin I₁	Blood platelets	Inhibition of aggregation and secretion
Inhibitory		
Epinephrine (α_2 receptors)	Blood platelets	Stimulation of aggregation and secretion
	Fat cells	Decreased lipid breakdown
Adenosine	Fat cells	Decreased lipid breakdown

Source: Berridge, 1985.

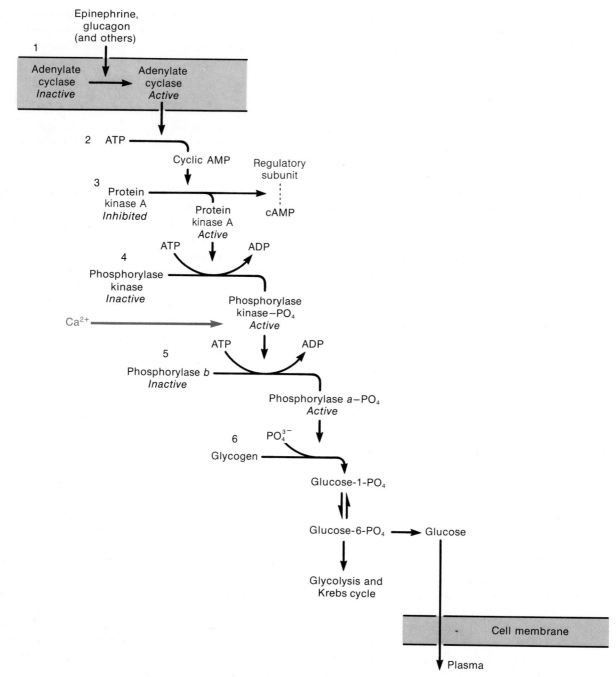

Figure 9-13 Epinephrine and glucagon stimulate breakdown of glycogen to glucose (glycogenolysis) in muscle and liver, respectively. Hormone binding to β-adrenoreceptors triggers a sequence of reactions in which several enzymes are converted from an inactive to active form. As a result of this enzyme cascade the original signal is greatly amplified. See text for discussion. [Adapted from Goldberg, 1975.]

enzyme, **phosphorylase kinase** (step 4), thereby activating it. Phosphorylase kinase-PO$_4$ in turn catalyzes phosphorylation of phosphorylase *b* to form the active form, called **phospho-rylase *a*** (step 5). It is the latter enzyme that cleaves glycogen with addition of PO$_4^{3-}$ to form glucose 1-phosphate (step 6). In cells, glucose 1-phosphate is readily converted to glucose 6-phosphate, which enters the glycolytic pathway or is dephosphorylated to glucose, which is transported across the plasma membrane into the bloodstream.

The cAMP-dependent protein kinase A that stimulates formation of phosphorylase *a* also acts in an indirect way to inhibit glycogen synthetase, the enzyme that catalyzes the polymerization of glucose into glycogen. Thus a hormone-stimulated increase in intracellular cAMP stimulates breakdown of glycogen and inhibits its synthesis. This synergistic effect is important, for it keeps the rise in glucose from driving by mass action the resynthesis of glycogen from glucose. Conversely, a decrease in cAMP inhibits glycogen breakdown and stimulates glycogen synthesis. This example illustrates that multiple cAMP-mediated effects can occur simultaneously within a single cell. Spotlight 9-2 describes the amplification of the hormone signal during glucose mobilization.

cGMP as a second messenger In addition to cAMP, most animal cells also can use **cyclic GMP** (cGMP) as a second messenger (see Figure 9-10). The intracellular concentration of cGMP is one-tenth or less than that of cAMP. The cGMP signaling pathway is not well understood, but it clearly differs from the cAMP pathway in some respects. Guanylate cyclase, which catalyzes the production of cGMP from the ATP analog GTP, can occur in two forms, one bound to the plasma membrane and one free in the cytoplasm. In contrast, adenylate cyclase is always bound to the plasma membrane. The two enzymes also differ in their responses to Ca^{2+}. Studies on isolated guanylate cyclase indicate that the enzyme is inactive at low concentrations of Ca^{2+} and becomes progressively more active as the Ca^{2+} concentration is increased. Isolated preparations of adenylate cyclase, on the other hand, are most active at low concentrations of Ca^{2+} and are inhibited by high concentrations. Moreover, the Ca^{2+} concentration providing optimal enzyme activity is lower for adenylate cyclase than for guanylate cyclase. In view of the difference in the responses of these two enzymes to Ca^{2+}, the relative concentrations of cAMP and cGMP can, in principle, be influenced by intracellular concentration of free Ca^{2+}. In addition, the greater dependence of cGMP synthesis on Ca^{2+} suggests that in some systems intracellular Ca^{2+} acts as a second messenger to stimulate the production of cGMP; in such a case, cGMP would, in fact, act as a third messenger. Like cAMP, cGMP activates a specific protein kinase, protein kinase G, which then phosphorylates effector proteins in the cell.

Hormonal stimulation of the same type of receptor can simultaneously induce changes in the cAMP and cGMP levels. For example, stimulation of the β-adrenoreceptors of the brain, lymphocytes, cardiac muscle, and smooth muscle simultaneously produces a rise in the level of cAMP and a drop in the level of cGMP. Conversely, stimulation of the muscarinic acetylcholine receptors in these tissues results in a drop in the level of cAMP but a rise in the level of cGMP. In some tissues, cAMP and cGMP exert opposing physiological actions. For example, the rate and strength of the heart beat are increased by an epinephrine-induced rise in cAMP, but decreased by an acetylcholine-induced rise in cGMP.

Inositol phospholipid signaling systems
In the early 1950s it was found that some extracellular signaling molecules stimulate the incorporation of radioactive phosphate into phosphatidylinositol (PI), a minor phospholipid in cell membranes. This finding led M. R. Hokin and L. E. Hokin (1953) to suggest that inositol phospholipids (phosphoinositides) play a role in hormone actions. Since then, inositol phospholipids have enjoyed periods of interest, neglect, controversy, and in the early 1980s acceptance as important second messengers in transducing many hormonal and other extracellular signals into a wide variety of cellular responses.

Why do you think calcium, rather than any other ion, has become the central intracellular messenger in biological systems?

Figure 9-14 outlines the chain of events linking extracellular signals to intracellular responses via the inositol phospholipid (IP) signaling system. Although not as well understood as the cAMP pathway, this system of lipid messengers has certain general features reminiscent of the cAMP cascade, which can be seen by comparing the Figures 9-14A and 9-11A. In both cases the membrane houses a receptor, a transducer G protein, and an amplifier enzyme, which catalyzes formation of second-messenger molecules from phosphorylated precursors. These second messengers in turn activate internal regulators, primarily protein kinases, which then activate various tissue-specific, kinase-specific effector molecules.

A closer look at Figure 9-14 reveals the distinguishing features of the IP pathway. Unlike the cAMP system, which has both stimulatory and inhibitory G proteins, the IP system only has a stimulatory G protein. Stimulation of this protein, tentatively called G_p, induces activation of phosphoinositide-specific *phospholipase C* (PLC), the amplifier enzyme in the IP pathway. (G_p is similar but not identical to G_s, which activates adenylate cyclase in the cAMP pathway.) PLC hydrolyzes phosphatidylinositol 4,5-biphosphate (PIP_2) into two major second messengers, **inositol trisphosphate** ($InsP_3$) and **diacylglycerol** (DAG). A remarkable feature of the IP system is that PIP_2, the precursor for production of the second messengers, is itself a constituent of the membrane. A phospholipid bearing three phosphate groups, PIP_2 is located primarily in the inner half of the lipid bilayer, where it can come in contact with membrane-bound phospholipase C (Figure 9-15). Once formed, the water-soluble IP_3 diffuses away from the membrane into the cytosol; the other second messenger, DAG, is lipid-insoluble and remains in the cytoplasmic half of the plasma membrane. These two second messengers subsequently follow their own pathways, but the two branches of the IP system sometimes collaborate in producing a cellular response. $InsP_3$ and DAG are rapidly metabolized and their degradation products are used to replenish PIP_2.

$InsP_3$ acts on intracellular calcium stores such as the endoplasmic reticulum (called the sarcoplasmic reticulum in muscle). Some $InsP_3$ is phosphorylated to form inositol 1,3,4,5-tetrakisphosphate ($InsP_4$), which enhances the entry of Ca^{2+} from the cell exterior into the cell through Ca^{2+} channels in the plasma membrane. The Ca^{2+} released by $InsP_3$ acts as another messenger and thus can be considered a third messenger in this system. For example, Ca^{2+} binds to

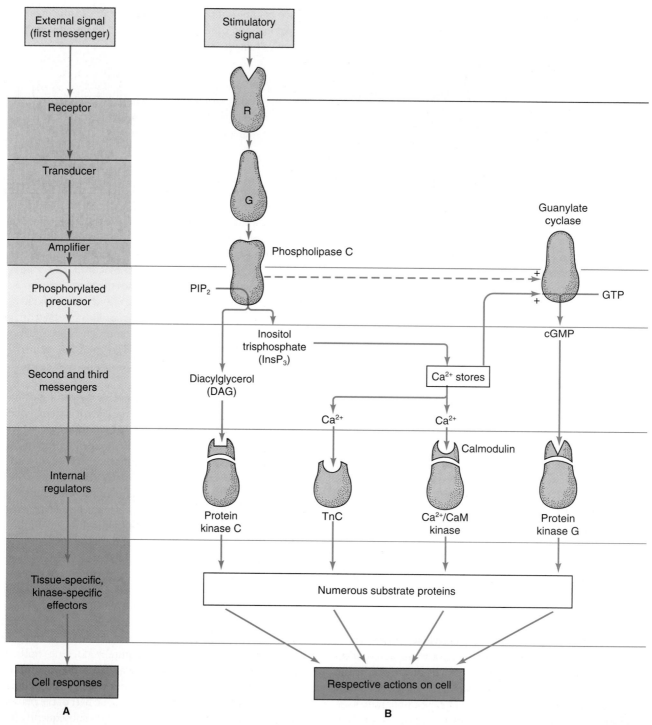

Figure 9-14 Binding of hormones by some G protein–linked receptors induces formation of the phospholipid-derived second messengers di-acylglycerol (DAG) and inositol trisphosphate (InsP₃). **(A)** Generic scheme of inositol phospholipid pathway, which is nearly identical with that of the cAMP pathway (see Figure 9-11A). **(B)** Condensed outline of the inositol phospholipid second-messenger system. The amplifier enzyme in this pathway is phosphoinositide-specific phospholipase C (PLC). The direct activation of guanylate cyclase by PLC (dashed pathway) is not fully established. Note that Ca²⁺ mobilized from intracellular stores may activate troponin C (TnC), form a complex with calmodulin (CaM) that activates Ca²⁺/calmodulin-dependent kinase (Ca²⁺/CaM kinase), promote activation of protein kinase C, or increase cGMP production by stimulating membrane-bound guanylate cyclase.

and activates troponin C (TnC) and calmodulin (CaM), as well as a number of other regulator and effector molecules (see Figure 9-14). As we'll discuss in Chapter 10, Ca²⁺/TnC stimulates muscle contraction directly. Ca²⁺/calmodulin may

act as an effector protein, or bind to and activate a number of enzymes and other effector proteins, of which the most studied is Ca²⁺/calmodulin kinase. These proteins induce various cellular responses through different mechanisms.

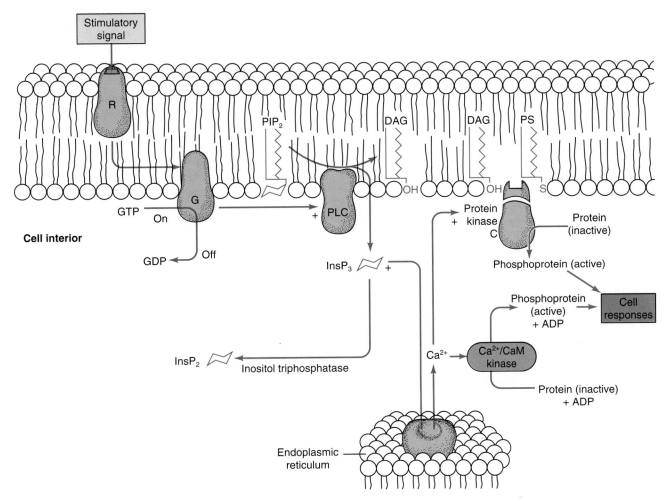

Figure 9-15 In the inositol phospholipid system, one second messenger acts in the membrane and the other in the cytosol. Hormone binding induces formation of the coupled GTP-activated G protein, which activates phosphoinositide-specific phospholipase C (PLC). PLC then catalyzes hydrolysis of membrane phosphatidylinositol 4,5-biphosphate (PIP$_2$) into diacylglycerol (DAG), which remains in the membrane, and inositol trisphosphate (InsP$_3$), which diffuses into the cytosol. DAG promotes activation of the membrane-bound protein kinase C; this activation also requires Ca^{2+} and phosphatidylserine (PS), another membrane phospholipid. InsP$_3$ promotes the liberation of Ca^{2+} from intracellular storage depots like the endoplasmic reticulum. The free Ca^{2+} has numerous regulatory functions, including stimulation of Ca^{2+}/calmodulin kinase (Ca^{2+}/CaM kinase). [Adapted from Berridge, 1985.]

The actions of DAG, the other second messenger in the IP system, occur in the plasma membrane, in which DAG molecules can move laterally by diffusion. DAG has two potential signaling roles. First, it can be cleaved to release arachidonic acid, a precursor in synthesis of the prostaglandins and other biologically active eicosanoids. Second, and more important, DAG activates membrane-bound protein kinase C by a mechanism analogous to activation of protein kinase A by cAMP. Although protein kinase C occurs in both the cytosol and in the inner portion of the cell membrane, it can only be activated when associated with the membrane. The activation of protein kinase C by DAG depends on Ca^{2+} and phosphatidylserine (PS), another phospholipid constituent of the membrane. Binding of DAG and PS to protein kinase C, located in the cytoplasmic half of the membrane, increases the affinity of the enzyme for Ca^{2+}; as a result, protein kinase C can be activated at the usual low concentrations of Ca^{2+} present in the

cytosol. Thus, the activation of protein kinase C requires two intracellular messengers, DAG and Ca^{2+}, both of which can be induced by the same extracellular signal.

Among the tissue-specific responses induced by hormones via the inositol phospholipid pathway are the following (Berridge, 1985):

- Breakdown of liver glycogen stimulated by vasopressin

- DNA synthesis in fibroblasts stimulated by growth factors

- Secretion of prolactin from anterior pituitary gland stimulated by thyrotropin-releasing hormone

Ca^{2+} signaling systems
The concentration of free Ca^{2+} in the cytosol can be increased in two ways: (1) release of Ca^{2+} from intracellular calcium stores such as the endoplasmic reticulum (called

SPOTLIGHT 9-2

AMPLIFICATION BY ENZYME CASCADES

The cAMP signaling pathway stimulated by binding of epinephrine or glucagon and leading to glycogen breakdown exemplifies the amplification of hormone action by enzyme cascades. The pathway between hormone binding and the breakdown of glycogen is rather complex (see Figure 9-13). It would be simpler, of course, for cAMP to activate the final enzyme in the sequence directly, thereby saving several steps in this multistep sequence. On the other hand, the large number of steps makes sense if we consider the need for amplification—that is, for producing a large effect in response to a few hormone molecules.

The diagram shows the steps in this pathway at which *biochemical amplification* occurs. Steps 1, 2, 4, and 5 involve an *activating reaction* that converts a catalytically inactive (or weakly active) molecule into an active enzyme. The result is a progressive amplification through four steps. In addition, each molecule of the final effector in this pathway, phosphorylase *a*, can itself convert many glycogen molecules into glucose 1-phosphate. If it is assumed, conservatively, that each activated enzyme molecule catalyzes the activation of 100 molecules in the next step, then the five amplification steps would produce an overall amplification of 10^{10}. That is, the interaction of one molecule of glucagon or epinephrine with its membrane receptor in liver or muscle cells can result in the mobilization of about 10,000,000,000 or more molecules of glucose.

The basal intracellular concentration of cAMP, the key to transmitting the signal from the membrane to the cytosol, is very low (10^{-12} to 10^{-8} M). Thus the hormone-induced production of only a small absolute number of cAMP molecules will represent a large percentage increase in cAMP concentration. Binding of just a few hormone molecules therefore can produce significant changes in cAMP levels and ultimately mobilize large amounts of glucose from glycogen in a short time.

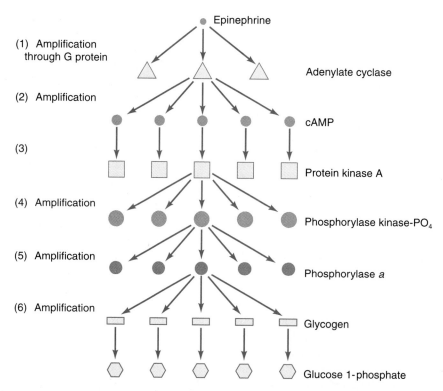

Enzyme cascades greatly amplify hormone action. In the pathway of hormonal stimulation of glycogenolysis, biochemical amplification occurs at several steps, so binding of one molecule of epinephrine or glucagon can lead to production of 10^{10} molecules of glucose 1-phosphate. Such a cascading of amplifying steps in the form of a series of enzyme-activating reactions can explain the extremely high potencies of many hormones. See Figure 9-13 for reaction sequence. [Adapted from H. D. Lodish et al., 1995.]

sarcoplasmic reticulum in muscle) and (2) influx of Ca^{2+} from the cell exterior through Ca^{2+} channels in the plasma membrane. As described in the previous section, $InsP_3$ stimulates release of calcium from intracellular calcium stores. The entry of Ca^{2+} from the cell exterior, through Ca^{2+} channels in the plasma membrane, has been shown to be stimulated by $InsP_4$, phosphorylation of the Ca^{2+} channel by cAMP-dependent kinase, electrical stimulation, or receptor activation itself, which we will discuss later.

In recent decades it has become clear that Ca^{2+} plays an important and ubiquitous role as both an intracellular regulatory agent and a messenger linking external signals to cellular responses. Two important characteristics of cells permit Ca^{2+} to function effectively in cellular regulation and signaling: (1) the ability of cells to increase and decrease the intracellular Ca^{2+} level over a wide concentration range and (2) the presence within cells of numerous proteins whose activity is modulated by the binding of Ca^{2+}. We'll first discuss these aspects of calcium's role in the cell and then look at how Ca^{2+} functions as a second messenger.

Modulation of intracellular Ca^{2+} concentration Most Ca^{2+} ions that enter the cell from the exterior are rapidly bound to anionic sites on protein molecules in the cytosol; only a small percentage of the ions remain ionized and free to diffuse. As a result, although the total Ca^{2+} content of most cells is about 1 mM (10^{-3} M), the concentration of free, ionized Ca^{2+} in the cytosol is maintained at extraordinarily low levels, usually below 10^{-7} M. (Note that unless indicated otherwise, references to intracellular levels of Ca^{2+} and other ions refer to free, unbound ions.) The advantage of this very low intracellular Ca^{2+} concentration is quite simple: influx of quite small amounts of Ca^{2+} from the extracellular space produces a very large increase in the concentration of free Ca^{2+} in the cytosol. This concept is illustrated by comparing the relative changes in the intracellular concentration of Ca^{2+} and Na^+ that result from the entry of equal quantities of these two ion species in response to a transient increase in the permeability of the plasma meKmbrane to both ions (Figure 9-16). Likewise, release of small amounts of Ca^{2+} from intracellular calcium stores causes a relatively large increase in the concentration of free Ca^{2+} in the cytosol. Thus, the cell maintains the cytosolic Ca^{2+} level extremely low, permitting it to increase as much as tenfold as a consequence of Ca^{2+} flow into the cell or out of the intracellular calcium stores.

Since the extracellular Ca^{2+} concentration is typically about 10^{-3} M, the electrochemical gradient favors entry of Ca^{2+} in cells. The cell has two primary mechanisms for removing excess Ca^{2+} from the cytosol, thereby keeping the free Ca^{2+} level low: primary and secondary active transport of Ca^{2+} across the plasma membrane to the exterior (see Chapter 4) and movement of Ca^{2+} ions into the endoplasmic reticulum via a Ca^{2+} pump in the reticulum membrane. Two additional mechanisms help keep the intracellular Ca^{2+} level from transiently becoming too high. First, various cytosolic proteins bind Ca^{2+} when the Ca^{2+} level increases and release Ca^{2+} when the level decreases. In effect, these proteins "buffer" the Ca^{2+} concentration, limiting perturbations in free Ca^{2+} levels, just as pH buffers limit perturbations in free H^+ levels. Second, when the cytosolic Ca^{2+} level becomes abnormally high, the mitochondria may import Ca^{2+} in exchange for H^+.

Certain technical advances have been essential in studying the physiological effects of changes in the intracellular Ca^{2+} concentrations. One of these advances was the discovery in 1963 of the jellyfish protein **aequorin**, which emits light when it complexes with Ca^{2+}. Since light can be measured with very sensitive instruments, injection of aequorin into cells has provided a means of detecting minute changes in the intracellular free Ca^{2+} level. More recently, calcium-sensitive dyes (e.g., arsenazo III) and calcium-sensitive fluorescent molecules (e.g., quin-2 and fura-2) have opened up new possibilities for sensitive optical measurement of Ca^{2+} levels within single living cells.

Ca^{2+}-binding proteins The other important feature typical of Ca^{2+}-mediated intracellular control and signaling is the presence of multiple Ca^{2+}-binding sites in certain enzymes and various regulatory proteins. These specialized binding sites have a very high affinity for Ca^{2+}, allowing tight binding of the cation at very low concentrations of free Ca^{2+}. The Ca^{2+}-binding sites in all these proteins consist of acidic amino acid residues, which are negatively charged and rich in oxygen atoms. The oxygen atoms, carrying full or

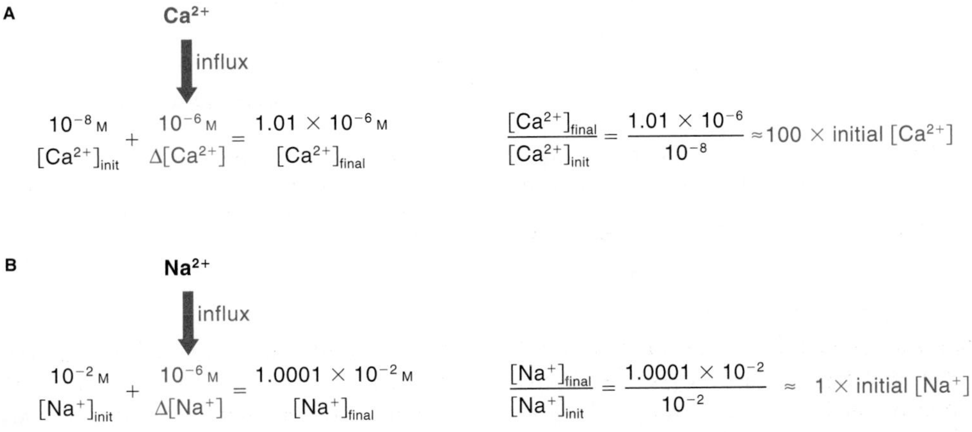

A

$$Ca^{2+} \downarrow \text{influx}$$

$$\underset{[Ca^{2+}]_{init}}{10^{-8}\,M} + \underset{\Delta[Ca^{2+}]}{10^{-6}\,M} = \underset{[Ca^{2+}]_{final}}{1.01 \times 10^{-6}\,M}$$

$$\frac{[Ca^{2+}]_{final}}{[Ca^{2+}]_{init}} = \frac{1.01 \times 10^{-6}}{10^{-8}} \approx 100 \times \text{initial } [Ca^{2+}]$$

B

$$Na^{2+} \downarrow \text{influx}$$

$$\underset{[Na^+]_{init}}{10^{-2}\,M} + \underset{\Delta[Na^+]}{10^{-6}\,M} = \underset{[Na^+]_{final}}{1.0001 \times 10^{-2}\,M}$$

$$\frac{[Na^+]_{final}}{[Na^+]_{init}} = \frac{1.0001 \times 10^{-2}}{10^{-2}} \approx 1 \times \text{initial } [Na^+]$$

Figure 9-16 The intracellular concentration of free Ca^{2+} is elevated many fold by influx of small amounts of Ca^{2+}. **(A)** In this example, the low initial intracellular concentration of Ca^{2+}, $[Ca^{2+}]_{init}$, of 10^{-8} M is raised 100-fold by a transient influx, $\Delta[Ca^{2+}]$, equivalent to a 10^{-6} M increment. **(B)** Since $[Na^+]_{init}$ is already 10^{-2} M, a 10^{-6} M increment, $\Delta[Na^+]$, produces virtually no change in the intracellular Na^+ concentration.

A

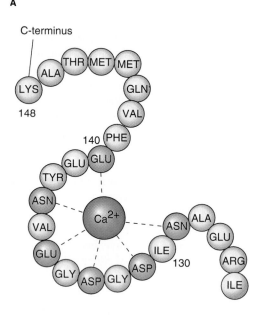

C

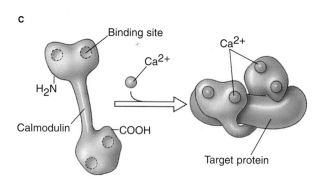

B

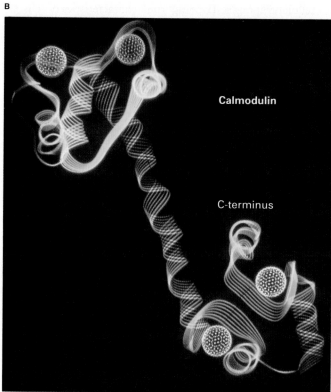

Figure 9-17 Calmodulin, a cytosolic protein with four Ca^{2+}-binding sites, forms the Ca^{2+}/calmodulin complex, an important intracellular regulator. **(A)** Amino acid sequence of the Ca^{2+}-binding site at the C-terminus of calmodulin. Each binding site contains aspartate, glutamate, and asparagine residues, shown in orange, whose side chains form ionic bonds with a Ca^{2+} ion, forming a loop in the backbone. Other binding sites also contain threonine and serine residues, whose side chain oxygen atoms also associate with the Ca^{2+} ion. **(B)** Model of the calmodulin molecule with four bound Ca^{2+} ions (blue spheres). **(C)** Diagram illustrating Ca^{2+}-induced conformational change in calmodulin. Calmodulin undergoes a conformation change when all four Ca^{2+}-binding sites are occupied. The resulting Ca^{2+}/calmodulin complex can bind to numerous target proteins, modulating their activity. [Part B courtesy Y. S. Babu and W. J. Cook; parts A and C adapted from Lodish et al., 1995.]

partial negative charges, occur in a loop of the peptide chain, so that six to eight oxygen atoms form a cavity of just the right size to harbor the positively charged calcium ion (Figure 9-17A). In fact, about 70% of the entire amino acid sequences of various Ca^{2+}-binding regulatory proteins are homologous.

Binding of Ca^{2+} to these proteins generally leads to a conformational change in the molecule. This change in conformation can produce an allosteric effect that alters the properties of the molecule. For example, binding of Ca^{2+} to troponin C, which is found only in striated muscle, causes a conformational change in the molecule that initiates a series of steps leading to contraction and is important in regulating contraction of vertebrate striated muscle. We'll discuss troponin C, the first Ca^{2+}-binding regulatory protein to be discovered, in detail in Chapter 10.

Calmodulin, a Ca^{2+}-binding protein closely related to troponin C, is present in relatively large amounts in every

eukaryotic tissue examined thus far. It functions as a multipurpose intracellular regulatory protein, mediating most Ca^{2+}-regulated processes. The single polypeptide chain of calmodulin, consisting of 148 amino acid residues, contains four Ca^{2+}-binding sites (Figure 9-17B). Binding of Ca^{2+} to all four sites produces a Ca^{2+}/calmodulin complex that can bind to and activate numerous enzymes and effector proteins (Figure 9-17C). For example, Ca^{2+}/calmodulin binds to the regulatory subunit of Ca^{2+}/calmodulin kinase. Once freed of its regulatory subunit, the catalytic subunit of Ca^{2+}/calmodulin kinase can phosphorylate serine and threonine residues on various effector proteins, which induce cellular responses (see Figure 9-14). Other enzymes and cellular processes regulated by Ca^{2+}/calmodulin are shown in Figure 9-18. Note that Ca^{2+}/calmodulin activates myosin light-chain kinase, a protein that regulates contraction in vertebrate smooth muscle; this function is somewhat analogous to that of troponin C in vertebrate striated muscle.

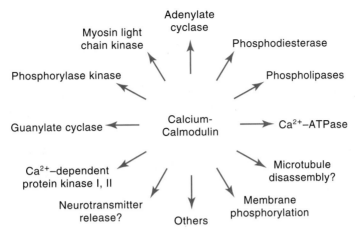

Figure 9-18 Calcium/calmodulin regulates many processes or enzymes within cells. Among these are adenylate cyclase and guanylate cyclase, which catalyze formation of the nucleotide second messengers. [Adapted from Cheung, 1979.]

Second-messenger role of Ca^{2+} Earlier we learned that Ca^{2+} acts as a third messenger in the inositol phospholipid system. Stimulation of other receptor systems leads to an influx of Ca^{2+}, which then can act as a second (and only) messenger, as illustrated in Figure 9-19. Various signals can activate Ca^{2+} second-messenger pathways. For example, activation of α-adrenoreceptors by epinephrine in the mammalian liver and salivary gland stimulates Ca^{2+} influx through opening of Ca^{2+} channels in the plasma membrane, whereas membrane depolarization causes opening of Ca^{2+} channels in muscle.

Membrane-enzyme signaling systems
Some cell-surface receptors seem to signal the cell directly through their intrinsic enzyme activity. Such receptors have a ligand-binding domain on the extracellular surface of the plasma membrane and a catalytic domain on the intracellular surface. Binding of an external signal to this type of receptor triggers a conformational change that causes the catalytic domain to become activated. The activated catalytic domain, in turn, induces further intracellular changes that result in the cellular responses.

To date, cell-surface receptors with intrinsic protein kinase or guanylate cyclase activity have been identified. The best-studied of these in animal cells are **receptor tyrosine kinases** (RTKs), which are known to bind insulin and a number of growth factors including platelet-derived growth factor (PDGF). When activated by external signal binding, RTKs transfer the phosphate group from ATP to the hydroxyl group on a tyrosine residue of selected proteins in the cytosol (Figure 9-20A). In all cases studied, RTKs also phosphorylate themselves when activated; this autophosphorylation enhances the activity of the kinase—an example of positive feedback regulation. Atrial natriuretic peptide (ANP) has been shown to activate a receptor guanylate cyclase (Figure 9-20B). A glance back at Figures 9-14 and 9-19 shows that membrane-bound guanylate cyclase can be activated by Ca^{2+} generated in other signaling

pathways. Since the receptor guanylate cyclase has a ligand-binding domain, it can be activated directly by hormone binding. The cGMP produced by activation of this receptor can function as a second messenger to mediate cellular responses as in other pathways.

Second-messenger networks
It is important to note that a single hormone can trigger several second-messenger systems by activating different types of receptors, even in the same cell. Binding of epinephrine to α- and β-adrenoreceptors in the mammalian salivary gland is an example of a *divergent pathway* in which the two second messengers—namely, intracellular free Ca^{2+} and cAMP, respectively—mediate different cell responses (Figure 9-21A). In mammalian liver, however, binding of epinephrine to α- and β-adrenoreceptors leads to the same cell response, an example of a *convergent pathway* (Figure 9-21B). In this case, both second messengers, Ca^{2+} and cAMP, activate phosphorylase kinase, which in turn stimulates glycogenolysis as discussed éarlier.

A more complicated example of second-messenger networks involves **serotonin** (5-hydroxytryptamine, 5-HT), a lipid-insoluble amine that functions as both a neurotransmitter and an endocrine hormone regulating gastric secretion and smooth muscle contraction in blood vessels. As shown in Figure 9-22, serotonin binds to several receptor subtypes, which are linked to various second-messenger pathways or ion channels; some of these converge, while others diverge. Like other lipid-insoluble hormones, serotonin binds to cell-surface receptors; however, its mode of action, which ultimately affects gene transcription, differs from that of most lipid-insoluble hormones (see Figure 9-8B).

As we saw earlier, activation of different receptors may stimulate or inhibit the same signaling system; for example, norepinephrine stimulates, but neuropeptide Y inhibits, the inositol phospholipid system. Another possible pattern is a single receptor capable of coupling to two

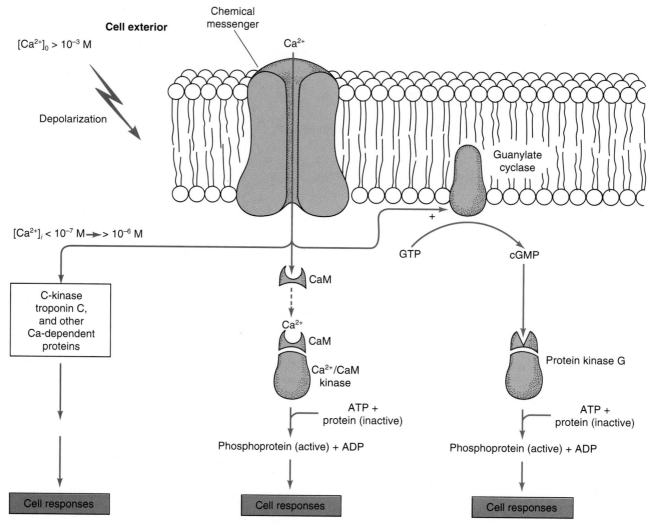

Figure 9-19 Stimulation of receptors that function as calcium-selective ion channels causes an influx of Ca²⁺, which acts as a second messenger. Either membrane depolarization or binding of a chemical messenger (e.g., extracellular hormone) can "open" an ion channel, permitting Ca²⁺ to move through the channel down its electrochemical gradient into the cytosol. The resulting local increase in cytosolic free Ca²⁺ from a resting level of $<10^{-7}$ M to $>10^{-6}$ M can activate several different intracellular signaling pathways, leading to various cell responses. CaM = calmodulin.

different G proteins, each with its own second-messenger system or both with the same second-messenger system. For example, somatostatin stimulates adenylate cyclase in a number of cell types through two different G proteins, one sensitive to and the other insensitive to pertussis toxin. Another example of a single receptor linked to several second-messenger pathways is the octopamine/tyramine receptor in *Drosophila*. Activation of this receptor inhibits adenylate cyclase via one G protein and activates phospholipase C via a different G protein, leading to an elevation in intracellular Ca²⁺. Interestingly, tyramine has a more potent effect on the adenylate cyclase pathway than octopamine, whereas octopamine has the greater effect on the phospholipase C pathway. Thus two agonists, differing in structure by a single hydroxyl group, can differentially couple this receptor to two second-messenger pathways.

 Why are there many more types of cell-surface receptor than G proteins? Is there more interaction between pathways of hormonal action outside or inside cells?

Changes in intracellular Ca²⁺ can have a multitude of effects, including modulation of other second-messenger systems. As discussed previously, Ca²⁺/calmodulin binds to and regulates numerous enzymes, including adenylate cyclase and guanylate cyclase, which form cAMP and cGMP, as well as the two phosphodiesterases that break down these second messengers. Conversely, in some cells, protein kinase A can phosphorylate some Ca²⁺ channels and alter their activity. Protein kinase A and Ca²⁺-activated kinases,

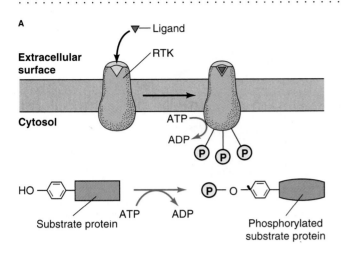

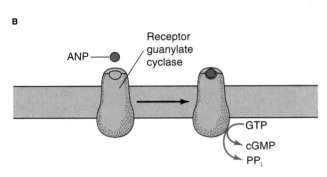

Figure 9-20 Some hormone receptors have intrinsic catalytic activity, which is stimulated by hormone binding. **(A)** Binding of ligand (e.g., insulin) to a receptor tyrosine kinase (RTK) activates catalytic activity in the cytosolic domain of the receptor. In some cases, the activated receptor may directly phosphorylate certain substrate proteins; in other cases, it binds a transducer protein that initiates a rather complicated signaling pathway. **(B)** The receptor for atrial natriuretic peptide (ANP) has guanylate cyclase activity. Hormone binding leads to production of the second messenger cGMP. [Adapted from Lodish et al., 1995.]

generated in different signaling pathways, frequently phosphorylate different sites on the same proteins. Within a signaling system such as the inositol phospholipid pathway, the two branches ($InsP_3$ and DAG) may interact, modulating the overall cellular response (see Figure 9-15).

Clearly, although intracellular signaling systems often are described as separate pathways, *in vivo* nothing could be farther from the truth. Because extensive interactions occur between many elements of the various signaling pathways, we cannot really understand their physiological roles by viewing them only as isolated pathways.

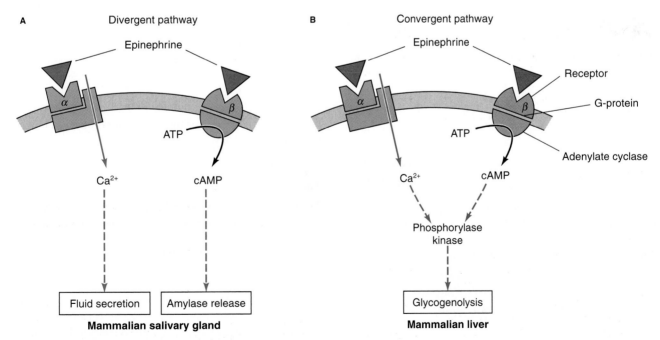

Figure 9-21 A single hormone may bind to different receptors, initiating convergent and/or divergent signaling pathways. Binding of epinephrine to α- and β-adrenoreceptors leads to increases in intracellular Ca^{2+} and cAMP, respectively. In the mammalian salivary gland **(A)** these two second messengers mediate divergent pathways, leading to different, independent end effects—fluid secretion and amylase secretion by secretory cells in the gland. In mammalian liver **(B)**, these two second

messengers both induce activation of phosphorylase kinase, which catalyzes the breakdown of glycogen to glucose (glycogenolysis) (see Figure 9-13). Thus, binding of the same hormone to different receptors triggers convergent pathways leading to the same end response. There is growing evidence that epinephrine is not unique in having multiple receptors in the same animal, or even in the same cell.

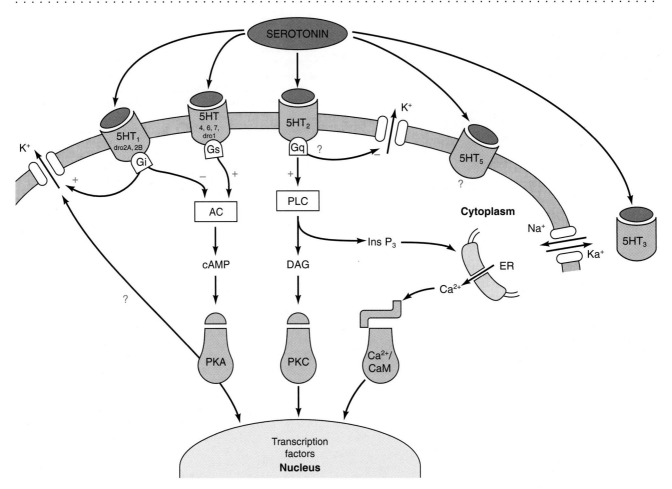

Figure 9-22 Serotonin binds to multiple receptors, which are linked to convergent and divergent second-messenger pathways. Binding of serotonin, also known as 5-hydroxytryptamine (5HT), to some receptors leads to production of cAMP, diacylglycerol (DAG), or inositol trisphosphate (InsP$_3$), which all can mediate the same cellular responses in cells of different tissues, or even in the same cells. The various receptors illustrated represent subclasses of the serotonin-receptor family (dro = *Drosophila*). G$_i$ = inhibitory G proteins; G$_s$ = stimulatory G proteins; G$_q$ = pertussis toxin–insensitive G proteins; AC = adenylate cyclase; PLC = phospholipase C; ER = endoplasmic reticulum; PKA = protein kinase A; PKC = protein kinase C; Ca^{2+}/CaM kinase = Ca^{2+}/calmodulin-dependent protein kinase. [Adapted from Saudou and Hen, 1994.]

PHYSIOLOGICAL EFFECTS OF HORMONES

As noted earlier, most hormones produce tissue-specific physiological effects. That is, a given hormone generally induces responses only in selected tissues, and may induce different responses in different tissues. This specificity in hormonal action depends partly on the restricted distribution of the components of hormone-triggered signaling pathways (especially receptors) and partly on the preferential expression of effector proteins in different tissues. In the following sections, we'll examine the physiological effects of four major categories of hormones.

Metabolic and Developmental Hormones

Several different hormones regulate metabolism and various developmental processes. Produced in various endocrine tissues, these hormones have diverse structures (e.g., steroids, catecholamines, peptides). Table 9-7 summarizes the properties of the major metabolic and developmental hormones.

Glucocorticoids and catecholamines

The adrenal gland, which is situated close to the kidney, is actually composed of two developmentally and functionally unrelated glandular tissues: an outer cortex, derived from nonneural tissue, and an inner medulla, derived from the neural crest (see Figure 8-12). As described in Chapter 8, the adrenal medulla synthesizes and secretes the catecholamines epinephrine and norepinephrine, which can bind to α- and β-adrenoreceptors. Stimulation of α-adrenoreceptors may cause a decrease in cAMP via a coupled inhibitory G protein (G$_i$); may trigger the inositol phospholipid pathway, leading to release of Ca^{2+} from intracellular stores; or may activate an associated Ca^{2+} channel, leading to an influx of extracellular Ca^{2+} into the cell. Stimulation of β-adrenoreceptors usually is coupled, via a stimulatory G protein (G$_s$), to an increase in cAMP. The catecholamines affect contraction of smooth muscle, induce vasoconstric-

TABLE 9-7
Metabolic and developmental hormones

Hormone	Tissue of origin	Structure	Target tissue	Primary action	Regulation
Glucagon	Pancreas (alpha cells)	Peptide	Liver, adipose tissue	Stimulates glycogenolysis and release of glucose from liver; promotes lipolysis	Low serum glucose increases secretion; somatostatin inhibits release
Glucocorticoids (e.g., cortisol)	Adrenal cortex	Steroid	Liver, adipose tissue	Stimulate mobilization of amino acids from muscle and gluconeogenesis in liver to raise blood glucose; increase transfer of fatty acids from adipose tissue to liver; exhibit anti-inflammatory action	Physiological stress increases secretion; biological clock via CRH and ACTH controls diurnal changes in secretion
Growth hormone (GH)	Adenohypophysis	Peptide	All tissues	Stimulates RNA synthesis, protein synthesis, and tissue growth; increases transport of glucose and amino acids into cells; increase lipolysis and antibody formation	Reduced plasma glucose and increased plasma amino acid levels stimulate release via GRH; somatostatin inhibits release
Insulin	Pancreas (beta cells)	Peptide	All tissues except most neural tissue	Increases glucose and amino acid uptake by cells	High plasma glucose and amino acid levels and presence of glucagon increase secretion; somatostatin inhibits secretion
Norepinephrine and epinephrine	Adrenal medulla (chromaffin cells)	Catecholamine	Most tissues	Increase cardiac activity; induce vasoconstriction; increase glycolysis, hyperglycemia, and lipolysis	Sympathetic stimulation via splanchnic nerves increase secretion
Thyroxine	Thyroid	Tyrosine derivative	Most cells, but especially those of muscle, heart, liver, and kidney	Increases metabolic rate, thermogenesis, growth, and development; promotes amphibian metamorphosis	TSH induces release

tion, and stimulate glycolysis and lipolysis. Their physiological effects are discussed in Chapter 8 and summarized in Table 8-2.

In this chapter, we focus on the hormones produced by cells of the adrenal cortex. When stimulated by ACTH, the adrenal cortex synthesizes and secretes a family of steroids derived from cholesterol (Figure 9-23). These hormones fall into three functional categories:

- Reproductive hormones

- Mineralocorticoids, which regulate kidney function

- Glucocorticoids, which have widespread actions, including mobilization of amino acids and glucose and anti-inflammatory actions

In this section, we discuss the glucocorticoids; the reproductive hormones and mineralocorticoids are covered in later sections.

Several adrenocortical hormones have **glucocorticoid** activity, including cortisol, cortisone, and corticosterone. Of these, cortisol is the most important in humans. The basal level of secretion of glucocorticoids is regulated via negative feedback by the hormones themselves on the CRH-secreting neurons of the hypothalamus and the ACTH-secreting cells of the anterior pituitary gland (Figure 9-24). The basal level of glucocorticoid secretion also undergoes a diurnal rhythm resulting from cyclic variation in CRH secretion, which appears to be influenced by an endogenous biological clock. Basal glucocorticoid levels in humans are maximal during the early hours of the morning prior to waking. This is adaptively useful because of the energy-mobilizing actions of these hormones. In addition to such endogenous regulation of secretion, the adrenal cortex is stimulated to secrete glucocorticoids in response to stress of various types (including starvation). Stress, acting through the nervous system, causes an elevation in ACTH and hence stimulation of the adrenal cortex.

 What are the advantages and disadvantages of evoking a stress response?

The glucocorticoids act on the liver, increasing the synthesis of enzymes that promote gluconeogenesis (synthesis of glucose from substances other than carbohydrates). Some of the newly synthesized glucose may be converted

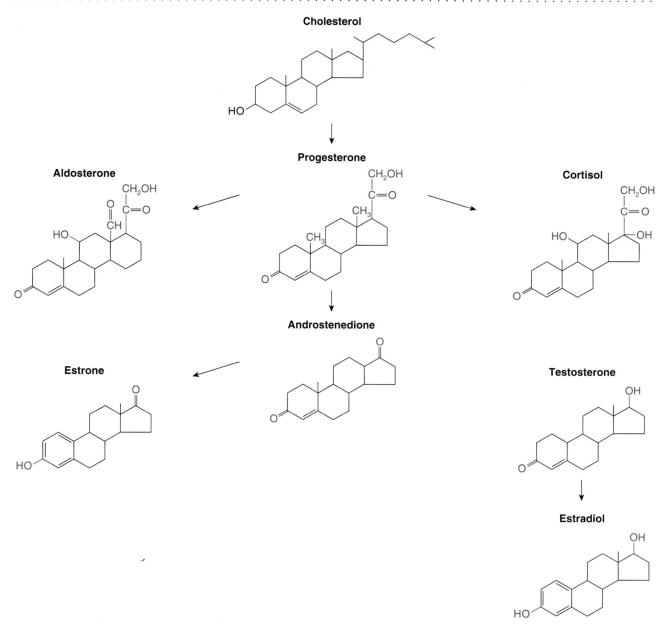

Figure 9-23 Cholesterol is the precursor for three major classes of hormones: mineralocorticoids, glucocorticoids, and reproductive hormones. Modifications to the cholesterol structure, shown in boldface, yield a large number of related steroid hormones and intermediates. (Some intermediates have been omitted in the synthetic pathway shown here.) Several steroid hormones have mineralocorticoid or glucocorticoid activity, but aldosterone and cortisol are the primary ones, respectively, in mammals. The adrenal cortex is the primary site for secretion of these hormones. The reproductive hormones (progesterone, testosterone, estrone, estradiol) are secreted mostly from the gonads, although they also are secreted from the adrenal cortex.

into glycogen, which is stored in the liver and muscle. Most of the newly produced glucose, however, is released into the circulation, causing a rise in blood glucose levels, but the glucocorticoids also act to reduce uptake of glucose into peripheral tissues such as muscle. At the same time, the uptake of amino acids by muscle tissues is decreased by glucocorticoids, and amino acids are released from muscle cells into the circulation. This release increases the quantity of amino acids available for deamination and conversion into glucose in the liver under glucocorticoid stimulation. This mechanism is especially important during starvation, the end result being the degradation of tissue proteins to

maintain adequate blood glucose to sustain energy production in critical tissues such as the brain. The glucocorticoids also stimulate mobilization of fatty acids from stores of fat in adipose tissue. These can be used as substrates for gluconeogenesis in the liver or metabolized directly in muscle to provide energy for contraction. All these actions thus increase the availability of quick energy to muscle and nervous tissue. The glucocorticoids have numerous other actions including stimulation of gastric secretion and inhibition of immune responses.

As discussed earlier, the glucocorticoids, like other lipid-soluble steroid hormones, bind to specific receptors in

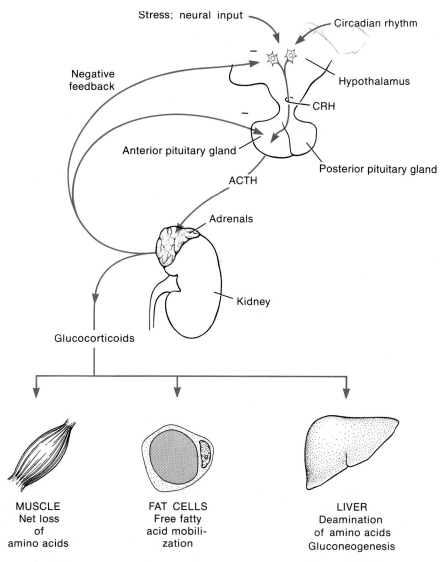

Stress; neural input

Circadian rhythm

Negative
feedback

Hypothalamus

CRH

Anterior pituitary gland

Posterior pituitary gland

ACTH

Adrenals

Kidney

Glucocorticoids

MUSCLE
Net loss
of
amino acids

FAT CELLS
Free fatty
acid mobili-
zation

LIVER
Deamination
of amino acids
Gluconeogenesis

Figure 9-24 Secretion of glucocorticoids, and hence their effects on target tissues, is regulated by neural stimuli and negative feedback. Neural stimuli induce release of corticotropin-releasing hormone (CRH) from hypothalamic neurosecretory cells. Subsequent release of adrenocorticotropin hormone (ACTH) from the anterior pituitary gland stimulates secretion of glucocorticoids by the adrenal cortex. These steroids produce an increase in blood glucose and liver glycogen by stimulating conversion of amino acids and fats to glucose. Negative feedback by the glucocorticoids to both the pituitary and the hypothalamus may limit ACTH release.

the cytosol, forming hormone-receptor complexes that enter the nucleus and regulate the transcription of specific genes (see Figures 9-8 and 9-9).

Thyroid hormones
The follicles of the thyroid tissue are stimulated by thyroid-stimulating hormone (TSH) to synthesize and release two major thyroid hormones—3,5,3′-triiodothyronine (T_3) and thyroxine (T_4)—from two iodinated tyrosine precursors (Figure 9-25). Iodine is actively accumulated by the thyroid tissue from the blood. The secretion of thyroid hormones is regulated by negative feedback of these hormones on the hypothalamic neurons that secrete TSH-releasing hormone (TRH) and on the TSH-secreting cells of the anterior pituitary gland (Figure 9-26). Superimposed on this regulation, however, is the stimulation of the hypothalamus by stress; a low skin temperature, for example, will stimulate the release of hypothalamic TRH.

The thyroid hormones act on the liver, kidney, heart, nervous system, and skeletal muscle, sensitizing these tissues to epinephrine and stimulating cellular respiration, oxygen consumption, and metabolic rate. The acceleration of metabolism stimulated by thyroid hormones leads to a rise in heat production. This is of major importance in the thermoregulation of many vertebrates (see Chapter 16).

The thyroid hormones also significantly affect the development and maturation of various mammalian vertebrate groups. The developmental effects of thyroid hormones occur only in the presence of growth hormone (GH), and vice versa. The synergistic actions of the thyroid

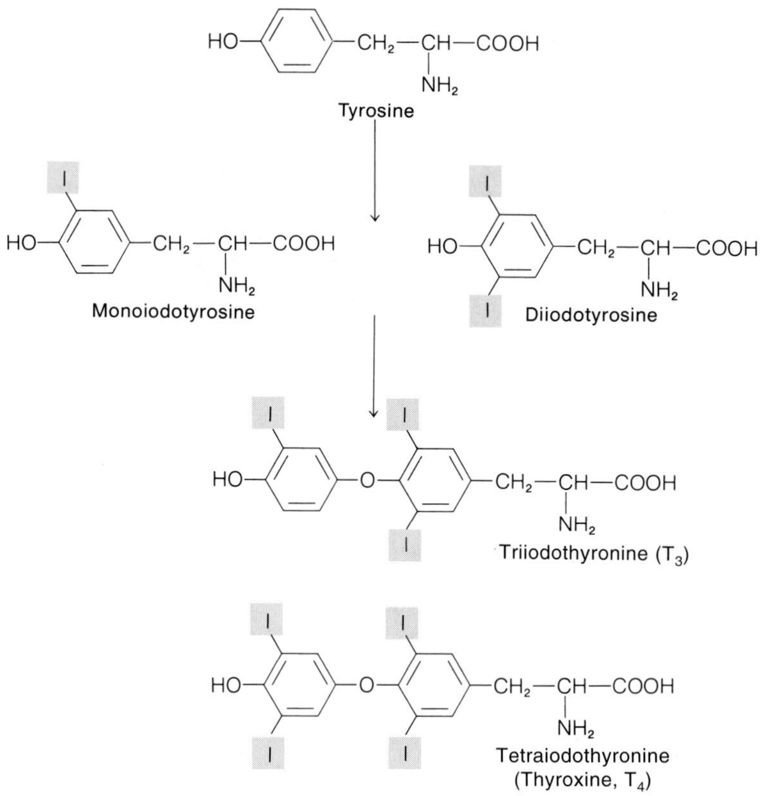

Figure 9-25 The thyroid hormones are produced from iodinated derivatives of the amino acid tyrosine. Condensation of the tyrosine derivatives yields 3,5,3'-triiodothyronine (T_3) and thyroxine (T_4); the two rings in each hormone are linked by an ether bond. T_3 is also produced by removal of one iodide from thyroxine.

hormones and growth hormone promote protein synthesis during development. Hypothyroidism resulting from the lack of dietary iodine during early stages of development in fish, birds, and mammals results in a deficiency disease (called **cretinism** in humans) in which somatic, neural, and sexual development are severely retarded, the metabolic rate is reduced to as little as about half the normal rate, and resistance to infection is reduced. Inadequate production of thyroid hormones, leads to excessive production of TSH due to decreased negative feedback to the hypothalamus and the anterior pituitary gland. The resulting overstimulation of the thyroid gland by TSH causes hypertrophy of the gland (**goiter**). Raising the dietary iodine level increases thyroid hormone production, thereby establishing normal feedback control on TSH production. Thus, the incidence of both cretinism and goiter have been reduced in areas where table salt is routinely "iodized" and the population is no longer dependent on natural trace amounts of iodine in food (mainly in seafood).

The thyroid hormones, like the steroid hormones, are lipid soluble and bind to specific receptors within the cytosol. Both types of hormones exert their effects by regulating the transcription of specific genes and ultimately the production of the proteins encoded by these genes. For this reason, the effects of these hormones develop slowly. For example, it may take up to 48 hours after a rise in blood levels of thyroid hormones before their effects are seen.

Insulin and glucagon

Insulin is secreted by the *beta cells* of the pancreatic **islets of Langerhans,** small patches of endocrine tissue scattered throughout the exocrine tissue of the pancreas. High blood glucose acts as the major stimulus to the pancreatic beta cells to secrete insulin (Figure 9-27). The release of insulin is also stimulated by glucagon, growth hormone, gastric inhibitory peptide (GIP, also known as glucose-dependent insulin-releasing peptide), epinephrine, and elevated levels of amino acids.

Insulin has important effects on carbohydrate, fat, and protein metabolism. With regard to carbohydrate metabolism, insulin has two major actions: increasing the rate of uptake of glucose into cells of liver, muscle, and adipose tissue and stimulating glycogenesis (polymerization of glucose to glycogen). As for lipid metabolism, insulin stimulates lipogenesis in liver and adipose tissue. In protein metabolism, insulin stimulates the uptake of amino acids into liver and muscles and the incorporation of amino acids into protein.

Diabetes mellitus in humans, which occurs in two major forms, is characterized by an absolute or relative deficiency of insulin. Type I diabetes mellitus is associated with a loss of pancreatic beta-cell mass, which leads to diminished or decreased insulin production and secretion (i.e., absolute insulin deficiency). Type II diabetes mellitus, on the other hand, is associated with defective insulin recep-

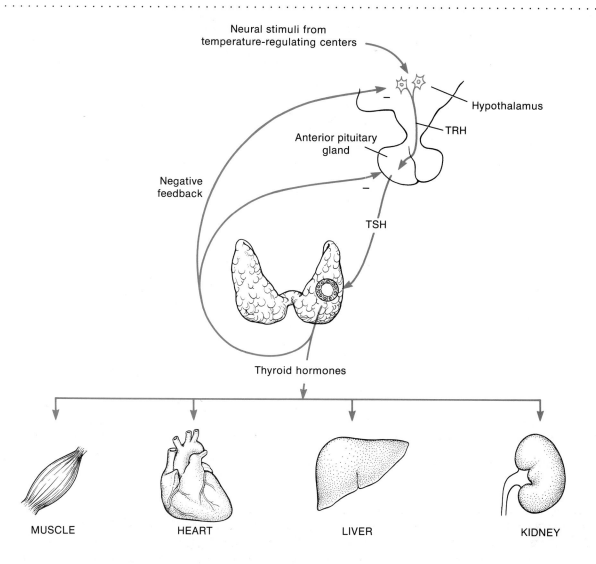

Neural stimuli from
temperature-regulating centers

Hypothalamus

TRH

Anterior pituitary
gland

Negative
feedback

TSH

Thyroid hormones

MUSCLE   HEART   LIVER   KIDNEY

Increased oxygen consumption and heat production

Figure 9-26 The thyroid hormones, which regulate metabolism in various tissues, are regulated by neural stimuli and negative feedback. A low skin temperature and stress stimulates TSH-releasing hormone (TRH) release from hypothalamic neurosecretory cells; TRH then stimulates secretion of thyroid-stimulating hormone (TSH) from the anterior pituitary gland. The thyroid responds by secreting the thyroid hormones, which cause increased metabolism in skeletal and cardiac muscle, liver, and kidney and hence lead to the metabolic generation of heat. Feedback inhibition by thyroid hormones apparently occurs at the levels of both the anterior pituitary gland and the hypothalamus. The follicle shown superimposed on the thyroid gland is drawn at a disproportionately large scale.

tors (i.e., relative insulin deficiency). Whatever the cause, insulin deficiency leads to **hyperglycemia** (high levels of blood glucose), **glycosuria** (spillover of excess glucose into the urine, which occurs when the blood glucose levels exceed the renal threshold for glucose), and a reduced ability to synthesize lipid and protein, which are broken down to supply energy because cells are deficient in glucose. In addition, mobilized fat particles that cannot be rapidly metabolized accumulate in the blood as **ketone bodies**. These are excreted in the urine but can also interfere with liver function. These disturbances in carbohydrate, lipid, and protein metabolism also produce a large number of complications in various organs (e.g., cataract and cardiovascular diseases).

 Why do diabetics lose weight, eat more, and produce more urine than non-diabetics?

Although the insulin receptor exhibits tyrosine kinase activity, the intracellular signaling pathway triggered by binding of insulin differs from that associated with other receptors of this type. Phosphorylation of various effector and regulatory proteins by the activated insulin receptor presumably mediates the various short-term and long-term effects of insulin. Insulin binding also induces the formation of peptide insulin mediators, which can inhibit adenylate

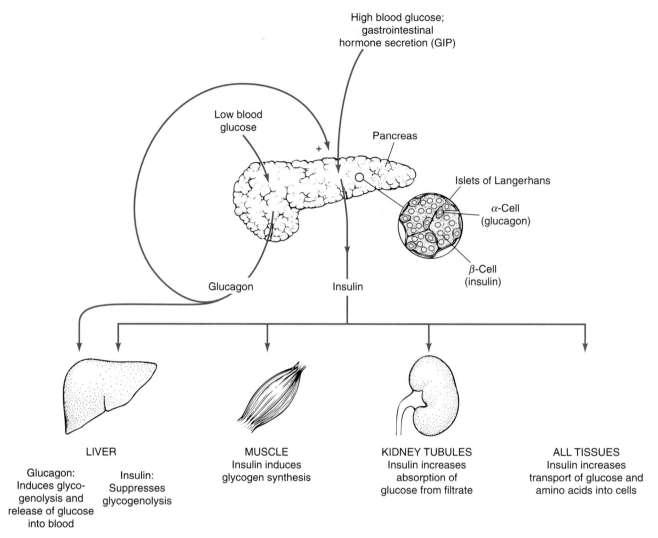

Figure 9-27 The pancreatic hormones insulin and glucagon play a major role in regulating blood glucose levels. High levels of blood glucose and glucagon and/or gastrointestinal hormones signaling food ingestion (e.g., gastrointestinal inhibitory peptide, GIP) stimulate the pancreatic β- cells to secrete insulin, which stimulates glucose uptake in all tissues. Glucagon, secreted by pancreatic α-cells, exerts an action that is antagonistic to that of insulin in the liver, where it stimulates glycogenolysis and glucose release. Insulin has several other effects.

cyclase and activate cAMP phosphodiesterase. This dual action has the effect of lowering intracellular cAMP levels.

Glucagon is secreted by the *alpha cells* of the pancreatic islets in response to **hypoglycemia** (low levels of blood glucose). This hormone has the opposite effects of insulin, stimulating glycogenolysis in the liver; it also stimulates lipolysis, providing lipids for gluconeogenesis (see Figure 9-27). The antagonistic actions of insulin and glucagon are important in maintaining an appropriate blood glucose level, so that adequate glucose is available for all tissues. Like epinephrine, which also promotes breakdown of glycogen, glucagon binds to receptors linked to the cAMP second-messenger pathway.

Growth hormone
The production and release of growth hormone (GH) in the anterior pituitary gland is under the direct control of GH-

releasing hormone (GRH) and GH-inhibiting hormone (GIH), otherwise known as somatostatin (see Table 9-2). In addition, the release of GRH and GIH is regulated by such factors as blood glucose levels (Figure 9-28). Reduced glucose levels, for example, indirectly stimulate release of growth hormone by increasing the secretion of GRH.

Growth hormone exerts both metabolic and developmental effects. Many of its diverse metabolic effects are opposite to those of insulin. For example, growth hormone induces the mobilization of stored fat for energy metabolism, whereas insulin induces the breakdown of stored fat. The fatty acids released from adipose tissue into the bloodstream in response to growth hormone are converted in the liver to ketone bodies for release into the circulation. Growth hormone also stimulates fatty acid uptake in muscle, further promoting the utilization of fatty acids as an energy source. By increasing the utilization of fatty acids, growth hormone helps conserve muscle glycogen stores (see Figure 9-28).

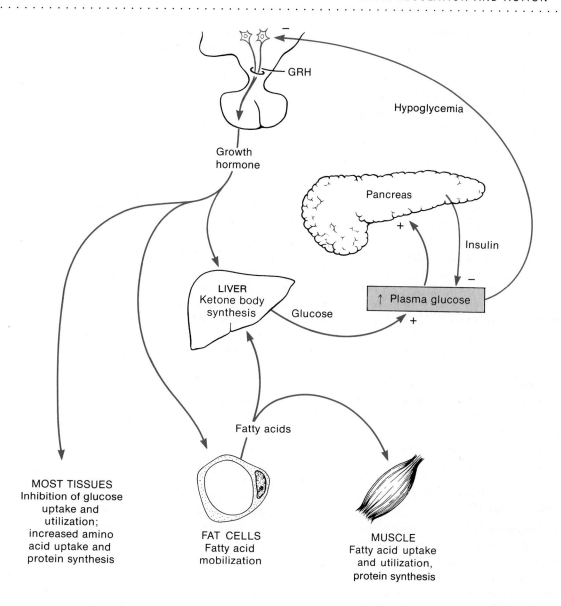

Figure 9-28 Many of the actions of growth hormone are antagonistic to those of insulin. Output of insulin from pancreatic beta cells occurs in response to high blood glucose, as after a meal. Growth hormone (GH) is released, usually several hours after a meal or after prolonged exercise, in response to insulin-induced hypoglycemia. Growth hormone causes lipolysis and fatty acid uptake by muscle tissue for energy and by the liver for ketone body synthesis. The GH-induced general depression of glucose uptake (except in the central nervous system) leads to a rise in plasma glucose, which then stimulates insulin secretion. The insulin stimulates glucose uptake into cells and thus counteracts GH-induced hyperglycemia.

In contrast to insulin, which causes a decrease in blood glucose levels, growth hormone causes an elevation of blood glucose. Therefore, growth hormone counteracts hypoglycemia, whereas insulin counteracts hypeglycemia. Growth hormone elevates blood glucose by three mechanisms: it stimulates gluconeogenesis from fat, blocks glucose uptake by tissues other than the nervous system, and promotes the utilization of fatty acids in place of glucose. Thus both glucagon, which stimulates glycogen breakdown in the liver, and growth hormone act to maintain adequate blood glucose levels. Growth hormone reaches its peak plasma level several hours after a meal, when immediate energy supplies (e.g., blood glucose, amino acids, and fatty acids) have begun to decrease. Furthermore, growth hormone stimulates insulin secretion both directly, through its action on the pancreatic beta cells, and indirectly, through its effect in elevating plasma glucose levels.

Growth hormone also stimulates RNA and protein synthesis, which may account for its developmental effects in promoting the growth of tissues—in particular, cartilage and subsequently bone. GH-stimulated tissue growth occurs by an increase in cell number (i.e., cell proliferation) rather than an increase in cell size. As noted above, thyroid hormones and growth hormone work synergistically to promote tissue growth during development. The growth-enhancing effects of growth hormone depend very much on the stage of development of the animal: the neonatal mammal is relatively insensitive to growth hormone but becomes more sensitive as it grows. GH not only stimulates proliferation of cells directly, it also stimulates the liver to produce other growth-promoting factors, called insulin-like growth factors, that also act directly on cells to promote growth. Disturbances in the secretion of growth hormone lead to several patterns of abnormal growth and development in humans:

- **Gigantism:** excessive size and stature caused by hypersecretion of growth hormone during childhood (before puberty)

- **Acromegaly:** enlargement of the bones of the head and of the extremities caused by hypersecretion of growth hormone beginning after maturity

- **Dwarfism:** abnormal underdevelopment of the body caused by insufficient secretion of growth hormone during childhood and adolescence

Little is known about the cell-surface receptors that bind growth hormone or the intracellular signaling pathways stimulated by hormone binding. However, application of growth hormone to tissues from young animals has been shown to inhibit adenylate cyclase activity and hence lead to a decrease in cAMP levels.

Hormones That Regulate Water and Electrolyte Balance

The major organs involved in regulation of water and electrolyte balance in vertebrates are the kidney, intestine, and bone, and in fish, the gills. Since epithelial cells are responsible for the uptake or excretion of water and electrolytes, most of the hormones that regulate water and electrolyte balance act on these epithelial tissues. The processes for maintaining water and electrolyte balance are described in more detail in Chapter 14. Here we consider the hormones that play a major role in regulating these processes (Table 9-8).

Antidiuretic hormone (ADH), also called **vasopressin,** regulates water turnover in the mammalian kidney. Secretion of this neurohormone from the posterior pituitary gland is stimulated by high blood osmolarity, acting on osmoreceptors in the anterior hypothalamus (see Figure 9-7). By increasing the water permeability of the kidney collecting duct, ADH stimulates resorption of water from the forming urine; the end result is a reduction in urine volume and increased water retention by the body. Increases in venous blood pressure, reflecting increases in blood volume, stimulate atrial stretch receptors in the heart; these then send an inhibitory signal to the hypothalamus that decreases ADH release and, therefore, enhances urine production, leading to a reduction in blood volume. ADH also enhances the release of ACTH and TSH from the anterior pituitary gland.

Mammals produce arginine vasopressin, but other vertebrates produce slightly different nonapeptides with similar actions, as noted earlier. Reptiles, fish, and birds produce a related peptide, called arginine vasotocin, which exerts effects similar to those of vasopressin and oxytocin (see Table 9-4). Like vasopressin, vasotocin promotes water resorption by the animal. In addition, this hormone may play a role in sexual behavior and is associated with expulsion of eggs from the oviduct in turtles (somewhat analogous to oxytocin action). Both vasopressin and vasotocin are known to cause smooth muscle contraction. ADH and its analogs exert their effects through the cAMP pathway.

The **mineralocorticoids,** in particular **aldosterone,** enhance resorption of sodium (and, indirectly, chloride) by the distal tubules and the collecting tubules of the kidney, thereby increasing the osmolarity of the blood. Aldosterone is one of the steroid hormones secreted by the adrenal cortex under the stimulation of ACTH (adrenal corticotrophic hormone). Secretion of aldosterone is stimulated by angiotensin II and high blood K^+ and is subject to negative feedback by the hormone on the CRH-secreting neurons of the hypothalamus and on the ACTH-secreting cells of the anterior pituitary gland (see Figure 9-6). The mineralocorticoids, like other steroid hormones, mediate their effects by binding to intracellular receptors and modifying gene expression.

Atrial natriuretic peptide (ANP) acts on the kidney to reduce sodium and, therefore, water resorption, leading to an increase in urine production and sodium excretion in the kidney. Thus the effects of this hormone counteract those of aldosterone and ADH. ANP is released by the atrium of the heart into the blood in response to an increase in venous pressure. Its mechanism of action is not clear.

As we saw earlier, Ca^{2+} plays a key role as a second messenger and regulatory agent in the cell. Thus careful regulation of the concentration of Ca^{2+} in the blood and the extracellular fluid is critical. This ion is actively absorbed through the intestinal wall into the plasma and is deposited in bone, the major depot for storage of Ca^{2+}. Elimination of Ca^{2+} from the body occurs through the kidney. The balance of these processes, which determines the blood Ca^{2+} concentration, is influenced by three hormones: parathyroid hormone, calcitonin, and calcitriol.

Parathyroid hormone (PTH), also known as **parathormone,** is secreted from the paired parathyroid glands in response to a drop in plasma Ca^{2+} levels. It acts to increase plasma Ca^{2+} by promoting Ca^{2+} mobilization from bone, increasing Ca^{2+} uptake from the forming urine in kidney tubules, increasing renal PO_4^{3-} excretion, and enhancing intestinal Ca^{2+} absorption (Figure 9-29). PTH works in conjunction with **calcitriol,** a steroid-like compound produced from vitamin D ingested with some foods and from vitamin D_3, which can be synthesized from cholesterol in the skin. Conversion of these precursors into calcitriol involves reactions in the liver and kidneys. The actions of calcitriol are similar to those of parathyroid hormone.

TABLE 9-8
Mammalian hormones involved in regulating water and electrolyte balance

Hormone	Tissue of origin	Structure	Target tissue	Primary action	Regulation
Antidiuretic hormone (ADH), or vasopressin	Neurohypophysis	Nonapeptide	Kidneys	Increases water resorption	Increased plasma osmotic pressure or decreased blood volume stimulates release
Atrial natriuretic peptide (ANP)	Heart (atrium)	Peptide	Kidneys	Reduces Na^+ and water resorption	Increased venous pressure stimulates release
Calcitonin	Thyroid (parafollicular cells)	Peptide	Bones, kidneys	Decreases release of Ca^{2+} from bone; increases renal Ca^{2+} and PO_4^{3-} excretion	Increased plasma Ca^{2+} stimulates secretion
Mineralocorticoids (e.g., aldosterone)	Adrenal cortex	Steroid	Distal kidney tubules	Promotes resorption of Na^+ from urinary filtrate	Angiotensin II stimulates secretion
Parathyroid hormone (PTH)	Parathyroid hormone	Peptide	Bones, kidneys, intestine	Increases release of Ca^{2+} from bone; with calcitriol increases intestinal Ca^{2+} absorption; decreases renal Ca^{2+} excretion	Decreased plasma Ca^{2+} stimulates secretion

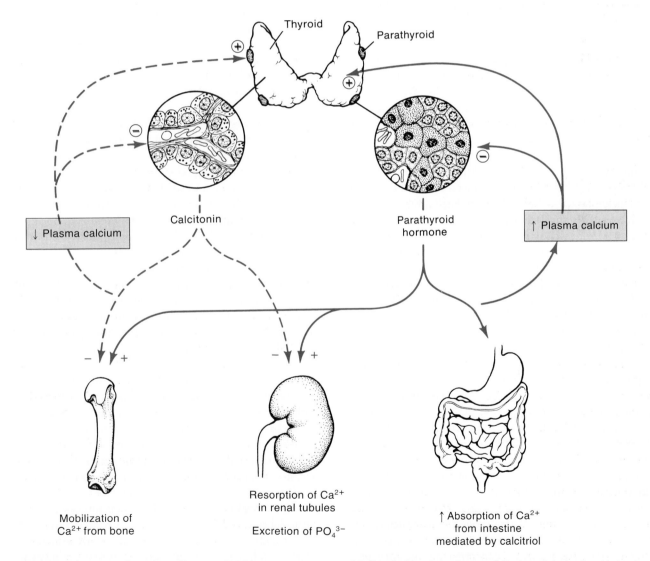

Figure 9-29 Calcitonin and parathyroid hormone (PTH) have opposite effects on plasma Ca^{2+} levels in mammals. Low levels of plasma Ca^{2+} stimulate the cells of the parathyroid glands to release PTH, which has several actions all tending to increase plasma Ca^{2+}. High concentrations of Ca^{2+} in the blood stimulate parafollicular cells in the thyroid gland to release calcitonin, which acts to increase plasma Ca^{2+}. Calcitriol, the active hormonal form of vitamin D, also increases intestinal absorption of Ca^{2+}.

TABLE 9-9
Important mammalian reproductive hormones

Hormone	Tissue of origin	Structure	Target tissue	Primary action	Regulation
Primary Sex Hormones					
Estradiol-17β (estrogens)	Ovarian follicle, corpus luteum, adrenal cortex	Steroid	Most tissues	Promotes development and maintenance of female characteristics and behavior, oocyte maturation, and uterine proliferation	Increased FSH and LH levels stimulate secretion
Progesterone	Corpus luteum, adrenal cortex	Steroid	Uterus, mammary glands	Maintains uterine secretion; stimulates mammary duct formation	Increased LH and prolactin levels stimulate secretion
Testosterone (androgens)	Testes (cells of Leydig), adrenal cortex	Steroid	Most tissues	Promotes development and maintenance of male characteristics and behavior and spermatogenesis	Increased LH level stimulates secretion
Other Hormones					
Oxytocin	Neurohypophysis	Nonapeptide	Uterus, mammary glands	Promotes smooth muscle contraction and milk ejection	Cervical distention and suckling stimulate release; high progesterone inhibits release
Prolactin (PL)	Adenohypophysis	Peptide	Mammary glands (alveolar cells)	Increases synthesis of milk proteins and growth of mammary glands; elicits maternal behavior	Continuous secretion of PL-inhibiting hormone (PIH) normally blocks release; increased estrogen and decreased PIH secretion permit release

Calcitonin is secreted from the parafollicular, or C, cells in the thyroid gland in response high plasma Ca^{2+} levels. It rapidly suppresses Ca^{2+} loss from bone, quickly countering the effects of PTH. Although calcitonin and PTH have opposing actions on bone metabolism, there is no feedback interaction between them. Each hormone, however, exerts negative feedback on its own secretion. The dominance of calcitonin prevents hypercalcemia and extensive dissolution of the skeleton. Essentially, then, bone acts as a large reservoir and buffer for Ca^{2+} and also PO_4^{3-}. The plasma Ca^{2+} and PO_4^{3-} levels are held within narrow limits by the opposing actions of PTH and calcitonin, which regulate the flux of these minerals between plasma and bone.

PTH and calcitonin are both peptide hormones and bind to cell-surface receptors. Little is known about the intracellular signaling pathways mediating their effects. Calcitriol is lipid soluble and presumably binds to an intracellular receptor.

Reproductive Hormones

In vertebrates several steroid hormones that affect reproduction (the estrogens, the androgens, and progesterone) are produced in the gonads (testis or ovary) and adrenal cortex of both sexes from cholesterol (see Figure 9-23). Cholesterol is first converted to **progesterone,** which is then transformed into the **androgens** (androstenedione and testosterone). These are then converted into the **estrogens,** of which estradiol-17β is the most potent. The steroid sex hormones, like other steroid hormones, bind to intracellular receptors and modify expression of specific genes. In ad-

dition to the steroid sex hormones, two peptide hormones produced in the pituitary gland function in parturition and lactation. Table 9-9 summarizes the properties of the steroid and peptide reproductive hormones.

The production and secretion of the steroid sex hormones in both males and females are promoted by follicle-stimulating hormone (FSH) and luteinizing hormone (LH), which are synthesized in the anterior pituitary (see Table 9-3). These tropic hormones are released from the anterior pituitary gland in response to the hypothalamic gonadotropin-releasing hormone (GnRH). The steroid sex hormones exert negative feedback on the GnRH-secreting neurons of the hypothalamus and on the anterior pituitary endocrine cells that produce FSH and LH.

Steroid sex hormones in males

The seminiferous tubules of the mammalian testes are lined with germ cells and Sertoli cells (Figure 9-30). Binding of FSH to receptors on Sertoli cells stimulates spermatogenesis in the germ cells after sexual maturity, either continuously or seasonally, depending on the species. The cells of Sertoli support development of the sperm and are responsible for synthesis of androgen-binding protein (ABP) and inhibin. Lying between the seminiferous tubules are interstitial cells, called **Leydig cells,** which produce and secrete sex hormones, particularly **testosterone.** Both testosterone itself and inhibin provide inhibitory feedback to the hypothalamic centers controlling GnRH production and hence diminish release of the gonadotropins FSH and LH from the anterior pituitary gland.

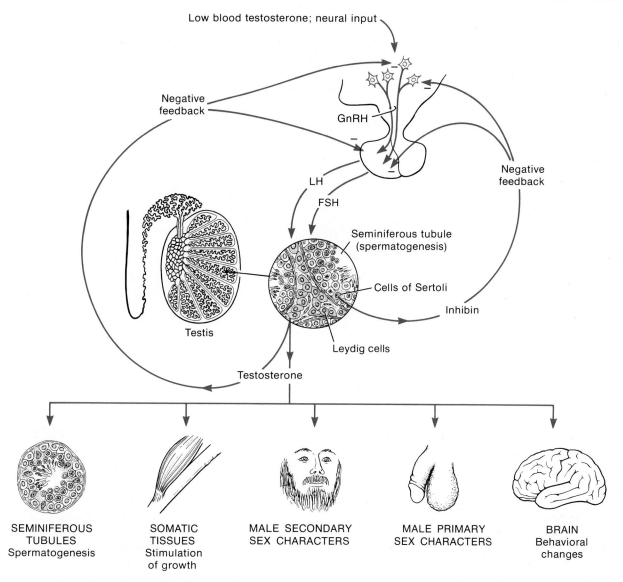

Figure 9-30 Testosterone, the primary sex hormone in males, has numerous actions and is regulated by neural stimuli and feedback control. A decrease in blood testosterone stimulates the secretion of gonadotropin-releasing hormone (GnRH), which promotes the release of follicle-stimulating hormone (FSH) and luteinizing hormone (LH). Some of the actions of testosterone are indicated at the bottom of the figure. High testosterone and inhibin, which also is secreted by the testes, inhibits FSH secretion both directly and indirectly.

The estrogens and androgens are important in both sexes in various aspects of growth, development, and morphologic differentiation, as well as in the development and regulation of sexual and reproductive behaviors and cycles. However, androgens predominate in the male, whereas estrogens predominate in the female. The androgens trigger development of the primary male sexual characteristics (e.g., the penis, vas deferens, seminal vesicles, prostate gland, epididymis) in the embryo and the secondary male sexual characteristics (e.g., the lion's mane, the rooster's comb and plumage, and facial hair in men) at the time of puberty. The androgens also contribute to general growth and protein synthesis—in particular, the synthesis of myofibrillar proteins in muscle, as evidenced by the greater muscularity of the males relative to the females in many vertebrate species.

Steroid sex hormones in females: regulation of the menstrual cycle

Unlike androgens, which stimulate prenatal differentiation of the embryonic male genital tract, the estrogens play no such role in early differentiation of the female tract. However, estrogens stimulate later development of primary sexual characteristics such as the uterus, ovary, and vagina. The estrogens also are responsible for development of the secondary female sex characteristics such as the breast and for regulation of reproductive cycles (Figure 9-31).

Simultaneous reproduction within an entire population can be of obvious survival value to a species. The gathering of large numbers of individuals of both sexes for mating, bearing young, and parenting the young during this period of high vulnerability can be timed to coincide

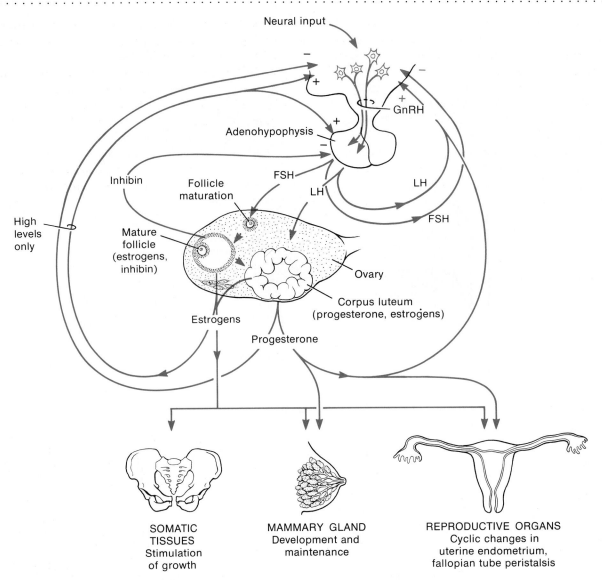

Figure 9-31 Estrogens and progesterone, the primary steroid sex hormones in females, mediate reproductive cycles and other effects under complex regulation. In mammals a decrease in progesterone and estrogen levels, as well as neural inputs, stimulates release of gonadotropin-releasing hormone (GnRH). This acts on the anterior pituitary gland, stimulating secretion of follicle-stimulating hormone (FSH), which promotes development of the primordial follicles in the ovary. Estrogens secreted by the follicles and by the interstitial cells eventually reach levels that stimulate release of luteinizing hormone (LH), which triggers ovulation and subsequent development of the corpus luteum. The corpus luteum secretes primarily progesterone and estrogens, which are needed to maintain pregnancy. Eventually, the high levels of FSH and LH, as well as progesterone, inhibit the activity of the hypothalamic neurosecretory cells, leading to a decrease in gonadotropin secretion. This prevents menstrual cycling during pregnancy.

with favorable weather and an adequate food supply. Moreover, the sudden appearance of large numbers of defenseless individuals of a species can have an overwhelming effect on even the most voracious of predators, permitting the survival of enough individuals of the new generation to assure survival of the species. In general, reproductive cycles arise from within the animal under the control of the neuroendocrine system, but these inner cycles are constrained by environmental signals such as the changes in day length that accompany the changing seasons.

 Why doesn't the mother reject the fertilized egg? Does the mother produce antibodies against the developing fetus?

Female mammals and birds are born with a full complement of **oocytes,** each of which becomes embedded in a follicle within the ovary and is capable of developing into one **ovum.** Most of the follicles and their oocytes degenerate early, but even before puberty some develop just short of yolk formation or maturation. In humans, about 400 ova are available for release between **menarche** (onset of menstruation) and **menopause**. Oogenesis in lower vertebrates occurs throughout life.

In mammalian females, the menstrual cycle is composed of the follicular phase and luteal phase (Figure 9-32, *left*). The **follicular phase** begins with FSH stimulating development of 15–20 ovarian follicles, fluid-filled cavities enclosed by a membranous sac of several cell layers, including the theca interna and ovarian granulosa. LH then stimulates the theca interna to synthesize and secrete androgens. FSH stimulates the production of an enzyme that

then converts the androgens to estrogens in the ovarian granulosa, leading to a substantial increase in estrogen levels. At *high* estrogen levels, characteristic of the time just prior to ovulation, estrogen activates the hypothalamus and anterior pituitary gland, producing a surge in release of FSH and LH, an example of positive feedback. This FSH accelerates maturation of the developing follicles; only one follicle completes its maturation and under the influence of LH ruptures at the surface of the ovary, releasing the ovum. The increase in estrogens during the follicular phase also stimulates proliferation of the **endometrium**, the tissue that lines the uterus.

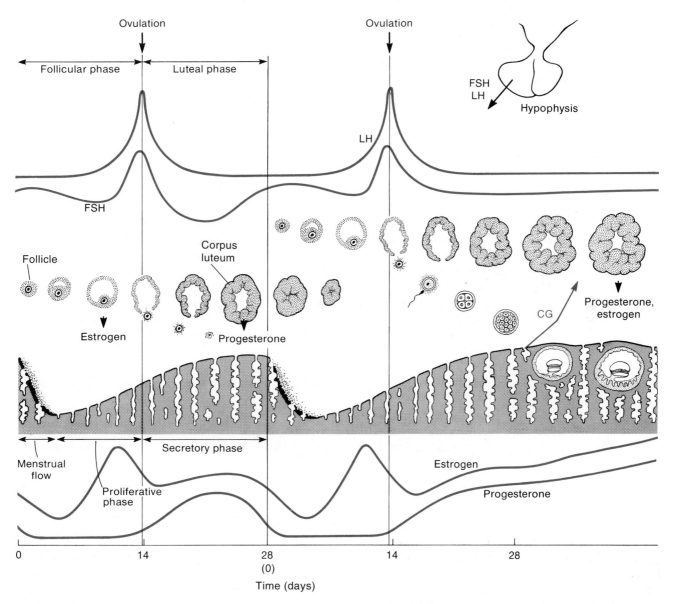

Figure 9-32 The primate menstrual cycle is regulated by periodic changes in the levels of the gonadotropins, estrogens, and progesterone. Before ovulation, follicle-stimulating hormone (FSH) promotes maturation of ovarian follicles, which secrete estrogen. *High* estrogen levels cause a surge of luteinizing hormone (LH), which triggers ovulation from one follicle. LH promotes development of the corpus luteum, and induces the corpus luteum to secrete progesterone and some estrogen. In the absence of implantation (*left*), the progesterone and estrogen levels peak and then fall, initiating menstruation. The subsequent decrease in estrogen, progesterone, and inhibin levels allows pituitary secretion of FSH and LH to increase again, thus initiating a new cycle. If implantation and pregnancy occur (*right*), secretion of chorionic gonadotropin (CG) by the placenta "rescues" the corpus luteum, which maintains secretion of estrogen and progesterone for the first two to three months of pregnancy in the human. Thereafter, the placenta itself secretes estrogens and progesterone. [Adapted from McNaught and Callander, 1975.]

During the **luteal phase,** which begins with ovulation, estrogen secretion declines and LH transforms the ruptured follicle into a temporary endocrine tissue, the **corpus luteum.** The corpus luteum secretes estrogen and progesterone, which exert negative feedback on GnRH release by the hypothalamus, leading to decreased secretion of FSH and LH. The ovarian hormone inhibin, which is released along with the ovum, acts on the anterior pituitary, inhibiting FSH (but not LH) release. Progesterone stimulates secretion of endometrial fluid by the endometrial tissue, preparing it for implantation of a fertilized ovum. In the absence of fertilization and implantation of an ovum, the corpus luteum degenerates after a 14 ± 1-day period (in humans), and secretion of estrogen and progesterone subsides. In humans and some other primates, this precipitates the **menses,** or shedding of the uterine lining. With the reduction in estrogen, progesterone, and inhibin levels, FSH and LH secretion by the pituitary increases again, initiating a new cycle.

If the released ovum is fertilized as it travels down the ciliated fallopian tube and the fertilized ovum becomes implanted in the endometrium of the placental mammal, the developing placenta begins to produce chorionic gonadotropin (CG) (see Figure 9-32, *right*). This hormone, whose action is similar to LH, induces further growth of the active corpus luteum, so that estrogen and progesterone secretion continues. The placenta begins secreting CG within about a day of implantation of the ovum and effectively takes over the gonadotropic function of the pituitary during early pregnancy by maintaining the corpus luteum. Pituitary FSH and LH are not secreted again until after parturition (birth of the fetus). In many mammals, including humans, the corpus luteum continues to grow and to secrete estrogen and progesterone until the placenta fully takes over the production of these hormones, at which time the corpus luteum degenerates. In other mammals, such as the rat, continued secretion by the corpus luteum, stimulated by prolactin, is essential to the maintenance of pregnancy throughout its term.

The durations of the follicular and luteal phases of the reproductive cycle vary among different mammalian groups. They are about equal in the primate menstrual cycle, but in the nonprimate mammals the luteal phase is much shorter. The number of cycles per year also varies among species. The human menstrual cycle of approximately 28 days normally occurs 13 times a year. Among nonprimate mammals, some have only one cycle per year (usually in spring); others, such as the laboratory rat, have multiple cycles throughout the year.

During gestation progesterone and estrogen, secreted from the corpus luteum or placenta, initiate growth of the mammary tissues in preparation for lactation. Prolactin and placental lactogen, a hormone produced in the placenta, also aid in preparing the mammary glands for lactation, but synthesis of milk is inhibited by progesterone during pregnancy. The negative feedback of estrogen and progesterone on the hypothalamus and anterior pituitary gland prevent release of FSH and LH during pregnancy, thereby preventing ovulation. Birth control pills contain small amounts of progesterone and estradiol or their synthetic analogs. Taken daily, these steroids mimic the earliest stages of pregnancy, preventing ovulation and also acting on the endometrium, thereby providing a highly effective means of avoiding conception.

Hormones involved in parturition and lactation

As pregnancy nears term, cervical distention stimulates release of oxytocin from the posterior pituitary gland (see Table 9-9). This hormone induces contractions of the smooth muscle in the uterine wall, which are critical to the normal birth process (**parturition**). Certain prostaglandins also may stimulate uterine contractions during childbirth. After parturition, a decrease in progesterone levels relieves inhibition of the milk-synthesizing machinery, permitting lactation to begin. Milk production is mediated by prolactin, along with the glucocorticoids, and the release of milk is induced by oxytocin. Both prolactin and oxytocin are released during suckling as a result of neural input to the hypothalamus arising from stimulation of the nipples.

Prostaglandins

The long-chain, unsaturated, hydroxy fatty acids called prostaglandins were first discovered in the 1930s in semi-

TABLE 9-10
Selected prostaglandins

Tissue of origin	Target tissue	Primary action	Regulation
Seminal vesicles, uterus, ovaries	Uterus, ovaries, fallopian tubes	Potentiates smooth muscle contraction and possibly luteolysis; may mediate LH stimulation of estrogen and progesterone synthesis	Introduced during coitus with semen
Kidney	Blood vessels, especially in kidneys	Regulates vasodilation or contraction	Increased angiotensin II and epinephrine stimulate secretion; inactivated in lungs and liver
Neural tissue	Adrenergic terminals	Blocks norepinephrine-sensitive adenylate cyclase	Neural activity increases release

nal fluid (see Figure 9-1B). They were thought to be produced by the prostate gland—hence the name. Since then, the prostaglandins in seminal fluid have been shown to be produced by the seminal vesicles. As noted earlier, prostaglandins are synthesized in membranes from arachidonic acid, which is produced by cleavage of membrane phospholipids by phospholipases (see Figure 9-15). They have now been found in virtually all mammalian tissues, in some cases acting locally as paracrine agents and in other cases acting on distant target tissues in a more classical endocrine fashion (see Figure 8-1). The 16 or more different prostaglandins identified to date fall into nine classes (designated PGA, PGB, PGC, . . . PGI). Some of them are converted to other biologically active prostaglandins. The prostaglandins undergo rapid oxidative degradation to inactive products in the liver and lungs.

The numerous prostaglandins have diverse actions on a variety of tissues, making it difficult to generalize about this group of hormones. Although they are lipid-soluble, prostaglandins bind to cell-surface receptors linked to the cAMP pathway. Many of their effects involve smooth muscle. The effects of some prostaglandins produced in selected tissues are shown in Table 9-10. For example, prostaglandins produced in the kidney act on the smooth muscle of blood vessels to regulate vasodilation and vasoconstriction. Prostaglandins are also involved in the function of blood cells, such as platelets, and in inflammatory responses. Aspirin acts as an anti-inflammatory agent by inhibiting prostaglandin synthesis.

HORMONAL ACTION IN INVERTEBRATES

Endocrine cells—in particular, neurosecretory cells—have been identified in all invertebrate groups, including the primitive hydroid coelenterates. In *Hydra*, for example, neurons secrete what is believed to be a growth-promoting hormone during budding, regeneration, and growth. This is perhaps not surprising since invertebrates account for the vast majority of animal species on earth, and their success is based at least in part on relatively sophisticated endocrine systems. Hormone actions have been studied in a limited number of invertebrate species, typically those with particularly accessible systems. Hormonal regulation of development in insects has been widely studied and will serve to illustrate general principles of hormone action in invertebrates.

Insects fall into two groups based on their pattern of development: hemimetabolous insects exhibit incomplete **metamorphosis,** and holometabolous insects exhibit complete metamorphosis. The life cycle of **hemimetabolous** insects—including the Hemiptera (bugs), Orthoptera (locusts, crickets), and Dictyoptera (roaches, mantids)—begins with development of the egg into an immature nymphal stage. The **nymph** eats and grows and undergoes several molts, replacing its old exoskeleton with a soft new one that expands to a larger size before hardening.

The stages between molts are termed **instars.** The final nymphal instar gives rise to the adult stage. The development of **holometabolous** insects—including the Diptera (flies), Lepidoptera (butterflies, moths), and Coleoptera (beetles)—is more complex. The egg develops into a **larva** (e.g., maggot, "worm," caterpillar), which grows through several instars. The larva is specialized for eating and therefore the insect stage that causes the major damage to many agricultural crops. The last larval instar molts to become a **pupa,** an outwardly dormant stage in which extensive internal reorganization takes place to give rise to the adult form. The adult, which shows little morphologic resemblance to the pupa or previous stages, is the reproductive stage, and in some species is not even equipped to feed.

The first experiments demonstrating probable endocrine control of insect development were done between 1917 and 1922 by S. Kopec, who ligated the last larval instars of a moth at various times during the instar. He found that when the ligature was tied before a certain critical period, the larva would pupate anteriorly to the ligature but remain larval posteriorly. Cutting the nerve cord had no effect, so he concluded that a circulating, pupa-inducing substance had its origin in a tissue located in the anterior portion of the larva. By testing various tissues, Kopec found that removal of the brain prevents pupation and that reimplantation of the brain allows it to proceed again. It was subsequently found that a neurohormone secreted by cells in the brain stimulates the prothoracic glands, the tissue that elaborates the molt-inducing hormone. Thus, ligating posterior to the thoracic glands after their activation by the brain-derived hormone, prevents pupation of the abdomen. Pupation can be initiated by implanting activated thoracic glands into the isolated abdomen.

The hardiness of insects makes them ideal subjects for experiments that demonstrate the humoral control of molting and metamorphosis. It is possible, for instance, to carry out extended **parabiosis** experiments in which two insects or two parts of one insect are joined so that they share a common circulation, exchanging body fluid (Figure 9-33). Windows made of thin glass make it possible to observe developmental changes in the tissues of the separated parts.

Five major hormones, three of them produced by neurosecretory cells, are now known to control development in insects (Table 9-11 and Figure 9-34):

- **Prothoracicotropic hormone (PTTH)** is a neurohormone produced by neurosecretory cells that have their cell bodies in the pars intercerebralis of the brain. PTTH appears to be a small protein with a molecular weight of about 5000.

- **Juvenile hormone** is synthesized and released from the **corpora allata,** which are nonneural paired glands somewhat analogous to the anterior pituitary gland. Several juvenile hormone homologs occur naturally in insects; they all have a modified fatty acid structure (Figure 9-35A).

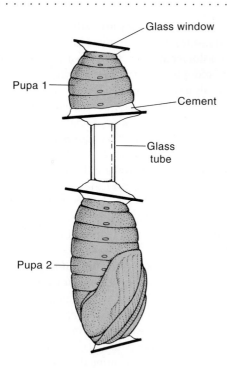

Figure 9-33 Parabiosis, the joining of body parts from different individuals, is a useful experimental method in insect endocrinology. Insect tissues readily survive such radical surgery as transection and decapitation. In this example, the abdomen of one pupa is joined to another pupa through a glass tube. Glass windows at either end permit visual inspection of the developing tissues.

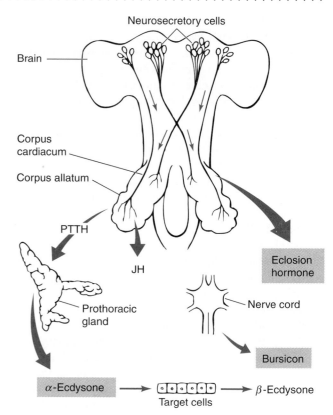

Figure 9-34 Of the five major insect developmental hormones, three are produced by neurosecretory cells and two by endocrine tissues. Neurosecretory cells in the brain synthesize prothoracicotropic hormone (PTTH) and eclosion hormone, which are stored in nerve terminals until their release into blood sinus spaces in the corpus cardiaca and corpus allata, two paired neurohemal organs. A third neurohormone, bursicon, is released primarily from nerve terminals in the nerve cord. The corpus allatum also contains nonneural cells that elaborate juvenile hormone (JH). Under the stimulus of PTTH, the prothoracic gland produces and secretes α-ecdysone, which is converted to the active molting hormone β-ecdysone. [Adapted from Riddiford and Truman, 1978.]

- **Ecdysone,** produced by the prothoracic glands, is synthesized from cholesterol. It is structurally similar to vertebrate steroid hormones but contains more hydroxyl groups (Figure 9-35B).

- **Eclosion hormone,** a peptide neurohormone, is released from neurosecretory cells whose terminals are in the **corpora cardiaca,** which are paired neurohemal organs immediately posterior to the brain.

- **Bursicon,** also a neurohormone, is produced by other neurosecretory cells in the brain and nerve cord. It is a protein with a molecular weight of about 40,000.

PTTH is shipped by axoplasmic transport along the axons of the neurosecretory cells to storage depots, or neurohemal organs, formed by the terminals of the axons (see Figure 9-34). The corpus cardiacum was thought to be the neurohemal organ that stores and releases PTTH, but more recent evidence from the tobacco hornworm moth, *Manduca sexta,* indicates that the axons of the PTTH-producing neurosecretory brain cells actually pass through the corpus cardiacum and end within the corpus allatum, which is located at the posterior end of the corpus cardiacum. Thus, the corpus allatum appears to be the site at which the neurosecretory endings release PTTH into the blood. It remains to be determined if this is true of all insects.

After its release into the blood, PTTH activates the prothoracic gland to synthesize and secrete the molt-inducing factor, α-ecdysone. Insects require cholesterol in their diets to synthesize this steroid hormone. It is now thought that α-ecdysone is a prohormone converted to the physiologically active form, 20-hydroxyecdysone (β-ecdysone), in several peripheral target tissues (see Figure 9-35B).

Juvenile hormone, acting in association with β-ecdysone, promotes the retention of the immature ("juvenile") characteristics of the larva, thereby postponing metamorphosis until larval development is completed. The presence of juvenile hormone in the early nymphal instar was demonstrated in the mid-1930s in experiments by V. B. Wigglesworth in which parabiotic coupling of the early instar to a final instar prevented the latter from becoming an adult. The circulating concentration of juvenile hormone is highest early in larval life, dropping to a minimum at the end of the pupal period (Figure 9-36). Metamorphosis to the adult stage occurs when juvenile hormone disappears from the circulation. The concentration then rises again in the reproductively active adult. In the males of some insect species, juvenile hormone promotes development of the

TABLE 9-11
Insect developmental hormones

Hormone	Tissue of origin	Structure	Target tissue	Primary action	Regulation
Bursicon	Neurosecretory cells in brain and nerve cord	Protein (MW ~40,000)	Epidermis	Promotes cuticle development; induces tanning of cuticle of newly molted adults	Stimuli associated with molting stimulate secretion
Ecdysone (molting hormone)	Prothoracic glands, ovarian follicle	Steroid	Epidermis, fat body, imaginal disks	Increases synthesis of RNA, protein, mitochondria, and endoplasmic reticulum; promotes secretion of new cuticle	PTTH stimulates secretion
Eclosion hormone	Neurosecretory cells in brain	Peptide	Nervous system	Induces emergence of adult from puparium	Endogenous "clock"
Juvenile hormone (JH)	Corpus allatum	Fatty acid derivative	Epidermis, ovarian follicles, sex accessory glands, fat body	In larva, promotes synthesis of larval structures and inhibits metamorphosis In adult, stimulates synthesis and uptake of yolk protein; activates ovarian follicles and sex accessory glands	Inhibitory and stimulatory factors from the brain control secretion
Prothoracicotropin (PITH)	Neurosecretory cells in brain	Small protein (MW ~5000)	Prothoracic gland	Stimulate ecdysone release	Various environmental and internal cues (e.g., photoperiod, temperature, crowding, abdominal stretch) stimulate release; JH inhibits release in some species

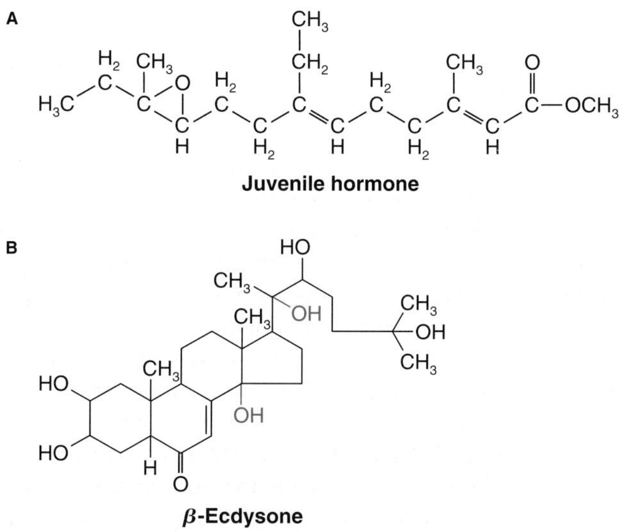

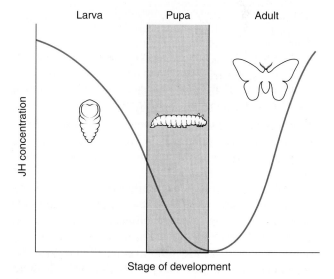

A

Juvenile hormone

B

β-Ecdysone

Figure 9-35 Juvenile hormone and β-ecdysone play key roles in regulating insect development. **(A)** The structure of juvenile hormone from the cecropia moth *Hyalophora cecropia*. This hormone promotes the retention of juvenile characteristics in larvae and induces reproductive maturation in adults. Several homologs of juvenile hormone occur naturally in insects. **(B)** The structure of β-ecdysone, the physiologically active molt-inducing hormone. The prohormone α-ecdysone, which lacks the hydroxyl group on C-20 (red), is synthesized from cholesterol in the prothoracic glands of insects. After its release, α-ecdysone is converted in certain target tissues into the active hormone β-ecdysone.

Figure 9-36 Normal progression through the insect life cycle depends on changes in the level of juvenile hormone. Metamorphosis of the juvenile larval form to the pupa occurs when the concentration of juvenile hormone falls below a certain threshold level. After the adult insect emerges and feeds, secretion of juvenile hormone begins again, regulating ovarian activity and stimulating development of male accessory organs. [Adapted from Spratt, 1971.]

accessory sexual organs; in many female insects, it induces yolk synthesis and promotes maturation of the eggs.

Thus, the normal development of an insect depends on precisely adjusted concentrations of juvenile hormone at each stage. The role of this hormone is somewhat analogous to that of thyroid hormones in the regulation of amphibian development. In both cases, disturbance of the relationship between hormone concentration and developmental stage leads to abnormal development. Because of its potency in preventing maturation in insects, juvenile hormone and its synthetic analogs are promising as potential nontoxic, ecologically sound means of combating insect pests, and one against which the insect would find it difficult to develop resistance.

During growth and development of insects, the epidermis undergoes conspicuous changes, including production of the **cuticle,** the chitinous, horny outer covering. Therefore, considerable attention has been given to the production of new cuticle, its tanning, and the shedding of the old cuticle during molting. PTTH, juvenile hormone, and β-ecdysone are all involved in the initiation of molting (Figure 9-37). Ecdysone, secreted by the prothoracic glands in response to stimulation by PTTH, acts on the epidermis to initiate production of the new cuticle, which begins with **apolysis,** the detachment of the old cuticle from the underlying epidermal cells. The epithelial cells then begin synthesizing the materials for the new cuticle, while the old cuticle is partially digested from beneath by enzymes in the molting fluid secreted by the epidermis. At high concentrations of juvenile hormone a larval-type new cuticle is formed, whereas at low concentrations levels an adult-type cuticle is produced and other events of metamorphosis ensue.

Two additional hormones, eclosion hormone and bursicon, are responsible for promoting the terminal phase of the molting process. Shedding of the cuticle of the pupae, termed **ecdysis,** is triggered by eclosion hormone at least in some holometabolous species. The pale, soft cuticle of a newly molted insect is expanded by respiratory movements of the insect to the next size before it hardens, or tans, under the influence of bursicon (see Figure 9-37).

Figure 9-38, on the opposite page, outlines the hormonal interactions that regulate metamorphosis of *Hyalophora cecropia,* a holometabolous insect. PTTH release initiates larval ecdyses (molts) and stimulates the prothoracic gland to secrete the molting hormone ecdysone. Growth continues through a series of instars, which remain larval as long as the concentration of juvenile hormone remains above a minimum. This process of growth and molting is usually completed in four or five instars, during which the concentration of juvenile hormone progressively declines. Once the juvenile-inducing effects of juvenile hormone are removed, the larva molts to the pupal stage. The pupa is the over-wintering stage of the cecropia moth, causing it to undergo obligatory diapause. Prolonged exposure to cold stimulates release of PTTH in the pupa, inducing the release of ecdysone; in the absence of juvenile hormone, ecdysone induces pupal development into the adult moth.

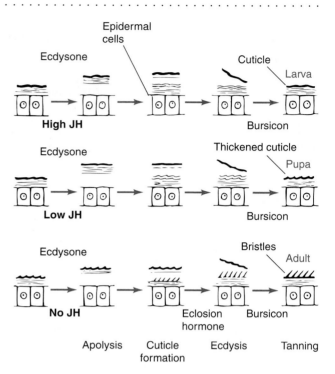

Figure 9-37 The cuticular changes involved in molts leading to larval, pupal, and adult stages are controlled by the level of juvenile hormone (JH). Ecdysone initiates production of new cuticle, beginning with detachment of the old cuticle (called apolysis). Shedding of the old cuticle (called ecdysis) is triggered by eclosion hormone. Although bursicon regulates hardening and darkening (tanning) of the new cuticle, the JH concentration determines whether it has larval, pupal, or adult characteristics (top to bottom). [Adapted from Riddiford and Truman, 1978.]

SUMMARY

The physiological and biochemical processes occurring in cells, tissues, and organs are controlled and coordinated in animals in large part by special blood-borne messenger molecules termed hormones, which are released from endocrine secretory tissues. In vertebrates these hormones fall into four chemical categories: (1) amines, (2) prostaglandins, (3) steroids, and (4) peptides and proteins. After a hormone is released from its site of origin, it circulates at low concentration in the bloodstream throughout the body. The selective actions of hormones on specific target tissues depends on the preferential distribution among tissues of hormone-specific receptors and various effector proteins that mediate hormone-induced cellular responses.

The secretion of hormones from endocrine tissues is stimulated either by hormones released from other edocrine tissues or by neurohormones released from specialized neurons; the latter form the basis for neuroendocrine reflexes. In addition, some endocrine tissues respond directly to conditions of the extracellular environment. The secretory activities of most endocrine tissues are modulated by negative feedback; that is, the increasing concentration of the hormone itself, or a response to the hormone by the target tissue (e.g., reduced blood glucose levels in the insulin loop), has an inhibitory effect on the synthesis or release of the

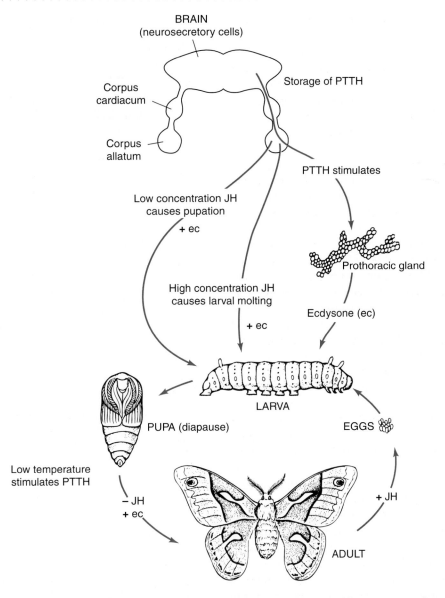

BRAIN
(neurosecretory cells)

Corpus
cardiacum

Corpus
allatum

Storage of PTTH

PTTH stimulates

Low concentration JH
causes pupation
+ ec

High concentration JH
causes larval molting
+ ec

Prothoracic gland

Ecdysone (ec)

LARVA

PUPA (diapause)

EGGS

Low temperature
stimulates PTTH

− JH
+ ec

+ JH

ADULT

Figure 9-38 Interactions of juvenile hormone and ecdysone regulate metamorphosis in holometabolous insects. This example shows the de-velopmental sequence of the cecropia moth. See text for discussion. [Adapted from Spratt, 1971.]

same hormone. Positive feedback occurs in some systems; occasionally hormone secretion is feedforward (i.e., not regulated by any consequences of the secreted hormone).

To exert their effect, all hormones must bind to specific receptors; this binding initiates intracellular mechanisms leading to a cellular response(s). Steroid hormones and thyroid hormones, being lipid soluble, enter cells freely and bind to receptor proteins present in the cytosol. The resulting hormone-receptor complexes translocate into the nucleus, where they bind to regulatory elements in the DNA thereby stimulating (and in a few cases inhibiting) transcription of specific genes. All other hormones bind to cell-receptors located in the plasma membrane of target cells. This binding triggers one or more intracellular signal-transduction pathways leading to the cell's responses.

In the cAMP signaling system, hormone binding activates a G protein, which then stimulates adenylate cyclase

to convert ATP into the second messenger cAMP. Binding of some hormones stimulates guanylate cyclase to produce the related second messenger cGMP, in most cases by a somewhat different sequence of steps. Once formed, cyclic nucleotide activates a specific protein kinase, which then phosphory-lates various effector proteins mediating cellular responses.

In the inositol phospholipid signaling system, hormone binding to a G protein–linked receptor activates phospho-inositide-specific phospholipase C, which then hydrolyzes PIP_2 into two major second messengers, $InsP_3$ and DAG. $InsP_3$ induces release of Ca^{2+} from intracellular stores. In addition, $InsP_3$ is converted to $InsP_4$, which promotes entry of Ca^{2+} from the cell exterior into the cell. The resulting increase in free cytosolic Ca^{2+} regulates the activity of a variety of cellular proteins. DAG, on the other hand, remains in the membrane and activates membrane-bound protein kinase C. This kinase then phosphorylates various effector proteins, which are responsible for the cellular responses.

In the Ca^{2+} signaling system, hormone stimulation of the receptor directly activates Ca^{2+} channels in the plasma membrane, thereby stimulating Ca^{2+} influx. Hormone-induced changes in intracellular Ca^{2+} levels regulate diverse cellular processes. In membrane–enzyme signaling systems, hormone binding activates intrinsic enzyme activity in the cytosolic domain of the receptor. The activated enzymes, in turn, induce intracellular responses.

Even in the same cell, a single hormone may bind to different cell-surface receptors linked to different second messengers, thereby inducing the same cell response (convergent pathways) or different responses (divergent pathways). A particular receptor may be coupled to two different G proteins, each linked to its own second-messenger pathway or both linked to the same second-messenger pathway. Other variations in signaling systems, some involving third messengers, also are possible. Clearly, intracellular signaling pathways typically interact with each other in various ways to control hormone-induced intracellular responses.

Although most hormones have multiple actions, they can usefully be grouped into several functional classes plus the very diverse prostaglandins. The production and secretion of several direct-acting hormones are regulated indirectly by hypothalamic releasing and inhibiting hormones and directly by tropic hormones produced in the anterior pituitary gland.

The following hormones have major roles in regulating metabolism and developmental processes: glucocorticoids and catecholamines, which are produced in the adrenal glands and affect energy metabolism; thyroid hormones, which regulate metabolic rate; insulin and glucagon, which are produced in the pancreas and have opposite effects on blood glucose levels; and growth hormone, which is produced in the anterior pituitary and works synergistically with the thyroid hormones to promote growth and development.

The reproductive hormones include the androgens (in males) and estrogens (in females), which promote development of sexual characteristics and gametes (sperm or oocytes). In females, progesterone acts to prepare the endometrium for implantation and helps prepare breast tissue for lactation; oxytocin stimulates uterine contractions during birth and milk ejection after birth; and prolactin promotes formation of milk and maternal behavior.

The hormones most responsible for regulating water and electrolyte balance are antidiuretic hormone (ADH), which increases water resorption in the kidney; mineralocorticoids, promotes Na^+ resorption in the kidney; atrial natriuretic peptide (ANP), which reduces Na^+ and water resorption in the kidney; parathyroid hormone and calcitriol (derived from vitamin D or cholesterol), which act to increase the plasma Ca^{2+} concentration; and calcitonin, which has the opposite action, decreasing the blood Ca^{2+} concentration.

REVIEW QUESTIONS

1. Give three examples of chemical regulation that do not involve the secretion of specific hormone molecules.

2. What criteria must be met before a tissue can be unequivocally identified as having an endocrine function?

3. Give examples of short-loop and long-loop negative feedback in the control of hormone secretion.

4. Discuss two examples that illustrate the intimate functional association of the nervous and endocrine systems.

5. Explain how it is possible that the actions of epinephrine and glucagon are similar but their actions are confined to different tissues.

6. How can a single second messenger (e.g., cAMP or $InsP_3$), induced by binding of different hormones, mediate different cellular responses in different tissues?

7. Explain how a small number of hormone molecules can elicit cell responses involving millions of times as many molecules.

8. What is the significance of protein phosphorylation in intracellular signaling systems?

9. How can a working muscle mobilize glycogen stores in the absence of epinephrine-induced glycogenolysis?

10. Describe two ways in which the concentration of free cytosolic Ca^{2+} becomes elevated. Discuss the role of Ca^{2+} as a second and third messenger.

11. Describe the similarities and differences that characterize four intracellular signaling systems described in this chapter.

12. Describe the interactions between cAMP and the inositol phospholipid pathway.

13. Describe the interrelations between Ca^{2+} and cAMP in the mammalian salivary gland and liver to illustrate convergent and divergent second-messenger pathways activated in response to a single first messenger such as epinephrine.

14. How do the glucocorticoids (also growth hormone and glucagon) combat hypoglycemia?

15. What role do the thyroid hormones play in amphibian development?

16. How does insulin produce its hypoglycemic effects?

17. What factors influence secretion of growth hormone?

18. Discuss the endocrine control of the menstrual cycle.

19. Explain how birth control pills prevent conception.

20. Discuss the role of juvenile hormone in the development and metamorphosis of an insect.

SUGGESTED READINGS

Alberts, B., D. Bray et al. 1994. *Molecular Biology of the Cell*. 3d ed. New York & London: Garland Publishing.

Hadley, M. E. 1992. *Endocrinology*. 3d ed. Englewood Cliffs, N.J.: Prentice Hall.

Raymond, J. R. 1995. Multiple mechanisms of receptor-G protein signaling specificity. *Am. J. Physiol.* 269 (Renal Fluid Electrolyte Physiol. 38):F141–F158.

Robb, S., T. R. Cheek et al. 1994. Agonist-specific coupling of a cloned *Drosophila* octopamine/tyramine receptor to multiple second messenger systems. *EMBO J.* 13:1325–1330.

Truman, J. W. 1992. The eclosion hormone system of insects. *Prog. Brain. Res.* 92:361–374.

CHAPTER
10
MUSCLES AND ANIMAL MOVEMENT

Animal movements—such as locomotion, eating, and copulation—are generated by three fundamentally different mechanisms: amoeboid movement, ciliary and flagellar bending, and muscle contraction. In addition, sound production and nearly all other forms of communication that do not rely on specific chemical signals are based on muscle contractions. Most muscles contract when neurons send signals to them, initiating a series of events that cause the muscles to become shorter. Muscle contractions are the most apparent and dramatic macroscopic signs of animal life, and they have excited the imagination of many people since ancient times. In the second century A.D., Galen hypothesized that "animal spirits" flow from nerves into muscles, inflating the muscles and increasing their diameter at the expense of their length, causing them to shorten.

Even as recently as the 1950s, it was suggested that muscles shorten because linear molecules of "contractile proteins" within the muscles are caused to shorten. The hypothesis stated that these molecules are helical in shape and that changes in the pitch of the helix produce changes in length. This hypothesis was short-lived, however, because the development of new techniques in the 1950s led to dramatic advances in our understanding of muscle function. Through evidence from electron microscopy, biochemistry, and biophysics, we have learned how the contractile mechanism of muscle is organized and how it produces shortening. It is also becoming clear how the process of contraction is initiated by electrical activity in the membrane of muscle fibers.

Muscles are classified, on both morphological and functional grounds, into two major types, **smooth muscle** and **striated muscle**. One muscle type—vertebrate striated muscle (primarily frog and rabbit skeletal muscle)—is the best understood, and in this chapter we will consider it in detail. Striated muscle itself can be subdivided into *skeletal muscle* and *cardiac muscle*. However, the mechanism by which all muscles contract is nearly identical, and the major differences between the classes are found in their cellular organization.

Recent research on comparative and integrative aspects of muscle function has uncovered unexpected diversity among skeletal muscles and revealed an elegant matching between the design of a muscle and its biological function. Several examples of this variation in organization are discussed in this chapter.

STRUCTURAL BASIS OF MUSCLE CONTRACTION

The general organization of skeletal muscle tissue is depicted in Figure 10-1. Muscles can move parts of an animal because each end of the muscle is attached to a bone or to some other structure, and when the muscle shortens, the physical relationship between the anchor points changes. Typically a muscle is anchored at each end by a tough strap of connective tissue called a **tendon**. Each muscle consists of long, cylindrical, multinucleate cells (or **muscle fibers**), which are arranged in parallel with one another. This arrangement allows all of the fibers in a muscle to pull in parallel with one another. Striated muscle fibers range from 5 to 100 μm in diameter, and may be many centimeters in length. (Consider the length of the calf muscle fibers of a professional basketball player.) One reason for this extraordinary size is probably that each fiber arises from many single embryonic muscle cells, called **myoblasts,** which fuse during embryonic development to form multinucleated units called **myotubes.** Each myotube contains many nuclei within a single plasma membrane and differentiates into an adult muscle fiber, sometimes called a myofiber. Each muscle fiber is, in turn, composed of numerous parallel subunits called **myofibrils,** which consist of longitudinally repeated units called **sarcomeres.** The sarcomere is the functional unit of striated muscle. The myofibrils of a muscle fiber are lined up with the sarcomeres in register, so the fiber looks banded, or striated, when it is observed with a light microscope. This banded appearance gave rise to the name striated muscle.

The structure of striated muscle fibers provides an elegant example of structure as the basis of function. The

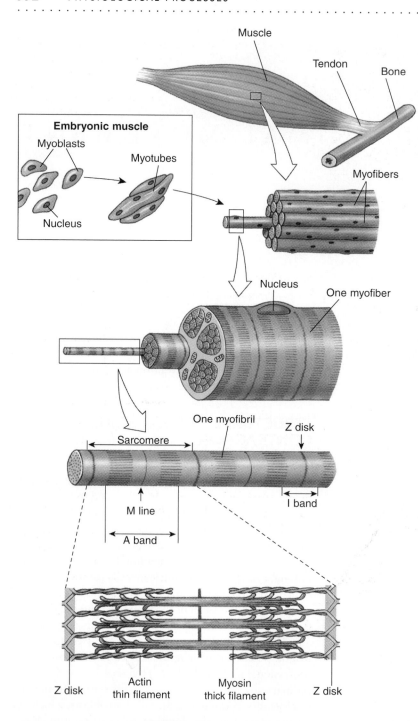

Figure 10-1 All vertebrate skeletal muscles are organized in a stereotyped hierarchy. The organ called a muscle consists of parallel multinucleate fibers, each of which contains many myofibrils. Muscles are attached to bones or other anchor points through tough connective tissue bands called tendons. Each muscle fiber is derived embryonically from a group of myoblasts that fuse to form myotubes. A myotube then synthesizes the proteins characteristic of muscle fibers and differentiates into its adult form. The myofibrils are made up of sarcomeres, arranged end-to-end. Each sarcomere contains thin filaments of actin and thick filaments of myosin, which interdigitate in a precise geometric relationship (see Figure 10-3). The thin filaments are anchored in regions called Z disks. [Adapted from Lodish et al., 1995.]

electron micrograph in Figure 10-2 shows a longitudinal section of several myofibrils. Each sarcomere is bounded at either end by a **Z line** (or **Z disk**) which contains α-actinin, one of the proteins found in all motile cells. Extending in both directions from the Z line of a myofibril are numerous **thin filaments** consisting largely of the protein **actin.** These thin filaments interdigitate with **thick filaments** made up of the protein **myosin.** Interdigitated thick and thin filaments make up the densest portion of the sarcomere, the **A band** (so named because this band is *anisotropic* and strongly polarizes visible light). The lighter portion in the center of the A band is called the **H zone,** which contains only thick filaments. In the middle of the H zone is the **M line,** which has been shown to contain enzymes that are important in en-

ergy metabolism (e.g., creatine kinase). The portion of the sarcomere between two A bands is called the **I band** (so named because this region is *isotropic* and does not polarize light).

If cross-sections are made through the various regions of a single sarcomere, the precisely arranged geometric relationship between thick and thin filaments is revealed (Figure 10-3). Only actin-containing thin filaments are seen when the section is made through an I band, and only myosin-containing thick filaments are seen in a section through an H zone. In the region of overlap, each myosin filament is surrounded by six thin filaments, and it shares these actin filaments with surrounding thick filaments. Each actin filament is surrounded by three myosin filaments.

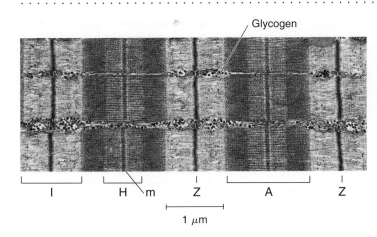

Glycogen

I   H   m   Z   A   Z

1 μm

Figure 10-2 The sarcomeres within different myofibrils are aligned in register producing the characteristic bands of skeletal muscle. Electron micrographs not only reveal these bands but also the sarcomere components giving rise to them. This electron micrograph of a longitudinal section through a frog skeletal muscle includes two complete sarcomeres of three myofibrils. The various bands and the Z disks (which appear as lines in longitudinal section) are labeled. Dark granules between fibrils are glycogen. [Courtesy of L. D. Peachey.]

When a section through a sarcomere is examined at high magnification with an electron microscope, small projections, called **cross-bridges,** are visible; these projections extend outward from the myosin filaments and make contact with the actin filaments during contraction (Figure 10-4A). The cross-bridges along the axis of the thick filament occur in groups of three; in each group, the cross-bridges are spaced about 14.3 nm apart and the angular displacement around the filament between successive cross-bridges is 120° (Figure 10-4B). This arrangement results in each thick filament having nine rows of cross-bridges, with adjacent cross-bridges in a row separated by 43 nm.

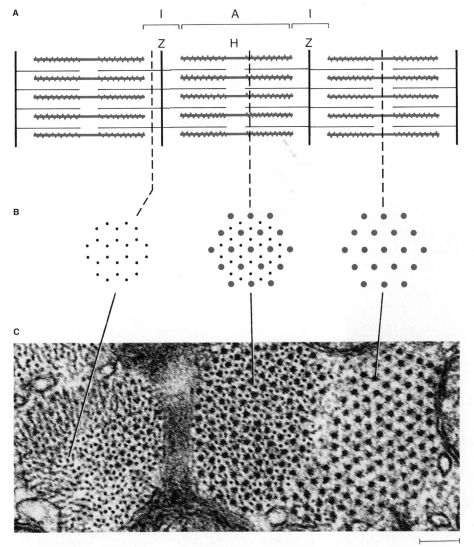

A   I   A   I

Z   H   Z

B

C

0.1 μm

Figure 10-3 Within a myofibril, thin filaments extending from the Z disks overlap with thick filaments in a precise geometric pattern. **(A)** Diagram of three sarcomeres, showing thick and thin myofilaments forming I, A, and H bands and Z disks. **(B)** Diagram of the geometric relation between thick and thin filaments in cross-sections made at different locations in a sarcomere. **(C)** Electron micrograph of a cross-section through myofibrils of a spider monkey extraocular muscle in which the sarcomeres of adjacent myofibrils are out of register so they can be matched with the profiles shown in part B. [Courtesy of L. D. Peachey.]

A

Crossbridges

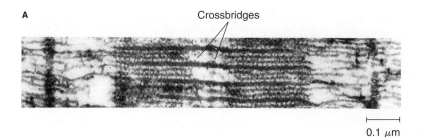

0.1 μm

Figure 10-4 Cross-bridges protrude from the myosin thick filaments toward the actin thin filaments. **(A)** Electron micrograph in which cross-bridges appear as faint projections extending from the myosin toward the actin filaments. **(B)** Diagram of the two-stranded helical arrangement of cross-bridges on the myosin thick filaments, drawn much larger than they are in part A. [Part A from Huxley, 1963; part B adapted from Murray, 1974.]

B

14.3 nm   42.9 nm

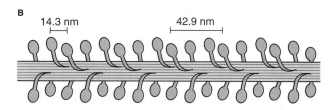

Myofilament Substructure

In the mid-nineteenth century, Wilhelm Kühne showed that different proteins can be extracted from a skeletal muscle that has been minced and then soaked in solutions containing various concentrations of salts. Nonstructural soluble proteins, such as myoglobin, are extracted by distilled water. Actin and myosin filaments are solubilized by highly concentrated salt solutions, which break the bonds holding actin and myosin monomeric proteins together in the filaments. Other proteins extracted under these conditions are discussed later in the chapter.

Our present knowledge of muscle contraction rests in part on analysis of the structure and composition of isolated actin and myosin filaments. Fragments of myofibrils that are several sarcomeres in length can be prepared by homogenizing fresh muscle in a laboratory blender. If the homogenization is carried out gently in a *relaxing* solution that contains magnesium, ATP, and a calcium-*chelating agent* such as EGTA, the formation of bonds between the myosin cross-bridges and the actin filaments is prevented. (EGTA and other calcium-chelating agents tightly bind Ca^{2+}, removing it from the solution.) As a result, the my-

A

0.1 μm

B

Actin   Troponin   Tropomyosin
        complex

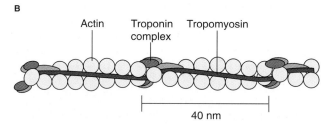

40 nm

Figure 10-5 Negatively stained individual isolated actin filaments can be visualized with an electron microscope at high magnification. **(A)** Electron micrograph of F-actin filaments. Note the two-stranded helical arrangement of the globular monomers. The specimen was prepared for microscopy by shadowing the actin filaments with a thin film of metal. **(B)** Diagram showing G-actin monomers in the two-stranded helix of F-actin. Intact thin filaments contain two other proteins—tropomyosin and troponin, a complex of three subunits. This structure has been deduced from electron micrographs (such as the one in part A) and from x-ray diffraction studies. [Part A courtesy of R. B. Rice; part B adapted from Ebashi et al., 1969.]

ofibrils fall apart into their constituent actin and myosin filaments.

An actin filament resembles two strings of beads twisted around each other into a two-stranded helix (Figure 10-5). Each "bead" in the string is a monomeric molecule of **G-actin,** so called because of its globular shape. The molecules of G-actin (approximate diameter of 5.5 nm) are polymerized to form the long two-stranded helix of **F-actin,** so called because of its filamentous appearance. Purified G-actin will polymerize *in vitro* to form F-actin filaments with the same physical structure as in muscle. The F-actin helix has a pitch of about 73 nm, so that its two strands cross over each other once every 36.5 nm. (This F-actin helix should not be confused with the far smaller α-helix found in other peptide chains.) In frog muscle, actin thin filaments are about 1 μm long and about 8 nm thick, and they are joined at one end to the material that constitutes the Z disk. Positioned in the grooves of the actin helix are filamentous molecules of the protein **tropomyosin.** At intervals of about 40 nm along the actin filament, a complex of globular protein molecules collectively called **troponin** is attached to tropomyosin (see Figure 10-5B). The role of troponin and tropomyosin in controlling muscle contraction is discussed later in this chapter.

A myosin molecule is composed of two identical heavy chains, which are long and thin, with an average length of 150 nm and a width of about 2 nm, and several much smaller light chains (Figure 10-6). One end of a myosin molecule forms a globular double "head" region, about 4 nm thick and 20 nm long. The long slender portion of the molecule constitutes its "neck" and "tail." The head region contains all of the enzymatic and actin-binding activity of the complete myosin molecule. The head is made up of the globular ends of the two heavy chains plus three or four (depending on the species) myosin light chains. The light chains are calcium-binding proteins. They differ among muscle types and influence the maximal velocity at which a muscle shortens. The α-helical portions of the heavy chains are twisted around each other, forming the neck and tail region of the molecule. When myosin molecules are treated with the proteolytic enzyme trypsin, they separate into two parts called **light meromyosin** (LMM) and **heavy meromyosin** (HMM). LMM constitutes the major part of the tail region, and HMM includes the globular head and the neck.

Myosin molecules will aggregate spontaneously *in vitro* to reconstitute myosin thick filaments when the ionic strength of a solution of myosin molecules is reduced. The first step in the formation of myosin filaments occurs when several myosin molecules aggregate with their tails overlapping and their heads pointing outward from the region of overlap and in opposite directions (Figure 10-7). The result is a short filament with a central region that lacks heads. This bare zone in the middle of thick filaments has implications for the contractile behavior of muscle, and we will return to it when we consider the length-tension relationship of a sarcomere. A filament grows as molecules of

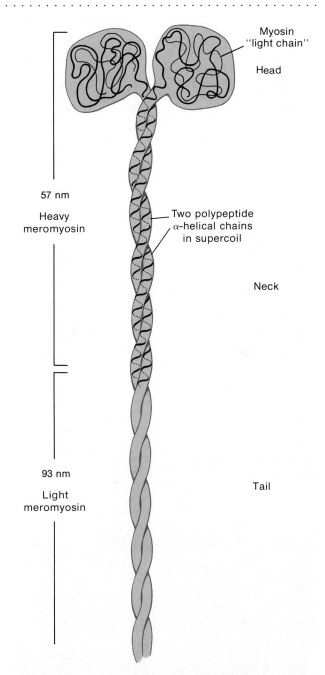

Figure 10-6 Each myosin molecule consists of a globular double head and a long thin tail. Myosin is separated into light and heavy meromyosin by digestion with the protease trypsin. Light meromyosin contains most of the tail; heavy meromyosin contains the globular heads and the beginning of the tail. [Adapted from Lehninger, 1993.]

myosin are added to each end, with their tails pointing toward the center of the filament and overlapping with the tails of previously added molecules. The head of each newly added myosin molecule projects laterally from the filament. Because the myosin molecules are added symmetrically to the two ends, the heads on each half of the filament are oriented opposite those of the other half (see Figure 10-7). Aggregation continues until for vertebrate myosin, the filament is about 1.6 μm long and about 12 nm thick. It is as yet unclear why filaments stop growing at that length.

A

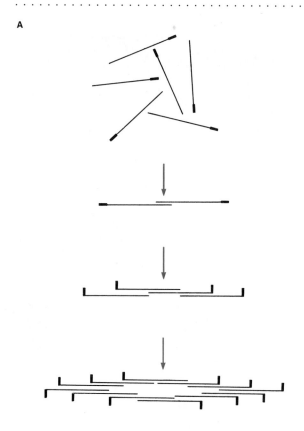

B

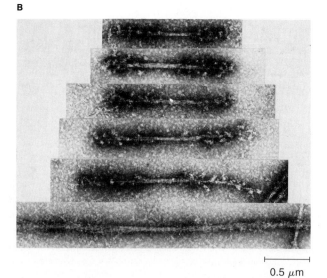

0.5 μm

Figure 10-7 Myosin molecules will polymerize spontaneously to reconstitute thick filaments with an organization identical to that found in muscle. **(A)** Diagram illustrating the spontaneous formation *in vitro* of thick filaments from a solution of myosin molecules. **(B)** Electron micrographs of myosin molecules assembled into thick filaments of various lengths. Note that the spontaneously formed filaments have a double-ended organization similar to the thick filaments in muscle. The diffuse dark staining at the ends of these thick filaments indicates the location of the myosin heads at both ends of the filaments; the lighter region at the center of each thick filament contains only the tails of myosin molecules. [Adapted from Huxley, 1969.]

Contraction of Sarcomeres: The Sliding-Filament Theory

The striations that define the sarcomeres were first observed with the light microscope well over a century ago. It was also observed that the sarcomeres change in length during stretch or contraction of a muscle, and that these changes correspond to the change in muscle length. Using a specially built interference light microscope, which permitted more accurate measurement of the sarcomeres, Andrew F. Huxley and R. Niedergerke in 1954 confirmed earlier reports that the A bands (myosin filaments) maintain a constant length when a muscle shortens, whereas the I bands and the H zone (zones where actin and myosin filaments do not overlap) become shorter. When a muscle is stretched, the A band again maintains a constant length, but the I bands and H zone become longer. That same year, Hugh E. Huxley and Jean Hanson reported that neither the myosin thick filaments nor the actin thin filaments seen in electron micrographs change their lengths when a sarcomere shortens or is stretched (Figure 10-8A). Instead, as the length of a sarcomere changes, it is the extent of *overlap* between actin and myosin filaments that changes.

Largely based on the observations described in the previous paragraph, H. E. Huxley and A. F. Huxley independently proposed the **sliding-filament theory** of muscle contraction. This theory states that during muscle contraction, sarcomeres shorten because the thin (actin) filaments actively slide along between the thick (myosin) filaments. The process pulls the actin filaments closer to the center of the sarcomere, and because the thin filaments are anchored in the Z disks, the sarcomeres become shorter (Spotlight 10-1, on page 360). When a muscle relaxes or is stretched, the overlap between thin and thick filaments is reduced, and the sarcomeres elongate. This theory differed radically from earlier hypotheses explaining muscle contraction, but it accounted for all of the data that had been observed up to that time.

One of the strongest pieces of evidence in support of the sliding-filament theory is the length-tension relation for a sarcomere. The length-tension curve relates the amount of overlap between actin and myosin filaments to the tension developed by an active sarcomere under that condition. According to the sliding-filament theory, each myosin cross-bridge that interacts with an actin filament generates force independently of all other cross-bridges and provides an increment of tension. Thus, the total tension produced by a sarcomere should be proportional to the number of cross-bridges that can interact with actin filaments, and this number should in turn be proportional to the amount of overlap between thick and thin filaments. The sliding-filament theory also predicts that no *active* tension (i.e., beyond that due to the elasticity of the muscle fiber) will develop if a sarcomere is stretched so far that there can no longer be any overlap between actin and myosin filaments.

 To test these predicted relationships between filament overlap and tension generated, single frog muscle fibers were stimulated to contract at different fixed sarcomere

A

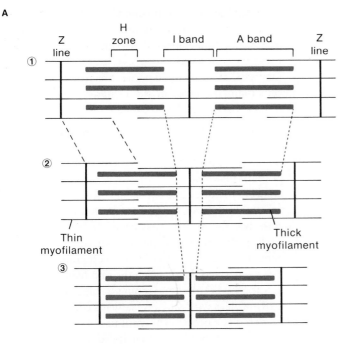

B

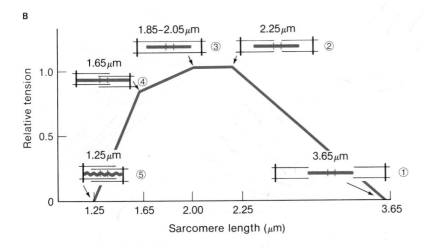

Figure 10-8 The sliding-filament hypothesis states that muscle fibers shorten when their actin and myosin filaments move past one another. **(A)** Relations of the myofilaments when two sarcomeres shorten. Note that the lengths of the thick filaments and of the thin filaments remain constant; only the amount by which thick and thin filaments overlap changes. When the filaments slide past one another, the I band narrows as the thin filaments slide toward the center of each A band, as was initially observed by Huxley and Niedergerke (1954). **(B)** Length-tension relation for a typical vertebrate sarcomere. The length and configuration of the sarcomere is depicted schematically near the curve at critical points. The active tension produced by the muscle is maximal when the overlap between thick and thin filaments allows the largest number of cross-bridges to be formed between actin and myosin. Tension drops off with increased length, because the thick and thin filaments overlap less and fewer cross-bridges can be formed. It also drops off with decreased length, because thin filaments begin to collide with one another, preventing further shortening. Skeletal muscle rarely operates over such a broad range of lengths, because the structure of the bones and joints limits the range of movement, so sarcomere length normally never departs significantly from the plateau region of this curve. [Adapted from Gordon et al., 1966.]

lengths. As the length of the sarcomeres is changed, so is the amount of overlap between the actin and myosin filaments (Figure 10-8B). The sarcomere length was adjusted with the aid of an electromechanical system that controlled the initial tension, allowing the sarcomeres to be held at any desired constant length. Then the tension generated when the fiber was stimulated to contract was measured and plotted as a function of sarcomere length. When the fiber was stretched until there was no overlap at all between thick and thin filaments, stimulation produced no tension beyond the passive elastic tension required to stretch the fiber to that length. When the fiber was held so that the actin filaments overlapped completely with the segment of the myosin filament bearing the cross-bridges, the tension generated was maximal. When the fiber was so short that the actin filaments in the two halves of the sarcomere collided, the fiber-generated tension decreased with further shortening. Tension decreased still further if the fiber was so short that the myosin filaments crumpled up against the Z disks.

In fact, it was possible to predict some of the properties of the length-tension curve before the experiments had been carried out. As noted above, the sliding-filament hypothesis assumes that the force generated by a sarcomere is proportional to the number of cross-bridges binding myosin filaments to actin filaments. It also assumes that cross-bridges are evenly distributed along each thick filament, except in the bare zone where no cross-bridges are present. (This second assumption has been experimentally verified.) From these assumptions and the dimensions of the filaments given in Figure 10-9A, it is possible to predict the shape of the sarcomere length-tension curve.

- *At what sarcomere length will the filaments be pulled beyond overlap and hence generate no force?* To determine the sarcomere length for a given amount of overlap between filaments of known lengths, imagine a very tiny ant trying to crawl along the filaments from the center of one Z disk to the center of the next

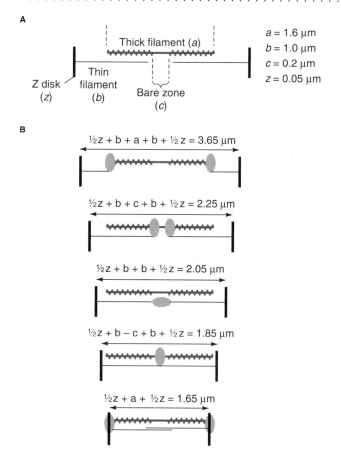

A

Thick filament (*a*)

Z disk (*z*) Thin filament (*b*) Bare zone (*c*)

$a = 1.6 \; \mu m$
$b = 1.0 \; \mu m$
$c = 0.2 \; \mu m$
$z = 0.05 \; \mu m$

B

$\tfrac{1}{2}z + b + a + b + \tfrac{1}{2}z = 3.65 \; \mu m$

$\tfrac{1}{2}z + b + c + b + \tfrac{1}{2}z = 2.25 \; \mu m$

$\tfrac{1}{2}z + b + b + \tfrac{1}{2}z = 2.05 \; \mu m$

$\tfrac{1}{2}z + b - c + b + \tfrac{1}{2}z = 1.85 \; \mu m$

$\tfrac{1}{2}z + a + \tfrac{1}{2}z = 1.65 \; \mu m$

Figure 10-9 If the lengths of thick and thin filaments are known, the sarcomere length can be predicted for various amounts of overlap. **(A)** Filament lengths measured from high-resolution electron micrographs of frog muscle fibers. *a*, length of thick filament; *b*, length of thin filament; *c*, length of bare zone that is free of cross-bridges at center of thick filament; *z*, thickness of Z disk. **(B)** Amounts of overlap between thick and thin fibers at inflections in the sarcomere length-tension curve. Each condition in this figure is matched to a point on the curve in Figure 10-8B. The colored ovals on each drawing indicate regions in which the interaction between filaments is critical for determining how much force can be generated by the sarcomere. The equation accompanying each drawing calculates the length of a single sarcomere for that condition of overlap. [Part B adapted from Gordon et al., 1966.]

Z disk. In a sarcomere that has been stretched just to the point where the filaments fail to overlap, the ant would have to cross one half of the thickness of a Z disk (0.025 μm), crawl along 1 actin thin filament (1.0 μm), step down to a thick filament and traverse it (1.6 μm), step up to a thin filament on the other side of the sarcomere and traverse it (1.0 μm), and finally cover the rest of the distance to the center of the second Z disk (0.025 μm). The total distance traveled would be 3.65 μm (Figure 10-9B, condition 1).

- *Why is there a plateau in maximal force between 2.05 and 2.25 μm?* When the sarcomere is 2.25 μm long, the ends of the thin filaments line up with the beginning of the bare zone on the thick filaments (where there are no cross-bridges). Thus all the cross-bridges on the thick filaments are optimally aligned to interact with

actin-binding sites on the thin filaments (Figure 10-9B, condition 2). As the sarcomere shortens farther, no more cross-bridges are added to the number that can interact with the thin filaments, so the force generated remains the same. The end of the plateau occurs when the thin filaments meet in the center of the sarcomere (Figure 10-9B, condition 3).

- *Why does the force fall as the sarcomere continues to shorten?* The sliding-filament hypothesis does not make quantitative predictions about muscle force below the point of maximal overlap, so it has been necessary to answer this question experimentally. From one perspective, the force might be expected to remain constant because all the cross-bridges overlap with actin sites on the thin filaments and can, at least in theory, generate force. However, two effects could reduce the force generated. First, when the thin filaments overlap at the middle of the sarcomere, binding of myosin cross-bridges with the thin filaments could be sterically hindered (Figure 10-9B, condition 4). Second, cross-bridges might bind with an inappropriate thin filament (one projecting from the Z disk at the other side of the sarcomere) and exert a force that pushes the Z disks apart, rather than pulls them together. Such a force would be considered negative and would need to be subtracted from the force generated by the normal cross-bridges.

- *Why does force decline steeply at 1.65 μm and fall to zero at about 1.25 μm?* The force generated declines steeply when the sarcomere is so short that the thick filaments contact the Z disks at both ends of the sarcomere (Figure 10-9B, condition 5). At this point, any further shortening of the sarcomere would require that the thick filaments be compressed. The actual slope of this decline and the length of the sarcomere at which no force can be produced cannot be predicted from the sliding-filament theory because these values would depend on the elastic modulus of the thick filaments and on how many cross-bridges are generating force.

- *How will changing the length of the filaments affect the shape of the length-tension curve?* Thin filaments in the muscles of mammals are about 1.2 μm long, (i.e., about 0.2 μm longer than thin filaments in frogs). From this value and using the calculations shown in Figure 10-9, we would predict that the plateau of the sarcomere length-tension graph of mammalian muscle would occur at sarcomere lengths between 2.45 and 2.65 μm and that force would fall to zero at 4.05 μm. The length of the thick filaments can also affect the properties of the length-tension curve. In all vertebrates, the thick filaments are about 1.60 μm long, but they are considerably longer in some invertebrates. Not only will the longer thick filaments change the shape of the sarcomere length-tension relationship, but they will also change the absolute force. Longer thick filaments can have more cross-bridges working in parallel and, hence, can generate higher forces (see Spotlight 10-1).

In the experiments that tested these predictions, it was crucial that the length measurements be made on small groups of sarcomeres located near the center of the muscle fiber and that the sarcomeres behave uniformly. Measurements made earlier, with less precise techniques, yielded rounded curves because the thick and thin filaments in many sarcomeres of a whole muscle—and, indeed, of a single fiber—varied in the amount that the thick and thin fibers overlapped at any given instant. A rounded curve would have failed to confirm the predictions of the sliding-filament theory and have seriously misled muscle physiologists.

Cross-Bridges and the Production of Force

Determining precisely how cross-bridges work is one of the greatest challenges facing current researchers studying muscle function. According to recent versions of the sliding-filament theory, the force driving muscle contraction arises when several different sites on the myosin head bind sequentially to sites on the actin filaments. All bonds between the head and the actin filaments are then broken, freeing the head for another cycle of sequential binding at a site farther along the actin filament. We will now consider these processes in detail.

Cross-bridge chemistry

Myosin cross-bridges must attach to binding sites on actin filaments in order to generate force, but cross-bridges must also be able to detach. If cross-bridges never detached from actin, a muscle could never relax. In addition, if cross-bridges could not detach, a muscle would be unable to shorten; attached cross-bridges would prevent filaments from sliding past one another, locking the muscle at one length. As a result, the cross-bridges must attach and detach from the thin filaments in a cyclic fashion.

Such a system poses a challenge to biochemists: How can myosin first attach to actin, so force can be generated, and then detach, so the filaments can continue sliding or the muscle can relax? The first hints about the chemistry of the interaction between myosin cross-bridges and actin filaments came from studies begun several decades ago on crude and purified extracts of muscle. Several interesting physical properties were exhibited by semipurified solutions of actin and myosin, which were extracted with concentrated salt solutions from freshly minced rabbit muscle and then precipitated with ammonium chloride.

When actin (A) and myosin (M) are mixed in the absence of ATP, they form a stable complex called **actomyosin** (AM). However, the addition of ATP to the solution causes rapid dissociation of the complex into actin and myosin-ATP:

$$AM + ATP = A + M\text{-}ATP$$

The observation that ATP is required for the dissociation of actomyosin (and for cross-bridge detachment) explains a phenomenon well-known to readers of detective novels. Following death, a human or other animal gradually becomes rigid and will hold the same position for hours or even for days. This condition, called **rigor mortis,** differs from muscle contraction, because in rigor mortis the muscles do not shorten. Instead, they simply maintain the same length for a long time. This rigidity occurs because when all of the ATP is used up after a cell dies, myosin binds to actin and cannot detach, producing rigidity.

When ATP binds to myosin, it is rapidly hydrolyzed to ADP and P_i, but the breakdown products unbind from myosin only very slowly. Thus the rate of ATP hydrolysis by myosin is very slow, and the rate-limiting step is the release of ADP and P_i from myosin. However, when actin binds to myosin, the release of ADP and P_i is greatly speeded up, presumably through an allosteric change in the conformation of myosin. This actin-induced effect greatly increases the rate at which myosin can hydrolyze ATP:

$$M\text{-}ATP \longrightarrow M\text{-}ADP\text{-}P_i \xrightarrow{\text{very slow}} M + ADP + P_i$$

$$M\text{-}ADP\text{-}P_i + A \xrightarrow{\text{fast}} AM + ADP + P_i$$

Because the binding of actin to the myosin-ADP-P_i complex releases energy, formation of actomyosin (AM) is kinetically favored. Combining these reactions produces a cycle of binding and unbinding (Figure 10-10). The net effect of one turn of the cycle is to split one molecule of ATP into ADP + P_i, liberating energy.

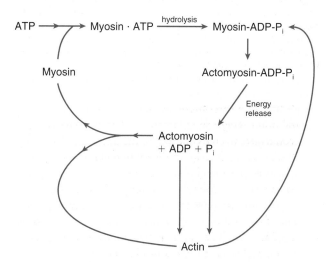

Figure 10-10 In the presence of ATP, myosin and actin filaments associate and dissociate in a cycle. ATP binds to actomyosin, splitting the molecule into actin and myosin. The myosin then acts as an ATPase, hydrolyzing ATP, but the release of the products ADP and P_i is slow unless actin rebinds to the myosin, increasing the rate of release. The net result of each turn of the cycle is the hydrolysis of one molecule of ATP, liberating energy, which can be used to generate force.

SPOTLIGHT 10-1

PARALLEL AND SERIES ARRANGEMENTS: THE GEOMETRY OF MUSCLE

Muscle has a highly organized, crystal-like geometry—all the way from the structure of the filaments to the organization of whole muscles. In this arrangement, some structural components are arranged in parallel with one another, while others are in series. These arrangements strongly affect the mechanics of muscle contraction.

Cross-bridges on one end of a thick filament are arranged in parallel with one another, but cross-bridges formed by the two ends of a filament oppose each other. Every cross-bridge extends from a thick filament to a thin filament independent of all other cross-bridges. Because of this arrangement, the forces produced by the cross-bridges along one thick filament are additive, like the force produced by each person in a tug of war or a current that moves through parallel resistors in a circuit. The force in one direction generated by a thick filament is equal to the force per cross-bridge times the number of cross-bridges on one half of the thick filament. What about the cross-bridges on the other half of the thick filament?

Hugh Huxley was the first to observe that the myosin monomers making up one half of a thick filament are assembled with their heads all pointed toward one Z disk, whereas those that make up the other half are oriented with their heads toward the other Z disk. This polarized configuration is crucial for the effective generation of force. Each set of cross-bridges exerts force on the thin filament directed toward the center of the sarcomere, so the force produced by the cross-bridges pull the Z disks to-gether. The force exerted by a thin filament on a thick filament is equal and opposite to the force exerted by the thick filament on the thin filament. The opposite polarity of the cross-bridges at the two ends of a thick filament means that a thin filament on one end of a sarcomere exerts a force onto that thick filament that is just balanced by the force of a matching thin filament on the other side. Hence the net force exerted *on* a thick filament by the surrounding thin filaments is zero, and the thick filament stays in the center of the sarcomere (part A of diagram). For example, if the cross-bridges on the right side of a sarcomere are generating a force of 100, then the cross-bridges on the left side must also be generating a force of 100, so the sum of the force on the thick filament is zero and the thick filament will stay in the middle of the sarcomere. However, if you were to attach a force transducer to one of the Z disks and measure the force generated by the cross-bridges, you would measure a force of 100, that is, the force generated by the bridges in one-half of the sarcomere.

What would happen if this polarity wasn't built into the thick filament? If all of the cross-bridges along a thick filament were lined up in the same direction, the thin filaments would exert a force in only one direction, and the thick filament would travel along the thin filaments toward one of the Z disks (part B of diagram). There would be a unidirectional net force on the thick filament (to the right in the diagram), and the thick filament would travel toward the right Z disk with unpredictable results. In this situation the sarcomere would not be able to generate force by shortening.

Some invertebrates have long thick filaments with many cross-bridges acting in parallel and the potential for generating greater force. However, this potential depends on the ratio of the number of cross-bridges to the total length of the filament, which in turn depends on the fraction of the length of each thick filament that is populated by cross-bridges (i.e., excluding the bare zones).

· ·

Energy transduction by cross-bridges

One of the major questions regarding the function of myosin cross-bridges is how chemical energy is transduced into mechanical energy by the cycle depicted in Figure 10-10. How do the cross-bridges generate a force between thick and thin filaments that causes the filaments to slide past one another? This question has been investigated in experiments with partially intact muscle fibers and in experiments *in vitro* with "skinned" muscle fibers (Spotlight 10-2, on page 364). Although various hypotheses have been proposed, the most widely accepted view is that a rocking or partial rotation of the actin-bound myosin head produces force (Figure 10-11A), and that this force is transmitted to the thick filament through the neck of the myosin molecule, which connects the head of the myosin molecule and the thick filament. In this hypothesis, the neck acts as a *cross-bridge link* between the myosin head and the thick filament, transmitting the force produced as the head rotates on the actin filament.

Studies of the mechanical properties of contracting muscle by Andrew F. Huxley and R. M. Simmons provided support for this view of cross-bridge function. They discovered that much of the elasticity that exists in series with the contractile components of muscle (see *Series Elastic Component* later in this chapter) resides in the cross-bridges themselves—presumably in the cross-bridge link. They hypothesized that as the myosin head rotates against the actin filament, the link is stretched elastically, which stores mechanical energy in the link (Figure 10-11B). According to their hypothesis, the rotation is produced when four sites on the myosin head (M_1 to M_4) interact in sequence with sites on the actin filament. These sites are ordered so that the actin-myosin affinity increases progressively from M_1 through M_4. Thus, after site M_1 attaches, it is energetically favorable for the head to rotate allowing site M_2 to attach, then M_3, and finally M_4 (left to right in Figure 10-11B).

The elasticity of the link would allow the rotational movement to occur without a single large and abrupt

Sarcomeres are arranged in series. Sarcomeres in a myofibril are arranged end to end (Z disk to Z disk), just as resistors can be placed in series in a circuit. When resistors are placed in series, the current through each resistor is the same as the current through every other resistor in the series. Similarly, the force generated by a series of sarcomeres is the same all along the chain of sarcomeres. Thus, although a chain of sarcomeres would have a tremendous number of cross-bridges in it, the force generated by the chain is determined by the force generated by any one sarcomere, and that force in turn is determined by the number of cross-bridges that are working in parallel on one half of the sarcomere.

However, because the sarcomeres are arranged in series, length changes and contraction velocities are additive. For instance, assume there are 1000 sarcomeres in series, each 2 μm long. If each sarcomere shortens by 0.1 μm, the whole series will shorten by 1000 $\times$ 0.1 μm = 100 μm. Similarly, if thin filaments move past a thick filament in each sarcomere at 10 $\mu m \cdot s^{-1}$, then the chain of sarcomeres will shorten at 2 $\times$ 1000 $\times$ 10 $\mu m \cdot s^{-1}$, or 20 $mm \cdot s^{-1}$. (Note: each Z disk in a sarcomere is moving toward the center of the sarcomere at 10 $\mu m \cdot s^{-1}$, so the overall shortening velocity of each sarcomere is twice that of each half sarcomere.) The amplification factor for length change demonstrates that to obtain very rapid shortening, it is necessary to have as many sarcomeres in series as possible.

Muscle fibers are in parallel. Each muscle fiber typically extends from one tendon to another and generates force between the tendons, independent of surrounding muscle fibers. Hence muscle fibers are arranged in parallel, and the force generated by each fiber is additive. One way to increase the force that can be generated by a muscle is simply to put more fibers in parallel. This mechanism occurs transiently when the nervous system recruits different numbers of muscle fibers to perform different activities.

The precision in the geometry of muscle makes it possible to calculate the force generated by a single cross-bridge if you

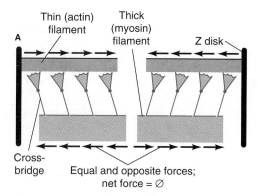

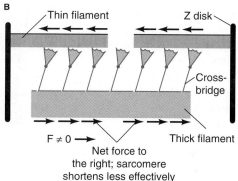

know the amount of force generated by a whole muscle (or even by a whole animal). For instance, consider a frog muscle with a 1 cm^2 cross-section that can generate 30 newtons of force. There are about 5 $\times$ 10^{10} thick filaments per square centimeter, and we know that thick filaments are arranged in parallel. So each thick filament must generate 6 $\times$ 10^{-10} newtons, or 600 piconewtons (pN) of force. There are about 150 cross-bridges at each end of a thick filament, so each cross-bridge must generate approximately 4 pN of force.

change in tension. Once it was stretched, the link would transmit its tension smoothly to the thick filament, generating force to shorten the sarcomere. One major piece of evidence supporting this hypothesis is the observation that the series elasticity of a muscle fiber is proportional to the amount of overlap between the thick and thin filaments; hence it is proportional to the number of attached cross-bridges. Moreover, sudden small decreases in fiber length are accompanied by very rapid recovery of tension, which presumably results from rotation of the cross-bridge heads into their more stable positions of interaction with actin sites (i.e., from the M_1 to the M_4 sites).

Not all details of cross-bridge function have yet been rigorously established. However, our present understanding of the sequence of events in cross-bridge function is summarized as follows (Figure 10-11C):

1. The head of each cross-bridge attaches to an actin filament by the first of a sequence of stable sites. It then goes to the second, third, and fourth sites in succession, forming successively stronger myosin-actin interactions with lower energy states.

2. This sequential binding causes the myosin head to rock or rotate, pulling on the link that connects the myosin head to the thick filament. The elasticity of the link allows the steplike rocking of the head to progress without sudden large changes in tension.

3. Tension in the link is transmitted to the myosin filament.

4. When the rotation of the head is complete, the myosin head dissociates from the actin filament and rotates back to its relaxed position.

The myosin head detaches from actin only if Mg^{2+}-ATP binds to the ATPase site of the head region. The ATP is then hydrolyzed, which is accompanied by a conformational change in the myosin head, leaving the head in an energized state, ready to rebind to a site on the actin that is a little

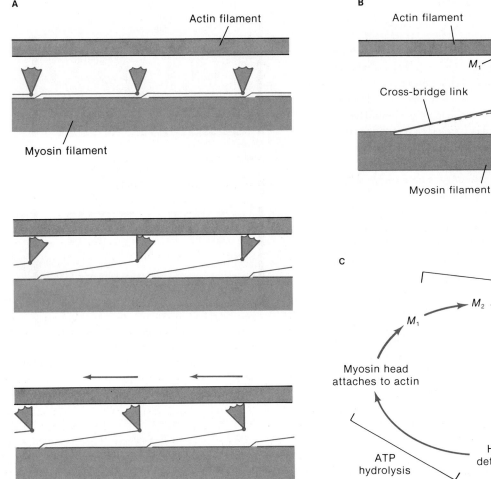

Figure 10-11 Thick and thin filaments slide past one another driven by changes in bonds between myosin cross-bridges and actin. **(A)** Sequence of events in the attachment of cross-bridges to actin filaments. In the relaxed state *(top)*, myosin heads are not bound to actin. Myosin heads attach to actin *(middle)* and through the formation of a series of bonds, the heads rotate, pulling on the actin filament *(bottom)* and causing it to slide past the myosin filament. Although the cross-bridges are shown acting in unison, they actually act asynchronously. **(B)** Modeling of cross-bridge function suggests that there are four separate binding sites on the myosin head. These sites, M_1 to M_4, interact in sequence (left to right) with sites on the actin filament. The rotation of the head produced in this manner causes the head to pull on the elastic cross-bridge link, stretching it. This tension is transferred to the actin filament, pulling it toward the left in this diagram and causing it to slide past the myosin filament. **(C)** Summary of cross-bridge cycle. Note that the myosin head detaches only if ATP binds. [Parts A and B adapted from Huxley and Simmons, 1971; part C adapted from Keynes and Aidley, 1981.]

farther along the thin filament. This cycle is repeated over and over, and the filaments slide past one another in small incremental steps of attachment, rotation, and detachment of the many cross-bridges on each thick filament.

MECHANICS OF MUSCLE CONTRACTION

Many of the mechanical properties of contracting muscle were elucidated before 1950, when the mechanism of contraction was not yet understood. It is useful to consider these classic findings and to attempt to explain them in terms of our current understanding of cross-bridge behavior.

The term *contraction* refers to the activation of muscles and the resultant generation of force. Muscle contrac-

tions have been categorized based on what happens to the length of active muscles. In an **isometric** (i.e., "same length") **contraction,** the length of a muscle is held fixed, preventing it from shortening (Figure 10-12A). The previous discussion of the length-tension relationship for a sarcomere was based exclusively on isometric contractions. Note that although no external shortening is permitted during an isometric contraction, there can be a very small amount of *internal* shortening (about 1%), which occurs when intracellular and extracellular elastic components—such as cross-bridge links and connective tissue that is in series with the muscle fibers—are stretched. In an **isotonic** (i.e., "same tension") **contraction,** the muscle shortens as force is generated (Figure 10-12B). Such contractions are the familiar ones that produce most movements. Contraction also may occur even when a muscle is lengthened by

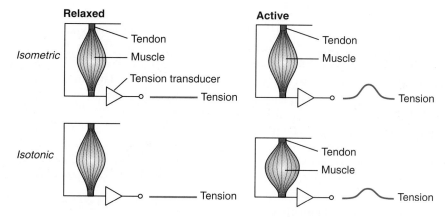

Figure 10-12 Muscles can contract isometrically or isotonically. In isometric contraction, the length of the muscle is held fixed, either by an experimenter or by the physical situation. For example, if you tried to pick up your car with your left arm, the contraction would be isometric, because the heavy weight of the car would prevent your arm muscles from shortening. In isotonic contraction, the muscle is allowed to shorten during the time when tension is being generated. Isotonic contractions of muscles move joints when we walk, run, and produce other movements.

externally applied force while it is generating force, for example if a heavy weight is suddenly applied to a contracting muscle.

Relation between Force and Shortening Velocity

For animals to move, muscles must shorten, so the relation between the production of force and the rate at which a muscle shortens (the so-called *force-velocity curve*) is crucial for understanding the design of a muscular system. Historically, to measure the force-velocity relation a muscle was attached to a lever with a weight attached to the other side of the fulcrum (Figure 10-13A). In more recent experiments, a motor controlled by a feedback circuit (a servomotor) replaces the weight, providing finer control. The system is arranged so there is a limit on how much the weight, or the servomotor, can stretch the muscle. In this setup, when the muscle is electrically stimulated, it starts to

contract, and when the force generated by the muscle becomes equal to the force exerted on the weight by gravity, the muscle begins to shorten at a constant velocity (i.e., the muscle begins to contract isotonically).

In the example depicted in Figure 10-13B, the maximal isometric tension that the muscle can produce is 100 g; that is, if the muscle contracts against a load of 100 g or more, it cannot shorten. If the load is less than 100 g, the muscle will shorten, at least a little. If a 50-g weight is attached, the muscle will shorten at a slow velocity; if lighter weights are attached, the muscle shortens faster. When it shortens against no weight at all, it shortens at its fastest speed, which is the maximal velocity of shortening, V_{max}. Plotting the force generated against the shortening velocity generates a hyperbolic curve that Archibald V. Hill, an important pioneer in muscle physiology, described in the 1930s by the following equation:

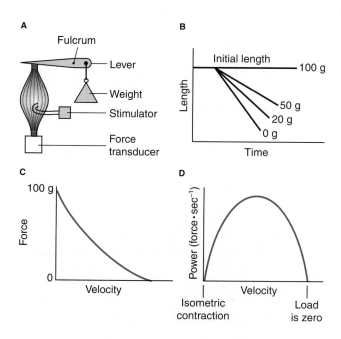

Figure 10-13 The force against which a muscle works and the velocity at which it shortens are reciprocally related. **(A)** Typical setup for measuring the relationship of force and velocity of a muscle. The muscle works against the weight that is hung on the other side of the fulcrum of the lever. When the stimulated muscle generates a larger force than the weight, it shortens and pulls down on the lever. A servomotor system could be used in this kind of experiment to provide finer control of the initial muscle length and the load. **(B)** The rate at which the muscle contracts as it works against four different loads: 100 g, 50 g, 20 g, and 0 g. When the load is smaller, so less force must be generated, the muscle contracts more rapidly. The maximal isometric tension of this muscle is 100 g, so when it works against the 100 g weight, it cannot shorten. At the start of the experiment, the length of the muscle would be set to optimize the overlap between thick and thin filaments in the sarcomeres. **(C)** Force-velocity curve based on the data shown in part B. At the maximal force of 100 g, the velocity of shortening is zero; that is, contraction is isometric. **(D)** Power-velocity curve calculated by multiplying the force and the velocity for each data point in part C. The power is zero if either the force or the velocity equals zero.

$$V \doteq \frac{b(P_0 - P)}{P + a} \qquad (10\text{-}1)$$

where V is the velocity of shortening; P, the force (or load); P_0, the maximal isometric tension of that muscle; b, a constant with dimensions of velocity; and a, a constant with dimensions of force (Figure 10-13C).

Equation 10-1 implies that as the load increases, the shortening velocity decreases. You probably are familiar with this principle from personal experience; you can lift a feather from a table much more rapidly than you can lift a heavy weight. Notice that the decline in force with increased velocity does not reflect a change in myofilament overlap. On the contrary, these experiments are purposely performed at the plateau of the length-tension relationship, so the number of cross-bridges that can interact with actin remains high and unchanged during shortening.

The relation between power and velocity is as important as the relation between force and velocity. For a fish to swim or a frog to jump, its muscles must generate mechanical power. The mechanical work performed by a muscle is the product of force times length change (ΔL). Mechanical power is given by

$$\text{power} = \frac{\text{work}}{\text{time}} = \frac{(\text{force})(\Delta L)}{\text{time}} = (\text{force})(\text{velocity})$$

Hence, multiplying the force generated by the muscle by the velocity at which it shortens yields the power production

under each condition. As shown in Figure 10-13D, the power generated is maximal at intermediate velocities. Power falls to zero if either the velocity of shortening or the force is zero.

As we will see later in this chapter, it is useful to describe force production or power production as V/V_{max}, where V is the velocity of shortening in a particular condition and V_{max} is the maximal velocity of shortening. Power production by the muscle shown in Figure 10-13 is maximal at a V/V_{max} of about $0.15-0.4$; this relation appears to hold for all muscles, no matter how fast their V_{max} is.

Effect of Cross-Bridges on Force-Velocity Relation

From the force-velocity curve described in the last section, we know that the force generated by a muscle drops as its speed of shortening increases. Recall that this relation is not caused by a change in the amount of overlap between thin and thick filaments—rather, it is observed at the maximal overlap. From our previous discussion of the role played by cross-bridges in isometric contraction, this fall in force with increased velocity could result if fewer cross-bridges are attached during rapid shortening, if each of the cross-bridges that are attached generate a smaller force, or both. Andrew Huxley's 1957 model of cross-bridge kinetics, although it has been superseded in some fine details, still provides the basic principles for understanding the overall mechanics and energetics of muscle contraction.

According to Huxley's model, cross-bridges are considered to be elastic structures that generate zero force

SPOTLIGHT 10-2

SKINNED MUSCLE FIBERS

One of the early advances that contributed significantly to muscle-cell physiology occurred when Albert Szent-Gyorgyi developed a procedure for isolating muscle fibers in which the intracellular structure remains intact, but the membrane no longer prevents free exchange of materials between the cytoplasm and the extracellular solution. This kind of preparation is called a "skinned" muscle fiber because its outer membrane has been entirely removed or rendered so leaky that it is functionally absent.

In Szent-Gyorgyi's procedure, muscle fibers are soaked for several days or weeks at a temperature below 0°C in a solution made up of equal parts of glycerin and water. Under these conditions the cell membrane becomes disrupted, and all soluble substances in the myoplasm are leached out, leaving intact the insoluble molecules that make up the contractile machinery. The glycerin in the solution prevents the formation of ice crystals, which could break up the structural organization of the fibers, and it also helps to solubilize the membranes. Storing the tissue at low temperature preserves the enzymes, but slows down catabolic processes that would cause the cells to digest themselves. These glycerin-extracted muscle fibers can be reactivated (i.e.,

made to contract and relax) if they are placed in appropriate conditions. In such fibers, an investigator can control the composition of the intracellular fluids without any interference from regulatory mechanisms that are normally present in an intact muscle fiber.

Another method of extracting some substances from cells, while leaving the insoluble proteins intact, employs nonionic detergents, such as the Triton X series. These agents, which are used at about 0°C, rapidly solubilize the lipid components of the cell membrane, allowing soluble metabolites to diffuse out of the cell and substances in the extracellular medium to diffuse rapidly into the cell. Fibers treated in this way are called "chemically skinned muscle fibers." This process requires only minutes, rather than days or weeks as is the case of glycerin extraction, thus saving much time.

A final way of producing skinned fibers is to actually dissect away the cell membrane using fine forceps. This process, which resembles removing the casing from a link sausage, requires great manual dexterity. With practice, however, a person can prepare fibers that remain structurally intact by this method.

Regardless of which method is used to prepare skinned muscle fibers, all of them allow experimenters to manipulate the chemical environment surrounding the contractile machinery of muscle fibers, expanding the opportunities to understand the molecular basis of contraction.

when they are at equilibrium. This behavior is similar to a piece of spring steel·projecting out from a surface. If the piece of steel is deformed by bending it, a restoring force is created to return it to its original position. Similarly, when a cross-bridge is bent toward or away from the Z disk, a restoring force is created that tends to bring it back to its original position; the magnitude of this force is proportional to the displacement of the cross-bridge from the equilibrium position (Figure 10-14A). If a cross-bridge bent toward the Z disk were attached to a thin filament when the restoring force was active, it would pull the Z disk toward the center of the sarcomere; this force is considered to be in the "positive" direction. By contrast, if a cross-bridge bent away from the Z disk were attached to a thin filament while the restoring force was active, it would push the Z disk away from the center of the sarcomere; this is viewed as a "negative" force.

Figure 10-14B illustrates how the forces generated by cross-bridge displacement cause movement of a thin filament. When the cross-bridge is at the equilibrium position (0), the force, F_0, is zero; when the cross-bridge is bent toward the Z disk, the force is positive (F_1 and F_2); the force is negative ($F_{1'}$ and F_3) when the cross-bridge is displaced away from the Z disk. The force generated by one thick filament is equal to $\Sigma n_i F_i$, the sum of the product of the number of attached cross-bridges at each displacement, n_i, and the force produced per cross-bridge at that displacement, F_i. As the velocity of shortening changes, the number of cross-bridges that are attached drops, and the displacement of the cross-bridges that are attached becomes smaller (Figure 10-14C). In addition, during rapid shortening some cross-bridges become attached when they are in a position that generates negative force. As a result of all of these changes, the net force produced during rapid shortening is lower than the force during slow shortening.

According to Huxley's theory, unattached cross-bridges are moved away from their neutral position by random thermal motion. If cross-bridges attached randomly to thin filaments, no force would be generated as the result of this thermal motion, because the number of cross-bridges generating negative force would equal the number generating positive force. However, cross-bridges can initially attach only when they are in a position that would generate positive force. Thus when a muscle is loaded maximally and is contracting isometrically, there will be a random distribution of cross-bridges that generate a positive force, and because all cross-bridges are generating positive force, the average force per cross-bridge will be positive and large.

If cross-bridges attach to thin filaments only when they are displaced to a position that produces positive force, how can cross-bridges generate negative force? During shortening, thin filaments move toward the center of the sarcomere, so cross-bridges that are attached at a sharp angle toward the Z disk (e.g., cross-bridge 2 in Figure 10-14B) are shifted closer to the equilibrium position, and their production of force is reduced by the movement of the thin filament. A cross-bridge that is attached at a very shallow an-

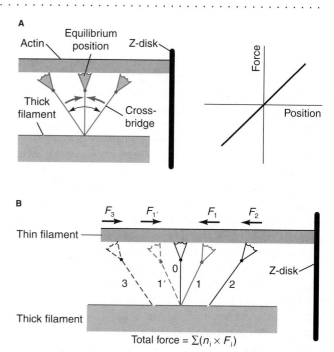

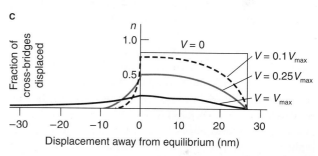

Figure 10-14 Cross-bridges generate force when they are moved away from their equilibrium position. **(A)** Relations between the position of the cross-bridge and the magnitude and direction of the force generated. At the equilibrium position, no force is generated. Displacement of the cross-bridge away from equilibrium in either direction generates a restoring force that tends to bring the cross-bridge back to equilibrium. **(B)** Cross-bridges attached to a thin filament at different positions. When displaced toward the Z disk (solid lines), they generate a positive force (e.g., F_2); when displaced away from the Z disk (dotted lines), they generate a negative force (e.g., F_3). The total force is the sum of the forces generated by all cross-bridges. Movement of the thin filament can change the displacement of some cross-bridges (e.g. 1 to 1'), causing them to exert negative ($F_{1'}$) rather than positive (F_1) force. **(C)** Fraction of the total number of cross-bridges that are attached and displaced. As the velocity at which thick and thin filaments slide past each other increases, fewer cross-bridges are attached, and the position (and force production) of the cross-bridges becomes more negative. At V_{max}, the net force generated by the cross-bridges equals zero, because the positive force generated by some cross-bridges equals the negative force generated by others. Conversely, when the muscle contracts isometrically ($V = 0$), the production of force is maximal because many cross-bridges are attached, and all of the attached cross-bridges are in a position that produces positive force.

gle (e.g., cross-bridge 1 in Figure 10-14B) can be dragged over to a position (1') that causes it to generate a negative force ($F_{1'}$). Of course, this effect could not go on indefinitely because such cross-bridges would produce more and more negative force, preventing further sliding of the thin filament. Each cross-bridge must detach, and the time it takes for a cross-bridge to detach is the key to what limits the maximal velocity of shortening:

Assuming that it takes a fixed time for cross-bridges to detach, then as the velocity at which the filaments slide past each other increases, more cross-bridges will be dragged to a position from which they generate a negative force before they can detach. There should, then, be a velocity at which the negative force generated by cross-bridges that have been dragged to the negative side of the equilibrium position will just balance the positive force generated by the attached cross-bridges on the positive side. At this point the net force generated by all attached cross-bridges is zero. Because the muscle cannot shorten any faster than this rate, this constitutes the maximal velocity of shortening, V_{max}. Thus at V_{max} some cross-bridges are attached, but the net force—or average force per cross-bridge—is zero. It follows that a muscle can have a fast V_{max} if its cross-bridges detach rapidly, allowing the cross-bridges to break their bond with the thin filaments before they generate large negative forces.

According to this model, two reasons explain the observed decrease in force as the velocity of shortening increases (see Figure 10-13D). First, the average force generated by the cross-bridges drops with velocity. Second, the total number of cross-bridges attached at any one time declines. The argument supporting this aspect of the model is based on chemical kinetics: As cross-bridges are dragged to positions in which they generate negative force, they detach faster, which causes fewer cross-bridges to be attached at higher velocities. At V_{max}, as few as 20% of the cross-bridges are thought to be attached.

REGULATION OF CONTRACTION

Up to this point we have considered only how cross-bridges on myosin thick filaments in a maximally activated muscle fiber bind and unbind to actin thin filaments, thereby generating force. Of course, if a muscle were "on" or activated all the time, we would be in a constant state of rigidity, unable to move, talk, or breathe. Thus, to perform useful work, muscles must turn on and off at the appropriate time. The mechanisms by which contraction is regulated—that is, turned on and off—are discussed in the following sections.

Role of Calcium in Cross-bridge Attachment

Although we now know that Ca^{2+} plays a crucial role in regulating the contractile activity of muscles, the evidence supporting this regulatory function of Ca^{2+} accumulated slowly. The earliest evidence for a physiological role of Ca^{2+} came from the work of Sidney Ringer and Dudley W. Buxton in the late nineteenth century. They found that an isolated frog heart stops contracting if Ca^{2+} is omitted from the bathing saline. (This observation marked the origin of *Ringer solution* and other physiological salines.) The possibility that Ca^{2+} participates in the regulation of muscle contraction was first tested in the 1940s when several researchers introduced various cations into the interior of muscle fibers. Of all the ions tested, only Ca^{2+} was found to produce contraction when it was present in concentrations similar to those normally found in living tissue. It was subsequently discovered that skeletal muscle fails to contract in response to stimulation if its internal calcium stores are depleted.

The concentration of Ca^{2+} ions is normally very low in the cytosol of muscle fibers—10^{-6} M or lower. Initial attempts to study the events of contraction in solution were foiled because it was impossible to maintain the Ca^{2+} concentration of experimental solutions as low as that in the cytosol. For example, even double-distilled water contains a higher concentration of Ca^{2+} than 10^{-6} M. The discovery of calcium-chelating agents, such as EDTA (ethylenediaminetetraacetic acid) and EGTA, overcame this obstacle to experimentally studying the effect of Ca^{2+} on contraction. The development of methods for preparing skinned muscle fibers, which lack the outer membrane, also facilitated research on the role of Ca^{2+} (see Spotlight 10-2).

The quantitative relation between the free cytosolic Ca^{2+} concentration in muscle fibers and contraction has been determined by exposing naked myofibrils (e.g., skinned muscle fibers) to solutions of different Ca^{2+} concentrations. Contraction of such preparations occurs only if the solutions also contain ATP, because ATP is required for muscle contraction (see Figure 10-11C). In experiments of this type, the myofibrils contract and generate tension only when Ca^{2+} (as well as ATP) is added to the surrounding solution; when the Ca^{2+} is removed, the myofibrils relax once again (Figure 10-15A). The amount of tension generated rises sigmoidally from 0 at a Ca^{2+} concentration of about 10^{-8} M to a maximum at about 5×10^{-6} M.

As we've seen already, force can be developed only when myosin cross-bridges bind to actin thin filaments, so anything that inhibits or facilitates this attachment will affect contraction. The key to how Ca^{2+} induces contraction lies in the two proteins—troponin and tropomyosin—associated with actin filaments (see Figure 10-5B). Troponin, a complex of several polypeptide chains, binds to tropomyosin about every 40 nm along the actin filament (Figure 10-16A). Troponin is the *only* protein in either the thin or the thick filaments of vertebrate striated muscle that has a high binding affinity for Ca^{2+}; each troponin complex binds four Ca^{2+} ions. When a myofibril is relaxed, tropomyosin occupies a position that interferes sterically with the binding of myosin heads to the actin filament (Figure 10-16B). When Ca^{2+} binds to troponin, the troponin molecule undergoes a change in conformation that moves tropomyosin out of the way, permitting myosin heads access to binding sites on the actin filament. Thus, when Ca^{2+} binds to troponin, it removes a constant inhibition of attachment between myosin cross-bridges and thin filaments. It is inferred from experimental results like those shown in Figure 10-15B that cross-bridges can bind to actin at concentrations of free Ca^{2+} above 10^{-7} M.

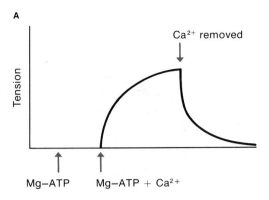

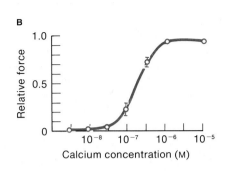

Figure 10-15 Free calcium ions regulate the state of muscle contraction. **(A)** Glycerin-extracted muscle fibers generate tension when they are exposed to Ca^{2+} and Mg-ATP. They relax when Ca^{2+} is removed, even if Mg-ATP is still present. **(B)** The force generated by a skinned muscle fiber varies with the concentration of Ca^{2+} in the surrounding medium. Force increases with increasing Ca^{2+} concentration , up to some maximum value. [Adapted from Hellam and Podolsky, 1967.]

As discussed earlier, the ATPase activity of myosin heads increases dramatically when the heads bind to actin. Because Ca^{2+} increases the binding of myosin heads, it would be expected to increase the ATPase activity of myosin. Indeed, the ATPase activity of skinned fibers has been shown to increase with the Ca^{2+} concentration of the surrounding solution (Figure 10-17A). The normal cycling of muscle contraction and relaxation depends on the presence of both ATP and Ca^{2+} in the cytosol of muscle fibers. This is demonstrated by the data in Figure 10-17B. In this experiment, when a glycerin-extracted muscle fiber was initially exposed to Ca^{2+} in the absence of Mg-ATP, no tension was generated. When Mg-ATP was added, tension developed, and that tension was then maintained even when Mg-ATP was removed; that is, rigor mortis set in. Once the muscle was in rigor, removal of Ca^{2+} had no effect because the lack of ATP caused all the attached cross-bridges to be frozen in place. Adding Mg-ATP back to the muscle that was in rigor and bathed in Ca^{2+}-free solution caused it to relax. Thus, both ATP and Ca^{2+} must be present if the thick and thin filaments are to interact effectively to produce tension.

The role of Ca^{2+} in regulating the actin-myosin interaction via troponin and tropomyosin applies to vertebrate skeletal and cardiac muscle. The role of Ca^{2+} differs in most other muscles. At least two other calcium-dependent regulatory mechanisms controlling actin-myosin interactions exist. In most invertebrate striated muscles, calcium initiates contraction by binding to the myosin light chains of the cross-bridge heads. Contraction of vertebrate smooth muscle and of nonmuscle actomyosin depends on a calcium-dependent phosphorylation of the myosin head, as described in the last section of this chapter.

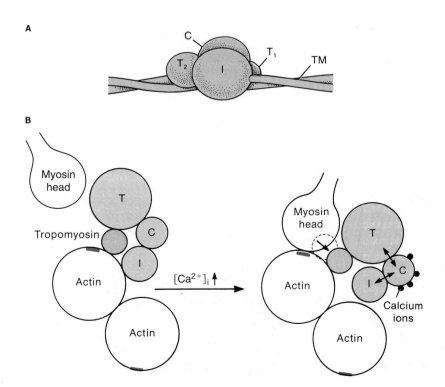

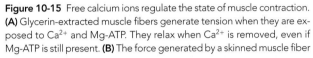

Figure 10-16 Troponin and tropomyosin regulate binding between myosin cross-bridges and actin thin filaments. **(A)** Three-dimensional model depicting the association of tropomyosin, (TM) and the troponin complex, containing subunits C, I, and T, with the actin filament. **(B)** Mechanism of Ca^{2+}-mediated regulation of actin-myosin interaction. When the concentration of Ca^{2+} is low *(left)*, the troponin complex binds with actin and tropomyosin, sterically preventing myosin cross-bridges from binding to actin. If the concentration of cytoplasmic Ca^{2+} increases, troponin C binds Ca^{2+}, changing the subunit affinities and causing the tropomyosin molecule to move away from the myosin-binding site on actin. Cross-bridges can then bind cyclically, and the thick and thin filaments can slide past one another until Ca^{2+} is removed from the troponin complex. [Part A adapted from Phillips et al., 1986; part B adapted from Ebashi et al., 1980.]

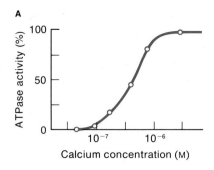

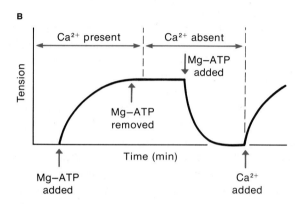

Figure 10-17 Free calcium modulates both the ATPase activity and the tension developed by glycerin-extracted muscle fibers. **(A)** The ATPase activity of myosin increases sigmoidally with the Ca^{2+} concentration of the surrounding solution, with a threshold of about 10^{-8} M. **(B)** Both Ca^{2+} and Mg-ATP are required for muscles to contract, but relaxation occurs only in the presence of Mg-ATP and the absence of Ca^{2+}. If Mg-ATP is removed once tension has developed, the fiber enters *rigor mortis* (flat part of the curve). *Rigor* is relieved only by removal of Ca^{2+} and addition of Mg-ATP. [Part A adapted from Bendall, 1969.]

Excitation-Contraction Coupling

Given what is known about cross-bridge attachment, sliding filaments, and the crucial role played by Ca^{2+}, it would seem likely that the regulation of contraction must include some mechanism that controls the concentration of free Ca^{2+} in the cytosol and couples the excitation of muscle with contraction. Recall from the discussion in Chapter 6 that when an action potential (AP) arrives at the neuromuscular junction, it triggers the release of acetylcholine from the motor neuron. The acetylcholine binds to postsynaptic receptor proteins, opening ion channels in muscle fibers. Current through these channels has a reversal potential that is more positive than the threshold of muscle fibers, so the synaptic potential at a neuromuscular junction can trigger all-or-none APs in the fiber. The AP that is initiated at the neuromuscular junction propagates away from the end-plate in all directions, exciting the entire membrane of the muscle fiber and setting in motion the sequence of events leading to contraction (Figure 10-18A).

At the neuromuscular junction, a single AP in the motor neuron can cause an AP in the postsynaptic muscle fiber, making this synapse quantitatively different from many neuron-neuron synapses. Whenever an AP is propagated in

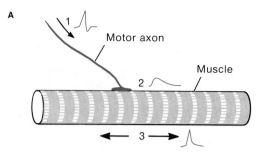

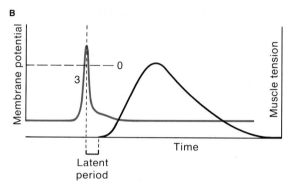

Figure 10-18 Muscle fibers contract when a postsynaptic potential at the neuromuscular junction causes a propagated action potential (AP) in the fiber. **(A)** An AP in a motor neuron (1) causes a postsynaptic potential in the muscle fiber (2), which gives rise to a propagated muscle AP (3). **(B)** The AP in the muscle fiber (colored trace) is followed, after a latent period, by a transient, all-or-none contraction (black trace), the muscle twitch.

a muscle fiber, it initiates a brief contraction, a *twitch*. Several milliseconds pass between the time that the AP is initiated and the time when the twitch begins (Figure 10-18B). During this latent period, **excitation-contraction coupling** occurs. The net effect of excitation-contraction coupling is to link the concentration of free Ca^{2+} in the cytosol to an AP in the plasma membrane of the muscle fiber. We examine the details of this critical process in the following sections.

Membrane potential and contraction

As illustrated in Figure 5-14, if some of the Na^+ ions in normal saline bathing excitable cells are replaced by K^+ ions, the membrane potential, V_m, will shift toward depolarization. When muscle fibers are suddenly depolarized in this way, they produce a transient contraction, which is called a **contracture** to differentiate it from a normal contraction. In the experiment depicted in Figure 10-19, a single frog muscle fiber was exposed to various concentrations of extracellular K^+ while the membrane potential and muscle tension were monitored. When the membrane was depolarized to about -60 mV, tension began to develop; with further depolarization, tension increased sigmoidally, reaching a maximum at about -25 mV.

This experiment demonstrates that the contractile system can produce graded contraction when the membrane

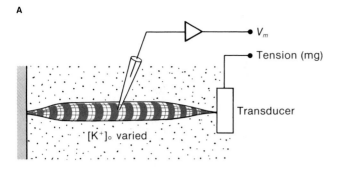

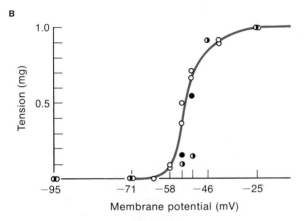

Figure 10-19 The tension developed by a muscle fiber varies with the membrane potential, V_m. **(A)** Setup for measuring membrane potential in and tension produced by an isolated muscle fiber as the concentration of KCl is varied in the extracellular solution. **(B)** The tension produced by the muscle fiber as a function of depolarization. Data points are plotted, and the colored curve shows the sigmoidal function that best fits the points. The threshold potential is about −60 mV. [Adapted from Hodgkin and Horowicz, 1960.]

is depolarized to different values. However, a single twitch in response to a single AP is typically an all-or-none event. How can these two observations be reconciled? During an AP in a muscle fiber, the membrane potential swings from a resting value of about −90 mV to an overshoot of about +50 mV. At the peak of the AP, the membrane potential is as much as 75 mV more positive than the potential required to give a maximal contracture. As a result, during an AP the membrane potential of the muscle fiber exceeds the value at which contraction—measured during steady depolarization—is fully activated. The twitch is all-or-none, because the AP is all-or none.

A potential difference across the surface membrane directly affects a region that extends at most only a fraction of a micrometer away from the inner surface of the membrane. As a result, no potential change across the surface membrane can directly exert any influence on the great bulk of the myofibrils in a typical skeletal muscle fiber, which is 50–100 μm in diameter. There must be something that couples depolarization of the surface membrane to activity of myofibrils deep within each muscle fiber. Electrotonic spread of local circuit currents produced by a propagated AP were experimentally ruled out because when currents of physiological magnitude were passed be-

tween two microelectrodes inserted into a muscle fiber they produced no contraction.

The hypothesis that Ca^{2+} might play a role in linking membrane potential and contraction was suggested relatively early. During the 1930s and 1940s, Lewis V. Heilbrunn argued for the importance of calcium in many cellular processes, including muscle contraction. He proposed that the contraction of muscle is controlled by intracellular changes in calcium concentration. We now know that this hypothesis is essentially correct, although it was widely rejected at first because of a fundamental misunderstanding about the nature of excitation-contraction coupling. It was assumed that calcium would have to enter the cytosol of the muscle fiber (also called the **myoplasm**) through the surface membrane to initiate contraction. A. V. Hill pointed out that the rate of diffusion of an ion or a molecule from the surface membrane to the center of a muscle fiber that is 25–50 μm in radius is several orders of magnitude too slow to account for the short observed latent period (about 2 ms) between an AP at the surface membrane and activation of the entire cross-section of the muscle fiber. From this logic, Hill correctly concluded that a *process* rather than a substance must couple the surface signal to myofibrils that lie deep within the muscle fiber. As we will see, it is the AP itself that is conducted deep into the cell interior, where it causes the release of intracellular Ca^{2+} from internal storage depots that surround the myofibrils. Elevation of the concentration of free Ca^{2+} in the myoplasm permits myosin cross-bridges to attach to the actin thin filaments and generate force.

T Tubules

Both anatomic and physiological evidence suggesting a mechanism of intracellular communication linking the surface membrane to the internal myofibrils came to light about 10 years after Hill's calculation. In 1958, Andrew F. Huxley and Robert E. Taylor studied the details of the process by stimulating the outside surface of single frog muscle fibers with tubular glass microelectrodes (Figure 10-20A). Their most significant findings were the following:

- Pulses of current that were too small to initiate a propagated AP but sufficient to depolarize the membrane under the pipette opening led to small local contractions (Figure 10-20B). Contractions occurred, however, only when the tip of the pipette was positioned directly over a Z disk.

- Contractions occurred only around the perimeter of the fiber and very close to the Z disk.

- Contractions spread further inward as the intensity of the stimulating current was increased.

- Contractions were limited to both half sarcomeres immediately on either side of the Z disk over which the electrode was positioned. In other words, there is *inward* spread, but no *longitudinal* spread, of the graded contraction.

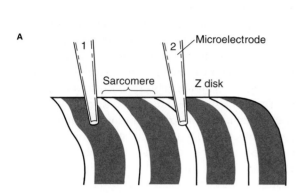

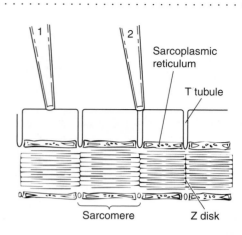

Figure 10-20 When frog muscles are stimulated by an extracellular micropipette, they can contract, but only when the pipette is placed in particular locations. **(A)** Experimental setup showing positioning of the tip of the stimulating pipette either in the center of a sarcomere (1) or directly over a Z disk (2). **(B)** Local contractions occur only if the opening of the stimulating pipette is lined up with the Z disk (2), placing it over the minute entrances to the T tubules, which are located in the plane of the Z disk. Stimulation in the middle of a sarcomere (1) produces little or no contraction.

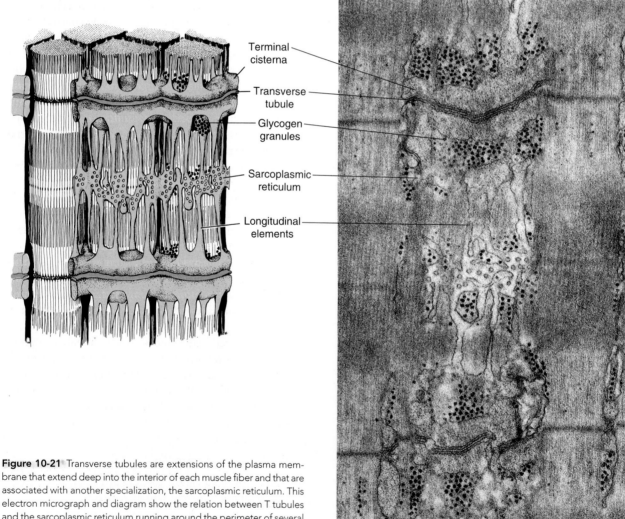

Figure 10-21 Transverse tubules are extensions of the plasma membrane that extend deep into the interior of each muscle fiber and that are associated with another specialization, the sarcoplasmic reticulum. This electron micrograph and diagram show the relation between T tubules and the sarcoplasmic reticulum running around the perimeter of several myofibrils in muscle from a frog. Notice that these structures may be buried deep within the interior of a single fiber, so the plasma membrane could be located as much as 50 μm from these structures. Dark spots in the electron micrograph are glycogen granules. [Adapted from Peachey, 1965.]

Electron microscopic studies of amphibian skeletal muscle performed about the same time provided an anatomic correlate of these physiological findings. Running around the perimeter of each myofibril at the level of the Z disk is a membrane-limited **transverse tubule** (or **T-tubule**) less than 0.1 μm in diameter, which is continuous with similar tubules surrounding neighboring myofibrils in the same sarcomere (Figure 10-21, on opposite page). The membrane of this anastomosing system of tubules is connected directly to the surface membrane of the muscle fiber, and the lumen of the T-tubule system is continuous with the solution on the outside of the fiber. This continuity was confirmed by the demonstration that ferritin or horseradish peroxidase—large protein molecules that produce an electron-opaque stain—appear in the lumen of the T tubules if a muscle fiber is exposed to these molecules for a period before the tissue is fixed for electron microscopy. These charged molecules are much too large to cross cell membranes, so the T tubules must be open to the extracellular space, arising as invaginations of the surface membrane.

The T-tubule system provides the anatomic link between the surface membrane and the myofibrils deep inside the muscle fiber. When Huxley and Taylor placed their stimulating pipette at the Z disks, over the entrance to a T tubule (see Figure 10-20), depolarizing current spread down the tubule and initiated contraction deep within the muscle fiber. If they produced hyperpolarizing current from their pipette instead, no contraction occurred. Comparative studies further strengthened the conclusion that T tubules carry excitation into muscle fibers. In crabs and lizards, the T tubules are located at the ends of the A bands, instead of at the Z disks (Figure 10-22). In these species, contraction is produced when a stimulating pipette is placed at the edge of an A band, rather than over a Z disk. From such results it has been concluded that T tubules, rather than the Z disks or any other part of the sarcomere, are most likely to transmit excitation into frog muscle fibers.

Further confirmation that T tubules play an important role in excitation-contraction coupling was obtained when the connection between the T tubules and the surface membrane was broken by osmotically shocking muscle fibers with a 50% glycerol solution. When the T tubules are disconnected from the surface membrane in this way, membrane depolarization no longer evokes a contraction; that is, physically uncoupling the T-tubule system from the membrane functionally uncouples the contractile system from the excitation process.

The inward spread of electrical signals down the T tubules was at first thought to occur by way of electrotonic conduction, but the spread of excitation to the center of a muscle fiber following depolarization of the membrane is reduced if tetrodotoxin is added to the bath or if the concentration of Na^+ is reduced in the extracellular fluid. Either treatment will reduce or eliminate sodium-based APs, suggesting that the APs characteristic of the surface membrane are actively carried deep into the muscle fiber by the membranes of the T tubules. In muscle fibers that do not produce APs (e.g., many arthropod muscles), the T tubules carry passive electrotonic signals into the fibers' interior, just as depolarization of the surface membrane is graded, rather than propagated and all-or-none.

Sarcoplasmic reticulum

Striated muscle fibers contain a second intracellular membrane system, the **sarcoplasmic reticulum** (SR), in addition to the T-tubule system. In frog muscle, the sarcoplasmic reticulum forms a hollow collar around each myofibril on either side of a Z disk and extends from one Z disk to the next as well (see Figure 10-21). The *terminal cisternae* of the sarcoplasmic reticulum make intimate contact with T tubules, which are sandwiched between the terminal cisternae of adjacent sarcomeres. When an AP is conducted into the T tubule it causes the release of Ca^{2+} ions stored in the neighboring sarcoplasmic reticulum.

How does Ca^{2+} get into the sarcoplasmic reticulum to begin with? When muscle fibers are broken up and their contents are fractionated, SR membranes form vesicles about 1 μm in diameter. If oxalate ions—which tightly bind Ca^{2+} forming highly insoluble calcium oxalate—are present in the solution in which the vesicles form, a calcium oxalate precipitate accumulates within the vesicles. This observation and others have been interpreted to mean that the SR membrane actively transports Ca^{2+} ions from the surrounding medium and concentrates it inside the vesicles. Electron micrographs of unfractionated muscle tissue that has been exposed to oxalate reveal calcium oxalate inside the terminal cisternae.

The calcium-sequestering activity of the sarcoplasmic reticulum is sufficiently powerful to keep the concentration of free Ca^{2+} in the myoplasm of resting muscle fibers below 10^{-7} M, which is sufficient to remove any Ca^{2+} bound to troponin. In other words, the sarcoplasmic reticulum is

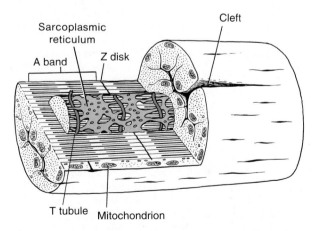

Sarcoplasmic reticulum

Cleft

A band

Z disk

T tubule   Mitochondrion

Figure 10-22 In crab muscle fibers, the T tubules are located at the edge of the A bands, rather than at the Z disks as in frog fibers. Stimulation with an extracellular micropipette produces a local contraction in crab fibers only when the tip of the pipette is placed near the edge of the A bands. Compared with frog muscle fibers, crab fibers have a larger diameter and contain deep clefts. [Adapted from Ashley, 1971.]

capable of driving the concentration of intracellular free Ca^{2+} so low that contraction is prevented. This ability of the sarcoplasmic reticulum to remove Ca^{2+} from the myoplasm depends on the activity of proteins within the SR membrane that bind and transport Ca^{2+} ions. In freeze-fracture electron micrographs, the membrane of the longitudinal elements of the sarcoplasmic reticulum contains many densely packed inclusions that have been associated with the calcium pump molecules. As in other active-transport systems, the calcium pump of the sarcoplasmic reticulum requires ATP as its energy source.

Under normal conditions, Ca^{2+} within the sarcoplasmic reticulum is bound to a protein called **calsequestrin.** As a result, the concentration of free Ca^{2+} within the sarcoplasmic reticulum remains relatively low, which reduces the gradient against which the pumps must work. In addition, because Ca^{2+} is stored bound to protein molecules, the sarcoplasmic reticulum can store a large amount of Ca^{2+}.

Combining the observation that Ca^{2+} ions are accumulated by the sarcoplasmic reticulum with what is known about the role Ca^{2+} plays in the interaction between thin and thick filaments, it seems likely that muscle contraction could be initiated when the sarcoplasmic reticulum releases Ca^{2+} into the myoplasm. The first direct evidence that the concentration of free Ca^{2+} rises in muscle fibers in response to stimulation came from a photometric method using the calcium-sensitive bioluminescent protein **aequorin** isolated from a species of jellyfish (see Figure 6-28). The chemistry of aequorin is complex, causing it to respond slowly to changes in the concentration of free Ca^{2+}. For this reason, recent experiments have employed fluorescent dyes that change their fluorescent properties when the amount of available free Ca^{2+} changes; these dyes respond much more rapidly than aequorin to changes in the Ca^{2+} concentration.

One calcium-indicator fluorescent dye used in such experiments is *furaptra*. In the absence of Ca^{2+} this dye fluoresces; that is, it emits light at a particular wavelength when illuminated with exciting light of a different wavelength. Because the intensity of furaptra fluorescence decreases as the Ca^{2+} concentration increases, this dye can be used to monitor changes in Ca^{2+} concentration within muscle fibers. For example, when a furaptra-loaded muscle fiber is stimulated electrically, the fluorescence of the dye first declines and then returns to its initial value (Figure 10-23). This observation has been interpreted to indicate that when the muscle is electrically stimulated, the amount of free Ca^{2+} in the myoplasm increases. A very small amount of the newly released Ca^{2+} binds to the furaptra, and the fluorescence of the dye declines. As the released Ca^{2+} is resequestered, Ca^{2+} unbinds from the dye, and the dye's fluorescence rises again.

All of this evidence indicates that contraction is activated when Ca^{2+} ions are released from the sarcoplasmic reticulum, and that this release somehow occurs when APs initiated at the surface membrane are transmitted into the depths of the muscle fiber by the T tubules. The anatomy of the T tubules and sarcoplasmic reticulum has suggested how this coupling happens. As noted above, each T tubule is located in close apposition to the terminal cisternae of the sarcoplasmic reticulum (see Figure 10-21). In fact, histologists have for decades called this portion of a muscle fiber the *triad*, because sections through the region consistently

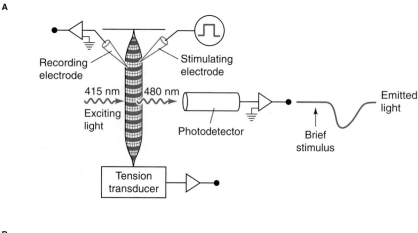

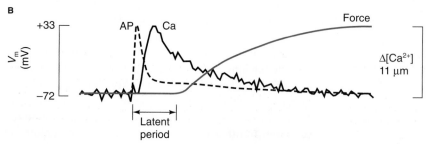

Figure 10-23 The amount of free Ca^{2+} and changes in its concentration in a muscle fiber can be measured using a calcium-sensitive fluorescent dye, such as furaptra. **(A)** In this experimental setup, a muscle fiber injected with the fluorescent dye is electrically stimulated and the subsequent changes in fluorescence, the membrane potential, and the production of tension by the fiber are recorded. **(B)** When the muscle fiber is stimulated, an AP is propagated along the surface membrane and recorded by the recording microelectrode. A short time later, the fluorescence signal from the calcium-sensitive dye inside the fiber indicates that the Ca^{2+} concentration within the fiber has increased, and even later, the tension transducer measures the production of tension by the fiber. Notice that the tension begins to rise only after the AP is over and the intracellular Ca^{2+} concentration is declining. [Part B courtesy of D. M. Baylor.]

revealed three associated tubes or sacks. Two of the sacks are always large, and they are located on either side of a much smaller tube or sack. We now know that the two sacks are two terminal cisternae of the sarcoplasmic reticulum and the smaller central tube is a T tubule. How can an AP in a T tubule be conveyed to the sarcoplasmic reticulum, causing it to release Ca^{2+}? Although the details are not yet completely understood, the general mechanism is now known.

Membrane receptors in triads In 1970, electron microscopy experiments by Clara Franzini-Armstrong revealed electron-dense molecules in the part of the SR membrane that lies adjacent to the T tubule. She called these molecules "feet;" they have more recently been named the *ryanodine receptor,* because they bind the drug ryanodine. Knox Chandler and his colleagues then proposed that these proteins represent Ca^{2+} channels and incorporated them into the "plunger model" for release of Ca^{2+} from the sarcoplasmic reticulum (Figure 10-24). In this model, depolarization of the T tubule causes a plug to be removed from the Ca^{2+} channels in the SR membrane, allowing Ca^{2+} to escape into the myoplasm driven down its steep electrochemical gradient. When the T-tubule membrane repolarizes, the plug is replaced, preventing further Ca^{2+} release.

The plunger model proposed an explanation of how the T-tubule membrane and the SR membrane might be coupled, but it failed to identify what caused the plunger to move in response to depolarization of the T-tubule membrane. Additional electron microscopy studies revealed a cluster of proteins in the T-tubule membrane directly across from each ryanodine receptor in the SR membrane. These T-tubule proteins, called *dihydropyridine receptors,* are voltage-sensitive. Because the ryanodine receptors extend most of the way across the cleft, a direct mechanical linkage between ryanodine receptors and dihydropyridine receptors was proposed (Figure 10-25). So the model proposes that when the T tubule depolarizes, the voltage-sensitive dihydropyridine receptor undergoes a conformational change and either mechanically dislodges the ryanodine receptor from calcium-selective channels in the SR membrane or forces the ryanodine receptor into a conformational change that opens the calcium channels.

Interestingly, only about half of the ryanodine receptors are associated directly with voltage-sensitive dihydropyridine receptors in the T-tubule membrane. This finding suggests that if there is a direct mechanical linkage between T tubules and the sarcoplasmic reticulum, only about half of the ryanodine receptors participate. It has been proposed that the other, unlinked ryanodine receptors are activated by the increase in myoplasmic free Ca^{2+} resulting from opening of the mechanically linked channels. Activation of these unlinked ryanodine receptors would in turn open more Ca^{2+} channels in the membrane of the sarcoplasmic reticulum. Such a mechanism, called Ca^{2+}-induced Ca^{2+} release, has been found in other tissues including the excitation-contraction coupling mechanism in cardiac muscle.

Time course of calcium release and reuptake The ability to measure rapid changes in the myoplasmic Ca^{2+} concentration, combined with the information about calcium binding by troponin and the kinetics of the calcium pumps in the SR membrane, has permitted the Ca^{2+} fluxes during muscle contraction and relaxation to be modeled. Such modeling suggests that when T tubules become depolarized, Ca^{2+} flows out of the sarcoplasmic reticulum for

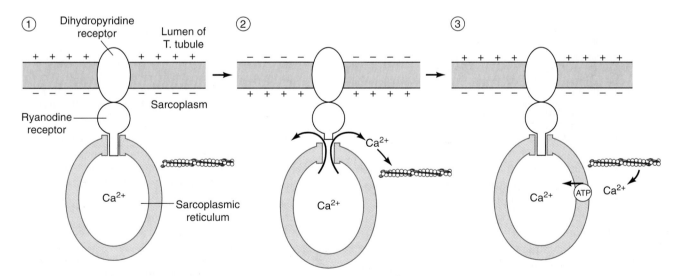

Figure 10-24 Depolarization of the T-tubule membrane indirectly causes calcium channels in the sarcoplasmic reticulum to open. When the membrane of the T tubule is at rest (1), calcium channels in the SR membrane are blocked by the "feet" of ryanodine receptors. When the T-tubule membrane depolarizes (2), voltage-sensitive dihydropyridine receptors convey the signal to the ryanodine receptors, and the "plungers" blocking calcium channels in the SR membrane are removed, allowing Ca^{2+} to flow out of the SR lumen into the myoplasm. The free Ca^{2+} binds to troponin, revealing cross-bridge binding sites on actin molecules. When the membrane potential returns to rest (3), the ryanodine receptors again block the calcium channels. Calcium pumps in the sarcoplasmic reticulum resequester Ca^{2+}, shifting the equilibrium of Ca^{2+} binding on troponin and concealing the cross-bridge binding sites. [Adapted from Berridge, 1993.]

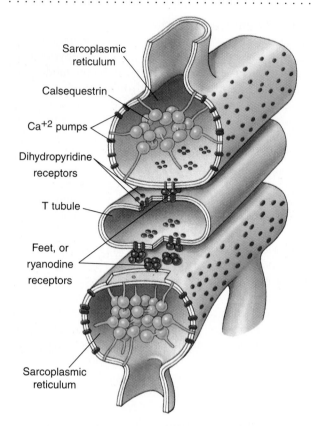

Figure 10-25 In an intact triad, several molecules contribute to the control of myoplasmic calcium. The voltage-sensitive dihydropyridine receptor and the ryanodine receptor work together linking depolarization of the T tubule to opening of calcium-selective channels in the SR membrane through which Ca^{2+} moves from the SR lumen into the myoplasm. The feet of the ryanodine receptors extend into the gap between the T tubules and sarcoplasmic reticulum. A calcium ATPase (calcium pump) in the SR membrane resequesters Ca^{2+} from the myoplasm, and calsequestrin inside the SR binds Ca^{2+}, reducing the concentration of free ionic Ca^{2+} inside the SR. [Adapted from Block et al., 1988.]

Labels on figure:
- Sarcoplasmic reticulum
- Calsequestrin
- Ca^{+2} pumps
- Dihydropyridine receptors
- T tubule
- Feet, or ryanodine receptors
- Sarcoplasmic reticulum

several milliseconds and then the calcium channels close. The mechanism for the closure is not fully understood.

Most of the Ca^{2+} that leaves the sarcoplasmic reticulum binds very quickly to troponin. The concentration of troponin in muscle fibers is about 240 μM, which represents a large buffer for Ca^{2+} ions. Thus, a very small amount of the released Ca^{2+} remains free in myoplasm, and it is only this small amount of unbound Ca^{2+} that is detected by a calcium-indicator dye in experiments such as the one shown in Figure 10-23. Both during and following the release of Ca^{2+} from the sarcoplasmic reticulum, the free Ca^{2+} in the myoplasm is pumped back into the SR lumen, lowering the myoplasmic level of free Ca^{2+}. As the concentration of free Ca^{2+} in the myoplasm becomes very low, Ca^{2+} bound to troponin is released back into the myoplasm and then is subsequently pumped back into the sarcoplasmic reticulum where it binds to calsequestrin.

Contraction-Relaxation Cycle

Starting with a relaxed skeletal muscle, the following sequence of events leads to contraction and then relaxation of a skeletal muscle fiber:

1. The surface membrane of the fiber is depolarized by an AP or, in some muscles, by synaptic potentials. In an animal, APs in skeletal muscle fibers are generated by synaptic potentials, so neuronal input is required for initiating contraction in skeletal muscle.

2. The AP is conducted deep into the muscle fiber along the T tubules.

3. In response to depolarization of the T-tubule membrane, voltage-sensitive dihydropyridine receptors in the T-tubule membrane undergo a conformational change that—through direct mechanical linkage to ryanodine receptors in the SR membrane—causes opening of Ca^{2+} channels in the SR membrane (see Figure 10-24, steps 1 and 2).

4. As Ca^{2+} flows out from the lumen of the sarcoplasmic reticulum, the free Ca^{2+} concentration of the myoplasm increases from a resting value of below 10^{-7} M to an active level of about 10^{-6} M, or higher, within a few milliseconds. The Ca^{2+} channels in the SR membrane then close.

5. Most of the Ca^{2+} that enters the myoplasm binds rapidly to troponin, inducing a conformational change in the troponin molecules. This conformational change causes a change in the position of the tropomyosin molecule, eliminating steric hindrance and allowing myosin cross-bridges to bind to actin thin filaments (see Figure 10-16B).

6. Myosin cross-bridges attach to the actin filaments and go through a series of binding steps that cause the myosin head to rotate against the actin filaments, pulling on the cross-bridge link (see Figure 10-11A,B). This pulling produces force on, and in some cases active sliding of, the thin filaments toward the center of the sarcomere, causing the sarcomere to shorten by a small amount (see Figure 10-8A).

7. ATP binds to the ATPase site on the myosin head causing the myosin head to detach from the thin filament. ATP is then hydrolyzed, and the energy of the hydrolysis is stored as a conformational change in the myosin molecule, which then reattaches to the next site along the actin filament as long as binding sites are still available, and the cycle of binding and unbinding is repeated (see Figure 10-11C). During a single contraction, each cross-bridge attaches, pulls, and detaches many times as it progresses along the actin filament toward the Z disk.

8. Finally, calcium pumps in the SR membrane actively transport Ca^{2+} from the myoplasm back into the SR lumen (see Figure 10-24, step 3). As the concentration of free Ca^{2+} in the myoplasm drops, Ca^{2+} bound to troponin is released, allowing tropomyosin again to inhibit cross-bridge attachment, so the muscle relaxes. The muscle remains relaxed until the next depolarization.

THE TRANSIENT PRODUCTION OF FORCE

Up to now, we have considered the mechanics of myosin cross-bridges that have been activated maximally. As we've seen already, there is a delay, or **latent period,** be-tween the action potential in a muscle fiber and the generation of force by the fiber (see Figures 10-18B and 10-23B). The latent period includes all of the time that is required for initiation of an action potential in the muscle fiber, propagation of the AP along the T tubules into the fiber, release of Ca^{2+} from the sarcoplasmic reticulum and diffusion of the Ca^{2+} ions to troponin molecules, binding of Ca^{2+} to troponin, activation of myosin cross-bridges and their binding to actin thin filaments, and generation of force. The time required for all of these processes is short: The delay from the peak of the AP to the first sign of tension in some muscles can be as short as 2 ms. Now we consider the mechanical properties of muscle fibers as they become activated and generate tension and then relax in the body.

Series Elastic Component

A muscle can be modeled as a contractile element that is arranged in parallel with one **elastic** component and in series with another elastic component as depicted in Figure 10-26A. (According to Hooke's law, the length of an object with ideal elasticity increases in proportion to the force applied.) The parallel elastic component in this model represents the properties of the plasma membrane of the muscle fibers and connective tissues that run in parallel with the muscle fibers. The **series elastic component,** also called the *series elastic elements,* represents tendons, connective tissues that link muscle fibers to the tendons, and perhaps the Z disks of the sarcomeres. An additional important constituent of the series elastic elements appears to be the myosin cross-bridge links themselves, which appear to undergo some stretch in response to tension (see Figure 10-11B). Representing all of the elastic components by only two components greatly simplifies the model, making it easier to manipulate mathematically while maintaining sufficient accuracy that the model can

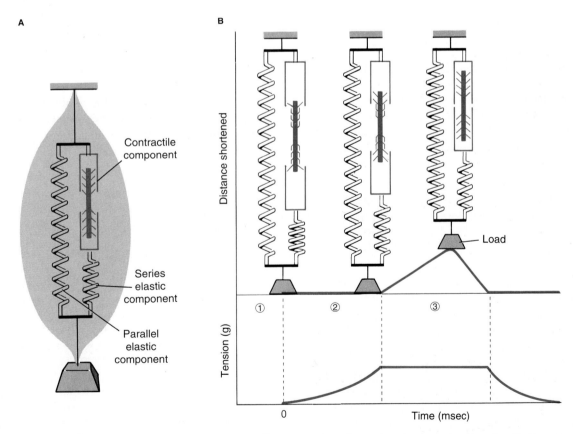

Figure 10-26 A muscle fiber, or an entire muscle, can be represented by a mechanical model that includes a contractile component and elastic components. **(A)** Mechanical model of a muscle consisting of a contractile component (the sarcomeres), in series with one elastic component (e.g., tendons) and in parallel with another elastic component (e.g., the outer membrane). **(B)** Role of series elastic component in muscle contraction. At the beginning of contraction in this model muscle, the weight rests on a surface (1). As the thick and thin filaments begin to slide past one another and tension increases, the series elastic components are stretched (1 ⟶ 2), but the length of the muscle has not yet changed; contraction up to this point (2) is isometric. Once the muscle generates tension that is equal to the weight of the load, the load is lifted and the contraction becomes isotonic (3). Note that as contraction progresses, the thick and thin filaments overlap increasingly and more cross-bridges become active. [Adapted from Vander et al., 1975.]

increase our understanding of the mechanics of muscle contraction.

As illustrated in Figure 10-26B, as the muscle becomes activated and the contractile component begins to shorten, the series elastic component must first be stretched before tension can be transmitted to the external load (steps 1 and 2). When the tension developed in the series elastic component equals the weight of the load, the muscle begins to shorten, and it lifts the load off of the surface (step 3). In steps 1 and 2, the contraction is isometric, whereas in going from step 2 to 3 it becomes isotonic as the load is finally lifted. If the load were sufficiently heavy that the muscle never produced tension equal to the weight of the load, the contraction would have remained isometric throughout. At maximal tension during an isometric contraction, the small shortening of the contractile component stretches the series elastic component by an amount equivalent to about 2% of the muscle length, even though the external length of the muscle does not change.

It takes time for the thin and thick filaments to slide past each other by cross-bridge activity as the series elastic component becomes stretched and tension builds up. Thus the series elastic component acts to slow the development of tension in the muscle and to smooth out abrupt changes in tension.

The Active State

During contraction, external shortening of a fiber and production of tension reach a maximum within 10 to 500 milliseconds, depending on the kind of muscle, the temperature, and the load. At first glance, this statement might suggest that the contractile mechanism is activated with a similar slowly rising time course. It is important, however, not to confuse the time it takes a *muscle* to develop tension with the time course of *cross-bridge* activity. Cross-bridges become activated and attach to thin filaments before the filaments begin to slide past one another. In addition, when the filaments slide, they must first take up the slack in the series elastic component before tension can be fully developed.

The state of the cross-bridges after activation, but before the muscle has had a chance to develop full tension, can be determined by application of *quick stretches* with a special apparatus. These stretches can be applied at various times after stimulation and before and during contraction. The rationale for quick-stretch experiments is that when the stretch is applied, it stretches the series elastic component, eliminating the time that is normally required for the contractile mechanism to take up the slack. The maneuver thus improves the time resolution of measuring the state of cross-bridge activity. The "internal" tension recorded by the sensing device during a quick stretch represents the tensile strength of the bonds between the thick and thin filaments, which depends on the holding strength of the cross-bridges at the instant of stretch. If the stretch applied is stronger than the holding strength of the cross-bridges, the cross-bridges will slip, and the filaments will slide past each

other. Thus, the loading during a quick stretch that is just necessary to make the thick and thin filaments slide apart approximates the load-carrying capacity of the muscle at the time of stretch. This tension should be proportional to the average number of active cross-bridges per sarcomere.

In the relaxed state, muscle has very little resistance to stretch other than the compliance of connective tissue, the sarcolemma, and other elastic components. Quick-stretch experiments revealed that after stimulation of a muscle, its resistance to stretch rises steeply and reaches a maximum at about the time when external shortening or tension in the unstretched muscle is just getting under way. After a brief plateau, the load-carrying ability decreases to the low level characteristic of the relaxed muscle.

The term **active state** is used to describe the increase in load-carrying ability (i.e., tension) of the muscle that is measured in quick-stretch experiments following a brief stimulation (Figure 10-27A). The active state corresponds to the formation of bonds between myosin cross-bridges and actin thin filaments and to the subsequent small internal shortening generated by the cross-bridges. Because cross-bridge activity is controlled by the concentration of free Ca^{2+} in the myoplasm, the time course of the active state is believed to approximately parallel the changes in myoplasmic Ca^{2+} concentration following stimulation. The brief increase in tension due to cross-bridge activity is called a twitch.

If stimulation of a muscle is prolonged, the active state persists. The prolongation of the active state by a barrage of high-frequency APs is called **tetanus**. In this state, the measurable external isometric tension can increase until it reaches the value of the active state measured by quick-stretch experiments (Figure 10-27B). The difference between twitches and tetanus are considered in the next section.

Twitches and Tetanus

The graph in Figure 10-27A raises a question: Why is the maximal external isometric tension produced by the muscle during a twitch so much lower than the internal tension associated with the active state? In other words, during a brief contraction, why does the muscle produce so much less tension than it is actually capable of producing?

During a single twitch, the active state is rapidly terminated by the calcium-sequestering activity of the sarcoplasmic reticulum, which efficiently removes Ca^{2+} from the myoplasm soon after it is released. Thus, the active state begins to decay before the filaments have had time to slide far enough even to stretch the series elastic component to a fully developed tension. For this reason, the tension of which the contractile system is capable cannot be realized in a single twitch.

Before the peak of the twitch tension, the contractile elements store potential energy in the series elastic elements by progressively stretching them. If a second AP follows the first before the sarcoplasmic reticulum can entirely remove the previously released Ca^{2+} from the myoplasm, the con-

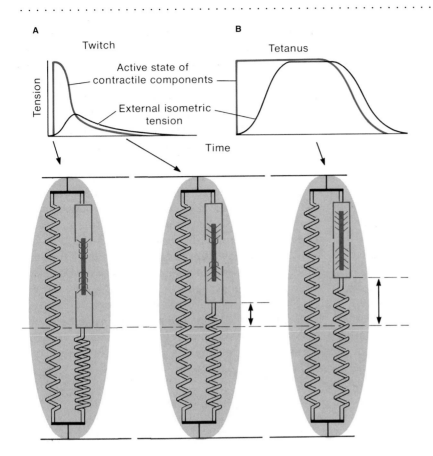

Figure 10-27 The time course of the active state differs from the time course of tension production. **(A)** The active state—as measured in quick-stretch experiments—develops rapidly in response to a brief stimulus. This kind of brief response is called a twitch. Measurable external isometric tension develops considerably more slowly and fails to reach the same tension that can be measured during a quick stretch. **(B)** In response to a prolonged stimulus, a tetanic contraction (tetanus) develops. In this case, the external isometric tension has time to reach the same value as the internal tension measured during quick-stretch experiments. [Adapted from Vander et al., 1975.]

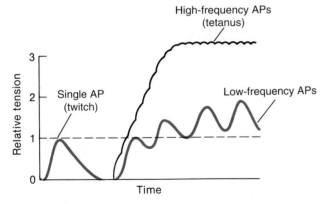

Figure 10-28 Twitches fuse when the stimulating APs arrive in rapid succession. A single AP produces a single twitch. If a set of low-frequency APs is conducted along the muscle fiber, each successive twitch begins before the muscle has time to return to its relaxed condition following the previous twitch. At a maximum frequency, the twitches fuse with one another, producing a long, strong contraction called tetanus.

by which individual twitches fuse can vary, reaching a maximal value in tetanus (Figure 10-28).

ENERGETICS OF MUSCLE CONTRACTION

In muscle contraction two major processes require the expenditure of energy. The most obvious one is the hydrolysis of ATP by myosin cross-bridges as they cyclically attach to and detach from actin thin filaments (see Figure 10-10). The other major process that uses energy is the pumping of Ca^{2+} released into the myoplasm during excitation-contraction coupling back into the sarcoplasmic reticulum against a Ca^{2+} concentration gradient (see Figure 10-24, step 3).

Biochemical studies have determined that 2 molecules of ATP are required to pump each Ca^{2+} ion into the sarcoplasmic reticulum. During a twitch, a certain amount of Ca^{2+} is released in a few milliseconds after the AP, and exactly that amount of Ca^{2+} must eventually be pumped back into the SR if the muscle fiber is to remain healthy. During tetanus (just as in a twitch), the calcium pumps immediately start to resequester the Ca^{2+} released during the first stimulus; however, they do not have time to remove all of it from the myoplasm, and each successive AP causes more Ca^{2+} to be released. The buildup of Ca^{2+} in the myoplasm keeps the troponin saturated with Ca^{2+} until the APs cease; at that point, the calcium pumps eventually can resequester all of the released Ca^{2+} back into the SR lumen. To

centration of Ca^{2+} remains high in the myoplasm, and the active state is prolonged. With the active state prolonged, isometric tension continues to increase with time until the tension produced by the internal shortening of the contractile components and the stretching of the series elastic component is just sufficient to cause cross-bridges to slip and prevent further shortening of the contractile components. The muscle has then reached full *tetanus tension*. Depending on the repetition rate of muscle APs, the amount

maintain the condition of tetanus, ATP is steadily hydrolyzed by both the myosin ATPase and the calcium pumps in the membrane of the SR.

ATP Usage by Myosin ATPase and Calcium Pumps

The relative consumption of ATP by myosin ATPase and the calcium pumps in the SR membrane was determined in experiments with tetanized frog muscle. In these experiments, the muscle preparations were stretched to different degrees producing different amounts of overlap between thick and thin filaments. As the muscle is stretched more and more, fewer cross-bridges can interact with actin, reducing both the amount of force that can be produced and the amount of ATP hydrolyzed by the myosin ATPase. (Remember that myosin can hydrolyze ATP on its own, but the resulting ADP and P_i are released much more slowly than they are when the myosin is bound to actin. As a result, if myosin cross-bridges have few available sites where they can bind to actin, the ATPase activity of the filaments will be low.) By contrast, stretching the muscle should have little or no effect on the rate at which Ca^{2+} is released from and resequestered by the sarcoplasmic reticulum, because these processes are mediated by membrane proteins whose activity is unrelated to the amount of myofilament overlap. Thus, as the muscles become increasingly stretched, the total ATPase activity will decline, and at the length where the myofilaments no longer overlap any ATPase activity measured will be due entirely to the calcium pumps.

Using this experimental approach, researchers found that the calcium pumps accounted for about 25%–30% of total ATPase activity during muscle contraction. It is generally assumed that this percentage is constant for all muscles; that is, muscles with a higher maximal contraction velocity also have faster calcium pumps in their SR. It is possible, however, that in the very fast sound-producing muscles discussed later in this chapter, pumping of Ca^{2+} may account for a larger fraction of the overall energy usage of the muscle.

Regeneration of ATP during Muscle Activity

As the previous discussion indicates, muscles use ATP exclusively to power muscle contraction. Yet early determinations of the overall ATP usage during muscle contraction produced a surprising result: The ATP concentrations in stimulated and unstimulated muscles (matched as closely as possible) were nearly identical. For many years this finding caused some muscle physiologists to hypothesize that muscles do not actually use ATP to power contractions. However, an alternative explanation, which turned out to be correct, was that in addition to ATP, muscle fibers contain a second high-energy molecule. Eventually this molecule was identified as **creatine phosphate,** also known as **phosphocreatine** (see Figure 3-39). Within muscle fibers, the enzyme creatine phosphokinase transfers a high-energy phosphate from creatine phosphate to ADP, regenerating ATP so quickly that the ATP concentration remains constant. Because of this reaction, accurately measuring the

amount of ATP hydrolyzed by the muscle is best done by measuring either the drop in the concentration of creatine phosphate or the rise in the concentration of P_i.

Beyond the technical issue of accurately measuring the rate of ATP hydrolysis, the creatine phosphokinase reaction is extremely important for effective muscle function. If a muscle runs out of ATP, it goes into rigor (see Figure 10-17B). It is, therefore, essential for survival that the concentration of ATP in muscles be buffered. During endurance activities, oxidative and anaerobic metabolism can generate ATP rapidly enough to maintain a sufficient ATP concentration to power muscle contraction. However, during high-intensity, short-duration activity (e.g., when an animal sprints to run down prey or to avoid becoming prey), the concentration of ATP within muscles is maintained constant by continuous rephosphorylation of ADP by the creatine phosphokinase reaction (Figure 10-29).

The concentration of creatine phosphate in muscle fibers (20–40 mM) is many times larger than the reserve of ATP (about 5 mM in muscle fibers). As a result, an animal can use the large reserve of high-energy phosphate in crea-

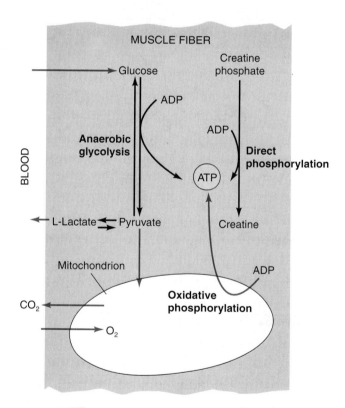

Figure 10-29 The ATP that provides energy for muscle contraction comes from several different sources. In direct phosphorylation, high-energy phosphates are transferred from creatine phosphate to ADP, regenerating ATP. The concentration of creatine phosphate in muscle fibers is very much higher than the concentration of ATP, so creatine phosphate effectively buffers the ATP concentration. Anaerobic glycolysis hydrolyzes glucose, rephosphorylating ADP in the process. Lactate accumulates as a by-product. Oxidative phosphorylation of ADP regenerates ATP, but is slower than the other two processes and requires O_2 to proceed. Red arrows indicate the transfer of materials into, out of, or between compartments in the fiber. Black arrows indicate chemical reactions. (See Chapter 3 for more details.)

tine phosphate to power muscle contraction until anaerobic and oxidative metabolism start to generate ATP, allowing it to locomote for a much longer time than it could just on its ATP. An animal's survival may depend on this extra source of energy. Moreover, the creatine phosphokinase reaction holds the ATP concentration nearly constant while supplying extra energy. The ATP concentration is stabilized because a large equilibrium constant greatly favors phosphorylation of ADP by creatine phosphate. Under most conditions, only the concentration of creatine phosphate falls within a muscle, while the concentration of ATP remains nearly constant.

FIBER TYPES IN VERTEBRATE SKELETAL MUSCLE

The muscular systems of animals perform a great variety of motor tasks, ranging from very fast movements that are over in 50–100 milliseconds to long-distance migrations and from sustained tetanic contractions to the rapid contractions of sound production, which occur at a frequency of several hundred per second. Even a casual observer notices the diversity in the external attributes of muscular systems such as wings, fins, and legs. There is an equally impressive diversity in the characteristics of the muscles themselves. To produce such a broad range of activities, muscles must be organized to perform very differently. Recent experiments have shown that the properties of the muscle often are found to be very well matched to the other components in a system, optimizing the system for its biological function. To appreciate how well a muscle is adapted to its biological role, it is necessary to examine its properties in light of the job it must perform.

Classification of Fiber Types

The skeletal muscles of vertebrates consist of muscle fibers of more than one type. Some contain a high proportion of tonic fibers specialized for slow steady contraction; these are most useful for maintaining postural tone. Other muscles contain a high percentage of twitch fibers, which are specialized for rapid movements of limbs. These different kinds of muscle fibers can be distinguished on the basis of biochemical, metabolic, and histochemical criteria.

Among the properties that distinguish the various fiber types are the following:

- The electrical properties of the membrane determine if a fiber will respond with an all-or-none twitch or with a graded contraction. If the membrane produces APs, the fiber will contract with all-or-none twitches.

- The maximal rate of contraction, V_{max}, is determined by the rate at which cross-bridges detach from the actin thin filaments (which in turn is determined by the nature of the myosin heavy chains).

- The time during which myoplasmic free Ca^{2+} remains elevated following an AP depends primarily upon the density of calcium-pump molecules in the membrane of the sarcoplasmic reticulum.

- The number of mitochondria and the density of a fiber's blood supply determine its rate of oxidative ATP production and hence its resistance to fatigue.

Based on these and other properties, four major groups of vertebrate skeletal muscle are recognized—tonic fibers and three types of twitch (or phasic) fibers (Table 10-1).

Tonic muscle fibers contract very slowly and do not produce twitches. They are found in the postural muscles of amphibians, reptiles, and birds, as well as in the muscle spindles and extraocular muscles of mammals. Tonic fibers normally do not produce APs and, in fact, they do not require an AP to spread excitation, because the innervating motor neuron runs the length of the muscle fiber and makes repeated synapses all along. In these muscle fibers, the myosin cross-bridges attach and detach very slowly, accounting for their extremely slow shortening velocity and ability to generate isometric tension very efficiently.

Slow twitch (or *type I*) fibers contract slowly and fatigue slowly; they are found in mammalian postural muscles. They are characterized by a slow-to-moderate V_{max} and slow Ca^{2+} kinetics. They generate all-or-none APs, so they contract in response to motor neuron input with all-or-none twitches. Like other twitch fibers, type I fibers typically have one or at most a few motor endplates; in

TABLE 10-1
Properties of twitch (phasic) fibers in mammalian skeletal muscles

Property	Slow oxidative (type I)	Fast oxidative (type IIa)	Fast glycolytic (type IIb)
Rate of contraction (V_{max})	Slow	Fast	Fast
Myosin ATPase activity	Low	High	High
Resistance to fatigue	High	Intermediate	Low
Capacity for oxidative phosphorylation	High	High	Low
Enzymes for anaerobic glycolysis	Low	Intermediate	High
Number of mitochondria	Many	Many	Few
Fiber diameter	Small	Intermediate	Large
Force per cross-sectional area	Low	Intermediate	High

Source: Adapted from L. Sherwood, 1993.

mammals all of the endplates on a single fiber are made by a single motor neuron. Slow twitch fibers are used both for maintaining posture and for moderately fast, repetitive movements. They fatigue very slowly for two reasons: They contain a large number of mitochondria and have a rich blood supply bringing plenty of oxygen, allowing them to depend on oxidative phosphorylation, and they use ATP at a relatively slow rate. They are also characterized by a reddish color (examples are the dark-colored meat of fish and fowl) because they contain a high concentration of the oxygen-storage protein **myoglobin** (see Chapter 13). Muscles that contain a high proportion of this type of fiber are often called *red muscle*.

Fast twitch oxidative (or *type IIa*) fibers have a high V_{max} and activate quickly. With their many mitochondria, they can produce ATP quickly by oxidative phosphorylation and thus fatigue slowly. These muscles are prominent in the flight muscles of wild birds; they are specialized for rapid repetitive movements, as in sustained, strenuous locomotion.

Fast twitch glycolytic (or *type IIb*) fibers contract rapidly and fatigue quickly. They have a high V_{max}, and they both activate and relax quickly because of their rapid Ca^{2+} kinetics. These fibers are generally recruited when very rapid contraction is required. Because these fibers contain few mitochondria and thus depend on anaerobic glycolysis to generate ATP, they fatigue rather quickly. A familiar example of this type of muscle is found in the white breast muscles of domestic fowl, which are never used for flying and cannot produce sustained activity. Ectothermic vertebrates, such as amphibians and reptiles, also make extensive use of glycolytic muscle fibers.

These categories are somewhat arbitrary, because some muscle fibers combine properties of different types. In addition, the absolute values for many of the parameters vary among species. For instance, the slow twitch fibers of a mouse have a faster V_{max} than do fast twitch oxidative fibers of a horse. Within a given muscle, however, the fiber types can be distinguished by their histological properties. For example, histochemical staining reveals differences in the properties of myosin ATPase in different fiber types (Figure 10-30). Another useful method for distinguishing fiber types is based on the abundance of oxidative enzymes such as succinic dehydrogenase.

Functional Rationale for Different Fiber Types

Although the properties of the various fiber types may seem quite different, they all are composed of the same basic building blocks and use the same mechanism for contraction. However, the various fiber types do differ in some molecular properties (e.g., length of myofilaments, detachment rate of myosin cross-bridges, and number of calcium pumps in the SR membrane) that can affect the overall contractile properties of a muscle, which is composed of many fibers.

What do animals gain by having different types of muscle fibers? Fast fibers obviously are necessary if an animal is

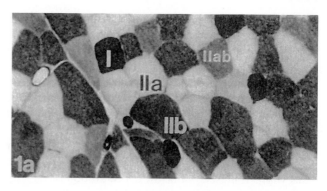

Figure 10-30 Histochemical staining for myosin ATPase activity reveals different types of fibers within a single muscle. This photograph of a section through a muscle from a horse contains slow oxidative (type I), fast oxidative (type IIa), and fast glycolytic (type IIb) fibers. Type IIab has intermediate properties. [Courtesy of L. Rome.]

to move fast, but then why have slow muscle fibers? A basic principle in muscle physiology is that there is always a trade-off between *speed* and *energetic cost*. Very fast muscles require a large amount of ATP. In contrast, slow muscles perform less rapidly, but they also require relatively little energy. To better understand this trade-off, it is useful to compare the energetic cost and mechanical abilities of fiber types that have different values of V_{max}.

The technique for measuring energy utilization by muscle that has the highest time resolution, and the one on which many conclusions about muscle energetics have been based, is to measure heat. The hydrolysis of ATP by muscles is an exothermic reaction, so as a result of the reaction, some heat is released. During a typical contraction, this heat warms up the muscle by a very tiny amount, about $0.001-0.01°C$. Very fast and very sensitive thermometers called thermopiles, which measure the heat without removing it from the muscle, can be used to monitor the heat generated by a muscle with very high resolution. In theory, the amount of ATP hydrolyzed by a muscle can be calculated by measuring the work done during contraction and comparing it with the enthalpy of ATP. Unfortunately, during contraction heat is absorbed and produced by many other chemical reactions and physical processes (e.g., stretching elastic elements) that are unrelated to the hydrolysis of ATP. For this reason, it is impossible to precisely relate heat production to the use of ATP. Nonetheless, measurements of heat have yielded considerable insight into how muscles use energy during contractions.

A muscle fiber's mechanical properties (i.e., force generation and power production) and energetic properties (i.e., rate of ATP use and efficiency) depend on both V and V/V_{max}. For a given rate of shortening, V, the force and mechanical power produced per cross-sectional area can be considerably higher in a fiber with a high V_{max} than it is in a slower fiber (Figure 10-31A,B); furthermore, the generation of force and power are maximal at intermediate values of V/V_{max}. It therefore takes fewer high-V_{max} fibers than low-V_{max} fibers to generate the same force or power.

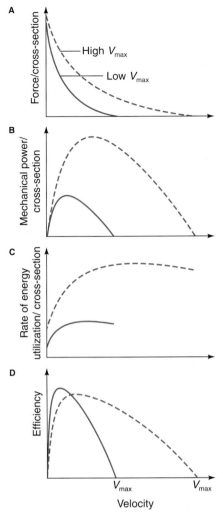

Figure 10-31 Force, power, rate of energy utilization, and efficiency vary as a function of shortening velocity. Fibers with a high V_{max} can generate more force (**A**) and mechanical power (**B**) than those with a low V_{max}, but they also use more energy at all shortening velocities (**C**). The efficiency of contraction is calculated as the power output divided by the energy used (**D**). Note that low-V_{max} fibers are more efficient at low shortening velocities, whereas high-V_{max} fibers are more efficient at higher rates of shortening. These curves were derived from heat, oxygen, and mechanical measurements of frog muscle contraction. [Adapted from Hill, 1964; Hill, 1938; and Rome and Kushmerick, 1983.]

It might, then, seem advantageous to have only muscle fibers with high values of V_{max}. There is, however, an energetic price paid for a high V_{max}. Measurements of heat liberated and high-energy phosphate hydrolyzed show that utilization of ATP is also a function of V and V/V_{max}. The rate at which ATP is hydrolyzed increases with increasing V/V_{max} up to a maximum and then decreases as V/V_{max} approaches 1 (Figure 10-31C). This increase in the rate at which ATP is hydrolyzed can be understood in terms of the Huxley model of cross-bridge function (see Figure 10-11). As the muscles shorten faster and faster, cross-bridges detach faster and hence go through ATP molecules faster. Notice in Figure 10-31C that the rate at which ATP is used is considerable higher in fibers with a high V_{max} than it is in fibers with a low V_{max} at all rates of shortening.

There is, thus, an adaptive balance between the mechanics and the energetics of contraction. From the combination of mechanical and energetic data the *efficiency* of muscle contraction (defined as the ratio between mechanical power output and energy utilization) can be calculated. Efficiency turns out also to be a function of V/V_{max} (Figure 10-31D). Fibers with a low V_{max} are more efficient at low rates of shortening; at higher rates of shortening, however, fibers with a high V_{max} are more efficient. As a consequence, if an animal is to produce both slow and rapid movements efficiently, it needs to have both kinds of fibers and to use the fibers appropriately to produce the two kinds of movements.

ADAPTATION OF MUSCLES FOR VARIOUS ACTIVITIES

The principles that determine the mechanical properties of muscles are illustrated in three very different kinds of motor activity: jumping of frogs, swimming of fish, and sound production by toadfish and rattlesnakes. We will consider each of these activities and the muscles that are used to produce them. The discussion will focus on three features of a working muscle:

- The amount of overlap between thick and thin filaments (where on its length-tension curve the muscle is working)

- The relative velocity of shortening, V/V_{max}, which determines the power and efficiency of the muscle

- The activation state of the muscle

In this section we will draw heavily from work byLawrence Rome and his colleagues, who have contributed significantly to our understanding of the comparative physiology of muscles.

Adaptation for Power: Jumping Frogs

When they jump, frogs move rapidly (in 50 to 100 milliseconds) from a crouched position, in which their potential and kinetic energy is zero, to an extended position, in which their potential and kinetic energy is high. Mechanical work must be done if the potential and kinetic energy of a body is to increase, and because that work must be performed in such a short time, the muscles that produce the jump must generate high power (i.e., work/unit time). In fact, the distance a frog jumps depends directly on how much power its muscles produce. Thus we might expect that a frog's jumping muscles would have properties that allow them to generate high power.

From our earlier discussion, we know that a muscle generating high power exhibits three properties: (1) it operates within the plateau of the sarcomere length–tension curve where maximal force is generated (see Figure 10-8B); (2) it shortens at a rate at which maximal power is generated (see Figure 10-13D); and (3) it becomes maximally activated before shortening begins. To determine whether a frog's jumping muscles actually have these properties,

G. Lutz and L. C. Rome observed frogs *(Rana pipiens)* as they jumped and also studied isolated frog muscles, integrating the results of the two kinds of experiments.

Length-tension relation

To examine the length-tension relation of muscles during a jump, Lutz and Rome measured the length and changes in length of the semimembranous muscle, a hip extensor. The measurements were made both while an intact frog jumped and in an isolated limb whose position was manipulated to match the shape of a jumping frog's leg (Figure 10-32). By plotting the length changes versus hip angle, they determined the *moment arm* of the muscle. (From physics, the moment arm is the distance separating a fixed fulcrum from a point at which a force is exerted that will tend to rotate a mass around the fixed point, as illustrated in the inset in Figure 10-32.) In this case, the moment arm—that is, the distance between the hip joint and the location where the muscle attaches to the pelvis—is crucial because it determines both the mechanical advantage of the muscle and the length changes it must make to produce any given change in the angle of the hip joint.

The length of sarcomeres in the hip flexor was determined when the hip was in the crouched position and the extended, jumping position. During a jump, the sarcomere length goes from 2.34 μm in the crouched position to 1.82 μm at the point of take-off. To determine where these lengths fall in the sarcomere length-tension relation, they were compared with a length-tension curve from a closely related species of frog *(Rana temporaria)* as shown in Figure 10-33A. The measured lengths of the hip flexor sarcomeres fell along the plateau of the sarcomere length-tension curve; thus, as expected for a power muscle, the fibers of the hip flexor operate very near their optimal power production during a jump. It has been calculated that this muscle generates at least 90% of its maximal tension throughout the jump. In addition, because of more subtle properties, if the initial sarcomere length were either longer or shorter than the measured length, the muscle would do less work.

A number of factors must be matched to produce this optimal behavior. The lengths of the myofilaments and the number of sarcomeres per muscle fiber must combine so that there is optimal overlap of the thick and thin filaments when the frog is in the crouched position. In addition, given the change in the angle of the hip joint that occurs during jumping, the moment arm of the hip joint must allow the muscle and its sarcomeres to undergo the appropriate changes in length.

Value of V/V$_{max}$

The V_{max} of the hip flexor muscle is about 10 muscle lengths per second, and it generates maximal power at 3.44 muscle lengths per second (Figure 10-33B). The mean rate at which the muscle shortens during a jump is 3.43 muscle lengths per second, that is, at a V/V_{max} of 0.32, precisely the rate at which the muscle produces maximal power. Thus the frog's muscles, joint configuration, and mass are all

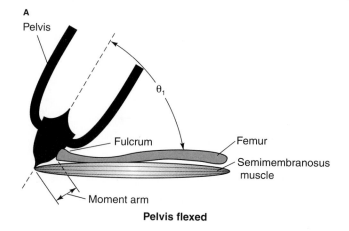

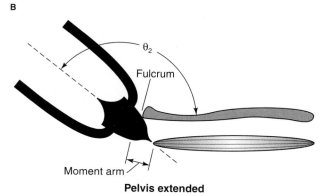

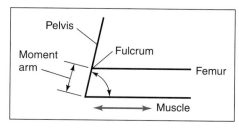

Figure 10-32 When the hip flexor muscle contracts, the hip joint rotates around the attachment point between the pelvis and the femur. **(A)** When the frog is in a crouched position, the angle of the hip joint (θ_1) is small and the extensor muscle (the semimembranous muscle) is relaxed. **(B)** When the frog jumps, the angle of the hip joint increases (θ_2), because contraction of the muscle acts through the moment arm indicated on the diagram and pulls on the pelvis. *Inset:* Schematic diagram illustrating the mechanical components contributing to jumping. [Adapted from Lutz and Rome, 1996.]

matched to allow the hip flexor muscle to shorten at a V/V_{max} appropriate for maximal generation of power.

State of activation

Even if the hip flexor muscle begins to contract at the optimal sarcomere length and shortens at an optimal rate allowing maximal power to be generated, it also must be maximally activated to generate maximal power. If the muscle started to shorten before it became fully activated, it would generate a force lower than the maximum possible at that velocity (i.e., the actual force would lie below the force-velocity curve) and the power would also be lower than the maximum. As discussed earlier, the time it takes for activation to occur depends on the rate at which Ca^{2+} is released

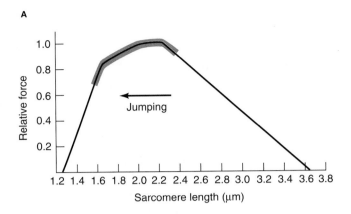

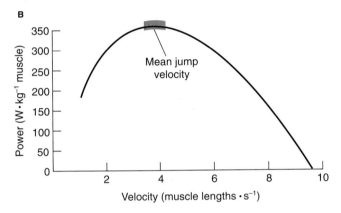

Figure 10-33 The mechanics of the hip flexor muscle of a frog operate optimally during a jump. **(A)** At the beginning of a jump, the sarcomeres in the semimembranous muscle are 2.34 μm long and they shorten to 1.83 μm during the jump (red portion of curve). Even at the shortest sarcomere length, the muscle still generates over 90% of its maximal tension. **(B)** At the velocity used during jumping, the muscle operates in the portion of the power curve (red) in which at least 99% of maximal power is generated. The velocity of shortening is expressed in terms of muscle-lengths per second to take into account the difference in length encountered among muscles. [Adapted from Lutz and Rome, 1994.]

and binds to troponin, and the rate of cross-bridge attachment. If the frog hip flexor is to become maximally activated before shortening begins, activation must occur rapidly and movement of the hip joint must be delayed until activation is complete (which will depend on the mass of the frog).

One way to determine if the flexor muscle shortens only after it becomes maximally activated would be to measure the force and power generated by the muscle in a frog as it jumped. However, such a measurement would require that force transducers be implanted into the frog to measure the behavior of a single muscle, a strategy that is not yet technically practical. In an alternative approach, the length of the hip flexor muscle and the electrical activity of fibers within the muscle are measured as carefully as possible in an intact frog, and these values are then reproduced in a muscle that has been removed from the frog's body.

This second approach was used in the experiment depicted in Figure 10-34. Electrical activity of the muscle in

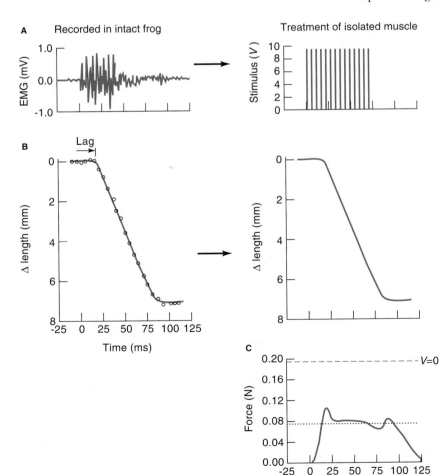

Figure 10-34 Electrical activity patterns and length changes recorded from muscles in an intact frog can be imposed upon isolated muscles to study the force generated. **(A)** Electromyogram recorded from the semimembranous muscle of a frog during a jump *(left)*, and the abstracted pattern of stimulation that was imposed on an isolated muscle *(right)*. **(B)** Rate at which the muscle shortened in an intact frog during jumping and the change in length imposed on the isolated muscle while it was being electrically stimulated as shown in part A. **(C)** Force produced by the isolated muscle (solid line) during the experimental manipulation mimicking the stimulation and shortening rate observed during a jump in an intact frog. The dashed line shows the isometric force produced by this muscle when $V = 0$, and the dotted line shows the force that would be expected in the same muscle contracting at the imposed shortening velocity during a force-velocity experiment similar to the one shown in Figure 10-31A. [Adapted from Lutz and Rome, 1994.]

an intact frog was measured by tiny electrodes implanted in the muscle; such electrodes record APs in muscle fibers, just as extracellular electrodes record APs in nerve bundles (see Spotlight 6-1). The record obtained from these electrodes is called an *electromyogram* (EMG). The APs in the fibers within a muscle are not synchronous, and the amplitude of the signal from any particular fiber depends on how close the fiber is to the electrode, so an EMG recording can appear very complex. However, the pattern of APs in the biggest units recorded by the EMG electrode can be abstracted from the record, and an isolated muscle can then be stimulated electrically using that pattern (see Figure 10-34A). In addition, the temporal pattern of length change in the hip flexor muscle was measured while an intact frog jumped, and this pattern of length change was imposed upon the isolated muscle while it was simultaneously being driven electrically (see Figure 10-34B). When treated in this way, the isolated hip flexor muscle generated the maximal force expected at the imposed shortening velocity (see Figure 10-34C), strongly suggesting that it is maximally activated during jumping. The implication of this result is that its molecular components of activation in this muscle are strikingly matched to its mechanical properties.

Diversity of Function: Swimming Fish

The study of muscles in fish has, for two reasons, been particularly useful in helping to elucidate how muscular systems are organized. First, fish display a wide diversity of movements that can be elicited readily and analyzed quantitatively. Second, different kinds of movements are powered by different muscle fiber types, which in fish are anatomically separated, permitting the activity of different fiber types to be monitored by electromyogram electrodes (Figure 10-35). (As noted earlier, the muscles in most vertebrates contain more than one type of fiber, making electrical monitoring of activity in one particular fiber type difficult or impossible.)

During the many movements of which fish are capable, the change in sarcomere length is roughly proportional to the curvature of the spine. While a carp is swimming at a velocity of 25 cm·s⁻¹, the curvature changes very little along most of its spine (Figure 10-36A), indicating that the length of sarcomeres along its body changes very little. In contrast, when the fish is startled—for example, by a loud sound—and produces an escape response, its spine curves, indicating that sarcomeres have shortened on one side of the body and lengthened on the other (Figure 10-36B). Notice the difference in time scale between swimming and the escape response. During steady swimming, one tailbeat takes about 400 ms, whereas in the escape response, the body of the fish changes from straight to highly curved in only 25 ms.

The muscles of a fish must, then, be able to generate both slow, low-amplitude movements and fast, high-amplitude movements. Earlier in this chapter, it was argued that muscles must be finely tuned to a particular activity in

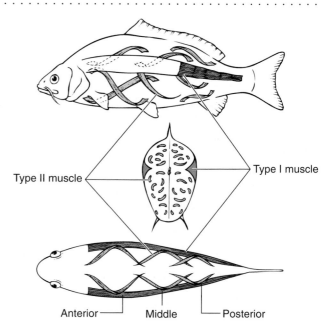

Figure 10-35 In fish, muscle fiber types are anatomically separate from one another, facilitating electromyographic monitoring of activity in specific fiber types. These diagrams show the arrangement of two muscle fiber types in a carp. Type I (slow oxidative) muscle fibers (in red) are found in red muscles, which lie in a thin layer just under the skin; the thickness of this layer has been enlarged in this drawing. These muscles run parallel to the body axis, so the change in sarcomere length during shortening is related to the curvature of the spine and to the distance separating the muscle layer from the spine. Type IIb (fast glycolytic) fibers (in gray) compose the white muscles, which are located deeper in the body. These muscles run helically, rather than parallel to the long axis of the body. Because of their anatomic arrangement, white muscles need to shorten only about 25% as much as red muscles do to produce a particular change in the curvature of the body. [Adapted from Rome et al., 1988.]

order to perform optimally, but these two kinds of behavior seem to require very different properties. Can muscles perform such different tasks, while at the same time the active fibers operate optimally? If they can, *how* do they do it?

Electromyograms recorded from fish while they either swam or responded to a loud sound revealed that different muscles, containing different fibers types, were active during the two very different behaviors. The separation of fiber types into distinctly different muscles in fish facilitated this conclusion. When a fish is swimming steadily, only *red muscles* are active; these are composed of slow oxidative (type I) fibers. In contrast, white muscles, composed of fast glycolytic (type IIa) fibers, are recruited to produce fast swimming or very rapid movements such as the escape response. A fish can produce very different kinds of movements well, because for each movement it uses muscles that are specialized to match the demands of the particular task. Let's examine the same three properties of these fish muscles that we considered for the hip flexor muscles of a frog.

Length-tension relation

The conclusion, stated earlier, that sarcomere length can be related directly to the curvature in the body of a fish is

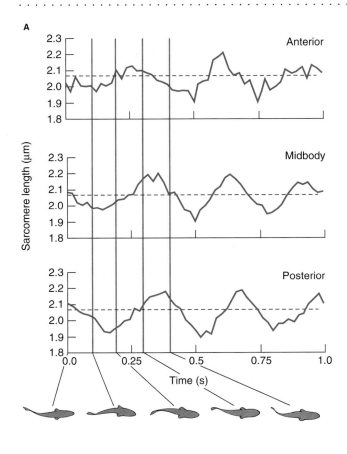

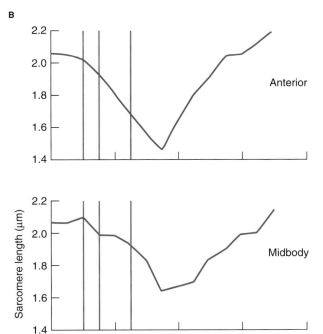

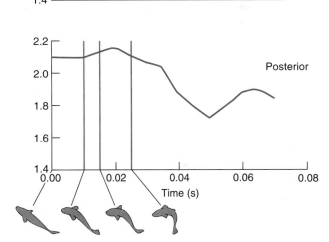

Figure 10-36 Swimming movements and the escape response in fish differ greatly in magnitude and time course, reflecting differences in changes of sarcomere length. Shown here are the calculated changes in sarcomere length within muscles located on one side of the anterior, midbody, and posterior of a carp engaged in two activities: **(A)** swimming steadily at 25 cm·s^{-1} or **(B)** escape response elicited by a loud sound. The changes in sarcomere length were calculated from the shape of the fish at each time-point in a video tape recording of the behavior. The shape of the fish body at selected time points is indicated in figures below the graph. Type I (slow oxidative) muscles are active during steady swimming, whereas type IIb (fast glycolytic) muscles produce the escape response. [Part A adapted from Rome et al., 1990a; part B adapted from Rome et al., 1988.]

based upon measurements of sarcomere length made in fish that had been frozen into the shapes assumed by living fish as they perform different behaviors. These measurements indicate that the sarcomeres in the red muscles of slowly swimming fishes repeatedly vary in length between 1.89 μm and 2.25 μm, centered around a sarcomere length of 2.07 μm (see Figure 10-36A). These values then need to be compared with the length-tension curve for fish sarcomeres to determine if the thick and thin filaments maintain optimal overlap while the sarcomeres undergo these changes in length. Electron microscopic examination of red and white muscles from carp reveal that the lengths of myofilaments in fish muscles are nearly identical with the lengths of these filaments in frog muscle. This finding indicates that the sar-

comere length-tension curve for frog muscle provides a good approximation of the same relation in carp. Comparison of the sarcomere lengths measured in swimming carp with the frog length-tension relation shows that in swimming the red muscles generate at least 96% of their maximal force (Figure 10-37A, *left*).

In the escape response, the fish moves rapidly, and its body curves dramatically. Recall that the red muscles in carp run parallel to the long axis of the fish, whereas the white muscles run helically (see Figure 10-35). Given the anatomic arrangement of the red muscles, the sarcomeres would have to shorten to a length of 1.4 μm to produce the escape response, whereas the sarcomeres in white muscles need shorten only to about 1.75 μm during this

A **Steady swimming**

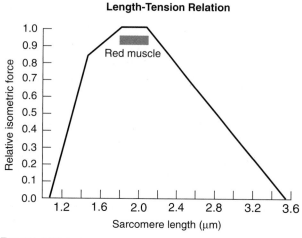

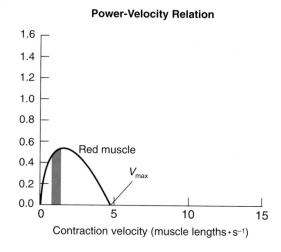

B **Escape response**

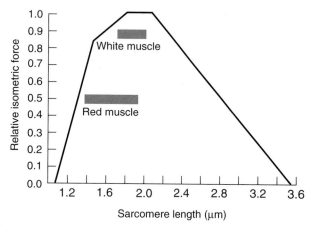

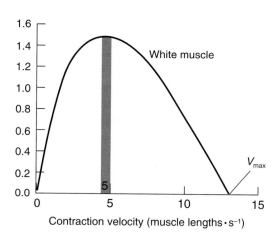

Figure 10-37 The properties of red and white muscles of fish make them optimally suited for different kinds of activities **(A)** The changes in the sarcomere length of red muscles during contraction coincide with the plateau of the sarcomere length-tension curve *(left)*. The colored bar indicates the lengths of red muscle sarcomeres during slow, steady swimming. In addition, the contraction velocities of red muscles during swimming correspond to values of V/V_{max} between 0.17 and 0.36, near the value at which red fibers produce maximal power *(right; colored region)*.

(B) Because of their anatomic arrangement, white fibers can produce the escape response at a more favorable region of the sarcomere length-tension curve than can red fibers *(left)*. In addition, the high V_{max} of white fibers allows them to generate power when they are shortening very rapidly *(right)*. Indeed, during the escape response the V/V_{max} for white muscles is 0.38, precisely at the peak of their power curve. [Adapted from Rome and Sosnicki, 1991.]

behavior. In other words, the mechanical advantage conferred by the anatomy of the white muscles allows them to produce a given change in the curvature of the spine with much less sarcomere shortening than would be required in red muscles. Thus, white muscles are much better suited to producing the escape response and generate about 85% of their maximal force during this behavior (see Figure 10-37B, *left*). When white muscle is used in less extreme movements (e.g., when a fish is swimming rapidly), the curvature of the backbone is not nearly as extreme, the sarcomeres shorten less, and the muscles generate nearly maximal force.

Because the fish uses different muscles to produce different movements, the myofilament overlap (sarcomere length) is never far from its optimal level, even in the most extreme movements. The lengths of the thick and thin filaments and the anatomic arrangement of the muscle fiber types combine to allow this optimization.

Value of V/V_{max}

In addition to their different anatomic arrangements, the red and white muscles of a carp have different values of V_{max}. The V_{max} of carp red muscle is 4.65 muscle lengths per second, whereas the V_{max} of carp white muscle is 12.8 muscle lengths per second, about 2.5 times higher. During steady swimming the red muscle shortens at a V/V_{max} of 0.17–0.36, which is near the value at which maximal power is generated (Figure 10-37A, *right*). At higher swimming speeds (higher values of V/V_{max}), a fish needs to generate greater mechanical power, but at these higher values the mechanical power output of the red muscle actually declines. In order to swim faster, a fish must activate white muscles as well.

In contrast to steady swimming, the escape response depends on activity in the white muscles. To power the escape response, the red muscles would have to shorten at 20 muscle lengths per second—four times faster than their V_{max}. White muscle in the anatomic orientation of the red muscles would also be unable to power the escape response, because the V_{max} of these muscles is only about 13 muscle lengths per second. However, the helical arrangement of the white muscles allows them to produce the escape response when they shorten at only about five muscle lengths per second, which corresponds to a V/V_{max} of about 0.38, the value at which the carp's white muscles produce the most power (Figure 10-37B, *right*).

Perhaps a fish would be better off with only white muscles. The white muscles could certainly power slow swimming. However, the high V_{max} of white muscles would mean that the V/V_{max} during slow swimming would be so low (0.01–0.03) that they would be extremely inefficient. Red muscles also can produce adequate power to generate slow swimming, and they do it much more efficiently than white muscles could. Thus, the anatomic arrangement and the V_{max} of the two kinds of muscle suits them to the particular behavior during which they are active. Fish need both kinds of muscles if they are to perform both slow swimming and fast escape responses optimally.

Kinetics of activation and relaxation

In considering the one-shot jump of frogs, our main concern was to determine whether the muscle becomes maximally activated during the early phase of shortening. The kinetics of muscle relaxation were essentially irrelevant. A fundamentally different problem is faced by animals during cyclical locomotion, such as swimming by fish. Swimming will be most efficient if muscles do not have to work against one another. For example, when muscles on one

side of the fish shorten, they will be most efficient in changing the shape of the fish—allowing it to push against the water—if the muscles on the other side of the body are relaxed. If muscles on both sides of the fish contracted maximally and simultaneously, the fish would be rigid and straight. Thus it is essential for each muscle to relax after it shortens, so that it offers no resistance when the contralateral muscle becomes active.

To better understand how the kinetics of activation and relaxation affects the generation of power during cyclical muscle contractions, Robert Josephson has applied the "workloop" technique to muscles. In this approach, muscles are driven by a servomotor system through the cyclical changes in length that occur during locomotion, and the investigator delivers a stimulus to the muscle at a particular time in the cycle. In this kind of experiment, the timing of the stimulus, the duration of the stimulus, the intrinsic activation and relaxation rate of the muscle, and the value of V_{max} for the muscle interact to determine how much power the muscle generates.

A useful way for quantifying these potentially complex interactions is to measure the amount of *net* work (force × change in length) the muscle generates during one cycle of shortening and lengthening (Figure 10-38A). Net work is graphically equivalent to the area contained within the force-length loop (Figure 10-38B). A muscle generates *positive work* only when it is shortening; thus positive work is equal to the area under the force-length curve during the shortening phase of a cycle. A muscle generates *negative work* when it is forcibly lengthened by a contralateral muscle (or a servomotor); thus negative work is equal to the area under the force-length curve during the lengthening phase of the cycle. The net work—the difference between the positive and negative work done during one cycle—is the area between the positive and negative legs of the force-

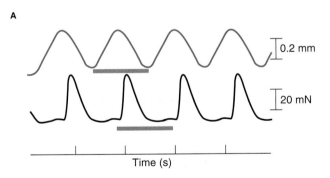
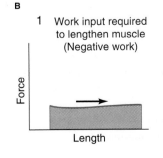
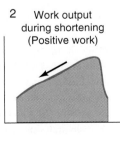
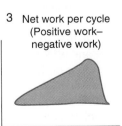

A

0.2 mm

20 mN

Time (s)

B

1 Work input required to lengthen muscle (Negative work)

2 Work output during shortening (Positive work)

3 Net work per cycle (Positive work– negative work)

Force

Length

Figure 10-38 Workloops graphically depict the net work done during cyclical muscle contractions. **(A)** The length (upper record) and tension (lower record) of a katydid flight muscle that is being driven to shorten and lengthen cyclically. Red bars indicate the duration of a single cycle. **(B)** The length-force relation during one complete cycle. In (1) the muscle is becoming longer because it is being stretched by an outside force; the shaded area under the curve represents the negative work done during this phase. In (2) the muscle shortens; the shaded area represents the positive work done during this phase. The net work (3) is the difference between the negative work and the positive work and equals the area encompassed by the length-force curve. [Adapted from Josephson, 1985.]

length curve for one cycle. For the muscle to generate net positive work, it must generate a greater force at each position during shortening than is required to stretch the muscle to that length. The net power generated in a cyclical contraction is expressed by:

$$(\text{positive work} - \text{negative work})_{cycle} \times \text{frequency of cycles}$$

It would seem that muscles might operate optimally if their fibers were fully activated during shortening (as in the frog) and could fully relax before they were forced to elongate by the activity of other muscles. If a muscle could be fully activated instantaneously and then relaxed instantaneously, the generation of force during shortening would be given by the force-velocity curve. There is, however, a problem. A muscle that was maximally activated throughout shortening and that then relaxed instantly at the end of shortening would be very energetically expensive for two reasons. First, such a muscle would have to pump Ca^{2+} back into its sarcoplasmic reticulum very rapidly, requiring a huge number of calcium pumps to be continuously active—a large expenditure of ATP. Second, instantaneous relaxation would require that cross-bridges detach very rapidly, but rapidly cycling cross-bridges use ATP much faster than slower cycling cross-bridges. A muscle with more modest rates of calcium pumping and cross-bridge cycling will be energetically less expensive, allowing it to work more efficiently. Efficiency of operation is of special concern for muscles, such as the swim muscles of an active fish, that are used almost continuously.

If a muscle relaxes slowly, allowing it to be metabolically efficient, the timing of stimulation becomes important. For a slowly relaxing muscle to be reasonably relaxed before it lengthens, stimulation must start during the lengthening phase and continue only into the very earliest part of the shortening phase; this stimulation pattern, however, will reduce the amount of work the muscle can do. Once again, there is a trade-off between two desirable features. In this case, it is the ability of the muscle to do work versus its metabolic efficiency. Workloop experiments have been performed on swimming fish to determine whether the swim muscles emphasize rapid relaxation, which is metabolically costly, or lower work output, which is less metabolically costly.

The basic experimental approach used in these studies was similar to that described for the frog hip flexor muscle. The electrical activity of muscles and changes in muscle length were determined in swimming fish. Then using the type of setup illustrated in Figure 10-34 isolated muscles were stimulated identically with the electromyogram, and the length of the muscles was controlled to match the changes measured during swimming. The force and power generated by the muscles under these conditions were determined, and the work done by the muscles then was determined by plotting the workloop. These experiments revealed that the posterior muscles do more net work than anterior muscles during slow steady swimming (Figure 10-39).

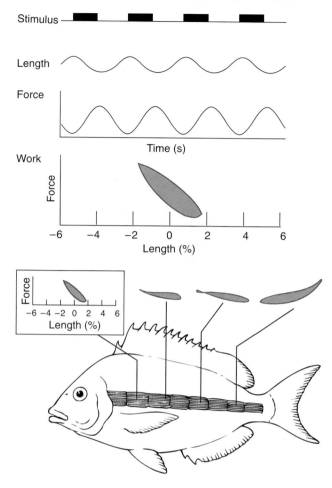

Figure 10-39 During slow, steady swimming, the posterior muscles do more net work than the anterior muscles. The area of workloops determined for muscles at several locations along the body of a scup indicates the net work performed by the various muscles. The scale of the workloops for each position along the body is the same. *Inset:* Typical stimulus, length, and force tracings used to plot workloops in these experiments. [Adapted from Rome et al., 1993.]

Muscles from different locations along the axis of the body receive different stimulus patterns and change in length by different amounts, which affects both the force generated and the power produced. The stimulus *duty cycle* (the percentage of one cycle during which the muscle is stimulated) is about 50% in the anterior part of the fish and falls to only about 25% in the posterior part of the fish. In addition, posterior muscles change in length much more than do anterior muscles during swimming. The combination of big changes in length and a short duty cycle permits the posterior muscles to generate a great deal of mechanical power. When isolated anterior muscles were exposed to the same set of conditions (i.e., stimulation pattern and length changes), they generated the same amount of power as the posterior muscles; it is the pattern of muscle *behavior* at the back of the fish that generates more power, not an intrinsic property of the muscle fibers themselves.

Examination of the workloops for red muscles during swimming indicates that the swim muscles operate with a slow activation rate and a slow relaxation rate. Stimulation to the posterior muscles, which generate the most power, begins during lengthening and ends just after the beginning of shortening, as predicted for this kind of muscle earlier in this section. As a result, the muscle must be relaxing through most of its power stroke, reducing the mechanical power it generates, but presumably also decreasing the energetic cost and perhaps thereby increasing the efficiency.

Adaptation for Speed: Sound Production

Some animals produce sound through mechanisms—for example, the movement of a column of air past a vibrating membrane or past the vocal cords—that are not directly coupled to muscle contraction. In other animals, however, sound is produced when muscles generate forces that directly cause structures, such as the swimbladder of a toadfish or the rattle on the tail of a rattlesnake, to vibrate. In these animals, the sound-producing (or *sonic*) muscles must undergo contraction-relaxation cycles at the frequency at which the sound is made, which is 10 to 100 times faster than most locomotory muscles operate.

In the last section, we saw that the swim muscles of fish have relatively slow rates of relaxation, allowing them to avoid the high energetic cost of excessive calcium pumping. When these swim muscles are experimentally stimulated at the high frequencies needed for sound production, they are unable to relax between stimuli and therefore contract tetanically (see Figure 10-28). If the sound muscles became tetanic in the same way, it would render the animal incapable of making any sound. Sonic muscles must then have unique properties that allow them to operate at the high frequencies associated with sound production, which sometimes exceed 80 cycles per second (hertz, or Hz).

Toadfish swimbladder

The male toadfish, *Opsanus tau*, produces a "boatwhistle" mating call ten to twelve times per minute for many hours to attract females to its nest. This tone is generated by oscillatory contractions of the muscles encircling the fish's gas-filled swimbladder, which is described in detail in Chapter 13. The steady swimming movements of toadfish occur at about 1–2 Hz, and their rapid escape response occurs at 5–10 Hz. In order to produce sound, however, the muscles of the swimbladder must contract and relax at frequencies of several hundred hertz. To understand the differences among these various types of muscles that allow them to function with such different time courses, the properties of each of their three kinds of muscle—swim, escape, and sonic—have been studied. The time course of many biological events is characterized by the **half-width,** which is the width on the temporal axis of the event when the measured variable is equal to half of its peak value. The half-width of a single twitch is 500 ms in red (swim) muscle, 200 ms in white (escape) muscle, and only 10 ms in swimbladder (sonic) muscle.

If a muscle is to activate and to relax rapidly, two conditions must be met. First, Ca^{2+}, the trigger for muscle con-

traction, must enter the myoplasm rapidly and be removed rapidly (Figure 10-40, steps 1 and 4). Second, myosin cross-bridges must attach to actin and generate force soon after the myoplasmic Ca^{2+} concentration rises (steps 2 and 3) and then detach and stop generating force soon after the Ca^{2+} concentration falls (steps 5 and 6). Myoplasmic free Ca^{2+} in the red and white muscles of toadfish rises and falls with typical kinetics, but the Ca^{2+} transient in the sonic muscles is the fastest ever measured for any fiber type (Figure 10-41A). Similarly, force measurements indicate that the sonic muscles both contract and relax about 50 times faster than the red muscles (Figure 10-41B).

The effect of the very fast Ca^{2+} transient measured in sonic muscles is most obvious during repeated stimulation. When red muscle is stimulated at 3.5 stimuli per second (3.5 Hz), Ca^{2+} lingers in the myoplasm so long that myoplasmic free Ca^{2+} cannot return to its resting value between stimuli. Indeed, the myoplasmic Ca^{2+} concentration remains above the threshold required for the generation of force for most of the time between stimuli, so a partially fused tetanus is produced (Figure 10-42A). By contrast, the swimbladder sonic muscle has such a fast Ca^{2+} transient, that even at 67 Hz, the myoplasmic Ca^{2+} concentration returns to its baseline value before every stimulus. Because the myoplasmic free Ca^{2+} concentration is below the threshold for the generation of force for much of the time between stimuli, only the first two twitches in the series are fused (Figure 10-42B). The production of an individual twitch in response to each stimulus is required for generating the oscillation of the swimbladder that produces sound.

The ability of a muscle to relax rapidly requires not only a brief Ca^{2+} transient in the myoplasm, but also the rapid release of bound Ca^{2+} from troponin (see Figure 10-40, step 5). Mathematical modeling indicates that if the time constant for the release of Ca^{2+} from troponin were the same in sonic fibers as it is in the fast white fibers of frog, twitches in sonic muscles would be longer than the ones that have been observed. Comparisons of the time course of force generation and of myoplasmic Ca^{2+} transients suggest that the release of Ca^{2+} from troponin in sonic muscles of the toadfish swimbladder must occur three times faster than it does in frog white fibers.

Finally, for force to drop quickly following the dissociation of Ca^{2+} from troponin, the myosin cross-bridges must detach rapidly from actin filaments. The Huxley model discussed earlier in this chapter suggests that the maximal velocity of shortening of a muscle, V_{max}, must be proportional to the rate at which cross-bridges detach from actin. Indeed, the V_{max} of swimbladder sonic muscle (about 12 muscle lengths per second) is exceptionally fast, being five times higher than toadfish red muscle and two and a half times higher than toadfish white muscle.

Experiments with toadfish indicate that sonic fibers have a number of adaptations that permit them to operate at very high frequencies. Ultrastructural and biochemical studies, for example, suggest that the short Ca^{2+} transient is achieved by an unusually high density of calcium channels

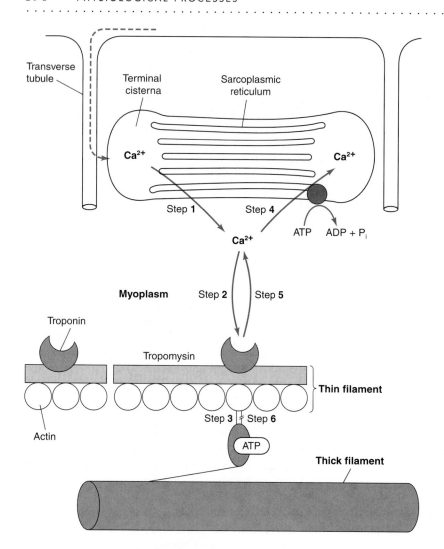

Figure 10-40 In muscles that operate at high frequencies, critical steps in the contraction-relaxation cycle must occur very rapidly. During activation, the stimulating signal (dashed arrow) is conducted down the T tubule and communicated into the sarcoplasmic reticulum, leading to opening of calcium channels in the SR membrane and Ca^{2+} outflow into the myoplasm (step 1). Binding of the free Ca^{2+} to troponin (step 2) relieves the tropomyosin-mediated inhibition of binding between actin and myosin. Myosin cross-bridges then attach to the actin filaments (step 3), and the thick and thin filaments slide past one another. During relaxation, calcium pumps in the SR membrane resequester Ca^{2+} (step 4). The resulting drop in myoplasmic Ca^{2+} favors release of bound Ca^{2+} from troponin (step 5), so that tropomyosin again is able to prevent attachment of cross-bridges (step 6).

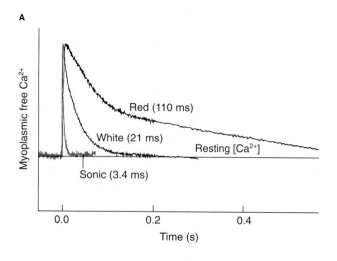

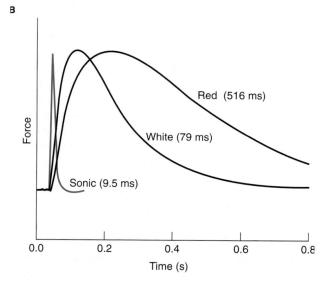

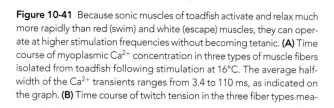

Figure 10-41 Because sonic muscles of toadfish activate and relax much more rapidly than red (swim) and white (escape) muscles, they can operate at higher stimulation frequencies without becoming tetanic. **(A)** Time course of myoplasmic Ca^{2+} concentration in three types of muscle fibers isolated from toadfish following stimulation at 16°C. The average half-width of the Ca^{2+} transients ranges from 3.4 to 110 ms, as indicated on the graph. **(B)** Time course of twitch tension in the three fiber types mea-sured under the same conditions as in part A. Sonic muscles both contract and relax much faster than do the red or white fibers. The average half-width of twitch tension ranges from 9.5 to 516 ms. Thus sonic muscle operates more than 50-fold faster than red muscle. The Ca^{2+} concentration and tension records are normalized to their maximal values for all fiber types. [Adapted from Rome et al., 1996.]

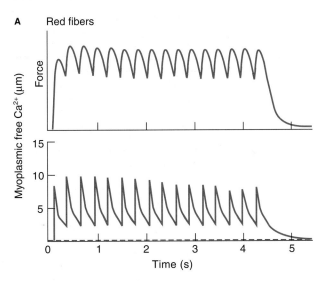

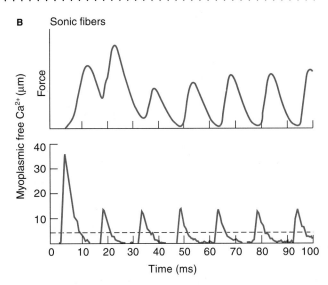

Figure 10-42 Red muscle from a toadfish contracts tetanically in response to relatively low-frequency stimulation, whereas sonic muscle produces individual twitches, even when stimulated at a much higher frequency. **(A)** Myoplasmic free Ca²⁺ in and force generated by a red fiber from a toadfish stimulated at 3.5 Hz. The threshold concentration of myoplasmic free Ca²⁺ necessary for the generation of force is shown by the dashed line on the Ca²⁺ trace. **(B)** Myoplasmic free Ca²⁺ in and force generated by a sonic fiber from a toadfish swimbladder stimulated at 67 Hz. Although the threshold Ca²⁺ concentration for activation (dashed line) is much higher in the swimbladder sonic fiber than in the red fiber, the Ca²⁺ transient is fast enough that the concentration falls below the threshold value between each stimulus. Notice the difference in time scale between parts A and B. [Adapted from Rome et al., 1996.]

in the SR membrane through which Ca²⁺ is released; an unusually high density of calcium pumps through which Ca²⁺ is resequestered; an increased concentration of certain calcium-binding proteins (e.g., troponin); and a fiber morphology such that the distance between the SR membrane and the myofilaments is particularly short, reducing the time required for diffusion. The rapid release of Ca²⁺ from troponin likely reflects decreased affinity of troponin for Ca²⁺. Finally, the rapid detachment of cross-bridges implies that myosin in sonic fibers also has special molecular properties.

These adaptations allow swimbladder sonic muscles to perform mechanical work at high operating frequencies. To emit continuous sound, sonic muscle must generate work to overcome frictional losses in the sound-producing system and to produce sound energy. Workloop experiments, similar to those described previously for the swim muscles of fish, show that swimbladder fibers can perform work at frequencies above 200 Hz at 25°C, the highest frequency for work production ever recorded in vertebrate muscle. By comparison, the highest frequency known for vertebrate locomotory muscles is 25–30 Hz, measured in mouse and lizard fast twitch muscles at 35°C.

Rattlesnakes

Rattlesnakes in the genus *Crotalus* also use special noise-making muscles, but as a warning to members of other species rather than to attract conspecifics for mating. Rattling is a loud and effective warning that renders these snakes, like many venomous animals, very conspicuous. Unlike the periodic toadfish boatwhistle, rattling can be sustained continuously for up to 3 hours. However, the rapidity with which the rattle-producing muscles contract

suggests that these muscles might have many features in common with the toadfish swimbladder sonic muscles.

Indeed, it has been found that the fibers that shake the rattle (shaker fibers) have a very rapid calcium transient, with a half-width of 4–5 milliseconds at 16°C, only 1–2 milliseconds slower than that of swimbladder sonic fibers (Figure 10-43A). However, at 16°C the half-width of the shaker muscle twitch is considerably longer than that of swimbladder (Figure 10-43B). The twitch of the shaker muscle is slower probably because its cross-bridges detach more slowly; the V_{max} of the shaker muscle is about 7 muscle lengths per second, only about half the V_{max} of the swimbladder muscle. In addition, Ca²⁺ may detach from troponin more slowly, but this feature has yet to be determined. The properties of rattlesnake shaker muscle suggest that a rapid calcium transient alone is not sufficient for the production of very rapid contractions. The release of Ca²⁺ from troponin and the detachment of cross-bridges from actin filaments must be unusually fast as well. For example, at 16°C shaker fibers can be stimulated up to only about 20 Hz before summation of force begins, with tetanic fusion occurring at about 50 Hz. (Rattling occurs at 30 Hz at 16°C.) In contrast, sonic fibers produce individual twitches at 67 Hz at 16°C.

However, many snakes are active at temperatures above 30°C, and they rattle at 90 Hz at a temperature of 35°C; at this temperature, the calcium transient and the twitch speed are even faster in the shaker muscle than in the swimbladder sonic muscle at 16°C (see Figure 10-43). Most likely, both the V_{max} of shaker fibers and the rate at which Ca²⁺ is released from troponin are higher at 35°C than at 16°C. At 35°C, shaker fibers can be stimulated at

A

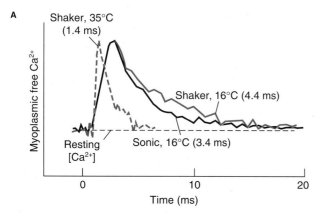

B

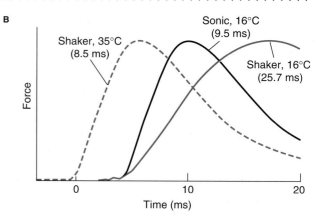

Figure 10-43 At 16°C, rattlesnake shaker muscles have a comparable calcium transient to toadfish sonic muscles, but their twitches last longer. **(A)** Myoplasmic free Ca²⁺ following stimulation of toadfish sonic fibers and rattlesnake shaker fibers at the indicated temperatures. The half-widths of the calcium transient, indicated in parentheses, are quite similar in sonic and shaker fibers at 16°C; however, at 35°C, a typical ambient temperature for snakes, the calcium transient in shaker fibers is much more rapid. **(B)** Time course of twitch tension in toadfish sonic fibers and rattlesnake shaker fibers measured under the same conditions as in part A. At 16°C, the half-width of twitch tension is nearly three times longer in shaker fibers than in sonic fibers, but at 35°C, the shaker fibers contract and relax considerably more rapidly. The Ca²⁺ concentration and force records are normalized to their maximal values. [Adapted from Rome, et al., 1996.]

100 Hz without complete tetanus, and they can perform work at 90 Hz. The similarities in the properties of toadfish swimbladder sonic muscles and rattlesnake shaker muscles suggest that in these species convergent evolution has arrived at similar solutions to the problems that are posed by high-frequency oscillatory contraction.

Energetic and spatial constraints on muscle operation at high frequencies
The rapid calcium kinetics that are required in muscle fibers that contract and relax rapidly could potentially demand a relative increase in the surface area (and volume) of the sarcoplasmic reticulum and in the number of mitochondria within each fiber. Any such increase must reduce the space that is available within each fiber for myofilaments, the structures that generate force. In toadfish sonic fibers, for example, the rate of Ca²⁺ uptake by the sarcoplasmic reticulum is about 50-fold greater than it is in red fibers, and about 30% of the entire volume of a sonic fiber is occupied by the sarcoplasmic reticulum. In addition, if a fast muscle is to operate continuously, as the rattlesnake shaker muscle does, it must use aerobic metabolism to generate enough ATP to fuel a high rate of calcium pumping; thus each fiber must contain many mitochondria, displacing even more myofilaments.

In other words, there must be a balance between features that allow a muscle to operate very rapidly and the amount of space left in each fiber for the contractile machinery—another example of a trade-off. If too much of the fiber volume is occupied by the components that support rapid calcium kinetics, the energetics of a fiber might be impressive, but there might be too few myofilaments to generate enough power to do the required work.

Vertebrate sonic muscles and some insect sonic muscles can operate at frequencies well above 100 Hz, but producing sounds does not require the production of high force,

and in many cases, the effort does not need to be sustained for long periods. The situation is quite different for flight by some insects, which depends on muscles that can operate at high frequency and produce considerable power. To power flight at wingbeat frequencies greater than 100 Hz, insects have evolved special muscle fibers that can produce high power at high frequencies. We discuss the properties of these high-power, high-frequency insect muscles in the next section.

 In our discussion of frog hip flexor muscles, fish swim muscles, and sound-producing muscles, we have encountered several trade-offs. What are some of these trade-offs, and what kinds of evolutionary forces might lead an organism to have "chosen" one of the features or the other?

High-Power, High-Frequency Muscles: Asynchronous Flight Muscles

Most species in the Hymenoptera (bees and wasps), Diptera (flies), Coleoptera (beetles), and Hemiptera (true bugs) contain flight muscles that are a notable exception to the rule that no more than one contraction is evoked by a single depolarization of the surface membrane. This unusual type of striated skeletal muscle does not exhibit a one-to-one relation between the arrival of motor impulses and the timing of individual contractions. These flight muscles are called **fibrillar muscles,** or more commonly **asynchronous muscles,** to distinguish them from muscles that contract in synchrony with each AP from a motor neuron. Although the timing of contraction is unrelated to the timing of neuronal input to an asynchronous muscle, a constant train of motor impulses and muscle depolarizations is required to maintain an asynchronous muscle in an active condition.

In some species of small insects, the wingbeat frequency (and the frequency of wing-muscle contractions) far exceeds the maximal maintained discharge rates of which axons are capable. Wingbeat frequency has been found to vary inversely with wing size. A tiny midge, for example, beats its wings at a frequency in excess of 1000 Hz, producing a high-pitched sound that can be perceived by the human ear.

How can contractions take place in asynchronous muscle fibers independent of the timing of the input stimulation? In some respects, asynchronous muscles are very similar to the more-common synchronous muscles. For example, activation of asynchronous muscle requires a sufficiently high concentration of free Ca^{2+} in the myoplasm, and there is some relation between neuronal input to the muscle and the myoplasmic Ca^{2+} concentration. In fact, as long as neuronal impulses continue to arrive at the neuromuscular junction, the myoplasmic Ca^{2+} concentration is maintained at a steady activating level. However, even if myoplasmic free Ca^{2+} is elevated, the active state is not initiated until the muscle is given a sudden stretch. The active state is terminated if tension on the muscle is released. This property has been demonstrated in glycerin-extracted asynchronous muscles. In the presence of a constant concentration of Ca^{2+} higher than 10^{-7} M, an extracted asynchronous muscle actively develops tension if a stretch is applied, and it oscillates repeatedly between contraction and relaxation if it is coupled to a mechanical system that has an appropriate resonant frequency (Figure 10-44).

The mechanics of flight differ considerably in insects possessing synchronous flight muscles from those possessing asynchronous muscles. In those insects with synchronous flight muscles (e.g., damselfly), the wings are elevated and depressed by simple lever mechanics (Figure 10-45A). Insects with asynchronous flight muscles have a more complex musculoskeletal arrangement in which the thorax can exist in either of two stable configurations. In these insects, contractions of the antagonistically arranged flight muscles change the shape of the thorax to generate only two wing positions—up or down (Figure 10-45B). Contraction of the *elevator muscles* pulls down the roof of the thorax, causing the wings to move up. As the roof of the thorax snaps down past an unstable "click" point (much as in a clicker toy), the *depressor muscles* are stretched and thereby activated, and the tension on the elevator muscles is suddenly released, thereby relaxing them. Contraction of the depressor muscles shortens the thoracic exoskeleton front to back, so as to expand the thorax dorsoventrally, causing the wings to move up. As the roof of the thorax moves back past the "click" position, the elevator muscles are stretched and thereby activated, and the tension on the depressor muscles is relieved, thereby relaxing them. This alternation between elevator and depressor muscle activity can continue as long as conditions are appropriate for the thick and thin filaments of the fibers to slide past one another and generate force.

When neuronal input to asynchronous flight muscles ceases, the muscle membrane repolarizes and the myoplasmic Ca^{2+} concentration drops, with the result that the cross-bridges are unable to attach to the actin filaments. If stretch is applied when myoplasmic Ca^{2+} is too low, it no

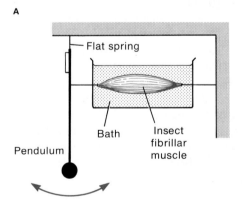

Figure 10-44 A glycerin-extracted asynchronous muscle will contract and relax repeatedly and rapidly if sufficient Ca^{2+} is present *and* it receives appropriate mechanical stretch. **(A)** In this experimental setup, the muscle is surrounded by a saline solution and mounted between a pendulum and a fixed surface. A tension transducer monitors the force generated by the muscle. Once the muscle is stimulated to contract once, it pulls on the pendulum, which in turn pulls on the muscle. The system creates a mechanical resonance between the muscle and the pendulum: First, the muscle moves the pendulum, and then the pendulum stretches the muscle, reactivating it. If the resonance frequency of the pendulum matches the requirements of the muscle, the muscle will continue to contract and relax rhythmically as long as the concentration of free Ca^{2+} in the saline is sufficiently high. **(B)** The ability of an asynchronous muscle to produce oscillations in the setup shown in part A depends on the Ca^{2+} concentration of the solution bathing the muscle. [Adapted from Jewell and Ruegg, 1966.]

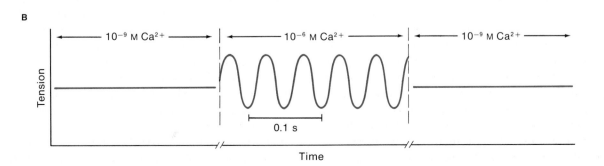

A
B

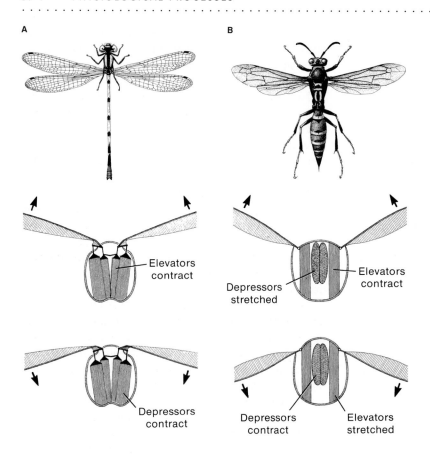

Figure 10-45 The mechanics of insect flight powered by synchronous flight muscles differ from the mechanics of flight powered by asynchronous muscles. **(A)** The synchronous flight muscles in a damselfly are arranged so that elevators (*middle*) and depressors (*bottom*) work vertically to raise and to lower the wings, respectively. In the transverse section through the thorax shown in these diagrams, the muscle fibers run dorsal to ventral. **(B)** The asynchronous flight muscles in the wasp work in two directions, forcing the roof of the thorax up and down between two stable positions. Contraction of the elevator muscles (*middle*), which run dorsal to ventral, pull the roof of the thorax down, raising the wings. Contraction of the depressor muscles (*lower*), which run from anterior to posterior (so they are shown in cross-section in the diagram), causes the roof of the thorax to arch upward, lowering the wings. The thorax is stable in either of these two configurations, but not in intermediate shapes. [From "The Flight Muscles of Insects" by D. S. Smith. Copyright © 1965 by Scientific American, Inc. All rights reserved.]

longer produces an active state, and flight movements cease. In essence, motor neuronal input to asynchronous muscles acts largely as an on-off switch, rather than as a regulator of the frequency of contraction. The frequency of contraction depends on the mechanical properties of the muscle and the mechanical resonance of the flight apparatus (thorax, muscles, wings). If the wings are clipped short, the wingbeat frequency increases, even though the frequency of incoming APs remains unchanged.

The force-velocity curves of insect asynchronous flight muscles are similar in shape to those of the synchronous muscles in vertebrates. Indeed, the workloop method of studying the mechanics of muscle contraction was initially developed on asynchronous muscles, and the data illustrated in Figure 10-38 were recorded on the flight muscle of a katydid.

With their novel mechanical arrangement, the asynchronous flight muscles of insects avoid many of the constraints that limit contractile frequency in most muscle fibers, making them capable of extraordinarily high-frequency contraction, even though the concentration of Ca^{2+} in the myoplasm changes slowly. As a result, these muscles do not require a large sarcoplasmic reticulum to pump Ca^{2+}, as do sound-producing muscles in vertebrates; nor do they require a large volume of mitochondria to generate ATP to power calcium pumps. For this reason, asynchronous fibers can devote more space to force-generating myofilaments, and mitochondria are required largely to supply ATP for the myosin ATPase.

? Asynchronous flight muscles have evaded many of the constraints imposed on the properties of the fastest, sound-producing vertebrate muscle fibers. What disadvantages accompany this functional arrangement? Why might so few species have developed this type of muscle? For example, birds also fly; why have they not hit upon this mechanism to facilitate flight?

NEURONAL CONTROL OF MUSCLE CONTRACTION

Effective animal movement requires that the contractions of many fibers within a muscle—and of many muscles within the body—be correctly timed with respect to one another. This coordination is generated within the nervous system, because the timing of most muscle contractions is controlled by the arrival of neuronal impulses at the neuromuscular junctions of the muscle. (Insect asynchronous muscles are one exception to this rule.) In addition to controlling the timing of contraction, the nervous system regulates the strength of contractions by selecting among different fiber types and by determining how many fibers will be active simultaneously. A motor system that was limited to all-or-none contractions of all skeletal muscles would produce a very limited repertoire of movement. Fine control of muscle contraction has been achieved in different organisms by various means during the course of evolution. Vertebrate and arthropod neuromuscular mechanisms lend themselves especially well to comparison because of the different mechanisms for the control of movement that have evolved in these two groups of animals.

Motor Control in Vertebrates

All vertebrate skeletal muscles are innervated by motor neurons whose cell bodies are located in the ventral horn of the gray matter of the spinal cord (or in particular locations in the brain). The axon of a motor neuron leaves the spinal cord by a ventral root, continues to the muscle by way of a peripheral nerve, and finally branches repeatedly to innervate skeletal muscle fibers. (The anatomy and organization of the vertebrate central nervous system are described in more detail in Chapter 11.) Depending on the muscle innervated, a single motor neuron may innervate only a few fibers or a thousand or more. Although a single motor neuron can innervate many muscle fibers, in vertebrates each muscle fiber receives input from only one motor neuron.

A motor neuron and the muscle fibers that it innervates form a **motor unit.** Vertebrate spinal motor neurons receive an enormous number of synaptic inputs from sensory neurons and from interneurons. In vertebrates these spinal motor neurons are the only means available for controlling contraction of the muscles, so they have been called "the final common pathway" of neuronal output. When an AP is initiated in a motor neuron as a conse-

quence of synaptic inputs, the membrane excitation spreads into all of its terminal branches, activating all of its endplates (see Figure 6-13). All vertebrate spinal α motor neurons produce the neurotransmitter acetylcholine (ACh); when the endplates of a α motor neuron are activated, acetylcholine is released onto all of the fibers in the neuron's motor unit. In twitch muscle fibers, the neuromuscular junction is typically sufficiently depolarized by a single incoming AP to bring the muscle fiber above threshold. As a result, a single AP in the motor neuron is usually sufficient to produce an AP, and a subsequent twitch, in these muscle fibers (Figure 10-46A). Each time a motor neuron fires an AP, all of the muscle fibers in its motor unit contract. Whether the contractions consist of single twitches or of sustained tetanic contractions depends on the frequency of the APs generated in the motor neuron by its synaptic input.

The tight coupling between the APs in a motor neuron and those in the twitch muscle fibers of its motor unit means that the amount by which tension can be modulated in a motor unit is very small, because there is no gradation between total inactivity and a twitch. If many APs occur in succession in the motor neuron, the result is partial tetanus unless the firing rate of the neuron is high enough to elicit a full, smooth tetanic contraction (see Figure 10-28). In vertebrates, the problem of increasing the overall muscle tension in a graded fashion is solved by recruiting increased numbers of active motor units, as well as by varying the average frequency at which the population of motor neurons fires.

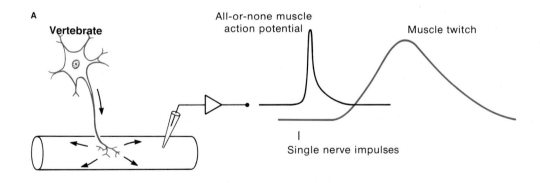

A

Vertebrate

All-or-none muscle action potential

Muscle twitch

Single nerve impulses

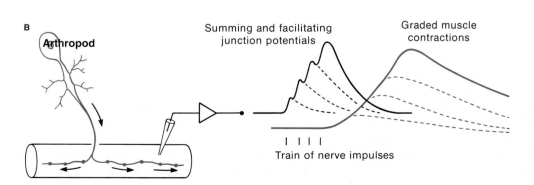

B

Arthropod

Summing and facilitating junction potentials

Graded muscle contractions

Train of nerve impulses

Figure 10-46 Most vertebrate muscles are made up of twitch fibers, whereas many invertebrate muscles are made up of fibers that produce graded contractions. **(A)** Vertebrate twitch muscle fibers produce an all-or-none twitch in response to each all-or-none AP in the membrane of the fiber. **(B)** Many arthropod muscle fibers, as well as vertebrate tonic muscle fibers, produce graded contractions in response to the overlapping postsynaptic potentials at multiple motor synapses distributed along the fiber length.

For example, if a small number of motor units in a muscle are maximally active, the muscle will contract with a small fraction of its total maximal tension. On the other hand, if all the motor neurons innervating the muscle are recruited to fire at a high rate, all the motor units are brought into a state of full tetanus, producing the maximal contraction of which the muscle is capable. In addition, many vertebrate muscles contain different types of fibers (see Figure 10-30), so the nervous system can modulate *which* fibers are active, as well as *how many* are active. Thus, the timing and force of a muscle's contraction is modulated by activity in the motor neurons innervating that muscle; by differentially activating motor neurons, the central nervous system can determine the strength and duration of muscle contractions.

The slow tonic muscle fibers of vertebrates (found primarily in amphibians and lizards) are unusual in that they receive multiterminal innervation—that is, the motor neuron makes many synapses along the length of each fiber. In these fibers, which lack all-or-none APs, the synaptic potentials produced by the broadly distributed neuromuscular junctions are sufficient to generate their graded contractions (Figure 10-46B). The tension produced by these muscles is strongly dependent on the frequency of motor neuron activity, and these tonic muscle fibers generally are found where slow, sustained contractions are required.

In mammals, only the extraocular muscles and intrafusal spindle fibers contain slow tonic fibers. As noted already, however, most skeletal muscles contain several types of twitch fibers. Typically, all the fibers within a single motor unit are of the same type. In addition, the properties of the innervating motor neuron are often matched to the properties of the muscle fibers. For instance, motor neurons innervating slow oxidative (type I) muscle fibers carry impulses at a lower frequency than do motor neurons innervating fast glycolytic (type IIb) fibers.

Motor Control in Arthropods

Arthropod nervous systems consist of a relatively small number of neurons compared with vertebrates, so a small number of motor units must generate the full range of contractions, from weak to strong, without relying on extensive recruitment of new motor units. Moreover, many types of arthropod muscles never produce APs, or do so only under certain conditions. In these muscles, as in the tonic muscle fibers of vertebrates, contraction is controlled by graded depolarization of the muscle fiber membrane rather than by the frequency of muscle APs. The pattern of neuronal control that has evolved under these constraints is quite different from the pattern of motor control in vertebrates.

Each vertebrate twitch muscle fiber is innervated at only one or two endplates, and postsynaptic potentials initiate APs that originate very near the endplate and are propagated along the muscle fiber. In contrast, crustacean skeletal muscle fibers, like vertebrate tonic fibers, receive many synaptic terminals located along the entire length of the muscle fiber, so no propagated AP is required to spread the

signal in the muscle fiber (see Figure 10-46). The synaptic potentials along the distributed neuromuscular junctions are summed, and the shorter the interval between excitatory synaptic potentials, the greater the depolarization of the muscle membrane. Because the coupling between membrane potential and tension is graded, each muscle fiber can produce a wide range of tension, instead of being limited to all-or-none twitches or tetanus, with no possibilities in between. For this reason, arthropod muscles function well over a large range of tensions with very few motor units, with the variability in tension produced by single fibers replacing the effect of recruitment in most vertebrate muscles. In some arthropod muscles, one motor neuron may innervate all—or at least most—of the fibers in the muscle.

The mechanical difference between twitch muscles and tonic muscles is based on the manner in which myoplasmic Ca^{2+} is controlled. In twitch muscles Ca^{2+} is released from the sarcoplasmic reticulum in an all-or-none manner in response to all-or-none APs, whereas in graded tonic muscles Ca^{2+} is released from the sarcoplasmic reticulum in a graded fashion, because the electrical signals that are conducted along the membrane are graded rather than all-or-none (Figure 10-47).

In many invertebrates the flexibility of motor control is further enhanced by *multineuronal innervation* of muscle fibers. Each muscle fiber receives synapses from several motor neurons, including one or two inhibitory neurons (Figure 10-48). The synaptic effects of inhibitory and ex-

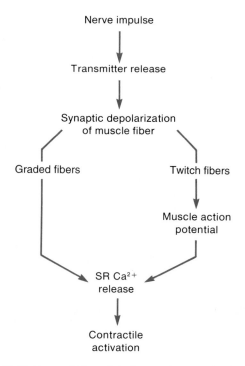

Figure 10-47 Neuronal APs are linked to muscle contraction through excitation-contraction coupling in the muscle fiber. In graded tonic muscle fibers *(left)*, the membrane potential varies based on summation and facilitation of synaptic potentials at synapses distributed all along the muscle fiber. In twitch muscle fibers *(right)*, the muscle fiber membrane carries APs, which give the contraction an all-or-none characteristic.

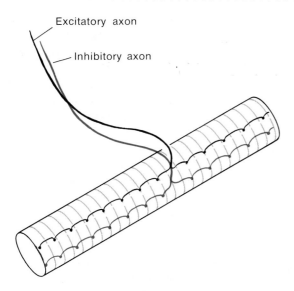

Figure 10-48 In contrast to vertebrate twitch fibers, each arthropod muscle fiber receives synaptic input from several motor neurons. There are usually several excitor axons, and there may be one or more inhibitors as well. Each motor neuron makes many synapses along the muscle fiber, producing a distributed neuromuscular junction. One motor neuron of each type is shown here for simplicity.

most arthropod muscles of several types of muscle fibers exhibiting different electrical, contractile, and morphologic properties. At one end of the spectrum are fibers with rapid all-or-none contractions, which resemble vertebrate twitch fibers. When a series of intracellular current pulses are delivered to such fibers, they produce a series of subthreshold depolarizations until the firing level is exceeded (Figure 10-49A). Once this occurs, the membrane responds with an all-or-none AP, which then elicits an all-or-none fast twitch. At the opposite end of the spectrum in crustacean muscle are fibers in which the electrical responses show little sign of regenerative depolarization, and the contractions are fully graded with the amount of depolarization (Figure 10-49C). Between these two extremes is a continuum of intermediate muscle-fiber types (Figure 10-49B). The differences in contractile behavior of these fiber types are correlated with morphologic differences. The slowly contracting fibers have relatively fewer T tubules and less sarcoplasmic reticulum than the rapidly contracting fibers. As in vertebrates, the properties of motor neurons innervating each type of muscle fiber are at least somewhat matched to the properties of the fibers themselves.

citatory motor axons directly sum at the level of the muscle fiber. In these systems, there typically is one excitatory neuron that produces exceptionally large excitatory synaptic potentials in the muscle fiber. This *fast* excitor axon can generate a strong contraction with less facilitation and summation than can a *slow* excitor axon, which must fire repeatedly at high frequency to produce similar levels of depolarization, and hence contraction, in the muscle fiber.

The variety and complexity of peripheral motor organization is increased still further by the presence in

 What are the advantages of the vertebrate and invertebrate patterns of motor neuron innervation? For example, what are the advantages in having muscle fibers receive input from several motor neurons, including inhibitory ones? What are the disadvantages? What are the advantages of having one single neuron that controls the activity of an entire motor unit? What are the disadvantages?

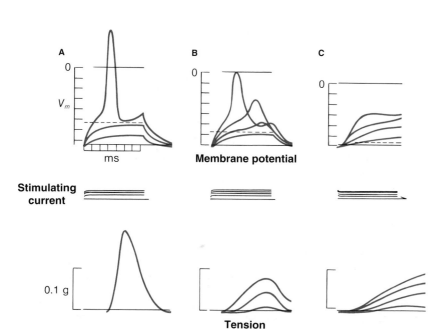

Figure 10-49 Crustacean muscles can contain fibers that differ greatly in their properties. Shown here are the membrane potentials *(top)* and tension generated *(bottom)* in three types of crustacean muscle fibers following intracellular stimulation *(middle)*. **(A)** All-or-none twitch fibers produce APs and fast twitches. **(B)** Intermediate graded fibers produce nonpropagating graded potentials and graded contractions. **(C)** Slow fibers produce only very small and slow depolarizations, and they contract very slowly. [Adapted from G. Hoyle, 1967, in *Invertebrate Nervous Systems,* C. A. G. Wiersma, ed. © 1967 by University of Chicago Press.]

CARDIAC MUSCLE

Cardiac muscle, the second type of striated muscle, shares many characteristics with skeletal muscle but differs in several important ways (Table 10-2). For example, vertebrate skeletal muscle fibers are individually innervated by an excitatory motor axon, whereas cardiac (ventricular) muscle fibers in most but not all vertebrates are innervated only diffusely by neurons of the sympathetic (excitatory) and parasympathetic (inhibitory) divisions of the autonomic system (see Chapter 11 for more discussion of the autonomic nervous system). The cardiac innervation is modulatory only and does not produce discrete postsynaptic potentials. Its actions are to increase or decrease the strength of spontaneous myogenic contractions, which are induced by the electrical activity within the pacemaker region of the heart (see Chapter 12). Another difference is that a cardiac muscle cell, or *myocyte,* contains one nucleus, whereas skeletal muscle cells are multinucleate. Cardiac muscle cells are connected electrically such that an AP initiated in the pacemaker region spreads rapidly, from muscle cell to muscle cell, through fast-conducting pathways to all muscle cells within the heart, ensuring that the atria and ventricles each contract as a unit.

Although the contractile mechanism of vertebrate ventricular muscle fundamentally resembles that of skeletal twitch muscle, their membrane APs differ. In contrast to the very short duration of the AP in skeletal muscle, the AP in cardiac muscle has a plateau phase that is hundreds of milliseconds long following the upstroke (see Figure 12-7). The long duration of the cardiac-muscle AP and the associated long refractory period of several hundred milliseconds prevents tetanic contraction and permits the muscle to relax, allowing the ventricle to fill with blood between APs. As a result of regularly paced, prolonged APs, the heart contracts and relaxes at a rate suitable for its function as a pump.

What functional advantage occurs because the ventricular AP lasts much longer than the AP in atrial muscle?

As in skeletal twitch muscle, contraction of cardiac muscle is activated by an increase in the cytosolic Ca^{2+} concentration. The rise in cytosolic Ca^{2+} depends both on influx across the plasma membrane and release from the sarcoplasmic reticulum. The cells of mammalian cardiac muscle possess an elaborate sarcoplasmic reticulum and system of T tubules (Figure 10-50). Membrane depolarization activates voltage-gated L-type calcium channels in the T tubules resulting in an inward flow of Ca^{2+} from the extracellular space. This small influx of Ca^{2+} triggers the release of a much larger pool of Ca^{2+} from the sarcoplasmic reticulum via calcium channels in the SR membrane, leading to contraction. Calcium is removed rapidly from the cytosol by calcium pumps in the SR membrane and by Na^+/Ca^{2+} exchange proteins in the sarcolemma.

TABLE 10-2
Characteristics of the major types of muscle fibers in vertebrates

Property/component	Striated muscle		Smooth (nonstriated) muscle	
	Skeletal	Cardiac	Multi-unit	Single-unit
Visible banding pattern	Yes	Yes	No	No
Myosin thick filaments and actin thin filaments	Yes	Yes	Yes	Yes
Tropomyosin and troponin	Yes	Yes	No	No
Transverse tubules	Yes	Yes	No	No
Sarcoplasmic reticulum	Well developed	Well developed	Very little	Very little
Mechanism of contraction	Sliding of thick and thin filaments past each other	Sliding of thick and thin filaments past each other	Sliding of thick and thin filaments past each other	Sliding of thick and thin filaments past each other
Innervation	Somatic nerves	Autonomic nerves	Autonomic nerves	Autonomic nerves
Initiation of contraction*	Neurogenic	Myogenic	Neurogenic	Myogenic
Source of Ca^{2+} for activation†	SR	ECF and SR	ECF and SR	ECF and SR
Gap junctions between fibers	No	Yes	No	Yes
Speed of contraction	Fast or slow depending on fiber type	Slow	Very slow	Very slow
Clear-cut relationship between length and tension	Yes	Yes	No	No

*Neurogenic muscles contract only when stimulated by synaptic input from a neuron. Myogenic muscles endogenously produce depolarizing membrane potentials, allowing them to contract independently of any neuronal input.

†SR, sarcoplasmic reticulum; ECF, extracellular fluid.

Source: Adapted from L. Sherwood, 1993.

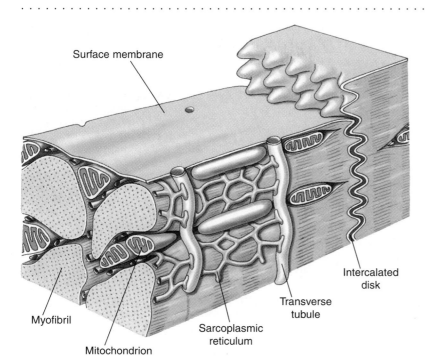

Surface membrane

Myofibril

Mitochondrion

Sarcoplasmic reticulum

Transverse tubule

Intercalated disk

Figure 10-50 Adult mammalian ventricular cardiac muscle has an extensive sarcoplasmic reticulum. The cells are connected electrically through intercalated disks, at which the extensive membranes of neighboring cells are joined by numerous gap junctions and desmosomes. [Adapted from Threadgold, 1967.]

The relative importance of the SR and the plasma membrane for Ca^{2+} regulation varies among species. Cardiac muscle of the frog has only a rudimentary reticulum and tubular system. The myocytes of the frog heart are much smaller than adult mammalian cardiac muscle fibers, and their relatively large surface-to-volume ratio reduces the need for an elaborate intracellular reticulum for the storage, release, and uptake of Ca^{2+}. Instead, most of the Ca^{2+} regulating contraction in amphibian heart cells enter through the surface membrane as a result of the membrane's increased calcium permeability during depolarization. Adult mammalian hearts, on the other hand, largely depend on calcium release from the sarcoplasmic reticulum.

As in skeletal muscle, ryanodine receptors mediate excitation-contraction coupling in mammalian cardiac muscle. Low concentrations of ryanodine (in the nanomolar range) lock the calcium channels in cardiac SR membrane in the open state (see Figure 10-24). The Ca^{2+} released from the sarcoplasmic reticulum following ryanodine treatment at low concentrations is removed from myocytes by Na^+/Ca^{2+} exchange across the sarcolemma. The end result is that SR calcium stores are reduced, the capacity to release Ca^{2+} from the SR is diminished, and cardiac contractility falls. Because the effects of ryanodine vary with the importance of the SR in regulating cardiac contraction, this drug has little effect on the contraction of the frog heart but a marked effect on the contraction of the adult rat heart.

The amount of tension that can be developed by a cardiac muscle depends on the amount of Ca^{2+} in the myoplasm. In the frog heart, when muscle cells are depolarized, Ca^{2+} flows into the cell because of the increased calcium permeability of the depolarized membrane. Because the influx of Ca^{2+} is voltage dependent, tension develops as a function of depolarization, with greater depolarization producing greater tension (Figure 10-51A). Reduction of the extracellular Ca^{2+} concentration leads to a weaker contraction for a given depolarization, because less Ca^{2+} enters the cell (Figure 10-51B). The intracellular Ca^{2+} concentration in cardiac muscle is determined not only by depolarization but also by a number of other factors including the action of catecholamines on the heart. The catecholamines epinephrine and norepinephrine that circulate in the blood or are released from neuron terminals activate α- and β-adrenoreceptors on the surface of cardiac cells. Stimulation of α-adrenoreceptors activates the inositol phospholipid second-messenger system (see Figures 9-14 and 9-15), resulting in increased release of Ca^{2+} from the sarcoplasmic reticulum. In contrast, β-adrenoreceptor stimulation activates the adenyl cyclase second-messenger system (see Figures 9-11 and 9-12), resulting in increased calcium flux across the sarcolemma. Thus stimulation of both types of adrenoreceptors augments cardiac contraction.

The time course of cardiac contraction is determined by the duration of the increase in cytosolic Ca^{2+} concentration and the cross-bridge cycling rate, both of which may be temperature dependent. Rapid cooling of the mammalian heart from 30 to 10°C produces a prolonged contracture because the reduction in temperature slows the calcium pump in the SR membrane and Na^+/Ca^{2+} exchange across the sarcolemma, increasing the duration of the calcium pulse. Animals living at low temperatures can maintain high heart rates because they have enhanced calcium release and removal mechanisms compared with mammalian hearts at the same low temperature. Some animals, such as carp, that use their muscles over a wide temperature range produce two different forms of myosin: a low-temperature (winter) form and a high-temperature (summer) form. These different forms of myosin allow a carp to maintain a reasonably stable time

A

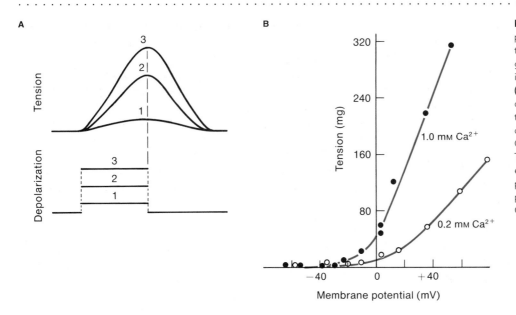

B

Figure 10-51 The larger the depolarization and the higher the extracellular Ca^{2+} concentration, the greater the tension developed in isolated frog ventricular muscle. **(A)** Tension (upper traces) developed at three voltage steps (lower traces). **(B)** Effect of the amount of depolarization and extracellular Ca^{2+} levels on tension developed. The tension was recorded at the end of each voltage step and is plotted against the membrane potential in millivolts. [Morad and Orkand, 1971.]

course of cardiac contraction throughout the year despite wide fluctuations in temperature.

SMOOTH MUSCLE

Muscle fibers are called "smooth" if they lack the characteristic striations produced by the organized groups of actin and myosin filaments that form sarcomeres. These least specialized muscle fibers have myosin that is similar to the myosin found in contractile nonmuscle cells. The myofilaments of smooth muscle are gathered into bundles of thick and thin filaments in dense bodies, or they are connected to attachment plaques located at the sarcolemma. The attachment plaques contain high concentrations of α-actinin, which is present in the Z disk of skeletal muscle, and the protein vinculin, which is not present in the Z disk. In smooth muscle fibers, vinculin binds to the α-actinin and anchors the actin filaments to the sarcolemma.

Vertebrate smooth muscles can be categorized into two categories: **single-unit muscles** and **multi-unit muscles** (see Table 10-2). In single-unit muscles, the individual muscle cells, which typically are small and spindle-shaped, are coupled with one another through electrically conducting gap junctions. If one or a few of the cells spontaneously depolarize, the rest of the cells depolarize, too, because excitation is passed through the gap junctions. As a result, a few cells in single-unit smooth muscles can generate contraction that moves throughout the entire muscle in a wave. Neurons synapse onto single-unit muscle cells and can *modulate* the rate and strength of contraction, but neuronal input is not required for contraction. Single-unit smooth muscle forms the walls of vertebrate visceral organs (e.g., alimentary canal, urinary bladder, ureters, and uterus). In contrast, in multi-unit muscles each cell acts independently and contracts only when it receives synaptic input from neurons. The muscles in the iris of the eye that regulate the diameter of the pupil are multi-unit muscles.

The innervation of smooth muscle differs significantly from that of skeletal muscle, which has discrete and intimate synaptic junctions between the motor terminal and the muscle fiber. In vertebrate smooth muscle, the neurotransmitter is released from many swellings, or varicosities, along the length of autonomic axons within the smooth muscle tissue. These do not form intimate junctions, but instead, the transmitter released at a given varicosity diffuses over some distance, encountering a number of the small, spindle-shaped smooth muscle cells along the way. Receptor molecules on the smooth muscle cells appear to be distributed diffusely over the cell surface. Smooth muscle of vertebrates is generally under autonomic and hormonal control and is generally not "voluntary," as is control of skeletal muscle (one exception to this rule may be the urinary bladder). An interesting feature of single-unit smooth muscle is the sensitivity of the membrane to mechanical stimuli. Stretching the muscle produces some depolarization, which in turn produces some contraction. As a result, muscle tension is maintained over a large range of muscle length. This response of smooth muscle to stretch accounts, at least in part, for the autoregulation seen in arterioles; that is, contraction of smooth muscle in the arteriolar wall in response to a rise in blood pressure maintains a reasonably constant blood flow in the peripheral tissues (see Chapter 12). Likewise, peristaltic movements along the intestinal tract rely on stretch-induced contractions of single-unit smooth muscles.

Smooth-muscle cells contract and relax far more slowly than striated muscle fibers and generally are capable of more sustained contractions. This is reflected in the duration and amplitude of the cytosolic Ca^{2+} pulse, which also initiates contraction in smooth muscle, and the slower process of excitation-contraction coupling, which is different from that in striated muscle. The slow release and uptake of Ca^{2+} in smooth-muscle cells is associated with a poorly developed sarcoplasmic reticulum, which is composed only of smooth flat vesicles close to the inner surface of the cell membrane. A highly developed sarcoplasmic reticulum like that of striated muscle fibers is unnecessary because the cells of smooth muscle are much smaller and therefore have larger surface-to-volume ratios than striated fibers. No point in the cytoplasm is more than a few micrometers distant from the surface membrane. Thus, the surface membrane of smooth-muscle cells can perform calcium-regulating functions similar to those of SR membranes in striated muscle.

In smooth-muscle cells, Ca^{2+} is constant' pumped outward across the surface membrane keeping internal Ca^{2+} levels very low. When the membrane is depolarized, it becomes more permeable to Ca^{2+} ions, permitting an influx of Ca^{2+}, which activates contraction. Relaxation occurs when the calcium permeability returns to its low resting level while the membrane pumps Ca^{2+} out of the cell. Large depolarizations of the membrane generate APs in which Ca^{2+} carries the inward current. Action potentials produce the greatest Ca^{2+} influx and thus evoke the largest contractions, because the tension generated is proportional to the intracellular level of Ca^{2+}.

Excitation-contraction coupling in smooth muscle occurs by several different mechanisms. As discussed earlier, regulation in striated muscle involves Ca^{2+} binding to troponin (see Figure 10-16), but smooth muscle lacks troponin. Rather, Ca^{2+} binds to **calmodulin** forming a Ca^{2+}/calmodulin complex (see Figure 9-17); this complex in turn binds to an elongated protein called **caldesmon**. In the absence of Ca^{2+}, caldesmon binds to actin thin filaments, restricting myosin-actin interactions and inhibiting muscle contraction. However, caldesmon that has been phosphorylated by protein kinase C cannot bind to thin filaments and therefore does not inhibit myosin-actin interactions. Thus phosphorylation of caldesmon or the binding of Ca^{2+}/calmodulin to caldesmon activates smooth-muscle contraction (Figure 10-52A).

A Caldesmon regulation of actin

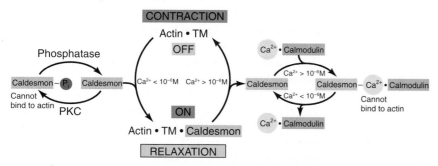

B Ca^{2+} binding to myosin light chains

D Regulation of myosin light chains by protein kinase C

Figure 10-52 Both actin- and myosin-dependent mechanisms control smooth-muscle contraction and relaxation. **(A)** Binding of caldesmon to the actin and tropomyosin (TM) of thin filaments prevents contraction. At cytosolic Ca^{2+} levels above 10^{-6} M, formation of the Ca^{2+}/calmodulin complex occurs. Binding of this complex to caldesmon releases it from thin filaments, allowing the muscle to contract. Phosphorylation of caldesmon by protein kinase C (PKC) also prevents it from binding to thin filaments and promotes contraction. **(B)** Binding of Ca^{2+} to the regulatory light chains of myosin allows actin-myosin interactions and promotes contraction. **(C)** Phosphorylation of the regulatory light chains by myosin LC kinase, which is activated by Ca^{2+}/calmodulin, also promotes muscle contraction. **(D).** Phosphorylation of the regulatory light chains by protein kinase C, at a site other than that acted upon by myosin LC kinase, inhibits myosin-actin interactions and causes smooth-muscle relaxation. [Adapted from Lodish et al., 1995.]

C Phosphorylation of myosin light chains

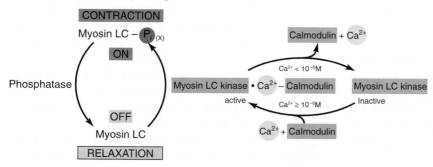

Three other mechanisms for regulating smooth-muscle contraction involve the *regulatory light chains* of myosin. In smooth muscle and some invertebrate muscle, binding of Ca^{2+} directly to the regulatory light chains induces a conformational change in the myosin head that allows it to bind to actin, so the muscle contracts (Figure 10-52B). Phosphorylation of the myosin light chains by *myosin light-chain (LC) kinase* also causes contraction in vertebrate smooth muscle (Figure 10-52C). Because myosin LC kinase is activated by Ca^{2+}/calmodulin, the rate of phosphorylation is also calcium-dependent. Phosphorylation of another site on the regulatory light chain of myosin by protein kinase C, however, induces a conformational change that prevents actin-myosin interactions, resulting in relaxation (Figure 10-52D). Thus phosphorylation of caldesmon by protein kinase C and of the myosin regulatory light chains by myosin LC kinase results in contraction, whereas phosphorylation of the regulatory light chains at another site by protein kinase C results in relaxation. The slowness in the actions of protein kinases, along with the slow changes in cytosolic Ca^{2+} levels, contributes to the slow rate of contraction seen in many smooth muscles.

Why is excitation-contraction coupling much more variable in smooth muscle than in skeletal and cardiac muscle?

Smooth-muscle contraction is modulated by a wide variety of stimuli, both neuronal and humoral, which can inhibit or activate contraction. All of these stimuli operate to influence cytosolic Ca^{2+} levels and/or the level of protein kinase C, myosin LC kinase, and muscle phosphatases. As in cardiac muscle, cytosolic Ca^{2+} levels in smooth muscle are regulated by hormonal activation of the adenyl cyclase and inositol phospholipid second-messenger systems. The diverse mechanisms that control smooth muscle result in the complex patterns of contraction observed in this type of muscle.

SUMMARY

Muscles are classified into two major types: striated and smooth. Because striated muscle has been so intensively studied, its structure is perhaps better understood than that of any other tissue. It appears striated when viewed with a light microscope because it contains a regular array of parallel myofilaments, organized into banded sarcomeres. A sarcomere contains an almost crystalline array of myosin and actin filaments. Myosin thick filaments interdigitate with and slide between the actin thin filaments. During muscle activity, the filaments slide past one another due to interactions between the actin filaments and cross-bridges that project from the myosin filaments.

The head of a myosin cross-bridge hydrolyzes ATP, a process that is significantly accelerated when the cross-bridge binds to actin. Hydrolysis of ATP, which requires Mg^{2+}, produces a conformational change that permits the cross-bridge to undergo a cycle of detachment, attachment, and rotation on the actin filament. After a myosin head detaches from actin, it can reattach farther along the actin filament if conditions still favor attachment. The force-generating rotation of the myosin head against the actin is believed to occur as multiple sites on the myosin head sequentially bind to the actin filament. As successive sites bind and unbind, the rotation of the head against the actin filament stretches the attachment of the myosin head to the myosin rod, which then pulls on the thin filament, causing it to slide past the myosin thick filament toward the center of the sarcomere. Because this process happens symmetrically at the two ends of the thick filaments, a sarcomere shortens symmetrically as the Z disks are pulled toward the center of the sarcomere. A number of muscle properties can now be understood in the light of this sliding-filament hypothesis of muscle contraction.

At rest, myosin heads cannot bind to actin sites because the associated protein tropomyosin sterically hinders binding. When a muscle fiber is depolarized, the AP propagates into the T tubules, which, through a series of steps, causes calcium channels in the SR membrane to open. As a result, Ca^{2+} is released from the sarcoplasmic reticulum into the myoplasm and then binds to troponin, a multisubunit globular protein complex that is attached to both actin and tropomyosin. When calcium binds to troponin, the protein undergoes a conformational change, which is transmitted to tropomyosin, thereby revealing myosin-binding sites on actin thin filaments. In this way, Ca^{2+} regulates contraction in vertebrate striated muscle. As the surface membrane repolarizes, the sarcoplasmic reticulum begins taking up Ca^{2+} again, removing it from troponin, terminating the active state of the muscle, and causing the muscle to relax as long as ATP is present. Although activation of smooth muscle also requires Ca^{2+}, the mechanisms involved differ from that in striated muscle.

The mechanical and energetic properties of different muscle fiber types exhibit remarkable diversity and adaptation to various activities. Muscular systems have evolved so that muscles operate at the optimal myofilament overlap to generate maximal force, and they shorten at the appropriate velocity (V/V_{max}) to generate maximal mechanical power with near-optimal efficiency. The kinetics of activation and relaxation tend to be relatively slow in locomotory muscles, minimizing the high energetic cost of calcium pumping by the sarcoplasmic reticulum. In contrast, vertebrate sound-producing muscles must operate at very high frequencies, and thus have very rapid calcium pumping. Although sonic muscles are accordingly very expensive energetically, they only make up a small fraction of any animal's muscle mass. Finally, insect asynchronous flight muscles are able to generate high mechanical power at very high frequency while avoiding the high energetic cost of calcium pumping, because they use stretch activation and shortening deactivation to achieve high-frequency modulation of force.

The control of muscle tension by the nervous system has evolved in several ways in different animal groups. Most vertebrate striated muscle fibers respond to impulses from a single motor neuron with all-or-none twitches, because they contract when the fibers themselves generate all-or-none APs. However, the muscle APs fuse into a steady tetanic contraction if the stimulating impulses occur with high enough frequency. Many arthropod striated muscle fibers (as well as vertebrate tonic fibers) contract in a graded, rather than in an all-or-none fashion, in response to graded, nonpropagated depolarizations at synapses that are distributed all along the muscle fiber. Most arthropod muscle fibers receive inhibitory innervation, in addition to synaptic input from several excitatory motor axons.

Vertebrate cardiac muscle is organized at the myofibrillar level much like striated skeletal muscle, but differs in several ways. The fibers of cardiac muscle consist of many short individual cells electrically coupled to one another via gap junctions. In skeletal muscle, in contrast, embryonic cells fuse into long multinucleated fibers, losing their integrity as individual cells. The ionic mechanisms of cardiac muscle are specialized for pacemaker activity in regions of the atria and for prolonged APs in the ventricle.

Multi-unit smooth muscle is composed of independently acting cells that require neuronal input to contract. Single-unit smooth muscle, which is more common in vertebrates, is composed of spindle-shaped cells that are electrically coupled to one another. Single-unit smooth muscle is present in the walls of vertebrate visceral organs; it can contract in response to stretch. All smooth muscle contains actin and myosin fibers, but they are not present in the organized manner characteristic of striated muscle. Smooth-muscle contraction is activated by Ca^{2+}, which enters the myoplasm primarily across the plasma membrane during depolarization, rather than from the sarcoplasmic reticulum as in striated muscle. This source of Ca^{2+} is feasible in smooth muscle because the contractions are slow and because the small cells have a large surface-to-volume ratio and small intracellular diffusion distances.

REVIEW QUESTIONS

1. Describe the organization and components of each of these structures: myofilaments, myofibrils, muscle fibers, and muscle.

2. What kinds of evidence led A. F. Huxley and H. E. Huxley to propose the sliding-filament hypothesis?

3. Draw a sarcomere and label its components.

4. Discuss the contributions of myosin, actin, troponin, and tropomyosin to contraction of striated muscle.

5. Predict the sarcomere length-tension graph of a muscle with the following filament dimensions: thick filament, 1.6 μm; bare zone, 0.4 μm; thin filament, 1.1 μm; Z disk, 0.05 μm.

6. Why do muscles become rigid several hours after an animal dies?

7. How is the force that causes the thick and thin filaments to slide past one another produced by the myosin cross-bridges?

8. When a muscle is shortening at V_{max}, what is the net force generated by its cross-bridges? What is the power produced?

9. Why does the velocity of shortening decrease as heavier loads are placed on a muscle?

10. Explain the steps by which Ca^{2+} regulates contraction of striated muscle fibers.

11. List the steps of muscle activation and of muscle relaxation.

12. How can depolarization of the surface membrane of a striated muscle fiber cause Ca^{2+} to be released from the sarcoplasmic reticulum? What molecules are involved?

13. What are the major processes in muscle function that require ATP?

14. What limits the tension that can be produced by a myofibril? By a muscle fiber? By a muscle?

15. What allows a muscle fiber to produce greater tension during tetanic contraction than during a single twitch?

16. Define mechanical power. Define efficiency. Why are mechanical power and efficiency equal to zero during isometric contractions and when a muscle shortens at V_{max}?

17. During locomotion what is the disadvantage of using too slow a muscle fiber (V_{max} is too low) to power a movement requiring a given shortening velocity? What is the disadvantage of using too fast a fiber (V_{max} is too high)? What is an optimal value of V_{max}?

18. Describe the features of the fish muscular system that enables it to produce both relatively slow movements with little backbone curvature as well as very fast movements with large backbone curvature.

19. Why must the muscles that produce sound relax very rapidly? What are the adaptations that permit the muscle to relax quickly?

20. Why is a large energetic cost associated with rapidly-relaxing muscles? How do insect asynchronous muscles avoid some of this cost?

21. What factors determine frequency of contraction in insect asynchronous muscle?

22. Discuss the major functional differences between skeletal, smooth, and cardiac muscle.

SUGGESTED READINGS

Alexander, R. McN., and G. Goldspink, eds. 1977. *Mechanics and Energetics of Animal Locomotion*. London: Chapman and Hall.

Alexander, R. McN. 1989. *Dynamics of Dinosaurs and Other Extinct Giants*. New York: Columbia University Press.

Alexander, R. McN 1992. *Exploring Biomechanics*. New York: Scientific American Library.

Ebashi, S., K. Maruyama, and M. Endo, eds. 1980. *Muscle Contraction: Its Regulatory Mechanisms*. New York: Springer-Verlag.

Huxley, H. E. 1969. *The mechanism of muscular contraction*. Science 164:1356–1365.

Josephson, R. E. 1993. Contraction dynamics and power output of skeletal muscle. *Annu. Rev. Phsyiol.* 55:527–546.

Lutz, G., and L. C. Rome. 1994. Built for jumping: the design of frog muscular system. *Science* 263:370–372.

McMahon, T. A. 1984. *Muscles, Reflexes, and Locomotion*. Princeton, N.J.: Princeton University Press.

Rome, L. C., R. P. Funke et al. 1988. Why animals have different muscle fibre types. *Nature* 355:824–827.

Stein, R. B. 1980. *Nerve and Muscle*. New York: Plenum.

Taylor, C. R., E. Weibel, and L. Bolis, eds. 1985. *Design and Performance of Muscular Systems*. Journal of Experimental Biology, Vol. 115. Cambridge: The Company of Biologists, Ltd.

Woledge, R. C., N. A. Curtin, and E. Homsher. 1985. *Energetic Aspects of Muscle Contraction*. New York: Academic Press.

BEHAVIOR: INITIATION, PATTERNS, AND CONTROL

Human beings have been studying—and making predictions about—animal behavior for as long as the species *Homo sapiens* has existed. Although our current interest in the origin and control of animal behavior stems at least in part from a search for models of our own behavior, our ancestors probably studied the topic to optimize hunting strategies and to minimize the chance that they themselves would become prey. In some respects, our curiosity about what animals do, and why they do it, may originate from a pressing need to know what that lion will do next. The complexity of this problem becomes evident when we consider all of the processes that contribute to behavior. How is information that will result in behavior collected by the various sensory organs? Where in the nervous system are decisions made, and where is coordinated action organized? How is activity in the nervous system translated into effective behavior? Understanding the function of the nervous and endocrine systems is a prerequisite for answering these and related questions about animal behavior.

All behavioral acts are ultimately generated by activity in motor neurons that causes muscles to contract. The behavior of an animal—the net movement caused by spatial and temporal patterns of muscle contraction—is constantly modified in response to stimulation from the environment. Some of these responses are simple and predictable reflexes. Other kinds of behavior depend strongly on information stored from past experience and are, therefore, less predictable to an observer who has no access to the animal's memories. The "hardware" underlying all behavior is composed of **neuronal networks,** or interconnecting circuits of neurons. Unlike electrical circuits, which are wired in a single way, neuronal networks are not "hard wired." Instead, they exhibit *plasticity,* which is the ability to be modified functionally, and even to some extent anatomically, in response to experience.

The simplest neuronal network is a **reflex arc,** in which sensory input is transmitted through some number of synapses to produce a motor output signal, which then causes muscles to contract. It is possible that the primordial reflex arc may have consisted of a **receptor cell** that directly innervated an **effector cell** (Figure 11-1A). For example, in the pharynx of the nematode *Caenorhabditis elegans,* single cells have been identified that probably serve both receptor and motor output functions. In even simpler organisms, sensory and motor functions are carried out in one cell (Spotlight 11-1).

In primitive animals, receptors and effectors may have been distributed throughout the organism, allowing each region of the body to respond to the environment relatively independently, without necessarily activating other regions. This kind of distributed nervous system is found in modern jellyfish and in the free-living coelenterate *Hydra.* However, as more-complex animals evolved, neurons became more numerous, circuits became more complex, and the nervous system became compacted into a central nervous system. Within the central nervous system, many neurons are located in close proximity, increasing the possibilities for interconnection. With many—even most—neurons being located in the central nervous system, receptors and effectors in the periphery are connected to the central nervous system by long axons.

A common simple reflex in modern animals is the **monosynaptic reflex arc** (Figure 11-1B), in which a sensory neuron (the receptor) synapses in the central nervous system onto a motor neuron that innervates a muscle (the effector). This type of reflex comprises three elements: a sensory neuron, a motor neuron, and muscle fibers. When the sensory neuron becomes sufficiently activated, it excites the motor neuron and, hence, the effector muscle. All through animal phylogeny, these elementary components of the reflex arc—sensory input pathways and motor neurons that synapse onto muscles—are characterized by features that have been conserved from the most primitive invertebrates to the most complex vertebrates.

Most reflex arcs contain more than one central synapse and are, therefore, **polysynaptic pathways.** Such a pathway includes at least one interneuron connecting the sensory and motor neurons (Figure 11-1C). In evolution, the number of interneurons increased enormously as animals became more complex, a development that allowed

A

Stimulus

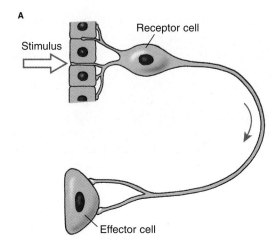

Receptor cell

Effector cell

B

Receptor cell

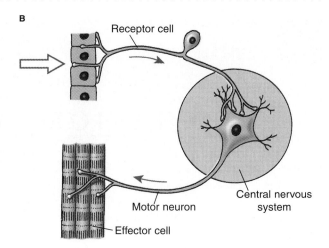

Central nervous system

Motor neuron

Effector cell

Figure 11-1 In simple reflex arcs, sensory receptors activate effector cells through a small number of synapses. (**A**) In this primitive reflex, the receptor cell directly innervates and activates an effector cell. Some chemoreceptor cells in the pharynx of the nematode *C. elegans* probably serve this function. (**B**) A monosynaptic reflex arc consists of a receptor neuron that synapses onto a motor neuron, which in turn activates muscle fibers. This kind of reflex is called monosynaptic because it includes only a single synapse within the central nervous system. (**C**) This more complicated reflex relies on several synapses in series. In parts B and C, the dashed circle encloses the part of the reflex arc that lies within the central nervous system.

C

Receptor cell

Afferent (sensory) neuron

Interneuron

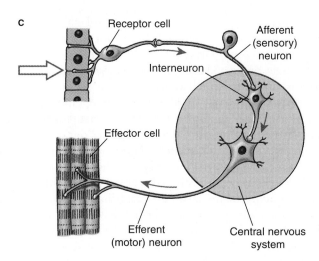

Effector cell

Efferent (motor) neuron

Central nervous system

behavioral complexity to increase dramatically in higher animals. Accumulated evidence strongly suggests that if a species has a large number of neurons interposed between input and output neurons, this situation—by itself—confers great potential for learning.

Although many properties of motor neurons, which are the *final common pathway* for motor output, have been conserved through evolutionary time, those of neuronal systems that process sensory information have not been so strongly conserved. Some elements of sensory transduction are common to many senses (see Chapter 7), but the properties of central neurons that process sensory signals have been exquisitely tuned by the circumstances of the species and can differ dramatically, depending on which sensory system has become most important in each species. For example, although both birds and bats fly—and thus encounter the same kind of environment—the way in which these animals encode information about their environment in sensory signals is quite different. As they fly in the dark, bats glean information about their surroundings by emitting sounds and listening for echos reflected by surfaces. Most birds rely heavily on vision. If a bird and a bat were to fly in the same area, the sensory systems of the two animals would represent their surroundings quite differently. Information about the distance between the bat and an object would be represented by the intensity of echoed sound or in its time of arrival; for the bird, information about distance would be represented by relative focal planes, relative position on the retina, and relative position of the two eyes. As a result, birds and bats use very different regions of the brain and mechanisms of information processing to interpret sensory signals from their environment. The auditory regions of the bat's brain are large and complex, whereas the visual region is small; in contrast, the visual regions of the bird's brain are large and complex.

Despite such differences across sensory modalities, there are general principles that apply to many sensory processing systems. For example, independent parameters of the stimulus—such as the color, size, and direction of motion of a visual stimulus—are processed in separate, parallel pathways. As information from each sensory modality is transmitted through the brain, the properties of the stimulus that are represented in each brain region become increasingly specific. In addition, most stimuli are organized systematically within each region of the brain, generating a map in which parts of the body, parts of the environment,

SPOTLIGHT 11-1

BEHAVIOR IN ANIMALS THAT LACK A NERVOUS SYSTEM

The behavior of multicellular animals depends on activity in the nervous system, but protozoa produce several interesting behaviors, too, although these single-celled creatures lack any neurons or muscles. Instead, events within the single cell serve the same functions as do sensory receptor cells, interneurons, motor neurons, and muscles. Consideration of the mechanisms that allow these apparently simple organisms to produce surprisingly complex behaviors can be a source of insight into the enormously conservative nature of evolution.

The ciliate *Paramecium* produces an *avoiding response* if it bumps into an object or if it is touched. A touch on the posterior

end of an individual organism causes it to swim forward more rapidly; a touch on its anterior end causes it to reverse direction (part A of the accompanying illustration). The direction in which a *Paramecium* swims depends on the direction in which its cilia are beating; a reversal in the direction of progress is caused by a reversal of the ciliary beat. What mechanism could account for such a reversal in response to mechanical stimulation?

Experiments in the laboratory of Roger Eckert revealed that the membrane of *Paramecium* includes stretch-activated ion channels that are differentially distributed: the channels in the anterior membrane are Ca^{2+} selective, whereas the channels in the posterior membrane are K^+ selective (see illustration).

Thus, in single-celled organisms, behavior is controlled by changes in membrane potential caused when ion fluxes across membranes change, when ion channels in the membrane open. Perhaps the properties of neurons and muscles can be regarded as an abstraction of some of the capabilities of these multifunctional "simple" organisms.

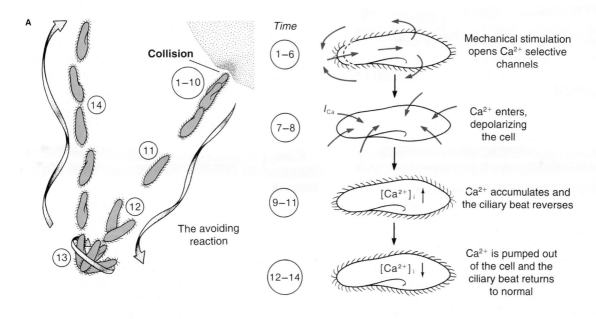

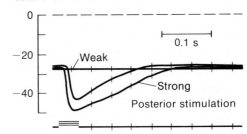

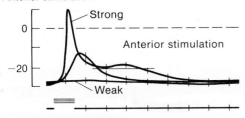

A *Paramecium* avoids objects with which it has collided by changing its direction and rate of swimming. (**A**) After it collides with an object, a *Paramecium* backs up, changes direction, and swims away in a new direction *(left)*. The passage of time is indicated by sequential numbers. Mechanical stimulation of the anterior end opens Ca^{2+}-selective channels *(right)*, increasing the internal free Ca^{2+} concentration, which in turn reverses the ciliary beat. (**B**) When the Ca^{2+} channels are opened by mechanical stim-

ulation to the anterior end, the membrane depolarizes and produces a graded, but weakly regenerative, change in the membrane potential that is associated with reversal of the ciliary beat. (**C**) Mechanical stimulation of the posterior end opens K^+-selective channels, producing a graded hyperpolarization of the membrane that accelerates the ciliary beat by an unknown mechanism. [Part A adapted from Grell, 1973; parts B and C adapted from Eckert, 1972.]

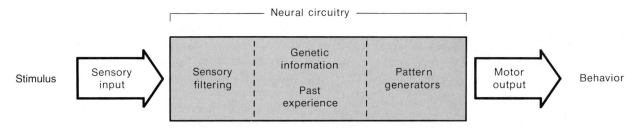

Figure 11-2 Information processing within the central nervous system can be divided into functionally related compartments. Sensory input enters from receptors and is filtered and processed. The sensory information is then integrated with other sensory information, with genetic programming, and with memories of past experience, eventually activating neurons that in turn generate appropriate motor output.

or some other attribute of the stimulus (e.g., the frequency of sounds) are arranged topographically in an orderly fashion with respect to one another. The original view that these maps are static has been replaced by the understanding that to some extent their topography can be regulated dynamically by use-dependent mechanisms. We consider these principles in detail later in this chapter.

Animals can change their behavior as a result of experience, which we call *learning,* and can store the information to be used in the future, which we call *memory.* Study of the physiological (and potentially the anatomic) mechanisms that underlie learning and memory has focused on changes at synapses. Recent insights into the cellular and molecular basis of learning and memory can now be incorporated into our view of how behavior is generated. Both learning and the production of complex behavior patterns must depend on the enormous amount of neuronal circuitry that lies between the relatively simple afferent sensory pathways and the efferent motor pathways. In higher organisms, most neurons are central interneurons. In Figure 11-2, this complex interface between sensory input and motor output is represented as a single box separating the input and output neurons. Indeed, this part of the nervous system remains a "black box" that is understood in only the most fragmentary sense. The outwardly observable relations between stimulus and response are the domain of behavioral psychology. In contrast, one goal of neuroscientists is to understand how the connecting neuronal circuitry works. One approach has been to study the nervous systems of relatively simple animals to determine what occurs in the time between the reception of a stimulus and the production of observable behavior. In this chapter, we examine some of the systems that are helping us to understand what the nervous system does when it processes sensory information and generates patterned responses.

EVOLUTION OF NERVOUS SYSTEMS

The evolution of the nervous system cannot be reconstructed directly from the fossil record in the same way that we can reconstruct the evolution of, say, the leg bones in vertebrates, because soft neuronal tissues leave little trace. However, examination of the neuronal organization of increasingly complex members of present-day phyla gives us some basis for speculating about the routes by which more complex nervous systems may have evolved from simpler ones. Until recently, in fact, comparing the structure and function of nervous systems in animals from different phyla was the only basis for inferring evolutionary relationships. It is now also possible to compare DNA sequences that have been conserved across species and to infer evolutionary relationships on the basis of how similar or different the sequences are in modern species. This method of **molecular phylogeny** provides information that complements more conventional analyses based on the study of whole organisms. Reconstruction of phylogenetic relationships based on a comparison of DNA sequences requires that the fossils or living organisms being compared have in common a particular molecule or molecules for which the nucleic acid coding sequence can be determined.

At the cellular level, the nervous system seems to have undergone surprisingly little modification in the course of evolution. The electrical and chemical properties of nerve cells in vertebrates and invertebrates are remarkably similar (see Chapters 5 and 6). Many general principles of neuronal function were derived from studies of the relatively simple nervous systems of invertebrates and lower vertebrates, which are more amenable to experiments than are the more complex nervous systems of higher vertebrates. In particular, neurons in many invertebrate species are large, accessible, and readily recognizable from animal to animal, making their activity relatively easy to record and to analyze. Recently, it has even become possible to do biochemical and molecular analyses of single neurons that have been identified in, and dissected from, invertebrate nervous systems.

The anatomically simplest nervous systems consist of very fine axons that are distributed in a diffuse network (Figure 11-3). Such **nerve nets** are commonly found among the coelenterates. These axons make synaptic contacts at points of intersection, and a stimulus applied to one part of an organism produces a response that spreads in all directions away from the point of stimulation. If the stimulus is repeated at brief intervals, conduction is *facilitated* and the signal spreads further. Very little is known about synaptic mechanisms in diffuse nerve nets, because the axons are extremely fine, making intracellular recording from them very difficult technically. However, even in the very simple nerve

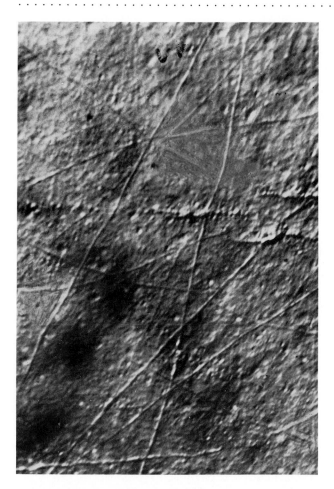

Figure 11-3 Axons in the nerve net of the jellyfish *Aurelia* can be seen on the subumbrellar surface when the organism is viewed in oblique light. The axons are arranged in a diffuse network that runs in all directions. They innervate muscles that cause the umbrella to contract. [Courtesy of A. Horridge.]

nets of coelenterates and ctenophores, there is evidence that the neurons are organized into reflex arcs.

A major early advance in the evolution of nervous systems was the organization of neurons into **ganglia,** or clusters of neuron cell bodies. Ganglia are found in the coelenterates and are common throughout most of the animal kingdom. A ganglion (Figure 11-4A, C) consists of neuronal somata that are organized around a mass of nerve processes (axons and dendrites) called a **neuropil.** This mode of organization permits extensive interconnections to be made among the neurons, at the same time allowing each neuron to produce a minimum number of **collateral processes**—that is, side branches arising from the axon or axons. A neuropil may appear to be a randomly arranged tangle of fine processes, but injection of markers or dyes into individual neurons (Figure 11-4B and D) reveals that the major features of each particular type of neuron are organized similarly from one individual animal to another, even though fine details may differ. Moreover, physiological evidence shows that connections within the neuropil are so orderly that identical synaptic interactions can be ob-

served between homologous neurons in different individuals of the same species.

Segmented invertebrates have nervous systems that are somewhat decentralized, consisting of a ganglion within each body segment. Each segmental ganglion usually serves the reflex functions of the segment in which it is located, plus perhaps those of one or more adjacent body segments. In addition, information is exchanged among ganglia through trunks of axons, called **connectives.** The result is a series of ganglia and connectives in the ventral nerve cord that is characteristic of annelids and arthropods (Figure 11-5; see also Figure 5-5). At the anterior end of the ventral nerve cord in these phyla, one or more relatively large clusters of neurons form a **brain**—or superganglion—which receives sensory information from the anterior end of the animal and which controls movements of the head. In addition, neurons with their somata in the brain often exert some control over other ganglia along the ventral nerve cord and can contribute to coordinating movements that require the body to work as a unit. This coalescence of neurons at the anterior end of the animal, where many sensory receptors are concentrated, is called *cephalization* and is a common feature of central nervous systems. (Not all brains are located at the anterior end of an organism; for example, at the posterior end of its ventral nerve cord, the leech has a tail brain that is even larger than the head brain.)

The segmental ganglia of annelids and arthropods have been used extensively for neuronal analysis because each ganglion contains a relatively small number of neurons; and, in many cases, several ganglia contain identical complements of neurons. Thus, in principle, an analysis of the interactions of neurons in one segment can be generalized to many—even to all—other segments of the nerve cord. For example, this approach has been useful in studying the nervous system of the leech (see Figure 5-5), in which the ganglia are highly similar to one another. Despite the simplicity of structure in its nervous system, a leech can perform such complex tasks as swimming to seek food or crawling to escape danger, which makes it particularly well suited for studies of the cellular basis of behavior.

The structure of the nervous system varies among the other invertebrate phyla. In contrast with worms and arthropods, whose body plans are segmentally structured and bilaterally symmetrical, an echinoderm typically has a nerve ring and radiating nerves arranged about its axis of symmetry. Perhaps as a result of their radial symmetry, echinoderms lack any brainlike ganglion. Mollusks have a nonsegmental nervous system consisting of several dissimilar ganglia that are connected by long nerve trunks.

Neurons in some molluskan species have contributed greatly to our understanding of neuronal interactions. The ganglia of opisthobranch mollusks (e.g., the sea hare *Aplysia*) and nudibranchs (e.g., *Tritonia*) contain a number of neurons with extraordinarily large cell bodies; in *Aplysia,* the somata of some neurons are more than 1 mm in diameter. Individual giant neurons can be recognized

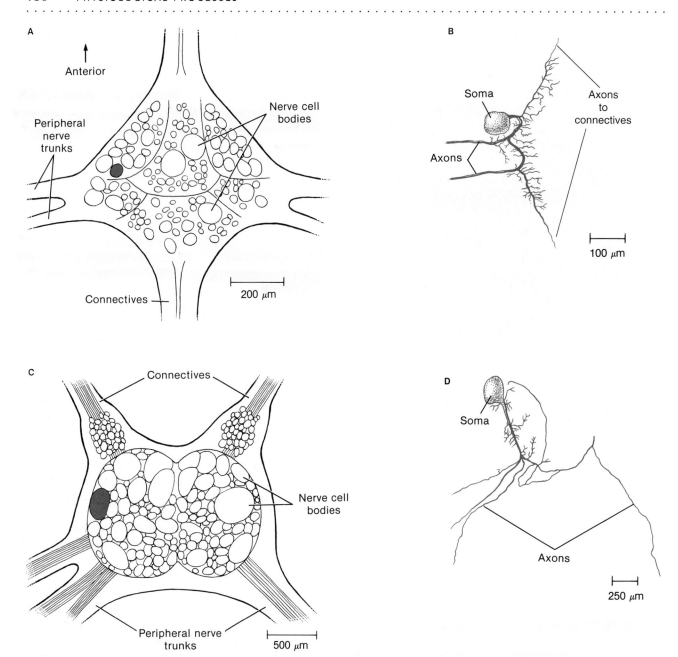

Figure 11-4 Neuronal somata are organized into ganglia in many species—including the higher vertebrates. This figure illustrates ganglia from members of two invertebrate phyla. In these ganglia, individual nerve cells can be readily identified from one specimen to another. (**A**) A segmental ganglion in the annelid leech *Hirudo*, showing the position of individual nerve cell bodies. The paired connectives at the top and bottom of the diagram contain axons that connect neurons in the ganglion with neurons in the ganglia of adjacent body segments. The peripheral nerve trunks emerging laterally carry motor and sensory axons to the viscera and muscles. (**B**) Mechanosensory neuron (shown in color in part A) that has been labeled by injecting it with an intracellular marker. This marker remains within the cell and diffuses into all of its branches. Note the numerous small branches on which synaptic contacts can be made with similar branches of other cells. The two large axons enter the peripheral nerve trunks at left and the two smaller axons enter the connectives. (**C**) Schematic diagram of the abdominal ganglion of the molluskan sea slug *Aplysia californica*. (**D**) Morphology of a neuron (shown in color part C) injected with an intracellular marker. This neuron sends axonal branches to all of the peripheral nerves shown at the bottom of part C. [Part A adapted from Yau, 1976; part B adapted from Muller, 1979; part C adapted from Kandel, 1976; part D adapted from Winlow and Kandel, 1976.]

visually from preparation to preparation, and they lend themselves well to long-term electrical recording, injection of experimental agents, and isolation for microchemical analysis. As in the nervous systems of annelids and arthropods, individual neurons in mollusks can be reliably identified on the basis of their location and the size of their so-

mata, enabling an experimenter to determine the properties that characterize a particular type of cell and to estimate the amount of variation that can occur from individual to individual.

The most complex nervous system known among the invertebrates belongs to the octopus. The brain alone is es-

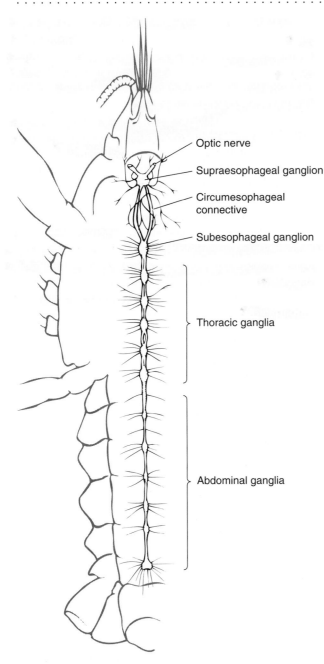

Figure 11-5 The ventral nerve cord of the crayfish *Astacus* illustrates the segmented arrangement of the nervous system in many invertebrates. There is one ganglion in each body segment. Connectives contain axons that run between ganglia, and the nerve roots that link the ganglia with structures in the periphery contain sensory and motor axons. Bilateral symmetry, another property typical of nervous systems in most phyla, also is illustrated in this diagram.

Labels in figure:
Optic nerve
Supraesophageal ganglion
Circumesophageal connective
Subesophageal ganglion
Thoracic ganglia
Abdominal ganglia

In general, the nervous systems of invertebrate animals, excluding the octopus, contain significantly fewer neurons than do those of the vertebrates; for this reason, invertebrate nervous systems are often referred to as "simple." However, superficial appearances can be deceiving, and the functional sophistication of even relatively simple nervous systems becomes apparent on closer examination. Later in this chapter, we will consider examples of the behavior of invertebrates and the neuronal circuitry that underlies the behavior.

From observation of the variety of nervous systems in the animal kingdom, several principles of evolution have emerged:

1. Nervous systems in all organisms are based on one type of cell, the neuron. Although neurons are sculpted into myriad shapes in the course of development, the mechanisms of electrical signaling within the cell and the nature of the chemical signals that allow information to be transmitted between cells are highly conserved across phylogeny.

2. The organization of the nervous system evolved through the elaboration of one fundamental pattern: the reflex arc. Just as the neuron is the basic structural unit of the nervous system, the reflex arc is its basic operating unit. In its simplest form, a reflex produces a stereotypical response to a particular sensory stimulus. In the simplest reflex arc, a monosynaptic reflex, a sensory neuron synapses directly onto a motor neuron, which synapses onto muscle cells. Coordinated contraction of particular muscles produces behavior.

3. Through evolution, there has been a trend toward gathering neurons into a central nervous system, connected to peripheral sensory receptors and muscles by long axons. The organization of these networks favors one-way conduction through neurons, from dendrites to the axon and, thence, to the axon terminals, although the biophysical properties of axons would enable them to conduct either toward or away from the soma.

4. Complicated organisms have more neurons than do simpler organisms, and their neurons are often concentrated in a brain, usually contained in the head.

5. As nervous systems became more complex through evolution, new structures were added on to older structures, rather than replacing them. As a result, structures that subserve new functions, which have arisen more recently in evolution, are often seen to be literally layered onto older, more primitive structures.

6. The size of particular regions in the brain of a species is usually related to the importance to that species of sensory input into, or motor control out of, those regions. For example, in animals that depend primarily on vision for survival, the regions of the brain that process visual information are typically larger than all other sensory areas. In nocturnal animals, other brain regions—for example, those that process hearing or olfaction, which are independent of light—are largest.

timated to contain 10^8 neurons—compare this number with the 10^5 neurons in the entire leech body. The neurons in the octopus brain are arranged in a series of highly specialized lobes and tracts that evidently evolved from the more dispersed ganglia of the lower mollusks. If number of neurons in any way correlates with intelligence, the octopus should be fairly smart, and behavioral studies have shown that, by invertebrate standards, the octopus is indeed quite intelligent.

ORGANIZATION OF THE VERTEBRATE NERVOUS SYSTEM

The vertebrate nervous system is organized into identifiable structural and functional regions (Figure 11-6), although neurons in many regions may work together to process incoming information and to generate appropriate behavior. For example, the system is divided into the *central nervous system* (CNS) and the *peripheral nervous system* (PNS). The central nervous system contains most of the neuronal cell bodies. It includes the cell bodies of all interneurons and of most neurons that innervate muscles and other effectors. The *peripheral nervous system* includes nerves, which are bundles of axons from sensory and motor neurons; ganglia that contain the cell bodies of some autonomic neurons; and ganglia that contain the cell bodies of most sensory neurons. (The retina is contained entirely *within* the central nervous system, unlike other sensory neurons.) Nerves are **afferent** if they carry information toward the brain and **efferent** if they carry information away from it. Many vertebrate nerves are **mixed nerves,** which include both afferent and efferent axons.

Efferent output from the central nervous system can be divided into two main systems: **somatic** and **autonomic.**

The somatic system is also referred to as the *voluntary system,* because motor neurons of this division control skeletal muscles and produce voluntary movements. The autonomic nervous system includes motor neurons that modulate the contraction of smooth and cardiac muscle and the secretory activity of glands. The autonomic nervous system thus controls the "housekeeping" systems, such as heartbeat, digestion, and temperature regulation. The term *autonomic* literally means "self-managed," and it was introduced when the relation between the autonomic nervous system and the more voluntary parts of the central nervous system was poorly understood. We now know that the apparently automatic responses regulated through the autonomic nervous system are integrated and controlled within the central nervous system, just like the responses produced consciously through more voluntary channels. The neurons of the autonomic system are themselves divided in two categories: **sympathetic** and **parasympathetic,** which differ from one another anatomically and functionally.

A striking characteristic of vertebrate nervous systems is their enormous redundancy; that is, there are a large number of *individual* neurons of every identifiable *type.*

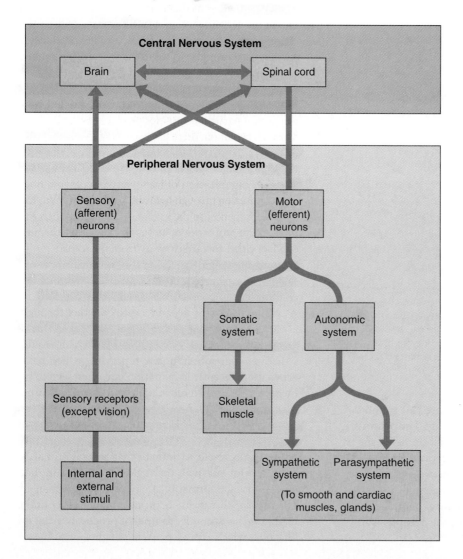

Figure 11-6 The vertebrate nervous system is organized into identifiable regions. The central nervous system consists of the brain and spinal cord. Information about the environment is brought to the central nervous system by sensory (afferent) neurons. The animal's response is generated and shaped by motor (efferent) neurons. Somatic motor neurons control the contraction of skeletal muscle. Autonomic neurons, divided into the sympathetic and parasympathetic systems, control the activity of smooth and cardiac muscles and of glands.

In an arthropod nervous system, a single motor neuron may control essentially all of the fibers in a particular muscle. In some cases, a single neuron may even, by itself, innervate more than one muscle of a limb. In contrast, in the vertebrates, each skeletal muscle is typically innervated by a pool of several hundred motor neurons. Each motor neuron in the pool controls a motor unit that typically consists of about 100 muscle fibers. (However, motor units may be much smaller than that or they may be much larger, in some cases containing more than 2000 fibers.) The motor neurons of each pool have common physiological properties, so information obtained from one motor neuron generally characterizes the whole pool. If this redundancy did not exist and if each neuron in the vertebrate central nervous system had unique properties, the enormous number of neurons would stymie any attempt to study the system. Instead, researchers have observed a fundamental orderliness in the system, which is thought to have arisen as natural selection conserved many neuronal properties through time while permitting duplication of individual units.

Major Divisions of the Central Nervous System

Although vertebrates, with their relatively large brains, display a great degree of cephalization, they retain a basic segmental organization in much of the central nervous system. Segmentation is particularly apparent in the organization of the spinal cord (Figure 11-7), but it can also be seen in the *cranial nerves*, which connect centers of the brain with structures in the head and the rest of body.

The spinal cord

In simpler vertebrates, the **spinal cord,** enclosed and protected by the vertebral column (Figure 11-8A), is a segmentally organized site of reflex action that can act independently of the brain but that also receives a large amount of input from higher centers. As vertebrates became more complex through evolution and the brain acquired more control over spinal function, the functional segmental organization of the spinal cord remained intact. The spinal cord is divided into regions—cervical, thoracic, lumbar, and sacral—named on the basis of their location. Within each region, the cord is divided into segments, each of

A

B

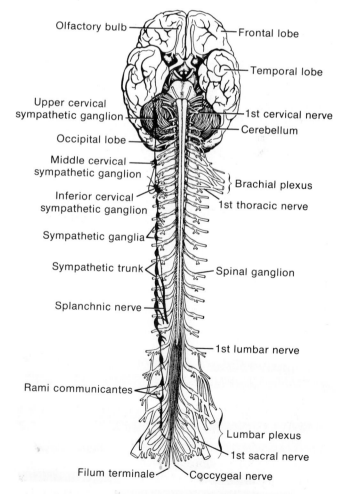

Figure 11-7 The structure of the vertebrate central nervous system is very elaborate in the head, but segmental organization is maintained in both the brain and the spinal cord. Diagrams of a ventral view of the frog (**A**) and the human (**B**) central nervous systems. Although segmentation is not apparent in many structures of the brain, it is the basis of organization within the spinal cord. It is also apparent in the organization of the cranial nerves. [Adapted from Wiedersheim, 1907; Neal and Rand, 1936.]

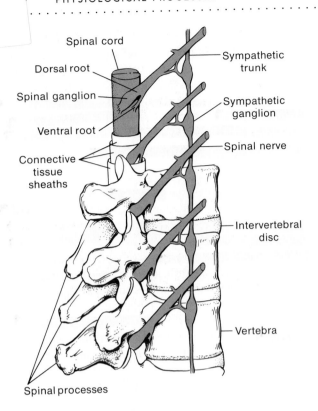

Spinal cord
Dorsal root
Spinal ganglion
Ventral root
Connective tissue sheaths
Spinal processes
Sympathetic trunk
Sympathetic ganglion
Spinal nerve
Intervertebral disc
Vertebra

Figure 11-8 Each spinal segment is connected to the periphery by a dorsal root, through which afferent signals enter, and a ventral root, through which efferent signals leave. (**A**) Relation between the vertebral column, the spinal cord, and the spinal nerves, which emerge between the vertebrae. Vertebrae are separated from one another by cushions of connective tissue, the intervertebral discs. Regions of the sympathetic nervous system that lie in the periphery are included in this diagram. Nervous tissue is shown in color. (**B**) Diagram of a cross-section through a single spinal segment. Neurons in a polysynaptic reflex arc that connect input from skin receptor endings to output onto skeletal muscles are shown in color. The pathway includes an interneuron in addition to the sensory receptor and the motor neuron. Note that the cell body of the sensory afferent neuron resides in a dorsal root ganglion outside the spinal cord. [Adapted from Montagna, 1959.]

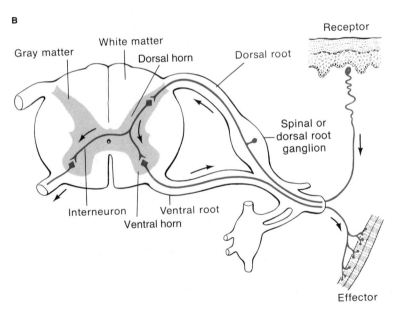

B

Gray matter
White matter
Dorsal horn
Dorsal root
Receptor
Spinal or dorsal root ganglion
Interneuron
Ventral root
Ventral horn
Effector

which receives information from, and sends information to, the periphery through the dorsal and ventral **spinal roots** (Figure 11-8B). Seen in cross section, the spinal cord is bilaterally symmetrical. Ascending (sensory) and descending (motor) axons are grouped around the outside surface of the cord and form well defined tracts. This outer region of the cord is called the **white matter,** named for the white appearance of the myelin that ensheathes many of the axons. The more centrally located **gray matter** of the spinal cord contains cell bodies and dendrites of interneurons and motor neurons, as well as the axons and presynaptic terminals of neurons that synapse onto these neurons. Most of the structures in the gray matter are not myelinated, so this central region lacks the shiny whiteness of the tracts. A fluid-filled central lumen, the **spinal canal,** is continuous with the fluid-filled cavities within the brain, which are called the **cerebral ventricles.** The **cerebrospinal fluid** in these cavities has a composition similar to that of blood plasma.

The regular organization of the spinal cord has facilitated its study by neurophysiologists. To a very large extent, afferent and efferent pathways are anatomically separated from one another (a pattern that was observed more than a century ago and that has been referred to as the "Bell-Magendie rule"). Afferent axons enter the central nervous system through the dorsal spinal roots (see Figure 11-8B); efferent nerve fibers carry information out of the central nervous system through the ventral roots. (There are minor exceptions to this rule. For example, fine unmyelinated sensory afferents enter the spinal cord through at least some of the ventral roots in the cat.) The cell bodies of spinal motor neurons are located in the ventral gray matter, called the **ventral horn,** on each side of the spinal cord, whereas the cell bodies of interneurons that receive sensory input are located in the dorsal gray matter, called the **dorsal horn.** Afferent axons that synapse onto sensory interneurons within the cord arise from sensory neurons whose cell bodies lie in the **dorsal root ganglia** outside the central nervous system. (There is one dorsal root ganglion on either side of each spinal segment.) The segregation of sensory and motor axons into the dorsal and ventral roots makes it possible to selectively stimulate the sensory input into (or motor output arising from) a single spinal segment. Alternatively, input to a segment or output from it can be selectively eliminated by transecting one of the spinal roots.

Many reflex connections are located in the spinal cord. Later in this chapter, we consider two spinal reflexes, the *stretch reflex* and the *pain-withdrawal reflex.* In addition, neuronal connections that produce patterned movements of locomotion—for example, walking, running, and hopping—are located in the spinal cord. Although signals from the brain can activate, suppress, or modulate behavioral patterns that arise from activity in spinal circuits, it appears that connections among spinal neurons are sufficient by themselves to generate complex and coordinated behaviors.

The brain

In all vertebrates, the brain contains many groups of neurons with specialized functions, such as the reception and processing of information from the eyes and the initiation of movements that require coordination of the whole body. In the higher vertebrates, the brain contains many more neurons than does the spinal cord, and it exerts very strong control over the rest of the nervous system.

Structure of the vertebrate brain The basic structure of the vertebrate brain is the same in all vertebrate classes (Figure 11-9). At the most caudal part of the brain, where the brain joins the spinal cord, the cord enlarges to form the **medulla** (also called the *medulla oblongata.*) The medulla contains centers that control respiration, the cardiovascular reflexes, and gastric secretion. It also contains groups of neurons that receive sensory information from several modalities and transmit it to other sensory and motor centers in the brain.

The **cerebellum,** which is dorsal to the medulla, consists of a pair of hemispheres that have a smooth surface in lower vertebrates and a very convoluted surface in higher vertebrates. The convolutions increase the surface area, providing space for many more neurons. The cerebellum has long been known to contribute to the production of

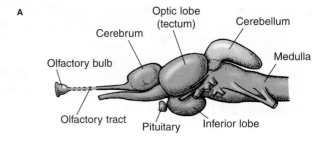

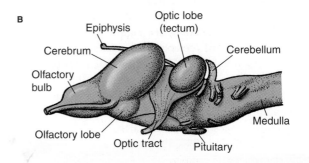

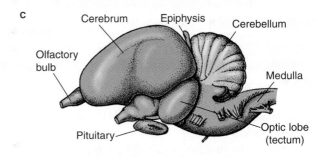

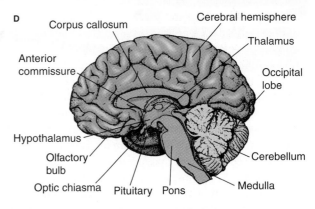

Figure 11-9 The brains of all vertebrate classes have certain structures in common, although the relative sizes of each region vary among species. (**A**) Lateral view of the brain of a fish. (**B**) Lateral view of the brain of a frog. (**C**) Lateral view of the brain of a bird. (**D**) Lateral view of a human brain that has been sectioned down the midline along the plane of bilateral symmetry. Note the gradual increase in the relative size of the cerebrum with evolutionary development. In contrast, the tectum (called the *superior colliculus* in mammals) maintained its relative size during evolution. The cerebellum, which has an important function in the coordination of movement, is highly developed in birds and mammals. [Adapted from Romer, 1955.]

motor output. It integrates information arriving from the semicircular canals and other proprioceptors and from the visual and auditory systems. These inputs are compared in the cerebellum, and the resulting output helps to coordinate the motor output responsible for maintaining posture, for orienting an animal in space, and for producing accurate limb movements. The relative size of the cerebellum differs greatly among species and exemplifies the principle that the size of a brain region is correlated with its relative importance in the behavior of each species. For example, a comparison of the relative sizes of the cerebellum in the brains of birds and of mammals (see Figure 11-9C, D), reveals that, typically, a bird's cerebellum is significantly larger than that of a mammal. This difference in size is thought to be related to the greater complexity of controlling the movements and orientation of an animal flying through space compared with that of controlling movement along the two-dimensional surface of the earth.

The cerebellum lacks any direct connection to the spinal cord, so it cannot control movements directly. Instead, it sends signals to regions of the brain that do directly control movement. In addition, the cerebellum has been found to participate in the *learning* of motor skills, and recent observations suggest that abnormalities in cerebellar neurons may contribute to the problems faced by autistic people. As we learn more about cerebellar function, it is possible that this part of the brain will be found to have a much broader role in the regulation of behavior than has been suspected.

Some of the structures of the midbrain act as relay stations, receiving information and transmitting it to other centers. In mammals, the **pons**, which means "bridge," consists of fiber tracts that interconnect many different regions of the brain. For example, they connect the cerebellum and the medulla with higher centers. The **tectum** *(optic lobe),* located in the midbrain, receives and integrates visual, tactile, and auditory inputs. Information from each of the sensory modalities that comes into the tectum is organized into a map that represents some feature of the environment. For example, in the map of visual input, locations in the environment that are close to one another are represented as being close to one another in the visual map. Maps from different sensory modalities are located in different layers of the tectum, but the maps are congruent with one another. For example, in the map of auditory inputs, the locations of sound sources are related to one another in the same way as objects in the visual world are organized in the visual map. In fishes and amphibians, the tectum exerts major control over body movements. Indeed, surgical removal of the cerebral hemispheres from a frog's brain does not substantially reduce the frog's behavioral capacity, but removal of the tectum incapacitates the animal. The tectum also serves important functions in reptiles and birds, but in mammals it serves primarily as a way station for signals on their way to higher centers.

The **thalamus** is a major coordinating center for sensory and motor signaling. It serves as a relay station for sensory input, providing initial information processing as well. Thalamic function can be modified by higher centers; for example, it receives an enormous number of inputs from the cerebral cortex. The **hypothalamus** includes a number of centers that control visceral functions, such as body temperature regulation, eating, drinking, and sexual appetite. Hypothalamic centers also participate in the expression of emotional reactions such as excitement, pleasure, and rage. Neuroendocrine cells in the hypothalamus control water and electrolyte balance and the secretory activity of the pituitary gland (see Chapter 9).

The anterior part of the brain contains structures related to its oldest and its newest features. The olfactory system forms the largest part of the anterior brain in many primitive vertebrates, suggesting that detecting food odors and interpreting chemical communication must have been powerful selective agents in the evolution of early vertebrates. Perhaps as a manifestation of its ancient importance, olfaction is the only sensory modality that is not processed through the thalamus but travels directly to the cerebrum. In lower vertebrates, the primitive **cerebrum** integrates olfactory signals and organizes motor responses to them. The large cerebral hemispheres that dominate the human brain evolved from this small cerebrum and its limited function.

Development of the vertebrate brain The segmental structure of the vertebrate central nervous system appears to be based on developmental processes that have been strongly conserved during evolution. The progenitor of the entire nervous system is the *neural tube,* a structure that arises from part of the outermost layer of a gastrula-stage embryo. The first step in the development of the brain is the formation of three expanded vesicles in the most anterior part of the neural tube; these vesicles are—from anterior to posterior—the forebrain, the midbrain, and the hindbrain (Figure 11-10). The center of the tube is a fluid-filled cavity, which is the precursor of the cerebral ventricles. In later stages, the three regions generate five subdivisions. Each subdivision grows as a result of cell division, particularly along the surface of the fluid-filled cavity, the **ventricular zone,** and of migration by cells away from this zone. Embryonic neurons usually migrate within their segment of origin, although some have been found to cross boundaries between regions. The linear and segmental organization that appears early in the development of the brain is ultimately obscured by distortions produced through unequal growth within the embryonic subdivisions. However, the linear organization of the primary vesicles is preserved to some extent in the organization of the pathways traversed in the adult brain by incoming and outgoing information.

Organization of the mammalian cerebral cortex In higher mammals, the **cerebral cortex**—the layer of cells that covers the cerebral hemispheres—is thrown into prominent folds that greatly increase the surface area and, hence, the total number of cortical neurons. This surface layer of gray

Neural tube   **Primary vesicle**   **Secondary vesicle**   **Adult structures**

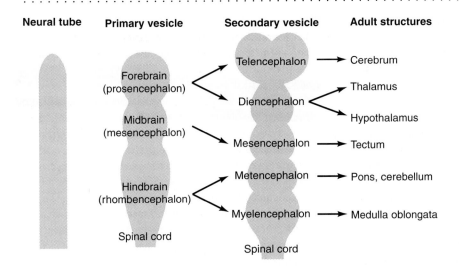

Telencephalon ⟶ Cerebrum

Forebrain
(prosencephalon)

Diencephalon ⟶ Thalamus

⟶ Hypothalamus

Midbrain
(mesencephalon) ⟶ Mesencephalon ⟶ Tectum

Metencephalon ⟶ Pons, cerebellum

Hindbrain
(rhombencephalon)

Myelencephalon ⟶ Medulla oblongata

Spinal cord

Spinal cord

Figure 11-10 The vertebrate brain begins as three linearly arranged enlargements of the anterior neural tube, which then become elaborated. Initially, the brain consists of the forebrain, the midbrain, and the hindbrain. Later, the forebrain divides into the telencephalon and the diencephalon; the hindbrain forms the metencephalon and the myelencephalon. At the right is a partial listing of adult structures that arise from the secondary vesicles. Sensory information is typically received in structures derived from the myelencephalon and metencephalon, is sent rostrally through structures derived from the mesencephalon and diencephalon, and is then sent on to the cerebral cortex, which develops from the telencephalon. Thus, although morphogenetic changes in the brain obscure its segmental origins, the pattern of information processing echoes early segmentation.

matter is organized into layers that are parallel to the surface, each having a recognizable pattern of input and output. In addition, it is organized into functional regions (Figure 11-11). Some areas of the cerebral cortex contain neurons that are purely sensory in function; that is, they receive sensory information, process it, and pass it on. Other regions are purely motor in function. In primitive mammals, such as rats, the sensory and motor areas account for almost all of the cortex. In contrast, the cerebral cortex of human beings and other higher mammals contains large regions that are neither clearly sensory nor clearly motor in function. These regions, such as the frontal cortex, are referred to as **association cortex,** and they are responsible for such complex functions as intersensory associations, memory, and communication.

The cortical regions that are purely sensory in function include the primary auditory, somatosensory, and visual cortical areas. (**Primary projection cortex** is the first location in the cortex to which information from a particular sensory modality is transmitted.) The amount of brain space allotted to each of the sensory modalities is related to the principal habits of an animal species. For example, the primary somatosensory cortex—which receives information from neurons that sense stimuli such as touch, temperature, and pain—is relatively much larger in rats and in tarsiers (a type of primitive primate) than it is in champanzees and human beings. The larger somatosensory cortex in rats and tarsiers correlates with their dependence on collecting and discriminating tactile information. On the other hand, all primates—including human beings,

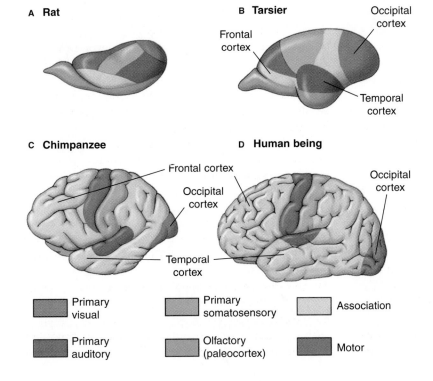

A   Rat

B   Tarsier

Occipital cortex

Frontal cortex

Temporal cortex

C   Chimpanzee

Frontal cortex

Occipital cortex

Temporal cortex

D   Human being

Occipital cortex

Primary visual

Primary auditory

Primary somatosensory

Olfactory (paleocortex)

Association

Motor

Figure 11-11 The mammalian cerebral cortex is composed of regions having specific functions. Side views of the cerebral cortex of four different mammals, showing the functional divisions within the cortex. Regions that have purely sensory or purely motor function are shaded. Unshaded regions serve "association" functions. Notice that the relative amount of association cortex increases from rat to human being. The frontal, temporal, and occipital regions are labeled on the three primate brains. In all drawings, the posterior pole of the brain points to the right.

panzees, and tarsiers—have much larger primary visual cortices than do rodents such as rats.

The size and location of cortical regions have been determined in electrophysiological experiments in which activity recorded from neurons within each site is correlated with the application of specific stimuli. Perhaps the most dramatic of these experiments have been performed on human beings. Many neurosurgical procedures are performed while the patients are awake and relatively alert. (The central nervous system itself has no pain receptors, so local anesthesia in the scalp and skull at the site of electrode entry allows the patient to remain comfortable and awake during the surgery.) In these procedures, the stimulation of neurons within a particular region of sensory cortex evokes sensations in the patients, who are able to tell the surgeon where in the periphery the sensation seems to arise and which modality is being stimulated. These experiments, conducted in the course of therapeutic surgery, have supported the hypothesis that *all* sensation occurs in the central nervous system, primarily in the sensory areas of the cerebral cortex.

The somatosensory cortex exemplifies the mapping of sensory input within the sensory cortex. This region of the cortex is organized so that different locations receive inputs from specific areas of the body. Sensations from areas of the body that are adjacent to one another are mapped so that they are adjacent to one another in the cor-

tex (Figure 11-12A). More neurons are dedicated to peripheral areas from which "important" information is received—for example, the face and the hands—than to other parts of the body. Indeed, in human beings, approximately half of the somatosensory cortex receives input from the face and hands, with the other half responsible for the remainder of the body surface. This remarkable dedication of resources to tactile information coming from the face and the hands correlates with the importance of these areas in our daily lives. For example, the human central nervous system requires detailed sensory information from the hands to carry out the fine manipulations that we perform with our hands when we do tasks and when we identify objects by touch. The map of sensory distribution, also called the *sensory homunculus*, differs in detail among species, reflecting the different roles played by particular parts of the body. In raccoons, for example, more sensory cortical area is dedicated to the forepaws than to any other part of the body, presumably because these animals manipulate their food and other objects extensively in their daily lives.

A further elaboration of the general principle that space in the brain is organized according to the importance of incoming sensory information is particularly well illustrated by the somatosensory cortex of the star-nosed mole. In this animal, 11 fleshy rays protrude from each side of the nose (Figure 11-13A, B), each covered with small swellings

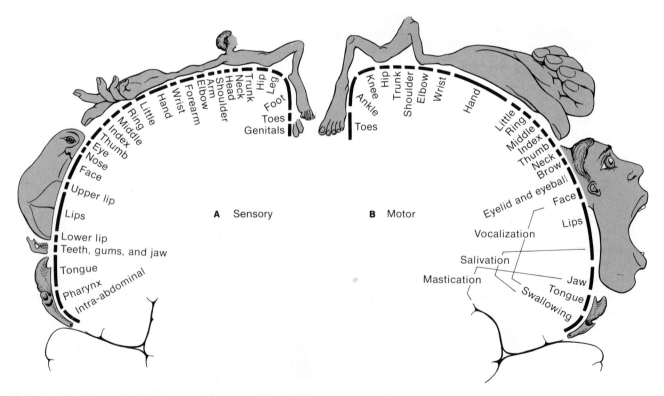

Figure 11-12 Both sensory and motor cortical regions are organized into topographical maps of the body. (**A**) Map of a transverse section through the somatosensory cortex, depicting the location of neurons and their corresponding peripheral projections—that is, the places in the periphery where the stimuli are subjectively "felt." Information coming from the face and hands occupies about half of the entire human somatosensory cortex. (**B**) Map of a transverse section through the human motor cortex, showing the distribution of cortical neurons that project to motor neurons of the brain and spinal cord, which in turn control activity in specific skeletal muscles.

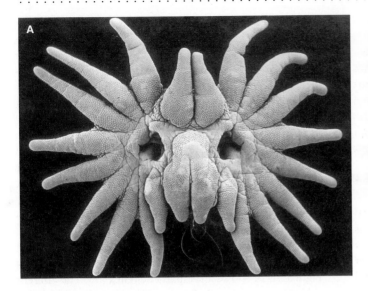

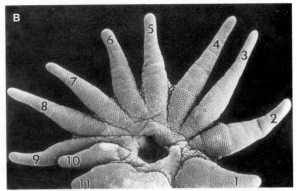

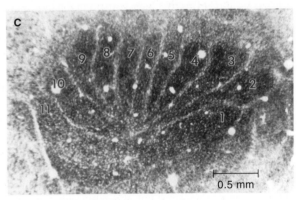

Figure 11-13 Sensory receptors in the elaborate nose of the star-nosed mole map onto specific regions of the sensory cortex. (**A**) The nose of a star-nosed mole, seen from the front. (**B**) A higher-magnification view of the fleshy rays surrounding the right nostril. (**C**) A brain section, parallel to the surface of the somatosensory cortex, that has been processed for cytochrome oxidase. The pattern of raylike stripes of darkly stained tissue matches the pattern of rays on the nose. Medial, up; posterior, right. [Courtesy of K. Catania.]

of the skin under which lie arrays of receptors and nerve endings. When it hunts for food, the animal waves these fleshy rays, bringing them into contact with the substrate. The projections from this remarkable organ were traced anatomically to the somatosensory cortex. Sensory input from the rays crosses the midline of the mole; and, in the contralateral somatosensory cortex, there are eleven wedge-shaped cellular regions, each of which receives input from one ray (Figure 11-13C). Apparently, each ray of the nose maps directly onto a stripe of neurons in the cortex, illustrating the principle that, in some cases, the nervous system organizes sensory input by sending information from specific sensory locations to morphologically distinct structures in the brain. Similar mapping has been reported for the projection from individual facial vibrissae (i.e., sensory hairs) of mice and rats onto specific sets of neurons, called *barrel fields*, in the somatosensory cortex. In species such as star-nosed moles or mammals that have prominent "whiskers" on their faces, the somatosensory cortex spatially sorts information from these readily identifiable sensory structures. Spatially sorting incoming sensory information may be common among all species, even though it is less readily observable in species that lack such spectacular sensory organs.

The **auditory cortex** of the temporal lobe and the **visual cortex** of the occipital lobe (see Figure 11-11) are both purely sensory. Direct electrical stimulation of these areas during neurosurgery evokes rudimentary auditory and vi-

sual sensations. The relative size of these areas also correlates with their importance to the animal. For example, the visual cortex in the tarsier occupies nearly a third of the entire cortical surface, whereas it takes up only a tiny fraction of the cortex in a rat (see Figure 11-11). Within these areas, the sensory world is arranged in a highly ordered fashion. In the primary visual cortex, for example, the two-dimensional world that stimulates the retina is mapped directly onto the two-dimensional surface of the cortex; that is to say, points lying near one another in the environment excite cells that lie near one another in the cortex. This *retinotopic mapping* was one of the first features of brain organization to be discovered, and it has guided exploration of other sensory systems. However, understanding how other sensory modalities are processed has presented challenges. The visual scene and the area of the cortical surface can be directly represented in two-dimensional maps, but mapping of the other senses requires more complex computations by the nervous system, as we will see.

The **motor cortex** is located adjacent to the somatosensory cortex, and it, too, exhibits topographical representation of the periphery (see Figure 11-12B). As in the somatosensory cortex, the spatial distribution of neurons in the motor cortex correlates with the location of the muscles that are controlled by the neurons. Neurons that are near neighbors in the motor cortex control muscles that are neighbors in the body. In addition, many neurons are dedicated to controlling muscles that make precise movements;

fewer neurons control muscles that make only large, imprecise movements. For example, the muscles that move the human fingers execute very detailed and finely tuned motions, and the amount of motor cortex dedicated to controlling the fingers is very large. In contrast, the movements of the toes are relatively simple and coarsely controlled, and the amount of motor cortex dedicated to the toes is correspondingly quite small. Interestingly, the fine details of motor maps are somewhat plastic; they can change to some extent, depending on use.

The control of movement arises from activity in the motor cortex, based on input from other areas of the cortex and the brain. The signal travels to the body by several pathways, including the **corticospinal tract,** which contains the axons of neurons that have somata in the motor cortex and that synapse in the spinal cord. (The convention of naming tracts according to the location of the somata and synapses is common in the vertebrate nervous system.) The number of synapses that separate neurons in the motor cortex from the spinal motor neurons varies among species. The pathway includes more synapses in lower vertebrates, and fewer in the higher vertebrates such as primates. In rabbits, for example, a signal from the motor cortex must be transmitted across several synapses in the spinal cord before it reaches the motor neurons. In cats, few synapses separate the motor cortex from the spinal segment containing the target motor neurons; but, when the signal reaches the appropriate spinal segment, it must traverse several interneurons before it reaches the spinal motor neurons. In primates, some cortical motor neurons synapse directly onto spinal motor neurons, an arrangement that is thought to confer special motor capacities on primates. However, neurons that directly connect the motor cortex with spinal motor neurons constitute only about 3% of the motor control neurons. Thus even in human beings, most motor control is achieved through less direct pathways.

Neurons controlling motor activity maintain a steady background of low-level synaptic input to the motor neurons. An increase in their activity synaptically activates motor neurons, which can produce forceful movements of the limbs. This behavior occurs naturally when a subject consciously generates a strong contraction of a muscle, which is why the system is called the "voluntary" system. Movements can also be evoked experimentally in anesthetized animals when clusters of neurons in the motor cortex are directly stimulated with weak electrical current.

The Autonomic Nervous System

Visceral function in vertebrates is regulated largely without conscious control and primarily by the **autonomic nervous system** (see Figure 11-6). As mentioned earlier, the autonomic nervous system is partitioned into two distinct divisions: the *sympathetic* and *parasympathetic* pathways. In general, these two branches act in continuous opposition to one another, and the balance between the two systems determines the state of the animal. When an animal is in a relaxed state or is sleeping without impinging stimuli, the parasympathetic pathways dominate, lowering the heart rate, diverting metabolic energy to housekeeping tasks such as digestion, and so forth. When an animal is threatened or frightened, causing it to prepare for emergency activity, increased activity in sympathetic neurons inhibits the housekeeping functions and enhances functions that can support intense physical activity. The heart rate becomes elevated, levels of glucose in the blood rise, and blood flow to skeletal muscles increases. These two states—deep sleep and alarm—are at opposite ends of a continuum. Most of the time, the two systems are nearly in balance; as a result, physiological variables such as heart rate are held at some intermediate value. A short list comparing some sympathetic and parasympathetic actions is given in Table 11-1.

The functional unit in both the sympathetic and the parasympathetic divisions is the autonomic reflex arc. Figure 11-14 shows a sympathetic reflex arc. The afferent side of an autonomic reflex arc is largely indistinguishable from a somatic reflex arc, although the sensory neurons are likely to respond to different stimuli—for example, the concentration of glucose in the blood or the oxygen tension in the tissues. The efferent side of an autonomic reflex arc is quite different from a somatic reflex arc. In both divisions of the autonomic nervous system, the motor output is carried by two neurons in series (Figure 11-15A). The soma of the first neuron, called the *preganglionic neuron,* is located in the central nervous system, and the soma of the second neuron, called the *postganglionic neuron,* lies in a *sympathetic chain ganglion* (also called a *paravertebral ganglion*). Postganglionic neurons lie entirely outside of the central

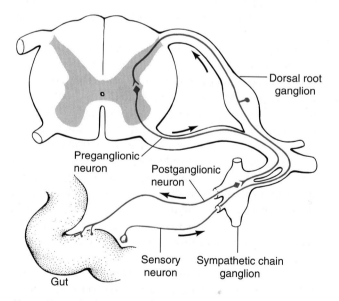

Figure 11-14 The functional unit of the autonomic nervous system is the autonomic reflex arc. This diagram illustrates a reflex arc of the sympathetic nervous system. Sensory input passes directly through the sympathetic chain ganglion, synapsing somewhere in the central nervous system. Motor output coming from higher centers synapses onto the preganglionic neuron, which in turn synapses onto postganglionic neurons in a sympathetic chain ganglion. Postganglionic neurons synapse onto the target organ. [Adapted from Montagna, 1959.]

TABLE 11-1
Opposing effects on target organs of the sympathetic and parasympathetic branches of the autonomic nervous system

Target tissue	Sympathetic effect	Parasympathetic effect
Eye		
Radial muscles of iris	Pupillary dilation	
Iris sphincter muscles		Pupillary constriction
Ciliary muscle (controlling thickness of lens)	Relaxation (focuses on distant objects)	Constriction (focuses on close objects)
Lacrimal (tear) glands		Stimulates production of tears
Salivary glands	Stimulates production of a small amount of viscous saliva ("dry mouth")	Stimulates production of a large amount of dilute saliva
Arterioles	Vasoconstriction, especially in vessels supplying skin	Little or no effect
Heart		
Pacemaker cells	Increases rate of heartbeat	Decreases rate of heartbeat
Ventricular contractile fibers	Increases force of contraction	Little or no effect
Lungs		
Bronchioles	Dilates bronchioles	Constricts bronchioles
Mucous glands	No effect	Stimulates secretion of mucus
Gastrointestinal tract		
Sphincter muscles	Contraction	Relaxation
Motility and tone of smooth muscles	Inhibits	Stimulates
Exocrine secretion of glands	Inhibits	Stimulates
Gallbladder	Inhibits contraction	Stimulates contraction
Liver	Increases glycogenolysis and, therefore, blood sugar	No effect
Urinary bladder	No effect	Contraction of muscles
Adrenal medulla	Stimulates secretion	No effect

nervous system, and they synapse onto the target cell of the autonomic reflex.

The location and properties of these postganglionic neurons depend on the autonomic branch to which they belong (see Figure 11-15). In the sympathetic branch, preganglionic neurons synapse onto postganglionic neurons in the sympathetic chain ganglia, which contain the cell bodies of the postganglionic neurons. The axons of the postganglionic neurons then extend to the target organs, which may lie quite far away from the ganglion. (An exception to this pattern is the *celiac ganglion*, which contains the somata of sympathetic postganglionic neurons innervating the stomach, liver, spleen, pancreas, kidney, and adrenal gland. It is located in the abdominal cavity.)

Preganglionic neurons of the parasympathetic nervous system, on the other hand, synapse onto postganglionic neurons in ganglia that lie near—or even in—the walls of the target organs. Hence, in the parasympathetic branch, axons of preganglionic neurons may be very long, and axons of postganglionic neurons are typically short. Preganglionic neurons of the sympathetic nervous system are located in the cervical, thoracic, and lumbar regions of the spinal cord. The cell bodies of parasympathetic preganglionic neurons lie in the head and in the sacral spinal cord.

The two divisions of the autonomic nervous system also differ pharmacologically. All preganglionic neurons are cholinergic; that is, their neurotransmitter is acetylcholine. However, the neurotransmitter of a postganglionic neuron depends on the division to which the neuron belongs. The neurotransmitter of parasympathetic postganglionic neu-

rons is acetylcholine; the neurotransmitter of sympathetic postganglionic neurons is norepinephrine. (Sympathetic postganglionic neurons also release a little epinephrine.)

Postganglionic neurons of the parasympathetic and sympathetic branches typically innervate the same target organs (Figure 11-15B) and typically exert opposite effects on their shared targets. For example, pacemaker activity of the heart is slowed down by acetylcholine released by parasympathetic postganglionic neurons, whereas it is accelerated by norepinephrine released by sympathetic postganglionic neurons. These actions of the parasympathetic and sympathetic systems are reversed in the digestive tract, where acetylcholine from parasympathetic neurons stimulates intestinal motility and digestive secretion, whereas norepinephrine from sympathetic neurons inhibits these functions.

Within the autonomic nervous system, there are two kinds of receptors for each of the neurotransmitters acetylcholine and norepinephrine. For each transmitter, the two types of receptors can be distinguished by drugs that act as agonists (i.e., mimic the natural transmitter) or as blockers. The two kinds of cholinergic receptors are called *nicotinic* and *muscarinic*. Nicotinic and muscarinic receptors were first distinguished from one another on the basis of physiological reactions to two substances—nicotine and muscarine—that are never produced naturally in the body. Nicotine, a plant alkaloid, acts as an agonist at some cholinergic synapses, including the synapses between pre- and postganglionic neurons in both divisions of the autonomic nervous system (see Spotlight 6-3). Muscarine,

A

Sympathetic

Parasympathetic

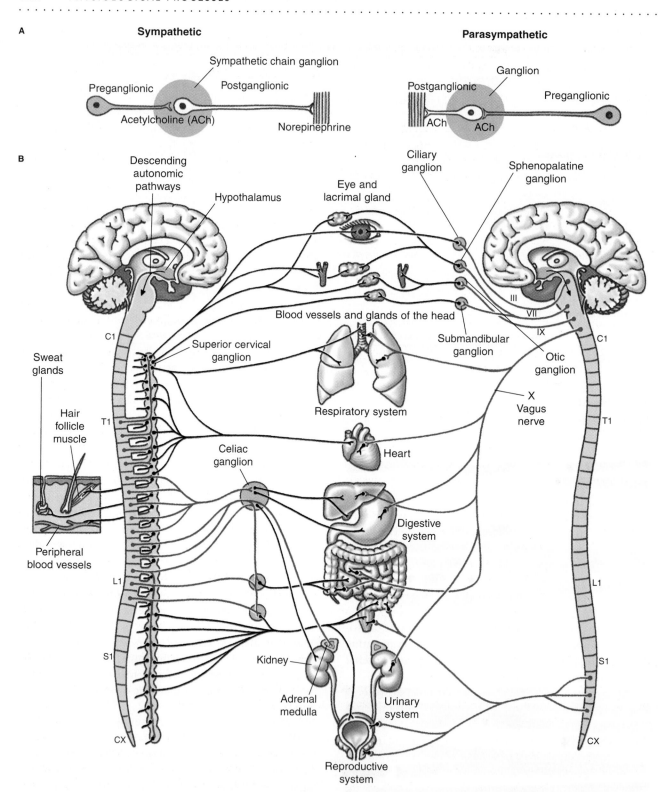

B

Figure 11-15 The two divisions of the autonomic nervous system share many targets. (**A**) Both the sympathetic and the parasympathetic divisions innervate their targets through a two-neuron chain. The soma of each preganglionic neuron lies within the central nervous system. The preganglionic neurons synapse in peripheral ganglia onto postganglionic neurons that contact the target organs. The two divisions differ pharmacologically. The preganglionic neurons of both divisions are cholinergic. The postganglionic neurons of the parasympathetic division also are cholinergic, but the postganglionic neurons of the sympathetic system are adrenergic, primarily using norepinephrine as their transmitter. Ab-

breviation: ACh, acetylcholine. (**B**) Locations and targets of pre- and postganglionic neurons in the sympathetic (*left*) and parasympathetic (*right*) branches of the autonomic nervous system. Preganglionic neurons are shown in color; postganglionic neurons are shown as black lines. This diagram illustrates the human autonomic nervous system, but the system is similar in most vertebrates. Abbreviations: C1, first cervical segment of the spinal cord; T1, first thoracic segment; L1, first lumbar segment; S1, first sacral segment; CX, coxal segment. In the parasympathetic nervous system, several pathways run in cranial nerves, which are indicated by roman numerals.

which is extracted from particular toxic mushrooms, acts as an agonist at other cholinergic synapses, including synapses made by parasympathetic postganglionic neurons onto their target cells. Curare (D-tubocurarine) blocks the action of acetylcholine on nicotinic receptors (including those at the endplate of skeletal muscle), whereas atropine blocks muscarinic receptors. Nicotinic and muscarinic receptors have completely different molecular structures and response mechanisms. Nicotinic receptors consist of protein complexes that bind the neurotransmitter and contain ion-selective channels within their structure. Muscarinic receptors are molecules with seven transmembrane domains, and they affect ion channels through the mediation of intracellular second messengers (see Chapter 6).

The adrenergic postganglionic neurons of the sympathetic division innervate two kinds of norepinephrine receptors, designated α and β receptors. Like the acetylcholine receptors, the adrenergic receptors are distinguished from one another by their pharmacology. Adrenergic α receptors are more sensitive to norepinephrine than to the drug *isoproterenol,* and they are selectively blocked by *phenoxybenzamine.* Adrenergic β receptors are more sensitive to isoproterenol than to epinephrine and are selectively blocked by *propranolol.* The two types of adrenergic receptors activate separate, but parallel, intracellular regulatory pathways.

Although all vertebrates have autonomic nervous systems, the two divisions are not well defined in all groups. For example, in teleost fishes, sympathetic and parasympathetic divisions are distinguishable; but, in cyclostome fishes, the autonomic nervous system is not partitioned into two divisions. The autonomic nervous system of amphibians appears to be essentially identical with that of mammals, but reptiles may not have such a clearly defined distinction between sympathetic and parasympathetic divisions. Comparative studies of the autonomic nervous system have been relatively rare, and further research may reveal some phylogenetic surprises.

ANIMAL BEHAVIOR

To understand how behavior can be initiated and controlled by the brain and other parts of the central nervous system, it is necessary to know the properties of neurons and the ways in which information is transmitted from one neuron to the next. We have already considered how individual neurons are linked into circuits. The simplest neuronal circuits are reflex arcs that depend on a small number of neurons, but the circuits that produce behavior can be immensely complex. To understand these more complicated circuits, we must consider several questions. First, what exactly about behavior do we want to explain? Second, can we construct this explanation from known circuits and their interactions? Third, do general principles emerge or is every behavior a "special case"? To suggest how clearly and completely these questions can currently be answered, in the remainder of this chapter we first consider

some complex behaviors and then turn to some properties of networks known to underlie behavior.

Basic Behavioral Concepts

A basic problem facing any animal is what to do (and not to do) in any given situation. Understanding the mechanisms that allow animals to make these choices is a major challenge. Behavioral scientists have taken two complementary approaches to meeting this challenge. The *neuroethological* approach has been to bring the animal into the laboratory and to observe behavior in this drastically simplified set of well defined circumstances. In the laboratory, there are no predators; the number and sex of conspecifics are controlled by the experimenter; there are unusual lights, smells, and sounds; and the animal is often limited to a relatively small and confined space. In some cases, the nervous system of the animal is exposed surgically to allow the experimenter to record from neurons while behavior is going on. Studies under these reduced conditions are useful, even necessary, for obtaining answers to circumscribed, well defined questions about behavior, but the results from such experiments can be difficult to translate into an understanding of how animals deal with challenges in their everyday lives. The *ethological* approach is to go into the field and observe the animal as it goes about its business in its normal environment. Observing animals in their normal state is an ancient practice, and our ancestors no doubt made such observations with dinner in mind. However, for the physiologist, these natural conditions raise severe problems that differ from those created by laboratory conditions. While observing an animal in its natural setting, it is difficult to determine the relative importance of all the different things that the animal does, and it is essentially impossible to record activity from the animal's nervous system. However, progress has been made in recognizing repeatable patterns in natural behaviors.

On the basis of observations made in the field and of many ingenious experiments, many workers have compiled detailed inventories of what animals do and how and where individual animals spend their time. These reports offer a wealth of information but few organizing principles. From considering such observations, ethologists Konrad Lorenz and Niko Tinbergen developed an interpretive framework that ultimately crystallized these data into a useful system. They recognized that the behavioral repertoire of an animal seemed to be constructed from elemental motor and sensory "units." They called the unit motor patterns **fixed action patterns** (also called *modal action patterns*), and the corresponding unit sensory elements they called **key stimuli** (also called *sign stimuli*). They were awarded a Nobel Prize in 1973 in recognition of the importance of their observations.

Fixed action patterns have six properties:

1. They are relatively complex motor acts, each consisting of a specific temporal sequence of components. They are not simple reflexes.

2. They are typically elicited by specific key stimuli rather than by general stimuli.

3. Fixed action patterns are normally elicited by an environmental stimulus; but, if an experimenter removes the stimulus after the animal's movements have begun, the behavior will usually continue to completion. This all-or-none property of fixed action patterns distinguishes them from reflexes, which typically require continued stimulation to maintain movement. It is as though the key stimulus is required to turn the pattern on; but, once begun, it plays out independently of further stimulation.

4. The stimulus threshold for fixed action patterns varies with the state of the animal, and the variation can be quite large. For example, immediately after an animal has copulated, it cannot be induced to copulate again without intense stimulation. The threshold for eliciting copulation then drops as time passes.

5. When they are presented with the appropriate stimulus, all members of a species (perhaps that are the same age or the same sex or both) will perform a given fixed action pattern nearly identically. Some fixed action patterns are common to many species of a genus. The properties of fixed action patterns are so reliable within some taxa that comparison of variations within these patterns has been used to deduce taxonomic relationships.

6. Fixed action patterns are typically performed in a recognizable form even by animals that have had no prior experience with the key stimuli. That is, these patterns are inherited genetically, although it is clear that in many species the patterns can change somewhat with experience. This property of heritability sparked a long-lived debate about the dichotomy between inherited and learned behaviors (i.e., "nature versus nurture") that still flares up periodically like an incompletely extinguished fire.

The stereotyped nature of muscle activity during fixed action patterns has made them extremely useful to physiologists interested in the cellular basis of motor activity during behavior. The ability to reliably elicit precisely the same muscle contractions over and over again by applying the appropriate key stimulus has provided a window into the working of the nervous system, particularly in simple animals. Some typical fixed action patterns are illustrated in Figure 11-16.

Key stimuli, which specifically stimulate an animal to perform a fixed action pattern, have also been called *releasers*, because they appear to "release" a prepatterned be-

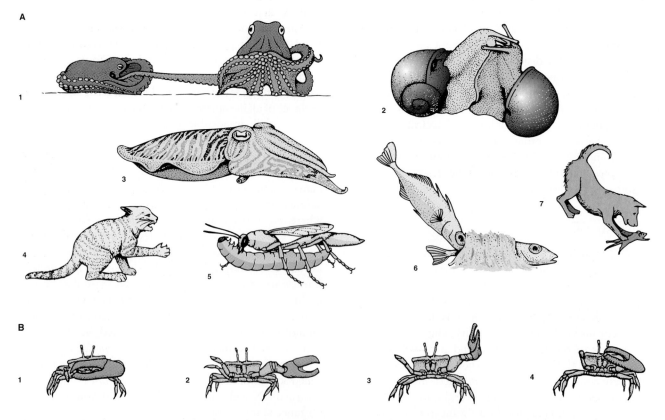

Figure 11-16 Fixed action patterns can be observed in most species. These behaviors appear to be genetically programmed, rather than learned. (**A**) Mating behavior in (1) octopus and (2) *Helix pomatia*. (3) Male *Sepia officinalis* in sexual display. (4) European wildcat striking with its paw. (5) A digger wasp capturing its prey. (6) Male three-spined stickleback quivering to stimulate a female to spawn. (7) The "mouse jump" carried out by a domestic dog. (**B**) The courtship behavior of a male fiddler crab. The crab attempts to attract a female by waving its large claw (1) forward, then (2) laterally, then (3) upward, and finally (4) forward again. [Part A, 2, 3, 5, and 6, adapted from Tinbergen, 1951, 1 adapted from Buddenbrock, 1956, 4 adapted from Lindemann, 1955, 7 adapted from Lorenz, 1954; part B adapted from Lorenz and Tinbergen, 1938.]

havioral response within an animal. To sort out the features of a stimulus that are essential to its effectiveness, ethologists have made models that are shaped and colored to mimic the suspected signal. The features are varied among the various models to determine which of them is most effective at eliciting the particular fixed action pattern. A classic example of a fixed action pattern that has been studied with the use of models is the aggressive response of a courting male three-spined stickleback fish. When a courting male sees another male stickleback, it displays aggressively and may even attack. Presenting courting male fish with models that differed in shape and in color revealed that the key stimulus for releasing this behavior is the red underside characteristic of male sticklebacks (Figure 11-17A). The shape of the model male fish was relatively unimportant, and the red underside was ineffective as a sign stimulus if the model was in any orientation other than horizontal (Figure 11-17B). Thus, it is not simply the color red but the red belly in the proper visual context that acts to release the aggressive behavior of the male fish. The discovery of these key stimuli suggested further experimental work, just as the discovery of fixed action patterns had. The highly specific nature of the releasing stimuli suggested that there might be adaptations within the stickleback's visual pathways that were specially tuned to important features, acting perhaps as sensory filters to admit only the appropriate parts of the stimulus. Indeed, the theme of specialized detectors has proved to be an important and powerful organizing concept in sensory physiology.

Fixed action patterns and key stimuli do not occur in isolation but are instead woven seamlessly into behavioral interchanges between animals. In courtship, caring for offspring, or aggressive encounters, animals use complex sets of movements to communicate their behavioral intent. In courtship, for example, a movement by one animal sends a signal to its partner. The partner may, in turn, move in a way that sends a signal back.

Genetic components of fixed action patterns and key stimuli have been used to probe how species-specific behavior is organized. An example of this approach has been the study of the wing movements that produce the chirps and trills in the courting song of a male cricket. Patterns of cricket songs are species-specific and are largely independent of environmental factors other than temperature. Moreover, the sound pattern produced by a species is directly related to the pattern of action potentials (APs) in motor neurons that control the sound-producing muscles. One set of studies examined sound production by males of two related species. One species produces a short trill (consisting of about two sound pulses), whereas the other species produces a long trill (consisting of about ten sound pulses). The trills of F_1 hybrids, the progeny produced by crossing individuals of the two species, consisted of an intermediate number of pulses (about four). Backcrosses to produce various genetic combinations demonstrated that the neuronal network producing the song pattern is under rigid genetic control, sufficiently precise to specify differences as small as the exact number of APs going to the chirp-producing muscles.

Similar experiments on vertebrate species have indicated that, in these higher organisms, too, some aspects of behavior are genetically controlled and that hybrid individuals perform an intermediate form of the behavior. Even quite complex behavioral patterns requiring pattern

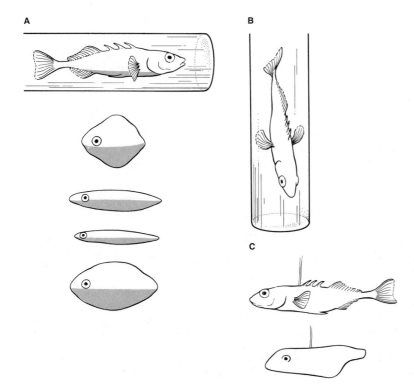

A   B   C

Figure 11-17 Studies of the aggressive behavior of a courting male stickleback fish have been facilitated by the use of models. Different models were used to study the aggressive behavior of the male fish. (**A**) Responses to these models indicated that the feature of another fish that releases aggressive behavior is the horizontal red underside of the intruding male. The models in part A were all effective releasers; that shown in part B was not. (**B**) A vertically oriented male fish elicited no response. (**C**) The shape of the model fish was unimportant. [Adapted from Tinbergen, 1951.]

recognition and appropriate responses can be genetically encoded. For example, the ability to orient with respect to stellar patterns, complete with time compensation to account for the Earth's rotation, is fully expressed by some species of birds, even if they have been hatched indoors and shielded from prior exposure to the sky. Under those conditions, the birds could never have learned from practice to orient properly in relation to the sky. At least some of the information required for proper orientation resides in the genetic material, and the nervous system of the bird is programmed to take off in the correct direction when it is presented with the appropriate visual input from the night sky.

It is reasonable to say that behavior in higher animals is based on both genetic and learned components. The relative contribution made by heredity and by experience varies greatly among different behaviors and among different species. Genetically programmed behavior seems to dominate in animals with simple nervous systems, but even very simple organisms exhibit the ability to learn from experience. Observations suggest that the more complex a nervous system is, the greater is its potential to learn, and this potential enables organisms not only to depart from, but also to enlarge upon, their limited repertoire of inherited, fixed patterns of behavior.

Examples of Behavior

A brief introduction to the fundamental units of behavior cannot begin to capture the remarkable range of behavioral capacities known in animals. Various animals have evolved specialized sensory and motor capabilities that allow them to produce many remarkable behaviors. We will consider some well known examples to demonstrate the complexity of the neuronal systems that underlie behavior.

Animal orientation

Many animals move predictably with respect to specific stimuli—a behavior called *orienting*. This behavior requires the integration of sensory input and the coordination of motor output, so it depends on the properties of the sensory receptor neurons, the connections within the central nervous system, and the muscles that cause the body of the animal to move. We now consider several examples of orientation and navigation to illustrate some of the mechanisms that underlie these ubiquitous behaviors and to exemplify the complex behavioral capabilities of even simple organisms.

A common nocturnal sight for many city dwellers is the scurrying of cockroaches as they abandon their snacks and disappear when the kitchen light is switched on. This behavior is an example of a **taxis**—a movement that is directed with respect to a stimulus. Scurrying cockroaches move *away* from a light stimulus, which is called *negative phototaxis* (Figure 11-18). An animal that turns *toward* the light is said to exhibit *positive phototaxis*. Jacques Loeb (1918) suggested early in the twentieth century that simple taxes occur when asymmetrical motor activation is caused by asymmetrical sensory input. According to this

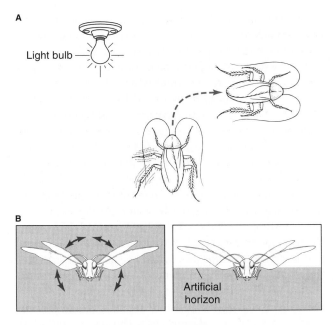

Figure 11-18 Sensory input can change behavior. (**A**) A cockroach runs away from light, a negative phototaxis. (**B**) A tethered locust flying in the dark tends to roll about its longitudinal axis. An artificial horizon stabilizes the insect's body position because visual input modulates the motor output to the flight muscles.

view, negative phototaxis would occur when the light impinging on one eye produced a strong **ipsilateral** motor output (i.e., on the same side as the eye), causing the animal to veer away from the source of the light. Positive phototaxis would occur if the light stimulus to an eye stimulated **contralateral** locomotor output (i.e., on the side away from the light), causing the animal to turn toward the source of light. Consistent with this hypothesis, it has been found that, if they are blinded in one eye, positively phototactic animals will orient so that the intact eye points away from the light.

Sensory information modulates behavior in other ways as well. For example, it can be used to correct for structural or functional asymmetries that arise in the central nervous system or in the structures (e.g., wings, legs, fins, etc.) that influence locomotion. A locust can continue to fly in a straight line even after one of its four wings is either partly or entirely removed, as long as it can use its eyes for orientation. In the dark, even an intact locust that has been tethered so that its behavior can be observed will roll about its longitudinal axis if it is induced to fly. The roll is due to slight asymmetries in the structure of the wings and in the centrally generated motor output. A tethered locust ceases to roll if it is provided with visual cues in the form of an artificial horizon (see Figure 11-18B); visual input allows the motor output to the wings to be corrected, stabilizing the insect's position. The visual input provides information to the central neurons that control flight, regulating the relative outputs to the left and right sets of wing muscles so as to hold the horizon seen by the animal in a horizontal orientation.

The importance of sensory feedback for orientation and locomotion in human beings is confirmed in our daily experience. For example, as noted in Chapter 1, the driver of a car continually makes minor steering adjustments: her eyes are the sensors in a feedback system in which her neuromuscular system, coupled to the car's steering mechanism, corrects any deviations that the car makes from the center of the lane. Deviations may be due to asymmetries in the driver's nervous and muscular systems or to irregularities or imperfections in the road and vehicle. When a person is deprived of this visual feedback, behavior changes. Blindfolded human subjects, either walking or driving a car in an open flat field, take a more or less circular course, with the size and direction (clockwise or counterclockwise) of the average circle differing among subjects. Similar turning biases are seen in animals at all phylogenic levels. Visual and other forms of sensory feedback compensate for these inherent locomotor biases, some of which may be due to congenital asymmetries in the function of the muscles and nervous system.

Many animals locate their prey from vibrations that are set up in the substrate by their prey. A spider is alerted to prey in its web, for example, by vibrations of the silken threads — vibrations that are detected by mechanoreceptor organs located in the spider's legs. Another group of arachnids, the nocturnal desert scorpions, use sand-borne vibrations produced by movements of their prey to locate and orient toward potential victims that are as far away as 0.5 m. At 15 cm or less, the scorpion can determine the distance, as well as the direction, of the source of vibration. Besides typical mechanoreceptor sensilla, a scorpion possesses a set of especially sensitive receptors on each of its eight legs (Figure 11-19). Although the sensitivity of these receptors to vibration is considerably less than that calculated for cochlear hair cells, it is nonetheless impressive. With the use of a finely tuned mechanical displacement stimulator, it has been shown that these receptors respond when the tarsal segment of the leg is displaced by less than 0.1 nm, which makes the receptors capable of detecting the direction in which sand-borne vibrations are traveling.

To orient accurately toward the source of vibration, a scorpion keeps all of its eight legs in contact with the substratum in a precisely spaced circle (see Figure 11-19B). In this arrangement, the leg closest to the source will be the first to sense the vibrational waves along the sandy substratum. Central neurons that receive the sensory inputs from the receptors appear to compare the timing of impulses that arise in the vibration receptors of all the legs. Those legs facing the stimulus source intercept the waves first, and those on the opposite side receive the stimulus about 1 ms later (vibrational waves travel 40–50 m·s⁻¹ in the sand). By integrating the timing of APs from the different legs, the central nervous system calculates the direction of the stimulus source, and it then produces the

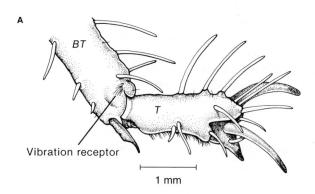

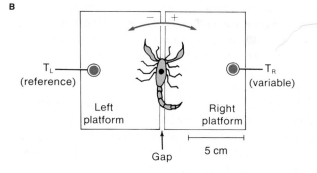

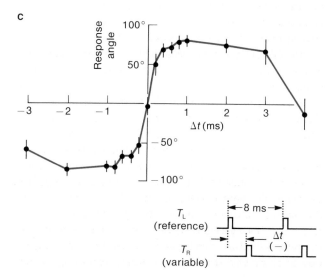

Figure 11-19 Sensory receptors on the legs of desert scorpions allow them to receive information about vibrations produced by potential prey animals and carried through the sand. Central neurons respond to the time at which information arrives from the eight legs and produce output that causes the scorpion to orient toward the source of vibration. (**A**) Location of one of the vibration receptor organs. Abbreviations: *BT,* basitarsus; *T,* tarsus. (**B**) Experimental setup to manipulate the timing of sensory input to the vibration receptors. (T_L and T_R, left and right transducers.) (**C**) Relation between the time at which sensory input from the various legs arrives in the central nervous system and the orientation angle assumed by the scorpion. The scorpion turns toward the legs from which the sensory input was first received. [Adapted from Brownell and Farley, 1979a, b.]

appropriate motor output allowing the scorpion to orient and respond in the direction of the source.

Many aquatic organisms can orient in relation to vibrational stimuli. For example, some surface-swimming insects and many fishes and amphibians sense reflections of waves that were initiated by their own swimming movements and that have bounced off nearby objects. Fishes and amphibians detect these reflected waves by using the hair cells of their lateral line system (see Chapter 7). Some mammals and a few birds have refined the mechanism of detecting reflected waves as a basis for orientation; they emit sound waves and detect the reflected sounds with the hair cells of their ears. This specialized orienting behavior is called **echolocation.**

Echolocation
The highly refined auditory mechanisms of mammals and birds have allowed the evolution of remarkable forms of acoustical orientation in which an animal emits high-frequency sound pulses and uses the return echoes to detect the direction, distance, size, and texture of objects in its environment. This sonarlike use of auditory signals is most highly developed in two groups of mammals: the microchiropteran bats and some cetaceans, such as porpoises and dolphins. Among birds, only two genera, the Asian cave swift *Collocalia* and the South American oil bird *Steatornis,* appear to use echolocation. These birds use their tongues to produce audible clicks for this purpose.

Near the end of the eighteenth century, the Italian naturalist Lassaro Spallanzani discovered that bats use echolocation. He became curious about the ability of bats to avoid obstacles even when flying in total darkness, whereas his pet owl required at least a dim light to do the same. After some false leads, he confirmed a previously published report by Louis Jurine, a Swiss surgeon, that plugging a bat's ears interferes with its ability to navigate in the dark. He further found that blinded bats could find their way home to their favorite belfry in the cathedral at Pavia. He observed that these blinded bats caught insects quite successfully: when they were caught, killed, and dissected, their stomachs were stuffed with insects that had been caught on the wing during their trip back to the belfry. At the time, little was known about the physics of sound, and Spallanzani overlooked the possibility that the bats themselves emitted sounds that were inaudible to human beings. As a result, he came to the erroneous conclusion that the bats navigated by detecting the echoes of sounds made by their own wing-beats and that they located their prey by homing in on the buzzing of insect wings.

It was not until 1938 that Donald Griffin and Robert Galambos, both students at Harvard University, used newly developed acoustical equipment to determine that bats emit ultrasonic cries and use the echoes of these sounds to "see in the dark" (Figure 11-20). Further studies by Griffin and his colleagues revealed the phenomenal echolocating capabilities of insectivorous bats. High-speed photography showed that, by using echolocation, a bat can capture two separate mosquitos or fruitflies in about 0.5 second. The fish-eating bats of Trinidad can even use echolocation to find and capture their underwater prey by detecting the ripples that are produced on the water's surface when a fish swims just under the surface.

Insectivorous bats capture an insect by three phases of acoustical orientation (Figure 11-20B). The "cruising" phase, which occurs during straight flight, consists of pulsed sounds separated by silent periods of at least 50 ms. Each pulse of sound is **frequency modulated** (FM), sweeping downward through a frequency spectrum somewhere between 100 and 20 kHz. (Because human beings cannot hear sounds above about 20 kHz, these sounds are called *ultrasonic.*) The second phase begins when the bat detects its prey. In this phase, pulses are produced at shorter intervals. The third, and final, phase consists of a buzzlike emission in which the intervals become even shorter, the duration of each pulse drops to about 0.5 ms, and the frequency of the sounds is reduced. Finally, the bat scoops up the insect with its wings or with the webbing between its hind legs, guiding the insect to its mouth.

The sounds produced by these bats are highly energetic, reaching intensities above 200 dynes·cm^{-2} close to the mouth of the bat. This sound is as intense as the noise made when a jet plane takes off and passes only 100 m overhead; it is 20 times as intense as the sound of a pneumatic jackhammer only several yards away. Nonetheless, the sound energy that returns as an echo reflected from a small object is very weak, because sound intensity, like other forms of radiating energy, drops off as the square of distance. This reduction in energy with distance holds both for the cries themselves and for the tiny fraction of the emitted energy that is reflected back from a small object, so a bat faces a formidable neuronal task in differentiating very weak and complex echoes from its own, far more powerful, emitted sounds.

In echolocating bats, a number of morphological and neuronal modifications assist in detecting the echoes. The snout is covered by complex folds, and the nostrils are spaced to produce a megaphone effect. The pinnae of the ears are very large to help capture echoes. The eardrum and ear ossicles are especially small and light, providing high fidelity at high sound frequencies. In the emission of sounds, muscles controlling the auditory ossicles contract briefly, reducing the sensitivity of the ear (this is a common feature in the ears of mammals). Blood sinuses, connective tissue, and fatty tissue isolate the inner ear from the skull, reducing direct transmission of sound from the mouth to the inner ear. Finally, the auditory centers of the brain occupy a very large fraction of the bat's relatively small brain. Many regions in the bat's brain receive auditory signals and, through the processes of neuronal computation, construct from these auditory cues a spatial representation of the external world. Similar mechanisms for processing auditory information have been carefully studied in the brain of the barn owl and are considered later in this chapter.

A

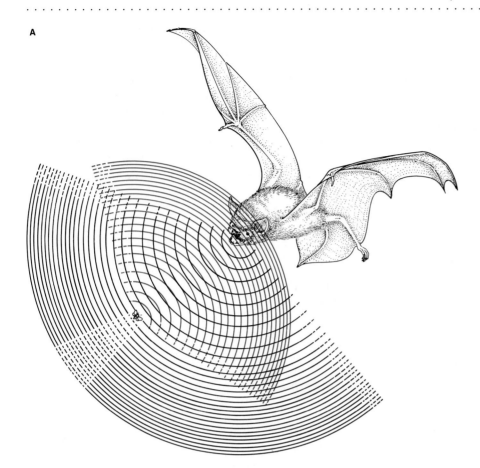

Figure 11-20 Bats find and capture their prey by using echolocation. (**A**) Portrait of a little brown bat, which has specialized structures of the snout and pinnae of the ears. The bat emits a frequency-modulated sound, which is reflected from objects in its environment, including the insects that are its prey. The spacing of the curves in this drawing indicate the changing frequency of the emitted and reflected sound over the duration of a sound pulse. Only a small fraction of the emitted sound (black curves) is reflected back to the bat (red curves), and only a small fraction of the reflected sound is intercepted by the bat. (**B**) Three phases of echolocation in the pursuit of an insect by the bat *Eptesicus*. In the initial phase, the cries are separated in time and they sweep downward through frequencies between 100 and about 40 kHz. When the bat detects an insect, the cries become more frequent, still scanning through about the same frequency band. At close range, the cries turn into a buzz, with a very short period between each cry and a smaller range of frequencies scanned. [Part B adapted from Simmons et al., 1979.]

B

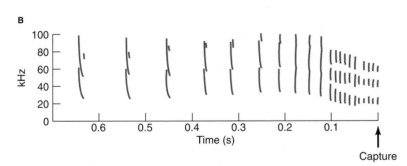

Animal navigation

Navigation over long distances is a striking extension of simple orienting responses. Many kinds of animals—from monarch butterflies to golden plovers and gray whales—migrate over long distances through unfamiliar territory. These navigational abilities have long been surrounded by an aura of mystery, because the animals typically rely on cues that people cannot detect. In addition, the study of navigation has been difficult because animals often use several different sensory cues to guide them, and this redundancy has made the experimental study of navigation confusing. Often, one system predominates when conditions for other systems are unfavorable; but, if conditions change, the animal can shift its strategy. Experimenters—who are inclined to vary only one parameter at a time—have found themselves mystified by an animal's continued ability to find its way when one of its sensory modalities has

been disabled. For example, it is now evident that, to varying degrees, birds use particular landmarks, other visual cues such as the plane of polarized light, odors, sounds, the position of the sun and stars, and even the Earth's magnetic field as they find their way back home or to some other location.

Animals from many phyla can navigate. Bees use the sun's position and the pattern of polarized light in the sky to keep track of the direction from their hive to food sources, and they can signal this direction to their hive mates by their "waggle dances" (Figure 11-21). Some birds can navigate over vast stretches of ocean that appear to be devoid of landmarks. When they are exposed to the night sky in a planetarium, some species of nocturnally migrating birds—for example, warblers—orient with respect to the projected patterns of stars. As time passes during the night, if a projected constellation is moved across the dome

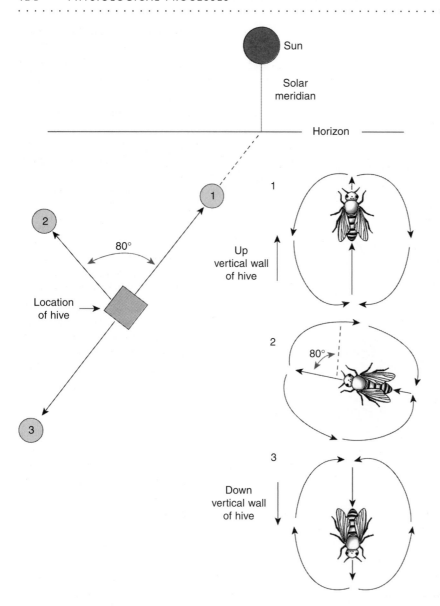

Figure 11-21 Bees can encode information about the location of a food source with respect to the hive by indicating the direction of the source related to the position of the sun. In this experiment, three potential food sources (numbered 1 to 3) were placed in very different locations. Scout bees that find a source indicate the direction of the food with respect to the hive by the direction in which they move on the vertical wall of the hive. The direction of the straight part of their "dance" shows the direction of the food source relative to the sun. [Adapted from Camhi, 1984.]

to mimic the Earth's rotation, the birds orient themselves in the "correct" direction with respect to the projected sky, continually compensating for the time-dependent shift in the position of constellations. Arbitrary changes in the position of the projected sky will cause the birds to change their orientation. It appears that in order to compensate for the rotation of the Earth relative to celestial references, birds, bees, and other celestial navigators refer to internal "clocks." This mechanism is called a *clock-compass*, and the underlying physiology is poorly understood. If a bee or a bird is put on a light-dark schedule with "dawn" and "dusk" shifted from natural lighting by several hours, it will enter the incorrect time into its internal clock-compass and will orient with a compass deviation equivalent to the artificial phase shift in its day-night cycle (Figure 11-22).

Animals can also depend on cues unavailable to human beings to guide orientation. For example, although it has long been suspected that some animals use the Earth's magnetic field for orientation and navigation, only recently has the hypothesis been supported by experimental evidence.

The behavior of homing pigeons is a good example. Homing pigeons deprived of familiar landmarks can still find their way home, even on an overcast day when it is impossible to detect the position of the sun. Under these conditions, they typically fly in circles for a few minutes and then head in the correct direction toward home. However, if small magnets are attached to their heads or if they are transported to release points in containers that make it impossible for them to detect Earth's magnetic field en route, they become disoriented. Local magnetic anomalies—for example, those caused by rich iron deposits—also disrupt the pigeons' ability to orient properly.

Cave salamanders of the genus *Eurycea* can find their way "home" in complete darkness, so it seemed possible that these animals, too, rely on Earth's magnetic field for navigation. In experiments to test their ability to use a magnetic field as a navigational cue, salamanders were trained to go to a particular place in a maze that was always oriented in the same way with respect to Earth's magnetic field. They were then tested in cross-shaped pas-

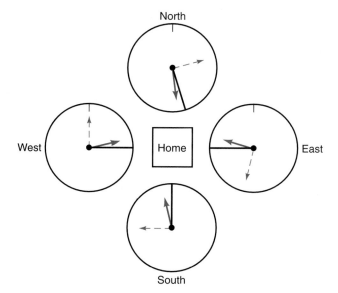

Figure 11-22 The clock-compass of homing pigeons can be manipulated by changing the timing of the pigeons' light cycle. Homing pigeons were trained to return home from each of four locations. Then half of the pigeons were maintained on an artificial cycle that was 6 hours ahead of the normal light-dark cycle. When the time-shifted pigeons were released from each of the test locations, they took off in a direction that was 90° counterclockwise to the correct orientation. The solid black line indicates the direction to the home cage. The solid red arrow indicates the average direction taken by pigeons that had been maintained on a normal light-dark cycle, and the dashed red arrow indicates the average direction taken by pigeons that were time shifted. There were approximately 10 pigeons maintained under each set of conditions. [Adapted from Camhi, 1984.]

sageways whose orientation to Earth's magnetic field could be manipulated. The results of the experiment confirmed that the salamanders can use Earth's magnetic field for orientation. This finding has significant implications concerning the mechanism by which animals can detect magnetic fields. Animals, such as birds, that move rapidly could be detecting the magnetic field directly. Alternatively, they could be detecting magnetic fields by sensing electrical currents induced in their saline body fluids as these conductors are moved through Earth's magnetic field (as described by Maxwell's laws of electricity and magnetism). Salamanders move very slowly compared with birds, and it is unlikely that electrical fields set up in their body fluids would be large enough to provide a basis for indirectly detecting magnetic fields. It therefore seems most likely that at least some animals are able to detect magnetic fields directly.

What sort of mechanism could underly a magnetic sense? This question cannot be answered with certainty. However, *magnetite*, a magnetic material of biological origin, has been found in or near the brains of several animals that appear to respond to magnetic fields—a finding that is both intriguing and suggestive. For example, a small black magnetite-containing structure lies between the brain and the skull of pigeons. Pelagic whales that may navigate by using Earth's magnetic field also have mag-

netite in parts of the cerebral cortex. Catastrophic strandings of these whales in unfamiliar inshore waters have been shown to correlate significantly with the occurrence of geomagnetic disturbances in the areas in which they swam astray.

The distribution and location of magnetite suggests that it may play a role in the detection of magnetic fields, but no actual receptor cells that transduce magnetic signals into neuronal signals have been identified. Some species of mud-dwelling bacteria may provide a model showing how orientation based on magnetite might work. These mud-dwelling bacteria contain particles of magnetite, and they have been found to orient to Earth's magnetic field. Bacteria that live in the Northern Hemisphere orient toward the North Pole, whereas those indigenous to the Southern Hemisphere orient toward the South Pole, and this difference has been shown to depend on the orientation of magnetite particles in the bacteria. The orientation of the magnetite enables the bacteria to swim downward into deeper mud, following the angle with which the magnetic field enters the Earth at each location. If the bacteria are placed in a drop of water within an artificial magnetic field, they collect at the appropriate side of the water drop; if the magnetic field is reversed, they swim to the opposite side of the drop.

The American eel *(Anguilla rostrata)* illustrates another navigational method based on geomagnetism that is available only to marine organisms. Larval eels migrate from spawning grounds in the Sargasso Sea to the Atlantic coast of North America, a distance of about 1000 km. When it was initially suggested that they employ the Earth's magnetic field for guidance, the response was skepticism because the density of the magnetic field is so low. However, the lateral-line system of eels contains extremely sensitive electroreceptors, and these electroreceptors are the basis for this ability. The movement of seawater in ocean currents acts as an enormous generator, because the salt water functions as a conductor moving through the Earth's magnetic field. The geoelectrical fields set up in the sea by ocean currents, such as the Gulf Stream, reach intensities of about $0.5 \, mV \cdot cm^{-1}$, or a 1.0 V drop over 20 km. The minute electrical currents produced by these small voltage gradients can apparently be detected by the eel's electroreceptors, as shown by training eels to reduce their heart rates when the surrounding electrical field changes. The heart rate of a trained eel drops when the surrounding electrical field changes by as little as $0.002 \, mV \cdot cm^{-1}$. Because the fields generated in the ocean are two or three orders of magnitude greater than this value, it is entirely possible that eels can orient with respect to the geoelectrical field.

This brief survey of animal orientation and navigation reveals the complexity of the behavior generated by the underlying neuronal systems. In view of these capabilities, it is daunting to begin a study of the neurons responsible for producing behavior, and, as we will see, the descriptions of the neuronal basis of behavior commonly lag far behind the descriptions of the behavior itself.

PROPERTIES OF NEURONAL CIRCUITS

Despite the extreme complexity of nervous systems, several generalizations can be made about their organization and function. First, neuronal circuits consist of specific connections between neurons, and the pattern of connectivity is essentially the same in all normal individuals of a species. These connections, which are typically established during embryonic development, are maintained and shaped by use throughout an organism's life. If a particular circuit remains unused for a long period, connections in the circuit become weaker, and there can be significant loss of function. Recent evidence indicates that the opposite also is true: within certain limits, repeated activity can increase the strength of existing connections.

The crucial importance of proper connections in producing behavior is illustrated by an experiment on a simple reflex connection in a frog (Figure 11-23). In this experiment, the consequences of altering neuronal connections were tested by surgically disconnecting the sensory fibers entering the spinal cord from one side and reconnecting them to the dorsal root of the opposite side. The severed frog neurons reconnected to the appropriate kind of central neurons, but contacted the wrong side of the spinal cord. In a normal frog, a noxious stimulus to a leg causes a reflex withdrawal of the leg. After neuronal connections had regenerated in this new configuration, a noxious stimulus to the leg on the surgically manipulated side evoked an inappropriate withdrawal of the other leg. The abnormal connections produced inappropriate behavior.

Connections within the nervous system also play an enormously powerful role in sensory perception. In the nineteenth century, Johannes Müller formulated the law of **nerve-specific energy,** stating that the modality of a sensation is determined not by the nature of the stimulus, but rather by the central connections of the nerve fibers activated by a stimulus. This notion is now universally accepted, and we know that the various senses, as well as the topographic distribution of sensory receptors, are represented by particular regions of the cortex (see Figure 11-12). Direct electrical stimulation of local areas in the somatosensory cortex evokes sensations in the subject's consciousness that are more or less similar to those produced by stimulation of the corresponding sense organ. Conversely, peripheral stimulation excites electrical signals at points in the somatosensory cortex that correspond to the specific region of the skin that was stimulated.

A second generalization about the nervous system is that the metabolic state, electrical properties, and summed total of all synaptic inputs impinging on each individual neuron determine how that neuron will respond at any given moment. Each active neuron, by virtue of its connections, in turn influences activity in other neurons.

A third generalization is that the complexity and variety of functions carried out by nervous systems arise from two levels of organization: (1) individual neurons can generate different kinds of signals and (2) neurons are orga-

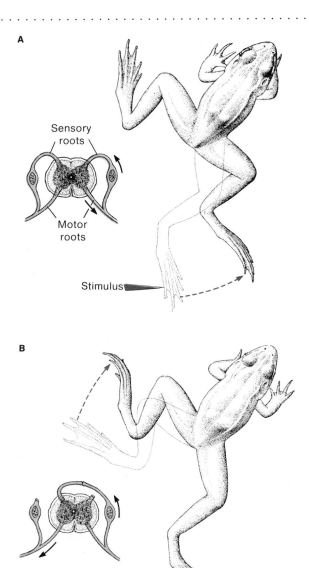

Figure 11-23 The "wiring diagram" of the nervous system is crucial for producing appropriate behavior. (**A**) In a normal frog, when a noxious stimulus is applied to a leg, the leg is reflexly withdrawn. (**B**) If the dorsal roots are cut and are caused to regrow into the contralateral spinal cord, the frog responds to a noxious stimulus on the surgically manipulated side by withdrawing the opposite, unstimulated leg. After the surgery, the sensory input from the right leg is connected to the motor neurons controlling the left leg. [From "The Growth of Nerve Circuits" by R. W. Sperry. Copyright © 1959 by Scientific American, Inc. All rights reserved.]

nized into an enormously complex and varied circuitry. The two basic kinds of signals—propagated, all-or-none APs and nonpropagated, graded synaptic and receptor potentials—are treated in Chapter 6. Synapses can be excitatory or inhibitory, strong or weak. In the end, all neuronal signals depend on the flow of ionic currents through ion channels, driven by electrochemical gradients.

Behavior is produced when sensory information is received by the central nervous system, processed, and used to generate motor output. Alternatively, ongoing activity in the central nervous system can, by itself, generate motor output—which happens, for example, when you decide to stop reading and close this book. To understand the entire process, it is necessary to understand how the central nervous system processes sensory information and how neuronal activity can lead to patterns of muscle contraction. We now consider the pieces of a complete input-output pathway that can generate behavior: sensory input, central processing, and motor output.

Pieces of the Neuronal Puzzle

In considering the complete network that underlies a particular behavior, we can recognize subcircuits whose properties affect the way that the entire network functions. **Sensory filter networks** transmit only certain features of complex sensory input, while blocking out other features. **Central pattern-generating networks** (CPGs) produce patterned motor output that generates more or less stereotyped movements. The output of some pattern generators—for example, those governing locomotion and respiration—is cyclic. The output of other pattern generators—such as those that control the movement of the tongue when frogs and toads capture their insect prey—is noncyclic. Superimposed on some central pattern-generating networks is a *motor command system* in which moment-to-moment changes in sensory input can modulate the motor output. Some behaviors require the participation of both a sensory filter network on the input side and a central pattern generator on the output side. An example is the feeding reflex of the frog, described in the next section. However, some of the simplest behaviors (e.g., the

knee-jerk reflex) are independent of either a sensory filter network or a motor control network.

Even a small number of neurons can be combined into circuits in a number of different ways. In fact, in higher vertebrates, it is common for a single neuron to receive thousands of presynaptic terminals from other neurons, some of them excitatory and some inhibitory. Each neuron may itself branch many times and innervate many other neurons. **Divergence**—the repeated branching of an axon—gives each neuron widespread influence on many postsynaptic neurons (Figure 11-24A). **Convergence** of inputs onto a single neuron (Figure 11-24B) allows a neuron to integrate signals from many presynaptic neurons. Most neurons, such as a mammalian spinal motor neuron, are rarely depolarized to threshold without considerable spatial and temporal summation of excitatory synaptic input. As a result, neurons typically produce APs only when there is more or less simultaneous activity in a number of excitatory presynaptic neurons, producing a convergence of inputs. There can be both divergence and convergence within a single network. For example, when information from the retina is transmitted to the brain, signals from each point at the center of the visual field are carried in several neurons, indicating divergence of information from single photoreceptors. In contrast, signals from the peripheral part of the visual field are combined over a wide area, indicating convergence of signals from many photoreceptors.

Neurons typically receive simultaneous barrages of excitatory and inhibitory synaptic inputs, which are integrated by the postsynaptic cell. Excitation of a neuron can

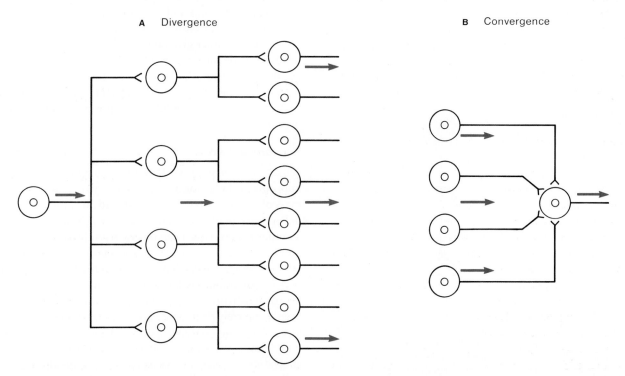

Figure 11-24 Information is carried through divergent and convergent pathways in the nervous system. (**A**) Divergence is the branching of one neuron, allowing it to innervate several others. (**B**) Convergence refers to the innervation of a single neuron by many presynaptic neurons.

be suppressed if its active inhibitory synapses outweigh its active excitatory synapses; so inhibitory synapses modulate the ease with which excitatory inputs can bring the neuron to threshold (Figure 11-25A). When many inhibitory synapses onto a neuron are active, a greater number of excitatory synapses must be activated to bring the postsynaptic neuron to the firing level. However, the net effect of inhibition depends on connections within the network (Figure 11-25B). Inhibition of activity in an inhibitory neuron—called **disinhibition**—can cause a net increase of excitation in the network.

Feedback is employed extensively in neuronal circuits. An example of positive feedback is shown in Figure 11-26A, in which a branch of a neuron in a hypothetical reverberating circuit excites an interneuron that feeds back to reexcite the initial neuron and keep it firing for extended periods. In theory, this neuron could continue to fire indefinitely once it has been excited by synaptic input. If the interneuron in a feedback circuit is inhibitory instead of excitatory (Figure 11-26B), it produces negative feedback, reducing the probability that the initial neuron will fire. A well known instance of negative neuronal feedback is that onto the motor neurons in the vertebrate spinal cord. Here, the α motor neurons make small branches that innervate short inhibitory interneurons, the **Renshaw cells** (Figure 11-27). Renshaw cells are excited each time the mo-

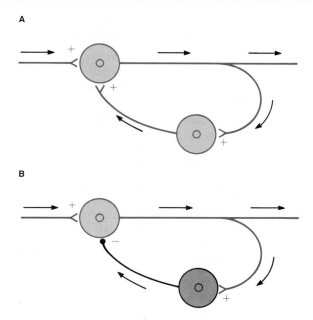

Figure 11-26 Neurons can be arranged in local feedback circuits. (**A**) Recurrent facilitation. If the interneuron is excitatory, it produces positive feedback and prolongs activity in the initial neuron. (**B**) Recurrent inhibition. If the interneuron is inhibitory, it produces negative feedback and limits activity in the initial neuron.

tor neurons fire, and they feed back, reducing the level of excitation within the motor neurons. The Renshaw cells respond to stimulation by producing a high-frequency train of impulses, which in turn cause inhibitory postsynaptic potentials in the motor neurons. We do not fully understand the functional significance of the Renshaw input onto motor neurons, but we do know that it limits motor neuron activity. Strychnine blocks the glycine-mediated synapses made by Renshaw cells onto the motor neurons—and probably other glycine-mediated synapses as well—producing convulsions, spastic paralysis, and death from paralysis of the respiratory muscles. These gruesome consequences of blocking inhibitory synapses demonstrate the importance of synaptic inhibition in normal function of the nervous system.

Sensory Networks

Sensory neuronal networks—the first step in generating appropriate behavioral responses—sort and refine the mass of information that is available to an animal. Individual sensory neurons respond to only a limited range of stimulus energy, which can be described by a tuning curve (Spotlight 11-2). This property of receptors and other properties of sensory networks combine to filter incoming sensory information. It is now clear that, in addition, sensory neuronal networks can magnify, amplify, add, subtract, and even completely reconfigure the original pattern of sensory input. The visual system is particularly well studied, so we will examine it in some detail to illustrate some general organizational principles in sensory systems. In addition, we will consider the auditory system of the barn owl as an example of how neurons can transform sensory signals.

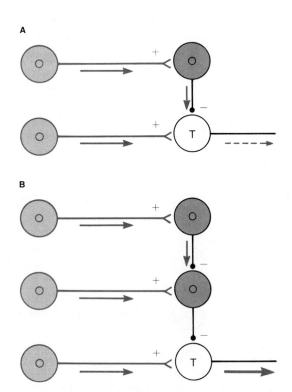

Figure 11-25 The net effect of inhibitory neuronal activity depends on the arrangement of the network. (**A**) Activity in the inhibitory neuron (gray), reduces the probability of APs (indicated by a dashed arrow) in the terminal cell, labeled T. (**B**) If two inhibitory neurons are arranged in series and if the second one is tonically excited or spontaneously active, excitation of the first inhibitor will reduce inhibitory input onto the terminal cell (T), increasing its net activation (indicated by a solid arrow).

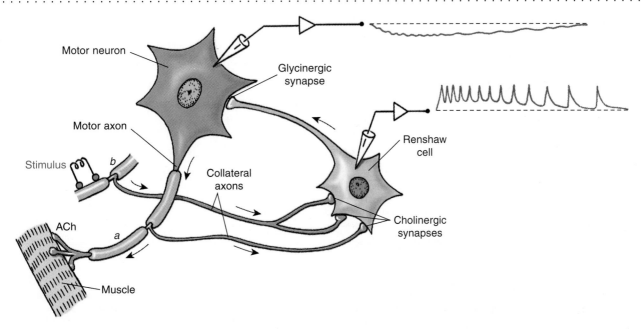

Figure 11-27 Renshaw cells feed back on and inhibit spinal α motor neurons. A simplified diagram of motor neuron axons with their collateral branches innervating a Renshaw cell. The Renshaw cells make glycinergic inhibitory synapses onto the motor neurons. The trace shown next to each recording electrode illustrates a typical record. In this case, the Renshaw cell was excited by an antidromic AP caused in motor neuron *b* by an electrical stimulus. Action potentials in the Renshaw cell caused an inhibitory postsynaptic potential in motor neuron *a*. These inhibitory postsynaptic potentials can be blocked by strychnine. [Adapted from Eccles, 1969.]

In its simplest form, sensory processing can be seen as the abstraction of features from the original information. A classic example of this is illustrated in Figure 11-28. Recordings from axons in the optic nerve of a frog show that some of the neurons respond only to certain features of the visual field. Some neurons are remarkably specific. For example, one kind of axon carries APs only when the photoreceptors to which it is connected are activated by a small object, such as a fly, that is moving against a light background containing stationary objects. These neurons do not respond if the entire scene moves or if the background light is simply turned on or off. Because frogs catch and eat moving insects, this class of neurons may tell the frog's brain that supper is available *now*, information that is likely to be more significant to the frog than are most of the other details in the visual scene. Anyone who tries to maintain frogs in captivity soon learns that a frog does not recognize a dead insect as food. In other words, a dead fly fails to activate the circuitry that triggers the frog's feeding response, apparently because it is not a small object moving against a stationary background—the distinguishing feature of a live fly.

This example raises the key question of where neuronal transformations of sensory information take place. In the frog, axons within the optic nerve are third-order neurons, meaning that information has crossed a minimum of two synapses to reach them. To discover where specialized detection takes place in the visual system, we need to look at the organization of the retina, beginning at the photoreceptors.

Lateral inhibition

A common feature of visual systems is that they enhance contrast in the scene, particularly at the boundaries that

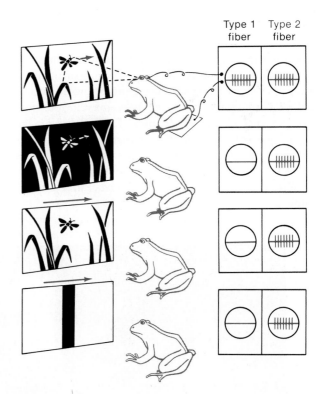

Figure 11-28 Some neurons in the frog retina respond specifically to stimuli that resemble a moving fly. A frog will respond behaviorally, by sticking out its tongue, to a small, dark object moving in its visual field (1). It fails to respond to movement of a light object against a dark background (2), to movement of the background against a stationary small object (3), or to other nonspecific visual stimuli (4). Some axons in the optic nerve (labeled Type 1 in the drawing) become active only in response to a small, dark, sharply outlined object moving across a lighter background. Other fibers (Type 2) are activated by a wide variety of visual stimuli, such as a moving background or a large moving bar. [Adapted from Bullock and Horridge, 1965.]

SPOTLIGHT 11-2

TUNING CURVES: THE RESPONSE OF A NEURON PLOTTED AGAINST THE PARAMETERS OF A STIMULUS

Recordings of activity made from individual neurons in sensory areas of the cortex indicate that each neuron responds to a range of stimuli, but it responds optimally to very specific parameters of a stimulus. Plots showing how the response of a sensory neuron changes with a parameter of the stimulus are called *tuning curves* (illustrated in the accompanying figure). Some neurons, such as neuron *c* in the figure, are very broadly tuned. Others, such as neuron *d* in the figure, are very narrowly tuned. The way in which information flows through a neuronal circuit depends heavily on the tuning curves of the neurons at each level of processing. For example, very narrowly tuned neurons act as filters, permitting only signals with particular properties to pass to the next level. In turn, the tuning curve of a central neuron depends on its pattern of synaptic input.

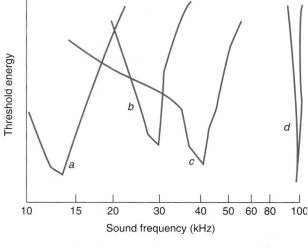

Tuning curves indicate the relations between the activity in a neuron and parameters of effective stimuli. This graph illustrates the range of sound frequency to which four primary auditory neurons (represented by *a, b, c,* and *d*) in the ear of the bat *Rhinolophus* respond. Each neuron is most sensitive to a particular frequency (the threshold energy that stimulates the neuron is lowest), but it can be stimulated by other frequencies within some range. Receptor *d* is very narrowly tuned, whereas receptor *c* is broadly tuned. Sounds outside the tuning curve of a neuron fail to activate it at normal energy levels. [Adapted from Camhi, 1984.]

separate objects. The neuronal mechanisms for emphasizing such differences probably evolved because small differences in stimulus energy can be important cues to an animal. You can observe contrast enhancement in your own visual system by looking at Figure 11-29. Each band in the figure appears to be lightest at its border with its darker neighbor and darkest at its border with its lighter neighbor. In fact, the actual luminosity of each band is uniform across its entire width, and the apparent difference in brightness within a band is an illusion.

What causes this illusion? It is a consequence of **lateral inhibition** at the level of the receptors. This phenomenon was studied in a series of experiments in the laboratory of H. K. Hartline at Rockefeller University in the mid-1950s, and the work received recognition by a Nobel Prize in 1967. In these experiments, the activity of a single ommatidium was recorded first when a bright stimulus light focused on only that ommatidium. You might suppose that adding more stimulus light to the whole eye would increase the number of APs produced by that single ommatidium.

Figure 11-29 Lateral inhibition enhances contrast between adjacent areas. Each band in this figure is actually uniformly dense from its left to its right border, but each appears to be lighter near its dark neighbor and darker near its light neighbor. You can see that each band is of uniform density by covering all of the bands but one with two pieces of paper.

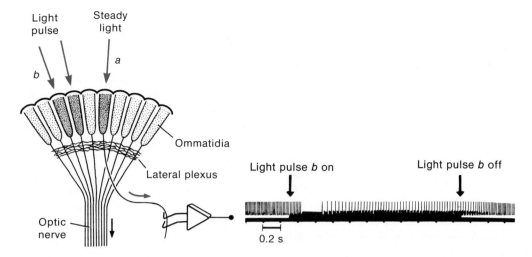

Figure 11-30 Lateral inhibition can be observed in the eye of the horse-shoe crab, *Limulus*. The output of a single ommatidium is decreased when neighboring ommatidia are stimulated. [Adapted from Hartline et al., 1956.]

However, adding more light by switching on the room lights *decreased* the firing rate of the single ommatidium. The diffuse light in the room stimulated the surrounding ommatidia and inhibited the test ommatidium. This phenomenon, lateral inhibition, has since been observed in many other visual systems, as well as in other sensory systems.

In another experiment, the source of lateral inhibition in the lateral eye of the horseshoe crab, *Limulus*, was shown to be neighboring photoreceptors (Figure 11-30). While the response of a single ommatidium to a steady light was recorded, a light pulse was presented to nearby ommatidia. The onset of light in adjacent receptor cells caused a reduction in the response of the ommatidium whose activity was being recorded. Inhibition between interacting units is completely reciprocal, and the amount of inhibition decreases with distance; inhibition is strongest between nearest neighbors. Lateral inhibition in the *Limulus* eye is mediated through the **lateral plexus,** a set of collateral branches from eccentric-cell axons that form inhibitory synapses on one another (see Chapter 7). Action potentials in eccentric-cell collaterals cause the release of inhibitory transmitter from synaptic terminals onto neighboring eccentric-cell axons. Because the inhibition exerted by a unit on its neighbors increases as the unit's activity increases, a strongly stimulated ommatidium will strongly inhibit neighboring, less strongly stimulated units. At the same time, the strongly stimulated unit receives weak inhibition from its neighbors. This interaction enhances the contrast in activity level between neighboring units that are exposed to different intensities of light (Figure 11-31). Contrast

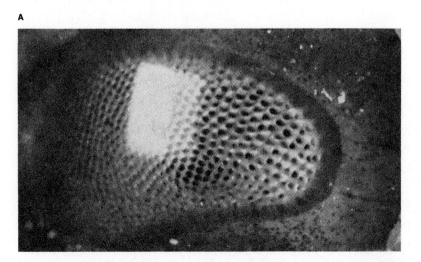

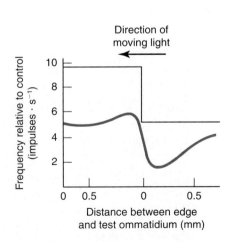

Figure 11-31 Contrast is enhanced most strongly at the boundaries between dim and bright regions. (**A**) In this experiment, the activity of a single ommatidium is recorded, while a strongly lit rectangular field is moved across the moderately lit compound eye. (**B**) Activity in the ommatidium is plotted against the position of the approaching bright edge. If all other ommatidia are masked, the output of the ommatidium will change in an abrupt, steplike manner as the bright rectangle reaches the ommatidium (black plot in part B). However, if the mask is removed, and the light border is allowed to pass over all the ommatidia, the output of the single ommatidium will generate a response similar to that represented by the red plot. [From "How Cells Receive Stimuli" by W. H. Miller, F. Ratliff, and H. K. Hartline. Copyright © 1961 by Scientific American, Inc. All rights reserved.]

enhancement is greatest for units that lie at the boundary separating a bright and a dim region, because lateral inhibition diminishes with distance. Thus, lateral inhibition sharpens the visual detection of edges by increasing contrast at the borders between areas of different luminosities. You experienced the effects of these interactions when you looked at Figure 11-29.

Visual processing thus begins in the very first neurons of the network. Processing of visual information by neurons farther along the chain continues to abstract and accentuate properties of borders and other features of visual stimuli.

Information processing in the vertebrate retina

The image of the world that falls onto the retina is a relatively accurate representation of the field viewed by that eye, limited only by the optics of the eye. The way in which the visual system transforms this raw material into a perceived image has been the subject of intense study at several levels of organization, ranging from the study of visual transduction (see Chapter 7) to a consideration of neurons in the brain that might recognize entire perceived objects. The visual system is the best studied of the sensory systems, probably because vision is so important to primates, including human beings. However, principles that have emerged from the study of vision apply to other sensory systems as well, which suggests that evolution may have settled on certain universal solutions to some problems commonly faced by neuronal networks in general. In this section, we examine some of what is known about the neuronal processes underlying visual perception.

The visual pathway of vertebrates begins in the retina and continues to the optic tectum in lower vertebrates (Figure 11-32A), or to the **lateral geniculate nuclei** and the **visual cortex** in birds and mammals (Figure 11-32B). The visual system can be viewed as a series of connected cellular plates (Figure 11-32C). The cells within each plate have properties in common. In the projections from one plate to the next, information both converges and diverges.

There is a significant amount of synaptic interconnection and, hence, potential image processing within the retina itself. The photoreceptors connect with **bipolar cells** that, in turn, contact *ganglion cells,* whose axons make up the optic nerve (Figure 11-33). The receptors are *first-order cells,* the bipolar cells are *second-order cells,* and the ganglion cells are *third-order cells* in the afferent pathway. This nomenclature is oversimplified, because there are two other types of retinal neurons—the **horizontal cells** and the **amacrine cells**—that are particularly important for mediating lateral interactions within the retina. The horizontal cells receive input from neighboring and moderately distant receptor cells, and they synapse onto bipolar cells. The amacrine cells interconnect bipolar cells and ganglion cells.

Studies combining intracellular recording techniques with the injection of fluorescent marker dyes have revealed the electrical activity typical of each retina cell type

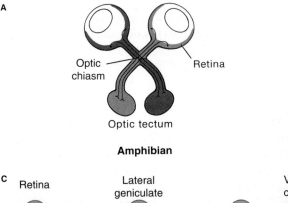

A

Optic chiasm

Retina

Optic tectum

Amphibian

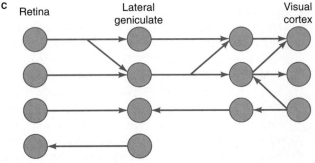

C   Retina

Lateral geniculate

Visual cortex

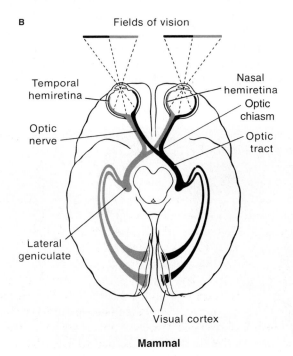

B

Fields of vision

Temporal hemiretina

Optic nerve

Lateral geniculate

Nasal hemiretina

Optic chiasm

Optic tract

Visual cortex

Mammal

Figure 11-32 Visual information is transmitted from the retina to the brain through layers of cells. (**A**) In an amphibian, the left and right sides of the optic tectum each receive projections from the entire field of view of the contralateral eye. (**B**) In a mammal, each side of the visual field is projected to the opposite side of the visual cortex. For example, the temporal half of the left retina and the nasal half of the right retina project to the left visual cortex. (**C**) The neurons that initially process visual information are organized in layers. The retina contains the first three layers, and the remainder are in the brain, in the lateral geniculate nuclei and in the cortex. Information converges and diverges between the layers, and it flows in both directions between the layers. [Part A from "Retinal Processing of Visual Images" by C. R. Michael. Copyright © 1969 by Scientific American, Inc. All rights reserved. Part B adapted from Noback and Demarest, 1972.]

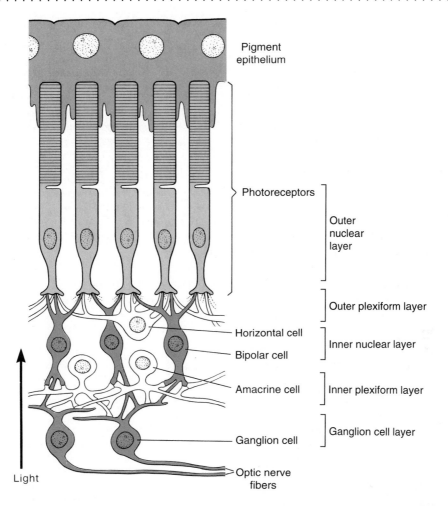

Pigment
epithelium

Photoreceptors

Outer
nuclear
layer

Outer plexiform layer

Horizontal cell

Bipolar cell

Inner nuclear layer

Amacrine cell

Inner plexiform layer

Ganglion cell

Ganglion cell layer

Optic nerve
fibers

Light

Figure 11-33 The function of the vertebrate retina is based on five major types of neurons. Photoreceptors receive light stimuli and transduce them into neuronal signals. Bipolar cells carry signals from photoreceptors to the ganglion cells, which send their axons into the central nervous system through the optic nerves. Horizontal and amacrine cells, which are located in the outer and inner plexiform layers, respectively, carry signals laterally. [From "Visual Cells," by R. W. Young. Copyright © 1970 by Scientific American, Inc. All rights reserved.]

(Figure 11-34). Vertebrate photoreceptor cells hyperpolarize when they are illuminated (see Chapter 7). They release synaptic transmitter continuously in the dark, and transmitter release is reduced when they hyperpolarize in response to illumination. Similarly, horizontal cells produce only graded hyperpolarizations in response to light (see Figure 11-34). Bipolar cells can produce graded potential changes of either polarity. A ganglion cell responds with the same polarity as do bipolar cells that innervate it. It becomes depolarized and fires APs when the bipolar cells synapsing on it depolarize, and it becomes hyperpolarized and ceases spontaneous firing when its bipolar inputs hyperpolarize. Amacrine cells respond transiently at the onset and offset of light in response to input from bipolar cells.

Bipolar cells typically connect more than one receptor to each ganglion cell, and they may also connect each receptor cell to several ganglion cells. Thus, convergence and divergence already exist between the first- and third-order cells of the visual system, but the amount depends on retinal location. In mammals, both convergence and divergence are minimal in the **fovea**, or *area centralis*, (the area

in the center of the retina on which visual images are sharply focused). This lack of convergence and divergence produces very high visual acuity based on one-to-one-to-one connections between cone photoreceptors, bipolar cells, and ganglion cells. (Cones are the majority of photoreceptors in the fovea.) Outside the fovea, each ganglion cell receives input from many receptor cells—primarily rods—conferring on these ganglion cells a greater sensitivity to dim illumination but a lower degree of visual acuity.

Structurally, the output of the retina is carried in the optic nerve by axons of ganglion cells, but how is the output organized? Understanding the information exported by ganglion cells hinges on the concept of a **receptive field,** an idea that was first proposed by Sherrington and was applied to visual processing by Hartline in the 1940s. A cell's receptive field is the area on the retina in which light stimuli affect the cell's activity. The receptive field of a ganglion cell is roughly centered on the cell and varies in size, depending on the degree to which photoreceptor and bipolar cells converge in the pathway to each ganglion cell. At the center of the fovea, a ganglion cell's

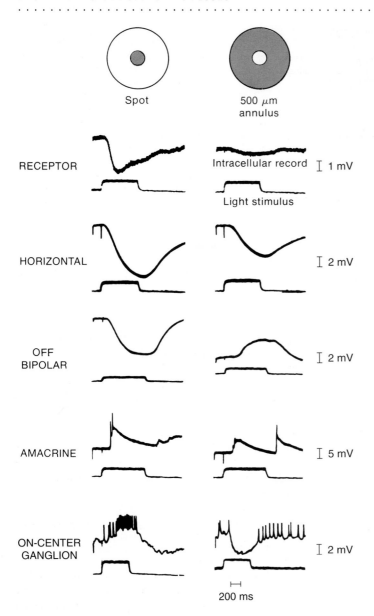

Spot

500 μm
annulus

RECEPTOR

Intracellular record   ⊥ 1 mV

Light stimulus

HORIZONTAL

⊥ 2 mV

OFF
BIPOLAR

⊥ 2 mV

AMACRINE

⊥ 5 mV

ON-CENTER
GANGLION

⊥ 2 mV

⊢⊣
200 ms

Figure 11-34 Each type of retinal neuron has a distinctive electrical response to light. Activity was recorded in each type of cell in response to a spot of light focused directly on the receptors in the region *(left)* and in response to an annulus of light surrounding the photoreceptors *(right)*. The duration of the stimulus is indicated in the lower trace on each record. In this example, the ganglion cell is activated by a light that shines on the center of its receptive field. Note that the bipolar and ganglion cells produce responses of opposite polarities to the spot and the annulus. This effect is believed to be due to lateral inhibition similar to that in *Limulus*. Notice that the *off* bipolar cell and the on-center ganglion cell shown in this figure would not be synaptically connected. See Figure 11-36 for a detailed depiction of how ganglion cell responses are related to signals in bipolar cells. [Adapted from Werblin and Dowling, 1969.]

receptive field extends over only one or a few photoreceptors; at the periphery of the retina, where convergence is great, the receptive field of a ganglion cell can be as large as 2 mm in diameter.

Each ganglion cell is spontaneously active in the dark, and the level of activity changes when a spot of light falls within its receptive field. Depending on which receptor cells are illuminated, the frequency of APs in a ganglion cell may increase if a small spot of light enters the cell's receptive field—an *on response*. Alternatively the frequency of APs may drop in response to a light—an *off response*. The receptive field of a ganglion cell is typically divided into a *center* and a *surround*, and the response of the cell depends on whether the center or the surround is being stimulated or both are (Figure 11-35). In an *on-center* ganglion cell, the frequency of APs increases when the center of its receptive field is illuminated (see Figure 11-35A). If a ring of light shines on the entire receptive field, with the center of the ring over the center of the field, activity in the cell drops. A

weaker *off* response is elicited by a spot of light that illuminates only a part of the field. The ring surrounding the center of the receptive field is called the *inhibitory surround* of the receptive field. An *off-center cell* exhibits the converse behavior, ceasing or reducing its activity when the center of its receptive field is illuminated and increasing its firing when the surround is illuminated.

The center-surround organization of receptive fields depends on lateral inhibition similar to that found in the compound eye of *Limulus*. Lateral interaction in the vertebrate retina takes place primarily through the activity of the horizontal cells in the **outer plexiform layer** (see Figure 11-33). The horizontal cells have extensive lateral processes and are interconnected with neighboring horizontal cells through electrotonic junctions. In addition, they make chemical synapses on bipolar cells and receive synaptic inputs from receptor cells. Light that falls in the surround of a ganglion cell's receptive field exerts its effects through lateral connections made by horizontal cells. Because hori-

A On-center ganglion cell   **B** Off-center ganglion cell

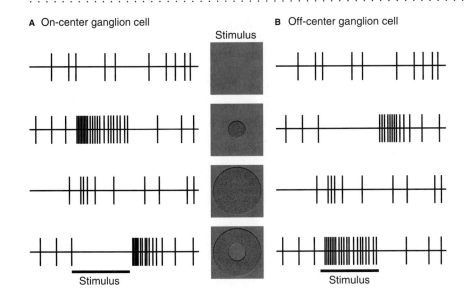

Stimulus

Stimulus

Stimulus

Figure 11-35 Retinal ganglion cells have either on-center responses or off-center responses to light stimuli. (**A**) Four recordings from a typical on-center retinal ganglion cell. Each record shows activity in the ganglion cell during a 2.5 second interval. Stimuli are shown in the middle of the figure. In the dark, APs in the cell are slow and more or less random. The lower three records show responses to a small spot, to a large spot that includes the center of the receptive field plus a surround, and to a ring that covers only the surround. (**B**) Responses of an off-center retinal ganglion cell to the same set of stimuli. [Adapted from Hubel, 1995.]

zontal cells form an extensive syncytial network, communicating with one another through low-resistance gap junctions, input from any receptor onto a horizontal cell produces a hyperpolarizing signal that spreads electrotonically in all directions away from the receptor. Every bipolar cell receives input from surrounding receptor cells by means of horizontal cells, and this input falls off with distance because the graded, hyperpolarizing potentials in horizontal cells decay as they spread electrotonically. The *indirect* input that a bipolar cell receives from outlying receptors through the horizontal cell network opposes the *direct* input that it receives from photoreceptors, providing the basis of the center-surround organization in retinal receptive fields. The local, direct-line pathway from photoreceptor through bipolar cell to ganglion cell produces the center response. The indirect pathway from photoreceptors through horizontal cells to bipolar cells and, thence, to ganglion cells mediates the response to the surround. These two pathways show how particular features of a stimulus can be extracted by even a relatively simple neuronal network.

The distinctive responses of on-center and off-center ganglion cells arise from their connections with two classes of bipolar cells: *on bipolar cells* and *off bipolar cells*. These two types of bipolar cells respond oppositely to synaptic input, both from receptors and from horizontal cells (Figure 11-36). The *off* bipolar cells become hyperpolarized by illumination of receptors, whereas the *on* bipolar cells become depolarized. In both types of bipolar cells, a light flashed onto the surround produces a response, mediated by horizontal cells, that is of the opposite electrical sign to that produced by illumination of the center. Each bipolar cell causes potential changes in its ganglion cell or cells that are of the same sign as the potential change occurring in the bipolar. Thus, ganglion cells innervated by *on* bipolars will have on-center receptive fields, whereas those innervated by *off* bipolars will have off-center fields. An on-center ganglion cell is excited by light in the center of its

receptive field because it receives direct synaptic input from *on* bipolar cells. It is inhibited by light in the surround of its receptive field, because horizontal cells that receive input from surrounding photoreceptors inhibit the *on* bipolar cells in the direct pathway from photoreceptors to the ganglion cell.

The responses of *on* and *off* bipolar cells depend on how the cells respond to the neurotransmitter released by photoreceptor cells and the different neurotransmitter released by horizontal cells. *On* bipolar cells are steadily hyperpolarized in the dark by transmitter that is steadily secreted from the partially depolarized receptor cells. When a light stimulus causes photoreceptors to hyperpolarize, their release of transmitter drops and *on* bipolar cells are allowed to depolarize. This depolarization causes the *on* bipolar cells to release an excitatory transmitter that depolarizes ganglion cells, increasing the frequency of APs in the ganglion cells. In contrast, the *off* bipolars have a different class of postsynaptic channels with different ionic selectivity and are steadily depolarized in the dark by neurotransmitter released by the photoreceptors. When light falls on the photoreceptors and they hyperpolarize, reduction in their release of neurotransmitter causes the *off* bipolar cells to hyperpolarize. This hyperpolarization is accompanied by a drop in transmitter release by the *off* bipolar cells, producing a hyperpolarization of postsynaptic ganglion cells.

In summary, the receptive field organization of the vertebrate retina depends on three basic features:

1. Two kinds of ganglion cells receive input from two corresponding kinds of bipolar cells. The connections produce on-center and off-center ganglion cell responses.

2. Receptors in the surround of the receptive field exert their effects through a network of electrically interconnected horizontal cells that synapse onto the two kinds of bipolar cells.

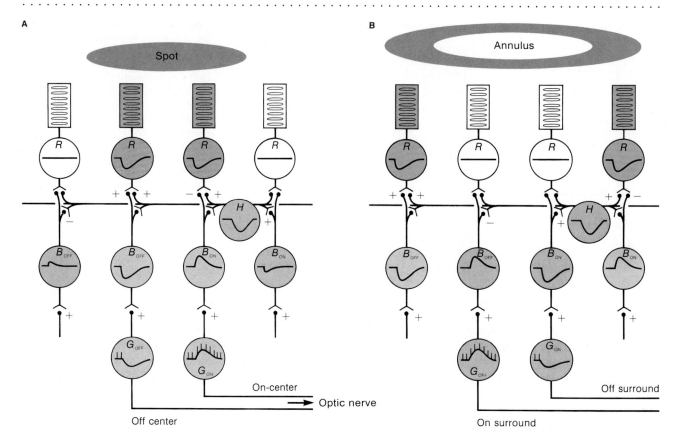

Figure 11-36 Connections within the retina produce the responses characteristic of on-center and off-center ganglion cells. Two kinds of bipolar cells, B_{ON}, and B_{OFF}, respond oppositely to direct input from the receptors, R, and to indirect input carried laterally by the horizontal cells, H. The *on* bipolar cells become depolarized during activation of the overlying receptor cells and are weakly hyperpolarized by lateral input from horizontal cells. The *off* bipolars behave oppositely. (**A**) Responses of bipolar cells and ganglion cells to a spot of light. (**B**) Responses of bipolar cells and ganglion cells to an annulus of light. Amacrine cells have been omitted from the diagram for simplicity. The direct pathway from photoreceptors to ganglion cells, G, is shown in color. The indirect, lateral pathway through horizontal cells is shown in gray. The plus and minus signs indicate synaptic transfer that conserves (+) or inverts (−) the polarity of the signal.

3. Direct input to bipolar cells from overlying receptors and indirect input to the cells through the horizontal cell network oppose each other and thereby produce the contrasting center-surround organization seen in both the on-center and the off-center ganglion cells.

The organization of the retina reveals several general principles that apply to the other parts of the central nervous system. First, nerve cells can signal each other electrotonically without APs *if the distances are small.* Nonspiking neurons can in fact convey more information more accurately than can all-or-none signals. Electrotonic signals are attenuated with distance, which limits the range of effects such as lateral inhibition. Second, reception of stimuli is not necessarily synonymous with depolarization. In some nerve cells (e.g., photoreceptors and some horizontal cells), hyperpolarization is the normal response to stimulation; it modulates synaptic transmission by causing a drop in the steady release of transmitter. Third, the postsynaptic response in a neuron cannot be predicted from the sign of the potential change in the presynaptic neuron. A cell can be either depolarized or hyperpolarized in response to hyper-

polarization of the presynaptic cell. The postsynaptic response depends on the ionic currents produced in the postsynaptic cell as a result of the modulated release of transmitter by the presynaptic neuron.

Information processing in the visual cortex

What happens to a retinal image after it has been transformed into an array of receptive field responses within the retina? Physically, the information is carried by axons to visual areas within the brain. The details of this pathway vary among species. In mammals and birds, the axons of retinal ganglion cells are routed either to the ipsilateral or to the contralateral side of the brain at the **optic chiasm,** the site where some axons cross the midline (see Figure 11-32B); whereas, in vertebrates more primitive than birds, all optic fibers are routed to the contralateral side at the chiasm (see Figure 11-32A). To some extent, the degree of crossing at the optic chiasm depends on how much overlap there is between the visual fields of the two eyes. In animals in which the visual field of one eye is entirely different from the visual field of the other eye, all retinal ganglion cell axons cross the midline. In mammals, ganglion cell axons synapse with fourth-order cells in the lateral geniculate nucleus of the thalamus. Lateral geniculate neurons send axons to synapse on fifth-order cortical neurons in the occipital cortex (see Figure 11-11) in an area called Area 17, also called *primary visual cortex* because it is the first region of the cortex in the pathway to receive visual information.

The pattern of synaptic relations within the lateral geniculate is based on the source and the nature of information carried by retinal ganglion cells, and it constitutes another step in the processing of visual input. Each lateral geniculate nucleus, or body, is composed of six layers of cells, stacked like a club sandwich that has been folded (Figure 11-37). The top four layers contain neurons with small somata, which are called *parvocellular* neurons, and the bottom two layers contain neurons with large somata, called *magnocellular* neurons. Input onto these neurons is tightly organized. Each lateral geniculate body receives information from only one half of the visual field (i.e., one of the two visual fields illustrated in Figure 11-32B), and cells in each layer receive input from only one retina. Each neuron in the lateral geniculate receives information from only one eye. Neurons in a given layer receive information from the same eye, and the layers alternate from one eye to the other, with the pattern of alternation changing between the fourth and fifth layers (see Figure 11-37). Across all layers, the topography of the corresponding retinal surface is preserved exactly, and the topography is kept in register among the layers. If we pass an electrode along the path indicated by a dashed line in Figure 11-37, we will encounter cells that respond to a light stimulus in precisely the same point in the visual field, but the eye of origin will switch from left to right as our electrode moves from one layer to the next.

Are there functional differences among the layers that receive information from each eye? Yes, the cells in each layer respond to particular properties of a stimulus, and the response varies from layer to layer. For example, in the monkey, cells within the four dorsal layers respond to the color of a stimulus, whereas cells of the two deepest layers do not. In contrast, the two deepest layers respond to movement, whereas the outer four layers do not. This spatial sorting of outputs from ganglion cells illustrates another principle of brain organization: information about a single stimulus is divided among parallel pathways. This pattern, called **parallel processing,** is a major theme of research into higher brain function. The receptive fields of neurons in the geniculate do not differ substantially from those of retinal ganglion cells. That is, they have a concentrically arranged center-surround organization of either the off-center or on-center type.

The difficult question of how the visual world is organized in the next visual projection area, Area 17, was extensively and insightfully analyzed by David Hubel and Torsten Wiesel in the 1960s, and they received the Nobel Prize in 1981 in recognition of the importance of their work. In their experiments, they recorded the activity of individual neurons in the brain of an anesthetized cat while a simple visual stimulus—such as a dot, circle, bar, or edge—was projected onto a screen positioned to cover the entire visual field of the cat (Figure 11-38A). The

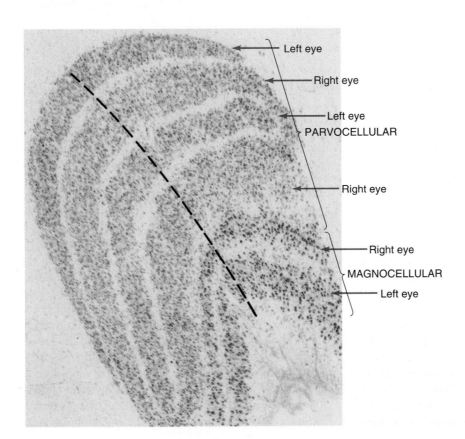

Left eye
Right eye
Left eye
PARVOCELLULAR
Right eye
Right eye
MAGNOCELLULAR
Left eye

Figure 11-37 Cells of the mammalian lateral geniculate nucleus are organized into layers, each of which receives information from only one eye. Histological section of the left lateral geniculate body of a macaque monkey; section is parallel to the face. The cells of the outer four layers have small somata and are called parvocellular. Cells in the deeper layers are magnocellular. In the left lateral geniculate, all cells receive information about the right visual field. In addition, the outermost layer receives input from only the left eye, whereas cells in the next layer receive input from only the right eye, and so forth. A recording electrode passed from one layer to the next would reveal that cells along the path indicated by the dashed line respond to precisely the same location in visual space, but the eye that received the information alternates. [Adapted from Hubel, 1995.]

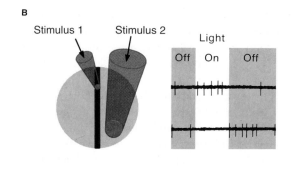

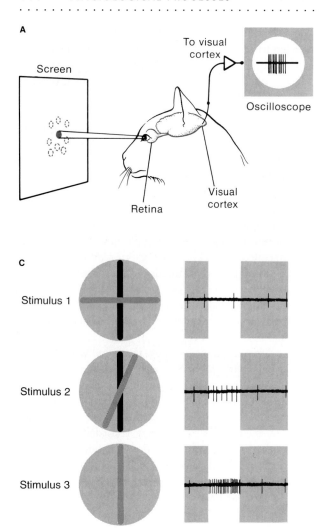

Figure 11-38 Neurons in Area 17 of a cat have very different receptive fields from those of retinal ganglion cells or lateral geniculate cells. (**A**) Experimental setup used to study the responses of cells in the visual cortex. An electrode is advanced through the cortex while light stimuli are projected onto a screen. (**B**) The receptive field of cortical simple cells is bar shaped. A spot of light anywhere along the *on* part of this receptive field (stimulus 1) produces a small excitation of the simple cell. A spot of light adjacent to the bar-shaped *on* region (stimulus 2) causes inhibition of APs in these tonically active cells. (**C**) Rotating a bar of light (red bar) across the receptive field of a simple cell produces maximum activity in the cell when the bar coincides completely with the *on* region of the cell's receptive field (stimulus 3) and partial excitation at other orientations (e.g., stimulus 2). [Part A from "Cellular Communication" by G. S. Stent. Copyright © 1972 by Scientific American, Inc. All rights reserved. Parts B and C from "The Visual Cortex of the Brain" by D. H. Hubel. Copyright © 1963 by Scientific American, Inc. All rights reserved.]

responses that they recorded from cortical neurons were correlated with the position, shape, and movement of the projected images. In retrospect, Hubel, Wiesel, and their collaborators made two important decisions in their experiments that allowed them to uncover order and regularity amid the enormous complexity of the visual brain. First, they chose to use more complex stimuli than just simple spots, and they asked which of these stimuli was most effective in eliciting a response in each neuron. Second, they recorded from many cells with each electrode penetration, allowing them to learn what neighboring cells had in common and how cells were grouped in the brain. These strategies allowed them to discover several different kinds of order among the interconnections of the visual cortex, and their discoveries have provided a model for examining other sensory systems.

Hubel and Wiesel's major discovery about the responses of cells in the visual cortex was that they responded to entirely different properties of stimuli than did retinal ganglion cells. Cortical cells responded most strongly to bars projected in different orientations. They called the two major classes of cells that they found *simple cells* and *complex cells,* based on the nature of their optimal stimulus.

They found that cells of each type were arranged systematically in space according to their optimal stimuli.

The receptive fields of simple cells are long and bar shaped, and the *on* region of the field has a straight border separating it from the *off* region (Figure 11-38B), rather than the circular border found for cells in the retina and in the lateral geniculate. As for retinal ganglion and geniculate cells, the receptive field of a simple cell lies in a fixed position on the retina and, hence, represents a particular part of the total visual field. There is some variation in the receptive fields of simple cells: some have a bar-shaped *on* region surrounded by an *off* region; for others, the receptive field consists of an *off* bar surrounded by an *on* region; and, for still others, it consists of a straight edge with an *off* region on one side and an *on* region on the other side.

A stimulus bar elicits maximal activity in a simple cell when it overlaps completely with the cell's *on* receptive field (Figure 11-38C). When the bar is rotated so it no longer aligns with the orientation of the receptive field, either it has no effect on the spontaneous activity of the simple cell or it inhibits the cell's activity. If the bar of light is displaced so that it falls just outside the *on* region, the cell is maximally inhibited. The orientation and the on-off boundaries differ from one simple cell to another; so, when a bar of light moves horizontally or vertically across the retina, it activates one simple cell after another as it enters one receptive field after another.

What makes simple cells respond specifically to straight bars or to borders of precise orientation and location? Hubel and Wiesel suggested—and recent experiments have confirmed—that each simple cell receives excitatory connections from lateral geniculate cells whose *on* centers are arranged linearly on the retina (Figure 11-39A). Simple cells that respond to borders, rather than to bars, are thought to receive inputs as shown in Figure 11-39B. A simple cell would receive maximal input when light fell on all of the receptors that activate the *on*-center fields of the ganglion cells and the geniculate cells on the pathway to that cell. Any additional illumination would fall on the inhibitory surround of the ganglion cells and could only reduce the response of the cortical cell.

Complex cells constitute the next level of abstraction in the processing of visual information. Complex cells are believed to be innervated by simple cells, which would make complex cells sixth-order cells in the hierarchy of visual information processing. Like simple cells, complex cells respond best to straight borders of specific angular orientation on the retina. Unlike the simple cells, however, complex cells do not have topographically fixed receptive fields. Appropriate stimuli presented within relatively large areas of the retina are equally effective at activating complex cells; as for simple cells, general illumination over the whole retina is not an effective stimulus. Some complex cells respond to bars of light of specific orientation (Figure 11-40A). Others give an *on* response to a straight border when the light is on one side and an *off* response when the light is on the other side. Still other complex cells respond optimally to a moving border that progresses in only one direction (Figure 11-40B). For these cells, movement in the other direction evokes either a weak response or no response at all. These receptive fields can be explained by a combination of synaptic inputs from simple cells. As a light-dark border moves through the receptive fields of the simple cells that synapse onto a complex cell, each simple cell excites the complex cell in turn as the light-dark border passes through the *on-off* borders in the receptive fields of the simple cells. This arrangement could produce directional sensitivity to movement of the *on-off* boundary (see Figure 11-40B). If the boundary moved so that sequential simple cells were illuminated, one simple cell after another would be excited, exciting the complex cell. When each simple cell became inhibited by the dark side of the moving edge, the next one in line would be excited. In contrast, if the boundary moved so that simple cells were sequentially exposed first to inhibition and only later to stimulation, one simple cell after another would inhibit the complex cell, counteracting any tendency for excitation caused by the bright side of the edge.

The properties of individual cortical cells suggest that they abstract features of the visual scene, such as edges, as

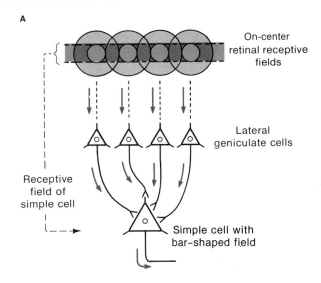

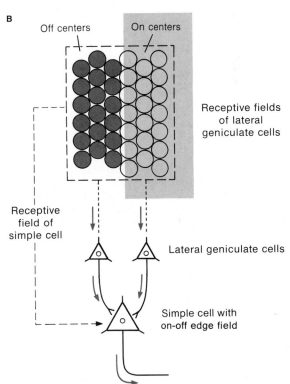

Figure 11-39 The responses of simple cells in the visual cortex arise from the pattern of their synaptic inputs. (**A**) The fixed, bar-shaped receptive field of a simple cell arises from convergence of outputs from ganglion and lateral geniculate cells whose circular on-center receptive fields are linearly aligned. (**B**) An on-off straight-border receptive field results from the convergence of off-center and on-center geniculate cells onto the simple cell.

a first step toward analysis and recognition. The spatial relations among visual cortical cells are correlated in an orderly fashion with their functional properties. In their experiments, Hubel and Wiesel discovered that cells adjacent to one another responded to similar features of a stimulus. When they penetrated the visual cortex with an electrode that was perpendicular to the cortical surface and recorded

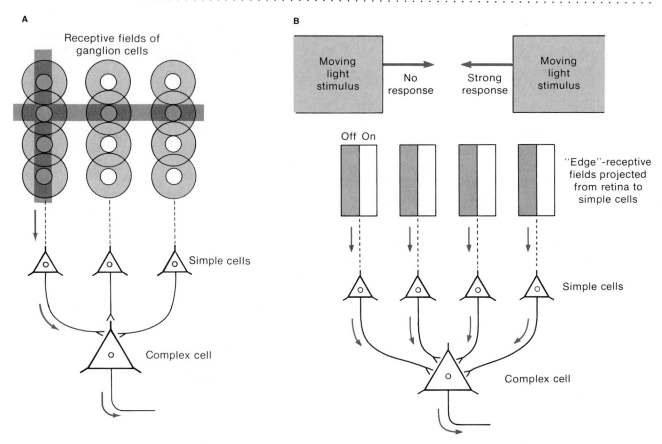

Figure 11-40 Responses in complex cells could be based on their pattern of input from simple cells. (**A**) Some complex cells respond to bars of light that have a specific angular orientation, but their location can be anywhere within a large receptive field. This pattern of response could be evoked by the convergence of many simple cells, each having similarly oriented bar-shaped receptive fields. In this example, the vertical bar of light stimulates one simple cell to fire because it falls on a row of ganglion-cell receptive fields that produce the bar-shaped receptive field of a simple cell. If the bar were moved to the right, it would excite another simple cell that synapses onto the same complex cell, producing activ-

ity in the complex cell. In contrast, a horizontal bar of light produces only a subthreshold response in the simple cells, and hence no signal is sent to the complex cell. (**B**) Some complex cells respond to edges of light moving in only one direction. This response pattern could be produced by the convergence of a population of simple cells, all of which are sensitive to light-dark edges of the same orientation. Excitation of the complex cell would occur if the edge were moved so that it illuminated the *on* side of the simple-cell receptive fields before it illuminated the *off* side. Movement in the opposite direction would produce only inhibition.

responses from cells that were located along that pathway, they discovered that the cells along each pathway responded to bars having the same orientation. When they moved the electrode laterally and made another penetration, they found a column of cells that responded optimally to a stimulus having a different orientation from that of the optimal stimulus for the neighboring column of cells. Each such set of cells is called a *cortical column*. In contrast, recording from cells along a track parallel to the surface revealed an amazingly regular change in the optimal stimulus orientation, with the preferred orientation shifting about 10 degrees each time the electrode advanced 50 µm. This result implies that the cells of the visual cortex are organized in columns according to a feature of their optimal stimulus and that this difference changes in an orderly fashion across the cortex (Figure 11-41A).

The columnar organization of cells with similar response properties had been seen earlier in the somatosensory cortex, where adjacent columns contained cells re-

sponding to touch or to the bending of a particular joint. However, the orderly array of orientation columns was only the first of the functionally based subdivisions found in the visual cortex. The next to be discovered related to the eye from which the visual signal came. By injecting one eye with a radioactive tracer molecule that is transported to the visual cortex, Hubel and his colleagues identified the projection pattern of each eye onto the cortical surface. These experiments revealed a second columnar system in which alternating columns represent one or the other eye (Figure 11-41B, but see Spotlight 11-3). Three-dimensional reconstructions of these columns, called **ocular dominance columns,** show their distribution across the cortical surface (Figure 11-41C).

These experiments revealed that the visual cortex is subdivided into small functional units that analyze the stimulus into its constituent features before passing it on to higher levels for further analysis. This modular organization is superimposed on a fundamental spatial map, which

SPOTLIGHT 11-3

SPECIFICITY OF NEURONAL CONNECTIONS AND INTERACTIONS

Notice in Figure 11-32B that the half of a visual image that falls on the *temporal* part (the side toward the ear) of one retina falls on the *nasal* part (the side toward the nose) of the other retina and vice versa. In human beings, the ganglion cells on the right side of each retina send their axons to the right side of the brain, and those on the left side send their axons to the left side of the brain. Thus the nasal half of the right retina and the temporal half of the left retina project to the left side of the brain.

David Hubel and Torsten Wiesel, in their studies of visual processing in the brain, found that some neurons in the right and left visual cortices have receptive fields in both retinas, and that these receptive fields are optically in register. That is, cortical cells that receive input from both retinas get information from both eyes from precisely the same small region of the visual field. These cortical cells receive extremely accurate neuronal projections from ganglion cells "seeing" the same part of the field, but located in the two retinas. These findings confirm the suggestion of Johannes Müller, made more than a century ago, that information originating from analogous receptors (i.e., those that "see" the same part of the visual field) on both right and left retinas converges on specific neurons in the brain. This high degree of morphological specificity underscores the precision with which synaptic contacts are established within the central nervous system.

The neurons of the visual cortex are arranged in a remarkably orderly manner. When a recording electrode is gradually advanced through the cortex in a path perpendicular to its surface, and encounters neurons in successively deeper layers, all of the neurons along the track have receptive field properties in common. For example, all of the cells may be simple cells with the same orientation.

The precise and orderly arrangements of neuronal connectivity present one of the greatest challenges to neurobiology: to discover the mechanisms that guide neurons to find functionally appropriate synaptic partners during embryonic development.

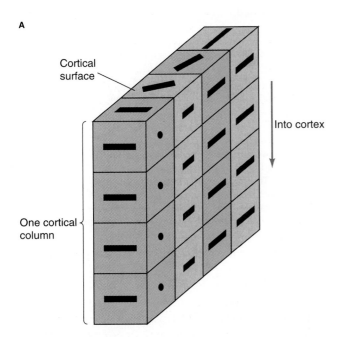

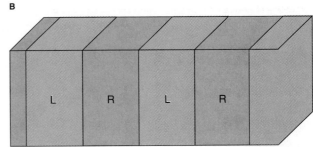

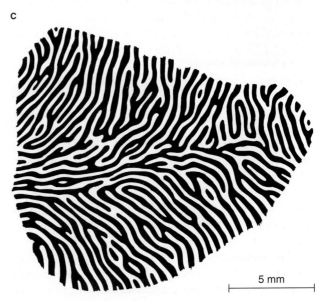

Figure 11-41 Neurons of the visual cortex are arranged in columns that are perpendicular to the cortical surface. (**A**) Diagram illustrating the organization of columns of cells that respond to orientation of stimuli. Columns are collections of cells for which the optimal stimulus orientation is the same for all cells in the column. Cells in adjacent columns have different optimal stimulus orientations, and this orientation varies systematically from column to column. (**B**) The eye that excites cortical neurons also alternates between neighboring columns. Red columns are excited by the left eye; gray columns by the right eye. (**C**) Simon LeVay's reconstruction of the ocular-dominance columns in part of Area 17. [Adapted from Hubel, 1995.]

Cortical surface

Into cortex

One cortical column

5 mm

persists through the layers of the visual cells. To understand the nature of the spatial map at the cortical level, experiments have been done to plot the visual field onto the cortex directly, with the use of a radioactive labeling technique (Figure 11-42). Radioactive 2-deoxyglucose was injected into an anesthetized monkey and then a complex, target-shaped stimulus was projected onto its retina. Active neurons take up more 2-deoxyglucose than do resting neurons, so cortical neurons that were activated by the stimulus would be expected to contain more radioactivity than their inactive neighbors. The pattern of radioactivity observed in the visual cortex revealed that, although the two-dimensional surface of the retina was completely represented on the cortical surface, the cortical pattern was not an exact replica of the spatial features of the stimulus on the retina. Instead, retinal regions representing the center of gaze (the fovea) were greatly magnified relative to those representing the peripheral view. This pattern corresponds to the difference in visual acuity across the retinal surface, as well as to differences in the convergence of primary photoreceptors onto subsequent layers of neurons. This distortion of the map in accord with the needs and habits of the animal is characteristic of all animals with well developed visual systems. For example, animals, such as rabbits, that live on large open plains have an elongated, horizontal region of specialization, called a **retinal streak,** that provides the greatest number of photoreceptors, and the least convergence, for receiving stimuli along the visual horizon.

All of the various levels of cortical organization must be combined to provide the next set of cortical cells with a complete picture of visual stimuli, and the manner in which this synthesis is accomplished is still the subject of intense research. For example, it now appears possible that some high-order visual neurons may be active only if a specific object (e.g., a face) enters their receptive field.

The visual cortex has taught physiologists several important principles about the organization of sensory networks. First, the visual system is organized hierarchically. At each level, cells require more complicated stimuli to excite them optimally, and this complexity arises from the convergence of cells having simpler receptive fields onto cells having more complicated receptive fields. Second, although convergence is apparent as we follow a stimulus into the system, the parallel analysis of distinct features of a stimulus requires divergence of information as well. The simultaneous analysis of different features of a stimulus, which occurs along parallel pathways, appears to be an important principle of functional organization. Third, the activity of cortical neurons in Area 17 results in abstractions of some features of the visual stimuli. Fourth, the visual cortex does not receive a simple one-to-one projection from the retina in either space or time. Instead, some regions in the visual field are expanded dramatically in their cortical representation, whereas others are compressed.

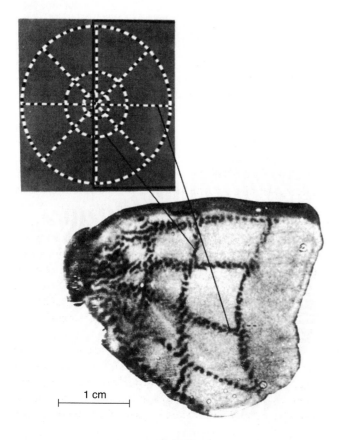

Figure 11-42 Visual space is represented on the surface of the visual cortex, but in a somewhat distorted form. This target-shaped stimulus with radial lines was centered on the visual fields of an anesthetized macaque monkey for 45 minutes after radioactive 2-deoxyglucose had been injected into the monkey's bloodstream. One eye was held closed. The cortex was removed, flattened, frozen, and sectioned. The lower picture shows a section parallel to the cortical surface. The roughly vertical lines of label represent the curved lines of the stimulus; the horizontal lines of label represent the radial lines in the right visual field. The lines are broken because only one eye was stimulated. This dotted pattern displays the ocular dominance columns. [Adapted from Tootell et al., 1982.]

1 cm

The organization of the visual system poses major challenges for developmental neurobiologists. How might the precision and complexity of synaptic connections in the visual system arise during embryonic development?

The auditory map of an owl brain

The retinotopic and somatotopic maps described previously are found in many levels of the brain as sensory information is transmitted through the nervous system. We can recognize these maps because, even in a distorted form, they mimic the spatial organization of objects in the outside world. The two-dimensional array of cells on the retinal surface produces a two-dimensional map of the surroundings, and the spatial relations in the surroundings are preserved as the image is projected onto the cells of the lateral geniculate and into the cortex. For other sensory systems, the nature of the possible central maps is not so obvious. In the auditory system, for example, the arrangement of hair cells along the cochlea is correlated with their sensitivity to particular frequencies of sound (see Chapter 7). If the spatial order of these hair cells were preserved in the projection of their axons to the brain, a brain map of sound frequency, a **tonotopic map,** would be the result. Indeed, tonotopic maps have been found in some auditory regions of the brain. However, it is not obvious how sorting sounds by frequency would help an animal acquire information about its environment. We know that human beings can locate the source of a sound in space, but knowing just the frequency of the sound would not help much in solving this problem.

How does an animal locate a sound in space? Information about where the source of a sound lies relative to a listener is encoded in the intensity of the sound and in the relation between the times at which the sound reaches the two ears. If a source is to the left of the animal, sounds reach the left ear first and arrive somewhat later at the right ear. The time separating the arrival of the sound at first one ear and then the other can be computed by the nervous system as an indication of where the sound originated. To understand how this is achieved, Eric Knudsen and Mark Konishi studied barn owls, birds that depend critically on locating the sources of sounds in darkness.

Barn owls have several characteristics that make them excellent animals in which to study the neuronal mechanisms that underlie sound localization. First, if light is available, owls use both vision and hearing to guide hunting, but they can capture mice in complete darkness, finding their prey only by listening to sounds (Figure 11-43). In addition, an owl cannot move its eyes within the orbits; instead it must move its whole head whether it is orienting to a sound or to a visible object, and this orienting response is quite accurate. Owls can point their heads to-

Figure 11-43 Barn owls can capture mice in total darkness. These images are from a film that was made by using only infrared illumination that the owl cannot see. The owl successfully captured the mouse in total darkness. [Courtesy of M. Konishi.]

ward the source of a sound with an accuracy of 1 to 2 degrees in both *azimuth* (lateral distance away from a point straight in front of the owl's head) and *elevation* (vertical distance away from a point straight in front of the owl's head).

To test its orienting ability, an owl was placed on a perch and sounds were generated by a speaker whose location could be varied over a hemisphere in space while remaining at a fixed distance from the bird (Figure 11-44A). The orientation of the owl's head was monitored as the owl oriented toward the sounds produced by the speaker. The orientation of the head in response to each sound was expressed in degrees of elevation and azimuth (Figure 11-44B). Careful behavioral observations indicated that the owl was using two kinds of cues in its orienting response: the intensity of sounds was used to determine the elevation of the target and their relative times of arrival at the two ears were used to determine the azimuth of the target.

To examine the role played by intensity cues, either the right or the left ear was plugged to attenuate the sounds, using plugs that either weakly or strongly reduced the sound intensity. The results of this experiment revealed that an owl consistently misdirected its gaze when one of its ears

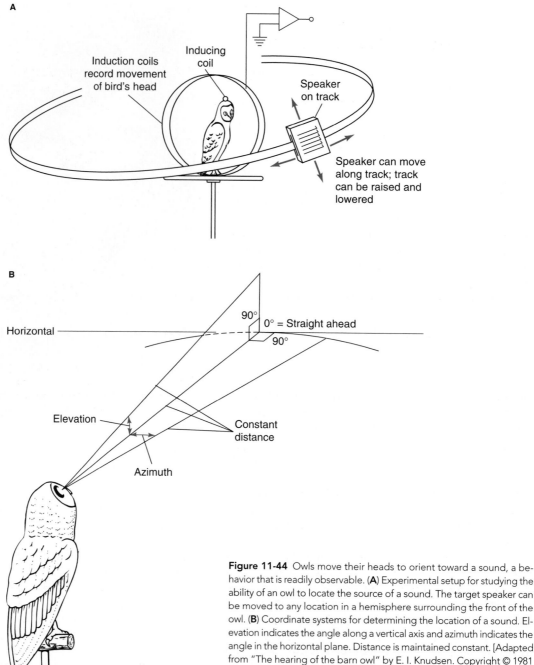

Figure 11-44 Owls move their heads to orient toward a sound, a behavior that is readily observable. (**A**) Experimental setup for studying the ability of an owl to locate the source of a sound. The target speaker can be moved to any location in a hemisphere surrounding the front of the owl. (**B**) Coordinate systems for determining the location of a sound. Elevation indicates the angle along a vertical axis and azimuth indicates the angle in the horizontal plane. Distance is maintained constant. [Adapted from "The hearing of the barn owl" by E. I. Knudsen. Copyright © 1981 by Scientific American, Inc. All rights reserved.]

was plugged (Figure 11-45A). With the right ear plugged, the owl oriented below the actual source and slightly to the left. With its left ear plugged, it oriented above the source and slightly to the right. In other words, when the sound was louder in the right ear, it seemed to the owl to be coming from above; whereas, when the sound was louder in the left ear, it seemed to be coming from below. The slight difference in the azimuth angle of orientation suggests that some information about the horizontal location is available from the intensity, but intensity cannot entirely account for orientation along that dimension.

How can interaural intensity differences allow an owl to discriminate the elevation of a sound source? The answer

lies—at least in part—in anatomy. The region around the openings of an owl's ears is made of stiff feathers, called the *facial ruff,* which form a surface that very effectively directs sounds into the ear canals like the fleshy pinna of the mammalian ear. When these feathers are removed, the external auditory canals of the owl are seen to be asymmetric (Figure 11-45B). The opening of the right ear is directed upward, whereas the opening of the left ear is directed downward. This arrangement could provide a basis for discriminating elevation from intensity cues. The importance of the facial ruff was revealed by removing these feathers. If it lacked its facial ruff, an owl was no longer able to identify the elevation of sound sources, although its estimates along

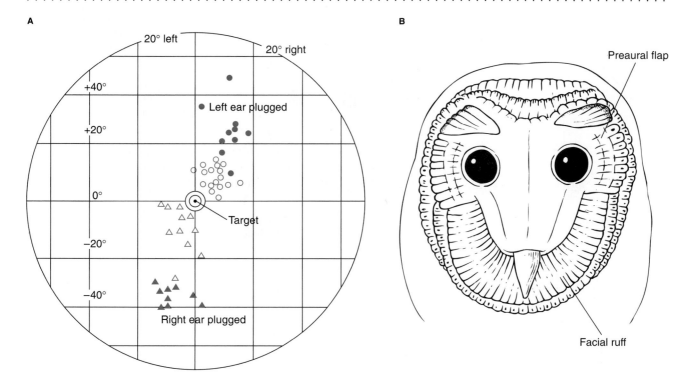

Figure 11-45 Plugging one of its ears caused an owl to make errors in locating the source of a sound. (**A**) A target was presented directly in front of the owl, which had either a hard (high attenuation, solid symbols) or a soft (less attenuation, open symbols) plug in one ear. Note that with the left ear plugged (circles) the owl judged the sound to be *above* its real location. With the right ear plugged, the owl made mistakes in the oppo- site direction. (**B**) Facial ruff showing the asymmetry in the auditory open-ings. The right ear canal points slightly upward, whereas the left ear canal points slightly downward. This small difference is amplified by the posi-tion of the feathers in the facial ruff. [Adapted from "The hearing of the barn owl" by E. I. Knudsen. Copyright © 1981 by Scientific American, Inc. All rights reserved.]

the horizontal axis remained as accurate without the facial ruff as with it. Thus, the facial ruff must amplify the direc-tional asymmetry of the ears and is essential for discrimi-nating differences in elevation among sound sources.

How does an owl locate sounds along the horizontal or azimuthal meridian? From behavioral experiments, it was clear that differences in the time at which sounds arrived at each of the ears were important for this discrimination. However, the relevant cue could have been either disparity in the onset (or offset) of the sound or ongoing disparity that occurred during the duration of sound (Figure 11-46). Disparity of onset (or offset) refers to the difference in the time at which a given signal first reaches each ear; the ear nearest the source receives the signal first. Disparity can also occur between the signals that are received at the two ears as a sound continues; just as the onset of a sound reaches the two ears at different times, identifiable features of the sound reach first one ear and then the other. These two types of disparity were independently varied by im-planting small speakers in an owl's ears. In response to dis-parity of onset, the owls failed to make "correct" head movements; whereas, in response to ongoing disparities ranging from 10 to 80 μs, owls made rapid head orienta-tions to the correct place in the azimuth corresponding to that time difference (Figure 11-46B). These experiments show that owls orient to sounds in space with remarkable accuracy. Elevation is judged from differences in the inten-

sity of sounds arriving at each ear, and azimuth is judged from the ongoing temporal disparity between sounds ar-riving at each ear.

How is information about a sound's location in space represented in the nervous system? The ears cannot di-rectly provide the brain with a representation of external space. Instead, as we have seen, an owl must compute the difference in intensity between sound signals sensed by its two ears to determine the elevation of a sound, and it must compute an ongoing evaluation of disparity between sound signals reaching its ears to determine the position of the sound in the azimuthal plane. How and where these comparisons are made and how the output is represented in the brain were discovered by Knudsen and Konishi in the late 1970s.

Knudsen and Konishi identified a collection of space-specific neurons in a midbrain nucleus. Each of these cells responds best to sound signals that are located at a partic-ular point in space, and each cell has a receptive field with an on-center, off-surround organization similar to that found in retinal ganglion cells (Figure 11-47A). Sounds that are located within the center region of the cell's receptive field (mean diameter ≈ 25°) excite the cell, whereas sounds in the surround of the receptive field inhibit its response. The neurons are arrayed in the nucleus to form a spatial map (Figure 11-47B) analogous to the retinotopic map de-rived from the retina and the somatotopic map derived

A

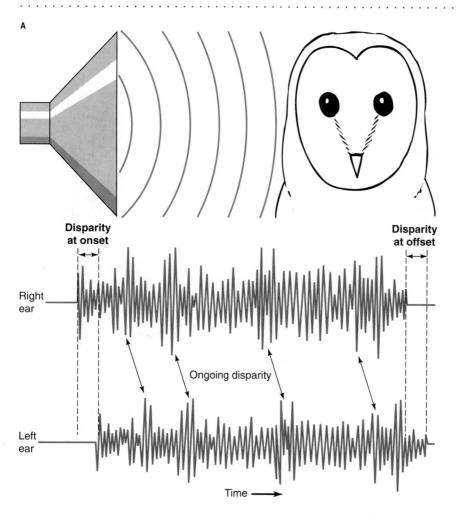

Disparity at onset

Disparity at offset

Right ear

Ongoing disparity

Left ear

Time →

Figure 11-46 Owls judge azimuthal location from the disparity between ongoing sounds that arrive at the ears. (**A**) Onset disparity occurs when a sound reaches one ear before the other. Ongoing disparity refers to the continued difference in sound waves as perceived by both ears. (**B**) Owls use the ongoing disparity between the sounds reaching the two ears to localize a signal accurately in space. The linear relation between azimuth and ongoing disparity between signals at the two ears suggests that this type of disparity is the relevant cue. [Adapted from "The hearing of the barn owl" by E. I. Knudsen. Copyright © 1981 by Scientific American, Inc. All rights reserved.]

B

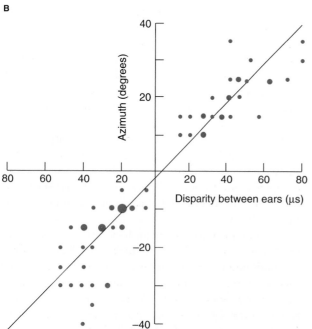

Azimuth (degrees)

Disparity between ears (μs)

Another feature common to this map and other brain maps is that the size of the receptive field for cells that receive information from directly in front of the animal is smaller than it is for cells that receive information from the sides of the animal. The area directly in front of the animal projects to a larger part of the nucleus and is therefore magnified compared with the area to the sides of the animal. This representation is reminiscent of the exaggerated representation in the visual cortex of the retinal fovea and of the large representation of the hands and face on the somatosensory cortex. In barn owls, the nucleus where these spatial fields are recorded is the *mesencephalicus lateralis dorsalis* (MLD), which is the avian homolog of the mammalian inferior colliculus. (The *inferior colliculus* is a major auditory center that lies just beneath the superior colliculus—the mammalian homolog of the optic tectum). The MLD nucleus passes a map of sound location in space to higher centers. Disparity between signals is sensed by neurons in nuclei that lie below the MLD in the midbrain. These neurons, which are called *coincidence detectors*, receive input from both ears, and their activity changes, depending on whether signals from the two ears arrive simultaneously or sequentially. The mechanisms by which

from the body surface. Cells at each point on the surface of the nucleus respond by firing APs in response to sounds at a particular point in space. Adjacent points in the nucleus respond to stimuli that are adjacent in space.

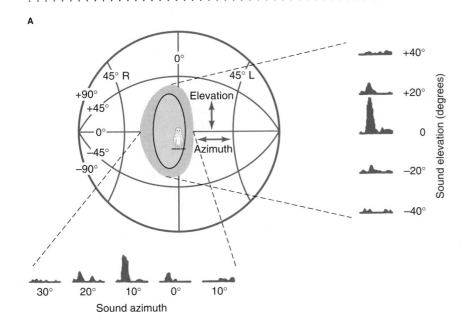

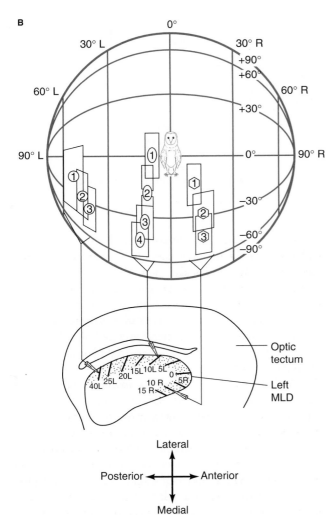

Figure 11-47 Auditory neurons in part of an owl's brain have spatially organized receptive fields. (**A**) Receptive field of a single cell showing the *on*-center (red) and *off*-surround (gray) plotted on a hemisphere. This cell responds most strongly to sounds at 0 degrees of elevation and 10 degrees to the right of center. Sounds that are 20 degrees away from this location stimulate the cell only weakly, and sounds that are 40 degrees away inhibit it. (**B**) Spatial auditory map in a barn owl's mesencephalicus lateralis dorsalis nucleus. Data for three electrode penetrations into the nucleus are shown. The location and orientation of each electrode track is shown on the *bottom* diagram, which depicts the nucleus as if it had been sectioned in a horizontal plane (orientation is indicated below the diagram). Neurons encountered along one track are numbered sequentially, and the receptive field of each neuron is shown. Neurons along one track respond to contiguous locations in space; and, as the electrode moved from one track to another, the azimuthal angle of the receptive fields (indicated on the diagram of the nucleus) changed smoothly. [Adapted from Knudsen and Konishi, 1978.]

sponse properties of neurons. Since then, similar computed maps have been found in the brains of bats, which, like owls, hunt by using auditory information. The spatial representation of sound in an owl's brain ultimately projects to the tectum, where it meets—and is congruent with—a map of space generated by the visual system. Adjacent layers of the tectum, then, are topographically correlated, with one processing information about sounds and the other processing information about visual input. This arrangement suggests that behavior can be organized more effectively if all of the sensory information about an object in space is first assembled at one location. The next problem in understanding the production of behavior is a consideration of where and how sensory information leads to a decision to act.

Motor Networks

The sensory side of the nervous system acquires and analyzes information about the outside world, which is essential for producing behavior that is matched to an animal's current circumstances. This information must then be passed on to neurons that are responsible for generating coordinated movement. Relatively little detail is known about the interface between the sensory and motor sides of this process, in part because investigators have worked independently at understanding either sensory or motor systems. However, in a few cases, this sensory-to-motor connection has been successfully explored, either in very simple reflexes of vertebrates or in more complex behavior of invertebrates.

We will consider motor control systems of increasing complexity, from those that produce simple reflex responses through networks controlling repetitive actions to

differences in sound intensity are computed by the owl's brain are still being investigated.

The barn owl's map of acoustic space was the first example of a brain map that is generated *de novo* from the re-complex networks that reveal general principles of central neuromotor organization. Motor patterns of different complexity display various amounts of flexibility. Fixed action patterns are relatively inflexible; they occur repeatedly with little variation, but many behaviors are extremely plastic. The animal can shape them to fit each new set of circumstances. One of the challenges for studies of motor control is to understand the neuronal activity that allows an organism to produce behavior that changes from moment to moment, as the situation changes.

Levels of motor control

The study of how neurons control muscle activity has mostly been focused either on animals with simple nervous systems or on repetitive actions by more complicated animals. The neuronal control of fixed action patterns has been a major topic of this work, because the all-or-none property of these behaviors suggests that a single neuronal decision must generate the behavior. This concept of a decision does not imply a conscious process, but rather that activation of a neuronal "switch" in the central nervous system is sufficient to initiate the behavioral pattern. Conceptually, this idea can be formalized as a hierarchical motor control system in which sensory input is used to select specific motor output. The lowest level of control is the motor neuron that connects to a muscle; activity in motor neurons is regulated by integrated neuronal input (Figure 11-48).

Initially, some physiologists believed that a short feedback loop between stretch receptors in leg muscles and the spinal motor neurons controlling those muscles might account for the motions of walking made by vertebrates. However, it has become clear that repetitive motor output—such as walking, swimming or flying—depends on activity in a central network that generates essential features of the motor pattern. The pattern of walking, swimming, or flying can be modified in response to sensory feedback and varies with features of the terrain or with water or wind currents in the environment. Finally, control is exerted from centers higher in the nervous system, whose decisions or commands are also influenced by sensory input. Notice that, in this control hierarchy, no strict chain of command is followed identically in every case. A wide variety of distinct environmental inputs can lead to related kinds of motor output, and feedback control operates at all levels of the system.

Simple reflexes

The simplest circuitry that controls the activity of skeletal muscles is the reflex arc. It takes only two kinds of neurons—muscle stretch receptors (also called 1a-afferent neurons) and spinal α motor neurons—connected together to produce the **myotatic reflex**, or *muscle stretch reflex* (Figure 11-49A). Because the basic form of this reflex requires only the synapse between afferent and efferent neurons, with no interposed interneurons, it is a monosynaptic reflex.

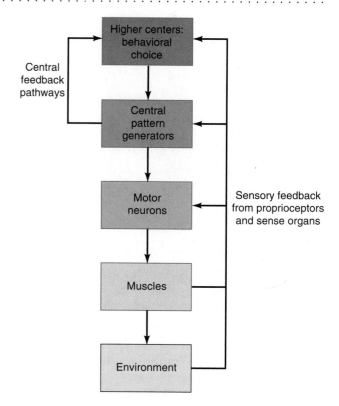

Figure 11-48 Motor control systems are arranged hierarchically. Neurons in the brain and nerve cord exert control over the entire motor side of the nervous system, generating decisions concerning motor output. These decisions modulate activity within sets of neurons, called central pattern generators, that activate motor neurons according to more or less preset patterns. The motor neurons provide the only pathway between the nervous system and the muscles, which in the final analysis cause behavior. Feedback occurs at all levels of the hierarchy, potentially shaping the output.

The sensory endings of stretch receptor neurons are located within each muscle, associated with sensory structures called **muscle spindle organs.** Each spindle organ contains a small bundle of specialized muscle fibers called **intrafusal fibers** to distinguish them from the majority of contractile fibers, which are called **extrafusal fibers.** Extrafusal fibers are the skeletal muscle fibers discussed in Chapter 10, and they are innervated by α motor neurons. The intrafusal fibers are small in mass and in number, and they do not contribute to the production of tension by the muscle. Instead, they participate in a feedback loop that regulates how sensitive the spindle organs are to stretch.

Muscle spindles lie parallel to the extrafusal fibers; so, if something happens to stretch the muscle (e.g., a weight is added to an isolated muscle or a joint bends, stretching the muscle that runs over the joint), the muscle spindles are stretched too. Stretching the central region of the muscle spindles increases the frequency of APs in the 1a-afferent axons. These afferent axons make excitatory synapses directly on the α motor neurons that control the muscle that contains their spindle organs; so, when the activity in the 1a-afferent axons increases, it tends to excite the motor

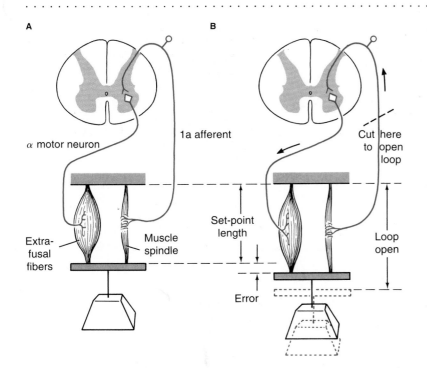

Figure 11-49 Only two kinds of neurons are required to produce the muscle stretch reflex. (**A**) The steady state of a muscle that is holding up a light weight. (**B**) If a heavier weight is added to the muscle, it stretches the muscle, activating stretch receptors, which synapse onto α motor neurons in the same spinal segment and cause the muscle to contract more forcefully. If the sensory axons were cut, there would be no feedback onto the motor neurons and the weight would cause the muscle to elongate (dashed lines). (**C**) Sequence of events that lead to production of the stretch reflex.

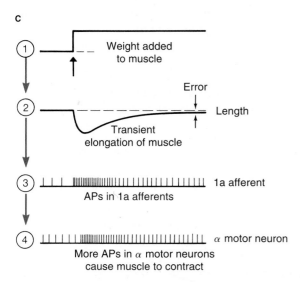

neurons, causing reflex contraction in the muscle that was stretched (Figure 11-49B, C).

The stretch receptors provide negative feedback because stretch of the muscle initiates neuronal activity that causes the muscle to contract, opposing the stretch. A familiar example of a stretch reflex is the knee-jerk reflex evoked when the tendon that crosses the knee cap (also called the **patella**) is tapped. Tapping the tendon produces a sudden stretch of the quadriceps muscle on the ventral surface of the thigh, activating the muscle and causing the knee joint to extend. The arclike nature of the reflex is revealed when the dorsal root into the appropriate segment of spinal cord is cut. Severing the dorsal root leaves all of the motor innervation intact but removes sensory input to the spinal segment. When the dorsal root is cut, the muscles innervated by the spinal segment go limp, even though their motor input is intact.

Notice that, when a muscle contracts under the influence of the stretch reflex, tension is removed from the muscle spindles. If nothing else happened, the 1a-afferents would then become silent; and, if the muscle were to be stretched a little more, the muscle spindles would be unable to respond unless their own length could be regulated. The intrafusal fibers—under the control of another set of motor neurons, the *γ-efferents*—regulate the length of the stretch receptors. When a muscle shortens, driven by its α motor neurons, activity in the γ-efferents also causes the intrafusal fibers to shorten, maintaining a constant tension in the spindle fibers. In this way, the γ-efferents allow the spindle fibers to maintain their sensitivity to muscle stretch through a wide range of muscle length.

Centrally generated motor rhythms

Locomotion and respiration typically consist of rhythmic movements produced by repetitive patterns of muscle contraction. Each phase of such a neuromotor cycle is both preceded and followed by characteristic activity in motor neurons. Bursts of activity are consistently related to one another in time. Logically, these repetitive acts could depend on moment-to-moment sensory input to the nervous system or on the autonomous motor output of pattern-generating networks that happens entirely independently of sensory input or on some combination of these two mechanisms (Figure 11-50). Regulation of repetitive motor output has been examined in many animal systems, and it appears that both mechanisms play a role. These experiments are typically carried out in *semi-intact preparations*—that is, in animals in which the nervous system has been exposed for recording but that can still carry out recognizable behaviors. In some behaviors, isolated nerve cords can

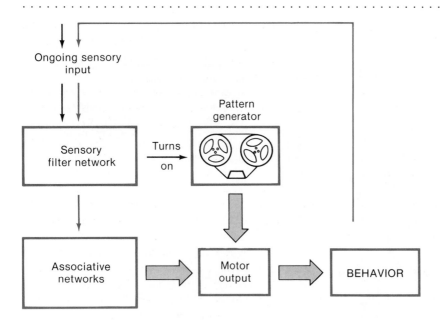

Figure 11-50 Motor output from the nervous system depends on a combination of sensory input and central pattern generation. Sensory input arises in part from the environment and in part from sensory receptors in the body of the animal. Central pattern generators—represented by a tape recorder in this diagram because these neurons produce the same pattern of output over and over again—play an important role in shaping behavior, but they provide only part of the input onto the motor neurons.

produce all features of a motor output pattern; and, although the concept of an isolated nerve cord *behaving* may seem strange, these behaviors can be studied either in semi-intact preparations or in isolated nerve cords, depending on which is most convenient.

Central motor patterns have been most clearly demonstrated in the nervous systems of some invertebrates—for example, in the neuromotor control of rhythmic locomotory movements. Grasshopper flight is controlled by mus-

cles that cause alternate up-and-down movements of the two pairs of wings, and these muscles receive the appropriate sequence of nerve impulses carried by several motor axons. (See Chapter 10 for more information on insect flight.) The patterns of activity in these motor neurons continue to occur with appropriate phase relations, even if sensory input from the muscles or joints of the wings is eliminated by cutting the sensory nerves (Figure 11-51A). This persistence suggests that the motor pattern may be largely generated

A

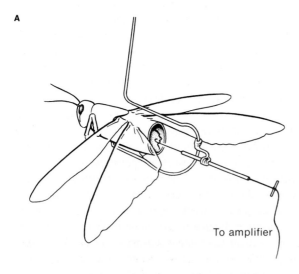

To amplifier

C

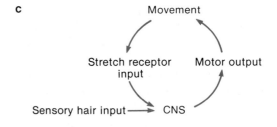

B

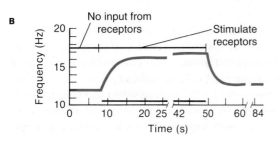

Figure 11-51 Both a central pattern generator and sensory feedback contribute to the production of grasshopper flight. (**A**) Experimental arrangement. An eviscerated grasshopper or locust is mounted so it can flap its wings when it is stimulated by air blowing on facial receptor hairs. Electrodes for recording motor output and for stimulating receptor nerves are fixed in place. (**B**) When sensory receptor neurons at the base of the wings are destroyed, the central pattern generator produces a low-frequency pattern. Electrically stimulating the receptor axons increases the frequency of the endogenous motor output. The time during which the receptor nerve was being stimulated is indicated by the black line. After the stimulation ceases, the rhythm returns to a low frequency. (**C**) Cyclic organization of flight motor output. External sensory input (e.g., a puff of air on hair receptors) stimulates the flight motor output. Wing movements activate stretch receptors that provide input stimulating the flight motor. Notice that this loop resembles the positive feedback loop illustrated in Figure 11-26. [Adapted from Wilson, 1964, 1971.]

within the central nervous system by a network of neurons that interact to coordinate the timing between contraction of different muscles.

Does sensory input play any role in the control of grasshopper flight, which seems to be driven by centrally programmed motor output? Sensory feedback from stretch receptors at the base of each wing is stimulated by the movement of the wings and can modify motor output, increasing the frequency, intensity, and precision of the rhythm. If these receptors are destroyed, the neuronal output to the wing muscles slows down to about half its normal frequency, although the phase relations among impulses in the different motor neurons are retained. The original frequency of the rhythm can be restored if the nerve roots containing the axons of the wing-joint receptors are stimulated electrically (Figure 11-51B). Interestingly, although the motor rhythm increases in frequency when it is provided with sensory input, the timing of the motor output is not closely related to the timing of impulses in the sensory nerves. Random stimulation of the wing-joint receptor axons can speed up the motor output, although sensory input is most effective if it occurs during a particular phase of the wingbeat cycle. Thus, proprioceptive feedback is not required for proper phasing of motor impulses to the flight muscles; but, when the central flight pattern generator becomes activated, sensory feedback reinforces its output (Figure 11-51C).

What turns the flight motor on and off? When a grasshopper jumps off the substrate to begin flight, hair receptors on its head are stimulated by the passing air. This specific sensory input initiates the output of the flight motor. When the insect alights, the flight central pattern generator is turned off by signals originating in mechanoreceptors of the foot (the foot is called the *tarsus* in insects).

Endogenous pattern-generating networks have been shown to exist in a number of invertebrate nervous systems. For example, the cyclic motor output to the abdominal swimmerets of the crayfish persists not only in an isolated nerve cord, but even in single, isolated, abdominal ganglia. This intrinsic rhythm is initiated and maintained by activity of "command" interneurons, whose somata are located in the supraesophageal ganglion of the brain. Although the bursting pattern in each abdominal ganglion requires maintained activity in one, or perhaps several, of the interneurons, there is no simple one-to-one relation between the firing pattern of these interneurons and the pattern of motor output to the swimmerets. The crucial interneurons appear to be providing a general level of excitation, which keeps the central pattern generator active.

One of the best studied rhythmic patterns is escape swimming in the nudibranch mollusk *Tritonia* (Figure 11-52A). This sea slug swims away from noxious stimuli by making alternating dorsal and ventral flexions of its body, which are produced by alternating contractions of dorsal and ventral flexor muscles. The central pattern is generated by interconnections of three types of neurons—a cerebral neuron (C2), the dorsal swim interneurons, and

the ventral swim interneurons—which synapse onto the flexion neurons (Figure 11-52B). The cerebral neuron (C2), the dorsal swim interneurons, and the ventral swim interneurons are linked by reciprocal connections many of which are a mixture of excitatory and inhibitory synapses. Reciprocal inhibitory synapses between neurons have been found in many central pattern generators that produce rhythmic outputs; the reciprocal inhibitory synapses in the central pattern generator for swimming in *Tritonia* have been shown to be necessary for generating swimming in this species. After the initial stimulus, the dorsal and ventral swim interneurons produce alternating bursts of neuronal activity, which activate the flexion neurons responsible for motor output. Intracellular recordings of activity in all five neuron types indicate that the swimming rhythm depends on both the membrane properties of individual neurons and their synaptic connections. Thus, the rhythm is neurogenic, produced by interactions between neurons. Recently, it has been demonstrated that synaptic strengths among the neurons of this network can be modulated during an episode of swimming to change the properties of the network, even while it is producing the swimming output.

Autonomous central neuronal control also exists to various degrees in vertebrates. Respiratory movements, which are driven by cells in the brain stem, persist in mammals when sensory input from the thoracic muscles is eliminated by cutting the appropriate sensory nerve roots. Toads in which all sensory roots, except those of the cranial nerves, have been cut still produce simple coordinated walking movements, although these movements are hard to discern, because loss of the myotatic reflex arc causes muscles to become flaccid. Motor output to the swimming muscles of sharks and lampreys continues in a normal, alternating pattern when segmental sensory input is eliminated. However, the intersegmental sequencing of motor output, which normally travels from the anterior to the posterior may be disrupted.

Walking movements have been investigated in cats that are supported on a treadmill after the brain stem has been transected above the medulla oblongata (called a *spinal cat* preparation). Such studies reveal that the walking sequence can occur without input from the brain. Moreover, a rudimentary walking rhythm has been seen to continue even after the dorsal roots have been transected, eliminating sensory input. Thus, even in vertebrates, some aspects of rhythmic movements are programmed into intrinsic connections between neurons within the spinal cord and hind brain, and they can continue even if sensory feedback and other sensory inputs are disrupted.

Central command systems

The stimulation of appropriate neurons in the central nervous system can elicit coordinated movements of various degrees of complexity. Electrical stimulation of one such **command system,** in the nerve cord of the crayfish, causes the animal to assume the defense posture, with open claws held high and body arched upward on extended forelegs.

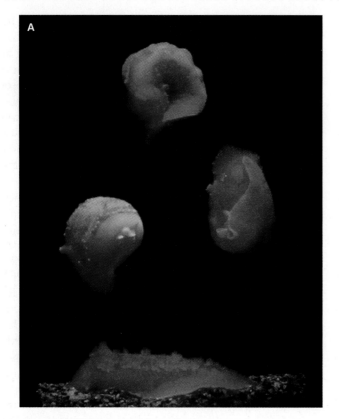

B

CPG

Flexible
neurons

C

C2

DSI

DFN

Stimulus

Figure 11-52 Swimming in the nudibranch mollusk *Tritonia* is controlled by a central pattern generator consisting of three types of neurons. (**A**) If a *Tritonia* is threatened (such as by a nudibranch-eating starfish), it rises off the substrate and swims by rhythmically contracting dorsal and ventral flexor muscles. (**B**) Three interconnected types of neurons act together to generate the swim motor pattern. An excitatory synapse is represented by a bar; an inhibitory synapse by a solid circle; a combination of the two symbols represents a multifunctional synapse. Membrane properties and synaptic interactions determine the swim motor pattern, which changes if these parameters change. (**C**) Recordings of activity in swim central pattern generating (CPG) neurons in an isolated brain after electrical stimulation of the pedal nerve. Abbreviations: C2, cerebral neuron; DSI, dorsal swim interneurons; VSI, ventral swim interneurons; DFN, dorsal flexion neurons; VFN, ventral flexion neurons. [Part A courtesy of P. Katz; parts B and C adapted from Katz et al., 1994.]

Appropriate sensory input excites this system through one specific interneuron, and this interneuron diverges broadly, producing excitation in some motor neurons and inhibition in others. Command systems in arthropods characteristically activate many muscles in a coordinated manner and produce reciprocal actions in a given body segment; that is, antagonists are inhibited while synergists are excited. Perhaps not surprisingly, the command interneurons that are most effective in eliciting a coordinated motor response are generally least easily activated by simple sensory input.

The discovery of this command neuron in the crayfish initially caused physiologists to hypothesize that a lot of an animal's behavior might be controlled by a small population of command neurons, each of which was responsible for producing and shaping a particular behavior. In this case, "choosing" among behaviors would depend on which command neurons were most active. However, further study of the neuronal basis of behavior suggests that most command functions arise within networks of neurons, in which all contributing neurons play an important role. To determine experimentally whether a neuron fills a command function, it is necessary to show that the activity of

the neuron is both *necessary* and *sufficient* for causing the particular motor output. That is, removing the neuron from the network must block, or greatly modify, the behavior (necessity) and activating only that neuron must produce the behavior (sufficiency).

When the necessity and sufficiency tests are carried out to determine the neuronal basis of many behaviors, three observations appear again and again. First, many neurons are multifunctional, functioning differently under different conditions. For example, some retinal bipolar cells have been found to carry signals from rod photoreceptors in dim light and from cone photoreceptors in bright light. There must be a shift in their connectivity pattern as the level of ambient light changes. Second, one neuron may belong to different levels of a hierarchical control system (see Figure 11-48). For example, one neuron in the *Tritonia* swim control network acts both in the central pattern generator for swimming and in the command system for escape. Third, because networks can be reconfigured, depending on the situation, there must be mechanisms that can modify neuronal connectivity. Anatomical connections may constrain the range of possible outputs for a set of neurons, but functional connections define their output at any given time. One of the best understood mechanisms for shifting neuronal networks among possible functional configurations is neuromodulation (see Chapter 6). Neuromodulators can cause changes in synaptic efficacy that dynamically reconfigure a collection of neurons into a new functional unit. Recognition that "anatomy is *not* destiny" in the nervous system has changed the way systems are analyzed. In this section, we consider two systems that have been analyzed in sufficient detail to provide examples of these three principles in the organization of command systems.

Many invertebrates escape from potential predators by using stereotypical movements. One well studied example is the crayfish, *Procambarus clarkii*, which has two types of escape response, depending on the location of the stimulus (Figure 11-53A). In each behavior, at least one giant axon is part of the control circuitry, a typical pattern in the neuronal control of many escape responses. Large axons carry signals rapidly, allowing an animal to escape faster. In the

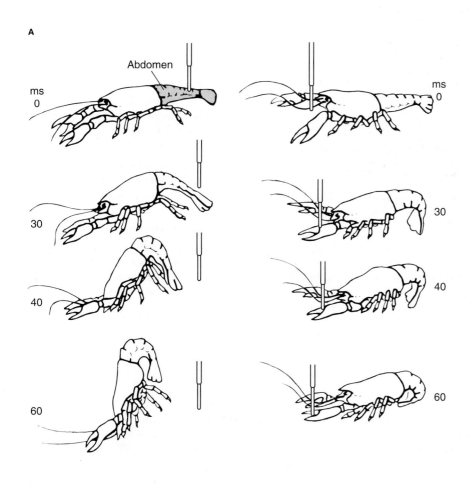

A

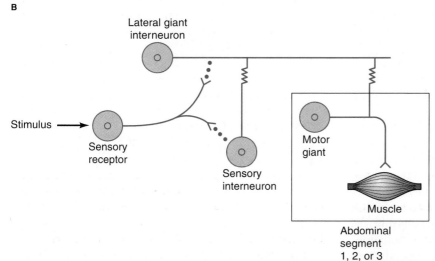

B

Figure 11-53 Tactile stimulation of giant interneurons cause crayfish to change their posture. (**A**) Stimulation of the abdomen *(upper left)* evokes an abdominal flexion that moves the crayfish upward and forward. This behavior is mediated by the lateral giant interneurons. Anterior stimulation at the antennae *(upper right)* evokes an abdominal flexion of a different form that propels the animal backward. This response is mediated by the medial giant interneurons. In both cases, the behavior moves the animal away from the stimulus. In the diagrams, the time after stimulation is indicated in milliseconds and time proceeds from top to bottom. (**B**) Simplified diagram of the circuit that mediates the crayfish escape response to touch on the abdomen. Sensory input is carried by chemical (triangles) and electrical (resistor symbols) synapses to the lateral giant neuron, which makes a rapid, electrotonic synapse onto the motor giant neuron. The motor giant neuron synapses onto the abdominal flexor muscles. The large size of the giant axons produces high conduction velocities, and the electrotonic synapses provide rapid communication between neurons. Electrically stimulating the lateral giant interneuron produces flexion only in abdominal segments 1 through 3. Compare this effect with the final posture of a crayfish that has been touched on the abdomen—flexion is pronounced in the anterior abdomen. [Adapted from Wine and Krasne, 1972, 1982.]

crayfish, there are two giant fibers: the medial giant interneuron, which controls flexion to propel the animal backward; and the lateral giant interneuron, which plays a key role in propelling the animal up and forward. The basic circuitry surrounding the lateral giant interneuron is shown in Figure 11-53B. The neuronal network surrounding the medial giant is quite different—on both the input and the output sides of the network—which explains why the behavior is so different when the crayfish is touched on its antenna.

The escape responses of crayfish illustrate some other features of motor control systems. First, if a crayfish is stimulated repeatedly, it fails to respond after about 10 minutes of stimulation; the response is said to *habituate*. Although habituation can occur at many different points in the network, it has been found that this behavior habituates because less neurotransmitter is released from the terminals of the sensory afferent neurons when stimuli are repeated for long periods. Second, the overall control of the crayfish tail includes a second, parallel pathway that can also elicit the tail-flip response. The fast flexor motor neurons, which are not giant motor neurons, produce more precise control of the tail flip, although it is neither as rapid nor as vigorous as the flip produced by the giant neurons. When the tail flip is initiated by the motor giant neurons, the second pathway is activated, too, although the slow pathway can operate by itself. Third, if the level of serotonin, a neuromodulator, changes in the crayfish, the response of the crayfish to a particular stimulus can change dramatically. An aggressive crayfish can become submissive, and vice versa. Thus neuromodulation must modify the connections between sensory and motor neurons.

The crayfish escape response is a typical fixed action pattern, and the neurons that control it illustrate several features of command systems outlined earlier. Perhaps the most important feature is the existence of multiple control points within the network, which offer several ways to initiate or to slightly alter the performance of the behavior. This flexibility within the constraints of a fixed action pattern has been a source of insight into the organization of behavior.

 What might be the evolutionary advantages and disadvantages of fixed action patterns and motor command systems? What would be the advantages and disadvantages of behavioral plasticity?

The recognition that synaptic neuromodulators can change the properties of a network has opened new avenues of thought. Central command systems, each of which was once believed to drive a single behavioral pattern to completion, must now be viewed as being plastic, with neurons taking on different synaptic relations, depending on

circumstances. The premier example of dynamic network assembly can be found in a collection of 30 large neurons that make up the **stomatogastric ganglion** (STG) of crustaceans. The esophagus and stomach of lobsters and crabs is a complex structure that is responsible for ingesting, storing, chewing, grinding, and filtering food (Figure 11-54). There are four functional regions of the stomatogastric system: the esophagus, the cardiac sac, the gastric mill, and the pylorus. The neurons of the STG control all of the muscular chambers responsible for ingestion and peristaltic movement of food. They also control the bony teeth that are responsible for chewing and grinding it. Because most neurons in the STG are motor neurons that innervate the muscles in the stomatogastric system, their intrinsic properties have been of direct interest in efforts to discover the functional architecture of each subnetwork that can be formed from this small set of neurons. The ganglion can be divided into three networks of neurons, which control muscles in the esophageal, gastric mill, and pyloric regions of the stomatogastric system. The esophageal, gastric, and pyloric networks can each generate patterns of rhythmic output that are independent of the other two (Figure 11-55A). The frequency of output from each of the networks is a characteristic of the network.

Input from modulatory neurons changes the behavior of these neurons drastically. For example, two electrically coupled neurons, called *PS neurons,* reconfigure the networks. When PS neurons fire, a valve between the esophagus and the stomach opens, and swallowing behavior is initiated. Then an entirely new rhythm begins, coordinating

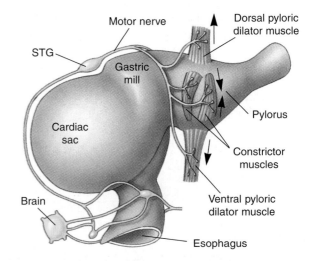

Figure 11-54 The stomatogastric nervous system controls activity in the esophagus, gastric mill, and pylorus of the lobster. The stomatogastric ganglion (STG, one of four ganglia in the system) contains only 30 neurons, most of which are motor neurons and all of which have been identified and characterized. The output of these neurons controls contraction of muscles that cause food to be swallowed, chewed, and moved to the rest of the digestive system. (Muscles that control the pylorus are shown. Constrictor muscles close the pylorus, preventing food from moving out. Dilator muscles open the pylorus, allowing food to move into the next segment of the digestive system. These muscles receive input from STG neurons.) [Adapted from Hall, 1992.]

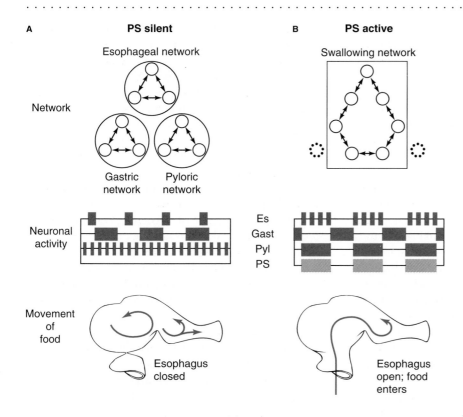

A PS silent

Esophageal network

Network

Gastric
network   Pyloric
network

Neuronal
activity

Movement
of
food

Esophagus
closed

B PS active

Swallowing network

Es
Gast
Pyl
PS

Esophagus
open; food
enters

Figure 11-55 Modulatory inputs to the stomatogastric ganglion change the neuronal outputs dramatically, reconfiguring subnetworks in the ganglion. (**A**) When the modulatory PS neurons are silent, neurons in the stomatogastric ganglion are organized into three separate subnetworks, the esophageal, gastric, and pyloric networks. Each of these subnetworks produces a rhythmic output, but the outputs are not temporally coordinated with one another. In this state, food is chewed and moved about within the cardiac sac and the pyloric cavity (indicated by red arrows); no food enters or leaves this part of the digestive tract. (**B**) When PS neurons are active, neurons of all three subnetworks are recruited into a new network in which their activity is coordinated to produce "swallowing" (indicated by the red arrow). Abbreviations Es, Gast, Pyl, and PS indicate activity of neurons in the esophageal, gastric, and pyloric subnetworks and of the PS neurons. [Adapted from Meyrand, 1994.]

all three parts of the STG system to produce a set of peristaltic waves that travel from the esophagus to the pylorus (Figure 11-55B). All other rhythms are inhibited during this behavior. When activity in the PS neurons ceases, yet another rhythm transiently appears, but eventually all of the neurons in the swallowing network return to their original activity patterns. The neurons that control swallowing behavior are called the *swallowing network*, which includes neurons that, in the absence of PS neuron activity, are active in the esophageal, gastric, or pyloric networks.

Some of the neuromodulators that control the activity of STG neurons have been identified. The biogenic amine serotonin and the neuropeptides proctolin and cholecystokinin all change the output pattern of at least some neurons in the STG.

The reconfiguration of a small pool of neurons into several functional networks suggests a new view of the neurons responsible for controlling motor output. Previous work has shown that a single anatomically defined network can produce different forms of output in response to neuromodulatory agents, but the stomatogastric system suggests that the composition of the network, too, can be plastic. Dynamic specification of many functional networks within a defined set of neurons offers a large increase in the number of possible ways that motor output can be controlled. Clearly, one challenge is to discover where the control of this mechanism resides and how it is regulated.

SUMMARY

All behavior is controlled by the motor output of the nervous system. Motor neurons are organized into diverse networks that may be somewhat plastic, allowing flexibility in behavioral responses. Understanding behavior at the neuronal level requires an understanding of how neurons interact to produce behavioral output.

In the course of evolution, the primitive, distributed, anatomically diffuse "nerve nets" characteristic of coelenterates became condensed into nerve cords and ganglia, which are seen even in some jellyfish. In segmented animals, the anterior end, initially specialized as the location of many sensory organs, became differentiated to contain a superganglion, or brain.

The most complicated nervous systems are found in vertebrates. These systems can be divided into the central and the peripheral nervous systems. All neurons in the nervous system are afferent neurons, efferent neurons, or interneurons, and most of the neurons in complex neuronal networks are interneurons. To a large extent, the connections in central networks appear to be preprogrammed genetically; but, during development and thereafter, they are sustained by and can be modified by use.

The integration of input and the production of subsequent activity in each neuron of a network depends primarily on two major sets of factors: (1) the organization of circuits and synapses formed by interacting neurons, and (2) the way in which individual neurons process or integrate incoming information to produce their own APs. The integrative properties of a neuron depend on the anatomy of the neuron, on its connections, and on the properties of its cell membrane and ion channels.

The study of the neuronal control of behavior has been aided by the identification of specialized behaviors called fixed action patterns. These highly stereotyped motor patterns are typically elicited by a particular stimulus, or key stimulus. Understanding the behavioral capacities of animals at the neuronal level is the goal of neuroethology.

Sensory neuronal networks sort and refine information available to the animal. They magnify, amplify, filter, and reconfigure the original sensory input. The mammalian visual system has taught us a great deal about how sensory systems function. Electrical recording from cells in the visual cortex indicates that individual central neurons are activated by, and extract, certain features of a stimulus, rather than generating a point-to-point representation of the peripheral input. In addition, studies of the visual system indicate that there is a heirarchical arrangement of neurons and that the specificity of sensory features evoking activity in the neurons increases with each level, until only specific features of the visual stimulus evoke responses at the higher levels. Some cells may be activated only by such complicated stimuli as a face. Studies in the barn owl have shown that a map of auditory space is computed from intensity and timing differences between sounds as they are received at the two ears. Such computed spatial maps are in register anatomically with other sensory maps, such as the retinotopic map from the visual system.

The simplest neuronal networks are monosynaptic reflex arcs, the most familiar of which is the stretch reflex of vertebrates. More complex behaviors include locomotory movements that are based in part on central "motor programs" that determine, for example, the sequence of muscle contractions that produce coordinated locomotory movements. Feedback from proprioceptive sensory neurons can exert an influence on the strength and the frequency of the motor output and, in most rhythmic motor activity, it also contributes to fine-tuning its coordination.

Muscle output is controlled by a hierarchiacal system. An example of the lowest level of control is the monosynaptic stretch reflex arc that is responsible for maintaining postural tone. At the next level are rhythmic motor patterns characteristic of walking, swimming, and crawling. Finally, the control of complex fixed action patterns resides at the top of the hierarchy. Attempts to understand the relations between levels of control have been most successful in the relatively simple motor systems of invertebrates. In these model systems, it is clear that a particular neuron may participate in several motor networks that function at different levels. Moreover, in some systems, neuromodulatory substances dynamically control the configuration of networks.

REVIEW QUESTIONS

1. Action potentials in all neurons are fundamentally alike; so how is the modality of input from the various sense organs recognized by the central nervous system?

2. Describe the general organization of the vertebrate brain and spinal cord.

3. Compare and contrast the sympathetic and the parasympathetic divisions of the autonomic nervous system. How do they differ anatomically? How do they differ functionally and biochemically?

4. It has been said that all sensation takes place in the brain. Explain what is meant by this statement.

5. How can an increased rate of firing in an inhibitory interneuron produce increased activity in other neurons?

6. What is the source of the continuous low-level synaptic input and slow tonic firing in vertebrate spinal α motor neurons?

7. Describe the organization of the vertebrate retina.

8. Each eye of a primate sees about the same field, but the right hemisphere of the brain "sees" the left half of the visual field, whereas the left hemisphere "sees" the right half of the field. How does this occur?

9. Why does the evening sky appear to have a lighter band outlining the silhouette of a mountain range?

10. What is meant by the "receptive field" of a cortical neuron?

11. How can the receptive field of a simple cell in the visual cortex be a bar or a straight edge, when cells of the lateral geniculate have circular receptive fields?

12. What would happen to your posture if all of your muscle spindles suddenly ceased to function?

13. How do γ-efferent fibers change the sensitivity of muscle spindles?

14. Discuss some of the general insights into neuronal organization that have resulted from studies of the retina and visual cortex.

15. The nervous system is sometimes compared to a telephone system or a computer. Discuss some properties of the nervous system that make this a good analogy and others that make it a poor analogy.

16. What evidence indicates that some complex behavior patterns are inherited and cannot be ascribed entirely to learning?

17. What supports the statement that the major factor responsible for differences in the functioning of different nervous systems is neuronal circuitry and not the properties of single nerve cells?

18. What are releasing stimuli and fixed action patterns? Give at least one example of each.

19. What is a central pattern generator? What are some of the properties of central pattern generators, and what roles do they play in the control of behavior? Describe some examples of central pattern generators.

20. What is a command neuron? What is a command system? Describe some examples of command systems.

21. How can one neuron play different roles in several central pattern generators?

SUGGESTED READINGS

Camhi, J. 1984. *Neuroethology.* Sunderland, MA: Sinauer. (An excellent textbook summarizing many subfields of this rapidly growing discipline.)

Carew, T. J., and C. L. Sahley. 1986. Invertebrate learning and memory: From behavior to molecule. *Ann. Rev. Neurosci.* 9:435–487. (A review of progress toward understanding this important form of plasticity in the nervous system.)

Dowling, J. 1987. *The Retina: An Approachable Part of the Brain.* Cambridge, MA: Belknap Press. (A description of the structural and functional organization of the vertebrate retina, written by a major contributor to our current knowledge of this remarkable organ.)

Ewert, J.-P. 1980. *Neuroethology.* Berlin: Springer-Verlag. (An introduction to a relatively new field: the study of the neuronal basis of behavior.)

Grillner, S., and P. Wallen. 1985. Central pattern generators for locomotion, with special reference to vertebrates. *Ann. Rev. Neurosci.* 8:233–261. (A review of the properties of central pattern generators, with emphasis on the CPG for swimming in lampreys.)

Gwinner, E. 1986. Internal rhythms in bird migration. *Scientific American* 254:84–92. (A biological approach to this otherwise apparently mysterious navigational ability.)

Hubel, D. 1995. *Eye, Brain, and Vision.* New York: Scientific American Library Paperbacks. (An exceedingly readable review of information processing in the visual system written by one of the most prolific and creative researchers in the field.)

Kandel, E., J. Schwartz, and T. Jessell. 1991. *Principles of Neural Science,* 3d ed. New York: Elsevier. (A giant compendium of information about the nervous system, with some emphasis on vertebrate—particularly mammalian—species.)

Knudsen, E. I. 1981. The hearing of the barn owl. *Scientific American* 245:113–125. (A very readable discussion of the remarkable auditory nervous system of this bird, including a description of some very creative physiological experimentation.)

Konishi, M. 1985. Birdsong: From behavior to neuron. *Ann. Rev. Neurosci.* 8:125–170. (A review of the neuronal basis of the production of bird songs, written by one of the most eminent experts on the avian brain.)

McFarland, D. 1993. *Animal Behaviour: Psychobiology, Ethology, and Evolution.* New York: Wiley. (A classic text covering the study of animal behavior.)

Nicholls, J. G., A. R. Martin, and B. G. Wallace. 1992. *From Neuron to Brain: A Cellular and Molecular Approach the Nervous System,* 3d ed. Sunderland, MA: Sinauer. (A very readable text describing both single-cell properties and circuitry.)

Nauta, W. J. H., and M. Feirtag. 1986. *Fundamental Neuroanatomy.* New York: W. H. Freeman and Company. (A comprehensive description of mammalian neuroanatomy.)

INTEGRATION OF
PHYSIOLOGICAL SYSTEMS

To this point we have discussed the basic principles of animal physiology (Chapters 1–4), followed by a discussion of the nervous, muscular, and
endocrine systems and the processes by which they
regulate physiological function (Chapters 5–11).
What remains to be discussed in Part III (Chapters
12–16) are the various regulated physiological systems that are involved in the day-to-day efforts of animals to acquire and store nutrients and energy, to expel wastes, to respond to changing environments, and
to reproduce.

Textbooks in animal physiology historically have
treated each of the regulated physiological systems of
animals more or less separately, with relatively little
focus on their mutual functional and structural inter-
dependencies. This approach persists both for convenience of discussion and because, to some extent, it
reflects the patterns of interests of biologists in particular animal systems. Physiologists usually identify
themselves as, for example, "cardiovascular physiologists;" few would stress the more integrated aspects
of their field by identifying themselves as, for exam-

ple, "energy transfer physiologists" studying the coordinated transport of nutrients, wastes, and heat between the environment and an animal's interior.

Further, because there are similarities between the
circulatory systems of all animals, it is convenient to
discuss individual systems in a single chapter. The division of physiological systems into units, useful in organizing a course or a book, however, has yielded
generations of students with the misimpression that
animals function as a series of loosely linked physiological systems that happen to be enclosed in a single
organism. For that reason we want to stress that animals operate as integrated systems that are responsive to, and constrained by, their surrounding environment. These interrelated systems act in a highly
coordinated fashion when faced with stresses that are
either environmental (temperature, pressure, etc.) or
biological (predation, disease, etc.).

The actual design and function of an individual
physiological system is modified by constraints placed
on it because it is part of a larger physiological network. Because these systems are highly mutually

dependent upon each other, environmental stresses may present conflicting demands upon individual systems. It is important to think about these interactions in terms of both space and time. Examples abound. Lung vital capacity in some snakes is initially reduced following ingestion of a large prey item because of space limitations in the visceral cavity. However, full lung capacity slowly returns as the meal is digested (interactions between respiration and digestion in space and time). A similar condition exists in humans after a large meal or during pregnancy. As another example, muscle power responds to physical training over time but this is not simply related to increased muscle mass. There must also be increased blood flow to the muscle which may require changes in the heart and respiration (interactions between locomotor and cardio-respiratory systems over time). In addition, the skeletal frame must be strengthened to withstand the increased stress placed on the body by this training.

Although we wish to emphasize the importance of taking an integrated view of the physiology of an animal, we realize that, at the same time, it is not practical to ask a student to learn simultaneously everything about all regulated physiological systems. Thus, we have divided the regulated systems into several different chapters. While each of these chapters focuses on a particular system and its functions, examples are used throughout that will emphasize the interactions between physiological systems and the way they respond in a coordinated manner to environmental change.

Chapters 12 through 14 of Part III discuss truly multi-function systems. The circulation (Chapter 12) is a means of distributing material between tissues, in particular oxygen, carbon dioxide, and various nutrients and excretory products. Acquisition of oxygen and the elimination of carbon dioxide is the subject of Chapter 13. The circulation and respiratory systems of animals both function together in homeostasis, for example, by regulating acid-base status and, in some systems, ionic and osmotic conditions within the animal (Chapter 14). Animals use a variety of mechanisms to acquire energy, ranging from filter feeding to predation, as described in Chapter 15. This chapter discusses the mechanics, control, and chemistry of food acquisition, digestion, and assimilation. The concluding chapter (Chapter 16) is, in many ways, a summary of the themes of the book, delving into the energetics of animals. Energy use in movement, reproduction, growth, and maintaining homeostasis is explored and placed in the overall objective of surviving to reproduce.

CIRCULATION

In animals 1 mm or less in diameter, materials are transported within the body by diffusion. In larger animals, however, adequate rates of material transport within the body can no longer be achieved by diffusion alone. In these animals, circulatory systems have evolved to transport respiratory gases, nutrients, waste products, hormones, antibodies, salts, and other materials among various regions of the body. Blood, the medium for transport of such materials, is a complex tissue containing many special cell types. It acts as a vehicle for most homeostatic processes and plays some role in nearly all physiological functions.

This chapter reviews the circulation of blood and how it is controlled to meet the requirements of the tissues. Most attention is given to the mammalian circulatory system, because it is the best known. Mammals are very active, predominantly aerobic, terrestrial animals, and their circulatory system evolved to meet their particular requirements. The mammalian system is only one of several types of circulation. All circulatory systems, however, comprise the following basic parts, which have similar functions in different animals:

1. A main *propulsive organ*, usually a heart, which forces blood around in the body
2. An *arterial system*, which can act both to distribute blood and as a pressure reservoir
3. *Capillaries*, in which transfer of materials occurs between blood and tissues
4. A *venous system*, which acts as a blood (volume) reservoir and as a system for returning blood to the heart

The arteries, capillaries, and veins constitute the *peripheral circulation.*

GENERAL PLAN OF THE CIRCULATORY SYSTEM

The movement of blood through the body results from any or all of the following mechanisms:

- Forces imparted by rhythmic contractions of the heart
- Elastic recoil of arteries following filling by cardiac contraction
- Squeezing of blood vessels during body movements
- Peristaltic contractions of smooth muscle surrounding blood vessels

The relative importance of each of these mechanisms in generating flow varies among animals. In vertebrates, the heart plays the major role in blood circulation; in arthropods, movements of the limbs and contractions of the dorsal heart are equally important in generating blood flow; in the giant earthworm, *Megascolides australis*, peristaltic contractions of the dorsal vessel are responsible for moving blood in an anterior direction and filling the lateral hearts, which pump blood into the ventral vessel for distribution to the body (Figure 12-1A). This worm, which can be up to 6 m in length, is divided into segments separated by membranous structures (septa). Tracer studies have shown that the anterior 13 segments, each of which contains two lateral hearts, have a rapid circulation, but the remaining segments, which lack lateral hearts, have a very sluggish circulation. Because of the peristaltic contractions of the dorsal vessel, the blood pressure is considerably higher in the dorsal vessel than in the ventral vessel (Figure 12-1B). In all animals, valves and/or septa determine the direction of flow, and smooth muscle surrounding blood vessels alters vessel diameter, thereby regulating the amount of blood that flows through a particular pathway and controlling the distribution of blood within the body.

Open Circulations

Many invertebrates have an *open circulation*, that is, a system in which blood pumped by the heart empties via an artery into an open fluid space, the **hemocoel,** which lies between the ectoderm and endoderm. The fluid contained within the hemocoel, referred to as **hemolymph,** or blood, is not circulated through capillaries but bathes the tissues directly. Figure 12-2 (A and B) illustrates the organization of the main vessels in the

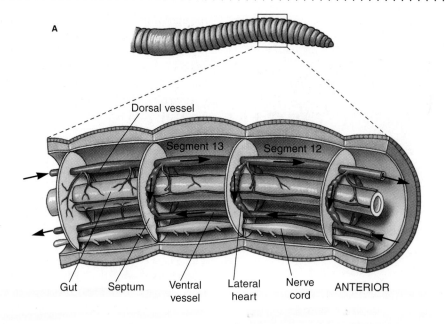

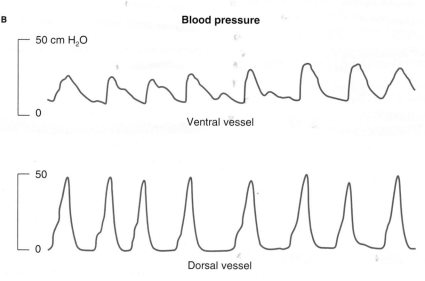

Figure 12-1 In the giant earthworm, *Megascolides australis,* peristaltic contractions of the dorsal vessel and pumping by the lateral hearts are both important in moving blood. **(A)** Blood flows from the dorsal vessel into the lateral hearts, present in the 13 anterior segments, and then is pumped into the ventral vessel. **(B)** Peak blood pressure is about twice as high in the dorsal vessel, because of its peristaltic contractions, than in the ventral vessel. [Adapted from Jones et al., 1994.]

open circulation of two groups of invertebrates. The hemocoel is often large and may constitute 20%–40% of body volume. In some crabs, for instance, blood volume is about 30% of body volume. Open circulatory systems have low pressures, with arterial pressures seldom exceeding 0.6–1.3 kilopascals (kPa), or 4.5–9.7 mm Hg (1 kPa = 7.5 mm Hg). Higher pressures have been recorded in portions of the open circulation of the terrestrial snail *Helix,* but these are exceptional. In snails, these high pressures are generated by contractions of the heart, whereas in some bivalve mollusks high pressures in the foot are generated by contractions of surrounding muscles rather than of the heart.

Animals with an open circulation generally have a limited ability to alter the velocity and distribution of blood flow. As a result, in bivalve mollusks and other animals that have an open circulation and use blood for gas transport, changes in oxygen uptake are usually slow and maximal rates of oxygen transfer low per unit weight. Nonetheless, such animals exert some control over both the flow and distribution of hemolymph; moreover, the blood is distributed throughout the tissues in many small channels in animals with an open circulation. In the absence of such features, even moderate rates of oxygen consumption would be impossible because of the large diffu-

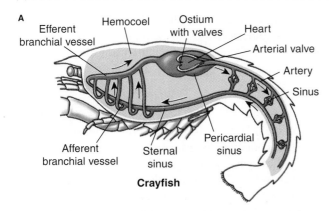

Crayfish

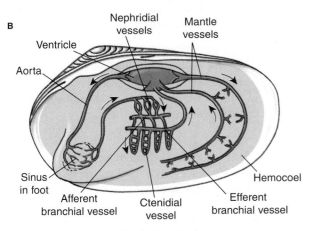

Bivalve mollusk

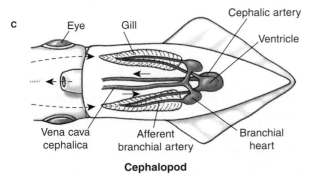

Cephalopod

Figure 12-2 Most invertebrates have an open circulation, but cephalopods have a closed circulation. The main blood vessels in the open circulation of crayfish **(A)** and bivalve mollusks **(B)** empty into a large surrounding space, the hemocoel, which makes up about 30% of total body volume. Compared with an open circulation, the closed circulation of cephalopods **(C)** is characterized by a higher blood pressure and more efficient delivery of oxygen. In all diagrams only the main blood vessels are shown. The arrows indicate blood flow.

sion distances for oxygen between the hemolymph and the active tissue.

Insects have an open circulation but do not depend on it for oxygen transport and thus can achieve high rates of oxygen transfer. They have evolved a tracheal system in which direct gas transport to tissues occurs through air-filled tubes that bypass the blood, which plays a negligible role in oxygen transport. Consequently, although insects have an open circulation, they have a large capacity for aerobic metabolism. The insect tracheal system is described in Chapter 13.

Closed Circulations

In a *closed circulation,* blood flows in a continuous circuit of tubes from arteries to veins through capillaries. All vertebrates and some invertebrates, such as cephalopods (octopuses, squids), have this type of circulation (Figure 12-2C). In general, there is a more complete separation of functions in closed circulatory systems than in open ones. The blood volume in the closed circulation of vertebrates typically is about 5%–10% of body volume, much smaller than that of open-circulation invertebrates. The extracellular volume in vertebrates, expressed as a percentage of body volume is similar to the hemocoel volume in invertebrates. The closed circulatory system of vertebrates is a specialized portion of their extracellular space.

In a closed circulation, the heart is the main propulsive organ, pumping blood into the *arterial system* and maintaining a *high blood pressure* in the arteries. The arterial system, in turn, acts as a pressure reservoir forcing blood through the capillaries. The capillary walls are thin, thus allowing high rates of transfer of material between blood and tissues by diffusion, transport, or filtration. Each tissue has many capillaries, so that each cell is no more than two or three cells away from a capillary. Capillary networks are in parallel, allowing fine control of blood distribution and, therefore, oxygen delivery to tissues. Animals with a closed circulation can increase oxygen delivery to a tissue very rapidly. For this reason, squid, unlike many other invertebrates, can swim rapidly and maintain high rates of oxygen uptake; that is, their closed circulation permits sufficient flow and efficient distribution of hemolymph to the muscles to support short bursts of high-level activity.

The blood is under sufficiently high pressure in a closed circulation to permit the **ultrafiltration** of blood in the tissues, especially the kidneys. Ultrafiltration refers to the separation of an ultrafiltrate, devoid of colloidal particles, from plasma by filtration though a semipermeable membrane (capillary wall) using pressure (blood pressure) to force the fluid through the membrane. Ultrafiltration occurs in most vertebrate kidneys, resulting in the net movement of a protein-free plasma from the blood into the kidney tubule. In general, all capillary walls are permeable and as pressures are high, so fluid slowly filters across the walls and into the space between cells. A **lymphatic system** has evolved in conjunction with the high-pressure, closed circulatory system of vertebrates to recover fluid lost to tissues from the blood. The extent of filtration depends largely on the blood pressure and the permeability of the capillary wall. Filtration across capillary walls can be decreased either by a reduction in the permeability of the capillary walls or in the pressure of the blood. For example, the vessels in some tissues have less permeable walls than those found in other tissues. And in the liver and lung, where permeability is high for other reasons, pressures are lower than those in the rest of the body.

Different pressures can be maintained in the **systemic** (body) and **pulmonary** (lung) circulations of mammals because the mammalian circulatory system is equipped with a completely divided heart (Figure 12-3). The right side of the heart pumps blood in the pulmonary circulation, and the left side pumps blood in the systemic circulation. This means, however, that the flows in the pulmonary and systemic circuits must be equal because blood returning from the lung is pumped around the body. In other vertebrates the heart is not completely divided and flow to the lung can be varied independently of body blood flow.

The *venous system* collects blood from the capillaries and delivers it to the heart via the veins, which are typically low-pressure, compliant structures in which large changes in volume have little effect on venous pressure. Thus, the venous system contains most of the blood and acts as a large-volume reservoir. Blood donors give blood from this reservoir, and since there is little change in pressure as the venous volume decreases, the volumes and flows in other regions of the circulation are not markedly altered.

THE HEART

Hearts are valved, muscular pumps that propel blood around the body. Hearts consist of one or more muscular chambers connected in series and guarded by valves or, in a few cases, sphincters (e.g., in some molluscan hearts), which allow blood to flow in only one direction. The mammalian heart has four chambers, two **atria** and two **ventricles**. Contractions of the heart result in the ejection of blood into the circulatory system. Multiple heart chambers permit stepwise increases in pressure as blood passes from the venous to the arterial side of the circulation (Figure 12-4).

 Why have so many species evolved a single large multi-chambered heart rather than a series of smaller hearts distributed throughout the circulation?

Vertebrate cardiac and skeletal muscle fibers are similar in many respects, except that the T-tubule system is less extensive in cardiac muscle cells of lower vertebrates and cardiac muscle cells are electrically coupled (see Chapter 10). Except for differences in the uptake and release of Ca^{2+}, the mechanisms of contraction of vertebrate skeletal and cardiac muscle are generally considered to be alike. The **myocardium** (i.e., heart muscle) consists of three types of muscle fiber, which differ in size and functional properties:

- The myocardial cells in the sinus node (or sinoatrial node) and in the atrioventricular node are often smaller than others, are only weakly contractile, are autorhythmic, and exhibit very slow conduction between cells.

- The largest myocardial cells, found in the inner surface of the ventricular wall, are also weakly contractile, but are specialized for fast conduction and constitute the system for spreading the excitation over the heart.

- The intermediate-sized myocardial cells are strongly contractile and constitute the bulk of the heart.

Electrical Activity of the Heart

A heartbeat consists of a rhythmic contraction (**systole**) and relaxation (**diastole**) of the whole muscle mass. Contraction of each cell is associated with an **action potential** (AP) in that cell. Electrical activity, initiated in the **pacemaker** region of the heart, spreads over the heart from one cell to another because the cells are electrically coupled via membrane junctions (see Chapter 4). The nature and extent of coupling determines the pattern by which the electrical wave of excitation spreads over the heart and also influences the rate of conduction.

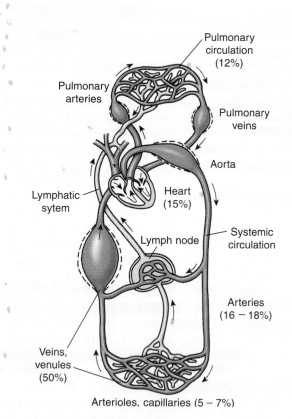

Figure 12-3 The closed circulation in mammals includes a fully divided heart, which permits different pressures in the pulmonary and systemic portions. This diagram illustrates the main components of the mammalian circulation, with oxygenated blood in the systemic system and the pulmonary system shown in red, and deoxygenated blood shown in blue. The associated lymphatic system (yellow) returns fluid from the extracellular space to the bloodstream via the thoracic duct. The percentages indicate the relative proportion of blood in different parts of the circulation. The lymphatic system and associated lymph nodes also play a key role in the immune response.

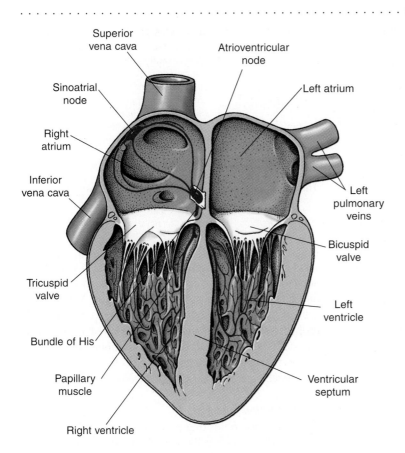

Superior
vena cava

Atrioventricular
node

Sinoatrial
node

Left atrium

Right
atrium

Inferior
vena cava

Left
pulmonary
veins

Bicuspid
valve

Tricuspid
valve

Left
ventricle

Bundle of His

Papillary
muscle

Ventricular
septum

Right ventricle

Figure 12-4 The multi-chambered mammalian heart permits the pressure to increase as blood moves from the venous to the arterial side. This cutaway view depicts the rear portion of the human heart with the impulse pathways shown in color. Impulses originate in the pacemaker, located in the sinoatrial node, spread to the atrioventricular node, from which they are transmitted to the ventricles. Pacemaker cells in some invertebrates are modified nerve cells; in other invertebrates and all vertebrates, they are usually described as modified muscle cells. [Adapted from E. F. Adolph, 1967. Copyright © 1967 by Scientific American, Inc. All rights reserved.]

Neurogenic and myogenic pacemakers

In vertebrate hearts, the pacemaker is situated in the **sinus venosus** or in a remnant of it called the **sinoatrial node** (see Figure 12-4). The pacemaker consists of small, weakly contractile, specialized muscle cells that are capable of spontaneous activity. These cells may be either neurons, as in the *neurogenic* pacemaker in many invertebrate hearts, or muscle cells, as in the *myogenic* pacemaker in vertebrate and some invertebrate hearts. Hearts are often categorized by the type of pacemaker and hence are called either neurogenic or myogenic hearts.

In many invertebrates, it is not clear whether the pacemaker is neurogenic or myogenic. Decapod crustaceans (shrimps, lobsters, crabs), however, do have neurogenic hearts. In these animals, the cardiac ganglion, situated on the heart, acts as a pacemaker. If the cardiac ganglion is removed, the heart stops beating, although the ganglion continues to be active and shows intrinsic rhythmicity. The cardiac ganglion consists of nine or more neurons (depending on the species), divided into small and large cells. The small cells act as pacemakers and are connected to large follower cells, which are all electrically coupled. Activity from the small pacemaker cells is fed into and integrated by the large follower cells and then distributed to the heart muscle. The crustacean cardiac ganglion is innervated by excitatory and inhibitory nerves originating in the central nervous system (CNS); these nerves can alter the rate of firing of the ganglion and, therefore, the *heart rate* (beats per minute).

Vertebrate, molluscan, and many other invertebrate hearts are driven by myogenic pacemakers. These tissues have been studied extensively in a variety of species. A myogenic heart may contain many cells capable of pacemaker activity, but because all cardiac cells are electrically coupled, the cell (or group of cells) with the fastest intrinsic activity is the one that stimulates the whole heart to contract and determines the heart rate. These pacemaker cells will normally overshadow those with slower pacemaker activity; however, if the normal pacemaker stops for some reason, the other pacemaker cells take over, producing a new, lower heart rate. Thus, cells with the capacity for spontaneous electrical activity may be categorized as pacemakers and latent pacemakers. In the event that a latent pacemaker becomes uncoupled electrically from the pacemaker, it may beat and control a portion of cardiac muscle, generally an entire chamber, at a rate different from that of the normal pacemaker. Such an **ectopic pacemaker** is dangerous because it will desynchronize the pumping action of the heart chambers.

Cardiac pacemaker potentials

An important characteristic of pacemaker cells is the absence of a stable resting potential. Consequently, the membranes of cells in pacemaker tissue undergo a steady depolarization, termed a **pacemaker potential,** during each diastole (Figure 12-5). As the pacemaker potential brings the membrane to the threshold potential, it gives rise to an all-or-none cardiac action potential. The interval between cardiac APs, which of course determines the heart rate, depends on the rate of depolarization of the pacemaker potential, as well as the extent of repolarization and the threshold potential for

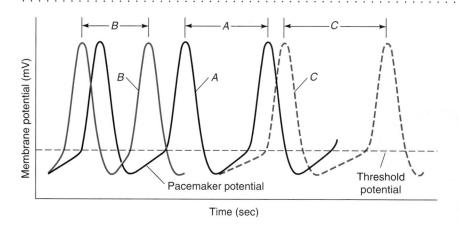

Figure 12-5 Pacemaker cells undergo spontaneous depolarization of the membrane, referred to as the pacemaker potential, triggering cardiac action potentials autorhymically (curve *A*). A more rapid depolarization increases the firing rate (curve *B*) and thus the heart rate, whereas a slower depolarization slows the firing rate (curve *C*) and heart rate.

the cardiac AP. A faster depolarization brings the membrane to a firing level sooner and thus increases the frequency of firing, leading to a faster heart rate, whereas a slower depolarization does the opposite (see Figure 12-5).

Pacemaker activity has its origin in time-dependent changes in membrane conductance. In the frog sinus venosus, the pacemaker depolarization begins immediately after the previous AP, when the potassium conductance of the membrane is very high. The potassium conductance then gradually drops, and the membrane shows a corresponding depolarization owing to the intracellular accumulation of potassium ions and a moderately high, steady conductance for sodium. The pacemaker depolarization continues until it activates the sodium conductance. The Hodgkin cycle then predominates to produce the rapid regenerative upstroke of the cardiac AP (see Chapter 5).

Acetylcholine, which is released from parasympathetic terminals of the vagus nerve (tenth cranial nerve), slows the heart by increasing potassium conductance of the pacemaker cells. This increased conductance keeps the membrane potential near the potassium equilibrium potential for a longer time, thereby slowing the pacemaker depolarization and delaying the onset of the next upstroke (Figure 12-6A). On the

other hand, norepinephrine released from sympathetic nerves accelerates the pacemaker depolarization potential, thus increasing the heart rate (Figure 12-6B). Although norepinephrine increases sodium and calcium conductance, this probably is not the main mechanism involved in speeding up the pacemaker rhythm. It is possible that norepinephrine decreases the time-dependent potassium efflux during diastole and thereby increases the rate of pacemaker depolarization.

Cardiac action potentials

The action potentials that precede contraction in all vertebrate cardiac muscle cells are of longer duration than those in skeletal muscle. The AP in skeletal muscle is completed and the membrane is in a nonrefractory stage before the onset of contraction; hence, repetitive stimulation and tetanic contraction are possible (Figure 12-7A). In cardiac muscle, by contrast, the action potential plateaus and the membrane remains in a refractory state until the heart has returned to a relaxed state (Figure 12-7B). Thus, summation of contractions cannot occur in cardiac muscle.

Cardiac APs begin with a rapid depolarization that results from a large and rapid increase in sodium conductance. This differs from the slow depolarization of the

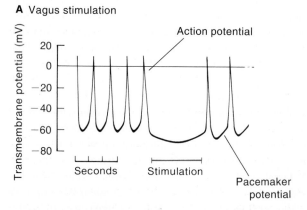

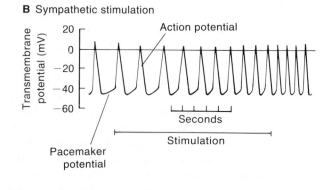

Figure 12-6 Parasympathetic stimulation via the vagus nerve and sympathetic stimulation have opposite effects on the pacemaker potential and heart rate. **(A)** Vagus stimulation produces a rise in diastolic (resting) transmembrane potential, a decrease in the rate of depolarization, and a decrease in the duration and frequency of the action potential. **(B)** Sympathetic stimulation produces an increase in the frequency of firing of the pacemaker cells. [Hutter and Trautwein, 1956.]

A Skeletal muscle

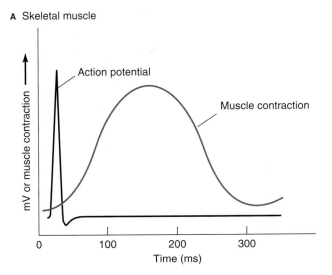

B Cardiac muscle

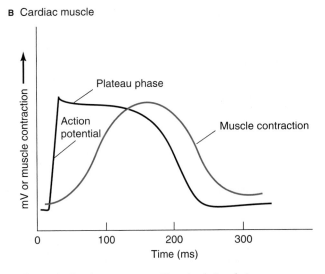

Figure 12-7 Action potentials in skeletal muscle are of very short duration **(A)**, whereas cardiac action potentials exhibit a prolonged repolarization, or plateau phase, during which cardiac muscle is refractory to stimulation **(B)**. For this reason, repetitive stimulation during a contraction and summation of contractions can occur in the skeletal muscle but not in heart muscle.

pacemaker potential, which is marked by a stable sodium conductance and decreasing potassium conductance. Repolarization of the cell membrane is delayed while the membrane remains depolarized in a so-called *plateau phase* for hundreds of milliseconds (see Figure 12-7B). The long duration of the cardiac AP produces a prolonged contraction, so that an entire chamber can fully contract before any portion begins to relax—a process that is essential for efficient pumping of blood.

The prolonged plateau of the cardiac AP results from maintenance of a high calcium conductance and a delay in the subsequent increase in potassium conductance (unlike the situation in skeletal muscle). The high calcium conductance during the plateau phase allows Ca^{2+} ions to flow into the cell, because the equilibrium potential for calcium is directed strongly inwards. This influx is especially important in lower vertebrates, in which a considerable proportion of the calcium essential for activation of contraction enters through the surface membrane. In birds and mammals, the surface-to-volume ratio of the larger cardiac cells is too small to allow sufficient entry of calcium to fully activate contraction. Therefore, most of the calcium is released—by depolarization of the T tubules (see Chapter 10)—from the extensive sarcoplasmic reticulum characteristic of the hearts of higher vertebrates. A rapid repolarization terminates the plateau phase, due to a fall in calcium conductance and an increase in potassium conductance.

The duration of the plateau and the rates of depolarization and repolarization vary in different cells of the same heart. The summation of these changes are recorded as the **electrocardiogram** (Figure 12-8). Atrial cells generally have an AP of shorter duration than ventricular cells. The duration of the AP in atrial or ventricular fibers from hearts of different species also varies. The duration of the AP is one factor correlated with the maximum frequency of the heartbeat; in smaller mammals, the duration of the ven-

tricular AP is shorter, thus heart rates generally are higher, than in larger mammals.

Because of the great diversity among the hearts of different invertebrate phyla, few generalizations can be made about the ionic mechanisms generating the cardiac AP of invertebrate hearts. The one widespread characteristic is participation of calcium. For instance, bivalve mollusk hearts have a calcium AP.

Transmission of excitation over the heart
Electrical activity initiated in the pacemaker region is conducted over the entire heart, depolarization in one cell resulting in the depolarization of neighboring cells by virtue of current flow through **gap junctions** (see Figure 4-35). These junctions between cells are located in regions of close apposition between neighboring myocardial cells, termed the **intercalated disk**. Adhesion of cells at intercalated disks is strengthened by the presence of desmosomes. The area of contact is increased by folding and interdigitation of membranes (Figure 12-9). The extent of infolding and interdigitation increases during development of the heart and also varies among species. Gap junctions are regions of low resistance between cells and allow current flow from one cell to the next across intercalated disks.

Although the junctions between myocardial cells can conduct in both directions, transmission is usually unidirectional because the impulse is initiated in and spreads only from the pacemaker region. There are usually several pathways for excitation of any single cardiac muscle fiber, since intercellular connections are numerous. If a portion of the heart becomes nonfunctional, the wave of excitation can easily flow around that portion, so that the remainder of the heart can still be excited. The prolonged nature of cardiac APs ensures that multiple connections do not result in multiple stimulation and a reverberation of activity in cardiac muscle. An AP initiated in the pacemaker region

A Electrocardiogram

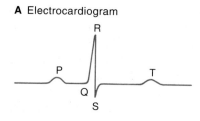

B Cardiac action potentials

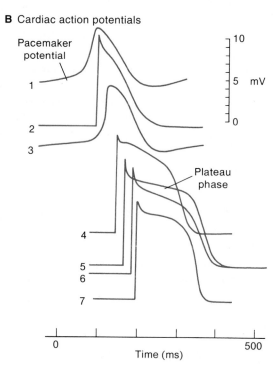

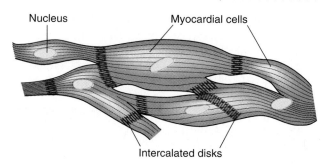

Figure 12-9 Electrical activity can spread throughout the heart because of the close apposition of myocardial cells at intercalated disks, which are densely packed with gap junctions. Shown here is a schematic diagram of myocardial cells from a mammalian heart. The folding and the interdigitation of membranes are characteristic of intercalated disks. Although desmosomes are present in these regions, thereby strengthening the adhesion between cells, they are not easily distinguished.

Figure 12-8 The electrocardiogram represents the summation of the electrical activity in various parts of the heart. **(A)** The major components of the electrocardiogram (ECG) reflect atrial depolarization (P), ventricular depolarization (QRS), and ventricular repolarization (T). **(B)** The amplitude, configuration, and duration of cardiac action potentials differ at various sites. The potential changes were recorded from the following sites: (1) sinoatrial node, (2) atrium, (3) atrioventricular node, (4) bundle of His, (5) Purkinje fiber in a false tendon, (6) terminal Purkinje fiber, and (7) ventricular muscle fiber. The numbers indicate the sequence in which the various sites fire. [Part B from Hoffman and Cranefield, 1960.]

ture branches into right and left bundles, which subdivide into Purkinje fibers that extend into the myocardium of the two ventricles. Conduction is slow through the nodal fibers (about 0.1 m·s^{-1}) but rapid through the bundle of His ($4-5$ m·s^{-1}). The bundle of His and the Purkinje fibers deliver the wave of excitation to all regions of the ventricular myocardium very rapidly, causing all the ventricular muscle fibers to contract together. As each wave of excitation arrives, the ventricular myocardial cells contract almost immediately, with the wave of excitation passing at a velocity of 0.5 m·s^{-1} from the internal lining of the heart wall (**endocardium**) to the external lining (**epicardium**). The functional significance of the electrical organization of the myocardium is its ability to generate separate, synchronous contractions of the atria and the ventricles. Thus, slow conduction through the atrioventricular node allows atrial contractions to precede ventricular contractions and also allows time for blood to move from the atria into the ventricles.

Because of the large number of cells involved, the currents that flow during the synchronous activity of cardiac cells can be detected as small changes in potential from points all over the body. These potential changes—recorded as the electrocardiogram—are a reflection of electrical activity in the heart and can be easily monitored and then analyzed. A P-wave is associated with depolarization of the atrium, a QRS complex with depolarization of the ventricle, and a T-wave with repolarization of the ventricle (see Figure 12-8A). The electrical activity associated with atrial repolarization is obscured by the much larger QRS complex. The exact form of the electrocardiogram varies with the species in question and is affected by the nature and position of recording electrodes, as well as by the nature of cardiac contraction. The electrocardiogram is valuable medically because it can be used to diagnose cardiac abnormalities.

As noted earlier, acetylcholine (ACh), released from cholinergic nerve fibers, increases the interval between APs in pacemaker cells and thus slows the heart rate (see

results in a single AP being conducted through all the other myocardial cells, and another AP from the pacemaker region is required for the next wave of excitation.

In the mammalian heart, the wave of excitation spreads from the sinoatrial node over both atria in a concentric fashion at a velocity of about 0.8 m·s^{-1}. The atria are connected electrically to the ventricles only through the **atrioventricular (AV) node;** in other regions the atria and ventricles are joined by connective tissue that does not conduct the wave of excitation from the atria to the ventricles (see Figure 12-4). Excitation spreads to the ventricle through small junctional fibers, in which the velocity of the wave of excitation is slowed to about 0.05 m·s^{-1}. The junctional fibers are connected to nodal fibers, which in turn are connected via transitional fibers to the **bundle of His;** this struc-

Figure 12-6A). This decrease in heart rate is sometimes referred to as a *negative chronotropic effect*. Parasympathetic cholinergic fibers in the vagus nerve innervate the sinus node and atrioventricular node of the vertebrate heart. As the heart rate slows, acetylcholine also reduces the velocity of conduction from the atria to the ventricles through the atrioventricular node. High levels of acetylcholine block transmission through the atrioventricular node, so that only every second or third wave of excitation is transmitted to the ventricle. Under these unusual conditions, the atrial rate will be two or three times that in the ventricle. Alternatively, high levels of acetylcholine may completely block conduction through the atrioventricular node *(atrioventricular block)*, giving rise to an ectopic pacemaker in the ventricle. The result is that the atria and ventricles are controlled by different pacemakers and contract at quite different rates, the two beats being uncoordinated. This would be devastating for a fish in which atrial contraction is very important for ventricular filling. It is not quite so devastating in mammals because atrial contraction only tops up the ventricles, they are filled mainly by the direct inflow of blood from the venous system through the relaxed atria.

The catecholamines epinephrine and norepinephrine have three distinct positive effects on heart function:

- Increased rate of myocardium contraction, or heart rate *(positive chronotropic effect)*

- Increased force of contraction of the myocardium *(positive inotropic effect)*

- Increased speed of conduction of the wave of excitation over the heart *(positive dromotropic effect)*

The effect of these catecholamines on the rate of contraction is mediated via the pacemaker, whereas the increased strength of contraction is a general effect on all myocardial cells. Norepinephrine also increases conduction velocity through the atrioventricular node. It is released from adrenergic nerve fibers that innervate the sinus node, atria, atrioventricular node, and ventricle, so that sympathetic adrenergic stimulation has a direct effect on all portions of the heart.

Coronary Circulation

The coronary circulation supplies nutrients and oxygen to the heart. The coronary supply to the heart is extensive and cardiac muscle has a much higher capillary density and more mitochondria than most skeletal muscles. There is also a high myoglobin content resulting in the typical red color of the heart. The blood pumped by the heart supplies nutrients to the inner spongy layer of the heart in many fish and amphibia as it flows through the heart, but even in these animals the coronary supply is necessary to deliver oxygen and other substrates to the outer, more dense, regions of the heart wall. In general hearts can use a wide variety of nutrients, including fatty acids, glucose, and lactate; the particular substrate used is determined largely by availability.

Because the heart primarily uses aerobic pathways to generate energy, it is very dependent on a continual oxygen supply. Thus a continual coronary flow is required to maintain cardiac performance. An increase in cardiac activity depends on increased metabolism, which in turn requires increased coronary flow. Adenosine is probably a key metabolite in maintaining the relationship between coronary flow and cardiac activity. Adenosine, which is formed from adenosine triphosphate (ATP) during cardiac metabolism, and other local metabolic factors cause dilation of coronary vessels and therefore increase coronary flow. Formation and release of adenosine increases with increased metabolism or during myocardial **hypoxia** (drop in oxygen level), leading to higher coronary flow. Sympathetic stimulation is a second, but less important, mechanism of increasing coronary flow. Circulating catecholamines increase cardiac contractility and cause coronary vasodilation mediated via β_1-adrenoreceptors.

 Mammalian hearts have an extensive coronary circulation throughout the myocardium, but in some fish hearts, the coronary circulation is restricted to epicardium, the external lining of the heart. What are the consequences of this differing organization to the nutrition of the heart?

Mechanical Properties of the Heart

The mechanical aspects of heart function relate to the changes in cardiac pressure and volume that lead to ejection of blood during each heartbeat. Now we'll examine these properties and determination of the work done by the heart.

Cardiac output, stroke volume, and heart rate

Cardiac output is the volume of blood pumped per unit time from a ventricle. In mammals it is defined as the volume ejected from the right or left ventricle, not the combined volume from both ventricles. The volume of blood ejected by each beat of the heart is termed the **stroke volume**. The mean stroke volume can be determined by dividing cardiac output by heart rate.

Stroke volume is the difference between the volume of the ventricle just before contraction *(end-diastolic volume)* and the volume of the ventricle at the end of a contraction *(end-systolic volume)*. Changes in stroke volume may result from changes in either end-diastolic or end-systolic volume. The end-diastolic volume is determined by four parameters:

- Venous filling pressure

- Pressures generated during atrial contraction

- Distensibility of the ventricular wall

- The time available for filling the ventricle

SPOTLIGHT 12-1

THE FRANK-STARLING MECHANISM

Otto Frank observed that the more the frog heart was filled, the greater the stroke volume. That is, increased venous return results in increased stroke volume. Frank derived a length-tension relationship for the frog myocardium and demonstrated that contractile tension increases with stretch up to a maximum and then decreases with further stretch. Ernest Starling, a dominant figure in many areas of physiology during the early 1900s, had come to similar conclusions as Frank. Although neither Starling nor Frank considered mechanical work, the increase in mechanical work from the ventricle caused by an increase in end-diastolic volume (or venous filling pressure) is termed the **Frank-Starling mechanism** (plot A). The curves derived from measuring work output from the ventricle at different venous filling pressures are known as **Starling curves** (plot B).

No single Starling curve, however, describes the relationship between venous filling pressure and work output from the ventricle. The mechanical properties of the heart are affected by a number of factors, including the level of activity in nerves innervating the heart and the composition of blood perfusing the myocardium. For instance, the relationship between ventricular work output and venous filling pressure is markedly affected by stimulation of sympathetic nerves innervating the heart.

Starling was a versatile researcher who along with William Bayliss discovered the hormone secretin. He coined the term *hormone* and defined their basic properties (see Chapter 9). Starling also made many contributions to our understanding of the circulation. In addition to the observations described by the Frank-Starling mechanism, he proposed the Starling hypothesis that the exchange of fluid between blood and tissues is due to the difference in the filtration and colloid osmotic pressures across the capillary wall. This hypothesis was subsequently verified largely by the work of E. Landis.

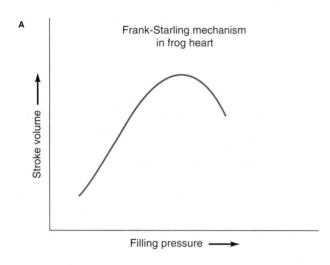

A — Frank-Starling mechanism in frog heart

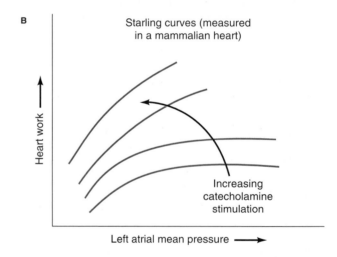

B — Starling curves (measured in a mammalian heart)

Increasing catecholamine stimulation

The end-systolic volume is determined by two parameters:

- The pressures generated during ventricular systole
- The pressure in the outflow channel from the heart (aortic or pulmonary artery pressure)

Increasing the venous filling pressure causes an increase in end-diastolic volume and results in an increased stroke volume from an isolated mammalian heart (Spotlight 12-1). End-systolic volume also increases, but not as much as end-diastolic volume. Thus, cardiac muscle behaves in a way similar to that of skeletal muscle in that stretch of the relaxed muscle within a certain range of lengths results in the development of increased tension during a contraction. Increases in arterial pressure also cause a rise in both end-diastolic and end-systolic volume with little change in stroke volume. In this instance, the increased mechanical work required to maintain stroke volume in the face of an elevated arterial pressure results from the increased stretch of cardiac muscle during diastole.

As noted above, epinephrine and norepinephrine released from sympathetic nerves or circulating in the blood increase the force of contraction of the ventricle; hence both the rate and the extent of ventricular emptying are increased by these catecholamines. The effects of cholinergic (i.e., vagus) nerve activity on the rate and force of ventricular power output during each beat are much less marked than the effects of adrenergic sympathetic nerves. This difference stems from the much more extensive innervation of the ventricles by adrenergic nerves than by cholinergic nerves.

The effects of sympathetic nerve stimulation and/or increased circulating levels of catecholamines represent a series of integrated actions. Stimulation of pacemaker cells leads to an increase in heart rate. Conduction velocity over the heart is increased to produce a more nearly synchronous beat of the ventricle. Both the rate of production of ATP and the rate of conversion of chemical energy to mechanical energy in ventricular cells increase, leading to an increase in ventricular work: the rate of ventricular emptying increases during systole, so that the same or a larger stroke volume is ejected in a much shorter time. This increased force of contraction is mediated by the action of catecholamines on both α- and β-adrenoreceptors (see Chapter 8 for more details). Thus, when adrenergic nerve stimulation increases the heart rate, the same stroke volume is ejected from the heart in a shorter time. So, although the time available for emptying and filling the heart is reduced as heart rate increases, stroke volume remains quite stable over a wide range of heart rates. For example, exercise in many mammals is associated with a large increase in heart rate with little change in stroke volume; only at the highest heart rates does the stroke volume fall (Figure 12-10). This situation occurs because, over a wide range of heart rates, increased sympathetic activity ensures more rapid ventricular emptying, and elevated venous pressures result in more rapid filling as heart rate increases.

There are limits, however, to which diastole can be shortened, determined by the maximum possible rate at which the ventricles can be filled and emptied as well as the nature of the coronary circulation. During contractions of the myocardium, coronary capillaries are occluded, so flow is very reduced during systole. Flow rises dramatically during diastole, but a decrease in the diastolic period tends to reduce the period of coronary blood flow. Catecholamines also cause coronary vasodilation and increase coronary flow. As we have noted, increases in cardiac output with exercise often are associated with large changes in heart rate and small changes in stroke volume in mammals (see Figure 12-10). Following sympathetic denervation of the heart, exercise results in similar increases in cardiac output,

but in this instance there are large changes in stroke volume rather than in heart rate. The increases in cardiac output are probably caused by an increase in venous return. The sympathetic nerves are not involved in increasing cardiac output per se but rather in raising heart rate and maintaining stroke volume, avoiding the large pressure oscillations associated with large stroke volumes and keeping the heart operating at or near its optimal stroke volume for efficiency of contraction. The sympathetic nerves thus play an important role in determining the relation between heart rate and stroke volume, but additional factors are involved in mediating the increase in cardiac output with exercise.

Changes in pressure and flow during a single heartbeat
Contractions of the heart cause fluctuations in cardiac pressure and volume as illustrated by the tracings in Figure 12-11A. The following sequence of events occurs during contraction of a mammalian heart (Figure 12-11B):

1. During diastole closed aortic valves maintain large pressure differences between the relaxed ventricles and the systemic and pulmonary aortas. The atrioventricular valves are open, and blood flows directly from the venous system into the ventricles.
2. When the atria contract, pressures rise in them and blood is ejected from them into the ventricles.
3. As the ventricles begin to contract, pressures rise in them and exceed those in the atria. At this point, the atrioventricular valves close, thus preventing backflow of blood into the atria, and ventricular contraction proceeds. During this phase, both the atrioventricular and the aortic valves are closed, so that the ventricles form sealed chambers and there is no volume change. That is, the ventricular contraction is *isometric*.
4. Pressures within the ventricles increase rapidly and eventually exceed those in the systemic and pulmonary aortas. The aortic valves then open, and blood is ejected into the aortas, resulting in a decrease in ventricular volume.
5. As the ventricles begin to relax, intraventricular pressures fall below the pressures in the aortas, the aortic valves close, and there is an isometric relaxation of the ventricle.

Once the ventricular pressures fall below those in the atria, the atrioventricular valves open, ventricular filling starts again, and the cycle is repeated. In the mammalian heart, the volume of blood forced into the ventricle by atrial contraction is about 30% of the volume of blood ejected into the aorta by ventricular contraction. Thus, ventricular filling is largely determined by the venous filling pressure, which forces blood from the veins directly through the atria into the ventricles. Atrial contraction simply tops off the nearly full ventricles with blood; but maximal cardiac output may be compromised if atrial contraction is impaired.

Contraction of cardiac muscle can be divided into two phases. The first is an isometric contraction during which

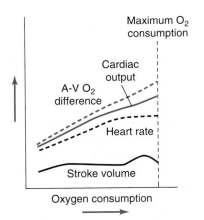

Figure 12-10 In humans and many other mammals, increased cellular oxygen needs during exercise are met in part by increasing heart rate rather than stroke volume, leading to higher cardiac output. At high levels of oxygen consumption, heart rate levels off and stroke volume increases and then decreases. In addition, extraction of oxygen from the blood in the capillaries increases during exercise, as indicated by the increase in the arterial-venous (A-V) O_2 difference. [Adapted from Rushmer, 1965b.]

A Changes in pressure and volume during heartbeat

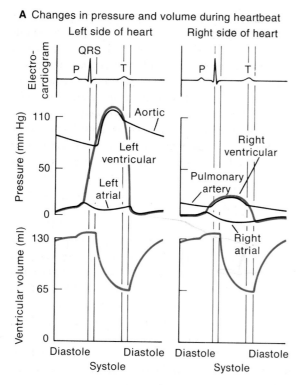

B Sequence of events in heartbeat

(1) **Mid-diastole**

(2) **Atrial contraction**

(3) **Isometric ventricular contraction**

(4) **Ventricular ejection**

(5) **Isometric ventricular relaxation**

Figure 12-11 During a single cardiac cycle, sequential contraction of the atria and ventricles and the opening and closing of valves produce characteristic changes in pressure and volume. **(A)** Changes in pressure and volume in the ventricles and aorta *(left)* and pulmonary artery *(right)* dur-ing a single cardiac cycle. **(B)** Sequence of events in contraction of mammalian heart. Black indicates contracted muscle; gray, relaxed muscle. See text for discussion. [Part A adapted from Vander et al., 1975.]

tension in the muscle and pressure in the ventricle increase rapidly. The second phase is essentially isotonic; tension does not change very much, for as soon as the aortic valves open, blood is ejected rapidly from the ventricles into the arterial system with little increase in ventricular pressure. Thus, tension is generated first with almost no change in length; then the muscle shortens with little change in tension. In other words, during each contraction, cardiac muscle switches from an isometric to an isotonic contraction.

Work done by the heart

It is a simple principle of physics that external work done is the product of mass times distance moved. In the present context, work can be calculated as the *change in pressure times flow.* Flow is directly related to the change in volume with each beat of the ventricle. With the pressure given in grams per square centimeter and the volume in cubic centimeters, pressure times volume equals grams times cubic centimeters divided by square centimeters, which equals grams times centimeters—the equivalent of mass times distance moved, or work. Thus, a plot of pressure times volume for a single contraction of a ventricle yields a *pressure-volume loop* whose area is proportional to the external work done by that ventricle.

Figure 12-12 illustrates pressure-volume loops for the right and left ventricles of a mammalian heart. The two

ventricles eject equal volumes of blood, but the pressures generated in the pulmonary circuit (right ventricle) are much lower; consequently, the external work done by the right ventricle is much less than that done by the left ventricle. As described in the previous section, blood is ejected from the ventricle when intraventricular pressures exceed the arterial pressure. If the arterial pressure is elevated, more external work must be done by the heart to raise the intraventricular pressure enough to maintain stroke volume at the original level. This, of course, means that there is an extra strain on the heart if blood pressure is high.

Not all energy expended by the heart will appear as changes in pressure and flow; some energy is expended to overcome frictional forces within the myocardium, and more is dissipated as heat. The external work done by the heart, expressed as a fraction of the total energy expended, is termed the *efficiency of contraction.* The external work done can be determined from measurements of pressure and flow and converted into milliliters of O_2 consumed. This, in turn, can be expressed as a fraction of the total O_2 uptake by the heart in order to measure the efficiency of contraction. In fact, not more than 10%–15% of the total energy expended by the heart appears as mechanical work.

Energy is expended to increase wall tension and raise blood pressure within the heart. According to **Laplace's law,** the relationship between wall tension and pressure in

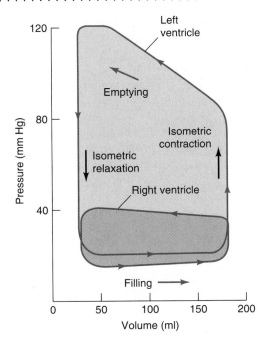

Figure 12-12 The area of a ventricular pressure-volume loop is proportional to the external work done by a ventricle in one cardiac cycle. Shown here are loops for the right and left ventricles of the mammalian heart. Once around a loop in a counterclockwise direction is equivalent to one heartbeat. Ventricular filling occurs at low pressure; pressure increases sharply only when the ventricles contract (the sharp upswing on the right-hand side of each loop). Ventricular volume decreases as blood flows into the arterial system, and ventricular pressure falls rapidly as the ventricle relaxes. Filling then begins again. Note that although the volume changes in both ventricles are similar, the pressure changes are much larger in the left ventricle than in the right one. Therefore, the left ventricle has a larger loop and hence does more external work than the right ventricle.

a hollow structure is related to the radius of curvature of the wall. If the structure is a sphere, then

$$P = \frac{2y}{R} \qquad (12\text{-}1)$$

where P is the **transmural pressure** (the pressure difference across the wall of the sphere), y is the wall tension, and R is the radius of the sphere. According to this relation, a large heart must generate twice the wall tension of a heart half its size to develop a similar pressure. Thus, more energy must be expended by larger hearts in developing pressure, and we might expect a larger ratio of muscle mass to total heart volume in these hearts. Hearts are not, of course, perfect spheres, but have a complex gross and microscopic morphology; nevertheless, Laplace's law applies in general. The energy expended in ejecting a given quantity of blood from the heart will depend on the efficiency of contraction, the pressures developed, and the size and shape of the heart.

The Pericardium

The heart is contained in a pericardial cavity and is surrounded by a connective-tissue membrane called the **pericardium.** The magnitude of the pressure changes within the pericardial sac depend on the rigidity of the pericardium and on the magnitude and rate of change of the heart volume. The membrane may be thin and flexible (compliant), in which case pressure changes within the pericardial cavity during each heartbeat are negligible. Or the pericardium may be quite rigid (noncompliant), in which case the intrapericardial pressure oscillates during each heartbeat.

The compliant pericardium enveloping the mammalian heart is formed of two layers, an outer fibrous layer and an inner serous layer. The serous layer is double, forming the inner lining of the pericardial space and the outer layer (epicardium) of the heart itself. In mammals, the serous layer secretes a fluid that acts as a lubricant, facilitating movement of the heart.

Crustaceans and bivalve mollusks have a noncompliant pericardium. In these animals, contractions of the ventricle reduce pressure in the pericardial cavity and enhance flow into the atria from the venous system (Figure 12-13). Thus, tension generated in the ventricular wall is utilized both to eject blood into the arterial system and to draw blood into the atria from the venous system.

The pericardium of elasmobranchs (sharks) and lungfishes also is noncompliant, whereas that of teleosts is compliant. The elasmobranch heart consists of four chambers—sinus venosum, atrium, ventricle, and **conus**—all contained within a rigid pericardium (Figure 12-14). The reduction in intrapericardial pressure that occurs during ventricular contraction in elasmobranchs produces a suction that helps expand the atrium and thereby increases venous return to the heart. If the pericardial cavity is opened, cardiac output is reduced; hence the increased venous return to the atrium caused by reduced pericardial pressure is

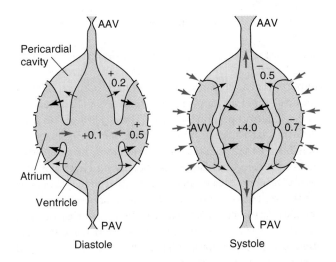

Figure 12-13 In the heart of the bivalve mollusk *Anodonta*, ventricular contraction not only ejects blood but also reduces pressure in the pericardial cavity, thus enhancing atrial filling. This occurs because of the noncompliant pericardium. Numbers are pressures in centimeters of seawater, which are expressed relative to ambient pressures. Large black arrows indicate the movements of the walls of contracting chambers; small arrows indicate movements of walls of relaxing chambers. The red arrows indicate the direction of blood flow. AAV, anterior aortic valve; PAV, posterior aortic valve; AVV, atrioventricular valve. [Adapted from Brand, 1972.]

Atrial contraction

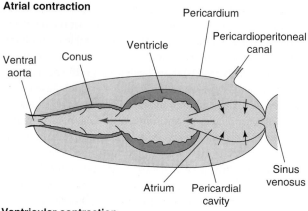

Ventricular contraction

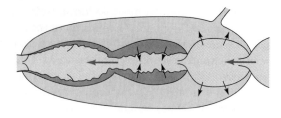

Conal contraction

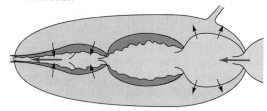

Figure 12-14 Because the elasmobranch heart is contained in a non-compliant pericardium, contractions of the ventricle reduce pressures in the pericardial cavity and assist atrial filling. In some elasmobranchs, fluid loss via the pericardioperitoneal canal during exercise, feeding, and coughing leads to an increase in heart size and stroke volume. Black arrows indicate direction of wall movement during muscle contraction or relaxation. Red arrows indicate direction of blood flow.

important in augmenting cardiac output. In some elasmobranchs, a *pericardioperitoneal canal* exists between the pericardial and peritoneal cavity. There is little or no fluid flow through this canal in resting fish, but during exercise, coughing, or feeding, the loss of fluid from the pericardial cavity via the canal causes an increase in heart size and stroke volume. This fluid is slowly replaced by plasma ultrafiltrate. Thus the thin, flexible pericardium of mammals, although protective, has little effect on cardiac output, whereas the more rigid pericardium of sharks, and the possible variations in pericardial fluid volume, can have a marked effect on cardiac output.

Vertebrate Hearts: Comparative Functional Morphology

The structure of the heart varies in different vertebrates, and a comparative analysis of vertebrate circulatory systems produces insights into the relationships between heart structure and function. Numerous cardiovascular differ-

ences distinguish air-breathing vertebrates from those that do not breathe air. Air-breathing vertebrates differ in the extent to which the systemic (body) and pulmonary (respiratory) circulations are separated.

The pulmonary circulation of birds and mammals is maintained at much lower pressures than is the systemic circulation. This is possible because they have two series of heart chambers in parallel. The left side of the heart ejects blood into the systemic circulation, and the right side ejects blood into the pulmonary circulation (see Figure 12-3). The advantage of a high blood pressure is that rapid transit times and sudden changes in flow can be readily achieved for blood passing through small-diameter capillaries. However, when the difference in pressure across a vessel wall (i.e., the transmural pressure) is high, fluid filters across the capillary wall; as a result, extensive lymphatic drainage of the tissues is necessary. In the mammalian lung, capillary flow can be maintained by relatively low input pressures, reducing the requirements for lymphatic drainage and avoiding the formation of large extracellular fluid spaces that could increase diffusion distances between blood and air and impair the gas transfer capacity of the lung. The advantage of a divided heart, like that of mammals, is that blood flow to the body and the lungs can be maintained by different input pressures. The disadvantage of a completely divided heart is that in order to avoid shifts in blood volume from the systemic to the pulmonary circuit, or vice versa, cardiac output must be the same in both sides of the heart, independent of the requirements in the two circuits.

In contrast, lungfishes, amphibians, reptiles, bird embryos, and fetal mammals have either an undivided ventricle or some other mechanism that allows the shunting of blood from one circulation to the other. These shunts usually result in the movement of blood from the right (respiratory, pulmonary) to the left (systemic) side of the heart during periods of reduced gas transfer in the lung. At such times, blood returning from the body, instead of being pumped to the lung, is shunted from the right to the left side of the heart and once again ejected into the systemic circuit, bypassing the lungs. In lungfishes, amphibians, and reptiles, flow to the lungs commonly is reduced during prolonged dives when gas transfer occurs across the skin and/or oxygen stores in the body are being used. Blood flow to the lungs is also reduced during development within the mother (mammals) or egg (birds), before the lungs become fully functional in gas exchange. Although a single undivided ventricle permits variations in the ratio of flows to the pulmonary and systemic circuits, the same pressures must be developed on both sides of the heart.

Water-breathing fishes

The heart of water-breathing fishes, including elasmobranchs and some bony fishes (teleosts), consists of four chambers in series. All chambers are contractile except the elastic bulbus of bony fish. A unidirectional flow of blood through the heart is maintained by valves at the sinoatrial and atrioventricular junctions and at the exit of the ventricle.

In elasmobranchs, the exit from the ventricle to the conus is guarded by a pair of flap valves, and there are from two to seven pairs of valves along the length of the conus depending on the species (see Figure 12-14). Conus length is variable among species; in general, more valves are found in those species with a longer conus. Just before ventricular contraction, all valves except the set most distal to the ventricle are open; that is, the conus and the ventricle are interconnected, but a closed valve at the exit of the conus maintains a pressure difference between the conus and the ventral aorta. During atrial contraction, both the ventricle and the conus are filled with blood. Ventricular contraction in elasmobranchs does not have an isovolumic phase, as in mammals, because at the onset of contraction blood is moved from the ventricle into the conus. Pressure rises in the ventricle and conus and eventually exceeds that in the ventral aorta. The distal valves open, and blood is ejected into the aorta. During conal contraction, which begins after the onset of ventricular contraction, the proximal valves close, preventing reflux of blood into the ventricle as it relaxes. Conal contraction proceeds relatively slowly away from the heart toward the aorta; each set of valves closes, in turn, to prevent backflow of blood.

As illustrated in Figure 12-15, blood pumped by the heart in typical water-breathing fish passes first through the gill (respiratory) circulation and then into a dorsal aorta that supplies the rest of the body (systemic circulation). Thus, unlike mammals, the respiratory and systemic circulations of fish are in *series* rather than in parallel, and the gill circulation is under higher pressures than the systemic circulation. The gills of fish are involved in ionic regulation as well as gas transfer and many of the functions of the mammalian kidney are located in the gills. The consequences of a high blood pressure in the fish gill on ionic and gas transfer is not clear.

Air-breathing fishes

Air-breathing has evolved in vertebrates many times, generally in response to hypoxic conditions, high water temperatures, or both. In general, air-breathing fish remain in water but rise to the surface to take in an air bubble to supplement oxygen supplies. Because the gill filaments usually collapse and stick together when exposed to air, they cannot be used for gas transfer in air. Hence, fish that have the ability to breathe air generally use structures other than the gills for this purpose, such as a portion of the gut or mouth, the swimbladder, or even the general skin surface.

Although the gills in air-breathing fish are not used for oxygen uptake from air, they are used for carbon dioxide excretion, as well as ionic and acid-base regulation. In many air-breathing fish, however, the gills are reduced in size, presumably to ameliorate oxygen loss from blood to water. The gills of the air-breathing teleost *Arapaima*, which is found in the Amazon River, are so small that only a fifth of oxygen uptake occurs across the gills even in water with normal oxygen levels. The bulk of oxygen uptake by this fish occurs via the swimbladder, which is highly vascularized and has many septa to increase surface area for exchange. In fact the gills of *Arapaima* are too small for the animal's oxygen requirements, and these fish die if denied access to air; that is, *Arapaima* is an obligate air-breathing fish.

Air-breathing fish have evolved a variety of blood shunts to permit changes in the distribution of blood to the

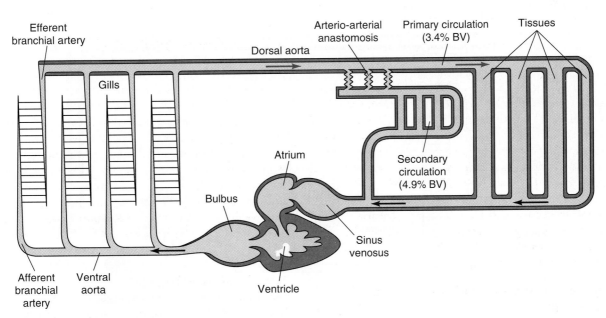

Figure 12-15 In a "typical" water-breathing teleost such as the trout, the respiratory circulation through the gills and the systemic circulation are in series. In the four-chamber, undivided heart, the pacemaker is in the sinus venosus. The ventricle ejects blood into the compliant bulbus and short ventral aorta. Blood flows through the gills into a stiff, long dorsal aorta. Most teleosts contain a low-hematocrit secondary circulation, which supplies nutrients but not much oxygen to the skin and gut. Black arrows indicate flow of deoxygenated blood; red arrows, flow of oxygenated blood. BV, body volume.

gills and the air-breathing organ. In the tropical freshwater teleost *Hoplerythrinus,* the posterior gill arches give rise to the coeliac artery, which perfuses the swimbladder and connects to the dorsal aorta by a narrow ductus. When the animal is breathing water, most of the cardiac output is directed to the first two gill arches and flows to the body. Following intake of air, the proportion of the blood flow to the posterior gill arches and therefore to the swimbladder increases, providing increased opportunity for oxygen uptake from the swimbladder.

 There are many more species of air-breathing fish in tropical than in temperate regions. Why?

The air-breathing fish *Channa argus* uses several mechanisms for achieving some separation of oxygenated and deoxygenated blood in the circulation. The most important mechanism is a division of the ventral aorta into two vessels, a posterior and an anterior ventral aorta. The anterior vessel supplies the first two gill arches and the air-breathing organ, whereas the posterior vessel supplies the posterior arches (Figure 12-16). The posterior arches are reduced in size, and the fourth arch is modified so that the afferent and efferent branchial arteries are in direct connection. Oxygenated blood is preferentially directed to the posterior arches, and the deoxygenated blood to the first two arches. This is achieved without division of the heart. The ventricle, however, is spongy (trabeculate), which may serve to prevent the mixing of blood in the ventricle, as has been suggested for the spongy heart of amphibians. In addition, the absence of sinoatrial valves in the *Channa* heart and the arrangement of the veins probably play an important role in preventing mixing of oxygenated and deoxygenated blood as these flows return to the heart in common vessels. Finally, muscular ridges on the wall of bulbus may prevent mixing of the oxygenated and deoxygenated flows when they are ejected from the heart. Once again this situation is similar to that seen in amphibians.

The division of the heart is more complete in the lungfishes (Dipnoi), which possess gills, lungs, and a pulmonary circulation. The African lungfish, *Protopterus,* has a partial septum in the atrium and ventricle and spiral folds in the bulbus cordis (Figure 12-17). This arrangement maintains the separation of oxygenated and deoxygenated blood in the heart. The anterior gill arches lack lamellae and oxygenated blood can flow from the left side of the heart directly to the tissues. Within the lamellae present in the posterior gill arches is a basal arterio-arterial connection that allows blood to bypass the lamellae when only the lung is in operation (e.g., during *estivation,* a state of torpor occurring in the summer). Blood from the posterior gill arches flows to the lungs or enters the dorsal aorta via a ductus. The ductus is richly innervated and is undoubtedly involved in controlling blood flow between the pulmonary artery and the systemic circulation. The initial segment of the pulmonary artery is muscular and is referred to as the *pulmonary vasomotor segment.* This vasomotor segment and the ductus probably act in a reciprocal fashion: when one constricts, the other dilates. The ductus in lungfish is analogous to the ductus arteriosus of fetal mammals, acting as a lung bypass when the lung is not functioning.

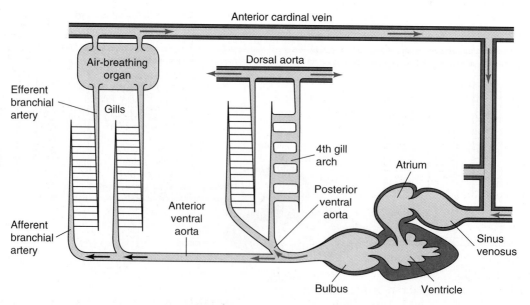

Figure 12-16 Even though the heart of the air-breathing teleost *Channa argus* is undivided, the flows of oxygenated and deoxygenated blood are partially separated. Deoxygenated blood (black arrows) preferentially flows through the first two gill arches and air-breathing organ, whereas oxygenated blood (red arrows) flows through posterior arches into the dorsal aorta. The fourth gill arch is modified so that the afferent and efferent branchial arteries are connected. Compare with Figure 12-15, which illustrates the circulation of more typical water-breathing teleosts. [Adapted from Ishimatzu and Itazawa, 1993.]

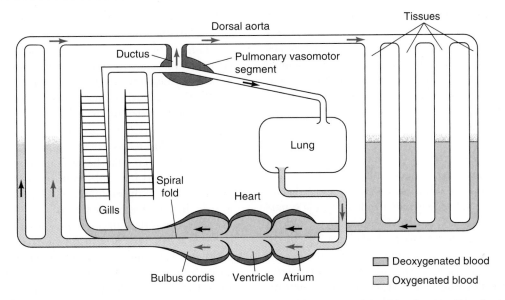

Figure 12-17 The circulation of the African lungfish, *Protopterus*, is marked by nearly complete separation of oxygenated blood (red arrows) and deoxygenated blood (black arrows). This separation is achieved by a septum dividing the atrial and ventricular chambers and a long spiral fold in the bulbus cordis. This fish possesses a lung and dis-tinct pulmonary circulation. The absence of lamellae in the anterior gill arches permits blood to flow directly to the systemic circulation via the dorsal aorta. The ductus and pulmonary vasomotor segment act recip-rocally to direct blood to the dorsal aorta or lungs depending on whether the fish is breathing in water or air. [Adapted from Randall, 1994.]

 When air-breathing vertebrates evolved, were the initial species moving out of water onto land large or small? Explain the reasons for your choice.

Amphibians

Amphibia have two completely separated atria, but a single ventricle. In the frog heart, the oxygenated and deoxygenated blood is separated even though the ventricle is undivided. Oxygenated blood from the lungs and skin is preferentially directed toward the body via the systemic arch, whereas deoxygenated blood from the body is directed toward the pulmocutaneous arch. This separation of oxygenated and deoxygenated blood is aided by a spiral fold within the *conus arteriosus* of the heart (Figure 12-18). Deoxygenated blood leaves the ventricle first during systole and enters the lung circulation. Pressures rise in the pulmocutaneous arch and become similar to those in the systemic arch. Blood then begins flowing into both arches, with the spiral fold partially dividing the systemic and pulmocutaneous flows within the conus arteriosus.

The volume of blood going to the lungs or body is inversely related to the resistance of the two circuits to flow. Immediately following a breath, the resistance to blood flow through the lung is low and blood flow is high; between breaths, resistance gradually increases and is associated with a fall in blood flow. These oscillations in pulmonary blood flow are possible because of the partial division of the amphibian heart. Although deoxygenated blood is directed toward the pulmocutaneous arch, the

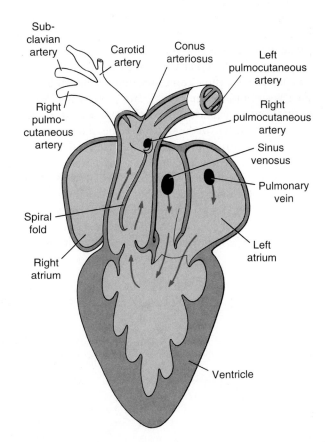

Figure 12-18 Even though the frog heart has a single ventricle, deoxygenated blood is directed to the lungs via the pulmocutaneous arch and oxygenated blood to the tissues via the systemic arch. This ventral view of the internal structure of the frog heart shows the position of the spiral fold, which aids in separating the two blood flows. [Adapted from Goodrich, 1958.]

ratio of pulmonary to systemic blood flow can be adjusted. That is, when the animal is not breathing, blood flow to the lungs can be reduced, so that most of the blood pumped by the ventricle is directed toward the body. When the animal is breathing, a more even distribution of flow to the lungs and body can be maintained. This distribution is possible only if the ventricle is not completely divided into right and left chambers (as it is in mammals).

Noncrocodilian reptiles
Most noncrocodilian reptiles, including turtles, snakes, and some lizards have a partially divided ventricle and right and left systemic arches. In these animals, the ventricle is partially subdivided by an incomplete muscular septum referred to as the *horizontal septum, Muskelleiste,* or *muscular ridge.* This horizontal septum separates the *cavum pulmonale* from the *cavum venosum* and *cavum arteriosum;* the latter two are partially separated by the *vertical septum* (Figure 12-19). The right atrium contracts slightly before the left atrium does and ejects deoxygenated blood into the cavum pulmonale across the free edge of the horizontal septum; ventricular contraction ejects this blood into the pulmonary artery. Oxygenated blood from the left atrium fills the cavum venosum and cavum arteriosum; from here the blood empties into the systemic arteries.

Measurements in turtles support the view that oxygenated blood from the left atrium passes into the systemic circuit, whereas deoxygenated blood from the right atrium passes into the pulmonary artery. Pulmonary artery diastolic pressure is often lower than systemic diastolic pressure; as a result, the pulmonary valves open first when the ventricle contracts. Thus, flow occurs earlier in the pulmonary artery than in the systemic arches during each cardiac cycle. In turtles, there may be some recirculation of arterial blood in the lung circuit; that is, there is a left-to-right shunt within the heart. The ventricle remains functionally undivided throughout the cardiac cycle, and the relative flow to the lungs and systemic circuits is determined by the resistance to flow in each part of the circulatory system. When the turtle breathes, resistance to flow through the

pulmonary circulation is low, and blood flow is high. When it does not breathe, as during a dive, pulmonary vascular resistance increases, but systemic vascular resistance decreases, resulting in a right-to-left shunt and a decrease in pulmonary blood flow. As in many other animals, there is a reduction in cardiac output associated with a marked slowing of the heart (**bradycardia**) during a dive.

The similarity in the pressures in the pulmonary and systemic outflow tracts in turtles, snakes, and some lizards indicates that their heart has a single ventricular chamber partially divided into subchambers even during systole (Figure 12-20A). In monitor lizards and related varanid lizards, however, the pulmonary outflows are at much lower pressures than the systemic outflows during systole (Figure 12-20B). The pressure in the cavum pulmonale, for instance, can be only a third of that in the cavum venosum during systole in *Varanus.* This pressure differential in veranid lizards is achieved by a pressure-tight contact between the muscular ridge (horizontal septum) and the wall of the heart during systole (Figure 12-21).

Crocodilian reptiles
Unlike other reptiles, crocodilian reptiles have a heart with a completely divided ventricle. The left systemic arch arises from the right ventricle; the right systemic arch, from the left ventricle. Close to the ventricles, the systemic arches are connected via the *foramen of Panizzae* (Figure 12-22A). The systemic arches also are joined by a short anastomosis caudal to the heart.

When a crocodilian reptile is breathing normally, the resistance to blood flow through the lungs is low, and pressures generated by the right ventricle are lower than those generated by the left ventricle during all phases of the cardiac cycle. In this case, blood is pumped by the left ventricle into the right systemic arch during systole, with the open aortic valve closing off the foramen of Panizzae (Figure 12-22B). There is a small reflux of blood into the left aorta from the right aorta via the anastomosis during systole. Because of this connection, pressures in the left systemic arch remain higher than the pressure in the right ven-

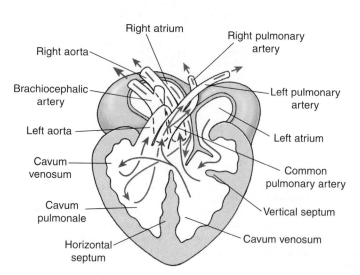

Right atrium
Right aorta
Right pulmonary artery
Brachiocephalic artery
Left pulmonary artery
Left aorta
Left atrium
Cavum venosum
Common pulmonary artery
Cavum pulmonale
Vertical septum
Horizontal septum
Cavum venosum

Figure 12-19 In noncrocodilian (chelonian) hearts, the ventricle is partially divided by the horizontal septum into the cavum venosum and ventral cavum pulmonale. The common pulmonary artery arises from the cavum pulmonale, whereas all of the systemic arteries arise from the cavum venosum. In this ventral view of the turtle heart, the arrows schematically indicate movement of oxygenated blood (red) and deoxygenated blood (black) but are not intended to represent the flow of separate bloodstreams through the heart. [Adapted from Shelton and Burggren, 1976.]

A

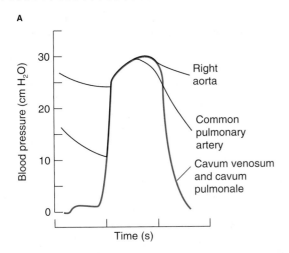

B

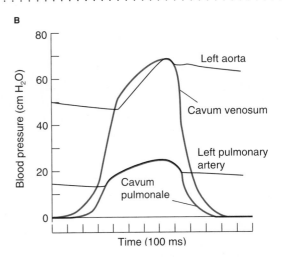

Figure 12-20 In turtles, the pressures in the systemic and pulmonary outflows are nearly identical during systole, whereas in veranid lizards they differ considerably. Shown are blood pressures measured simultaneously at the indicated sites during a single heartbeat in **(A)** a turtle, *Chrysemys scripta*, and **(B)** a monitor lizard, *Varanus exanthematicus*. [Part A from Shelton and Burggren, 1976; part B from Burggren and Johansen, 1982.]

tricle; consequently, the valves at the base of the left systemic arch remain closed throughout the cardiac cycle (Figure 12-22C). All blood ejected from the right ventricle passes into the pulmonary artery and flows to the lungs. Thus, the crocodilian reptile is functionally the same as the mammal in that there is complete separation of systemic and pulmonary blood flow.

Crocodilian reptiles, however, have the added capacity to shunt blood from the pulmonary to the systemic circuit. This P ⟶ S shunt is achieved by active closure of a valve at the base of the pulmonary outflow tract towards the end

of systole. Under some experimental circumstances peak right ventricular pressure becomes equal to left ventricular pressure and exceeds left systemic pressure. As a result, the valves at the base of the left systemic arch open, and blood from the right ventricle is ejected into the systemic circulation during late systole (Figure 12-22D,E). In this case, a portion of the deoxygenated blood returning to the heart from the body via the right atrium is recirculated in the systemic circuit. Exactly when the P ⟶ S shunt operates normally in the animal is not clear. The role of the foramen of Panizzae also remains enigmatic; it is only open during

A   **Diastole**

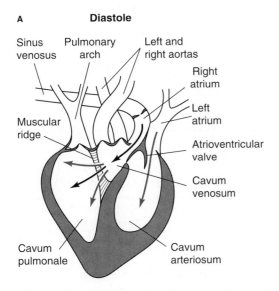

B   **Systole**

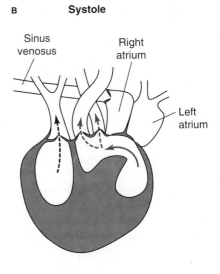

Figure 12-21 In veranid lizards, a pressure-tight separation between the cavum pulmonale and cavum venosum occurs during systole. **(A)** During diastole, the muscular ridge only partially separates the cavum venosum and cavum pulmonale. Thus, oxygenated blood (red arrows) remaining in the cavum venosum from the preceding systole is washed into the cavum pulmonale by deoxygenated blood (black arrows). The cavum arteriosum is filled with oxygenated blood. Separation between the cavum arteriosum and the cavum venosum is provided by at least one atrioventricular valve. **(B)** During systole, the muscular ridge is pressed tight against the outer heart wall, forming a pressure-tight barrier. Deoxygenated blood remaining in the cavum venosum from the preceding diastole is mixed with oxygenated blood from the cavum arteriosum and flushed into the aortic arches. Deoxygenated blood with an admixture of oxygenated blood is expelled from the cavum pulmonale into the pulmonary arch. With no connection between the cavum venosum and cavum pulmonale, different pressures can develop in the outflow tracts. [Adapted from Heisler et al., 1983.]

A Diastole

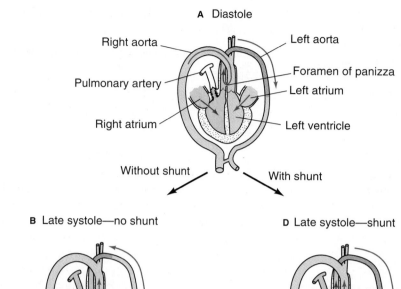

Figure 12-22 Under some conditions, a P ⟶ S shunt operates during late systole in crocodiles. These schematic diagrams and pressure and flow tracings illustrate what happens during the cardiac cycle with and without the shunt. See text for discussion. [Adapted from Jones, 1995.]

B Late systole—no shunt

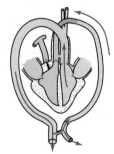

D Late systole—shunt

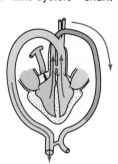

C No P→ S shunt

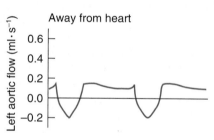

E P→ S shunt

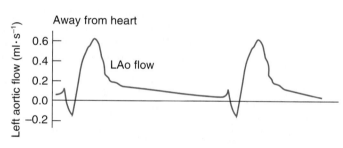

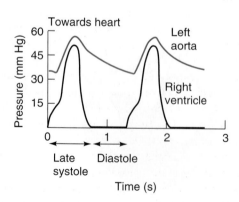

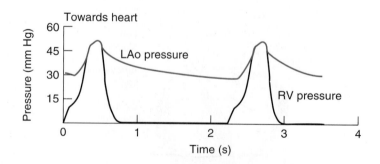

diastole, allowing flow between the aortic arches as the heart relaxes.

Mammals and birds

The heart in both mammals and birds, which consists of four chambers, is in fact two hearts beating as one. The heart originates as two separate tubes that join together dur-

ing development to form the multi-chambered heart of the postnatal animal. The right side pumps blood to the lung, the left side pumps blood around the body. Blood returning from the lungs enters the left atrium, passes into the left ventricle, and is then ejected into the body circulation. Blood from the body collects in the right atrium, passes into the right ventricle, and is pumped to the lungs (see Figure 12-3).

Valves prevent backflow of blood from the aorta to the ventricle, the atrium, and the veins. These valves are passive and are opened and closed by pressure differences between the heart chambers. The atrioventricular valves (bicuspid and tricuspid valves) are connected to the ventricular wall by fibrous strands (see Figure 12-4). These strands prevent the valves from being everted into the atria when the ventricles contract and intraventricular pressures are much higher than those in the atria. The walls of the ventricle, especially the left chamber, are thick and muscular. The inner surface of the ventricular muscle, or myocardium, is lined by an endothelial membrane, the endocardium. The ventricular myocardium is covered by the epicardium.

Mammalian fetus At birth, mammals shift from a placental to a pulmonary circulation, a process that involves several central cardiovascular readjustments. The lungs of the mammalian fetus are collapsed, presenting a high resistance to blood flow. In the fetus, the pulmonary artery is joined to the systemic arch via a short, but large-diameter, blood vessel, the **ductus arteriosus** (Figure 12-23). Heart function in the mammalian fetus exhibits three important features:

- Most of the blood ejected by the right ventricle is returned to the systemic circuit via the ductus arteriosus.

- Blood flow through the pulmonary circulation is greatly reduced.

- A marked right-to-left (P $\longrightarrow$ S) shunt operates; that is, blood flows from the pulmonary to the systemic circuit.

At birth, the lungs are inflated, reducing the resistance to flow in the pulmonary circuit. Blood ejected from the right ventricle passes into the pulmonary vessels, resulting in an increased venous return to the left side of the heart. At the same time, the placental circulation disappears, and the resistance to flow increases markedly in the systemic circuit. Pressures in the systemic circuit rise above those in the pulmonary circulation; if the ductus arteriosus fails to close after birth, this pressure difference results in a left-to-right (S $\longrightarrow$ P) shunt with blood flowing from the systemic to the pulmonary circuit. Generally, however, the ductus arteriosus becomes occluded, and blood flow through the ductus does not persist.

If the ductus arteriosus remains open after birth, blood flow to the lungs exceeds systemic flow, because a portion of the left ventricular output passes via the ductus into the pulmonary artery and to the lung. In these circumstances, systemic flow is often normal, but pulmonary flow may be twice the systemic flow, and cardiac output from the left ventricle may be twice that from the right ventricle. The result is a marked hypertrophy of the left ventricle. The work done by the left ventricle during exercise is also much greater than normal, and the capacity to increase output is limited. As a result, the maximum level of exercise is much reduced if the ductus arteriosus remains open after birth. Furthermore, this condition increases the blood pressure in the lungs, leading to a greater fluid loss across lung capillary walls and to possible pulmonary congestion. These problems only become detrimental when the left ventricle has become enlarged. An open ductus arteriosus is readily and easily correctable by surgery.

Fetal blood is oxygenated in the placenta and mixed with the blood returning from the lower body via the inferior vena cava, a vein that in turn empties into the right atrium (see Figure 12-23). In the interatrial septum is a hole, the *foramen ovale*, that is covered by a flap valve; oxygenated blood returning via the inferior vena cava is directed into the left atrium through the foramen ovale. Oxygenated blood is then pumped from the left atrium into the left ventricle and ejected into the aorta, whence it flows to the head and upper limbs. Deoxygenated blood returning to the right atrium via the superior vena cava is preferentially directed toward the right ventricle, whence it flows into the systemic circuit via the ductus arteriosus. At birth, pressures in the left atrium exceed the pressure in the right atrium; as a result, the foramen ovale closes, but its position is later indicated by a permanent depression.

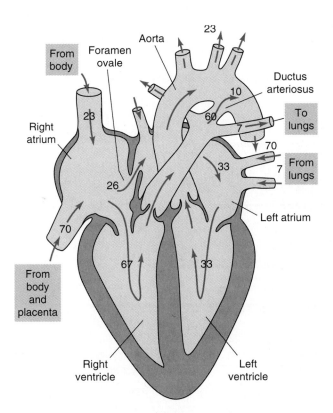

Figure 12-23 In the mammalian fetal heart, most of the blood ejected from the right ventricle returns to the systemic circulation via the ductus arteriosus. Oxygenated blood returning from the placenta is shunted from the right to the left atrium through the foramen ovale and then pumped into the aorta. After birth, the ductus arteriosus normally closes, so systemic and pulmonary circulations become separated. The numbers refer to the percentage of the combined cardiac output from the right and left ventricles that flows to and from different regions of the body.

Bird embryo A network of blood vessels, forming the *chorioallantois*, lies just under the shell of bird eggs. Oxygen diffusing across the eggshell is taken up by blood passing through the chorioallantois. The oxygenated blood leaving the chorioallantois and deoxygenated blood from the head and body enter the right atrium of the bird embryo heart. Oxygenated blood from the chorioallantoic circulation passes from the right into the left atrium through several large and numerous small holes in the interatrial septum. The oxygenated blood is then pumped into the left ventricle and ejected into the aorta, whence it flows to the head and body. After the young bird hatches, the holes in the interatrial septum close, completely separating the pulmonary and systemic circulations.

HEMODYNAMICS

As we have noted, contractions of the heart generate blood flow through the vessels—arteries, capillaries, and veins—that form the circulatory system. Before examining the properties of these vessels in detail, it is necessary to discuss the general patterns of blood flow in these vessels and the relationship between pressure and flow in the circulatory system. The laws describing the relationships between pressure and flow apply to both open and closed circulatory systems.

In vertebrates and other animals with a closed circulation, the blood flows in a continuous circuit. Since fluids are incompressible, blood pumped by the heart must cause flow of an equivalent volume in every other part of the circulation. That is, at any one time the same number of liters per minute flows through the arteries, the capillaries, and the veins. Furthermore, unless there is a change in total blood volume, a reduction in volume in one part of the circulation must lead to an increase in volume in another part.

The velocity of flow at any point is related not to the proximity of the heart but to the *total cross-sectional area* of that part of the circulation—that is, to the sum of the cross sections of all capillaries or arteries at that point in the circulation. Just as the velocity of water flow increases where a river narrows, so in the circulation the highest velocities of blood flow occur where the total cross-sectional area is smallest (and the lowest velocities occur where the cross-sectional area is largest). The arteries have the smallest total cross-sectional area, whereas the capillaries have by far the largest. Thus, the highest velocities occur in the aorta and pulmonary artery in mammals; then velocity falls markedly as blood flows through the capillaries, but it rises again as blood flows through the veins (Figure 12-24). The slow flow of blood in capillaries is of functional significance, because it is in capillaries that the time-consuming exchange of substances between blood and tissues takes place.

Laminar and Turbulent Flow

In many smaller vessels of the circulation, blood flow is streamlined. Such continuous **laminar flow** is characterized by a parabolic velocity profile across the vessel (Figure 12-25A). Flow is zero at the wall and maximal at

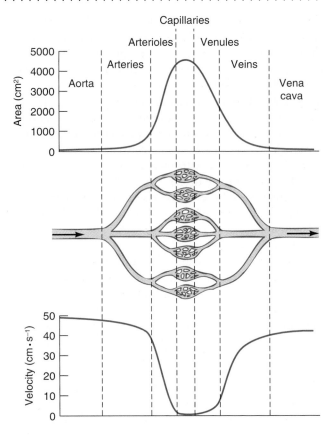

Figure 12-24 Blood velocity is inversely proportional to the cross-sectional area of the circulation at any given point. Blood velocity is highest in the arteries and veins and lowest in the capillaries; the converse is true for the cross-sectional area. [Adapted from Feigl, 1974.]

A Continuous laminar flow

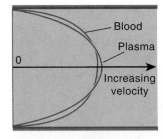

B Pulsatile laminar flow

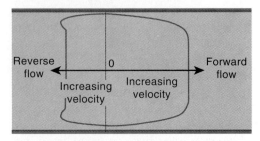

Figure 12-25 Blood flow through smaller vessels approximates continuous laminar flow, but in large elastic arteries pulsatile laminar flow is observed. As shown in these velocity profiles, the flow rate is higher towards the center of the vessel. **(A)** The presence of red blood cells flattens the profile of blood compared with that of plasma. **(B)** Pulsatile flow is marked by a flat profile and reversal of flow during each heartbeat.

the center along the axis of the vessel. A thin layer of blood adjacent to the vessel wall does not move, but the next layer of fluid slides over this layer, and so on, each successive layer moving at an increasingly higher velocity, with the maximum at the center of the vessel. A pressure difference supplies the force required to slide adjacent layers past each other, and viscosity is a measure of the resistance to sliding between adjacent layers of fluid. An increase in viscosity will require a larger pressure difference to maintain the same rate of flow, as discussed later.

The pulsatile laminar flow characteristic of large arteries has a more complex velocity profile than the continuous laminar flow characteristic of smaller vessels. In large arteries, blood is first accelerated and then slowed with each heartbeat; in addition, since the vessel walls are elastic, they expand and then relax as pressure oscillates with each heartbeat. Close to the heart, the direction of flow reverses each time the aortic valves shut. The end result is that the velocity across large arteries has a much flatter profile than the velocity across more peripheral blood vessels and the direction of flow oscillates (Figure 12-25B).

In **turbulent flow** fluid moves in directions not aligned with the axis of the flow, thus increasing the energy needed to move fluid through a vessel. Laminar flow is silent; turbulent flow noisy. In the bloodstream, turbulence causes vibrations that produce the sounds of the circulation. Detection of these sounds with a stethoscope can localize the points of turbulence. Blood pressure measurement with a sphygmomanometer depends on hearing the sounds associated with blood escaping past the pressure cuff as blood pressure rises during systole. Sounds can be heard in vessels when blood velocity exceeds a certain critical value and in heart valves when they open and close.

Although turbulent flow is uncommon in the peripheral circulation, it does occur in some situations. The **Reynolds number** *(Re)* is an empirically derived value that indicates whether flow will be laminar or turbulent under a particular set of conditions. A high Reynolds number indicates flow will be turbulent, whereas a low number indicates flow will be laminar. The *Re* is directly proportional to the flow rate, $\dot{Q}$ (in milliliters per second), and density, ρ, of the blood, and inversely proportional to the inside radius of the vessel, r (in centimeters), and viscosity, η, of the blood:

$$Re = \frac{2\dot{Q}\rho}{\pi r \eta} \qquad (12\text{-}2)$$

The ratio of viscosity to density (η/ρ) is the kinematic viscosity. The larger the kinematic viscosity, the less the likelihood that turbulence will occur. The relative viscosity, and therefore the kinematic viscosity, increases with **hematocrit** (volume of red blood cells per unit volume of blood), so that the presence of red blood cells decreases the occurrence of turbulence in the bloodstream.

In general, blood velocity is seldom high enough to create turbulence in undivided vessels with smooth walls, except during the very high blood flows associated with strenuous exercise. The highest flow rates in the mammalian circulation are in the proximal portions of the aorta and pulmonary artery, and turbulence may occur distal to the aortic and pulmonary valves at the peak of ventricular ejection or during backflow of blood as these valves close. In general, flow will be turbulent in portions of the circulation where vessel walls are smooth and the vessels are undivided only if *Re* is greater than about 1000, a value seldom observed. Small back eddies may form at arterial branches and, like the back eddies in rivers, can become detached from the main flow regime, being carried downstream as small discrete regions of turbulence. These eddies can form in the circulation when the *Re* is as low as 200.

Relationship between Pressure and Flow

Flow will occur between two sites where there is a difference in potential energy, which can be measured as a difference in pressure. Thus differences in pressure between two points in a flow path establish a pressure gradient and therefore the direction of flow—from high to low pressure. (An exception is a fluid at rest under gravity, where pressure increases uniformly with depth but flow does not occur.) When the heart contracts, the potential energy (pressure) in the ventricle increases. Pressures generated by heart contractions are dissipated by the flow of blood, because energy is used to overcome the resistance to flow through the vessels. For this reason, blood pressure falls as the blood passes from the arterial to the venous side of the circulation (Figure 12-26).

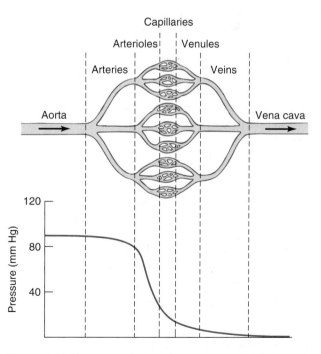

Figure 12-26 The pressure (potential energy) generated during each cardiac contraction is dissipated in overcoming the resistance to flow provided by the vessels. Because resistance is highest in the arterioles, the major pressure decrease occurs in this region of the circulation. [Adapted from Freigl, 1974.]

Role of kinetic energy

Energy is expended in setting the blood into motion, but once in motion flowing blood has inertia; that is, fluids in motion possess kinetic energy. In static fluids, potential energy is measured in terms of pressure; in fluids in motion, potential energy is measured in terms of both pressure and kinetic energy. As we'll see, however, kinetic energy generally makes a negligible contribution to the flow rate of blood. The kinetic energy per milliliter of fluid is given by $\frac{1}{2}(\rho v^2)$, where ρ is the density of the fluid and v is the velocity of flow. If the velocity is measured in centimeters per second and density in grams per millimeter, then kinetic energy has the units of dynes per square centimeter, the same as pressure.

The maximum velocity of blood flow occurs at the base of the aorta in mammals and is about 50 cm·s^{-1} at the peak of ventricular ejection, and the density of blood is about 1.055 g·ml^{-1}. Thus the kinetic energy of the blood in the aorta during peak ejection is calculated as $\frac{1}{2} \times 1.055 \times 50^2$, or 1 mm Hg. This is small compared with peak systolic transmural pressures of around 120 mm Hg. Blood velocity is low in the ventricle but accelerates as blood is ejected into the aorta; that is, blood gains kinetic energy as it leaves the ventricle. Pressure is converted into kinetic energy as blood is ejected from the heart, and this conversion accounts for most of the small drop in pressure that occurs between the ventricle and the aorta. Kinetic energy is highest in the aorta. In the capillaries the velocity is about 1 mm·s^{-1} and kinetic energy, therefore, is negligible.

Poiseuille's law

The relationship between pressure and continuous laminar flow of fluid in a rigid tube is described by **Poiseuille's law,** which states that the flow rate of a fluid, $\dot{Q}$, is directly proportional to the pressure difference, $P_1 - P_2$, along the length of the tube and the fourth power of the radius of the tube, r, and inversely proportional to the tube length, L, and fluid viscosity, η:

$$\dot{Q} = \frac{(P_1 - P_2)\pi r^4}{8L\eta} \qquad (12\text{-}3)$$

As $\dot{Q}$ is proportional to r^4, very small changes in r will have a profound effect on $\dot{Q}$. A doubling of vessel diameter, for instance, will lead to a 16-fold increase in flow if the pressure difference ($P_1 - P_2$) along the vessel remains unchanged.

Although Poiseuille's equation applies to steady flows in straight rigid tubes, it has been used, with some limitations discussed later, to analyze the relationship between pressure and flow in small arteries (arterioles), capillaries, and veins, even though these are not "rigid" tubes. Blood pressure and flow are pulsatile, and blood is a complex fluid consisting of plasma and cells. Since the blood vessel walls are not rigid, the oscillations in the pressure and flow of blood are not in phase; consequently, the relationship between the two is no longer accurately described by Poiseuille's law.

The extent of the deviation of the relationship between pressure and flow from that predicted by Poiseuille's law is indicated by the value of a nondimensional constant α:

$$\alpha = r\frac{\sqrt{2\pi nf\rho}}{\eta} \qquad (12\text{-}4)$$

where ρ and η are the density and viscosity of the fluid, respectively; f is the frequency of oscillation; n is the order of the harmonic component, and r is the radius of the vessel. If α is 0.5 or less, the relationship between pressure and flow is described by Poiseuille's equation. Because the value for α in the small terminal arteries and veins is about 0.5, this equation can be used to analyze the relationship between pressure and flow in this portion of the circulation. In contrast, the values of α for the arterial systems of mammals and birds range from 1.3 to 16.7, depending on the species and the physiological state of the animal. Thus, Poiseuille's law is not applicable to this portion of the circulation.

There have been only a few studies of the *in vivo* microcirculation due to the difficulty of measuring blood flow and pressure in capillaries. In those tissues where the relationship between pressure and flow in the microcirculation has been measured, it has been found to be nonlinear, indicating that Poiseuille's equation does not accurately describe the microcirculation. There are two reasons for this: the capillaries branch with collateral pathways that may open and close, and they are so small that red blood cells are deformed as they squeeze through the capillaries.

Resistance to flow

Because it is often difficult or impossible to measure the radii of all vessels in a vascular bed, we designate $8L\eta/\pi r^4$, the inverse of the term in Poiseuille's law (equation 12-3), as the resistance to flow, R, which is equal to the pressure difference ($P_1 - P_2$) across a vascular bed divided by the flow rate, $\dot{Q}$:

$$R = \frac{P_1 - P_2}{\dot{Q}} = \frac{8L\eta}{\pi r^4} \qquad (12\text{-}5)$$

The resistance to flow in the peripheral circulation is sometimes expressed in **peripheral resistance units** (PRUs), with 1 PRU being equal to the resistance in a vascular bed when a pressure difference of 1 mm Hg results in a flow of 1 ml·s^{-1}.

Blood flow through a vessel increases with increased pressure difference along a vessel and decreased resistance to flow, which is inversely proportional to the fourth power of the radius of the vessel. As pressure increases in an elastic vessel, so does the radius; as a result, flow increases as well. Let us consider a blood vessel with a constant pressure differential along its length but operating at two pressure levels:

Example 1: input pressure 100 mm Hg and outflow pressure 90 mm Hg; Δ = 10 mm Hg

Example 2: input pressure 20 mm Hg and outflow pressure 10 mm Hg; Δ = 10 mm Hg

The flow rate in this vessel will be much greater at the higher pressure (example 1) if the vessel is distensible, simply because the radius will be increased and the resistance to flow reduced.

Viscosity of blood

According to Poiseuille's law, the flow of blood is inversely related to its viscosity. Plasma has a viscosity relative to water of about 1.8; the addition of red blood cells increases the relative viscosity, so that mammalian and bird blood at 37°C have a relative viscosity between 3 and 4. Thus, owing largely to the presence of red blood cells, blood behaves as though it were three or four times more viscous than water. This characteristic means that larger pressure gradients are required to maintain the flow of blood through a vascular bed than would be needed if the vascular bed were perfused by plasma alone. However, blood flowing through small vessels behaves as if its relative viscosity were much reduced. In fact, in vessels less than 0.3 mm in diameter, the relative viscosity of blood decreases with diameter and approaches the viscosity of plasma. This phenomenon, called the **Fahraeus-Lindqvist effect,** is explained later.

As we saw earlier, the velocity profile across a vessel with continuous laminar fluid flow is parabolic, as is seen with plasma (see Figure 12-25A). The maximum velocity is twice the mean velocity, which can be determined by dividing the flow rate by the cross-sectional area of the tube. The rate of change in velocity is maximal near the walls and decreases toward the center of the vessel. In flowing blood, red cells tend to accumulate in the center of the vessel, where the velocity is highest but the rate of change in velocity between adjacent layers smallest. This accumulation leaves the walls relatively free of cells, so that fluid flowing from this area into small side vessels will have a low level of red blood cells and consist almost entirely of plasma. Such a process is referred to as **plasma skimming.**

The accumulation of red blood cells in the center of a bloodstream means that blood viscosity is highest in the center and decreases toward the walls. This differential in viscosity between the center and the walls of the bloodstream will alter the velocity profile of blood compared with that of plasma. The effect of this viscosity difference is a slight increase in blood flow at the walls and a slight reduction in flow at the center; that is, the parabolic shape of the velocity profile is flattened somewhat (see Figure 12-25A).

The hematocrit (percent volume of rbc's in blood) in small vessels is smaller than that in larger ones. In small vessels the boundary layer of plasma occupies a larger portion of the vessel lumen at a given flow than in larger vessels. This axial flow of red blood cells in small vessels means that the greatest change in velocity occurs in the plasma layers close to the walls and explains why the apparent viscosity of blood flowing in these small vessels approaches that of plasma. Thus the Fahraeus-Lindqvist effect can be explained in terms of the reduced hematocrit seen in small vessels. This decrease in the apparent viscosity of blood, which occurs in arterioles, reduces the energy required to drive blood through the microcirculation.

In very small vessels—those with a diameter of approximately 5 to 7 μm—further decreases in diameter lead to an inversion of the Fahraeus-Lindqvist effect, namely, an increase in the apparent viscosity of blood. In such vessels, the red blood cell completely fills the lumen and is distorted as it passes through. Because the red blood cell membrane is not firmly anchored to underlying structures, it can move over its own cell contents, acting somewhat like a tank tread as it moves along the walls of the vessel. Deformation of red blood cells in small vessels leads to a complex flow of erythrocyte membrane and surrounding fluid as the cells squeeze through the narrow lumen.

If flow is laminar but pulsatile, as in arteries, the velocity profile is flattened even more than with continuous laminar flow (see Figure 12-25B). Thus blood velocity is constant across much of the vessel and drops sharply near the walls. In turbulent flow, blood moves in various directions in relation to the axis of flow, so there is little accumulation of red blood cells in the center of the vessel. As a result, the blood viscosity and velocity of flow changes little across the vessel.

 Antarctic teleost fish operate at temperatures close to or even below 0°C. What effects might this have on pressure and flow in the circulation of these fish? What modifications may have evolved to compensate for these low temperatures?

Compliance in the circulatory system

A further consideration in analyzing the relationship between pressure and flow in the circulation is that blood vessels contain elastic fibers that enable them to distend. Vessels are not, in fact, the straight, rigid tubes to which Poiseuille's law applies. Rather, as pressure in a vessel increases, the walls are stretched and the volume of the vessel increases. The ratio of change in volume to change in pressure is termed the **compliance** of the system. The compliance of a system is related to its size and the elasticity of its walls. The greater the initial volume and elasticity of the walls, the greater will be compliance of the system.

The venous system is very compliant; that is, small changes in pressure produce large changes in volume. For this reason, the venous system can act as a *volume reservoir,* because large changes in volume have little effect on venous pressure (and therefore on the filling of the heart during diastole or capillary blood flow). The arterial system, which overall is less compliant than the venous system, acts as a *pressure reservoir* in order to maintain capillary blood flow. Nevertheless, the portions of the arterial system near the heart are elastic in order both to dampen the oscillations in pressure generated by contractions of the heart and to maintain flow in distal arteries during diastole.

In summary, a large number of factors affect the relationship between pressure and flow in the circulation. Velocity of flow depends on the total cross-sectional area of the circulation; it is highest in the arteries and veins and lowest in the capillaries because the sum of the cross-sectional areas of all the capillaries is higher than that of the arteries or veins (see Figure 12-24). Contractions of the heart generate pressure and flow. The highest pressures in the circulation are found in the ventricles and vessels leading from the heart. Pressures are dissipated as energy is lost overcoming the resistance to flow in the vessels. Changes in kinetic energy are reflected in only very small changes in blood pressure as the blood changes velocity. There are only small pressure drops through the arterial and venous systems, even though blood flow is high, because the vessels are large and resistance to flow is small. The largest pressure drop is seen across the arterioles because at this point in the circulation the flow is high and the vessels are small and have a high resistance (see Figure 12-26). The pattern of the flow of blood through this high-resistance pathway reduces the apparent viscosity of the blood (Fahraeus-Lindqvist effect) and therefore the resistance to flow; even so the largest drop in blood pressure occurs in the arterioles. The capillaries are even smaller than arterioles, but flow is much lower in each capillary; therefore, the pressure drop across the capillaries is much smaller than that across the arterioles.

THE PERIPHERAL CIRCULATION

Blood pumped from the left ventricle of the mammalian heart carries oxygenated blood via the arterial system to capillary beds in the tissues, where the oxygen is exchanged for carbon dioxide. The venous system returns the deoxygenated blood to the right atrium (see Figure 12-3). Although all blood vessels share some structural features, the vessels in various parts of the peripheral circulation are adapted for the functions they serve.

Figure 12-27 illustrates the structure of various sized arteries and veins. A layer of endothelial cells, called the **endothelium**, lines the lumen of all blood vessels. In larger vessels, the endothelium is surrounded by a layer of elastic and collagenous fibers, but the walls of capillaries consist of a single layer of endothelial cells. Circular and longitudinal smooth muscle fibers may intermingle with or surround the elastic and collagenous fibers. The walls of larger blood vessels comprise three layers:

- **Tunica adventitia:** the limiting fibrous outer coat
- **Tunica media:** middle layer consisting of circular and longitudinal muscle

- **Tunica intima:** inner layer, closest to the lumen, composed of endothelial cells and elastic fibers

The boundary between the tunica intima and the tunica media is not well defined; the tissues blend into one another. Owing to increased muscularization, arteries have a thickened tunica media, and the larger arteries close to the heart are more elastic, with a wide tunica intima. The thick walls of larger blood vessels require their own capillary circulation, termed the **vasa vasorum.** In general, arteries have thicker walls and much more smooth muscle than veins of similar outside diameter. In some veins, muscular tissue is absent.

Arterial System

The arterial system consists of a series of branching vessels with walls that are thick, elastic, and muscular—well suited to deliver blood from the heart to the fine capillaries that carry blood through the tissues. Arteries serve four main functions, as illustrated in Figure 12-28:

1. Act as a conduit for blood between the heart and capillaries
2. Act as a pressure reservoir for forcing blood into the small-diameter arterioles
3. Dampen oscillations in pressure and flow generated by the heart and produce a more even flow of blood into the capillaries
4. Control distribution of blood to different capillary networks via selective constriction of the terminal branches of the arterial tree

Arterial blood pressure, which is finely controlled, is determined by the volume of blood the arterial system contains and the properties of its walls. If either is altered, the pressure will change. The volume of blood in the arteries is determined by the rate of filling via cardiac contractions and of emptying via arterioles into capillaries. If cardiac output increases, arterial blood pressure will rise; if capillary flow increases, arterial blood pressure will fall. Normally, however, arterial blood pressure varies little, because the rates of filling and emptying are evenly matched (i.e., cardiac output and capillary flow are evenly matched).

Blood flow through the capillaries is proportional to the pressure difference between the arterial and venous systems. Because venous pressure is low and changes little, arterial pressure exerts primary control over the rate of capillary blood flow and is responsible for maintaining adequate perfusion of the tissues. Arterial pressure varies among species, generally ranging from 50 to 150 mm Hg. Pressure differences are small along large arteries (less than 1 mm Hg), but pressure drops considerably along small arteries and arterioles because of increasing resistance to flow with decreasing vessel diameters.

The oscillations in blood pressure and flow generated by contractions of the heart are dampened in the arterial system, because of the elasticity of arterial walls. As blood

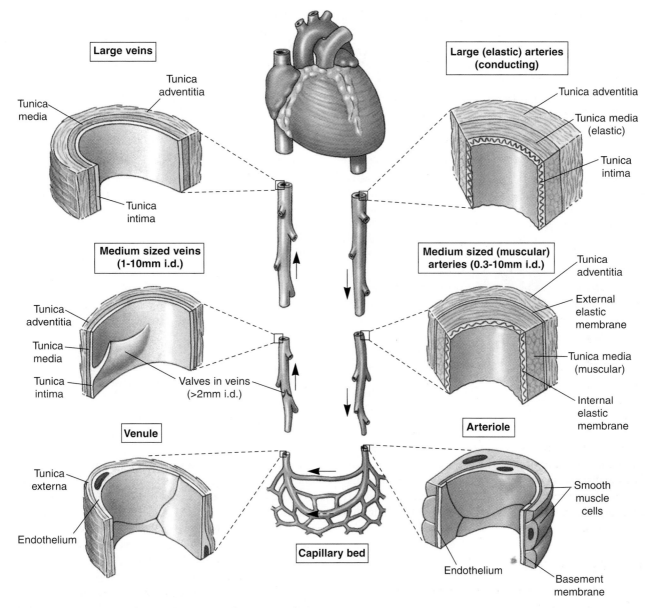

Figure 12-27 In the mammalian peripheral circulation, blood flows from the heart via progressively smaller arteries, then through the microcirculation, and finally back to the heart via progressively larger veins. A layer of endothelial cells, the endothelium, lines the lumen of all vessels. In larger vessels, the endothelium is surrounded by a muscle layer (tunica media) and outer fibrous layer (tunica adventitia). [Adapted from Martini and Timmons, 1995.]

is ejected into the arterial system, pressure rises and the vessels expand. As the heart relaxes, blood flow to the periphery is maintained by the elastic recoil of the vessel walls, resulting in a reduction in arterial volume (see Figure 12-28). If arteries were simply rigid tubes, pressures and flow in the periphery would exhibit the same stops and starts that occur at the exit of the ventricle during each heartbeat. Although elastic, arteries become progressively stiffer with increasing distension. As a result, they are easily distended at low pressures, but then resist further expansion at high pressures. The response of arterial walls to distension is similar in a wide variety of animals, reflecting similar structural and functional characteristics (Figure 12-29).

According to Laplace's law, the wall tension required to maintain a given transmural pressure within a hollow structure increases with increasing radius (see equation 12-1). Elastic vessels thus are unstable and tend to balloon; that is, since they cannot develop high wall tension as pressure increases, they tend to bulge out. In blood vessels, this instability is prevented by a collagen sheath that limits their expansion. Ballooning of a blood vessel (**aneurism**) can occur, however, if the collagen sheath breaks down.

In general, the elasticity of the arterial wall, as well as the thickness of the muscular layer, decreases with increasing distance from the heart. That is, farther from the heart, the arteries become more rigid and serve primarily as blood conduits. For example, the aorta of a dog becomes

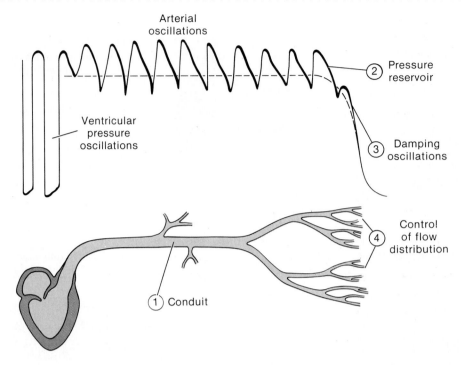

Figure 12-28 The systemic arterial system functions as a conduit and pressure reservoir; it also smoothes out pressure oscillations and controls distribution of flow to capillaries. The conduit function (1) is served by the vascular channels along which blood flows toward the periphery with minimal frictional loss of pressure. The distensible walls and high outflow re- sistance of arteries account for the pressure-reservoir function (2) and damping of oscillations in pressure and flow (3). Controlled hydraulic re- sistance in the peripheral vascular beds controls the distribution of blood to the various tissues (4). [Adapted from Rushmer, 1965a.]

progressively stiffer and its diameter decreases with in- creasing distance from the heart (Figure 12-30). In a whale, the aortic arch at the exit from the heart is very elastic and has a large diameter, but the arterial system beyond the aor-

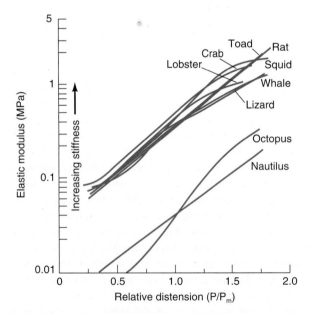

Figure 12-29 The elastic properties of arteries are surprisingly similar in a wide variety of animals, with nautilus and octopus being notable ex- ceptions. This similarity is reflected in plots of elastic modulus versus rel- ative distension, expressed as pressure (P) divided by the resting blood pressure (P_m) of the species. [Adapted from Shadwick, 1992.]

tic arch narrows rapidly and becomes much more rigid than that of a dog. The elastic whale aortic arch expands with each heartbeat, accommodating about 50%–75% of the stroke volume; the remainder flows into the portion of the arterial system downstream of the aortic arch. The change in ventricular volume with each heartbeat can be as much as 35 liters in a large whale, with a heart rate of around 12–18 per minute.

The extent of elastic tissue in arteries varies depending on the particular function of each vessel. In fishes, for ex- ample, blood pumped by the heart is forced into an elastic bulbus and a ventral aorta (see Figure 12-15). The blood then flows through the gills and passes into a dorsal aorta, the main conduit for the distribution of blood to the rest of the body. A smooth, continuous flow of blood is required in the gill capillaries for efficient gas transfer. The bulbus, the ventral aorta, and the afferent branchial arteries lead- ing to the gills are very compliant and act to smooth and maintain flow in the gills in the face of large oscillations produced by contractions of the heart. The dorsal aorta, which receives blood from the gills, is much less elastic than the ventral aorta. If the dorsal aorta were more elastic than the ventral aorta, there would be a rapid rush of blood through the gills during each heartbeat. This rush would in- crease rather than decrease the oscillations in flow through the gills. In this example, then, to ensure a steady blood flow through the gill capillaries, the major compliance must be placed before, not after, the gills to dampen the oscil- lations in flow through the gills. The ventral aorta must

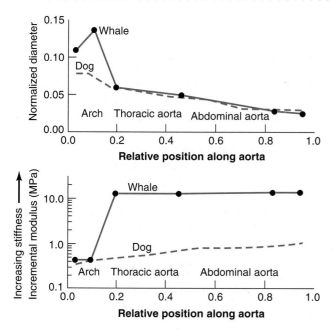

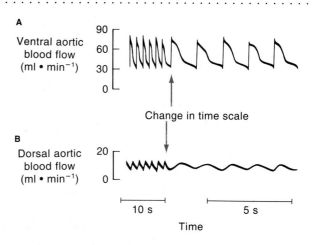

Figure 12-31 Blood flow is more pulsatile in the ventral aorta (**A**) than in the dorsal aorta (**B**) of fishes. The elasticity of the bulbus and ventral aorta help to dampen oscillations in pressure and flow. The recordings shown are from cod. [Jones et al., 1974.]

Figure 12-30 The arterial system of dogs and whales becomes stiffer and smaller in diameter with distance from the heart. In whales, there is an abrupt decrease in diameter and increase in stiffness between the aortic arch and the thoracic artery. [Adapted from Gosline and Shadwick, 1996.]

be elastic and the large-volume dorsal aorta relatively stiff to achieve a smoothing of flow through the gills (Figure 12-31).

 How might the arterial system of an invertebrate with an open circulation vary in structure and function from the vertebrate arterial system?

Blood pressure

Blood pressures reported for the arterial system are usually transmural pressures (i.e., the difference in pressure between the inside and outside, across the wall of the blood vessel). The pressure outside vessels is usually close to ambient, but changes in the extracellular pressure of tissues can have a marked effect on transmural pressure and therefore on vessel diameter and consequently blood flow. For example, contractions of the heart raise pressure around coronary vessels and result in a marked reduction in coronary flow during systole. Breathing in is associated with a reduction in thoracic pressure and thus raises transmural pressure in veins leading back to the heart, increasing venous return to the heart. During a heartbeat cycle, the maximum arterial pressure is referred to as systolic pressure and the minimum as diastolic pressure; the difference is the **pressure pulse.**

Transmural pressures are typically given in millimeters of mercury; both the systolic and diastolic pressures generally are indicated with a slash between them (e.g., 120/

80 mm Hg). Blood is 12.9 times less dense than mercury, so a blood pressure of 120 mm Hg is equal to $120 \times 12.9 = 1550$ mm (155 cm) of blood. In other words, if the blood vessel were suddenly opened, the blood would squirt out to a maximum height of 155 cm above the cut. To convert pressures in millimeters of mercury to kilopascals (kPa), multiply by 0.1333 kPa (e.g., 120 mm Hg $\times$ 0.1333 = 16 kPa).

The oscillations in pressure produced by the contractions and relaxations of the ventricle are reduced at the entrance to capillary beds and nonexistent in the venous system. Heart contractions cause small oscillations in pressure within capillaries. The pressure pulse travels at a velocity of $3-5 \text{ m} \cdot \text{s}^{-1}$. The velocity of the pressure pulse increases with decrease in artery diameter and increasing stiffness of the arterial wall. In the mammalian aorta, the pressure pulse travels at $3-5 \text{ m} \cdot \text{s}^{-1}$ and reaches $15-35 \text{ m} \cdot \text{s}^{-1}$ in small arteries.

Peak blood pressure and the size of the pressure pulse within the mammalian and avian aorta both *increase with distance from the heart* (Figure 12-32). This pulse amplification can be large during exercise. There are three possible explanations for this rather odd phenomenon. First, pressure waves are reflected from peripheral branches of the arterial tree; the initial and reflected waves summate; and, where peaks coincide, the pressure pulse and peak pressure are greater than where they are out of phase. If the initial and reflected waves are 180° out of phase, the oscillations in pressure will be reduced. It has been suggested that the heart is situated at a point where initial and reflected waves are out of phase, thus reducing peak arterial pressure in the aorta close to the ventricle. As distance from the heart increases, the initial and reflected pressure waves move into phase, and a peaking of pressures is observed in vessels of the periphery. Second, the decrease in elasticity and diameter of arteries with distance from the heart may cause an increase in the magnitude of the pressure pulse. Third, the pressure pulse is a complex waveform, consisting

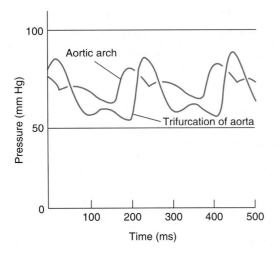

Figure 12-32 In the aorta of mammals and birds, the peak blood pressure and pressure pulse both increase with distance from the heart. Shown are simultaneous recordings of a rabbitís blood pressure in the aortic arch (2 cm from the heart) and at trifurcation of the aorta (24 cm from the heart). Note that the mean pressure is slightly less at trifurcation of the aorta than in the aortic arch near the heart. [Adapted from Langille, 1975.]

of several harmonics. Higher frequencies travel at higher velocities, and it has been suggested that the change in waveform of the pressure pulse with distance is due to summation of different harmonics. This third explanation is open to question, as the distances are too small to allow summation of harmonics.

Effect of gravity and body position on pressure and flow
When a person is lying down, the heart is at the same level as the feet and head, and pressures will be similar in arteries in the head, chest, and limbs. Once a person moves to a sitting or standing position, the relationship between the head, heart, and limbs changes with respect to gravity, and the heart is now a meter above the lower limbs. The result is an increase in arterial pressure in the lower limbs and a decrease in arterial pressure in the head. The height of the column of blood simply results in a higher blood pressure due to gravity.

Gravity has little effect on capillary flow, which is determined by the arterial-venous pressure difference rather than the absolute pressure. That is, gravity raises arterial and venous pressure by the same amount and therefore does not greatly affect the pressure gradient across a capillary bed. Because the vascular system is elastic, however, an increase in absolute pressure expands blood vessels, particularly the compliant veins. Thus, pooling of blood tends to occur, particularly in veins, in different regions of the body as an animal changes position with respect to gravity. This effect is related solely to the elasticity of blood vessels and would not occur if the blood flowed in rigid tubes.

The problems of pooling and maintaining capillary flow are acute in species with long necks. For instance, when the giraffe is standing with its head raised, its brain is about 6 meters above the ground and over 2 meters

above the heart (Figure 12-33A). If the arterial pressure of blood perfusing the brain is to be maintained at around 98 mm Hg, aortic blood pressure must be 195–300 mm Hg near the heart. Aortic blood pressures greater than 195 mm Hg have been recorded in an anesthetized giraffe whose head was raised about 1.5 meters (Figure 12-33B). Arterial pressures in the legs of the giraffe are even higher than aortic pressures; to prevent the pooling of blood, the giraffe has large quantities of connective tissue surrounding the leg vessels. As the giraffe lowers its head to the ground, arterial blood pressure at the level of the heart is reduced considerably, thus maintaining a relatively constant blood flow to the brain. The wide variation in aortic pressure as the giraffe moves its head position could lead to extensive pooling of blood (head raised) or decreased flow (head lowered) in arterioles other than those of the head. Pooling most likely is prevented by **vasoconstriction** of these peripheral vessels when the head is raised. Conversely, when the head is lowered, extensive **vasodilation** of arterioles leading to capillary beds other than those

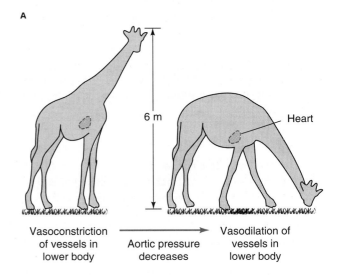

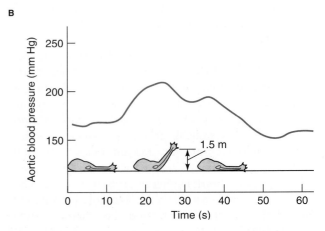

Figure 12-33 As animals with long necks raise and lower their heads, the cardiovascular system must adjust to maintain blood flow to the brain and avoid pooling of blood in the lower part of the body. See text for discussion. [Adapted from White, 1972.]

in the head probably maintains flow despite the lower aortic pressure.

The ability of the giraffe to regulate pressure and flow in peripheral vessels other than those to the head is particularly crucial for kidney function. If the kidney tubule were subjected to the enormous changes in blood pressure associated with the raising and lowering of the giraffe's head, the rate of glomerular filtration would be chaotic. Each time the animal lifted its head, the large increase in arterial blood pressure would lead to a very high rate of ultrafiltrate formation in the kidney; this in turn would require that fluid be reabsorbed at an equally high rate. In the absence of any appropriate controls, the giraffe could lower its head to drink and then lose any fluid gained as it was filtered through the kidney when the head was raised. Thus, the giraffe must have mechanisms for adjusting peripheral resistance to flow in various capillary beds as it swings its head from ground level to drink to a height of 6 m to eat. Similar problems must have been or are faced by a number of other animals with long necks, like dinosaurs and camels.

Pooling of blood, with changes of position with respect to gravity, is not a problem for animals in water, because the density of water is only slightly less than that of blood, whereas air is much less dense than blood. In water, the hydrostatic pressure increases with depth, and effectively matches the increase in blood pressure due to gravity; thus transmural pressure does not change, so the blood does not pool. Clearly, the circulatory problems faced by tall terrestrial dinosaurs were very different from those of aquatic dinosaurs.

Velocity of arterial blood flow
Blood flow and the oscillations in flow with each heartbeat are greatest at the exit to the ventricle, decreasing with increasing distance from the heart (Figure 12-34). At the base of the aorta, as noted earlier, flow is turbulent and reverses during diastole as closure of the aortic valves creates vor-

tices in the blood ejected into the aorta during systole. In most other parts of the circulation, flow is laminar, and oscillations in velocity are damped by the compliance of the aorta and proximal arteries.

Mean velocity in the aorta—the point of maximal blood velocity—is calculated as about 33 cm·s^{-1} in humans, based on a cross-sectional area of about 2.5 cm^2 and cardiac output of about 5 L·min^{-1}. If we assume that maximal velocity in a vessel is twice the mean velocity (valid only if the velocity profile is a parabola), then the maximal velocity of blood flow in the human aorta would be 66 cm·s^{-1}. If cardiac output is increased by a factor of 6 during heavy exercise, maximal velocity is raised to 3.96 m·s^{-1}. In contrast, the pressure pulse associated with each heartbeat travels through the circulation at 3–35 m·s^{-1}; thus, the pressure pulse travels faster than the flow pulse.

Venous System

The venous system acts as a conduit for the return of blood from the capillaries to the heart. It is a large-volume, low-pressure system consisting of vessels with a larger inside diameter than the corresponding arteries (see Figure 12-27). In mammals, about 50% of the total blood volume is contained in veins (see Figure 12-3). Venous pressures seldom exceed 11 mm Hg (1.5 kPa), roughly 10% of arterial pressures. The walls of veins are much thinner, contain less smooth muscle, and are less elastic than arterial walls; venous walls also contain more collagen than elastic fibers. As a result, the walls of veins are easily stretched and exhibit much less recoil than occurs in arteries. The large diameter and low pressure of veins permits the venous system to function as a *storage reservoir for blood*. If venous pressures were high, then according to Laplace's law (see equation 12-1), very high wall tensions would develop, requiring the walls to be very strong to prevent them from tearing.

In the event of blood loss, venous blood volume, not arterial volume, decreases in order to maintain arterial pressure and capillary blood flow. The decrease in the venous blood reservoir is compensated for by a reduction in venous volume. The walls of many veins are covered by smooth muscle innervated by sympathetic adrenergic fibers. Stimulation of these nerves cause vasoconstriction and a reduction in the size of the venous reservoir. This reflex allows some bleeding to take place without a drop in venous pressure. Blood donors actually lose part of their venous reservoir; the loss is temporary, however, and the venous system gradually expands as blood is replaced due to fluid retention.

Venous blood flow
Blood flow in veins is affected by a number of factors other than contractions of the heart. Contraction of limb muscles and pressure exerted by the diaphragm on the gut both result in the squeezing of veins in those parts of the body. Because veins contain *pocket valves* that allow flow only

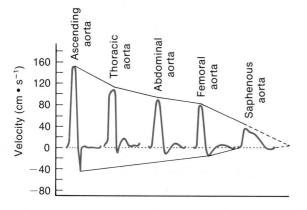

Figure 12-34 The maximal velocity of arterial blood and oscillations in flow decrease progressively with distance from the heart. A backflow phase is observed in the large arteries; in the ascending aorta, it is probably related to a brief reflux of blood through the aortic valves. These tracings were obtained from a dog's arteries. Oscillatory flow is damped out entirely in the capillaries. [Adapted from McDonald, 1960.]

toward the heart, this squeezing augments the return of blood to the heart. Thus venous return to the heart increases during exercise, as skeletal muscle contractions squeeze veins and drive blood towards the heart. The increase in venous return will increase cardiac output. Activation of this skeletal-muscle venous pump is associated with increased activity in sympathetic fibers innervating the venous smooth muscle, increasing smooth-muscle tone. This increase in venous tone ensures that the skeletal-muscle pump increases venous pressure and therefore return to the heart, rather than simply distending another part of the venous system. In the absence of skeletal muscle contraction, there may be considerable pooling of blood in the venous system of the limbs.

Breathing in mammals also contributes to the return of venous blood to the heart. Expansion of the thoracic cage reduces pressure within the chest and draws air into the lungs; this pressure reduction sucks blood from the veins of the head and abdominal cavity into the heart and large veins situated within the thoracic cavity. In sharks contractions of the ventricle reduce pressure in the pericardial cavity, so blood from the venous system is sucked into the atrium (see Figure 12-14).

Peristaltic contractions of the smooth muscle of **venules,** the small vessels joining capillaries to veins, can promote venous flow toward the heart. Such peristaltic activity has been observed in the venules of the bat wing.

Blood distribution in veins

Venous smooth muscle also aids in regulating the distribution of blood in the venous system. When a person shifts from a sitting position to a standing position, the change in the relative positions of heart and brain with respect to gravity activates sympathetic adrenergic fibers that innervate limb veins, causing contraction of venous smooth muscle and thereby promoting the redistribution of pooled blood. Such venoconstriction is inadequate, however, to maintain good circulation if the standing position is held for long periods in the absence of limb movements, as when soldiers stand immobile during a review. Under such circumstances, venous return to the heart, cardiac output, arterial pressure, and flow of blood to the brain are all reduced, which can result in fainting. Similar problems affect bedridden patients who attempt to stand after several days of inactivity and astronauts returning to Earth after a long period of weightlessness. In these instances other control systems involving baroreceptors (pressure receptors) and arterioles may be disrupted as well. In the absence of body changes that shift the relative positions of the heart and brain with respect to gravity, the system of corrections falls into disrepair, and the result is the pooling of blood. The reflex control of venous volume is normally reestablished with use.

The organization of the venous system is influenced by the degree of support offered by the medium. There was an extensive reorganization of the venous system as vertebrates moved into air and lost the support of water. As mentioned previously, the effects of gravity on blood distribution are not important in aquatic animals because the densities of water and blood are not very different. For this reason, pooling of blood due to gravity does not occur in aquatic animals. Because of the large difference between the density of air and blood, pooling became an immediate problem with the evolution of terrestrial forms. The required changes in the venous system are in addition to those required to maintain separation of oxygenated and deoxygenated blood through the heart.

Although the effects of gravity are minimal in aquatic animals, venous return to the heart is exacerbated by swimming in fish. As the fish moves forward, blood will collect in the tail due both to inertia and to compression waves that pass down the body associated with the swimming movements of the fish. To diminish these problems most veins returning to the heart pass down the center of the fish's body. Some fish also have an accessory caudal heart in the tail, which aids in propelling blood toward the central heart (Figure 12-35). The flow of water over the pectoral regions of some fish may reduce hydrostatic pressure in that region, so that venous return to the heart is promoted with increased swimming speed.

Countercurrent exchangers

Countercurrent exchangers are a common feature in the design of animals (see Spotlight 14-2). In many animals arteries and veins run next to each other with the blood flows

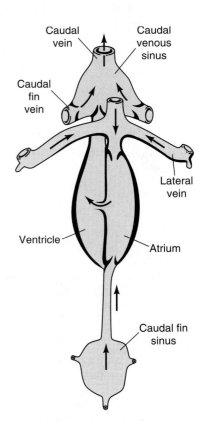

Figure 12-35 Some fish have a caudal heart located in the tail, which aids in returning deoxygenated blood to the central heart. The walls of the heart contain skeletal muscle and beat rhythmically. [Adapted from Kampmeier, 1969.]

moving in opposite directions (i.e., countercurrent blood flow). In many such instances, especially if the vessels are small, there will be exchange of heat between the countercurrent blood flows. Because heat is transferred much more easily than gas, it is possible to have heat exchange with little gas transfer. Countercurrent heat exchangers are common in the limbs of birds and mammals and are used to regulate the rate of heat loss via the limbs.

A countercurrent arrangement of small arterioles and venules is referred to as a *rete mirabile*. Before entering a tissue, an artery divides into a large number of small capillaries that parallel a series of venous capillaries leaving the tissue. The "arterial" capillaries are surrounded by "venous" capillaries, and vice versa, forming an extensive exchange surface between inflowing and outflowing blood. These retial capillaries serve to transfer heat or gases between arterial blood entering a tissue and venous blood leaving it. In humans, this type of countercurrent exchanger is found only in the kidney. Tuna have a large number of retia mirabile, which are used to regulate the temperature of the brain, muscles, and eyes (see Figures 16-22 and 16-23). The rete mirabile leading to the physoclist swimbladder of other fish such as the eel function as a carbon dioxide countercurrent exchanger (see Figure 13-59).

Capillaries and the Microcirculation

Most tissues have such an extensive network of capillaries that any single cell is not more than three or four cells away from a capillary. This is important for the transfer of gases, nutrients, and waste products, because diffusion is an exceedingly slow process. Capillaries are usually about 1 mm long and 3–10 μm in diameter, just large enough for red blood cells to squeeze through. Large leukocytes, however, may become lodged in capillaries, stopping blood flow. The leukocytes are either dislodged by a rise in blood pressure or migrate slowly along the vessel wall until they reach a larger vessel and are swept into the bloodstream.

Microcirculatory beds

Figure 12-36 illustrates the vessels composing a microcirculatory bed. Small terminal arteries subdivide to form **arterioles,** which in turn subdivide to form **metarterioles** and subsequently capillaries, which then rejoin to form venules and veins. The arterioles are invested with smooth

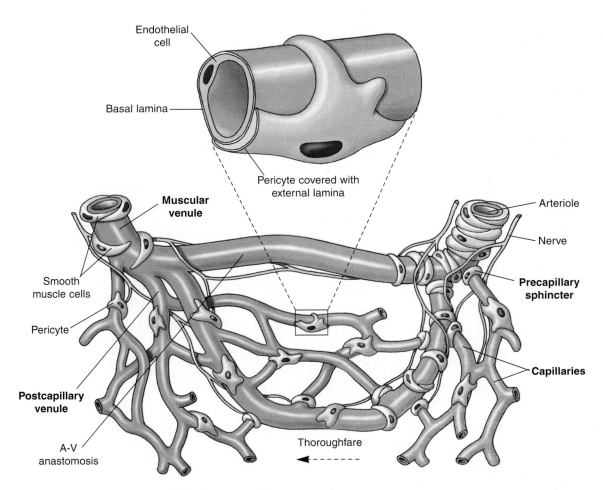

Figure 12-36 A microcirculatory bed consists of small arteries (arterioles), capillaries, and venules. Capillaries, which consist of a single layer of endothelial cells surrounded by a basement membrane and have occasional contractile pericyte cells wrapped around them. Direct flow from the arterial to venous system can occur via the thoroughfare channel, but most blood flows through the network of capillaries. The precapillary sphincter helps regulate flow into the capillary bed. Also see Figures 12-27 and 12-37.

muscle that becomes discontinuous in the metarterioles and ends in a smooth muscle ring, the *precapillary sphincter.* Capillary walls, which are completely devoid of connective tissue and smooth muscle, consist of a single layer of endothelial cells surrounded by a basement membrane of collagen and mucopolysaccharides. The capillaries are often categorized as arterial, middle, or venous capillaries, the latter being a little wider than the other two types. A few elongated cells with the ability to contract, called *pericyte cells,* are found wrapped around capillaries. The venous capillaries empty into pericytic venules, which in turn join the muscular venules and veins. The venules and veins are valved, and the muscle sheath appears after the first post-capillary valve. Even though the walls of capillaries are thin and fragile, they require only a small wall tension to resist stretch in response to pressure because of their small diameter (see equation 12-1).

The innervated smooth muscle of the arterioles and, in particular, the smooth muscle sphincter at the junction of arteries and arterioles control blood distribution to each capillary bed. Most arterioles are innervated by the sympathetic nervous system; a few arterioles (e.g., those in the lungs) are innervated by the parasympathetic nervous system. Different tissues have varying numbers of capillaries open to flow and show some variation in the control of blood flow through the capillary bed. In some tissues, opening and closing of the precapillary sphincters, which are not innervated and appear to be under local control, alter blood distribution within the capillary bed. In other tissues, however, most, if not all, of the capillaries tend to be open (e.g., in the brain) or closed (e.g., in the skin) for considerable periods. All capillaries combined have a potential volume of about 14% of an animal's total blood volume. At any one moment, however, only 30%–50% of all capillaries are open, and thus only 5%–7% of the total blood volume is contained in the capillaries.

Material transfer across capillary walls
Transfer of substances between blood and tissues occurs across the walls of capillaries, pericytic venules, and, to a lesser extent, metarterioles. The endothelium composing the capillary wall is several orders of magnitude more permeable than epithelial cell layers, allowing substances to move with relative ease in and out of capillaries. However, the capillaries in various tissues differ considerably in permeability. These permeability differences are associated with marked changes in the structure of the endothelium. Based on their wall structure, capillaries are classified into three types (Figure 12-37):

- Continuous capillaries, which are the least permeable, are located in muscle, nervous tissue, the lungs, connective tissue, and exocrine glands.

- Fenestrated capillaries, which exhibit intermediate permeability, are found in the renal glomerulus, intestines, and endocrine glands.

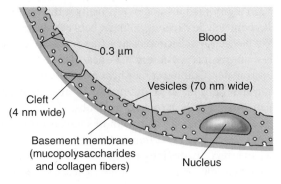

A Continuous capillary

0.3 μm

Blood

Vesicles (70 nm wide)

Cleft (4 nm wide)

Basement membrane (mucopolysaccharides and collagen fibers)

Nucleus

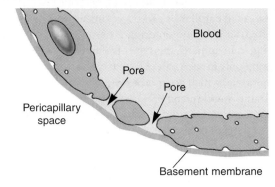

B Fenestrated capillary

Blood

Pore

Pore

Pericapillary space

Basement membrane

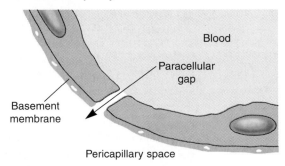

C Sinusoidal capillary

Blood

Paracellular gap

Basement membrane

Pericapillary space

Figure 12-37 Differences in the structure of the capillary endothelium define three types of capillaries, which are found in characteristic tissues. Shown here are portions of the endothelial wall. **(A)** Continuous capillary with 4-nm clefts, a complete basement membrane, and numerous vesicles. **(B)** Fenestrated capillary with pores through a thin portion of the wall, few vesicles, and a complete basement membrane. **(C)** Sinusoidal capillary with large paracellular gaps extending through the discontinuous basement membrane. In general, continuous capillaries are the least permeable and sinusoidal capillaries are the most permeable.

- Sinusoidal capillaries, which are the most permeable, are present in the liver, bone marrow, spleen, lymph nodes, and adrenal cortex.

In the *continuous capillaries* of skeletal muscle, which have been studied extensively, the endothelium is about 0.2–0.4 μm thick and is underlain by a continuous basement membrane (see Figure 12-37A). The endothelial cells are separated by *clefts*, which are about 4 nm wide at the narrowest point. Most of the cells contain large numbers of pinocytotic vesicles about 70 nm in diameter. Most of these vesicles are associated with the inner and outer membranes of the endothelial cells; the rest are located in the cell matrix.

Substances can move across the wall of continuous capillaries either through or between the endothelial cells. Lipid-soluble substances diffuse through the cell membrane, whereas water and ions diffuse through the water-filled clefts between cells. In addition, at least in brain capillaries, there are transport mechanisms for glucose and some amino acids. Large macromolecules can move across many capillary walls, but exactly how they are transferred is not always clear. Some evidence indicates that the numerous vesicles in endothelial cells play a role in transferring substances across the capillary wall. For example, electron microscopic studies have shown that when horseradish peroxidase is placed in the lumen of a muscle capillary, it first appears in vesicles near the lumen and then in vesicles close to the outer membrane, but never in the surrounding cytoplasm. This finding suggests that material is packaged in vesicles and shuttled through the endothelial cells. Supporting this concept of vesicle-mediated transport is the observation that endothelial cells of brain capillaries contain fewer vesicles and are less permeable than endothelial cells from other capillary beds. The reduced permeability of brain capillaries, however, is also considered to result from the tight junctions between endothelial cells. Another possibility is suggested by microscopic observations of capillaries in the rat diaphragm in which vesicles have been observed to coalesce, forming pores through the endothelial cells. Conceivably, then, substances diffuse through pores created by coalescence of nonmobile vesicles, rather than being packaged in vesicles that then move across the cell.

The continuous capillaries in the lung are less permeable than those in other tissues. In these less-permeable capillaries, the pressure pulse may play a role in augmenting movement of substances (e.g., oxygen) through the endothelium. As pressure rises, fluid is forced into the capillary wall, but as pressure drops, fluid returns to the blood. This tidal flushing of the capillary wall should enhance mixing in the endothelial barrier and effectively augment transfer.

In the capillaries of the renal glomerulus and gut, the inner and outer plasma membranes of endothelial cells are closely apposed and perforated by *pores* in some regions, forming a fenestrated endothelium (see Figure 12-37B). Not surprisingly, these *fenestrated capillaries* are permeable to nearly everything except large proteins and red blood cells. The kidney ultrafiltrate is formed across such an endothelial barrier. The basement membrane of fenestrated endothelia normally is complete and may constitute an important barrier to the movement of substances across fenestrated capillaries. These capillaries contain only a few vesicles, which probably play no role in transport.

The endothelium in *sinusoidal capillaries* is characterized by *large paracellular gaps* that extend through the basement membrane and an absence of vesicles in the cells (see Figure 12-37C). Liver and bone capillaries always contain large paracellular gaps, and most substance transfer across these capillaries occurs between the cells. As a result, the fluid surrounding the capillaries in liver has much the same composition as plasma.

The clefts, pores, and paracellular gaps through which substances can freely diffuse across capillary walls are about 4 nm wide, but only molecules much smaller than 4 nm can get through, indicating the presence of some further sieving mechanism. The diameter of these openings varies within a single capillary network and usually is larger in the pericytic venules than in the arterial capillaries. This is of functional significance because blood pressure, which is the filtration force for moving fluid across the wall, decreases from the arterial to the venous end of the capillary network. Inflammation or treatment with a variety of substances (e.g., histamine, bradykinin, and prostaglandins) increases the size of the openings at the venous end of the capillary network, making it very permeable.

Capillary pressure and flow

The arrangement of arterioles and venules is such that all capillaries are only a short distance from an arteriole, so that pressure and flow are fairly uniform throughout the capillary bed. Transmural pressures of about 10 mm Hg have been recorded in capillaries (Figure 12-38). High pressures inside a capillary result in the filtration of fluid from the plasma into the interstitial space. This filtration pressure is opposed by the plasma *colloid osmotic pressure*, which results largely from the higher concentration of proteins in the blood than in the interstitial fluid. Because of their large size, these plasma proteins are retained in the blood and not transported across the capillary wall.

To visualize the relationship of these two pressures, consider the schematic situation depicted in Figure 12-39. Generally, blood pressure is higher than the colloid osmotic pressure at the arterial end of a capillary bed, so fluid moves into the interstitial space (area 1). The blood pressure steadily decreases along the length of the capillary, while the colloid osmotic pressure remains constant. Once the blood pressure falls below the colloid osmotic pressure, fluid in the interstitial space is drawn back into the blood by osmosis (area 2). Thus the net movement of fluid at any point along the capillary is determined by two factors: (a) the *difference* between blood pressure and colloidal osmotic pressure and (b) the *permeability* of the capillary wall, which tends to increase toward the venous end.

This concept is sometimes referred to as the *Starling hypothesis*, after its initial proponent, Ernest Starling (1866–1927), whose prolific research also included studies on the relationship between ventricle work output and venous filling pressure (see Spotlight 12-1). In most capillary beds, the net loss of fluid at the arterial end is somewhat greater than the net uptake at the venous end of the

A

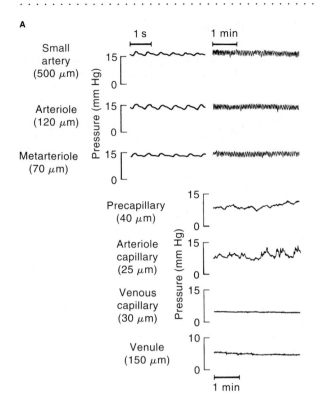

B

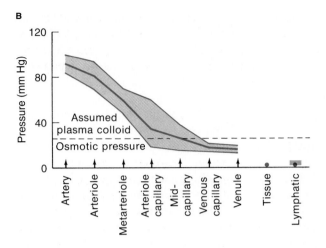

Figure 12-38 The pressure pulse is reduced and the mean blood pressure decreases as blood flows through a capillary bed. **(A)** Blood pressure tracings recorded throughout capillary bed of the frog mesentery. The pressure is smoothed and falls as blood flows across the capillary bed. **(B)** Plot of blood pressure versus location in circulation of the subcutaneous layers of the bat wing. Shaded area represents ±1 SE (standard error) from the average values indicated by thick line. Also plotted are typical tissue and lymphatic pressures for comparison. [Part A from Weiderhielm et al., 1964; part B from Weiderhielm and Weston, 1973.]

capillary. The fluid, however, does not accumulate in the tissues, but is drained by the lymphatic system and returned to the circulation. Thus, there typically is a circulation of fluid from the arterial end of capillary bed into the interstitial spaces and back into the blood across the venous end of the capillary bed or via the lymphatic system. Because of this bulk flow of fluid, the exchange of gases, nutrients, and wastes between blood and tissues exceeds that expected by diffusion alone.

Net filtration of fluid across capillary walls will result in an increase in tissue volume, termed **edema**, unless the excess fluid is carried away by the lymphatic system. In the kidney, capillary pressure is high and filtration pressures exceed colloid osmotic pressures; hence, an ultrafiltrate is formed in the kidney tubule, eventually to form urine. The kidney is encapsulated to prevent swelling of the tissue in the face of ultrafiltration. In most other tissues, there is only a small *net* movement of fluid across capillary walls and tissue volume remains constant. A rise in capillary pressure, owing to a rise in either arterial or venous pressure, will result in increased loss of fluid from the blood and tissue edema. In general, though, arterial pressure remains fairly constant to prevent large oscillations in tissue volume. A drop in colloid osmotic pressure can result from a loss of protein from the plasma by starvation or excretion or by increased capillary wall permeability, leading to movement of plasma proteins into the interstitial space. If filtration pressure remains constant, a decrease in colloid osmotic pressure will also result in an increase in net fluid loss to the tissue spaces.

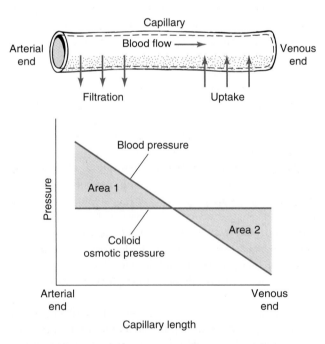

Figure 12-39 Net fluid flow across the capillary wall depends on the difference between the blood pressure and the colloid osmotic pressure of the extracellular fluid. At the arterial end of the capillary, blood pressure exceeds colloid osmotic pressure and fluid is filtered from the plasma into the extracellular space (area 1). At the venous end, the reverse is true and fluid is drawn back into the plasma from the extracellular space (area 2). Area 1 is somewhat larger than area 2 in most capillary beds; that is, there is a small net loss of fluid from the circulation to the extracellular space. In general, this tissue fluid is drained and returned to the bloodstream via the lymphatic system.

Why should lying down and raising the legs reduce ankle swelling in humans? Swollen ankles are not a common malady in giraffes. Why not?

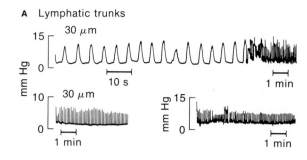

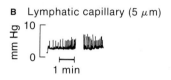

Figure 12-40 Pressures in the lymphatic system are similar to those in the venous system. These recordings are from lymphatic trunks **(A)** and lymphatic capillaries **(B)** in the wing of an unanaesthetized bat. They were obtained by micropuncture without prior surgical intervention. [Weider-hielm and Weston, 1973.]

THE LYMPHATIC SYSTEM

Lymph—a transparent, slightly yellow or sometimes milky fluid—is collected from the interstitial fluid in all parts of the body and returned to the blood via the lymphatic system. Because this fluid contains many white blood cells but no red blood cells, it is nearly colorless, making the lymphatic vessels difficult to see. As a result, even though the lymphatic system was first described about 400 years ago, it has not been nearly as extensively studied as the cardiovascular system.

The lymphatic system begins with blind-ending *lymphatic capillaries* that drain the interstitial spaces. These lymphatic capillaries join to form a treelike structure with branches reaching to all tissues. The larger lymphatic vessels resemble veins and empty via a duct into the blood circulation at a point of low pressure. In mammals and many other vertebrates, the lymph vessels drain via a **thoracic duct** into a very low pressure region of the venous system, usually close to the heart (see Figure 12-3). The lymphatic system serves to return to the blood the excess fluid and proteins that filter across capillary walls into the interstitial spaces. Large molecules, particularly fat absorbed from the gut and probably high-molecular-weight hormones, reach the blood via the lymphatic system.

The walls of lymphatic capillaries consist of a single layer of endothelial cells. The basement membrane is absent or discontinuous, and there are large paracellular gaps between adjoining cells. This feature has been demonstrated by microscopic observation of horseradish peroxidase or china ink particles passing through lymphatic capillary walls. Because lymphatic pressures are often slightly lower than the surrounding tissue pressures, interstitial fluid passes easily into lymphatic vessels. The vessels are valved and permit flow only away from the lymphatic capillaries. The larger lymphatic vessels are surrounded by smooth muscle and, in some instances, contract rhythmically, creating pressures of up to 10 mm Hg and driving fluid away from the tissues (Figure 12-40). The vessels also are squeezed by contractions of the gut and skeletal muscle and by general movements of the body, all of which promote lymph flow. Fats are taken up from the gut by the lymphatic system rather than directly into the blood. Folds in the gut wall, called villi, each contain a lymphatic vessel (**central lacteal**) into which fats and fat-soluble nutrients (e.g., vitamins A, D, E, and K) pass from the lumen of the gut (see Chapter 15). The lacteal is "milked" of its milky fat-containing lymph by contractions of the gut, pushing the lymph forward and eventually, via the thoracic duct, into the blood. Lymph vessels are innervated, but it is not clear what type of innervation exists nor what function these nerves have.

Lymph flow is variable, 11 ml·h^{-1} being an average value for resting humans. This is $\frac{1}{3000}$ of the cardiac output during the same time period. Nevertheless, although it is small, lymphatic flow is important in draining tissues of excess interstitial fluid. If lymph production exceeds lymph flow, severe edema can result. In the tropical disease filariasis, larval nematodes, transmitted by mosquitoes to humans, invade the lymphatic system causing blockage of lymph channels; in some cases, lymphatic drainage from certain parts of the body is blocked totally. The consequent edema can cause parts of the body to become so severely swollen that the condition has come to be called *elephantiasis* because of the resemblance of the swollen, hardened tissues to the hide of an elephant.

Reptiles and many amphibians have **lymph hearts,** which aid in the movement of fluid within the lymphatic system. Bird embryos have a pair of lymph hearts located in the region of the pelvis; these hearts persist in the adult bird of a few species. Mammals lack these structures for moving lymph. Frogs have not only multiple lymph hearts but also a very large-volume lymph space, which serves as a reservoir for water and ions and as a fluid buffer between the skin and underlying tissues. The large lymph volume in amphibians is derived from both plasma filtration across capillaries and the diffusion of water across the skin. The ratio of lymph flow to cardiac output is much higher in toads (approximately 1:60) than in mammals (approximately 1:3000), and toad lymph hearts, although having a much smaller stroke volume, can beat at rates higher than the blood heart.

Fish appear to either lack or have only a very rudimentary lymphatic system, although they have a secondary circulation that in the past was described as a lymphatic system. This secondary circulation, which has a low hematocrit, is connected to the primary circulation via arterio-arterial anastomoses and drains into the primary venous system near the heart (see Figure 12-15). The secondary circulation supplies nutrients but not much oxygen to the skin and gut but is not generally distributed to other parts of the body. The skin exchanges gases directly with the surrounding water. Because of its narrow distribution it is unlikely that the secondary circulation fulfills the lymphatic function of maintaining tissue fluid balance. It is not clear how fish maintain tissue fluid balance, but the absence of lymphatics seems to be related to the fact that fish live in a medium of similar density to their own bodies.

CIRCULATION AND THE IMMUNE RESPONSE

Both the circulatory and lymphatic systems are involved in the body's defense against infection. The crucial players in the immune response are **lymphocytes,** a type of white blood cell (**leukocyte**). The unique characteristic of lymphocytes is their ability to "recognize" foreign substances

(antigens) including those on the surface of invading pathogens, virus-infected cells, and tumor cells. There are two main types of lymphocytes: B lymphocytes (B cells) and T lymphocytes (T cells). The latter are subdivided into helper T (T_H) cells and cytotoxic T (T_C) cells. Lymphocytes are aided by other leukocytes, particularly neutrophils and macrophages. Under certain conditions, both neutrophils and macrophages can engulf microorganisms and foreign particulate matter by **phagocytosis.** These phagocytic cells also produce and release various cytotoxic factors and antibacterial substances.

The immune response consists of recognizing the invader, then marking and destroying it. Recognition is carried out exclusively by lymphocytes, whereas destruction can be effected by both lymphocytes and phagocytic cells (**phagocytes**). The lymphocyte recognition system must be able to discriminate between natural constituents of the body and foreign invaders, that is, to distinguish between self and nonself. Failure to recognize self leads to autoimmune diseases, some of which can be fatal.

Lymphocytes respond in three ways to an invasion by a pathogen (Figure 12-41). B cells develop into plasma cells, which secrete **antibodies** that bind to the pathogen, mark-

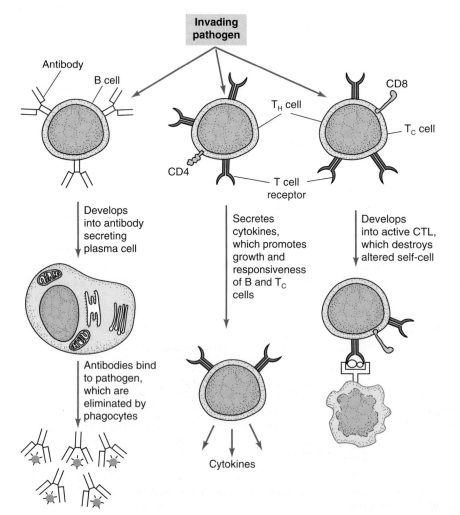

Figure 12-41 Three types of lymphocytes—B cells, helper T (T_H) cells, and cytotoxic T (T_C) cells—respond in different ways to antigen. Membrane-bound antibody on B cells and T-cell receptors on T cells recognize and bind antigen specifically. T_H and T_C cells can be distinguished by the presence of membrane molecules called CD4 and CD8. See text for discussion. [Adapted from Kuby, 1997.]

ing the cell for degradation by phagocytes. T_C cells can recognize tumor cells and pathogen-infected tumor cells; upon recognizing such cells, T_C are stimulated to mature into active cytotoxic T lymphocytes (CTLs), which destroy the altered self-cells. Recognition of antigen by T_H cells stimulates them to secrete *cytokines,* which in turn promote the growth and responsiveness of B cells, T_C cells, and macrophages, thereby increasing the strength of the immune response to a pathogen.

Leukocytes circulate in both the blood and lymph. Large numbers of lymphocytes are present in *lymph nodes,* which are located along the lymphatic vessels (see Figure 12-3). These nodes filter the lymph and help bring antigen into contact with lymphocytes. To get to tissues that have been invaded by pathogens, leukocytes must be able to leave the lymphatic and circulatory systems, a process termed **extravasation.** Normally, of course, leukocytes are swept along in the bloodstream and do not pass across vessel walls. At sites of infection, however, inflammatory signals are produced that induce the synthesis and activation of adhesive proteins on the blood side of the endothelium. As leukocytes roll past an inflamed vascular endothelium, P-selectin on the blood-facing surface binds to and slows the passing leukocytes (Figure 12-42). This interaction stimulates the leukocytes to produce integrin receptors (e.g., LFA-1), which then bind with intracellular adhesion molecules (ICAMs) on the surface of the endothelium. As

a result of these and other interactions, the cells adhere to the endothelium. Once firmly adhered, the leukocytes can move between the endothelial cells and migrate into the infected tissue.

REGULATION OF CIRCULATION

Regulation of circulation hinges on *controlling arterial blood pressure* so that three central priorities can be fulfilled:

- Delivering an adequate supply of blood to the brain and heart

- Supplying blood to other organs of the body, once the brain and heart supply is assured

- Controlling capillary pressure so as to maintain tissue volume and the composition of the interstitial fluid within reasonable ranges

The body employs a variety of receptors for monitoring the status of the cardiovascular system. In response to sensory inputs from these receptors, both neural and chemical signals induce appropriate adjustments to maintain an adequate arterial pressure. In this section, we first discuss regulatory features affecting the heart and main vessels, and then focus on the microcirculation.

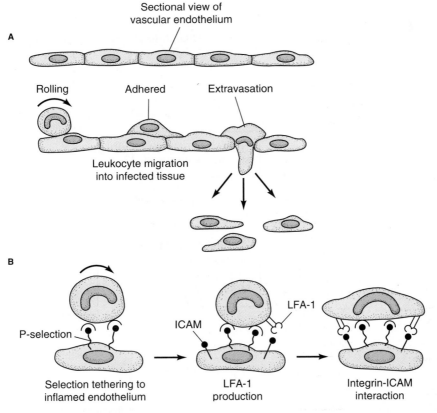

Figure 12-42 Leukocytes migrate from the circulation into tissues at sites of inflammation. **(A)** Overview of leukocyte adherence to and extravasation across inflamed vascular endothelium. **(B)** Some of the interactions between cell-surface molecules that cause leukocytes to adhere to an inflamed endothelium. [Adapted from Kuby, 1997.]

Control of the Central Cardiovascular System

Baroreceptors monitor blood pressure at various sites in the cardiovascular system. Information from baroreceptors, along with that from **chemoreceptors** monitoring the CO_2, O_2, and pH of the blood, is transmitted to the brain. Muscle contraction or changes in the composition of the extracellular fluid of muscles activate afferent fibers embedded in muscle tissue, and this in turn causes changes in the cardiovascular system. In addition, inputs from cardiac **mechanoreceptors** and from a variety of **thermoreceptors** lead to reflex effects on the cardiovascular system.

In mammals, the integration of these sensory inputs occurs in a collection of brain neurons referred to as the **medullary cardiovascular center,** located at the level of the medulla oblongata and pons. The medullary cardiovascular center also receives inputs from other regions of the brain, including the medullary respiratory center, hypothalamus, amygdala nucleus, and cortex. The output from the medullary cardiovascular center is fed into sympathetic and parasympathetic autonomic motor neurons that innervate the heart and the smooth muscle of arterioles and veins, as well as to other areas of the brain such as the medullary respiratory center.

Stimulation of *sympathetic nerves* increases the rate and force of contraction of the heart and causes vasoconstriction; the result is a marked increase in arterial blood pressure and cardiac output. In general, the reverse effects follow stimulation of *parasympathetic nerves,* the end result being a drop in arterial blood pressure and cardiac output. The medullary cardiovascular center can be divided into two functional regions, which have opposing effects on blood pressure:

- Stimulation of the *pressor center* results in sympathetic activation and a rise in blood pressure.

- Stimulation of the *depressor center* results in parasympathetic activation and a drop in blood pressure.

In general terms, various sensory inputs affect the balance between pressor and depressor activity: some activate the pressor center and inhibit the depressor center; others have the reverse effect. Thus, the various inputs that converge on the medullary cardiovascular center are modified and integrated. The result is an output that activates the pressor or the depressor center and produces cardiovascular changes in response to changing requirements of the body or disturbances to the cardiovascular system. Figure 12-43 presents an overview of this central circulatory control in mammals.

Arterial baroreceptors

Baroreceptors, which are widely distributed in the arterial system of vertebrates, show increased rates of firing with increases in blood pressure. Unmyelinated baroreceptors have been localized in the central cardiovascular system of amphibia, reptiles, and mammals. These unmyelinated baroreceptors only respond to pressures *above* normal, initiating reflexes that reduce arterial blood pressure and thus protect the animal from damaging increases in blood pressure. Myelinated baroreceptors, which have been found only in mammals, respond to blood pressures *below* normal, thus protecting the animal from prolonged periods of

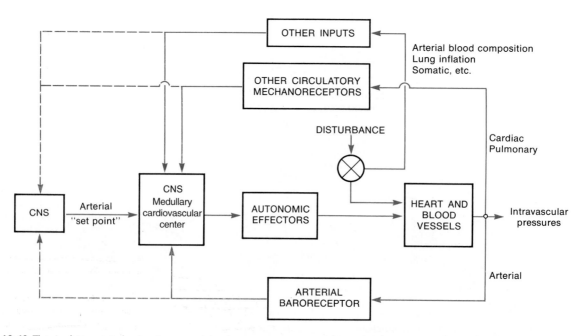

Figure 12-43 The circulatory control system in mammals involves a number of negative-feedback loops. Various receptors monitor changes in the state of the cardiovascular system, sending inputs to medullary cardiovascular center. After integrating these inputs and comparing with the arterial set point, this center sends signals via autonomic nerves to maintain an appropriate arterial blood pressure. The arterial set point is altered by inputs from other areas of the brain, which are in turn influenced by a variety of peripheral inputs (dashed lines). [Adapted from Korner, 1971.]

reduced blood pressure. Baroreceptors in the mammalian **carotid sinus** have been studied much more extensively than those in the aortic arch or subclavian, common carotid, and pulmonary arteries. In mammals there appear to be only minor quantitative differences between the baroreceptors of the carotid sinus and the aortic arch. Birds have aortic arch baroreceptors.

The carotid sinus in mammals is a dilation of the internal carotid at its origin, where the walls are somewhat thinner than in other portions of the artery. Buried in the thin walls of the carotid sinus are finely branched nerve endings that function as baroreceptors. Under normal physiological conditions, there is a resting discharge from these baroreceptors. An increase in blood pressure stretches the wall of the carotid sinus, causing an increase in discharge frequency from the baroreceptors. The relationship between blood pressure and baroreceptor impulse frequency is sigmoidal, the system being most sensitive over the physiological range of blood pressures (Figure 12-44). In addition, the baroreceptor discharge frequency is higher when pressure is pulsatile than when it is constant. The carotid sinus baroreceptors are most sensitive to frequencies of pressure oscillation between 1 and 10 hertz. Since the arterial pressure increases and decreases with each heartbeat, this frequency range is within the normal physiological range of arterial pressure oscillations. Similar observations have been made on the relationship between discharge frequency and pressure for pulmocutaneous arterial baroreceptors in the toad (Figure 12-45). Sympathetic efferent fibers terminate in the arterial wall near the carotid sinus baroreceptors; stimulation of these sympathetic fibers increases discharge of these baroreceptors. Under normal physiological conditions

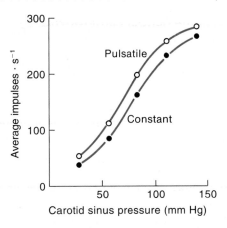

Figure 12-44 The discharge frequency of baroreceptors increases with pressure in a sigmoid fashion. These receptors are most sensitive within the physiological range of pressures and when blood flow is pulsatile. These values were recorded from a multifiber preparation of the carotid sinus nerve and plotted against the mean pressure in the carotid sinus during pulsatile or constant flow. [Adapted from Korner, 1971.]

these efferent neurons may be utilized by the central nervous system (CNS) to control sensitivity of the receptors.

Signals from baroreceptors in response to increased blood pressure are relayed through the medullary cardiovascular center to autonomic motor neurons, leading to a reflex reduction in both cardiac output and peripheral vascular resistance (Table 12-1). The reduction in cardiac output results from both a drop in the heart rate and the force of cardiac contraction. The culmination of the various autonomic effects is a decrease in arterial blood pressure. But as the arterial pressure decreases, so does the baroreceptor

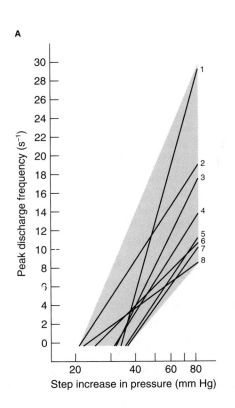

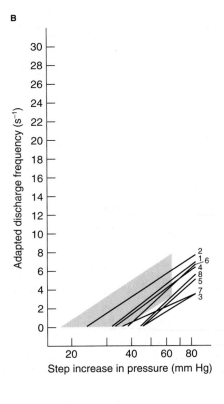

Figure 12-45 Baroreceptors are very sensitive to changes in pressure. The effect of step increases in pressure on the discharge frequency of pulmocutaneous baroreceptors in the toad are plotted **(A)** immediately after the pressure increase and **(B)** 45 seconds later. Each numbered black line represents one observation corresponding to the pressure increase shown on the horizontal axis. The rapid initial peak response is greater than the response 45 seconds after the pressure increase. [Adapted from Van Vilet and West, 1994.]

TABLE 12-1
Reflex effects observed during changes in carotid sinus pressure

Autonomic effector	Carotid sinus pressure*	
	Increased	Decreased
Cardiac vagus	++++	−
Cardiac sympathetic	−	+++
Splanchnic bed		
Resistance vessels	−−	++
Capacitance vessels	−−	++
Renal bed	≈0	+
Muscle bed		
Resistance vessels	−−	++++
Capacitance vessels	−	+
Skin		
Resistance vessels	−	++
Capacitance vessels	?0	0
Adrenal catecholamines	≈0	++
Antidiuretic hormone	?	++

*A + means increased autonomic effect; a −, decreased autonomic effect; and 0, no autonomic effect.

Source: Korner, 1971.

discharge frequency, causing a reflex increase in both cardiac output and peripheral resistance, which tends to increase arterial pressure. Thus the baroreceptor reflex of the carotid sinus is a negative feedback loop that tends to stabilize arterial blood pressure at a particular set point. The set point may be altered by interaction with other receptor inputs or may be reset centrally within the medullary cardiovascular center by inputs from other regions of the brain (see Figure 12-43).

Arterial chemoreceptors

Arterial chemoreceptors, which are located in the carotid and aortic bodies, are particularly important in regulating ventilation (see Chapter 13), but they also have some effect on the cardiovascular system. These chemoreceptors respond with an increase in discharge frequency to an increase in CO_2 or to decreases in O_2 and pH of the blood perfusing the carotid and aortic bodies. An increase in discharge frequency results in peripheral vasoconstriction and a slowing of the heart rate *if* the animal is not breathing (e.g., during submersion). Cardiac output is reduced while birds and mammals dive; peripheral vasoconstriction then ensures the maintenance of arterial blood pressure and therefore brain blood flow in the face of this reduction in cardiac output.

Peripheral vasoconstriction can cause a rise in arterial pressure, which then evokes reflex slowing of the heart by stimulation of the systemic baroreceptors. Nevertheless, stimulation of arterial chemoreceptors results in a slowing of the heartbeat even when arterial pressure is regulated at a constant level. Chemoreceptor stimulation thus has a direct effect on heart rate, as well as an indirect effect via

changes in arterial pressure resulting from peripheral vasoconstriction. Not surprisingly, there are many interactions between the control systems associated with both the respiratory and the cardiovascular system. For example, the discharge pattern from stretch receptors in the lungs has a marked effect on the nature of the cardiovascular changes caused by chemoreceptor stimulation. If the animal is breathing normally, changes in gas levels of the blood will cause one set of reflex changes; if, however, the animal is not breathing, chemoreceptor stimulation results in quite a different series of cardiovascular changes, as we will see in the later discussion on diving.

Cardiac receptors

Both mechanoreceptive and chemoreceptive afferent nerve endings are located in various regions of the heart. Information on the state of the heart collected by these receptors is transmitted via the spinal cord to the medullary cardiovascular center and other regions of the brain. In addition, stimulation of some cardiac receptors causes hormone release either directly from the atria or from other endocrine tissues within the body. Stimulation of cardiac receptors evokes a series of reflex responses including changes in heart rate and cardiac contractility and, under extreme conditions, the pain that can be associated with a heart attack.

Atrial receptors The atrial walls contain many mechanoreceptive afferent fibers, which are classified into three types. Myelinated *A-type* and *B-type* afferent fibers are embedded in the atria. A-fiber afferents respond to heart-rate changes and appear to relay information on heart rate to central cardiovascular control centers. B-fiber afferents respond to increases in the rate of filling and volume of the atria. Increases in venous volume result in an increase in venous pressure, which in turn raises the atrial filling and consequently the discharge frequency of the B-fibers. This increased activity is processed by the central cardiovascular centers leading to two major effects, one on the heart and one on the kidney. Stimulation of atrial B-fibers leads to a faster heart rate mediated via increased activity in the sympathetic outflow to the sinus node of the heart. Stimulation of these afferent fibers also causes a marked increase in urine excretion (**diuresis**), probably mediated by a decrease in antidiuretic hormone (ADH) levels in the blood. Thus, there is a negative feedback loop for regulating blood volume. An increase in blood volume raises venous pressure and atrial filling; this stimulates atrial B-fibers, leading to inhibition of ADH release from the pituitary. The resulting fall in blood ADH levels leads to diuresis and therefore a reduction in blood volume.

The third type of atrial mechanoreceptor comprises unmyelinated C-type afferent fibers innervating the junction of the veins and atria. Stimulation of these C-fiber afferents receptors affects both heart rate and blood pressure. If heart rate is low, distension of this region results in an increase in heart rate, whereas if heart rate is high, stimulation results in a fall in heart rate. Stimulation of C-fibers also causes a fall in blood pressure. Both myelinated and unmyelinated sympathetic fibers innervate the atria. Atrial contraction and atrial distension reflexly stimulate these fibers, causing an increase in heart rate.

The atrial wall also contains stretch-sensitive secretory cells that produce **atrial natriuretic peptide (ANP)**. This hormone, which is released into the blood upon stretch of these cells, has several endocrine effects. As its name indicates, ANP causes an increase in urine production and sodium excretion, thereby effectively reducing blood volume and therefore blood pressure. ANP inhibits release of renin by the kidney and production of aldosterone by the adrenal cortex. It thus diminishes the renin-angiotensin-aldosterone system, which stimulates sodium resorption and an increase in blood volume (see Chapter 14). In addition to these actions, ANP inhibits release of ADH and acts directly on the kidney to increase water and sodium excretion. ANP has been demonstrated to have a depressor effect, reducing both cardiac output and blood pressure. In addition ANP antagonizes the pressor effect of angiotensin.

Atrial natriuretic peptide belongs to a family of natriuretic peptides (A-, B-, C-, and V-type natriuretic peptide) sharing a common 17-amino-acid ring structure linked by a disulfide bridge. Since the initial investigations of ANP in the early 1980s, natriuretic peptides have been found in a wide variety of tissues, including the central nervous system. In many instances they may have an autocrine or paracrine function. For example, receptors for natriuretic hormone have been located in both the atria and ventricles of the hearts of several vertebrates. Binding of locally released natriuretic hormone to these receptors may reduce contractility, indicating a paracrine function within the heart.

When mammals dive under water, the thorax is compressed, resulting in an increase in venous pressure. What might be the physiological consequences of diving under water?

Ventricular receptors The endings of both myelinated and unmyelinated sensory afferent fibers are embedded in the ventricle. The myelinated fibers are mechanoreceptive and chemoreceptive, with separate endings for each modality. The mechanoreceptive endings are stimulated by interruption of coronary blood flow. The chemoreceptive endings are stimulated by substances like **bradykinin**. At low stimulation levels, these fibers cause increased sympathetic outflow and decreased vagal outflow to the heart, raising cardiac contractility as well as blood pressure. At higher stimulation levels, these fibers are necessary for the perception of pain in the heart. Myelinated afferent fibers are much less numerous than unmyelinated C-fiber afferent endings in the left ventricle. Stimulation of the C-fiber af-

ferents at low levels causes peripheral vasodilation and a reduction in heart rate. Increased stimulation of these fibers causes stomach relaxation and, at even higher frequencies, results in vomiting.

Skeletal muscle afferent fibers
Somewhat surprisingly, most nerves innervating skeletal muscle contain more afferent fibers than efferent fibers. The afferent fibers can be subdivided into four broad groups. Groups I and II are sensory fibers from muscle spindles and Golgi tendon organs; these seem to play little or no role in the control of the cardiovascular system. In contrast, stimulation of group III fibers, which are myelinated "free nerve endings," or group IV fibers, which are unmyelinated sensory endings, appears to have cardiovascular effects. These fibers are activated by either mechanical or chemical stimulation, with most fibers responding to only one modality. Mechanical stimulation may be due to contraction, squeezing, or stretching of the muscle. The changes in the extracellular fluid associated with muscle contraction also are thought to stimulate chemoreceptive muscle afferent fibers and evoke cardiovascular changes. Large changes in pH and osmotic pressure raise the activity of group IV fibers, but it is not clear if the pH or osmotic changes occurring *in vivo* are adequate to mediate cardiovascular effects.

Electrical stimulation of muscle afferents can result in either an increase or a decrease in arterial blood pressure, depending on the fibers being stimulated or the frequency of stimulation of a particular group of afferent nerves. At low frequencies, stimulation of some afferent fibers results in a fall in arterial blood pressure, whereas stimulation of the same fibers at high frequencies results in a rise in blood pressure. Electrical stimulation of afferent nerves from muscles usually causes a change in heart rate in the same direction as the change in blood pressure; that is, if blood pressure is elevated, so is heart rate, and vice versa. In those instances where electrical stimulation of muscle afferents causes an increase in heart rate and cardiac output, there is also a change in the distribution of blood in the body. Blood flow to the skin, kidney, gut, and inactive muscle is reduced, thus augmenting flow to the active muscles.

The cardiovascular response evoked by muscle contraction has been shown to disappear following dorsal root section, so the response is presumably reflex in origin, resulting from stimulation of afferent fibers in the muscle. The response varies depending on whether muscle contraction is isometric (static exercise) or isotonic (dynamic exercise). Static exercise is associated with an increase in arterial blood pressure with little change in cardiac output, whereas dynamic exercise results in a large increase in cardiac output with little change in arterial blood pressure. The sensory inputs from muscle afferent fibers are processed in the central cardiovascular center, leading to stimulation of the autonomic nerves innervating the heart and vessels, the efferent arm of the reflex arcs.

Control of the Microcirculation

Capillary blood flow adjusts to meet the demands of the tissues. If the requirements change suddenly, as in skeletal muscle during exercise, then capillary flow also changes. If requirements for nutrients vary little with time, as in the brain, then capillary flow also varies little. The regulation of capillary flow can be divided into two main types, nervous control and local control.

Nervous control of capillary blood flow

Nervous control serves to maintain arterial pressure by adjusting resistance to blood flow in the peripheral circulation. The vertebrate brain and heart must be perfused with blood at all times. An interruption in the perfusion of the human brain rapidly results in damage. Nervous control of arterioles ensures that only a limited number of capillaries will be open at any moment, for if all capillaries were open, there would be a rapid drop in arterial pressure and blood flow to the brain would be reduced. The nervous control of capillary flow operates under a priority system. If arterial pressure falls, blood flow to the gut, liver, and muscles is reduced to maintain flow to the brain and heart. Most arterioles are innervated by sympathetic nerves, which release norepinephrine at their endings. Some arterioles, however, are innervated by parasympathetic nerves, which release acetylcholine at their endings.

Sympathetic stimulation and circulating catecholamines

Binding of the catecholamine norepinephrine to α-adrenoreceptors in the smooth muscle of arterioles usually causes vasoconstriction and therefore a decrease in diameter of the arterioles. This decrease in diameter causes an increase in resistance to flow, thus reducing blood flow through that capillary bed. The generalized effect of sympathetic stimulation is peripheral vasoconstriction and subsequent rise in arterial blood pressure. This overall response is mediated by binding of norepinephrine from nerve endings to α-adrenoreceptors in vascular smooth muscle, resulting in an increase in smooth-muscle tension.

Stimulation of β-adrenoreceptors in arterial smooth muscle, however, often results in relaxation of the muscle and an increase in diameter of the arterioles (i.e., vasodilation), thereby decreasing the resistance to flow and increasing the blood flow through that capillary bed. Because β-adrenoreceptors are rarely located near nerve endings, they usually are stimulated by circulating catecholamines. Catecholamines are released into the bloodstream from adrenergic neurons of the autonomic nervous system and from chromaffin cells in the adrenal medulla. Circulating catecholamines are dominated by epinephrine released from the adrenal medulla (see Chapter 8). Epinephrine reacts with both α- and β-adrenoreceptors, causing vasoconstriction and vasodilation, respectively. Although α-adrenoreceptors are less sensitive to epinephrine, when activated they override the vasodilation mediated by β-adrenoreceptors. The result is that high levels of circulating epinephrine cause vasoconstriction and thus an increase in peripheral resistance via α-adrenoreceptor stimulation. At lower levels of circulating epinephrine, however, β-adrenoreceptor stimulation dominates, producing an overall vasodilation and a decrease in peripheral resistance. Even at levels of epinephrine that produce vasodilation, it causes a rise in arterial blood pressure by stimulating β-adrenoreceptors in the heart, causing a marked increase in cardiac output.

The β-adrenoreceptors can be divided into two subgroups: β_1-adrenoreceptors, which are stimulated by both circulating catecholamines (epinephrine) and adrenergic nerve stimulation (norepinephrine) and β_2-adrenoreceptors, which respond only to circulating catecholamines. In the peripheral circulation, only β_2-adrenoreceptors are present, whereas β_1-adrenoreceptors are found in the heart and coronary circulation, where both circulating catecholamines and neurally released norepinephrine can have a marked effect.

We can summarize these effects as follows:

- Stimulation of sympathetic nerves generally causes peripheral vasoconstriction and a rise in arterial blood pressure.

- An increase in circulating catecholamines causes a decrease in peripheral resistance, with a rise in arterial pressure because of concomitant stimulation of the heart and a rise in cardiac output.

The response in any vascular bed depends on several things: the type of catecholamine, the nature of the receptors involved, and the relationship between stimulation of the receptors and the change in muscle tone. Although stimulation of α-adrenoreceptors usually is associated with vasoconstriction and that of β-adrenoreceptors with vasodilation, this is not invariably the case. An additional complicating factor is that not all sympathetic fibers are adrenergic. In some instances, they may be cholinergic, releasing acetylcholine from their nerve endings. Stimulation of sympathetic cholinergic nerves causes vasodilation in the vasculature of skeletal muscle.

The action of catecholamines is extensively modulated by a variety of substances, including neuropeptide Y and adenosine. **Neuropeptide Y,** first isolated from porcine brain in 1982, is structurally related to mammalian pancreatic polypeptide and peptide YY. Neuropeptide Y is widespread throughout the animal kingdom and, so far, has been identified in many vertebrates and in insects. Neuropeptide Y is co-localized with norepinephrine in sympathetic ganglia and adrenergic nerves; it also is found in many nonadrenergic fibers. The atrial and ventricular myocardium and the coronary arteries are surrounded by nerve fibers that contain neuropeptide Y. In addition, it appears that myocardial cells themselves can synthesize and secrete neuropeptide Y. In general, neuropeptide Y decreases coronary blood flow and the contraction of cardiac muscle by reducing the level of inositol triphosphate ($InsP_3$), an intracellular second messenger (see Chapter 9).

Neuropeptide Y appears to ameliorate those actions of catecholamines on the heart and coronary circulation mediated via InsP$_3$. The role of neuropeptide Y in the peripheral circulation is less well understood, but it appears to ameliorate the increase in blood pressure resulting from norephinephrine-induced peripheral vasoconstriction mediated by α-adrenoreceptors.

ATP, as well as neuropeptide Y, is stored and co-released with catecholamines. ATP and its breakdown product, adenosine, act to inhibit release of catecholamines. Adenosine is released by many tissues during hypoxia but has only a paracrine or autocrine action because of rapid inactivation. Hypoxia tends to promote catecholamine release into the blood from chromaffin tissue, but this action is modulated by the local release of adenosine.

Parasympathetic stimulation   Arterioles in the circulation to the brain and the lungs are innervated by parasympathetic nerves. These nerves contain cholinergic fibers, which release acetylcholine from their nerve endings when stimulated. In mammals, parasympathetic nerve stimulation causes vasodilation in arterioles. Some parasympathetic neurons release ATP and other purines from their endings. Some of these **purinergic** neurons may participate in the control of capillary blood flow. ATP, for instance, causes vasodilation.

Local control of capillary blood flow
Tissues require a basal capillary blood flow to supply nutrients and O$_2$ and to remove waste products. Active tissues have greater requirements and thus capillary blood flow must increase during activity. In addition to nervous control of the central cardiovascular system, various mechanisms control the microcirculation at the local level. For instance, if a vessel is stretched by an increase in input pressure, the vascular smooth muscle responds by contracting, opposing any increase in vessel diameter. This tendency to maintain vessel diameter within narrow limits prevents large changes in resistance to flow and therefore maintains a relatively constant basal flow through the capillary bed. Local heating of a tissue, which may accompany inflammation, is associated with a marked vasodilation, whereas a reduction in temperature causes a vasoconstriction. Thus an ice pack can reduce the blood flow and, therefore, the swelling associated with damage to a tissue.

Numerous compounds also influence capillary blood flow within a tissue. These can be grouped into three types: compounds produced by the vascular endothelium; various vasocontrictors and vasodilators released from other cells; and metabolites associated with increased activity.

Endothelium-produced compounds   The endothelium is not merely a barrier between blood and the underlying tissues, but an active tissue, producing many compounds. Some of these, such as nitric oxide, endothelin, and prostacyclin, affect vascular smooth muscle and, therefore, capillary blood flow.

Nitric oxide is produced and released continuously by the vascular endothelium, causing relaxation of vascular smooth muscle. *Nitric oxide-mediated vasodilator tone* regulates blood flow and pressure in mammals and perhaps other vertebrates. Observation of endothelium-dependent vascular relaxation led to the discovery of endothelium-derived relaxing factor (EDRF). It is now known that this phenomenon results largely from the generation and release of nitric oxide, which activates guanyl cyclase, leading to an elevation of the intracellular second messenger cGMP (cyclic 3', 5' guanosine monophosphate). This compound in turn mediates muscle relaxation.

A family of enzymes, the nitric oxide synthases, oxidize L-arginine to nitric oxide and L-citrulline in the endothelium. Several nitric oxide synthases are calcium dependent, and calcium entry into endothelial cells has been shown to cause the production and release of nitric oxide and relaxation of surrounding smooth muscle. The finding that some calcium channels in the endothelium are stretch sensitive suggests that nitric oxide production in response to vessel stretch may be due to increased calcium entry into the endothelium. A variety of chemicals (e.g., acetylcholine, ATP, and bradykinin) stimulate release of nitric oxide, as does hypoxia, pH change, and increased vessel shear stress. There is evidence of increased production of nitric oxide with increasing pressure associated with each heartbeat.

Nitric oxide synthases have been found in a wide variety of animals including horseshoe crabs, the blood-sucking bug *Rhodnius*, lampreys, and man. Nitric oxide has been shown to have many functions other than maintaining vasodilator tone, so its presence in animals without vascular tone or in nonvascular tissues is not surprising. For example, nitric oxide released in the central nervous system by stimulation of *N*-methyl-D-aspartate receptors is involved in modulation of synaptic activity. Nitric oxide may also be involved in nonspecific defense reactions, the relaxation of nonvascular smooth muscle in the gastrointestinal and genito-urinary tracts, and the regulation of the release of some hormones. In addition, nitric oxide released by endothelial cells, platelets, and leukocytes modulates both cell adhesion and aggregation and inhibits thrombosis.

The vascular endothelium releases endothelins and prostacyclin, as well as nitric oxide. *Endothelins* are small vasoconstrictive proteins, containing 21 amino acid residues. Prostacyclin causes vasodilation and acts as an anticoagulant. It thus functions as an antagonist of the prostaglandin *thromboxane A$_2$*, which promotes blood clotting and causes vasoconstriction.

Inflammatory and other mediators   Thromboxane A$_2$ is formed in the plasma from arachidonic acid released by platelets when they bind to damaged tissues. Although thromboxane levels increase in a damaged tissue and cause vasoconstriction, local injury in mammals is accompanied by a marked vasodilation of vessels in the region of the damage, due largely to the local release of histamine. Histamine is released, not from endothelial cells, but from some connective tissue and white blood cells in injured tissues. Antihistamines ameliorate, but do not completely remove, this inflammatory response. Another group of potent vasodilators, **plasma kinins,** also are activated in damaged tissues. Tissue damage results in the release of proteolytic enzymes that split kininogen, an α_2-globulin, into kinins. Hypoxia also stimulates formation of kinins.

Among the vasocontrictors that act on arterioles are norepinephrine released from sympathetic nerves and **angiotensin II.** Angiotensin is formed, primarily in the lungs, from angiotensinogen, which circulates in the blood (see Chapter 14). Finally, **serotonin** acts as a vasoconstrictor or vasodilator, depending on the vascular bed and on the dose level. It is found in high concentration in the gut and blood platelets.

Histamine, bradykinin, and serotonin cause an increase in capillary permeability. As a result, large proteins and other macromolecules tend to distribute themselves more evenly between plasma and interstitial spaces, reducing the colloid osmotic pressure difference across the capillary wall. Filtration thus increases, and tissue edema occurs. On the other hand, norepinephrine, angiotensin II, and vasopressin tend to promote absorption of fluid from the interstitial fluid into the blood. This absorption can be achieved by reducing filtration pressure and/or the permeability of the capillaries.

> **?** The pulmonary circulation has high levels of the angiotensin-converting enzyme and is involved in catecholamine metabolism. Why are these functions located in the pulmonary circulation?

Metabolic conditions associated with activity When activity in a tissue increases, there must be a concomitant increase in blood flow. Local control of capillary flow ensures that the most active tissue has the most dilated vessels and therefore the most blood flow. The degree of dilation depends on local conditions in the tissue, and those conditions associated with high levels of activity generally cause vasodilation. The term **hyperemia** means increased blood flow to a tissue; **ischemia** means the cessation of flow. **Active hyperemia** refers to the increase in blood flow that follows increased activity in a tissue, particularly skeletal muscle.

Active tissues, metabolizing aerobically, are marked by a decrease in O_2 and an increase in CO_2, H^+, various other metabolites (e.g., adenosine, other ATP breakdown products), and heat. Extracellular K^+ also rises in skeletal muscle following exercise. All of these activity-related metabolic changes, as well as nitric oxide and prostacyclin, have been shown to cause vasodilation and a local increase in capillary blood flow. That is, *the most active tissue has the most dilated vessels and therefore the highest blood flow.*

Although low O_2 levels, indicative of tissue activity, cause vasodilation and increased blood flow in systemic capillaries, the lung capillary bed exhibits the opposite behavior. That is, low O_2 in the lung causes local vasoconstriction rather than vasodilation. The functional significance of this difference relates to the direction of gas transfer. In the lung capillaries, O_2 is taken up by the blood, and thus blood flow should be greatest in the regions of high O_2. In systemic capillaries, however, O_2 leaves the blood for delivery to the tissues, and the highest blood flow should be to the area of greatest need, which is indicated by regions of low O_2.

If blood flow to an organ is stopped by clamping the artery or by a powerful vasoconstriction, there will be a much higher blood flow to that organ when the occlusion is removed than there was before the occlusion. This phenomenon is termed **reactive hyperemia.** Presumably during the ischemic period (a period of no blood flow), O_2 levels are reduced, and CO_2, H^+, and other metabolites build up and cause a local vasodilation. The result is that when the occlusion is removed, blood flow is much higher than normal.

CARDIOVASCULAR RESPONSE TO EXTREME CONDITIONS

In the previous sections, we've described the general organization of the circulation and its regulation under usual conditions. The cardiovascular system responds in characteristic ways during exercise, diving, and hemorrhage to meet the physiological challenge of these extreme conditions.

Exercise

Regulation of the cardiovascular system during exercise is clearly a complex process involving central neural control mechanisms, peripheral neural reflex mechanisms (especially those involving skeletal muscle afferent fibers), and local control. Many cardiovascular changes seen during exercise can occur in the absence of neural mechanisms, indicating the importance of local control systems in increasing blood flow to active skeletal muscles. The central neural control mechanisms and reflexes from muscle afferent mechanoreceptive and chemoreceptive inputs, however, clearly play a role, the exact form varying with the nature of the exercise. For example, the reflex effect on the cardiovascular system of muscle afferent stimulation depends on the nature of the exercise:

- *Isometric contractions* of muscles tend to raise blood pressure with little effect on cardiac output.

- *Isotonic contractions* raise cardiac output but cause little change in arterial blood pressure.

During exercise, blood flow to skeletal muscle is increased in proportion to the level of activity of the muscle. The increase in flow to a muscle may be as much as twenty times; at the same time, transfer of oxygen from the blood to muscle may increase threefold, resulting in a sixtyfold increase in oxygen utilization by the muscle. Active hyperemia is primarily responsible for increasing blood flow to muscle; the resulting decrease in peripheral resistance leads to an increase in cardiac output mediated by sympathetic nerves. At the same time, there is a reduction in flow to the gut, kidney, and, at high levels of exercise, the skin (Figure 12-46). Cardiac output can increase up to ten times above the resting level owing to large increases in heart rate and small changes in stroke volume. Much of the increase in cardiac output can be accounted for by a decrease in peripheral resistance to about 50% of the resting value and by an increase in venous return to the heart due to both the pumping action of skeletal muscles on veins and the increase in breathing associated with exercise.

The increased sympathetic, but decreased parasympathetic, activity in nerves innervating the heart has the effect of increasing both heart rate and the force of contraction, so as to maintain stroke volume at a relatively constant level. In fact, stroke volume increases by about 1.5 times during exercise in mammals, despite the large increase in heart rate and the associated reduced time available for filling and emptying. Following sympathetic stimulation, however, blood is ejected more rapidly from the ventricles with each beat, maintaining stroke volume at higher heart rates. The relative role of changes in stroke volume and heart rate in generating the increase in cardiac output with exercise varies among animals. In fish, for example, the changes in stroke volume are much greater than the changes in heart rate, whereas in birds there are very large

changes in heart rate and little change in stroke volume during exercise.

Exercise is associated with only small changes in arterial blood pressure, pH, and gas tensions. The oscillations in P_{CO_2} and P_{O_2} with breathing are somewhat larger, as is the arterial pressure pulse. The increased pressure pulse is dampened to some extent because of increased elasticity of the arterial walls due to a rise in circulating catecholamines. It is probable that arterial chemoreceptors and baroreceptors play only a minor role in the cardiovascular changes associated with exercise. Motor neurons that innervate skeletal muscle are activated by higher brain centers in the cortex at the onset of exercise (see Chapter 10); it is possible that this activating system also initiates changes in lung ventilation and blood flow. Proprioceptive feedback from muscles may also play a role in increasing lung ventilation and cardiac output (see Chapter 13). A number of other changes augment gas transfer during exercise; for example, red blood cells are released from the spleen in many animals, increasing the oxygen carrying capacity of the blood. Thus, exercise is responsible for a complex series of integrated changes that lead to delivery of adequate oxygen and nutrients to the exercising muscle.

 What are the effects of temperature on exercise performance and capacity in humans and in fish?

Diving

Many air-breathing vertebrates can remain submerged for prolonged periods. During submersion for any period, all air-breathing vertebrates stop breathing, so the animal must rely on available oxygen stores in the blood (see Chapter 13). The cardiovascular system is adjusted to meter out the limited oxygen store to those organs—brain, heart, and some endocrine structures—that can least withstand **anoxia.**

Much of the information on the responses to submersion has been collected from studies of animals forced to dive, sometimes simply by holding an animal's head under water. Because naturally occurring dives vary considerably in depth, duration, and exercise level, information obtained on forced dives is not always directly applicable to natural dives. Whales and dolphins spend their lives in the water going to the surface to breathe, whereas seals may spend considerable time on land out of water. Other animals may spend most of their time on land and dive only occasionally. Oxygen stores vary in animals, so metabolism may be completely aerobic during some dives but largely anaerobic during others.

Figure 12-47 illustrates the typical cardiovascular changes that occur when a seal dives and remains submerged. In mammals, but not in other vertebrates, stimulation of the facial receptors that inhibit breathing cause a marked bradycardia. Although the initial pressurization of

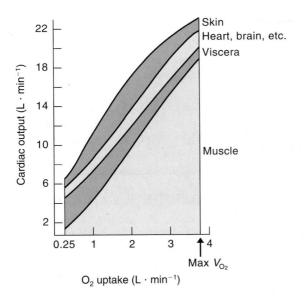

Figure 12-46 During exercise, total cardiac output increases and blood flow shifts to the active muscles. Shown is the approximate distribution of cardiac output at rest and at different levels of exercise up to the maximal oxygen consumption (Max V_{O_2}) in a normal young man. The progressive reduction in the absolute blood flow and percentage of cardiac output distributed to the viscera (splanchnic region and kidneys) augments muscle blood flow. Even skin is constricted during brief periods of exercise at high oxygen consumption. [Adapted from Rowell, 1974.]

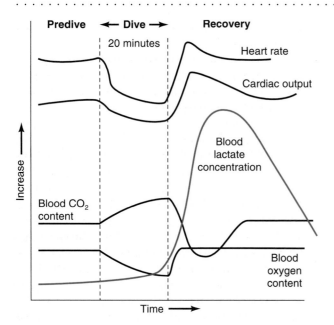

Figure 12-47 The cardiovascular system undergoes numerous adjustments when a seal dives. Heart rate, cardiac output, and blood O_2 content decrease during a dive, but blood CO_2 content increases. During the recovery period after a dive, blood lactate increases greatly; the other parameters first overshoot and then gradually return to predive values.

the lung can lead to a transient increase in blood O_2 and CO_2 levels, the continued utilization of O_2 during the dive results in a gradual fall in blood O_2 and rise in blood CO_2 levels. This fall in blood O_2 stimulates the arterial chemoreceptors and, in the absence of lung stretch-receptor activity, causes peripheral vasoconstriction and a reduction in heart rate and cardiac output; thus blood flow to many tissues is reduced so as to maintain flow to the brain, heart, and some endocrine organs.

The absence of lung stretch-receptor activity is due to the absence of breathing and the compression of the lung as the animal descends in the water column. The increase in peripheral resistance results from a marked rise in sympathetic output and involves constriction of fairly large arteries. Reductions in blood flow to the kidney have been recorded in Weddell seals during a dive. In some instances, blood flow to muscle decreases, but this depends on the level of exercise associated with the dive and the species. Sometimes arterial pressure rises during a dive, causing stimulation of arterial baroreceptors; in such dives bradycardia is maintained by a rise in both chemoreceptor and baroreceptor discharge frequency. The bradycardia is caused by an increase in parasympathetic and, to some extent, a decrease in sympathetic activity in fibers innervating the heart.

It has been shown in the seal that the generation of the diving bradycardia can involve some form of associative learning. In some trained seals, bradycardia occurs before the onset of the dive and, therefore, before the stimulation of any peripheral receptors. This psychogenic influence on heart rate can have a marked effect on the change in heart rate during a dive in many animals. In general, if heart rate

is low before a dive, there may be little or no change in heart rate during the dive. If the heart rate is high, then there may be a marked bradycardia due to wetting the face and a decrease in lung stretch-receptor activity.

The "water" receptors present in birds are not directly involved in the cardiovascular changes associated with submersion. A decrease in heart rate is not observed either in submerged ducks breathing air through a tracheal cannula or in submerged ducks following carotid body denervation (Figure 12-48). Thus, activation of the "water" receptors causes suspension of breathing (**apnea**); the subsequent drop in blood P_{O_2} and pH and the rise in P_{CO_2} result in stimulation of chemoreceptors, which then reflexly cause the cardiovascular changes.

Stimulation of lung stretch receptors in mammals modifies the reflex response initiated by chemoreceptor stimulation. In the absence of breathing, and hence stimulation of lung stretch receptors, different reflex responses are elicited by chemoreceptor stimulation than when the animal is breathing. In the absence of breathing, lung inflation tends to suppress the reflex cardiac inhibition and peripheral vasoconstriction caused by stimulation of arterial chemoreceptors. As a submerged animal rises in the water column, the lung becomes inflated, possibly activating stretch receptors in the lung and causing cardiac acceleration. When the animal is breathing, stimulation of arterial chemoreceptors results in a marked increase in lung ventilation. In this case, low blood O_2 and/or high blood CO_2 levels cause peripheral vasodilation. This vasodilation leads to an increase in cardiac output to maintain arterial pressure in the face of increased peripheral blood flow. Thus, the hypoxia (low oxygen) caused by cessation of breathing during a dive is associated with bradycardia and a reduc-

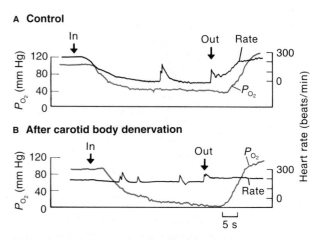

Figure 12-48 The usual decrease in heart rate (bradycardia) that occurs in submerged ducks depends on an intact carotid body innervation. Tracings show heart rate and oxygen tension (P_{O_2}) in the brachiocephalic artery during a period of submergence of the head in water indicated by the in and out arrows. **(A)** Control six-week-old duck with all nerves intact. **(B)** The same duck three weeks after denervation of the carotid bodies. [Jones and Purves, 1970a.]

tion in cardiac output. In contrast, hypoxia that occurs when the animal is breathing (e.g., at high altitude) is associated with an increase in heart rate and cardiac output.

Hemorrhage

Normally stimulation of arterial and atrial baroreceptors inhibits vasopressin release, as well as sympathetic outflow to the peripheral circulation. Hemorrhage reduces both venous and arterial blood pressure, reducing the discharge frequency of both atrial and arterial baroreceptors. This releases the baroreceptive inhibition of sympathetic outflow causing constriction of both arteries (vasoconstriction) and veins (venoconstriction), and an increase in cardiac output. The peripheral vasoconstriction and increased cardiac output raises arterial blood pressure, while the venoconstriction maintains venous return to the heart.

Hemorrhage-induced reduction in baroreceptor inhibition also promotes vasopressin release. In addition, there is an increase in renin/angiotensin/aldosterone activity, resulting from the fall in blood pressure and the associated decreased renal blood flow. Both vasopressin and aldosterone reduce urine formation, thereby conserving plasma volume. There is a marked stimulation of thirst and this helps to restore plasma volume. The reduced renal blood flow promotes kidney production of erythropoietin, which stimulates red blood cell production by the bone marrow. Thus lost red blood cells are replaced by increased production in the days (week) following the hemorrhage. The liver is also stimulated to increase the production of plasma proteins. The increase in production of erythrocytes and plasma proteins, along with the reduction in urine production and increased drinking rate, restores the blood to its original state.

SUMMARY

Circulatory systems can be divided into two broad categories—those with open and those with closed circulations. In open circulatory systems, transmural pressures are low, and blood pumped by the heart empties into a space in which blood bathes the cells directly. In closed circulatory systems, blood passes via capillaries from the arterial to the venous circulation. Transmural pressures are high, and fluid that has slowly leaked across capillary walls into the extracellular spaces is subsequently returned to the circulation via a lymphatic system.

The heart is a muscular pump that ejects blood into the arterial system. Excitation of the heart is initiated in a pacemaker, and the pattern of excitation of the rest of the muscle mass is determined by the nature of the contact between cells. The junctions between muscle fibers in the heart are of low resistance and allow the transfer of electrical activity from one cell to the next.

The initial phase of each heart contraction is isometric; this is followed by an isotonic phase in which blood is ejected into the arterial system. Cardiac output is dependent on venous inflow, and in mammals, changes in cardiac output are associated with changes in heart rate rather than in stroke volume.

Blood flow is generally streamlined (continuous laminar), but because the relationship between pressure and flow is complex, Poiseuille's law applies only to flow in smaller arteries and arterioles.

The arterial system acts as a pressure reservoir and a conduit for blood between the heart and capillaries. The elastic arteries dampen oscillations in pressure and flow caused by contractions of the heart, and the muscular arterioles control the distribution of blood to the capillaries. The venous system acts both as a conduit for blood between capillaries and the heart and as a blood reservoir. In mammals, 50% of the total blood volume is contained in veins.

Capillaries are the site of transfer of material between the blood and tissues. Only 30%–50% of all capillaries are open to blood flow at any particular time, but no capillary remains closed for long, because they all open and close continuously. Capillary blood flow is controlled by nerves that innervate smooth muscle around arterioles. Changes in the composition of blood and extracellular fluid in the region of a capillary bed cause the vessels either to constrict or to dilate, thereby altering blood flow.

The walls of capillaries are generally an order of magnitude more permeable than other cell layers. Material is transferred between blood and tissues by passing either through or between the endothelial cells that form the capillary wall. Endothelial cells contain large numbers of vesicles that may coalesce to form channels for the movement of material through the cell. In addition, some endothelial cells have specific carrier mechanisms for transferring glucose and amino acids. The size of the gaps between cells varies between capillary beds; brain capillaries have tight junctions, whereas liver capillaries have large gaps between cells.

Arterial pressure is regulated via central control mechanisms to maintain capillary blood flow, which can be further adjusted locally to meet the requirements of particular tissues. Arterial baroreceptors monitor blood pressure and reflexly alter cardiac output and peripheral resistance to maintain arterial pressure. Atrial and ventricular mechanoreceptors monitor venous pressure and derivatives of the heart contraction to ensure that activity of the heart is correlated to blood inflow from the venous system and blood outflow into the arterial system. Arterial chemoreceptors respond to changes in the pH and gas levels of the blood. All these sensory receptors feed information into the medullary cardiovascular center, where the inputs are integrated to ensure an appropriate response of the circulatory system to changing requirements of the animal, as during exercise. Natriuretic peptides, vasopressin, and the renin-angiotensin-aldosterone system operate in conjunction with neural reflexes to maintain blood volume following a drink or following hemorrhage.

In general, stimulation of sympathetic nerves innervating vascular smooth muscle causes peripheral vasoconstriction and a rise in arterial blood pressure, whereas an increase in circulating catecholamines (especially epinephrine) causes a decrease in peripheral resistance accompanied by a rise in arterial pressure due to a concomitant rise in cardiac output. The vascular endothelium releases various compounds (e.g., nitric oxide, endothelin, and prostacyclin) that cause localized vasoconstriction or vasodilation, thereby adjusting blood flow to tissue needs. Inflammatory mediators, including histamine and kinins, act to increase blood flow to sites of tissue injury. Finally, as aerobic metabolism in a tissue increases, there is a local increase in capillary blood flow, termed active hyperemia. This assures that the most active tissues normally have the highest capillary blood flow.

REVIEW QUESTIONS

1. Describe the properties of myogenic pacemakers.
2. Describe the transmission of excitation over the mammalian heart.
3. Describe the changes in pressure and flow during a single beat of the mammalian heart.
4. Discuss the factors that influence stroke volume of the heart.
5. What is the nature and function of the nervous innervation of the mammalian heart?
6. What is the effect on cardiac function of a rigid versus a compliant pericardium?
7. What is the functional significance of a partially divided ventricle in some reptiles?
8. Discuss the changes in circulation that occur at birth in the mammalian fetus.
9. Discuss the applicability of Poiseuille's equation to the relationship between pressure and flow in the circulation.
10. What are the functions served by the arterial system?
11. Describe the factors that determine capillary blood flow.
12. Describe the location of various baroreceptors and/or mechanoreceptors in the mammalian circulatory system and their role in cardiovascular regulation.
13. Compare and contrast the cardiovascular responses to breathing air low in oxygen with those associated with diving in mammals.
14. Describe the cardiovascular changes associated with exercise in mammals.
15. What are the consequences of raising or lowering arterial blood pressure for cardiac function and for exchange across capillary walls?
16. Discuss the relationship between capillary structure and organ function, comparing that found in different organs of the body.
17. Describe the ways in which substances are transferred between blood and tissues across capillary walls.
18. What are the functions served by the venous system?
19. Describe the effects of gravity on blood circulation in a terrestrial mammal. How are these effects altered if the animal is in water?
20. Define Laplace's law. Discuss the law in the context of the structure of the cardiovascular system.
21. Discuss the role of the lymphatic system in fluid circulation. Discuss how and why this role may vary in different parts of the body.

SUGGESTED READINGS

Bundgaard, M. 1980. Transport pathways in capillaries: in search of pores. *Ann. Rev. Physiol.* 42:325–326.

Crone, C. 1980. Ariadne's thread: an autobiographical essay on capillary permeability. *Microvasc. Res.* 20:133–149.

Heisler, N., ed. 1995. Mechanisms of Systemic Regulation: Respiration and Circulation. *Adv. Comp. Environ. Physiol.*, Vol. 21.

Hoar, W. S., D. J. Randall, and A. P. Farrell, eds. 1992. *Fish Physiology*. Vol XIIIA & B. New York: Academic Press.

Johansen, K., and W. Burggren, eds. 1985. *Cardiovascular Shunts*. (Alfred Benzon Symposium 21.) Copenhagen: Munksgaard.

Kooyman, G. L. 1989. *Diverse Divers*. Zoophysiology. Vol. 23. New York: Springer-Verlag.

Kuby, J. 1997. Leukocyte migration and inflammation. In *Immunology*, 3d ed. New York: W. H. Freeman.

Lewis, D. H., ed. 1979. Lymph circulation. *Acta Physiol. Scand.*, Suppl. 463.

Radomski, M. W., and E. Salas. 1995. Biological significance of nitric oxide. 4th Int. Congress. *Comp. Physiol. Biochem. Physiol. Zool.* 68:33–36.

Schmidt-Nielsen, K. 1972. *How Animals Work*. New York: Cambridge University Press.

Van Vilet, B. N., and N. H. West. 1994. Phylogenetic trends in the baroreceptor control of arterial blood pressure. *Physiol. Zool.* 67(6):1284–1304.

13

GAS EXCHANGE AND
ACID-BASE BALANCE

Only 200 years ago Antoine Lavoisier showed that animals utilize oxygen and produce carbon dioxide and heat (Spotlight 13-1). This process was later shown to take place at the level of the mitochondria (see Chapter 3). Animals obtain oxygen from the environment, using it for cellular respiration. The carbon dioxide generated is eventually liberated into the environment. For cellular respiration to proceed, there must be a steady supply of oxygen, and the waste product carbon dioxide must be continually removed. If carbon dioxide accumulates in the body, pH falls and the animal dies. Although the transport of oxygen and carbon dioxide occur in opposite directions, both processes have many elements in common. If gas transport is impaired, animals die due to lack of oxygen rather than accumulation of carbon dioxide, because oxygen is required for metabolism to continue and carbon dioxide is the product of aerobic metabolism. Air contains about 21% oxygen, but almost no carbon dioxide, the remainder being mostly nitrogen. Carbon dioxide added to the environment by animals is removed by photosynthetic bacteria, plants, and algae, which produce oxygen. This cycling of O_2 and CO_2 is part of the vast interdependency that exists between plants and animals.

In this chapter we review the transport of O_2 and CO_2 in the blood and the systems that have evolved in animals to facilitate the movement of these two gases both between the environment and the blood and between the blood and the tissues. The main focus is on systems found in vertebrates, particularly mammals, because these have been investigated most thoroughly. A number of systems that transport O_2 between the environment and tissues are of particular interest, including the one that moves oxygen into the swimbladder of fish against gradients that can be several atmospheres. This is described at the end of this chapter as an example of one the many intriguing problems of gas transfer in animals.

GENERAL CONSIDERATIONS

Oxygen and carbon dioxide are transferred passively from the environment across the body surface (i.e., skin or special respiratory epithelium) by diffusion. Relevant physical laws regarding the behavior of gases, along with some of the terminology used in respiratory physiology, are reviewed in Spotlight 13-2. To facilitate the rate of gas transfer for a given concentration difference, the surface area of the respiratory epithelium should be as large as possible and diffusion distances as small as possible.

The O_2 requirements and CO_2 production of an animal increase as a function of mass, but the rate of gas transfer across the body surface is related primarily to surface area. The surface area of a sphere increases as the square of its diameter, whereas the volume increases as the cube of its diameter. In very small animals the distances for diffusion are small, and the ratio of surface area to volume is large. For this reason, diffusion alone is sufficient for the transfer of gases in small animals, such as rotifers and protozoa, which are less than 0.5 mm in diameter. Increases in size result in increases in diffusion distances and reductions in the ratio of surface area to volume. Large surface-area-to-volume ratios are maintained in larger animals by the elaboration of special areas for the exchange of gases. In some animals the whole body surface participates in gas transfer, but in large, active animals there is a specialized respiratory surface. This surface is made up of a thin layer of cells, the *respiratory epithelium*, which is 0.5 to 15 μm thick. This surface comprises the major portion of the total body surface. In humans, for instance, the respiratory surface area of the lung is between 50 and 100 m^2, varying with age and lung inflation; the area of the rest of the body surface is less than 2 m^2.

Gas transfer between the environment and eggs, embryos, many larvae, and even some adult amphibians occurs by simple diffusion. Boundary layers of fluid low in oxygen and high in carbon dioxide are found whenever gas transfer occurs by diffusion alone. The thickness of this hypoxic (low-oxygen) layer increases with animal size, oxygen uptake, and decreasing temperatures. Stagnation of the medium close to the gas-exchange surface is avoided, in most animals, by the movement of air or water by breathing. A circulatory system has evolved in larger animals to transfer oxygen and carbon dioxide by the flow

SPOTLIGHT 13-1

EARLY EXPERIMENTS ON GAS EXCHANGE IN ANIMALS

Poul Astrup (1915–) and John Severinghaus (1922–), two prominent scientists in the field of gas exchange, described many of the most significant experiments leading to our present understanding of gas transfer in animals in their book *The History of Blood Gases, Acids and Bases* published in 1986. Studies of gas exchange in animals began as an extension of Robert Boyle's (1627–1691) work in the 17th century on the properties of air. He showed that both animals and flames died in a vacuum, indicating that something in air was required both to maintain life and to keep a candle burning.

Joseph Priestley (1733–1804), who lived near a brewery, was fascinated by the large volumes of gas produced during the brewing process. Continuing Boyle's experiments in modified form, Priestly heated various chemicals, collected the gases produced over water or mercury, and then determined if mice could live in the gases. He noticed that a mouse lived longer and the flame burned brighter in the gas produced by heating mercuric oxide than in the gas produced from other chemicals. He also observed that mice lived longer if plant material was present in their containers. Priestley's observations caused Benjamin Franklin to remark that the practice of cutting down trees near houses should cease as plants were able to restore air, which is spoiled by animals. Thus Priestley demonstrated that plants, as well as certain chemicals when heated, could produce some gas that keeps animals and flames alive. He thought this gas could absorb *phlogiston*, something that was released when material

was burned. According to this theory, coal contained a great deal of phlogiston which was released into the air during combustion leaving behind ash. That is, when substances were burned they lost phlogiston and, therefore, lost weight.

Antoine Lavoisier (1743–1794), however, found that phosphorus gained weight when burned in air and that some other substances, when heated in air, gained weight but did not do so if heated in a vacuum. In other words, something in air was consumed when some substances were heated. This was the end of the phlogiston theory. Lavoisier called the substance that was consumed during burning and was required to keep animals alive *oxygen*, from Greek words meaning "to form acid."

Lavoisier repeated some experiments of Henry Cavendish (1731–1810), who found that the *inflammable gas* evolved when metals are added to acid can combine with oxygen to form water. Lavoisier named this gas *hydrogen*, from Greek words meaning "to form water." He also repeated and extended some of Priestley's experiments and found that if mercuric oxide was heated with coal, *fixed air* (carbon dioxide) was formed. Fixed air had been described earlier by Joseph Black (1728–1799), who produced it by adding acid to chalk.

Expired air was known to contain some fixed air, and Lavoisier made the next large step. He realized that both burning coal and animals consumed oxygen and produced heat and carbon dioxide. He then measured oxygen uptake and heat production in animals and found that the amount of heat produced relative to oxygen uptake was about the same for animals and burning coal, although the rates of these processes were much slower in animals.

Lavoisier was also a tax collector. Such people are generally not held in high repute and this brilliant scientist was no exception: he was sent to the guillotine in 1794.

---

of blood between the tissues and the respiratory epithelium. Blood flows through an extensive capillary network and is spread in a thin film just beneath the respiratory surface, thereby reducing diffusion distances required to distribute the contained gases. The gases are transported between the respiratory surface and the tissues by bulk flow of blood in the circulatory system. Gases diffuse between blood and tissues across the capillary wall. Once again, to facilitate gas transfer, the area for diffusion is large, and the diffusion distance between any cell and the nearest capillary is small. **Graham's law** states that the rate of diffusion of a substance along a given gradient is inversely proportional to the square root of its molecular weight (or density). Because oxygen and carbon dioxide molecules are of similar size, they diffuse at similar rates in air; they are also utilized (O_2) and produced (CO_2) at approximately the same rates by animals. It can therefore be expected that a transfer system that meets the oxygen requirements of an animal will also ensure adequate rates of carbon dioxide removal.

Figure 13-1 schematically illustrates the components of the gas-transfer system in many animals, which involves four basic steps:

1. Breathing movements, which assure a continual supply of air or water to the respiratory surface (e.g., lungs or gills)
2. Diffusion of O_2 and CO_2 across the respiratory epithelium
3. Bulk transport of gases by the blood
4. Diffusion of O_2 and CO_2 across capillary walls between blood and mitochondria in tissue cells

The capacity of each of these steps is matched because natural selection tends to eliminate metabolically costly unutilized capacities. This matching of capacities in a chain of linked events has been referred to as *symmorphosis*. Presumably the capacities of elements in a chain will be determined by the capacity of the rate-limiting step. Capacities in a chain of events, however, are not always matched, and symmorphosis draws

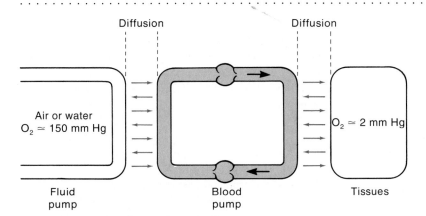

Figure 13-1 The gas-transport system of a vertebrate consists of two pumps and two diffusion barriers alternating in series between the external environment and the tissues. [Adapted from Rahn, 1967.]

attention to these apparently uneconomic design features. One explanation for over- or under-capacity is that a single element can be a link in several chains; thus its capacity may be appropriate for one chain of events, but be in excess for another, explaining the apparent excess capacity.

The rate of flux of gases varies enormously among animals, from 0.08 ml·g^{-1}·h^{-1} in an earth worm to 40 ml·g^{-1}·h^{-1} in a hovering hummingbird. The concentrations of aerobic enzymes (e.g., cytochrome oxidase) and cristae area per mitochondrion both increase with metabolic rate. The hummingbird and some insects, however, may have reached the upper limit for rates of oxygen utilization by animals. Clearly mitochondrial volume and density in muscles cannot be increased indefinitely without compromising the capacity of the muscles to contract; that is, there must be some relationship between structures that supply energy (mitochondria) and structures that use energy (myofilaments). The space occupied by mitochondria never exceeds 45% of the total volume in muscle, even in mammals, birds, and insects, the animals with the highest oxygen uptakes. There must also be limits to mitochondrial design in terms of the number of cristae per unit mitochondrial volume, the ultimate miniaturization being determined by the minimum volume required by the enzymes involved in energy production. It would seem that hummingbirds, and perhaps some other small mammals and a few insects, may have approached these limits of design determining the maximum rates of oxygen uptake.

Insects are usually much smaller than the smallest birds and mammals. Some large insects seem to have been displaced by small birds, rather like the monoplane displacing the biplane before World War II. Vertebrate miniaturization may be limited by the nature of their gas-transfer systems. Insects have a tracheal system that exchanges gases directly between the medium and the tissues, permitting high rates of oxygen uptake in very small animals.

> **?** What are the advantages and disadvantages of a tracheal system, compared with a blood circulation system, in the transfer of gases between the environment and the tissues?

OXYGEN AND CARBON DIOXIDE IN BLOOD

In considering the movement of oxygen and carbon dioxide between the environment and the cells, we will first discuss how these gases are transported in the blood, rather than starting with either the environment or the cell. We take this approach because the mechanisms by which oxygen and carbon dioxide are carried in the blood affects their transfer between the environment and the blood and the blood and the tissues.

Respiratory Pigments

Once oxygen diffuses across the respiratory epithelium into the blood, it combines with a **respiratory pigment** that gives the characteristic color to the blood. The best known respiratory pigment, **hemoglobin**, is red. By binding oxygen, the respiratory pigment increases the O$_2$ content of blood. In the absence of a respiratory pigment, the O$_2$ content of blood would be low. The **Bunsen solubility coefficient** of oxygen in blood at 37°C is 2.4 ml O$_2$ per 100 ml of blood per atmosphere of oxygen pressure. Therefore, the concentration of O$_2$ in physical solution (i.e., not bound to a respiratory pigment) in human blood at a normal arterial P_{O_2} will be only 0.3 ml O$_2$ per 100 ml blood, or 0.3 vol % O$_2$. In fact, the total O$_2$ content of human arterial blood at a normal arterial P_{O_2} is 20 vol %. The 70-fold increase in content is due to the combination of oxygen with hemoglobin. In most animals using hemoglobin as a respiratory pigment, the O$_2$ content in physical solution is only a small fraction of the total O$_2$ content of the blood. The Antarctic icefish is an exception among the vertebrates; the blood of this fish lacks a respiratory pigment and therefore has a low O$_2$ content. It compensates for the absence of hemoglobin with an increased blood volume and cardiac output, but its rate of O$_2$ uptake is reduced compared with that of species from the same habitat that have hemoglobin. Low temperatures probably are a factor in the evolution of fishes lacking hemoglobin. Low temperatures are associated with low metabolic rates in poikilotherms, and oxygen, like all gases, has a higher solubility at low temperatures.

Respiratory pigments are complexes of proteins and metallic ions, and each one has a characteristic color. The

SPOTLIGHT 13-2

THE GAS LAWS

Over 300 years ago, Robert Boyle determined that at a given temperature the product of pressure and volume is constant for a given number of molecules of gas. **Gay-Lussac's law** states that either the pressure or the volume of a gas is directly proportional to absolute temperature if the other is held constant. Combined, these laws are expressed in the equation of state for a gas:

$$PV = nR\,K$$

where P is pressure, V is volume, n is number of molecules of a gas, R is the universal gas constant ($0.08205\ L \cdot atm \cdot K^{-1} \cdot mol^{-1}$, or $8.314 \times 10^{7}\ ergs \cdot {}^{\circ}K^{-1} \cdot mol^{-1}$, or $1.987\ cal \cdot K^{-1} \cdot mol^{-1}$), and K is absolute temperature. For accurate use, the equation should be modified by using van der Waals constants.

The equation of state for a gas indicates that equal volumes of different gases at the same temperature and pressure contain equal numbers of molecules **(Avogadro's law)**. One mole of gas occupies approximately 22.414 liters at 0°C and 760 mm Hg. Because the number of molecules per unit volume is dependent on pressure and temperature, the conditions should always be stated along with the volume of gas. Gas volumes in physiology are usually reported as being at body temperature, atmospheric pressure, and saturated with water vapor (BTPS); at ambient temperature and pressure, saturated with water vapor (ATPS); or at standard temperature and pressure (0°C, 760 mm Hg) and dry, or zero water vapor pressure (STPD).

Gas volumes measured under one set of conditions (e.g., ATPS) can be converted to another (e.g., BTPS) by using the equation of state for a gas. For example, the volume of air expired from a mammalian lung at a body temperature of 37°C (273 + 37 = 310 K) is often measured at ambient room temperature, say 20°C (273 + 20 = 293 K). The drop in temperature will reduce the expired gas volume. A gas in contact with water will be saturated with water vapor. The water vapor pressure at 100% saturation varies with temperature. Expired air is saturated with

water, but as temperature decreases, water will condense, and this condensation will also reduce the expired gas volume. If the barometric pressure is 760 mm Hg, and the water vapor pressure at 37° and 20°C is 47 mm Hg and 17 mm Hg, respectively, then a measured gas volume at 20°C of 500 ml is converted to BTPS expired volume as follows:

$$500\ ml \times \frac{(760 - 17)}{(760 - 47)} \times \frac{(273 + 37)}{(273 + 20)} = 551\ ml$$

Thus, under the conditions stated above, a gas volume of 551 ml within the lung is reduced to 500 ml following exhalation because of the drop in gas temperature and the condensation of water.

Dalton's law of partial pressure states that the partial pressure of each gas in a mixture is independent of other gases present, so that the total pressure equals the sum of the partial pressures of all gases present. The partial pressure of a gas in a mixture will depend on the number of molecules present in a given volume at a given temperature. Usually, oxygen accounts for 20.94% of all gas molecules present in dry air; thus, if the total pressure is 760 mm Hg, the partial pressure of oxygen, P_{O_2}, will be $760 \times 0.2094 = 159$ mm Hg. But air usually contains water vapor, which contributes to the total pressure. If the air is 50% saturated with water vapor at 22°C, the water vapor pressure is 18 mm Hg. If the total pressure is 760, the partial pressure of oxygen will be $(760 - 18) \times 0.2094 = 155$ mm Hg. If the partial pressure of CO_2 in a gas mixture is 7.6 mm Hg and the total pressure is 760 mm Hg, then 1% of the molecules in air are CO_2.

Gases are soluble in liquids. The quantity of gas that dissolves at a given temperature is proportional to the partial pressure of that gas in the gas phase **(Henry's law)**. The quantity of gas in solution equals $\alpha\,P$, where P is the partial pressure of the gas and α is the **Bunsen solubility coefficient,** which is independent of P. The Bunsen solubility coefficient varies with the type of gas, the temperature, and the liquid in question, but is constant for any one gas in a given liquid at constant temperature. The Bunsen solubility coefficient for oxygen decreases with increases in ionic strength and temperature of water.

---

color of a respiratory pigment changes with its O_2 content. Thus, hemoglobin, which is bright red when it is loaded with O_2, becomes a dark maroon-red when deoxygenated. Vertebrate hemoglobin, except that of cyclostomes, has a molecular weight of 68,000 and contains four iron-containing porphyrin prosthetic groups, called **heme**, associated with globin, a tetrameric protein (Figure 13-2A). The globin molecule consists of two dimers, $\alpha_1\beta_1$ and $\alpha_2\beta_2$, each of which is a tightly cohering unit. The two dimers are more loosely connected to each other by salt bridges, except that the two β chains do not touch. Oxygenation alters these bridges, leading to conformational changes in the hemoglobin molecule. Hemoglobin can be dissociated

into four subunits of approximately equal weight, each containing one polypeptide chain and one heme group. **Myoglobin,** a respiratory pigment that stores O_2 in vertebrate muscles, is equivalent to one hemoglobin subunit and exhibits considerable sequence homology with the hemoglobin α chain.

In a hemoglobin molecule, iron in the ferrous state (Fe^{2+}) is bound into the porphyrin ring of the heme, forming coordinate links with the four pyrrole nitrogens (Figure 13-2B). The two remaining coordinate linkages are used to bind the heme group to an O_2 molecule and to the imidazole ring of a histidine residue in the globin (Figure 13-2C). If O_2 is bound, the molecule is referred to

A

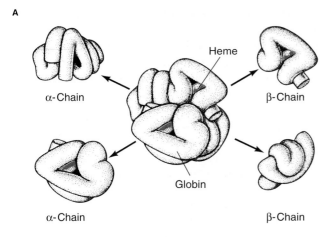

α-Chain

Heme

β-Chain

Globin

α-Chain

β-Chain

B

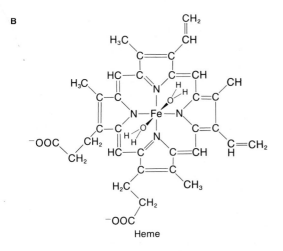

Heme

C

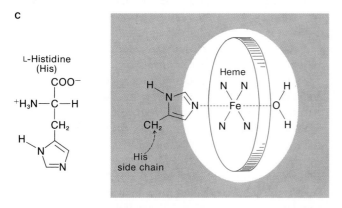

L-Histidine
(His)

Heme

His
side chain

Figure 13-2 Hemoglobin, the main respiratory pigment in vertebrates, consists of four globin protein subunits, each containing one heme molecule. **(A)** Schematic diagram of hemoglobin molecule, showing relationship of the α and β chains. Two of the four heme units (red) are visible in the folds formed by the polypeptide chains. **(B)** Structure of heme, formed by the combination of ferrous ion (Fe^{2+}) and protoporphyrin IX. **(C)** Schematic diagram of heme in a pocket formed by the globin molecule. The side chain of a histidine (His) residue in globin acts as an additional ligand for the iron atom in heme. When oxygen binds, it displaces the remaining H_2O ligand. [Adapted from McGilvery, 1970.]

as **oxyhemoglobin;** if O_2 is absent, it is called **deoxyhemoglobin.** Binding of O_2 to hemoglobin to form oxyhemoglobin does not oxidize ferrous to ferric iron. Oxidation of the ferrous iron in hemoglobin to the ferric state produces **methemoglobin,** which does not bind O_2 and therefore is nonfunctional. Although formation of methemoglobin occurs normally, red blood cells contain the enzyme methemoglobin reductase, which reduces methemoglobin to the functional ferrous form. Certain compounds (e.g., nitrites and chlorates) act either to oxidize hemoglobin or to inactivate methemoglobin reductase, thereby increasing the level of methemoglobin and impairing oxygen transport.

The affinity of hemoglobin for carbon monoxide is about 200 times greater than its affinity for oxygen. As a result, carbon monoxide will displace oxygen and saturate hemoglobin, even at very low partial pressures of carbon monoxide, causing a marked reduction in oxygen transport to the tissues. Hemoglobin saturated with carbon monoxide is called **carboxyhemoglobin.** The effect of such saturation on oxidative metabolism is similar to that of oxygen deprivation, which is why the carbon monoxide produced by cars or improperly stoked coal or wood stoves is so toxic. Even the levels found in city traffic can impair brain function owing to partial anoxia.

Hemoglobin is found in many invertebrate groups, but others possess different respiratory pigments, including **hemerythrin** (Priapulida, Brachiopoda, Annelida), **chlorocruorin** (Annelida), and **hemocyanin** (Mollusca, Arthropoda). Many invertebrates do not have a respiratory pigment. Hemocyanin, a large, copper-containing respiratory pigment, has many properties similar to those of hemoglobin, binding oxygen when the partial pressure is high and releasing it when the partial pressure is low. Hemocyanin binds oxygen in the ratio of 1 mol of O_2 to approximately 75,000 g of the respiratory pigment. In comparison, 4 mol of O_2 bind to 68,000 g of hemoglobin when it is completely saturated. Unlike hemoglobin, hemocyanin is not packaged in cells and is not associated with high levels of carbonic anhydrase in the blood. In its oxygenated form, it is light blue; in its unoxygenated form, it is colorless.

Oxygen Transport in Blood

Each hemoglobin molecule can combine with four oxygen molecules, each heme combining with one molecule of oxygen. The extent to which O_2 is bound to hemoglobin varies with the partial pressure of the gas, P_{O_2}. If all sites on the hemoglobin molecule are occupied by O_2, the blood is 100% saturated and the oxygen *content* of the blood is equal to its oxygen capacity. A millimole of heme can bind a millimole of O_2, which represents a volume of 22.4 ml of O_2. Human blood contains about 0.9 mmol of heme per 100 ml of blood. The oxygen capacity is therefore $0.9 \times 22.4 = 20.2$ vol %. The oxygen content of a unit volume of blood includes the O_2 in physical solution as well as that combined with hemoglobin, but in most cases the O_2 in physical solution is only a small fraction of the total O_2 content.

Because the oxygen capacity of blood increases in proportion to its hemoglobin concentration, the oxygen content commonly is expressed as a percentage of the oxygen capacity, that is, the *percent saturation*. This makes it possible to compare the oxygen content of blood of different hemoglobin content. *Oxygen dissociation curves* describe the relationship between percent saturation and the partial pressure of oxygen.

The oxygen dissociation curves of myoglobin and lamprey hemoglobin are hyperbolic, whereas the oxygen dissociation curves of other vertebrate hemoglobins are sigmoid (Figure 13-3). This difference occurs because myoglobin and lamprey hemoglobin have a single heme group, but other hemoglobins have four heme groups. The sigmoid shape of the dissociation curves exhibited by hemoglobins having several heme groups results from *subunit cooperativity;* that is, oxygenation of the first heme groups facilitates oxygenation of subsequent heme groups. The steep portion of the curve corresponds to oxygen levels at which at least one heme group is already occupied by an oxygen molecule, increasing the affinity of the remaining heme groups for oxygen. As a hemoglobin molecule is oxygenated, it goes through a conformational change from the *tense (T) state* to the *relaxed (R) state.* Oxygenation is associated with changes in the tertiary structure near the hemes that weaken or break connections between the $\alpha_1\beta_1$ and $\alpha_2\beta_2$ dimers, leading to a large change in the quaternary structure from the T to the R state. These conformational changes also produce changes in the dissociation of acidic side chains, so that protons (H^+ ions) are released as hemoglobin is oxygenated.

An important property of respiratory pigments is that they combine reversibly with O_2 over the range of partial pressures normally encountered in the animal. At low P_{O_2}, only a small amount of O_2 binds to the respiratory pigment; at high P_{O_2}, however, a large amount of O_2 is bound. Because of this property, the respiratory pigment can act as an oxygen carrier, loading at the respiratory surface (a region of high P_{O_2}) and unloading at tissues (a region of low P_{O_2}). In some animals, the predominant role of a respiratory pigment may be to serve as an oxygen reservoir, releasing O_2 to the tissues only when O_2 is relatively unavailable. In many animals at rest, the venous blood entering the lung or gills is around 70% saturated with oxygen; that is, most of the oxygen bound to hemoglobin is not removed during transit through the tissues. During exercise, when the oxygen demand by the tissues is increased, this venous reservoir of oxygen is tapped and venous saturation may drop to 30% or less.

Hemoglobins that have high oxygen affinities are saturated at low partial pressures of oxygen, whereas hemoglobins with low oxygen affinities are completely saturated only at relatively high partial pressures of oxygen. The affinity is expressed in terms of the P_{50}, the partial pressure of oxygen at which the hemoglobin is 50% saturated with oxygen; the lower the P_{50}, the higher the oxygen affinity. As the curves in Figure 13-3 demonstrate, myoglobin has a much higher oxygen affinity than hemoglobin. Variations in oxygen affinity among hemoglobins are related to differences in the protein globin, not to differences in the heme group. Each α and β chain of the globin molecule consists of between 141 and 147 amino acids, depending on the chain and the hemoglobin in question. The amino acid sequences of both the α and the β chains from different hemoglobins exhibit many similarities, but there are some differences. Although most amino acid substitutions are neutral, some have a marked impact on function. For example, a genetic defect resulting in substitution of valine for glutamic acid in position 6 of the β chain causes human hemoglobins to form large polymers that distort the erythrocyte into a sickle shape, giving rise to *sickle cell anemia.* Because these sickle cells cannot pass through small blood vessels, oxygen delivery to tissues is impaired. Individuals with both normal and sickle cell hemoglobins suffer only mild debilitation but have greater resistance to malaria, thus ensuring the continuation of the sickle cell gene in the population. Certain amino acids in globin bind various ligands, and substitution of these residues can cause changes in the oxygen affinity of hemoglobin.

The rate of oxygen transfer to and from blood increases in proportion to the difference in P_{O_2} across an epithelium. A hemoglobin with a high oxygen affinity facilitates the movement of O_2 into the blood from the environment because O_2 is bound to hemoglobin at low P_{O_2}; i.e., O_2 entering the blood is immediately bound to hemoglobin, so O_2 is removed from solution and P_{O_2} is kept low. Thus, a large difference in P_{O_2} is maintained across the respiratory epithelium—and therefore a high rate of oxygen transfer into the blood—until hemoglobin is fully saturated. Only then does blood P_{O_2} rise. Hemoglobin with a high oxygen affin-

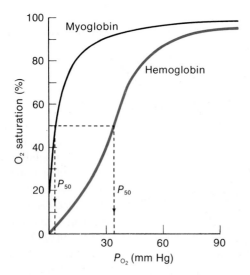

Figure 13-3 Hemoglobins with multiple heme groups have sigmoid oxygen dissociation curves, whereas myoglobin with only a single heme group has a hyperbolic dissociation curve. Lamprey hemoglobin, with a single heme group, has a dissociation curve similar to that of myoglobin. P_{50}, the partial pressure at which a respiratory pigment is 50% saturated with oxygen, is a measure of its oxygen affinity.

ity, however, will not release O_2 to the tissues until the P_{O_2} is very low. In contrast, a hemoglobin with a low oxygen affinity will facilitate the release of O_2 to the tissues, maintaining large differences in P_{O_2} between blood and tissues and a high rate of oxygen transfer to the tissues. Thus, a hemoglobin of high oxygen affinity favors the uptake of O_2 by the blood, whereas a hemoglobin of low oxygen affinity facilitates the release of O_2 to the tissues. From a functional viewpoint, therefore, hemoglobin should have a low O_2 affinity in the tissues and a high O_2 affinity at the respiratory surface. In light of this, it is highly significant that the oxygen affinity of hemoglobin is affected by changes in chemical and physical factors in the blood that favor oxygen binding at the respiratory epithelium and oxygen release in the tissues.

The hemoglobin-oxygen affinity is labile and dependent on the conditions within the red blood cell. For instance, the hemoglobin-oxygen affinity is reduced by the following:

- Elevated temperature

- Binding of organic phosphate ligands including 2,3-diphosphoglycerate (DPG), ATP, or GTP

- Decrease in pH (increase in H^+ concentration)

- Increase in CO_2

The hemoglobin molecule has a much higher affinity for ligands when it is in the T, or deoxygenated, state.

Increases in H^+ concentration (decreases in pH) cause a reduction in the oxygen affinity of hemoglobin, a phenomenon termed the **Bohr effect,** or Bohr shift (Figure 13-4). Carbon dioxide reacts with water to form carbonic acid and reacts with $-NH_2$ groups on plasma proteins and hemoglobin to form carbamino compounds. Thus an increase in P_{CO_2} causes a reduction in the oxygen affinity of hemoglobin in two ways: by decreasing blood pH (Bohr effect) and by promoting the direct combination of CO_2 with hemoglobin to form carbamino compounds. Therefore, when CO_2 enters the blood at the tissues it facilitates the unloading of O_2 from hemoglobin, whereas when CO_2 leaves the blood at the lung or gill, it facilitates the uptake of O_2 by the blood. The oxygen dissociation curve for myoglobin, unlike that for hemoglobin, is relatively insensitive to changes in pH.

Hemocyanins from the Dungeness crab, *Cancer magister,* and some other invertebrates exhibit a Bohr shift similar to that of hemoglobin (Figure 13-5). But hemocyanins from several gastropods and from the horseshoe crab, *Limulus,* show a greater oxygen affinity with a decrease in pH. This phenomenon, referred to as a *reverse Bohr effect,* may facilitate oxygen uptake during periods of low oxygen availability when prolonged reductions in blood pH occur in these animals.

As noted above, the binding of organic phosphate compounds to hemoglobin reduces the oxygen affinity of most vertebrate hemoglobins, except those from cyclostomes, crocodiles, and ruminants. The dominant erythrocytic organophosphate differs among species. For instance, mammalian erythrocytes contain high levels of 2,3-diphosphoglycerate (DPG); indeed, hemoglobin and DPG are nearly equimolar in human erythrocytes. DPG binds to specific amino acid residues in the β chains of deoxyhemoglobin, but DPG binding decreases with increasing pH. Increases in DPG levels accompany reductions in blood O_2 or hemoglobin concentrations, increases in pH, or both. Low blood O_2 levels may result from a climb to a higher

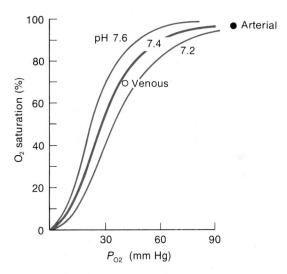

Figure 13-4 The oxygen affinity of hemoglobin decreases with decreasing pH. Because of this phenomenon, called the Bohr effect, changes in blood P_{CO_2}, which influence blood pH, indirectly affect hemoglobin-oxygen affinity. Shown are experimental blood oxygen dissociation curves in humans at three pH values. The P_{O_2} values of mixed venous and arterial blood are indicated. [Adapted from Bartels, 1971.]

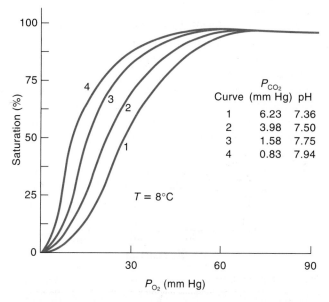

Figure 13-5 Some hemocyanins, like hemoglobin, exhibit a Bohr shift. Blood oxygen dissociation curves for the crab *Cancer magister* shown here indicate that the hemocyanin from this crab shows a Bohr shift. [Unpublished data supplied by D. G. McDonald.]

altitude, as both barometric pressure and the partial pressure of O_2 in air decrease with altitude. The resultant DPG rise in humans in response to high altitude is completed in 24 hours with a half-time of about 6 hours. At elevations of 3000 m, the DPG concentration in erythrocytes is 10% greater than it is at sea level. The low O_2 levels at altitude result in a decrease in blood O_2 levels, which stimulates breathing. The resulting increase in **ventilation** (i.e., exchange of air between the lungs and ambient air) reduces CO_2 levels in the blood and raises blood pH, which increases hemoglobin-oxygen affinity. The elevation in DPG at altitude offsets the effects of reduced CO_2 levels and maintains hemoglobin-oxygen affinity close to that at sea level.

In the erythrocytes of some vertebrates, other phosphorylated compounds are present in higher concentration than DPG, and consequently they play a more important role in modulating the oxygen affinity of hemoglobin than does DPG. In most fishes ATP and/or GTP has this function, whereas inositol pentaphosphate ($InsP_5$) is the dominant erythrocytic organophosphate in birds. In the Amazonian fish *Arapaima gigas,* ATP is the dominant erythrocytic organophosphate in the young aquatic form, but $InsP_5$ is dominant in the obligate air-breathing adult.

Phosphorylated compounds in the erythrocyte not only affect the oxygen affinity of hemoglobin but also increase the magnitude of the Bohr effect and may affect subunit interaction. It appears that in mammals the functional significance of increased DPG levels is to maintain hemoglobin-oxygen affinity under hypoxic (low-oxygen) conditions, as at high altitude. In contrast, hypoxia reduces erythrocytic organophosphate levels in fishes. In these animals, however, hypoxia is often associated with a decrease in blood pH (**acidosis**), rather than the increase in pH (**alkalosis**) seen in mammals at altitude. The effect of the reduction in ATP (or GTP) in fishes is to offset the effects of this hypoxia-associated acidosis, thereby maintaining blood-oxygen affinity. Thus in a functional sense the effects of changing erythrocytic organophosphate levels are similar in both fishes and mammals; in both instances the result is to maintain hemoglobin-oxygen affinity.

Reaction velocities for the binding of oxygen to hemoglobin are rapid and usually do not limit rates of oxygen transfer. The rate at which oxygen can bind to hemoglobin, however, also depends on the hemoglobin concentration. The higher the hemoglobin concentration the more oxygen bound per unit time. The more oxygen bound per unit time the longer the persistence of a large diffusion gradient across the respiratory epithelium for oxygen and, therefore, the higher the rate of oxygen transfer.

The presence of a respiratory pigment also increases the transfer of oxygen through the blood, because the oxygenated pigment co-diffuses with oxygen down the concentration gradient. That is, a gradient exists for both oxygen and the oxygenated pigment in the same direction

through the solution; the gradient for the deoxygenated pigment is in the reverse direction from that for oxygen and the oxygenated pigment. Hence, the oxygenated pigment diffuses in the same direction as oxygen, whereas the deoxygenated pigment diffuses in the reverse direction. Thus, a pigment such as hemoglobin may facilitate the mixing of gases in the blood, and myoglobin may play a similar role within tissues.

 Why are so many different organic phosphates used to modulate hemoglobin-oxygen affinity in vertebrates?

In some fishes, cephalopods, and crustaceans, an increase in CO_2 or a decrease in pH causes not only a reduction in the oxygen affinity of hemoglobin but also a reduction in oxygen capacity, which is termed the **Root effect,** or Root shift (Figure 13-6). In those hemoglobins showing a Root shift, low pH reduces oxygen binding to hemoglobin, so that even at high P_{O_2}, only some of the binding sites are oxygenated; that is, 100% saturation is never achieved.

An increase in temperature exacerbates problems of oxygen delivery in poikilothermic aquatic animals such as fishes. A rise in temperature not only reduces oxygen solubility in water but also decreases the oxygen affinity of hemoglobin, making oxygen transfer between water and blood more difficult. Unfortunately, this decrease in affinity occurs at a time when tissue oxygen requirements are increasing, also as a result of the rise in temperature.

It is generally assumed that a particular hemoglobin has evolved to meet the special gas-transfer and H^+ buffering requirements of the animal. Differences in the properties of hemoglobins are due to variation in the amino acid sequence of the peptide chains of the globin portion of the

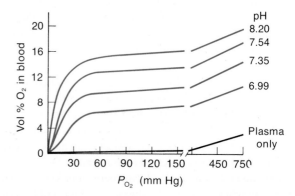

Figure 13-6 Reductions in pH decrease the blood oxygen capacity (Root effect) in hemoglobins from some teleost fishes. These oxygen equilibrium curves of eel blood were obtained at 14°C with the pH from 6.99 to 8.20. The bottom line describes the O_2 content of plasma. [Adapted from Steen, 1963.]

molecule, the heme portion being the same in all hemoglobins. Not only do hemoglobins vary among species, but they also may change during development. In humans, for example, several genes encode β-like globin chains, and the relative expression of these chains differs during prenatal and postnatal life (Figure 13-7). Human fetal hemoglobin, which contains γ chains, rather than adult β chains, has a higher O_2 affinity than adult hemoglobin. The higher O_2 affinity of fetal hemoglobin enhances oxygen transfer from mother to fetus. As the proportion of fetal hemoglobin decreases and adult hemoglobin increases following birth, the oxygen affinity of the blood decreases (Figure 13-8). Other mammals exhibit similar differences between fetal and adult hemoglobins.

It is important to remember that in most animals hemoglobin is contained within red blood cells, but the values of blood parameters usually refer to conditions in the plasma, not in the red blood cell. Differences in these parameters exist between the inside and outside of cells, including red blood cells. For example, the pH of mammalian arterial blood at 37°C is usually 7.4. This is the pH of arterial blood *plasma;* the pH inside the red blood cell is less, about 7.2 at 37°C.

Carbon Dioxide Transport in Blood

Carbon dioxide diffuses into the blood from the tissues, is transported in the blood, and diffuses across the respiratory surface into the environment. Carbon dioxide reacts with water to form carbonic acid, a weak acid, which dissociates into bicarbonate and carbonate ions:

$$CO_2 + H_2O \rightleftharpoons H_2CO_3 \rightleftharpoons H^+ + HCO_3^-$$
$$HCO_3^- \rightleftharpoons H^+ + CO_3^{2-}$$

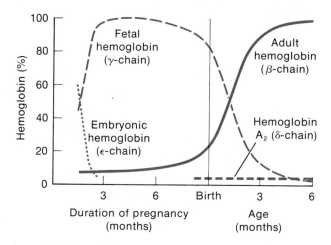

Figure 13-7 Hemoglobins change during development in humans. The relative amounts of the various hemoglobin β-like chains synthesized in the fetus changes during the course of pregnancy. Fetal hemoglobin, which contains two α and two γ chains, has a higher oxygen affinity than adult hemoglobin ($\alpha_2\beta_2$). [Adapted from Young, 1971.]

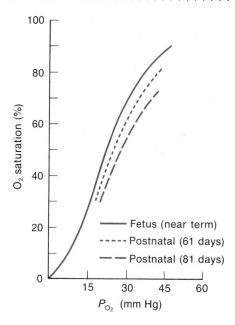

Figure 13-8 In humans, the oxygen affinity of blood decreases for about three months after birth as the fetal hemoglobin is replaced by adult hemoglobin (see Figure 13-7). These blood oxygen dissociation curves were determined at a pH of 7.40. [Adapted from Bartels, 1971.]

Carbon dioxide also reacts with hydroxyl ions to form bicarbonate:

$$H_2O \rightleftharpoons H^+ + OH^-$$
$$CO_2 + OH^- \rightleftharpoons HCO_3^-$$

The proportion of CO_2, HCO_3^-, and CO_3^{2-} in solution depends on pH, temperature, and the ionic strength of the solution. In mammalian blood at pH 7.4, the ratio of CO_2 to H_2CO_3 is approximately 1000:1, and the ratio of CO_2 to bicarbonate ions is about 1:20. Bicarbonate is, therefore, the predominant form of CO_2 in the blood at normal blood pH. The carbonate content is usually negligible in birds and mammals; in poikilotherms, however, with their low temperature and high blood pH, the carbonate content may approach 5% of the total CO_2 content of the blood, but bicarbonate is still the predominant form of CO_2.

Carbon dioxide also reacts with —NH_2 groups on proteins and, in particular, hemoglobin to form carbamino compounds.

$$\text{protein} —NH_2 + CO_2 \rightleftharpoons H^+ + \text{protein} —NHCOO^-$$

The extent of carbamino formation depends on the number of available terminal NH_2 groups, and it increases with blood pH and increasing CO_2 levels. The terminal NH_2 groups of both the α and β chains of mammalian, bird, and reptile hemoglobins are available for carbamino formation. The terminal NH_2 group of the α chain of fish and amphibian hemoglobins, however, is acetylated and therefore not available for carbamino formation. Because organophosphates bind to some of the same amino acids that are involved in carbamino formation, organophosphate binding reduces carbamino formation. However, high pH reduces organophosphate binding and so augments carbamino formation by making more NH_2 groups available. Because fish erythrocytes often have high organophosphate levels as well as acetylated α chains, fish rely less on carbamino formation for CO_2 transport than mammals.

The sum of all forms of CO_2 in the blood—that is, molecular CO_2, H_2CO_3, HCO_3^-, CO_3^{2-}, and carbamino compounds—is referred to as the *total CO_2 content* of the blood. The CO_2 content varies with P_{CO_2}, and the relationship can be described graphically in the form of a CO_2 dissociation curve (Figure 13-9). As P_{CO_2} increases, the major change is in the bicarbonate content of the blood. The formation of bicarbonate is, of course, pH dependent. The relationships between plasma HCO_3^- concentration and plasma pH at three values of P_{CO_2} are shown graphically in Figure 13-10. A decrease in pH at constant P_{CO_2} is associated with a fall in bicarbonate. The pH of red blood cells is less than that of plasma, but P_{CO_2} is in equilibrium across the cell membrane. Therefore, bicarbonate levels are lower in erythrocytes than in plasma. Erythrocytes usually constitute less than 50% of the blood volume (i.e., plasma volume is greater than erythrocyte volume), and the bicarbonate concentration is higher in plasma than in erythrocytes; it thus follows that most of the bicarbonate in the blood is in plasma.

Transfer of Gases to and from the Blood

As CO_2 is added to the blood in the tissues and removed at the respiratory surface, the levels of CO_2, HCO_3^-, and carbamino compounds all change. Carbon dioxide both enters and leaves the blood as molecular CO_2 rather than as bicarbonate ion because CO_2 molecules diffuse through membranes much more rapidly than HCO_3^- ions. In the tissues, CO_2 enters the blood and either is hydrated to form

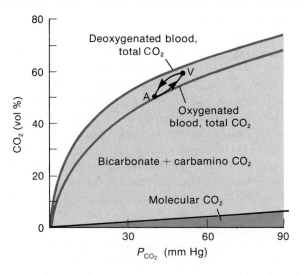

Figure 13-9 The total CO_2 content of blood increases with P_{CO_2}, but only the volume of molecular CO_2 increases linearly. Note that at a given P_{CO_2} oxygenated blood contains less CO_2 than deoxygenated blood (Haldane effect). A and V refer to arterial and venous blood levels, respectively.

HCO_3^- or reacts with—NH_2 groups of hemoglobin and other proteins to form carbamino compounds. The reverse process occurs when CO_2 is unloaded from the blood. The largest change is in the HCO_3^- concentration; changes in the levels of CO_2 and carbamino compounds usually represent less than 20% of total carbon dioxide excretion.

The reaction of CO_2 with OH^- to form HCO_3^- is slow and has an uncatalyzed time course of several seconds. But in the presence of the enzyme **carbonic anhydrase**, this reaction approaches equilibrium in much less than a second. Although plasma has a higher total CO_2 content than red blood cells, most of the CO_2 entering and leaving the plasma does so via erythrocytes, because carbonic anhydrase is present in red blood cells but not in the plasma. Therefore, formation of HCO_3^- ions in the tissues and CO_2 in the lungs occurs predominantly in red blood cells; once

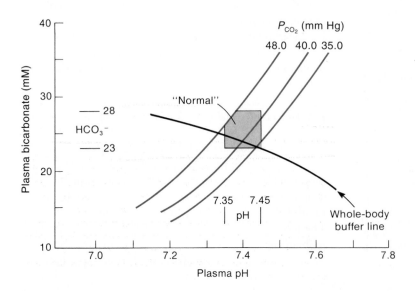

Figure 13-10 The pH, bicarbonate concentration, and P_{CO_2} in human plasma are interrelated and normally fall within quite narrow limits (indicated by red box)., However, when blood P_{CO_2} is altered *in vivo* by hyper- or hypoventilation, then plasma pH and bicarbonate are altered beyond the normal range, as indicated by the whole-body buffer line. [Adapted from Davenport, 1974.]

formed, HCO_3^- ions and CO_2 subsequently are transferred from or into the plasma.

On entering the blood from the tissues, CO_2 diffuses into red blood cells, and HCO_3^- is formed rapidly in the presence of carbonic anhydrase (Figure 13-11A). As the HCO_3^- level within erythrocytes rises, HCO_3^- ions move from the cells into the plasma. Electrical balance within the cells is maintained by anion exchange; as HCO_3^- ions leave the red blood cells, there is a net influx of Cl^- ions from the plasma into the cells, a process called the **chloride shift.** Red blood cells, unlike many other cells, are very permeable to both Cl^- and HCO_3^- because the membrane has a high concentration of a special anion carrier protein, the *band III protein.* This transport protein binds Cl^- and HCO_3^- and transfers them in *opposite* directions through the erythrocyte membrane. The anion exchange is passive and depends on concentration gradients to drive the process,

which can occur in either direction, bicarbonate flowing out of the erythrocyte in the tissues and into the erythrocyte at the respiratory surface (Figure 13-11B). Band III protein is present in all vertebrate erythrocytes except those of lamprey and hagfish. In these animals bicarbonate stays within the red blood cell and there is no anion transfer between the erythrocyte and the plasma.

A second reason why most of the CO_2 entering or leaving the blood passes through the red blood cells is that oxygenation of hemoglobin (Hb) causes H^+ release, thereby acidifying the cell interior; conversely, deoxygenation results in the binding of H^+ to Hb. Thus O_2 binding to Hb at the respiratory surface facilitates the formation of CO_2, whereas release of O_2 from Hb in the tissues facilitates the formation of HCO_3^- (Figure 13-12). As a result, changes in pH associated with the transfer of CO_2 into or out of the blood are minimized because of proton binding to and

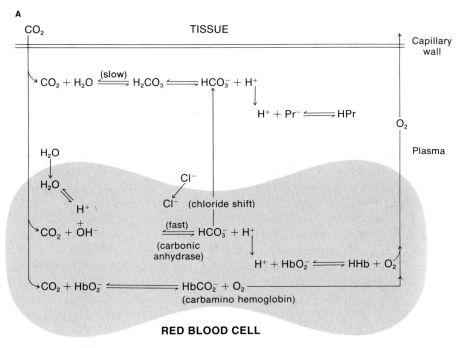

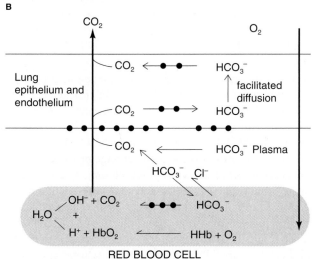

Figure 13-11 Most of the carbon dioxide entering the blood in the tissues and leaving the blood in the lungs passes through red blood cells. **(A)** Carbon dioxide produced in the tissues rapidly forms bicarbonate (HCO_3^-) in the red blood cell because the hydration reaction is catalyzed by carbonic anhydrase present in the cell. Bicarbonate leaves the erythrocyte in exchange for chloride, and excess protons are bound by deoxygenated hemoglobin (Hb). **(B)** These reactions are reversed in lungs. Oxygen entering the red blood cell displaces protons from Hb, and carbon dioxide enters the plasma. Carbonic anhydrase (indicated by solid circles) in the membrane of the lung endothelial cells converts some of the plasma bicarbonate to carbon dioxide. Movement of carbon dioxide across the respiratory surface is augmented by the diffusion of bicarbonate and its conversion back to carbon dioxide at the outer surface, a process termed facilitated diffusion.

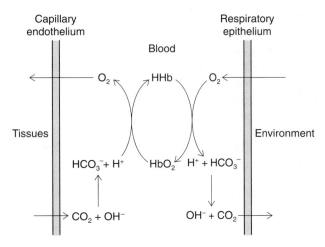

Figure 13-12 The pH changes associated with the changes in blood P_{CO_2} in the tissues and respiratory surface are offset by binding and release of H^+ ions by deoxygenated and oxygenated blood. For example, transfer of CO_2 into the blood in the tissues causes a decrease in pH due to formation of bicarbonate; concomitant deoxygenation of hemoglobin frees proton acceptors, which bind the excess H^+ ions. The opposite reactions occur at the respiratory epithelium.

proton release from hemoglobin as it is deoxygenated and oxygenated, respectively.

For example, as the P_{CO_2} increases in the tissues, the subsequent formation of HCO_3^- or carbamino compounds liberates H^+ ions. At the same time, release of oxygen forms deoxyhemoglobin, which binds protons. As deoxygenation proceeds, however, more proton acceptors become available on the hemoglobin molecule. In fact, complete deoxygenation of saturated hemoglobin, releasing 1 mol O_2, results in the binding of 0.7 mol of H^+ ions. Thus, when the ratio of CO_2 production to O_2 consumption (called the **respiratory quotient**) is 0.7, the transport of CO_2 can proceed without any change in blood pH. (As discussed in Chapter 16, the respiratory quotient depends on the type of diet.) Even when the respiratory quotient is 1, the additional 0.3 mol H^+ is buffered by blood proteins, including hemoglobin, and blood undergoes only a small change in pH. At a given P_{CO_2}, deoxyhemoglobin binds more protons, thereby facilitating HCO_3^- formation, and reacts with CO_2 to form carbamino hemoglobin more easily than does oxyhemoglobin. As a result, the total CO_2 content of deoxygenated blood at a given P_{CO_2} is higher than that of oxygenated blood (see Figure 13-9). Thus, deoxygenation of hemoglobin in the tissues reduces the change in P_{CO_2} and pH as CO_2 enters the blood; this is termed the **Haldane effect.**

In the lungs, two mechanisms are available for transfer of CO_2 from the blood. As noted already, carbonic anhydrase is absent from plasma, and thus the interconversion of CO_2 and HCO_3^- occurs at the slow, uncatalyzed rate in plasma. (Any carbonic anhydrase liberated by the breakdown of red blood cells is excreted via the kidney.) In the endothelial cells of lung capillaries, however, carbonic anhydrase is embedded in the cell surface, accessible to plasma CO_2 and HCO_3^-. Therefore, the conversion of HCO_3^- to

CO_2 can occur at the catalyzed rate in plasma as blood perfuses the lung capillaries (see Figure 13-11B). In addition, oxygenation of hemoglobin acidifies erythrocytes in the lung capillaries, facilitating the conversion of HCO_3^- to CO_2, which then diffuses into the plasma and across the lung epithelium. The resulting decrease in erythrocyte bicarbonate levels results in the influx of HCO_3^- ions from the plasma accompanied by the outward movement of Cl^- ions. The relative quantities of HCO_3^- converted to CO_2 in the erythrocytes and the plasma of the blood perfusing the respiratory epithelium is influenced by the extent of proton production associated with hemoglobin oxygenation and the amount of carbonic anhydrase activity in the walls of the respiratory epithelium. In teleost fish, for instance, the plasma perfusing the gills is not exposed to carbonic anhydrase. In these animals, most excretion of CO_2 occurs through the red blood cells and is tightly coupled to O_2 uptake through proton production by oxygenation of hemoglobin.

Carbonic anhydrase activity is also found on the endothelial surfaces of a number of systemic capillary beds, including those in skeletal muscle. In these capillaries, formation of HCO_3^- catalyzed by carbonic anhydrase can occur in the absence of red blood cells. Thus some of the CO_2 transferred into the blood in skeletal muscle does not pass through erythrocytes. Carbonic anhydrase also facilitates carbon dioxide transfer, referred to as facilitated diffusion of CO_2 (see Figure 13-11B), which results from the simultaneous diffusion through the epithelium of bicarbonate and protons, the latter also augmented by release from buffers. Carbonic anhydrase catalyzes the rapid interconversion of CO_2 and HCO_3^- in this process of facilitated diffusion, with CO_2 entering and leaving the cell.

There are at least seven forms of carbonic anhydrase, designated CA-I through CA-VII. All are similar in structure and catalyze the interconversion of carbon dioxide and bicarbonate. Carbonic anhydrase I (CA-I) and carbonic anhydrase II (CA-II), present in human red blood cells, have a molecular weight of about 29,000, containing about 260 amino acid residues. CA-II, an extremely efficient catalyst of the carbon dioxide-bicarbonate hydration-dehydration reactions, is found in a wide variety of tissues, including the brain, eye, kidney, cartilage, liver, lung, pancreas, gastric mucosa, skeletal muscle, and anterior pituitary, as well as red blood cells. This form is involved in a wide variety of functions, augmenting the supply of bicarbonate and/or protons for a number of cellular and metabolic processes. A few humans exhibit an inherited CA-II deficiency, the pattern of inheritance being autosomal recessive. Although these individuals have no detectable CA-II, they have normal levels of CA-I in their red blood cells. CA-II deficiency not only compromises the gas-exchange process but also produces many other symptoms including metabolic acidosis, renal tubular acidosis, and sometimes mental retardation. In addition, because CA-II is involved in production of protons needed for bone resorption in osteoclasts, its absence results in osteoporosis, often associated with multiple bone fractures. The wide range of symptoms associated with inherited CA-II deficiency reflects the large number of functions in which CA-II plays a role in augmenting proton and/or bicarbonate delivery.

The rate of movement of CO_2 and O_2 into or out of the red blood cell is determined by the diffusion distance and the diffusion coefficient of these substances through the cell. The diffusion difference and hence the rate of erythrocyte oxygenation might be expected to be related to cell size, which varies considerably among vertebrates. For instance, the amphibian *Necturus* has erythrocytes that are 600 times the volume of erythrocytes from a goat. Earlier studies demonstrated that small erythrocytes are oxygenated faster than larger cells *in vitro* (Figure 13-13), but this finding may have little relevance *in vivo*. Recent experiments using a whole-blood thin-film technique, which is analogous to the *in vivo* situation, have shown that oxygen uptake rates are independent of cell size. The explanation for this probably lies in the flattened shape of erythrocytes. If the large flat surface of the cells faces the respiratory medium as they pass single file through the respiratory capillaries, then their diffusion distances may be quite similar even though volumes of cells are very different. Thus the *in vitro* results are probably not applicable to the *in vivo* situation.

Excretion of CO_2 is considered to be limited by the rate of bicarbonate-chloride exchange across the erythrocyte membrane. The surface-to-volume ratio of erythrocytes, as well as the transport capacity for bicarbonate-chloride exchange mediated by band III protein, may be important in determining rates of carbon dioxide excretion. To see the interrelationship of these parameters, let's compare trout and human erythrocytes (Table 13-1). Red blood cells from trout are larger and have a much higher concentration of band III protein in their membranes than do red blood cells from humans. The higher concentration of band III protein presumably compensates for the increased cellular volume and offsets, at least to some degree, the effects of a lower body temperature in trout, compared with humans, on anion-exchange rates. Even so, anion exchange is slower across trout red blood cells at 15°C than across human red blood cells at 38°C. However, transit times for erythrocytes

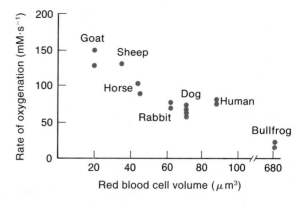

Figure 13-13 Small erythrocytes are oxygenated faster than large cells *in vitro*. However, cell size probably is unrelated to oxygenation rates *in vivo*. [From Holland and Forster, 1966.]

through the gills is longer than that in the lungs, allowing more time for anion exchange across the red blood cell.

Despite these considerations, it is still not clear why different species have evolved red blood cells of such different sizes. Those animals with large red blood cells also have large cells generally. Thus cell size may have been selected for reasons other than gas transfer and may be largely unrelated to gas-transfer rates. For instance, triploid salmon, whose red blood cells are 1.5 times the size of those of their diploid cousins but have the same hemoglobin concentration, are able to swim just as fast as their diploid cousins, indicating that the efficiency of gas transfer is comparable.

It's important to remember that *in vivo* gas transfer is a dynamic process that takes place as blood moves rapidly through capillaries. Rates of diffusion, reaction velocities, and steady-state conditions for gases in blood must all be taken into account in analyzing the process. For instance, a Bohr shift (e.g., decrease in hemoglobin-oxygen affinity with decrease in pH) would have little importance if it occurred after the blood had left the capillaries that supply an active tissue. The Bohr shift, in fact, occurs very rapidly, having a half-time at 37°C of 0.12 seconds in human red blood cells. Although a reduction in temperature always decreases the velocities of reactions involved in gas transfer in a species, these velocities do not vary and are not modulated to regulate gas-transfer rates at constant temperature. Concentration changes, however, are used to adjust gas-transfer rates over hours or days. For example the oxygen content of the blood depends on the concentration of hemoglobin, which is increased in many vertebrates in response to hypoxia. Rapid changes in gas-transfer rates in vertebrates are achieved either by adjusting the breathing rate and volume and/or by adjusting the flow rate and distribution of blood in both tissues and the respiratory surface.

REGULATION OF BODY pH

Animals have a body pH that is on the alkaline side of neutrality; that is, there are fewer hydrogen than hydroxyl ions in the body. The concentrations of hydrogen and hydroxyl ions are very low in aqueous solutions because water is

TABLE 13-1
Comparison of bicarbonate-chloride exchange system in trout and human erythrocytes

Property	Trout	Human
Cell surface (cm²)	2.67×10^{-6}	1.42×10^{-6}
Band III molecules per cell	8×10^6	1×10^6
Band III molecules per cm²	30×10^{11}	7×10^{11}
Half-time for Cl⁻ ion exchange (seconds):		
0/C	3.42	17.2
10/C	1.29	2.32
15/C	0.81	0.89
38/C	—	0.05

Source: Romano and Passow, 1984.

only weakly dissociated. Human blood plasma at 37°C has a pH of 7.4, or a hydrogen ion activity of 40 nanomoles per liter (1 nM = 10^{-9} M). Normal function can be maintained in mammals at 37°C over a blood plasma pH range of 7.0–7.8, that is, between 100 and 16 nM H$^+$. This is, in fact, a rather large percentage deviation from the normal H$^+$ concentration of 40 nM compared with the much lower tolerance of variations in Na$^+$ or K$^+$ levels in the body. It is important, however, to bear in mind that the absolute changes in concentration are small, as are the actual concentrations of H$^+$ ions in the body.

Blood pH in vertebrates is midway between the pK of the carbon dioxide/bicarbonate and ammonia/ammonium reactions (Figure 13-14A). Most cell membranes are not very permeable to HCO$_3^-$ and NH$_4^+$ ions but are very permeable to CO$_2$ and NH$_3$. Some membranes have a relatively low permeability to NH$_3$, but these are the exceptions rather than the rule. A body pH that is midway between these pKs ensures adequate rates of excretion by diffusion of the two major end products of metabolism, namely carbon dioxide and ammonia. Because these pKs vary with temperature, so does the pH of blood, ensuring adequate rates of excretion over a range of temperatures (Figure 13-14B).

Changes in body pH alter the dissociation of weak acids and thus the ionization of proteins. The net charge on proteins determines enzyme activity and subunit aggregation, influences membrane characteristics, and contributes to the osmotic pressure of body compartments. Osmotic pressure is affected because the charge on proteins is a major contributor to the total fixed charge within cells. A change in the fixed charge will alter the Donnan equilibrium of ions and therefore could affect the osmotic pressure. Any differences in osmotic pressure between body compartments disappear rapidly because membranes are permeable to water, and water movement will cause changes in the volume of various body compartments.

Thus animals regulate their internal pH, in the face of a continual metabolic release of hydrogen ions, to stabilize volume and regulate enzyme activity. Cells also undergo changes in pH either as a result of, or to regulate and control, cellular functions. For example pH plays a central role in such things as sea urchin sperm activation and the stimulation of glycolysis in frog muscle by insulin. Cells also undergo changes in pH as a result of external influences. For example, cells become acidotic during hypoxia because of an imbalance between proton production resulting from hydrolysis of ATP to ADP and proton consumption by NAD in those tissues subjected to anaerobic metabolism.

Hydrogen Ion Production and Excretion

Hydrogen ions are produced through metabolism or ingested in foods (e.g., citric acid in oranges) and then excreted on a continuing basis. The largest pool and flux is due, usually, to the metabolic production of CO$_2$, which at the pH of the body reacts with water to form H$^+$ and

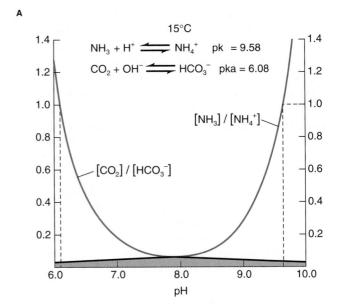

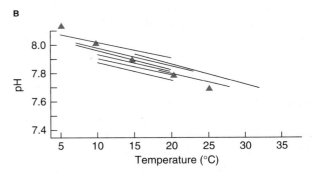

Figure 13-14 In vertebrates, the plasma pH is midway between the pKs of the ammonia/ammonium and carbon dioxide/bicarbonate reactions. **(A)** The effect of varying pH on the [CO$_2$]/[HCO$_3^-$] and [NH$_3$]/[NH$_4^+$] ratios in trout plasma at 15°C. The dashed lines mark the pH values at which the ratios equal 1 (i.e., the pK values). **(B)** Effect of temperature on plasma pH for several fishes. Red triangles are the calculated pH values at which the CO$_2$/HCO$_3^-$ and NH$_3$/NH$_4^+$ ratios are equal at various temperatures. Thus plasma pH is maintained at levels that ensure both NH$_3$ and CO$_2$ excretion. [Adapted from Randall and Wright, 1989.]

HCO$_3^-$ ions (see Figure 13-11A). At the respiratory surface HCO$_3^-$ is converted to CO$_2$, which is then excreted (see Figure 13-11B). Thus if the production and excretion of CO$_2$ are balanced, the overall effect of CO$_2$ flux on body pH will be zero. If CO$_2$ excretion is less than production, so that CO$_2$ accumulates, the body will be acidified; if the reverse occurs, the body pH will rise. Terrestrial vertebrates, however, can vary the rate of CO$_2$ excretion to maintain body pH.

Ingestion of meat usually results in a net intake of acid, whereas ingestion of plant food often results in a net intake of base. Generally there is a small net production of hydrogen ions as a result of diet and metabolic activity. Thus the overall effect of food ingestion and metabolism is a small continual production of acid. The pH of the body is maintained by excreting this acid via the kidney in terrestrial vertebrates or across regions of the body surface, such as the gills of fish or the skin of frogs. Changes in blood pH can also occur in response to acid movement between compartments. For example, following a heavy meal, the production of large volumes of acid in the stomach can produce an *alkaline tide* in the blood owing to transfer of acid from the blood into the stomach. In a similar manner, the production of large volumes of alkaline pancreatic juices can result in an *acid tide* in the blood.

 Crocodiles may eat a whole mammal as a single meal! What might be the changes in pH of the various body compartments as digestion proceeds?

As discussed in Chapter 3, the relationship between pH and the extent of dissociation of a weak acid, HA, is described by the **Henderson-Hasselbalch equation:**

$$pH = pK' + \log \frac{[A^-]}{[HA]}$$

When the pH of a solution of a weak acid is equal to the pK' of the acid, then 50% of the acid is in the undissociated form, HA, and 50% is in the dissociated form, $H^+ + A^-$. At 1 pH unit above the pK, the ratio of the undissociated to dissociated form is 10% to 90%, whereas at 2 pH units above the pK, the ratio becomes 1% to 99%. The Henderson-Hasselbalch equation can be rewritten for the CO_2/HCO_3^- acid-base pair as

$$pH = pK' + \log \frac{[HCO_3^-]}{\alpha P_{CO_2}}$$

where P_{CO_2} is the partial pressure of CO_2 in blood, α is the Bunsen solubility coefficient for CO_2, $[HCO_3^-]$ is the concentration of bicarbonate, and pK' is the apparent dissociation constant. The term "apparent" is used because this pK' is a lumped value for the combined reactions of CO_2 with water and the subsequent formation of bicarbonate, and is not a true pK. We can see from this equation that changes in pH will affect the ratio of HCO_3^- to P_{CO_2}, and vice versa. The pK' of the CO_2/HCO_3^- reaction is about 6.1, and the pK of the HCO_3^-/CO_3^{2-} reaction is around 9.4. At the pH of the body, about 95% of the CO_2 is in the form of HCO_3^-, the remainder being carbon dioxide and carbonic acid; the amount of CO_3^{2-} is negligible.

Weak acids have their greatest buffering action when pH = pK. Because the pK of plasma proteins and hemoglobin is close to the pH of blood, these compounds are important physical buffers in the blood. The CO_2/HCO_3^- pair, with an apparent pK' below the pH of the blood, is of less importance than either hemoglobins or proteins in providing a physical buffer system. The importance of the CO_2-bicarbonate system is that an increase in breathing can rapidly increase pH by lowering CO_2 levels in the blood, and that HCO_3^- can be excreted via the kidney to decrease blood pH. Although bicarbonate is not an important chemical buffer in living systems, it is often referred to as a buffer because the CO_2-to-bicarbonate ratio can be adjusted by excretion in order to regulate pH. The most important true buffers in the blood are proteins, especially hemoglobin. Phosphates are also significant buffers in many cells.

The importance of buffers in ameliorating pH changes can be seen by considering the effects of acid infusion on mammalian blood. About 28 mmol of hydrogen ions must be added to the blood to reduce pH from 7.4 to 7.0. In fact, only 60 nmol (about 0.2%) are required to change the pH of an aqueous solution to this extent; in blood, however, the bulk of the added 28 mmol of H^+ is buffered by conversion of HCO_3^- to CO_2 (18 mmol), hemoglobin (8.0 mmol), plasma proteins (1.7 mmol), and phosphates (0.3 mmol). Thus nearly 500,000 times as many hydrogen ions are buffered as are required to cause the pH to change from 7.4 to 7.0.

Clearly, if lung ventilation is reduced so that CO_2 excretion drops below CO_2 production, body CO_2 levels will rise and pH will fall. This decrease in body pH is referred to as **respiratory acidosis**. The reverse effect, that is a rise in pH due to increased lung ventilation, is termed **respiratory alkalosis**. The word "respiratory" is used to differentiate these pH changes from those caused by changes in metabolism or kidney function. For example, anaerobic metabolism results in net acid production, which reduces body pH; such changes are referred to as **metabolic acidosis.**

Body fluids, like other solutions, are electroneutral; that is, the sum of the anions equals the sum of the cations. The normal electrolyte status of human plasma is illustrated in Figure 13-15. The sum of bicarbonate, phosphates, and protein anions is referred to as the *buffer base*. The remaining cations and anions are referred to as strong ions (i.e., those completely dissociated in physiological solutions); the difference between the sum of strong cations and the sum of strong anions is referred to as the *strong ion difference* (SID) and is a reflection of the magnitude of the buffer base. Because a change in blood pH usually results in a change in the buffer base, SID also must change to maintain electrical neutrality. In this situation, the change in SID usually involves either sodium or chloride, since these are the major ions in the blood. For example, a reduction in bicarbonate must be associated with an increase in chloride or a reduction in sodium. Conversely, a change in the ratio of sodium to chloride will be associated with a change in the buffer base and therefore blood pH. Vomiting the stomach contents results in chloride loss and a reduction in blood chloride levels; as a consequence, bicarbonate levels are increased along with blood pH without any change in P_{CO_2}; this is referred to as **metabolic alkalosis**. Vomit originating from the duodenum, rather than the stomach, however, results in the loss of more bicarbonate than chloride, causing a metabolic acidosis.

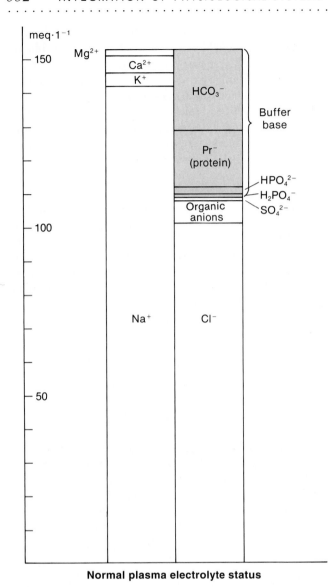

Normal plasma electrolyte status

Figure 13-15 All body fluids are electroneutral, containing equal numbers of positive and negative charges. This diagram shows the equivalent concentrations (meq $\cdot$ L^{-1}) of the major electrolytes in human plasma at normal pH. The concentration of the buffer base (the nonrespiratory acid-base displacement) depends on pH. Thus a pH increase or decrease that changes the buffer base concentration must be accompanied by a corresponding change in the concentration of one or more strong ions, usually sodium or chloride. [Adapted from Siggaard-Andersen, 1963.]

Hydrogen Ion Distribution between Compartments

Cell membranes separating the intracellular and extracellular compartments and layers of cells between two body compartments are much more permeable to carbon dioxide than to either hydrogen or bicarbonate ions. The permeability of most cell membranes to H$^+$ ions, although usually low, is often greater than that for K$^+$, Cl$^-$, and HCO$_3^-$ ions; a notable exception is the erythrocyte membrane, which is very permeable to HCO$_3^-$ and Cl$^-$ ions, but not very permeable to H$^+$ ions. Red blood cells and cells in the collecting duct of the mammalian kidney have

high levels of band III protein in their plasma membranes, but other cells do not. As discussed previously, band III protein mediates the exchange of HCO$_3^-$ for Cl$^-$ ions at high rates. Thus, although all cell membranes are permeable to CO$_2$, only a few membranes can transfer HCO$_3^-$ at high rates via the band III anion-exchange mechanism.

An increase in extracellular P_{CO_2} causes an increase in both bicarbonate and hydrogen ion concentration, thereby creating gradients for CO$_2$, HCO$_3^-$, and H$^+$ across the cell membrane. In cells that are very permeable to CO$_2$ but not very permeable to H$^+$ or bicarbonate, such a situation leads to rapid movement of CO$_2$ into the cell; as the CO$_2$ is converted to HCO$_3^-$, the intracellular pH falls sharply. Acidification associated with increased P_{CO_2} often occurs much more rapidly in the intracellular compartment than in the extracellular compartment because carbonic anhydrase, which catalyzes the conversion of CO$_2$ to HCO$_3^-$, is present inside cells but not always in the extracellular fluid. Even when P_{CO_2} remains elevated, intracellular pH slowly returns to the initial level due to the slow extrusion of acid (or uptake of base) across the cell membrane (Figure 13-16A). The rise in intracellular pH is such that if the P_{CO_2} level is returned to the original value, cell pH will be higher than the initial value; that is, there is a small overshoot in pH.

As noted earlier, most cell membranes are much more permeable to molecular ammonia, NH$_3$, than to ammonium ions, NH$_4^+$. If NH$_4$Cl levels in the extracellular fluid increase, ammonia penetrates the cell much more rapidly than ammonium ions. The result is, of course, that ammonia levels in the cell are increased much more rapidly too. Ammonia equilibrates across the membrane and combines with hydrogen ions to form ammonium ions within the cell, thus raising cell pH (Figure 13-16B). After reaching a maximum, pH starts to fall during prolonged NH$_4$Cl exposure because of a slow passive influx of NH$_4^+$ along with other acid-base regulating mechanisms in the membrane. The return of the external NH$_4$Cl level to the original value results in a sharp fall in intracellular pH as NH$_3$ diffuses out of the cell. However, because of the accumulation of intracellular NH$_4^+$, cell pH falls below the initial level, but slowly returns to the initial level as NH$_4^+$ diffuses out of the cell.

These mechanisms of pH adjustment are activated by either a reduction in intracellular pH or an increase in extracellular pH. In mammalian cells acid extrusion is reduced to low levels if extracellular pH falls below 7.0 or intracellular pH rises above 7.4. If an acid is injected into a cell, it is extruded from the cell at rates that increase in proportion to the decrease in cell pH. Although a portion of the H$^+$ efflux may be related to H$^+$ diffusion out of the cell, some of the efflux is coupled to sodium influx. This coupling of sodium and proton transport could be due to either a cation-exchange mechanism in the membrane or an electrogenic proton pump that increases membrane potential, thereby providing an electrochemical gradient for diffusion of Na$^+$ ions through sodium-selective channels. For exam-

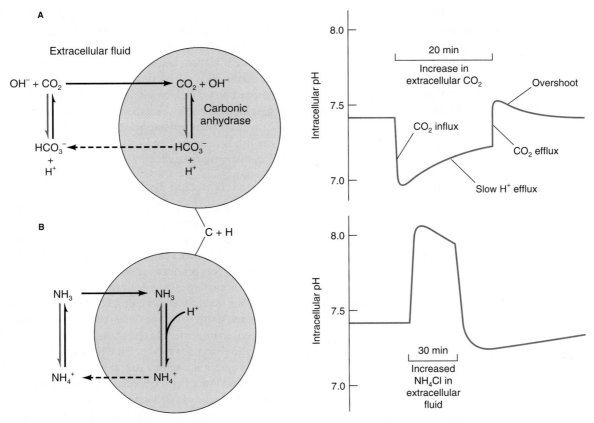

Figure 13-16 Changes in extracellular carbon dioxide and ammonium chloride levels cause changes in intracellular pH of tissue cells. **(A)** If CO_2 levels in the extracellular fluid are suddenly increased, CO_2 diffuses rapidly into the cell, forming bicarbonate and causing a sharp fall in intracellular pH. A subsequent slow efflux of H^+ ions (dashed line) leads to a gradual rise in the intracellular pH. **(B)** If extracellular NH_4Cl levels rise sharply, NH_3 diffuses rapidly into the cell and combines with hydrogen ions to form ammonium ions, which diffuse slowly across the cell membrane (dashed line). As a result, the intracellular pH increases.

ple, some cells can actively pump protons out via a proton ATPase in the membrane; this proton efflux can result in a sodium influx. Often acid extrusion is accompanied by chloride efflux, presumably in exchange for extracellular HCO_3^-, which has been shown to be required for pH regulation by cells. For instance, the drug SITS (4-acetamido-4'-isothiocyanostilbene-2,2'-disulfonic acid), which blocks chloride–bicarbonate exchange in erythrocytes, also inhibits pH regulation in other cells.

Thus both proton-exchange and anion-exchange mechanisms in the cell membrane play an important role in adjusting intracellular pH. An acid load in the cell is accompanied by H^+ efflux coupled to Na^+ influx and by HCO_3^- influx coupled to Cl^- efflux. The movement of HCO_3^- into the cell is equivalent to movement of H^+ out of the cell because HCO_3^- ions that enter the cell are converted to CO_2, releasing hydroxyl ions and increasing pH. The CO_2 so formed, leaves the cell and is converted to bicarbonate, releasing protons. This cycling of CO_2 and HCO_3^-, referred to as the **Jacobs-Stewart cycle**, functions to transfer H^+ ions from the cell interior in the face of an intracellular acid load, such as that generated by anaerobic metabolism (Figure 13-17).

In most vertebrate red blood cells, unlike most other cells, hydrogen ions are passively distributed across the

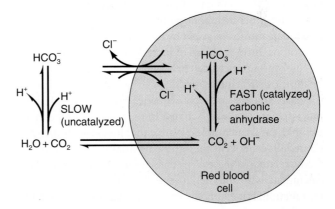

Figure 13-17 The Jacob-Stewart cycle is the cycling of carbon dioxide and bicarbonate to transfer acid between the extracellular and intracellular compartments. In a red blood cell, depicted here, the cycle generally operates to transfer acid from the plasma to the cell interior. Because carbonic anhydrase is present only inside cells, the slow, uncatalyzed interconversion of CO_2 and HCO_3^- in the extracellular fluid determines the rate of acid transfer.

membrane, and the membrane potential maintains a lower pH inside the red blood cell than in the plasma. A sudden addition of acid to the plasma (e.g., following anaerobic production of H^+) results in a fall in erythrocyte pH. The acid is transferred from the plasma to the interior of the erythrocyte not by diffusion of H^+ ions but by bicarbonate–chloride exchange (see Figure 13-17). The addition of H^+ to the plasma causes P_{CO_2} to increase due to the conversion of HCO_3^- to CO_2, which then diffuses into the red blood cell and is converted to HCO_3^-, thereby reducing intracellular pH. Bicarbonate then diffuses out of the cell via the chloride-bicarbonate exchange mechanism. Thus in erythrocytes, the Jacobs-Stewart cycle operates to transfer acid from the plasma to the cell interior.

Factors Influencing Intracellular pH

Intracellular pH will be stable if the rate of acid loading, from metabolism or from influx into the cell, is equal to the rate of acid removal. Any sudden increase in cell acidity will be counteracted by the various mechanisms discussed in previously:

- Buffering by physical buffers (e.g., proteins and phosphates) located within the cell

- Reaction of HCO_3^- with H^+ ions, forming CO_2, which then diffuses out of the cell.

- Passive diffusion or active transport of H^+ ions from the cell

- Cation-exchange (Na^+/H^+) or anion-exchange (HCO_3^-/Cl^-) mechanisms, or both, in the cell membrane

In addition, the generation of protons through metabolism may be modulated by pH. Many enzymes are inhibited by low pH, so that the inhibition of glycolysis (and possibly some other metabolic pathways) at low pH may serve to regulate intracellular pH by reducing the net production of protons during periods of increased acidity in cells.

In some instances, cell pH may be modulated to control or limit other cellular functions. It is not always clear if these pH changes are a consequence of, or are regulating, the associated cellular activity. In many cells intracellular pH (pH_i) and calcium levels are either inversely or directly related. In other cells they are sequentially, rather than directly, related; in these cases changes in pH_i may modulate calcium activity and therefore many of its actions on cellular function. For example, when frog eggs are fertilized, intracellular calcium levels increase transiently, followed by a sustained increase in pH_i. There is some evidence to indicate that this alkalinization of the cell may prolong the action of elevated calcium.

In a few cases the regulation of intracellular pH (pH_i) has a clear effect on cellular function. For example, the erythrocytes of many teleosts have a Na^+/H^+ exchanger and a HCO_3^-/Cl^- exchanger in the membrane. The hemoglobin in these animals exhibits a Root shift, that is, a decrease in blood oxygen capacity as blood pH falls (see Figure 13-6). Clearly, this effect would impair oxygen transport by erythrocytes during periods of metabolic acidosis in the absence of some countervailing mechanism. In

fact, catecholamines released into the blood during periods of metabolic acidosis activate the erythrocyte Na^+/H^+ exchanger, which moves H^+ out of and Na^+ ions into the cell. In fish with a large muscle mass, burst swimming results in a marked acidosis. This drop in plasma pH, if transferred to the red blood cell, would impair oxygen binding to hemoglobin and reduce the ability of the fish to swim aerobically. This does not happen because erythrocytic pH_i in these fish is regulated and remains high during the acidosis following burst-swimming activity.

Factors Influencing Body pH

A stable body pH requires that acid production be matched to acid excretion. In mammals, this symmetry is achieved by adjusting the excretion of CO_2 via the lungs and excretion of acid or bicarbonate via the kidneys, so that acid excretion balances production, which is largely determined by the metabolic requirements of the animal. The collecting duct of the mammalian kidney has A-type (acid-excreting) and B-type (base-excreting) cells, the activity of which can be altered to increase acid or base excretion. In aquatic animals, the external surfaces have the capacity to extrude acid in ways similar to that seen in the collecting duct of the mammalian kidney (see Chapter 14). For instance, the skin of frogs and gills of freshwater fish have a proton ATPase, which excretes protons, on the apical surface of the epithelium. Fish gills also have a HCO_3^-/Cl^- exchange mechanism. If these mechanisms are inhibited by drugs, body pH is affected.

Temperature can have a marked effect on body pH. The dissociation of water varies with temperature, and the pH of neutrality (i.e., $[H^+] = [OH^-]$) is 7.00 only at 25°C. The dissociation of water decreases, and the pH of neutrality (pN) therefore increases, with a decrease in temperature. At 37°C, pN is 6.8, whereas at 0°C, it is 7.46. Human plasma at 37°C has a pH of 7.4, so it is slightly alkaline. At pN, the ratio of OH^- to H^+ concentrations is 1. This ratio increases with increasing alkalinity; at pH 7.4 at 37°C it is about 20. Most animals maintain almost the same alkalinity in many of their tissues relative to pN independent of the temperature of their bodies (Figure 13-18). Fishes at 5°C have a plasma pH of 7.9–8.0; turtles at 20°C, a plasma pH of about 7.6; and mammals at 37°C, a plasma pH of 7.4. Thus, all have a similar relative alkalinity and the same ratio of OH^- to H^+ ions (about 20) in plasma. Tissues are generally less alkaline than plasma; for example, the pH_i of erythrocytes is about 0.2 pH units less than plasma, and the pH_i of muscle cells is about 7.0.

Temperature also has a marked effect on the pK' of plasma proteins and the CO_2/HCO_3^- system, the pK' increasing as temperature decreases. According to the Henderson-Hasselbalch equation, changes in pK' will cause changes in pH or in the dissociation of weak acids. However, the temperature-induced changes in plasma pH (see Figure 13-18) offset the temperature-dependent changes in the pK' of plasma proteins, so that

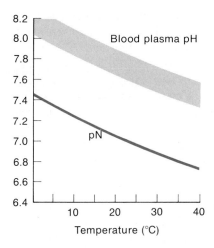

Figure 13-18 The pH at neutrality (pN) and plasma pH decrease with increasing temperature, but the relationship between the two is constant in most animals. In this graph, the effect of temperature on plasma pH in various turtles, frogs, and fishes is compared with the change in pN. [Adapted from Rahn, 1967.]

the extent to which the plasma proteins dissociate remains constant.

Because the pK' of the CO_2 hydration-dehydration reaction changes less with temperature than does blood pH, animals must adjust the ratio of CO_2 to HCO_3^- in the blood. In general, it appears that as temperature falls, air-breathing, poikilothermic vertebrates keep bicarbonate levels constant but decrease molecular CO_2 levels. In aquatic animals, on the other hand, CO_2 levels remain the same and bicarbonate levels increase as temperature drops. This process results in the same adjustment of the CO_2-to-bicarbonate ratio and hence pH in both aquatic and air-breathing vertebrates. The important point is that if body pH changes with temperature in the same way as the pK' of proteins, then the Henderson-Hasselbalch equation predicts the charge on proteins should remain unchanged. If there is little or no change in the net charge on proteins, function will be retained over a wide range of temperatures.

The ability of the body to redistribute acid between body compartments is of functional significance because some tissues are more adversely affected by changes in pH than others. The brain is particularly sensitive, whereas muscle can and does tolerate much larger oscillations in pH. As a result, the brain has extensive, if poorly understood, mechanisms for regulating the pH of the cerebrospinal fluid (CSF). In the face of a sudden acid load in the blood, hydrogen ions are taken up by the muscles, reducing oscillations in the blood and protecting the brain and other more sensitive tissues. Hydrogen ions are then slowly released into the blood from muscle and excreted either via the lungs as CO_2 or via the kidney in acid urine. Thus, when there is a sudden acid load in the body, the muscles can act as a temporary H^+ reservoir, thus reducing the magnitude of the oscillations in pH in other regions of the body.

GAS TRANSFER IN AIR: LUNGS AND OTHER SYSTEMS

The previous sections considered the properties of oxygen and carbon dioxide and described how these gases are carried in the blood and the effect they have on body pH. In this section, we examine the ways in which O_2 and CO_2 are transferred between air and blood. Focus is placed on the vertebrate *lung*, but other gas-transfer systems also are considered. In the next section gas transfer between water and blood across gills is discussed.

The structure of a gas-transfer system is influenced by the properties of the medium as well as the requirements of the animal. For example, the lungs of mammals have a very different structure from the gills of fish and are ventilated in a different manner. This dissimilarity exists because, although the density and viscosity of water are both approximately 1000 times greater than that of air, water contains only one-thirtieth as much molecular oxygen. Moreover, gas molecules diffuse 10,000 times more rapidly in air than in water. Thus, in general, air breathing consists of the reciprocal movement of air into and out of the lungs, whereas water breathing consists of a unidirectional flow of water over the gills (Figure 13-19A). The design objectives of fish gills are to minimize diffusion distances in water, creating a thin layer of water over the respiratory surface. These variations in the environment, in the structure of the respiratory apparatus, and in the nature of ventilation result in differences in the partial pressures of gases in the blood and tissues of air-breathing and water-breathing animals, particularly in P_{CO_2} (Figure 13-19B).

Functional Anatomy of the Lung

The vertebrate lung, which develops as a diverticulum of the gut, consists of a complex network of tubes and sacs, the actual structure varying considerably among species. The sizes of terminal air spaces become progressively smaller in the lungs of amphibians, reptiles, and mammals (in that order), but the total number of air spaces per unit volume of lung becomes greater. The structure of the amphibian lung is variable, ranging from a smooth-walled pouch in some urodeles to a lung subdivided by septa and folds into numerous interconnected air sacs in frogs and toads. The degree of subdivision is increased in reptiles, and increases even more in mammals, the total effect being an increase in respiratory surface area per unit volume of lung. In general, the area of the respiratory surface in mammals increases with body weight and the rate of oxygen uptake (Figure 13-20). Teleost fishes typically have a smaller respiratory surface area than mammals of equivalent body weight.

The mammalian lung consists of millions of blind-ending, interconnected sacs, termed **alveoli**. The **trachea** subdivides to form **bronchi**, which branch repeatedly, leading eventually to terminal **bronchioles** and finally respiratory bronchioles, each of which is connected to a terminal spray of alveolar ducts and sacs (Figure 13-21). The total

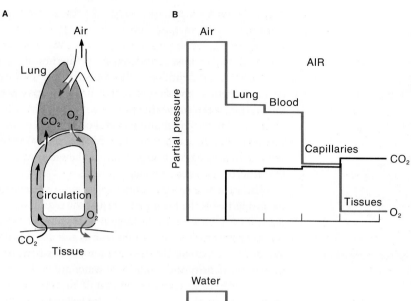

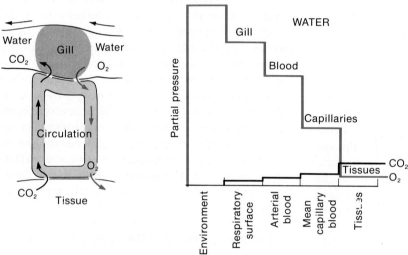

Figure 13-19 The different gas-transfer systems in air-breathing and water-breathing animals are associated with characteristic distribution of respiratory gases in the blood and tissues. **(A)** Schematic diagrams of O_2 and CO_2 flows in air-breathing and water-breathing animals. **(B)** Relative values of P_{O_2} and P_{CO_2} in the inhalant medium, blood, and tissues in air-breathing *(top)* and water-breathing *(bottom)* animals.

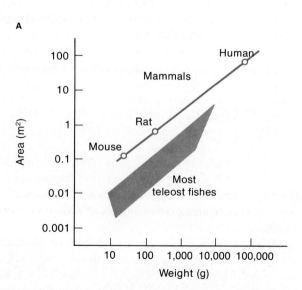

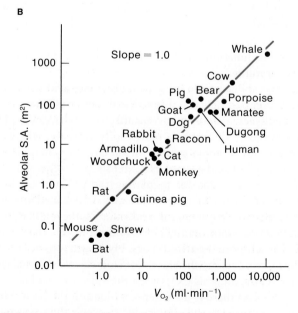

Figure 13-20 Respiratory surface area increases with size. **(A)** Relationship between respiratory surface area and body weight in selected mammals and teleost fishes. **(B)** Relationship between alveolar surface area (S. A.) and oxygen uptake in mammals. [Part A adapted from Randall, 1970; part B from Tenney and Temmers, 1963.]

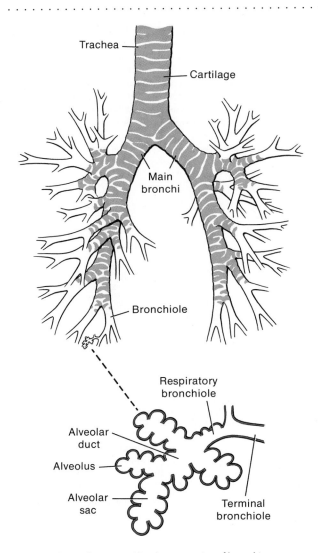

Figure 13-21 In the mammalian lung, a series of branching, progressively smaller ducts deliver air to the respiratory portion, consisting of terminal and respiratory bronchioles and alveolar ducts and sacs. Gas transfer occurs across the respiratory epithelium shown in red.

keeps the lungs clean (see Chapter 8). In the respiratory portions of the lung, smooth muscle replaces cartilage. Contraction of this smooth muscle can have a marked effect on the dimensions of the airways in the lungs.

Small mammals have a higher resting O_2 uptake rate per unit body weight than large mammals because of their greater alveolar surface area per unit body weight. This increase in respiratory surface area is achieved by a reduction in the size but an increase in the number of alveoli per unit volume of lung. In humans the number of alveoli increases rapidly after birth, the adult complement of about 300 million being attained by the age of eight years; subsequent increases in respiratory area are achieved by increases in the volume of each alveolus. The resting O_2 uptake rate per unit weight is higher in children than in adults; once again there is a correlation between uptake per unit weight and alveolar surface area per unit body weight.

The diffusion barrier in mammals is made up of an aqueous surface film, the epithelial cells of the alveolus, the interstitial layer, endothelial cells of the blood capillaries, plasma, and the wall of the red blood cell (Figure 13-22B). Several cell types compose the lung epithelium. Type I cells, the most abundant, constitute the major part of the lung epithelium. They are squamous epithelial cells, having a thin platelike structure, a single cell extending into the two adjacent alveoli with the nucleus tucked away in a corner. Type II cells are characterized by a laminated body within the cell and have surface villi; type II cells produce surfactants, discussed later. Type III cells are mitochondrion-rich cells with a brush border. These rare cells appear to be involved in NaCl uptake from lung fluid. In addition to these cells, a number of *alveolar macrophages* wander over the surface of the respiratory epithelium. It is generally assumed, but not demonstrated, that the coefficient of diffusion for gases does not vary in the lungs of different animals, the only structural variables being lung area and diffusion distance between air and blood.

The following terms are used to describe different types of breathing and lung ventilation:

- **Eupnea**—normal, quiet breathing typical of an animal at rest.

- **Hyperventilation** and **hypoventilation**—increase and decrease, respectively, in the amount of air moved in or out of the lung by changes in the rate and/or depth of breathing, such that ventilation no longer matches CO_2 production and blood CO_2 levels change

- **Hyperpnea**—increased lung ventilation due to increase in breathing in response to increased CO_2 production (e.g., during exercise)

- **Apnea**—absence of breathing

- **Dyspnea**—labored breathing associated with the unpleasant sensation of breathlessness

- **Polypnea**—increase in breathing rate without an increase in the depth of breathing

cross-sectional area of the airways increases rapidly as a result of extensive branching, although the diameter of individual air ducts decreases from the trachea to the terminal bronchioles. The terminal bronchioles, the respiratory bronchioles, the alveolar ducts, and the alveolar sacs constitute the respiratory portion of the lung. Gases are transferred across the thin-walled alveoli found in the regions distal to the terminal bronchioles, termed **acini**. The airways leading to the terminal bronchioles constitute the nonrespiratory portion of the lung. Alveoli in *adjoining acini* are interconnected by a series of holes, the **pores of Kohn**, allowing the collateral movement of air, which may be a significant factor in gas distribution during lung ventilation (Figure 13-22A).

Air ducts leading to the respiratory portion of the lung contain cartilage and a little smooth muscle and are lined with cilia. The epithelium secretes mucus, which is moved toward the mouth by the cilia. This "mucus escalator"

A

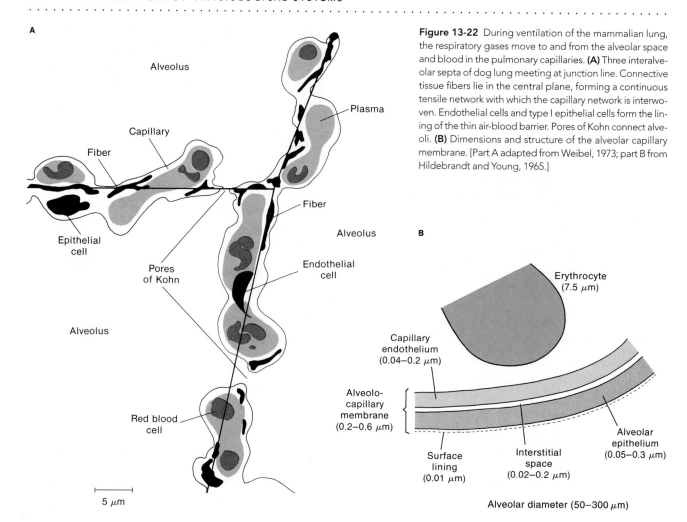

Figure 13-22 During ventilation of the mammalian lung, the respiratory gases move to and from the alveolar space and blood in the pulmonary capillaries. **(A)** Three interalveolar septa of dog lung meeting at junction line. Connective tissue fibers lie in the central plane, forming a continuous tensile network with which the capillary network is interwoven. Endothelial cells and type I epithelial cells form the lining of the thin air-blood barrier. Pores of Kohn connect alveoli. **(B)** Dimensions and structure of the alveolar capillary membrane. [Part A adapted from Weibel, 1973; part B from Hildebrandt and Young, 1965.]

Air exchanged between the alveoli and the environment must pass through a series of tubes (trachea, bronchi, nonrespiratory bronchioles) not directly involved in gas transfer. At the end of exhalation the air contained in these tubes will have come from the alveoli and will be low in oxygen and high in carbon dioxide. This air will be the first to move back into the alveoli at the next breath. At the end of inhalation the nonrespiratory tubes will be filled with fresh air and this volume will be the first to be exhaled with the next breath. Thus this volume is not involved in gas transfer and, therefore, is referred to as the *anatomical dead space volume*. Some air may be supplied to nonfunctional alveoli, or certain alveoli may be ventilated at too high a rate, increasing the volume of air not directly involved in gas exchange. This volume of air, termed the *physiological dead space*, is usually greater than, but includes, the anatomical dead space (Spotlight 13-3).

The amount of air moved in or out of the lungs with each breath is referred to as the **tidal volume**. The amount of fresh air moving in and out of the alveolar air sacs equals the tidal volume minus the anatomical dead-space volume, and is referred to as the **alveolar ventilation volume**. Only this gas volume is directly involved in gas transfer. The lungs are not completely emptied even at maximal expira-

tion, leaving a **residual volume** of air in the lungs. The maximum volume of air that can be moved in or out of the lungs is referred to as the **vital capacity** of the lungs. These and other terms used to describe various volumes and capacities associated with lung function are illustrated in Figure 13-23.

The O_2 content is lower and the CO_2 content is higher in alveolar gas than in ambient air because only a portion of the lungs' gas volume is changed with each breath. Alveolar ventilation in humans is about 350 ml, whereas the functional residual volume of the lungs exceeds 2000 ml. During inspiration the ducts leading to the alveoli elongate and widen, causing an increase in acinar volume. During breathing, air moves in and out of the acinus and may also move between adjacent alveoli through the pores of Kohn. Mixing of gases in the ducts and alveoli occurs by diffusion and by convection currents caused by breathing (Figure 13-24). In the alveolar ducts, O_2 diffuses toward the alveoli and CO_2 away from them. Partial pressures of O_2 and CO_2 are probably fairly uniform across the alveoli, because diffusion is rapid in air and the distances involved are small. The partial pressures of gases within the alveoli oscillate in phase with the breathing movements, the magnitude depending on the extent of tidal ventilation.

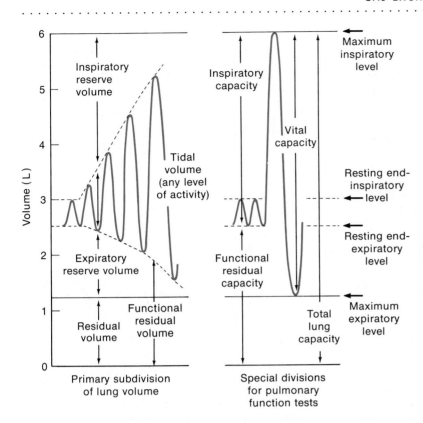

Figure 13-23 Numerous terms are used to describe various volumes and capacities associated with lung function. The tidal volume is the volume of air typically moved in and out of the lung, whereas the vital capacity is the maximum volume.

What might be the advantages for gas transfer of either vibrating a lung or having a tuned lung that vibrated during breathing?

The O_2 and CO_2 levels in alveolar gas are determined by both the rate of gas transfer across the respiratory epithelium and the rate of alveolar ventilation. Alveolar ventilation depends on breathing rate, tidal volume, and anatomical dead-space volume. Variations in the magnitude of the anatomical dead space will alter gas tensions in the alveolus in the absence of changes in tidal volume.

Thus, artificial increases in anatomical dead space, produced in human subjects breathing through a length of hose, result in a rise in CO_2 and a fall in O_2 in the lungs. As discussed in a later section, these changes activate chemoreceptors, leading to an increase in tidal volume. In animals with long necks (e.g., the giraffe and trumpeter swan), the tracheal length and therefore the anatomical dead space is greater than in those with short necks (Figure 13-25). In order to maintain adequate gas tensions in the lungs, long-necked animals have increased tidal volumes.

Breathing rate and tidal volume vary considerably in animals. Humans breathe about 12 times per minute and have a tidal volume at rest of about 10% of total lung volume. Such relatively rapid, shallow breathing produces small

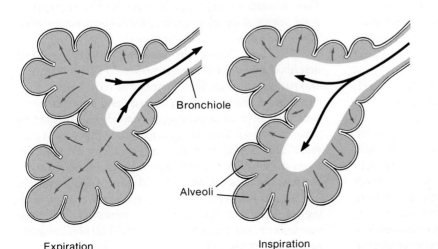

Figure 13-24 The direction of air flow (large arrows) in the respiratory portions of the lung changes during inspiration and expiration, but diffusion of oxygen (small arrows) is always towards the alveoli walls.

Expiration

Inspiration

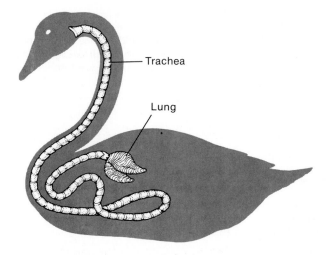

Figure 13-25 The extremely long trachea of the trumpeter swan results in a large anatomical dead-space volume. For comparison, see Figure 13-29 illustrating the length of the human trachea. [From Banko, 1960.]

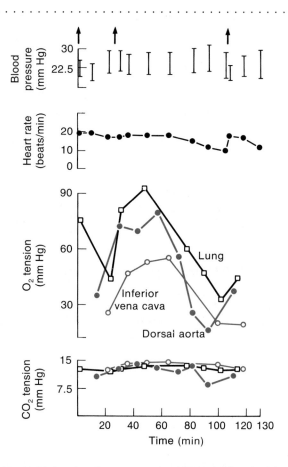

Figure 13-26 Breathing frequency tends to vary inversely with tidal volume and magnitude of oscillations in P_{O_2}. In *Amphiuma*, an aquatic, air-breathing amphibian that breathes infrequently, tidal volume and changes in P_{O_2} are large. Shown here are plots of blood pressure, heart rate, P_{O_2}, and P_{CO_2} in a 515-g *Amphiuma* during two breathing-diving cycles. Vertical arrows indicate when the animal surfaced and ventilated its lungs. Note that blood pressure, heart rate, and P_{CO_2} are nearly constant between breaths, whereas P_{O_2} shows large, slow oscillations in the lung and blood. [Adapted from Toews et al., 1971.]

oscillations in P_{O_2} in the lung and blood. In contrast, the exclusively aquatic but air-breathing amphibian *Amphiuma*, which lives in swamp water, rises to the surface of the water about once each hour to breathe; its tidal volume, however, is more than 50% of its lung volume. This large tidal volume, coupled with infrequent breathing, produces large, slow oscillations in P_{O_2} in the lung and blood, which are more or less in phase with the breathing movements (Figure 13-26). *Amphiuma* is preyed on by snakes and is most vulnerable when it rises to breathe. Because it lives in water of low oxygen content, aquatic respiration is not a suitable alternative. The hazard of being eaten while surfacing to breathe may have influenced the evolution of its very low breathing rate, its large tidal and lung volume, and its ability to make cardiovascular adjustments that help maintain O_2 delivery to the tissues in the face of widely oscillating blood gas levels. Carbon dioxide levels in *Amphiuma* do not oscillate in the same way as oxygen because carbon dioxide is lost across the skin and is not so dependent on lung ventilation.

In summary, O_2 and CO_2 levels in alveolar gas are determined by ventilation and the rate of gas transfer. Ventilation of the respiratory epithelium is determined by breathing rate, tidal volume, and anatomical dead-space volume. The nature and extent of ventilation also influences the magnitude of oscillations in O_2 and CO_2 in the blood during a breathing cycle.

Pulmonary Circulation

The lung, like the heart, receives blood from two sources. The major flow is of deoxygenated blood from the pulmonary artery that perfuses the lung, taking up O_2 and giving up CO_2; this is termed the *pulmonary circulation*. A second, smaller supply, the *bronchial circulation*, comes from the systemic (body) circulation and supplies the lung tissues with O_2 and other substrates for growth and maintenance. Our discussion here is confined to the pulmonary circulation.

In birds and mammals, pressures in the pulmonary circulation are lower than in the systemic circulation. This pressure difference reduces filtration of fluid into the lung. An extensive lymphatic drainage of lung tissues also helps ensure that no fluid collects in the lung (see Chapter 12). These features are important because any fluid that collects at the lung surface increases the diffusion distance between blood and air and reduces gas transfer.

Blood flow through the pulmonary circulation is best described as *sheet flow*—that is, flow of a liquid between two parallel surfaces. This contrasts with the laminar flow characteristic of flow through a tube and flow in the systemic circulation. The pulmonary capillary endothelium resembles two parallel surfaces, joined by pillar-like structures, with blood flowing between them. As pressure increases, the parallel surfaces move apart, leading to an increase in the thickness of the blood sheet. That is, pressure increases the thickness of the blood sheet rather than spreading out the flow in other directions. The mean arterial pressure in the human lung is about 12 mm Hg, oscillating between 7.5 mm Hg and 22 mm Hg with each contraction of

SPOTLIGHT 13-3

LUNG VOLUMES

The alveolar ventilation volume, V_A, equals the difference between the tidal ventilation volume, V_T, and the dead-space volume, V_D:

$$V_A = V_T - V_D$$

If f denotes breathing frequency, the volume of air moved in and out of the lung each minute, $V_A f$, is called the *alveolar minute volume*, or *respiratory minute volume*, symbolized as $\dot{V}_A$. The dot over the V indicates a rate function.

The anatomical dead space, V_{Danat}, is the volume of the non-respiratory portion of the lung; the physiological dead space, $V_{Dphysiol}$, is the volume of the lung not involved in gas transfer. If the partial pressure of CO_2 in expired air is denoted by P_ECO_2, the partial pressure of CO_2 in alveolar air by P_ACO_2, and the partial pressure of CO_2 in inspired air by P_ICO_2, then

$$P_ECO_2 \times V_T = (P_ACO_2 \times V_A) + (P_ICO_2 \times V_D)$$

But $V_A = V_T - V_D$, so substituting into this equation, we obtain

$$P_ECO_2 \times V_T = P_ACO_2(V_T - V_D) + (P_ICO_2 \times V_D)$$

and

$$P_ECO_2 \times V_T = (P_ACO_2 \times V_T) - (P_ACO_2 \times V_D) + (P_ICO_2 \times V_D)$$

By rearrangement

$$(P_ACO_2 \times V_D) - (P_ICO_2 \times V_D) =$$
$$(P_ACO_2 \times V_T) - (P_ECO_2 \times V_T)$$

$$V_D(P_ACO_2 - P_ICO_2) = V_T(P_ACO_2 - P_ECO_2)$$

$$V_{Dphysiol} = V_T \frac{(P_ACO_2 - P_ECO_2)}{(P_ACO_2 - P_ICO_2)}$$

But P_ICO_2 approaches zero, and P_ACO_2 is the same as the partial pressure of CO_2 in arterial blood, P_aCO_2. So the last expression can be written as follows:

$$V_{Dphysiol} = V_T \frac{(P_aCO_2 - P_ECO_2)}{P_aCO_2}$$

Thus the physiological dead space of the lungs can be calculated from measurements of tidal volume, V_T, and the CO_2 partial pressures in arterial blood, P_aCO_2, and expired air, P_ECO_2.

---

the heart. In the vertical (upright) human lung, arterial pressure is just sufficient to raise blood to the apex of the lung; hence flow is minimal at the top and increases toward the base of the lung (Figure 13-27). Blood is distributed more evenly to different parts of the horizontal lung.

The pulmonary vessels are very distensible and subject to distortion by breathing movements. Small vessels within the interalveolar septa are particularly sensitive to changes in alveolar pressure. The diameter of these thin-walled collapsible capillaries is determined by the transmural pressure

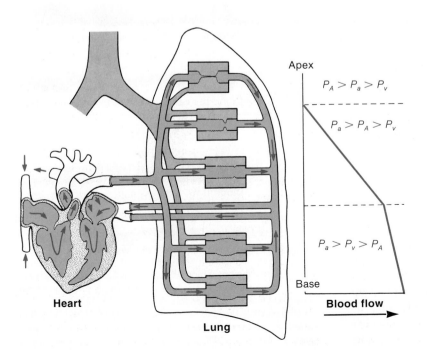

Figure 13-27 In the upper portion of the vertical lung, the diameter of alveolar capillaries, and hence blood flow through them, depends on the difference between the arterial pressure, P_a, and the alveolar pressure, P_A. In this schematic diagram of blood flow in the vertical human lung, the boxes represent the condition of vessels in the interalveolar septum in different portions of the lung. At the apex of the lung, P_A often exceeds P_a; as a result the capillaries collapse and blood flow ceases. P_v is the venous pressure. [Adapted from West, 1970.]

Apex

$P_A > P_a > P_v$

$P_a > P_A > P_v$

$P_a > P_v > P_A$

Base

Blood flow

Heart

Lung

(arterial blood pressure within capillaries, P_a, minus alveolar pressure, P_A). If the transmural pressure is negative (i.e., $P_A > P_a$), these capillaries collapse and blood flow ceases. This collapse may occur at the apex of the vertical human lung, where P_a is low (see Figure 13-27). If pulmonary arterial pressure is greater than alveolar pressure, which in turn is greater than pulmonary venous pressure, then the difference between arterial and alveolar pressure will determine the diameter of capillaries in the interalveolar septa and, in the manner of a sluice gate, control blood flow through the capillaries. Venous pressure will not affect flow into the venous reservoir as long as alveolar pressure exceeds venous pressure. Flow in the upper portion of the vertical lung is probably determined in this way by the difference between arterial blood pressure and alveolar pressure. Arterial blood pressure (and therefore blood flow) increases with distance from the apex of the lung.

In the bottom half of the vertical lung, where venous pressure exceeds alveolar pressure, blood flow is determined by the difference between arterial and venous blood pressures. This pressure difference does not vary with position, although both the arterial and venous pressures increase toward the base of the lung. This increase in absolute pressure results in an expansion of vessels and, therefore, a decrease in resistance to flow. Thus, flow increases toward the base of the lung, even though the arterial-venous pressure difference does not change (see Figure 13-27). The position of the lungs with respect to the heart is therefore an important determinant of pulmonary blood flow. The lungs surround the heart, thus minimizing the effect of gravity on pulmonary blood flow as an animal changes from a horizontal to vertical position. This close proximity of lungs and heart within the thorax also has significance for cardiac function: the reduced pressures within the thorax during inhalation aid venous return to the heart. This is often referred to as the *thoraco-abdominal pump*.

Even though the mammalian pulmonary circulation lacks well-defined arterioles, both sympathetic adrenergic and parasympathetic cholinergic fibers innervate the smooth muscle around the pulmonary blood vessels and bronchioles. The pulmonary circulation, however, has much less innervation than does the systemic circulation and is relatively unresponsive to nerve stimulation or injected drugs. Sympathetic nerve stimulation or the injection of norepinephrine causes a slight increase in resistance to blood flow, whereas parasympathetic nerve stimulation or acetylcholine has the opposite effect.

Reductions in either oxygen levels or pH cause local vasoconstriction of pulmonary blood vessels. The vasoconstrictor response to low oxygen, which is the opposite to that observed in systemic capillary networks, ensures that blood flows to the well-ventilated regions of the lung. Poorly ventilated regions of the lung will have low alveolar oxygen levels, causing a local vasoconstriction and therefore a reduction in blood flow to that area of the lung. Alternatively, a well-ventilated area of the lung will have high alveolar oxygen levels, so the local blood vessels will be di-

lated and blood flow will be high. Although pulmonary hypoxic vasoconstriction is important in directing blood flow to well-ventilated regions of the lung, it leads to problems when animals are exposed to general hypoxia, as may occur at high altitudes (see later section).

Cardiac output to the pulmonary circuit is identical to cardiac output in the systemic circuit in mammals and birds. In amphibians and reptiles, with a single or partially divided ventricle that ejects blood into both the pulmonary and the systemic circulation, the ratio of pulmonary to systemic blood flow can be altered. In turtles and frogs, there is a marked increase in blood flow to the lung following a breath due to pulmonary vasodilation. During periods between breaths in the frog *Xenopus*, pulmonary blood flow decreases, but systemic blood flow is hardly changed possibly because the ventricle is undivided (Figure 13-28). These animals breathe intermittently, and variable blood flow to the gas exchanger, independent of blood flow to the rest of the body, permits some control of the rate of oxygen use from the lung store and rapid renewal of blood oxygen stores during ventilation. In addition, cardiac work is reduced during apnea.

Ventilation of the Lung

The mechanism of lung ventilation varies considerably among animals. These variations reflect differences in the functional anatomy of the lungs and associated structures. First we will see how the mammalian lung is ventilated and then consider ventilation in birds, reptiles, frogs, and invertebrates.

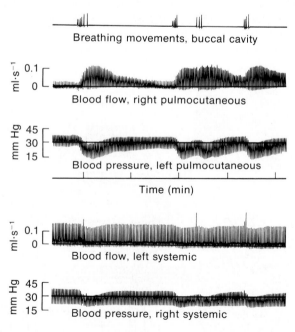

Figure 13-28 Pulmonary blood flow typically increases, whereas systemic flow remains constant, following breathing in turtles and frogs. These traces from the frog *Xenopus* record pressure changes in the buccal cavity produced by lung-ventilating movements of the buccal floor (*upper*), as well as the corresponding flow and pressure in the arterial arches. [From Shelton, 1970.]

Mammals

The lungs of mammals are elastic, multi-chambered bags, which are suspended within the **pleural cavity** and open to the exterior via a single tube, the trachea (Figure 13-29). The walls of the pleural cavity, often referred to as the **thoracic cage,** are formed by the ribs and the **diaphragm.** The lungs fill most of the thoracic cage, leaving a low-volume pleural space between the lungs and thoracic wall; this space is sealed and fluid filled. Because of their elasticity, isolated lungs are somewhat smaller than they are in the thoracic cage. *In situ* this elasticity creates a pressure below atmospheric in the fluid-filled pleural space. The fluid in the pleural cavity provides a flexible, lubricated connection between the outer lung surface and the thoracic wall. Fluids are essentially incompressible, so when the thoracic cage changes volume, the gas-filled lungs do too. If the thoracic cage is punctured, air is drawn into the pleural cavity and the lungs collapse—a condition known as **pneumothorax.**

When intact lungs are filled to various volumes and the entrance is closed with the muscles relaxed, then alveolar pressure varies directly with lung volume. At low pulmonary volumes, alveolar pressure is less than ambient pressure owing to the resistance of the thorax to collapse, whereas at high pulmonary volumes, alveolar pressure exceeds ambient pressure because of the forces required to expand the thoracic cage. If lung volume is large, then once the mouth and glottis are opened, air will flow out of the lungs because the weight of the ribs will reduce pulmonary volume. At some intermediate volume, V_r, alveolar pressure in the relaxed thorax is equal to ambient pressure (Figure 13-30).

During normal breathing, the thoracic cage is expanded and contracted by a series of skeletal muscles, the di-

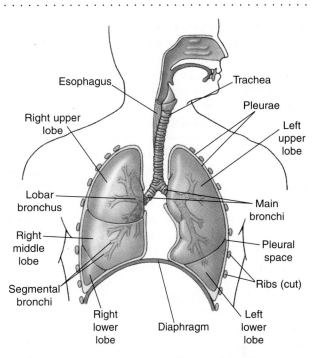

Figure 13-29 In mammals, the lungs fill most of the thoracic cavity, formed by the ribs and diaphragm. The right lung has three lobes, and the left lung, two lobes, in humans. The low-volume pleural space between the lungs and thoracic wall is fluid filled and sealed.

aphragm, and the external and internal intercostal muscles. Contractions of these muscles are determined by activity of motor neurons controlled by the respiratory center within the medulla oblongata, which we discuss later. The volume of the thorax increases as the ribs are raised and moved outward by contraction of the external intercostals and by contraction (and therefore the lowering) of the diaphragm

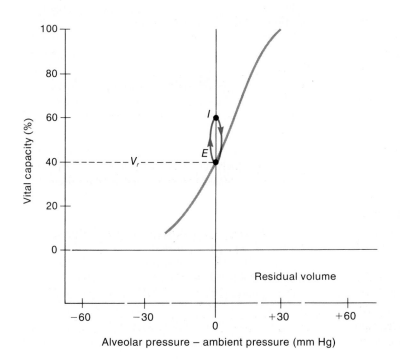

Figure 13-30 During quiet breathing with the thoracic muscles relaxed, the alveolar and ambient pressures are equal between breaths. This plot shows the relationship between lung volume and pressure within the thorax when muscles are relaxed, but the glottis is closed. V_r is the lung volume when alveolar pressure is the same as ambient pressure and the lung-chest system is relaxed. The points *I* and *E* represent the pressure and volume of the system following inspiration and expiration during quiet breathing.

(Figure 13-31A). Contractions of the diaphragm account for up to two-thirds of the increase in pulmonary volume. The increase in thoracic volume reduces alveolar pressure, and air is drawn into the lungs. Relaxation of the diaphragm and external intercostal muscles reduces thoracic volume, thereby raising alveolar pressure and forcing air out of the lungs (Figure 13-31B). During quiet breathing, pulmonary volume between breaths is at an intermediate value, V_r, at which the alveolar and ambient pressures are equal (see Figure 13-30). Under these conditions exhalation often is passive, simply due to relaxation of the diaphragm and external intercostals. With increased tidal volume, expiration becomes active, owing to contraction of the internal intercostal muscles, which further reduces thoracic volume until it drops below V_r at the end of expiration.

Birds

In birds, gas transfer takes place in small air capillaries (10 μm in diameter) that branch from tubes called **parabronchi** (Figure 13-32). The functional equivalent of mammalian alveolar sacs, parabronchi are a series of small tubes extending between large *dorsobronchi* and *ventrobronchi*, both of which are connected to an even larger

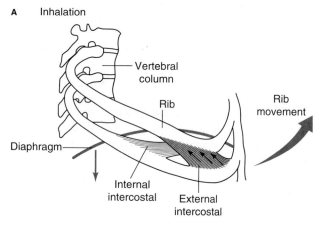

A Inhalation

Vertebral column

Rib

Rib movement

Diaphragm

Internal intercostal

External intercostal

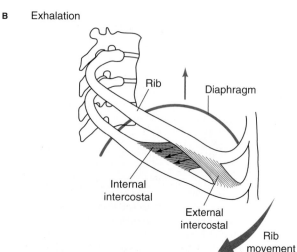

B Exhalation

Rib

Diaphragm

Internal intercostal

External intercostal

Rib movement

Figure 13-31 The volume of the thorax increases during inhalation **(A)** and decreases during exhalation **(B)** in mammals due to movement of the ribs and diaphragm.

tube, the *mesobronchus,* which joins the trachea anteriorly (Figure 13-33A). The parabronchi and connecting tubes form the lung, which is contained within a thoracic cavity. A tight horizontal septum closes the caudal end of the thoracic cage. The ribs, which are curved to prevent lateral compression, move forward only slightly during breathing; as a result, the volume of the thoracic cage and lung changes little during breathing. The large flight muscles of birds are attached to the sternum and have little influence on breathing. Although there is no mechanical relation between flight and respiratory movements in birds, "in phase" flight and breathing movements may result from synchronous neural activation of the two groups of muscles involved.

How, then, is the avian lung ventilated? The answer lies in the associated air-sac system connected to the lungs (see Figure 13-32). As these air sacs are squeezed, air is forced through the parabronchi. The system of air sacs, which extend as diverticula of the airways, penetrates into adjacent bones and between organs, reducing the density of the bird. Of the many air sacs, only the thoracic (cranial) and abdominal (caudal) sacs show marked changes in volume during breathing. Volume changes in the air sacs are achieved by a rocking motion of the sternum against the vertebral column and by lateral movements of the posterior ribs. Air flow is bidirectional in the mesobronchus, but unidirectional through the parabronchi (Figure 13-33B). During inspiration, air flows into the caudal air sacs through the mesobronchus; air also moves into the cranial air sacs via the dorsobronchus and the parabronchi. During expiration, air leaving the caudal air sacs passes through the parabronchi and, to a lesser extent, through the mesobronchus to the trachea. The cranial air sacs, whose volume changes less than that of the caudal air sacs, are reduced somewhat in volume by air moving from the cranial sac via the ventrobronchi to the trachea during expiration.

Oxygen diffuses into the air capillaries from the parabronchi and is taken up by the blood. The air in the parabronchi is changed during both inspiration and expiration, enhancing gas transfer in the bird lung. The unidirectional flow is achieved not by mechanical valves but by *aerodynamical valving.* The openings of the ventrobronchi and dorsobronchi into the mesobronchus show a variable, direction-dependent resistance to air flow. The structure of the openings is such that eddy formation, and therefore resistance to flow, varies with the direction of air flow.

Reptiles

The ribs of reptiles, like those of mammals, form a thoracic cage around the lungs. During inhalation, the ribs are moved cranially and ventrally, enlarging the thoracic cage. As this expansion reduces the pressure within the cage below atmospheric pressure, and the nares and glottis are open, air flows into the lungs. Relaxation of muscles that enlarge the thoracic cage releases energy stored in stretching the elastic component of the lung and body wall, al-

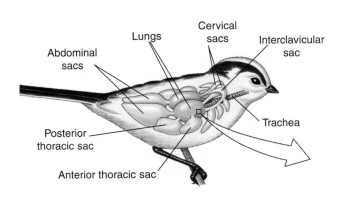

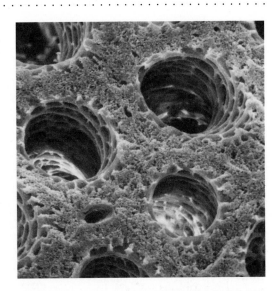

Figure 13-32 In bird lungs, gas exchange occurs in air capillaries extending from parabronchi, small tubelike structures that are the functional equivalent of alveoli in mammals. The parabronchi *(right)* and connecting tubes form the lung. During breathing, volume changes occur in the associated air sacs, not in the thoracic cage and lungs. [Photograph courtesy of H. R. Duncker.]

lowing passive exhalation. Although reptiles do not possess a diaphragm, pressure differences between the thoracic and abdominal cavities have been recorded, indicating at least a functional separation of these cavities.

In tortoises and turtles, the ribs are fused to a rigid shell. The lungs are filled by outward movements of the limb flanks and/or the plastron (ventral part of the shell) and by forward movement of the shoulders. The reverse process results in lung deflation. As a result, retraction of limbs and head into the shell leads to a decrease in pulmonary volume.

Frogs

In frogs, the nose opens into a buccal cavity, which is connected via the glottis to paired lungs. The frog can open and close its nares and glottis independently. Air is drawn into the buccal cavity with the nares open and glottis closed; then the nares are closed, the glottis is opened, and the buccal floor raised, forcing air from the buccal cavity into the lungs (Figure 13-34). This lung-filling process may be repeated several times in sequence. Expiration also may be a step process, the lungs releasing air in portions to the buccal cavity. Expiration may not be complete, so that some of the air from the lung is mixed with ambient air in the buccal cavity and then pumped back into the lungs. That is, a mixture of pulmonary air, presumably low in O_2 and high in CO_2, is mixed with fresh air in the buccal cavity and returned to the lungs. The reason for this complex method of lung ventilation is not clear, but it may be directed toward reducing oscillations in CO_2 levels in the lungs in order to stabilize and regulate blood P_{CO_2} and control blood pH.

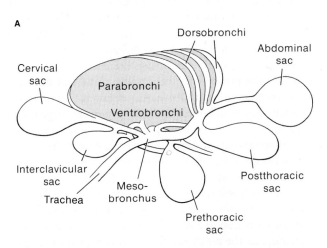

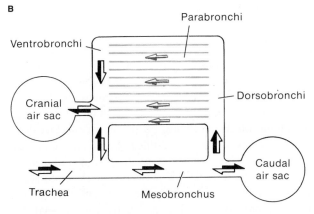

Figure 13-33 Squeezing of airs sacs forces air through the parabronchi in bird lungs. **(A)** The avian bronchial tree and associated air sacs. The air sacs of the cranial group (cervical, interclavicular, and prethoracic sacs) depart from the three cranial ventrobronchi, whereas the air sacs of the caudal group (postthoracic and abdominal sacs) are connected directly to the mesobronchus. **(B)** Schematic diagram of air flow through the bird lung. Flow in the parabronchi is unidirectional. Solid arrows represent flow during inspiration; open arrows, flow during expiration. [Adapted from Scheid et al., 1972.]

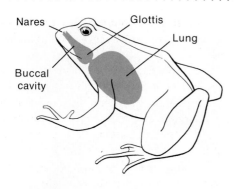

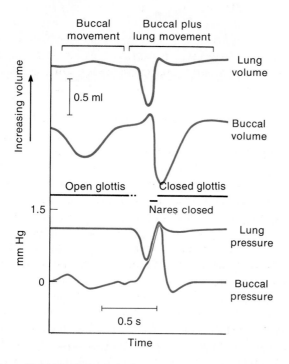

Figure 13-34 Ventilation in the frog is a stepwise process. Shown here are pressure and volume changes in the buccal cavity and lung of a frog during buccal movements alone with the glottis closed and during buccal and lung movements with the glottis open and the nares closed (i.e., lung filling). [Adapted from West and Jones, 1975.]

Invertebrates

Invertebrates exhibit a variety of gas-transfer mechanisms. Ventilation does not occur in some invertebrates, which rely only on diffusion of gases between the lung and the environment. In spiders, which have paired ventilated lungs on the abdomen, the respiratory surface consists of a series of thin, blood-filled plates that extend like the leaves of a book into a cavity guarded by an opening (spiracle). The spiracle can be opened or closed to regulate the rate of water loss from these "book lungs." Snails and slugs have ventilated lungs that are well-vascularized invaginations of the body surface, the mantle cavity. The volume change that the snail lung is capable of undergoing enables the animal to emerge from and withdraw into its rigid shell. When the snail retracts into its shell, the lungs empty, a situation similar to that seen in tortoises.

In aquatic snails, the lungs serve to reduce the animal's density.

Pulmonary Surfactants

The lung wall tension depends on the properties of the alveolar wall and the surface tension at the liquid–air interface. *Surface tension* is the force that tends to minimize the area of a liquid surface, causing liquid droplets to form a sphere. It also makes a surface film resistant to stretch, so that work must be done to stretch a liquid surface. Because the alveoli are so compliant, the surface tension of their liquid lining contributes about 70% of the lung wall resistance to stretch. If the liquid lining was just water, the alveolar wall tension would be much higher than in fact it is, and large forces would be required to inflate the lung and to separate membranes glued together by surface tension. The explanation for the relatively low surface tension of the liquid lining the lungs is the presence of **surfactants**, lipoprotein complexes that bestow a very low surface tension on the liquid-air interface. Lung surfactants not only reduce the effort associated with breathing but also help prevent collapse of alveoli.

Surfactants are produced by type II cells within the alveolar lining and have a half-life of about 12 hours in mammals. The predominant lipid in these lipoprotein complexes is dipalmitoyl lecithin. The lipoprotein film is stable, the lipid forming an outer monolayer firmly associated with the underlying protein layer. Synthesis of surfactants requires cortisol, and their release can be stimulated by sighing. Surfactants are found in the lungs of amphibians, reptiles, birds, and mammals, and they may be present in some fishes that build bubble nests.

The small dimensions of the fragile alveolar sacs create mechanical problems that might cause them to collapse. To understand why alveolar collapse is a problem, and how surfactants counteract it, consider a tiny bubble that is alternately inflated and then deflated. As discussed in Chapter 12, Laplace's law states that the pressure differential between the inside and outside of a bubble is proportional to $2y/R$, where y is the wall tension per unit length and R the radius of the bubble. If two bubbles have a similar wall tension but a different radius, the pressure in the small bubble will be higher than that in the large bubble. As a result, if the bubbles are joined, the small bubble will empty into the large bubble (Figure 13-35A,B).

A somewhat similar situation exists in the lung. We can consider the alveoli as a number of interconnected bubbles. If the wall tension is similar in alveoli of different size, the small alveoli will tend to collapse, emptying into the larger alveoli. This normally does not occur in the lung for two reasons: surrounding tissue helps prevent overexpansion of alveoli, and the properties of the alveolar surfactant lining are such that wall tension increases when the surface film is expanded and decreases when it is compressed. This occurs because the film expands as alveolar volume increases, so the surfactant spreads out and therefore is less effective in lowering surface tension (Figure 13-35C). The result of this effect is to minimize pressure differences between large and small alveoli, thus reducing the chance of collapse. Alveoli fold as their volume decreases, and the regions between the folds come to have a thick layer of surfactant. The very low surface tension of this thick surfactant layer permits easy inflation of collapsed and folded alveoli. If only water were present in the folds, large forces would be required to separate the layers and inflate the alveoli.

In mammals, surfactants appear in the fetal lung prior to birth, thereby reducing the forces required to inflate the lungs of the newborn. Newborns who produce no lung surfactants cannot inflate their lungs at birth without assistance. This condition, referred to as *newborn respiratory distress syndrome,* occurs primarily in premature babies. Assistance can be given to the baby by forcing air into the lungs, using positive pressure ventilation, and by surfactant replacement. In addition, pregnant women who are likely to have a premature birth can be given an injection of cortisol during gestation to stimulate surfactant production in the fetus.

Heat and Water Loss across the Lung

Increases in lung ventilation not only increase gas transfer but also result in more loss of heat and water. Thus, the evolution of lungs has involved some compromises. Air in contact with the respiratory surface becomes saturated with water vapor and comes into thermal equilibrium with the blood. Cool, dry air entering the lung of mammals is humidified and heated. Exhalation of this hot, humid air results in considerable loss of heat and water, which will be proportional to the rate of ventilation of the lung surface. Many air-breathing animals live in very dry environments, where water conservation is of paramount importance. It is therefore not surprising that these animals in particular have evolved means of minimizing the loss of water.

The rates of heat and water loss from the lung are intimately related. As air is inhaled, it is warmed and humidified by evaporation of water from the nasal mucosa. Because the evaporation of water cools the nasal mucosa, a temperature gradient exists along the nasal passages. The nose is cool at the tip, increasing in temperature toward the glottis. As the moist air leaving the lung is cooled, water condenses on the nasal mucosa, since the water vapor pressure for 100% saturation decreases with temperature. Thus the cooling of exhalant air in the nasal passages results in the conservation of both heat and water. The blood circulation to the nasal mucosa is capable of supplying water to saturate the inhalant air, but the temperature gradients established by water evaporation and air movement are not destroyed by the circulation.

The structure of the nasal passages in vertebrates is variable, and to some extent it can be correlated with the ability of animals to regulate heat and water loss. Humans have only a limited ability to cool exhaled air, which is saturated with water vapor and is at a temperature only a few degrees below core body temperature. Other animals have longer and narrower nasal passages for more effective water conservation, as we will discuss in Chapter 14.

Poikilotherms such as reptiles and amphibians, whose body temperatures adjust to the ambient temperature, exhale air saturated with water at temperatures about 0.5–1.0°C below body temperature. Pulmonary air temperatures and body surface temperatures are often slightly below ambient because of the continual evaporation of water. In some reptiles, however, body temperature is maintained above ambient. In the iguana, heat and water loss is

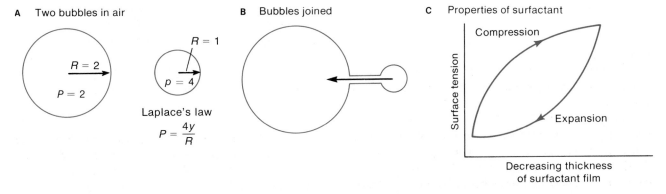

Figure 13-35 The presence of surfactant in the lungs helps prevent alveolar collapse. **(A)** Laplace's law states that the pressure (*P*) in a bubble decreases with increased radius (*R*) if the wall tension (*y*) remains constant. Thus, if two bubbles have the same wall tension but the radius of one is twice that of the other, the pressure in the small bubble is two times that of the large bubble. The equation is written $4y/R$ rather than $2y/R$ because the bubble in air has an inner and an outer surface. **(B)** If the bubbles are joined, the small bubble with the higher pressure collapses into the large bubble with the lower pressure. **(C)** The tendency of small alveoli to collapse into larger alveoli in the lung is ameliorated by a surfactant lining. As the surfactant film expands with the alveolus, the thickness of the film decreases and the surface tension increases. Because the surface tension is a major component of the wall tension, this effect tends to minimize pressure differences between alveoli of different sizes, thereby stabilizing them.

controlled in a manner similar to that observed in mammals. In addition, this lizard conserves water by humidifying air with water evaporated from the excretory fluid of the nasal salt glands. The rate of water loss is closely correlated with lung ventilation and, therefore, oxygen uptake. Reptiles generally have much lower oxygen requirements than mammals or birds, and so their rate of water loss is much less.

Gas Transfer in Bird Eggs

The shells of bird eggs have fixed dimensions but contain an embryo whose gas-transfer requirements increase by a factor of 10^3 between laying and hatching. Thus, the transfer of O_2 and CO_2 must take place across the shell at ever-increasing rates during development while the dimensions of the transfer surface (eggshell) do not change. Gases diffuse through small air-filled pores in the eggshell and then through underlying membranes, including the chorioallantoic membrane (Figure 13-36A). The chorioallantoic circulation is in close apposition to the eggshell and increases with the development of the embryo. Several factors contribute to the increase in gas-transfer rates during development in the bird's egg: development of an underlying circulation in the chorioallantoic membrane, an increase in blood

flow and volume, an increase in hematocrit and blood oxygen affinity, and an increase in the P_{O_2} difference across the eggshell (Figure 13-36B). The eggshell, once produced, does not change during the development of the embryo.

Water is lost from the egg during development, causing a gradual enlargement of an air space within the egg. The volume of this air cell is as much as 12 ml at hatching in the chicken egg. Just before they hatch, birds ventilate their lungs by poking their beaks into the air cell. Blood P_{CO_2} is initially low in the embryo, but it gradually rises to about 45 mm Hg just before hatching (Figure 13-36C). This pressure is maintained after hatching, thus avoiding any marked acid-base changes when the bird switches from the shell to its lungs for gas exchange.

The shell and underlying membranes, representing the barrier between ambient air and embryonic blood, can be divided into an outer gas phase (the air cell) and an inner liquid phase. At sea level the outer gas phase represents about 30%–40% of the total diffusive resistance to oxygen transfer, 85% of that to carbon dioxide, and 100% of that to water vapor. Eggs at altitude are exposed to a reduction in both oxygen and total gas pressure. The rate of diffusion of gases increases with a reduction in total pres-

Figure 13-36 During development of a bird embryo, gas transfer increases across the eggshell even though the shell structure does not change. **(A)** Diagram of the diffusion pathway between air and chick embryo blood across the eggshell in the region of the air cell. **(B)** Plot of P_{O_2} versus incubation age in the air cell and allantoic venous blood. **(C)** Plot of P_{CO_2} versus incubation age in the air cell and allantoic venous blood. There is no P_{CO_2} difference between air-cell gas and allantoic venous blood, whereas there is a P_{O_2} that increases during development of a chick embryo. [Adapted from Wangensteen, 1972.]

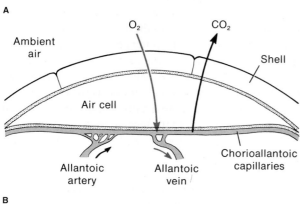

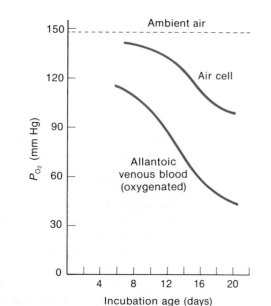

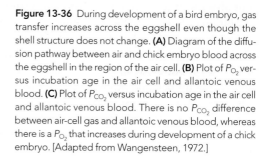

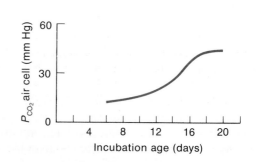

sure. The reduced oxygen pressure at altitude is partially offset by increased rates of oxygen diffusion in the gas phase; despite this, eggs become hypoxic at altitude. If eggs are kept in a hypoxic environment for a period of time, more capillaries develop in the chorioallantoic membrane, increasing oxygen diffusing capacity and offsetting the effects of altitude on oxygen transfer across the eggshell. Because carbon dioxide and water vapor also diffuse more rapidly at the reduced pressures associated with altitude, eggs at altitude also have a reduced blood P_{CO_2} and lose water more rapidly than those at sea level. Thus conditions affecting diffusion rates have a marked effect on CO_2 and water loss by eggs, and rates of water loss increase markedly in eggs exposed to reduced pressures. The properties of the shell are determined by the adult when the egg is laid. It appears that some birds can reduce the effective pore area of their eggs when acclimated to altitude.

Insect Tracheal Systems

The system that insects have evolved for transferring gases between the tissues and the environment differs fundamentally from that found in air-breathing vertebrates. The insect **tracheal system** takes advantage of the fact that oxygen and carbon dioxide diffuse 10,000 times more rapidly in air than in water, blood, or tissues. Tracheal systems consist of a series of air-filled tubes that penetrate from the body surface to the cells, acting as a pathway for the rapid movement of O_2 and CO_2, thereby avoiding the need for a circulatory system to transport gases between the respiratory surface and the tissues. These tubes, or tracheas, are invaginations of the body surface; thus their wall structure is similar to that of the cuticle. Except in a few primitive forms, the tracheal entrances, called **spiracles**, can be adjusted to control air flow into the tracheas, regulate water loss, and keep out dust. The bug *Rhodnius*, for example, dies in three days if its spiracles are kept open in a dry environment. The tracheas branch everywhere in the tissues; the smallest, terminal branches, or **tracheoles**, are blind-ending and poke between and into individual cells (without disrupting the cell membrane), delivering O_2 to regions very close to the mitochondria. Air sacs commonly are located at various intervals throughout the tracheal system; these sacs enlarge tracheal volume and therefore oxygen stores, and sometimes reduce the specific gravity of organs, either for buoyancy or for balance.

Tracheal ventilation

Diffusion of gases, even in air, is a slow process. Much more rapid transfer of oxygen and carbon dioxide can be achieved by the mass movement of gases, or **convection**. Larger insects usually have some mechanism for generating air flow in the bigger tubes of their tracheal system. The air sacs and tracheal tubes are often compressible, allowing changes in tracheal volume. Some larger insects ventilate the larger tubes and air sacs of the tracheal system by alternate compression and expansion of the body wall, particularly the abdomen. Different spiracles may open and close during different phases of the breathing cycle, thereby controlling the direction of air flow. In the locust, for instance, air enters through the thoracic spiracles but leaves through more posterior openings. Tracheal volume in insects is highly variable; it is 40% of body volume in the beetle *Melolontha* but only 6%–10% of body volume in the larva of the diving beetle *Dytiscus*. Each ventilation results in a maximum of 30% of tracheal volume being exchanged in *Melolontha* and 60% in *Dytiscus*. Not all insects ventilate their tracheal system; in fact, many calculations have shown that diffusion of gases in air is rapid enough to supply tissue demands in many species. To augment gas transfer, ventilation of trachea occurs in larger insects and, during high levels of activity, in some smaller insects.

In many insects the spiracles open and close, resulting it what is referred to as the insect *discontinuous ventilation cycle* (DVC). The DVC can be divided into three phases: an open phase, a closed phase, and an intermediate flutter phase when the spiracle oscillates rapidly between the open and closed states. Oxygen utilization and carbon dioxide production by the tissues occurs during all phases, oxygen being supplied from stores in the tracheal system when the spiracles are closed. Pressure in the tracheal system falls during the closed phase because oxygen levels decrease more rapidly than carbon dioxide levels increase. Carbon dioxide levels in the endotracheal space rise slowly during the closed phase because most of the carbon dioxide produced by metabolism is stored in the tissues. Thus during the flutter phase and at the onset of the open phase, gases move into the trachea both by bulk flow down a pressure gradient and by diffusion. Carbon dioxide and water diffuse from the endotracheal space during the open phase and even during the flutter phase, but not during the closed phase (Figure 13-37).

In theory, discontinuous ventilation can reduce water loss associated with respiration. The generation of low oxygen levels in the endotracheal space during the closed phase ensures high rates of oxygen diffusion into the tracheal space during the open phase compared with rates of water loss. The functional significance of the flutter phase in determining the rates of gas and water transfer is not clear, but it may enhance gas mixing in the tracheal space. In some instances, however, the role of discontinuous ventilation in water conservation appears to be of little significance. Many xeric species, which require little water, do not show discontinuous ventilation. For example, the lubber grasshopper does not display discontinuous ventilation during desiccation even though it is capable of doing so. In this case only about 5% of total water loss is via the tracheal system, so perhaps it is not surprising that the pattern of ventilation is not changed during desiccation. In water-stressed cockroaches, cuticular water fluxes are more than twice that of loss through the spiracles, and closure of pore structures in the cuticle can conserve water during periods of desiccation. Thus it is not clear why

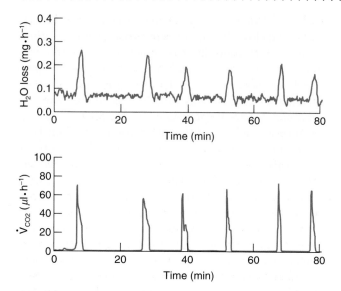

Figure 13-37 Some insects exhibit discontinuous ventilation as the result of opening and closing of the spiracles. These traces of water and carbon dioxide loss from an alate (would-be queen) harvester ant show that respiratory water loss is concentrated in the spiracle open phase associated with carbon dioxide excretion. The spiracles are closed between pulses of CO_2 excretion. The background cuticular water loss rates can be seen between the open phases. [From Lighton, 1994.]

many insects have adopted a pattern of discontinuous ventilation of their tracheal system. Although this mechanism reduces water loss in relation to oxygen uptake, the savings may not always be of much significance to the animal. The functional significance of the flutter phase is an even greater enigma.

 What are some of the problems faced by insects at high altitude? Are these similar to those faced by bird's eggs?

Gas exchange across tracheolar walls

Gases are transferred between air and tissues across the walls of the tracheoles. These walls are very thin, with an approximate thickness of only 40 to 70 nm. The tracheolar area is very large, and only rarely is a particular insect cell more than three cells away from a tracheole. The tips of the tracheoles, except in a few species, are filled with fluid, so that oxygen diffusing from the tracheoles to the tissues moves through the fluid in the tracheoles, the tracheolar wall, the extracellular space (often negligible), and the cell membrane to the mitochondria. This diffusion distance can be altered in active tissues either by an increase in tissue osmolarity, which causes water to move out of the tracheoles and into tissues or by changes in the activity of an ion pump, which results in the net flow of ions and water out of the tracheoles. As fluid is lost from the tracheoles, it is replaced by air, so that oxygen can more rapidly diffuse into the tissues (Figure 13-38). Insect flight muscle has the highest recorded

O_2 uptake rate of any tissue, with O_2 uptake increasing 10- to 100-fold above the resting value during flight. In general, more active tissues have more tracheoles, and in larger insects the tracheal system is more adequately ventilated.

Modified tracheal systems

There are many modifications of the generalized tracheal system just described. Some larval insects, for example, rely on cutaneous respiration, the tracheal system being closed off and filled with fluid. Some aquatic insects have a closed, air-filled tracheal system in which gases are transferred between water and air across *tracheal gills*. The gills are evaginations of the body that are filled with tracheas, the air of which is separated from the water by a 1-μm-thick membrane. Since this tracheal system is not readily compressible, it allows the insect to change depth under water without impairment of gas transfer.

Many aquatic insects, such as mosquito larvae, breathe through a *hydrofuge* (water-repellent) siphon that protrudes above the surface of the water; others take bubbles of air beneath the surface with them. The water bug *Notonecta* carries air bubbles that cling to velvetlike hydrofuge hairs on its ventral surface when submerged, and the water beetle *Dytiscus* dives with air bubbles beneath its wings or attached to its rear end. When such insects dive, gases are transferred between the bubble and the tissues via the tracheal system; gases can also diffuse, however, between the bubble and the water.

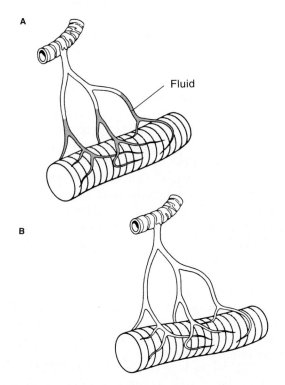

Figure 13-38 In resting muscle fibers, the terminal parts of tracheoles contain fluid **(A)**, but in active fibers, air may replace this fluid **(B)**, thereby increasing diffusion of oxygen into the muscle. [Adapted from Wigglesworth, 1965.]

Gas exchange in such "bubble-breathing" bugs thus involves diffusion across both tracheolar walls and bubble walls. The rate of O_2 transfer between water and the interior of a bubble will depend on the oxygen gradient established and the area of the air-water interface. In a pond the oxygen in the surface water is in equilibrium with the ambient air above the surface. Because surface water mixes with the deeper water, the P_{O_2} in the pond water will be in equilibrium with the air and will not vary with depth if the pond is well mixed and no oxygen is removed by aquatic animals or added through photosynthesis by aquatic plants. An air bubble transported to depth by a water bug or beetle will be compressed by hydrostatic pressure; as a result, gas pressure within the bubble will rise, exceeding that in the water. For every 10 m of depth, pressure in a bubble increases by approximately 1 atm.

If we consider a bubble just below the surface, the oxygen content of the bubble will decrease owing to uptake by the animal; this will establish an O_2 gradient between the bubble and the water (assuming the water is in gaseous equilibrium with air), so oxygen will diffuse into the bubble from the water. As P_{O_2} in the bubble is reduced, the nitrogen partial pressure, P_{N_2}, will increase; if the bubble is just below the surface, the pressure will be maintained at approximately atmospheric pressure. Nitrogen will therefore diffuse slowly from the bubble into the water (Figure 13-39). (Because of the high solubility of CO_2 in water, CO_2 levels in the bubble are always negligible.) If the bubble is taken to depth, however, the pressure will increase by 0.1 atm for every meter of depth, increasing both P_{O_2} and P_{N_2} and speeding the diffusion of both N_2 and O_2 from the bubble into the water. The bubble will gradually get smaller and eventually disappear as nitrogen leaves it. Thus, the life of the bubble depends on the insect's metabolic rate, the initial size of the bubble, and the depth to which it is taken. The bubble collapses because nitrogen is lost from it as the insect uses the oxygen. It has been calculated that up to seven times the initial bubble O_2 content diffuses into the bubble from the water and is therefore available to the insect before the bubble disappears. It is possible that aquatic air-breathing vertebrates like the beaver may take advantage of oxygen diffusing from water into gas bubbles trapped under ice. These

A

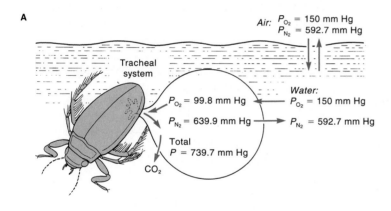

B

At a depth of 1 m, bubble life
is short because of rapid N_2 loss

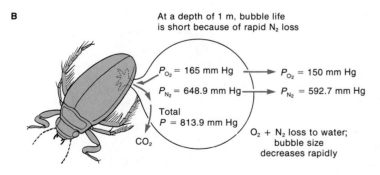

C

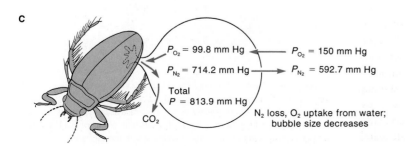

Figure 13-39 Some aquatic insects carry air bubbles when they dive. Under water, gas exchange occurs between the bubble and the insect's tracheal system, and between the bubble and water. The direction of gas flow will depend on the partial pressures of O_2, CO_2, and N_2 and the total pressure (P) in bubbles under water. **(A)** Conditions at start of descent. **(B)** Condition in bubble immediately after being taken to depth of 1 m. **(C)** Condition sometime later at same depth. Arrows indicate diffusion of gas molecules. Note that the sum of the gas partial pressures in the water phase (and in the atmosphere) always equals 742.7 mm Hg.

A

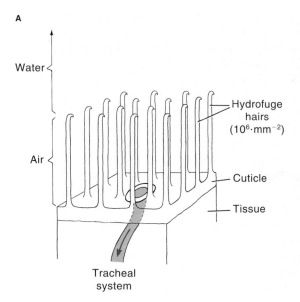

B

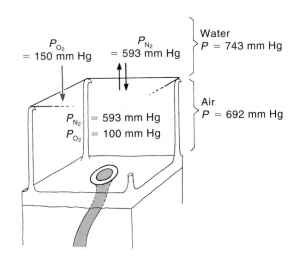

Figure 13-40 Hydrofuge hairs on the surface of some insects and insect eggs have an incompressible air space that acts as a gill under water. **(A)** Schematic diagram of plastron with protruding hydrofuge hairs. Oxygen diffuses from water into the air space contained within the plastron and then into the animal via the tracheal system. Typically, there are about 10^6 hairs per mm²; only a few are depicted here. **(B)** Partial pressures of oxygen and nitrogen in the air and water phases.

animals exhale under water and the air bubble produced rests under the ice and gains oxygen from the water; later the animals can inhale the rejuvenated air.

If the bubbles were noncollapsible, the insect would not need to surface, because oxygen would continue to diffuse from the water via the bubble into the tracheal system and thence to the tissues. In some insects (e.g., *Aphelocheirus*) a thin film of air trapped by hydrofuge hairs, called a **plastron,** in effect provides a noncollapsible bubble (Figure 13-40A). The plastron can withstand pressures of several atmospheres before collapsing. In the small air space, N_2 is presumably in equilibrium with the water, P_{O_2} is low, and oxygen therefore diffuses from water into the plastron, which is continuous with the tracheal system (Figure 13-40B).

GAS TRANSFER IN WATER: GILLS

Gills of fish and crabs are usually ventilated with a unidirectional flow of water (see Figure 13-19B). Tidal flow of water, similar to that of air in the lung, would be costly because of the high density and viscosity of water; thus the energetic cost of reversing the direction of flow of water is simply too high. The lamprey and sturgeon are exceptions to the rule that water flow through gills is unidirectional. The mouth of the parasitic lamprey is often blocked by attachment to a host. The gill pouches, although connected internally to the pharyngeal and mouth cavities, are ventilated by tidal movements of water through a single external opening to each pouch (Figure 13-41). This unusual method of gill ventilation is clearly associated with a parasitic mode of life. The ammocoete larvae of lampreys are not parasitic and maintain a unidirectional flow of water over their gills, typical of aquatic animals in general. Water

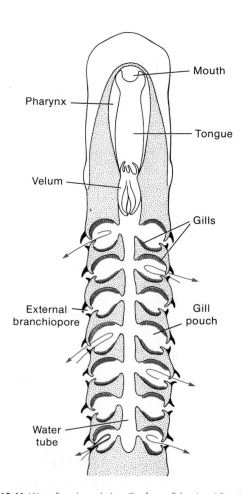

Figure 13-41 Water flow through the gills of most fishes is unidirectional, but in the adult lamprey, water moves in and out of each gill pouch via the external branchiopore. Shown here is a longitudinal transverse section through the head of an adult lamprey. Arrows mark the direction of water flow. The valves of the external branchiopore move in and out with the oscillating water flow.

flow through the mouth and gills of sturgeon is normally unidirectional, but if the animal has its mouth in mud while searching for food, it can generate a tidal flow of water through slits in gill coverings.

Flow and Gas Exchange across Gills

Blood flow through the fish gills can be described as sheet flow; that is, as pressure increases, the thickness, but not the other dimensions, of the blood sheet increases (Figure 13-42). In this respect, circulation through the gills is similar to the pulmonary circulation. The flow of blood relative to the flow of water in aquatic animals can be either concurrent or countercurrent, or some combination of these two arrangements (Figure 13-43). The advantage of a countercurrent over a concurrent flow of blood and water is that a larger difference in P_{O_2} can be maintained across the exchange surface, thus allowing more transfer of gas. A countercurrent flow is most advantageous if the values for O_2 content × flow (capacity/rate) are similar in blood perfusing and water flowing over the gills. If the capacity/rate values for blood and water differ considerably, then countercurrent flow has little advantage over concurrent flow. For example, if water flow were very high in relation to blood flow, there would be little change in P_{O_2} in the water as it flowed over the gills, and the mean P_{O_2} difference across the gills would be similar in concurrent and countercurrent arrangements of flow. Although the O_2 content of fish blood is generally much higher than that of water, the flow rate of water across gills is much higher than the flow rate of blood. Thus the capacity/rate values are similar in the blood perfusing and water flowing over the gills in most fishes, and countercurrent flow is typical.

Because water has a much lower oxygen content than air, water-breathing animals require a much higher venti-

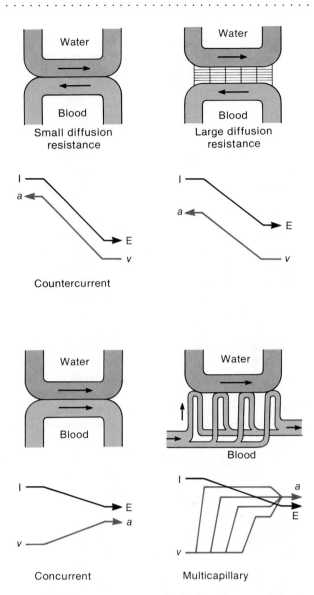

Figure 13-43 Various arrangements for the flows of water and blood at the respiratory surface are found in aquatic animals. Relative changes in P_{O_2} in water and blood are indicated below each diagram. I, inhalant; E, exhalant; a, arterial blood; v, venous blood.

lation rate to achieve a given oxygen uptake than do air-breathing animals. This requirement, combined with the much greater density of water compared with air, makes oxygen extraction from the environment a more costly exercise in water. This is offset somewhat by gills having a unidirectional, rather than tidal, flow of water. Water has a much higher heat capacity than air, and heat transfer is more rapid than gas transfer, so blood leaving the gills of a water-breathing animal is usually in thermal equilibrium with the environment. A few fish have some warm tissues; this is only possible because of a countercurrent blood supply to selected tissues. The countercurrent blood supply acts as a heat exchanger, reducing heat loss from the tissue and warming it to above ambient temperatures. Tuna, for instance, have warm muscles, eyes, and brains.

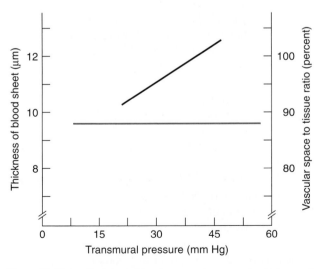

Figure 13-42 In gills, a rise in blood pressure increases the thickness of the blood sheet but not its height or length. In this plot, based on measurements in gill lamellae of the lingcod, *Ophiodon elongatus,* the black line shows the thickness of the blood sheet, and the red line the vascular space to tissue ratio, a measure of the height and length of the blood sheet. [From Farrell et al., 1980.]

What are the differences in the design of a countercurrent heat exchanger and countercurrent oxygen exchanger? Is it possible to design a gas exchanger that does not exchange heat, or a heat exchanger that does not exchange gases?

Flow of water over the gills of teleost fishes is maintained by the action of skeletal muscle pumps in the buccal and opercular cavities. Water is drawn into the mouth, passes over the gills, and exits through the *opercular* (gill-covering) clefts (Figure 13-44). Valves guard the entrance to the buccal cavity and opercular clefts, maintaining a unidirectional flow of water over the gills. The buccal cavity changes volume by raising and lowering the floor of the mouth. The *operculum* (gill covering) swings in and out, enlarging and reducing the size of the opercular cavities. Changes in volume in the two cavities are nearly in phase, but a pressure differential is maintained across the gills throughout most of each breathing cycle. The pressure in the opercular cavity is slightly below that in the buccal cavity, resulting in a unidirectional flow of water across the gills throughout most, if not all, of the breathing cycle.

Many active fish, such as tuna, "ram-ventilate" their gills, opening their mouths so as to ventilate the gills by the

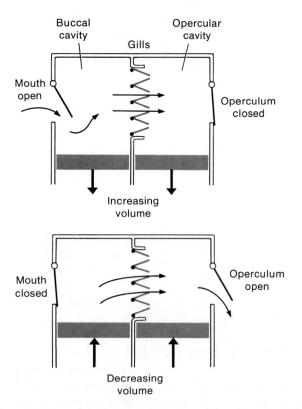

Figure 13-44 Unidirectional flow of water through the gills in teleost fish is achieved by sequential opening and closing of the mouth and operculum and by a small pressure differential between the buccal and opercular cavities. In this schematic diagram of gill ventilation, small arrows indicate water flow and large arrows indicate movement of the floor of the mouth.

forward motion of the body while swimming. The remora, a fish that attaches itself to the body of a shark, ventilates its own gills only when the shark stops swimming; normally, it relies on the forward motion of its host to ventilate its gills.

Functional Anatomy of the Gill

The details of gill structure vary among species, but the general plan is similar. The gills of teleost fishes are taken to be representative of an aquatic respiratory surface. The four *gill arches* on either side of the head separate the opercular and buccal cavities (Figure 13-45A). Each arch has two rows of filaments, and each filament, flattened dorsoventrally, has an upper and a lower row of *lamellae* (Figure 13-45B,C). The lamellae of successive filaments in a row are in close contact. The tips of filaments of adjacent arches are juxtaposed, so that the whole gill forms a sieve-like structure in the path of water flow. The gills are covered by mucus secreted from mucous cells within the epithelium. This mucous layer protects the gills and creates a boundary layer between the water and the epithelium.

Water flows in slitlike channels between neighboring lamellae (see Figure 13-45C,D). These channels are about 0.02–0.05 mm wide and about 0.2–1.6 mm long; the lamellae are about 0.1–0.5 mm high (Figure 13-46A, on page 556). As a result, the water flows in thin sheets between the lamellae, which represent the respiratory portion of the gill, and diffusion distances in water are reduced to a maximum of 0.01–0.025 mm (half the distance between adjacent lamellae on the same filament).

Gill lamellae are covered by thin sheets of epithelial cells, which are joined by tight junctions (Figure 13-46B, on page 556). The inner lamellar wall is formed by *pillar cells,* which occupy about 20% of the internal volume of the lamella. The pillar cells are associated with an extensive collagen network, which prevents the lamellae from bulging even though they are subjected to high blood pressures. Blood flows as a sheet in the spaces between the pillar cells, the flow being described by sheet-flow dynamics as in the lung. The diffusion distance between the center of the red blood cell and the water is between 3 and 8 μm, much larger than the diffusion distance across the mammalian lung epithelium (see Figure 13-22B). The total area of the lamellae is large, varying from 1.5 to 15 $cm^2 \cdot g^{-1}$ of body weight, depending on the size of the fish and on whether it is generally active or sluggish.

Fish gills normally are important in ion regulation and carry out many of the functions of the mammalian kidney. Ion exchange in gills is mediated by at least two types of cells, as discussed in Chapter 14. Because of the metabolic cost of this ion transport, oxygen consumption by gill tissue may be 10% or more of the total oxygen uptake of the fish.

When exposed to air, gills collapse and become nonfunctional, so a fish out of water usually becomes hypoxic, hypercapnic, and acidotic. A few fish and crabs can breathe air, generally using a modified swimbladder, mouth, gut,

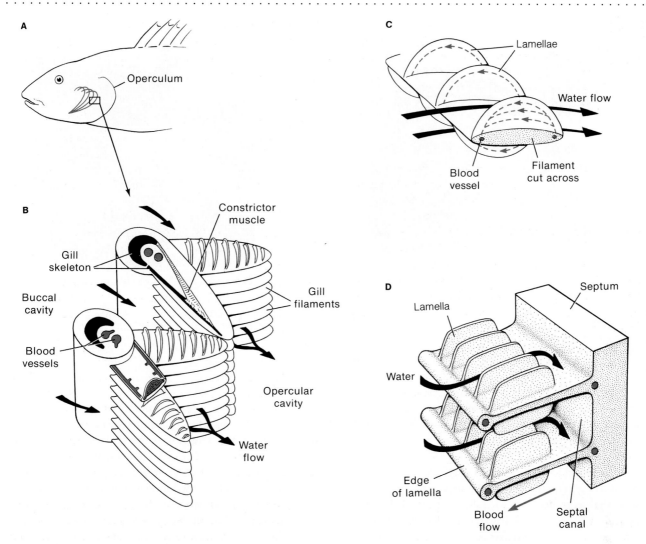

Figure 13-45 The general structure of gills is similar in all fishes, although minor variations are found among species. **(A)** Position of the four gill arches beneath the operculum on the left side of a teleost fish. **(B)** Enlarged view of part of two gill arches showing the filaments of adjacent rows touching at their tips. Also shown are the blood vessels that carry the blood before and after its passage over the gills. **(C)** Part of a single fila-

ment with three secondary folds (lamellae) on each side. The flow of blood (red arrows) is in the opposite direction to that of the water (black arrows). **(D)** Part of the dogfish gill. As in teleost fish, the flow of blood is in the opposite direction to that of the water. [Parts A–C adapted from Hughes, 1964; part D adapted from Grigg, 1970.]

or branchial cavity for this purpose (see *Air-Breathing Fishes* in Chapter 12). Air-breathing crabs usually show a decrease in oxygen consumption as well as a decrease in body carbon dioxide levels when they move from air to water breathing. The purple shore crab, *Leptograspus variegatus,* however, shows no change in body oxygen content as it moves between air and water and may regulate body carbon dioxide and therefore pH levels by adjusting the ratio of air to water breathing. Thus this crab truly is amphibious.

REGULATION OF GAS TRANSFER AND RESPIRATION

Because the regulation of the rate of O_2 and CO_2 transfer has been studied most extensively in mammals, this section focuses on mammalian regulation of gas transfer. The

movement of O_2 and CO_2 between the environment and mitochondria in mammals is regulated by altering lung ventilation and the flow and distribution of blood within the body. Here we place emphasis on the control of breathing; Chapter 12 presents details of the control of the cardiovascular system.

Ventilation-to-Perfusion Ratios

Energy is expended in ventilating the respiratory surface with air or water and in perfusing the respiratory epithelium with blood. The total cost of these two processes is difficult to assess, but probably amounts to 4%–10% of the total aerobic energy output of an animal, depending on the species in question and the physiological state of the animal. Thus, gas transfer between the environment and cell accounts for a considerable proportion of the total energy

A

Secondary lamellae

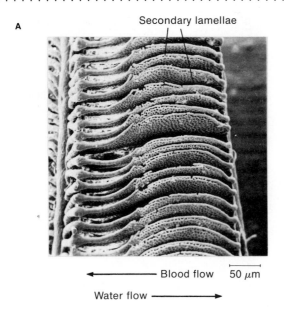

←——— Blood flow   50 μm

Water flow ———→

B

Water

Mucus

Epithelial
cell

Pillar cell

Basement
membrane

2 μm

Blood plasma

Figure 13-46 Water flows between gill lamellae, which are covered by a thin epithelial layer. **(A)** Scanning electron micrograph of a plastic cast of the vasculature of a trout gill filament, showing several lamellae. **(B)** Trans- verse section through a trout gill lamella showing components of the water-blood barrier. [Courtesy of B. J. Gannon.]

output of the animal and represents a significant selective pressure in favor of the evolution of mechanisms for the close regulation of ventilation and perfusion in order to conserve energy.

The rate of blood perfusion of the respiratory surface is related to the requirements of the tissues for gas transfer and to the gas-transport capacities of the blood. To ensure that sufficient oxygen is delivered to the respiratory surface to saturate the blood with oxygen, the rate of ventilation, $\dot{V}_A$, must be adjusted in accord with the perfusion rate, $\dot{Q}$, and the gas content of the two media so that the amount of oxygen delivered to respiratory surface equals that taken away in the blood. The oxygen content of arterial blood in humans normally is similar to that of air. The $\dot{V}_A/\dot{Q}$ ratio, therefore, is about 1 in humans (Figure 13-47A). Water, however, contains only about one-thirtieth as much dis- solved oxygen as an equivalent volume of air at the same P_{O_2} and temperature. Thus, in fishes, the ratio of water flow, $\dot{V}_G$, over and blood flow, $\dot{Q}$, through the gills is between 10:1 and 20:1 (Figure 13-47B), much higher than the $\dot{V}_A/\dot{Q}$ ratio in air-breathing mammals. Based on the difference in the oxygen content of water and air, the $\dot{V}_G/\dot{Q}$ ratio in fishes might be expected to be 30:1. However, it is lower than this because the oxygen capacity of the blood of lower vertebrates is often only half that of mammalian blood.

Any changes in the oxygen content of the inhalant medium will affect the $\dot{V}_A/\dot{Q}$ ratio. In order to maintain a given rate of oxygen uptake, a decrease in P_{O_2} of inhalant air or water must be compensated for by an increase in ventilation and hence an increase in the ventilation-to- perfusion ratio. Conversely, an increase in the inhalant P_{O_2} is accompanied by a decrease in ventilation if the rate of oxygen uptake remains the same.

The ventilation-to-perfusion ratio must be maintained over each portion of the respiratory surface as well as over the whole surface. The pattern of capillary blood flow can change in both gills and lungs, changing the distribution of blood over the respiratory surface. The distribution of air or water must reflect the blood distribution. Perfusion of an alveolus without ventilation is as pointless as ventilat- ing an alveolus without blood perfusion of that same alve- olus. Although such extreme situations are unlikely to oc- cur, the maintenance of too high or too low a blood flow or ventilation rate will result in energetically inefficient gas transfer per unit of energy expended. For efficient gas transfer, the optimal ventilation-to-perfusion ratio should be maintained over the whole respiratory surface. This op- timal maintenance does not preclude differential rates of blood perfusion over the respiratory surface, but requires only that the flows of blood and inhalant medium be matched.

The efficiency of gas exchange is diminished if some of the blood entering the lungs or gills either bypasses the res- piratory surface or perfuses a portion of the respiratory sur- face that is inadequately ventilated (Figure 13-48). The magnitude of such *venous shunts*, expressed as a percent- age of total flow to the respiratory epithelium, can be cal- culated from the arterial and venous O_2 content, assuming an ideal arterial O_2 content. In the lung, for instance, blood is almost in equilibrium with alveolar gas tensions. If these tensions and the blood oxygen dissociation curves are known, the expected ideal O_2 content of arterial blood can be determined. Let us assume that this ideal content is 20 ml of O_2 per 100 ml of blood (20 vol %) and the measured values for arterial and venous blood are 17 and 5 vol %, respectively. This reduction in measured arterial

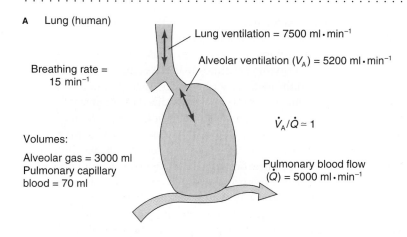

A   Lung (human)

Lung ventilation = 7500 ml·min^{-1}

Alveolar ventilation (V_A) = 5200 ml·min^{-1}

Breathing rate = 15 min^{-1}

$\dot{V}_A/\dot{Q} \approx 1$

Volumes:

Alveolar gas = 3000 ml
Pulmonary capillary blood = 70 ml

Pulmonary blood flow ($\dot{Q}$) = 5000 ml·min^{-1}

Figure 13-47 The ventilation-to-perfusion ratio in fish gills is much higher than in the human lung. Approximations of volumes and flows in the human lung and trout gill are shown; actual values may vary considerably.

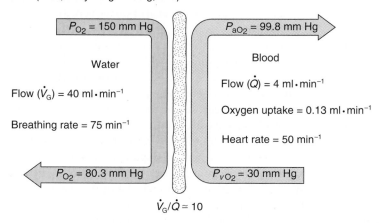

B   Gill (trout, body weight 200 g, 8°C)

P_{O_2} = 150 mm Hg

P_{aO_2} = 99.8 mm Hg

Water

Blood

Flow ($\dot{V}_G$) = 40 ml·min^{-1}

Flow ($\dot{Q}$) = 4 ml·min^{-1}

Breathing rate = 75 min^{-1}

Oxygen uptake = 0.13 ml·min^{-1}

Heart rate = 50 min^{-1}

P_{O_2} = 80.3 mm Hg

P_{vO_2} = 30 mm Hg

$\dot{V}_G/\dot{Q} \approx 10$

O_2 content from the ideal situation can be explained in terms of a venous shunt, oxygenated arterial blood (20 vol %) being mixed with venous blood (5 vol %) in the ratio of 4:1 to give a final arterial O_2 content of 17 vol %; that is, 20% of the blood perfusing the lung is passing

through one or more venous shunts. This is an extreme example to illustrate a point; in most cases, venous shunts are very small.

Flows of blood and inhalant medium (air or water) are regulated to maintain a near optimal ventilation-to-perfu-

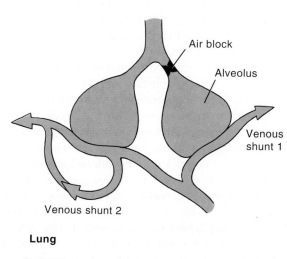

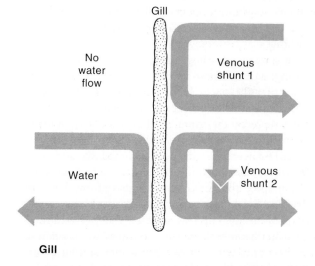

Air block

Alveolus

Venous shunt 1

Venous shunt 2

Lung

No water flow

Gill

Venous shunt 1

Water

Venous shunt 2

Gill

Figure 13-48 The efficiency of gas transfer in the lung and gills is decreased when blood flows to a portion of the respiratory surface without adequate ventilation (shunt 1) or because blood does not flow close enough to the respiratory epithelium (shunt 2). Blood flow is regulated to avoid the development of such venous shunts in the lung and gills.

sion ratio over the surface of the respiratory epithelium under a variety of conditions. In general terms, $\dot{Q}$ is regulated to meet the requirements of the tissues; $\dot{V}_A$ and $\dot{V}_G$ are regulated to maintain adequate rates of O_2 and CO_2 transfer. Such mechanisms as hypoxic vasoconstriction of blood vessels help to maintain optimal ventilation-to-perfusion ratios in different parts of the respiratory surface. As discussed earlier, low alveolar oxygen levels cause a vasoconstriction in lung vessels, thereby reducing blood flow to poorly ventilated, and therefore hypoxic, regions and increasing blood flow to well-ventilated regions of the respiratory surface. Blood perfusion of the respiratory surface tends to be less well distributed in resting animals. Blood pressure rises with exercise and blood is distributed more evenly under these conditions, resulting in a more even ventilation-to-perfusion ratio over the respiratory surface.

Neural Regulation of Breathing

The integration of breathing movements in all air-breathing vertebrates results from the central processing of many sensory inputs. The central processor consists of a *pattern generator,* determining the depth and amplitude of each breath, and a *rhythm generator,* controlling breathing frequency. Several sensory inputs adjust ventilation to maintain adequate rates of gas transfer and blood pH. Other inputs integrate breathing movements with feeding, talking and singing, or other body movements. Certain sensory inputs may cause coughing or swallowing reflexes, which protect the respiratory epithelium from environmental hazards. Other inputs function to optimize breathing patterns to minimize energy expenditure.

Medullary respiratory centers

As noted earlier, the mammalian lung is ventilated by the action of the diaphragm and muscles between the ribs (see Figures 13-29 and 13-31). These muscles are activated by spinal motor neurons and the phrenic nerve, which receive inputs from groups of neurons that constitute the **medullary respiratory centers.** The control of respiratory muscles can be very precise, allowing extremely fine control of air flow, as is required for such complex actions in humans as singing, whistling, and talking, as well as simply breathing. Microsections of the neonatal rat brain stem indicate that the *pre-Botzinger complex* in the ventral medulla is capable of generating the respiratory rhythm and may represent the central rhythm generator that maintains breathing rhythm in the adult. Rhythmic activity is enhanced by neurons in the pons and medulla, and some neurons just anterior to the medulla cause prolonged inspiration in the absence of rhythmic drive from the pons.

In 1868 Ewald Hering and Josef Breuer observed that inflation of the lungs decreases the frequency of breathing. (Breuer later became an early proponent of psychoanalysis and collaborated with Sigmund Freud in producing a book on hysteria.) The **Hering-Breuer reflex** is abolished by cutting the vagus nerve. Inflation of the lung stimulates pulmonary stretch receptors in the bronchi and/or bronchioles, which have a reflex inhibitory effect, via the vagus nerve, on the medullary inspiratory center (nucleus tractus solitarius) and therefore on inspiration. Thus the medulla contains a central rhythm generator that drives the pattern generator within the medullary respiratory center to cause breathing movements. This system is modified by inputs from other areas of the brain and from various peripheral receptors.

The medullary respiratory center contains **inspiratory neurons,** whose activity coincides with inspiration, and **expiratory neurons,** whose activity coincides with expiration. The respiratory rhythm was once considered to arise from reciprocal inhibition between inspiratory and expiratory neurons, with reexcitation and accommodation occurring within each set of neurons. But several lines of evidence indicate this model of the central rhythm generator is not tenable, and more recent studies suggest that respiratory rhythm depends primarily on the activity of inspiratory neurons.

Inspiratory neuronal activity, recorded from either the phrenic nerve or some individual neurons in the medulla, shows a rapid onset, a gradual increase, and then a sharp cutoff with each burst of activity associated with inhalation. This neuronal activity results in a contraction of the inspiratory muscles and a decrease in intrapulmonary pressure (Figure 13-49A). Increased blood CO_2 levels cause the progressive growth of inspiratory activity to increase more rapidly (Figure 13-49B). Thus, the rate of rise of inspiratory activity is increased by inputs from chemoreceptors, resulting in a more powerful inspiratory phase. The "off switch" of inspiratory neurons occurs once activity in the neuron has reached a threshold level. Expansion of the lung stimulates pulmonary stretch receptors, whose activity reduces the threshold for the inspiratory off switch (Figure 13-49C). Thus the pulmonary stretch receptors, through their action on inspiratory neurons, prevent overexpansion of the lung.

The interval between breaths is determined by the interval between bursts of inspiratory neuronal activity, which is related to the level of activity in the previous burst and in afferent nerves from pulmonary stretch receptors. In general, the greater the level of inspiratory activity (i.e., the deeper the breath), the longer the pause between inspirations. The result is that the ratio of inspiratory to expiratory duration remains constant in spite of changes in the length of each breathing cycle. This ratio is affected by the level of activity of the pulmonary stretch receptors. If, for example, the lung empties only slowly during expiration, the pulmonary stretch receptors will remain active while the lung remains inflated; the continued activity of the stretch receptors will prolong the duration of expiration and increase the time available for exhalation. The neuronal mechanisms causing phasic activation of inspiratory neurons are poorly understood, as is the nature of the central rhythm generator, possibly located in the pre-Botzinger complex in the ventral medullary region of the brain.

Exhalation often is a largely passive process, which does not depend on activity in expiratory neurons. This is

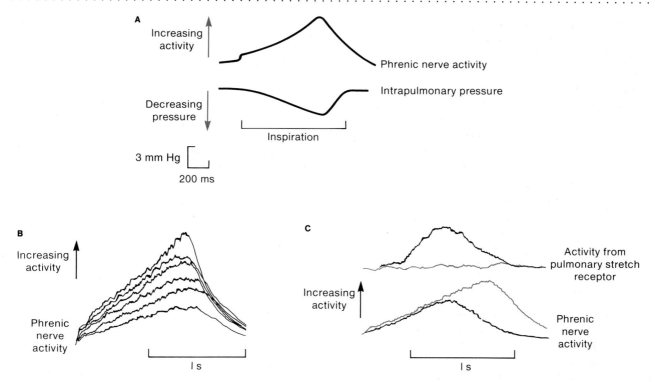

Figure 13-49 Phrenic nerve activity, stimulated by increasing alveolar P_{CO_2}, induces inspiration. **(A)** Relationship between phrenic nerve activity and intrapulmonary pressure during inspiration. Note the sudden onset, gradual rise, and then "off switch," or termination, of inspiratory activity. **(B)** Effect of increasing alveolar P_{CO_2} levels (P_ACO_2) on discharge in the phrenic nerve. Recordings were made at P_ACO_2 ranging from 28.5 mm Hg (bottom trace) to 60 mm Hg (top trace). The higher the P_ACO_2, the more rapid the rise of phrenic nerve activity during inspiration. **(C)** Effect of increasing activity from pulmonary stretch receptors on activity in the phrenic nerve. In the absence of stretch receptor activity, the switch off in phrenic nerve activity is delayed (red traces). An increase in receptor activity results in an earlier termination of activity in the phrenic nerve but does not affect the rate of increase in phrenic activity before the switching off (black traces).

especially true during quiet, normal breathing. Expiratory neurons are active only when inspiratory neurons are quiescent, and then they show a burst pattern somewhat similar to, but out of phase with, that of inspiratory neurons. Inspiratory neuronal activity inhibits expiratory activity, showing the dominance of inspiratory neurons in the generation of rhythmic breathing. In the absence of inspiratory activity, expiratory neurons are continually active. Inspiratory neuronal activity, however, imposes a rhythm, via inhibition, on expiratory neurons.

Fish, birds, and awake mammals usually breath rhythmically and continuously, whereas amphibians and reptiles often show episodic breathing, with pauses between episodes of rhythmic breathing. Recent studies of the bullfrog brain stem have shown that these episodic patterns of breathing are an intrinsic property of the brain stem and do not depend on sensory feedback. The *nucleus isthmi* in the bullfrog brain stem is involved not only in the integration of chemoreceptor input but also appears to be essential for the maintenance of episodic breathing. In sleeping mammals episodic breathing appears to be the result of the interaction between peripheral and central components of the control system. During sleep, central respiratory drive is reduced in mammals, and breathing is maintained by input from peripheral chemoreceptors. A breathing period increases oxygen and decreases carbon dioxide levels in the

blood, reducing peripheral chemoreceptor input to the respiratory center. Breathing stops until oxygen levels fall sufficiently to increase chemoreceptor drive enough to initiate breathing again. This results in the periodic breathing typical of many sleeping mammals. In the awake mammal central respiratory drive is sufficient to maintain continuous rhythmic breathing.

Factors affecting the rate and depth of breathing
Several types of receptors respond to stimuli that influence ventilation, causing reflex changes in the rate and/or depth of breathing. Among the stimuli affecting ventilation are changes in O_2, CO_2, and pH; emotions; sleep; lung inflation and deflation; lung irritation; variations in light and temperature; and the requirements for speech. These influences are integrated by the medullary respiratory centers. Breathing can also, of course, be controlled by conscious volition.

In most, if not all, animals, changes in O_2 and CO_2 lead to reflex changes in ventilation. The chemoreceptors involved have been localized in only a few groups of animals. Chemoreceptors monitor changes in O_2 and CO_2 in arterial blood in the **carotid bodies** and **aortic bodies** of mammals, in the carotid body of birds, and in the carotid labyrinth of amphibians. In teleost fish chemoreceptors located in the gills respond to reductions in O_2 levels in the water and the blood. In all cases, the chemoreceptors are innervated by branches of the ninth (glossopharyngeal) or tenth (vagus) cranial nerve.

Mammals and probably other air-breathing vertebrates also have central chemoreceptors, located in the medulla, that drive ventilation in response to decreases in the pH of the cerebrospinal fluid (CSF), usually caused by elevations in P_{CO_2}. Stimulation of this system is required to maintain normal breathing: if body P_{CO_2} falls, or is held at a low level experimentally, breathing will cease. These central chemoreceptors have little ability to respond to falling O_2 levels; the peripheral chemoreceptors have this role and are important in increasing ventilation during periods of hypoxia.

The carotid and aortic bodies of mammals receive a generous blood supply and have a high oxygen uptake per unit weight (Figure 13-50A). These arterial chemoreceptors consist of a number of lobules, or "glomoids," that surround very convoluted capillaries. The blood vessels can be divided into small and large capillaries and arteri-ovenous shunts. The arterioles are innervated by both sympathetic and parasympathetic postganglionic efferents. Each lobule consists of several glomus (type I) cells covered

by sustentacular (type II) cells. The glomus cells, thought to be the actual receptors, are small ovoid cells with a large nucleus and many dense-core vesicles, or granules (Figure 13-50B). These cells are interconnected by synapses and often possess cytoplasmic processes of different lengths. They are innervated by afferent fibers of the glossopharyngeal nerve and possibly preganglionic sympathetic efferents. A single nerve fiber may innervate 10 or 20 glomus cells. A glomus cell may be either presynaptic or postsynaptic, or both (reciprocal), with respect to a nerve fiber. A single nerve fiber may be postsynaptic (afferent) to one glomus cell with a presynaptic connection (efferent) to a neighboring glomus cell or even another region of the same glomus cell. Many glomus cells lack innervation but are synaptically connected to other glomus cells in the lobule. A few glomus cells may be innervated by sympathetic efferent fibers.

The chemoreceptors in the carotid and aortic bodies are stimulated by decreases in blood O_2 and pH and increases in blood CO_2. It is possible that the observed response to increasing CO_2 is due to changes in pH within these receptors rather than to changes in CO_2 *per se*. The result of chemoreceptor stimulation is to recruit new fibers and increase the

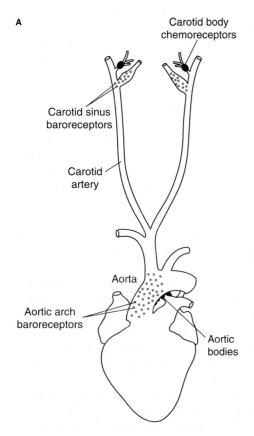

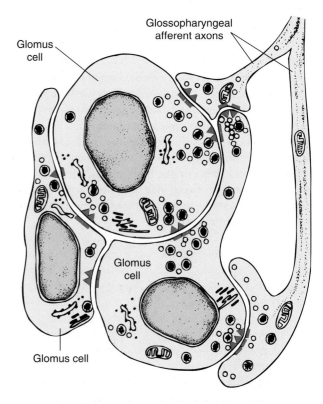

Figure 13-50 In mammals, chemoreceptors in the carotid and aortic bodies monitor blood gas levels and pH. **(A)** Diagram showing the location of carotid and aortic body chemoreceptors and carotid sinus and aortic arch baroreceptors (small red dots) in the dog. The baroreceptors help regulate arterial blood pressure (see Chapter 12). **(B)** Small portion of the rat carotid body, which consists of several lobules containing glo-

mus cells. These are connected by synapses and innervated by glossopharyngeal afferent fibers. Some regions of afferent nerve endings are presynaptic to the glomus cell, some postsynaptic, and some form reciprocal synapses. ▲, presynaptic regions. [Part A adapted from Comroe, 1962; part B adapted from McDonald and Mitchell, 1975.]

firing rate in the afferent nerves innervating glomus cells. The chemoreceptors adapt to changing arterial CO_2 levels. The carotid body chemoreceptors show a much larger response to pH and/or CO_2 changes than the aortic body chemoreceptors. Stimulation of these chemoreceptors leads to an increase in lung ventilation, mediated via the medullary respiratory center. The actual increase, in response to a given decrease in arterial P_{O_2}, depends on the blood CO_2 level, and vice versa (Figure 13-51). Efferent activity to the carotid body modulates the response. Increased sympathetic efferent activity constricts arterioles in the carotid body via an α-adrenergic mechanism, thereby reducing blood flow, which in turn increases the chemoreceptor discharge and lung ventilation. Nonsympathetic efferent activity in the carotid nerve reduces the response of the carotid body to changes in arterial blood P_{O_2} and P_{CO_2} and/or pH. Increases in temperature and osmolarity also stimulate the arterial chemoreceptors, and stimulation of the carotid nerve causes increased ADH release. Thus the carotid body chemoreceptors may play a role in osmoregulation as well as in the control of breathing and circulation.

As mentioned previously, mammals and possibly other air-breathing vertebrates have central chemoreceptors that are necessary for normal breathing. These H^+-sensitive receptors are located in the region of the medullary respiratory center and are stimulated by a decrease in the pH of the CSF. The CSF of mammals, and possibly of other vertebrates, is very low in protein and is essentially a solution of NaCl and $NaHCO_3$, with low but closely regulated levels of K^+, Mg^{2+}, and Ca^{2+}. The CSF is also poorly buffered; therefore small changes in P_{CO_2} have a marked effect on CSF pH. Because the blood–brain barrier is relatively impermeable to H^+, the central H^+-sensitive chemoreceptors are insensitive to changes in blood pH. However, changes in blood P_{CO_2} cause corresponding changes in the P_{CO_2} of the CSF, and these in turn result in changes in the pH of the CSF. An increase in P_{CO_2} leads to a decrease in the pH of the CSF; subsequent stimulation of the H^+-sensitive receptors causes reflex increases in breathing (Figure 13-52). Prolonged changes in P_{CO_2} result in the adjustment of the pH of the CSF by changes in HCO_3^- levels.

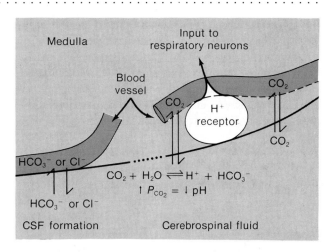

Figure 13-52 Central H^+-sensitive receptors are influenced by the pH of cerebrospinal fluid (CSF) and by arterial P_{CO_2}. Carbon dioxide molecules diffuse readily across the walls of the brain capillaries and alter CSF pH, but there is a barrier to other molecules. An increase in P_{CO_2} causes a decrease in the CSF pH; the resulting stimulation of H^+ receptors reflexively increases breathing. Across some capillary walls, exchange of HCO_3^- and Cl^- helps to maintain a constant pH of the CSF in the face of a prolonged change in P_{CO_2}.

In mammals and other air-breathing vertebrates, carbon dioxide rather than oxygen levels dominate in the control of breathing. In aquatic vertebrates, however, oxygen is the major factor in the control of breathing. In fact, fishes exposed to high oxygen levels will reduce breathing to the extent that there is a marked increase in P_{CO_2} in the blood. Two factors account for this difference. First, oxygen concentration is much more variable in the aquatic environment than in air. Second, oxygen is much less soluble than carbon dioxide in water; as a result, if ventilation is adequate to deliver oxygen to the gills, it will also be adequate to remove carbon dioxide from the blood. Under most conditions ventilation does not limit carbon dioxide excretion in aquatic animals. Only in the rare condition of very high oxygen levels in the water is ventilation reduced sufficiently to curtail carbon dioxide excretion in fish.

The lungs contain several types of receptors that help regulate inflation and prevent irritation of the respiratory surface. We saw earlier that stimulation of pulmonary stretch receptors prevents overinflation of the lung (Hering-Breuer reflex). In mammals increased CO_2 levels reduce the inhibitory effects of these pulmonary stretch receptors on the medullary respiratory center, thereby increasing the depth of breathing and lung ventilation. It is not clear whether the CO_2-sensitive receptors found in the lungs of birds are pure CO_2 chemoreceptors or CO_2-sensitive mechanoreceptors, as observed in mammals. Increased CO_2 in the lungs of birds, however, has a greater effect on sensory discharge from the lung than that observed in mammals.

In addition to pulmonary stretch receptors, a variety of irritant receptors are present in the lung. Stimulation of these receptors by mucus and dust or other irritant particles causes reflex *bronchioconstriction* and coughing. A third

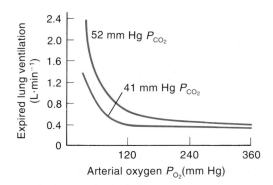

Figure 13-51 Lung ventilation rates increase with a decrease in arterial P_{O_2} and an increase in arterial P_{CO_2}. The relationships shown here are from measurements in the duck. [From Jones and Purves, 1970.]

group of receptors in the lung is positioned close to the pulmonary capillaries in interstitial spaces; these are called **juxtapulmonary capillary receptors,** or type J receptors. These receptors were previously termed "deflation receptors," but their natural stimulus appears not to be lung deflation but an increase in interstitial volume, as seen, for example, during pulmonary edema. Stimulation of type J receptors elicits a sensation of breathlessness. Violent exercise probably results in a rise in pulmonary capillary pressure and an increase in interstitial volume, which could cause stimulation of type J receptors and therefore breathlessness.

RESPIRATORY RESPONSES TO EXTREME CONDITIONS

Variations in the levels of respiratory gases, diving by air-breathing animals, and exercise all induce respiratory responses. Let's see how animals adjust to these extreme conditions.

Reduced Oxygen Levels (Hypoxia)

Aquatic animals are subjected to more frequent and rapid changes in oxygen levels than are air-breathing animals. Both mixing and diffusion are more rapid in air compared with water, so regions of local hypoxia develop more often in aquatic environments. Although photosynthesis can cause very high oxygen levels during the day in some aquatic environments, oxygen consumption by both biological and chemical processes can produce localized hypoxic regions. The changes in oxygen levels in water may or may not be accompanied by changes in carbon dioxide.

Many aquatic animals can withstand very long periods of hypoxia. Some fishes (e.g., carp) overwinter in the bottom mud of lakes where the P_{O_2} is very low. Many invertebrates also bury themselves in mud with a low P_{O_2} but high nutritive content. Some parasites live in hypoxic regions, such as the gut, during one or more phases of their life cycle. Limpets and bivalve mollusks close their shells during exposure at low tide to avoid desiccation, but as a consequence are subject to a period of hypoxia. Many of these animals utilize a variety of anaerobic metabolic pathways to survive the period of reduced oxygen availability. Others also adjust the respiratory and cardiovascular systems to maintain oxygen delivery in the face of reduced oxygen availability. For instance, aquatic hypoxia causes an increase in gill ventilation in many fish, as a result of stimulation of chemoreceptors on the gills. The increase in water flow offsets the reduction in oxygen content and maintains delivery of oxygen to the fish. In fishes, such as tuna, that ram ventilate their gills by swimming forward with their mouth open, the size of the gap increases with hypoxia to increase water flow over the gills.

Compared with aquatic environments, oxygen and carbon dioxide levels are relatively stable in air, and local regions of low oxygen or high carbon dioxide are rare

and easily avoided. There is, of course, a gradual reduction in P_{O_2} with altitude, and animals vary in their capacity to climb to high altitudes and withstand the accompanying reduction in ambient oxygen levels. The highest permanent human habitation is at about 5800 m, where the P_{O_2} is 80 mm Hg compared to about 155 mm Hg at sea level. Many birds migrate over long distances at altitudes above 6000 m, where atmospheric pressures would cause severe respiratory distress in many mammals. High altitudes are associated with low temperatures as well as low pressures, and this also has a marked effect on animal distribution.

A reduction in the P_{O_2} of ambient air results in a decrease in blood P_{O_2}, which in turn stimulates the carotid and aortic body chemoreceptors, causing an increase in lung ventilation in mammals. The rise in lung ventilation then leads to an increase in CO_2 elimination and a decrease in blood P_{CO_2}. The decrease in blood P_{CO_2} causes a reduction in P_{CO_2} and therefore an increase in pH of the CSF. Decreases in blood P_{CO_2} and increases in CSF pH tend to reduce ventilation, thereby attenuating the hypoxia-induced increase in lung ventilation. If, however, hypoxic conditions are maintained, as occurs when animals move to high altitude, both blood and CSF pH are returned to normal levels by the excretion of bicarbonate. This process takes about one week in humans. Thus, as CSF pH returns to normal, the reflex effects of hypoxia on ventilation predominate; the result is a gradual increase in ventilation as the animal acclimatizes to altitude. The response to prolonged hypoxia may also involve modulation of the effects of CO_2 on the carotid and aortic bodies to reset these chemoreceptors to the new lower CO_2 level at high altitude.

 What effect would blocking the carbonic anhydrase activity in erythrocytes have on the ventilatory responses of humans observed at altitude?

As mentioned earlier, low oxygen levels cause a local vasoconstriction in the pulmonary capillaries in mammals, producing a rise in pulmonary arterial blood pressure. This response normally has some importance in redistributing blood away from poorly ventilated, and therefore hypoxic, portions of the lung. When animals are subjected to a general hypoxic environment, however, the increase in the resistance to flow through the whole lung can have detrimental effects. Some mammals that live at high altitudes exhibit a reduced local pulmonary vasoconstriction in response to hypoxia; this is probably a genetically determined acclimation. Humans residing at high altitudes are usually small and barrel chested, and have large lung volumes. Lung development is oxygen insensitive, but the growth of limbs is reduced under hypoxic conditions. The high lung to body ratio enables these people to maintain oxygen uptake under hypoxic conditions. Pulmonary blood pressures are high, and there is often hypertrophy of the right ventricle. High pulmonary pressures produce more even distribution of blood in the lung, and so augment the diffusing capacity for oxygen.

Long-term adaptations also occur during prolonged exposure to hypoxia. Most vertebrates respond by increasing the number of red blood cells and the blood hemoglobin content—and therefore the oxygen capacity of the blood. A reduction in blood oxygen levels stimulates production of the hormone *erythropoietin* in the kidney and liver. Erythropoietin acts on the bone marrow to increase production of red blood cells *(erythropoiesis)*. Under hypoxic conditions, the levels of hemoglobin-binding organophosphates (e.g., DPG) changes, thus altering the oxygen affinity of hemoglobin. In humans, a climb to high altitude is accompanied by an increase in DPG levels and a reduction in the hemoglobin-oxygen affinity. The increasing DPG levels offset the effects of high blood pH on hemoglobin-oxygen affinity. The high blood pH results from hyperventilation in response to the low O_2 availability.

Hypoxia due to travel to high altitude also results in systemic vasodilation and an increase in cardiac output. The higher cardiac output lasts only a few days and returns to normal or drops below normal as O_2 supplies to tissues are restored by the compensatory increases in ventilation and blood hemoglobin levels. Exposure to hypoxia stimulates a proliferation of capillaries in tissues, ensuring a more adequate oxygen delivery to the tissues. The gills of fishes and amphibians are larger in species exposed to prolonged periods of hypoxia. Similar enlargement of the respiratory surface apparently does not occur in mammals. These processes augment the transfer of oxygen, its transport in the blood, and its delivery to the tissues, but they take from several hours to days or weeks to reach completion.

Increased Carbon Dioxide Levels (Hypercapnia)

In many animals, an increase in blood P_{CO_2} results in an increase in ventilation. In mammals the increase is proportional to the rise in the CO_2 level in the blood. The effect is mediated by modulation of the activity of several receptors that send messages to the medullary respiratory center. These receptors include the chemoreceptors of the aortic and carotid bodies and the mechanoreceptors in the lungs, but the response is dominated by the central H^+ receptors (see Figure 13-52). Correction of CSF pH, in the face of altered P_{CO_2} levels, is very important in the return of ventilation to normal.

A marked increase in ventilation occurs almost immediately in response to a rise in CO_2. The increase is maintained for long periods in the presence of increased CO_2, but ventilation eventually returns to a level slightly above the volume that prevailed before hypercapnia. This return to a value only slightly greater than the initial ventilation level is related to increases in levels of plasma bicarbonate and CSF bicarbonate, with the result that pH returns to normal even though the raised CO_2 levels are maintained.

Diving by Air-Breathing Animals

Many air-breathing vertebrates live in water and dive for varying periods of time. Dolphins and whales rise to the surface to breathe but spend most of their life submerged. The time between breaths varies with the diver, but is around 10–20 minutes for many diving vertebrates (Table 13-2). The elephant seal dives regularly to depths of 400 m, subjecting itself to a pressure of over 40 atm at the bottom of the dive. These pressures would crush the thoracic cage of man. There are reports of sperm whales diving to nearly 2000 m, and staying submerged for over an hour. These are of course maximum estimates; most dives are much shorter and to less depth.

Diving mammals and birds are, of course, subjected to periods of hypoxia during submergence. The mammalian entral nervous system (CNS) cannot withstand anoxia and must be supplied with oxygen throughout the dive. Diving animals solve the problem by utilizing oxygen stores in the lungs, blood, and tissues (Figure 13-53). Many diving animals have high hemoglobin and myoglobin levels, and their total oxygen stores generally are larger than those in non-diving animals. To minimize depletion of available stores, oxygen is preferentially delivered to the brain and the heart during a dive; blood flow to other organs may be reduced, and these tissues may adopt anaerobic metabolic pathways. There is a marked slowing of heart rate (bradycardia) and a reduction of cardiac output during a prolonged dive or if the animal is forcibly submerged in an experimental setting (see Figure 12-47). Air-breathing animals that spend prolonged periods submerged at sea must have sufficient oxygen stores to sustain aerobic metabolism, because they cannot tolerate the high accumulation of lactic acid that results from anaerobic metabolism. During prolonged dives, metabolic rates and thus oxygen needs often are reduced in such animals (e.g., elephant seals).

Some diving animals, such as the Weddell seal, exhale before diving, thus reducing the oxygen store in their

TABLE 13-2
Total oxygen stores, mean dive time, and mean dive depth in diving vertebrates

Species	O_2 stores $(ml \cdot kg^{-1})$	Mean drive time (minutes)	Mean depth of dive (meters)
Leatherback turtle	20	11	—
Penguins*	58	6	100
Weddell seal	60	15	100
Northern elephant seal	—	20	400
Human[†]	20	2	shallow

*The O_2 stores are for a King penguin; the dive time and depth are for an Emperor penguin.

[†]Leatherback turtles have similar oxygen stores as humans but can dive for much longer times because of their lower rate of oxygen use.

Source: Adapted from Kooyman, 1989.

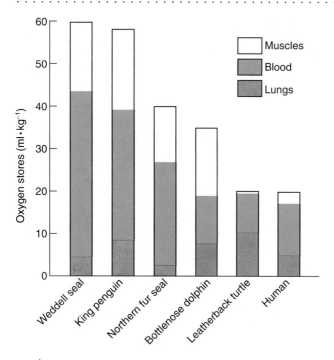

Figure 13-53 Air-breathing animals draw on oxygen stores in the lungs, blood, and tissues (especially muscles) when submerged. The generalized total oxygen stores (expressed in ml $O_2 \cdot kg^{-1}$ body weight) of the major taxa of marine divers are compared with humans. The distribution of the total O_2 stores in the lungs (gray), blood (red), and muscle (white) varies somewhat among species. [Adapted from Kooyman, 1989.]

lungs. During a deep dive, the increase in hydrostatic pressure results in lung compression. In those animals that reduce lung volume before a dive, air is forced out of the alveoli as the lungs collapse and is contained within the trachea and bronchi, which are more rigid but less permeable to gases. If gases remained in the alveoli, they would diffuse into the blood as pressure increased. At the end of the dive, the partial pressure of nitrogen in the blood would be high, and a rapid ascent would result in the formation of bubbles in the blood the equivalent of decompression sickness, or the "bends," in humans. Thus exhalation before diving reduces the chances of the bends occurring. Since only about 7% of a Weddell seal's total oxygen stores are in the lungs, pre-dive exhalation appears to be a reasonable trade-off.

Receptors that detect the presence of water and that inhibit inspiration during a dive are situated near the glottis and near the mouth and nose (depending on the species). The decrease in blood O_2 levels and increase in CO_2 levels that occur during a dive do not stimulate ventilation because inputs from the chemoreceptors of the carotid and aortic bodies are ignored by the respiratory neurons while the animal is submerged.

During birth, a mammal emerges from an aqueous environment into air and survives a short period of anoxia between the time the placental circulation stops and the time air is first inhaled. The respiratory and circulatory responses of the fetus during this period are similar in several respects to those of a diving mammal.

Exercise

Exercise increases O_2 utilization, CO_2 production, and metabolic acid production. Cardiac output increases to meet the higher demands of the tissues. Even though the transit time for blood through the lung capillaries is reduced, nearly complete gas transfer still occurs (Figure 13-54). Ventilation volume increases in order to maintain gas tensions in arterial blood in the face of increased blood flow. The increase in ventilation in mammals is rapid, coinciding with the onset of exercise. This initial sudden increase in ventilation volume is followed by a more gradual rise until a steady state is obtained both for ventilation volume and oxygen uptake (Figure 13-55). When exercise is terminated, there is a sudden decrease in breathing, followed by a gradual decline in ventilation volume. During exercise, O_2 levels are reduced and CO_2 and H^+ levels raised in venous blood, but the mean P_{O_2} and P_{CO_2} in arterial blood do not vary markedly, except during severe exercise. The oscillations in arterial blood P_{O_2} and P_{CO_2} associated with each breath increase in magnitude, although the mean level is unaltered.

Exercise covers a range from slow movements up to maximum exercise capacity. The phrase *moderate exercise* refers to exercise above resting levels that is aerobic, with only minor energy supplies derived from anaerobic glycolysis. *Severe exercise* refers to exercise in which oxygen uptake is maximal and further energy supplies are derived from anaerobic metabolism. *Heavy exercise* is a term sometimes used to denote the exercise level between moderate and severe exercise.

The onset of exercise involves many changes in lung ventilation and the cardiovascular system, as well as muscle contraction. In the initial stages, during the transition from rest to exercise, the animal is not in a steady state and part of the energy supply is derived from anaerobic

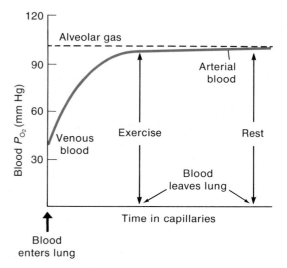

Figure 13-54 Blood P_{O_2} rapidly reaches near equilibration with alveolar P_{O_2} even during exercise. Although blood flow increases, and therefore blood spends less time in the lung capillaries, during exercise, increased ventilation allows equilibration to occur. [Adapted from West, 1970.]

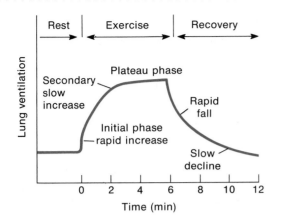

Figure 13-55 An increase in lung ventilation is one of several adjustments to meet the increased oxygen demand during exercise. Typical changes in lung ventilation during exercise and recovery in humans are depicted.

processes. If the exercise level is moderate and sustained, the animal moves into a new steady state typical of that exercise level, with increased lung ventilation, cardiac output, blood flow to the exercising muscles, and oxygen uptake. The relationship between lung ventilation and oxygen uptake is linear during moderate exercise, the slope of the relationship varying with the type of exercise.

A number of receptor systems, some not yet identified, appear to be involved in the respiratory responses to exercise. Contractions of muscles stimulate stretch, acceleration, and position mechanoreceptors in muscles, joints, and tendons. Activity in these receptors reflexively stimulates ventilation, and this system probably causes the sudden changes in ventilation that occur at the beginning and end of a period of exercise. The increase in ventilation varies with the group of muscles being stimulated. Leg exercise, for example, results in a larger increase in ventilation than arm exercise; the same is true for bicycle exercise versus exercise on a treadmill. It has also been suggested that changes in neural activity in the brain and spinal cord leading to muscle contraction may also affect the medullary respiratory center, causing an increase in ventilation.

Muscle contraction generates heat and raises body temperature, thereby increasing ventilation via action on temperature receptors in the hypothalamus. The exact response elicited by stimulation of the hypothalamus depends on the ambient temperature. The increase in ventilation is more pronounced in a hot environment. Since the rise and fall of temperature that follow exercise and subsequent rest are gradual, they would appear to account for only slow changes in ventilation during exercise.

In the absence of exercise, large changes in carbon dioxide and oxygen are required to produce equivalent changes in ventilation. It would seem that the chemoreceptors in both the aortic and the carotid bodies and in the medulla are probably not directly involved in the ventilatory responses to exercise, because mean P_{O_2} and P_{CO_2} levels in arterial blood do not change very much during exercise. However,

the sensitivity of these receptors may increase during exercise, so that relatively small changes in gas partial pressures can cause an increase in ventilation. In this regard, it is significant that catecholamines, which are released in increased quantities during exercise, increase the sensitivity of medullary receptors to changes in carbon dioxide.

Threshold levels of carbon dioxide are required to drive ventilation during exercise, as in resting conditions. Exercising sheep connected to an external artificial lung to maintain low P_{CO_2} and high P_{O_2} levels in their blood do not breathe. Ventilation in the intact mammal increases in proportion to the CO_2 delivery to the lung, but the location of any receptors involved is not known. There are chemical changes in exercising muscle, and these may play a role in reflexively stimulating ventilation via muscle afferent fibers.

Ventilation increases more during severe exercise than during moderate exercise, and the relationship between ventilation and oxygen uptake during severe exercise is no longer linear but becomes exponential. This large increase in ventilation is probably driven by the same mechanisms as in moderate exercise, with the added stimulation of a marked metabolic acidosis and high circulating catecholamine levels.

SWIMBLADDERS: OXYGEN ACCUMULATION AGAINST LARGE GRADIENTS

Fish are denser than the surrounding water and must generate upward hydrodynamic forces if they are to maintain position in the water column and not sink to the bottom. They can generate lift by swimming and using their fins and body as hydrofoils. The minimum speed below which sufficient lift cannot be generated is about $0.6 \ \text{m} \cdot \text{s}^{-1}$ for Skipjack tuna; thus these fish must, and in fact do, swim continually to maintain position in the water column. Other fish hover like a helicopter, using their pectoral fins to maintain position. In both cases there is an energetic cost to maintaining position that can be reduced by incorporation of a buoyancy device.

To avoid expending energy to maintain lift, many aquatic animals maintain a neutral buoyancy, compensating for a dense skeletal structure, by the incorporation of lighter materials in specialized organs. These "buoyancy tanks" may be NH_4Cl solutions (squids), lipid layers (many animals, including sharks), or air-filled swimbladders (many fishes). Ammonium chloride and lipid floats have the advantage of being essentially incompressible, not changing volume with the changes in hydrostatic pressure that accompany vertical movement in water. These float structures, however, are not much lighter than the other body tissues and so must be large if the animal is to achieve neutral buoyancy. Swimbladders are less dense and can be much smaller than NH_4Cl and lipid floats, but they are compressible and change in volume, thus changing the buoyancy of the animal with changes in depth.

Hydrostatic pressure increases by approximately 1 atm for every 10 m of depth. If a fish is swimming just below the surface and suddenly dives to a depth of 10 m, the total pressure in its swimbladder doubles from 1 to 2 atm and the bladder volume is reduced by one-half, thus increasing the density of the fish. The fish will now continue to sink because it is more dense than water. Similarly, if the fish rises to a shallower depth, its swimbladder will expand, decreasing the fish's density, so that it continues to rise. Although the low density of swimbladders is an advantage, they are essentially unstable because of the volume changes they undergo with changes in depth. One means of preventing volume changes is for gas to be removed or added as the fish ascends or descends, respectively. Many fishes do have mechanisms for increasing or decreasing the amount of gas in the swimbladder in order to maintain a constant volume over a wide range of pressures.

Fishes with swimbladders spend most of their time in the upper 200 m of lakes, seas, and oceans. The pressure in the bladder will range from 1 atm at the surface to about 21 atm at 200 m. Gases dissolved in water are generally in equilibrium with air, and neither the partial pressure nor the gas content in water will vary with depth, because water is virtually incompressible (Figure 13-56). The swimbladder gas in most fishes consists of O_2, but in some species the swimbladder is filled with CO_2 or N_2. If the fish dives to a depth of 100 m, O_2 is added to the swimbladder to maintain buoyancy. The aquatic environment is the source of this O_2, which is moved from the surrounding water to the swimbladder against a pressure difference—in this example, a difference of nearly 10 atm (water P_{O_2} = 0.228 atm; bladder P_{O_2} = 10 atm). To understand how this occurs, let us review the structure of the swimbladder.

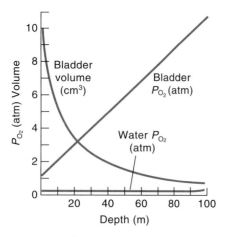

Figure 13-56 The volume of the swimbladder decreases and bladder P_{O_2} increases as a fish descends. Hydrostatic pressure increases by approximately 1 atm every 10 m. In this example, oxygen is assumed to be the only gas present and is neither added to nor removed from the bladder. Fishes can maintain constant density only by maintaining constant bladder volume, which is achieved by addition of oxygen to the bladder with increasing depth. Note the increasing P_{O_2} difference between water and bladder with depth. Oxygen must be moved from water into the swimbladder against this increasing P_{O_2} gradient.

 What are the problems of using a swimbladder for buoyancy at great depths. Would you expect to find deep-sea fishes with swimbladders?

The Rete Mirabile

The teleost swimbladder is a pouch of the foregut (Figure 13-57). In some fishes, there is a duct between the gut and bladder; in others, the duct is absent in the adult. The bladder wall is tough and impermeable to gases, with very little leakage even at very high pressures, but the wall expands easily if pressures inside the bladder exceed those surrounding the fish. Those animals capable of moving oxygen into the bladder against a high pressure gradient have a **rete mirabile**. The rete consists of several bundles of capillaries (both arterial and venous) in close apposition, so that there is countercurrent blood flow between arterial and venous blood. It has been calculated that eel retia have 88,000 venous capillaries and 116,000 arterial capillaries containing about 0.4 ml of blood. The surface area of contact between the venous and arterial capillaries is about 100 cm². Blood passes first through the arterial capillaries of the rete, then through a secretory epithelium (gas gland) in the bladder wall, and finally back through the venous capillaries in the rete. The arterial blood and the venous blood in the rete are separated by a distance of about 1.5 μm.

The rete structure allows blood to flow into the bladder wall without a concomitant large loss of gas from the swimbladder. Blood leaving the secretory epithelium at high P_{O_2} passes into the venous capillaries. The partial pressure of oxygen decreases in both arterial and venous capillaries with distance from the secretory epithelium. The P_{O_2} difference between arterial and venous blood at the end of the rete distal to the swimbladder is small compared with the P_{O_2} difference between the environment and the swimbladder, reducing the loss of oxygen from the swimbladder. It was thought that the reason oxygen levels dropped in the rete was because of the diffusion of oxygen from venous to arterial capillaries, the rete acting as a countercurrent exchanger (see Spotlight 14-2). H. Kobayashi, B. Pelster, and P. Scheid (1993), however, were unable to detect any significant transfer of oxygen across the rete. The P_{O_2} does fall in the blood flowing away from the gas gland because oxygen binds to hemoglobin, not because of any loss of oxygen to arterial blood entering the rete. Exactly how and why this occurs will be discussed later.

Oxygen Secretion

The rete structure reduces gas loss from the swimbladder, but *how is oxygen secreted into the swimbladder?* First, consider the relationship between P_{O_2}, oxygen solubility, and oxygen content. Oxygen is carried in blood bound to hemoglobin and in physical solution. If oxygen is released from hemoglobin into physical solution, P_{O_2} will increase. The release of oxygen from hemoglobin can be caused by a reduc-

A

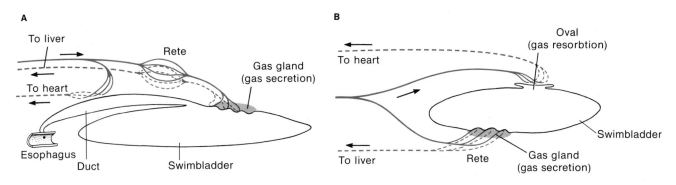

B

Figure 13-57 Two main types of swimbladder are found in fish. **(A)** A physostome swimbladder (e.g., from the eel, *Anguilla vulgaris*) is connected to the outside via a duct to the esophagus. **(B)** A physoclist swim-bladder (e.g., from the perch, *Perca fluviatilis*) lacks a duct. Gas enters and leaves the bladder via the blood. [Adapted from Denton, 1961.]

tion in pH via the Root-off shift (Figure 13-58). An increase in ionic concentration reduces oxygen solubility and also results in an increase in P_{O_2}, as long as the oxygen content in physical solution remains unchanged. Thus, an increase in blood P_{O_2} can be achieved by releasing oxygen from hemoglobin or increasing the ionic concentration of the blood.

The cells of the gas gland have few mitochondria and negligible Krebs cycle activity. For this reason, even in an oxygen atmosphere, glycolysis in the secretory epithelium (gas gland) of the swimbladder yields two lactate molecules and two protons for each glucose molecule. The pentose phosphate shunt, however, is active in the gas gland, producing carbon dioxide via decarboxylation of glucose without oxygen consumption. The production of carbon dioxide, lactate, and protons by gas-gland cells results in (1) a decrease in pH, which causes the release of oxygen from hemoglobin (Root-off shift) and (2) an increase in ionic concentration and therefore a reduction in oxygen solubility (sometimes termed the "salting-out effect"). Both changes cause the P_{O_2} in the secretory epithelium to increase more

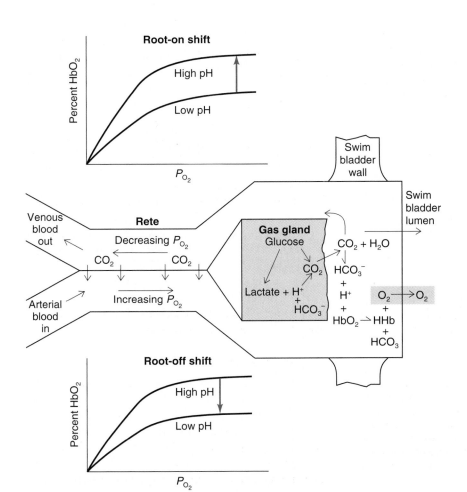

Figure 13-58 Anaerobic metabolism of glucose to lactate and CO_2 in the gas gland, located in the wall of the fish swimbladder, leads to a decrease in erythrocyte pH and release of oxygen from hemoglobin. As a result, the P_{O_2} in the blood flowing through the gas gland becomes greater than the P_{O_2} in the lumen of the swimbladder, so oxygen diffuses into the lumen. Root-off shift, leading to increase in P_{O_2}, occurs on arterial side of the rete, whereas Root-on shift, leading to decrease in P_{O_2}, occurs on the venous side.

than that in the swimbladder, so that oxygen diffuses from blood into the gas space of the swim bladder (see Figure 13-58). The salting-out effect will also reduce the solubility of other gases, such as nitrogen and carbon dioxide, and may explain the high levels of these gases sometimes observed in swim bladders.

Let's return now to the situation in the rete. As discussed earlier, erythrocytes are not very permeable to H^+ ions, so the drop in pH in the gas gland is transferred into the red blood cells by CO_2, which crosses cell membranes with ease (Figure 13-59). Acid produced in the gas gland reacts with HCO_3^-, probably taken up from the plasma, producing CO_2. Thus blood leaving the gas gland and entering the venous capillaries of the rete has a high CO_2 content. As the high-CO_2 venous blood flows through the rete, CO_2 diffuses into the arterial blood flowing towards the gas gland. This raises the pH of the venous blood, which in turn increases oxygen binding by hemoglobin (Root-on shift); as more oxygen is bound, the P_{O_2} in the venous blood falls as it flows away from the gas gland (see Figure 13-58). On the arterial side of the rete, the entering CO_2 lowers blood pH, which drives oxygen from the hemoglobin (Root-off shift), thereby raising the blood P_{O_2}. Thus the P_{O_2} changes in the rete result from loading and unloading of hemoglobin with oxygen, with the rete serving as a countercurrent exchanger for carbon dioxide and not oxygen. In fact the rete has a relatively low oxygen permeability.

The gas gland and associated rete enable fish to transfer oxygen into the swimbladder even though the bladder may contain oxygen at several atmospheres pressure. The bladder wall is slightly permeable to gases so there is a continual loss of gas that increases with depth (bladder pressure). Gases, therefore, must be secreted continually to maintain volume in the face of this loss. Eels migrating at depth across the oceans enlarge their rete and gas gland and decrease the permeability of the bladder wall, enabling them to maintain bladder volume at higher pressures. The permeability of the bladder wall is decreased by an increase in its thickness due to increased deposition of guanine. Eels turn from yellow to silver as a result of these guanine deposits. This occurs as they leave the rivers and begin their migration across the oceans.

SUMMARY

At the level of the mitochondria, the number of oxygen molecules that an animal extracts from the environment and utilizes is approximately the same as the number of carbon dioxide molecules it produces and releases into the environment. In very small animals, gases are transferred between the surface and the mitochondria by diffusion alone, but in larger animals a circulatory system has evolved for the bulk transfer of gases between the respiratory surface and the tissues.

Respiratory surfaces are characterized by large surface areas and small distances for diffusion between the inhalant medium and the blood to facilitate gas transfer. Breathing movements assure a continual supply of oxygen and prevent stagnation of the medium close to the respiratory epithelium. The design of the respiratory surface and the mechanism of breathing are related to the nature of the medium (i.e., gills in water, lungs in air).

Bulk transport of O_2 and CO_2 in the blood is augmented by the presence of a respiratory pigment (e.g., hemoglobin). The pigment not only increases the oxygen-carrying capacity of the blood, but also aids the uptake and release of O_2 and CO_2 at the lungs and tissues.

The rate of gas transfer across a respiratory surface depends on the ratio of ventilation rate of the respiratory surface to blood flow, $\dot{V}_A/\dot{Q}$, as well as on the absolute ventilation volume and cardiac output. These factors are closely regulated to maintain adequate rates of gas transfer to meet the requirements of the tissues. The control system, which has been studied extensively only in mammals, consists of a number of mechanoreceptors and chemoreceptors that feed information into a central integrating region, the medullary respiratory center. This center, through a variety

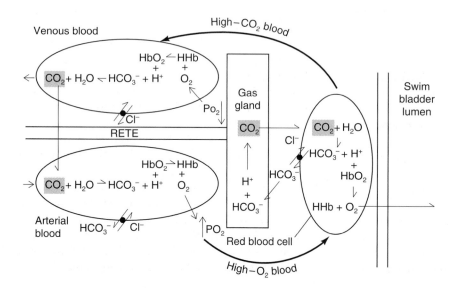

Figure 13-59 The rete associated with the gas gland acts as a countercurrent exchanger for carbon dioxide. Venous blood coming from the gas gland is high in CO_2, which diffuses into the arterial side of the rete, lowering pH and causing a Root-off shift (see Figure 13-58) and increased P_{O_2} in the arterial blood entering the gland. The CO_2 is recycled through the rete, further increasing P_{O_2} in the arterial blood and decreasing it in the venous blood.

of effectors, causes appropriate changes in breathing and blood flow to maintain rates of O_2 and CO_2 transfer at a level sufficient to meet the requirements of metabolism.

Animals regulate body pH in the face of continual production and excretion of H^+ ions. Production of H^+ varies with the metabolic requirements of the animal; H^+ excretion via the lungs and kidney is adjusted to match production. Buffers, particularly proteins and phosphates, ameliorate any oscillations in body pH due to an imbalance between acid production and excretion. The muscle tissues are utilized as a temporary H^+ ion reservoir, thereby further protecting more sensitive tissues such as the brain from wide swings in pH until the excess H^+ can be excreted from the body. Intracellular pH is adjusted by modulation of Na^+/H^+ and HCO_3^-/Cl^- exchange mechanisms located in the cell membrane.

Insects have evolved a tracheal system that takes advantage of the rapid diffusion of gases in air and avoids the necessity of transporting gases in the blood. The tracheal system consists of a series of air-filled, thin-walled tubes that extend throughout the body and serve as diffusion pathways for O_2 and CO_2 between the environment and the cells. In some large active insects, the tracheal system is ventilated.

Bird eggs and fish swimbladders present interesting problems in gas transfer. A bird's egg contains an embryo whose oxygen must be transferred across a shell of fixed dimensions, the transfer requirements increasing a thousandfold between laying and hatching. Gas tensions in the fish swimbladder often exceed that in the blood by several orders of magnitude, but the design of the blood supply and gas gland is such that gases move from the blood into the swimbladder.

REVIEW QUESTIONS

1. Calculate the percent change in volume when dry air at 20°C is inhaled into the human lung (temperature = 37°C).
2. Define the following terms: (a) oxygen capacity, (b) oxygen content, (c) percent saturation, (d) methemoglobin, (e) Bohr effect, and (f) Haldane effect.
3. Describe the role of hemoglobin in the transfer of oxygen and carbon dioxide.
4. Describe the effects of gravity on the distribution of blood in the human lung. What effect does alveolar pressure have on lung blood flow?
5. Compare and contrast ventilation of the mammalian lung and the bird lung.
6. What is the functional significance of the presence of surfactants in the lung?
7. How have insects avoided the necessity of transporting gases in the blood?
8. The number and dimensions of air pores in eggshells are constant for a given species. What effect would doubling the number of pores have on the transfer of oxygen, carbon dioxide, and water across the eggshell?

9. Discuss the role of the rete mirabile in the maintenance of high gas pressures in the fish swimbladder.
10. How is oxygen moved into the swimbladder of teleost fishes?
11. Describe the structural and functional differences between gills and lungs.
12. Why is the ventilation-to-perfusion ratio much higher in water-breathing than in air-breathing animals?
13. Describe the role of central chemoreceptors in the control of carbon dioxide excretion.
14. What is the importance of the Hering–Breuer reflex in the control of breathing?
15. Describe the processes involved in the acclimation of mammals to high altitude.
16. What is the effect on intracellular pH of elevating extracellular NH_4Cl levels at either low or high extracellular pH?
17. Describe the role of the CO_2-bicarbonate systems in pH regulation in mammals.
18. Explain the consequences of the localization of the enzyme carbonic anhydrase within the red blood cell and not in the plasma.
19. Describe the possible mode of operation of the medullary respiratory center.
20. Discuss the interaction between gas transfer and heat and water loss in air-breathing vertebrates.

SUGGESTED READINGS

Asrtrup, P., and J. W. Severinghaus. 1986. *The History of Blood Gases, Acids and Bases.* Copenhagen: Munksgaard.

Brauner, C. J., and D. J. Randall. 1996. The interaction between oxygen and carbon dioxide movements in fishes. *Comp. Biochem. Physiol.* 113A: 83–90.

Dejours, P. 1988. *Respiration in Water and Air.* Amsterdam: Elsevier.

Diamond, J. 1982. How eggs breathe while avoiding desiccation and drowning. *Nature* 295:10–11.

Evans, D. H. 1993. *The Physiology of Fishes.* Boca Raton, Fla.: CRC Press.

Euler, C. von. 1980. Central pattern generation during breathing. *Trends Neurosci.* 3:275–277.

Heisler, N., ed. 1995. *Mechanisms of Systemic Regulation: Respiration and Circulation.* Advances in Comparative and Environmental Physiology, Vol. 21. Berlin: Springer-Verlag.

Hochachka, P. W., and G. N. Somero. 1983. *Strategies of Biochemical Adaptation.* 2d ed. Princeton, N.J.: Princeton University Press.

Jensen, F. B. 1991. Multiple strategies in oxygen and carbon dioxide transport by haemoglobin. In A. J. Woakes, M. K. Grieshaber, and C. Bridges, eds., *Physiological Strategies for Gas Exchange and Metabolism.* Society of Experimental Biology Seminar Series, Vol. 41. Cambridge: Cambridge University Press.

Kobayashi, H., B. Pelster, and P. Scheid. 1993. Gas exchange in fish swimbladder. In P. Scheid, ed., *Respiration in Health and Disease: Lessons from Comparative Physiology.* Stuttgart: Gustav Fischer.

Krogh, A. 1968. *The Comparative Physiology of Respiratory Mechanisms.* New York: Dover.

Milvaganam, S. E. 1996. Structural basis for the Root effect in haemoglobin. *Nature Struct. Biol.* 3:275–283.

Nikinmaa, M. 1990. Vertebrate red blood cells: adaptations of function to respiratory requirements. In S. D. Bradshaw, W. Burggren, H. C. Heller, S. Ishii, H. Langer, G. Neuweiler, and D. J. Randall, *Zoophysiology,* Vol. 28. New York: Springer-Verlag.

Perutz, M. F. 1996. Cause of the Root effect in fish heamoglobins. *Nature Struct. Biol.* 3:211–212.

Rahn, H. 1966. Aquatic gas exchange theory. *Resp. Physiol.* 1:1–12.

Richter, D. W., K. Ballanyi, and S. Schwarzacher 1992. Mechanism of respiratory rhythm generation. *Curr. Opin. Neurobiol.* 2:788–793.

Roos, A., and W. F. Boron. 1981. Intracellular pH. *Physiol. Rev.* 61:296–434.

Schmidt-Nielsen, K. 1972. *How Animals Work.* Cambridge: Cambridge University Press.

Weber, R.E. 1992. Molecular strategies in the adaptation of vertebrate hemoglobin function. In S. C. Wood, R. E. Weber, A. R. Hargens, and R. W. Millard, *Physiological Adaptations in Vertebrates: Respiration, Circulation and Metabolism.* New York: Marcel Dekker.

West, J. B. 1974. *Respiratory Physiology: The Essentials.* Baltimore: Williams and Wilkins.

Zhu, X. L., and W. S. Sly. 1990. Carbonic anhydrase IV from human lung. *J. Biol. Chem.* 15:8795–8801.

IONIC AND OSMOTIC BALANCE

The unique physical and chemical properties of water undoubtedly played a major role in the origin of life, and all life processes take place in a watery milieu (see Chapter 3). Water is, in fact, indispensable for all biochemical and physiological processes. Indeed, the physicochemical nature of Earth life is largely a reflection of the special properties of water. The presence of water here on Earth made it possible for life as we know it to arise several billion years ago in a shallow, salty sea. Extracellular fluids surrounding living cells to this day reflect, to some extent, the composition of the primeval sea in which life evolved (Table 14-1).

The ability to survive in a variety of osmotic environments was achieved in the more advanced animal groups by the evolution of a stable internal environment, which acts to buffer the internal tissues against the vagaries and extremes of the external environment. Thus, the ability to maintain a suitable internal environment in the face of osmotic stress (something that tends to disturb the ionic and osmotic homeostasis of the animal) has played a most important role in animal evolution. There are two main reasons for this. First, animals are restricted in their geographic distribution by environmental factors, one of the most important being the osmotic nature of the environment. Second, geographic dispersal followed by genetic isolation is an important mechanism for the divergence of species in the process of evolution. If, for example, the arthropods and the vertebrates had not evolved mechanisms for regulating their extracellular compartments, they would have been unsuccessful in their invasion of the osmotically hostile freshwater and terrestrial environments. In the absence of competition from terrestrial arthropods and vertebrates, other groups would have evolved with greater diversity to fill the vacant terrestrial niches, and the living world would be quite different from the one we know.

In this chapter, we consider the osmotic environment, osmotic exchange between the animal and its environment, and mechanisms used by various animals to cope with environmental osmotic extremes. The movement of water and solutes across cell membranes and multicellular ep-

ithelial layers has been covered, along with other cellular mechanisms, in Chapter 4. That discussion forms an essential background for understanding the osmoregulatory processes in organs such as the kidney, gill, and salt gland covered in this chapter. Toward the chapter's end we discuss the closely related problem of elimination of toxic nitrogenous wastes produced during the metabolism of amino acids and proteins.

PROBLEMS OF OSMOREGULATION

One of the requirements in the regulation of the internal environment is that appropriate quantities of water be retained. Another major requirement for cell survival is the presence, in appropriate concentrations, of various solutes (e.g., salts and nutrient molecules) in the extracellular and intracellular compartments (Table 14-2). Some tissues require an extracellular ionic environment that is an approximation of seawater—namely, fluid high in sodium and chloride and relatively low in the other major ions, such as potassium and the divalent cations. For many marine invertebrates, the surrounding seawater itself can act as the extracellular medium; for most of the more complex forms, the internal fluids are in near ionic equilibrium with seawater. In contrast, the extracellular fluids of vertebrates, with the exception of the hagfishes, have an ionic concentration that is about one third that of seawater with much of the magnesium sulfate removed and some of the chloride replaced by bicarbonate anion (see Table 14-1). This presumably reflects the freshwater origin of most vertebrates, including marine teleost fishes. The extracellular fluids of marine teleosts are much more dilute than seawater, and these fishes maintain both an ionic and osmotic difference between their body fluids and seawater. Elasmobranchs, on the other hand, maintain an ionic difference but only minor osmotic differences; the high levels of urea in the body of elasmobranchs brings the osmolarity to slightly above that of seawater.

The intracellular environment of most animals is low in sodium but high in potassium, phosphate, and proteins

TABLE 14-1
Composition of extracellular fluids of representative animals*

	Habitat*	Osmolarity (mosM)	Ionic concentrations (mM)							
			Na$^+$	K$^+$	Ca^{2+}	Mg^{2+}	Cl$^-$	SO$_4^{2-}$	HPO$_4^{2-}$	Urea
Seawater†		1000	460	10	10	53	540	27		
Coelenterata										
Aurelia (jellyfish)	SW		454	10.2	9.7	51.0	554	14.6		
Echinodermata										
Asterias (starfish)	SW		428	9.5	11.7	49.2	487	26.7		
Annelida										
Arenicola (lugworm)	SW		459	10.1	10.0	52.4	537	24.4		
Lumbricus (earthworm)	Ter.		76	4.0	2.9		43			
Mollusca										
Aplysia (sea slug)	SW		492	9.7	13.3	49	543	28.2		
Loligo (squid)	SW		419	20.6	11.3	51.6	522	6.9		
Anodonta (clam)	FW		15.6	0.49	8.4	0.19	11.7	0.73		
Crustacea										
Cambarus (crayfish)	FW		146	3.9	8.1	4.3	139			
Homarus (lobster)	SW		472	10.0	15.6	6.7	470			
Insecta										
Locusta	Ter.		60	12	17	25				
Periplanta (cockroach)	Ter.		161	7.9	4.0	5.6	144			
Cyclostomata										
Eptatretus (hagfish)	SW	1002	554	6.8	8.8	23.4	532	1.7	2.1	3
Lampetra (lamprey)	FW	248	120	3.2	1.9	2.1	96	2.7		0.4
Chondrichthyes										
Dogfish shark	SW	1075	269	4.3	3.2	1.1	258	1	1.1	376
Carcharhinus	FW		200	8	3	2	180	0.5	4.0	132
Coelacantha										
Latimeria	SW		181	51.3	6.9	28.7	199			355
Teleostei										
Paralichthys (flounder)	SW	337	180	4	3	1	160	0.2		
Carassius (goldfish)	FW	293	142	2	6	3	107			
Amphibia										
Rana esculenta (frog)	FW	210	92	3	2.3	1.6	70			2
Rana cancrivora	FW	290	125	9			98			40
	80% SW	830	252	14			227			350
Reptilia										
Alligator	FW	278	140	3.6	5.1	3.0	111			
Aves										
Anas (duck)	FW	294	138	3.1	2.4		103		1.6	
Mammalia										
Homo sapiens	Ter.		142	4.0	5.0	2.0	104	1	2	
Lab rat	Ter.		145	6.2	3.1	1.6	116			

*The osmolarity and composition of seawater vary, and the values given here are not intended to be absolute. The composition of body fluids of osmoconformers will also vary, depending on the composition of the seawater in which they are tested.

†SW = seawater; FW = freshwater; Ter. = terrestrial.

Sources: Schmidt-Nielsen and Mackay, 1972; Prosser, 1973.

(Table 14-3). There are only minor and transient osmotic differences between the intracellular and extracellular fluids of animals. Thus the cell membrane maintains ionic, but not osmotic, differences between the intracellular and extracellular fluids, whereas the epithelium surrounding the body often maintains both ionic and osmotic differences between animals and their environments. In most multicellular animals, the entire body surface is not usually involved in ionic and osmotic regulation; rather, this regulation is effected by a specialized portion of the body surface such as the gills of fish or some internal structure like the

salt gland of elasmobranchs or the kidney of mammals. The rest of the body surface, with the exception of the lining of the gut, is relatively impermeable to ions and water.

Animals require nutrients and oxygen to maintain metabolism, and as a result of metabolism, they produce waste products. Cell membranes that are permeable to oxygen are also permeable to water, and energy must be expended to maintain the ionic and osmotic balance of the animal. An animal cannot reduce osmotic and ionic problems by sealing itself off from the environment because nutrients must be acquired and waste products excreted. Some animals en-

TABLE 14-2
Major inorganic ions of tissues

Ion	Distribution	Major functions
Na$^+$	Main extracellular cation	Is major source of extracellular osmotic pressure Provides potential energy for transport of substances across cell membranes Provides inward current for membrane excitation
K$^+$	Main cytosolic cation	Is source of cytosolic osmotic pressure Establishes the resting potential Provides outward current for membrane repolarization
Ca^{2+}	Low concentration in cells	Regulates exocytosis and muscle contraction Is involved in "cementing" cells together Regulates many enzymes and other cell proteins; acts as second messenger.
Mg^{2+}	Intra- and extracellular	Acts as cofactor for many enzymes (e.g., ATPases)
HPO$_4^{2-}$; HCO$_3$	Intra- and extracellular	Buffers H$^+$ concentration
Cl$^-$	Main extracellular anion of tissues	Is counterion for inorganic cations

cyst themselves, but this is feasible only if their metabolic rate is very much reduced. Brine shrimp larvae, for example, can survive in a state of suspended animation for many years with little or no growth; when in this state, they can be placed in water and revived. This is only possible because energy turnover is very reduced during encystment, limiting nutrient utilization and waste-product accumulation. Most animals, however, do not exist in a state of suspended animation and must ingest nutrients at a high rate and cope with the associated osmotic and ionic problems.

The waste products generated during metabolism are often toxic and cannot accumulate to high levels in the body without serious consequences. Thus the cellular environment must be freed of these toxic by-products of metabolism. In the smallest aquatic organisms, this purification happens simply by diffusion of the wastes into the surrounding water. In animals that have circulatory sys-

tems, the blood typically passes through excretory organs, generally termed *kidneys*. In terrestrial animals, the kidneys not only play an important role in the removal of organic wastes but are also the primary organs of osmoregulation.

 The rate of water turnover is different in a whale, a human, a shrimp, and a snake. How are they different and why?

A number of mechanisms are employed to handle osmotic problems and regulate the differences (1) between intracellular and extracellular compartments and (2) between the extracellular compartment and the external environment. These are collectively termed *osmoregulatory mechanisms*, a term coined in 1902 by Rudolf Hober to refer to

TABLE 14-3
Electrolyte composition of the human body fluids

Electrolytes	Serum (meq·kg^{-1} H$_2$O)	Interstitial fluid (meq·kg^{-1} H$_2$O)	Intracellular fluid (muscle) (meq·kg^{-1} H$_2$O)
Cations			
Na$^+$	142	145	10
K$^+$	4	4	156
Ca^{2+}	5		3
Mg^{2+}	2		26
Totals	153	149	195
Anions			
Cl$^-$	104	114	2
HCO$_{3-}$	27	31	8
HPO$_4^{2-}$	2		95
SO$_4^{2-}$	1		20
Organic acids	6		
Protein	13		55
Totals	153	145	180

Note: Some of the ions contained within cells are not completely dissolved within the cytosol, but may be partially sequestered within cytoplasmic organelles. Thus, the true free Ca^{2+} concentration in the cytosol is typically below the overall value given in the table for intracellular Ca^{2+}. Failure of anion and cation totals to agree reflects incomplete tabulation.

the regulation of osmotic pressure and ionic concentrations in the extracellular compartment of the animal body. The evolution of efficient osmoregulatory mechanisms had extraordinarily far-reaching effects on other aspects of animal speciation and diversification. The various adaptations and physiological mechanisms evolved by animals to cope with the rigors of the osmotic environment form especially fascinating examples of the resourcefulness of evolutionary adaptation. This is the theme of an excellent book by the late Homer Smith entitled *From Fish to Philosopher*.

Although there may be hourly and daily variations in osmotic balance, an animal is generally in an osmotic steady state over the long term. That is, on the average, the input and output of water and of salts over an extended period are equal. Water enters terrestrial animals with their food and drink. For animals living in a freshwater environment, water enters the body primarily through the respiratory epithelium—the gill surfaces of fish and invertebrates, and the integument of amphibians and many invertebrates. Water leaves the body in the urine, in the feces, and by evaporation through the lungs and integument, the outer covering.

The problem of osmotic regulation does not end with the intake and output of water. If that were so, osmoregulation would be a relatively simple matter: A frog sitting in freshwater that is far more dilute than its body fluids would merely have to eliminate the same amount of water as leaked in through its skin, and a camel would just stop urine production between oases. Osmoregulation also involves the maintenance of favorable solute concentrations in the extracellular compartment. Thus, the frog immersed in hypotonic pond water is faced not only with the need to eliminate excess water, but also with the problem of retaining salts, which tend to leak out through the skin, because the skin in amphibians is generally more permeable than that in the other vertebrate classes.

The osmotic exchanges that take place between an animal and its environment can be divided into two classes (Figure 14-1):

- *Obligatory osmotic exchanges,* which occur mainly in response to physical factors over which the animal has little or no physiological control

- *Regulated osmotic exchanges,* which are physiologically controlled and serve to aid in maintaining internal homeostasis.

Regulated exchanges generally serve to compensate for obligatory exchanges. The flux of a substance across a membrane is determined by its concentration gradient, the surface area of the membrane involved, the thickness of the membrane (i.e. the diffusion distance), and the permeability of the membrane. The same factors influence both obligatory and regulated exchanges. In the next section, we consider obligatory exchange and then, in the following sections, various mechanisms of regulated exchange.

OBLIGATORY EXCHANGE OF IONS AND WATER

The integument, respiratory surfaces, and other epithelia in contact with the surrounding milieu act as the barriers to obligatory exchange between an organism and its environment. The various factors that contribute to the obligatory exchange are outlined next.

Gradients Between the Animal and the Environment

The greater the difference between the concentration of a substance in the external medium and that in the body fluids, the greater the tendency for net diffusion in the direction of the lower concentration. Thus, although a frog immersed in a pond tends to take up water from its hypotonic environment, a bony fish in seawater is faced with the problem of losing water to the surrounding hypertonic seawater. Similarly, a marine fish with a lower NaCl content than that of seawater faces a continual diffusion of salt into the body, whereas a freshwater fish faces a continual loss of salt. The rate of transfer depends on the size of the gradient and the permeability and area of the animal's surface.

Surface-to-Volume Ratio

The volume of an animal varies with the cube of its linear dimensions, but its surface area varies with the square of its linear dimensions. Therefore, the surface-to-volume ratio is greater for small animals than for large animals. It fol-

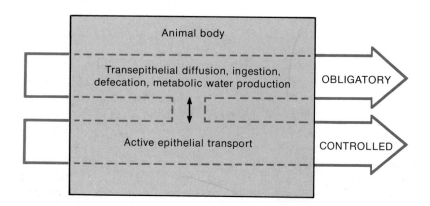

Figure 14-1 Two major classes of osmotic exchange—obligatory and controlled—occur between an animal and its environment. Obligatory exchanges occur in response to physical factors over which the animal has little short-term physiological control. Controlled exchanges are those that the animal can vary physiologically to maintain internal homeostasis. Substances entering the animal by either path can leave by the other path.

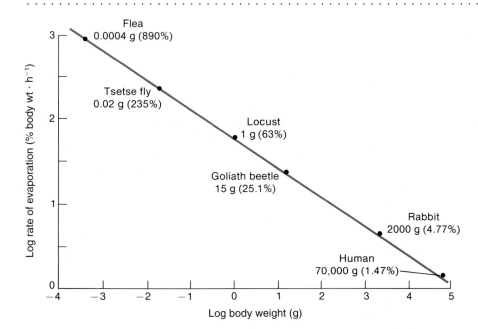

Figure 14-2 Small animals dehydrate more rapidly than large animals because of their high surface-to-mass (and thus surface-to-volume) ratios. This log-log plot shows the amount of water, as percentage of body weight, that is lost per hour under hot desert conditions versus body weight. [Edney and Nagy, 1976.]

lows that the surface area of the integument, through which water or a solute can exchange with the environment, is greater relative to the water content of a small animal than it is for a large animal. This means that for a given net rate of exchange across the integument (in moles per second per square centimeter), a small animal will dehydrate or hydrate more rapidly than a larger animal of the same shape (Figure 14-2).

Permeability of the Integument

The integument acts as a barrier between the extracellular compartment and the environment. Movement of water across the integument occurs through cells (*transcellular*) and between cells (*paracellular*). Pure phospholipid bilayers, however, are not very permeable to water, and transcellular movement of water across biological membranes depends on the presence of water channels. For example, erythrocytes swell or shrink rapidly in response to changes in the osmotic strength of the extracellular fluid because of the presence of a 28-kDa protein, appropriately called **aquaporin**. It appears that water channels in membranes are formed of a tetramer of identical aquaporin molecules. The role of aquaporin as a water channel was demonstrated in experiments with frogs eggs and oocytes, which are not very permeable to water and thus do not swell much when placed in pond water. When mRNA encoding the aquaporin protein was injected into frog oocytes, they became very water permeable and swelled when placed in water. Membrane permeability to water is presumably related to the concentration of aquaporin water channels within the phospholipid bilayer. Tight junctions between cells reduce the permeability of the paracellular pathway to water. The absence of aquaporin water channels in the membrane reduces the transcellular water permeability.

The permeability of the integument to water and solutes varies among animal groups. Amphibians generally

have moist, highly permeable skins, through which they exchange oxygen and carbon dioxide and through which water and ions move by passive diffusion. Amphibian skin compensates for loss of electrolytes by active transport of salts from the aquatic environment into the animal. Fish gills are necessarily permeable, as they engage in the exchange of oxygen and carbon dioxide between the blood and the aqueous environment. The gills, like the frog skin, also engage in active transport of salts. The volume of blood perfusing the gills of fish has been shown to decrease as respiratory demand drops and to rise in response to increased oxygen need. This reduction in blood perfusion of the gills effectively limits osmotic transfer through the gill epithelium during periods of low oxygen uptake. Thus, as oxygen transfer increases so does osmotic and ionic exchange across the gills.

In contrast, reptiles, some desert amphibians, birds, and many mammals have relatively impermeable skins and thus generally lose relatively little water through this route. In fact, the skin of some mammals (e.g., cow hide) is so impermeable that it can be used to carry water or even wine. The low permeability of the integument of terrestrial animals is maintained in those species that have secondarily become marine or aquatic, such as pond insects and marine mammals.

Not all vertebrates, however, have a relatively impermeable integument. Many amphibians, as well as mammals that perspire, can become dehydrated at low humidity because of water loss through the integument. Animals with highly permeable skin are simply not able to tolerate very hot, dry environments. Most frogs stay close to water. Toads and salamanders can venture a bit farther, but they also are limited to moist woods or meadows not far from puddles, streams, or bodies of water in which they can replenish their supply of body water. These animals also minimize water loss by behavioral strategies, avoiding

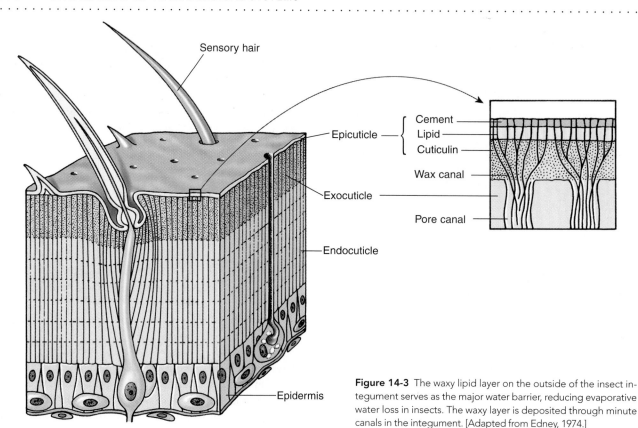

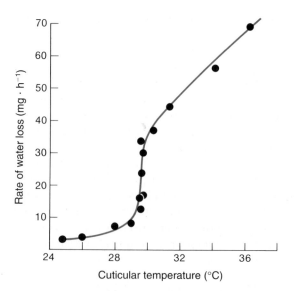

Figure 14-3 The waxy lipid layer on the outside of the insect integument serves as the major water barrier, reducing evaporative water loss in insects. The waxy layer is deposited through minute canals in the integument. [Adapted from Edney, 1974.]

desiccation by staying in cool, damp microenvironments during hot, dry times of day. The desert toads, *Chiromantis xerampelina* and *Phyllomedusa sauvagii*, have extremely low evaporative loss of water from their skin because it is covered by a secreted wax coating. These toads also excrete uric acid rather than ammonia or urea (see the later section *Excretion of Nitrogenous Wastes*).

Frogs and toads are endowed with a large-volume lymphatic system and an oversized urinary bladder in which they store water until needed. When these animals wander from a body of water or during periods of low rainfall, water will move osmotically from the lumen of the bladder into the partially dehydrated interstitial fluid and blood. The epithelium of the bladder, like the amphibian skin, is capable of actively transporting sodium and chloride from the bladder lumen into the body to compensate for the loss of salts that accompanies excessive hydration during times of plentiful water. Thus, the anuran bladder serves a dual function as a water reservoir in times of dehydration and as a source of salts during times of excessive hydration. The high water permeability of amphibian skin is used to advantage to take up water from hyposmotic sources such as puddles. Many anurans have specialized regions of skin on the abdomen and thighs, termed the *pelvic patch*, that when immersed can take up water at a rate of three times the body weight per day. The permeability of amphibian skin is controlled by the hormone arginine vasotocin (AVT) or more simply *vasotocin*; like the mammalian hormone vasopressin, or antidiuretic hormone (ADH), vasotocin enhances water permeability. The outer layers of toad skin

contain minute channels that draw water by capillary action to moisten the skin, conserving internal water during cutaneous evaporation.

Because insects have a waxy cuticle that is highly impermeable to water, their evaporative water loss is much lower than in many other animal groups (Table 14-4). The wax is deposited on the surface of the exoskeleton through

Figure 14-4 The rate of water loss from insects is much higher at temperatures above the melting point of the waxy layer covering the cuticle. The sharp break in this plot of water loss versus cuticular temperature in a cockroach corresponds to the melting point of the waxy cuticle. [From Beament, 1958.]

TABLE 14-4
Evaporative water loss of representative animals under desert conditions

Species	Water loss ($mg \cdot cm^{-2} \cdot h^{-1}$)	Remarks*
Antropods		
Eleodes armata (beetle)	0.20	30°C; 0 % r.h.
Hadrurus arizonensis (scorpion)	0.02	30°C; 0 % r.h.
Locusta migratoria (locust)	0.70	30°C; 0 % r.h.
Amphibian		
Cyclorana alboguttatus (frog)	4.90	25°C; 100 % r.h.
Reptiles		
Gehrydra variegata (gecko)	0.22	30°C; dry air
Uta stansburiana (lizard)	0.10	30°C
Birds		
Amphispiza belli (sparrow)	1.48	30°C
Phalaenpitus nutalllii (poorwill)	0.86	30°C
Mammals†		
Peromyscus eremicus (Cactus mouse)	0.66	30°C
Oryx beisa (African oryx)	3.24	22°C
Homo sapiens	22.32	70 kg; nude, sitting in sun; 35°C

*r.h. stands for relative humidity. Where not indicated, relative humidity is not available.

†The cactus mouse and African oryx are desert animals and employ various water-conservation measures. Thus their evaporative water loss is much less than that of humans.

Source: Hadley, 1972.

fine canals that penetrate the cuticle (Figure 14-3). The importance of the waxy layer for water retention by insects has been demonstrated by measuring the rate of water loss at different temperatures. In Figure 14-4, we see that there is a sudden jump in the rate of water loss coincident with the melting point of the wax coating. The major route of water loss in terrestrial insects is via the tracheal system, which consists of air-filled tracheoles that penetrate the tissues. As long as the tracheoles are open to the air, water vapor can diffuse out while oxygen and carbon dioxide diffuse down their respective gradients. The entrances to the tracheoles are guarded by valvelike spiracles that are closed periodically by the spiracular muscles, conserving water. The importance of this mechanism in water conservation in insects, however, has been questioned (see Chapter 13 for further discussion).

Feeding, Metabolic Factors, and Excretion

Water and solutes are taken in during feeding. Those end products of digestion and metabolism that cannot be used by the organism must be eliminated. Carbon dioxide diffuses into the environment from the respiratory surfaces. Al-

though water is another end product of cellular metabolism, it is produced in small enough quantities that its elimination is no problem (Table 14-5). In fact, this so-called **metabolic water** is the major source of water for many desert dwellers. Osmotic problems are posed by the inevitable production of nitrogenous end products of metabolism (e.g., ammonia and urea) and by the ingestion of salts, because water is required for their elimination from the body.

The diet may include excess water or excess salts. A seal feeding on marine invertebrates with an osmolarity similar to seawater ingests a relatively high quantity of salt relative to water but requires water to excrete the salt load. If the seal feeds on marine teleost fish, which are more dilute than seawater, the ingested salt load is much less. The seal burns fat to produce both energy and water when eating marine invertebrates but stores fat when eating fish. The burning of fat produces the water required to excrete the salt load associated with eating marine invertebrates (see Table 14-5). Thus the seal becomes fat when eating fish but gets thin eating marine invertebrates.

In terrestrial animals, the regulation of plasma ion concentrations and the excretion of nitrogenous wastes are accompanied by unavoidable losses of body water. A number of physiological adaptations tend to minimize the loss of water associated with these important functions of excretory systems. Among terrestrial invertebrates, insects are highly effective in conserving water in the course of eliminating nitrogenous and inorganic wastes. The extent to which ions are reabsorbed in the insect rectum or eliminated with the feces is regulated according to the osmotic condition of the insect. This is illustrated by an experiment in which locusts were allowed to drink either pure water or a concentrated saline solution containing NaCl and KCl ($450 \ mosm \cdot L^{-1}$). The salt concentration of the rectal fluid after the insect drank saline was several hundred times higher than after it drank pure water, whereas the salt concentration of the hemolymph increased by only about 50% after drinking saline (Table 14-6).

The kidney is the chief organ of osmoregulation and nitrogen excretion in most terrestrial vertebrates, especially mammals, which have no other provision for the excretion of salts or nitrogen. The kidneys of birds and mammals utilize *countercurrent multiplication* to produce a

TABLE 14-5
Production of metabolic water during oxidation of foodstuffs

	Foodstuff		
	Carbohydrates	Fats	Proteins
Grams of metabolic water per gram of food	0.56	1.07	0.40
Kilojoules expended per gram of food	17.58	39.94	17.54
Grams of metabolic water per kilojoule expended	0.032	0.027	0.023

Source: Edney and Nagy, 1976.

TABLE 14-6
Ionic regulation in locusts*

Fluid	Concentration (mean values in meq·L⁻¹)		
	Na	K	Cl
Saline for drinking	300	150	450
Hemolymph			
With water to drink	108	11	5
With saline to drink	158	19	569
Rectal fluid			
With water to drink	1	22	5
With saline to drink	405	241	569

*Desert locusts were given strong saline or pure water to drink. When they drank saline, the ionic concentrations in the hemolymph rose, but not to the level of the saline. Ionic concentrations in their rectal fluid became higher than those in saline.

Source: Edney and Nagy, 1976.

hyperosmotic urine, which is more concentrated than the plasma. This specialization, centered on a hairpin-like bend in the kidney tubules, called the **loop of Henle,** has undoubtedly been of major importance in allowing birds and mammals to exploit dry terrestrial environments. The loop of Henle reaches its highest degree of specialization in desert animals such as the kangaroo rat and Australian hopping mice, which can produce a urine of up to 9000 mosm·L⁻¹. In birds, the countercurrent organization of the loop of Henle is less efficient, perhaps because the avian kidney contains a mixture of "reptilian-type" tubules, which lack the loop of Henle, and "mammalian-type" tubules, which contain this specialized structure. The highest osmolarities determined in avian urine (in the salt-marsh Savannah sparrow) have been around 2000 mosm·L⁻¹. Reptiles and amphibians, whose kidneys are not organized for countercurrent multiplication, are unable to produce a hyperosmotic urine. As an adaptive consequence, some amphibians, when faced with dehydration, are able to cease urine production entirely during the period of osmotic stress.

Temperature, Exercise, and Respiration

Because of its high heat of vaporization, water is ideally suited for the elimination of body heat by evaporation from epithelial surfaces. During evaporation, those water molecules with the highest energy content enter the gaseous phase and thus take with them their thermal energy. As a result, the water left behind becomes cooler. The importance of water in temperature regulation leads to conflicts and compromises between physiological adaptation to environmental temperatures and osmotic stresses in terrestrial animals.

Desert animals, faced with both high temperatures and a meager water supply, are especially hard pressed as they must avoid becoming overheated and yet avoid losing large quantities of body water. In some instances desert mammals and birds will let their body temperature rise above 40°C rather than expend water for evaporative cooling. Strenuous exercise generates heat owing to muscle metabolism and must be compensated by a high rate of heat dissipation. This compensation can be accomplished best by evaporative cooling over respiratory surfaces (e.g., the lungs, air passages, and tongue) or by evaporative water loss through the skin. In some very active mammals, body temperature rises during exercise, but brain temperature remains normal due to a countercurrent heat exchanger in the nasal region that cools the brain blood supply. Even during basal conditions (no exercise beyond breathing), the nature of the respiratory mechanism of many terrestrial animals leads to the loss of water through the respiratory surfaces. The nose of mammals plays an important role in reducing water loss through this pathway.

As we have noted, respiratory surfaces are, by their very nature, a major avenue for water loss in air-breathing animals. The internalization of the respiratory surfaces in a body cavity (i.e., the lung) reduces evaporative loss in terrestrial vertebrates. Even within the lung, however, ventilation of the respiratory epithelium by unsaturated air will cause evaporation of the moisture wetting the epithelial surface. Such evaporative loss of water is enhanced in birds and mammals because their body temperature generally is higher than the ambient temperature. The same holds for those reptiles and amphibians that raise their body temperature by behavioral strategies. In such animals, the warmer expired air contains more water than the cooler inspired air as the water-holding capacity of air increases with temperature (Figure 14-5).

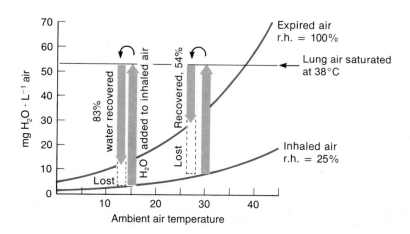

Figure 14-5 Water loss associated with respiration depends on the relationship between the body temperature and the inhaled-air temperature, as well as on the relative humidity of the inspired air. As unsaturated air is warmed in the lungs, its picks up moisture until it is saturated (gray bars). During exhalation, the air is cooled in the nasal passages, so that much of the water is recovered (red bars). The data here are for the kangaroo rat when the inhaled air is at 25% relative humidity (r.h.) and 15°C (left bars) or 30°C (right bars). Clearly, the amount of water recovered is greater and hence the amount lost is less when the inhaled air is at the lower temperature. Indeed, under these climatic conditions, the kangaroo rat exhales air at 13°C (lower than ambient!). [Adapted from Schmidt-Nielsen et al., 1970.]

The respiratory loss of water is minimized through a mechanism first discovered in the nose of the desert-dwelling kangaroo rat, *Dipodomys merriami*, by Knut Schmidt-Nielsen. This mechanism, termed a *temporal countercurrent system*, retains most of the respiratory water vapor by condensing it on cooled nasal passages during expiration. Air entering the nasal passages is warmed to about 37–38°C and humidified by heat and moisture absorbed from the tissues of the nasal passages, trachea, and bronchi (Figure 14-6A). The nasal passages are cooled by this evaporative water loss and the flow of cool air through the nasal passages. The tissue temperature is lowest at the tip of the nose and increases along the nasal passages towards the lung. The nose has a large blood supply to maintain the delivery of water to humidify the incoming air. The blood supply does not warm the nose because it is arranged in a countercurrent fashion, so that warm blood entering the nasal region is cooled by cold blood leaving the nose.

During exhalation, the process of heat exchange between air and nasal tissues is reversed. The warm expired air is cooled to somewhat above ambient as it passes back out through the nasal passages, which had been cooled by the same air during inhalation. As the expired air gives up some of its heat to the tissues of the nasal passages, most of the acquired moisture condenses on the cool nasal epithe-lium (Figure 14-6B). Mammals, including humans, employing this mechanism to humidify inhalant air have "cold" noses, which can be wet or even occasionally drip. With the next inhalation, this condensed moisture again contributes to the humidification of the inspired air, and the cycle is repeated, most of the vapor being recycled within the respiratory tract.

The nose, therefore, plays an important role in reducing the loss of water and heat from the body. The importance of the nose in cooling expired air can be detected easily by placing your hand in front of your nose and mouth and breathing out via your mouth and via your nose; the temperature difference is usually obvious. Because there is little cooling of air expired through the mouth, , the loss of water and heat is greater when breathing out via the mouth (e.g., when the nose is clogged due to a cold) compared with expiration through the nose. If air flow through the nasal passages is bypassed by placing a tube in the trachea, as during human or animal surgical operations, heat and water loss may increase; and surgical patients must be given additional food and water to compensate for the increased water loss. The increased water loss from the trachea can contribute to post-surgical sore throat, a common problem.

A similar mechanism for trapping exhaled moisture occurs in numerous birds and lizards. Where salt glands drain

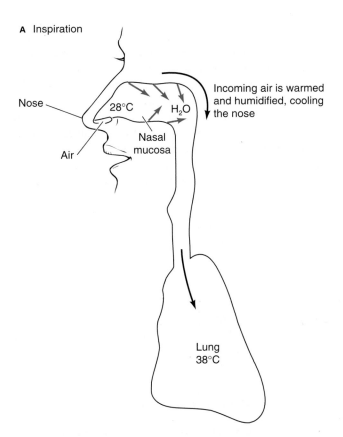

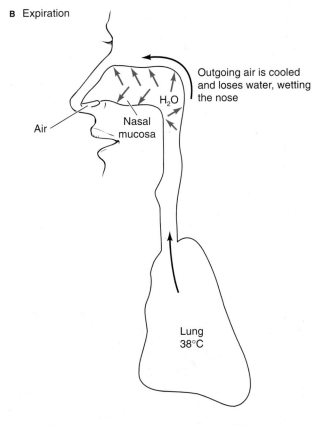

Figure 14-6 Temporal countercurrent exchange in the respiratory systems of many vertebrates acts to conserve body heat and body water. **(A)** During inspiration, cool air (e.g., at 28°C) is warmed and humidified as it flows toward the lungs, removing heat and water from the nasal passages. **(B)** During expiration, the same air loses most of the heat and water it gained earlier, as it warms and deposits water on the cooled nasal passages on its way out. Small red arrows indicate direction of heat and water movement; long arrows indicate direction of air flow.

into the nasal passages, as they do in iguanas, the water in the excreted salt solution enters the incoming air during inspiration and is largely conserved by being condensed during exhalation. Occassionally all the water from the salt solution is evaporated, leaving salt deposits around the nose of these seawater-drinking lizards. This will occur, only however, if the animal's temperature is above that of the environment.

Water loss via the lungs is small for mammals living in hot humid climates and large for those living in cold dry climates. The rate of ventilation and the ventilation pattern (e.g., breathing through either the mouth or nose) also affect the rate of water loss via the lungs. The problem of respiratory water loss is much less in animals with body temperatures similar to the ambient temperature; in this case the air has only to be humidified at ambient temperature. A reptile with a low metabolic rate and, therefore, a low ventilation rate and with a body temperature equal to the ambient temperature will have only minimal rates of water loss via the lung. This gives reptiles an advantage over mammals in regions where water is scarce.

 What is the effect of exercise on water flux in freshwater and marine teleosts?

OSMOREGULATORS AND OSMOCONFORMERS

Animals that maintain an internal osmolarity different from the medium in which they are immersed have been termed **osmoregulators.** An animal that does not actively control the osmotic condition of its body fluids and instead conforms to the osmolarity of the ambient medium is termed an **osmoconformer.** Table 14-1 reveals these two extremes of adaptation. Most vertebrates, with the notable exception of elasmobranchs and hagfish, are strict osmoregulators, maintaining the composition of the body fluids within a small osmotic range. Although some osmotic differences do exist among vertebrate species, the blood of vertebrates is **hyposmotic** (or slightly hyperosmotic, as in sharks) to seawater and significantly hyperosmotic to freshwater. This is true, as well, of fishes that migrate between freshwater and saltwater environments; these employ endocrine mechanisms to meet the changing osmotic stresses accompanying environmental change.

Many terrestrial invertebrates also osmoregulate to a large degree. Aquatic, brackish-water, and marine invertebrates are, of course, exposed to various environmental osmolarities. Marine invertebrates, as a rule, are in osmotic balance with seawater, and the ionic concentrations in their body fluids generally parallel the seawater in which the species live. This similarity has allowed the use of seawater as a physiological saline in studies of the tissues of marine species. For example, some large neurons removed

from marine invertebrates continue to function for many hours when placed in seawater. Ionic concentrations in the body fluids of freshwater and terrestrial invertebrates differ from seawater; in these animals the body fluids are invariably more dilute than seawater but considerably more concentrated than freshwater.

Some aquatic invertebrates are *strict osmoregulators* like the vertebrates, some are *limited osmoregulators,* and some are *strict osmoconformers.* These classes are illustrated in Figure 14-7, in which the osmolarity of the extracellular compartment is plotted against the osmolarity of the aqueous environment. As the osmolarity of the environment changes, the osmolarity of a strict osmoconformer changes by an equal amount, paralleling the line that describes internal-external osmolar equality. In contrast, a strict osmoregulator maintains a constant internal osmolarity over a large range of external osmolarities, so as to produce a horizontal plot parallel with the abscissa. Limited osmoregulators regulate over a limited range of osmolarities and conform at other environmental osmolarities.

 Do strict osmoconformers and strict osmoregulators really exist?

Osmoconformers display a high degree of *cellular osmotic tolerance,* whereas osmoregulators maintain strict *extracellular osmotic homeostasis* in the face of the large environmental differences in electrolyte concentration. In osmoregulating animals, the internal tissues are generally not able to cope with more than minor changes in extracellular osmolarity and must depend entirely on osmotic regulation of the extracellular fluid to maintain cell volume. The cells of osmoconformers, on the other hand, can cope with high plasma osmolarities by increasing their intracellular osmolarities, thereby maintaining cell volume. This is

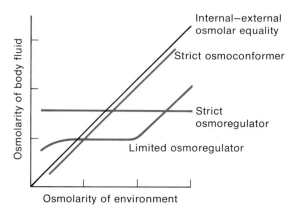

Figure 14-7 Aquatic animals can be classified into three groups based on the relationship between the osmolarity of their body fluids and that of the environment. In these plots of internal versus external osmolarity, the behavior of a strict osmoconformer parallels the line representing equality of internal-external osmolarity (black line).

achieved by increasing the concentration of intracellular organic **osmolytes,** which are substances that by their presence in high concentrations act to increase intracellular osmolarity. The use of such substances reduces the need to maintain osmotic pressure with inorganic ions, which could give rise to other problems (e.g., lower enzyme efficiency). In some marine vertebrates and invertebrates, organic osmolytes also are present in the blood and interstitial fluids, as well as inside cells, so that both extracellular and intracellular osmolarity are brought close to that of seawater. The best-known examples of such organic osmolytes are *urea* and *trimethylamine oxide,* both utilized by various marine elasmobranchs, the primitive coelacanth fish *Latimeria,* and the crab-eating, brackish-water frog *Rana cancrivora* of Southeast Asia (see Table 14-1, page 572).

OSMOREGULATION IN AQUEOUS AND TERRESTRIAL ENVIRONMENTS

Animals face quite distinct osmotic problems in aqueous and terrestrial environments. In this section, we first discuss osmoregulation by water-breathing animals and then consider air-breathing animals. Figure 14-8 presents an overview of water and salt exchange in various osmoregulating animals.

Water-Breathing Animals

Many aquatic animals find themselves and all their respiratory surfaces immersed in water. The osmolarities of aqueous environments range from a few milliosmoles per liter in freshwater lakes to about 1000 mosm·L^{-1} in ordinary seawater, or even more in landlocked salt seas. Intermediate environments, such as brackish bogs, marshes, and estuaries, have salinities ranging between these extremes. As a rule, the body fluids (i.e., interstitial fluids and blood) tend away from the environmental osmotic extremes. **Euryhaline** aquatic animals can tolerate a wide range of salinities, whereas **stenohaline** animals can tolerate only a narrow osmotic range. In this section, we consider the nature of the osmotic problems faced by freshwater and marine animals and their mechanisms for dealing with them.

Freshwater animals

The body fluids of freshwater animals, including invertebrates, fishes, amphibians, reptiles, and mammals, are generally hyperosmotic to their aqueous surroundings (see Table 14-1). Freshwater vertebrates have blood osmolarities in the range of 200 to 300 mosm·L^{-1}, while the osmolarity of freshwater is generally much less than 50 mosm·L^{-1}. Because they are hyperosmotic to their aqueous surroundings, freshwater animals face two kinds of osmoregulatory problems:

- They are subject to swelling by movement of water into their bodies owing to the osmotic gradient

- They are subject to the continual loss of body salts to the surrounding medium, which has a low salt content.

Thus freshwater animals must prevent the net gain of water and net loss of salts, which they accomplish by several means.

One way to avoid a net gain of water is production of a dilute urine. Among closely related fishes, for example, those that live in freshwater produce a more copious (i.e., plentiful and hence dilute) urine than their saltwater relatives (see Figure 14-8). The useful salts are largely retained by reabsorption into the blood from the ultrafiltrate in the tubules of the kidney, and thus a dilute urine is excreted. Nonetheless, some salts pass out in the urine, so there is a potential problem of gradually washing out biologically important salts such as KCl, NaCl, and $CaCl_2$. Lost salts are replaced, in part, from ingested food. An important specialization for salt replacement in freshwater animals is active transport of salt from the external dilute medium across the epithelium into the interstitial fluid and blood. This activity is accomplished across transporting epithelia such as those in the skin of amphibians and in the gills of fishes. In fishes and many aquatic invertebrates, the gills act as the major osmoregulatory organs, having many of the functions located in the kidneys of mammals.

Freshwater animals have remarkable abilities to take up salts from their dilute environment. Freshwater fishes are able, for example, to extract Na^+ and Cl^- ions with their gills from water containing less than 1 mM NaCl, even though the plasma concentration of the NaCl exceeds 100 mM (Figure 14-9A). Thus, the active transport of NaCl in the gills takes place against a concentration gradient in excess of 100-fold. The mechanism of sodium reabsorption appears to be similar in the gills of freshwater fishes, frog skin, turtle bladder, and the mammalian kidney. In all cases the cells of these epithelia are joined together by tight junctions. Transport of Na^+ into these cells is dependent on an electrogenic proton ATPase, which actively transports protons out of the cells into the environment. The mechanism of sodium reabsorption is discussed in detail later.

In some freshwater animals, including fishes, reptiles, birds, and mammals, water uptake and salt loss are minimized by an integument having low permeabilities to both salts and water. As a general rule, animals living in freshwater refrain from drinking fresh water, reducing the need to expel excess water.

Marine animals

In general, the intracellular and extracellular body fluids of marine invertebrates and the ascidians (primitive chordates) are close to seawater both in osmolarity (**isosmotic**) and in the plasma concentrations of the individual major inorganic salts (see Table 14-1). Such animals therefore need not expend much energy in regulating the osmolarity of their body fluids. A rare example of a vertebrate whose plasma is also isosmotic to the environment is the hagfish. It differs from most marine invertebrates, however, in that it does regulate the concentrations of individual ions. In particular, blood Ca^{2+}, Mg^{2+}, and SO_4^{2-} are maintained at

Type of animal	Blood concentration relative to environment	Urine concentration relative to blood	Osmoregulatory mechanisms
Marine elasmobranch	Isotonic	Isotonic	Does not drink seawater / Hypertonic NaCl from rectal gland
Marine teleost	Hypotonic	Isotonic	Drinks seawater / Secretes salt from gills
Freshwater teleost	Hypertonic	Strongly hypotonic	Drinks no water / Absorbs salt with gills
Amphibian	Hypertonic	Strongly hypotonic	Absorbs salts through skin
Marine reptile	Hypotonic	Isotonic	Drinks seawater / Hypertonic salt-gland secretion
Desert mammal	—	Strongly hypertonic	Drinks no water / Depends on metabolic water
Marine mammal	Hypotonic	Strongly hypertonic	Does not drink seawater
Marine bird	—	Weakly hypertonic	Drinks seawater / Hypertonic salt-gland secretion / Weakly hypertonic urine
Terrestrial bird	—	Weakly hypertonic	Drinks fresh water

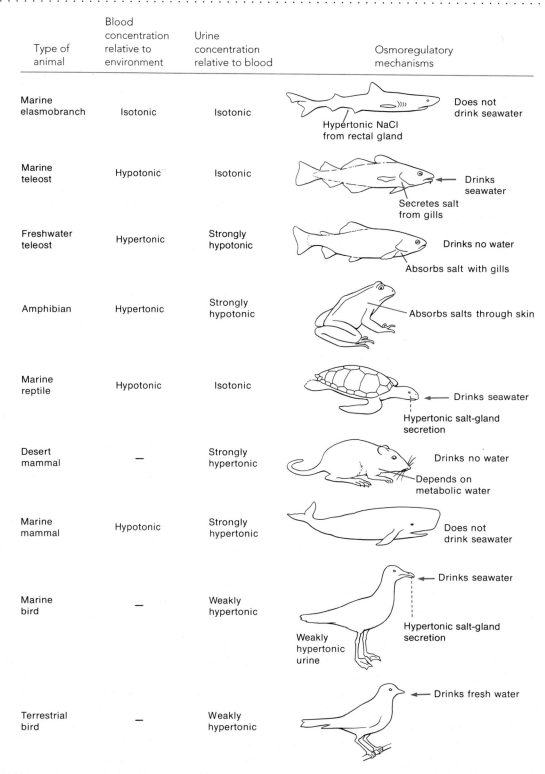

Figure 14-8 Animals living in different environments exhibit various osmoregulatory mechanisms. The active exchange of water and salt in some vertebrates is illustrated here. Passive loss of water through the skin, lungs, and alimentary tract is not indicated.

significantly lower concentrations than they are in seawater, whereas Na⁺ and Cl⁻ are higher. Since various functions of excitable tissues such as nerve and muscle of vertebrates are especially sensitive to the concentrations of Ca^{2+} and Mg^{2+}, the regulation of these divalent cations may have evolved to accommodate the requirements of neuromuscular function.

Like the hagfish, the cartilaginous fishes such as sharks, rays, and skates, as well as the primitive coelacanth *Latimeria*, have plasma that is approximately isosmotic to

A

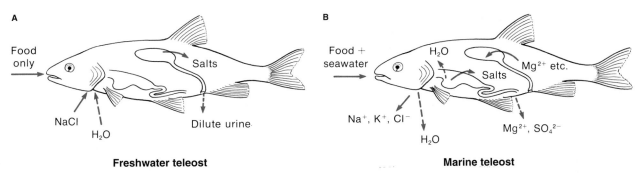

Freshwater teleost

B

Marine teleost

Figure 14-9 Salt and water exchange is in the opposite direction in freshwater and marine teleosts. **(A)** Freshwater teleosts prevent the net gain of water and loss of salt by copious excretion of a dilute urine from which most of the salt has been reabsorbed. **(B)** Marine teleosts face the opposite osmotic problems, namely, avoiding a water deficit and excess of salts. They achieve this by drinking seawater and then eliminating salts by several routes. Solid arrows indicate active processes; broken arrows, passive processes. Note the active role of the gills in salt transport in both groups. [Adapted from Prosser, 1973.]

seawater. They differ, however, in that they maintain far lower concentrations of electrolytes (i.e., inorganic ions), making up the difference with organic osmolytes such as urea and trimethylamine oxide (TMAO). High urea concentrations tend to cause the breakup of proteins into constituent subunits, whereas TMAO has the opposite effect, canceling the effect of urea and stabilizing protein structure in the face of high urea levels. In the elasmobranchs and coelacanths, excess inorganic electrolytes such as NaCl are excreted via the kidneys and also by means of a special excretory organ, the *rectal gland*, located at the end of the alimentary canal.

The body fluids of marine teleosts (modern bony fishes), like those of most higher vertebrates, are hypotonic to seawater, so there is a tendency for these fishes to lose water to the environment, especially across the gill epithelium. To replace the lost volume of water, they drink saltwater. Most of the net salt uptake is due to drinking seawater rather than salt uptake across the body surface or gills. By absorption across the intestinal epithelium, 70% to 80% of the ingested water enters the bloodstream, along with most of the NaCl and KCl in the seawater. Initially the ingested seawater is diluted by about 50% by diffusional uptake of salts across the esophagus. Active salt uptake occurs in the small intestine, via a Na/2Cl/K cotransporter across the apical membrane and then by active transport via a Na^+/K^+ ATPase across the basolateral membrane. Left behind in the gut and expelled through the anus are most of the divalent cations such as Ca^{2+}, Mg^{2+}, and SO_4^{2-}. The excess salt absorbed along with the water is subsequently eliminated from the blood by active transport of Na^+, Cl^-, and some K^+ across the gill epithelium into the seawater, and by secretion of divalent salts by the kidney (see Figure 14-9B). The urine is isotonic to the blood, but rich in those salts (especially Ca^{2+}, Mg^{2+}, and SO_4^{2-}) that are not secreted by the gills. The net result of the combined osmotic work of gills and kidneys in the marine teleost is a net retention of water.

The gills of marine teleosts, as might be expected, are organized differently from those of freshwater fish. In ma-

rine teleosts, the gill epithelium contains specialized cells, called **chloride cells,** that mediate transport of NaCl from the blood into the surrounding water. The mechanism of this transport, which makes it possible for these fishes to live in saltwater, is described in a later section.

 What is, and how would you estimate, the energetic cost of osmoregulation. Would you expect the cost to be less if freshwater and marine fish were placed in an isosmotic solution?

Some teleost species—for example, the salmon of the Pacific Northwest and eels in eastern North America and Europe—are able to maintain a more-or-less constant plasma osmolarity even though they migrate between marine and freshwater environments. Such fish undergo a physiological adaptation that enable them to maintain a more-or-less constant ionic composition in both environments. Some of the physiological changes that occur when teleosts migrate from freshwater to seawater begin before the animal enters seawater. Eels, for instance, reduce the permeability of the integument, changing from yellow to silver in the process. Likewise, the gill reorganization that characterizes adaptation of salmon to seawater begins as the fish migrate down river to the ocean. We'll discuss the adaptation of migrating teleosts in detail later in this chapter.

To summarize, freshwater animals tend to take in water passively and to remove it actively through the osmotic work of kidneys (vertebrates) or kidney-like organs (invertebrates). They lose salts to the dilute environment and replace them by actively absorbing ions from the surrounding fluids into their bodies through skin, gills, or other actively transporting epithelia. Marine fish, on the other hand, lose water osmotically through the gills or through the integument, if it is permeable. To replace lost water, marine fish drink seawater and actively secrete the excess salt ingested with the seawater back into the environment. This process takes place through active transport in extrarenal osmoregulatory organs such as gills and the rectal gland.

Air-Breathing Animals

Animals in a terrestrial environment can be thought of as submerged in an ocean of air rather than water. Unless the humidity of the air is high, animals having a water-permeable epithelium will be subject to dehydration very much as if they were submerged in a hypertonic medium such as seawater. Dehydration would be avoided if all epithelial surfaces exposed to air were totally impermeable to water. The evolutionary process has not found this to be a feasible solution to the problem of desiccation, since an epithelium that is impermeable to water (and thus dry) will have limited permeability to oxygen and carbon dioxide, and will thus be unsuited for the respiratory needs of a terrestrial animal. As a consequence, air-breathing animals are subject to dehydration through their respiratory epithelia. Air-breathing animals utilize various means to minimize water loss into the air by this route and others (see Figure 14-8).

Marine reptiles (e.g., iguanas, estuarine sea turtles, crocodiles, sea snakes) and marine birds drink seawater to obtain a supply of water but, like marine teleosts, are unable to produce a concentrated urine that is significantly hyperosmotic to their body fluids. Instead, they are endowed with glands specialized for the secretion of salts in a strong hyperosmotic fluid. These salt glands are generally located above the orbit of the eye in birds and near the nose or eyes in lizards. Brackish-water crocodiles were long suspected of using extrarenal means of excreting salts, and eventually salt glands were discovered in the tongue of this reptile. Although neither reptilian nor avian kidneys are capable of producing a very hypertonic urine, the salt glands of marine reptiles and birds secrete a sufficiently concentrated salt solution to enable them to drink saltwater even though their kidneys are unable to produce urine more concentrated than seawater (Figure 14-10A). Salt glands compensate in these groups for the inability of their kidney to produce a urine that is strongly hypertonic relative to body fluids. Marine mammals, which lack salt glands or similar specializations, avoid drinking seawater, get their water entirely from their food intake and its subsequent metabolism, and depend primarily on their kidneys for maintaining osmotic balance.

Human beings, like other mammals, are not equipped to drink seawater. The human kidney can remove up to about 6 g of Na^+ from the bloodstream per liter of urine produced. Because seawater contains about 12 g·L^{-1} of Na^+, imbibing seawater will cause a human being to accumulate salt without adding a physiologically equivalent amount of water (Figure 14-10B). Stated differently, to excrete the salt ingested with a given volume of seawater, the human kidney must pass more water than is contained in that volume. This, of course, will lead rapidly to dehydra-

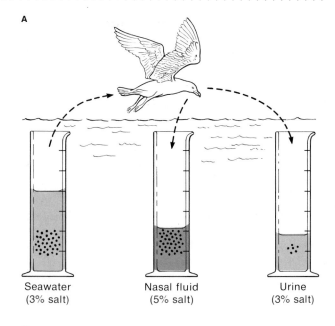

A

Seawater
(3% salt)

Nasal fluid
(5% salt)

Urine
(3% salt)

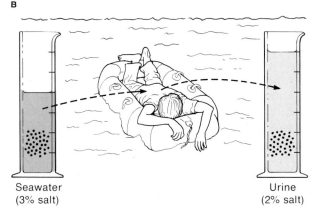

B

Seawater
(3% salt)

Urine
(2% salt)

Figure 14-10 Marine reptiles and birds drink seawater to obtain water, whereas most mammals become dehydrated if they drink seawater. **(A)** When marine birds drink seawater, they secrete NaCl via the salt glands, thereby eliminating 80% of the ingested salt along with only 50% of the ingested water. As a result, these birds can produce a hypotonic urine without dehydrating. **(B)** When humans and other mammals, who lack salt glands, drink seawater, they cannot concentrate the urine sufficiently to conserve water while eliminating the ingested salt. Like terrestrial mammals, marine mammals cannot drink seawater; these animals use various water-conservation mechanisms to survive. [From "Salt Glands" by K. Schmidt-Nielsen. Copyright © 1959 by Scientific American, Inc. All rights reserved.]

tion. Thus humans stranded at sea will die unless they have access to freshwater. They cannot replace lost water by drinking seawater. If they do so, it will only make the problem worse. Humans require a constant source of fresh drinking water to excrete accumulated salts and metabolic waste products.

Mammals cannot drink seawater, and yet marine mammals such as pinnipeds (e.g., sea lions, seals) and cetaceans (e.g., porpoises, whales) live in the ocean, even though they do not have extrarenal salt-secreting organs like the salt glands of birds and reptiles. Camels and many other mammals can survive in the desert. Thus, unlike humans, many mammals can survive in habitats where drinking water is not available. Joseph Priestley (1733–1804), who was the first to isolate many gases, including oxygen, observed that he could keep mice alive without water for three or four months, in a cage on a shelf above the kitchen fireplace in his home in Yorkshire, England. That is, he observed that mice could survive for many months without drinking water. Another small mammal, the kangaroo rat, *Dipodomys merriami*, a native of the American Southwest, has become a classic example of how small mammals survive without drinking water in the arid conditions of the desert. Let's see how these mammals, as well as certain terrestrial arthropods, survive in the absence of freshwater.

Desert-living mammals

The survival strategies practiced by the kangaroo rat exemplify a variety of osmoregulatory adaptations characteristic of many small desert mammals (Figure 14-11). The kangaroo rat and other desert mammals are faced with a physiological double jeopardy—excessive heat and near absence of free freshwater. Water regulation and temperature regulation are, of course, closely related, since one important means of channeling excess heat out of the body into the environment is by evaporative cooling. Since evaporative cooling is at odds with water conservation, most desert animals cannot afford this method and have devised means of circumventing it. The kangaroo rat, like many desert mammals, avoids much of the daytime heat by remaining in a burrow during daylight hours and coming out only at night. This nocturnal lifestyle is an important and widespread behavioral adaptation to desert life. Not only does the cool burrow reduce the animal's temperature load, but it reduces respiratory water loss. The nasal countercurrent mechanism for conserving respiratory moisture depends, of course, on the ambient temperature in the burrow being significantly lower than the 37°C to 40°C characteristic of the core temperatures of birds and mammals. If the rodent ventures out of its cool burrow into air close to its own temperature, its respiratory water loss will rise

abruptly, since the cooling properties of the nasal epithelium will be reduced. Desert mammals also generally avoid heat-generating exercise during the day, when removal of excess heat from the body is slowed by the higher ambient temperature. Because of its efficient kidneys, the kangaroo rat excretes a highly concentrated urine, and rectal absorption of water from the feces results in essentially dry fecal pellets.

By using all these adaptations for desert survival, the kangaroo rat greatly reduces its potential water loss. In spite of this extreme osmotic economy, the small amount of lost water must, of course, be replaced, or the animal will eventually dry up. Since the kangaroo rat eats dry seeds that contain only a trace of free water, is not known to drink, and in fact survives quite well in the near absence of free water, it must have a cryptic source of water. This, it turns out, is the metabolic water noted earlier (see Table 14-5). The exquisite conservation of water by the kangaroo rat allows it to survive primarily on the water produced during the oxidation of foods, so that over the long term water gains equal water losses (Table 14-7).

Unlike the kangaroo rat, camels are too large to hide from the hot desert sun in burrows. When deprived of drinking water, camels do not sweat but allow their body temperature to rise rather than losing water by evaporative cooling during the heat of the day. During the cooler night, the camel's body temperature drops and increases only slowly the next day because of the animal's large body mass

TABLE 14-7
Sources of water gain and loss by the kangaroo rat

Gains		Losses	
Metabolic water	90%	Evaporation + perspiration	70%
Free water in "dry" food	10%	Urine	25%
Drinking	0%	Feces	5%
	100%		100%

Source: Schmidt-Nielsen, 1972.

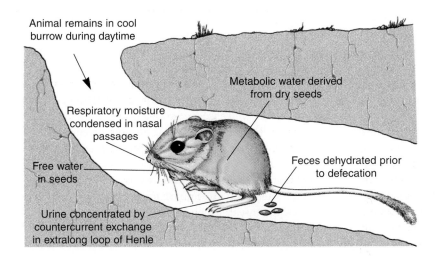

Figure 14-11 The water-conserving strategies of the kangaroo rat are characteristic of many small desert dwellers.

Animal remains in cool burrow during daytime

Metabolic water derived from dry seeds

Respiratory moisture condensed in nasal passages

Free water in seeds

Feces dehydrated prior to defecation

Urine concentrated by countercurrent exchange in extralong loop of Henle

A

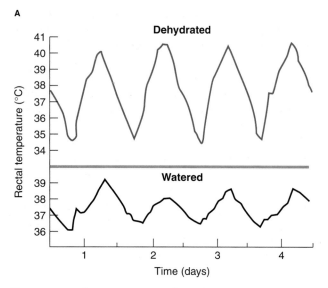

Dehydrated

Watered

B

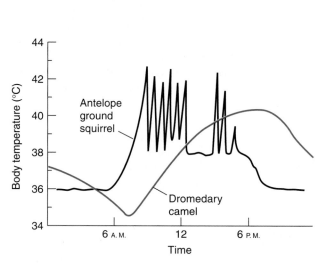

Figure 14-12 When water is scarce, large desert animals such as the camel exhibit a large, but slow, increase in body temperature during the day, whereas smaller animals heat up rapidly when exposed to the sun. **(A)** The daily temperature fluctuation in a well-watered camel and in a dehydrated one. When the camel is deprived of drinking water, the daily fluctuation may increase to as much as 7°C. This has a great influence on the use of water for temperature regulation. **(B)** Diagrammatic representation of the daily patterns of body temperature in a large and a small mammal subjected to heat stress under desert conditions. Small animals must enter burrows periodically to avoid overheating. [Part A from Schmidt-Nielsen, 1963; part B from Bartholomew, 1964.]

and thick fur, which acts as a heat shield. Nevertheless, the body temperature of a dehydrated camel may vary from 35°C at night to 41°C during the day (Figure 14-12A). This strategy of heating during the day and cooling at night is impossible in small rodents, whose body temperatures oscillate much more rapidly than in the larger camel (Figure 14-12B). Because of their small size, desert rodents heat up rapidly in the sun and must return to their burrow to cool down. The camel also reduces heating by orienting to give minimal surface exposure to direct sunlight. The camel, like other desert animals, produces dry feces and concentrated urine. When water is not available the camel does not produce urine but stores urea in the tissues. The camel can tolerate not only dehydration but also high urea levels in the body. When water becomes available, these ships of the desert can rehydrate by drinking 80 liters in 10 minutes.

Marine mammals
Marine mammals face problems similar to those of desert animals because they live in an environment without available drinking water. Water, water everywhere and not a drop to drink! The physiological responses of marine mammals, although different in detail, are generally similar to those of desert mammals. The emphasis is on water conservation. They are endowed, as are other mammals, with highly efficient kidneys capable of producing a very hypertonic urine. Seals have a characteristic labyrinth-like proliferation of epithelial surfaces in the nasal passages which reduces water loss via breathing. Whales and dolphins have a blow hole rather than the typical mammalian nose. These animals have large lung tidal volumes. The velocity of air

flow through the blow hole is high because both inspiration and expiration are rapid, and large volumes of gas are moved with each breath. It is possible that the expansion of air passing through the blow hole of a whale also cools the air, resulting in water condensation in the region of the blow hole that can be used to wet inspired air. This would reduce water loss via ventilation.

 What are the osmoregulatory problems faced by the young camels and whales? Can you suggest possible solutions to these problems?

A remarkable example of water retention in a marine mammal faced with desert-like problems of water conservation occurs in a recently weaned baby elephant seal. After being abandoned by its mother, the baby seal goes for 8–10 weeks without food or water. During this time the baby seal's only source of water is that derived from the oxidation of its body fat. It weighs about 140 kg at the time of weaning and loses only about 800 g of water per day, of which less than 500 g is lost through respiration. This economy is ascribed both to its nasal countercurrent heat exchanger and to a slowing of metabolic rate, which allows it to stop breathing for 40 minutes and then alternate with 5 minutes of deep breathing. The ability to suspend breathing is, of course, not uncommon for marine mammals such as the elephant seal, which can dive for prolonged periods. The ability to conserve water is also seen in the adult elephant seal. The large males spend up to three months on the beach and when on land do not drink or eat. The females suckle their young for about four weeks on the beach and then the pup is abandoned when the female goes to sea for four months. She returns for about a month to molt and then goes to sea for another six months. While at sea, the female elephant seal does not drink but relies on the water in fish in her diet and metabolic water to maintain her water requirements.

Terrestrial arthropods

Certain terrestrial arthropods have the ability to extract water vapor directly from the air, even, in some species, when the relative humidity is as low as 50% (Table 14-8). To date, this poorly understood ability has been demonstrated only in certain arachnids (ticks, mites) and in a number of wingless forms of insects, primarily larvae. Those species that exhibit this ability live in habitats devoid or nearly devoid of free water. The ability to remove water from the air is all the more remarkable in these arthropods because it normally occurs even when the vapor pressure of the hemolymph exceeds that of the air, which it does at all values of relative humidity below 99%.

TABLE 14-8
Critical equilibrium humidities for some arthropods that can extract water from the vapor phase

	Relative humidity
Arachnida	
Ixodes ricinus	92.0
Rhipicephalus sanguineus	84.0–90.0
Insecta	
Thermobia domestica	45.0
Tenebrio molitar larvae	88.0

Note: At relative humidities below critical, the animal is unable to extract moisture from the air.

Source: Edney and Nagy, 1976.

The water vapor pressure associated with a solution decreases with increasing ionic content, so highly concentrated salt solutions will absorb water from air. Insects take advantage of this by creating very concentrated solutions that can absorb water from air. In insects that extract water from air, the site of entry is often the rectum, which reduces the water content of fecal matter to remarkably low levels. As water is removed from the feces, the latter can take on new water from the air, if the water vapor pressure is high enough and if the air has access to the rectal lumen. In ticks, tissues in the mouth have been implicated in the uptake of water vapor. Here it appears that the salivary glands secrete a highly concentrated solution of KCl that in turn absorbs water from the air.

OSMOREGULATORY ORGANS

The osmoregulatory capabilities of metazoans depend to a great extent on the properties of *transport epithelia* located in gills, skin, kidneys, and gut. The highly specialized epithelial cells composing these epithelia differ from all other cell types in being anatomically and functionally *polarized*. The *apical surface* (sometimes referred to as the mucosal or luminal surface) of an epithelial cell faces a space that is continuous with the external world (the sea, the pond, the lumen of the gut, the lumen of a kidney tubule, etc.). The other side of the epithelial cell, the *basal surface* (sometimes referred to as the serosal surface) generally bears deep basal clefts and faces the internal compartment containing extracellular fluid. This internal compartment is the one that contains all the other cells of the remaining body tissues. These exist, so to speak, in their own private "pond," composed of the extracellular fluid in which they are bathed. The proper composition of this internal pond depends on the osmoregulatory work and barrier functions performed by epithelial cells.

The excretion of nitrogenous wastes varies among species depending on water availability. The nature of the nitrogenous end product varies and many different organs are involved in the excretion of ammonia, urea, and/or uric acid. In freshwater fish, for example, ammonia is usually the major nitrogenous end product and the gills are the major site of excretion. In mammals, by contrast, the major nitrogenous end product is urea and the kidneys the site of excretion. Because the excretion of nitrogenous end products is variable and not organ specific, it is discussed at the end of this chapter.

The mechanisms for transporting substances across epithelia were discussed in Chapter 4 and the same basic cellular machinery is used in all excretory or osmoregulatory organs. For example, similar salt-excreting cells are found in the nasal gland of birds and reptiles, the mammalian kidney, the rectal gland of elasmobranchs, and the gills of marine teleost fishes. Not only are the cells similar but their activities are regulated by similar arrays of hormones. The detailed functioning of organs with similar cellular structure may be different because of the gross organization of the organ. The capabilities of transport epithelia are greatly enhanced in osmoregulatory organs by their anatomic organization, as is exquisitely evident in the kidneys of mammals. Here, in addition to a high degree of cellular differentiation for transepithelial transport, the epithelium is organized into tubules arranged so as to enhance the transport efficiency of the tubular epithelium. This combination of cell function and tissue organization has produced a marvelously efficient osmoregulatory and excretory organ. The next several sections describe and compare the operation of various types of osmoregulatory organs found in different animals.

MAMMALIAN KIDNEY

The mammalian kidney is the osmoregulatory organ of which we have the most complete understanding, thanks to intensive research over the past four to five decades. The mammalian kidney performs certain functions that in lower vertebrates are shared by other organs such as the skin and bladder of amphibians, the gills of fishes, and the salt glands of reptiles and birds. Thus, the mammalian kidney is not representative of all vertebrate kidneys, which are organized somewhat differently in different groups of vertebrates.

Anatomy of the Mammalian Kidney

The gross anatomy of the mammalian kidney is shown in Figure 14-13. Each individual normally has two kidneys, one located on each side against the dorsal inner surface of the lower back, outside the peritoneum. In view of their small size (about 1% of total body weight in humans), the kidneys receive a remarkably large blood flow, equivalent to about 20%–25% of the total cardiac output. The kidney filters the equivalent of the blood volume every 4–5 minutes. The outer functional layer, the cortex, is covered by a tough capsule of connective tissue. The inner functional layer, the medulla, sends papillae projecting into the pelvis. The pelvis gives rise to the **ureters**, which empty into the urinary bladder. The urine leaves the bladder during **micturition** (urination) via the **urethra,** which leads to the end of the penis in males and into the vulva in females.

Human adults produce about a liter of slightly acidic (pH approximately 6.0) urine each day. Urine production rates vary diurnally, being high during the day and low at night, reflecting the time course of water intake and production of metabolic water. Urine contains water and other by-products of metabolism, such as urea, as well as NaCl, KCl, phosphates, and other substances that are present at concentrations in excess of the body's requirements. The objective is to maintain a more-or-less constant body composition; hence the volume and composition of urine reflects the volume of fluid taken in and the amount and composition of ingested food. The actual volume of urine produced is determined by the volume of water ingested plus the water produced during metabolism minus evaporative water loss via the lungs and sweating and, to a lesser extent, that lost with the feces. When voided, urine is normally clear and transparent, but after a rich meal the urine may become alkaline and slightly turbid. The smell and color of urine is determined by the diet. For example, consumption of methylene blue will give urine, which typically is yellow, a distinctive blue color, and consumption of asparagus will completely change the more usual, slightly aromatic odor of urine.

The release of urine is accomplished by the simultaneous contraction of the smooth muscle of the bladder wall and the relaxation of the skeletal muscle sphincter around the opening of the bladder. As the bladder wall is stretched by gradual filling of the bladder, stretch receptors in the wall of the bladder generate nerve impulses that are carried by sensory neurons to the spinal cord and brain, producing the "associated" sensation of fullness. The sphincter can then be relaxed by inhibition of motor impulses, allowing the smooth muscle of the bladder wall to contract under autonomic control and empty the contents. The presence of a bladder allows the controlled release of stored urine

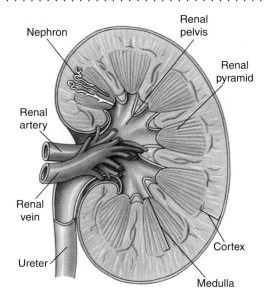

Figure 14-13 The functional units of the mammalian kidney, called nephrons, are arranged in a radiating fashion within the renal pyramids. The distal end of each nephron within a pyramid empties into a collecting duct, which passes through a papilla into a calyx. The renal calyces drain in a central cavity termed the renal pelvis. The urine passes from the pelvis into the ureter, which takes it to the urinary bladder. In this cross-sectional drawing, only one nephron is depicted, although each pyramid contains many nephrons.

rather than a continual dribble paralleling the flow of urine from the kidney into the bladder. Such controlled release is used by some animals to mark out their territory.

The functional unit of the mammalian kidney is the **nephron** (Figure 14-14), an intricate epithelial tube that is closed at its beginning but open at its distal end. Each kidney contains numerous nephrons, which empty into **collecting ducts.** These ducts combine to form papillary ducts, which eventually empty into the renal pelvis. At the closed end, the nephron is expanded—somewhat like a balloon that has been pushed in from one end toward its neck—to form the cup-shaped **Bowman's capsule.** The lumen of the capsule is continuous with the narrow lumen that extends through the renal tubule. A tuft of capillaries forms the renal **glomerulus** inside Bowman's capsule. This remarkable structure is responsible for the first step in urine formation. An **ultrafiltrate** of the blood passes through the single-cell layer of the capillary walls, through a basement membrane, and finally through another single-cell layer of epithelium that forms the wall of Bowman's capsule. The ultrafiltrate accumulates in the lumen of the capsule to begin its trip through the various segments of the renal tubule, finally descending the collecting duct and eventually into the renal pelvis.

The wall of the renal tubule is one cell layer thick; this epithelium separates the lumen, which contains the ultrafiltrate, from the interstitial fluid. In some portions of the nephron, these epithelial cells are morphologically specialized for transport, bearing a dense pile of microvilli on their luminal, or apical, surfaces and deep infoldings of their

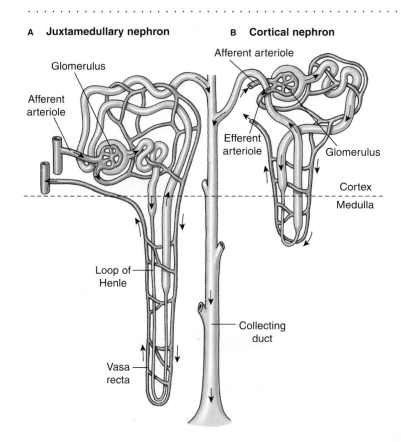

A Juxtamedullary nephron   **B Cortical nephron**

Glomerulus

Afferent arteriole

Afferent arteriole

Efferent arteriole

Glomerulus

Cortex

Medulla

Loop of Henle

Collecting duct

Vasa recta

Figure 14-14 The mammalian nephron is a long tubular structure, which is closed at its beginning in Bowman's capsule but open at its terminus, where it empties into a collecting duct. The renal tubule and collecting duct are shown in yellow; the vascular elements in red or blue. Juxtamedullary nephrons **(A)** have a long loop of Henle, which passes deep into the renal medulla and is associated with a vasa recta. The blood first passes through the capillaries of the glomerulus and then flows through the hairpin loops of the vasa recta, which plunges into the medulla of the kidney along with the loop of Henle. The more common cortical nephrons **(B)** have a short loop of Henle, only a small portion of which enters the medulla, and lack a vasa recta. In these nephrons, the blood passes from the afferent arteriole to the glomerular capillaries and then leaves the nephron via the efferent arteriole.

basal membranes (Figure 14-15). The epithelial cells are tied together by leaky tight junctions, which permit limited paracellular diffusion between the lumen and interstitial space surrounding the renal tubule.

The nephron can be divided into three main regions: the proximal nephron, the loop of Henle, and the distal nephron. The proximal nephron consists of Bowman's capsule and the **proximal tubule.** The hairpin loop of Henle comprises a descending limb and an ascending limb. The latter merges into the **distal tubule,** which joins a collecting duct serving several nephrons. The number of nephrons per kidney varies from several hundred in lower vertebrates to many thousands in small mammals, and a million or more in humans and other large species.

The loop of Henle, found only in the kidneys of birds and mammals, is considered to be of central importance in concentrating the urine. Vertebrates that lack the loop of Henle are incapable of producing a urine that is hyperosmotic to the blood. In mammals, the nephron is so oriented that the loop of Henle and the collecting duct lie parallel to each other (see Figure 14-14). The glomeruli are found in the renal cortex, and the loops of Henle reach down into the papillae of the medulla; thus the nephrons are arranged in a radiating fashion within the kidney (see Figure 14-13). Nephrons can be divided into two groups:

- *Juxtamedullary nephrons,* which have their glomeruli in the inner part of the cortex and long loops of Henle that plunge deeply into the medulla (see Figure 14-14A)

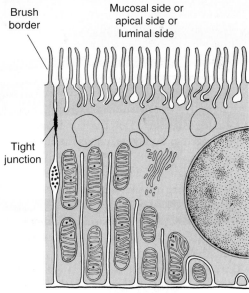

Brush border

Mucosal side or apical side or luminal side

Tight junction

Serosal side or basolateral side or blood side

Figure 14-15 The cells of the proximal tubule are specialized for the transport of salts and other substances from the luminal (apical) side to the serosal (blood) side. The apical membrane facing the lumen is thrown into fingerlike projections (microvilli), greatly increasing its surface area. This surface is referred to as a brush border. Mitochondria are concentrated near the basolateral (serosal) surface, which is thrown into deep basal clefts. These features allow the concentration of salts in the renal interstitium by active transport of salts across the basal membrane.

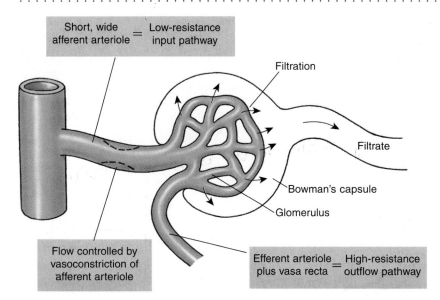

Figure 14-16 Blood pressure in the renal glomerulus is high because of the low-resistance input pathway (afferent arteriole) and the high-resistance output pathway. Regulation of glomerular blood pressure, which influences the filtration rate, is largely through modulation of the diameter of the afferent arteriole.

- *Cortical nephrons,* which have their glomeruli in the outer cortex and relatively short loops of Henle that extend only a short distance into the medulla (see Figure 14-14B)

The anatomy of the renal circulatory system is important in the function of the nephron. The renal artery subdivides to form a series of short *afferent arterioles,* one of which supplies each nephron (see Figure 14-14). The glomerular capillaries within Bowman's capsule are subjected to somewhat higher pressures than other capillaries because of the low-resistance input pathway and high-resistance output pathway (Figure 14-16). The capillaries of the glomerulus recombine to form an *efferent arteriole.* Unlike most other vessels, which would join to form veins, the efferent arteriole in juxtamedullary nephrons then subdivides again to form a second series of capillaries surrounding the loop of Henle. Thus the blood, on leaving the glomerulus located in the cortex, enters the efferent arteriole and is carried into the medulla in a descending and subsequently ascending loop of anastomosing (interconnecting) capillaries before leaving the kidney via a vein. The hairpin loops, which parallel the loops of Henle of juxtamedullary nephrons, are referred to as the **vasa recta** (see Figure 14-14A). Flow in the efferent arteriole is less than that in the afferent arteriole because around 10% of the blood is filtered across the Bowman's capsule. For humans this amounts to about a liter of filtrate formed every 10 minutes. Clearly urine flow rate is much less, so much of the initial filtrate formed in the Bowman's capsule is reabsorbed into the blood across the kidney tubule.

Urine Production

Three main processes contribute to the ultimate composition of the urine (Figure 14-17):

- *Glomerular filtration* of plasma to form an ultrafiltrate in the lumen of the Bowman's capsule

- *Tubular reabsorption* of approximately 99% of the water and most of the salts from the ultrafiltrate leaving behind and concentrating waste products such as urea

- *Tubular secretion* of a number of substances via active transport in nearly all instances

Formation of the ultrafiltrate is the initial step in urine production; reabsorption and secretion occur along the length of the renal tubule. In addition to these processes, the excretion of nitrogenous wastes, discussed at the end of the

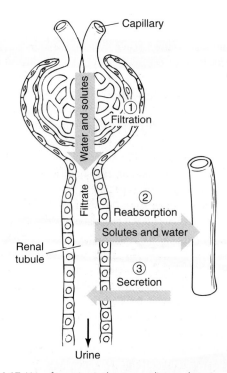

Figure 14-17 Urine formation in the mammalian nephron involves three main processes. Filtration, the initial step, takes place in Bowman's capsule, followed by reabsorption and secretion, which occur along the renal tubule. The final product of these processes is a hypertonic urine, whose composition differs from that of blood.

chapter, involves the synthesis of certain excreted products in the tubular cells and lumen.

Glomerular filtration

The glomerular ultrafiltrate contains essentially all the constituents of the blood except for blood cells and nearly all blood proteins. Filtration in the glomerulus is so extensive that 15%–25% of the water and solutes are removed from the plasma that flows through it. The glomerular ultrafiltrate is produced at the rate of about 125 ml·min⁻¹, or about 180 L·day⁻¹, in human kidneys. When this number is compared to the normal intake of water, it is evident that unless most of the glomerular ultrafiltrate is subsequently reabsorbed into the bloodstream, the body would be quickly dehydrated, thus much of the ultrafiltrate must be reabsorbed.

 What is the advantage of filtering the equivalent of the blood volume every 4 or 5 minutes and then reabsorbing most of the salts and water? Why not simply excrete toxic waste by a process of secretion?

The process of ultrafiltration in the glomerulus (Figure 14-18) depends on three factors: (1) the net hydrostatic pressure difference between the lumen of the capillary and the lumen of Bowman's capsule, which favors filtration; (2) the colloid osmotic pressure, which opposes filtration; and (3) the hydraulic permeability (sievelike properties) of the three-layered tissue separating these two compartments. The net pressure gradient results from the sum of the hydrostatic pressure difference between the two compartments and the colloid osmotic pressure difference. The latter arises because of the separation of proteins during the filtration process. In humans, the proteins remaining in the capillary plasma produce an osmotic pressure difference of about −30 mm Hg, and the hydrostatic pressure difference (capillary blood pressure minus the back pressure in the lumen of the Bowman's capsule) is about +40 mm Hg (Table 14-9). The result is a net filtration pressure of only about +10 mm Hg. This small pressure differential acting on the high permeability of the glomerular sieve produces a phenomenal rate of ultrafiltrate formation by the millions of glomeruli in each human kidney. It is important to note that the filtration process in the kidney is entirely passive, depending on hydrostatic pressure that derives its energy from the contractions of the heart. In lower vertebrates such as the salamander, the blood pressure in the glomerular capillaries is much lower than in humans, but the net filtration pressure is not that much less than in the human kidney, because of a lower intracapsular and osmotic pressure in the salamander (see Table 14-9).

TABLE 14-9
Balance sheet of pressures (in mm Hg) involved in glomerular ultrafiltration as illustrated in Figure 14-20

	Salamandar	Man
Glomerular capillary pressure	17.7	55
Intracapsular pressure	−1.5	−15
Net hydrostatic pressure	16.2	40
Colloid osmotic pressure	−10.4	−30
Net filtration pressure	5.8	10

Source: Pitts, 1968; Brenner et al., 1971.

Fluid filtered from the blood into the Bowman's capsule must cross the capillary wall, then the basement membrane, and finally the inner layer of the capsule. The glomerulus consists of fenestrated capillaries, which contain many large pores and are about 100 times more permeable than the continuous capillaries found in other parts of the body (see Figure 12-37). The basement membrane contains collagen for structural purposes and negatively charged glycoproteins that repel albumin and other negatively charged proteins. The hydraulic properties of the glomerular apparatus depend primarily on the sievelike properties of *filtration slits*, formed by a rather remarkable assemblage of fine cellular processes termed *pedicels*. These extend from larger processes of *podocytes* ("foot cells"), the cells composing the visceral layer of Bowman's capsule (Figure 14-19A). The pedicels are aligned in an array covering the endothelium (vascular epithelium) of glomerular capillaries. These fingerlike processes interdigitate so as to leave very small

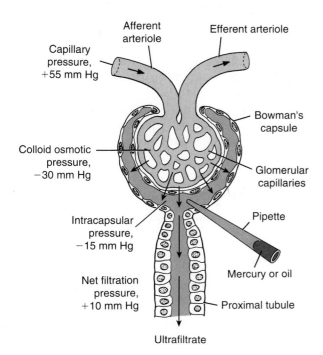

Figure 14-18 The net hydrostatic pressure affecting the glomerular filtration is determined by the sum of the various forces indicated at the left. Samples of the glomerular filtrate can be obtained by insertion of a micropipette, as shown on the right. The mercury in the pipette is pushed to the tip by pressure before penetration of the capsule. A sample is then sucked into the calibrated tip for subsequent microanalysis. [Adapted from Hoar, 1975.]

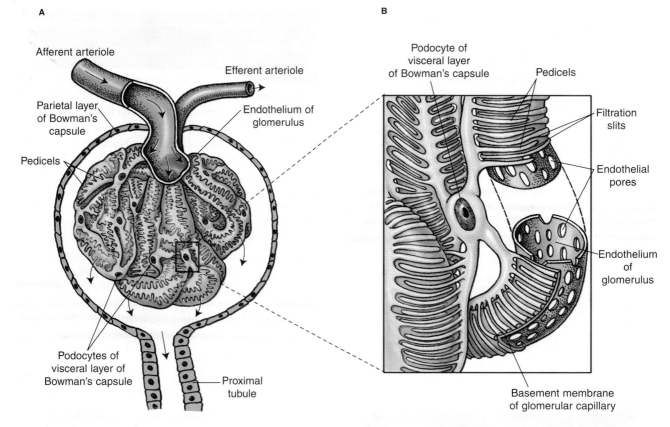

Figure 14-19 The inner (visceral) surface of Bowman's capsule is specialized for filtering the blood in the glomerular capillaries **(A)** Overview of glomerulus. The podocytes composing the visceral layer have long processes, termed pedicels, which cover the vascular epithelium. **(B)** Enlargement of the portion of part A that is enclosed in a box. Substances pass from the blood through the endothelial pores, across the basement membrane, and then through the filtration slits between pedicels.

spaces, the filtration slits, between them (Figure 14-19B). The filtrate, driven by the net pressure drop across the endothelium, passes through the pores formed in the walls of the glomerular capillaries and then through the filtration slits. The three-layered membrane separating the capillary lumen and lumen of Bowman's capsule acts as a molecular sieve, excluding almost all proteins from the ultrafiltrate based mainly on molecular size, but also on shape and charge (Table 14-10). There is a bulk flow of water through the sieve carrying with it ions, glucose, urea, and many other small molecules.

The kidneys are perfused by 500–600 ml of plasma per minute, or 20%–25% of the cardiac output, yet constitute less than 1% of the body weight. This preferential perfu-

TABLE 14-10

Relation between the molecular size of a substance and the ratio of its concentration in the filtrate appearing in Bowman's capsule to its concentration in the plasma, [filtrate]/[filtrand]

Substance	Mol. wt.	Dimensions (in nanometers)		[filtrate] [filtrand]
		Radius from diffusion coefficient	Dimensions from x-ray diffraction	
Water	18	0.11		1.0
Urea	6	0.16		1.0
Glucose	180	0.36		1.0
Sucrose	342	0.44		1.0
Insulin	5,500	1.48	54 / 8	0.98
Myoglobin	17,000	1.95	88 / 22	0.75
Egg albumin	43,500	2.85	54 / 32	0.22
Hemoglobin	68,000	3.25		0.03
Serum albumin	69,000	3.55	150 / 36	<0.01

Source: Pitts, 1968.

sion takes place in a relatively low-resistance vascular bed within the kidney. A high renal blood pressure is the result of the relatively direct arterial supply; because arteries and arterioles are large in diameter and short in length, the loss of pressure due to friction is minimized. The efferent arterioles (those taking the blood away from the glomeruli) are of smaller diameter and along with the capillaries of the vasa recta, constitute the major resistance of the renal vascular bed, ensuring high pressures within the glomeruli.

As noted already, the glomerular filtration rate depends largely on the net filtration pressure and the permeability of Bowman's capsule. The net filtration pressure depends on the blood pressure (glomerular capillary pressure), the intracapsular pressure, and the colloid osmotic pressure of the blood plasma (see Table 14-9). Under normal conditions, the colloid osmotic pressure and intracapsular pressure do not vary. The colloid osmotic pressure of plasma can be elevated during dehydration, and the intracapsular pressure can be increased by the presence of kidney stones obstructing the renal tubules; in both cases glomerular filtration rate will be reduced. In contrast, the seepage of plasma through burned skin can lower the colloid osmotic pressure of plasma, which in turn could increase the glomerular filtration rate. These examples, however, are exceptions rather than the rule.

Although blood pressure and cardiac output normally increase during exercise, these changes have little effect on the glomerular filtration rate in mammals because of regulatory processes that control blood flow to the kidney. This regulation is achieved by modulating the resistance to flow in the afferent arteriole leading to each nephron and depends on a number of interrelated mechanisms involving both paracrine and endocrine secretions as well as neural control.

Several intrinsic mechanisms provide autoregulation of the glomerular filtration rate. First, an increase in blood pressure will tend to stretch the afferent arteriole, which would be expected to increase the flow to the glomerulus. The wall of the afferent arteriole, however, responds to stretch by contraction, thus reducing the diameter of the arteriole and therefore increasing the resistance to flow. This myogenic mechanism thus reduces variations in flow to the glomerulus in the face of oscillations in blood pressure. Second, cells in the **juxtaglomerular apparatus (JGA),** which is located where the distal tubule passes close to the Bowman's capsule between the afferent and efferent arterioles, secrete substances that modulate renal blood flow.

The juxtaglomerular apparatus is composed of three types of cells (Figure 14-20):

- Modified distal-tubule cells, which form the *macula densa* and may monitor the osmolarity and flow of fluid in the distal tubule

- Specialized vascular cells, called *granular cells*, located between the afferent and efferent arterioles

- Secretory *juxtaglomerular cells*, modified smooth-muscle cells that are located primarily in the wall of the afferent arteriole

Under certain conditions, the juxtaglomerular cells release the hormone **renin**, which indirectly affects blood pressure

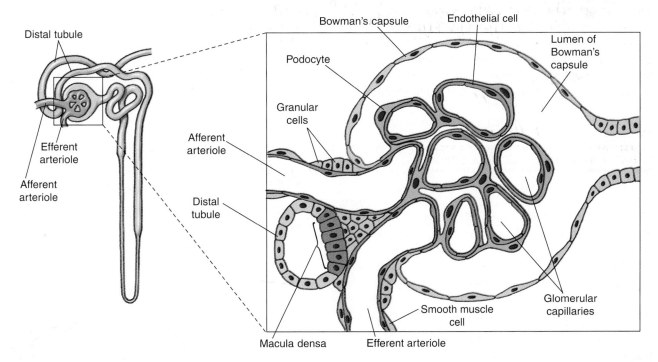

Figure 14-20 The juxtamedullary apparatus plays a key role in controlling blood flow through the glomerulus. This structure is composed of several cell types including modified distal-tubule cells, which constitute the macula densa; secretory juxtaglomerular cells in the wall of the afferent arteriole; and granular cells. [Adapted from Sherwood, 1993.]

and, therefore, renal blood flow as described below. The juxtaglomerular apparatus also releases various substances that act in a paracrine fashion to cause vasoconstriction or vasodilation of the afferent arteriole in response to increased or decreased flow, respectively, through the distal tubule. Thus myogenic and juxtaglomerular feedback-control mechanisms work together to autoregulate the glomerular filtration rate over a wide range of blood pressures.

In addition to these autoregulatory mechanisms, the glomerular filtration rate is subject to extrinsic neural control. The afferent arterioles are innervated by the sympathetic nervous system. Sympathetic activation causes vasocontriction of afferent arterioles and a reduction in glomerular filtration. This response, which overrides any autoregulation, occurs when there is a sharp drop in blood pressure, for example, as a result of extensive blood loss. The reduction in filtration rate helps to restore blood volume and pressure to normal. Conversely, an elevation in blood pressure reduces sympathetic vasoconstriction and enhances glomerular filtration, decreasing blood pressure and volume.

Sympathetic activation can also cause contractions of elements within the glomerulus, closing off portions of the filtering capillaries and effectively reducing the area available for filtration. The podocytes also are contractile, and when they contract, the number of filtration slits decreases. Thus contraction of either or both of these elements can effectively reduce the hydraulic permeability of Bowman's capsule. In the past, the hydraulic permeability of the glomerular membrane was thought to change only in disease states that caused the membrane to become leaky. It is now apparent that normal regulation of the glomerular filtration rate can involve changes in the hydraulic permeability of the glomerular membrane.

A reduction in renal blood pressure, a fall in solute delivery to the distal tubule, and/or activation of the sympathetic innervation induces release of the hormone renin from the secretory juxtaglomerular cells located in the wall of the afferent arteriole that carries blood into the glomerular capillaries in Bowman's capsule. Renin is a proteolytic enzyme whose release leads to increased levels of **angiotensin II** in the blood. This hormone has several actions, one of which is to cause general vasoconstriction (constriction of arterioles), which raises the blood pressure, thereby increasing both renal blood flow and the rate of glomerular filtration. Angiotensin II may also cause constriction of the efferent arterioles, raising glomerular blood pressure and increasing filtration. Angiotensin II also stimulates release of the steroid aldosterone from the adrenal cortex and vasopressin from the posterior pituitary. The role of these hormones in promoting the tubular reabsorption of salts and water is discussed later.

Tubular reabsorption

As the glomerular filtrate makes its way through the nephron, its original composition is quickly modified by re-

absorption of various metabolites, ions, and water. The human kidneys produce about 180 liters of filtrate per day, but the final volume of urine is only about 1 liter. Thus, over 99% of the water is reabsorbed. Of the 1800 g of NaCl typically occurring in the original filtrate, only 10 g (or less than 1%) appear in the urine of persons consuming 10 g of NaCl per day. Varying amounts of many other filtered solutes are also reabsorbed from the tubular lumen. In addition, some substances are secreted into the tubular fluid. The **renal clearance** of a substance is a measure of the extent to which it is reabsorbed or secreted in the kidneys, as explained in Spotlight 14-1.

To understand the relationship between clearance and reabsorption, let's consider glucose. A healthy mammal exhibits a plasma glucose clearance of 0 ml · min^{-1}. That is, even though the glucose molecule is small and is freely filtered by the glomerulus, normally it is completely reabsorbed by the epithelium of the renal tubule (Figure 14-21). Glucose is fully reabsorbed because its loss in the urine would mean a loss of chemical energy to the organism. Usually glucose appears in the urine only when the glucose concentration in the blood plasma, and hence in the glomerular filtrate, is very high. Figure 14-21 reveals that there is a maximum rate (milligrams per minute) at which glucose can be removed from the tubular urine by reabsorption. This transfer maximum, or Tm, is about 320 mg · min^{-1} in humans. Below plasma glucose levels of about 1.8 mg · ml^{-1}, all the glucose appearing in the glomerular filtrate is reabsorbed. At about 3.0 mg · ml^{-1}, the carrier mechanism is fully saturated, so that any additional amount of glucose appearing in the filtrate will be passed out in the urine. The arterial plasma glucose concentration in humans is normally held at about 1 mg · ml^{-1} by an endocrine feedback loop involving insulin. Since this level is well below the Tm for glucose, normal urine contains essentially no glucose. Because the high plasma glu-

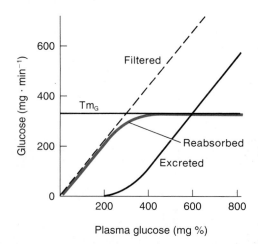

Figure 14-21 The concentration of glucose in the glomerular filtrate (broken line) is proportional to the plasma glucose concentration. The renal tubules are capable of reabsorbing the glucose by active transport (colored line) at rates up to 320 mg · min^{-1} (Tm$_G$). Glucose entering the filtrate in excess of this rate is necessarily excreted via the urine (black line).

SPOTLIGHT 14-1

RENAL CLEARANCE

The **renal clearance** of a plasma-borne substance is the volume of blood plasma from which that substance is "cleared" (i.e., completely removed) per unit time by the kidneys. A substance that is freely filtered into the nephron along with water, but that is neither reabsorbed nor secreted in the kidney, permits the calculation of the **glomerular filtration rate** (GFR) merely by dividing the amount of the nontransported substance appearing in the urine by the concentration of that substance in the plasma. One such molecule is **inulin** (*not* insulin), a small starchlike carbohydrate (molecular weight 5000). Since the inulin molecule is neither reabsorbed nor secreted by the renal tubule, the *inulin clearance* is identical with the rate at which the glomerular filtrate is produced—that is, the GFR, generally given in milliliters per minute.

If we know the GFR and the concentration of a freely filtered substance in the plasma (thus also its concentration in the ultrafiltrate), we can easily calculate whether the substance undergoes a net reabsorption or net secretion during the passage of the ultrafiltrate along the renal tubule. Thus, if less of the substance appears in the urine than was filtered in the glomerulus, it must have undergone some reabsorption in the tubule. This is true for water, NaCl, glucose, and many other essential constituents of the blood. If, however, the quantity of a substance appearing in the urine over a period of time is greater than the amount that passed into the nephron because of glomerular filtration, it can be concluded that this substance is actively secreted into the lumen of the tubule. Unfortunately, the *clearance technique* is of limited usefulness in studies of renal function, since it indicates only the *net* output of the kidney relative to input and fails to provide insight into the physiological details.

In renal clearance studies, a test substance such as inulin is first injected into the subject's circulation and allowed to mix to uniform concentration in the bloodstream. A sample of blood is removed, and the plasma concentration of inulin, P, is determined from the sample. The rate of appearance of inulin in the urine is determined by multiplying the concentration of inulin in the urine, U, by the volume of urine produced per minute, V. The amount of inulin appearing in the urine per minute (VU) must equal the rate of plasma filtration (GFR) multiplied by the plasma concentration of inulin:

$$\frac{VU}{(GFR)P} = \frac{\text{amt. inulin appearing in urine} \cdot \text{min}^{-1}}{\text{amt. inulin removed from blood} \cdot \text{min}^{-1}} = 1$$

In this special case, the substance used, inulin, is freely filtered and unchanged by tubular absorption or secretion. Therefore, the GFR and the clearance, C, of the substance are equal. Substituting C for GFR gives, for inulin,

$$\frac{VU}{CP} = 1$$

so the renal clearance is given by

$$\frac{VU}{P} = C = \text{renal clearance (ml} \cdot \text{min}^{-1})$$

If the amount of a substance, x, appearing in the urine per minute deviates from the amount of x present in the volume of plasma that is filtered per minute, this will be reflected in a value of C_x that differs from the inulin renal clearance, C. For example, if the inulin clearance of a subject, and hence the GFR, is 125 ml $\cdot$ min^{-1} and substance x exhibits a clearance of 62.5 ml $\cdot$ min^{-1}, then

$$\frac{VU_x}{P_x} = C_x = 62.5 \text{ ml} \cdot \text{min}^{-1} = 0.5 \text{ (GFR)}$$

In this case, a volume of plasma *equivalent* to half that filtered each minute is cleared of substance x. Stated differently, only half the amount of substance x present in a volume of blood plasma equal to the volume filtered each minute actually appears in the urine per minute.

There are two possible reasons why the renal clearance for a substance would be less than the GFR. First, it may not be freely filterable. For example, filtration of a substance may be hindered by its binding to serum proteins, by its large molecular size, or by some other factor. Second, a substance may be freely filtered, but it may be reabsorbed in the kidney tubules, thus reducing the amount that appears in the urine. As a matter of fact, most molecules below a molecular weight of about 5000 are freely filtered, but many of these are either partially reabsorbed or partially secreted (see Table 14-10). The extent of reabsorption or secretion can be gauged by the renal clearance of a substance. Reabsorption reduces the renal clearance to below the GFR. Tubular secretion, however, will cause more of a substance to appear in the urine than is carried into the tubule by glomerular filtration.

cose levels typical of diabetes mellitus exceed the reabsorption ability of the renal tubule, diabetics commonly have glucose in their urine.

The details of tubular function vary from species to species. Our knowledge of the changes in urinary composition along different portions of the nephron is based to a large extent on the technique of micropuncture, first developed by Alfred Richards and his coworkers in the 1920s. A glass capillary micropipette is used to remove a minute sample of the tubular fluid from the lumen of the nephron. The osmolarity of the sample (expressed as milliosmoles per liter) is then determined by measuring its freezing point. The lower the freezing point, the higher its osmolarity. The stopped-flow perfusion technique, a modification of

Richards' original technique, can be used to isolate a portion of the lumen and analyze its action on injected samples of a defined solution in vitro (Figure 14-22).

Microchemical methods are now used to determine the concentrations of individual ion species in the sample. In a technique developed more recently, a given segment of renal tubule is dissected from the kidney and perfused *in vitro* with a defined test solution; analysis of the perfusate provides insight to the movement of substances across the isolated tubule segment (Figure 14-23). The results of numerous studies using these techniques have detailed the roles of various portions of the nephron in the reabsorption of salts and water, which are summarized in Figure 14-24.

The *proximal tubule,* which initiates the process of concentrating the glomerular filtrate, is most important in active reabsorption of salts. In this segment, about 70% of the Na$^+$ is removed from the lumen by active transport, and a nearly proportional amount of water and certain other solutes, such as Cl$^-$, follow passively. So about 75% of the filtrate is reabsorbed before it reaches the loop of Henle.

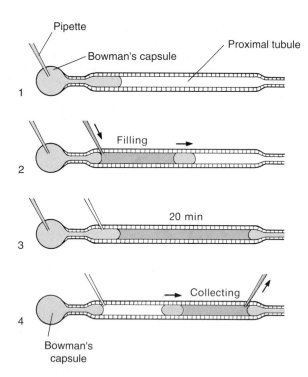

Figure 14-22 The stopped-flow perfusion technique is used to study the function of various portions of the renal tubule *in vitro.* A micropipette is inserted into Bowman's capsule (1), and oil (color) then is injected until it enters the proximal tubule. Perfusion fluid (gray) is injected through a second pipette into the middle of the column of oil, forcing a droplet ahead of it (2). The tubule is full when the droplet reaches the far end of the tubule (3). After about 20 minutes, the perfusion fluid is collected by the injection of a second liquid behind the oil remaining near the glomerulus (4). The ability of the tubule segment to reabsorb or secrete substances can be determined by comparing the composition of the perfusate before and after it is injected. [From "Pumps in the Living Cell" by Arthur K. Solomon. Copyright © 1962 by Scientific American, Inc. All rights reserved.]

The result is a tubular fluid that is isosmotic with respect to the plasma and interstitial fluids. Stopped-flow perfusion experiments revealed that when the NaCl concentration inside the tubule is decreased, the movement of water also decreases. This result is just the opposite of what would be expected if the outward movement of reabsorbed water occurred by simple osmotic diffusion, and it indicates that water transport is coupled to active sodium transport (see Chapter 4). The actual pumping of Na$^+$ takes place at the basolateral (serosal) surface of the epithelial cells of the proximal tubule, just as it does in frog skin and gallbladder epithelia. In amphibians, this active transport leaves the tubular lumen about 20 mV negative relative to the fluid surrounding the nephron. This potential difference probably accounts for the passive net diffusion of chloride out of the proximal tubule as the counterion for sodium. In the proximal tubule, NaHCO$_3$ is the major solute reabsorbed proximally and NaCl the major solute reabsorbed distally.

At the most distal portion of the proximal tubule (where it joins the thin descending limb of the loop of Henle), the glomerular filtrate is already reduced to one fourth of its original volume. As a result of the reduction in the volume of tubular fluid, substances that are not actively transported across the tubule or that do not passively diffuse across it are four times as concentrated toward the end of the proximal tubule than in the original filtrate. In spite of this great reduction in the volume of tubular fluid, the fluid at this point is isosmotic relative to the fluid outside the nephron, having an osmolarity of about 300 mosm · L^{-1}. It is interesting to note that the active transport of NaCl alone can account for the major changes in volume of the fluid along the proximal tubule and for the increased concentration of urea and many other filtered substances.

The proximal tubule is ideally structured for the massive reabsorption of salt and water. Numerous microvilli at the luminal border of the tubular epithelial cells form the so-called **brush border** (see Figure 14-15). These projections greatly increase the absorptive surface area of the membrane, thereby promoting diffusion of salt and water from the tubular lumen into the epithelial cell.

Glucose and amino acids are also reabsorbed in the proximal tubule by a sodium-dependent mechanism and are not normally present in the ultrafiltrate beyond the proximal tubule. Carriers on the apical membrane cotransport sodium and glucose or amino acids from the ultrafiltrate into the cell. The uptake process, which is uphill for glucose and amino acids, depends on the sodium electrochemical gradient created by the Na$^+$/K$^+$ ATPase in the basolateral membrane of the tubular cell. Once in the tubular cell, glucose and amino acids diffuse into the blood.

Phosphates, calcium ions, and other electrolytes normally found in the blood are reabsorbed up to that required by the body, and any excess is excreted. Parathyroid hormone modulates the reabsorption of phosphates and calcium by the kidney. Parathyroid hormone stimulates kidney 1α,25-hydroxylase activity, which in turn stimulates

R A Date for return

{ Gratuity
{ Tax

Flowers for
    speakers

Prayer in memory...

Ask Chris Andrews

Mari Ramos   Patty
Nicki Ornelas

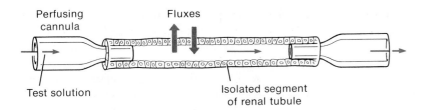

Figure 14-23 Perfusion of a dissected segment of renal tubule and chemical analysis of the perfusate permits determination of the fluxes of ions across the tubular wall *in vitro*.

the production of calcitriol, the active form of vitamin D. Calcitriol released into the blood stimulates calcium reabsorption and phosphate excretion from the kidney, as well as calcium absorption from the gut and release from bone (see Figure 9-29).

The *descending limb* and *thin segment of the ascending limb of the loop of Henle* are made up of very thin cells containing few mitochondria and no brush border. *In vitro* perfusion studies have demonstrated that there is no active salt transport in the descending limb. Moreover, this segment exhibits very low permeability to NaCl and low permeability to urea but is permeable to water. As discussed later, this differential permeability plays an important role

in the urine-concentrating system of the nephron. The thin segment of the ascending limb has also been shown by perfusion experiments to be inactive in salt transport, although it is highly permeable to NaCl. Its permeability to urea is low, and to water, very low. Again, this differing permeability plays a key role in the urine-concentrating mechanism of the nephron.

The *medullary thick ascending limb* differs from the rest of the loop of Henle in that it exhibits active transport of NaCl outward from the lumen to the interstitial space (see Figure 14-24). This portion, along with the rest of the ascending limb, has a very low permeability to water. As a result of NaCl reabsorption, the fluid reaching the distal

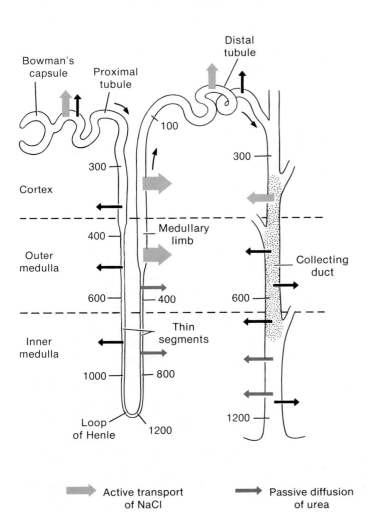

Figure 14-24 The movement of ions, water, and other substances in and out of the filtrate along the renal tubule determines the composition of the urine. In this schematic diagram, the fluxes of NaCl, water, and urea are shown in different portions of the mammalian renal tubule. The numbers indicate the filtrate tonicity in milliosmoles per liter. The relative rates of active transport of NaCl are indicated by the size of the arrow. The permeability of the stippled portion of the collecting duct is regulated by antidiuretic hormone (ADH). [Adapted from Pitts, 1959.]

tubule is somewhat hyposmotic relative to the interstitial fluid. The importance of salt reabsorption by the thick ascending tubule is discussed later in the section on the urine-concentrating mechanism.

The movement of salt and water across the *distal tubule* is complex. The distal tubule is important in the transport of K^+, H^+, and NH_3 into the lumen and of Na^+, Cl^-, and HCO_3^- out of the lumen and back into the interstitial fluid. As salts are pumped out of the tubule, water follows passively. The transport of salts in the distal tubule is under endocrine control, and is adjusted in response to osmotic conditions.

Because the *collecting duct* is permeable to water, water flows from the dilute urine in the duct into the more concentrated interstitial fluid of the renal medulla (see Figure 14-24). This is the final step in the production of a hyperosmotic urine. The water permeability of the duct is variable and controlled by antidiuretic hormone (ADH). Thus the rate at which water is absorbed is under delicate feedback control. The collecting duct reabsorbs NaCl by active transport of sodium. The inner medullary segment of the collecting duct, toward its distal end, is highly permeable to urea. The significance of this will become clear in our later discussion of the countercurrent mechanism that concentrates the urine in the collecting duct.

Now that we have summarized the movement of water, ions, and glucose out of the tubular filtrate, let's examine sodium reabsorption in the nephron more closely. In the proximal tubule and ascending limb of the loop of Henle, sodium is transferred across the apical membrane of the tubular epithelium via cotransporters and then is actively transported into the blood via a Na^+/K^+ ATPase (Figure 14-25A). The electrochemical gradient for Na^+ between the ultrafiltrate and the blood favors the diffusion of Na^+ through channels in the apical membrane from the ultrafiltrate into the tubular cells. Sodium is also exchanged for a proton via an electrically neutral Na^+/H^+ exchanger; in this case, the downhill movement of Na^+ energizes the uphill movement of H^+ into the lumen (Figure 14-25B). Farther along in the distal tubule and collecting duct, sodium reabsorption is coupled to secretion of protons into the urine by acid-secreting cells, which are involved primarily in pH regulation. These cells are described in detail later.

Sodium chloride represents more than 90% of the osmotic activity of the extracellular fluid. Because reabsorption of salt results in the reabsorption of water, the amount of salt in the body is an important determinant of the volume of the extracellular fluid (ECF). If ECF volume is large, then blood pressure tends to rise. The converse is true for a reduction in ECF, for example, as a result of blood loss. Thus blood pressure is an indication of blood volume, which in turn is a reflection of the salt content of the body. When cells of the macula densa portion of the juxtaglomerular apparatus sense a decrease in blood pressure and/or solute delivery to the distal tubule, they stimulate release of renin from the juxtaglomerular cells in the walls of the afferent arteriole (see Figure 14-20). As outlined in

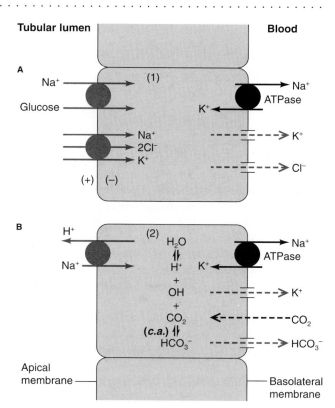

Figure 14-25 Several transport systems are involved in reabsorption of Na^+ in the proximal tubule and ascending limb of the loop of Henle in the mammalian kidney. **(A)** Sodium passively crosses the apical membrane via Na/2Cl/K and glucose/Na^+ cotransporters. A Na^+/K^+ ATPase in the basolateral membrane actively removes Na^+ from the cell into the blood; K^+ and Cl^- exit via ion channels down their concentration gradient. **(B)** The movement of Na^+ down its electrochemical gradient into the cell also energizes the outward movement of protons via an electrically neutral Na^+/H^+ exchanger. CO_2 in the blood diffuses into the cell, where carbonic anhydrase (*c.a.*) ensures a high rate of proton delivery to the exchanger. A basolateral sodium pump transports Na^+ from the cell into the blood. K^+ and HCO_3^- exit via ion channels down their electrochemical gradient.

Figure 14-26A, renin acts to cause a rise in blood levels of angiotensin II and subsequently aldosterone; the latter promotes sodium reabsorption from the filtrate.

Renin, a proteolytic enzyme, cleaves angiotensinogen, a glycoprotein molecule that is manufactured in the liver and is present in the α_2-globulin fraction of plasma proteins. Cleavage of angiotensinogin releases a 10-residue peptide, angiotensin I. Angiotensinogen-converting enzyme (ACE) then removes two additional amino acids to form the 8-residue peptide, angiotensin II (Figure 14-26B). Much of the formation of angiotensin II occurs during the passage of blood through the lungs. Angiotensin II stimulates the secretion of aldosterone from the adrenal cortex and also causes a general vasoconstriction, which raises blood pressure. Removal of the amino-terminal aspartic acid residue from angiotensin II yields angiotensin III, which also causes secretion of aldosterone from the adrenal cortex but to a lesser extent than angiotensin II.

Like other steroid hormones, **aldosterone** diffuses across the cell membrane and binds to cytoplasmic recep-

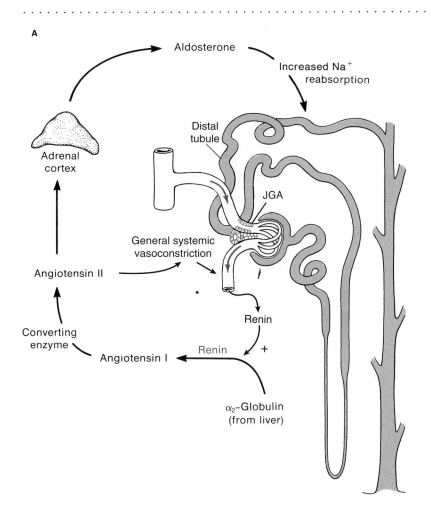

A

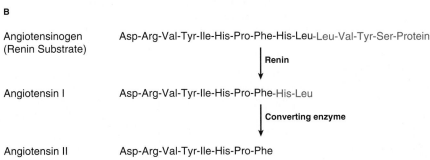

Figure 14-26 The renin-angiotensin system plays an important role in controlling sodium reabsorption in the mammalian kidney. **(A)** Renin is liberated by secretory cells in the juxtaglomerular apparatus (JGA) in response to decreased pressure in the afferent arteriole and to low Na⁺ concentration in the distal tubule. Circulating renin leads to an increase in the titer of angiotensin II and aldosterone. Aldosterone stimulates Na⁺ reabsorption from the filtrate in the renal tubule. **(B)** Renin is a proteolytic enzyme that cleaves angiotensinogen, an α_2-globulin, yielding angiotensin I. Another proteolytic enzyme then removes the two carboxyl-terminal residues to give angiotensin II.

tors in target cells, leading to an increase in transcription of specific genes and ultimately synthesis of the encoded proteins (see Figure 9-9). Aldosterone acts on the cells of the tubular epithelium to increase sodium reabsorption but does so without affecting water permeability. Three mechanisms have been proposed to account for the aldosterone-induced increase in sodium reabsorption across tubular epithelial cells (Figure 14-27):

1. Sodium-pump hypothesis: Increased activity of Na⁺/K⁺ ATPase in the basolateral membrane, perhaps due to changes in membrane structure that enhance ATPase activity as well as to increased synthesis of the pump protein.

2. Metabolic hypothesis: Increase in the production of ATP to power the sodium pump, perhaps due to aldosterone-stimulated increase in fatty acid metabolism.

3. Permease hypothesis: Increased permeability of the apical membrane to Na⁺ ions, presumably due to an increase in the number of sodium channels in the membrane.

Quite possibly, all three mechanisms operate in tubular cells stimulated by aldosterone.

Increased circulating levels of angiotensin II also increase the synthesis of **vasopressin,** also called **antidiuretic hormone (ADH),** in the hypothalamus and its release from the posterior pituitary (see Figures 9-5 and 9-7). Vaso-

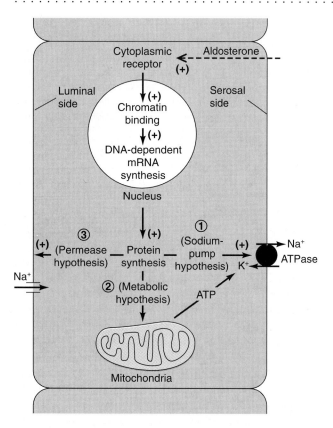

Figure 14-27 Aldosterone, a steroid hormone that stimulates gene expression, increases sodium reabsorption in the kidney. Depicted here are three mechanisms proposed to explain the effect of aldosterone: an increase in the activity of the Na⁺/K⁺ ATPase directly (sodium-pump hypothesis) or indirectly by increasing ATP levels (metabolic hypothesis), and an increase in the activity of sodium channels (permease hypothesis). [Adapted from M. E. Hadley, 1992.]

Tubular secretion

The nephron has several distinct systems that secrete substances by transporting them from the plasma into the tubular lumen. Most thoroughly investigated are the systems for secretion of K^+, H^+, NH_3, organic acids, and organic bases. Although the number of secretory mechanisms and transport molecules must be limited, nevertheless, the nephron is capable of secreting innumerable "new" substances, including drugs and toxins, as well as endogenous, naturally occurring molecules. How is the nephron able to recognize and transport all these diverse substances? The answer seems to reside in the role of the vertebrate liver in modifying many such molecules so that they can react with the transport systems located in the wall of the nephron. These secretory mechanisms are important because they remove potentially dangerous substances from the blood. In the liver, many of these substances, along with normal metabolites, are conjugated with glucuronic acid or its sulfate. Both these classes of conjugated molecules are actively transported by the system that recognizes and secretes organic acids. Since they are highly polar, these conjugated molecules, once deposited by the transport machinery in the lumen of the nephron, cannot readily diffuse back across the wall of the nephron into the peritubular space and from there into the blood, and so these substances are excreted in the urine.

 A few marine teleost fish species have aglomerular kidneys. What might have been the selective forces operating on the evolution of such structures?

pressin acts through cyclic AMP to increase the water permeability of the principal cells in the distal tubule and collecting duct, increasing the number of water channels in the apical membrane and, therefore, promoting water reabsorption. Aldosterone, unlike vasopressin, does not act through cyclic AMP, but does act with vasopressin to enhance both sodium and water reabsorption by the kidney.

Atrial natriuretic peptide (ANP), released from the atrium of the heart into the blood in response to an increase in venous pressure, causes an increase in urine production and sodium excretion. It thus has the opposite effect of the renin-angiotensin system on the kidney. ANP inhibits the release of vasopressin and renin and the production of aldosterone from the adrenal gland. ANP acts directly on the kidney to reduce sodium and, therefore, water reabsorption (see Chapter 12).

 Diving to depth raises venous pressure in humans. What effect might this have on kidney function? What differences might be expected in the regulation of kidney function between humans and whales?

Normally, most of the potassium ions, which are freely filtered at the glomerulus, are reabsorbed from the filtrate in the proximal tubule and the loop of Henle due to the presence of a Na/2Cl/K cotransport system in the apical membrane and Na⁺/K⁺ ATPase in the basal membrane (see Figure 14-25A). Potassium channels in the basal membrane allow potassium to be recycled across the basal membrane. The rate of active reabsorption in the proximal tubule and loop of Henle continues unabated even when the level of K^+ in the blood and filtrate rises to high levels in response to excessive intake of this ion. However, the distal tubule and collecting duct are able to secrete K^+ into the tubular filtrate to achieve homeostasis in the face of a high body load of potassium. The secretion of K^+ involves the active transport of K^+ from the interstitial fluid into the tubular cell by the usual Na⁺/K⁺ ATPase in the basolateral membrane and subsequent leakage of cytosolic K^+ through potassium channels in the apical membrane into the tubular fluid (Figure 14-28). The latter is electronegative with respect to the cytosol, so K^+ can simply diffuse down its electrochemical gradient from inside the renal tubular cell into the lumen.

Tubular lumen   **Blood**

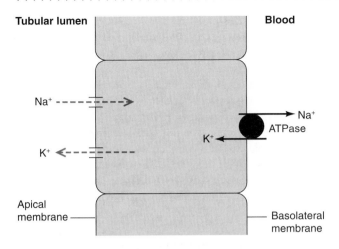

Apical membrane ———   ——— Basolateral membrane

Figure 14-28 In the distal tubule and collecting duct, K^+ can be secreted into the tubular filtrate. A Na^+/K^+ ATPase in the basolateral membrane actively transports K^+ into the tubular epithelium; it then passively moves down its electrochemical gradient via potassium channels in the apical membrane into the lumen.

The rate of potassium secretion (and sodium reabsorption) by these mechanisms is stimulated by aldosterone, which is released in response to elevated plasma potassium levels as well as reduced sodium levels. Reduced potassium levels directly stimulate the adrenal glands, whereas reduced blood sodium levels stimulate the adrenals via activation of the renin-angiotensin system. Stimulation of sodium reabsorption is therefore coupled to potassium secretion through the action of aldosterone; one cannot be corrected without affecting the other. Release of aldosterone in response to low blood sodium levels will enhance sodium reabsorption but could also lead to abnormally low blood potassium levels due to enhanced potassium secretion and excretion.

Because high extracellular potassium levels may cause cardiac arrest and convulsions, excess K^+ ions must be quickly removed from the plasma. Insulin is released in response to high potassium levels and stimulates potassium uptake by cells, especially fat cells. Potassium is then slowly released from these cells and removed by the somewhat slower renal mechanisms. Thus, like aldosterone, insulin release can also lead to low plasma potassium levels.

Regulation of pH by the Kidney

As discussed in detail in Chapter 13, the carbon dioxide/bicarbonate buffering system is primarily responsible for determining the pH of the extracellular space in mammals. This system involves three reactions:

(1) $CO_2 + H_2O \rightleftharpoons H_2CO_3 \rightleftharpoons HCO_3^- + H^+$

(2) $CO_2 + OH^- + H^+ \rightleftharpoons HCO_3 + H^+$

(3) $HOH \rightleftharpoons OH^- + H^+$

Reaction (1) occurs very slowly at body temperatures, but reaction (2) is catalyzed by the enzyme carbonic anhydrase

and is therefore rapid. Two factors have the most effect on the CO_2/HCO_3^- system in mammals: excretion of CO_2 via the lung and excretion of acid via the kidney. The ratio of lung ventilation to CO_2 production largely determines the CO_2 concentration of the body. For example, when lung ventilation is reduced, CO_2 levels increase and the blood pH drops as hydrogen ions and bicarbonate ions accumulate (see Figure 13-10). Changes in breathing can adjust carbon dioxide excretion and, therefore, modulate body pH in the short term (see Chapter 13). The excretion of acid (H^+ ions) in the urine is ultimately responsible for maintaining the plasma HCO_3^- concentration in mammals. Acid excretion across the skin of amphibia or the gills of fish supplements or takes over the role of acid excretion by the kidney in these animals.

The concentration of HCO_3^- in mammalian plasma is around $25 \times 10^{-3} \; mol \cdot L^{-1}$, whereas the H^+ concentration is around $40 \times 10^{-9} \; mol \cdot L^{-1}$. The concentrations of bicarbonate and protons in the glomerular ultrafiltrate are similar to those in the plasma; that is, the filtrate contains large quantities of bicarbonate but a very low concentration of protons. Yet urine has a pH of around 6.0 and contains little or no bicarbonate. Thus acid is added to the filtrate and most, if not all, of the bicarbonate is removed in the process of urine formation. At pH 6 the urine still has a very low concentration of protons, and the change in H^+ concentration alone, as the filtrate flows down the tubule, would not be sufficient to maintain body pH in the face of continual metabolic production of acid. In fact most of the acid added to the urine is buffered by either phosphate or ammonia.

Because protons are added to the tubular filtrate along the entire length of the tubule, the filtrate becomes progressively more acidic. In the proximal tubule and loop of Henle, protons are secreted via a H^+/Na^+ exchanger discussed earlier (see Figure 14-25B). The distal tubule and the collecting duct contain cells, referred to as A-type cells, that have a proton ATPase in the apical membrane and a chloride/bicarbonate exchange system in the basolateral membrane. (This anion exchanger is similar to the band 3 protein in the red blood cell membrane.) These cells also contain high levels of carbonic anhydrase, so that intracellular carbon dioxide is rapidly hydrated forming bicarbonate ions and protons; the protons are transported across the apical membrane and the bicarbonate ions move across the basal membrane. The secreted protons can react with bicarbonate in the ultrafiltrate to form carbon dioxide and water, which can diffuse back into the cell. Thus the secretion of protons from the A-type cell can result in the net uptake of bicarbonate into the blood through the cycling of carbon dioxide (Figure 14-29A). Clearly the A-type cell is an acid-secreting cell.

Removal of protons from an A-type cell makes the intracellular potential more negative, thereby enhancing sodium reabsorption from the filtrate. The intracellular sodium level is kept low by the activity of a Na^+/K^+ ATPase in the basolateral membrane, which transports Na^+ from the cell into

A

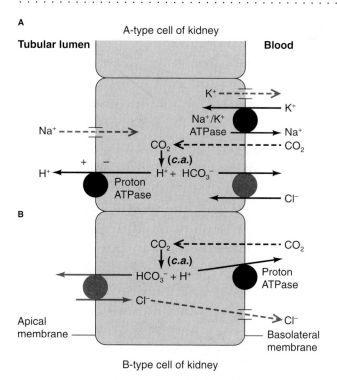

Figure 14-29 Body pH in mammals can be modulated by regulating the relative activity of acid-secreting (A-type) cells and base-secreting (B-type) cells in the distal tubule and collecting duct of the kidney. **(A)** A-type cells pump protons into the lumen via an apical H^+ ATPase, acidifying the filtrate; the resulting increase in the potential across the apical membrane favors reabsorption of Na^+. **(B)** B-type cells use the H^+ ATPase in the basolateral membrane to pump protons into the blood, accompanied by the reabsorption of Cl^-. Both cell types contain carbonic anhydrase (c.a.), which rapidly forms H^+ and HCO_3^- ions from CO_2 diffusing into the cell from the blood.

the extracellular flood. The basolateral membrane of the A-type cell also contains K^+ channels, and K^+ is cycled through this membrane by the Na^+/K^+ ATPase. Thus acidification of the filtrate by A-type cells is coupled to sodium reabsorption.

The distal tubule and collecting duct also contain base-secreting cells, called B-type cells. These cells have a chloride/bicarbonate exchanger in the apical membrane; this exchanger differs from the band 3-type protein found in the basal membrane of the A-type cell. As illustrated in Figure 14-29B, B-type cells contain carbonic anhydrase and secrete bicarbonate into the lumen of the tubule in exchange for chloride. Protons and chloride ions move across the basolateral membrane via a proton ATPase and chloride channels.

A mammal can regulate its body pH by altering the activity of these A-type and B-type cells. The activity of A-type cells and, therefore, acid secretion increase during acidosis, whereas increased B-type cell activity and bicarbonate secretion are associated with alkalosis. Changes in the activity of the A-type cells involve alteration in both the proton ATPase activity in the apical membrane and the number of bicarbonate/chloride exchangers present in the basal membrane.

Proton secretion by renal tubular cells reduces the pH of the ultrafiltrate, thereby increasing the gradient against which protons are transported. Thus the ability to secrete protons decreases with filtrate pH; when the pH of the filtrate drops below 4.5, acid secretion stops. If the ultrafiltrate is buffered, however, more protons can be secreted across the tubular epithelium without a drop in pH sufficient to inhibit the proton pump. The ultrafiltrate is buffered by bicarbonate, phosphates, and ammonia. Acid secreted into the ultrafiltrate reacts with bicarbonate to form carbon dioxide, with HPO_4^{2-} to form $H_2PO_4^-$, or with NH_3 (ammonia) to form NH_4^+ (ammonium) ions (Figure 14-30). The tubular membrane is essentially impermeable to both phosphates and ammonium ions. Phosphates are filtered from the blood in the glomerulus, whereas ammonia diffuses from the blood across the tubular cells into the lumen, where it is converted to ammonium ions. Both phosphates and ammonium ions are trapped in the filtrate and then excreted from the body. The bicarbonate, phosphate, and ammonia buffer systems compete for protons secreted into the filtrate. Phosphate levels depend on diet, with excess phosphate being filtered into the ultrafiltrate. Thus the capacity of the phosphate buffer system (i.e., the number of protons that it can bind) depends on what the animal eats and is independent of the acid-base requirements of the animal. Body pH is not generally regulated by selection of appropriate foods.

Under acidotic conditions, plasma bicarbonate levels often fall; as a result, bicarbonate levels in the filtrate are reduced and less is available to act as a buffer. Under such conditions, ammonia is a major vehicle for elimination of excess acid. Ammonia is produced within the renal tubular cells by enzymatic deamination of amino acids, especially glutamine (see Figure 14-30). In its nonpolar, un-ionized

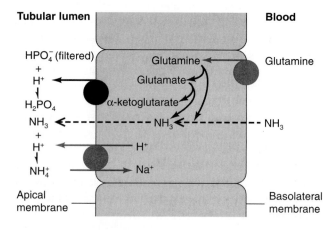

Figure 14-30 Buffering of the renal filtrate by HPO_4^- and NH_4^+ permits greater secretion of protons. The phosphate ions in the lumen arise by filtration, whereas the ammonium ions arise by passive diffusion of NH_3 from the blood across the tubular cells or by intracellular breakdown of glutamine. Glutamine (and other amino acids) enter the tubular cells via basolateral transporters and is deaminated, yielding NH_3, which diffuses across the apical membrane into the lumen. Because the membrane is largely impermeable to both $H_2PO_4^-$ and NH_4^+, both ions are trapped in the urine and excreted.

form, ammonia freely diffuses across the cell membrane into the lumen where it reacts with protons forming NH_4^+ ions. Because the highly polar NH_4^+ is impermeant, it traps both nitrogen atoms and protons in the urine, thus serving as a vehicle for their excretion. If acidotic conditions continue in the body for a few days, ammonia production by the tubular epithelium increases, NH_4^+ concentration in the filtrate rises, and acid excretion by the kidney is increased. The secretion of ammonia is highly adaptive. Mammals that have entered a state of metabolic acidosis (excess acid production) show dramatic increases in ammonia production and secretion, as this is the body's major adaptive long-term mechanism for correcting an acid load.

 What changes in the pattern of acid-base regulation occurred with the evolution of air-breathing vertebrates?

Urine-Concentrating Mechanism

The urine of birds and mammals becomes concentrated by osmotic removal of water from the filtrate in the collecting ducts as they course through the renal medulla. There is a clear-cut correlation between the architecture of the vertebrate kidney and its ability to manufacture a urine that is hypertonic relative to the body fluids. Kidneys capable of producing a hypertonic urine (i.e., those of mammals and birds) all have nephrons featuring the loop of Henle. Moreover, the ability of a mammal to concentrate the urine is directly related to the length of the loops of Henle in its kidneys. The loops of Henle are longest in desert dwellers, such as the kangaroo rat; these longer loops produce larger overall gradients in osmolarity from renal cortex to medulla, thus permitting more efficient osmotic extraction of water from the collecting duct. In general, the longer the loop and the deeper it extends into the renal medulla, the greater the concentrating power of the nephron. Thus, desert mammals have both the longest loops of Henle and the most hypertonic urine.

In addition to this correlation between the anatomy and concentrating ability of the nephron, the **tonicity** of the interstitial fluid progressively increases toward the deeper regions of the renal medulla (Figure 14-31) for reasons discussed later. These findings led B. Hargitay and Werner Kuhn to propose in 1951 that the loop of Henle acts as a **countercurrent multiplier** (Spotlight 14-2). Though a very attractive and plausible hypothesis, it was initially hard to test because of the difficulty of sampling the intratubular fluid in the thin loop of Henle. Determinations of the melting point of the fluid in slices of frozen kidney and subsequently *in situ* perfusion experiments of segments of the loop provided experimental support for the countercurrent hypothesis.

These studies showed that the fluid entering the descending limb of the loop of Henle from the proximal

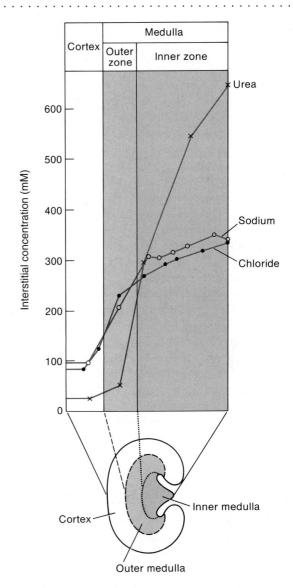

Figure 14-31 Solute concentrations in the interstitium of the mammalian kidney progressively increase from the cortex into the depths of the medulla. Shown here are the interstitial concentrations (in millimoles per liter) of urea, sodium, and chloride at different depths. Note that most of the increase in urea concentration occurs across the inner medulla, whereas most of the increase in NaCl concentration occurs across the outer medulla. Since the osmotic contributions of Na^+ and Cl^- sum, the total osmotic contributions of NaCl and urea are about equal deep within the medulla. [Adapted from Ulrich et al., 1961.]

tubule is isosmotic with respect to the extracellular fluid at that point (i.e., the outer portion of the renal medulla), having a concentration of about 300 $mosm \cdot L^{-1}$ (see Figure 14-24). The concentration of the fluid gradually increases as it makes its way down the descending limb toward the hairpin turn in the loop, where its concentration reaches 1000–3000 $mosm \cdot L^{-1}$ in most mammals. At this point, too, it is nearly isosmotic relative to the surrounding extracellular fluid in the deep portion of the renal medulla. This increase in the osmolarity of the tubular fluid flowing down the descending limb occurs because the wall of the descending limb is relatively permeable to water, but far less

SPOTLIGHT 14-2

COUNTERCURRENT SYSTEMS

In 1944 Lyman C. Craig published a method of concentrating chemical compounds based on the countercurrent principle. This method has proved useful in many industrial and laboratory applications. As in many other instances, human ingenuity turns out to be a reflection of Nature's inventiveness; countercurrent mechanisms have been found to operate in a variety of biological systems, including the vertebrate kidney, the gas-secreting organ of swim bladders and gills of fishes, and the limbs of various birds and mammals that live in cold climates.

The principle can be illustrated with a hypothetical countercurrent multiplier that employs an active-transport mechanism much like the one that operates in the mammalian kidney. The model shown in part A of the accompanying figure consists of a tube bent into a loop with a common dividing wall between the two limbs. A NaCl solution flows into one limb of the tube and then out the other. Let us assume that, within the common wall separating the two limbs of the tube, there is a mechanism that actively transports NaCl from the outflow limb to the inflow limb of the tube, without any accompanying movement of water. As bulk flow carries the fluid along the inflow limb, the effect of NaCl transport is cumulative, and the salt concentration becomes pro-

gressively higher. As the fluid rounds the bend and begins flowing out the other limb, its salt concentration progressively falls as a result of the cumulative effect of outward NaCl transport along the length of the outflow limb. By the time it reaches the end of that limb, its osmolarity is slightly lower than that of the fresh fluid beginning its inward flow in the other limb. This establishes a salt gradient along the tube.

This example resembles the loop of Henle in principle but not in detail. The loop of Henle has no common wall dividing the two limbs; nevertheless, the limbs are coupled functionally through the interstitial fluid, so that the NaCl pumped out of the ascending limb can diffuse the short distance toward the descending limb and cause osmotic reabsorption of water from that limb. Several important points should be noted about countercurrent systems such as the loop of Henle and the simple model illustrated here.

First, the standing concentration gradient set up in both limbs is due to both the continual movement of fluid through the system and the cumulative effect of transfer from the outflow limb to the inflow limb. The gradient would disappear if either fluid movement or transport across the membrane were to cease.

Second, the difference in concentration from one end to the other of each two limbs of the countercurrent multiplier is far greater than the difference across the partition separating the limbs at any one point (part B of the figure). As a consequence, the countercurrent multiplier can produce greater concentration changes than would be attained by a simple transport epithelium without the configuration of a countercurrent system. The longer the multiplier, the greater the concentration differences that can be attained.

Third, the multiplier system can work only if it contains an asymmetry. In the model in part A, there is an *active* energy-requiring net transport of NaCl in one direction across the partition. A *passive* countercurrent system, such as one used to conserve heat, does not require the expenditure of energy (part C of the figure). In the extremities of birds and mammals that inhabit cold climates, for example, a temperature differential exists between the arterial and venous flow of blood, because the blood is cooled as it descends into the leg. As a result of this asymmetry and the countercurrent arrangement of the vessels, the arterial blood gives up some of its heat to the venous blood leaving the leg, thereby reducing the amount lost to the environment.

A

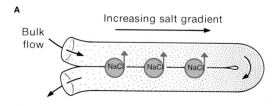

B

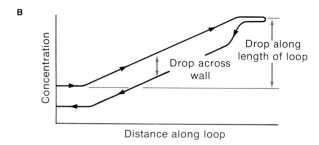

C

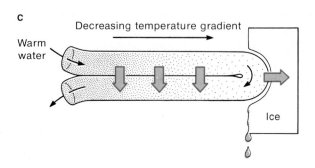

Active countercurrent systems require the expenditure of energy, whereas passive ones do not. **(A)** Model of an active system in which a salt solution flows through a U-shaped tube with a common dividing wall. The active transport of NaCl from the outflow to the inflow limb constitutes an asymmetry necessary for the multiplier system to work. **(B)** A plot of salt concentration along the two limbs. Note that the concentration difference across the wall at any point is small relative to the total concentration difference along the length of the loop. The length of the loop as well as the efficiency of transport across the wall will determine the overall concentration gradient along the entire length of the loop. **(C)** Model of a passive system in which warm water flows down the input limb and gives up part of its heat to cooler water flowing in the opposite direction in the outflow limb. Some heat is lost to the heat sink represented by the ice, but much more of the heat is conserved by passive transfer from the inflow to the outflow limb.

permeable to NaCl or urea. Thus the osmotic loss of water allows the tubular fluid to approach osmotic equilibrium with the interstitial fluid around the hairpin turn of the loop. As the tubular fluid flows up the ascending limb, it undergoes a progressive loss of NaCl (but not water). Most of the NaCl is actively transported across the wall of the thick segment of the ascending limb, although there is some passive loss of NaCl across the thin segment. Both the thin and the thick segments of the ascending limb are relatively impermeable to water.

The functional asymmetry between the descending and the ascending limbs of the loop of Henle, together with the countercurrent principle, accounts for the observed interstitial *corticomedullary osmotic gradient* of NaCl and urea represented by the gray wedge in Figure 14-32. The interstitial osmotic gradient is believed to be established by a combination of features that include the active transport of NaCl from the ascending thick segment and selective passive permeabilities to water, salt, and urea along specific segments of the nephron.

Recall that the descending limb of the loop of Henle has high-water, low-urea, and low-salt permeability, whereas the ascending limb has low-water, low-urea, and high-salt permeability. As shown in Figure 14-32 (step 1), NaCl is actively transported out of the tubular fluid in the thick segment of the ascending limb of the loop of Henle and in the distal tubule. The mechanism of this active salt transport is similar to salt secretion in the rectal gland of sharks and in the gills of saltwater teleosts (discussed in later section) except that in the kidney salt is transferred from the tubular lumen into the blood. The loss of NaCl from these segments and its addition to the surrounding interstitium leads to the osmotic loss of water (step 2) from the distal tubule and from the salt-impermeant descending limb in the cortex and outer medulla.

Because of the net loss of water and salt from the filtrate in the loop of Henle and the distal tubule, the filtrate entering the collecting duct has a high urea concentration. The renal tubule up to this point is largely impermeable to urea, but as the collecting duct passes into the depths of the medulla, it becomes highly permeable to urea. As a result, urea leaks out down its concentration gradient (step 3), raising the interstitial osmolarity of the inner medulla. The resulting high interstitial osmolarity draws water from the

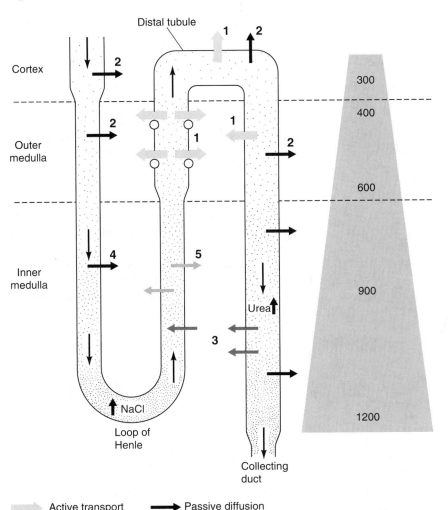

Figure 14-32 The steady-state corticomedullary osmotic gradient in the renal interstitium depends on differing permeabilities and active salt transport in different segments of juxtaglomerular nephrons, as well as on the anatomic layout of the nephrons and their circulatory supply (the vasa recta, not shown). The gray wedge depicts the osmotic gradient in the extracellular fluid with the small numbers indicating the total osmolarity. Active transport of NaCl from the ascending thick limb of the loop of Henle and distal tubule (step 1) is largely responsible for the interstitial osmolarity in the cortex and outer medulla. The high osmolarity of the inner medulla depends largely on the passive diffusion of urea from the lower collecting duct (step 3), the only portion of the nephron highly permeable to urea. Some urea reenters the filtrate in the thin limb of the loop of Henle, where the urea level is relatively low, leading to recycling of urea (thin red arrow). See text for further discussion of the various transport steps depicted. [Adapted from Jamison and Maffly, 1976.]

descending limb of the loop of Henle (step 4), producing a very high intratubular solute concentration at the bottom of the loop. As the highly concentrated tubular fluid then flows up the highly salt-permeable thin segment of the ascending limb, NaCl leaks out (step 5) down its concentration gradient. The lower collecting duct is the only section of the nephron with a high urea permeability. The high osmolarity of the inner medullary interstitium thus depends largely on the passive accumulation there of urea by the countercurrent mechanism of the nephron. If the ascending limb were as permeable to urea as is the collecting duct, this accumulation would not occur. If NaCl were not actively removed (with water following passively), urea would not become concentrated in the collecting duct, and the high medullary accumulation of urea would not take place either.

It is interesting that the interstitial medullary urea gradient is established largely by passive means, although the active transport of NaCl is an essential component of the system and accounts for most of the metabolic energy expenditure necessary to set up the NaCl and urea gradients. The result of this combination of cellular specialization and tissue organization is a standing corticomedullary gradient of urea and NaCl in which the osmolarity becomes progressively higher with distance into the depths of the renal medulla, both inside the tubule and in the peritubular interstitium. This gradient is responsible for the final osmotic loss of water from the collecting ducts into the interstitium and the consequent production of a hyperosmotic urine.

A countercurrent feature in the organization of vasa recta, the blood vessels around the nephron, is essential in maintaining the standing concentration gradient in the interstitium. Blood descends from the cortex into the deeper portions of the medulla in capillaries that form looplike networks around each juxtamedullary nephron and then ascends toward the cortex (see Figure 14-14A). In this circuit, the blood takes up salt and gives up water osmotically as the surrounding interstitial fluid becomes increasingly hyperosmotic. Thus the osmolarity of the blood increases as it descends via the vasa recta into the medullary depths (Figure 14-33). The reverse occurs as the blood returns to the cortex and encounters an interstitium of progressively lower osmolarity. As a result, there is little net change in blood osmolarity during the circuit through the vasa recta, although the water and solutes removed from the glomerular filtrate in its passage through the nephron are carried away by the blood. However, this represents only a small percentage of the large volume of blood that perfuses the kidney.

An important consequence of the countercurrent organization of the vasa recta is that it allows a high rate of renal blood flow (essential for effective glomerular filtration) without disrupting the corticomedullary standing gradient of NaCl and urea concentration. As the blood leaves the glomerulus and moves down the vasa recta into the medulla, it passively accepts interstitial NaCl and urea as it encounters ever-increasing interstitial osmolarities. NaCl

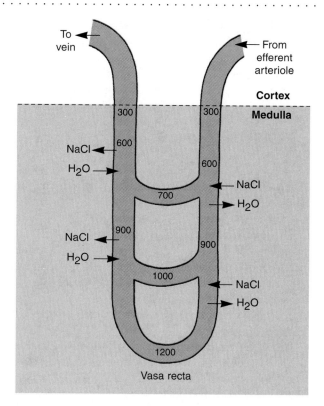

Figure 14-33 The countercurrent arrangement of the vasa recta helps maintain the interstitial corticomedullary osmotic gradient. This schematic diagram of the vasa recta indicates the passive fluxes of NaCl and water and the osmolarity of the blood at various points. Note that the blood osmolarity is the same at the beginning and end of the vasa recta.

and urea in the blood reach their peak concentrations as the blood traverses the loop of the vasa recta in the depths of the medulla. On ascending back toward the cortex, the excess NaCl and urea diffuse back into the interstitium, staying behind as the blood leaves the kidney. But before leaving the kidney, the blood, in fact, regains some of the water it lost during glomerular filtration. This happens because the colloid osmotic pressure of the blood plasma is elevated during ultrafiltration.

Control of Water Reabsorption

The tubular fluid is concentrated by the osmotic removal of water as it passes down the collecting duct into the hyperosmotic depths of the renal medulla (see Figure 14-32). This concentration provides a means of regulating the amount of water passed in the urine. The rate at which water is osmotically drawn out across the wall of the collecting duct from the urine into the interstitial fluid depends on the water permeability of the wall of the collecting duct. Vasopressin, also known as antidiuretic hormone (ADH), regulates the water permeability of the collecting duct and thereby controls the amount of water leaving the animal via the urine. The higher the level of ADH in the blood, the more permeable the epithelial wall of the collecting duct, and hence the more water is drawn out of the urine as it passes down the duct toward the renal pelvis. The effect of ADH on water reabsorption from the duct is shown in Figure 14-34.

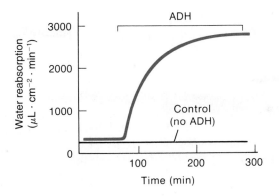

Figure 14-34 Antidiuretic hormone (ADH) increases the water permeability of portions of the collecting duct (see Figure 14-35). The data shown are from a perfusion experiment in which the fluid perfused through the duct and in the external bath were held constant at 125 mosM and 290 mosM, respectively. In the absence of ADH little water was reabsorbed from the perfusate, but application of ADH caused a dramatic increase in reabsorption. [Adapted from Grantham, 1971.]

The blood ADH level is a function of the osmotic pressure of the plasma and blood pressure. The neurosecretory cells that produce ADH have their cell bodies in the hypothalamus and their axon terminals in the neurohypophysis (posterior pituitary gland). These osmotically sensitive cells respond to increased plasma osmolarity by increasing the rate at which ADH is released into the bloodstream from their axon terminals, thereby increasing the blood level of ADH and reabsorption of water from the collecting duct

(Figure 14-35). If, for example, the osmolarity of the blood is increased as a result of dehydration, the activity of the neurosecretory neurons is increased, more ADH is released, the collecting ducts become more permeable, and water is osmotically drawn from the urine at a higher rate. This process results in the excretion of a more concentrated urine and the conservation of body water.

The hypothalamic cells that produce and release ADH receive inhibitory input from arterial and atrial baroreceptors that respond to increases in blood pressure. Hemorrhage, for example, results in a fall in blood pressure, reducing the activity of these baroreceptors (see Figure 12-44); the resulting decreased inhibitory input to the ADH-producing cells in the hypothalamus leads to increased release of ADH and reduced loss of water in the urine, thus helping to restore blood volume. Conversely, any factor that raises the venous blood pressure (e.g., an increase in blood volume due to dilution by ingested water) will inhibit the ADH-producing hypothalamic cells, causing an increased loss of body water via the urine. The ingestion of drinks containing ethyl alcohol inhibits the release of ADH and therefore leads to copious urination and an increase of plasma osmolarity beyond the normal set-point level. Some degree of dehydration results, and this contributes to the uncomfortable feeling of a hangover.

The action of mammalian ADH and the related peptide, arginine vasotocin, of nonmammalian vertebrate species, is not limited to the kidney. If these antidiuretic hor-

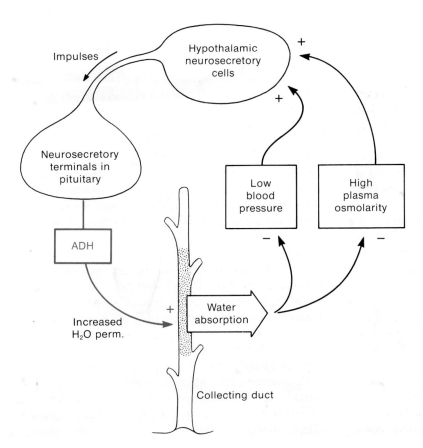

Figure 14-35 The osmolarity of the blood is under feedback regulation by the action of antidiuretic hormone on the collecting duct. Antidiuretic hormone (ADH) increases the water permeability of the stippled region, enhancing the rate of osmotic removal of water from the urine. The increased recovery of water counteracts worsening of the conditions that stimulate ADH secretion.

mones are applied to frog skin and toad bladder, they increase the water permeability of those epithelia.

To summarize the mechanisms we've discussed, the formation of urine in the mammalian kidney begins with the concentration of the glomerular filtrate into a hyperosmotic fluid in the proximal tubule. About 75% of the salt and water are removed from the filtrate in osmotically equivalent amounts as it passes through the proximal tubule, leaving urea and certain other substances behind. As the filtrate moves through the loop of Henle and the distal tubule, there is little net change in its osmolarity, but the countercurrent mechanism sets up a standing concentration gradient in the medullary interstitium along the length of the loop. This gradient provides the basis for the osmotic removal of water from the urine as it makes its way down the collecting duct within the medulla. Interestingly, this process takes place without active transport of water at any place along the nephron.

An animal can experience osmotic stress owing to changes in temperature or salinity and owing to the ingestion of food and drink. Perturbations in the osmotic state of the body fluids are minimized through feedback mechanisms by which the osmoregulatory organs adjust their activity so as to maintain the internal status quo. These feedback control mechanisms may be neural, endocrine, or a combination of the two. In mammals, adjustments in the volume and concentration of the urine are the primary means for maintaining osmotic homeostasis. In response to osmotic stress and other signals, mammals can regulate several aspects of urine formation including (1) the glomerular filtration rate, (2) the rate at which salts and water are absorbed from the lumen of the renal tubule, (3) secretion of unwanted substances, and (4) the rate at which water is osmotically drawn out of the pre-urine in the collecting duct.

NONMAMMALIAN VERTEBRATE KIDNEYS

In the kidneys of the marine hagfishes (class Cyclostomata), the nephrons possess glomeruli but no tubules, so the Bowman's capsules empty directly into collecting ducts. The kidneys are used largely to excrete divalent ions (e.g., Ca^{2+}, Mg^{2+}, and SO_4^{2-}) and carry out little or no osmoregulation. Thus, the extracellular fluids of the most primitive living vertebrate, the hagfish, are relatively similar to seawater in concentration of major salts, and their plasma is essentially isotonic relative to seawater (see Table 14-1).

As a general rule, the kidneys in freshwater teleosts have larger glomeruli and more of them than do those in their marine relatives. Because their bodies are hypertonic to the environment and water diffuses into their bodies, freshwater teleosts maintain water balance by producing large volumes of dilute urine. The kidney nephrons in certain marine teleosts have neither glomeruli nor Bowman's capsules. In such **aglomerular** kidneys, the urine is formed entirely by secretion because there is no specialized mecha-

nism for the production of a filtrate. These fish are hypotonic to their environment and so lose water continually across the skin and gills. Their problem is water conservation and they produce only small volumes of urine. Little urea is formed and ammonia is excreted across the gills.

Amphibians and reptiles appear incapable of producing a hypertonic urine (i.e., of higher osmolarity than the plasma), because they lack the countercurrent system of the loop of Henle that is necessary to produce urine of significantly greater osmolarity than the plasma. Only mammals and birds are known to have a renal countercurrent organization, and thus only these animals, apparently, have their plumbing so organized as to allow osmotic countercurrent multiplication. The avian kidney contains a mixture of reptilian-type and mammalian-type nephrons. That is, some avian nephrons lack a loop of Henle, and in some birds the loop is oriented perpendicular to the collecting duct, producing a less efficient concentrating mechanism.

The elasmobranch *Raja erinacea* (a skate) has been shown to have a complex renal tubule organization that has the anatomic requisites for countercurrent multiplication. However, the skate nephron is functionally quite different from the mammalian nephron. As we have seen, the mammalian kidney excretes urea and retains water to produce a hypertonic urine. The elasmobranch kidney, in contrast, retains urea (which is used as an osmolyte) and does not produce a concentrated urine. Tubular bundles constitute the countercurrent system in elasmobranch kidneys. These tubular bundles have been described in the kidneys of marine elasmobranchs, which have high levels of urea in their tissues and reabsorb urea from the kidney ultrafiltrate. Freshwater stingrays, on the other hand, do not reabsorb filtered urea and their kidneys lack tubular bundles, indicating that the bundles are the site of urea reabsorption. Thus, the function of the countercurrent organization of the elasmobranch nephron may be to conserve urea.

EXTRARENAL OSMOREGULATORY ORGANS IN VERTEBRATES

As indicated in the previous section, many vertebrates rely on extrarenal osmoregulatory organs to maintain osmotic homeostasis. First, we consider specialized glands for excreting salt found in some animals, and then see how fish gills are used for osmoregulation.

Salt Glands

Elasmobranchs, marine birds, and some reptiles possess glands that secrete salt by cellular mechanisms similar to sodium reabsorption in the mammalian kidney.

Elasmobranch rectal gland

Marine elasmobranchs, although slightly hypertonic to seawater, have a much lower NaCl content than seawater. As a result there is a continual influx of NaCl into the body of the animal. The excess salt is removed largely by the rectal gland, which produces a concentrated salt solution and is the major (perhaps the only) extrarenal site for secretion of excess NaCl by marine elasmobranchs. The gland functions to regulate extracellular volume by controlling the amount of NaCl in the body.

The rectal gland consists of a large number of blind-ending tubules that drain into a duct, which opens into the intestine near the rectum. The fluid produced by the gland can have a slightly higher salt concentration than seawater but is isosmotic to the plasma of the fish. The blood of elasmobranchs is also slightly hyperosmotic to seawater but has a much lower salt concentration, the osmolarity of the blood being made up by high concentrations of urea and trimethylamine oxide (TMAO). Elasmobranchs are able to tolerate high levels of urea, which normally causes the dissociation of multi-subunit enzymes, thus inhibiting their activity. In contrast, TMAO promotes the association of subunits, thereby counteracting the effect of urea. Urea and TMAO do not appear in the rectal gland fluid, only NaCl.

Formation of the secreted fluid in the rectal gland does not involve filtration of the blood; rather NaCl is secreted into the lumen of the tubule and water follows. The cells of the tubule wall of the rectal gland consist of a single type of cell, a salt-secreting cell similar to the chloride cell found in the gills of marine teleosts. This cell has a very extensive, folded, basolateral membrane, whose surface area is much larger than that of the apical (mucosal) membrane. The basolateral (serosal) membrane contains high concentrations of a Na$^+$/K$^+$ ATPase, which pumps Na$^+$ out and K$^+$ into the cell; the K$^+$, however, cycles back out through the many potassium channels also present in the basolateral membrane (Figure 14-36). The activity of the Na$^+$/K$^+$ ATPase generates a large sodium gradient across the basal membrane, which drives NaCl uptake via a Na/2Cl/K cotrans-

port system also in the basolateral membrane. Thus as Na$^+$ and K$^+$ cycle across the basal membrane, the Cl$^-$ level inside the cell rises above that in the tubular lumen; eventually Cl$^-$ exits via chloride channels in the apical (mucosal) membrane moving down its concentration gradient. The overall effect is the movement of Cl$^-$ from the serosal (blood) side of the tubular wall into the lumen. This creates an electrical potential, with the serosal side positive and the lumen negative; the resulting electrochemical gradient for sodium permits diffusion of Na$^+$ from the serosal side through paracellular pathways into the lumen. Water follows the transport of NaCl and is distributed passively across the tubular wall, but the wall is impermeable to urea and TMAO. Thus the rectal gland produces a solution that has a much higher NaCl concentration than the blood but is isosmotic with the blood.

The hearts of dogfish sharks contain a natriuretic peptide hormone that stimulates chloride secretion in perfused rectal glands. Although circulating natriuretic peptide levels have not yet been measured in elasmobranchs, it is possible that natriuretic peptides released from the heart into the circulation stimulate secretion by the rectal gland, reducing extracellular volume. The appropriate stimulus for the release of the natriuretic peptide would seem to be a rise in venous pressure, that is, filling pressure of the heart. In fact, the heart of a teleost fish, the rainbow trout, has been shown to contain natriuretic peptide hormone, which is released into the circulation by increased venous pressure.

 Do you think dinosaurs had salt glands?

Salt gland in birds and reptiles

In 1957, Knut Schmidt-Nielsen and his coworkers, investigating the means by which marine birds maintain their osmotic balance without access to freshwater, discovered that the nasal salt glands secrete a hypertonic solution of NaCl. It was found in those early studies that if cormorants or gulls are administered seawater by intravenous injection or by stomach tube, the increase in the plasma salt concentration leads to a prolonged nasal secretion of fluid with an osmolarity two to three times that of the plasma. Salt glands have subsequently been described in many species of birds and reptiles, especially those subjected to the osmotic stress of a marine or desert environment. These species include nearly all marine birds, ostriches, the marine iguana, sea snakes, and marine turtles, as well as many terrestrial reptiles. Crocodilians have a similar salt-secreting gland in the tongue.

The salt glands of birds and some reptiles occupy shallow depressions in the skull above the eyes. In birds, the salt gland consists of many lobes about 1 mm in diameter, each of which drains via branching secretory tubules and a central canal into a duct that, in turn, runs through the beak and empties into the nostrils (Figure 14-37A,B). Active

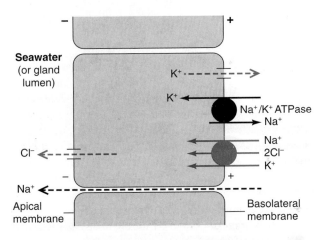

Figure 14-36 Salt-secreting cells present in the rectal gland of sharks, the bird and reptile nasal gland, and the gills of marine teleosts all use the same basic mechanism for transporting salt from the blood. Operation of the Na$^+$/K$^+$ ATPase and Na/2Cl/K cotransporter in the basolateral membrane results in net movement of Cl$^-$ from the blood into the gland tubules or seawater in the case of fish. The transmembrane potential created by this movement increases the sodium electrochemical gradient sufficiently so that Na$^+$ can diffuse via paracellular channels even against a high concentration gradient.

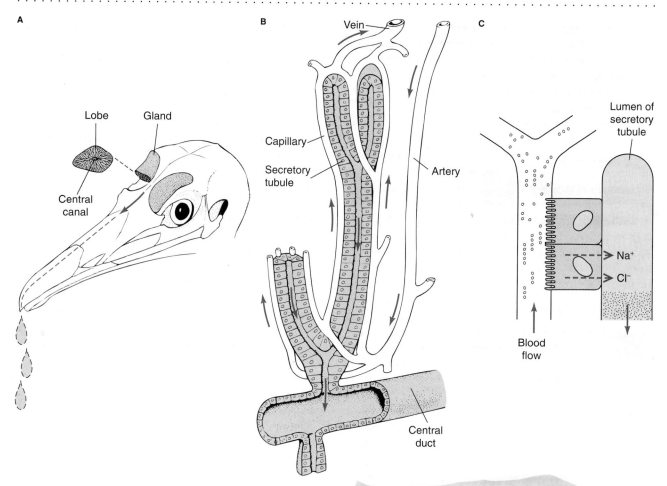

Figure 14-37 Marine birds maintain osmotic balance by excretion of a concentrated salt solution from glands located above the orbit. **(A)** The avian salt gland consists of a longitudinal arrangement of many lobes, which drain via a central canal into a duct that carries secretions to the nasal passages. **(B)** Each lobe consists of tubules and capillaries arranged radially around a central canal. Single tubules are surrounded by capillaries in which blood flows counter to the flow of secretory fluid in the tubule. This countercurrent flow facilitates the transfer of salt from the blood to the tubule, since the uphill gradient of salt concentration between capillary and tubule lumen is thereby minimized at each point along the length of the tubule. **(C)** The secretory cells constituting the tubular wall transport NaCl from the blood into the lumen via the mechanism depicted in Figure 14-36. These cells have a brush border and contain many mitochondria. [Part A adapted from Schmidt-Nielsen, 1960; part B from "Salt Glands," by K. Schmidt-Nielsen. Copyright © 1959 by Scientific American, Inc. All rights reserved.]

secretion takes place across the epithelium of the secretory tubules, which consists of characteristic salt-secreting cells. These have a profusion of deep infoldings in the basolateral membrane and are heavily laden with mitochondria. As in many other transport epithelia, adjacent cells are tied together by junctions, which preclude the massive leakage of water past the cells, from one side of the epithelium to the other. These cell junctions, however, are not as tight as those binding cells of the frog skin together but are leaky allowing the paracellular movement of ions, as in the rectal gland.

The formation of fluid in the nasal gland, as in the rectal gland, does not include filtration of the blood. The absence of filtration can be deduced from the failure of small filterable molecules (e.g., inulin or sucrose) that are injected into the bloodstream to appear in the gland fluid. High concentrations of a Na$^+$/K$^+$ ATPase have been demonstrated in the basolateral membrane of the tubular cells. Application of ouabain to the basal surface of the epithelium blocks salt transport. Since this inhibitor does not pass across epithelia and can block the pump only by direct contact with the ATPase, the sodium-transport mechanism appears to operate in the basal membrane of the epithelial cells, as it does in the rectal gland. Increased salt secretion is associated with increased Na$^+$/K$^+$ ATPase activity in the salt gland. The Na$^+$/K$^+$ ATPase also occurs to some extent in the apical membrane of the bird nasal gland. The basal membrane of the salt-gland epithelium also contains a Na/2Cl/K cotransporter and potassium channels, and the apical membrane contain chloride channels. The net result is movement of NaCl from the blood across the epithelium into the lumen of the salt gland (Figure 14-37C).

As we saw earlier, the salt solution produced by the elasmobranch rectal gland is isosmotic to plasma; in contrast, the fluid produced by the nasal gland is hyperosmotic to plasma. In both cases, the gland fluid has a high salt concentration, but the osmolarity of the blood of elasmobranchs is much higher than that of birds and reptiles. It is not clear how the solution produced by the nasal glands of birds and reptiles is concentrated. It is possible that the initial solution in the apex of the tubule is isosmotic to plasma and becomes more concentrated as it passes down the tubule. The cells of the secretory epithelium of a single tubule become larger, with deeper paracellular channels, towards the base of the tubule, indicating that the fluid may become more concentrated towards the base of the tubule. Those birds that can produce the most concentrated salt solutions have the largest secretory cells, with long paracellular channels between cells. In addition, the avian salt gland and its blood flow are organized as a countercurrent system and this might aid in concentration of the salt solution. The capillaries are so arranged that the flow of blood is parallel to the secretory tubules and occurs in the direction opposite to the flow of secretory fluid (see Figure 14-37B). This flow maintains a minimal concentration gradient from blood to tubular lumen along the entire length of a tubule; it thereby minimizes the concentration gradient for uphill transport from the plasma to the secretory fluid.

The salt gland is not always active but responds to a salt load and/or expansion of the extracellular space. When birds drink seawater, water will diffuse from the body into the gut because seawater has a higher osmolarity than the body fluids. At the same time NaCl will diffuse from the seawater in the gut into the body. Thus the initial effect of drinking saltwater is to reduce extracellular volume while increasing NaCl levels in extracellular fluid and the blood (Figure 14-38A). The salt level in the gut thus will drop because of salt loss to the body and diffusion of water from the body into the gut. After a while the osmolarity of the gut fluids will fall below that of the body, so that water movement between the body into the gut will reverse, that is, water moves into the body, following the movement of salt and expanding extracellular volume. The initial reduction in extracellular volume inhibits nasal fluid production immediately after drinking seawater. The subsequent elevation of both extracellular volume and salt content, acts as a strong stimulus for salt secretion, thus there is often a short delay between drinking saltwater and secretion from the nasal gland. Since the solution secreted from the salt gland is more concentrated than the seawater taken in, the bird ends up gaining osmotically free water, as illustrated in Figure 14-38B.

Regulation of the secretory activity of the avian salt gland involves both parasympathetic neural control and neuroendocrine control through the pituitary (Figure 14-39). Osmoreceptors in the hypothalamus respond to an increase in plasma tonicity by a sensory discharge. This response, together with input from extracranial osmoreceptors and/or volume receptors, activates parasympathetic cholinergic neurons that innervate the salt

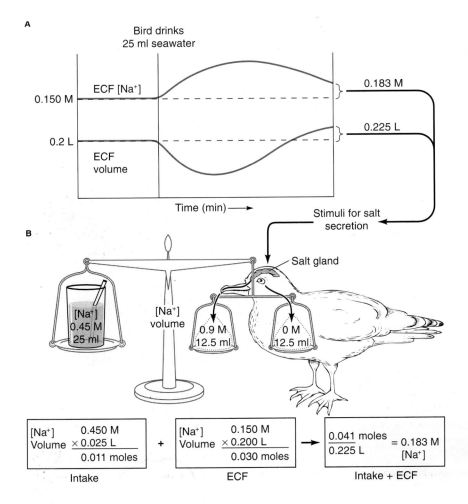

Figure 14-38 Because the salt-gland secretion is more concentrated than seawater, birds that drink seawater gain osmotically free water. Before drinking seawater, the gull in this example has an extracellular fluid (ECF) volume of 0.2 liters and an ECF [Na⁺] of 0.15 M; thus the ECF contains 0.03 moles of Na⁺. The bird then drinks 0.025 liters of seawater with a [Na⁺] of 0.45 M, so it ingests 0.011 moles of Na⁺. Initially the ECF volume decreases and ECF [Na⁺] increases **(A)** because Na⁺ moves from the seawater in the gut into the ECF (down its concentration gradient), while water moves into the gut until osmotic equilibrium is established between the ECF and gut. The initial decrease in ECF volume inhibits salt-gland secretion. As the ECF [Na⁺] rises, water from the gut moves back into the ECF. When both the ECF volume and [Na⁺] are above their base levels, the salt gland is stimulated **(B)**. If the secretion has a [Na⁺] of 0.9 M (twice as concentrated as the ingested seawater), then the gull can secrete all the ingested salt in half the volume. In this example, the gull has a net profit of 12.5 ml of osmotically free water; this water can be used to excrete other ions (and molecules) via the kidneys, which continue to filter actively. [Adapted from unpublished material courtesy of Dr. Maryanne Hughes.]

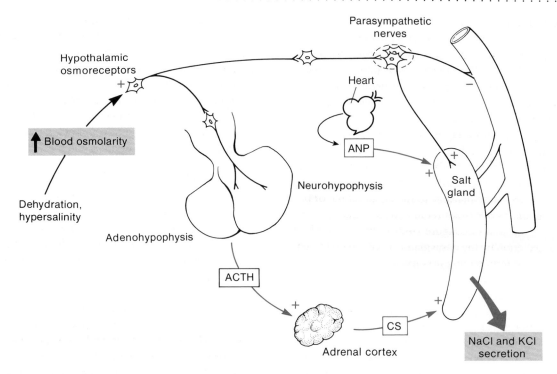

Figure 14-39 The increase in avian salt-gland secretion in response to a rise in blood osmolarity and decrease in blood pressure is mediated by direct and indirect mechanisms. Stimulation of osmotically sensitive neurons in the hypothalamus and sensory input from peripheral osmoreceptors activate direct parasympathetic pathways to the salt gland and to the vessels supplying blood to the gland. Atrial natriuretic peptide (ANP), re-leased from the heart in response to low venous pressure, also directly stimulates secretion. Pituitary secretion of ACTH in response to increasing blood osmolarity indirectly promotes salt secretion by stimulating release of corticosterone (CS) from the adrenal cortex. This hormone acts directly on the gland making it responsive to blood tonicity.

gland. Acetylcholine liberated from the terminals of those neurons not only stimulates the secretion of salt but also enhances secretion by causing vasodilation and thus increased blood flow to the secretory tissue. Acetylcholine acts on muscarinic receptors in the secretory cells of the gland, triggering the inositol phospholipid intracellular signaling system that leads to a rise in cytosolic calcium levels (see Figure 9-14). The increased intracellular calcium levels activate chloride and potassium channels in the plasma membrane of the secretory cells. A variety of other agents can stimulate secretion by increasing cAMP levels, which, in turn, activate chloride channels. The end result of an intracellular increase in the levels of inositol phosphate and/or cAMP is salt secretion.

Secretion is also stimulated by adrenocorticosteroids and by prolactin. Although direct neural control is most important in making short-term adjustments to osmotic stress, corticosterone is required for maintaining the responsiveness of the salt gland. For instance, when an animal's adrenal cortex, the source of corticosteroids, is removed, the infusion of a high-tonicity salt solution is no longer effective in stimulating salt-gland secretion (Figure 14-40). But if corticosterone is then injected into the experimental animal, salt-gland function is retained. Atrial natriuretic peptide (ANP), secreted from the heart in response to increased venous pressure, also stimulates secretion by acting directly on the secretory cells of the bird salt gland. This hormone

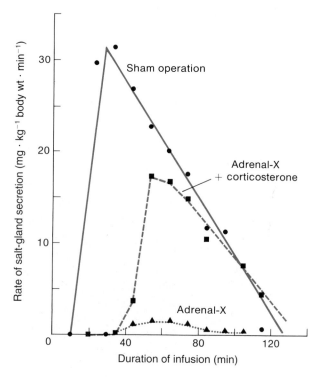

Figure 14-40 The responsiveness of the avian salt gland to high blood osmolarity is dependent on corticosterone. Two days after adrenalectomy (adrenal-X), 10% NaCl was infused into the blood of animals with (black squares) and without (black triangles) corticosterone replacement therapy. [Adapted from Thomas and Phillips, 1975.]

causes a transient increase in salt secretion, presumably reducing blood volume and, therefore, venous pressure.

In addition, increased blood osmolarity stimulates secretion of AVT, an ADH-like neurohypophyseal peptide. Although AVT has no effect on salt-gland secretion, it reduces glomerular filtration and water and salt excretion in the bird's kidneys, which are capable of producing urine that is only slightly hypertonic to plasma. The action of AVT on the kidney, combined with salt excretion via the salt gland, results in water retention by the body and a decrease in blood osmolarity. As in mammals, low blood pressure and/or solute delivery to the distal tubule of bird kidneys stimulates release of renin and subsequent formation of angiotensin II (see Figure 14-26). Angiotensin II inhibits salt secretion by the nasal gland via action on the central nervous system, having no direct effect on the gland.

The reason that birds and reptiles can drink seawater and survive is because, unlike mammals, they have a nasal gland that can excrete hypertonic salt solutions. Mammals have salt-secreting cells located in the thick ascending limb of the loop of Henle that are similar to those found in the nasal gland of birds and the rectal gland of elasmobranchs. In addition, these mammalian cells seem to be controlled by the same array of hormones, namely natriuretic peptides and the renin-angiotensin system. In mammals, however, these salt-secreting cells are not arranged in a way that permits production of a hypertonic salt solution that can be excreted from the body. Thus organization at the organ, as well as the cellular and molecular levels, is important in determining the ability of an animal to survive in a range of environments.

Fish Gills

The epithelial surface area of a gill must be large if it is to function efficiently as an organ for respiratory gas exchange. Although this feature makes gills an osmotic liability for animals such as fishes, which are out of osmotic equilibrium with their aqueous environment, it does make gills suitable as organs for osmoregulation. Thus, the gills of numerous aquatic species, vertebrate and invertebrate, are active not only in gas exchange but also in such diverse functions as ion transport, excretion of nitrogenous wastes, and maintenance of the acid-base balance. In teleost fishes, for example, the gills play the central role in coping with osmotic stress.

The structure of a teleost gill is illustrated in Figure 14-41. The epithelium separating the blood from the external water consists of several cell types including mucous cells, chloride cells, and pavement cells (Figure 14-42). The epithelium of the lamellae consists mostly of flat pavement cells no more than $3-5\ \mu m$ thick, but containing some mitochondria. These are clearly best suited for respiratory exchange, acting as minimal barriers for diffusion of gases. The epithelium covering the gill filaments also contains chloride cells, which are more columnar in shape and several times thicker from base to apex than the pavement cells. Chloride cells are deeply invaginated by infoldings of the basolateral membrane and are heavily laden with mitochondria and with enzymes related to active salt transport. Pavement cells and chloride cells are joined by tight junctions limiting the paracellular movement of water and ions.

Secretion of salt in seawater

Chloride cells were first described in 1932 by Ancel Keys and Edward Willmer, who ascribed to them the transport of chloride because they exhibit histochemical similarities to cells that secrete hydrochloric acid in the amphibian stomach, and because it had already been shown that the gill of marine teleosts is the site of extrarenal excretion of chloride (and sodium). Subsequent histochemical studies confirmed the presence of high levels of chloride in these

[margin handwritten notes: AVT; function of gills; Function of gills in teleost; How the various cells of gills in teleost play a role]

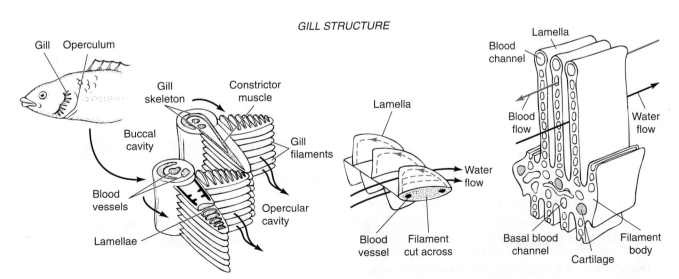

GILL STRUCTURE

Figure 14-41 Fish gills function as both respiratory and osmoregulatory organs. These drawings show a portion of the teleost gill in increasing magnification. In addition to gas exchange between the blood and water, Na$^+$ can move in and out of the blood in the lamellae. Black arrows indicate water flow; red arrow and dotted lines indicate blood flow.

A

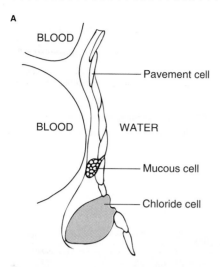

BLOOD

Pavement cell

BLOOD   WATER

Mucous cell

Chloride cell

Figure 14-42 The epithelium of the teleost gill is composed largely of pavement cells interspersed with a few mucous and chloride cells. **(A)** Drawing of lamellar epithelium showing typical distribution of pavement, mucous, and chloride cells. The chloride cells tend to be at the base of the lamellae. **(B)** Electron micrograph of a freshwater teleost chloride cell with adjacent pavement cells. **(C)** Electron micrograph of dogfish gill mucous cell containing many large mucous granules. [Electron micrographs courtesy of Jonathan Wilson.]

B

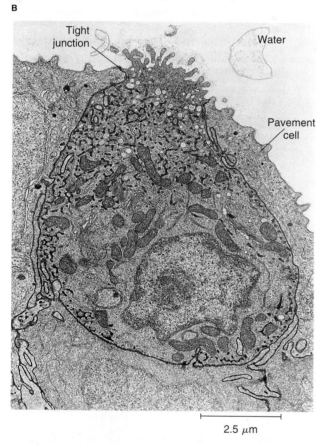

Tight junction

Water

Pavement cell

2.5 μm

C

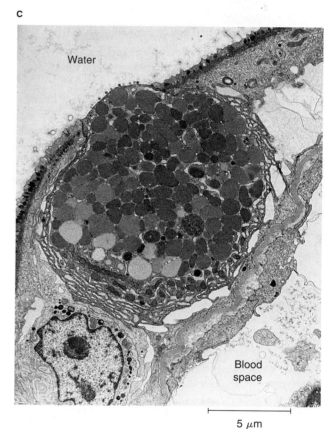

Water

Blood space

5 μm

cells, especially near the pit that develops on the apical (mucosal or external) border of these cells in fishes that have become adapted to high salinities.

The mechanism of salt transport by chloride cells is similar to that of the salt-secreting cells illustrated in Figure 14-36. Thus, chloride cells have high levels of a Na^+/K^+ ATPase associated with Na/2Cl/K cotransporters in the basolateral membrane and a chloride channel in the apical membrane. Each chloride cell is associated with an accessory cell (distinct from a pavement cell), and Na^+ diffuses from blood to seawater through the less-tight paracellular channel between the chloride and accessory cell. In

the case of marine teleosts, the secretion of salt occurs against an osmotic gradient and there is no movement of water following the movement of salt. Thus the shark rectal gland, avian nasal gland, seawater teleost gill, and the thick ascending limb of the loop of Henle in the mammalian kidney tubule all appear to contain salt-secreting cells that transport NaCl by the same basic mechanism (see Figures 14-25A and 14-36). In the mammalian kidney, however, the direction of salt transport is into the blood rather than into the environment, as in the other cases.

Since chloride cells were first characterized and associated with the transport of Cl^- across the gills of marine teleosts, they have been found to also mediate the exchange of other ions including Ca^{2+}. For instance, Ca^{2+} in the water is taken up via calcium channels in the apical membrane of the chloride cells and then actively transported into the blood via a Ca^{2+} ATPase, present at high levels in the basolateral membrane.

Uptake of salt in freshwater

Pavement cells in gills of freshwater fish

The pavement cells in the gills of freshwater fish appear to have a proton ATPase and sodium channels in the apical membrane. The proton ATPase is presumably electrogenic and pumps protons out of the gills, generating a potential that draws Na^+ into the cell, a mechanism similar to that demonstrated in frog skin and the mammalian kidney (see Figure 14-29A). A clear relationship between the activity of the proton pump and the apical membrane potential has yet to be demonstrated in fish gills. A Na^+/K^+ ATPase in the basolateral membrane pumps Na^+ out of the cell into the blood, and K^+ cycles through potassium channels in the membrane. Thus a proton ATPase appears to energize Na^+ uptake across the apical membrane, whereas a Na^+/K^+ ATPase moves sodium across the basolateral membrane of the gills of freshwater fish.

Chloride cells

The gill epithelium of freshwater fish also possess chloride cells, which mediate uptake of Ca^{2+} from the water. These cells, which differ from the chloride cells in marine teleost fish, have an anion transporter on the apical membrane and high levels of proton ATPase within the cell. They may be involved in the uptake of Cl^-, as well as Ca^{2+}, by freshwater fish.

Physiological adaptation in migrating fish

In species (e.g., salmon and eels) that regularly migrate between seawater and freshwater, the gill epithelium changes so as to adapt to changes in environmental salinity. These fishes actively take up NaCl in freshwater and actively excrete it in saltwater by the mechanisms described above. The physiological adaptation of the gills involves the synthesis and/or destruction of molecular components of the epithelial transport systems and changes in the morphology and number of the chloride cells. When fishes that are able to tolerate a wide range of salinities are transferred between freshwater and saltwater, it can take up to a few days for physiological adaptation to the new environment to occur and for the animal to regain osmotic homeostasis. It is now known that osmoregulatory adaptation is mediated by endocrine hormones that influence epithelial differentiation and metabolism. The steroid hormone cortisol and growth hormone stimulate the changes in gill structure associated with the transition from freshwater to seawater, whereas prolactin stimulates the changes in gill structure that accompany the reverse transition.

First, let's consider what happens when fish migrate from freshwater to saltwater (Table 14-11, part A). When the fish are in freshwater, the proton ATPase in pavement cells is active. But when they move from freshwater to seawater, the proton ATPase is down-regulated because Na^+

TABLE 14-11

Physiological adaptation accompanying movement of fish to water of differing salinity

(A) Freshwater → saltwater
1. Proton ATPase that powers active uptake of NaCl is down-regulated.
2. Increased Na^+ influx into body raises plasma Na^+, stimulating an increase in plasma cortisol and growth hormone levels.
3. Hormones induce an increase in the number of chloride cells and an elaboration of their basolateral membrane, resulting in increased infoldings.
4. As a result, Na^+/K^+ ATPase activity and NaCl secretion increases.
5. Plasma Na^+ levels return to normal.

(B) Saltwater → freshwater
1. Low external sodium closes paracellular gaps between chloride and accessory cells, so NaCl efflux falls rapidly.
2. Plasma prolactin levels increase.
3. Hormone causes the number of chloride cells to decrease and the apical pits to disappear.
4. As a result Na^+/K^+ ATPase activity falls.
5. Up-regulation of proton ATPase returns fish to the freshwater condition.

uptake is no longer required. The influx of sodium from seawater causes plasma Na^+ levels to rise, which in turn stimulates secretion of cortisol (Figure 14-43A). Cortisol, along with growth hormone, induces an increase in the number of typical seawater chloride cells. As a result of these changes, the gill Na^+/K^+ ATPase activity and salt secretion increase (Figure 14-43B). In salmon, cortisol release begins while the fish is moving down river, thus pre-adapting the fish for a seawater existence. This process is termed *smolting*, and the end product is a *smolt*, a fish ready for seawater transfer. The increase in plasma Na^+ that occurs when the fish enters seawater causes additional release of cortisol and starts the changes that will allow the fish to live in seawater. After reaching the ocean, it normally takes about a week for the elevated plasma Na^+ to return to normal levels, similar to those of freshwater fish (see Figure 14-43A).

smolting

When a marine teleost moves from seawater to freshwater, more-or-less opposite changes occur, adapting the fish to low-salt water (see Table 14-11, part B). Initially, the paracellular gaps in the gill epithelium close, reducing salt loss. A rise in the plasma prolactin level stimulates changes in the chloride cells so that the Na^+/K^+ ATPase activity falls. Finally, up-regulation of the proton ATPase allows the uptake of salt necessary for survival in freshwater.

What are the design conflicts that exist in the gills, a structure used for both gas exchange and ionic regulation?

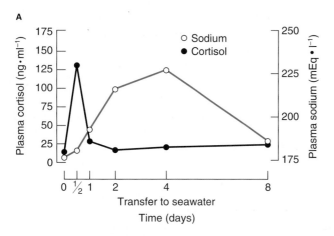

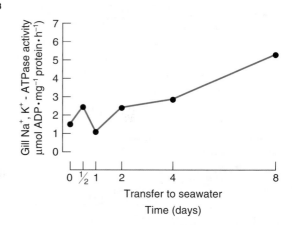

Figure 14-43 Cortisol plays a major role in inducing the physiological adaptation that occurs when coho salmon are transferred from freshwater to seawater. **(A)** Initially after a fish moves into seawater, the plasma sodium level begins to rise, stimulating secretion of cortisol. **(B)** The spike in plasma cortisol levels mediates a number of changes in the gills in-

cluding an increase in the gill Na^+/K^+ ATPase activity. As this activity increases, sodium secretion from the gill rises; thus, after several days in seawater, plasma sodium returns to values close to those observed in freshwater fish.

INVERTEBRATE OSMOREGULATORY ORGANS

In general, invertebrate osmoregulatory organs employ mechanisms of filtration, reabsorption, and secretion similar in principle to those of the vertebrate kidney to produce a urine that is significantly different in osmolarity and composition from the body fluids. Insects and possibly some spiders are the only invertebrates known to produce a concentrated urine. These mechanisms are used to differing extents in various organs in different groups of animals. That there has been convergent evolution of physiological mechanisms in nonhomologous organs underscores the utility of these mechanisms.

Filtration-Reabsorption Systems

Several lines of evidence indicate that filtration of plasma, similar in principle to that which occurs in the Bowman's capsule of vertebrates, underlies the formation of the primary urine in both mollusks and crustaceans. For example, when the nondigestible polysaccharide inulin is injected into the bloodstream or coelomic fluid, it appears in high concentrations in the urine. (This also occurs in mammals.) Since it is unlikely that such substances are actively secreted, they must enter the urine during a filtration process in which all those molecules below a certain size pass through a sievelike membrane of tissue. During the reabsorption of water and essential solutes, these polymers remain behind in the urine.

As in vertebrates, the normal urine of some invertebrates contains little or no glucose, even though substantial levels occur in the blood. However, studies with several mollusks have shown that when the blood glucose is elevated by artificial means (e.g., by injection), glucose appears in the urine. In each species, glucose appears in the urine at a characteristic threshold concentration of blood glucose; the urine glucose concentration rises linearly with

blood glucose concentration beyond the threshold level. This behavior parallels that in the mammalian kidney (see Figure 14-21) and probably results from saturation of the transport system by which glucose filtered into the tubular fluid is reabsorbed from the filtrate into the blood. Once the transport system is saturated, the "spillover" of glucose in the urine is proportional to its concentration in the blood. More conclusive evidence is obtained with the drug phlorizin, which is known to block active glucose transport. When phlorizin is administered to mollusks and crustaceans, glucose appears in the urine even at normal blood glucose levels. The most reasonable explanation for this effect is that glucose enters the urine as part of a filtrate and remains in the urine when the reabsorption mechanism is blocked by phlorizin.

Further support for the filtration-reabsorption mechanism comes from analyses of tubular fluids near suspected sites of filtration indicating that their composition is similar to that of the plasma. Finally, the rate of urine formation in some invertebrates has been found to depend on the blood pressure. This relationship is consistent with a filtration mechanism, but the change in blood pressure may also produce a change in the circulation to the osmoregulatory organ.

The site of primary urine formation by filtration is known in only a few invertebrates. In a number of marine and freshwater mollusks, filtration takes place across the wall of the heart into the pericardial cavity, and the filtrate is conducted to the "kidney" through a special canal. Glucose, amino acids, and essential electrolytes are reabsorbed in the kidney. In the crayfish, the major organ of osmoregulation is the so-called *antennal gland* (Figure 14-44). Part of that organ, the coelomosac, resembles the vertebrate glomerulus in ultrastructure. Micropuncture measurements have shown that the excretory fluid that collects in the coelomosac is produced by ultrafiltration of the blood. The antennal gland of Crustacea is clearly involved

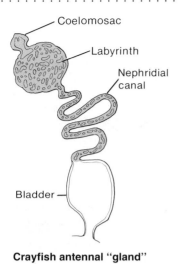

Crayfish antennal "gland"

Figure 14-44 Osmoregulation in some invertebrates depends on filtration-reabsorption organs that differ structurally from the mammalian kidney but are functionally analogous. In this schematic diagram of the crayfish antennal gland, the osmotically active tissues are shown in color. Filtration of the blood produces the initial excretory fluid, which then is modified by selective reabsorption of various substances. [Adapted from Phillips, 1975.]

in regulating the concentration of ions (e.g., Mg^{2+}) in the hemolymph.

Since the final urine in mollusks and crustaceans differs in composition from the initial filtrate, there must be either secretion of substances into the filtrate or reabsorption of substances from the filtrate. The reabsorption of electrolytes is well established in freshwater species, for the final urine has a lower salt concentration than either the plasma or the filtrate. Glucose must also be reabsorbed, since it is present in the plasma and in the filtrate but is either absent or present at very low concentrations in the final urine.

It is interesting that the filtration-reabsorption type of osmoregulatory system has appeared in at least three phyla (Mollusca, Arthropoda, Chordata) and perhaps more. This kind of system has the important advantage that all the low-molecular-weight constituents of the plasma are filtered into the ultrafiltrate in proportion to their concentration in the plasma. Such physiologically important molecules as glucose and, in freshwater animals, such ions as Na^+, K^+, Cl^-, and Ca^{2+} are subsequently removed from the filtrate by tubular reabsorption, leaving toxic substances or unimportant molecules behind to be excreted in the urine. This process avoids the need for active transport into the urine of toxic metabolites or, for that matter, unnatural, man-made substances of a neutral or toxic nature encountered in the environment. Thus, an advantage of the filtration-reabsorption system is that it permits the excretion of unknown and unwanted chemicals taken in from the environment without the necessity for a large number of distinct transport systems.

A disadvantage of the filtration-reabsorption osmoregulatory system is its high energetic cost for the organism.

The filtering of large quantities of plasma requires the active uptake of large quantities of salts, either in the excretory organ itself or in other organs, such as gills or skin. In frog skin, for example, it has been shown that 1 mol of oxygen must be reduced in the synthesis of ATP for every 16–18 mol of sodium ions transported. In freshwater clams, the cost of maintaining sodium balance amounts to about 20% of the total energy metabolism. In marine invertebrates, however, the filtration-reabsorption system proves metabolically less expensive, since salt conservation is much less of a problem.

Secretory-Reabsorption Systems

Insects can survive in both freshwater and arid terrestrial environments; given their often large surface-to-volume ratios, the osmotic demands placed on these insects can be extreme. The locust, for example, has a large capacity to regulate the ionic strength of the hemolymph (blood). During dehydration the hemolymph volume may decrease by up to 90%, but its ionic composition is maintained. In addition, when this insect is given solutions to drink that range in osmotic strength from that of seawater to that of tapwater, hemolymph osmotic pressure changes by only 30%. This capacity to regulate hemolymph composition depends on a secretory-type osmoregulatory system.

In broad outline, the osmoregulatory system of locusts and other insects consists of **Malpighian tubules** and the hindgut (ileum, colon, and rectum). The closed ends of the long, thin Malpighian tubules lie in the hemocoel (the blood-containing body cavity); the tubules empty into the alimentary canal at the junction of the midgut and hindgut (Figure 14-45). The secretion formed in the tubules passes into the hindgut, where it is dehydrated and passes into the rectum and voided as a concentrated urine through the anus. The presence of a tracheal system for respiration in insects (described in Chapter 13) diminishes the importance of an efficient circulatory system. As a consequence, the Malpighian tubules do not receive a direct, pressurized arterial blood supply, as the mammalian nephron does. Instead, they are surrounded with blood, which is at a pressure essentially no greater than the pressure within the tubules. Since there is no significant pressure differential across the walls of the Malpighian tubules, filtration cannot play a role in urine formation in insects. Instead, the urine must be formed entirely by secretion, with the subsequent reabsorption of some constituents of the secreted fluid. This process is analogous to the formation of urine by secretion in the aglomerular kidneys of marine teleosts. The serosal surface of the Malpighian tubule exhibits a profusion of microvilli and mitochondria, a specialization often associated with a highly active secretory epithelium.

What are the limitations placed on the osmoregulatory system in animals having a low-pressure open circulation?

A

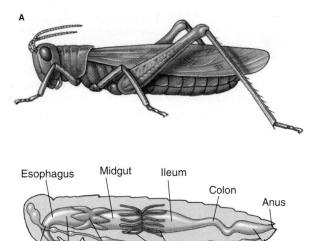

B

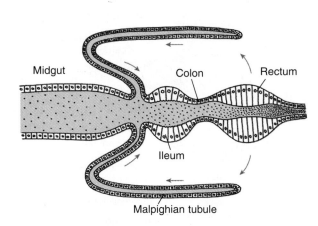

Figure 14-45 Osmoregulation in insects involves a secretory-reabsorption mechanism. **(A)** External side view and cross-sectional view of locust. **(B)** Simplified diagram showing relation of Malpighian tubule to gut of the locust. The pre-urine is produced by secretion into the lumen of the Malpighian tubules, which lie in the blood-containing hemocoel. The pre-urine flows into the rectum, where it is concentrated by the reabsorption of water; although ions also are reabsorbed, the excreted urine is hypertonic to the hemolymph. The arrows indicate the circular pathway of water and ion movement. The insect body contains numerous Malpighian tubules, even though only two are depicted.

pre-urine into the hindgut

The details of urine formation by tubular secretion differ among different insects, but some major features seem to be common throughout. KCl and, to a lesser extent, NaCl are transported from the hemocoel into the tubular lumen, along with such waste products of nitrogen metabolism as uric acid and allantoin. It appears that the transport of K⁺ is the major driving force for the formation of the pre-urine in the Malpighian tubules, with most of the other substances following passively. This has been concluded from the following observations:

K⁺ = driving force of pre-urine

- The pre-urine is isotonic or slightly hypertonic relative to the hemolymph.

- The pre-urine has a high K⁺ concentration in all insects.

- The rate of pre-urine formation is a function of K⁺ concentration in the fluid surrounding the tubule, higher K⁺ concentrations producing more rapid pre-urine accumulation.

- The formation of pre-urine is largely independent of the Na⁺ concentration of the surrounding fluid.

characteristics of pre-urine

Although K⁺ is osmotically the most important substance actively transported, there is evidence that active transport plays an important role in the secretion of uric acid and other nitrogenous wastes.

The pre-urine formed in the Malpighian tubules is relatively uniform in composition from one species to another, and in each species it remains isotonic relative to the hemolymph under different osmoregulatory demands. The fluid formed in the Malpighian tubules passes into the hindgut, where several important changes in its composition occur. In the hindgut, water and ions are removed in amounts that maintain the proper composition of the hemolymph. Thus, it is in the hindgut that the composition of the final urine is determined. The water and ions removed from the urine by the hindgut are transferred through intimate connections to the lumen of the Malpighian tubules. These substances are thereby retained and recycled in the Malpighian tubule-hindgut circuit (see Figure 14-45B).

The most complete study of the osmoregulatory behavior of the hindgut has been done with the desert locust *Schistocerca*. The serosal surface of the ileum and rectum is a highly specialized secretory epithelium (Figure 14-46). When a solution similar to hemolymph is injected into the hindgut of this insect, water, K⁺, Na⁺, and Cl⁻ are absorbed into the surrounding hemolymph. Evidence from electrical measurements suggests that the ions are transported actively with water following. An electrogenic chloride pump and potassium channels in the apical membrane appear to mediate KCl uptake from the hindgut lumen into the cells of the gut lining. Sodium uptake from the lumen is coupled to amino acid uptake and/or ammonium ion excretion. The KCl then moves from the cell into the hemolymph via appropriate channels in the basolateral membrane, while sodium is removed from the cell into the hemolymph by a Na⁺/K⁺ ATPase (Figure 14-47). Acid is excreted into the lumen of the hindgut by a proton ATPase. The locust hindgut is capable of removing a large amount of ions plus water, leaving behind an excess of

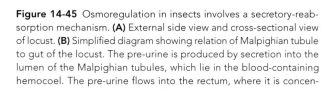

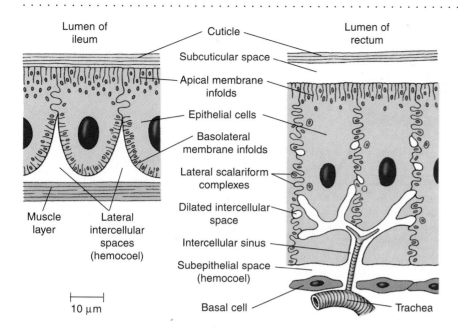

Figure 14-46 The hindgut of insects is specialized for transport of water and ions from the lumen to the surrounding hemocoel. Shown here are the ultrastructural organization and gross dimensions of the epithelium of the ileum and rectum in the locust, both of which are involved in reabsorption. Note the extensive infolding of the apical membrane and extensive lateral intercellular spaces. [Adapted from Irvine et al., 1988.]

ions and waste products so that the excreted urine is hypertonic with an osmolarity up to four times that of the blood.

In the mealworm *Tenebrio,* the urine-to-blood osmolarity ratio can be as high as 10, which is comparable to the concentrating ability of the most efficient mammalian kidneys. It has been suggested that uphill transport of water in *Tenebrio* and some other species results from a countercurrent arrangement of the Malpighian tubules, the perirectal space, and the rectum (Figure 14-48). Water is drawn

osmotically from the rectum into the Malpighian tubules because of the KCl gradient produced by active transport. The direction of bulk flow in these compartments is such that the osmotic gradient along the length of the complex is maximized, with the absolute osmolarities highest toward the anal end of the rectum. This gradient may allow the concentrations near the anal end to exceed those of the hemolymph by several times.

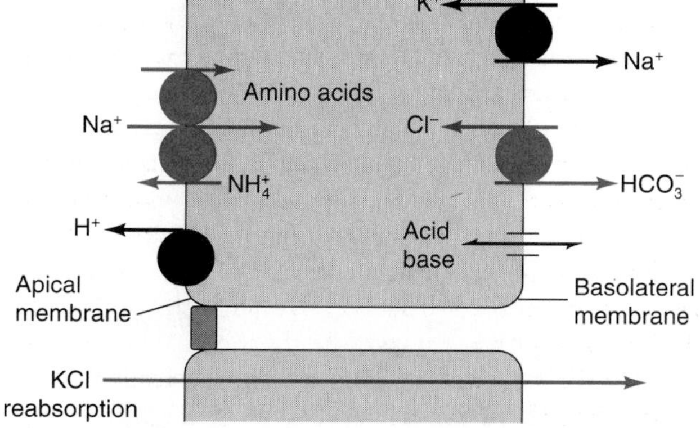

There is evidence for the feedback regulation of osmolarity among the invertebrates, especially in insects. The bug *Rhodnius* becomes bloated after sucking blood from a mammalian host. Within 2–3 minutes, the Malpighian tubules increase their secretion of fluid by more than a thousand times, producing a copious urine. Artificially bloating the insect with a saline solution does not produce such diuresis in an unfed *Rhodnius.* It has also been found that isolated Malpighian tubules immersed in the hemolymph of unfed individuals remain quiescent, but if immersed in the hemolymph of a recently fed *Rhodnius* they produce a copious secretion. A factor that stimulates the secretion of these tubules can be extracted from the neural tissue containing the cell bodies or axons of neurosecretory cells, primarily those of the metathoracic ganglion. Thus, it appears that these cells release a diuretic hormone in response to a factor present in the ingested blood. The only identified neurohumor that stimulates the diuretic action of the neurosecretory cells is serotonin. Similar findings in other insect species suggest that diuretic and antidiuretic hormones produced in the nervous system regulate the secretory activity of the Malpighian tubules or the reabsorptive activity of the rectum. In earthworms, removal of the anterior ganglion results in the retention of water and a concomitant decrease in plasma osmolarity. Injection of homogenized brain tissue reverses these effects, suggesting a humoral mechanism.

Figure 14-47 Ions are transported in and out of locust rectal cells by numerous mechanisms. The primary effect is the reabsorption of KCl and water and excretion of ammonia and acid.

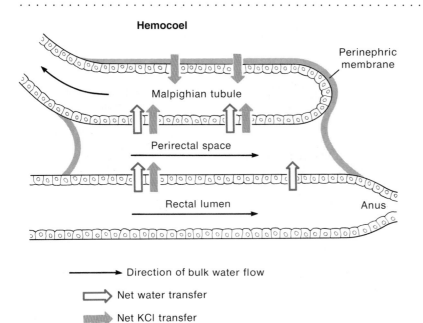

Hemocoel

Perinephric membrane

Malpighian tubule

Perirectal space

Rectal lumen

Anus

→ Direction of bulk water flow

⟹ Net water transfer

⟹ Net KCl transfer

Figure 14-48 The countercurrent arrangement of the water-extraction apparatus of the rectum of the mealworm beetle, *Tenebrio*, probably accounts for the high urine-concentrating ability of this organism. Most of the water and KCl entering the rectal lumen is recycled into the Malpighian tubules. See text for further discussion. [Adapted from Phillips, 1970.]

EXCRETION OF NITROGENOUS WASTES

When amino acids are catabolized, the amino group is released or transferred to another molecule for removal or for reuse. Unlike the atoms of the carbon skeleton of amino acids, which can be oxidized to CO_2 and water, the amino group must either be salvaged for the resynthesis of amino acids or be excreted dissolved in water to avoid a toxic rise in the plasma concentration of nitrogenous wastes. Elevated body ammonia levels have several deleterious effects on metabolism and amino acid transport; also, NH_4^+ can substitute for K^+ in ion-exchange mechanisms, leading to convulsions, coma and eventually death. Thus, in most animals, there is a close link between osmoregulatory functions and processes involved in the elimination of excess nitrogen. In those animals faced with a limited water supply, this relation gives rise to a serious problem—namely, the inevitable conflict between conserving water, on the one hand, and preventing toxic accumulation of nitrogenous wastes, on the other. As we will see, animals have evolved excretory strategies appropriate to their water economies.

Animals generally excrete most excess nitrogen as *ammonia, urea,* or *uric acid* (Figure 14-49). Lesser quantities of nitrogen are excreted in the form of such compounds as creatinine, creatine, or trimethylamine oxide; very small quantities of amino acids, purines, and pyrimidines also may be excreted. The three primary nitrogenous excretory compounds differ in their properties, so that in the course of evolution some animal groups have found it more opportune to produce one or the other of these forms for excretory purposes during all or part of their life cycles (Figure 14-50).

Ammonia is more toxic than urea or uric acid and must be kept at low levels in the body. Because excretion of ammonia occurs by diffusion, a large volume of water is re-quired to maintain the concentration of ammonia in the excretory fluid below that in the body, which is necessary for diffusion to occur. This means that about 0.5 liters of water is needed to excrete 1 g of nitrogen in the form of ammonia. Urea is less toxic than ammonia and requires only 0.05 liters of water to excrete 1 g of nitrogen as urea, that is, only 10% of the water required to excrete the equivalent amount of nitrogen as ammonia. Urea synthesis, however, consumes ATP; therefore, if plenty of water is available and ammonia levels in the body can be kept low enough, excreting nitrogenous waste as ammonia saves energy. Even less water is required to excrete uric acid, with only 0.001 liters required for uric acid excretion per gram of nitrogen, or only 1% of that needed for ammonia excretion. Uric acid is only slightly soluble in water and is excreted as a white pasty precipitate, characteristic of bird feces. The low solubility of uric acid has adaptive significance in that, in its precipitated form, uric acid contributes nothing to the tonicity of the "urine" or feces.

Thus, in general, water availability normally determines the nature and pattern of nitrogenous excretion. Aquatic animals normally excrete ammonia across their gills, whereas terrestrial animals normally excrete urea or

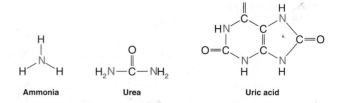

Ammonia   Urea   Uric acid

Figure 14-49 Most excess nitrogen is eliminated in the form of ammonia, urea, and/or uric acid. Of these common nitrogenous excretory products, the amount of water utilized in excretion of 1 g of nitrogen is greatest for ammonia, which is highly soluble, and least for uric acid, which is relatively insoluble. Note the differences in number of nitrogen atoms per molecule.

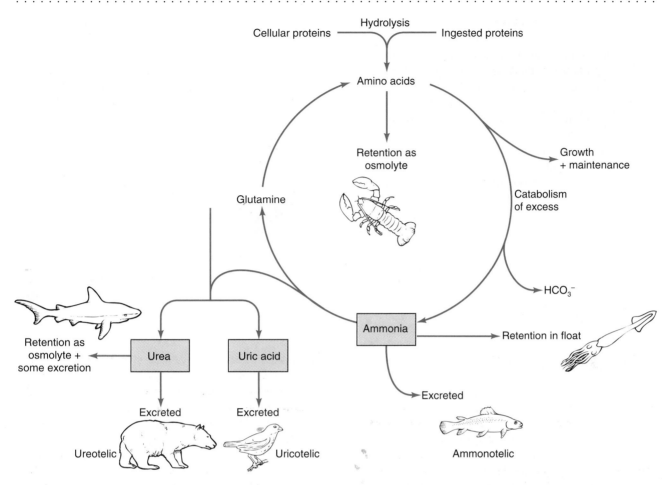

Figure 14-50 Although exceptions occur, water availability correlates with the predominant nitrogen excretory product (pink boxes) found in different animals. This general overview of nitrogen metabolism and excretion in animals shows the points at which they differ. Based on the main nitrogen excretory product used by an animal, it can be classified as ammonotelic, ureotelic, or uricotelic. In some animals certain nitrogenous compounds act as osmolytes, which are substances used to adjust body osmolarity. [Adapted from Wright, 1995.]

uric acid via their kidney. That is, the type of excretion is generally related to habitat: Terrestrial birds excrete about 90% of their nitrogenous wastes as uric acid and only 3%–4% as ammonia, but semiaquatic birds, such as ducks, excrete only 50% of their nitrogenous wastes as uric acid and 30% as ammonia. Mammals excrete most of their waste as urea. Frog tadpoles are aquatic and excrete ammonia; after they metamorphose into the terrestrial adult form, they excrete urea. Avian embryos produce ammonia for the first day or so, and then switch to uric acid, which is deposited within the egg as an insoluble solid and thus has no effect on the osmolarity of the precious little fluid contained in the egg. Lizards and snakes have various developmental schedules for switching from the production of ammonia and urea to the production of mainly uric acid. In species that lay their eggs in moist sand, the switch to uric acid production occurs late in development, but before hatching. The switch to uric acid production is a kind of biochemical metamorphosis that prepares the organism for a dry, terrestrial habitat. It is evident, however, that there are overlaps of different excretory products in animals in similar environments.

Ammonia-Excreting (Ammonotelic) Animals

Most teleosts and aquatic invertebrates are **ammonotelic;** that is, they excrete their nitrogenous wastes primarily as ammonia, producing little or no urea. As just noted, most land animals convert nitrogenous wastes to either uric acid or urea to save water. Interesting exceptions are the terrestrial isopod (sowbug), as well as some terrestrial snails and crabs; these animals eliminate a significant portion of their nitrogen waste by ammonia volatilization.

Cell membranes are generally permeable to un-ionized ammonia (NH_3) but not very permeable to ammonium (NH_4^+) ions. Ammonia excretion is due largely to the passive diffusion of un-ionized ammonia. In most teleosts, nearly all ammonia is excreted as NH_3. The associated excretion of H^+ and carbon dioxide acidifies the water next to the gill surface, trapping NH_3 as the largely impermeant NH_4^+ and enhancing ammonia excretion. Some membranes, however, have a low permeability to NH_3, as well as NH_4^+. Membranes of *Xenopus* eggs and those of the cells of the thick ascending limb of the loop of Henle in the mammalian kidney are examples of structures having a low NH_3 permeability.

Amino groups of various amino acids are transferred, with the aid of a transaminase enzyme, to glutamate, which is then deaminated to form ammonium ions and α-ketoglutarate in the liver. In the liver, glutamate is also converted to **glutamine,** which is much less toxic than ammonia and crosses membranes easily but is not normally excreted in any quantity. Although mammals excrete most nitrogenous waste as urea, they excrete small amounts of ammonia in the urine. The less toxic glutamine, rather than ammonia, is released from the mammalian liver into the blood and taken up by the kidney. The glutamine is then deaminated in the cells of the kidney tubules, and ammonia is liberated into the tubular fluid. The excreted ammonia can take up a proton to form the NH_4^+, which cannot diffuse back into the tubular cells and thus leaves the body via the urine (see Figure 14-30). Since ammonia in both its free and ionized forms is highly toxic, it makes good sense that glutamine, which is nontoxic, should act as the amino-group carrier through blood and tissues until its deamination in the ammonotelic kidney.

A blood concentration of only $0.05 \; mmol \cdot L^{-1}$ ammonia is toxic to most mammals, causing convulsions and death. Similar acute toxic effects have been observed in many other animals, including birds, reptiles, and fish. Mexican guano bats are unusual among mammals in that they can withstand very high levels of ammonia (1800 ppm) in the atmosphere of the caves in which they live. This level is sufficient to kill humans, so enter guano bat caves with care! The toxicity of NH_3 is due in part to the elevation of pH it produces, which causes changes in the tertiary structure of proteins. Ammonia also interferes with some ion-transport mechanisms, because NH_4^+ substitutes for K^+ in some cases. Ammonia can also effect brain blood flow and some aspects of synaptic transmission, particularly glutamate metabolism.

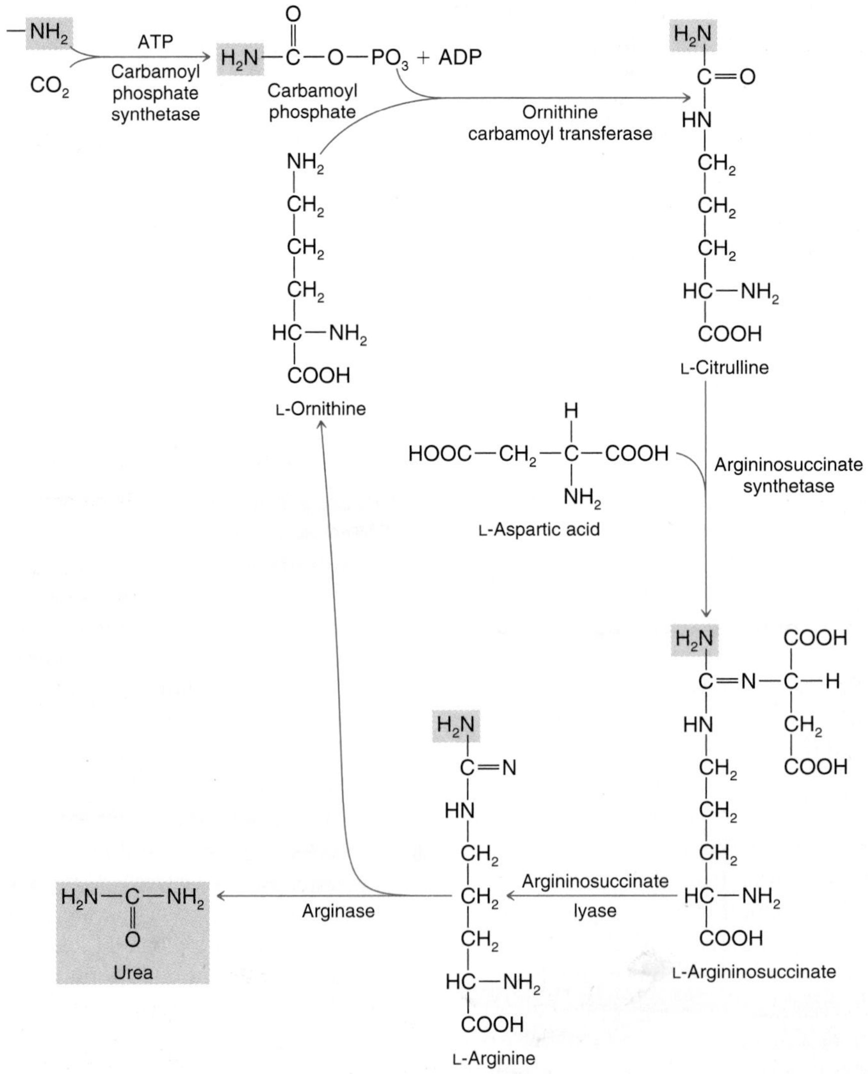

Figure 14-51 Urea is formed by the ornithine-urea cycle in all vertebrates except teleosts. Because ATP is required for the first step, ureotelic animals consume more energy in the excretion of nitrogen than do other animals.

What are the effects of variations in the pH of body compartments on the transfer and distribution of ammonia and ammonium ions within the body?

Some squid, shrimps, and tunicates sequester NH_4^+ in high concentration in specialized acidified chambers that act as a float making the animal more buoyant. In the floats of these marine animals, NH_4^+ is substituted for heavier Ca^{2+}, Mg^{2+}, and SO_4^{2-} ions (see Figure 14-50). Ammonium levels in the floats are very high and the tissues that make up the float must be resistant to the toxic actions of ammonia. Ammonia levels in other regions of the body are relatively low.

Urea-Excreting (Ureotelic) Animals

Ureotelic animals excrete most nitrogen wastes as urea, which is quite soluble in water, is far less toxic than ammonia, and requires much less water for excretion. Moreover, urea contains two nitrogen atoms per molecule. Ureotelic animals utilize one of two pathways for urea formation. With the exception of most teleost fishes, vertebrates synthesize urea primarily in the liver via the **ornithine-urea cycle** (Figure 14-51). In this cycle, two $-NH_2$ groups and a

molecule of CO_2 are added to ornithine to form arginine. The enzyme arginase, which is present in relatively large quantities in these animals, then catalyzes removal of the urea molecule from arginine, regenerating ornithine.

The Lake Magadi tilapia, *Oreochromis alcalicus grahami,* is a completely aquatic freshwater teleost fish; unlike most teleosts, it excretes all nitrogenous waste as urea. The high pH of Lake Magadi (about 10) impairs ammonia excretion to an extent that leads to ammonia accumulation and death in other fish. *Oreochromis alcalicus grahami* can live in Lake Magadi because it converts ammonia to urea via the ornithine-urea cycle, thereby avoiding ammonia toxicity. Elasmobranch fish use urea produced from ammonia via the ornithine-urea cycle to increase their body osmolarity; they also excrete most of their nitrogenous waste as urea across their gills. So not all aquatic animals excrete ammonia.

Most teleosts and many invertebrates utilize the so-called **uricolytic pathway** in which urea is produced from uric acid that either arises from a transamination via aspartate or is produced during nucleic acid metabolism. In this pathway, uric acid is converted first to allantoin and allantoic acid with the help of the enzymes uricase and allantoinase, respectively, and then to urea using allantoicase (Figure 14-52). During evolution most mammals have lost

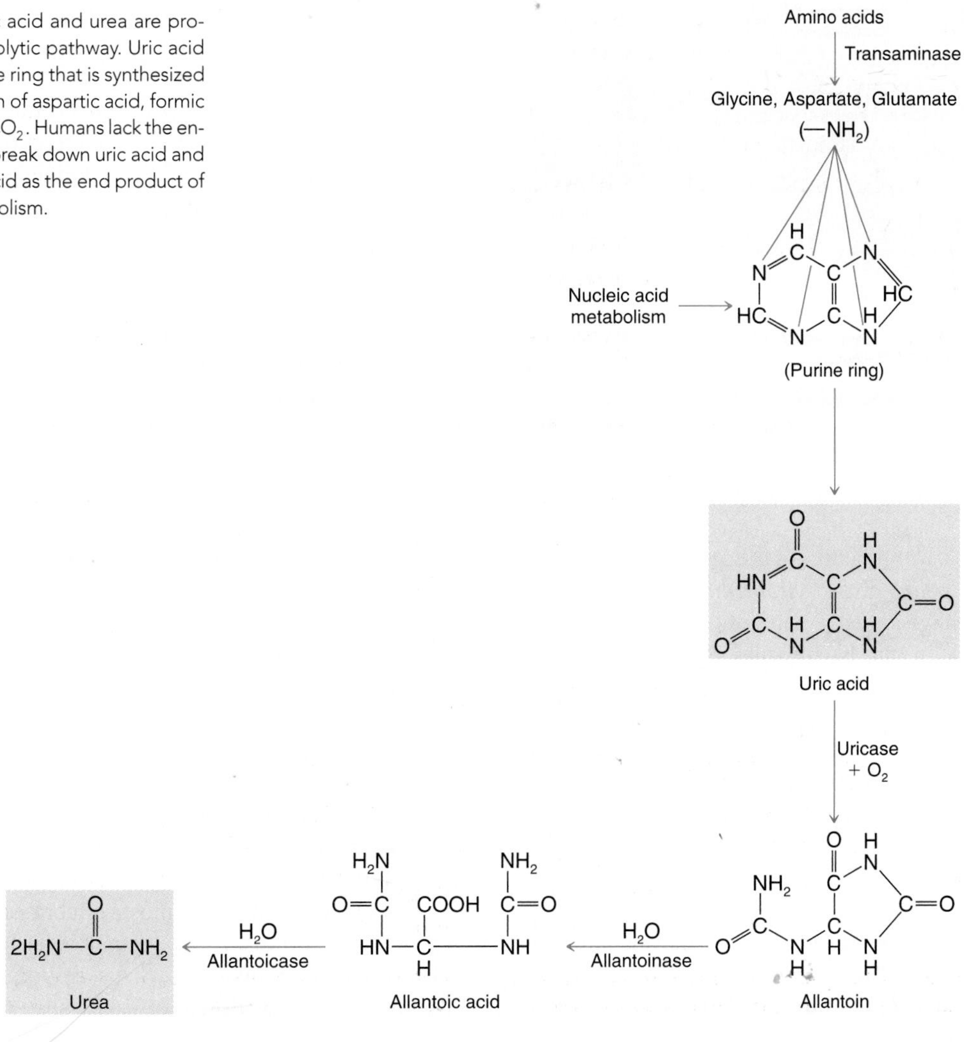

Figure 14-52 Uric acid and urea are produced via the uricolytic pathway. Uric acid arises from a purine ring that is synthesized by a complex union of aspartic acid, formic acid, glycine, and CO_2. Humans lack the enzymes needed to break down uric acid and thus excrete uric acid as the end product of nucleic acid metabolism.

the ability to produce allantoicase and allantoinase; hominoid primates also cannot synthesize uricase and thus excrete uric acid as an end product of nucleic acid metabolism. Uric acid excretion is normally only about 1% by weight of urea excretion in humans. Should uric acid production or intake increase, however, uric acid levels in the blood may rise because excretion may be compromised due to low uric acid solubility, especially if urine volume is small. The low solubility may also result in the precipitation of uric acid crystals when levels of uric acid rise in the blood, which in turn causes the painful condition called gout.

Urea crosses membranes either through aqueous pores or via specialized membrane transporters, as lipid membranes are not very permeable to urea. Specific urea transporters are thought to be present in elasmobranchs and several other vertebrates, including mammals. Urea transporters are widely distributed and in some instances may be involved in rapid urea transport to stabilize cell volume in the face of an osmotic shock.

Uric Acid-Excreting (Uricotelic) Animals

Uricotelic animals—birds, reptiles, and most terrestrial arthropods—excrete nitrogen chiefly in the form of uric acid or the closely related compound guanine. Uric acid and guanine have the advantage of carrying away four nitrogen atoms per molecule. The nitrogen atoms incorporated into uric acid ultimately arise from the breakdown of the amino acids glycine, aspartate, and glutamine (see Figure 14-52). Because these animals lack uricase, they cannot break down uric acid. Thus the catalysis of nitrogenous molecules is terminated at uric acid, which largely precipitates because of its low solubility and is excreted as the end product, requiring very little urinary water. In general, uricotelic animals are adapted to conditions of limited availability of water.

Uric acid is transported from the blood into the cells of the renal tubule via a urate-anion exchanger or via a urate uniporter. It then moves from the cells into the lumen of the tubule down an electrochemical gradient and is excreted in the urine. Urate transport competes with *para*-aminohippurate transport in the kidney tubule of birds but not in reptiles.

Two unusual amphibians are the arid-land toads *Chiromantis xerampelina* and *Phyllomedusa sauvagii*. These toads not only have an extremely low evaporative water loss from their skin but, like reptiles, excrete nitrogen as uric acid rather than ammonia or urea as most other amphibians do. The low solubility of uric acid causes it to precipitate readily in the cloaca, and allows these toads, like reptiles and birds, to minimize the volume of urine necessary to eliminate their excess nitrogen.

SUMMARY

The extracellular environment in many marine and nonmarine animals broadly resembles dilute seawater. This similarity may have had its origin in the shallow, dilute primeval seas that are believed to have been the setting for the early evolution of animal life. The ability of many animals to regulate the composition of their internal environment is closely correlated with their ability to occupy ecological environments that are osmotically at odds with the osmotic requirements of their tissues. Osmoregulation requires the exchange of salts and water between the extracellular environment and the external environment to compensate for obligatory, or uncontrolled, losses and gains. The transport of solutes and water across epithelial layers is fundamental to all osmoregulatory activity. The obligatory exchange of water depends on (1) the osmotic gradient that exists between the internal and external environments, (2) the surface-to-volume ratio of the animal, (3) the permeability of the integument, (4) the intake of food and water, (5) the evaporative losses required for thermoregulation, and (6) the disposal of digestive and metabolic wastes in urine and feces.

Marine and terrestrial animals are faced with dehydration, whereas freshwater animals must prevent hydration by uncontrolled osmotic uptake of water. Marine birds, reptiles, and teleosts replace lost water by drinking seawater and actively secreting salt through secretory epithelia. Freshwater fishes do not drink water; they replace lost salts by active uptake. Birds and mammals are the only vertebrates that secrete a hypertonic urine. Many desert species in addition utilize mechanisms for minimizing respiratory water loss.

Most vertebrate kidneys utilize filtration, reabsorption, and secretion to form urine. A countercurrent mechanism present in mammalian and avian kidneys allows production of a hypertonic urine. The filtration of the plasma in the glomerulus is dependent on arterial pressure. Crystalloids and small organic molecules are filtered, leaving blood cells and large molecules behind. Salts and organic molecules such as sugars are partially reabsorbed from the glomerular filtrate in the renal tubules, and certain substances are secreted into the tubular fluid. A countercurrent multiplier system that includes the collecting duct and the loop of Henle sets up a steep extracellular concentration gradient of salt and urea that extends deep into the medulla of the mammalian kidney. Water is drawn osmotically out of the collecting duct as it passes through high medullary concentrations of salt and urea toward the renal pelvis. Endocrine control of the water permeability of the collecting duct determines the volume of water reabsorbed and retained in the circulation. The final urine, then, is the product of filtration, reabsorption, and secretion. These processes allow the urinary composition to depart strongly from the proportions of substances occurring in the blood.

The formation of urine follows the same major outline in all or most vertebrates and invertebrates. A pre-urine is formed that contains essentially all the small molecules and ions found in the blood. In most vertebrates and in the crustaceans and mollusks, this formation is accomplished by ultrafiltration; in insects, by the secretion through the epithelium of the Malpighian tubules of KCl, NaCl, and phosphate, with water and other small molecules, such as amino acids and sugars, following passively by osmosis and diffusion down their concentration gradients. The pre-urine is subsequently modified by the selective reabsorption of ions and water and, in some animals, by secretion of waste substances into the lumen of the nephron by the tubular epithelium.

Birds and reptiles can drink seawater, excreting the salt load through a nasal salt gland. Elasmobranchs excrete salt via a rectal gland, which is made up of salt-secreting cells similar to those found in the thick ascending limb of the loop of Henle in the mammalian kidney, the avian and reptilian salt gland, and the chloride cell in the gills of marine teleosts. The hormonal regulation of the activity of these cells is also similar in sharks, birds, reptiles, and mammals. The gills of teleost fishes and many invertebrates perform osmoregulation by active transport of salts, the direction of transport being inward in freshwater fish and outward in marine fish.

The nitrogen produced in the catabolism of amino acids and proteins is concentrated into one of three forms of nitrogenous waste, depending on the osmotic environment of different animal groups. Ammonia, highly toxic and soluble, requires large volumes of water for dilution and subsequent excretion across the gills of teleosts. Uric acid is less toxic and poorly soluble; it is excreted as a semisolid suspension via the kidney of birds and reptiles. Urea is the least toxic and its excretion requires a moderately small amount of water. Mammals largely convert their nitrogenous wastes into urea, which is excreted in the urine; elasmobranchs use urea as an osmotic agent in their blood and excrete most excess nitrogen as urea across their gills.

REVIEW QUESTIONS

1. How has the development of osmoregulatory mechanisms affected animal evolution?
2. What factors influence obligatory osmotic exchange with the environment?
3. Explain why respiration, temperature regulation, and water balance in terrestrial animals are closely inter-related. Give examples.
4. Describe three anatomic or physiological mechanisms used by insects to minimize dehydration in dry environments.
5. How do marine and freshwater fishes maintain osmotic homeostasis?
6. Name and describe the three major processes used by the vertebrate kidney to achieve the final composition of the urine.
7. What factors determine the rate of ultrafiltration in the glomerulus?
8. What is meant by the renal clearance of a substance?
9. If the intratubular fluid in the loop of Henle remains nearly isosmotic relative to the extracellular fluid along its path, and is even slightly hypotonic on leaving the loop, in what way is the final urine made hypertonic in the mammalian kidney?

10. Explain why the consumption of 1 liter of beer will lead to a greater urine production than an equal volume of water.
11. What role does the kidney play in the regulation of blood pressure?
12. Discuss the role of the kidney in the control of plasma pH.
13. How do insects produce concentrated, hypertonic urine and excrement?
14. In the course of evolution, terrestrial organisms have come to excrete mainly uric acid and urea rather than ammonia. What are the adaptive reasons for such a change?
15. Explain why gulls can drink seawater and survive but humans cannot.
16. After the injection of inulin into a small mammal, the plasma inulin concentration was found to be $1 \text{ mg} \cdot \text{ml}^{-1}$, the concentration in the urine $10 \text{ mg} \cdot \text{ml}^{-1}$, and the urine flow rate through the ureter $10 \text{ ml} \cdot \text{h}^{-1}$. What was the rate of plasma filtration and the clearance in milliliters per minute? How much water was reabsorbed in the tubules per hour?
17. What evidence is there that the mammalian nephron employs tubular secretion as one means of eliminating substances into the urine?
18. Why is a countercurrent system more efficient in physical transport and transfer than a system in which fluids in opposed vessels flow in the same direction?
19. What are the similarities and differences between the elasmobranch rectal gland and the bird salt gland?

SUGGESTED READINGS

Braun, E. J., and D. H. Thomas. 1991. *Integrative aspects of osmoregulation in birds.* Symposium 38:2103–2146. Acta XX Congressus Internationalis Ornithologici. New Zealand Ornithological Congress Trust Board.

Gupta, B. L., R. B. Moreton, J. L. Oschman, and B. J. Wall. 1977. *Transport of Ions and Water in Animals.* London: Academic Press.

Krogh, A. 1939. *Osmotic Regulation in Aquatic Animals.* Cambridge: Cambridge University Press.

Kultz, D., K. Jurss, and L. Jonas. 1995. Cellular and epithelial adjustments to altered salinity in the gill and opercular epithelium of a cichlid fish (*Oreochromis mossambicus*). *Cell. Tissue Res.* 279:65–73.

Larsen, E. H. 1991. Chloride transport by high-resistance heterocellular epithelia. *Physiol. Rev.* 71:235–283.

Pitts, R. F. 1974. *Physiology of the Kidney and Body Fluids.* 3d ed. Chicago: Year Book Medical Publishers.

Phillips, J. E., et al. 1994. Mechanisms of acid-base transport and control in locust excretory system. *Physiol. Zool.* 67:95–119.

Riordan, J. R., B. Forbush, III, and J. W. Hanrahan. 1994. The molecular basis of chloride transport in shark rectal gland. *J. Exp. Biol.* 196:405–418.

Rodriguez-Boulan, E., and W. J. Nelson, eds. 1993. Epithelial and Neuronal Cell Polarity and Differentiation. *J. Cell Sci.* Suppl. 17. Cambridge, UK: Company of Biologists.

Schmidt-Nielsen, K. 1972. *How Animals Work.* Cambridge: Cambridge University Press.

Schmidt-Nielsen, K. 1981. Countercurrent systems in animals. *Sci. Am.* 244:118–128.

Smith, H. W. 1953. *From Fish to Philosopher.* Boston: Little, Brown.

Wood, C. M., and T. J. Shuttleworth, eds. 1995. *Cellular and Molecular Approaches to Fish Ionic Regulation.* San Diego: Academic Press.

Wright, P. A. 1995. Nitrogen excretion: three end products, many physiological roles. *J. Exp. Biol.* 198:273–281.

ACQUIRING ENERGY: FEEDING, DIGESTION, AND METABOLISM

In order to grow, maintain itself, and reproduce, every animal requires both raw materials and energy. These materials and the energy used in their metabolism come from food, but what actually constitutes a food item varies greatly between animals, ranging from individual molecules absorbed across the general body surface to living prey swallowed whole. Regardless of its origin, which can be plant, animal, or inorganic sources, food is used as material for production of new tissue, for the repair of existing tissue, and for reproduction. Food also serves as an energy source for ongoing processes, such as movement and metabolism.

The chemical energy contained within food ultimately is derived from the sun (see Figure 3-3). Chlorophyll-containing plants are photosynthetic, **autotrophic** (self-nourishing) organisms that harness radiant energy to synthesize complex carbon compounds from simple precursors—CO_2 and H_2O. These compounds are repositories of chemical energy that can be released and utilized through coupled reactions to drive energy-consuming processes in living tissue. Almost all organisms are **heterotrophic**, depending on energy-yielding carbon compounds derived from the ingestion of other plants or animals, and ultimately on the photosynthesizers, which gather in the sun's energy. The exception of the relatively recently discovered "deep-sea vent" invertebrates, deriving their nutrition from mineral-rich vent waters, only highlights the normal dependency of animal life on the energy of the sun.

The flow of energy from the sun through a photosynthetic autotroph to a molecule of ATP in a heterotrophic animal is shown in Figure 15-1. Monosaccharides such as glucose are synthesized by green plants from CO_2 and H_2O. These elementary carbon compounds occur at the beginning of the **food chain**, which represents a series of organisms linked together by the fact that each "link" of the chain serves as a food item for the next. Each group of organisms represents a **trophic level**. In a short food chain with only two trophic levels, green plants are eaten by a large heterotroph, such as an elephant. This heterotroph, having no natural predators except humans, is at the end of that food chain until it dies and is consumed by bacteria

and carrion-eating scavengers. In a longer chain, a representative succession would be phytoplankton > zooplankton > small fish > medium fish > large fish, and the nutrient flow is generally more complex (see Figure 3-3).

Usable material and free energy are lost in passing from one trophic level to another in a food chain. The grain produced in a 1-hectare (ha; 2.47-acre) field of wheat contains more material and energy directly available for human consumption than it does if that same grain is used as cattle feed, converted to beef, and then consumed by humans. For example, a 1-ha grain field produces on average 5 times more protein than does a hectare devoted to beef production, while a hectare of legumes produces 10 times more. A cow must be fed more than 20 kg of plant protein to produce just 1 kg of protein for human consumption. Humans are at the top trophic level of that food chain. At each level of feeding, digestion, and incorporation along the food chain, there is considerable energy loss due to the energetic cost of tissue maintenance and of food digestion and reassembly into new molecules to be incorporated into tissue. Consequently, a shorter food chain generally conserves greater amounts of photosynthetically captured energy for the top consumer than a long one does, if the efficiency of transfer from each trophic level to the next one is approximately equal.

We will now consider how various animals acquire food items.

FEEDING METHODS

Obtaining food adequate in both quantity and quality occupies much of the routine behavior of most animals. Certainly an animal's physiology and morphology are the result of natural selection that favors effective acquisition of energy from food while avoiding becoming someone else's meal. The complexity and sophistication of the nervous and muscular systems, for example, attest to the power of the selective forces acting on the organism. As these vary, so do the variety of methods by which animals feed. Sessile (non-mobile) bottom-dwelling species commonly resort to surface absorption, filter feeding, or trapping. Mobile

Sessile's modes of feeding
↓
Mobile mode

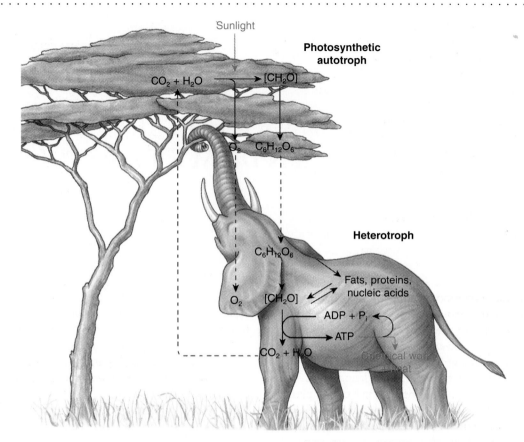

Sunlight

Photosynthetic autotroph

$CO_2 + H_2O$ → $[CH_2O]$

O_2　$C_6H_{12}O_6$

Heterotroph

$C_6H_{12}O_6$

Fats, proteins, nucleic acids

O_2　$[CH_2O]$

$ADP + P_i$

ATP

$CO_2 + H_2O$

Chemical work Heat

Figure 15-1 Two trophic levels occur in this generalized flow diagram of chemical energy through a food chain. The flowchart begins with the photosynthetic formation of high-energy-content molecules (sugars) from low-energy-content raw materials (CO_2 and H_2O) in plants. Oxidation of carbon compounds yields free energy coupled to the synthesis of high-energy compounds, such as ATP, used as common energy currency in metabolism. Chemical energy content is at its peak following the photosynthetic production of sugars. When the plant material is consumed by a heterotroph, some chemical energy is converted to heat and thus is lost as a direct source of energy for driving biological processes. This food chain has only two trophic levels, but most food chains have many more intervening levels.

animals follow a more active sequence, which in the extreme of many carnivores (meat eaters) includes searching, stalking, pouncing, capturing, and killing.

Food Absorption through Exterior Body Surfaces

The feeding method that is least dependent on specialized capture and digestive organs involves absorption of nutrients directly across the body wall. Certain protozoans, endoparasites (animals that live within other animals), and aquatic invertebrates are able to take up nutrient molecules from the surrounding medium directly through their soft body wall. Endoparasites such as parasitic protozoans, tapeworms, flukes, and certain mollusks and crustaceans are surrounded by host tissues or by alimentary canal fluids, both of which are high in nutrients. Tapeworms, which may be many meters long, lack even a rudimentary digestive system. Tapeworms evolved from a primitive flatworm that lacked a body cavity (i.e., was acoelomic). However, some endoparasites appear to have secondarily lost the digestive apparatus present in their ancestors. For example, parasitic crustaceans, which belong to the cirripeds (barnacle group), lack an alimentary canal, but they appear to have evolved from nonparasitic ancestors possessing a gut.

Some free-living protozoans and invertebrates derive part of their nutrients by direct surface uptake from the surrounding medium. Small molecules such as amino acids are taken up from dilute solution by transport mechanisms (described in Chapter 4), against what can often be a huge concentration gradient. In some of these organisms, larger molecules or particles are taken up by a bulk process such as phagocytosis, which is described next.

Endocytosis

Endocytosis represents a more active form of "feeding" than passive absorption directly across the body wall. Like direct nutrient absorption, however, it occurs at the local cellular rather than tissue or organismal level. Endocytosis includes two processes, phagocytosis ("cell eating") and pinocytosis ("cell drinking"). In **phagocytosis**, pseudopod-like protuberances extend out and envelope relatively large nutrient particles. **Pinocytosis** occurs when a smaller particle binds to the cell surface and the plasma membrane invaginates (folds inward) under it, forming an **endocytotic cavity.** Whether captured by phagocytosis or pinocytosis, the morsel is then engulfed in a membrane-enclosed vesicle that pinches off from the bottom of the cavity.

The vesicle (or **food vacuole** in Protozoa) fuses with **lysosomes**, organelles containing intracellular digestive enzymes, whereafter it is called a **secondary vacuole**. After digestion, the contents of the vacuole pass through the vacuole wall into the cytoplasm. The remaining undigested material is excreted externally by exocytosis, essentially a reverse process of pinocytosis. Feeding by pinocytosis and phagocytosis is familiar in protozoans such as *Paramecium*, but also occurs in the lining of the alimentary canals and other tissues of many multicellular animals.

Filter Feeding

Many aquatic animals use **filter feeding**, also called **suspension feeding**, to capture food. Food items (usually phytoplankton or zooplankton) are carried to specialized entrapment devices either on the body surface or within it.

Most marine filter feeders are small, sessile animals, such as sponges, brachiopods, lamellibranchs, and tunicates. Food items are carried along on water currents that either occur naturally or are generated by the movements of body parts of the filter-feeding animal itself, such as cilia or flagella. Brachiopods respond behaviorally to currents, rotating on their pedal stalks to present the most efficient hydrodynamic orientation for capturing the water current. A number of other sessile animals located in moving water make use of **Bernoulli's effect** (i.e., a drop in fluid pressure as fluid velocity increases) to increase the rate of water flowing through the entrapment sites, at no energy cost to themselves. An example of such passively assisted filter feeding is seen in sponges (Figure 15-2). The flow of water across the large terminal opening causes a drop in pressure (Bernoulli's effect) outside the osculum. As a result, water is drawn out of the sponge through the osculum, and is drawn in through the numerous **ostia** (mouth-like openings) in the body wall. The drop in pressure is facilitated by the shape of the sponge's exterior, which causes the water over the osculum to flow with greater velocity than the water flowing past the ostia. Food particles, swept into the ostia of the sponge along with the water, are engulfed by **choanocytes**, the flagellated cells lining the body cavity. The flagella of the choanocytes also create internal water currents within spongocoel, the hollow water-filled interior. Some sponges living in moving water "pump" a volume of water equivalent to up to 20,000 times their body volume per day.

Mucus, a sticky mixture of mucopolysaccharides, often plays an important role in filter feeding. Waterborne microorganisms and food particles are trapped in a layer of mucus that covers a ciliated epithelium. The mucus is then transported to the oral parts by beating cilia. The cilia propel water through sessile animals not only to capture suspended food but also to aid in respiration. This is of greatest importance in still water. In mollusks such as the mussel, *Mytilus*, the cilia on the surface of the ctenidium draw a stream of water through the inhalant siphon, passing the water between the gill filaments (Figure 15-3). These cilia are also responsible for keeping mucus traveling down

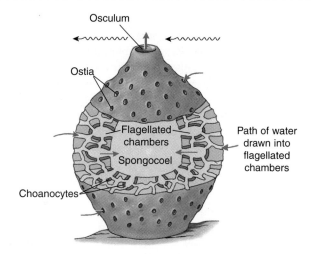

Figure 15-2 Water flows in an organized fashion through syconoid sponges. In this diagrammatic section, red arrows indicate water flow. A significant proportion of this flow results from the reduction in hydrostatic pressure at the osculum due to the Bernoulli effect, produced by the transverse water currents (black arrows) flowing over the osculum at elevated velocity. Water flow is also generated from the activity of flagellated choanocytes that line the flagellated chambers (and give them their name). The choanocytes are found in the regions of the flagellated chambers lined in red. Water enters the sponge through the ostia, passes through the flagellated chambers, and ends up in the inner cavity, the spongocoel. Nutrients are then taken up into individual cells by endocytosis. [Adapted from Hyman, 1940; Vogel, 1978.]

along the filaments (i.e., 90° to the water flow) to the tip of the gill, where it travels in a special groove under ciliary power toward the mouth in a rope-like string of mucus. Sand and other indigestible particles are sorted out and rejected (presumably on the basis of texture), passing out with the water leaving the exhalant siphon.

Non-sessile animals filter-feed by various mechanisms. A number of fishes are planktivorous, using modified gill

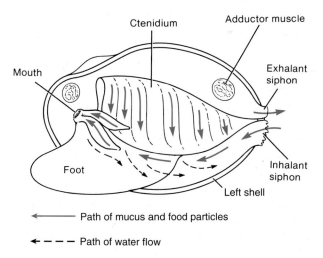

Path of mucus and food particles

Path of water flow

Figure 15-3 Lamellibranch mollusks employ ciliary feeding. Side view of generalized lamellibranch with right valve removed. Red arrows show paths of inhaled water with food particles in the inhalant siphon and, over the surfaces of the gills (ctenidium). After passing down the gills, sand and other indigestible materials are eliminated out the exhalant siphon, while food particles are passed by cilia to the mouth.

rakers to strain food out of the flow of water passing through the mouth and over the gills. Juveniles of the paddlefish, *Polyodon spathula,* swim rapidly and continuously both to ventilate their gills and to filter out food items (see Chapter 16). Filter feeding is also very common in amphibian larvae. In *Xenopus laevis,* the South African clawed toad, the branchial chamber contains gills bearing branchial filter plates that entrap suspended organic material. The material becomes entrapped in mucus, which is then swept by cilia into the esophagus to be swallowed. In *Xenopus,* branchial respiration and food ingestion may present functional conflicts. As the gill filter plates load up with suspended food items, the resistance to water flow through the gills rises sharply. Indeed, in larval *Xenopus,* gill ventilation decreases in proportion to food density in inspired water, presumably maintaining a constant rate of food ingestion. Increased cutaneous and pulmonary respiration apparently can compensate for the lack of branchial gas exchange when optimal conditions for filter feeding result in reduced branchial water flow.

The largest filter feeders are the baleen whales, such as the right whale. Horny **baleen plates** bear a fringe of parallel filaments of hair-like keratin that hang down from the upper and lower jaws and act as strainers analogous to the gill rakers of fishes or larval amphibians (Figure 15-4A). These whales swim with jaws open into schools of pelagic crustaceans such as krill, engulfing vast numbers suspended in tons of water. As the jaws close, the water is squeezed back out through the baleen strainers with the help of the large tongue, and the crustaceans, left behind inside the mouth, are swallowed. Clearly, filter feeding can be a very effective form of food capture and can support an animal of huge dimensions.

Birds such as flamingos also use filter feeding to capture small animals and other morsels they find in the muddy bottoms of their freshwater habitat (Figure 15-4B). The flamingo and the right whale exhibit remarkable convergent evolution: they both have a deep-sided lower jaw, a recurved rostrum, fibrous-fringe filters suspended from the upper jaw, and a large, fleshy tongue. Both feed by filling the mouth cavity with water, and then using the tongue as a piston to force water out through the filters, trapping and retaining waterborne food particles.

Fluid Feeding

Fluid feeding involves a variety of structures and mechanisms, including piercing and sucking, and cutting and licking.

Piercing and sucking

Feeding by piercing a prey or food item and sucking fluids from it occurs among the platyhelminths, nematodes, annelids, and arthropods. Leeches, among the annelids, are true bloodsuckers, using an anticoagulant in their saliva to prevent clotting in their prey's blood. In fact, the anticoagulants in leeches have been chemically isolated and are being used clinically. Leeches themselves are still used for medicinal purposes to reduce swelling by removing extracellular fluid following cosmetic and other forms of surgery. Some free-living flatworms seize their invertebrate prey by wrapping themselves around it. They then penetrate the body wall with a protrusible pharynx that sucks out the victim's body fluids and viscera. Penetration by the pharynx and liquefaction of the victim's tissues are facilitated by proteolytic enzymes secreted by the muscular pharynx.

Large numbers of arthropods feed by piercing and sucking. Most familiar and irksome of these to humans are mosquitoes, fleas, bedbugs, and lice, which can be vectors of disease. The majority of sucking arthropods victimize animal hosts. However, especially among the Hemiptera (true bugs) are species that pierce and suck plants, from which they draw sap. Sucking insects generally possess fine piercing mouthparts in the form of a **proboscis** (Figure 15-5A). Often, the two **maxillae** are shaped so that they make up two canals that run to the tip of the proboscis (Figure 15-5B and C). One of these, the dorsal canal, is the passage for blood or sap sucked from the host. The other, the ventral canal, carries saliva, containing anticoagulants or enzymes, from the salivary glands into the host. Sucking occurs by the action of a muscular pharynx. After feeding, most insects are able to fold the proboscis back out of the way.

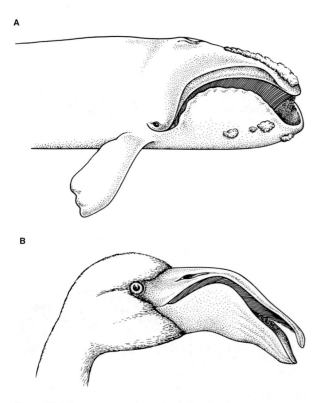

A

B

Figure 15-4 Convergent evolution has led to filter-feeding mechanisms in both **(A)** the black right whale, *Eubalaena glacialis,* and **(B)** the lesser flamingo, *Phoeniconaias minor.* Both the baleen in the whale's mouth and the fringe along the edge of the flamingo's bill act as strainers. [Adapted from Milner, 1981.]

Fluid feeders like mosquitoes or ticks can ingest a huge amount of fluid relative to their body mass in a very short period of time. While this represents an abundance of food, what type of short-term physiological imbalances could the ingestion of blood, sap, or other fluids pose for the fluid feeder?

Seizing of Prey

Predators use various types of mouthparts and other appendages to capture and masticate animals and plants. Often toxins are used to further immobilize items of prey.

Jaws, teeth, and beaks

Although no true teeth occur among the invertebrates, various invertebrates have beak-like or tooth-like chitinous structures for biting or feeding. Invertebrates such as the preying mantis and the lobster also have anterior limbs modified for prey capture (Figure 15-6). Spiders and their relatives have needle-like mouthparts for injection of venom, while cephalopods like the octopus has a sharp, tearing beak. Among the vertebrates, hagfishes, sharks, bony fishes, amphibians, and reptiles have pointed teeth, mounted on the jaws or palate, that aid in holding, tearing, and swallowing prey.

The teeth of non-mammalian vertebrates are usually non-differentiated, with a single tooth type found throughout the mouth. One notable exception is found among the poisonous snakes, such as vipers, cobras, and rattlesnakes, which have modified teeth, called **fangs,** that they use to inject venom (Figure 15-7). These fangs either are equipped with a groove that guides the venom or are hollow, very much like a syringe. In rattlesnakes, the fangs fold back against the roof of the mouth, but extend perpendicularly when the mouth is opened to strike at prey. A snake's jaws are held together with an elastic ligament that allows them

Cutting and licking

Numerous invertebrates and a few vertebrates feed by cutting the body wall of a prey item and then licking, or sponging the body fluids that leak from the cut. The blackfly and related biting flies have mouthparts with a sharpened mandible for cutting, and a large, sponge-like labium for transferring the body fluid (usually blood) to the esophagus. Among the chordates, a few phyletically ancient fishes (lampreys, hagfishes) use rasp-like mouths to make large, circular flesh wounds on the host. They feed on the blood created from these wounds. Vampire bats use their teeth to make puncture wounds in cattle, from which they lick oozing blood. The saliva of these bats contains an anticoagulant, as well as an analgesic to prevent the host from feeling the effects of the bite, at least until the bat has finished feeding.

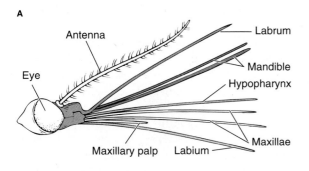

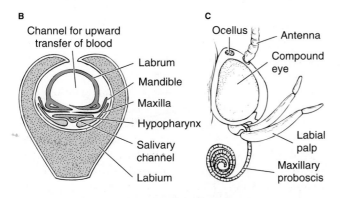

Figure 15-5 Sucking insects use tubular mouthparts for feeding. **(A)** lateral view of the head of a mosquito, with the mouthparts separated for identification. **(B)** Cross-section of the assembled mouthparts of the mosquito, showing separate channels for blood to move superiorly to the mouth, and for saliva to move inferiorly to the wound. **(C)** Lateral view of the head of a moth, showing the sucking mouthparts curled up between feeding bouts. [Adapted from Rupert and Barnes, 1994.]

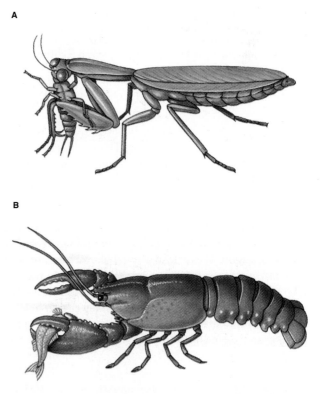

Figure 15-6 The anterior limbs of arthropods are often modified for seizing prey and holding it while the mouthparts tear off small pieces to be swallowed. **(A)** A mantispid insect (order Neuroptera). **(B)** Lobsters (order Decapoda) have one claw modified for tearing and one modified for crushing.

A

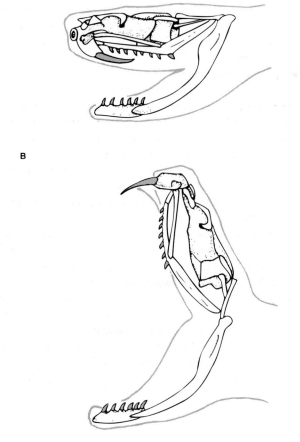

B

A

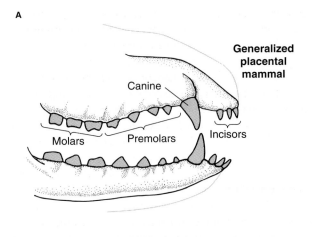

B

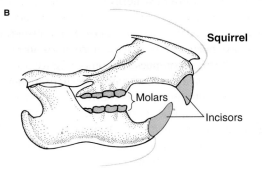

C

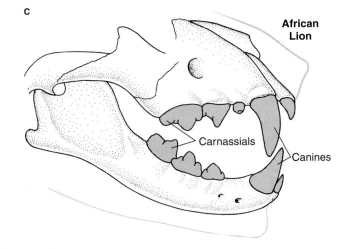

D

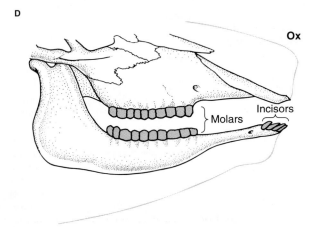

Figure 15-7 Rattlesnakes have modified teeth, known as *fangs*, which they use to inject venom into their prey. These side views of a rattlesnake skull show **(A)** a non-striking position, with the jaws only partially open and the hinged fangs folded into the roof of the mouth, and **(B)** a striking position, in which the jaws are open wide and the fangs extended. The extraordinary flexibility of the lower jaws allows the snake to swallow prey whole after injecting it with deadly venom. [Adapted from Parker, 1963.]

to spread apart during swallowing. This enables the snake to swallow animals larger than the diameter of its head (see Figure 15-7). Swallowing prey whole is relatively common, and very evident in prey capture and consumption in snakes.

Mammals use their teeth for seizing and masticating their prey. Their teeth have developed very different shapes during evolution (Figure 15-8). Chisel-like **incisors** are used for gnawing, especially by rodents and rabbits. In the elephants (and before them, mammoths), the incisors are modified into a pair of *tusks*. Pointed, dagger-like **canines** are used by the carnivores, insectivores, and primates for piercing and tearing food. In some groups, like the wild pigs

Figure 15-8 Mammalian dentition is specialized for food type. **(A)** Teeth of a generalized placental mammal, showing the major divisions of dentition. **(B)** Squirrel, showing incisors enlarged for gnawing. **(C)** African lion, with carnassial teeth modified for shearing bone and tendon. **(D)** Ox, with extensive molars for grinding plant material. [Adapted from Romer, 1962; Cornwall, 1956.]

and walruses, the canines are elongated as tusks, which are used for prying and fighting. Most complex and interesting in their form are the molars of some herbivorous groups such as cattle, including oxen, pigs, hippopotamuses, and horses and zebras. These teeth, which are used in a side-to-side grinding motion, are composed of folded layers of enamel, cement, and dentine, all of which differ in hardness and in rate of wearing. Because the softer dentine wears rather quickly, the harder enamel and cement layers form ridges that enhance the effectiveness of the molars for chewing grass and other tough vegetation. Many mammals, such as the cats (the domestic cat and the great cats, such as the lion) use limbs equipped with sharp claws to supplement the teeth as food-capturing structures.

Instead of teeth, birds have horny beaks, in a multitude of shapes and sizes, evolved to adapt to each species' unique food sources and methods of obtaining them. For instance, beaks may have finely serrated edges, sharp, hook-like upper bills, or sharp, wood-pecking points (Figure 15-9). Seed-eating birds eat their food whole (perhaps after removing the outer hull), but may grind the swallowed seed in a muscular **crop** or **gizzard** containing pebbles that act like "millstones." Raptorial birds (hawks, eagles), endowed with excellent vision and flight mobility, capture prey with their talons as well as their beaks.

Toxins

A large number of animals from different phyla use **toxins** either to subdue prey or to fend off predators. Most of these toxins act at synapses in the nervous system. Surprisingly simple animals can use sophisticated arrays of venom-producing cells. Among the coelenterates (hydras, jellyfish, anemones, corals) for example, there is extensive use of **nematocysts** (stinging cells). Concentrated in large numbers on the tentacles, the nematocysts inject paralytic toxins into prey and immobilize it while the tentacles carry it to the mouth (Figure 15-10). Many nemertine worms paralyze their prey by injecting venom through a stiletto-like proboscis. Venoms are also used by annelids, gastropod mollusks (including one species of octopus), and a wide variety of arthropods.

Among the last group, scorpions and spiders are most notorious for their toxins, which are usually highly specific chemicals that bind to specific receptor types. After grabbing its prey with its large chelae (pincer-like organs), a scorpion will arch its tail and then plunge its sting into its prey (Figure 15-11). The scorpion then injects the victim with a poison containing a **neurotoxin** that interferes with the proper firing of nerve impulses. Spider poisons also contain neurotoxins. The venom from the black widow spider contains a substance that induces massive release of

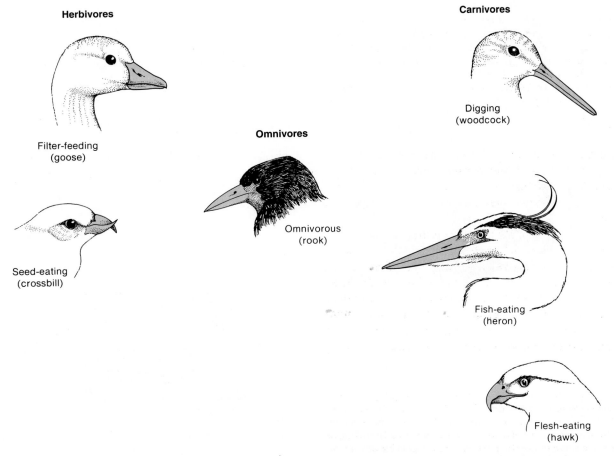

Figure 15-9 Bird beaks are adapted to suit herbivory, omnivory, and carnivory. [Adapted from Marshall and Hughes, 1980.]

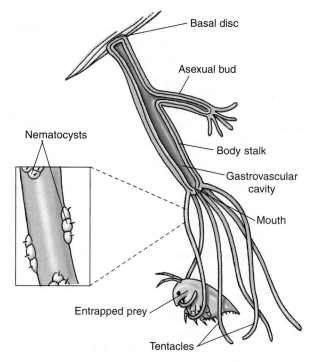

Figure 15-10 Tentacles bearing stinging nematocysts dangle from around the mouth of the hydra. Small prey items (generally zooplankton) are stung, paralyzed, and then transferred to the mouth for ingestion. [Adapted from Rupert and Barnes, 1994.]

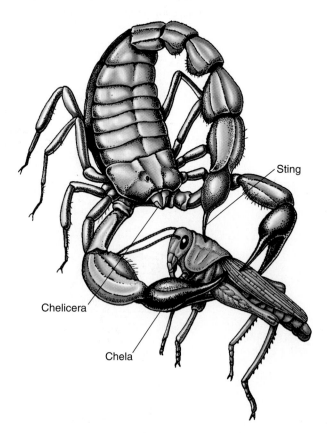

Figure 15-11 The scorpion *Androctonus* captures prey and then injects poison to subdue it. As the prey item is grabbed by the chela, the tail is arched over the scorpion's head to bring the sting into position. The prey is then impaled by the sting, which injects the rapidly acting toxin. [Adapted from Jennings, 1972.]

neurotransmitter at the motor endplate in muscle. A neurotoxin, **α-bungarotoxin** (see Spotlight 6-3), found in the venom of the cobra-like krait, binds to nicotinic acetylcholine (ACh) receptors, thereby blocking neuromuscular transmission in vertebrates. The venoms of various species of rattlesnake contain hemolytic (blood cell–destroying) substances.

Toxins, although highly effective, are generally expensive to produce. Usually carefully measured doses of toxins are delivered during a bite or sting. Toxins also must be specially stored before administration to avoid self-poisoning. Toxins are generally proteins and, as such, are rendered harmless by the proteolytic enzymes of the predator's digestive system when it ingests its poisoned prey.

Herbivory and Grazing to Collect Food

Herbivores often have mouth parts specialized for feeding on plant material. Many gastropods use a rasp-like structure termed a **radula** to scrape algae from rock surfaces or to rasp through vegetation (Figure 15-12). Vertebrate herbivores have bony plates (some fish and reptiles) or teeth primarily in the form of **molars** with wide flat surfaces mod-

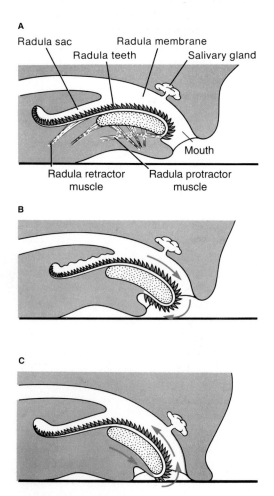

Figure 15-12 The head of a gastropod mollusk contains a powerful radula. **(A)** The rasp-like radula, revealed in sagittal section, is used for grazing on vegetation. **(B)** Protraction of the radula. **(C)** Retraction of the radula. [Adapted from Rupert and Barnes, 1994.]

ified for grinding plant material. Plants (especially some grasses) contain relatively large amounts of silicates, and can be tremendously abrasive. Consequently, the molars of herbivores often are coated in especially tough enamel to resist wear. Alternatively, some herbivores such as small rodents (microtines) have continuously growing, rootless teeth.

OVERVIEW OF ALIMENTARY SYSTEMS

Alimentary systems play an essential role in providing nourishment through digesting and absorbing food, and removing from the body indigestible materials and toxic by-products of digestion. The most primitive "alimentary system" is the plasma membrane of unicellular organisms, in which microscopic food particles are engulfed, undigested, by endocytosis directly into the cell itself. Once in the cell, food particles undergo **intracellular digestion** by acids and enzymes. More complex multicellular animals rely primarily on **extracellular digestion** carried out by true alimentary systems.

From an anatomical perspective, there are myriad designs of alimentary systems. However, from a physiological perspective, alimentary systems fall into one of three categories on the basis of how they process food in a "reactor," or site of chemical digestion. So-called batch reactors are blind tubes or cavities that receive food and eliminate wastes in a pulsed fashion; that is, one batch is processed and eliminated before the next one is brought in

(Figure 15-13, left). Coelenterates, for example, have a blind tube or cavity, the **coelenteron,** which opens only at a "mouth" that serves also for the expulsion of undigested remains. In all phyla higher than the flatworms, ingested material passes through a hollow, tubular cavity—the **alimentary canal**—extending through the organism and open at both ends. Processing goes on continuously, rather than in pulses, with new food being ingested while older food is still being processed. Some alimentary canals can be modeled as **ideal continuous-flow, stirred-tank reactors,** in which food is continually added and mixed into a homogeneous mass, and the products of digestion are continuously eliminated, overflowing from the reactor (Figure 15-13, middle). An example of such a reactor is the forestomach of ruminants. The third way of processing food is in a **plug-flow reactor,** in which a **bolus** (a discrete plug or collection) of food is progressively digested as it winds its way through a long, tube-like digestive reactor (Figure 15-13, right). Unlike the stirred-tank reactor, its composition varies according to its position along the reactor tube. The small intestine of many vertebrates functions as a plug-flow reactor. It is important to recognize that many animals combine features of both continuous- and plug-flow reactors. As you will see below, in many animals chemical digestion begins in the stomach, configured as a continuous-flow, stirred-tank reactor, and then continues on into the small intestine, configured as a plug-flow reactor.

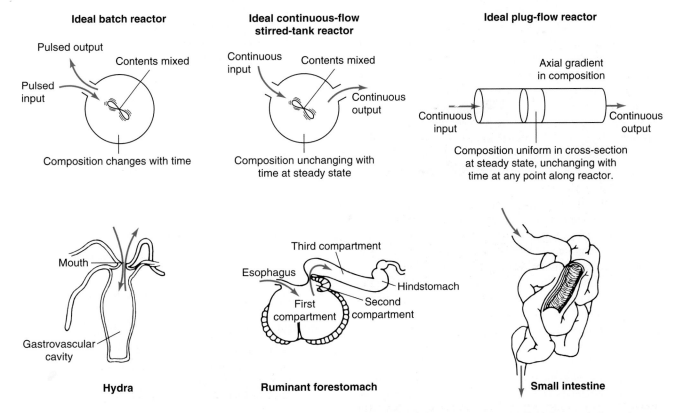

Figure 15-13 Digestive systems are functionally classified according to the type of chemical reactor they form. *(Left)* Batch reactors are found in simple organisms like *Hydra*. *(Middle)* Ruminants have a continuous-flow, stirred-tank reactor in the form of a forestomach. *(Right)* The small intestine of many vertebrates acts as a plug-flow reactor, which may function in addition to the stomach. [From Hume, 1989; adapted from Penry and Jumars, 1987.]

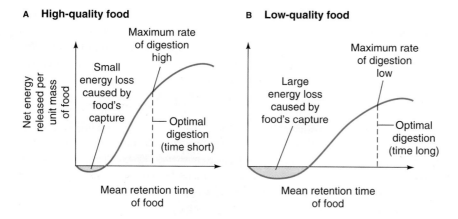

A High-quality food

B Low-quality food

Figure 15-14 The quality of food will greatly influence the time required for digestion in a continuous-flow digestive reactor. **(A)** A high-quality food requires minimal energy to capture and eat and, once eaten, is quickly digested to release large amounts of energy. The maximum rate of digestion occurs at the point of the curved line with the steepest slope. **(B)** A low-quality food requires considerable energy to capture and eat, and takes a long period of digestion to yield only low amounts of energy. [Adapted from Hume, 1989; Sibly, 1981.]

It is critically important that the design of the alimentary canal and the reactors it contains matches well with the quality of the food that the animal routinely eats. A high-quality food can release maximum amounts of energy with minimal time spent in the digestive reactor, whatever its type (Figure 15-14). A lower-quality food, on the other hand, requires a longer period of digestion to release its energy. This in turn requires longer periods spent in the reactor and longer transit times through the alimentary canal. As also indicated in Fig. 15-14, the amount of energy spent in capturing a particular food must also be factored into consideration of food quality.

A generalized alimentary canal, or digestive tract, is illustrated in (Figure 15-15). The lumen of this alimentary canal is topologically external to the body. Sphincters and other devices guard the entrance to and exit from the canal, preventing uncontrolled exchange between the lumen and the external environment. Ingested material is subjected to various mechanical, chemical, and bacterial treatments as it passes through this canal, and **digestive juices** (primarily enzymes and acids) are mixed with the ingested material at appropriate regions in the alimentary canal. As the ingested material is first mechanically broken down and then chemically digested, nutrients undergo **absorption** and are then transported into the circulatory system. Undigested, non-absorbed material is stored briefly until it, along with bacterial remains, is expelled as feces by the process of **defecation.**

The overall tubular organization of the alimentary canal is efficient because it allows ingested material to travel in one direction, passing through different regions that can then be specialized for particular digestive tasks. For example, the alimentary canal near the point of ingestion is often specialized for acid secretion, while more distant regions are alkaline. This regional specialization allows both acid and base secretion to occur at the same time and to permit different types of digestive action.

In general, alimentary canals can be divided on a structural and functional basis into four major divisions (see Figure 15-15): (1) headgut, (2) foregut, (3) midgut and (4) hindgut. These regions are specialized for (1) receiving ingested material, (2) conducting, storing, and digesting ingested material, (3) digesting and absorbing nutrients, and (4) absorbing water and defecating. Representative alimentary canals from the different invertebrate and vertebrate classes are illustrated in Figures 15-16 and 15-17, respectively.

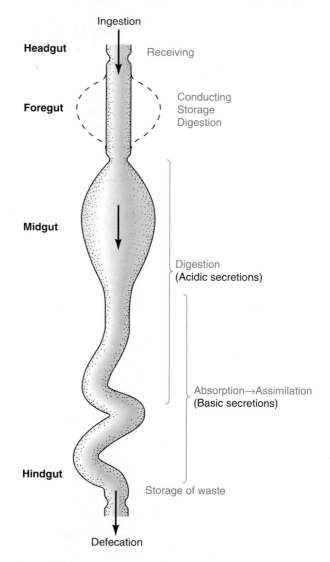

Figure 15-15 A digestive tract with one-way passage of food allows simultaneous operation of sequential stages in the processing of food and reduces mixing of digested and undigested matter. The dashed outline represents a "crop," a storage region found in some animals.

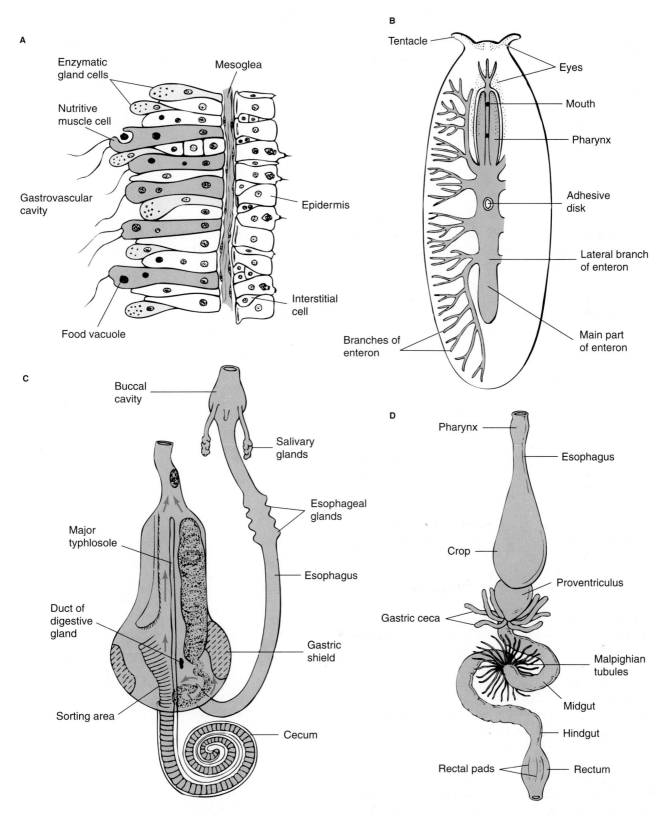

Figure 15-16 Digestive systems of invertebrates show great variation, ranging from simple to highly complex. **(A)** Section through body wall of *Hydra,* a coelenterate. The epithelial lining of the coelenteron includes phagocytosing cells (called *nutritive muscle cells*) and gland cells that secrete digestive enzymes. **(B)** Digestive system of a polyclad flatworm. **(C)** Digestive system of a prosobranch gastropod mollusk. Arrows show ciliary currents and the rotation of the mucous mass. **(D)** Digestive system of the cockroach *Periplaneta.* The proventriculus (or gizzard) contains chitinous teeth for grinding food. [Part C from Rupert and Barnes, 1994; part D from Imms, 1949.]

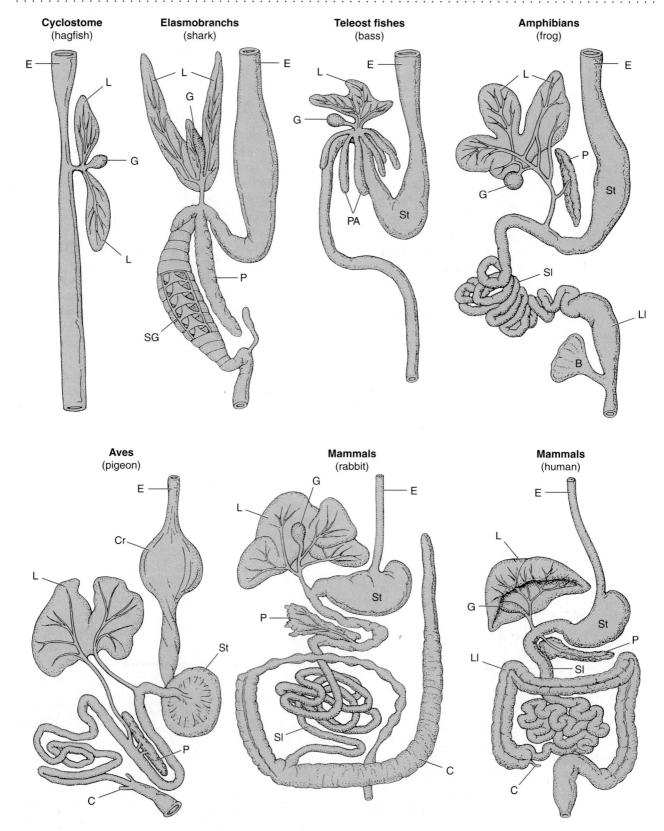

Figure 15-17 The tube-like digestive system of vertebrates has a basic organizational plan, with common elements of esophagus, stomach, intestine, and colon. B, bladder; C, cecum; Cr, crop (gizzard); E, esophagus; G, gallbladder; L, liver; LI, large intestine; P, pancreas; PA, pyloric appendices; SG, spiral gut; SI, small intestine; St, stomach. [From Florey, 1966; adapted from Stempell, 1926.]

Headgut: Food Reception

The **headgut** is the anterior (cranial) region of the alimentary canal, providing an external opening for food entry (see Figure 15-15). It consists of organs and structures for feeding and swallowing, including the mouthparts, buccal cavity, pharynx, and associated structures such as bills, teeth, tongue, and salivary glands. Where a common pathway exists leading to both the alimentary canal and the passageway (e.g., the trachea) ending in the organ of internal gas exchange, there may be additional sphincter- or valve-like structures that control and divert the flow of ingested material and inspired water or air into their respective channels.

Other than in small-particle feeders such as coelenterates, flatworms, and sponges, the headgut of most metazoans has **salivary glands,** the secretions of which aid ingestion and the mechanical (and often chemical) digestion of food. The primary function of the salivary secretion, **saliva,** is lubrication to assist swallowing. The lubrication is provided in many cases by a slippery mucus of which the chief constituent is a type of mucopolysaccharide named **mucin.** The saliva often contains additional agents, such as digestive enzymes, toxins, and anticoagulants (in blood-lapping or blood-sucking animals such as vampire bats and leeches). (See Chapter 8 for a discussion of salivary glands.)

Tongues, an innovation of the chordates, assist in the mechanical digestion and swallowing of food. In some animals tongues are used to grasp food. They are also used in chemoreception, bearing gustatory receptors called *taste buds* (see Figure 7-16A). Snakes use their forked tongues to take olfactory samples from the air and the substratum, retracting the tongue to wipe the samples in Jacobson's organ, which consists of a pair of richly innervated chemosensory pits located in the roof of the buccal cavity. Jacobson's organs are found in other reptiles and some amphibians.

Foregut: Food Conduction, Storage, and Digestion

In most species the **foregut** consists of an **esophagus,** a tube that leads from the oral region to the digestive region of the alimentary canal, and a **stomach** (see Figure 15-15).

Esophagus

The esophagus conducts food from the headgut to the digestive areas, usually the stomach (see below). In chordates and some invertebrates, the esophagus conducts the bolus, or mass of chewed food mixed with saliva, by **peristaltic movement** (see Chapter 11) from the buccal cavity or pharynx. In some animals, this conducting region contains a sac-like expanded section, the **crop,** which is used to store food before digestion. The presence of a crop, generally found in animals that feed infrequently, allows quantities of food to be stored for digestion at a later time. Leeches, for example, feed very infrequently, with weeks or months between feeding periods. However, they ingest large quantities of blood at a "sitting," storing the blood for many weeks and digesting it in small amounts between their rare feedings. In some animals crops are also used to ferment or digest foods for purposes other than their immediate digestion. Parent birds prepare food in this way to be regurgitated for their nestlings.

Stomach

In vertebrates and some invertebrates digestion takes place primarily in the stomach and the midgut. The stomach serves as a storage site for food, and in many species begins the initial stages of digestion. In most vertebrates, for example, the stomach initiates protein digestion by secreting the enzyme pepsinogen (later converted to pepsin) and hydrochloric acid, which provides the highly acidic environment required for pepsin activation. Contraction of the muscular walls of the stomach also provides mechanical mixing of food, saliva, and stomach secretions.

Stomachs are classified as monogastric or digastric, according to the number of chambers they possess. A **monogastric stomach** consists of a single strong muscular tube or sac. Vertebrates that are carnivorous or omnivorous characteristically have a monogastric stomach (Figure 15-18). Instead of a stomach, some invertebrates, such as insects (see Figure 15-16D), have outpouchings termed **gastric ceca** (singular, **cecum**), which are lined with enzyme-secreting cells, as well as phagocytic cells that engulf partially digested food and continue the process of digestion. In these alimentary systems the processes of digestion and absorption are completed in the ceca, and the remainder of the alimentary canal is concerned primarily with water and electrolyte balance and defecation.

Some birds have a tough, muscular gizzard, or crop, or both (see Figure 15-17). Sand, pebbles, or stones are swallowed and then lodge in the gizzard, where they aid in the grinding of seeds and grains. The proventriculus of insects and the stomach of decapod crustaceans, comparable to the bird's gizzard, contain grinding apparatuses for chewing swallowed food. Some fish such as mullets also have gizzards. On the other hand, some fish and larval toads lack stomachs altogether, with material from the esophagus entering into what is functionally the midgut.

Multichambered **digastric stomachs** (Figure 15-19) are found in the mammalian suborder Ruminantia (deer, elk, giraffe, bison, sheep, cattle, etc.). Somewhat similar digastric stomachs occur outside this suborder, in particular in the suborder Tylopoda (camel, llama, alpaca, vicuaña). Microorganisms in the first division of the stomach carry out **fermentation,** the anaerobic conversion of organic compounds to simpler compounds, yielding energy as ATP. All of the above-named groups carry out **rumination,** in which partially digested food is regurgitated (transported back to the mouth) for remastication (additional chewing). This process allows the ruminant (a gazelle on the open savanna, for example) to swallow food hastily while grazing and then to chew it more thoroughly later when at rest in a place of relative safety from predators. After the regurgitated food is chewed, it is swallowed again. This time it passes into the second division of the digastric stomach and

A

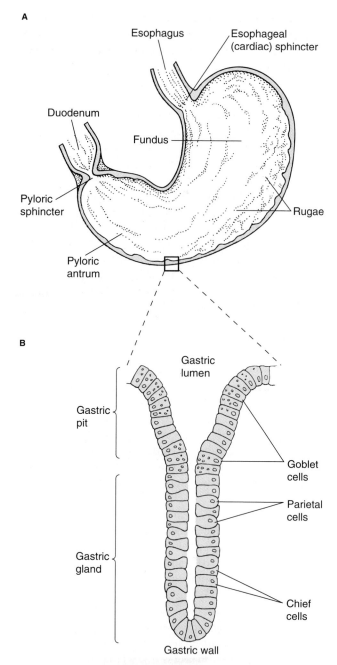

B

Figure 15-18 The monogastric stomach is a single chamber lined with a specialized epithelium. **(A)** Major parts of the mammalian stomach. **(B)** Detail of fundic, or gastric, glands lining a single gastric pit. The inner layer of the stomach is lined with thousands of gastric pits, into which the gastric glands open and dump their digestive juices. The epithelium of the gastric gland contains chief (pepsinogen-secreting) and parietal (HCl-secreting) cells as well as goblet (mucus-secreting) cells.

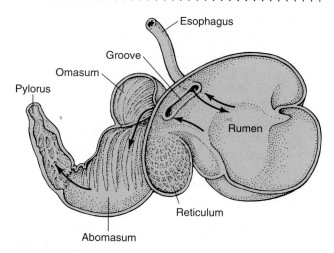

Figure 15-19 The digastric stomach of ruminants has multiple chambers for storing and digesting food materials. This sheep stomach, characteristic of the ruminants, has two divisions made up of four chambers. The rumen and reticulum make up the fermentative division. The omasum and abomasum (true stomach) make up the digestive division.

act as a fermentation vat that receives grazed vegetation. Bacteria and protozoans in these chambers thrive on the vegetation, causing extensive digestive breakdown by fermentation of carbohydrates to butyrate, lactate, acetate, and propionate. These products of fermentation, along with some peptides, amino acids, and short-chain fatty acids, are absorbed into the bloodstream from the rumen fluid. Symbiotic microorganisms grown in the rumen, along with undigested particles, are passed into the omasum (absent in the Tylopoda) and then into the abomasum. Only the latter secretes digestive enzymes and is homologous to the monogastric stomach of non-ruminants.

Fermentation in the stomach is not limited to ruminating animals. It is found in other animals in which the passage of food in the stomach is delayed, allowing the growth of symbiotic microorganisms in a zone anterior to the digestive stomach, as in the kangaroo and the crops of galliform (chicken-like) birds.

Midgut: Chemical Digestion and Absorption

In vertebrates, the **midgut** is the major site for the chemical digestion of proteins, fats, and carbohydrates. Once digested to their component molecules, these materials are then absorbed in the midgut and transported away from the alimentary canal in the blood. As food is ready to pass on from the vertebrate stomach, it is released into the midgut through the **pyloric sphincter**, which relaxes as the peristaltic movements of the stomach squeeze the acidic contents into the duodenum, the initial segment of the **small intestine** (see Figure 15-18A). Digestion continues in the small intestine, generally in an alkaline environment.

General structure and function of the midgut
Among the vertebrates, carnivores have shorter and simpler intestines than do herbivores, reflecting the shorter time required to digest meat than vegetation. For example, a tad-

begins the second stage of digestion. In this stage **hydrolysis** takes place with the assistance of digestive enzymes secreted by the stomach lining.

The digastric stomach of the Ruminantia (see Figure 15-19) has four chambers, separated into two divisions. The first division consists of the **rumen** and **reticulum** chambers; the second division comprises the **omasum** and the **abomasum** (true stomach). The rumen and reticulum pole, which is almost always herbivorous, has a longer intestine than the adult frog, which is carnivorous.

The vertebrate midgut or small intestine is typically divided into three distinct portions. The first, rather short, section is the **duodenum,** the lining of which secretes mucus and fluids and receives secretions carried by ducts from the **liver and pancreas.** Next is the **jejunum,** which also secretes fluid and is involved in digestion and absorption. The most posterior section, the **ileum,** acts primarily to absorb nutrients digested previously in the duodenum and jejunum, although some secretion occurs from the ileum.

As just noted, the secretory functions of the vertebrate duodenal epithelium are supplemented by secretions from the liver and pancreas. The cells of the liver produce **bile salts,** which are carried in the **bile fluid** to the duodenum through the **bile duct.** Bile fluid has two important functions. It emulsifies fats, and it helps neutralize acidity introduced into the duodenum from the stomach. The pancreas, an important exocrine organ described in Chapter 9, produces **pancreatic juice,** which contains many of the proteases, lipases, and carbohydrases essential for intestinal digestion in vertebrates. Pancreatic juice is released into the **pancreatic duct** and, like bile, is important in neutralizing gastric acid in the intestine.

The intestine of most animals contains large numbers of bacteria, protozoans, and fungi. These multiply, contributing enzymatically to digestion, and are usually, in turn, digested themselves. An important function of some intestinal symbionts is the synthesis of essential vitamins.

The midgut region varies greatly not only in structure, but also in function in different animal groups. In many invertebrates, especially those with extensive ceca and diverticula (blind outpouchings of the alimentary canal), the intestine serves no digestive function. In some air-breathing fishes (e.g., the weather loach, *Misgurnus anguillicaudatus*), the midgut is modified into a gas exchange organ where O_2 from gulped air is exchanged with CO_2 from the cells, with residual gas then being expelled out the anus.

Intestinal epithelium

The vertebrate small intestine has adaptations at every anatomical level, from its gross anatomy to the organelles of individual cells, all designed to amplify the surface area available for absorption of nutrients. In humans, the lumen of the small intestine has a gross cylindrical surface area of only about 0.4 m², or about 7–8 pages of this book. However, because of the enormous elaboration of absorptive surfaces provided by this hierarchy of structures, the true area is increased at least 500 times, to a total of 200 to 300m², or about the size of a doubles tennis court. Since the rate of absorption is generally proportional to the area of the apical surface membrane of the cells lining the epithelium, this huge increase in surface area greatly aids absorption of digested substances from the fluid within the intestine. We will now examine this remarkable system of valleys and peaks, peninsulas and inlets.

The general organization of the vertebrate small intestine is shown in Figure 15-20A. The outermost layer is the **serosa,** which is the same tissue that covers the visceral organs of the abdomen. The serosa overlies an outer layer of

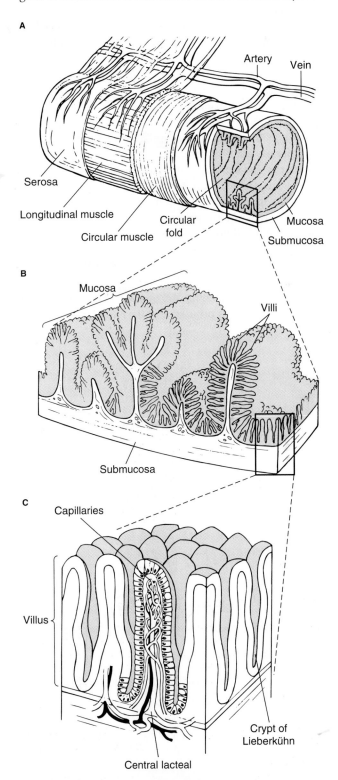

Figure 15-20 The anatomy of the small intestine is dominated by specializations for increased surface area. **(A)** Overall plan. **(B)** Intestinal folds of the mucosa are covered by **(C)** finger-like villi. [From "The Lining of the Small Intestine," by F. Moog. Copyright © 1981 by Scientific American, Inc. All rights reserved.]

longitudinal smooth muscle, while an inner layer of circular smooth muscle surrounds the epithelial layer, which consists of the submucosa (a layer of fibrous connective tissue) and the mucosa (or mucous membrane). Projecting into and encircling the lumen of the small intestine are numerous folds of the mucosa, variously called the folds of Kerckring or the circular folds (Figure 15-20A and B). In addition to increasing surface area, these folds serve to slow the progress of food through the intestine, allowing more time for digestion. At the next anatomical level are the finger-like villi (Figure 15-20B and C), which line the folds, standing about 1 mm tall. Each villus sits in a circular depression known as the crypt of Lieberkühn (see Figure 15-20C). Within each villus is a network of blood vessels—arterioles, capillaries and venules—and a network of lymph vessels, the largest of which is the central lacteal. Nutrients taken up from the intestine are transferred into these blood and lymph vessels for transport to other tissues; the central lacteal can, in addition, take up larger particles.

The villi are lined with the actual absorptive surface of the small intestine, the cells of the digestive epithelium (Figure 15-21). The epithelium consists of goblet cells interspersed among columnar absorptive cells (Figure 15-21A). The absorptive cells proliferate at the base of the villus and

steadily migrate toward its tip, where they are sloughed off at the rate of about 2×10^{10} cells per day in the human intestine, meaning that the entire midgut lining is replaced every few days.

The next level in the hierarchy of absorptive adaptations is found at the apical surface of each absorptive cell, where there are striated structures called microvilli, which collectively form the brush border (Figure 15-21B and D). There are up to several thousand microvilli per cell (about 2×10^5 per square millimeter); each standing 0.5 to 1.5 μm tall and about 0.1 μm wide. The membrane of the microvillus is continuous with the plasma membrane of the epithelium and contains actin filaments that form crossbridge links with myosin filaments present at the base of each microvillus (Figure 15-21C). Intermittent actin-myosin interaction produces rhythmic motions of the microvilli, which might help mix and exchange the intestinal chyme (semifluid mass of partially digested food) near the absorptive surface.

The surfaces of the microvilli are covered by the glycocalyx, a meshwork up to 0.3 μm thick comprising acid mucopolysaccharides and glycoproteins (Figure 15-21C). Water and mucus are trapped within the interstices of the glycocalyx. The mucus is secreted by the mucous (goblet)

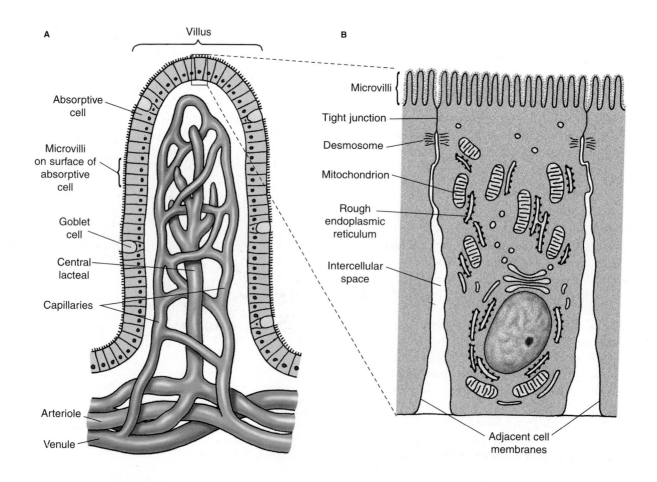

A Villus

Absorptive cell

Microvilli on surface of absorptive cell

Goblet cell

Central lacteal

Capillaries

Arteriole

Venule

B

Microvilli

Tight junction

Desmosome

Mitochondrion

Rough endoplasmic reticulum

Intercellular space

Adjacent cell membranes

cells, named for their shape, that occur among the absorptive cells (see Figure 15-21A).

Adjacent absorptive cells are held together by desmosomes (Chapter 4). Near the apex, the **zonula occludens** encircles each cell, making a tight junction with its neighbors (Figure 15-21B). The tight junctions are especially tight in this epithelium, so that the apical membranes of the absorptive cells effectively form a continuous sheet of apical membrane, without breaks between cells. Because of the virtual impermeability of the tight junctions, all nutrients must pass across this membrane and through the absorptive cell cytoplasm to get from the lumen to the blood and lymph vessels within the villi. Little, if any, paracellular passage occurs.

Hindgut: Water and Ion Absorption and Defecation

The **hindgut** serves to store the remnants of digested food (see Figure 15-15). From this material is absorbed inorganic ions and excess water for return to the blood. In vertebrates, this function is carried out primarily in the latter portion of the small intestine and in the large intestine. In some insects, the feces within the rectum are rendered almost dry by a specialized mechanism for removing water from the rectal contents (Chapter 14). The hindgut also functions as the major site for bacterial digestion of intestinal contents through the action of the bacterial flora found in herbivorous reptiles, birds, and most herbivorous mammals.

In many species it is the hindgut that consolidates undigested material and bacteria growing in the hindgut into **feces**. The feces pass into the cloaca or rectum and are then expelled through the **anus** in the process of **defecation** (see below).

The hindgut is also the site of **hindgut fermentation** in many animals (Figure 15-22). The colon acts as a modified plug-flow reactor in most large animals that are hindgut fermenters (e.g., horses, zebras, tapirs, sirenians, elephants, rhinos, and marsupial wombats). In smaller hindgut fermenters the tremendously enlarged cecum acts as a continuous-flow, stirred-tank reactor (rabbits, many rodents, hyraxes, howler monkeys, koalas, and brushtail and ringtail opossums).

The hindgut terminates in a **cloaca** in many vertebrates, including hagfish, lungfish, *Latimeria*, elasmobranchs, adult amphibians, reptiles, birds, and a few mammals (monotremes, marsupials, some Insectivora, a few rodents). The cloaca aids in urinary ion and water resorption in those species in which the ureters terminate in the cloaca rather than in external genitalia.

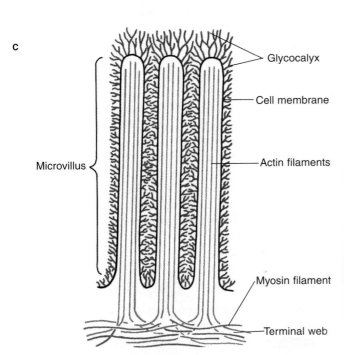

C

Glycocalyx

Cell membrane

Microvillus

Actin filaments

Myosin filament

Terminal web

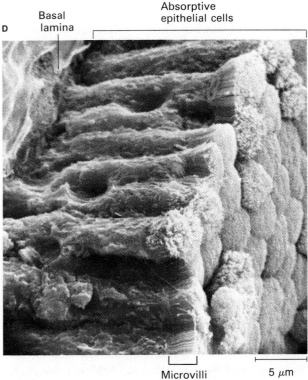

D

Basal lamina

Absorptive epithelial cells

Microvilli

5 μm

Figure 15-21 The lining of the mammalian small intestine has a complex microanatomy specialized for absorption and secretion. The luminal surface is shown in color. **(A)** A villus covered with the mucosal epithelium, which consists primarily of absorptive cells and occasional goblet cells. **(B)** An absorptive cell. The luminal, or apical, surface of the absorptive cell bears a brush border of microvilli. **(C)** The microvilli consist of evaginations of the surface membrane, enclosing bundles of actin filaments. **(D)** Scanning electron micrograph of a group of absorptive cells from the human small intestine, showing the brush border. [Parts A–C from "The Lining of the Small Intestine," by F. Moog. Copyright © 1981 by Scientific American, Inc. All rights reserved. Part D from Lodish et al., 1995.]

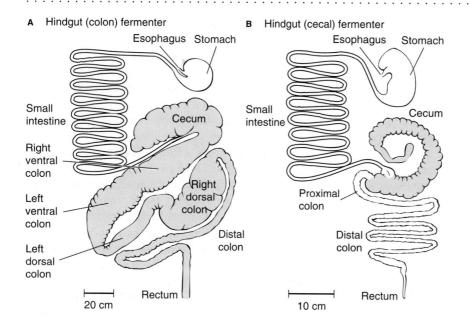

A Hindgut (colon) fermenter

B Hindgut (cecal) fermenter

Figure 15-22 The digestive tract of a colon fermenter has an enlarged colon compared with that of a cecal fermenter, which has an enlarged cecum. **(A)** Digestive tract of the horse *Equus caballus*, a colon fermenter. Site of fermentation is shown in red. **(B)** Digestive tract of the rabbit *Oryctolagus cuniculus*, a cecal fermenter. [Adapted from Stevens, 1988.]

 Several species of air-breathing catfishes swallow air and extract oxygen from the gas bubble across the wall of a specially modified intestine. How would you design an alimentary canal that could serve the dual purposes of digestion-absorption and gas exchange?

Dynamics of Gut Structure—Influence of Diet

Research over the past decade has changed our traditional view of the gut as a relatively static set of organs and tissues. In fact, we now know that gut size and structure are quite dynamic, responding to changes in both energy demand and food quality in most animals, be they carnivores or herbivores. Most responsive is overall gut size. House wrens *(Troglodytes aedon),* induced to increase food intake through exposure to combinations of lowered ambient temperature and enforced exercise over several months, responded by increasing the overall length of the small intestine by about one-fifth. Efficacy of nutrient uptake increases as a result. The mass of the empty stomach of the ground squirrel *(Spermophilus tridecemlineatus)* increases three- to fourfold within a few months after rousing from hibernation. Although reptiles have a much lower rate of metabolism than birds and mammals (see Chapter 16), some reptiles appear to remodel their gut in response to food intake much more rapidly than has been shown for birds and mammals, sometimes within a few hours or days. In the Burmese python *(Python molurus)*, anterior small intestine mass increases by over 40% over fasting levels within 6 hours of a large meal (25% of body mass), and reaches double the fasting mass two days after a meal. These changes are due largely to proliferation of the mucosal rather than serosal layer. Associated with these morphological changes were increases in capacity for amino acid uptake that ranged from 10–24 times the fasting values.

Even when overall length and diameter are not affected by dietary changes, the "microstructure" in terms of villi may change, resulting in alterations in absorptive surface area. These changes can lead to an overall increase in nutrient absorption when an animal's energy demands are great, as well as aid in slowing the passage of digesting food, to enhance extraction of nutrients. This latter situation is particularly evident in cecal fermenters.

Dietary adjustments can similarly alter the cellular and macromolecular makeup of the gut. Research over the last few decades by investigators, including Jared Diamond and William Karasov, has shown that most intestinal membrane transporters are regulated by dietary levels of their substrates. Increasing substrate levels stimulates an increase in concentration and/or activity of transporters for glucose, fructose, some nonessential amino acids, and peptides. The proliferation of transporters appears to be matched to the level of nutrient intake so as to provide no more than the uptake capacity necessary.

It is important to emphasize that an expansion of gut surface area or nutrient transporter proteins carries with it significant metabolic cost in support of this new macro- or microstructure. Consequently, most changes in gut structure appear to be completely reversible, to reduce the metabolic cost of maintaining the gut during periods when food resources are scarce.

MOTILITY OF THE ALIMENTARY CANAL

The ability of the alimentary tract to contract and propel ingested material along its length, a characteristic called **motility,** is important to digestive function for:

1. Translocation of food along the entire length of the alimentary canal and the final expulsion of fecal material

2. Mechanical treatment by grinding and kneading to help mix in digestive juices and convert food to a soluble form

3. Mixing of the contents so that there is continual renewal of material in contact with the absorbing and secreting surfaces of the epithelial lining

Muscular and Ciliary Motility

Motility can be achieved by two different mechanisms—muscular motility and ciliary motility. **Muscular motility,** in which transport is achieved by muscle contraction of the walls of the alimentary canal, is the only mechanism found in arthropods and chordates. In chordates, motility is achieved strictly by smooth muscle fibers, but in many arthropods motility is achieved by striated fiber contraction. Muscular mechanisms permit handling of harder and larger pieces of food. Ciliary motility, in which cilia lining the digestive tract generate currents of fluid within, is the only mechanism used to translocate food along the alimentary canals of annelids, lamellibranch mollusks, tunicates, and cephalochordates. However, ciliary motility is used in conjunction with muscular mechanisms in echinoderms and most mollusks.

Peristalsis

The alimentary musculature is made up of smooth muscle tissue in all animal groups other than arthropods, where it comprises striated muscle. The arrangement of the musculature in vertebrates consists of an inner **circular** layer and an outer **longitudinal** layer (Figure 15-23; see also Figure 15-20A). The contraction of the circular layer coordinated with relaxation of the longitudinal layer produces an active constriction with an elongation. Active shortening of the longitudinal layer with relaxation of the circular layer produces distension. **Peristalsis** occurs as a traveling wave of constriction produced by contraction of circular muscle and is preceded along its length by a simultaneous contraction of the longitudinal muscle and relaxation of the circular muscle (Figure 15-24). This pattern of contraction "pushes" the luminal contents in the direction of the peristaltic wave. Mixing of the luminal contents is achieved primarily by a process called **segmentation,** which consists of rhythmic contractions of the circular muscle layer that occur asynchronously along the intestine at various points without participation of the longitudinal muscle.

Swallowing in vertebrates involves the integrated movements of muscles in the tongue and pharynx, as well as peristaltic movements of the esophagus, which are under direct neural control of the medulla oblongata of the brain. These actions propel a bolus to the stomach. **Regurgitation** occurs when peristalsis takes place in the reverse direction,

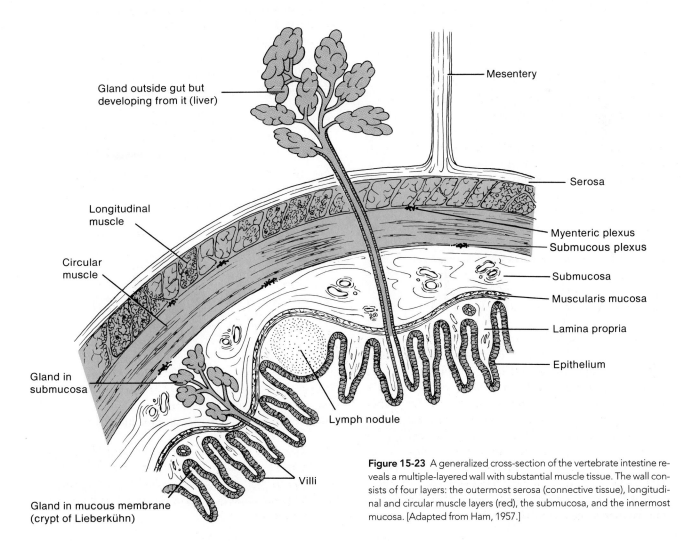

Gland outside gut but developing from it (liver)

Longitudinal muscle

Circular muscle

Gland in submucosa

Gland in mucous membrane (crypt of Lieberkühn)

Villi

Lymph nodule

Mesentery

Serosa

Myenteric plexus

Submucous plexus

Submucosa

Muscularis mucosa

Lamina propria

Epithelium

Figure 15-23 A generalized cross-section of the vertebrate intestine reveals a multiple-layered wall with substantial muscle tissue. The wall consists of four layers: the outermost serosa (connective tissue), longitudinal and circular muscle layers (red), the submucosa, and the innermost mucosa. [Adapted from Ham, 1957.]

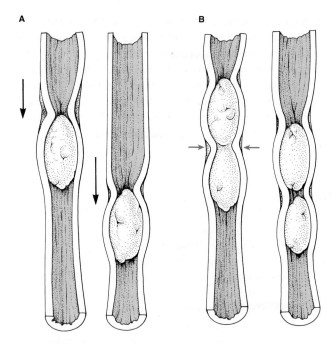

Figure 15-24 Coordinated contraction of the gastrointestinal tract propels material through its lumen. **(A)** Peristalsis occurs as a traveling wave of contraction of circular muscle preceded by relaxation. This produces longitudinal movement of the bolus. **(B)** Segmentation occurs as alternating relaxations and contractions, primarily of circular muscle. The result is a kneading and mixing of the intestinal contents.

moving the luminal contents back into the buccal cavity. Ruminants regularly use regurgitation to bring up the unchewed food for further chewing, and other vertebrates use it during emesis (vomiting).

Normal peristalsis in the vertebrate stomach occurs with the ring of contraction only partially closed. Consequently there is a mixing action in which the contents are squeezed backward (opposite to the direction of the wave) centrally through the partially open ring and forward peripherally in the direction of peristalsis as the partially closed ring of contraction moves from the cardiac to the pyloric end of the stomach.

Control of Motility

The coordinated contractions of circular and longitudinal smooth muscle layers that provide alimentary canal motility in vertebrates are regulated by a combination of distinct mechanisms.

Intrinsic control

The smooth muscle tissue in the wall of the alimentary tract is **myogenic**—that is, capable of producing an intrinsic cycle of electrical activity that leads to muscle contraction without external neural stimulation. This cycle occurs as rhythmic depolarizations and repolarizations called the **basic electric rhythm** (BER). This rhythm consists of spontaneous slow waves of depolarization that progress slowly along the muscle layers (Figure 15-25). Some of these slow waves give rise to action potentials (APs) produced by an inward current carried by calcium ions. These calcium "spikes" lead to contractions of the smooth muscle cells in which they occur. The amplitude of the slow-wave BER is modulated by local influences such as stretching of the muscle tissue. Such stretching would occur when a chamber of the alimentary canal is stretched by contents in its lumen. Another influence on contraction is chemical stimulation of the mucosa by substances in the chyme.

Extrinsic (neural, hormonal) control

Intrinsic patterns of the BER are modulated by locally released gastrointestinal **peptide hormones** (Table 15-1; also,

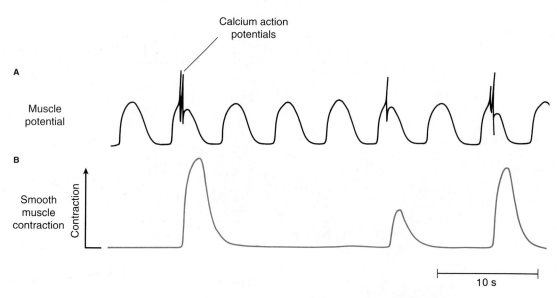

Figure 15-25 Electrical and mechanical (contraction) activity are coordinated in the cat jejunum. **(A)** The slow basic electric rhythm evident as oscillations in the muscle potential occasionally gives rise to calcium action potentials at their peaks. **(B)** Calcium action potentials elicit contractions of the smooth muscle in which they occur. [Adapted from Bortoff, 1985.]

see Spotlight 9-1). Thus, a chemical stimulant in the chyme can cause the release of a local hormone, and this, in turn, can modulate the motility of the muscle tissue.

In addition to local stimuli, intestinal motility is influenced by diffuse innervation from the sympathetic, parasympathetic, and peptidergic (purinergic) divisions of the autonomic nervous system (see Chapter 9). Sympathetic and parasympathetic postganglionic neurons form networks dispersed throughout the smooth muscle layers (Figure 15-26). The parasympathetic network made up of cholinergic neurons is divided into the **myenteric plexus** and the **submucosal plexus.** These plexi, which receive their parasympathetic input primarily via branches of the vagus nerve, mediate excitatory actions (i.e., increased motility and gastrointestinal secretion) of the digestive tract. In contrast, the innervation from the sympathetic division is

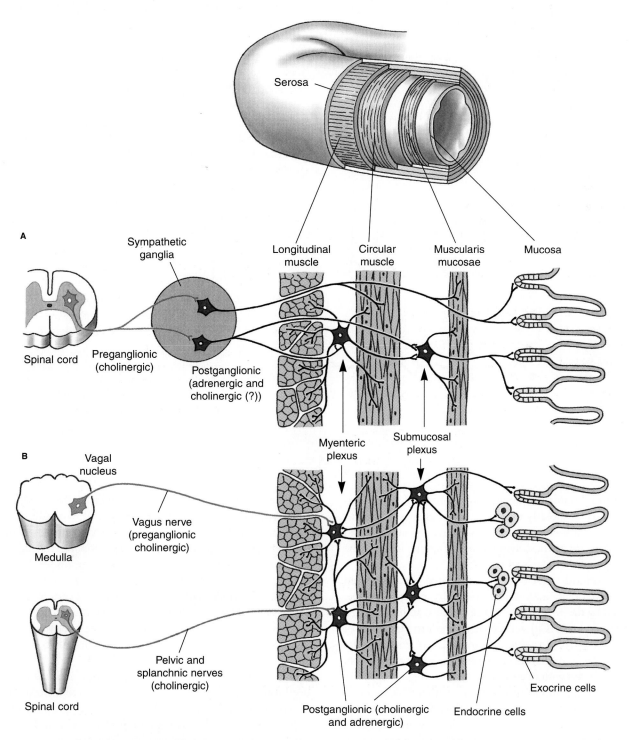

Figure 15-26 The gastrointestinal tract has rich sympathetic and parasympathetic innervation. **(A)** Efferent sympathetic innervation. **(B)** Parasympathetic innervation. All nerve endings on the gastrointesti- nal target tissues (muscle, glands) are postganglionic. [Adapted from Davenport, 1977.]

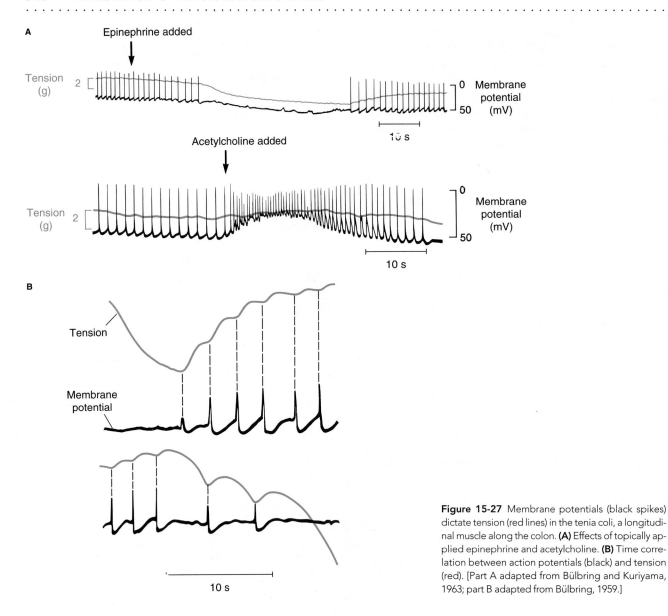

Figure 15-27 Membrane potentials (black spikes) dictate tension (red lines) in the tenia coli, a longitudinal muscle along the colon. **(A)** Effects of topically applied epinephrine and acetylcholine. **(B)** Time correlation between action potentials (black) and tension (red). [Part A adapted from Bülbring and Kuriyama, 1963; part B adapted from Bülbring, 1959.]

primarily inhibitory. Postganglionic neurons of the sympathetic division directly innervate all the tissues of the gut wall as well as neurons of the myenteric and submucosal plexi. Activity of these sympathetic efferents inhibits the motility of the stomach and intestine.

The smooth muscle cells are inhibited (i.e., prevented from developing action potentials) by norepinephrine, released by the sympathetic nerve endings, and are excited by acetylcholine (ACh), released in response to the activity of the parasympathetic nerves (Figure 15-27A). Each impulse associated with excitation produces an increment of tension, which subsides with cessation of impulses (Figure 15-27B). Evidence of the importance of the smooth muscle innervation in maintaining tone is found in Hirschsprung's disease (also known as *congenital megacolon*), in which there is a congenital absence of ganglion cells in the wall of the rectum. Lacking smooth muscle tone, the colon becomes greatly extended, which can lead to recurrent fecal impactions.

The peristaltic movements described in the previous section are coordinated by the intrinsic BER, with the local participation of the myenteric plexus. This contrasts with the peristaltic movements of the swallowing reflex, in which the movements of the esophagus are under direct control of the central nervous system.

Smooth muscle in the alimentary canal of vertebrates is also regulated by non-adrenergic, non-cholinergic neurons that release a variety of peptides and purine nucleotides. In the nearly three decades since this was first discovered, aminergic neurons have been identified that release ATP, 5-HT, dopamine, GABA, while peptidergic neurons have been found that release enkephalins, vasoactive intestinal polypeptide (VIP), substance P, bombesin/gastrin-releasing peptide, neurotensin, cholecystokinin (CCK), and neuropeptide Y/pancreatic polypeptide. This host of transmitter substances allows very fine control over the numerous interacting functions of the alimentary canal.

TABLE 15-1
Action of some enzymes secreted in the mouth, stomach, pancreas, and small intestine of mammals

Enzyme	Site of action	Substrate	Products of action
Mouth			
Salivary α-amylase	Mouth	Starch	Disaccharides (few)
Stomach			
Pepsinogen → pepsin	Stomach	Proteins	Large peptides
Pancreas			
Pancreatic α-amylase	Small intestine	Starch	Disaccharides
Trypsinogen → trypsin	Small intestine	Proteins	Large peptides
Chymotrypsin	Small intestine	Proteins	Large peptides
Elastase	Small intestine	Elastin	Large peptides
Carboxypeptidases	Small intestine	Large peptide	Small peptides (oligopeptides)
Aminopeptidases	Small intestine	Large peptide	Oligopeptides
Lipase	Small intestine	Triglycerides	Monoglycerides, fatty acids, glycerol
Nucleases	Small intestine	Nucleic acids	Nucleotides
Small intestine			
Enterokinase	Small intestine	Trypsinogen	Trypsin
Disaccharidases	Small intestine*	Disaccharides	Monosaccharides
Peptidases	Small intestine*	Oligopeptides	Amino acids
Nucleotidases	Small intestine*	Nucleotides	Nucleosidases, phosphoric acid
Nucleosidases	Small intestine*	Nucleosides	Sugars, purines, pyrimidines

*Intracellular

GASTROINTESTINAL SECRETIONS

The alimentary canal of most animals produces both endocrine and exocrine secretions. In fact, the alimentary canal in many animals has been described as the "largest endocrine and exocrine gland of the body." As explained in Chapters 8 and 9, hormones are produced in the alimentary canal by cells of endocrine glands and are liberated directly into the bloodstream, acting as messengers to receptor molecules in target tissues, which usually include other tissues of the alimentary canal.

Exocrine gastrointestinal secretions usually consist of aqueous mixtures of substances rather than a single species of molecule. Exocrine tissues of the alimentary canal include the salivary glands, secretory cells in the stomach and intestinal epithelium, and secretory cells of the liver and pancreas. The **primary secretions** of the exocrine glands of the alimentary canal enter the acinar lumen of the gland, and then generally become secondarily modified in the gland's secretory duct. This **secondary modification** can involve further transport of water and electrolytes into or out of the duct to produce the final secretory juice, as illustrated in the salivary gland (Figure 15-28) and described in detail in Chapter 8.

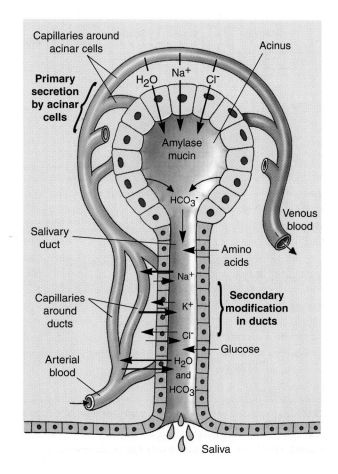

Figure 15-28 Formation of saliva in the mammalian salivary gland depends on active transport and osmosis. The acinar cells transport electrolytes from their basal sides into the acinus and secrete mucin and amylase by exocytosis, with water flowing into the lumen by osmosis. As the salivary fluid moves down the duct, it undergoes modification by active transport across the epithelium of the duct. [Adapted from Davenport, 1985.]

Exocrine Secretions of the Alimentary Canal

There are large variations in the composition of the secretions from different regions of the alimentary canal. However, these mixtures usually consist of some combination of water, ions, mucus, and enzymes.

Water and electrolytes

The exocrine glands of the alimentary canal typically secrete large quantities of water-based fluids bearing digestive enzymes and other chemicals into the alimentary canal lumen (Figure 15-29). Most of this water is reabsorbed in the distal portions of the gut.

In aqueous solution, the mucus produced in the goblet cells of the stomach and intestine (see Figures 15-18 and 15-21) provides a slippery, thick lubricant that helps prevent mechanical and enzymatic injury to the lining of the gut. The salivary glands and pancreas secrete a thinner mucoid solution.

Secretion of inorganic constituents of digestive fluids generally occurs in two steps. First, water and ions are secreted into the lumen of the gland either by passive ultrafiltration due to a hydrostatic pressure gradient across the luminal epithelium, or by active (energy-requiring) processes from the interstitial fluid bathing the basal portions of the acinar cells. The latter is believed usually to entail active transport of ions by these cells, which is then followed by the osmotic flow of water into the acinus. There is subsequent secondary modification of this ultrafiltrate by active or passive transport across the epithelium lining the ducts as the fluid passes along the exocrine ducts toward the alimentary canal.

Bile and bile salts

The vertebrate liver does not produce digestive enzymes. However, it does secrete **bile,** a fluid essential for digestion of fats. Bile consists of water and a weakly basic mixture of cholesterol, lecithin, inorganic salts, bile salts, and bile pigments. The **bile salts** are organic salts composed of bile acids manufactured by the liver from cholesterol and conjugated with amino acids complexed with sodium (Figure 15-30). The **bile pigments** derive from biliverdin and bilirubin, which are products of the breakdown of hemoglobin spilled into plasma from old, ruptured red blood cells. Bile produced in the liver is transported via the hepatic duct to the **gallbladder,** where it is concentrated and stored. Water is removed osmotically, following active transport of Na^+ and Cl^- from the bile across the gallbladder epithelium.

Bile serves numerous functions important to digestion. First, its high alkalinity is important in the terminal stages of digestion because it buffers the high acidity provided by the gastric juice secreted earlier in the digestive process.

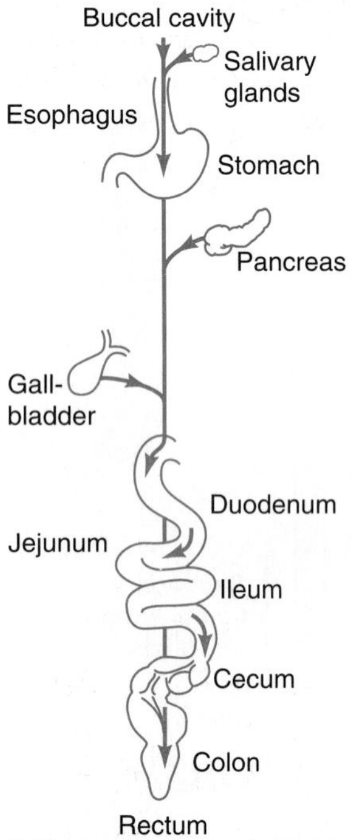

Region	Secretion	Daily amount (L)	pH	Composition*
Buccal cavity — Salivary glands	Saliva	1+	6.5	Amylase, bicarbonate
Stomach	Gastric juice	1–3	1.5	Pepsinogen, HCl, rennin in infants, "intrinsic factor"
Pancreas	Pancreatic juice	1	7–8	Trypsinogen, chymotrypsinogen, carboxy- and aminopeptidase, lipase, amylase, maltase, nucleases, bicarbonate
Gallbladder	Bile	1	7–8	Fats and fatty acids, bile salts and pigments, cholesterol
Duodenum	'Succus entericus'	1	7–8	Enterokinase, carboxy- and aminopeptidases, maltase, lactase, sucrase, lipase, nucleases

*Excluding mucus and water, which together make up some 95% of the actual secretion.

Figure 15-29 Important digestive secretions occur at all points along the human alimentary canal. The approximate volume and pH of each secretion is shown on the right.

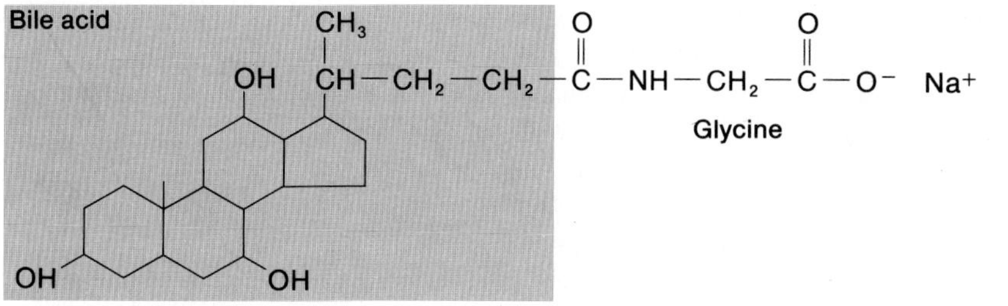

Figure 15-30 Sodium glycholate is the mammalian bile salt. Cholic acid (colored area) is conjugated with the amino acid glycine and sodium.

Second, bile salts facilitate enzymatic fat digestion by breaking down fat into microscopic droplets that collectively have a much higher surface area. The ability of the bile salts to disperse fatty, water-insoluble substances derives from their amphipathic nature. That is, the bile salt molecule contains a lipid-soluble bile acid together with a water-soluble amino acid. Thus, it acts as a detergent for the emulsification of fat droplets, dispersing them in aqueous solution for more effective attack by digestive enzymes. Ultimately, the bile salts are removed from the large intestine by highly efficient active transport and returned to the bloodstream. The bile salts then become bound to a plasma carrier protein and are returned to the liver to be recycled. Bile salts also disperse lipid-soluble vitamins for transport in the blood.

Third, bile fluid contains waste substances removed from the blood by the liver, such as hemoglobin pigments, cholesterol, steroids, and drugs. These substances are either digested or excreted in the feces.

Digestive enzymes

An animal must first digest food before it can be used for tissue maintenance and growth and as a source of chemical energy. **Digestion** is primarily a complex chemical process in which special **digestive enzymes** catalyze the hydrolysis of large foodstuff molecules into simpler compounds that are small enough to cross cell membranes of the intestinal barrier. For example, starch, a long-chain polysaccharide, is degraded to much smaller disaccharides and monosaccharides; proteins are hydrolyzed into polypeptides and then into tripeptides, dipeptides, and amino acids.

All digestive enzymes carry out hydrolysis, adding H^+ to one residue and OH^- to the other (Figure 15-31). Hydrolysis of the anhydrous bonds frees the constituent residues (e.g., monosaccharides, amino acids, monoglycerides) from which the polymer is formed, making them small enough for absorption from the alimentary canal into the circulating body fluids and for subsequent entry into cells to be metabolized.

Digestive enzymes, like all enzymes, exhibit substrate specificity and are sensitive to temperature, pH, and certain ions (see Chapter 3). Corresponding to the three major types of foodstuffs are three major groups of digestive enzymes: proteases, carbohydrases, and lipases.

Proteases Proteases are proteolytic enzymes, categorized as either **endopeptidases** or **exopeptidases**. Both types of enzymes attack peptide bonds of proteins and polypeptides (Figure 15-31A, Table 15-2). They differ in that endopeptidases confine their attacks to bonds well within (*endo*, "within") the protein molecule, breaking large peptide chains into shorter polypeptide segments. These shorter

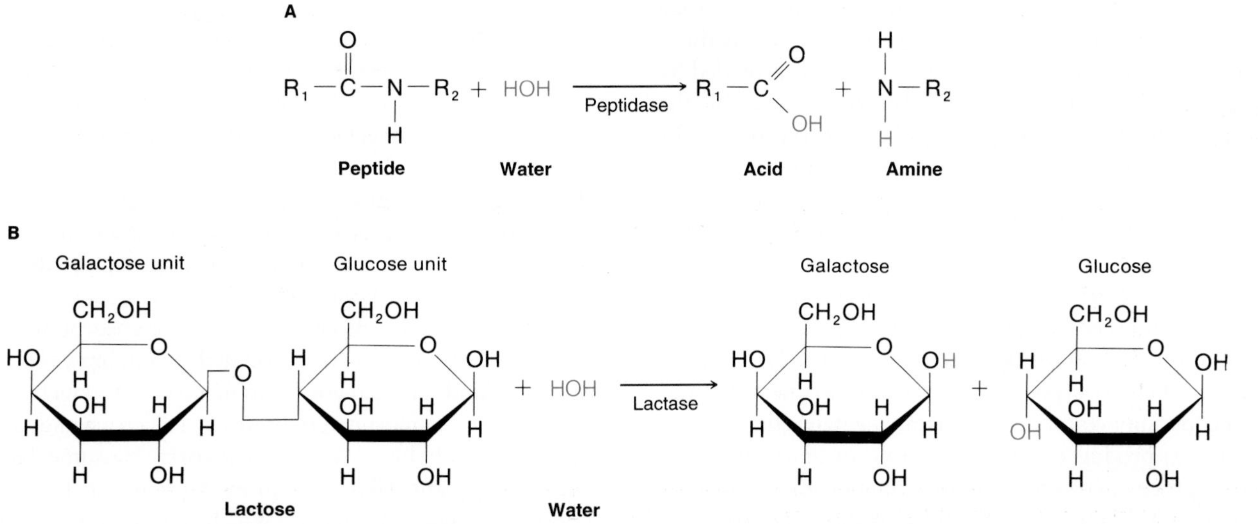

Figure 15-31 Peptides (**A**) and disaccharides (**B**) are broken down by hydrolysis. Under enzyme catalysis, a molecule of water is added to the two residues as shown, breaking the covalent bond holding the residues together.

TABLE 15-2
Some gastrointestinal peptide hormones

Hormone	Tissues of origin	Target tissue	Primary action	Stimulus to secretion
Gastrin	Stomach and duodenum	Secretory cells and muscles of stomach	HCl production and secretion; stimulation of gastric motility	Vagus nerve activity; peptides and proteins in stomach
Cholecystokinin (CCK)*	Upper small intestine	Gallbladder	Contraction of gallbladder	Fatty acids and amino acids in duodenum
		Pancreas	Pancreatic juice secretion	
Secretin*	Duodenum	Pancreas, secretory cells, and muscles of stomach	Water and NaHCO₃ secretion; inhibition of gastric motility	Food and strong acid in stomach and small intestine
Gastric inhibitory peptide (GIP)	Upper small intestine	Gastric mucosa and musculature	Inhibition of gastric secretion and motility	Monosaccharides and fats in duodenum
Bulbogastrone	Upper small intestine	Stomach	Inhibition of gastric secretion and motility	Acid in duodenum
Vasoactive intestinal peptide (VIP)*	Duodenum	Stomach, intestine	Increase of blood flow; secretion of thin pancreatic fluid; inhibition of gastric secretion	Fats in duodenum
Enteroglucagon	Duodenum	Jejunum, pancreas	Inhibition of motility and secretion	Carbohydrates in duodenum
Enkephalin*	Small intestine	Stomach, pancreas, intestine	Stimulation of HCl secretion; inhibition of pancreatic enzyme secretion and intestinal motility	Basic conditions in stomach and intestine
Somastostatin*	Small intestine	Stomach, pancreas, intestines, splanchnic arterioles	Inhibition of HCl secretion, pancreatic secretion, intestinal motility, and splanchnic blood flow	Acid in lumen of stomach

*Peptides marked with an asterisk are also found in central nervous tissue as neuropeptides. Neuropeptides not listed here, but identified in both brain and gut tissue, include substance P, neurotensin, bombesin, insulin, pancreatic polypeptide, and ACTH.

segments provide a much greater number of sites of action for the exopeptidases. The exopeptidases attack only peptide bonds near the end (*exo*, "outside") of a peptide chain, providing free amino acids, plus dipeptides and tripeptides. Some proteases exhibit marked specificity for particular amino acid residues located on either side of the bonds they attack. Thus, the endopeptidase **trypsin** attacks only those peptide bonds in which the carboxyl group is provided by arginine or lysine, regardless of where they occur within the peptide chain. The endopeptidase **chymotrypsin** attacks peptide bonds containing the carboxyl groups of tyrosine, phenylalanine, tryptophan, leucine, and methionine.

In mammals, protein digestion usually begins in the stomach by the action of the gastric protease **pepsin**. There are different forms of this enzyme, but the most powerful form functions best at a low pH value of around 2. The action of pepsin is aided by secretion of gastric HCl and results in the hydrolysis of proteins into polypeptides and some free amino acids. In the mammalian intestine, several proteases produced by the pancreas continue the proteolytic process, yielding a mixture of free amino acids and small peptide chains. Finally, proteolytic enzymes intimately associated with the epithelium of the intestinal wall hydrolyze the polypeptides into **oligopeptides**, which con-

sist of residues of two or three amino acids, and then further break these down into individual amino acids.

Carbohydrases Carbohydrases can be divided functionally into polysaccharidases and glycosidases. **Polysaccharidases** hydrolyze the glycosidic bonds of long-chain carbohydrates such as cellulose, glycogen, and starch. The most common polysaccharidases are the **amylases,** which hydrolyze all but the terminal glycosidic bonds within starch and glycogen, producing disaccharides and oligosaccharides. The **glycosidases,** which occur in the glycocalyx attached to the surface of the absorptive cells (see Figure 15-21C), act on disaccharides such as sucrose, fructose, maltose, and lactose by hydrolyzing the remaining alpha-1,6 and alpha-1,4 glycosidic bonds. This breaks these sugars down into their constituent monosaccharides for absorption (see Figure 15-31B). Amylases are secreted in vertebrates by the salivary glands and pancreas and in small amounts by the stomach, and in most invertebrates by salivary glands and intestinal epithelium. Many herbivores consume large amounts of plant cell walls, containing cellulose, hemicellulose, and lignin. Cellulose, which is in greatest abundance, consists of glucose molecules polymerized via beta-1,4 bonds. **Cellulase,** an enzyme that digests cellulose and hemicellulose, is produced by symbiotic microorganisms in the gut of host animals as diverse as termites and cattle, which themselves are incapable of producing cellulase. In termites, cellulase is liberated into the intestinal lumen by the symbiont and functions extracellularly to digest the ingested wood. In cattle, the symbiotic microbes take up cellulose molecules (from ingested grass, etc.), digesting them intracellularly and passing some digested fragments into the surrounding fluid. These symbiotic gut bacteria, in turn, multiply and are themselves subsequently digested. Were it not for these symbiotic microorganisms, cellulose (the major nutritional constituent of grass, hay, and leaves) would be unavailable as food for grazing and browsing animals. Only a few animals, such as the shipworm *Toredo* (a wood-boring clam), *Limnoria* (an isopod), and the silverfish (an insect), can secrete cellulase without the help of symbionts.

Lipases Fats are water-insoluble, which presents a special problem for their digestion. Fats must undergo a special, two-stage treatment before they can be processed in the aqueous contents of the digestive tract. First, fats are **emulsified**—that is, they are rendered water-soluble by dispersing them into small droplets through the mechanical churning of the intestinal contents produced by segmentation (see Figure 15-24). The process of emulsification is aided by the chemical action of detergents such as bile salts and the phospholipid **lecithin** under conditions of neutral or alkaline pH. Bile salts have a hydrophobic, fat-soluble end and a hydrophilic, water-soluble end. Lipid attaches to the hydrophobic end, while water attaches to the hydrophilic end, dispersing the fat in the water-based fluid of the digestive tract. The overall effect is comparable to making mayonnaise, in which salad oil is dispersed in vinegar and egg yolk.

The second step, in vertebrates, is the formation of **micelles** (see Figure 2-16), aided by bile salts. Micelles are exceedingly small spherical structures formed from molecules which have polar hydrophilic groups at one end and nonpolar hydrophobic groups at the other end and which are assembled so that their polar ends face outward into the aqueous solution. The lipid core of each micelle is about 10^{-6} times the size of the original emulsified fat droplets, greatly increasing the surface area available for pancreatic lipase digestion. Enzymatic degradation then results from the action of intestinal lipases (in invertebrates) or pancreatic lipases (in vertebrates), producing fatty acids plus monoglycerides and diglycerides. In the absence of sufficient bile salts, fat digestion by the lipase is incomplete, and undigested fat is allowed to enter the colon.

Proenzymes Certain digestive enzymes, in particular proteolytic enzymes, are synthesized, stored, and released in an inactive molecular form known as a **proenzyme**, or **zymogen**. Proenzymes require activation, usually by hydrochloric acid in the lumen of the gastric gland, before they can carry out their degradative functions. Initial packaging of the enzyme in an inactive form prevents self-digestion of the enzyme and its tissue container while it is stored in zymogen granules. The proenzyme is activated by the removal of a portion of the molecule, either by the action of another enzyme specific for this purpose or through a rise in ambient acidity. Trypsin and chymotrypsin are good examples of enzymes originally constituted as proenzymes. The proenzyme **trypsinogen**, a 249-residue polypeptide, is inert until a 6-residue segment is cleaved from the NH_2-terminal end. This cleavage is achieved either by the action of another trypsin molecule or by **enterokinase**, an intestinal proteolytic enzyme. Trypsin also activates **chymotrypsinogen** through hydrolytic active form, **chymotrypsin**.

Other digestive enzymes In addition to the major classes of digestive enzymes just described, there are others playing a less important role in digestion. **Nucleases, nucleotidases, and nucleosidases,** as their names imply, hydrolyze nucleic acids and their residues. **Esterases** hydrolyze esters, which include those fruity-smelling compounds characteristic of ripe fruit. These and other minor digestive enzymes are not essential for nutrition, but they enhance the efficient use of ingested food.

Control of Digestive Secretions

Among vertebrates, the primary stimulus for secretion of digestive juices in a given part of the digestive tract is the presence of food there or, in some instances, elsewhere in the tract. The presence of food molecules stimulates chemosensory endings, which leads to the reflex activation of autonomic efferents that activate or inhibit motility and exocrine secretion. Appropriate food molecules also directly stimulate epithelial endocrine cells by contact with their receptors, causing reflex secretion of gastrointestinal hormones into the local circulation. These reflexes permit secretory organs outside the alimentary tract proper (the liver and pancreas, for example) to be properly coordinated with the need for digestion of food passing along the digestive tract. Gastrointestinal secretion is largely under the control of **gastrointestinal peptide hormones** secreted by endocrine cells of the gastric and intestinal mucosa. Several of these hormones turn out to be identical with neuropeptides that act as transmitters in the central nervous system. This suggests that the genetic machinery for producing these biologically active peptides has been put to use by cells of both the central nervous system and the gastrointestinal tract. Some gastrointestinal hormones are listed in Table 15-2.

Often ignored in the control of digestive secretion in animals is the role of cognition or thought processes. Cephalic influences such as mental images of food as well as learned behaviors also stimulate digestive secretion, at least in mammals (Spotlight 15-1). However, none of these neural and hormonal mechanisms regulating secretion is under simple voluntary control.

The characteristics of digestive secretion (rate of secretion, quantity of secretion) depends on several interacting features, including: (1) whether secretion is neurally or hormonally controlled, (2) where in the alimentary canal

SPOTLIGHT 15-1

BEHAVIORAL CONDITIONING IN FEEDING AND DIGESTION

The experiments of the Russian physiologist Ivan Pavlov figure prominently in the histories of both psychology and physiology. Pavlov, nearly a century ago, demonstrated reflexive secretion of saliva in dogs. A dog was given food following the sounding of a bell. Normally a dog will salivate in response to the sight or taste of food, but not in response to a bell. However, after several presentations of the bell (conditioned stimulus) together with food (unconditioned stimulus), the bell alone elicited salivation. This was the first recognition of a **conditioned reflex.** These experiments became important for the development of theories of animal behavior and psychology. In the context of this chapter, Pavlov's experiments demonstrated that some secretions of the digestive tract are under **cephalic control** (i.e., con-

trolled by the brain). Thus, in vertebrates, neural control of digestive secretions consists of two categories. In the first, secretomotor output to gland tissue occurs by an unconditioned reflex elicited directly by food in contact with chemoreceptors. In the second, secretomotor output is evoked indirectly by **association** of a conditioned stimulus with an unconditioned stimulus.

Another example of cephalic control of secretions is the reflexive secretion of salivary and gastric fluids evoked by the sight, smell, or anticipation of food. This reaction is based on past experience (i.e., associative learning). Closely related to this is the discovery that some animals exhibit one-trial avoidance learning of noxious food. Thus, a meal will be rejected even before it is tasted if it looks or smells like something previously sampled that proved to be noxious. Insect-eating birds have been found to avoid a particular species of bad-tasting insect prey on the basis of a one-trial experience with that prey. Examples of avoidance of noxious foods by one-trial learning have also been described in several mammalian species.

---

secretion occurs, and (3) how long food is normally present in the region being stimulated. For example, salivary secretion is very rapid and entirely under involuntary neural control, gastric secretions are under hormonal as well as neural control, and intestinal secretions are slower and are primarily under hormonal control. As in other systems, neural control predominates in rapid reflexes, whereas endocrine mechanisms are involved in reflexes that develop over minutes or hours.

Compared with vertebrates, very little is known about the control of digestive secretions in the invertebrates. Filter feeders evidently maintain a steady secretion of digestive fluids while they continuously feed. Other invertebrates secrete enzymes in response to the presence of food in the alimentary canal, but the precise control mechanisms have yet to be intensively studied. The formidable variety of invertebrate types further precludes generalizing about their digestive systems.

Salivary and gastric secretions

Mammalian saliva contains water, electrolytes, mucin, amylase, and antimicrobial agents such as lysozyme and thiocyanate (see Figure 15-28). In the absence of food, the salivary glands produce a slow flow of watery saliva. Secretion of saliva is stimulated by the presence of food in the mouth, or, indeed, by any mechanical stimulation of tissues within the mouth, via cholinergic parasympathetic nerves to the salivary glands. Cognitive awareness of food has an identical effect (see Spotlight 15-1). The amylase in saliva mixes with the food during chewing and digests starches. The mucin and watery fluid conditions the food bolus to help it slide smoothly toward the stomach by the peristaltic movements of the esophagus.

A major secretion of the stomach lining is **hydrochloric acid** (HCl), which is produced by the **parietal,** or **oxyntic,** cells located in the gastric mucosa. The secretion of HCl is stimulated by:

- Vagal motor discharges.

- The action of the gastric hormone gastrin, in conjunction with histamine, a local hormone with paracrine actions synthesized in the mast cells of the gastric mucosa. (Both hormones are required for HCl secretion because they bind to different receptors on the parietal cell membrane, both of which must be filled for HCl secretion to occur.)

- Secretatgogues in food, such as caffeine, alcohol, and the active ingredients of spices.

The secreted HCl helps break the peptide bonds of proteins, activates some gastric enzymes, and kills microorganisms that enter with the food. In some animals, the amount of H^+ used to produce secreted HCl is so great that blood and other extracellular fluids may actually become alkalotic for hours or days after ingestion of a large meal. This so-called **alkaline tide** can result in a rise in blood pH of 0.5 or even 1.0 pH unit in crocodiles, snakes, and other predators that have large, infrequent meals.

The parietal cells produce a concentration of hydrogen ions in the gastric juice 10^6 times greater than in plasma (Figure 15-32). They do this with the aid of the enzyme **carbonic anhydrase,** which catalyzes the reaction of water with carbon dioxide:

$$CO_2 + H_2O \underset{\substack{\text{carbonic} \\ \text{anhydrase}}}{\rightleftharpoons} H_2CO_3$$

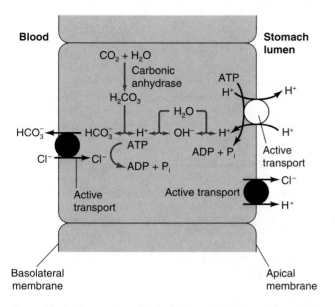

Figure 15-32 The secretion of hydrochloric acid (HCl) by gastric parietal cells employs primary active transport of H^+ produced from the breakdown of water. A Cl^-/HCO_3^- exchange pump resides on the basolateral membrane and a Cl^- channel on the apical membrane.

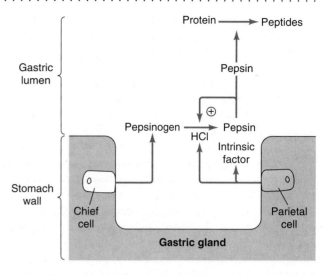

Figure 15-33 The powerful proteolytic enzyme pepsin is secreted in an inactive form (pepsinogen), which is then activated by HCl. The chief (zygomatic) cell secretes the pepsinogen, while the parietal cell secretes the HCl, as well as intrinsic factor.

The HCO_3^- resulting from the dissociation of H_2CO_3 is exported from the parietal cell into the plasma in exchange for Cl^- via a HCO_3^-/Cl^- antiporter in the basolateral membrane. The imported Cl^- diffuses to the apical membrane, where it exits via a Cl^- channel and enters the lumen of the gastric gland. The H^+ created by carbonic anhydrase is actively secreted by the apical cell membrane into the lumen of the gastric gland. These processes of importing and exporting ions allow the parietal cell to maintain a constant pH while providing the stomach with a highly acidic solution.

Pepsin is the major enzyme secreted by the stomach. This proteolytic enzyme is secreted in the form of the proenzyme **pepsinogen** by the exocrine cells called **chief cells,** or **zygomatic cells** (Figure 15-33). Chief cell secretion is under vagal control and is also stimulated by the hormone **gastrin,** which arises from the gastric wall (Figure 15-34). The inactive pepsinogen, of which there are several variants, is converted to the active **pepsin** by a low-pH-dependent cleavage of a part of the peptide chain. Pepsin, an endopeptidase, selectively cleaves inner peptide bonds that occur adjacent to carboxylic side groups of large protein molecules.

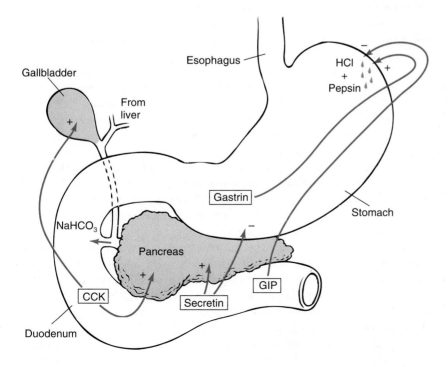

Figure 15-34 Vertebrate gastrointestinal hormones influence secretion and mechanical activity of the digestive tract. Gastrin from the lower stomach stimulates the flow of both HCl and pepsin from stomach secretory cells as well as the churning action of the muscular walls. Gastrin is secreted in response to intragastric protein, stomach distension, and input from the vagus nerve. Gastric inhibitory peptide (GIP), liberated from the small intestine in response to high levels of fatty acids, inhibits these activities. Neutralization and digestion of the chyme is accomplished by pancreatic secretions stimulated by cholecystokinin (CCK), which also induces contraction of the gallbladder, liberating the fat-emulsifying bile into the small intestine. CCK is secreted in response to the presence of amino and fatty acids in the duodenum. Secretin stimulates pancreatic secretion, but inhibits gastric activity. The plus and minus signs indicate stimulation and inhibition, respectively.

Goblet cells in the lining of the stomach secrete a gastric mucus containing various mucopolysaccharides. The mucus coats the gastric epithelium, protecting it from digestion by pepsin and HCl. HCl can penetrate the layer of mucus but is neutralized by alkaline electrolytes trapped within the mucus.

In some young mammals, including bovine calves (but not human infants), the stomach secretes **rennin,** an endopeptidase that clots milk by promoting the formation of calcium caseinate from the milk protein casein. The curdled milk is then digested by proteolytic enzymes, including rennin.

Gastric secretion in mammals occurs in three distinct phases: the cephalic, the gastric, and the intestinal phase. In the **cephalic phase,** gastric secretion occurs in response to the sight, smell, and taste of food, or in response to conditioned reflexes (see Spotlight 15-1). This phase is mediated by the brain (hence the term *cephalic*) and is abolished by section of the vagus nerve. In the **gastric phase,** mediated by the hormone gastrin and the compound **histamine,** the secretion of HCl and pepsin is stimulated directly by the presence of food in the stomach, which stimulates both chemoreceptors and mechanoreceptors. The **intestinal phase** is controlled by gastrin, as well as the hormones **secretin, vasoactive intestinal peptide** (VIP), and **gastric inhibitory peptide** (GIP) (see Table 15-2). GIP, for instance, is liberated by endocrine cells in the mucosa of the upper small intestine in response to the entry of fats and sugar into the duodenum (see Table 15-2 and Figure 15-34).

Understanding how the gastric phase of secretion is regulated was determined by use of the "Heidenhain pouch," which is a denervated pouch surgically constructed within an animal from part of the stomach. Secretions from the pouch are directed outside the body wall, where they can be collected and their volume measured. The pouch's only contact with the rest of the stomach is indirect, through the circulation. Since it is not innervated, the pouch can exhibit no cephalic phase secretion. However, it does secrete gastric juice in response to food placed into the stomach proper. The researchers correctly interpreted this finding as evidence that a hormonal messenger is released into the bloodstream when food is in the stomach. The hormone was named *gastrin* (see Table 15-2) and was later found to be a polypeptide. Gastrin is secreted from endocrine cells of the pyloric mucosa of the stomach in response to gastric chyme containing protein and to distension of the stomach. It stimulates stomach motility by binding to smooth muscle, and it induces a strong secretion of HCl and moderate secretion of pepsin by binding to secretory cells in the stomach lining. When the pH of the gastric chyme drops to 3.5 or below, gastrin secretion slows, and at pH 1.5 it stops. As already noted, secretion of histamine by the gastric mucosa also stimulates secretion of HCl, as does mechanical distension of the stomach.

The intestinal phase of gastric secretion is more complex (see Figure 15-34). As food enters the duodenum of the small intestine, partially digested proteins in acidic chyme directly stimulate the duodenum's mucosa to secrete **enteric gastrin** (also called **intestinal gastrin**). Enteric gastrin has the same action as stomach gastrin, stimulating the gastric glands to increase their rate of secretion. In humans, at least, the intestinal phase is thought to play a relatively small role in overall regulation of gastric secretion.

The secretion of gastric juices can be reduced both by the absence of stimulating factors and by reflex inhibition. The **enterogastric reflex,** which inhibits gastric secretion, is triggered when the duodenum is stretched by chyme pumped from the stomach, and when this chyme contains partially digested proteins or has particularly low pH. Gastric secretion can also be inhibited by strong activation of the sympathetic nervous system. Action potentials in the sympathetic nerves terminating in the stomach release norepinephrine, which inhibits both gastric secretion and gastric emptying.

Intestinal and pancreatic secretions

The epithelium of the mammalian small intestine secretes **intestinal juice,** or **succus entericus,** which is a mixture of two fluids. **Brunner's glands** in the first part of the duodenum between the pyloric sphincter and the pancreatic duct secrete a viscous, enzyme-free, alkaline mucoid fluid that enables the duodenum to withstand the acidic chyme coming from the stomach until it can be neutralized by the alkaline pancreatic and biliary secretions coming from the pancreatic duct. A thinner, enzyme-rich alkaline fluid arises in the crypts of Lieberkühn (see Figure 15-20) and mixes with duodenal secretions. The secretion of intestinal juice is regulated by several hormones, including secretin, gastric inhibitory peptide (GIP), and gastrin, and additionally is under neural control. Distension of the wall of the small intestine elicits a local secretory reflex. Vagal innervation also stimulates secretion.

The large intestine secretes no enzymes. However, it does secrete a thin alkaline fluid containing bicarbonate and potassium ions plus some mucus that binds the fecal matter together.

In addition to its endocrine secretion of insulin from the islets of Langerhans (see Chapter 9), the pancreas contains exocrine tissue that produces several digestive secretions that enter the small intestine through the pancreatic duct. The pancreatic enzymes, including alpha-amylase, trypsin, chymotrypsin, elastase, carboxypeptidases, aminopeptidases, lipases, and nucleases, are delivered in an alkaline, bicarbonate-rich fluid that helps neutralize the acid chyme formed in the stomach. This buffering is essential, since the pancreatic enzymes require a neutral or slightly alkaline pH for optimum activity.

Exocrine secretion by the pancreas is controlled by the peptide hormones produced in the upper small intestine. Acid chyme reaching the small intestine from the stomach stimulates the release of secretin and VIP, both produced by endocrine cells in the upper small intestine (see Table 15-2). These peptides are transported in the blood, reaching the duct cells of the pancreas and stimulating them to produce its thin bicarbonate fluid. Peptide hormones only weakly stimulate secretion of pancreatic enzymes, however. Gastrin secreted from the stomach lining also elicits a small flow of pancreatic juice in anticipation of the food that will enter the duodenum.

Secretion of pancreatic enzymes is elicited by another upper intestinal hormone—the peptide **cholecystokinin** (see Table 15-2)—secreted from epithelial endocrine cells in response to fatty acids and amino acids in the intestinal chyme. Cholecystokinin is now known to be identical with pancreozymin, and so both are currently referred to as *cholecystokinin (CCK)*. It stimulates pancreatic secretion of enzymes as well as contraction of the smooth muscle wall of the gallbladder, forcing bile into the duodenum (see Figure 15-34).

The neuropeptides somatostatin and enkephalin have also been identified in endocrine cells of the upper intestinal mucosa in vertebrate guts. Both hormones have a variety of actions on gastrointestinal function. Somatostatin, which normally acts through paracrine effects, inhibits gastric acid secretion, pancreatic secretion, and intestinal motility, as well as blood flow. The enkephalins inhibit gastric acid secretion, stimulate pancreatic enzyme secretion, and inhibit intestinal motility.

The composition of pancreatic secretions can be modified in some species by the content of the diet. Thus, a diet high in carbohydrates over several weeks will result in an increase in the amylase content of pancreatic enzymes. Similar correlations have been noted between protein and proteases, and fat and lipases.

 What effect would a diminished flow of secretions from the pancreatic duct have on digestion? Why does blockage of the pancreatic duct of mammals potentially lead to rapid death?

ABSORPTION

The breakdown products of digestion (amino acids from proteins, sugars from carbohydrates, etc.) are transported from the gut to the animal's tissues and cells. In a unicellular organism, the products of digestion leave the food vacuole to enter the surrounding cytoplasm. In a multicellular animal, these products must be transported across the **absorptive epithelium** into the circulation, and then move from the blood into the tissues.

Digestion products are absorbed mainly via the microvilli of the apical membrane of the absorptive cell (see Figure 15-21). The digestive and absorptive mechanisms of the microvilli include the glycocalyx, digestive enzymes intimately associated with the membrane, and specific intramembrane transporter proteins. In the basolateral membranes other mechanisms transfer these substances out of the absorptive cell into the interstitial fluid and eventually into the general circulation.

Nutrient Uptake in the Intestine

The carbohydrate-rich filaments composing the glycocalyx covering the microvilli arise from, and are continuous with, the surface membrane of the microvillus itself. The filaments of the glycocalyx appear to be the carbohydrate side chains of glycoproteins embedded in the membrane. Further, the brush border (microvilli plus glycocalyx) has been found to contain digestive enzymes for the final digestive stages of various small foodstuff molecules. These enzymes are membrane-associated glycoproteins having carbohydrate side chains protruding into the lumen. The enzymes found associated with the brush border include disaccharidases, aminopeptidases, and phosphatases. Thus, some of the terminal stages of digestion are carried out at the absorptive cell membrane, close to the sites of uptake from the lumen into the absorptive cells.

Several transfer processes are involved in absorption. These include passive diffusion, facilitated diffusion, cotransport, countertransport, and active transport (see Chapter 4), and endocytosis. The type of transfer mechanism used depends on the type of molecule being transported during the absorption process.

Simple diffusion
Simple diffusion can take place across the lipid bilayer (providing the diffusing substance has a high lipid solubility) or through water-filled pores. Substances that diffuse across the brush border membrane of the intestine include fatty acids, monoglycerides, cholesterol, and other fat-soluble substances. Substances that pass through water-filled pores include water, certain sugars, alcohols, and other small, water-soluble molecules. For nonelectrolytes, net diffusion rate is proportional to their chemical concentration gradient. For electrolytes, it is proportional to the electrochemical gradient. In passive diffusion, net transfer is always "downhill," using the energy of the concentration gradient.

 Can you design an experiment that would indicate the metabolic cost of taking up a nutrient by active transport compared with one taken up by passive diffusion?

Carrier-mediated transport
The absorption of monosaccharides and amino acids presents two problems. First, these molecules are hydrophilic because of their —OH groups, because of charges they may bear, or because of both. Second, they are too large to be carried through water-filled pores by solvent drag or simple diffusion. These problems are overcome by **carrier-mediated transport** across the absorptive cell membrane (Figure 15-35). For example, sugars such as fructose are carried down their concentration gradient by facilitated diffusion, a process in which a hydrophilic, lipid-insoluble substance diffuses down its chemical gradient with the help

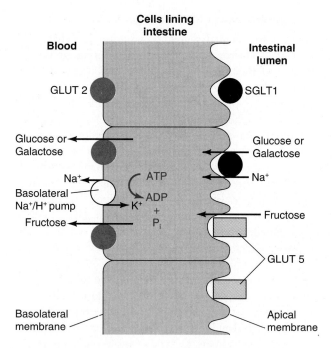

Figure 15-35 Facilitated transport proteins (GLUT5, GLUT2) are central to this model of sugar transport across the cells lining the small intestine. SGLT1 is an integral membrane protein that couples the transport of Na^+ to that of glucose or galactose. Fructose passes through the brush border by way of the GLUT5 transport protein. Sugar uptake is powered by sodium and electrical gradients across the membrane. Then, sugars are transported down their concentration gradients via the GLUT2 transport proteins on the basolateral membrane. A basolateral Na^+/K^+ pump pumps out Na^+, creating the gradient that powers the whole process. [Adapted from Wright, 1993.]

of specific protein channels located in the membranes. This process is powered by coupling sugar transport to the sodium gradients and electrical gradients across the plasma membrane. SGLT1 is the integral membrane protein that couples the transport of Na^+ with glucose across the brush border. In this model, GLUT5 is the brush border fructose transporter, and GLUT2 is the basolateral membrane transporter for fructose as well as glucose and galactose (not shown).

Some monosaccharides are taken up into the absorptive cells by a related mechanism, **hydrolase transport,** in which a glycosidase attached to the membrane hydrolyzes the parent disaccharide (e.g., sucrose, maltose) and also acts as, or is coupled to, the transfer mechanism of the monosaccharide into the absorptive cell.

Once through the epithelium, sugar and amino acid molecules enter the blood by diffusion into the capillaries within the villi. Upon reaching other tissues of the body, sugars and amino acids are transferred by the same types of active transport and facilitated diffusion mechanisms into other body cells.

Active transport

In the mammalian intestine, the sodium-driven transport of amino acids into the absorptive cells takes place via four

separate and non-competing cotransport systems. Each system transports just one of four categories of amino acids:

1. The 3 dibasic amino acids (lysine, arginine, and histidine) having two basic amino groups each
2. The diacidic amino acids (glutamate and aspartate), having two carboxyl groups each
3. A special class consisting of glycine, proline, and hydroxyproline
4. The remaining neutral amino acids

Yet another separate transport system exists for dipeptides and tripeptides. Once inside the cell, dipeptides and tripeptides are cleaved into their constituent amino acids by intracellular peptidases. This has the advantage of preventing a concentration buildup of the oligopeptides within the cell, so there is always a large inwardly directed gradient promoting their inward transport.

Special handling of lipids

The digestion products of fats, monoglycerides, fatty acids, and glycerol diffuse through the brush border membrane and are reconstructed within the absorptive cell into triglycerides. They are collected together with phospholipids and cholesterol into tiny droplets termed **chylomicrons,** about 150 μm in diameter (Figure 15-36). Chylomicrons are coated with a layer of protein, and are loosely contained in vesicles formed by the Golgi apparatus. They are subsequently expelled by exocytosis through fusion of these vesicles with the basolateral membrane of the absorptive cell.

Endocytosis

Transport of sugars and amino acids across the basolateral membranes occurs by facilitated transport, as noted earlier. Some oligopeptides are taken up by absorptive cells through endocytosis. In newborn mammals this process is responsible for the uptake in the intestine of immunoglobulin molecules derived from the mother's milk that escape digestion. Once inside the absorptive cell, nutrients pass through the basolateral membranes of the absorptive cell (see Figure 15-36) into the interior of the villus and then move from the interstitial fluid into the circulatory system.

Blood Transport of Nutrients

From the interstitial fluid of the villus, digestion products enter the blood or the lymphatic circulation (see Figure 15-36). Fishes have relatively simple lymphatic vessels, but these are well developed in all other vertebrates. In humans, about 80% of the chylomicrons, for example, enter the bloodstream via **lymph** carried in the **lymphatic system,** a modified ultrafiltrate of blood plasma, while the rest enter the blood directly. The pathway into the lymphatic system begins with the blind central lacteal of the villus (see Figure 15-21A). In humans **lymph** is returned to the circulation via the **thoracic lymph duct.** Sugars and amino acids primarily enter the capillaries of the villus, which are drained by venules that lead into the **hepatic portal vein.** This vein takes the blood from the intestine directly to the

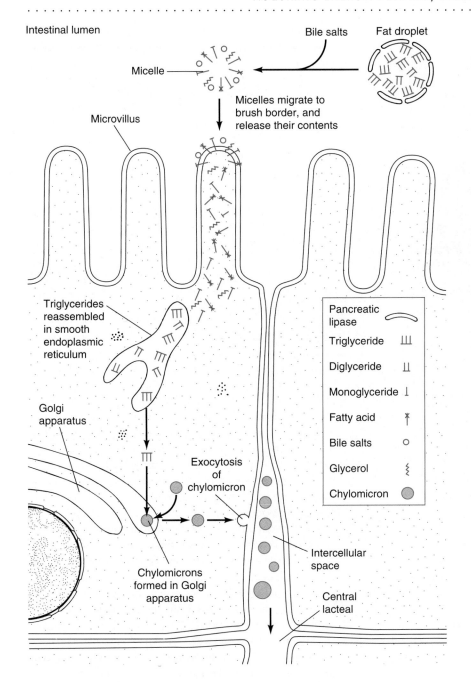

Intestinal lumen

Bile salts

Fat droplet

Micelle

Micelles migrate to brush border, and release their contents

Microvillus

Triglycerides reassembled in smooth endoplasmic reticulum

Golgi apparatus

Exocytosis of chylomicron

Chylomicrons formed in Golgi apparatus

Intercellular space

Central lacteal

Pancreatic lipase

Triglyceride

Diglyceride

Monoglyceride

Fatty acid

Bile salts

Glycerol

Chylomicron

Figure 15-36 Lipids are transported from the intestinal lumen through absorptive cells and into the interstitial space. Hydrolytic products of triglyceride digestion—monoglycerides, fatty acids, and glycerol—form micelles with the bile salts in solution. Micelles transport these materials to the brush border, where their contents enter the absorptive cell by passive, lipid-soluble diffusion across the microvillus membrane. Within the cell they are resynthesized into triglycerides in the smooth endoplasmic reticulum and, together with a smaller amount of phospholipids and cholesterol, are stored in the Golgi apparatus as chylomicrons—droplets about 150 μm in diameter. These then leave the basolateral portions of the cell by exocytosis.

liver. There, under the influence of insulin, much of the glucose is taken up into **hepatocytes,** and in these cells it is converted to glycogen granules for storage and subsequent release into the circulation after being converted back into glucose. The hormonal regulation of glycogen breakdown, sugar metabolism, fat metabolism, and amino acid metabolism is discussed in Chapter 9.

Water and Electrolyte Balance in the Gut

In the process of producing and secreting their various digestive juices, the exocrine tissues of the alimentary canal and its accessory organs pass a great deal of water and electrolytes into the lumen of the alimentary canal. In humans, this can normally amount to over 8 liters per day (Figure 15-37), or about 1.5 times the total blood volume.

Clearly, this quantity of water, not to mention the electrolytes contained within it, cannot be lost from the body with the feces. In fact, nearly all secreted water and electrolytes, along with ingested water, is recovered by uptake in the intestine. Although water is reabsorbed throughout the intestine, most of the reabsorption takes place in the lower part of the small intestine.

The cells in the alimentary canal responsible for water uptake are bound together by tight junctions near their apical borders (see Figure 15-21B), nearly obliterating free paracellular pathways. Tracer studies using deuterium oxide, D_2O, indicate that water leaves the intestinal lumen through channels in the absorptive cell membrane that occupy only 0.1% of the epithelial surface. Flux studies employing isotopically labeled solutes indicate that these

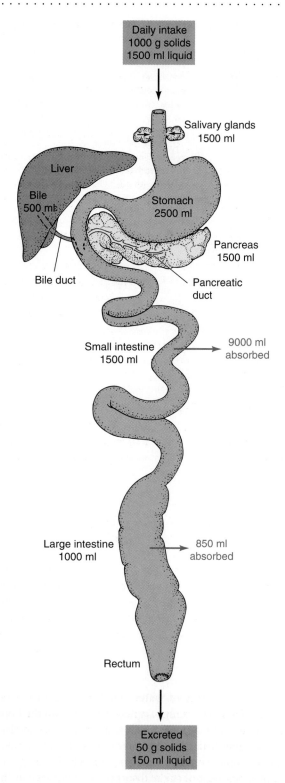

Figure 15-37 Fluid fluxes occur all along the length of the human alimentary canal. Volumes vary with the condition and body mass of the subject. Values in black, in milliliters, are the amounts of fluid entering the alimentary tract, and those in red are the amounts reabsorbed from the lumen. [Adapted from Madge, 1975.]

Because osmotic pressure is the motive force leading to net water movement from the intestinal lumen to the interior of the villus, this movement is entirely passive. In fact, there is no evidence of active transport of water in any living organism—animal, plant, or microbe. The osmotic gradient driving water from the lumen into the villus is set up primarily by the active transport of substances from the lumen into the villus, in particular the transport of salt, sugar, and amino acids. The elevated osmotic pressure within the villus that results from this active transport, especially in the lateral clefts of the epithelium (see Figures 4-48 and 4-49), draws water osmotically from the absorptive cell. This water is then replenished by water entering osmotically across the apical membrane from the lumen.

Most of the absorption of water and electrolytes across the absorptive cell epithelium occurs at or near the tips of the villi. The greater proportion of water absorption at the villus tip results from an elevated concentration of Na^+ near the upper end of the villus lumen, which decreases with increasing distance from the villus tip. There are two reasons for this concentration gradient. First, most of the active absorption of Na^+ takes place across absorptive cells located at the tip of each villus. Chloride follows, and NaCl accumulation is therefore greatest at the blind upper end of the villus lumen. Second, the organization of circulation within the villus leads to a further concentration of NaCl at the upper end of the villus lumen because of a countercurrent mechanism (see Spotlight 12-2). Arterial blood flowing toward the tip of the villus picks up Na^+ and Cl^- from NaCl-enriched blood leaving the villus in a descending venule. The "short-circuiting" of NaCl in this manner recirculates and concentrates it in the villus tip, promoting osmotic flow of water from the intestinal lumen into the villus.

The absorption of Na^+ and Cl^- into the villus is enhanced by high concentrations of glucose and certain other hexose sugars in the intestinal lumen, which stimulate sodium-sugar cotransport.

Excessive uptake of water from the lumen across the intestinal wall results in abnormally dry lumen contents (and hence constipation). This situation is normally prevented by an inhibitory action on electrolyte and water uptake by some of the gastrointestinal hormones. Gastrin acts indirectly to inhibit water absorption from the small intestine, while secretin and CCK reduce the uptake of Na^+, K^+, and Cl^- in the upper jejunum. Bile acids and fatty acids also inhibit the absorption of water and electrolytes.

Unlike water, Ca^{2+} requires a special active transport mechanism for absorption from the gut. The calcium ion is first bound to a calcium-binding protein found in the microvillus membrane and is then transported as a complex into the absorptive cell by an energy-consuming process. From the absorptive cell the Ca^{2+} then passes into the blood. The presence of calcium-binding protein is regulated by the hormone calcitriol, formerly known as *1,25-dihydroxy-vitamin D₃*. The release of Ca^{2+} from the absorptive cell into the blood is accelerated by parathyroid hormone.

channels exclude water-soluble molecules with molecular weights exceeding $200 \text{ g} \cdot \text{mol}^{-1}$. Smaller solute molecules are carried passively along with the water by **solvent drag** as it flows down its osmotic gradient through hydrated channels.

Vitamin B_{12}, which has a molecular weight of $1357\,g\cdot mol^{-1}$, is the largest water-soluble essential nutrient taken up intact across the intestinal lumen in the region of the distal ileum. This highly charged cobalt-containing compound is associated with food protein, to which it is bound as a coenzyme. In the process of absorption, B_{12} transfers from the dietary protein to a mucoprotein known as **intrinsic factor** (or **hemopoietic factor**) that is produced by the H^+-secreting parietal cells of the stomach. Since B_{12} is essential for the synthesis and maturation of red blood corpuscles, **pernicious anemia** occurs when B_{12} absorption is prevented by interference with its binding to intrinsic factor. Some tapeworms "steal" B_{12} in the intestine of the host by producing a compound that removes it from intrinsic factor, making it unavailable to the host but available to the tapeworm.

NUTRITIONAL REQUIREMENTS

Whatever the form of food capture, ingestion, and digestion, all animals must acquire an appropriate variety and amount of nutritive substances, as we will now consider. **Nutrients** are substances that serve as sources of metabolic energy and as raw material for growth, repair of tissues, and production of gametes. Nutrients also include essential trace elements such as iodine, zinc, and other metals that may be required in extremely small quantities. There is wide variation between the nutritional needs of different species. Within a species, nutritional needs vary according to phenotypic differences in body size and composition and activity, and also with age, sex, and reproductive state. A gravid (egg-bearing) or pregnant female may require more nutrients than a male, while a male producing sperm may have greater nutritive needs than one that is not producing sperm. Regardless of reproductive state, a small animal requires more food for energy per gram of body weight than does a larger animal, because its metabolic rate per unit body weight is higher. Similarly, an animal with a high body temperature requires more food, to satisfy greater energy needs, than does an animal with lower body temperature. (The energetics of temperature, size, and other factors are discussed in Chapter 16.)

Energy Balance

A **balanced nutritional state** exists when an animal has sufficient food intake of all nutrients necessary for long-term growth and maintenance. The nutritional requirements include (1) sufficient sources of energy to power all body processes, (2) enough protein and amino acids to maintain a positive nitrogen balance (i.e., to avoid net loss of body proteins), (3) enough water and minerals to compensate for their loss or incorporation into body tissues, and (4) those essential amino acids and vitamins not synthesized within the body.

Energy balance requires that caloric intake over a given period of time equal the number of calories consumed for tissue maintenance and repair and for work (metabolic and otherwise), plus the production of body heat in birds and mammals. Thus,

$$
\begin{aligned}
\text{caloric intake} &= \text{caloric output} \\
&= \text{calories consumed by tissues} \\
&\quad + \text{heat produced}
\end{aligned}
$$

Insufficient intake of calories can be temporarily offset by using stores of fat and carbohydrates or proteins within tissues, but this produces a resultant loss of body weight. Conversely, caloric intake in excess of what is required for energy balance will result in increased storage of body fat, as in the large fat stores accumulated before long migrations in migratory birds or laid down by mammals before onset of hibernation.

Animals differ in their abilities to synthesize the substances fundamental to maintenance and growth. Thus, for a given animal species, certain cofactors (Zn, I etc.) or building blocks (amino acids, etc.) essential for important biochemical reactions or for the production of tissue molecules may be required from food sources simply because those substances cannot be produced by the animal itself. Such items are known as **essential nutrients**.

Nutrient Molecules

A wide variety of molecules serve as nutrient molecules, including water, proteins and amino acids, carbohydrates, fats and lipids, nucleic acids, inorganic salts, and vitamins.

Water

Of all the constituents of animal tissue, none is more pervasively important to living tissue than water. This unique and marvelous substance can constitute 95% or more of the weight of some animal tissues. It is replenished in most animals by drinking (see Chapter 12) and by ingestion with food. Some marine and desert animals depend almost entirely on "metabolic water"—water produced during the oxidation of fats and carbohydrates—to replace water lost by evaporation, defecation, and urination (see Chapter 16).

Proteins and amino acids

Proteins are used as structural components of tissues and as enzymes. They can also be utilized as energy sources if first broken down to amino acids (see Chapter 3). The proteins of animal tissues are composed of about 20 different amino acids. The ability to synthesize amino acids differs among species. Those amino acids that cannot be synthesized by an animal, but are required for synthesis of essential proteins, are the so-called **essential amino acids** for that animal. Recognition of this requirement has been of enormous economic significance in the poultry industry. The rate of growth of chickens at one time was limited by too small a proportion of a few essential amino acids in the grain diet they were provided. Supplementing the diet with these amino acids allowed full utilization of the other amino acids present in the feed, greatly increasing the rate of protein synthesis and hence the rate of poultry growth and egg laying. Microbiologists artificially induce this limiting condition by genetically engineering microbes that require a specific amino acid (e.g., lysine) not normally found in their environment. Thus, the microbes will grow only in an environment enriched with the amino acid, serving as a safeguard preventing their spread through normal populations.

Carbohydrates

Carbohydrates are used primarily as immediate (glucose 6-phosphate) or stored (glycogen) sources of chemical energy. However, they may also be converted to metabolic intermediates or to fats (see Chapter 3). Conversely, proteins and fats can be converted by most animals into carbohydrates. The major sources of carbohydrate are the sugars, starches, and cellulose found in plants and the glycogen stored in animal tissues.

Lipids

Lipid (fat) molecules are especially suitable as concentrated energy reserves. Each gram of fat provides over 2 times as much caloric energy as a gram of protein or carbohydrate. Consequently, lipids can store significantly more chemical energy per unit volume of tissue. Fat is commonly stored by animals for periods of **caloric deficit,** as during hibernation, when energy expenditure exceeds energy intake. Lipids are

TABLE 15-3
Some mammalian vitamins

Vitamin (designation in letter or name)	Major dietary sources; solubility*	Uptake; storage	Function in mammals[†]	Deficiency symptoms
Carotene (A)	Egg yolk, green or yellow vegetables, fruits; FS	Absorbed from gut with aid of bite; stored in liver	Formation of visual pigments; maintenance of epithelial structures; important in fetal development	Night blindness, skin lesions, birth defects
Calciferol (D₃)	Fish oils liver; FS	Absorbed from gut; little storage	Enhancement of calcium absorption from gut; bone and tooth formation	Rickets (defective bone formation) in children, osteomalacia in adults
Tocopherol (E)	Green leafy vegetables, meat, milk, eggs, butter; FS	Absorbed from gut; stored in adipose and muscle tissues	In humans, maintenance of red cells; antioxidant. In other mammals, maintenance of pregnancy	Increased fragility of red blood cells, muscular dystrophies, abortion, muscular wastage
Napthoquinone (K)	Synthesis by intestinal flora, liver, green leafy vegetables; FS	Absorbed from gut; little storage, excreted in feces	Enabling of prothrombin synthesis by liver	Failure of coagulation
Thiamine (B₁)	Brain, liver, kidney, heart, whole grains, nuts, beans, potatoes	Absorbed from gut; stored in liver, brain, kidney	Formation of cocarboxylase enzyme involved in decarboxylation (Krebs cycle)	Stoppage of CH_2O metabolism at pyruvate, beriberi, neuritis, heart failure
Riboflavin (B₂)	Milk, eggs, lean meat, liver, whole grains; WS	Absorbed from gut; stored in kidney, liver, heart	Flavoproteins in oxidative phosphorylation	Photophobia, fissuring of the skin
Niacin	Lean meat, liver, whole grains; WS	Absorbed from gut; distributed to all tissues	Coenzyme in hydrogen transport (NAD, NADP)	Pellagra, skin lesions, digestive disturbances, dementia
Cyanocobalamin (B₁₂)	Liver, kidney, brain, fish, eggs, bacterial synthesis in gut; WS	Absorbed from gut; stored in liver, kidney, brain	Nucleoprotein synthesis; formation of erythrocytes	Pernicious anemia, malformed erythrocytes
Folic acid (folacin, pteroylglutamic acid)	Meats; WS	Absorbed from gut; utilized as taken in	Nucleoprotein synthesis; formation of erythrocytes	Failure of erythrocytes to mature, anemia
Pyridoxine (B₆)	Whole grains, traces in many foods; WS	Absorbed from gut; half appears in urine	Coenzyme for amino and fatty acid metabolism	Dermatitis, nervous disorders
Pantothenic acid	Many foods; WS	Absorbed from gut; stored in all tissues	Constituent of coenzyme A (CoA)	Neuromotor, cardiovascular disorders
Biotin	Egg yolk, tomatoes, liver, synthesis by flora of GI tract; WS	Absorbed from gut	Protein and fatty acid synthesis; CO_2 fixation; transamination	Scaly dermatitis, muscle pains, weakness
Ascorbic acid (C)	Citrus; WS	Absorbed from gut; little storage	Vital element for collagen and ground substance; antioxidant	Scurvy (failure to form connective tissue)

*FS = fat-soluble; WS = water-soluble

[†]Most vitamins have numerous functions. Listed are a mere sampling.

also important in certain tissue components such as plasma membranes and other membrane-based organelles of the cell and the myelin sheaths of axons. The fatty molecules or lipids include fatty acids, monoglycerides, triglycerides, sterols, and phospholipids.

Nucleic acids

Although nucleic acids are essential for the genetic machinery of the cell, all animal cells appear to be capable of synthesizing them from simple precursors. Thus, the intake of intact nucleic acids is not necessary from a nutritional perspective.

Inorganic salts

Some chloride, sulfate, phosphate, and carbonate salts of the metals calcium, potassium, sodium, and magnesium are important constituents of intra- and extracellular fluids. Calcium phosphate occurs as **hydroxyapatite** $[Ca_{10}(PO_4)_6(OH)_2]$, a crystalline material that lends hardness and rigidity to the bones of vertebrates and the shells of mollusks. Iron, copper, and other metals are required for redox reactions (as cofactors) and for oxygen transport and binding (hemoglobin, myoglobin). Many enzymes require specific metal atoms to complete their catalytic functions. Animal tissues need moderate quantities of some ions (Ca, P, K, Na, Mg, S, and Cl) and trace amounts of others (Mn, Fe, I, Co, Cu, Zn, and Se).

Vitamins

Vitamins are a diverse and chemically unrelated group of organic substances that generally are required in small quantities primarily to act as cofactors for enzymes. Some vitamins important in human nutrition are listed in Table 15-3, along with their diverse functions. Detailed nutritional vitamin requirements are known primarily for domesticated animals grown for their meat, eggs, or other products. Very little is known about the vitamins involved in the metabolism of lower vertebrates and especially invertebrates.

The ability to synthesize different vitamins differs between species, and those essential vitamins that an animal cannot produce itself must be obtained from other sources, primarily from plants but also from dietary animal flesh or from intestinal microbes. Ascorbic acid (vitamin C) is synthesized by many animals, but not by humans, who acquire it mainly from citrus fruits. Scurvy, a condition of ascorbic acid deficiency in humans, was common on board ships before the British admiralty instituted the use of citrus fruit — especially limes — to supplement the diet of the crews. Their use of limes led to the general term *limey* to describe the English. Humans also are unable to produce vitamins K and B_{12}, which are produced by intestinal bacteria and then absorbed for distribution to the tissues. Fat-soluble vitamins such as A, D_3, E, and K are stored in body fat deposits. Water-soluble vitamins such as ascorbic acid are not stored in the body, however, and so must be ingested or produced continually to maintain adequate levels.

SUMMARY

All heterotrophic organisms acquire carbon compounds of moderate to high energy content from the tissues of other plants and animals. The chemical energy contained in these compounds originally was converted from radiant energy into chemical energy trapped in sugar molecules by photosynthesizing autotrophs. Subsequent synthetic activity by autotrophs and heterotrophs converts these simple carbon compounds into more complex carbohydrates, fats, and proteins.

Animals obtain food in many different ways, including absorption through the body surface in some aquatic or marine species, endocytosis in microorganisms, filter feeding, mucus trapping, sucking, biting, and chewing. Once ingested, the food may be temporarily stored, as in a crop or rumen, or immediately subjected to digestion. Digestion consists of the enzymatic hydrolysis of large molecules into their monomeric building blocks. In multicellular animals, this takes place extracellularly in an alimentary canal. Digestive hydrolysis occurs only at low-energy bonds, most of the chemical energy of foodstuffs being conserved for intracellular energy metabolism once the products of digestion have been assimilated into the animal's tissues. Stepwise intracellular oxidations by coupled reactions then lead to a controlled release of chemical bond energy and material for cell growth and functioning.

Digestion in vertebrates begins in a region of low pH, the stomach, and proceeds to a region of higher pH, the small intestine. Proteolytic enzymes are released as proenzymes, or zymogens, which are inactive until a portion of the peptide chain is removed by digestion. This procedure avoids the problem of proteolytic destruction of the enzyme-producing cells that store and secrete the zymogen granules containing the proenzyme. Other exocrine cells secrete digestive enzymes (e.g., carbohydrases and lipases), mucin, or electrolytes such as HCl or $NaHCO_3$.

The motility of the vertebrate digestive tract depends on the coordinated activity of longitudinal and circular layers of smooth muscle. Peristalsis occurs when a ring of circular contraction proceeds along the gut preceded by a region in which the circular muscles are relaxed. The parasympathetic innervation stimulates motility, whereas the sympathetic innervation inhibits motility.

The motility of smooth muscle as well as the secretion of digestive juices is under fine neural and endocrine control. All gastrointestinal hormones are peptides, and many of them also function as neuropeptides in the central nervous system, where they act as transmitters or short-range neurohormones. Both direct activation by food in the gut and neural activation stimulate the endocrine cells of the gastrointestinal mucosa that secrete peptide hormones. These hormones are active in either stimulating or inhibiting the activity of the various kinds of exocrine cells in the gut that produce digestive enzymes and juices.

Digestion products are taken up by the absorptive cells of the intestinal mucosa and transferred to the lymphatic and circulatory systems. The absorptive surface, in effect consisting of a continuous membrane sheet formed by the apical membranes of the myriad absorptive cells joined by tight junctions, is greatly increased in area by virtue of microvilli, the microscopic evaginations of the apical membrane. The absorptive cells cover larger finger-like villi that reside on convoluted folds and ridges in the wall of the intestine, further increasing the surface area.

The process of terminal digestion takes place in the brush border, formed by the microvilli and the glycocalyx, that covers the apical membrane. Here short-chain sugars and peptides are hydrolyzed into monomeric residues before membrane transport takes place. Transport of some sugars can occur by facilitated diffusion, which requires a membrane transport protein but no metabolic energy. Most sugars and amino acids require energy expenditure for adequate rates of absorption. An important transport mechanism for these substances is cotransport with Na^+, utilizing a common membrane protein and the potential energy of the electrochemical gradient driving Na^+ from the lumen into the cytoplasm of the absorptive cell. Endocytosis plays a role in the uptake of small polypeptides and, rarely, of larger proteins, such as immunoglobulin in newborn animals. Fatty substances enter the absorptive cell by simple diffusion across the cell membrane.

Water and electrolytes enter the alimentary canal as constituents of digestive juices, but these quantities are nearly all recovered by active uptake of solutes by the intestinal mucosa. Active transport of solutes from the intestinal lumen results in the passive osmotic movement of water from the lumen into the cells and eventually back into the bloodstream. Without such recycling of electrolytes and water, the digestive system would impose a lethal osmotic load on the animal.

REVIEW QUESTIONS

1. Define the terms *digestion, absorption, assimilation,* and *nutrition.*
2. In what way is Bernoulli's effect significant to the feeding of a sponge?
3. Cite two unrelated examples of proteins produced specifically for the purpose of obtaining and utilizing food.
4. What is an essential amino acid?
5. Explain why it would be inadvisable for the digestive system to fragment amino acids, hexose sugars, and fatty acids into still smaller molecular fragments, even though doing so might facilitate absorption.
6. Explain why proteolytic enzymes fail to digest the exocrine cells in which they are produced and stored before release.
7. Give several examples of symbiotic microorganisms in alimentary canals, and explain how they benefit the host.
8. State two adaptive advantages of the digastric stomach. Since it has three or four chambers, why is it referred to as *digastric?*

9. Explain how bile aids the digestive process even though it contains little or no enzyme.
10. Outline the autonomic innervation of the intestinal wall, explaining the organization and functions of sympathetic and parasympathetic innervation.
11. How is HCl produced and secreted into the stomach by parietal cells?
12. Compare and contrast endocrine and exocrine systems. What do they have in common?
13. What is meant by secondary modification of an exocrine secretion?
14. Describe the roles of gastrin, secretin, and cholecystokinin in mammalian digestion.
15. Why are some gastrointestinal hormones also classified as neuropeptides? Give examples.
16. Explain what is meant by the cephalic, the gastric, and the intestinal phases of gastric secretion. How are they regulated?
17. How are amino acids and some sugars transported against a concentration gradient from intestinal lumen into epithelial cells?
18. Why is the countercurrent principle important in the removal of the water from the intestinal lumen?
19. How is pernicious anemia related to intestinal function?

SUGGESTED READINGS

Chivers, D. J., and P. Langer. 1994. *The Digestive System in Mammals: Food, Form and Function.* New York: Cambridge University Press.

Davenport, H. W. 1985. *Physiology of the Digestive Tract.* 4th ed. Chicago: Year Book Medical Publishers.

Diamond, J. M. 1991. Evolutionary design of intestinal nutrient absorption: enough but not too much. *News. Physiol. Sci.* 6:92–96.

Johnson, L. R. 1991. *Gastrointestinal Physiology.* 4th ed. St. Louis: Mosby-Year Book.

Karasov, W. H. 1987. Nutrient requirements and the design and function of guts in fish, reptiles, and mammals. In P. Dejours, L. Bolis, C. R. Taylor, and E. R. Weibel, eds., *Comparative Physiology: Life in Water and on Land,* pp. 181–191. Fidia Research Series, IX. Liviana, Padua, Italy.

Karasov, W. H. 1990. Digestion in birds: chemical and physiological determinants and ecological implications. In M. Morrision, C. J. Ralph, J. Verner, and J. R. Jehl, eds., *Studies in Avian Foraging: Theory, Methodology, and Applications. Stud. Avian Biol.* 13:391–415. Kansas: Cooper Ornithological Society.

Mayer, E. A., X. P. Sun, and R. F. Willenbucher. 1992. Contraction coupling in colonic smooth muscle. *Ann. Rev. Physiol.* 54:395–414.

Stevens, C. E. 1988. *Comparative Physiology of the Vertebrate Digestive System.* Cambridge, Eng.: Cambridge University Press.

CHAPTER
16

USING ENERGY: MEETING ENVIRONMENTAL CHALLENGES

Animals require food from which they derive chemical energy to perform work, to maintain their structural integrity, and, ultimately, to reproduce. In Chapters 3 and 15, you learned that animals degrade large, organic compounds to transfer some of their chemical energy to special "high energy" molecules (e.g., ATP). These molecules are subsequently used to drive endergonic reactions. Thus, animals eventually use the chemical energy of foodstuffs to produce electrical, ionic, and osmotic gradients and muscle contraction. The more effectively an animal captures and uses the energy resources available in its environment, the better able it is to compete with other members of its species, and the greater is the fitness of the species in an evolutionary sense.

This chapter explores the various factors that affect energy expenditure by animals and considers in particular the relation between metabolic rate and body temperature, body size, locomotion, and reproduction. In many respects, it is properly the last chapter of the book, because it integrates the animal and its physiology into its environment.

THE CONCEPT OF ENERGY METABOLISM

The term *metabolism*, in its broadest sense, is the sum total of all the chemical reactions occurring in an organism (see Chapter 3). Because the rate of a chemical reaction increases with temperature, the metabolic activity of an animal is closely linked to its own body temperature. Low body temperatures preclude high metabolic rates because of the temperature dependence of enzymatic reactions. High metabolic rates, on the other hand, with their high rates of heat production, may lead to overheating and attendant deleterious effects on tissue function, especially in hot climates. In cold climates, excessive heat loss can lower body temperature to dangerously low levels at which a further drop in temperature leads to reduced body heat production, in a vicious circle of reduced metabolic heat production and continued cooling. Thus, body temperature is a vital variable that touches on all aspects of animal func-

tion, which must be maintained in the presence of temperature fluctuations in the environment. Some animals continuously maintain body temperatures elevated above that of the environment; whereas, in others, body temperature is less tightly regulated or even unregulated.

Animals (like machines) are much less than 100% efficient in their energy conversions, so a large fraction of metabolic energy is in the form of heat produced as a by-product of the release of free energy during exergonic reactions such as those occurring in muscle contraction. This metabolic heat is comparable to the waste heat produced by a gasoline engine in converting chemical energy into mechanical work. Yet, in many animals, heat is not "wasted" in the usual sense of the word in that metabolic heat production is used to raise the temperature of the animal's tissues to levels that significantly enhance the rate of biochemical reactions.

Body mass also affects an animal's energy expenditure. Smaller animals tend to have higher mass-specific metabolic rates than do larger animals. Thus, body mass affects many different physiological processes and the performance of most physiological systems. Like body mass, muscular activity affects the rate of energy expenditure. The consumption of metabolic fuel in a hummingbird hovering over a nectar-laden flower is far greater than that in that same bird roosting at night.

Finally, the act of reproduction may constitute a major commitment of acquired and stored energy. In some animals, the release of gametes results in relatively little energy loss; whereas, in others, a huge percentage of the annual energy intake is used in producing eggs or sperm and caring for young.

Metabolic pathways fall into two major categories:

1. **Anabolism,** which requires energy and is associated with repair, regeneration, and growth, is the assembly of simple substances into more complex molecules required by the organism. Although it is difficult to measure anabolic metabolism quantitatively, a positive nitrogen balance (i.e., net incorporation of nitrogen) in

an organism is one index of anabolism. That is, anabolic activity leads to a net incorporation of nitrogen-containing molecules through protein synthesis, rather than a net loss due to protein breakdown.

2. **Catabolism,** in contrast, is the breaking down of complex, energy- or material-rich molecules into simpler ones. In catabolism, the degradation of complex molecules into simpler ones is accompanied by the release of chemical energy. Some of this energy is stored as high-energy phosphate compounds, such as ATP, which are subsequently used to power cellular activities (see Figure 3-70). Simpler metabolic intermediates, such as glucose or lactate, can serve as energy storage compounds in that they are substrates for additional exergonic reactions (see Figure 3-79).

In the absence of external work or the storage of chemical energy, *all energy released during metabolic processes appears eventually as heat.* This simple fact makes it possible to use heat production as an index of *energy metabolism,* provided the organism is in a thermal steady state with its environment. The conversion of chemical energy into heat is measured as the *metabolic rate*—heat energy released per unit time. Although heat production is a useful measure of metabolic rate, there are other common and traditional measures such as oxygen consumption. Currently, nuclear magnetic resonance (NMR) is being used to directly (and noninvasively) characterize the metabolism of high-energy phosphate groups taking place within animal tissues.

Measurements of metabolic rates are useful not only to physiologists, but also to ecologists, animal behaviorists, evolutionary biologists, and many others because metabolic rates can be used to calculate the energy requirements of an animal. To survive over the long term, an animal must take in as much energy in the form of energy-yielding foodstuff molecules as the total of all the energy that it releases and stores. Measurements of metabolic rate at different ambient (environmental) temperatures may provide information about the heat-conserving or heat-dissipating mechanisms of an animal. Measurements of metabolic rate during different types of exercise help us to understand the energy cost of such activities. How metabolically expensive is it, for example, simply to stay alive, to be big or small, or to fly, swim, run, or walk a given distance?

The metabolic rate of an animal varies with the kinds and intensity of processes taking place. These processes include tissue growth and repair; chemical, osmotic, electrical, and mechanical internal work; and external work for locomotion and communication (Figure 16-1).

In addition to body and environmental temperatures, body mass, reproductive status, and activity, other factors that affect metabolic rate are time of day, season of year, age, sex, shape, stress, type of food being metabolized, and reproductive status. Consequently, metabolic rates of different animals can be meaningfully compared only under carefully chosen and closely controlled conditions, which are considered in the next section.

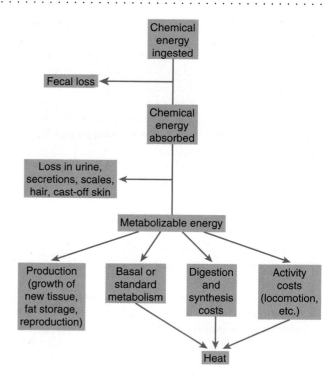

Figure 16-1 Chemical energy from foodstuffs must be taken in and utilized by an animal. Part of the potential chemical energy ingested is unabsorbed and is degraded by intestinal flora or passed out in fecal matter. Of the chemical energy absorbed, some is lost in urine and other secretions, and some appears directly as heat resulting from exergonic metabolic reactions (chemical, electrical, or mechanical work). Any energy left over is conserved in the anabolic buildup of tissues.

MEASURING METABOLIC RATE

Physiologists intending to measure metabolic rate recognize several different normal metabolic levels, or states, that can affect their measurements.

Basal and Standard Metabolic Rates

The **basal metabolic rate** (BMR) is the stable rate of energy metabolism measured in mammals and birds under conditions of minimum environmental and physiological stress (namely, at rest with no temperature stress) and after fasting has temporarily halted digestive and absorptive processes. Ambient temperature affects body temperature in almost all animals other than birds and mammals. Because the minimum metabolic rate varies with the body temperature, it is necessary to measure the equivalent of the basal metabolic rate at a controlled, specified body temperature at which the animal is not expending additional metabolic energy to warm or cool itself. For that reason, the **standard metabolic rate** (SMR) is defined as an animal's resting and fasting metabolism *at a given body temperature.* Interestingly, the SMR of some ectotherms depends on their previous temperature history, owing to metabolic compensation or thermal acclimation, which is described later.

Basal and standard metabolic rates are useful measurements for comparing baseline metabolic rates both between and within species. However, they give little information about the metabolic costs of normal activities carried out by the animals, because the conditions under which the BMR and the SMR are measured differ greatly from natural conditions—that is, for these measurements, the animal is in an unnaturally controlled and quiet state. The term that best describes the metabolic rate of an animal in its natural state is called its **field metabolic rate** (FMR), which is the *average* rate of energy utilization as the animal goes about its normal activities, which may range from complete inactivity during resting to maximum exertion when chasing prey (or being preyed upon).

Metabolic Scope

The range of metabolic rates of which an animal is capable is called its **aerobic metabolic scope,** defined as the ratio of the maximum sustainable metabolic rate to the BMR (or the SMR) determined under controlled, resting conditions. This dimensionless number (e.g., 5, 7, or 14) indicates the increase in an animal's maximal energy expenditure (usually measured as oxygen consumption) over and above the amount that it expends under resting conditions. Commonly, metabolic rate increases from 10 to 15 times with activity in many animals. It should be noted, however, that, because sustained activity is normally powered by aerobic metabolism, this type of measurement does not take into account the contribution of anaerobic processes to activity; such processes build up an oxygen debt and hence are not sustainable.

The concept of metabolic scope associated with activity applies to all animals, regardless of their mode of locomotion. In fishes swimming in flow tanks, for example, faster swimming can be stimulated by increasing the rate of water flow. These experiments indicate that metabolic scope varies with body size. The ratio of active to standard metabolism in salmon, for example, increases from less than 5 in young specimens weighing 5 g to more than 16 in those weighing 2.5 kg. This general relation is compounded when making interspecific comparisons between animals with different modes of activity. Although smaller than a 5-g larval salmon, flying insects, especially those that sustain high body temperatures during flight, can exhibit ratios of as much as 100. These ratios are probably the highest in the animal kingdom.

Studies on metabolic scope have inherent complexities and pitfalls. As mentioned, there may be a significant contribution from anaerobic metabolism, leading to a buildup of **oxygen debt,** especially during short periods of exertion (Figure 16-2). White muscles in some vertebrates are specially adapted to develop an oxygen debt through anaerobic metabolism and are therefore particularly suitable for short-term spurts of intense activity. This component of total metabolism may not be detected in short-term metabolic measurements, because the aerobic breakdown of anaerobic products can be delayed. For this reason, it is best to make measurements of metabolic scope only during sustained activity at a constant level of exertion.

An additional practical problem in determining metabolic scope is that the maximum exertion in an apparatus designed by the experimenter may not produce the animal's actual maximum possible exertion, because the animal's cooperation and motivation may not be maximized.

Finally, measurements of metabolic scope can be skewed by variations in SMR. Very low SMRs under conditions such as sleep and torpor can lead to very high estimates of metabolic scopes.

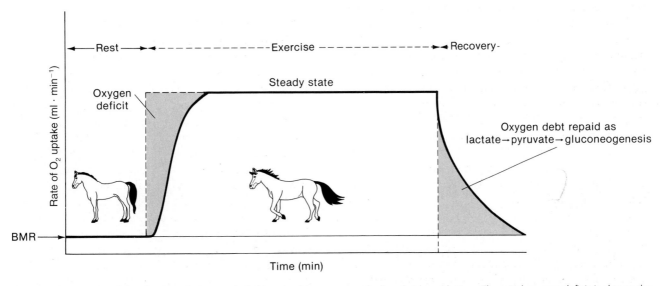

Figure 16-2 An oxygen deficit develops during a period of sustained, intense activity. Active muscle tissue with anaerobic capabilities can amass an oxygen deficit, which is subsequently paid off in the form of a delayed oxidation of an anaerobic product such as lactic acid. As a result, an elevated metabolic rate continues after cessation of activity but gradually subsides with time. The initial oxygen deficit is due to the use of preexisting stores of high-energy phosphagens amassed during rest. Replenishment of these stores is included in the repayment of the oxygen debt.

Would a similarly sized factorial increase in aerobic metabolism during activity have the same energetic implications in an animal with a very high BMR (e.g., a bird) and one with a very low SMR (e.g., a cockroach)?

Direct Calorimetry

If no physical work were being performed and chemical synthesis were occurring, all the chemical energy released by an animal in carrying out its metabolic functions would ultimately flow from the animal as heat. This is formalized in **Hess's law** (1840), which states that the total energy released in the breakdown of a fuel to a given set of end products is always the same, irrespective of the intermediate chemical steps or pathways used. The metabolic rate of an organism is therefore effectively determined by measuring the amount of energy released as heat over a given period. Such measurements are made in a calorimeter, and the method is called *direct calorimetry*. The animal, usually unrestrained and minimally disturbed, is placed in a well-insulated chamber for the measurement. Heat lost from the animal is determined from the rise in temperature of a known mass of water used to trap that heat. The earliest and simplest calorimeter was that devised in the 1780s by Antoine Lavoisier and Pierre de Laplace, in which the heat given off by an animal in a chamber melted ice packed around that chamber. The heat loss was calculated from the mass of collected water and the latent heat of melting ice. In one type of modern calorimeter, water flows through coiled copper pipes in the measuring chamber. The total heat lost by the subject is the sum of the heat gained by the water plus the latent heat present in the water vapor of the expired air and of evaporated skin moisture. To measure

this latent heat, the mass of the water vapor is determined by passing the air through sulfuric acid, which absorbs the water. The energy content of each gram of water absorbed is 2.45 kJ (0.585 kcal), the latent heat of vaporization of water at 20°C. The results are generally reported in calories or kilocalories per hour (see Spotlight 16-1 for a discussion of the numerous, and sometimes confusing, ways in which energy units are expressed).

Although simple in principle, direct calorimetry can be rather cumbersome in practice. The technique may be too imprecise for animals having very low metabolic rates, and very large animals require a calorimetry chamber of impractical dimensions. Consequently, direct calorimetry has been used most often for birds and small mammals with high metabolic rates. Another disadvantage of direct calorimetry is that an animal's behavior (and therefore its metabolism) is unavoidably altered because of the restrictions imposed by the measurement conditions.

Indirect Calorimetry: Measurement from Food Intake and Waste Excretion

Metabolic rate can be estimated from a "balance sheet" that sums energy gain and compares it with energy loss. Living organisms obey the laws of energy conservation and transformations that were initially derived for nonliving chemical and physical systems (Chapter 3). We could, therefore, in principle determine the metabolic rate of an animal in an energy steady state by using the following formulation:

$$\begin{aligned} &\text{rate of chemical energy intake} \\ &\quad - \text{rate of chemical energy loss} \\ &\qquad = \text{metabolic rate (heat production)} \quad (16\text{-}1) \end{aligned}$$

Total energy intake over a given period equals the chemical energy content of ingested food over that same period.

SPOTLIGHT 16-1

ENERGY UNITS (OR WHEN IS A CALORIE NOT A CALORIE?)

The most commonly used unit of measurement of heat is the *calorie*, abbreviated "cal," which is defined as the quantity of heat required to raise the temperature of 1 g of water 1°C. This heat quantity varies slightly with temperature, and so the calorie is more precisely the amount of heat required to raise the temperature of 1 g of water from 14.5°C to 15.5°C.

Because a calorie is a very small quantity of heat relative to many biological processes, a more practical unit of heat energy is the *kilocalorie* (1 kcal = 1000 cal). Unfortunately, confusion has

arisen over the use of the popular term *Calorie* (note the capital C), which designates 1000 calories (cal). In fact, if the label of a soda can states that it contains "125 Calories," almost invariably this means that it contains 125 *kilo*calories. The use of calorie and kilocalorie have persisted largely because they are familiar to most people.

According to the International System of Units (Système International d'Unités, SI), heat is defined in terms of work, and the unit of measurement is the *joule* (J). Again, the more useful version is the *kilojoule* (1 kJ = 1000 J). Thus, 1 cal = 4.184 J, and 1 kcal = 4.184 kJ. If we assume a respiratory quotient (R_Q) of 0.79, which is a typical value, 1 liter of oxygen used in the oxidation of substrate will release 4.8 kcal, or 20.1 kJ, of heat energy.

Power is the amount of energy expended per unit time, and has the SI units of *watts* (W), with 1 W = 1 J·s^{-1}. Conversion tables for units of energy are given in Appendix 3.

Energy loss is the unabsorbed chemical energy that remains in the feces and urine produced by the animal over the same period. The energy content of either food or wastes can be obtained from the heat of combustion of these materials in a **bomb calorimeter.** In this method, the material to be tested is first dried and then placed inside an ignition chamber enveloped in a jacket containing a known amount of water. The material is burned to ash (using no additional fuel) with the aid of oxygen gas. The resulting heat is captured in the surrounding water jacket. The amount of energy released from the burning of the test substance is then determined from the increase in temperature of the surrounding water. The energy released from the burned material is equivalent to that which would be released if all of this material were to be passed through aerobic metabolic pathways.

When using the balance-sheet approach to energy metabolism, one must contend with variables that are difficult to control. For example, not all the energy extracted from food is available for the metabolic needs of the animal. Depending on the food type, a variable fraction may actually be digested and absorbed in the digestive tract (Chapter 15). One must correct for this fraction when calculating total energy intake. Another complicating factor is that energy can be obtained during the period of measurement from an animal's tissue reserves (e.g., stored fat). Because these energy reserves can be exhausted, the animal will eventually lose weight, signaling a nonsteady state (a violation of one of the assumptions of this technique).

The balance-sheet method does not measure BMR, SMR, or RMR (resting metabolic rate in a fed, thermoregulating, inactive animal), which are best measured by more direct methods, considered next.

Indirect Measures of Metabolic Rate

Indirect measures of metabolic rate depend on the measurement of some variable other than heat production related to energy utilization. The energy contained in food molecules becomes available for use by an animal when those molecules or their products are subjected to oxidation, as described in Chapter 3. In aerobic oxidation, the amount of heat produced is related to the quantity of oxygen consumed. Thus, measurements of oxygen uptake ($\dot{M}_{O_2}$) and carbon dioxide production ($\dot{M}_{CO_2}$), expressed as moles of gas per hour, can be used to calculate metabolic rate.* **Respirometry** is the measurement of an animal's respiratory exchange—that is, its $\dot{M}_{O_2}$ and $\dot{M}_{CO_2}$. In *closed system respirometry,* an animal is confined to a closed, water- or air-filled chamber in which the amounts of oxygen consumed and carbon dioxide produced are monitored for a given time period. Oxygen consumption is revealed by

successive determinations of the decreasing amount of oxygen dissolved in the water or present in the air contained in the chamber. These measurements can be obtained very conveniently with the aid of an oxygen electrode and the appropriate electronic circuitry. The partial pressure of oxygen of the water or air directly determines the signal produced in the electrode. In gaseous phase (that is, an air-filled respirometer), O_2 can be measured by a mass spectrometer or an electrochemical cell in addition to an oxygen electrode. In water or air, CO_2 can be determined with a CO_2 electrode, but the complex chemistry of CO_2 dissolved in water (see Chapter 13) makes the interpretation of these values more complex. In gases, CO_2 can be accurately measured with a CO_2 electrode, an infrared device, a gas chromatograph, or a mass spectrometer. Generally, O_2 is easier to measure than CO_2, so $\dot{M}_{O_2}$ is more commonly reported than $\dot{M}_{CO_2}$ as a measure of metabolic rate.

All of these methods of gas analysis allow mass-flow analytical techniques in which the flow of gas or water into and out of a chamber is monitored, and the difference in gas concentrations or partial pressures is used to calculate respiratory exchange. Such systems employ *flow-through* or *open* respirometry. Importantly, the chambers in which the animals reside must be well stirred so that the gas exiting the chamber is in equilibrium with that throughout the chamber. Open system respirometry can also be carried out on animals fitted with breathing masks, a method especially useful in wind-tunnel tests of flying animals or on animals running on treadmills.

Closed and open respirometry can be combined in a single experiment, as illustrated in Figure 16-3. Such combined systems are frequently used in partitioning total gas exchange into pulmonary, branchial, and cutaneous locations in such animals as amphibian larvae and air-breathing fishes that simultaneously employ water and air breathing.

The determination of metabolic rate from O_2 consumption rests on important assumptions:

1. The relevant chemical reactions are assumed to be aerobic. This assumption usually holds for most animals at rest, because energy available from anaerobic reactions is relatively minor except during vigorous activity. However, anaerobiosis is important in animals that live in oxygen-poor environments, as do gut parasites and invertebrates that dwell in deep lake-bottom muds. Oxygen consumption would be an unreliable index of metabolic rate in such animals and would underestimate the true metabolic rate.

2. The amount of heat produced (i.e., energy released) when a given volume of oxygen is consumed is assumed to be constant irrespective of the metabolic substrate. This assumption is not precisely true: more heat is produced when 1 liter of O_2 is used in the breakdown of carbohydrates than when fat or protein is the substrate. However, the error resulting from this assumption is no greater than about 10%. Unfortunately, it is generally difficult to precisely identify the

---

* Oxygen uptake and carbon dioxide production are also commonly expressed as volume of gas, $\dot{V}_{O_2}$ and $\dot{V}_{CO_2}$, respectively. This is less desirable than expressing these values as molar amounts, which by definition are completely independent of measurement temperature and atmospheric pressure; $\dot{V}_{O_2}$ or $\dot{V}_{CO_2}$ reported in a paper can accurately be converted into other units such as $\dot{M}_{O_2}$ or $\dot{M}_{CO_2}$ only if the author has reported temperature and pressure, which is not always done.

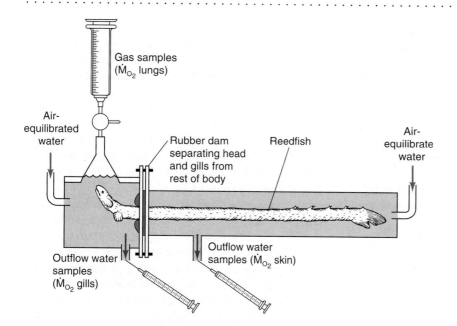

Figure 16-3 Open and closed respirometery can be combined in a single experiment to measure an animal's gas exchange partitioning between various sites. In this experiment on the reedfish *Calamoicthys calabaricus*, two independent open systems are used to determine branchial and cutaneous aquatic oxygen uptake. A third, closed respirometery system includes the air in the funnel above the head, from which the animal breathes air. Gas samples taken after an air breath are used to calculate aerial oxygen consumption. [Adapted from Sacca and Burggren, 1982.]

substrate(s) being oxidized in correcting for differences in caloric yield.

3. The O_2 stores in the body are small, so the minute-to-minute oxygen consumption from air or water flowing over the gas-exchange organs fairly accurately represents the metabolic rate. (Note that the ability to store CO_2 in body tissues is much greater than the ability to store O_2, so the minute-to-minute elimination of CO_2 is a much less reliable indicator of metabolic rate.)

A final important method for measuring metabolic rate employs *isotopic techniques*. These techniques first came to prominence in the measurement of water fluxes in animals: deuterium- or tritium-labeled water is injected into an animal, and the specific activity in serial blood or other body fluid samples is determined. The decline in specific activity with time indicates the loss of labeled water and therefore the outward water flux. The use of isotopes was then extended to the measurement of CO_2 production as a measure of metabolic rate. In essence, isotopes of oxygen and hydrogen are injected into an animal. The subsequent decline in the about of isotopic oxygen (^{18}O) in body water is related to the rates of CO_2 loss through exhalation and through water loss, the latter measured by the disappearance of deuterium- or tritium-labeled water. Although the numerous assumptions required must be validated for each experimental setting, the great advantage of the technique is that it can be employed on intact, unrestrained, normally behaving animals. The many studies by Ken Nagy and his colleagues have shown the usefulness of this technique in measuring *field metabolic rate*.

Respiratory Quotient

To translate the amount of oxygen consumed into equivalent heat production, we must know the relative amounts of carbon and hydrogen oxidized. The oxidation of hydrogen atoms is hard to determine, however, because metabolic water (i.e., that produced by oxidation of hydrogen atoms available in foodstuffs), together with other water, is lost in the urine and from a variety of body surfaces at a rate that is irregular and determined by unrelated factors (e.g., osmotic stress and ambient relative humidity). It is more practical to measure, along with the O_2 consumed, the amount of carbon converted into CO_2, as explained earlier. As noted in Chapter 13, the ratio of the volume of CO_2 produced to the volume of O_2 removed from within a given time is called the *respiratory quotient* (R_Q):

$$R_Q = \frac{[\text{rate of } CO_2 \text{ production}]}{[\text{rate of } O_2 \text{ consumption}]} \quad (16\text{-}2)$$

Under resting, steady-state conditions, the R_Q in Table 16-1 is characteristic of the type of molecule catabolized

TABLE 16-1

Heat production and respiratory quotient for the three major foodstuff types

	Heat production (kJ)			R_Q $\left(\dfrac{\text{Liter } CO_2}{\text{Liter } O_2}\right)$
	Per gram of foodstuff	Per liter of O_2 consumed	Per liter of CO_2 produced	
Carbohydrates	17.1	21.1	21.1	1.00
Fats	38.9	19.8	27.9	0.71
Proteins (to urea)	17.6	18.6	23.3	0.80

(carbohydrate, fat, or protein). Thus, the R_Q reflects the proportions of carbon and hydrogen in the food molecules.

The following examples illustrate how the R_Q of the major food types may be calculated from a formulation of their oxidation reactions:

- **Carbohydrates.** The general formula of carbohydrates is $(CH_2O)_n$. In the complete oxidation of a carbohydrate, oxygen is used *in effect* only to oxidize the carbon to form CO_2. Upon complete oxidation, each mole of a carbohydrate produces n moles of both H_2O and CO_2 and consumes n moles of O_2. The R_Q for carbohydrate oxidation is thus 1. The overall catabolism of glucose, for example, may be formulated as

$$C_6H_{12}O_6 + 6\ O_2 \rightleftharpoons 6\ CO_2 + 6\ H_2O$$

$$R_Q = \frac{6 \text{ volumes of } CO_2}{6 \text{ volumes of } O_2}$$
$$= 1$$

- **Fats.** The R_Q characteristic of the oxidation of a fat such as tripalmitin may be calculated as follows:

$$2\ C_{51}H_{98}O_6 + 145\ O_2 \rightleftharpoons 102\ CO_2 + 98\ H_2O$$

$$R_Q = \frac{102 \text{ volumes of } CO_2}{145 \text{ volumes of } O_2}$$
$$= 0.70$$

Because different fats contain different ratios of carbon, hydrogen, and oxygen, they differ slightly in their R_Qs.

- **Proteins.** The R_Q characteristic of protein catabolism presents a special problem because proteins are not completely broken down in oxidative metabolism. Some of the oxygen and carbon of the constituent amino acid residues remains combined with nitrogen and is excreted as nitrogenous wastes in urine and feces. In mammals, the excreted end product is urea, $(NH_2)_2CO$; in birds, it is primarily uric acid, $C_5H_4N_4O_2$. To obtain the R_Q, it is therefore necessary to know the amount of ingested protein as well as the amount and kind of nitrogenous wastes excreted. The oxidation of carbon and hydrogen in the catabolism of protein typically produces

$$R_Q = \frac{96.7 \text{ volumes of } O_2}{77.5 \text{ volumes of } CO_2}$$
$$= 0.80.$$

It is routinely assumed in making deductions from R_Q that (1) the only substances metabolized are carbohydrates, fats, and proteins; (2) no synthesis takes place alongside breakdown; and (3) the amount of CO_2 exhaled in a given time equals the CO_2 produced by the tissues in that interval. These assumptions are not strictly true, so caution must be exercised in using R_Q values at rest and in postabsorp-

tive (fasting) states. Under such conditions, protein utilization is negligible, and carbohydrate utilization is minor, so the animal is considered to be metabolizing primarily fat. From Table 16-1, it can be seen that the oxidation of 1 g of mixed carbohydrate releases about 17.1 kJ (4.1 kcal) as heat. When 1 liter of O_2 is used to oxidize carbohydrate, 21.1 kJ (5.05 kcal) of heat is obtained; the value for fats is 19.87 kJ (4.7 kcal) and for protein (metabolized to urea), 18.6 kJ (4.46 kcal). A fasting aerobic animal presumed to be metabolizing mainly fats produces about 20.1 kJ (4.80 kcal) of heat for every liter of oxygen consumed.

Another term often used to describe the ratio of $\dot{M}_{O_2}$ to $\dot{M}_{CO_2}$ is the *respiratory exchange ratio* (R_E), which is a measure of the *instantaneous* relation between $\dot{M}_{O_2}$ and $\dot{M}_{CO_2}$ as measured from gas exiting the respirometer or face mask. When CO_2, for example, is temporarily being stored in body tissues rather than eliminated from the body (as during a period of submergence in a diving animal), the apparent $\dot{M}_{CO_2}$ is lower than is actually occurring at the tissue level. Under these conditions, the R_E will be lower than the R_Q until a new CO_2 steady state is reached in body tissues and CO_2 is once again eliminated at the same rate at which it is produced by cellular respiration.

Energy Storage

Although animals continually expend metabolic energy, most do not ingest food steadily. Consequently, they do not strike a moment-to-moment balance between food intake and energy expenditure. As food is taken in bursts (i.e., in discrete meals), the animals immediate energy requirements are exceeded. However, the excess is stored for later use, primarily as fats and carbohydrates.

Protein is not an ideal storage material for energy reserves because nitrogen is a relatively scarce commodity and is generally the limiting factor in growth and reproduction; it would be wasteful to tie up valuable nitrogen in energy reserves. Fat is the most effective form of energy storage, because oxidation of fat yields 38.9 kJ·g^{-1} (9.3 kcal·g^{-1}), nearly twice the yield per gram for carbohydrate or protein (Table 16-1). This efficiency is of great importance in animals such as migrating birds or insects, in which economy of weight and volume is of the essence. Not only is the energy yield per gram of carbohydrate lower than that of fats, but carbohydrates are stored in a bulky hydrated form, with as much as 4 to 5 g of water required per gram of carbohydrate, whereas fats are stored in a dehydrated state. Nonetheless, some carbohydrates are important in energy storage. Glycogen, a branched, starchlike carbohydrate polymer, is stored as granules in skeletal muscle fibers and liver cells of vertebrates. Muscle glycogen can be rapidly converted into glucose for oxidation within the muscle cells during intense activity, and liver glycogen is used to maintain blood glucose levels. Glycogen is broken down directly into glucose 6-phosphate, providing fuel for carbohydrate metabolism more directly and rapidly than does fat. Thus, carbohydrates tend to be used to power short-term increases in metabolism—during activity, for example. Fats, which cannot be directly metabolized anaerobically, are metabolized aerobically in response to longer-term demands for energy and during fasting when carbohydrate stores have been depleted.

Specific Dynamic Action

Max Rubner reported in 1885 that a marked increase in metabolism accompanies the processes of digestion and assimilation of food independently of other activities. He gave this phenomenon the rather awkward name **specific dynamic action** (SDA). Since then, SDA has been documented in all five vertebrate classes, as well as in invertebrates including crustaceans, insects, and mollusks. Generally, an animal's oxygen consumption and heat production increases within about 1 hour after a meal is eaten, reaching a peak some 3–6 hours later and remaining elevated above the basal value for several hours (Figure 16-4). In fish, amphibians, and reptiles with an SDA equivalent to a doubling or tripling of metabolic rate, there are also attendant large increases in heart rate and cardiac output and a temporary redistribution of blood toward the gut. Similar cardiovascular changes of lesser magnitude occur in animals with less prominent SDA responses (e.g., in humans).

The mechanism of SDA is not clearly understood, but apparently the work of digestion (and the concomitant increase in metabolism of the tissues of the gastrointestinal tract) is responsible for only a small part of the elevated metabolism. A more likely explanation for this rise in metabolic rate may be that certain organs, such as the liver, expend extra energy processing recently absorbed nutrients for entry into metabolic pathways. The extra energy consumed by such processes is lost as heat. The increase in heat

production differs, depending on the ingested food materials. The magnitude of the increased metabolic rate ranges from 5% to 10% of total energy of ingested carbohydrates and fats and from 25% to 30% of that of proteins.

Specific dynamic action probably accounts for some of the variation in metabolic rate reported by different researchers for a single species. Very different metabolic rates could be obtained, depending on whether the animals measured were in a postabsorptive state or were in some part of the SDA response. Consequently, basal metabolism must be measured only during the postabsorptive state so as to minimize any contribution of SDA.

BODY SIZE AND METABOLIC RATE

Body size is one of the more important physical characteristics that affects an animal's physiology. The study of how both anatomical and physiological characteristics change with body mass is called scaling. Changes in body size introduce changes that are not always simple and proportional (i.e., geometric). For example, the doubling of the height of an animal while retaining the same body proportions is accompanied by a fourfold increase in surface area and an eightfold increase in mass. The consequences of nongeometric scaling for the functional anatomy and physiology of the animal are immediately evident. You can carry out a thought experiment by imagining a mouse scaled up to the size of an elephant while retaining its mouselike body proportions. Clearly, the imaginary enlarged mouse has different proportions from those of an elephant, and its relatively slender legs would probably collapse under the weight of the very massive body. After all, for each doubling in height of the imaginary mouse, the mass increases by a factor of eight (height cubed) while the cross-sectional area of leg bone increases by a factor of only four (height squared). These same scaling factors make a small mouse capable of jumping many times its own body length without harm, whereas an elephant is essentially "earthbound."

Changes in body mass have great effects on an animal's metabolic rate. Consider, for example, the respiratory and metabolic requirements of a tiny water shrew during diving compared with those of a submerged whale. Although both whales and water shrews normally dive, a whale can hold its breath and remain under water far longer than a shrew. The reason stems from the general principle that *small animals must respire at higher rates per unit body mass than large animals*. In fact, there is an inverse relation between the rate of O_2 consumption per gram of body mass and the total mass of the animal. Thus, a 100 g mammal consumes far more energy per unit mass per unit time than does a 1000 g mammal. The nonproportionality of the basal metabolic rates of mammals ranging from very small to very large is illustrated by the well known "mouse to elephant" curve (Figure 16-5A). A similar relation holds not only for other vertebrate groups, but throughout the animal and plant kingdom. Few biological principles are so widely applicable.

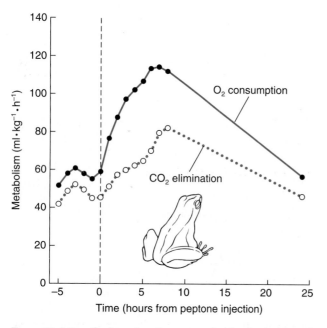

Figure 16-4 Specific dynamic action occurs after feeding in the toad *Bufo marinus*. Specific dynamic action was induced by injecting peptone (a mixture of amino acids produced from chemically digested meat protein) into the animal's stomach. [Adapted from Wang et al., 1995.]

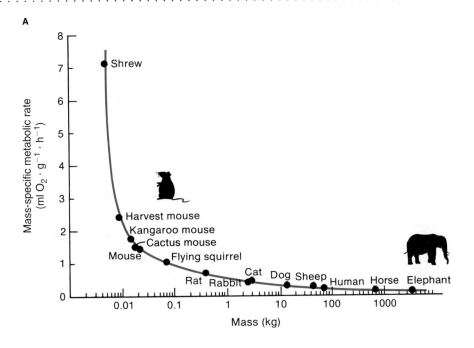

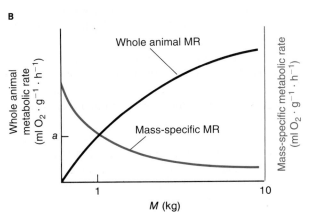

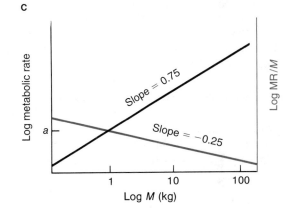

Figure 16-5 Mass-specific metabolic rate (MR) in mammals declines as body mass (M) increases. **(A)** The mouse-to-elephant curve, with metabolic intensity (mass-specific metabolic rate) given as O_2 consumption per unit mass plotted against body mass. Note the logarithmic scale of body mass. **(B)** Generalized relations between overall metabolic rate and body mass (black curve) and between metabolic intensity and body mass (red curve). **(C)** Log-log plots of part B. The MR and log MR plots in parts B and C cross at M – 1 kg. [Part A adapted from Schmidt-Nielsen, 1975.]

The inverse relation between metabolic rate and body mass applies within species as well as between species. Thus, a small human being, cockroach, or fish tends to have a higher metabolic rate per unit mass per unit time than does a larger member of that species. However, this relation is often difficult to demonstrate *within* a species, where the overall range of body mass may be quite small compared with that *between* species and where other factors such as sex, nutrition, and season may exert compounding effects.

Metabolic rate is a power function of body mass, as described by the simple relation

$$\text{MR} = aM^b \qquad (16\text{-}3)$$

in which MR is the basal or standard metabolic rate, M is the body mass, a is the intercept of the log-log regression

line (and differs between species), and b is an empirically determined exponent that expresses the rate of change of MR with change in body mass.

Mass-specific metabolic rate, also termed *metabolic intensity*, is the metabolic rate of a unit mass of tissue (i.e., amount of O_2 consumed per kilogram per hour). It is determined by dividing both sides of equation 16-3 by M:

$$\frac{\text{MR}}{M} = \frac{aM^b}{M} = aM^{(b-1)} \qquad (16\text{-}4)$$

The relation described by equation 16-3 is shown in color in Figure 16-5B. Because it is often more convenient to work with straight-line rather than curved plots (e.g., for statistical analysis) equations 16-3 and 16-4 are often put into their logarithmic form. Thus, equation 16-3 becomes

$$\log MR = \log a + b(\log M) \qquad (16\text{-}5)$$

and equation 16-4 becomes

$$\frac{\log MR}{M} = \log a + (b - 1) \log M \qquad (16\text{-}6)$$

These logarithmic equations are plotted in Figure 16-5C. See Appendix 2 for a discussion of logarithmic equations.

Notice the difference in how whole-animal metabolic rate (black plots) and mass-specific metabolic rate (red plots) change with changes in body mass. These graphs show that overall metabolic rate rises with increasing body mass, whereas the mass-specific metabolic rate (metabolic rate of a unit mass of tissue) *decreases* with increasing body mass. This principle first emerges in the mouse-to-elephant plot in Figure 16-5A.

The value of exponent b lies close to 0.75 for many different taxonomic groups of vertebrates and invertebrates and even holds true for various unicellular taxa (Figure 16-6). The exponential relation between body size and metabolic rate has been attracting the attention of physiologists since it was first recognized more than a century ago. There have been many attempts at giving a rational explanation for this nearly universal logarithmic relation between body mass and metabolism. In 1883, Max Rubner proposed an attractive theory known as the *surface*

hypothesis. Rubner reasoned that the metabolic rate of birds and mammals that maintain a more or less constant body temperature should be proportional to body surface area because the rate of heat transfer between two compartments (i.e., warm animal body and cool environment) is proportional, all else being equal, to their area of mutual contact (Spotlight 16-2). The surface area of an object of *isometric shape* (i.e., nonvarying proportions) and uniform density varies as the 0.67 (or $\frac{2}{3}$) power of its mass. This is because mass increases as the cube of linear dimension, whereas surface area increases only as the square. As noted, this relation holds true for a series of animals of different mass only if body proportions remain constant. This provision is generally satisfied only by adult individuals of different size within a species, because they tend to obey the principle of **isometry**—namely, proportionality of shape regardless of size. In this case, it follows that the surface area must vary as the 0.67 power of body mass. However, the principle of isometry is *not* followed in individuals of different size belonging to related but *different* species. Instead, they tend to follow the principle of **allometry**—namely, systematic changes in body proportions with increasing species size. An example of allometry was alluded to earlier when we compared the proportions of an elephant with those of a mouse. In a comparison of surface-to-mass relations in mammals of different species ranging from mice to whales, surface areas were found to be proportional to the 0.63 power of body mass (Figure 16-7).

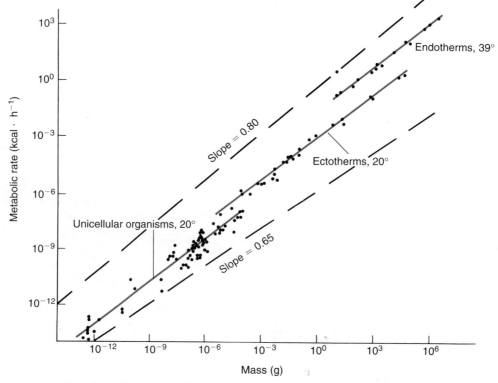

Figure 16-6 A wide variety of groups of animals (including unicellular organisms) show the same general relation between metabolic rate and body mass. Metabolic rate is related to body mass by similar exponents in all three groups for which data are presented here. All three solid lines represent slopes (exponents in the allometric equation) of 0.75. The vertical position of each group on the graph is related to coefficient a in equation 16-3. [From Hemmingsen, 1969.]

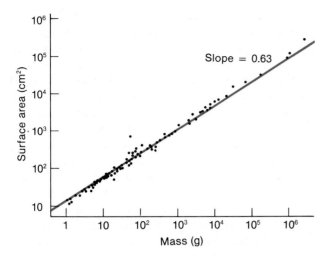

Figure 16-7 Body surface area of mammals ranging from mice to whales is very closely correlated with body mass. The slope of the line gives an exponent of 0.63 rather than the 0.67 predicted by the isometric (i.e., proportional) scaling. The allometric (i.e., disproportional) scaling arises from the fact that, with increasing species size, there is a progressive relative thickening of body structures (i.e., bones, muscles, etc.), so a large species has relatively less surface area than would be predicted from isometric scaling. Recall the relative proportions of mouse and elephant. [From McMahon and Bonner, 1983.]

The surface hypothesis of Rubner gained support over the years from numerous findings that metabolic rate in animals maintaining a constant body temperature is *approximately* proportional to body surface area. An especially close correlation can be seen in comparing the metabolic rates of adult guinea pigs (all of the same species), which were found to be proportional to body mass raised to the 0.67 power (Figure 16-8A) or, assuming isometry of shape,

proportional to the surface area of the individual. Recall that isometry—and, hence a 0.67, power relating surface area to body mass—is characteristic of adult individuals of the *same* species.

In spite of the logical attractiveness of the surface hypothesis, it is not without flaws. True, the difference in metabolic intensity between large and small homeotherms may indeed be an adaptation to the more rapid loss of heat from a smaller animal owing to surface-to-volume relations, the small animal having more surface area per unit mass. Nonetheless, several contradictory observations raise serious concerns about the surface hypothesis. First, when metabolic rates of individuals of *different* species of mammals are plotted against body mass, the exponent relating metabolic rate to body mass is found to be approximately 0.75 (Figure 16-8B). The 0.75 power relating metabolic rate to body mass was first discovered by Max Kleiber (1932) and is often referred to as **Kleiber's law.** The exponent 0.75 is significantly higher than that predicted by the surface hypothesis (see Figure 16-8B); recall that the surface area of mammalian individuals taken from various species of differing size is proportional to the body mass raised to the 0.63 power (see Figure 16-7). Thus, in comparing different species, the differences in metabolic rate clearly cannot be predicted simply on the basis of differences in body surface area.

Another flaw of the surface hypothesis arises from the simple observation that the metabolic rates of animals whose body temperatures vary with that of their surroundings (such as fish, amphibians, reptiles and most invertebrates) exhibit nearly the same relation to body mass as the metabolic rates of animals that actively maintain a constant, high body temperature (i.e., birds and mammals;

A

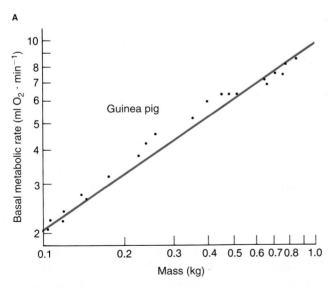

B

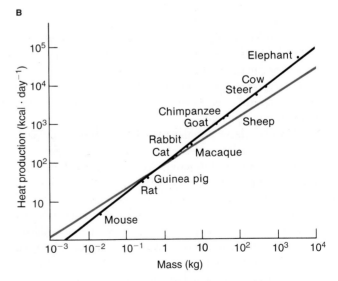

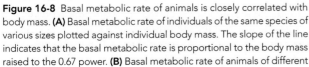

Figure 16-8 Basal metabolic rate of animals is closely correlated with body mass. **(A)** Basal metabolic rate of individuals of the same species of various sizes plotted against individual body mass. The slope of the line indicates that the basal metabolic rate is proportional to the body mass raised to the 0.67 power. **(B)** Basal metabolic rate of animals of different

species plotted against body mass. The black line through the points has a slope of 0.75. The red line has the slope of 0.63 predicted by the surface law. The discrepancy is statistically significant, indicating that metabolic rate is not strictly related to surface area in mammals. [Part A from Wilkie, 1977; black line in part B from Kleiber, 1932.]

SPOTLIGHT 16-2

THE REYNOLDS NUMBER: IMPLICATIONS FOR BIG AND SMALL ANIMALS

The energy expended in propelling an animal through a fluid medium (water or air) depends in part on the *flow pattern* set up in the medium. The flow pattern is determined not only by the density and viscosity of the medium, but also by the dimensions and velocity of the animal. Osborn Reynolds combined these four factors in a dimensionless ratio that relates *internal forces* (proportional to density, size , and velocity) to *viscous forces*. This is the so-called **Reynolds number** *(Re),* calculated as

$$Re = \frac{\rho V L}{\mu}$$

where ρ is the density of the medium, V is the velocity of the body, L is an appropriate linear dimension, and μ is the viscosity of the medium. Thus, when a body moves through a fluid such as water or air, the flow pattern depends on its *Re*. The larger the object or the higher its speed in water, the higher is its *Re*. The same object moving at the same speed in air as in water would be characterized by a lower *Re* in air (about 15 times lower) because of the much lower density of air.

An *Re* below about 1.0 characterizes movement in which the object produces a purely laminar pattern of flow in the fluid passing over its surface. If *Re* is above about 40, turbulence begins to appear in the wake of the object. As *Re* exceeds about 10^6, the fluid in contact with the object becomes turbulent. At this point, the energy needed to increase the velocity further rises steeply.

The velocity at which turbulence appears is higher for a streamlined object such as a dolphin than for an unstreamlined object such as a human scuba diver with protruding tanks, and so forth. Because the value of a streamlined form is in reducing turbulence, it has no advantage for small organisms operating at very low Reynolds numbers, in that such organisms do not experience turbulence.

To a small organism such as a bacterium, spermatozoan, or ciliate, the watery medium appears far more viscous than it does to a human being. The viscosity encountered by a paramecium swimming through water has been compared to the viscosity that would be experienced by a person swimming through a pool of honey (which, indeed, is difficult to imagine). This is another example of an allometric scaling effect. Viscous effects are proportional to surface area, which rises with the square of animal length. Inertial effects, due to momentum of the moving animal, are proportional to mass, which rises with the cube of length. Thus, the movements of small organisms are dominated by viscous effects, whereas those of large animals are dominated by inertial effects.

The relative importance of these two factors in the "coasting" of objects of different size can be illustrated by forcing a tiny toothpick (low *Re*) floating on water to a given velocity, say $0.1 \, \mathrm{m \cdot s^{-1}}$, and then doing the same to a large floating log (high *Re*) of similar physical proportions but larger size. When it is let go, the little toothpick abruptly comes to rest due to the drag exerted on it by the viscosity and cohesion of the water. In contrast, the massive log coasts for many seconds after its release because its far greater momentum (based on its mass) overcomes the drag (based on its surface area). Similarly, a paramecium comes to an abrupt halt if it stops its rapidly beating cilia, whereas a whale coasts with little loss of velocity between the slow thrusts of its flukes.

---

see Figure 16-6). There is no self-evident reason why the metabolic rate of animals with variable body temperature should be causally related through heat loss to body surface area. Relatively little or no metabolic energy is expended to warm animals when they are in temperature equilibrium with the environment.

Scaling effects are also evident at the cellular level. There is a correlation between differences in metabolic intensity of animals of differing sizes and the number of mitochondria per unit volume of tissue. The cells of a small mammal contain more mitochondria and mitochondrial enzymes in a given volume of tissue than do the cells of a large mammal. Because mitochondria are sites of oxidative respiration, this correlation comes as no surprise. However, we are still left with the problem of how metabolic intensity is functionally related to body size.

The question of why large animals have lower metabolic rates per volume of tissue than those of small animals and the functional reasons for the allometric relations that exist between metabolic rate (as well as other variables) and ani-

mal size have been considered extensively. McMahon and Bonner (1983) pointed out that the *cross-sectional area* rather than the surface area of the body (or rather of its parts) more closely resembles the scaling of metabolic rate to body mass, because the cross-sectional area of any body part in a series of animals of increasing size should be proportional to the 0.75 power of body mass, owing to allometric principles that require an elephant's leg to be proportionately thicker than a mouse's leg. Remember that metabolic rate bears the same (0.75) power relation to body mass in a wide range of animals (see Figures 16-6 and 16-8B).

Although the allometry of metabolic rate is well documented, comparative physiologists have yet to "prove" definitively why this relation exists, and both experiments and thought on the subject continue. However, there is no doubt of the physiological implications of allometry to animals. Small animals with proportionately higher metabolic rates must spend more of their time looking for resources and may also be more susceptible to temporary shortages of metabolic substrates or oxygen.

? In *Gulliver's Travels* (written by Jonathan Swift, 1667–1745), Gulliver travels to a land peopled with giants (the Brobdingnagians) and to a land peopled with tiny persons (Lilliputians). Knowing what you do now about allometry, what physiological and structural problems might be faced by each population that Gulliver visited?

TEMPERATURE AND ANIMAL ENERGETICS

Few environmental factors have a larger influence on animal energetics than temperature. Those animals whose body temperature fluctuates with that of the environment experience corresponding temperature-induced changes in metabolic rate, whereas those that can maintain a constant body temperature in fluctuating environmental temperatures have to expend metabolic energy to do so.

Temperature Dependence of Metabolic Rate

Chemical reaction rates, especially those of enzymatic reactions, are highly temperature dependent. Therefore, tissue metabolism and, ultimately, the life of an organism depend on maintenance of the internal environment at temperatures compatible with metabolic reactions facilitated by enzymes. When we consider the effect of temperature on the rate of a reaction, it is useful to obtain a *temperature quotient* by comparing the rate at two different temperatures. A temperature difference of 10 Celsius degrees has become a standard (if arbitrary) span over which to determine the temperature sensitivity of a biological function. The so-called Q_{10} is calculated by using the **van't Hoff equation:**

$$Q_{10} = (k_2/k_1)^{10/(t_2 - t_1)} \qquad (16\text{-}7)$$

where k_1 and k_2 are rates of reaction (rate constants) at temperatures t_1 and t_2, respectively. The beauty of the Q_{10} concept is that it can be applied both to simple processes such as single enzymatic reactions and to complex processes such as running and growing. To relate the van't Hoff equation to metabolic rate, consider the following form of the van't Hoff equation:

$$Q_{10} = (MR_2/MR_1)^{10/(t_2 - t_1)} \qquad (16\text{-}8)$$

in which MR_1 and MR_2 are the metabolic rates at temperature t_1 and t_2, respectively. For temperature intervals of precisely 10 degrees, the following simpler form of equation 6-8 can be used:

$$Q_{10} = \frac{MR_{(t + 10)}}{MR_t} \qquad (16\text{-}9)$$

[handwritten annotations: "higher temp" pointing to $MR_{(t+10)}$, "lower temp" pointing to MR_t]

where MR_t is the metabolic rate at the lower temperature, and $MR_{(t + 10)}$ is the metabolic rate at the higher temperature.

The Q_{10} of a given enzymatic reaction depends on the particular temperature range being considered, so it is important, when citing a Q_{10} value, to clearly indicate the range of temperatures (i.e., t_1 and t_2) for which it was determined. As a rule of thumb, chemical reactions (and such physiological processes as metabolism, growth, locomotion, etc.) have Q_{10} values of about 2 to 3, whereas purely physical processes (such as diffusion) have lower temperature sensitivities (i.e., closer to 1.0).

The temperature effect on enzymes results in the metabolic rate of an animal increasing exponentially with body temperature, as described by the equation

$$\frac{MR}{M} = k10^{b_1 t} \qquad (16\text{-}10)$$

where MR is metabolic rate and M is body mass (making MR/M the metabolic intensity in kilocalories per kilogram per hour), k and b_1 are constants, and t is temperature in degrees Celsius. Because enzyme action largely dictates metabolic rate, we can see this same relation when looking at the effect of body temperature on oxygen consumption in animals that do not maintain body temperature at a constant value (Figure 16-9A). Again, it is useful to transform the relation into a logarithmic one to produce a linear plot. Thus, equation 16-10 becomes

$$\frac{\log MR}{M} = \log k + b_1 t \qquad (16\text{-}11)$$

Now the coefficient b_1 gives the slope of the line—that is, the rate of increase in log MR/M per degree (Figure 16-9B).

The metabolic rates in most animals with variable body temperature increase two- to threefold for every 10 degree (Celsius) increase in ambient temperature, in accordance with what would be predicted for the Q_{10} of enzymes. Yet, the metabolic rates of some ectotherms exhibit a remarkable temperature independence. For example, some intertidal invertebrates that experience large swings in ambient temperature with the ebb and flow of the tides have metabolic rates with a Q_{10} very close to 1.0, so the rate of metabolism changes very little with temperature changes as large as 20 degrees. These animals appear to possess enzyme systems with extremely broad temperature optima, which prevents their inactivation during environmental temperature swings. Such enzyme systems may be due to a staggering of the temperature optima of sequential enzymes in a reaction, such that a drop in the rate of one step in a sequence of reactions "compensates" for an increased rate of other steps in the sequence. The "biological clocks" of animals also are temperature insensitive with Q_{10}s of 1. Otherwise, the "time" kept by these "clocks" would be utterly dependent on an animal's body temperature, and even a fever in a mammal or bird would throw off its body rhythms.

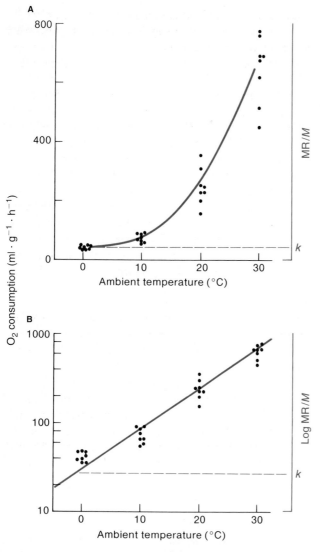

Figure 16-9 The oxygen consumption of the tiger moth caterpillar increases sharply as its body temperature increases. **(A)** Geometric coordinates. **(B)** Semilog coordinates. The generalized ordinates are shown in color at the right in reference to equations 16-11 and 16-12. The constant k is obtained by extrapolating the metabolic rate to a body temperature of 0°C and is the proportionality factor in equations 16-11 and 16-12. [From Scholander et al., 1953.]

Thermal acclimation—enzymatic mechanisms

Environmental heat or cold elicits compensatory changes in physiology and, in some cases, morphology in many species. These changes help an individual organism to cope with the temperature stress. An animal that cannot escape the winter cold (e.g., a pond-dwelling teleost fish in a temperate environment) will gradually undergo, in the course of several weeks, a whole suite of compensatory biochemical adaptations to low temperature. As noted in Chapter 1, the overall change that the animal undergoes in the natural setting is termed *acclimatization*. We will confine ourselves here to a more restricted concept, *acclimation*, which refers to the specific physiological change(s) developed over time in the laboratory in response to varying

a single environmental condition such as temperature. (Recall that evolutionary *adaptation* refers to evolutionary changes over thousands of generations of a species).

Enzymatic acclimation

Acclimation occurs in individual tissues as well as in whole animals. For instance, at a given experimental temperature, winter acclimated frogs and summer acclimated frogs have different contractile properties of skeletal muscles and different heart rates. Similarly, nerve conduction persists at low temperatures in cold-acclimated fishes, but it is blocked at these same temperatures in warm-acclimated ones. How can this be explained? It is reasonable to suppose that enzymatic reactions have been affected. In Figure 16-10, the plots of O_2 consumption against temperature for frogs acclimated at 5°C and 25°C show differing slopes. That is, the net respiratory processes in the two acclimation groups exhibit different Q_{10}s, suggesting that there has been a modification in the temperature sensitivity of enzyme activity. A change in the rate of enzymatically controlled reactions can indicate a change either in the molecular structure of one or more enzymes or in some other factor that affects enzyme kinetics.

In some instances of acclimation, however, thermal compensation appears to result simply from a change in the quantity of an enzyme rather than its characteristics. This is indicated in experiments in which the plot relating a metabolic function to the test temperature exhibits displacement without a change in slope (Figure 16-11). Because the Q_{10} of the process remains unchanged but the activity is higher at every temperature in the cold-acclimated group, the acclimation appears to have led to an increase in the number of enzyme molecules without any change in the kinetics of the enzymes. The particular time course of acclimation depends on the rate at which enzyme type or concentration can be modified.

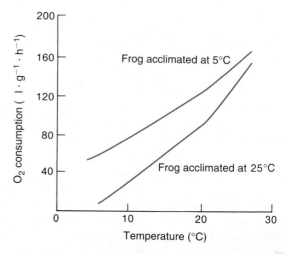

Figure 16-10 At any given measurement temperature, oxygen consumption in frogs acclimated at 5°C is greater than oxygen consumption in frogs acclimated at 25°C. This phenomenon minimizes the disruptive effects of temperature change in these and other ectotherms.

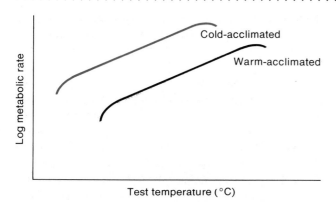

Figure 16-11 Temperature acclimation greatly influences temperature effects on metabolic rate. Generalized plot of log metabolic rate against test temperature in a cold-acclimated individual and in a warm-acclimated one. The similarity of slope in the two plots indicates identical Q_{10}s.

Homeoviscous membrane adaptations

The cell membrane, which is composed largely of a bilayer of lipids with embedded proteins, is very sensitive to temperature change. Low temperatures can cause the membrane to enter a gel-like phase with very high membrane lipid viscosity, whereas high temperatures can cause the membrane to become "hyperfluid" with very little viscosity. Either situation can cause increasingly disruptive changes in physical properties as temperatures move away from optimal values for a particular animal. The many functions of the cell membrane, which range from forming a physical barrier to general solute diffusion to facilitating transmembrane movement of specific solutes, can be in jeopardy if the membrane lipid viscosity becomes too high or too low. We can imagine the effect of temperature on lipid viscosity by recalling that room temperatures lie below the melting point of a cooking grease but above the melting point of a cooking oil. The difference between the oil and the grease lies in the degree of hydrogenation of the carbon backbone. The greater the proportion of unsaturated (i.e., double unhydrogenated) carbon–carbon bonds of a lipid's fatty acid molecules, the lower its melting point. At temperatures above the melting point, the lipid is less viscous, or "oily"; below the melting point, it is more viscous, or "waxy."

Part of the acclimatization of ectothermic animals to cold or hot environments is that the membrane lipids become more saturated during acclimatization to warmth and less saturated during acclimatization to cold, helping to stabilize the form of the lipids and thus the cellular functions that spring from them. This phenomenon is called **homeoviscous adaptation,** referring to adaptations at the molecular level through natural selection that help minimize temperature-induced differences in viscosity.

Unfortunately, there is no simple measure of membrane fluidity. Most often used as an index is the steady-state fluorescence anisotropy (a measure of the lack of symmetry of a molecule or structure). 1,6-Diphenyl-1,3,5-hexatriene

(DPH) is a commonly used probe of membrane fluidity. A high fluorescence anisotropy indicates a high degree of lipid polarization and membrane order attendant with a low membrane viscosity. Figure 16-12 shows changes in DPH polarization of the basolateral membranes of enterocytes isolated from rainbow trout. Initially, acute temperature changes are accompanied by changes in membrane polarization and fluidity. However, with time, homeoviscous adaptation of the membrane lipids results in a lipid polarization and membrane viscosity after acclimation at 5°C that are similar to those after acclimation at 20°C.

As Hazel (1995) argues, homeoviscous adaptation is a powerful paradigm for explaining acclimation and adaptation in animals with variable body temperatures but cannot be the only explanation. Some animals become fully acclimated to temperature change with moderate or even no homeoviscous adaptation in lipid membrane properties. Altered expression of membrane proteins and proliferation of mitochondrial and sarcoplasmic reticular membranes, along with homeoviscous adaptation of membrane lipids, present a picture of cell membranes as dynamic structures that change in complex ways to retain their function despite temperature change.

Finally, there are also regional differences in lipid properties, including melting point, in some mammals. In the limbs, which may be subjected to near-freezing temperatures, the tissue lipids are less saturated and so have a lower melting point than the fats in the body core. At 37°C, the fats in the limbs are much "oilier" than the waxier fats of

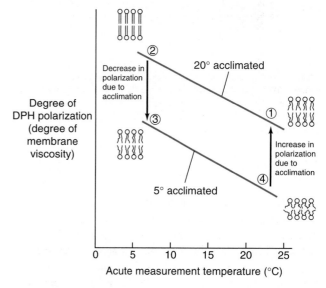

Figure 16-12 Homeoviscous adaptation maintains relatively constant membrane lipid properties in rainbow trout enterocytes. After an initial measurement at 25°C (point 1), a warm-acclimated (20°C) trout is rapidly cooled to 5°C. Initially, the membranes of its enterocytes become more polarized and more viscous (point 2), but as homeoviscous adaptation occurs the lipid membranes become less polarized and regain their fluidity (point 3). Similarly, if a cold-acclimated trout at a measurement temperature of 5°C is rapidly warmed to 25°C and measured, the lipid membranes are initially highly depolarized (point 4) but become more polarized with acclimation (point 1). [Adapted from Hazel, 1995.]

the warmer body regions. The low-viscosity oils extracted from the limbs of slaughtered cattle are marketed as neat's-foot oil, a penetrating leather preservative and softener.

> | ? | Intertidal animals in temperate climates can experience changes in ambient temperature of as much as 50 degrees (Celsius scale) in the summer as they are alternately covered by cool water and then exposed to the hot sun. How would the concept of thermal acclimation apply to these animals, and what type of physiological or biochemical adaptations or both would you expect them to have to help them cope?

Determinants of Body Heat and Temperature

The temperature of an animal depends on the amount of heat (calories) contained per unit mass of tissue (see Spotlight 16-2). Because tissues consist primarily of water, the **heat capacity** of tissues between 0°C and 40°C approximates $1.0 \ \text{cal} \cdot °C^{-1} \cdot g^{-1}$. It follows that the larger the animal, the greater its body heat content at a given temperature. The rate of change of body heat depends on (1) the rate of *heat production* through metabolic means, (2) the rate of external *heat gain*, and (3) the rate of *heat loss* to the environment (Figure 16-13). We can state that

body heat = heat produced + (heat gained − heat lost)
= heat produced + heat transferred

Thus, body heat, and hence the body temperature of an animal, can be regulated by changes in the rate of heat production and heat transfer or exchange (i.e., heat gained minus heat lost).

Numerous factors affect the rate of body heat production. Behavioral mechanisms such as simple exercise cause an increase in heat production by elevating metabolism. The activation of autonomic mechanisms leading to release of hormones can produce accelerated metabolism of energy reserves. Acclimatization mechanisms, which are slower than the other two processes, often lead to an elevation in basal metabolism and the associated heat production.

The total heat content of an animal is determined by the metabolic production of heat and the *thermal flux* between the animal and its terrestrial surroundings, as shown in Figure 16-13. The relation between these factors can be represented as

$$H_{tot} = H_v + H_c + H_r + H_e + H_s$$

in which H_{tot} is the total heat, H_v is the heat produced metabolically, H_c is the heat lost or gained by conduction and convection, H_r is the net heat transfer by radiation, H_e is the heat lost by evaporation, and H_s is the heat stored in the body. Heat leaving the animal has negative (−) value,

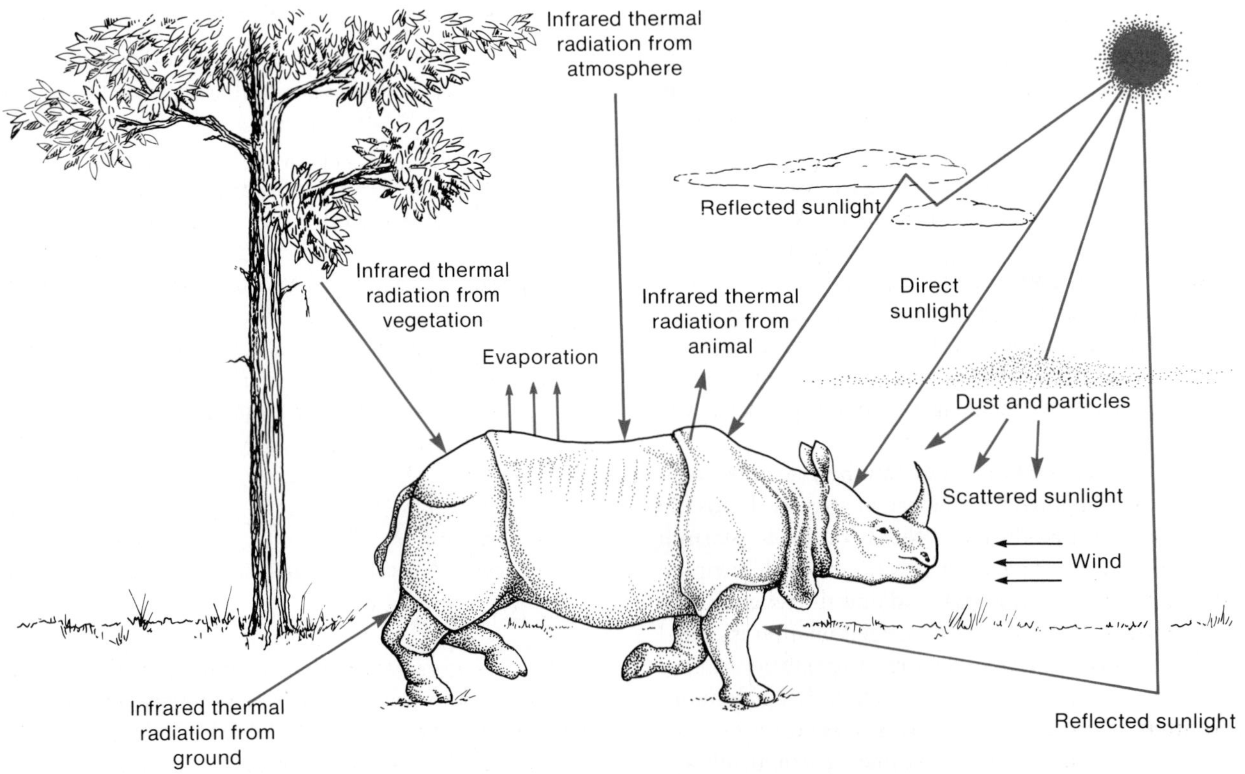

Figure 16-13 Heat is transferred between an animal and its environment in numerous ways. Infrared thermal radiation and direct and reflected sunlight transfer heat into the animal, whereas radiation and evaporation transfer heat out to the environment. [Adapted from Porter and Gates, 1969.]

whereas heat entering the body from the environment has positive (+) value. Animals can lose heat by conduction, convection, radiation and evaporation. Let us now consider each of these key terms.

Conduction The transfer of heat between objects and substances that are in contact with each other is conduction. It results from the direct transfer of kinetic energy of the motion from molecule to molecule, with the net flow of energy being from the warmer to the cooler region. The rate of heat transfer through a solid conductor of uniform properties can be expressed as

$$Q = \frac{kA^{(t_2 - t_1)}}{l} \qquad (16\text{-}12)$$

in which Q is the rate of heat transfer (in joules per centimeter per second) by conduction; k is the *thermal conductivity* of the conductor; A is the cross-sectional area (in square centimeters); and l is the distance (in centimeters) between points 1 and 2, which are at temperatures t_1 and t_2, respectively. Conduction is not limited to heat flow within a given substance; it may also be between two phases, such as the flow of heat from skin into the air or water in contact with the body surface.

Convection The transfer of heat contained in a mass of a gas or liquid by the movement of that mass is *convection*. Convection may result from an externally imposed flow (e.g., wind) or from the changes in density of the mass produced by heating or cooling of the gas or fluid. Convection can accelerate heat transfer by conduction between a solid and a fluid, because continuous replacement of the fluid (e.g., air, water, or blood) in contact with a solid of a different temperature maximizes the temperature difference between the two phases and thus facilitates the conductive transfer of heat between the solid and the fluid.

Radiation The transfer of heat by electromagnetic **radiation** takes place without direct contact between objects. All physical bodies at a temperature above absolute zero emit electromagnetic radiation in proportion to the fourth power of the absolute temperature of the surface. As an example of how radiation works, the sun's rays may warm a black body to a temperature well above the temperature of the air surrounding the body. A dark body both radiates and absorbs more strongly than does a more reflective body having a lower *emissivity*. For temperature differences between the surfaces of two bodies of about 20 Celsius degrees or less, the net radiant heat exchange is approximately proportional to the temperature difference.

Evaporation Every liquid has its own *latent heat of vaporization*, which is the amount of energy required to change it from a liquid to a gas of the same temperature—that is, to evaporate. The energy required to convert 1 g of water into water vapor is relatively high, approximately 585 cal.

Many animals dissipate heat by allowing water to be evaporated from body surfaces.

Heat storage Heat storage leads to an increase in temperature of the heat-storing mass. The larger the mass, or the higher its specific heat, the smaller its rise in temperature (in °C) for a given quantity of heat (in joules) absorbed. Thus, a large animal that has a small surface-to-mass ratio tends to heat up more slowly in response to an environmental heat load than does a small animal that has a relatively high surface-to-mass ratio. This follows from the simple fact that heat exchange with the environment must take place through the body surface.

The rate of heat transfer (kilocalories per hour) into or out of an animal also depends on several factors. Changing the value of any one of them alters heat flow across the body surface in the direction of the temperature gradient:

- *Surface area* per gram of tissue decreases with increases in body mass, providing small animals with a high heat flux per unit of body weight (as already noted). Animals can sometimes control their apparent surface area by changing posture (e.g., by extending limbs or drawing them close to the body).

- *Temperature difference* between the environment and the animal's body has a large effect by altering the *temperature gradient* (i.e., change in temperature per unit distance) for heat transfer. The closer an animal maintains its temperature to the ambient temperature, the less heat will flow into or out of its body.

- *Specific heat conductance* of the animal's surface varies with the nature of the body surface. Animals with high heat conductances in surface tissues are typically close to the temperature of their surroundings, with some exceptions, such as the elevation of body temperature when an animal basks in sunlight. Animals that actively maintain a constant body temperature (birds, mammals) have feathers, fur, or blubber that decrease the heat conductance of their body surfaces. An important feature of fur and feathers is that they trap and hold air, which has a very low thermal conductivity and therefore further retards the transfer of heat. Such insulation spreads out the temperature difference between the body core and the animal's surroundings over a distance of several millimeters or centimeters so that the temperature gradient is less steep and thus the rate of heat flow is reduced.

Most animals have body temperatures similar to those of their environments. Animals breathing water can maintain only parts of the body above ambient temperature because oxygen transfer is slower than heat transfer and water contains little oxygen but has a high specific heat, and so delivery of oxygen to the respiratory surface inevitably removes all heat produced by metabolism. Air-breathing animals, on the other hand, can obtain a sufficient amount of oxygen from a small volume of air and can heat that air to high temperature. They have heat "to spare" to raise body temperature. Air, unlike water, has a high O_2 content and a low specific heat. Thus, air-breathing animals can raise body temperature above ambient temperature, whereas water-breathing animals cannot. Some animals can keep parts of their bodies (e.g., dog legs, bird feet, tuna muscle, etc.) at different temperatures because of countercurrent heat exchangers (see next section, *Temperature classifications of animals*). Thus, water-breathing animals such as tuna can maintain regions of their bodies above ambient temperatures; but, each time the blood perfuses the respiratory surface, its temperature approaches ambient temperature because heat loss exceeds heat production. Air-breathing animals can have an elevated body temperature because it is theoretically possible for heat production to exceed heat loss. To maintain high body temperature requires a high rate of heat production because of obligatory heat loss through respiration.

Animals use several different mechanisms to regulate the exchange of heat between themselves and the environment:

- *Behavioral control* includes moving to a part of the environment where heat exchange with the environment favors attaining optimal body temperature. For instance, a desert ground squirrel retires to its burrow during the midday heat; a lizard suns itself to gain heat by radiation from its surroundings, raising its body temperature well above ambient temperature. Animals also control the amount of surface area available for heat exchange by adjusting their postures.

- *Autonomic control* of blood flow to the vertebrate skin affects the temperature gradient and, hence, the heat flux at the body surface (Figure 16-14). For example, the activation of piloerector muscles increases the extent of fluffing of pelage and plumage, which increases the effectiveness of insulation by increasing the amount of trapped, unstirred air (Figure 16-15A). Sweating and salivation during panting cause evaporative cooling.

- *Acclimatization* includes long-term changes in pelage or subdermal, fatty-layer insulation, as well as changes in the capacity for autonomic control of evaporative heat loss through sweating. Acclimatization can also include the capacity for metabolic heat generation, as in finches, for example.

Temperature Classifications of Animals

It should be clear to you now that animals deal with variation in the thermal characteristics of their environment in a variety of ways. The "traditional" scheme used by comparative physiologists to classify thermoregulatory modes of animals is based on the stability of body temperatures. When exposed to changing air or water temperatures in the laboratory, **homeotherms** (or **homoiotherms**) maintain

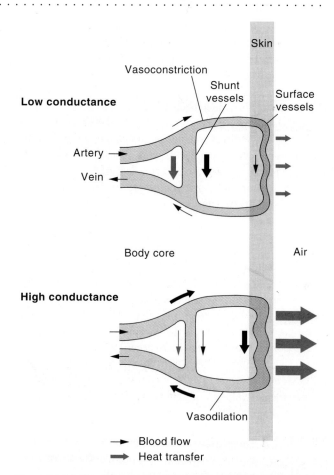

Figure 16-14 Blood flow to the skin helps regulate the heat conductance of the body surface. Vasomotor control of peripheral arterioles shunts the arterial blood either to the skin or away from it. In response to environmental cold, peripheral blood vessels vasoconstrict, shunting blood away from the surface of an endotherm. In response to high temperatures, the blood is diverted to the skin, where it approaches temperature equilibrium with the environment. In ectotherms, cutaneous blood flow is often increased through peripheral vasodilation to absorb heat from the environment.

body temperatures above ambient temperatures and regulate their body temperatures within a narrow physiological range by controlling heat production and heat loss (Figure 16-16). In most mammals, the normal physiological range for core body temperature is typically from 37°C to 38°C; whereas, in birds, it is closer to 40°C. Some vertebrates other than birds and mammals and some invertebrates also can control their body temperatures in this manner, although such control is often limited to periods of activity or rapid growth in these organisms.

Poikilotherms are those animals in which body temperature tends to fluctuate more or less with the ambient temperature when air or water temperatures are varied experimentally. The colloquial terms "warm blooded" for homeotherms and "cold blooded" for poikilotherms are unsatisfactory because many poikilotherms can become quite warm. For example, a locust sustaining flight in the equatorial sun or a lizard running across the sand at midday in a hot desert may have blood temperatures exceeding those of warm blooded mammals.

A

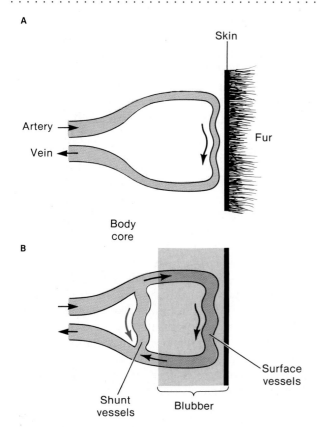

B

Figure 16-15 Fur and blubber act as heat insulation. **(A)** Fur is outside the skin and circulation, and its insulating properties can be changed rapidly only by flattening or fluffing through pilomotor control. **(B)** Because blubber is located under the skin and is supplied with blood vessels, its insulating value can be regulated by shunting the blood through vasomotor control to the surface or away from the surface below the blubber.

Early comparative physiologists considered all fishes, amphibians, reptiles, and invertebrates to be poikilotherms, because all of these animals were thought to lack the high rates of heat production found in birds and mammals. Several difficulties with the homeotherm-poikilotherm classification scheme became apparent with the completion of

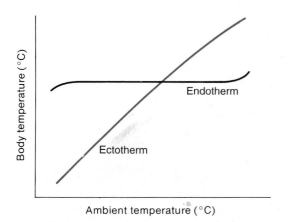

Figure 16-16 Homeotherms maintain body temperature as ambient temperature changes, whereas the body temperatures of poikilotherms more closely track ambient temperature.

more field studies (especially with the use of radio telemetry of body temperature). For example, some deep-sea fishes have more stable body temperatures than do many higher vertebrates because these fishes live in specialized environments that are thermally very stable. Many so-called poikilotherms (e.g., lizards) are able to regulate their body temperatures quite well in their natural surroundings by controlling heat exchange with their environment, although this ability is ultimately limited by the availability of heat in the environment. Moreover, numerous birds and mammals are now known to allow their body temperatures to vary widely, either regionally in the body or in the whole body over time.

These inconsistencies have led to a more widely applicable temperature classification scheme based on the *source* of body heat. In this scheme, *endothermic animals* generate their own heat, and *ectothermic animals* rely almost entirely on environmental sources of heat. (It should be emphasized that the concepts of homeothermy versus poikilothermy as well as endothermy versus ectothermy are idealized extremes, and most organisms are not at these extremes.)

Endotherms are animals that generate their own body heat through heat production as a by-product of metabolism, typically elevating their body temperatures considerably above ambient temperatures. Most produce heat metabolically at high rates, and many have relatively low thermal conductivity because of good insulation (fur, feathers, fat), which enables them to conserve heat in spite of a high temperature gradient between body and environment. Mammals and birds exemplify animals that regulate their temperatures within relatively narrow limits and are therefore said to be *homeothermic endotherms*. A few large fishes (sharks and larger tuna) and some flying insects are termed *regional heterothermic endotherms* because they maintain regions of their body above ambient temperatures, sometimes for short periods of time under specific circumstances, as in flying insects. Because endotherms (all birds and mammals plus many terrestrial reptiles and a number of insects) maintain their body temperatures well above ambient temperatures in cold climates, they have been able to invade habitats that are too cold for most ectotherms. Endotherms keep warm at considerable metabolic cost: the metabolic rate of an endotherm at rest is usually at least five times that of an ectotherm of equal size and body temperature.

Ectotherms produce metabolic heat at comparatively low rates—rates normally too low to allow for endothermy. Often, ectotherms have low rates of metabolic heat production and high **thermal conductances**—that is, they are poorly insulated. As a result, heat derived from metabolic processes is quickly lost to cooler surroundings. Accordingly, heat exchange with the environment is much more important than metabolic heat production in determining an ectotherm's body temperature. On the other hand, the high thermal conductance allows ectotherms to absorb heat readily from their surroundings. *Behavioral temperature regulation* is the principal means by which ectotherms regulate their body temperatures. (That reptiles regulate body temperature is well known). Behavioral temperature regulation can be demonstrated in the laboratory by placing animals in thermal gradients and monitoring the preferred body temperatures. Alternatively, animals can be placed in a "shuttle box," which consists of one chamber that is well below the preferred body temperature and a connected chamber that is well above the preferred body temperature. The animal will "shuttle" back and forth, keeping its body temperature at a level between those of the two thermal environments.

Field observations combined with radio telemetry of body temperature indicate the considerable thermoregulatory ability of reptiles through behavioral as well as physiological means (e.g., shunting blood to the skin to cool or warm). Many ectotherms needing to change body temperature behave in a way that facilitates heat absorption from the environment or helps the animal unload heat to the environment (or minimizes heat uptake from the environment). A lizard or a snake may bask in the sun with its body oriented for maximal warming until it achieves a temperature suitable for efficient muscular function. Small ectotherms in hot environments (lizards, ants) often elevate their bodies to the extent that their legs allow to avoid the hottest temperatures immediately adjacent to the sand or rock over which they are moving.

In general, the most effective thermoregulatory action taken by ectotherms is movement into a suitable **microclimate** in the environment. A burrow under a rock, for example, is often much more moderate in temperature than the surface temperature (Figure 16-17). Intertidal zones in the tropics often appear to be devoid of invertebrate life during the heat of the day, but that same habitat may be teeming with life at night as animals emerge from their microclimates underneath rocks and in burrows.

Heterotherms are those animals capable of varying degrees of endothermic heat production, but they generally do not regulate body temperature within a narrow range. They may be divided into two groups, *regional* and *temporal* heterotherms. *Temporal heterotherms* constitute a broad category of animals whose temperatures vary widely over time. Monotremes (egg-laying mammals) such as the echidna are temporal heterotherms (Figure 16-18), as are other mammals and birds in torpor and hibernation. Temporal heterothermy is also shown by many flying insects, pythons, and some fishes, which can raise the temperature of their bodies (or regions of their bodies, as in the fishes) well above ambient temperature by virtue of heat generated as a by-product of intense muscular activity. Some insects prepare for flight by exercising their flight muscles for a time to raise their temperatures before takeoff.

Some species of small mammals and birds have accurate temperature control mechanisms and so are basically homeothermic. Yet, they behave like temporal heterotherms because they allow their body temperatures to undergo daily cyclical fluctuations, having endothermic temperatures during periods of activity and lower temperatures during periods of rest. In hot environments, this flexibility gives certain large animals, such as camels, the ability to absorb great quantities of heat during the day and to give it off again during the cooler night. Certain tiny endotherms, such as hummingbirds, must eat frequently to support their

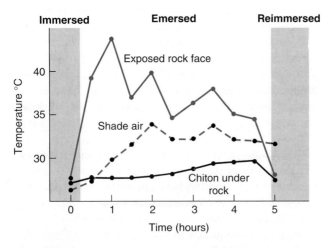

Figure 16-17 Microclimates under rocks afford ectotherms protection from harsh thermal environments. The tropical chiton *Chiton stokesii* can live in an intertidal zone that reaches lethally high temperatures during the day by seeking much cooler microclimates under rocks. Temperatures in this illustration were recorded from the exposed face of the chiton-bearing rock, in shaded air, and in the space beneath the attached foot of a chiton hidden under a rock. [Adapted from McMahon et al., 1991.]

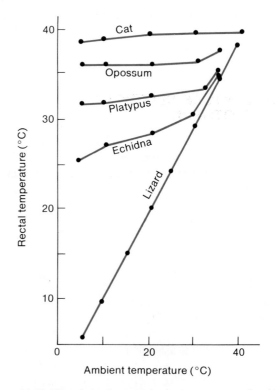

Figure 16-18 The relation between body temperature and ambient temperature differs in various animals. The cat is a strict homeotherm, maintaining body temperature independent of ambient temperature, whereas monotremes (platypus and echidna) are temporal heterotherms. The lizard is a strict heterotherm. [Adapted from Marshall and Hughes, 1980.]

high daytime metabolic rate. To avoid running out of energy stores at night when they cannot feed, they enter into a state of **torpor** during which they allow metabolic rate to decline and body temperature to drop toward ambient temperature. Even some large endotherms resort to a long winter torpor with reduced body temperature to save energy (see *Hibernation and winter sleep* later in this chapter).

Regional heterotherms are generally ectotherms that can achieve high *core* (i.e., deep-tissue) *temperatures* through muscular activity, while their peripheral tissues and extremities approach the ambient temperature. As mentioned earlier, examples include mako sharks, tuna, and many flying insects. Elevated temperatures generally allow higher metabolic rates than would be achieved at ambient environmental temperatures. Fishes that are regional heterotherms depend on *countercurrent heat exchangers*. Heat is conserved in the body core by a specialized parallel arrangement of incoming arteries and outgoing veins (see *Temperature relations of heterotherms*) that, in the case of heat exchangers, facilitates heat transfer between blood vessels and retains heat in the body core. Some large billfin fishes (e.g., marlin) use specialized ocular muscle called "heater tissue" to elevate brain temperature (see *Thermogenesis* later in this chapter). Another special example of regional heterothermy is seen in the scrotums of some mammals, including canines, cattle, and human beings, which hold the testes outside the body core to keep them at a slightly lower temperature. The scrotum shortens in cool air, drawing the testes against the warmer body, and lengthens as its temperature rises. These actions regulate testicular temperature and in particular prevent overheating of the testes, which has a harmful effect on sperm production.

TEMPERATURE RELATIONS OF ECTOTHERMS

Ectotherms occupy a wide variety of environments—both hot and cold. A few very specialized environments have highly stable temperatures, varying no more than a degree or two throughout the year. Examples are the shallow marine waters under the Arctic and Antarctic ice, the deep regions of the seas, the air deep within the interior of many caves, and the microenvironments within deep groundwater. Normally, however, almost all environments show significant long- or short-term variation in temperature. Such variation is maximized in terrestrial temperate environments, some of which may have daytime summer surface temperatures of nearly 40°C and nighttime winter temperatures of −40°C. Most animals occupy microhabitats with less extreme temperature fluctuations. However, some degree of thermal stress is inherent in living in most environments, and a wide variety of mechanisms have evolved through natural selection to sustain animal life.

Ectotherms in Freezing and Cold Environments

Because the body temperature of many ectotherms depends to a considerable extent on the ambient temperature, freezing is a threat to those species living in environments with ambient temperatures below freezing. The formation of ice crystals within cells is usually lethal because, as the crystals grow in size, they rupture and destroy the cells. No animal is known to survive complete freezing of its tissue water, but some come close. Certain beetles can withstand freezing temperature because the extracellular fluid contains a substance that accelerates nucleation (the process of crystal formation). Consequently, the extracellular fluids freeze more readily than the intracellular fluids. As ice forms in the extracellular fluid, the unfrozen extracellular fluid becomes more concentrated with solutes. This process draws water out of the cells, lowering the intracellular freezing point. As the temperature drops further, the process continues and produces further depression of the freezing point of the remaining intracellular water. The freshwater larvae of the midge *Chironomus*, which survive repeated freezing, yield some unfrozen liquid at temperatures as low as −32°C. If ice crystals form and grow within cells, they damage the tissue by breaking the cells. In contrast, ice crystals that form outside the cells do little damage. The adaptiveness of this process therefore lies in crystal formation in the extracellular space where little tissue damage is caused. Red blood cells, yeast, sperm, and other cell types also can withstand freezing damage, provided intracellular ion concentrations do not rise above those levels that cause damage to cell organelles. As K. B. Storey and J. M. Storey and their colleagues (1992) have found, a few vertebrates, primarily anuran amphibians, also can withstand freezing. Both ice-nucleating proteins that initiate and control the formation of extracellular ice and cryoprotectants ("antifreezes") are employed to survive freezing environments.

Some animals can undergo *supercooling*, in which the body fluids can be cooled to below their freezing temperature, yet remain unfrozen because ice crystals fail to form. Ice crystals will not form if they have no nuclei (mechanical "seeds," so to speak) to initiate crystal formation. Thus, certain fishes dwelling at the bottom of Arctic fjords live in a continually supercooled state and normally do not freeze. If they brush against frozen ice on the water surface, however, ice crystallizes rapidly throughout their bodies and they die immediately. Thus, survival for these fish depends on their remaining well below the surface where ice is absent.

The body fluids of some cold-climate ectotherms contain *antifreeze substances*. For example, the body fluids of a number of arthropods, including mites and various insects, contain glycerol, the concentration of which typically increases in the winter. Glycerol, acting as an antifreeze solute, lowers the freezing point to as low as −17°C. The tissues of larvae of the parasitic wasp *Brachon cephi* can withstand even lower temperatures; they have been supercooled to −47°C without ice crystal formation. The blood of the antarctic ice fish *Trematomus* contains a glycoprotein antifreeze that is from 200 to 500 times as effective as an equivalent concentration of sodium chloride in preventing ice crystal formation. The glycoprotein lowers the

Glycerol acts as an anti-freeze

Critical thermal maxima relate to physiological process

temperature at which the ice crystals enlarge, but it does not lower the temperature at which they melt.

For many animals living in cool (but not freezing) environments, survival requires maintaining adequate metabolism at the very low levels of enzyme activity characteristic of low temperatures. Many animals living in cold environments have enzymes that show maximal activity at temperatures many degrees below those of homologous enzymes in animals living in warmer environments. Figure 16-19 shows strong evidence of thermal adaptation in the Michaelis-Menten constant (K_m) of pyruvate for A$_4$-lactate dehydrogenase. The K_m value in *Termatomus centronotus*, a fish that lives in water that is almost always at $-1.9°C$, is far higher than that in fishes and other vertebrates that inhabit more eurythermic environments. Even within a single species (barracuda), fish that inhabit cooler waters have a K_m for pyruvate that is higher than that in individuals of the same species that inhabit warmer waters, allowing the cooler fish to maintain a relatively higher metabolism for their body temperature than that of the warmer fish.

Ectotherms in Warm and Hot Environments

Heat exchange with the environment is closely related to body surface area, so the temperature of a small ectotherm (which has a relatively large surface area) rises and falls rapidly as environmental temperatures undergo daily fluctuations. All ectotherms have a **critical thermal maximum**, a temperature above which long-term survival is not possible. Generally, this is determined by measuring the temperature at which 50% mortality occurs in a population.

The critical thermal maximum varies enormously, depending on the organism. Some thermophilic bacteria can thrive at temperatures above 90°C, although almost all metazoans have critical thermal maxima below 45°C. The physiological causes for a critical thermal maximum are varied. An *ultimate upper limit* is the temperature at which proteins are denatured, though proteins such as enzymes usually fail to function at levels well below the temperature

ultimate upper limit

of denaturation. Often, critical thermal maxima relate to a breakdown in some critical physiological process. For example, most tissue functions in many ectotherms are handicapped by a decreased affinity of the respiratory pigment for oxygen at the upper limit of body temperature. At 50°C, the blood of a chuckwalla *(Sauromalus)* cannot achieve more than 50% O$_2$ saturation of arterial blood, which prevents vigorous activity by the animal. At slightly lower temperatures (47–48°C) that prevent higher arterial oxygen saturation levels, the desert iguana *Dipsosaurus* continues to be active. Above 43°C, the iguana pants, much as a dog does, to increase heat loss through evaporative cooling.

Most ectotherms never experience such extreme temperatures. Yet, even in temperate climates, many experience general environmental temperatures that are high enough to require active responses to prevent unacceptably elevated body temperatures. Many ectotherms expose their bodies to sun or to shade to absorb more or less heat, respectively, from the environment. The effectiveness of such *behavioral thermoregulation* is enhanced by the high heat conductance of ectotherms. Certain reptiles, however, additionally use nonbehavioral (i.e., physiological) means to control the rates at which their bodies heat and cool. For example, the diving Galapagos marine iguana *Amblyrhynchus* (Figure 16-20) can permit its body temperature to rise at about twice the rate at which it drops by regulating both heart rate and flow of blood to its surface tissues. When the iguana wishes to warm up, it basks in the sun and simultaneously diverts cooler core blood to the surface (see Figure 16-20A). The net effect is a large difference between body and environmental temperature. The increased blood flow increases the skin's heat conductance and speeds absorption of heat into the animal. Increased pumping of blood accelerates the removal of heat from surface tissues to deeper tissues. During the iguana's prolonged feeding dives in the cool ocean, loss of body heat is slowed by a reduction in blood flow to the surface tissues and by a general

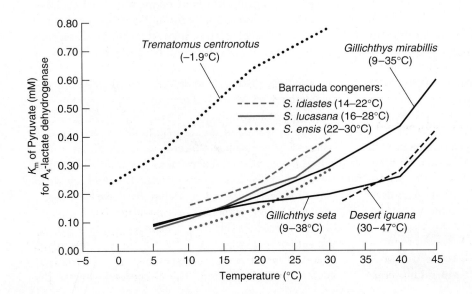

Figure 16-19 Higher K_ms of pyruvate for A$_4$-lactate dehydrogenase have been selected for animals living in colder environments. This relation holds both between species (*Trematomus* vs. *Gillichthys*) and within species (barracuda). [Adapted from Somero, 1995.]

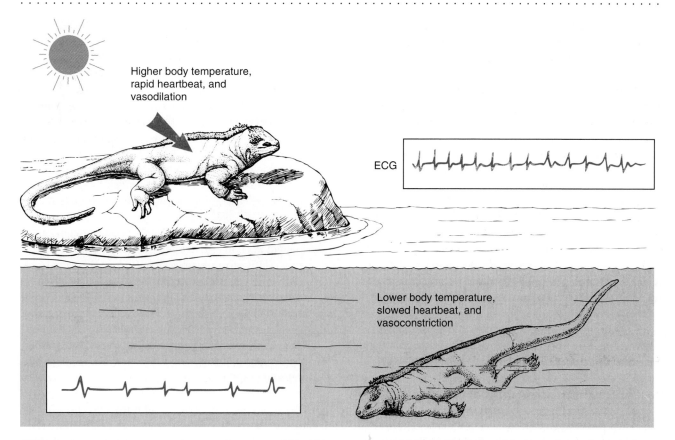

Figure 16-20 The Galapagos marine iguana heats and cools at different rates, indicating an active regulation of heat exchange with its environments. **(A)** On land, the basking iguana absorbs heat from the sun's rays. Vasodilation of cutaneous blood vessels and rapid heartbeat (as recorded in the electrocardiogram, ECG) assure heating of the blood and efficient circulation, which quickly distributes the heat throughout the body. Underwater heat loss is retarded by a slowed heartbeat and vasoconstriction in cutaneous blood vessels, both of which minimize the flow of blood to the skin.

slowing of the circulation. This is apparent in experiments that demonstrate a hysteresis (asymmetrical response) of the heart rate relative to body temperature during a rise and fall in body temperature (Figure 16-21). Essential physical concepts underlying this tactic include not only the difference between the rate of heat convection and the rate of heat conduction, but also the differing heat capacities of air and water. Because of its far higher heat capacity, water can remove heat by conduction from the surface of the iguana much faster than air can; thus, it is especially important that circulation to the skin be slowed during diving.

Similar dissimilarities between rates of heating and cooling, which indicate active regulatory processes, have been observed in amphibians and arthropods.

Costs and Benefits of Ectothermy: A Comparison with Endothermy

Early comparative physiologists assumed that ectothermy was inferior to endothermy as a way of life. The endothermal vertebrates (primarily birds and mammals) were viewed as more complex recent evolutionary arrivals than the primarily ectothermal "lower vertebrates" (fishes, amphibians, and lizards). More recently, however, the term "lower vertebrates" has fallen out of favor as we realize that they are as highly adapted for their way of life as are birds and mammals for theirs. Indeed, the endotherms and ectotherms do represent different life-styles, the former representing a fast, high-energy way of life and the latter representing a slower, low-energy approach. Many of the

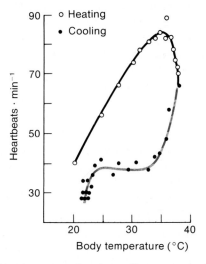

Figure 16-21 A hysteresis in the relation of heart rate to body temperature during heating, and then cooling, in water can be demonstrated in the marine iguana. The heart rate rises steeply during heating but drops back still more steeply with cooling. [From Bartholomew and Lasiewski, 1965.]

anatomical and functional properties of ectothermic vertebrates are adaptations facilitating a life of modest energy requirements. Those modest requirements allow some reptiles, amphibians, and fishes, for example, to exploit terrestrial niches unavailable to birds and mammals. Small salamanders with low metabolic rates live in great abundance in the cool litter on forest floors of New England. The total biomass of such salamanders is estimated to exceed that of the forest's birds and mammals. Body size is often a critical factor in the advantages of ectothermy in certain environments. Because few ectotherms elevate their body temperatures above ambient temperatures, they do not experience the increased loss of body heat that occurs with decreasing body size (resulting from increased surface-to-volume ratio). Thus, ectotherms can function with much smaller body masses than endotherms. Shrews and hummingbirds are unusually small endotherms, but many ectotherms, such as frogs and salamanders and most invertebrates, are much smaller.

Thus, in considering the "costs and benefits" of ectothermy relative to endothermy in animals, the following generalizations can be made:

1. Because their body temperatures are generally closer to ambient temperatures, ectotherms generally live at a lower metabolic rate. As a consequence, ectotherms can "invest" a larger proportion of their "energy budget" in growth and reproduction. Ectotherms require less food, and so they can spend less time foraging and more time quietly avoiding predators. They also need less water, because they lose less by evaporation from their typically cooler surfaces, and they need not be massive for the purpose of reducing surface-to-volume ratio.

2. The benefits in item 1 are balanced by certain costs, among which is the inability of ectotherms to regulate their body temperatures (unless their environments permit behavioral thermoregulation). For example, a lizard can elevate its temperature by basking only if there is sufficient solar radiation, which limits the times of day and seasons of the year when such activity is possible. Other costs of ectothermy are that a low rate of aerobic metabolism limits the duration of bursts of high activity and an oxygen debt develops during anaerobic respiration. Such factors have been evoked to argue that large dinosaurs must have been endotherms.

3. The biological costs and benefits of being an endotherm are the inverse of those of being an ectotherm. Because of their high rates of aerobic respiration and their elevated temperatures, endotherms maintain elevated body temperatures and can generally sustain longer periods of intense activity. Thus, the endotherms can be thought of as "high-rolling big spenders" in energetic terms compared with the energetically more modest ectotherms, which are characterized by lower energy intake and lower energy expenditure. Another advantage of endothermy is that the constancy of body temperature allows enzymes to function more efficiently over a relatively narrow range.

4. Endothermic animals can do certain things on a bigger, faster scale, but they do so at a price. The field metabolic rates (daily costs of survival in their natural habitats) of endotherms are as much as 17 times as high as the field metabolic rates of ectotherms. The price paid by the endotherms for their high metabolic rate includes the requirement that they take in much larger amounts of food and water daily. Thus, a 300 g rodent needs 17 times as much food per day as does a 300 g lizard living in the same habitat and having the same diet of insects. The high rate of respiratory gas exchange makes endotherms susceptible to dehydration in hot, dry climates. High body temperature relative to ambient temperature makes very small body mass problematic because of surface-to-volume considerations that cause a small animal to lose heat faster than a larger one. Because such a large quantity of energy is consumed by an endotherm to elevate and maintain body temperature, only a relatively small percentage of energy can be budgeted for growth and reproduction.

It is apparent, then, that ectothermy and endothermy constitute a metabolic dichotomy affecting far more than just body temperature. Indeed, the implications of these two types of energy economies also extend to such areas as activity, physiology, behavior, and evolution. Both ectothermy and endothermy have their respective advantages and disadvantages. Ectothermy is always mechanistically less complex than endothermy. Some thermoregulating terrestrial ectotherms are capable of regulating their body temperatures with precision and at levels as much as 30 Celsius degrees higher than the air temperature. Whereas endotherms typically maintain a relatively constant temperature set point, some thermoregulating ectotherms can vary the temperature set point, depending on activity requirements; such regulation allows body temperature to fall during periods of rest and rise prior to activity, as in the basking behavior of lizards. This has the advantage of fuel economy, very much like lowering and raising the thermostatic temperature setting in a home according to the temperature needs of its inhabitants.

Endothermy and ectothermy also offer animals differing advantages in different climates. In the tropics, ectotherms such as reptiles compete successfully with, or even outcompete, mammals both in the abundance of species and in numbers of individuals. This competitive success is thought to be due in part to (1) the warm, relatively even tropical climate, allowing reptiles to expand into nocturnal activity rhythms, whereas tropical mammals tend to be diurnal in habit; and (2) the greater energy economy enjoyed by the ectotherms, because they need not expend energy to elevate body temperature. The metabolic energy thus saved by tropical ectotherms can be diverted to reproduction and to other uses that promote species survival. In moderate and cold climates, ectotherms are necessarily more sluggish, are thus less competent as predators, and are generally less abundant than mammals are in those climates. Endotherms have a significant competitive edge over ectotherms in the cold because their tissues are kept warm. Generally speaking, the farther from the equator, the higher the prevalence of terrestrial endothermy. In polar regions, for example, there are no reptiles and almost no insects, and only a few genera of amphibians and insects occupy subpolar arctic environments.

Almost all mammals are endothermic, with the exception of hibernators, marsupials (e.g., opossums), and the monotremes (e.g., the echidna), which are heterothermic. Without concern for whether these heterothermic mammals evolved from endothermic ancestors or have retained the original ancestral condition, what environmental selection pressures lead to the *persistence* of heterothermy in marsupials and monotremes?

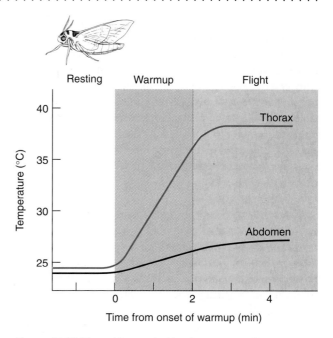

Figure 16-22 The sphinx moth *Manduca sexta* undergoes a preflight thermogenesis. Shivering of the thoracic flight muscles causes a steep increase in thoracic temperature prior to flight [Adapted from Heinrich, 1974.]

TEMPERATURE RELATIONS OF HETEROTHERMS

Heterotherms are animals intermediate between pure ectotherms and endotherms. As mentioned earlier in this chapter (in *Temperature classifications of animals*), certain insects and fishes are heterotherms. Some flying insects, including locusts, beetles, cicadas, and arctic flies, can be considered both temporal and regional heterotherms because, when preparing to fly, they raise the core temperatures in their thoracic parts to more or less regulated levels. At moderate ambient temperatures, these insects are unable to take off and fly without prior warmup because their flight muscles contract too slowly to produce sufficient power for flight at temperatures much below 40°C. When inactive, however, these insects behave strictly as ectotherms. After such an insect is aloft, its flight muscles produce enough heat to maintain elevated muscle temperatures, and the insect even employs heat-dissipating mechanisms to prevent overheating. These flying insects generally have quite a large mass; and some, such as bumblebees, butterflies, and moths, are covered with heat-insulating "hairs" or scales. To warm up, these insects activate their large thoracic flight muscles, which are among the most metabolically active tissues known. The activation is such that antagonistic muscles work against one another, producing heat without much wing movement other than small, rapid vibrations akin to *shivering*. Flight is finally initiated when the thoracic temperature has reached the temperature that is maintained during flight, about 40°C (Figure 16-22).

Like endotherms in general, endothermic flying insects face problems of regulating their body temperature when their environments have large temperature gradients. At am-

bient temperatures approaching 0°C, convective heat loss is generally so rapid that flight temperatures cannot be maintained. High ambient temperatures, on the other hand, place the insect in danger of overheating. Thus, at ambient temperatures above 20°C, the hovering sphinx moth *Manduca sexta* prevents thoracic overheating by regulating the flow of warm blood to the abdomen. The flow of heat from the active thorax to the relatively inactive and poorly insulated abdomen increases the loss of heat to the environment through the body surface and especially through the tracheal system.

An interesting and somewhat unusual example of true shivering thermogenesis in an insect is found in honeybee swarms, in which individual honeybees regulate the swarm's core temperature by muscle contraction in the form of shivering movements together with alterations in swarm structure. At low ambient temperatures (e.g., 5°C), the swarm packs more tightly, restricting the free flow of air into and out of the swarm to that needed for respiration. Through shivering activity, the core of the swarm can be maintained as high as 35°C. In warm weather, in contrast, the swarm loosens, providing ventilatory passages for air flow so that the core temperature exceeds the outside temperature by only a few degrees.

Another example of muscle-generated heat in a heterothermal species is found in the brooding female Indian python as it elevates its body temperature with shivering thermogenesis so as to provide warmth for the group of eggs around which it coils itself. In the laboratory, the rate of occurrence of muscle contractions was found to increase with declining ambient temperature, and the increase in contractions was accompanied by an increased difference between the ambient and body temperatures.

Unlike terrestrial ectothermic vertebrates, which can bask in the sun to warm up, marine ectotherms cannot obtain radiant energy as a source of underwater heat, owing to the high infrared absorption of water. As a result, fishes can rise above ambient temperature only through intensive metabolic activity. Many teleosts are strictly ectothermic, operating at core temperatures close to ambient temperatures. However, as already mentioned, some fishes, such as tunas, have specializations for generating and retaining sufficient heat to raise the temperature of body muscle, brain, and eyes some 10 degrees or more above their surroundings. These fishes can therefore be classified as regional heterotherms. The large mass (and hence small surface-to-volume ratio) of some of these fishes helps them attain a relatively constant muscle temperature. In these fishes, the retention of heat in the body core depends crucially on the organization of the vascular system. Unlike ectothermic fishes, which have a centrally located aorta and postcardinal vein (Figure 16-23A), heterothermic fishes, (e.g., tunas and lamnid sharks such as the mako) have

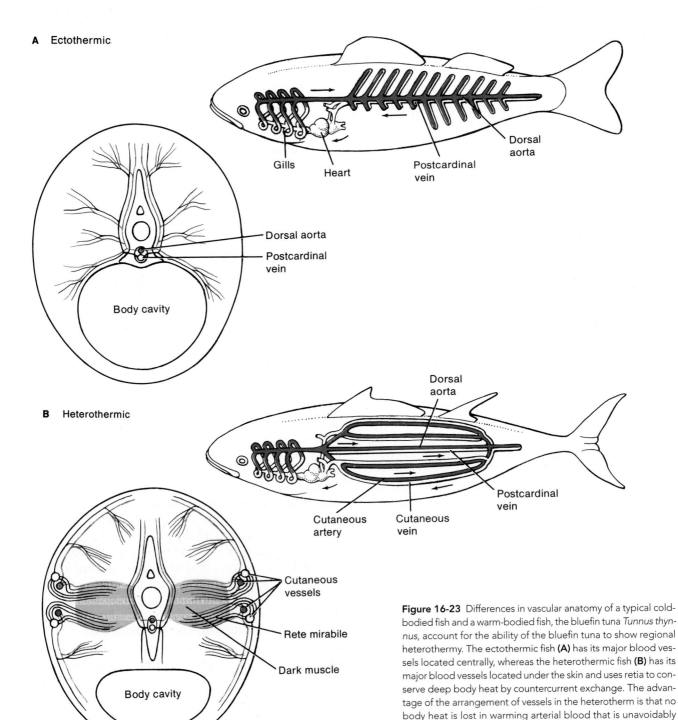

Figure 16-23 Differences in vascular anatomy of a typical cold-bodied fish and a warm-bodied fish, the bluefin tuna *Tunnus thynnus*, account for the ability of the bluefin tuna to show regional heterothermy. The ectothermic fish **(A)** has its major blood vessels located centrally, whereas the heterothermic fish **(B)** has its major blood vessels located under the skin and uses retia to conserve deep body heat by countercurrent exchange. The advantage of the arrangement of vessels in the heterotherm is that no body heat is lost in warming arterial blood that is unavoidably cooled while passing through the gills. [Adapted from F. G. Carey, 1973.]

major blood vessels (lateral cutaneous arteries and veins) located under the skin (Figure 16-23B). Blood is delivered to the deep red muscles through a *rete mira-bile* that acts as a heat-exchange system (Figure 16-24). Arterial blood, which is rapidly and unavoidably cooled during passage through the extensively perfused respiratory tissues of the gills and through the surface vessels, passes from the cool periphery into the warm deeper muscle tissue through a rete of fine arteries that intermingle with small veins carrying warm blood away from the muscles. This constitutes a **countercurrent heat-exchange system** such that the cool arterial blood passing from the surface toward the core picks up heat from the warm

venous blood leaving the muscle tissue and passing toward the periphery. This permits the retention of heat in the deep red-muscle tissue and minimizes heat loss to the surroundings.

Two anatomical features permit heterothermic fishes to maintain their swimming muscles at a temperature suitable for vigorous muscular activity while the temperature of the surface tissues approaches that of the surrounding water. First, the red (dark) aerobic swimming muscles are located relatively deep in the core of a fish's body. Second, escape to the periphery (skin, gills, etc.) of the heat produced during muscular activity is retarded by the countercurrent arrangement of the vessels composing the rete.

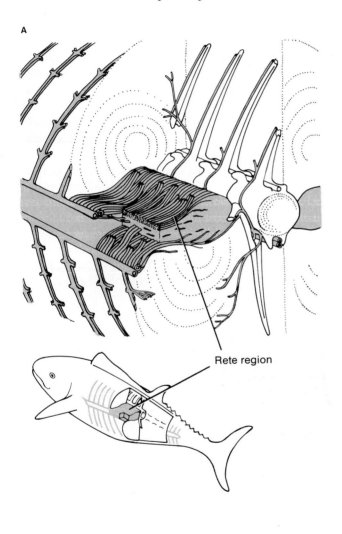

A

Rete region

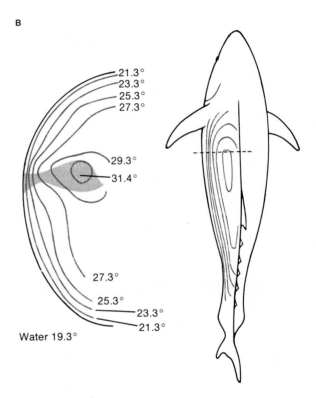

B

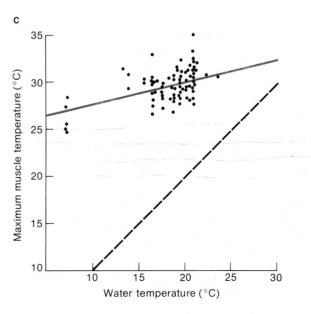

C

Figure 16-24 The bluefin tuna controls regional body temperature with a countercurrent arterial-venous heat-exchange rete. The rete (shown in color) helps the tuna retain heat produced in active deep muscles. **(A)** Enlargement of the rete area. **(B)** Isotherms *(left)*, plotted at 2 Celsius degree intervals, show temperature distribution in cross section *(right)*. **(C)** Maximum muscle temperatures of bluefins caught in waters of different temperatures. The dashed line indicates equality between body temperatures and water temperatures. [From Carey and Teal, 1966.]

Another important factor is that these regional heterotherms swim continuously, so the red muscle never cools down to the ambient temperature. One of the implications of regional heterothermy is the energy savings in the cool tissues while the temperatures of only certain tissues, such as the swimming muscles, are being elevated.

TEMPERATURE RELATIONS OF ENDOTHERMS

In homeothermic endotherms (most mammals and birds), body temperature is closely regulated by homeostatic mechanisms that regulate rates of heat production and heat loss so as to maintain a relatively constant body temperature independent of environmental temperatures. As mentioned earlier, core temperatures are maintained nearly constant between 37°C and 38°C in mammals and about 40°C in birds. The temperatures of peripheral tissues and extremities are held less constant and are sometimes allowed to approach environmental temperatures. Basal heat production for different homeotherms of a given size is about the same, and the basal metabolic rate can be 10 times as high as the standard metabolic rate of ectotherms of comparable size measured at similar body temperatures. This elevated basal metabolism, in conjunction with heat-conserving and heat-dissipating mechanisms, allows homeotherms to maintain constant body temperatures as much as 30 Celsius degrees or more above ambient temperatures.

Mechanisms for Body Temperature Regulation

Endotherms use a wide variety of both physiological and behavioral mechanisms to maintain body temperature within a narrow range. Before these mechanisms are considered, however, the concept of thermal neutral zone must be introduced.

Thermal neutral zone

The degree of thermoregulatory activity that homeotherms require to maintain a constant core temperature increases with increasing extremes of environmental temperature. At moderate temperatures, the basal rate of heat production balances heat loss to the environment. Within this range of temperatures, termed the **thermal neutral zone** (Figure 16-25), an endotherm does not need to expend energy to maintain its body temperature; it can regulate its body temperature by adjusting the rate of heat loss through alterations in the thermal conductance of the body surface. These adjustments include vasomotor responses (see Figures 16-14 and 16-15), postural changes to alter exposed areas of surface, and regulation of the insulating effectiveness of the pelage by raising or lowering hairs or feathers. Thus, within this range, fur or feathers are fluffed by pilomotor muscles in the skin to provide a thicker layer of stagnant air; at the upper end of this range, fur or feathers are held closer to the skin. The "goose bumps" of human beings are a vestige of the pilomotor control of a long-lost pelage.

As the ambient temperature decreases, an endotherm will eventually reach its *lower critical temperature* (LCT; see Figure 16-25), below which the basal metabolic rate becomes insufficient to balance heat loss despite these manifold adjustments in thermal conductance. Below this temperature, an endotherm must increase heat production above basal levels to offset heat loss (i.e., by thermogenesis, described in the next subsection). Heat production then rises linearly with decreasing temperature below the lower critical temperature, in what is termed the *zone of metabolic regulation* (see Figure 16-25). If the environmental temperature drops below the zone of metabolic regulation, compensating mechanisms fail, the body cools, and the metabolic rate drops. Many animals tolerate varying degrees of **hypothermia** during their normal rest period (including human beings during sleep). However, if an animal's body temperature falls below its normal values, the animal enters a state of hypothermia (see Figure 16-25). If this condition persists, the animal grows progressively cooler and, because cooling only further lowers the metabolic rate, the animal soon dies.

The thermal neutral zone lies entirely below the normal body temperature, T_b (37–40°C), as shown in Figure 16-25.

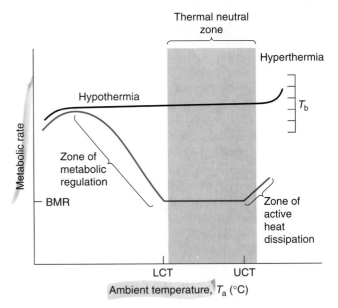

Figure 16-25 The resting metabolic rate of an endothermic homeotherm (red plot) is higher at extremes of ambient temperature. The thermal neutral zone extends from the lower (LCT) to the upper critical temperature (UCT). Above and below this range, the metabolic rate must rise to either increase thermogenesis in the zone of metabolic regulation or increase active dissipation of heat by evaporative cooling if body temperature, T_b (black plot) is to remain essentially constant. Within the thermal neutral zone, body temperature is regulated entirely by changing the heat conductance of the body surface, which requires essentially no change in metabolic effort. At ambient temperatures below the LCT, thermogenesis is unable to replace body heat at the rate at which it is lost to the environment, and hypothermia sets in. At ambient temperatures above the UCT, heat production and gain exceed the rate of heat loss, and hyperthermia occurs.

To see why, consider that heat loss by passive mechanisms cannot be increased further beyond the *upper critical temperature,* because the surface insulation is minimal (i.e., it cannot be made lower) at that temperature. Any further increase in ambient temperature, T_a, above that temperature will therefore cause a rise in body temperature, unless active heat-dissipating mechanisms such as sweating or panting are brought into play. Without evaporative heat loss, temperatures above the thermal neutral zone lead to **hyperthermia**, because the heat produced by basal metabolism cannot escape passively from the body as fast as it is being produced. Regardless of ambient temperature, a living animal is continuously producing some heat and, unless this heat is dissipated, the body temperature must continue to rise. (Sauna and hot-tub enthusiasts should bear this in mind.)

Why does the metabolic rate rise linearly with temperature below the lower critical temperature, along a line that extrapolates to zero at an ambient temperature equal to body temperature, as shown in Figure 16-25? This is explained by considering **Fourier's law of heat flow:**

$$Q = C(T_b - T_a) \qquad (16\text{-}13)$$

in which Q is the rate of heat loss from the body (in calories per minute) and C is the thermal conductance (see Spotlight 16-2). Because T_b is constant, Q varies linearly with the ambient temperature. The thermal conductance determines the slope of the plot below the neutral zone; the better the insulation (i.e., the lower C is), the shallower the slope and the less heat must be produced metabolically at low temperatures.

The extrapolated intercept with zero is at T_b because, if $T_a = T_b$, $C(T_b - T_a) = 0$. With $Q = 0$, there is no net heat loss. We know that the metabolic rate does not normally drop below the basal metabolic rate. When $T_a = T_b$, body temperature must be above the neutral zone because there

is no gradient for heat loss, so the animal will tend to warm up. The animal must cool by some means other than heat conduction. The only means of cooling when T_a lies above the upper critical temperature is by evaporation.

Thermogenesis

When the ambient temperature drops below the lower critical temperature, an endothermic animal responds by generating large amounts of additional heat from energy stores, thereby preventing a decrease in the core temperature. There are two primary means of extra heat production other than exercise: shivering and nonshivering **thermogenesis**. Both processes convert chemical energy into heat by a normal energy-converting metabolic mechanism that is adapted to primarily produce heat. Essentially all the chemical-bond energy released in this process is fully degraded to heat rather than to chemical or mechanical work.

Shivering is a means of using muscle contraction to liberate heat. Shivering thermogenesis occurs in some insects as well as in endothermic vertebrates. The nervous system activates groups of antagonistic skeletal muscles so that there is little net muscle movement other than shivering. The activation of muscle causes ATP to be hydrolyzed to provide energy for contraction. Because the muscle contractions are inefficiently timed and mutually opposed, they produce no useful physical work, but the chemical energy released during contraction appears as heat.

In *nonshivering thermogenesis,* enzyme systems for the metabolism of fats are activated throughout the body, so conventional fats are broken down and oxidized to produce heat. Very little of the energy released is conserved in the form of newly synthesized ATP. A specialization found in a few mammals for fat-fueled thermogenesis is **brown fat**, also called *brown adipose tissue* (BAT). Generally found as small deposits in the neck and between the shoulders (Figure 16-26), brown fat is an adaptation for rapid,

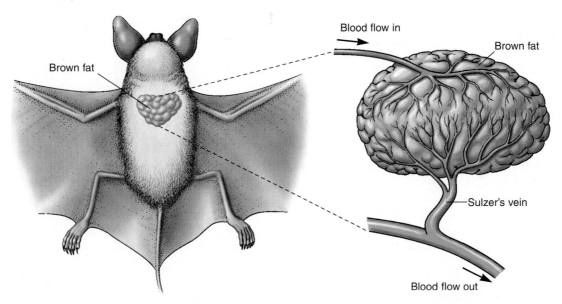

Figure 16-26 Brown-fat deposits are found between the scapulae in bats and many other mammals. The detail shows the special vasculariza-tion of this tissue. During brown-fat oxidation, this tissue is detectable as a warm region by its infrared emission.

massive heat production. This fat contains such extensive vascularization and so many mitochondria that it is brown (owing largely to mitochondrial cytochrome oxidase) rather than white. In brown fat, oxidation takes place within the fat cells themselves, which are richly endowed with fat-metabolizing enzyme systems. In ordinary body fat, deposits must first be reduced to fatty acids, which enter the circulation and eventually are taken up by other tissues, where they are oxidized.

Nonshivering thermogenesis in fat (including brown fat) is activated by the sympathetic nervous system through the release of norepinephrine, which binds to receptors on the adipose cells of brown-fat tissue. Through a second-messenger mechanism, described in Chapter 9, this signal leads to thermogenesis by two mechanisms. In the first of these mechanisms, normal ATP utilization for cellular processes rises in these fat cells in response to the sympathetic signal, accounting for part of the increased heat production. Through processes such as ion pumping by the plasma membrane, ATP is hydrolyzed to produce work and heat. In the second mechanism, ATP production is uncoupled during respiratory chain oxidation. The resynthesis of ATP from ADP and P_i is normally coupled to the movement of protons (H^+) down their electrochemical gradient from intermembrane space into mitochondria across the inner mitochondrial membrane. Thermogenesis in brown fat is characterized by the appearance in the inner mitochondrial membrane of uncoupling proteins, which provide a pathway for protons to leak across this membrane without the energy of their downhill movement being harnessed for the phosphorylation of ADP to ATP. Once inside the mitochondrion, the protons oxidize substrate oxygen to produce water and heat, or else further utilization of metabolic energy is required to subsequently pump them into the intermembrane space and eventually out of the mitochondria.

Brown fat heats up significantly during thermogenesis. This newly produced heat is rapidly dispersed to other parts of the body by blood flowing through the extensive vasculature of the brown fat tissue. Nonshivering thermogenesis is especially pronounced during arousal of hibernating or torpid mammals, when it supplements shivering to facilitate rapid warming. One consequence of acclimation to cold by mammals is an increase in brown fat deposits, which allows for a gradual changeover from shivering to nonshivering thermogenesis at low ambient temperatures. The acclimatory increase in brown-fat thermogenesis is mediated by the thyroid hormones. Brown fat is also present in some mammalian infants, including human infants, where it is generally located in the region of the neck and shoulders, the spine, and the chest. Because an infant is relatively small and inactive at birth, deposits of brown fat provide an important and rapid means of warming if the infant is threatened with temperature reduction.

Another example of tissues specialized for heat production are the **heater tissues** formed from modified eye muscles in billfishes. B. A. Block and her colleagues (1994) have extensively investigated these tissues, which have an enormous capacity to generate heat (as high as $250 \text{ W} \cdot \text{kg}^{-1}$). Heater cells, which lack myofibrils and sarcomeres, produce heat through the release of Ca^{2+} from internal cytoplasmic stores. The released Ca^{2+} ions then stimulate catabolic processes and mitochondrial respiration (Figure 16-27).

Endothermy in cold environments

Endotherms adapted to cold environments have necessarily evolved a number of mechanisms, both temporary and permanent, that help them retain body heat. For example, an animal sensing heat loss in a windy place will fluff its fur or feathers and move to a more sheltered area. This reduces

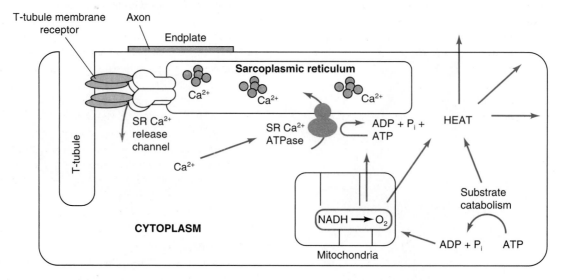

Figure 16-27 The metabolism of heater tissues in billfish is specialized for heat production. This model for nonshivering thermogenesis in the skeletal muscle of billfish heater tissue shows how exogenously stimulated release of Ca^{2+} into the cytoplasm stimulates mitochondrial respiration and the associated release of heat. After stimulation, a T-tubule receptor activates a sarcoplasmic reticulum (SR) Ca^{2+} channel, releasing Ca^{2+} stored in the sarcoplasmic reticulum. The increased cytoplasmic Ca^{2+} ions then trigger the release of energy in the form of heat from ATP previously manufactured in the mitochondria. [Adapted from Block, 1994.]

convection and the dissipation of body heat by the wind. More enduring responses to cold include the thick layers of insulation in many arctic animals, in the form of subcutaneous fat or a thicker pelage or plumage. The insulating effectiveness of pelages in arctic and subarctic animals changes with both season and latitude to match insulation qualities with insulation needs. In addition, animals living in the temperate zone exhibit seasonal variations by shedding old fur or feathers and growing new bodily cover, thereby providing thick insulation during the winter, yet preventing overheating during the summer.

The specific conductances of homeotherms vary over a large range and decrease with body size (Figure 16-28). Larger animals have lower specific heat conductances owing to their generally thicker coats of fur or feathers. In addition, they face smaller heat loss in cold climates because of their relatively smaller surface areas. Thus, one adaptation of endotherms to cold latitudes is an increase in body size. As the surface-to-volme ratio becomes smaller, pelage becomes thicker and conductance decreases. With increased insulation, the lower critical temperature of a homeotherm decreases and the thermal neutral zone extends to lower temperatures (Figure 16-29). An exception is that small animals and immature animals often have feathers or fur that is less conductive per unit thickness, as evident in the fluffy feathers of young chicks, for example.

Blubber, a fatty tissue typically found under the skin in cetaceans, is a good insulator because, like air, blubber has a lower thermal conductivity than does water, which is the main constituent of nonfatty tissues. In addition, fatty tissues are metabolically very inactive and require little perfusion by blood, which would ordinarily carry heat to be lost at the body surface. In whales, the outermost regions of the thick blubber layer are always at a temperature near that of the surrounding water.

An important means of controlling heat loss from the surface is the diversion of blood flow to or away from the skin (see Figure 16-14). Vasoconstriction of arterioles leading to the skin keeps warm blood from perfusing cold skin and conserves the heat of the body core. An interesting advantage of blubber over pelage in the control of heat loss is illustrated in Figure 16-15B, which reminds us that fur is located outside the body proper, whereas blubber is contained

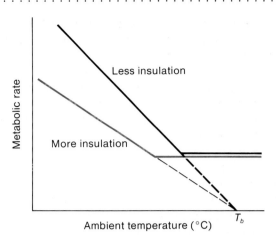

Figure 16-29 The decline in metabolic rate in endotherms as ambient temperatures fall depends on the extent of the animal's insulation. A decrease in insulation (i.e., an increase in conductance) raises the lower critical temperature and makes the slope of increasing metabolism steeper. The slope, however, still extrapolates to body temperature at zero metabolic rate.

within the body and is supplied with blood vessels. Thus, whereas the insulating properties of fur remain unaffected by circulatory adjustments to the area, the insulating properties of blubber depend on whether blood flow to the surface is restricted or not. Hence, the more blood flow is diverted away from the vessels within the blubber, the higher the effective thickness of the insulating layer. Conversely, the greater the blood flow into the blubber, the lower the effective thickness of the insulating layer. This ability to regulate heat transfer through blubber allows a marine mammal to facilitate the loss of excessive body heat by shunting its surface blood effectively to the outer regions of the insulating layer of blubber during periods of intense activity in warmer waters or when lying on land in warm air.

Countercurrent heat exchange

Effective locomotion requires that the limbs of endotherms not be mechanically hindered by a massive layer of insulation. The flukes and flippers of cetaceans and seals and the legs of wading birds, arctic wolves, caribou, and other cold-weather homeotherms require blood to nourish cutaneous tissue and limb muscles used in locomotion. The well vascularized limbs are potential major avenues of body heat loss because they are thin and have large surface areas.

Excessive heat loss from these appendages can be reduced drastically by countercurrent heat exchange. Countercurrent exchange systems have already been discussed in the context of oxygen and carbon dioxide exchange (Chapter 13). Arterial blood, originating in the animal's core, is warm. Conversely, the venous blood returning from peripheral tissues may be very cold. As blood flows from the core, it enters arteries in the limb that lie next to veins that carry blood returning from the extremity. As the arteries and veins pass each other, the warm arterial blood gives up heat to the returning venous blood and thus becomes successively cooler as it enters the extremity. By the time it reaches the periphery, the arterial blood is precooled to

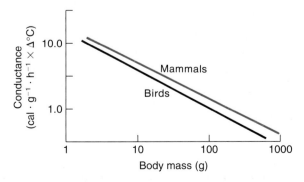

Figure 16-28 Thermal conductance decreases exponentially as body mass increases.

within a few degrees of the ambient temperature, and little heat is lost. Conversely, the returning venous blood is warmed by the arterial blood, so it is nearly at core temperature as it flows into the core. The advantage of such a system is that heat exchange with the environment is restricted without a reduction in blood flow and the oxygen and nutrients that the blood carries. A similar situation was described for heterothermic fishes (see Figures 16-23 and 16-24). Another highly evolved example of countercurrent heat exchange is found in the mirabile of the flipper of the porpoise. Here, the artery carrying warm blood flowing toward the extremity is completely encased in a circlet of veins carrying cold blood back from the extremity. Birds and arctic land mammals also use countercurrent exchange to minimize heat loss from their extremities in cold climates, and to some extent this mechanism is also present in the extremities of human beings. As a result, the extremities of cold-climate endotherms are maintained at temperatures that are far below the core temperature and often approach ambient temperature (Figure 16-30). The efficiency of the countercurrent heat exchangers can generally be regulated by vasomotor control in which blood flow is shunted past the heat exchanger network by parallel vessels.

Endothermy in hot environments—dissipation of body heat

In very hot, dry climates, large animals have the advantages of relatively low surface-to-mass ratios and large heat capacities. Camels, well known for their ability to tolerate heat, have not only a large body mass, but also a thick pelage that helps insulate them from external heat. Low surface-area-to-volume ratio and thick pelage retard the absorption of heat from the surroundings. Furthermore, because of its large mass and the high specific heat of tissue water, the camel, as well as other large mammals, can absorb relatively large quantities of heat for a given rise in body temperature. These features also result in a slow loss of heat during the cool hours of the night. Thus, the large mass acts as a heat buffer that, by reducing rates of both absorption and loss of heat, minimizes extreme temperature fluctuations. A dehydrated camel can also tolerate an elevation of its core temperature by several degrees, further increasing its heat-absorbing capacity. Large amounts of heat gradually accumulated during daytime hours are subsequently dissipated in the cool of the night. In preparation for the next onslaught of daytime heat, the dehydrated camel allows its core temperature to drop several degrees below normal during the night. As a consequence, the camel starts the day with a heat deficit, which allows it to absorb an equivalent amount of additional heat during the hot part of the day without reaching harmful temperatures. This practice, called **limited heterothermy,** allows the camel to tolerate the extreme daytime desert heat without using much water for evaporative cooling.

Limited temporal heterothermy is also practiced by the antelope ground squirrel *(Ammospermophilus leucurus),* a diurnal desert mammal. Because of its small mass, the antelope ground squirrel cannot continuously gain heat for several hours in the hot sun, and its small surface-to-mass ratio would lead to rapid heating. Instead, this desert mammal exposes itself to high environmental temperatures for only about 8 minutes at a time. It then returns to its burrow, where its stored heat escapes into the cool underground air. By allowing its temperature to drop a bit below normal before returning to the hot desert floor, it is able to extend its stay a few minutes without lethal overheating.

The temperature of the body surface is an important factor affecting heat loss to the environment, because it de-

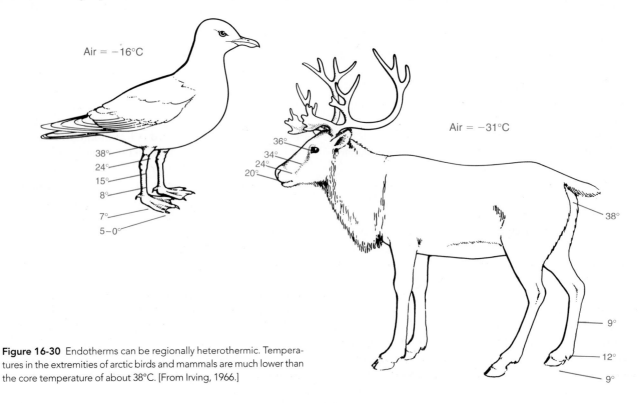

Figure 16-30 Endotherms can be regionally heterothermic. Temperatures in the extremities of arctic birds and mammals are much lower than the core temperature of about 38°C. [From Irving, 1966.]

termines the temperature gradient, $T_b - T_a$, along which heat will flow. Heat can be lost by conduction, convection, and radiation (see Spotlight 16-2) as long as the ambient temperature is below the body surface temperature. The closer the surface temperature is to the core temperature in an endotherm, the higher the rate of heat loss through the surface to cooler surroundings. Heat is transferred from the core to the surface primarily by the circulation; the rate of heat loss to the environment is regulated by the flow of blood to surface vessels (see Figures 16-14 and 16-15).

Endotherms thus use various "heat windows" to regulate the loss of body heat, opening or shutting them by regulating blood flow. These heat windows permit the loss of heat by radiation, conduction, and, in some cases, evaporative cooling. An example of such temperature-regulating windows can be seen in the thin, membranous, and lightly furred ears of rabbits, with their extensively interconnecting arterioles and venules. Another example is seen in the horns of various mammals; in goats and cattle, the horns are highly vascularized by a network of blood vessels that, under conditions of heat load, vasodilate and act as radiators of heat. Similarly, the legs and snout, having large surface-to-volume ratios, are used as thermal windows for the dissipation of heat by regulation of the rate of blood flow through the arterioles serving the skin of the appendages. Some mammals living under conditions of intense solar radiation or high temperatures have certain areas of the body surface exceptionally lightly furred or even naked to facilitate heat loss by radiant, evaporative, or conductive means. Such areas generally include the axilla (armpit), groin, scrotum, and parts of the ventral surface. Some of these areas, such as the udder and scrotum, carry additional temperature sensors that are used to detect changes in air temperature with minimal interference from the core temperature. By this means, the animal can anticipate changing temperature loads and make the corresponding adjustments in advance.

Variations in posture or body orientation also can affect rate of heat absorbance or loss. For example, the guanaco, a medium-sized camel-like inhabitant of the Andes, has very densely matted hair on its back and a lighter covering of fur on its head and neck and the outer sides of its legs. The inner sides of the upper thighs and the underside are nearly naked, acting as thermal windows covering nearly 20% of the body surface. By adjusting the posture and orientation of its body with respect to solar radiation and cooling breezes, the guanaco can adjust the degree to which its thermal windows are open or shut, permitting a fivefold change in thermal conductance. This posturally controlled flexibility in surface insulation permits a variability in heat transfer across the surface of endotherms that is independent of surface-to-mass ratio.

Evaporative cooling

The evaporation of 1 g of water requires 2448 J (585 cal) of energy. Consequently, evaporation is the most effective means of removing excess body heat, *assuming that there is sufficient water available* to "waste" in this fashion. Certain reptiles

and birds and some mammals take available body water (saliva and urine) or standing water from the environment and spread it on various body surfaces, allowing it to evaporate at the expense of body heat. Animals with naturally moist skin, such as amphibians, may have a body temperature lower than ambient temperature because of evaporative cooling, though this is not an effect that has been selected for.

Some vertebrates use sweating or panting to produce evaporative cooling. In *sweating,* found in some mammals, sweat glands in the skin actively extrude water through pores onto the surface of the skin (see Chapter 8). Sweating is under autonomic control. Although it is a mechanism for evaporative cooling, sweating can persist in the absence of evaporation when the relative humidity of air is very high. Water will continue to be secreted from sweat glands even if the humidity is too high for evaporation to keep up with the rate of sweating, leading not only to elevated body temperature, but also to elevated water (and salt) loss.

In *panting,* mammals and birds use the respiratory system to lose heat by evaporative cooling (see also Chapter 14 on osmoregulation in desert environments). Panting mammals breathe through the mouth instead of through the nose. Heat is carried away in exhalant air because the dimensions of the mouth are such that exhalant air retains the heat absorbed in the lungs. As noted earlier, the nasal passages and their vascularization are effective in many mammalian species in retaining both water and body heat. Mammals also hyperventilate to increase heat loss. A change in alveolar ventilation, however, will result in a change in blood P_{CO_2} and blood pH. This situation is avoided during panting by a disproportionate increase in dead-space ventilation (i.e., flow through the mouth and trachea) *without* an increase in ventilation of the alveolar respiratory surface (Figure 16-31). Breathing rate is increased,

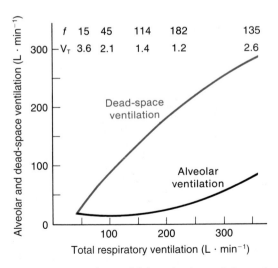

Figure 16-31 Panting induces a shift from alveolar ventilation to alveolar and dead-space ventilation. As the total respiratory ventilation (*abscissa*) increases in the panting ox, the dead-space ventilation (flow through the mouth and trachea) increases steadily. The alveolar ventilation, however, does not increase until the total ventilation exceeds about 200 L·min⁻¹. In extreme panting, the respiratory frequency (*f*) decreases as tidal volume (V_T) increases (figures at top of graph). [Adapted from Hales, 1966.]

but tidal volume is reduced. Overheated canines and birds pant by inhaling through the nose and exhaling through the mouth, exposing the tongue and other mouth structures to encourage further water evaporation and therefore heat loss (Figure 16-32). Panting produces a one-way air flow over the nonrespiratory surfaces of the nose, trachea, bronchi, and mouth, causing evaporation without causing stagnation of saturated air in these passages. The amount of respiratory work required in panting is lower than it might seem, because a panting animal causes its respiratory system to oscillate at its resonant frequency, thereby minimizing muscular effort. Panting is accompanied by increased secretions from the salivary glands of the nose, secretions that are under autonomic control. Most of the water that is not evaporated by panting is swallowed and conserved.

Because evaporation from skin or respiratory epithelium is the most effective means of ridding the body of excess heat, there is a close link between water balance and temperature control in hot environments (see Chapter 14). In hot, dry, desertlike environments, animals can be faced with the choice of either overheating or desiccating. Dehydrated mammals conserve water by reducing evaporation caused by panting or sweating and thus allow the body temperature to rise. Having a small heat capacity, a small mammal exposed to the desert heat will, in the absence of thermoregulatory water, undergo a rise in temperature that is far more rapid and more threatening than it would be for a larger animal. To survive, small mammals must either drink water or stay out of the heat.

The reciprocity of water conservation and heat dissipation in a small desert animal can be illustrated by considering the water balance and temperature control in the kangaroo rat. To conserve water, this animal uses a temporal countercurrent heat-exchange system in which the nasal epithelium is cooled during inspiration by the inhalant air. During exhalation, most of the moisture picked up by the air in the warm, humid respiratory passages is conserved by its condensation on the cool nasal epithelium. However, this mechanism also recycles body heat and requires that the inhaled air be cooler than the body core. As a consequence, the kangaroo rat is confined to its cool burrow during the hot times of day. If inhaled air were at or above body temperature, the kangaroo rat's loss of respiratory moisture would increase. Although the evaporative loss of water would help cool the animal, it would also seriously alter its water balance.

The importance of water in the control of body temperature in a large desert mammal has been long known, as illustrated by simple and now classic observations made of camels. The camels were either allowed to drink freely or were subjected to periods of dehydration during which water was withheld for several days. Rectal temperatures were highest in daytime and lowest at nighttime. These fluctuations were minimal when the camels were allowed to drink but still large in comparison with a water-drinking human being. The temperature swings became even more exaggerated during periods of dehydration, when reserves of body water dwindled, leaving less for heat storage and for thermoregulation by sweating.

Thermostatic Regulation of Body Temperature

Homeothermic endotherms use a system of body temperature control that is similar to the mechanized thermostatic control found in a laboratory temperature bath (see Figure 1-4) or a home heating system. In the water bath, a *temperature comparator* compares the water temperature, T_w, detected by a temperature sensor with a *set-point temperature*, T_{set}. If T_w is below T_{set}, the *thermostat* closes the circuit that activates the production of additional heat until $T_w = T_{set}$, after which the thermostat contacts open and heat production ceases. The cycle is repeated as T_w drops again. This analogy is especially apt in the zone of metabolic regulation (see Figure 16-25), in which heat production increases with decreasing ambient temperature.

Both homeothermic endotherms and homeothermic ectotherms also use nonmetabolic means to regulate body temperature. The regulation of body temperature, T_b, is not fully understood even after decades of research on the topic but appears to work along principles of negative feedback (see Spotlight 1-1). Most animals have not one but many temperature sensors in various regions of the body. Furthermore, to maintain T_b at about T_{set}, homeothermic animals can call on several heat-producing and heat-exchanging mechanisms, so the thermostat controls heat-

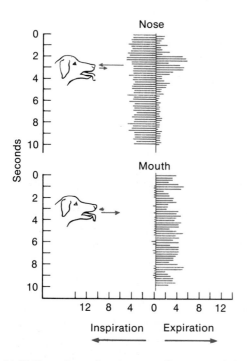

Figure 16-32 The pathway of respiratory gas flow varies with the extent of panting in the dog. *(Top)* Air flow through the nose of a panting dog. Horizontal lines extending to the left of the vertical midline indicate inspiration; to the right, expiration. Mean inhaled and exhaled volumes are indicated by vectors placed adjacent to the dog's nose. *(Bottom)* Air flow through the mouth of a panting dog. Inspiration through the mouth is virtually zero; expiration through the mouth carries most of the air taken in through the nose. [From Schmidt-Nielsen et al., 1970.]

conserving and heat-loss mechanisms as well as heat production. This means of control is analogous to the microprocessor-controlled heating and cooling system of the futuristic "smart house" in which the thermostat, in addition to cycling the furnace and air conditioner, controls the position of window shades, window opening and closing, the conductance of the wall and roof insulation, and so forth. Furthermore, control of thermogenesis in the homeotherm is not all-or-none, like the turning on and off of a furnace. Instead, the rate of heat production by metabolic means is graded according to need. The colder the temperature sensors become (within limits), the higher the rate of thermogenesis. Engineers call this *proportional control*, because heat production and conservation are more or less proportional to the difference of $T_b - T_{set}$.

The hypothalamus—the mammal's "thermostat"

Mammalian body temperature can vary widely (as much as 30 Celsius degrees) between the periphery and the body core, with the extremities undergoing far more variation than the core. Temperature-sensitive neurons and nerve endings exist in the brain, the spinal cord, the skin, and sites in the body core, providing input to thermostatic centers in the brain. Although a mammal may have several thermoregulatory centers, the most important one, considered to be the body's "thermostat," is located in the hypothalamus (see Figure 9-5). It was discovered by Henry G. Barbour in 1912 in the course of a series of experiments in which a small temperature-controlled probe was implanted in different parts of the rabbit brain. The probe evoked strong thermoresponses only when it was used to heat or cool the hypothalamus. Cooling the hypothalamus produced an increase in metabolic rate and a rise in T_b, whereas heating it evoked panting and a drop in T_b. This experiment is analogous to changing the temperature of a thermostat in your home by holding a lighted match nearby. As the thermostat is warmed above its set-point temperature, it shuts down the furnace, allowing the room temperature to drop below the set point. An apparatus for controlling hypothalamic temperature and measuring a homeotherm's response to changes in that temperature is shown in Figure 16-33.

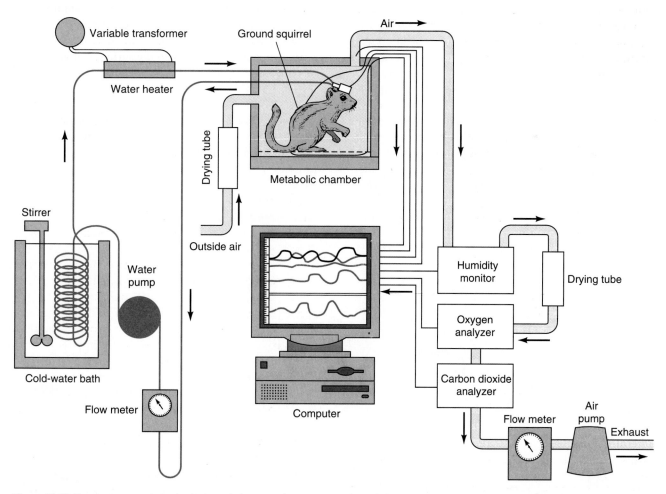

Figure 16-33 Temperature regulation by the hypothalamus can be measured by experimentally changing hypothalamic temperature. The apparatus measures the temperature sensitivity of the hypothalamus and thermoregulatory response to changes in hypothalamic temperature, which is altered by means of a water-perfused thermode implanted in the hypothalamus. Metabolic rate and evaporative water loss are measured by analyzing the effluent air for water, O_2, and CO_2 content. The metabolic chamber is at constant temperature [From "The Thermostat of Vertebrate Animals" by H. C. Heller, L. I. Crawshaw, and H. T. Hammel. Copyright © 1978 by Scientific American, Inc. All rights reserved.]

Experimental procedures like Barbour's have shown that the mammalian hypothalamic thermostat is highly sensitive to temperature. Variation of mammalian brain temperature of only a few Celsius degrees seriously affects the functioning of the brain, so it is not surprising to find the major thermoregulatory center of mammals located there. Neurons that are highly temperature sensitive are located in the anterior part of the hypothalamic thermostat. Some of these neurons show a sharply defined increase in firing frequency with increased hypothalamic temperature (Figure 16-34). These neurons are believed to activate heat-dissipating responses such as vasodilation and sweating. Others show a decrease in firing frequency with increase in temperature above a certain value. Still other neurons increase their firing frequency when the brain temperature drops below the set-point temperature. They appear to control the activation of heat-producing responses (e.g., shivering, nonshivering thermogenesis, brown-fat metabolism) and heat-conserving (e.g., pilomotor) responses.

In addition to the information about its own temperature generated by these thermosensitive neurons, the hypothalamus receives neural input from thermoreceptors in other parts of the body. All this thermal information is integrated and used to control the output of the thermostat. Neural pathways leaving the hypothalamus make connections with other parts of the nervous system that regulate heat production and heat loss. Some of these pathways are activated by high temperatures signaled from peripheral and spinal thermoreceptors and by the hypothalamic temperature-sensitive neurons. The efferent pathways activate increased sweating and panting, as well as a lowered peripheral vasomotor tone that produces increased blood flow to the skin. Conversely, body cooling leads to thermogenesis and increased peripheral vasomotor tone. These same responses can be elicited without cooling the whole body by simply cooling the neurons of the hypothalamus.

Thus, experimentally lowering hypothalamic temperature in a dog leads to elevated metabolic heat production by shivering. On the other hand, warming the dog's hypothalamus elicits the heat-dissipating response of panting.

A rise in core temperature of only 0.5 Celsius degrees in most mammals causes such extreme peripheral vasodilation that the blood flow to the skin can increase several times above normal. In human beings, this response produces a flushed appearance to the skin. The effect of elevated core temperature on peripheral vasodilation and, hence, skin temperature is illustrated in Figure 16-35, which shows that the skin temperature of a rabbit's ear rose very sharply from less than 15°C to about 35°C at the point at which the rabbit's core temperature exceeded 39.4°C. Because the temperature of the ear reached a maximum, it can be assumed that the vessels of the ear dilated fully as soon as the core temperature exceeded this limit.

The effect of the hypothalamic thermostat on such peripheral heat-exchange mechanisms is about 20 times as great in some mammals as reflexive adjustments initiated by peripheral temperature sensors. This hypothalamic "override" is significant in light of the importance of carefully regulated brain temperature. Without dominance of the hypothalamic thermostat, an internally overheated animal exercising in a cold environment would fail to activate heat-dissipating blood flow to the surface capillaries, and its core temperature would continue to rise to dangerous levels.

In some homeotherms, especially small animals subject to rapid cooling at low ambient temperatures, the set-point temperature of the hypothalamic thermostat changes with ambient temperature, presumably because ambient deviations are sensed by peripheral receptors. Thus, in the kangaroo rat, a sudden drop in ambient temperature is quickly followed by a rise in set-point temperature. This rise causes an increase in metabolic heat production in anticipation of increased heat loss to the environment.

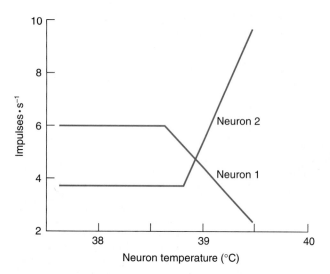

Figure 16-34 Different hypothalamic neurons show different temperature-activity patterns in a rabbit. Neuron 1 exhibits a linear decrease at a temperature above 38.4°C, whereas neuron 2 shows a steep linear increase at a temperature above 38.7°C. [Adapted from Hellon, 1967.]

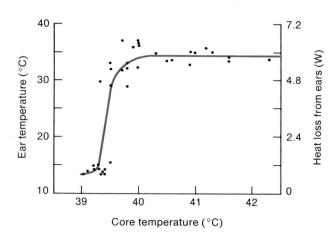

Figure 16-35 Ear heat loss rises suddenly as core temperature increases in a rabbit at an ambient temperature of 10°C. The core temperature was raised by forcing the rabbits to run on a treadmill. As temperature rose to above 39.5°C, blood flow to the ears increased, raising ear temperature and heat loss (given in watts). [From Kluger, 1979.]

The relations between thermoregulatory responses controlled by the hypothalamic centers and those controlled by the core temperature are illustrated in Figure 16-36. Small deviations in core temperature from the set point produce only peripheral vasomotor and pilomotor responses (black plot) that in effect alter the thermal conductance of the body. These small deviations in core temperature usually result from moderate variations within a range of ambient temperatures corresponding to the thermal neutral zone (see Figure 16-25). When the core temperature is forced out of this range by more extreme ambient-temperature deviations or by exercise, passive thermoregulatory responses no longer suffice, and the hypothalamic centers institute active measures—that is, thermogenesis or evaporative heat loss (red plots in Figure 16-36).

Nonmammalian thermoregulatory centers
Thermostatic control of body temperature has received less attention in birds than in mammals, perhaps because the manner of control seems to be more complex in birds. The region of the hypothalamus that serves as the thermoregulatory center in mammals is virtually insensitive to temperature changes in those birds tested (mainly pigeons). The spinal cord was found to be a site of central temperature sensing in pigeons, penguins, and ducks, but core receptors outside the central nervous system are the dominant temperature receptors in birds. The temperature sensors in the core presumably signal the avian hypothalamic thermostat, which in turn integrates the input and activates the thermoregulatory effectors.

Fishes and reptiles, like birds and mammals, have a temperature-sensitive center in the hypothalamus. Heating the hypothalamus with an implanted thermode leads to hyperventilation in the scorpion fish; cooling leads to slower ventilatory movements. Peripheral cooling also produces similar ventilatory responses. Because the fish's rate of metabolism varies with body temperature, a rise in temperature leads to an increased need for oxygen. The temperature-determined adjustment in the rate of respiration is adaptive in that it anticipates changes in respiratory need and serves to minimize fluctuations in blood oxygen. The

reptilian response to cooling of the hypothalamus is to engage in *thermophilic* (i.e., heat-seeking) *behavior,* whereas heating of the hypothalamus elicits *thermophobic* (i.e., heat-avoiding) *behavior.*

Experiments by S. C. Wood and his colleagues (1991) have drawn some intriguing links between behavioral thermoregulation and hypoxic exposure. As an alternative to increasing convective O_2 supply to tissues by increasing ventilation and cardiac output during hypoxic exposure, a wide variety of both vertebrates and invertebrates respond to hypoxia by reducing oxygen demand by temporarily selecting a lower preferred body temperature. Thus, vertebrates such as mice, toads, fishes, and lizards in an experimental thermal gradient will move toward a cooler region when made hypoxic (Figure 16-37). Invertebrates such as spiders and crayfish and even unicellular organisms such as *Amoeba* also respond in this way.

Fever

An interesting feature of the hypothalamic thermoregulatory center is its sensitivity to certain chemicals collectively termed **pyrogens** (fever-producing substances). There are two general categories of pyrogens, based on their origins. *Exogenous pyrogens* are endotoxins produced by gram-negative bacteria. These heat-stable, high-molecular-weight polysaccharides are so potent that a mere 10^{-9} g of purified endotoxin injected into a large mammal causes an elevation of body temperature. *Endogenous pyrogens,* on the other hand, arise from the animal's own tissues and, unlike those of bacterial origin, are heat-labile proteins. Leukocytes release endogenous pyrogens in response to circulating exogenous pyrogens produced by infectious bacteria. Thus, it appears that exogenous pyrogens cause a rise in body temperature indirectly by stimulating the release of endogenous pyrogens that act directly on the hypothalamic center. This idea is supported by evidence that the hypothalamus is more sensitive to direct application of endogenous pyrogens than to exogenous ones.

The sensitivity of the hypothalamic temperature-sensing neurons to these pyrogenic molecules leads to an

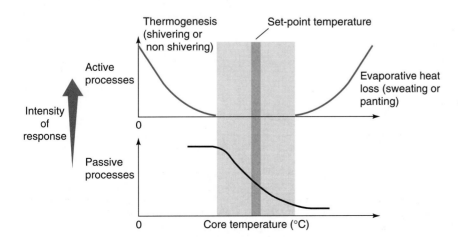

Figure 16-36 The degree of an animal's thermoregulatory response is greatest at body temperatures above or below normal core temperature. Within a range (light gray area) of the set-point temperature (dark gray area), regulation of body temperature is only through control of heat conductance to the environment by varying the peripheral blood flow or the insulating effectiveness of fur or feathers (black plot). Above and below this range, these passive measures are exhausted, and active thermogenesis *(left)* or evaporative heat loss *(right)* develops (red plots).

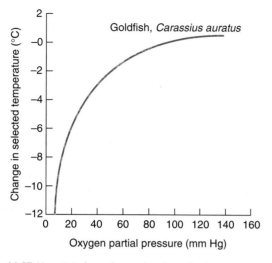

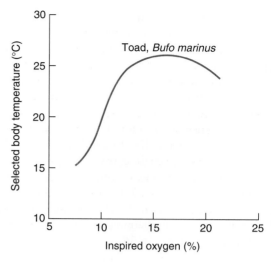

Figure 16-37 Hypoxia induces changes in selected body temperature in goldfish (*Carassius auratus*) and toads (*Bufo marinus*). [Adapted from Wood, 1991.]

elevation in the set point to a higher temperature than normal. The result is that the body temperature rises several degrees, and the animal experiences a **fever.** Anesthetics and opiates such as morphine, in contrast with pyrogens, cause a lowering of the set-point temperature and hence a drop in body temperature. The adaptive significance of endogenous pyrogens and of their production of fever in homeotherms may relate to the bacteriostatic effects of elevated body temperature.

Pyrogenic bacteria elevate body temperature in some ectotherms as well as in endotherms. In a classic experiment by H. A. Bernheim and M. G. Kluger (1976), body temperatures were monitored in a desert iguana under

laboratory conditions simulating a desert environment before and after administration of pyrogenic bacteria (Figure 16-38). In response to the fever-producing bacteria, the lizards positioned themselves more frequently in radiantly heated zones of the artificial environment, effectively raising their temperatures to unusually high levels (i.e., fever). This behavioral response and the fever temperatures it produced conferred protection against bacterial infection. This protection is thought to take two forms: (1) the antiviral and antitumor agent *interferon* is more effective at higher temperatures, and (2) elevated temperatures also diminish the growth of some microbes.

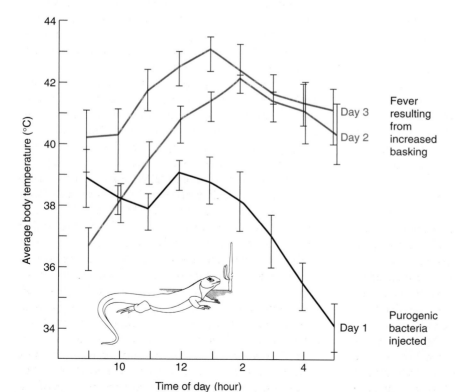

Figure 16-38 An ectotherm responds to injections of pyrogenic bacteria with a fever. Like other lizards, the desert iguana, *Dipsosaurus dorsalis*, regulates its body temperature behaviorally by adjusting its location and posture with respect to radiant heat from the sun or hot objects such as dark rocks. After infection by pyrogenic bacteria, the lizards raised their body temperatures above normal levels by increased basking behavior. The graph shows the increase in body temperature with successive days after injection of the bacteria. [Adapted from Bernheim and Kluger, 1976.]

Thermoregulation during activity

The energy efficiency of muscle contraction is only about 25%. For every joule of chemical energy converted into mechanical work, 3 J of energy is degraded to heat. During exercise, this extra heat, added to the heat produced by basal metabolism, will cause a rise in body temperature above the set-point temperature unless it can be dissipated to the environment at the same rate at which it is produced. Most of the excess heat manages to be transferred to the environment, but a rise in core temperature of homeotherms does occur during exercise, indicating incomplete removal of the excess heat. The rise in temperature is moderately useful in two respects: (1) it increases the difference $T_b - T_a$ and thereby increases the effectiveness of the heat-loss processes by increasing the gradient for heat loss, and (2) it leads to an increased rate of metabolic reactions, including those that support physical activity. However, the core temperature can rise to dangerously high levels during heavy exercise in warm environments, and this excess heat has to be dealt with.

The level to which the core temperature rises in homeotherms is proportional to the rate of muscular work. During light or moderate exercise in cool environments, body temperature rises to a new level and is regulated at that level as long as the exercise continues. Thus, temperature appears to remain under the control of the body's thermostat. The rise in temperature proportional to the level of exercise appears to be a consequence primarily of an increase in the error signal, $T_b - T_{set}$, of the thermostatic feedback control by the hypothalamus. The error signal is the *difference* between the thermostat's set point and the actual core temperature. The greater this difference (i.e., the greater the error signal), the greater the activation of heat-loss mechanisms. Thus, the rate of heat dissipation increases as the core temperature rises above the set point, and a new equilibrium becomes established between heat production and heat loss. During heavy exercise, especially in warm environments, the heat-dissipating mechanisms are not able to balance heat production until body temperature rises several degrees, increasing the $T_b - T_a$ difference. Thus, elevations of 4 to 5 Celsius degrees in core temperature are commonly observed in human beings after strenuous, sustained running and in race horses, greyhounds, and sled dogs after racing.

The rise in T_b (and in the error signal as T_b rises above T_{set}) is kept small by the high sensitivity of the feedback control of heat-dissipating mechanisms. For example, a small increase in T_b above the set-point temperature produces a strong and steep increase in the rate of sweating (Figure 16-39). The effectiveness of heat loss is affected by humidity of the surrounding air—the higher the humidity, the less effective the heat loss (as any person familiar with either desert or high-humidity environments can testify). The heat-loss mechanisms are initiated by vigorous exercise even before peripheral body temperature has undergone any significant increase. For example, in human beings, an increased rate of sweating begins within 2 seconds after on-

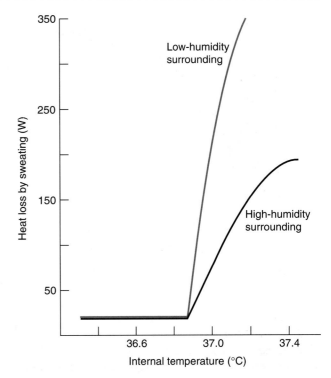

Figure 16-39 Rate of sweating in humans increases sharply as body temperature approaches about 37°C. Core temperature was elevated by exercise or by elevating the ambient temperature. [Adapted from Benzinger, 1961.]

set of heavy physical work, even though there is no detectable increase in skin temperature in that time. However, core blood temperature shows a detectable rise in temperature within 1 second after exercise has begun. Apparently, the onset of sweating, nearly concurrent with the onset of neural activity underlying exercise, results from the reflex activation of sweating by central temperature receptors. The set points for heat loss are lower in well trained athletes, especially in warm weather.

A special countercurrent heat exchanger to prevent overheating of the brain during such strenuous exercise as running is employed by certain groups of hoofed mammals (e.g., sheep, goats, and gazelles) and carnivores (e.g., cats and dogs). This system, the **carotid rete** (Figure 16-40), uses cool venous blood returning from respiratory passages to remove heat from hot arterial blood traveling toward the brain. In these animals, most of the blood to the brain flows through the external carotid artery. At the base of the skull, the carotid anastomoses into hundreds of small arteries that form a vascular rete, the vessels of which rejoin just before passage into the brain (see Figure 16-40). These arteries pass through a large sinus of venous blood, the **cavernous sinus**. This venous blood is significantly cooler than the arterial blood because it has come from the walls of the nasal passages, where it was cooled by respiratory air flow. Thus, the hot arterial blood flowing through the rete gives up some of its heat to the cooler venous blood before it enters the skull. As a result, brain temperature may be from 2 to 3 Celsius degrees lower than the core temperature of the

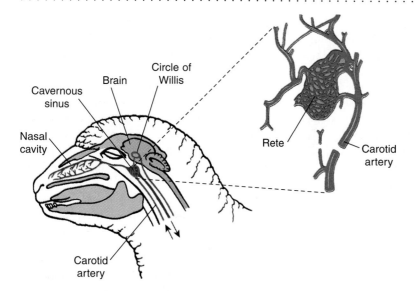

Figure 16-40 The sheep has a carotid rete for countercurrent cooling of carotid blood. The carotid rete, found in sheep and some other mammals, is shown in red. A network of small arteries acts as a heat exchanger for blood supplying the brain. Cool venous blood returning from the nasal cavity bathes the carotid rete contained in the cavernous sinus, removing heat from arterial blood flowing to the circle of Willis and then to the brain. [Adapted from Hayward and Baker, 1969.]

body. Although sustained running in hot surroundings inevitably places a heat load on these animals, the most serious and acute consequence of overheating—spastic brain function—is thereby prevented. This system of cooling is most effective when the animal is breathing hard during strenuous exercise.

DORMANCY: SPECIALIZED METABOLIC STATES

Dormancy is a general term for reduced body activities, including reduced metabolic rate. It often includes heterothermy. Dormancy can be variously classified according to its depth (in reference both to ability for arousal and to decrease in T_b) and its duration, and it includes *sleep, torpor, hibernation, winter sleep,* and *estivation.* Sleep has been the most thoroughly investigated (probably because it is the only state of dormancy experienced by people). The remaining four categories are less well understood than sleep; however, in homeotherms, all appear to be manifestations of physiologically related processes.

Sleep

Studied intensively in human beings and other mammals, sleep entails extensive adjustments in brain function. In mammals, slow wave sleep is associated with a drop in both hypothalamic temperature sensitivity and body temperature, as well as changes in respiratory and cardiovascular reflexes. During rapid-eye-movement (REM) sleep, hypothalamic temperature control is suspended. Although there may be a variety of triggers of sleep, in mammals there is evidence of sleep-inducing substances that build up during wakefulness, accumulating in extracellular fluids of the central nervous system. The identity and mode of action of these substances is under investigation. The time course and extent of sleep varies greatly among animals. Seals resting on pack ice sleep for only a few minutes at a time before

rousing to scan the ice for approaching polar bears. Human beings and many other mammals sleep for hours at a time. Many of the big carnivores (e.g., lions and tigers) sleep for as long as 20 hours a day, especially after a meal.

Torpor

The lower the T_b, the lower the basal metabolism and the lower the rate of conversion of energy stores, such as fatty tissues, into body heat. Thus, it is generally advantageous to allow body temperature to decrease during periods of nonfeeding and inactivity. Small endotherms, because of their high rates of metabolism, are subject to starvation during periods of inactivity when they are not feeding. During those periods, some animals enter a state of **torpor,** in which temperature and metabolic rate subside. Then, before the animal becomes active, its body temperature rises as a result of a burst of metabolic activity, especially through shivering or oxidation of brown-fat stores or both (if a mammal). Daily torpor is practiced by many terrestrial birds. The hummingbird is a classic example, allowing its body temperature to fall from a daytime level of about 40°C to a nighttime level as low as 13°C (in the rufous hummingbird when a low ambient temperature permits). Several species of small mammals also undergo torpor (e.g., shrews), but large mammals have too much thermal mass to cool down quickly for short periods of torpor.

Hibernation and Winter Sleep

A period of deep torpor, or winter dormancy, **hibernation** lasts for weeks or even several months in cold climates. It is entered into through slow wave sleep and is devoid of rapid-eye-movement sleep. Hibernation is common in mammals of the orders Rodentia, Insectivora, and Chiroptera, which can store sufficient energy reserves to survive the periods of nonfeeding. Many hibernators arouse periodically (as often as once a week or as infrequently as every 4–6 weeks) to empty their bladders and defecate.

During hibernation, the hypothalamic thermostat is reset to as low as 20 Celsius degrees or more below normal. At ambient temperatures between 5°C and 15°C, many hibernators keep their temperatures as little as 1 degree above ambient temperature. If the air temperature falls to dangerously low levels, the animal increases its metabolic rate to maintain a constant low T_b or becomes aroused.

Thermoregulatory control is not suspended during torpor and hibernation—it merely continues with a lowered set point and reduced sensitivity (gain), as in slow wave sleep. In the hibernating marmot, for example, experimental cooling of the anterior hypothalamus with an electronically controlled, implanted probe increases metabolic production of body heat. The increase in heat production is proportional to the difference between the set-point temperature and the actual hypothalamic temperature. The set-point temperature drops about 2.5 Celsius degrees within a day or two as the animal enters a deeper state of hibernation.

As might be expected, body functions are greatly slowed with the lowered body temperature characteristic of torpor and hibernation. The effect of reduced body temperature on the metabolic rate (expressed as $\dot{V}_{CO2}$) of golden-mantled ground squirrels is shown in Figure 16-41. In conjunction with a decrease in metabolism, blood flow in hibernating mammals is typically reduced to about 10% of prehibernation values, although the head and brown-fat tissue receive a much higher blood flow than do other tissues. The cardiac output decreases to only a small percentage of the normal rate. This retardation is accomplished by a drastic slowing of the rate of heartbeat, with stroke volume remaining essentially unchanged. As a result of reduced respiratory exchange, the blood of many hibernators becomes more acidic. This acidosis may further lower enzyme activity because of the departure from the pH optimum of metabolic enzymes.

The rate of arousal from hibernation is often much higher than the rate of entry into hibernation. Thus, in the ground squirrel, the transition to the torpid state is completed within 12 to 18 hours (Figure 16-42), whereas arousal requires less than 3 hours. The speed at which this midsized mammal arouses depends on rapid heating initiated by intensive oxidation of brown fat, accompanied by shivering. This frequently leads to a large surge in metabolic rate, as evident in Figure 16-42.

Although many small endotherms undergo a daily cycle of torpor, their high rates of metabolism preclude extended periods of torpor in the form of hibernation because, even in the hibernating state, they would quickly consume stored energy reserves with little remaining for the metabolically expensive process of arousal. All true hibernators are midsized mammals weighing at least several hundred grams and large enough to store sufficient reserves for extended hibernation. There are no true hibernators among large mammals. Bears, which were once thought to hibernate, in fact simply enter a "winter sleep" in which body temperature drops only a few degrees, and they remain curled up in a protected microhabitat such as a cave or hollow log. With its large body mass and low rate of heat loss, a bear can store sufficient energy reserves to enter winter sleep without dropping body temperature. Bears are able to wake and become active quickly at any point during the winter, making it dangerous to encounter a bear even if it is in winter sleep. Typically, however, bears stay in winter sleep for long periods, retaining metabolic wastes in their bodies and even bearing their cubs. Winter sleep, with its relatively high body temperature, does not offer the same degree of energy savings as deep hibernation, but a fall in body temperature of even a few degrees saves energy.

Why are there no large hibernators? First, they have less need to save fuel, because their normal basal metabolic rates are low relative to their fuel stores owing to the allometry of metabolism and fuel storage. Second, because of the large mass and relatively low rate of metabolism, a prolonged metabolic effort would be required to raise body temperature from a low level near ambient temperature to normal body temperature. It has been calculated, for example, that a large bear would require at least 24–48 hours to warm up to 37°C from a hibernating temperature of 5°C. Warming of such a large mass would also be energetically very expensive.

Estivation

The poorly defined term **estivation**, which has been called "summer sleep," refers to a dormancy that some species of both vertebrates and invertebrates enter in response to high ambient temperatures or danger of dehydration or both. Land snails such as *Helix* and *Otala* become dormant during long periods of low humidity after sealing the entrance to the shell by secreting a diaphragm-like operculum that retards loss of water by evaporation. Many land crabs similarly spend dry seasons in an inactive state at the bottom of their burrows. Well known as estivators are African lungfish, *Protopterus*. These air-breathing fish survive periods of drought in which their ponds dry up by estivating in the semidry bottom until the next rainy season floods the area. The lungfish seals itself inside a "cocoon," in which a

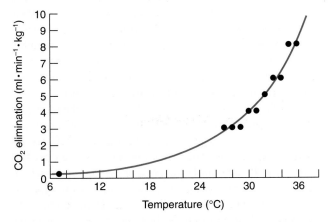

Figure 16-41 Both experimentally induced hypothermia and natural hibernation reduce metabolism in the golden-mantled squirrel. Open symbols, experimentally induced hypothermia; asterisk, unanesthetised animals awake and hibernating. [From Milsom, 1992.]

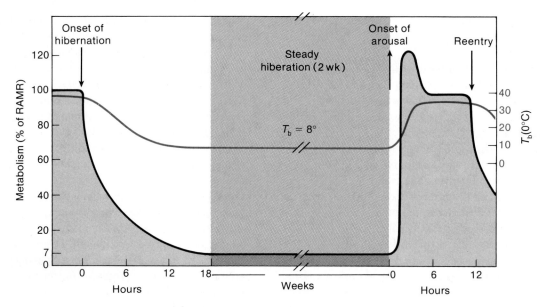

Figure 16-42 Metabolism increases briefly during an episode of arousal from hibernation in a ground squirrel. The squirrel was kept in a chamber having a temperature of 4°C. The period of steady-state hibernation is shaded in color, and the body temperature, T_b, is graphed in red. Metabolism is graphed in black. At onset of hibernation, the set point for body temperature is depressed. Metabolism decreases, allowing T_b to drop to 1–3 Celsius degrees above T_a throughout hibernation. Arousal occurs when the set-point temperature climbs to 38°C, and a strong surge of metabolic heat production raises T_b to the new set-point level. Abbreviation: RAMR, resting average metabolic rate. [From Swan, 1974.]

small tube leads from the fish's mouth to the exterior to allow ventilation of the lungs. Interestingly, chemical estivation-inducing factors in the plasma of estivating lungfish produce a torporlike state when injected into mammals. Some small mammals, such as the Columbian ground squirrel, spend the hot late summer inactive in their burrows, with their core temperatures approaching the ambient temperature. This state is probably similar physiologically to hibernation, but it differs in seasonal timing.

 How can you explain the finding that the injection of proteins isolated from the blood plasma of an estivating lungfish induces sleep in mice?

ENERGETICS OF LOCOMOTION

Early in this chapter, we considered the basal rate of metabolism characteristic of the resting animal. Additional energy over and above the basal rate is expended when the animal is active—that is, producing movement with its muscles. The most readily quantified type of muscle activity in most animals is simple *locomotion*. Because it is required for finding food and mates and for escaping predators, locomotion is also one of the more important types of routine activity. We now turn to an examination of the metabolic cost of animal locomotion.

Animal Size, Velocity, and Cost of Locomotion

The *metabolic cost of locomotion* is the amount of energy required to move a unit mass of animal a unit distance and is usually expressed in units of kilocalories per kilogram per kilometer. This energy is taken as that expended above and beyond that which is expended under basal conditions of rest. Measurements of O_2 consumption and CO_2 production associated with locomotion are generally made while the animal is running on a treadmill, swimming in a flow tank, or flying in a wind tunnel. The measured rate of gas exchange is then translated into rate of energy conversion.

Relations between the *net* work done in the locomotion of an animal and the *gross* energy conversion powering the underlying muscle activity are complicated by several factors, not all of which are sufficiently well understood to be discussed here. Nonetheless, we know that a significant percentage of muscular effort during locomotion does not contribute directly to the production of forward motion. Some muscle contraction holds limb joints in their proper articulating positions. Another large percentage of muscle work is performed in an elongating muscle to counteract gravity, to absorb shocks, and to finely tune the movements of limbs during contraction of antagonists. The comparative energetics of animal locomotion are further complicated by the well established inverse relation between the force produced by a contracting muscle and its rate of shortening (i.e., muscle length or sarcomere length per second; see Figure 10-13). The higher the rate of cross-bridge cycling, the higher the metabolic cost of shortening the muscle by a given distance. Small animals show higher rates of limb stride, tail beat, or wing motion. Thus, small animals employ higher rates of muscle shortening (and hence cross-bridge cycling) to achieve a given velocity of locomotion than do larger animals. For that reason, they must convert correspondingly larger amounts of metabolic energy to produce a given amount of force per unit cross section of contractile tissue in moving their limbs.

Several generalizations can be made in relating the overall energy cost of locomotion to the size and velocity of an animal. Locomotion is a metabolically expensive process. The rate of oxygen consumption in excess of the basal metabolic rate increases linearly with the velocity (Figure 16-43A). It is noteworthy, however, that the increase in energy utilization per unit weight for a given increment in speed is less for larger animals than for smaller ones. This can be seen in the different slopes of the plots in Figure 16-43A. When the cost of locomotion is plotted as energy utilization per gram of tissue per kilometer against body mass, it is again apparent that *larger animals expend less energy to move a given mass a given distance* (Figure 16-43B). The lower energy efficiency of small animals during locomotion may, to a limited degree, be due to the greater drag that they experience (to be discussed shortly), but this explanation certainly does not suffice for terrestrial animals moving at low and moderate speeds through air, where drag is negligible. More likely, the lower energy efficiency is related to the lower efficiency of rapidly contracting muscle.

The relation between velocity and the cost of locomotion is complex. As velocity of running increases in quadripedal mammals, for example, the metabolic cost of traveling a given distance initially decreases (Figure 16-44). This is because nonlocomotory expenses account for a progressively smaller fraction of the total energy expended. However, as velocity continues to increase, animals that swim, fly, or run all begin to experience an increase in the cost of locomotion as they generate near maximal locomotory velocities. Figure 16-45 illustrates this phenomenon in cephalopods (e.g., squids and nautilus), for which a typical U-shaped curve describes the cost of locomotion at various locomotory velocities. The cost of locomotion initially decreases sharply as velocity increases but then begins to rise at higher velocities.

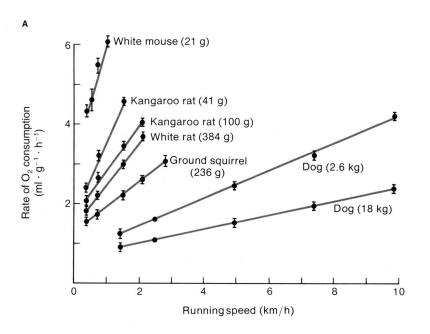

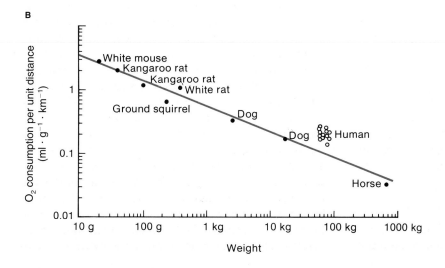

Figure 16-43 Metabolic rate during locomotion depends both on body size and on running speed. **(A)** Relation between rate of oxygen consumption and velocity of running in mammals of different sizes. The slope of each plot represents the cost of transporting a unit mass over a unit distance. **(B)** Log-log plot of metabolic cost of transporting 1 g a distance of 1 km in running mammals of different sizes. The cost of basal metabolism was subtracted before plotting the values. Data are from slopes of plots in part A. Values for tetrapods lie close to a straight line. [From Taylor et al., 1970.]

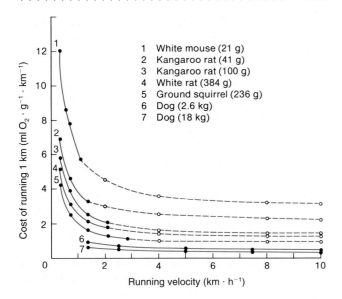

Figure 16-44 The energetic cost of transporting a unit of body mass by running decreases as body size increases in mammals. The cost of running 1 km drops and levels off with increasing velocity. Dashed proportions are extrapolated. [From Taylor et al., 1970.]

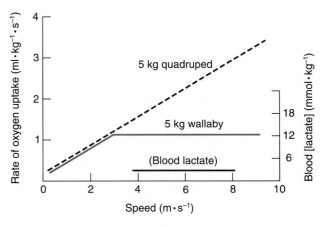

Figure 16-46 Hopping, bipedal animals such as wallabys and kangaroos can increase velocity with no increase in oxygen consumption. Similar-sized quadrapeds and wallabys initially show a linear increase in oxygen consumption as velocity increases. As wallabys switch to bipedal locomotion, however, their rate of oxygen consumption does not increase with further increases in velocity. That blood lactate similarly stays level as velocity increases indicates that the higher speeds are not achieved by an increase in anaerobic metabolism. [Adapted from Baudinette, 1991.]

A noted exception to the U-shaped, cost-velocity relation typical of many running animals is found in hopping bipedal animals, especially kangaroos and wallabies. At slow velocities, oxygen consumption increases linearly in both a wallaby and a typical quadriped of similar size (Figure 16-46). At moderate to high velocities, however, wallabies steadily increase velocity without increasing oxygen consumption—a seemingly impossible achievement. They do so by using their powerful hind legs as springs; the hind legs store much of the kinetic energy expended in elevating the animal's body mass during leg extension.

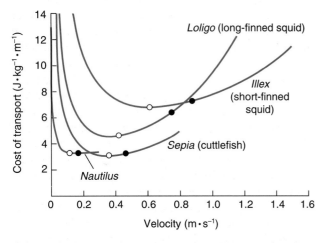

Figure 16-45 The metabolic cost of transport in cephalopods shows the "U" shape typical of many flying, swimming, and running animals. Both very slow and very rapid locomotion are relatively expensive. All cephalopods plotted weigh approximately 0.6 kg. [From O'Dor and Webber, 1991.]

Physical Factors Affecting Locomotion

The metabolic cost of moving a given mass of animal tissue over a given distance also depends on the purely physical factors of inertia and drag.

Inertia is the tendency of a mass to resist acceleration, whereas *momentum* refers to the tendency of a moving mass to sustain its velocity. These concepts are closely related, and effects due to both properties are often lumped under the term **inertial effects.**

Every object possesses both inertia and momentum proportional to its mass. The larger the animal, the greater its inertia, and the greater its momentum when it is in motion. The high inertial forces that must be overcome during the acceleration of a large animal account for a significant utilization of energy during the period of acceleration (Figure 16-47A). Small animals, like small cars or small airplanes, require less energy to accelerate to a given velocity. Likewise, they need less energy to decelerate. Therefore, a small animal starts and stops abruptly at the beginning and end of a locomotory effort, whereas a large animal accelerates more slowly after locomotion begins and slows down more gradually as locomotion ends (Figure 16-47B). Similarly, in terrestrial animals, the limbs are engaged in back-and-forth movements during running. The limbs are subject to inertial forces related to their mass as they accelerate and decelerate during locomotion. The limbs of a large animal exhibit greater inertia and momentum than do those of a small animal.

Because animals do not move in a vacuum, the energetics of sustained locomotion are affected by the physical properties of the gas or liquid through which they move. **Drag** is the force exerted in the opposite direction of an animal's movement and is caused by the viscosity and density of the gas or liquid environment through which the animal

A

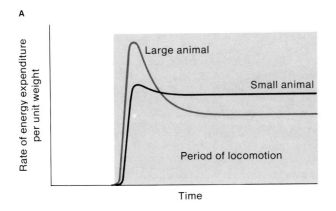

B

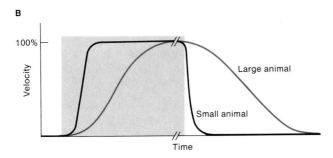

Figure 16-47 Body mass affects both rate of energy expenditure and acceleration during locomotion. **(A)** Rate at which energy is used per unit mass during onset and maintenance of locomotion (shaded region) in a large and a small animal of similar type. **(B)** Velocity of a small and a large animal during acceleration and deceleration at the beginning and end of a period of locomotion (shaded region).

moves. The drag produced in a given medium depends on the velocity, surface area, and shape of an object. For an object of a given shape, drag is proportional to the surface area. Because larger animals have lower surface-to-mass ratios, they experience less fluid drag per unit mass than do smaller animals, for whom overcoming drag is energetically more costly. Once it is under way, a larger animal expends less energy per unit mass to propel itself at a given velocity than does a smaller animal of similar type (see Figure 16-47A). Drag is also proportional to the *square* of an animal's velocity, meaning that the energy required to overcome drag and propel the animal at faster speeds increases with velocity.

These effects are far more pronounced in water than in air because water, having the higher viscosity and density, produces far more drag on a moving object than air does. Drag is of major importance in swimming and flying because of the high viscosity of water faced by swimmers and the high velocity experienced by flyers. Drag is of little importance in running, because the velocities attained in running are low, as is the viscosity of air. These relations are quantified in the Reynolds number (see Spotlight 16-2).

Aquatic, Aerial, and Terrestrial Locomotion

Animals have evolved myriad ways of moving in water, on land, and in air. Despite this diversity, each mode of loco-

motion is similarly constrained by the environment in which it is employed and by the laws of physics.

Swimming

Animals that swim in water need to support little or none of their own weight. Many have flotation bladders or large amounts of body fat that enable them to suspend themselves at a given depth with little expenditure of energy. However, although the high density of water allows them to be neutrally buoyant, it also produces high drag. This hindrance to objects moving through a fluid has led to a convergence of body forms among marine mammals and fishes. The streamlined, fusiform body shape is wonderfully developed in most sharks, teleost fishes, and dolphins. The reasons are evident enough on intuitive grounds, but they can be understood more clearly in relation to flow pattern.

The ease with which an object moves through water depends in part on the flow pattern of the water. The fluid at the immediate surface of the object moves at the same velocity as the object, whereas the fluid at a great distance is undisturbed. If the transition in fluid velocity is smoothly continuous as the fluid progresses away from the object's surface, then *laminar flow* (Chapter 12) occurs at the boundary layer—the layer of unstirred fluid immediately adjacent to the object's surface (Figure 16-48A). In contrast, *turbulent flow* results when there are sharp gradients and inconsistencies in the velocity of flow. Because of conservation of energy, pressure and velocity are reciprocally related in a given fluid system, and the higher the velocity of fluid at a given site, the lower the pressure. Thus, strongly differing flow rates around an object cause eddy currents owing to secondary flow patterns set up between regions of high and regions of low pressure. Moreover, the higher the viscosity of the medium or the higher the relative motion between the object and the surrounding fluid, the greater the shear forces produced and, hence, the greater the tendency toward turbulence. Because its production dissipates energy as heat, turbulence retards the efficient conversion of metabolic energy into propulsive movement.

Long, streamlined shapes promote laminar flow with minimal eddy current formation. Fishes and marine mammals such as seals, porpoises, and whales are admirably streamlined, exhibiting nearly turbulence-free passage through water even at high speeds. Flying birds are similarly streamlined in flight. An additional factor reducing turbulence in these animals is the compliance (deformability) of the body surface. A high body compliance damps small perturbations in the pressure of the water flowing over the body surface and thereby lessens the local variations in water pressure that give rise to energy-dissipating turbulence.

The speed of a swimming animal is proportional to the *power-to-drag* (thrust-to-drag) ratio. Power developed by contracting muscle is directly proportional to muscle mass, and, if we assume that muscle mass increases in proportion to total body mass, power (thrust) rises in proportion to body mass. On the other hand, for a large swimming animal, total drag increases but drag *per unit mass* decreases,

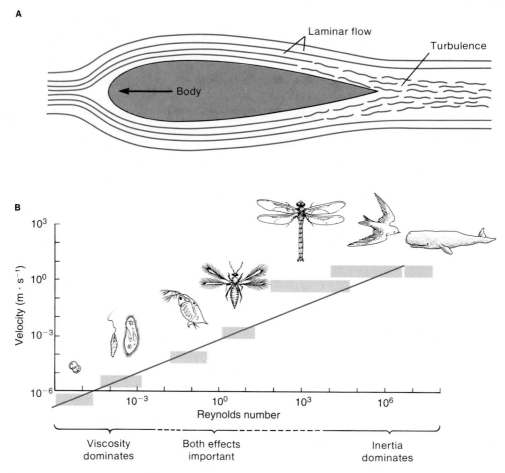

Figure 16-48 Velocity and fluid dynamics are highly dependent on body mass in both flying and swimming animals. **(A)** Fluid flow around a body moving through water. Movement through a fluid can create turbulence owing to uneven fluid pressures. Laminar flow occurs where pressure gradients are minimal. The larger the body and the less viscous the fluid, the higher the velocity before turbulence occurs. **(B)** Log of animal size plotted against log of the respective Reynolds numbers (*Re*s) at cruising velocities. Small animals that move slowly have small *Re*s because viscous forces predominate at small dimensions. Larger animals move rapidly with high *Re*s because inertial forces predominate at large dimensions. [Part B from Nachtigall, 1977.]

for a given velocity, as body mass increases. This is because, if the shape remains constant, the surface and cross-sectional areas (which determine the drag) increase as a function of some linear dimension squared, whereas body mass (which determines the power available) increases as a function of that linear dimension cubed. Thus, a large aquatic animal can develop power out of proportion to its drag forces; therefore, it is able to attain higher swimming velocities than those of a smaller animal of the same shape. Large fishes and mammals swim faster than their smaller counterparts. Because of the high drag forces developed in water and because drag increases as the square of velocity, aquatic animals can reach the speeds of a bird in flight only if they are much larger and more powerful than the bird.

Flying

Unlike water, air offers little buoyant support, so all flyers must overcome gravity by utilizing principles of aerodynamic *lift*. Although the effects of drag increase with speed, there is still less need for streamlining among birds than among fishes because of the low density of air. Thus, thanks to the relatively low drag forces developed, birds can achieve much higher speeds than those of fishes. The production of propulsive force, which drives the bird forward, and lift, which keeps it aloft, is accomplished simultaneously during the downstroke of the bird wing (Figure 16-49). The wing is driven downward and forward with an angle of attack that pushes the air both downward and backward, creating an upward and forward *thrust*. The components of lift and forward propulsion overcome the bird's weight and drag, respectively.

The body shapes of fishes and birds demonstrate the great differences that exist between the physical properties of water and those of air, as well as the biological divergence that results from adaptation to these two dissimilar media. When a bird is gliding, its elongated, extended wings that take the form of an airfoil produce excellent lift; but, if the bird were in water, they would obviously generate far too much drag. Thus, the wings of penguins are modified to serve as short paddles and are folded against the body while the birds are coasting under water. Because drag forces are much higher in water than in air, only

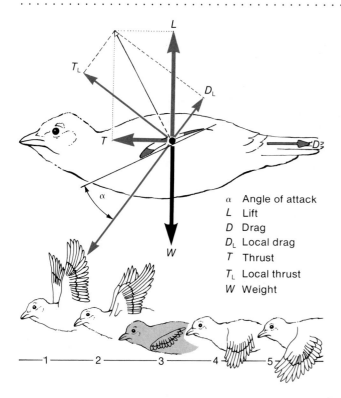

Figure 16-49 The downstroke of a bird wing develops forces in several directions. *(Top)* the wing in stage 3 of the flight cycle shown at the bottom. Red arrows illustrate forces relating to the wingbeat; black arrow illustrates force relating to the body. Induced drag equals the drag produced as a consequence of lift production. Induced thrust is complementary to induced drag. [From Nachtigall, 1977.]

α Angle of attack
L Lift
D Drag
D_L Local drag
T Thrust
T_L Local thrust
W Weight

medium-sized to large animals can coast in water. In contrast, only very small flying animals, smaller than a dragonfly, are unable to coast (glide) in air. Small insects such as flies and mosquitoes must continually beat their wings to maintain headway, because they have very little momentum.

Running
When swimming, flying, and running are compared with respect to the energy cost of moving a given body mass a given distance (Figure 16-50), it is apparent that terrestrial locomotion (i.e., running) is the most costly, whereas swimming is the least expensive. A swimming fish expends less energy in locomotion than a bird does in flying through the air because, as already noted, the fish is close to neutral buoyancy, whereas a bird must expend energy to stay aloft. But why is running less efficient than either flying or swimming?

Running differs from swimming and flying in the way limb muscles are used, and this difference accounts for the low work efficiency of running. When a biped or quadruped animal runs, its *center of mass* (CM) rises and falls cyclically with the gait. The rise in the CM occurs when the foot and leg extensors push the body up and forward, and the fall occurs as gravity inexorably tugs at the body, bringing it back to earth between locomotory extensions. Effi-

ciency is lost because the antigravity extensor muscles that contract to propel the CM upward and forward must also break the fall of the CM that occurs before the next stride. To control the fall, the extensor muscles must expend energy to resist lengthening as they slow the rate of descent of the body in preparation for the next cycle. This technically unproductive use of muscle energy to counteract the pull of gravity is said to produce "negative work." You may be familiar with such negative work carried out by your legs' extensor muscles while hiking *down* a steep trail.

In short, running or walking is less efficient than flying, swimming, or bicycling because the muscles must be used for deceleration (negative work) as well as acceleration (positive work). One reason that a person riding a bicycle is so efficient (and thus why people can cycle many times faster and farther than they can run) is that the CM does not rise and fall, which means that more muscular energy can be transferred into forward velocity.

Elastic energy storage in elastic elements of the limbs appears to be especially important in running and hopping animals. Consider the bounding of a kangaroo. The greater the height achieved during the hop, the greater is the speed of descent and, when the legs strike the ground, the more energy is transferred to the elastic elements in the limbs and the greater the force of elastic recoil of the limbs when they subsequently extend at the beginning of the next hop. Not many terrestrial animals actually hop, but the concept of elastic energy storage is important when considering changes in gait (e.g., walking to trotting to galloping in a horse). By changing gait at appropriate speeds, land animals enhance their locomotory efficiency and avoid potentially injurious forces on legs. For example, consider a pony made to trot on a treadmill at a speed at which it would normally gallop, or to gallop when it would normally trot, or to trot when it would normally walk. In all cases, it expends more energy than if allowed to change its gait naturally. Optimum gaits result from the relative amounts of energy stored in the elastic elements of the body, such as tendons, when performing the different gaits. For instance, little energy is stored when an animal walks; somewhat more is stored when it trots. When the animal is galloping, its entire trunk is involved in elastic storage. At least half the negative work done in absorbing kinetic energy during the stretching of an active muscle appears as heat; the remainder is stored in stretched elastic elements such as muscle cross-bridges, sarcoplasmic reticulum, and Z-lines of muscle cells and tendons. Only the elastically stored energy is available for recovery on the rebound, and only from 60% to 80% of that is recovered on its release. The energy converted into heat is not available for conversion into mechanical work in living tissue.

Locomotory energetics of ectotherms versus endotherms
You might think that, on simple energetic grounds, terrestrial endotherms and ectotherms of the same size will expend the same net metabolic energy to run at a given velocity. This reasonable assumption is almost, but not

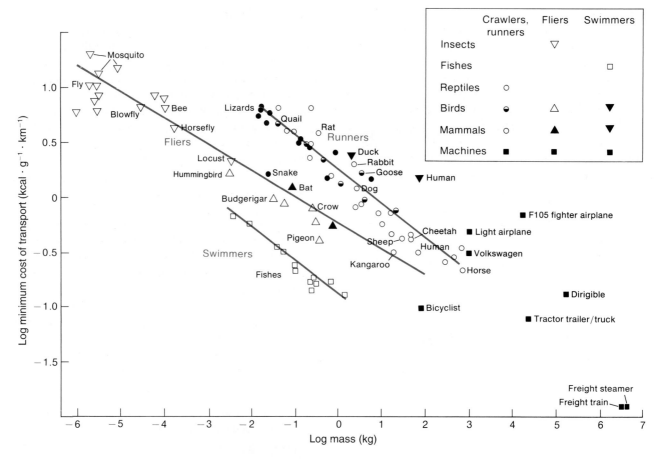

Figure 16-50 Cost of transport is more closely related to the kind of locomotion than it is to the kind of organism. Cost is given in kilo-calories per gram per kilometer for animals as well as machines. [From Tucker, 1975.]

entirely, correct. When O_2 consumption is plotted against running velocity for a lizard and a mammal of similar size, the aerobic parts of the plots for both exhibit rather similar slopes. Thus, when movement begins, a similar increment in metabolic energy expenditure is required for a similar incremental increase in velocity for both the mammal and the lizard. The difference between the two animals lies in the lower y-intercept of the lizard's plot relative to its standard metabolic rate while at rest. The reason for the differences between the resting metabolic rates and the y-intercepts is not completely certain, but these differences may represent the "postural cost" of locomotion—this cost being higher for a mammal than for a lizard.

As noted earlier, the rate of O_2 consumption rises linearly with increasing velocity of locomotion. This relation is true for ectotherms as well as for endotherms. An endotherm of a given mass typically has a basal metabolic rate about 6 to 10 times that of an ectotherm of similar mass. In both groups, a similar relation exists between the basal rate and the maximum metabolic rate that can be achieved with intense exercise. That is, the *factorial scope for locomotion* exhibited by both groups is about the same. Thus, in response to intense exercise, endotherms can achieve a maximum rate of O_2 consumption of as much as tenfold that of ectotherms of similar size, and so an en-

dotherm of a given size can achieve a higher rate of activity while undergoing aerobic metabolism than a similar-sized ectotherm can.

The locomotor speed at which the maximum rate of aerobic respiration is reached is termed the **maximum aerobic velocity** (MAV). As an animal exceeds its maximum aerobic velocity, the additional activity is supported entirely by anaerobic metabolism, which leads to glycolytic production of lactic acid. As lactic acid production progresses, an oxygen debt develops (see *Metabolic scope,* earlier in this chapter). Anaerobic metabolism is also associated with muscle fatigue (due to progressive depletion of chemical energy stores) and metabolic acidosis, which if extreme can disrupt tissue metabolism. Because of these two consequences, anaerobic metabolism is unsuitable for sustained activity. Thus, only locomotion below the maximum aerobic speed can be sustained by animals of either group. Because endotherms are capable of far higher rates of aerobic metabolism than are ectotherms, they are generally capable of higher rates of sustained locomotor activity.

Clearly, the implications for ectothermy and endothermy are not limited to mechanisms of temperature control but are also of great importance to the kinds of activity that an animal can undertake. The metabolic differences between ectotherms and endotherms determine, for example, how far and how fast they can travel. This is not to say that ectotherms cannot achieve rates of activity and speeds of locomotion as high as those of endotherms. However, because locomotor activity in excess of the maximum aerobic velocity relies on prodigious rates of anaerobic metabolism, high rates of locomotion can be sustained by ectotherms for only brief periods. This limitation can be observed in ectothermic vertebrates such as some species of lizards and frogs, in which a rapid burst of activity, seldom lasting more than a few seconds, takes the animal quickly to a new resting or hiding place when disturbed. In some fishes, more than 50% of body mass is glycolytic white muscle fibers specialized for very short bursts of locomotor activity. The disadvantage that ectotherms have in sustaining high rates of activity is offset by their more modest energy requirements, which enable them to spend more time hiding and less time looking for a meal.

 Why can fleas leap hundreds of times their own body length, whereas most medium-sized mammals can jump only a few body lengths, and large mammals do not even attempt to leap? Consider both the physical surroundings of the animals and their own structural makeup.

BODY RHYTHMS AND ENERGETICS

Most animals strive to maintain some sort of constancy in their interior milieu. Although variations in body temperature, metabolic rate, intracellular pH, and body energy content, for example, may of necessity fluctuate widely in response to environmental constraints and demands, most animals have a preferred range for these and other physiological variables. Despite the evolution of mechanisms that help achieve this relative constancy, almost all animals also show an innate, usually subtle, rhythmic variation in these variables. These variations occur on a daily, tidal, lunar, or other basis and can usually be linked to a rhythmic change in the animal's environment.

Early experiments in biological rhythms by **chronobiologists** concentrated on endotherms, where variations were found in body temperature and metabolic rate. It has been known for centuries, for example, that the body temperature of human beings living on typical activity-sleep cycles falls by a half degree or so during the early morning hours (about 3:00–5:00 A.M.), only to rise again at about normal waking time. In fact, virtually all animals and plants show some type of rhythmic variation in metabolism or some other physiological variable. Biological rhythms are so innate to animals that even individual cells in cell culture will show rhythms in rate of cell division. The cell does not have to be a particularly complex cell—a daily rhythm occurs, for example, in prokaryotic cells such as nitrogen-fixing cyanobacteria.

Circadian Rhythms

Biological rhythms lasting from milliseconds (at the cellular level) to years (at the whole-animal level) have been identified in a wide variety of animals. Most rhythms (and certainly many of the most prominent and well studied rhythms) relate to daily cycles, which are called **circadian rhythms.** A true circadian rhythm, which is endogenously generated, can be distinguished from a physiological or other variable that just happens to track daily changes in environment by the use of four different criteria.

First, a circadian rhythm shows *persistence,* remaining for at least several days or weeks in an animal that has been removed from the natural environment and placed in a laboratory setting with constant environmental conditions (constant temperature, constant light or constant dark, etc.). A true circadian rhythm will persist in the animal, manifest most often as a continuation of that animal's normal daily cycles, the innate period being near 24 hours. Although any of a variety of physiological or behavioral characteristics can be measured, one of the most frequently measured is general activity. Figure 16-51 shows a typical apparatus for measuring the locomotor activity of a rodent. When the animal runs in the exercise wheel, the activity is recorded, either directly on a chart recorder or on a

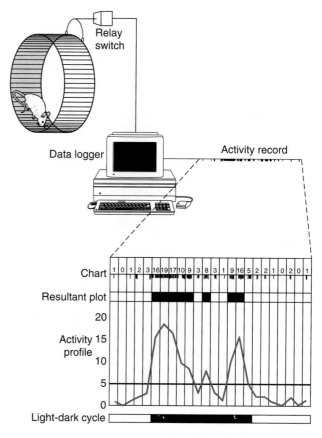

Figure 16-51 Circadian rhythms can be recorded by monitoring spontaneous activity. In this example, a rodent runs in an exercise wheel, which activates a switch closing an electric circuit every time it turns. The resulting electrical record of activity can be written directly on a chart recorder or, more conveniently, logged in a computer for later analysis.

computer. The apparatus can be modified by replacement of the exercise wheel with some other form of activity-measuring device to record activity in birds, fishes, or virtually any other animal. Figure 16-52 shows activity levels in a blinded house sparrow. For the first two weeks of the record, the sparrow was in a light-dark cycle. Even though the bird was unable to see, its circadian rhythm nonetheless persisted, with a free-running period several minutes longer than 24 hours.

The second characteristic feature of circadian rhythms is that they tend to be largely *body-temperature independent*. We have seen that metabolism and physiological processes that derive from and contribute to metabolism have a temperature quotient, Q_{10}, of between 2 and 3. Yet, a rise in temperature typically causes very little or no increase in the cycling of the circadian rhythm; in some animals, an increase in body temperature may actually slow down the circadian rhythm.

Circadian rhythms are also characterized by the fact that they can be made *conditionally arrhythmic*—that is, a certain set of environmental temperatures, lighting regimes,

oxygen levels, and so forth, can disrupt the normal circadian rhythm. Often, there is a threshold level of temperature below which the rhythm is finally disrupted. The effects of light are more graded. In the mosquito, forexample, a circadian activity rhythm is evident when the light phase of the light-dark cycle consists only of low light, but this rhythm gradually diminishes as light intensity increases.

A final characteristic feature of circadian rhythms is that they can be *entrained*. If an animal is placed in total darkness, for example, the length of its circadian rhythm remains close to 24 hours but is generally slightly shorter or longer, resulting in a progressive "creep" in activity cycles that is quite obvious in a long-term record of an animal's activity (see Figure 16-52). However, if the animal in darkness is then given a new lighting regime with a periodicity of slightly longer or shorter than 24 hours, the animal's activity will be entrained to the new lighting regime. Reentrainment is not instantaneous but takes place in a series of transients that manifest the inability of the internal clock to be shifted more than a certain amount each cycle. Through the use of new light cues, activity cycles can be advanced or delayed to the point of being in a completely opposite phase to that of the original circadian rhythm. In the experiment with the blinded house sparrow (see Figure 16-52), the feathers were removed from the top of its head. This allowed light to penetrate through the skull to the brain, parts of which are light sensitive, and the sparrow's activity was then entrained to the light-dark cycle. As the feathers grew in, the entrainment began to be lost and the circadian rhythm lengthened, but removal of the feathers for a second time returned the entrained activity rhythm. In a final experiment, injection of dye under the scalp on top of the skull blocked light penetration to the brain, and the circadian activity rhythm began to drift once again.

Light is the most commonly effective **zeitgeber,** or environmental entrainment factor. However, environmental temperature, food availability, and interactions with other animals of the same or different species also may be zeitgebers that exert an effect on metabolism, activity, and other basic facets of animal life.

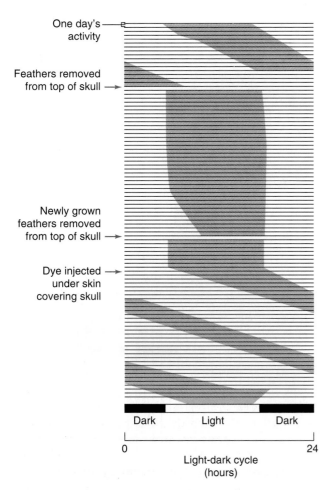

One day's activity

Feathers removed from top of skull →

Newly grown feathers removed from top of skull →

Dye injected under skin covering skull →

Dark   Light   Dark

0   24

Light-dark cycle (hours)

Figure 16-52 Circadian rhythms can be entrained by light or other cues. In this experiment with a blinded house sparrow in a light-dark cycle of lighting, even the small amount of light reaching the brain through the top of the skull is sufficient to entrain a daily activity rhythm. [Adapted from Menaker, 1968.]

 If the most basic biochemical reactions of cells are temperature sensitive, how can an inherent "biological clock" controlling circadian rhythms itself be temperature independent?

Noncircadian Endogenous Rhythms

With the circadian rhythm as the "standard," endogenous biological rhythms can be classified into **infradian rhythms,** less than a day in length, and **ultradian rhythms,** greater than a day in length.

Infradian cycles are usually related to aspects of cell function. In fact, some 400 distinct infradian rhythms in cell function have been identified to date. These infradian cycles greatly affect animal energetics, but the effects are more difficult to measure than are changes occurring on a daily or longer basis. Many infradian rhythms, such as those related to certain aspects of cell division, have not yet been correlated with any type of rhythmic environmental change. In some cases, external environment may have little or no role; in other cases, we have probably just failed to identify the environmental factor that entrains the rhythm.

Ultradian rhythms are very common in animals. The influences of the moon through both its light and its production of ocean tides greatly affects the physiology of many intertidal animals. **Circatidal rhythms,** which correlate with tidal cycles, are generally 12.4 hours in length. Many intertidal animals show such rhythms, which have many of the same characteristics (except for their length) as circadian rhythms (Figure 16-53). **Circalunar rhythms** correlate with the 29.5 day lunar cycle and affect reproduction in many animals. **Circannual cycles** correlate with the 365 day earth year and are most evident in the often strong, seasonal cycles that affect everything from fur color to hibernation to migration of many animals. Note that all of the *circ-* rhythms are endogenous rhythms—they persist if external cues are removed, and they are entrainable.

Temperature Regulation, Metabolism, and Biological Rhythms

Many animals show circadian or other rhythms in body temperature. In endotherms, considerable amounts of energy are used in maintaining constant body temperature, either directly through thermogenesis or indirectly by powering physiological mechanisms that regulate heat loss or gain or both. In ectotherms, body temperature directly affects the animal's metabolism. Thus, for both endoderms and ectotherms, circadian and other rhythms affecting body temperature also affect an animal's energy metabolism. Because the consequences of rhythms on temperature regulation and metabolism are inseparable, we will consider them together.

Vertebrate endotherms
Circadian rhythms in body temperature have been identified in most birds and mammals. There is a relatively strong scaling effect, with smaller animals showing larger circadian variation in body temperature. Thus, human beings weighing from 50 to 80 kg show a daily variation of about 0.6 Celsius degree, whereas the far smaller shrews, deer mice, and hummingbirds, each weighing only a few grams, may show daily temperature fluctuations of as much as 20 degrees. The large daily variation in these small endotherms is probably due to the fact that they frequently enter into a nightly (or, in some cases, daily) torpor. The large daily swing in body temperature probably relates to the greater metabolic cost of maintaining body temperature in very small endotherms. Consequently, the smaller the animal, the greater will be the energy savings from allowing body temperature to fall by several degrees.

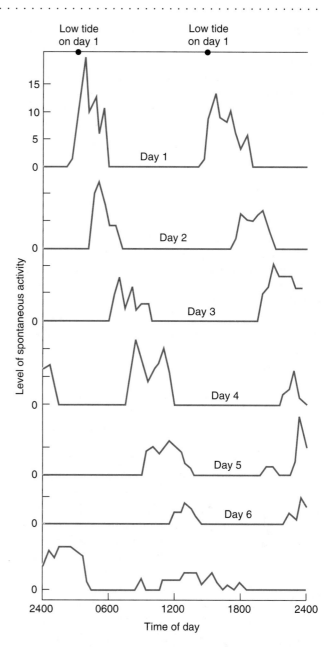

Figure 16-53 A fiddler crab kept in constant dark with no tide continues to show a tidal activity rhythm. There are two low tides each day, which occur on average 50 minutes later each day. The crab's activity continues to coincide with these tide cycles. [Adapted from Palmer, 1973.]

What is the root cause of a circadian body temperature rhythm in endotherms? Because body temperature in an endotherm is a function of its heat production and heat loss and gain from the environment, it follows that one or more of these factors must show rhythmic change to account for daily or other variations in body temperature. Relatively few studies have examined the total heat production and heat conduction budgets of an endotherm in the context of circadian or other rhythms. However, experiments in human subjects have simultaneously measured the circadian rhythms in body temperature, heat conductance, and heat production. These data show that changes in heat production (i.e., changes in metabolic rate) account for about one-quarter of the 0.6 degree daily swing in core temperature, with about three-quarters of the change resulting from changes in heat conductance between the core and the environment.

Some endotherms modify the amplitude, but not the periodicity, of circadian body temperature rhythms when exposed to stresses ranging from temperature extremes to inadequate food or water or both. In a classic, late-1950s study on endotherm thermoregulation, K. Schmidt-Nielson and his colleagues studied African camels *(Camelus dromedarius)* in the Sahara desert of Algeria. Well hydrated and fed camels showed a circadian body temperature rhythm with an amplitude of about 2 Celsius degrees, but this increased to about 6 degrees when the camels were unable to drink water. The maximum core temperature (in late afternoon) was increased, whereas the minimum core temperature (in early morning) decreased. These changes are presumably to reduce water loss through evaporative cooling in the day and to decrease heat loss because of a reduced thermal gradient to the cooler nighttime surroundings. Both reduce the need for metabolic heat production. Birds such as kestrels and pigeons, which also show circadian body temperature rhythms, similarly showed a larger daily core temperature range, mainly owing to a lower nighttime core temperature.

Field observations alone may be insufficient to identify circadian rhythms in regulated body temperature because many endotherms also show strong daily rhythms in activity levels. Depending on the animal's ability to dissipate metabolically produced heat, a rhythmic rise in body temperature could be solely the result of increased locomotor activity (which itself is a manifestation of circadian rhythms). However, two lines of evidence indicate that there is usually a distinct intrinsic rhythm in body temperature independent of activity: (1) temperature rhythms persist in animals in the laboratory in which activity levels have been corrected for or controlled, and (2) temperature rhythms persist in human beings over several days of complete bed rest. In reality, circadian rhythms in body temperature are often imposed on activity rhythms with a similar time component, amplifying the daily range of body temperature.

Ultradian rhythms in body temperature are best exemplified by the hibernators, which, as noted earlier, lower their core temperature by 20–35 Celsius degrees for periods of weeks or months, punctuated by brief periods of arousal. These rhythms can persist for at least four years in hibernating golden-mantled ground squirrels *(Citellus)* isolated at birth from any light or temperature zeitgebers. In hibernating bats, circadian rhythms in body temperature and metabolism can be measured at the new, much lower mean core temperature typical of hibernation. This emphasizes the generally temperature independent nature of the biological clock responsible for circadian rhythms. However, the circadian rhythm eventually disappears as hibernation continues. Rodents such as the 13-lined ground squirrel *(Spermophilus tridecemlineatus)* show no evidence

of a continuing circadian rhythm in O_2 consumption with the onset of hibernation.

Some mammals show little or no evidence of a circadian rhythm in either body temperature or metabolic rate. Such animals tend to be those that live in environments with very stable conditions of temperature, light, food availability, and so forth. Fossorial (burrow-dwelling) pocket gophers and moles, for example, live in constant darkness with little temperature variation and show no metabolic circadian rhythms. It is unclear what advantage there would be to a significant body temperature and metabolic rhythm in such animals. Mammals such as voles, which have a herbivorous diet with high bulk, feed nearly constantly to derive sufficient energy. These animals similarly show little or no metabolic rhythmicity.

Vertebrate ectotherms

All ectotherms by definition rely on external sources of heat to elevate body temperature. However, both the behavioral and the physiological adjustments made to regulate body temperature are modified by circadian rhythms in preferred body temperatures. Because metabolic rate is closely related to body temperature, circadian rhythms in O_2 consumption and CO_2 production are closely correlated with body temperature changes.

Fishes have long been known to show circadian rhythms in activity, body temperature, and metabolic rate. In many cases, daily periods of activity correspond to the highest body temperatures and metabolic rates. In a classic study, J. R. Brett (1971) examined lake-dwelling sockeye salmon, *Oncorhyncus nerka,* monitoring position in the water column (Figure 16-54). During the day, these salmon stayed in deep, cold water, and presumably both body temperature and metabolism reflected this low wa-

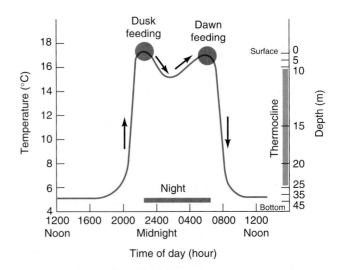

Figure 16-54 Vertical migration in sockeye salmon *(Onchorhynchus nerka)* shows a strong diurnal rhythm. These fish rise through the water column to feed at dusk. They remain near the surface to feed at dawn before descending to cold, deep water during the day. [Adapted from Brett, 1971.]

ter temperature. As dusk approached, the fish rose to the surface to feed, which brought them through the thermocline into water of about 17°C, where they remained for a bout of dawn feeding before descending to cold water for the day. This overall pattern of activity allows the sockeye salmon to conserve energy by having a cold-induced, low metabolic rate during periods of inactivity. Such circadian rhythms of vertical migration related to feeding are very common in pelagic fishes. When they take animals through temperature gradients, then metabolic rate will similarly show daily variations. But are there true circadian rhythms in metabolic rate or are these changes merely reflecting body temperature changes? In fact, daily rhythms in oxygen consumption persist in conditions of constant temperature and light in many fish species.

Studies on circadian rhythms in temperature regulation and metabolism in amphibians are few. Circadian rhythms in preferred body temperature and activity have been found in the aquatic salamander *Necturus maculosus* but not in the toad *Bufo boreas,* the larvae of the frog *Rana cascadae,* or the salamander *Plethodon cinereus.* Many species of toads and frogs show pronounced activity patterns related to feeding, predation, and so forth, but we currently know very little about how many of these patterns are cued by the external environment and how many are due to innate circadian rhythms—that is, are controlled by "biological clocks."

Many reptiles show a circadian rhythm in preferred body temperature. Because resting metabolism tracks body temperature in these animals, metabolic rate correspondingly ranges up and down in the course of a 24-hour period. Again, establishment of whether these thermoregulatory and metabolic rhythms are true innate rhythms requires monitoring animals in constant environmental conditions. In fact, in the lizard *Sceloporus occidentalis* and other species, a circadian rhythm in preferred body temperature persists for several days under constant light conditions in the laboratory. Similarly, circadian rhythms in oxygen consumption are found in some species of the lizard *Lacerta.*

Invertebrates

Invertebrates from many different phyla have been identified as having some degree of thermoregulatory capacity, primarily through behavioral means but also by physiological means (e.g., the metabolic heat production in flying insects considered earlier). Daily or other rhythms in preferred body temperatures have been observed in numerous Arthropoda—among them crayfish, shrimp, and a variety of insects, including silkmoths, bees, and wasps. Oxygen consumption tracks body temperature in these animals, so metabolic rate similarly shows daily cycles if an animal changes its body temperature. Daily rhythms in oxygen consumption have also been observed in earthworms, amphipods, seapens, and mollusks. Intertidal animals may show combinations of circadian, tidal, and lunar metabolic cycles.

With few exceptions, it is unknown whether these daily rhythms are endogenous or are exogenously triggered.

Male American silkmoths, *Hyalophora cecropia,* show endogenous rhythms of endothermic warming in preparation for flight; these rhythms are unaffected by ambient temperature. Honey bees *(Apis mellifera)* show very prominent circadian rhythms in oxygen consumption when kept in constant darkness, with oxygen consumption increasing from 20 to 30 times over resting levels during the periods that correspond to daylight activity and foraging. The fruit fly *Drosophila* similarly shows an increase in metabolic rate during the period corresponding to daylight when kept in constant darkness.

Unicellular organisms

The presence of true circadian or other rhythms in unicellular organisms is of particular interest to chronobiologists. Although the study of more complex animals has implicated the brain, pineal gland, and other tissues as the site of a biological clock, the presence of true biological rhythms in unicellular organisms indicates that all the necessary components for a clock can be found among the cellular organelles.

Unicellular organisms exhibit rhythms in rates of photosynthesis, oxidative metabolism, bioluminescence, cell division, growth, phototaxis, and vertical migration—to name but a few variables. The first definitive demonstration of an innate circadian rhythm was made in 1948 on the unicellular alga *Euglena gracilis,* which shows phototactic rhythms (migration to light). Since that time, a variety of other eukaryotic cells, including *Paramecium,* have been shown to exhibit true circadian rhythms that persist under constant conditions and can be entrained. More recently, circadian rhythms have been found in the much simpler prokaryotic cells of cyanobacteria.

The knowledge that individual cells show circadian cycles in cell division has been put to use in designing more effective chemotherapy for human cancer patients. Different chemotherapeutic agents act on different phases of the cell division cycle, which in human beings has a circadian rhythm. By timing the administration of the drug to the vulnerable period of the cell cycle (typically, two o'clock in the morning rather than during the usual business hours), as much as a 10-fold increase in effectiveness has been achieved. Moreover, undesirable and deleterious side effects of the treatment are also greatly reduced.

ENERGETICS OF REPRODUCTION

Reproduction is the ultimate goal of all organisms, and the evolution of virtually all specializations can be linked directly or indirectly to improvement in an animal's reproductive fitness. Given the overall importance of reproduction, it is not surprising that this process accounts for a considerable proportion of an animal's energy budget. Exactly what proportion depends on many factors, including mode of reproduction, body size, whether an animal is ectothermic or endothermic, and so forth. We begin by considering the different patterns of energy investment in reproduction that have evolved.

Patterns of Energetic Investment in Reproduction

Natural selection of reproductive structures and processes has resulted in a wide variety of reproductive patterns. The most favorable mode for a species is that which maximizes the reproductive value of the offspring by raising the largest number possible to sexual maturity. By the mid-1960s, several groups of ecologists and evolutionary biologists had recognized that, collectively, animals had evolved one of two general patterns of energy investment in reproduction. That is, they recognized that a given amount of energy to be invested in producing offspring can be partitioned, or "spent," in two different ways. These two patterns were called *r*-selection and *K*-selection, with the letters *r* and *K* coming from the logistics equation that models the growth rate of continually reproducing animal populations. (Consult a text in ecology or evolution for further details on the logistics equation and the growth of animal populations.)

r Selection — "smaller and more"

In the first pattern of energetic investment in reproduction, as exhibited by r-*selected animals,* an animal produces offspring that, at the start of their development, are very small. By virtue of each offspring having a very small energy content (and thus a small energy cost to the parents), the parent(s) can produce much larger numbers of offspring. Females of some species of sea urchins, for example, release as many as 100,000,000 eggs in a spawning. Among vertebrates, pelagic fishes may similarly release huge numbers of fertilized eggs; for example, the mackerel *Scomber scombrus* releases tens of thousands of eggs in a spawning. These are extreme examples, but most invertebrates and many ectothermic vertebrates produce dozens or more offspring in a single breeding episode.

The "trade off" in producing large numbers of small offspring is that the parent is less able to afford parental care to each of so many offspring. Most r-selected animals simply release their offspring into the environment to fend for themselves. Because they are small and vulnerable, relatively few survive to reproduce. The probability of a mackerel spawn surviving to reproductive age is 0.000006.

K Selection — "larger and fewer"

The second pattern of energetic investment in reproduction is exhibited by K-*selected animals.* These animals produce relatively large offspring — offspring that have a high energy content and represent a large energetic investment by the parents. As a consequence, the number of offspring produced is far smaller than that produced by r-selected animals. Mammals and birds, for example, tend to produce offspring that, at birth or hatching, are at least a few percent of the body mass of the mother and may be much larger. Litters or egg clutches rarely consist of more than 8 to 10 offspring, because of the huge energy cost of producing them. However, because numbers are small, the offspring are more manageable, and K-selected animals usually invest additional energy in parental care (see *Parental care as an energy cost of reproduction* later in this chapter).

Because the offspring of *K*-selected animals tend to be large and cared for during their early development, their chances of successfully reaching reproductive age are far greater. Contrast the six-in-a-million chance of a mackerel surviving with that of a bird or mammal, whose chances may approach 50% or higher.

The classification of animals as *r*- or *K*-selected, however, is not absolute, and many species show characteristics of both. For example, many Cichlid fishes produce hundreds of small larvae (an "*r*" characteristic). However, they then show "*K*" characteristics by investing huge amounts of time and energy in parental care — females may even stop feeding themselves for weeks to protect the young.

 Which population of animals — *r*-selected or *K*-selected — would experience a greater proportion of juvenile mortality if the availability of energy in the form of food were to become severely limited?

Allometry and the energy cost of reproduction

Like virtually all other aspects of an animal's physiology, allometric scaling affects the energy cost of reproduction. M. Reiss (1989) assembled data on the energy cost of reproduction within taxa that ranged from spiders to salamanders to mammals. These data show that, in general, larger animals invest relatively less energy in their offspring than do smaller animals. The value of the exponent in the allometric equation relating energy cost of reproduction to body mass ranges from a low of 0.52 in ducks and geese to 0.95 in hoverflies, with mammals showing a range of about 0.69–0.83.

If we consider both invertebrates and vertebrates, *within* a species, larger females devote relatively more energy to reproduction than do smaller females. One can see this clearly in mammals: a large, fat rodent, feline, or canine produces a larger litter than does a thin female with few energy reserves. Yet, in species with a sexual dimorphism in which males are larger than females, the data suggest that larger males expend relatively less energy on reproduction than do smaller males.

The "Cost" of Gamete Production

Reproduction begins with the production of gametes (eggs and sperm). The energy cost of gamete production varies greatly. Gamete production is a costly business in most invertebrates, with half or even more of the total energy assimilated being diverted into the gametes. Usually, the female makes the largest energy expenditure in producing yolky eggs that will nurture the growing embryos. Sperm production usually requires a lower energy expenditure. However, some males expend disproportionately large amounts of energy on sperm production. The testes and associated accessory organs in the male cricket (*Acheta domesticus),* for example, account for 25% of the animal's body mass. The spermatophore (the packet containing sperm that is passed to the female during copulation) that it produces is about 2.5% of its body mass, and it can produce two or more spermatophores per day.

Although the energy cost of gamete production may be high, very few invertebrates expend energy on protecting and nurturing their offspring after they are produced. Notable exceptions are found among the arthropods. Female scorpions carry their newly hatched offspring on their backs. For insects such as ants and bees, the entire social structure is organized on the care and nurturing of the offspring (see next subsection).

Ectothermic vertebrates, like invertebrates, spend a large proportion (as much as half) of their total energy budget on gamete production and reproduction. Sperm production is relatively inexpensive for males of all species, but egg production can be energetically costly. As already mentioned, many fishes produce large clutches of eggs. Different species of oviparous (egg-laying) amphibians produce egg clutches ranging from 4 to 15,000 eggs. Estimates of the energy costs are few; but, in the salamander *Desmognathusochrophaeus,* about 48% of the female's energy is used in a combination of egg production and parental care. In the lizard *Uta stansburiana,* reproductive behavior of males consumes 32% of their energy during spring, and the females expend as much as 83% of their energy (26% for increased metabolic costs and 57% for producing eggs) on reproduction. Female crocodiles, alligators, and, especially, brooding snakes such as pythons similarly invest large amounts of energy in combined egg production and parental care.

The energy cost of gamete production is highly variable among endothermic animals. In birds, some of which produce large numbers of relatively large, yolky eggs, estimates of the cost of egg laying range from less than 10% to more than 30% of the animal's total energy budget. Domestic chicken hens (white leghorn), which have been subjected to human selection to maximize efficiency of egg production, expend about 15% to 20% of their energy on egg production. The energy cost of producing sperm in roosters, however, is negligible. Similarly, the cost of sperm production in mammals is virtually negligible. Almost all mammals produce very small numbers of very small eggs and, again, their energy cost is negligible. However, as we will see next, almost all mammals invest a considerable amount of energy in protecting the developing offspring after fertilization.

Parental Care as an Energy Cost of Reproduction

Many animal species do not exhibit parental care but, in those species that do, it can be a major energy cost. Parental costs fall into two categories. In the first category is the cost of the actual transfer of material from a parent to the developing offspring. One of the best examples of this is mammalian **lactation,** in which large amounts of milk rich in lipids and carbohydrates are secreted to nourish the newborns. Typically, lactating mammals expend as much as 40% of their energy on producing milk. In dairy cattle, which have been selected for copious milk production, as much as 50% of total energy may be used for milk production. Animals other than mammals also produce secretions for their developing offspring. One or both parents of many bird species regurgitate semidigested food into the mouths of offspring. Although this practice is not as energetically costly as producing the same weight of milk, it is a real cost in that the regurgitated food might otherwise be digested and assimilated into the parent's own body. Doves produce "crop milk," a viscous fluid formed from the initial digestion of material temporarily stored in the crop and then given to their offspring. Certain species of viviparous and ovoviviparous amphibians and reptiles produce uterine secretions ("uterine milk") that nourishes the embryos until birth. Embryos of viviparous caecilian (apodan) amphibians use specialized dentition to scrape away and ingest the lining of the oviduct. The female of the poison-dart frog *Dendrobates pumilio* returns to the small ponds holding her larval offspring and deposits unfertilized eggs for her larvae to eat. Among the invertebrates, ants, bees, and wasps can expend large amounts of their total energy in gathering raw materials and producing honey or equivalent substances for the nourishment of the developing animals in the colony.

The second type of cost of reproduction associated with parenting is the metabolic cost of behaviors specifically associated with parenting. Complex parental care is evident in numerous invertebrate taxa, including mollusks *(Octopus),* polychaete worms, and the social insects (ants, bees, and wasps), which have elaborate behaviors of nest building, brooding, and so forth. Among the vertebrates, parental care is found in all classes and is widespread in birds and mammals.

Unfortunately, hard data on the energy cost of parental care are difficult to obtain, because such behaviors often include such parental activities as brooding, foraging, grooming, and so forth. Although such behaviors clearly require metabolic energy expenditure on the part of the parent, they are often complex, not easily duplicated in the laboratory, and not easily separated from on-going energy costs not related to parenting. (Some might say that the estimated $200,000 that it takes to raise and educate a child in the United States is an indirect cost of human parenting.)

ENERGY, ENVIRONMENT, AND EVOLUTION

In the introduction to this book as well as the introduction to this part of the book, we describe the interdependency of numerous physiological systems. As Donald Jackson (1987) commented in considering the problems faced by interacting physiological systems, "A disturbance to one part reverberates throughout the organism, and produces responses, compromises and adaptations of various functions." The resolution of potential conflicts between the differing demands of networked physiological systems must be made in the context of space and time. A conflict between the demands of two physiological systems can be tolerated for short periods of time but must eventually be resolved by some other appropriate physiological action. Moreover, different environmental stresses carry with them very different senses of urgency. Figure 16-55 indicates the very different amounts of time during which an absence of oxygen, water, and food and an excess of body heat can be tolerated by an organism. In most animals, a physiological conflict that denies an animal food or water, for example, can be tolerated far longer than a physiological conflict that denies an animal oxygen. Often then, physiological conflicts are resolved on the basis of which of two resulting conditions is the least threat to homeostasis. These tolerances also vary greatly between species: a human being can survive without oxygen for only a few minutes; a turtle can survive for several hours; and some simple metazoans survive without oxygen indefinitely and may actually be killed by oxygen.

The field of animal physiology is only now beginning to focus on interactions between different physiological systems, rather than on the isolated characters of the individual systems. This integrated approach will by necessity draw in systems-level physiology, ecological physiology, environmental physiology, and evolution.

A salmon swimming up a stream and traversing a series of water falls is sending large quantities of blood through almost all of the respiratory surfaces of its gills in an attempt to acquire oxygen and eliminate carbon dioxide. At the same time, it produces huge quantities of urine. Why is urine production so elevated, and what physiological systems have been thrown into conflict?

This concluding chapter has integrated the animal and its physiology into its environment. Environmental constraints place limitations and demands on design and function. Animals living in water have very different shapes from those of terrestrial animals. Drag forces are much greater in water than in air, and so aquatic animals are much more streamlined. Animals in water have a density that is similar to that of the environment, but this is not the case in air. Gravity has an important effect on the circulation in terrestrial animals not seen in aquatic animals. In terrestrial animals, blood tends to pool in veins, and there are many mechanisms to ensure adequate venous return to the heart. The giraffe must have a strong, fibrous skin around the lower part of its limb to prevent blood from pooling in veins in its legs. This problem does not occur in fishes and other aquatic animals. However, in these animals, motion results in strong forces on the body surface that could interfere with venous return; so, at least in fishes, most large veins travel through the center of a fish's body.

Survival of an individual animal often depends on the allocation and rate of use of available energy. Different an-

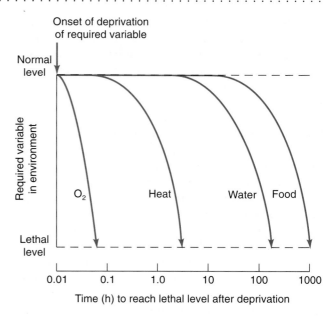

Figure 16-55 The rate of decline in cellular metabolic rate toward lethally low levels differs greatly among species after the abrupt removal of oxygen, food, or water or in failure to eliminate metabolically produced heat. Note that the time scale is logarithmic (1000 hours is about 42 days). Different species have very different tolerances for oxygen deprivation, starvation, and dehydration. For example, the brain cells of human beings begin to die within minutes of oxygen deprivation, whereas some species of turtles can survive for days or even months in the complete absence of oxygen. [Adapted from Jackson, 1987.]

imals adopt different strategies. Mammals, for example, have high rates of energy turnover that require them to seek food continually. Reptiles, on the other hand, have much lower energy turnover rates and can survive on much less energy. Different environments favor different strategies at different times. For example, reptiles seem to have the advantage in the water and in food-scarce desert environments during the day, but mammals seem to have the edge during the cooler nights. Mammals spend energy on maintaining a high body temperature and therefore need more food but can function in the cold nights.

Success of an individual animal is measured by the genetic legacy left by that animal—that is, by surviving to reproduce. Reproduction occurs when the animal is mature and conditions are favorable for survival of the young. For example, during periods of decreased energy availability, unfavorable for reproduction, the animal may enter a state of suspended animation, as in insect diapause or mammalian hibernation. There is a marked reduction in energy expenditure during these periods. Essentially, the animal is cutting its energetic costs during the "bad times" to balance energy input and output. There is a marked increase in energy expenditure during exercise. Animals often migrate to avoid certain environments where, for example, food is short and temperatures are low. The cost of migration varies with the nature of the environment, with flying being much less costly per unit distance than walking and running on land or swimming in water. The results of the process of animal evolution have given us numerous examples of adaptions for survival in a multitude of different habitats. These examples are variations in the organization of a series of basic component parts that make up the vast panoply of life.

SUMMARY

The utilization of chemical energy in tissue metabolism is accompanied by the inexorable production of heat as a low-grade energy by-product. The total energy liberated in the conversion of a higher energy compound into a lower energy end product is independent of the chemical route taken. In addition, a given class of food molecules consistently liberates the same amount of heat and requires the same amount of O_2 when oxidized to H_2O and CO_2. These characteristics of energy metabolism make it possible to use either the rate of heat production or the rate of O_2 consumption (and CO_2 production) as a measure of metabolic rate. The respiratory quotient—the ratio of CO_2 production to O_2 consumption—is useful in determining the proportions of carbohydrates, proteins, and fats metabolized, each of which has a different characteristic energy yield per liter of O_2 consumed.

The basal metabolic rate and the standard metabolic rate are related to body size—the smaller the animal, the higher the metabolic rate per unit mass of tissue (termed metabolic intensity). Although there is a fairly good correlation between metabolic intensity and body surface-to-volume ratio, suggesting that metabolic rate is determined by heat balance mechanisms, this correlation may be incidental. Similar correlations are seen in ectotherms that are in temperature equilibrium with the environment and in endotherms that steadily lose heat to the environment.

Dependence of enzymatic reactions and metabolic rate on tissue temperature is described by the Q_{10}, the ratio of metabolic rate at a given temperature to the metabolic rate at a temperature 10 Celsius degrees lower. This ratio typically lies between 2 and 3.

Endotherms are animals that generate most of their own body heat, allowing them to elevate their core temperatures above that of the environment. Ectotherms obtain most of their body heat from their surroundings, and some elevate their temperatures by various behavioral means, such as basking. Poikilothermy, homeothermy, and heterothermy refer to varying degrees of body temperature control.

Ectotherms show a variety of methods for survival in temperature extremes. Some species cope with subzero temperatures by using "antifreeze" substances or by supercooling without ice crystal formation, but no animals have been shown to survive freezing of water within cells. Other ectotherms elevate body temperature by shivering or nonshivering muscle contraction at certain times or in certain parts of their bodies. Such heat production is used by some insects and large fishes to warm locomotor muscles to optimal operating temperatures. Heat absorption or heat loss to the environment is regulated in some ectothermic species by changes in blood flow to the skin. In this way,

heat absorbed from the sun's rays can be quickly transferred by the blood from the body surface to the body core during heating or, conversely, core heat can be conserved in a cold environment by restricted circulation to the skin.

Endotherms subjected to cold environments conserve body heat by increasing the effectiveness of their surface insulation. They do this by decreasing peripheral circulation, increasing fluffiness or thickness of pelage or plumage, or adding fatty insulating tissue. In cold-climate endotherms, heat is also conserved by countercurrent heat-exchange mechanisms in the circulation to the limbs. Within the thermal neutral zone of ambient temperatures, changes in surface conductance compensate for changes in ambient temperature. Below this temperature zone, thermogenesis compensates for increased heat loss to the environment. Thermogenesis occurs by shivering or by nonshivering oxidation of substrates, as well as by exercise, specific dynamic action, activity of the Na^+-K^+ ATPase, and other activities.

At ambient temperatures above the thermal neutral zone, endotherms actively dissipate heat by means of evaporative cooling, either by sweating or by panting. The use of water places an osmotic burden on desert dwellers. Most small desert inhabitants, subject to rapid changes in body temperature, minimize such changes by usually remaining in cool microenvironments to avoid daytime heat. Large desert mammals, buffered against rapid temperature changes by more favorable surface-to-volume ratios and large thermal inertia, can conserve water that they would otherwise use for cooling by slowly absorbing heat during the day without reaching lethal body temperatures; they can then rid themselves of heat during the cool night. The brain is specially protected from overheating in some mammals by a highly developed carotid rete in which cool venous blood from the nasal epithelium removes heat from arterial blood heading toward the brain.

Body temperature in endotherms and some ectotherms is regulated by a neural thermostat sensitive to differences between the actual temperature of neural sensors and the thermostatic set-point temperature. Differences result in a neural output to thermoregulatory effectors for corrective heat loss or heat gain. Fever develops when the set-point temperature is raised by the cellular action of endogenous pyrogens, which are protein molecules released by leukocytes in response to exogenous pyrogens produced by infectious bacteria.

Ordinary sleep, torpor, hibernation, winter sleep, and estivation are all neurophysiologically and metabolically related forms of dormancy. During periods when food intake is necessarily absent or restricted, small- to medium-sized homeotherms allow their temperatures to drop in accord with a lowered set-point temperature of the body thermostat. By lowering body temperature to within a few degrees of ambient air, the homeotherm conserves energy stores. Oxidation of brown fat and shivering thermogenesis are used to produce rapid warming at the termination of torpor or hibernation.

The energetics of locomotion is also related to body size. The smaller the animal, the higher the metabolic cost of transporting a unit mass of body tissue over a given distance. The Reynolds number *(Re)* of a body moving through a liquid or gaseous medium is the ratio of the relative importance of inertial and viscous forces in the medium. Small animals swim with a low *Re* and large animals with a high *Re* because, with increasing size, viscosity plays a lesser role and inertia a greater one.

The rate of energy utilization during different kinds of locomotion typically increases with velocity. By changing gait from walking to running, hopping, or trotting, and so forth, terrestrial animals increase efficiency. Increased efficiency is achieved when the energy of falling at the end of the stride is elastically stored for release during the next stride, as in a hopping kangaroo.

Metabolic rate in many animals shows distinct endogenous rhythms that may be manifest in locomotor activity or changes in body temperature (in ectotherms). These rhythms may be circadian (daily), infradian (shorter than a day), or ultradian (longer than a day). Circadian rhythms are characterized by their persistence in the absence of environmental cues, are temperature independent, are conditionally arrhythmic, and are entrainable by zeitgebers such as light.

Reproduction requires a significant energy expenditure for many organisms. The two general patterns in which energy expended in the reproductive effort may be partitioned are *r*-selection and *K*-selection. *r*-Selected animals produce large numbers of very small offspring and offer no parental care. The low rate of survival is offset by the large number of offspring produced. *K*-Selected animals produce small numbers of large offspring. Owing in part to parental care, survival is high. Energy costs of reproduction to parents include the production of gametes; the cost of providing nutrition, as in mammalian lactation; and the cost of behaviors constituting parental care.

REVIEW QUESTIONS

1. Define ectotherm, endotherm, poikilotherm, homeotherm, basal metabolic rate, standard metabolic rate, and respiratory quotient.
2. Explain why the rate of heat production can be used to measure metabolic rate accurately.
3. Why can respiratory gas exchange be used as a measure of metabolic rate?
4. Why is the surface hypothesis inadequate as an explanation for the high metabolic intensity of small animals?
5. Why is the locomotion of a small aquatic animal affected more strongly by the viscosity of the medium than is that of a large animal?
6. What factors affect the flow pattern of fluid around a swimming animal? What factors minimize turbulence?
7. Why does riding a bicycle for 10 km require less energy than running the same distance at the same speed?

8. Why are the giant ants and other photographically scaled up insect monsters of old grade-B science fiction movies anatomically untenable?
9. Give examples of the effect of body size on the metabolism and locomotion of animals.
10. The potency of some medications depends on metabolic factors. Explain why it might be risky to give a 100 kg person 100 times the dose of a drug proved effective in a 1.0 kg guinea pig.
11. Give examples of low-temperature adaptations of some ectotherms and some endotherms.
12. What are some of the factors that determine the limits of the thermal neutral zone of a homeotherm?
13. What thermoregulatory mechanisms are available to a homeotherm at temperatures below and above the thermal neutral zone?
14. Explain and give examples of the relations that exist between water balance and temperature regulation in a desert animal.
15. Describe the integration of peripheral and core temperatures in the thermostatic control of temperature in a mammal.
16. Describe two naturally occurring situations in which the set-point temperature of the hypothalamic thermostat is changed and the body temperature correspondingly changes.
17. Explain the mechanism of heat production in two different kinds of thermogenesis.
18. What is the role of countercurrent heat exchange in porpoises, arctic mammals, tuna fishes, and sheep?
19. What are the sphinx moth's two major means of regulating thoracic temperature?
20. What means does the marine iguana use to speed the elevation of its body temperature and then retard cooling during diving?
21. How would you distinguish between (1) a true circadian rhythm, (2) a metabolic rhythm, and (3) a rhythm that is triggered by rhythmic changes in environmental cues?
22. Compare and contrast reproduction in *r*- and *K*-selected animals. What are the advantages of each?

SUGGESTED READINGS

Blake, R., ed. 1991. *Efficiency and Economy in Animal Physiology.* Cambridge: Cambridge University Press. (This book considers, in an evolutionary framework, the efficiency and metabolic costs of various types of animal locomotion.)

Block, B. A. 1994. Thermogenesis in muscle. *Annu. Rev. Physiol.* 56:535–577. (This comprensive review describes the biochemical and cellular mechanisms behind specialized heater tissues in endothermic animals.)

Carrey, C., ed. 1993. *Life in the Cold: Ecological, Physiological and Molecular Mechanisms.* Boulder: Westview Press. (This book presents a collection of multilevel reviews on hibernation, torpor, and other mechanisms by which animals survive life in the cold.)

Chadwick, D. J., and K. Ackrill, eds. 1995. *Circadian Clocks and Their Adjustment.* New York: Wiley. (Reviews of genetic, molecular, and neural bases of biological clocks controlling circadian rhythms.)

Cossins, A. R., and K. Bowler. 1987. *Temperature Biology of Animals.* London: Chapman and Hall. (This book focuses on the cellular aspects of thermoregulation in animals.)

Edmunds, L. N., Jr. 1988. *Cellular and Molecular Bases of Biological Clocks.* Berlin: Springer-Verlag. (This monograph explores the phenomenon of annual rhythmicity in a wide range of animals and plants.)

Heinrich, B. 1993. *The Hot-Blooded Insects: Strategies and Mechanisms of Thermoregulation.* Cambridge, Mass.: Harvard University Press. (This book explores the physiological and biochemical mechanisms of endothermy among the insects.)

Jones, J. H., and S. L. Lindstedt. 1993. Limits to maximal performance. *Annu. Rev. Physiol.* 55:547–569. (The metabolic and physiological factors that limit animal locomotor performance are the focus of this review.)

McMahon, T. A., and J. T. Bonner. 1983. *On Size and Life.* New York: Scientific American Books. (This delightfully illustrated book examines how scaling and allometry pervade the world around us.)

McNeil, R. A. 1992. *Exploring Biomechanics: Animals in Motion.* New York: Scientific American Library. (McNeil provides a comprehensive treatment of the anatomy and biomechanics of animal locomotion.)

Peters, R. H. 1983. *The Ecological Implications of Body Size.* Cambridge: Cambridge University Press. (This book is valuable not only for its lucid descriptions of allometry, but also for the extensive tabulated data in its numerous appendices.)

Ruben, J. 1995. The evolution of endothermy in mammals and birds: from physiology to fossils. *Annu. Rev. Physiol.* 995:69–95. (In this comprehensive review, Ruben speculates on the evolutionary processes leading to endothermy in vertebrates.)

Schmidt-Nielsen, K. 1983. *Scaling: Why is Animal Size So Important?* New York: Cambridge University Press. (This now-classic book provides a wealth of information on how anatomy and physiology are affected by body size and why physiologists should care.)

Trayhurn, P., and D. G. Nicholls, eds. 1986. *Brown Adipose Tissue.* London: E. Arnold. (Neural control mechanisms, biochemistry, metabolism, physiology, and anatomy of brown fat are all covered in this book.)

Woakes, A. J., and W. A. Foster, eds. 1991. The comparative physiology of exercise. *J. Exp. Biol.* 160. (This entire volume consists of a series of brief review papers describing the physiological implications of, and adapations for, exercise.)

APPENDIXES

Appendix 1: SI Units

Basic SI units

Physical quantity	Name of unit	Symbol for unit
Length	meter	m
Mass	kilogram	kg
Time	second	s
Electric current	ampere	A
Temperature	kelvin	K
Luminous intensity	candela	cd

SI multipliers

Multiplier	Prefix	Symbol
10^{12}	tera	T
10^9	giga	G
10^6	mega	M
10^3	kilo	k
10^2	hecto	h
10	deka	da
10^{-1}	deci	d
10^{-2}	centi	c
10^{-3}	milli	m
10^{-6}	micro	μ
10^{-9}	nano	n
10^{-12}	pico	p

Derived SI units

Physical quantity	Name of unit	Symbol for unit	Definition of unit
Acceleration	meter per second squared	$m \cdot s^2$	
Activity	1 per second	s^{-1}	
Electric capacitance	farad	F	$A \cdot s \cdot V^{-1}$
Electric charge	coulomb	C	$A \cdot s$
Electric field strength	volt per meter	$V \cdot m^{-1}$	
Electrical resistance	ohm	Ω	$V \cdot A^{-1}$
Entropy	joule per kelvin	$J \cdot K^{-1}$	
Force	newton	N	$kg \cdot m \cdot s^2$
Frequency	hertz	Hz	s^{-1}
Illumination	lux	lx	$lm \cdot m2$
Luminance	candela per square meter	$cd \cdot m^{-2}$	
Luminous flux	lumen	lm	$cd \cdot sr$
Power	watt	W	$J \cdot s^{-1}$
Pressure	newton per square meter	$N \cdot m^2$	
Voltage, potential difference	volt	V	$W \cdot A^{-1}$
Work, energy, heat	joule	J	$N \cdot m$

Appendix 2: Logs and Exponentials

Straight line equations:

If a straight line describes a plot of x against y, then b is the value of the intercept of the line on the ordinate and a is the slope of the line. The relationship between x and y is:

$$y = ax + b$$

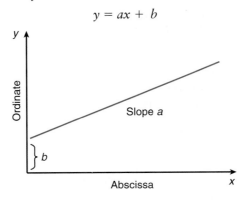

Exponential equations:

In biological systems there is often an exponential relationship between values, described by the equation:

$$y = b \cdot a^x$$

The logarithmic form of this equation is:

$$log\ y = log\ b + x\ log\ a$$

Using semi-log graph paper (linear abscissa) $log\ y$ can be plotted against x, giving a straight line to determine the values of slope a and intercept b.

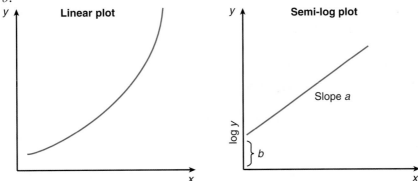

(Examples of the use of straight line and exponential equations may be found throughout the book, especially in chapters 3 and 16.)

Use of logarithm terms:

In the equation $y = 10^x$, x is the logarithm of y. That is, x is the power to which 10 must be raised in order to yield y. For example, the logarithm of 10 is 1, and the logarithm of 100 is 2. The equation

$$y = \frac{a}{b}$$

can be tranformed into a logarithmic equation:

$$log\ y = log\ \frac{a}{b} = log\ a - log\ b$$

just as multiplication can be transformed into the addition of logarithms. A convenient identity that is useful in calculating equilibrium potentials using the Goldman equation is:

$$log\ \frac{a}{b} = -\ log\ \frac{b}{a}$$

This identity is true because $log\ \frac{a}{b} = log\ a - log\ b$ and $log\ \frac{b}{a} = log\ b - log\ a$. Notice that $log\ a - log\ b = -\ (log\ b - log\ a)$, which proves the identity.

Appendix 3: Conversions, Formulas, Physical and Chemical Constants, Definitions

Units and conversion factors

To convert from	to	multiply by
angstroms	inches	3.937×10^{-9}
	meters	1×10^{-10}
	micrometers (μm)	1×10^{-4}
atmospheres	bars	1.01325
	dynes per square centimeter	1.01325×10^{6}
	grams per square centimeter	1033.23
	torr (= mmHg; 0°C)	760
	pounds per square inch	14.696
	pascals	1.013×10^{5}
bars	atmospheres	0.9869
	dynes per square centimeter	1×10^{6}
	grams per square centimeter	1019.716
	pounds per square inch	14.5038
	torr (= mmHg; 0°C)	750.062
	pascals	10^{5}
calories	British thermal units	3.968×10^{-3}
	ergs	4.184×10^{7}
	foot-pounds	3.08596
	kilocalories	10^{-3}
	horsepower-hours	1.55857×10^{-6}
	joules	40184
	watt-hours	1.1622×10^{-3}
	watt-seconds	4.184
ergs	British thermal units	9.48451×10^{-11}
	calories	2.39×10^{-8}
	dynes per centimeter	1
	foot-pounds	7.37562×10^{-8}
	gram-centimeters	1.0197×10^{-3}
	joules	1×10^{-7}
	watt-seconds	1×10^{-7}
grams	daltons	6.024×10^{23}
	grains	15.432358
	ounces (avdp)	3.52739×10^{-2}
	pounds (avdp)	2.2046×10^{-3}
inches	angstroms	2.54×10^{8}
	centimeters	2.54
	feet	8.333×10^{-2}
	meters	2.54×10^{-2}
joules	calories	0.239
	ergs	1×10^{7}
	foot-pounds	0.73756
	watt-hours	2.777×10^{-4}
	watt-seconds	1

Units and conversion factors (Continued)

To convert from	to	multiply by
liters	cubic centimeters	10^3
	gallons (US, liq)	0.2641794
	pints (US, liq)	2.113436
	quarts (US, liq)	1.056718
lumens	candle power	7.9577×10^{-2}
lux	lumens per square meter	1
meters	angstroms	1×10^{10}
	micrometers (μm)	1×10^6
	centimeters	100
	feet	3.2808
	inches	39.37
	kilometers	1×10^{-3}
	miles (statute)	6.2137×10^{-4}
	millimeters	1000
	yards	1.0936
newtons	dynes	10^5
pascals	bars	10^{-5}
	atmospheres	9.87×10^{-6}
	dynes per square centimeter	10
	grams per square centimeter	1.0197×10^{-2}
	torr (= mmHg; 0°C)	7.52×10^{-3}
	pounds per square inch	1.450×10^{-4}
watts	British thermal units per second	9.485×10^{-4}
	calories per minute	14.3197
	ergs per second	1×10^7
	foot-pounds per minute	44.2537
	horsepower	$1.341\ 10^{-3}$
	joules per second	1

Temperature conversions

$$°C = 5/9 \,(°F - 32)$$
$$°F = 9/5 \,(°C) + 32$$
$$0 \, K = -273.15°C = -459.67°F + 32$$
$$0°C = 273.15 \, K = 32°F$$

Useful formulas

Electric potential	$E = IR = q/C$	Electrostatic force of attraction	$F = \dfrac{q_1 q_2}{\varepsilon r^2}$

Electric potential
$$E = IR = q/C$$
E = electric potential (voltage)
I = current
R = resistance
q = charge
C = capacitance

Power
$$p = w/t$$
w = work
t = time

Electric power
$$p = RI^2 = EI$$
E = electric potential

Work
$$W = RI^2 t = EIt = Pt$$

Pressure
$$P = \text{force (f)/unit area}$$

Weight
$$W = mg$$
m = mass
g = acceleration of gravity

Force
$$f = ma$$
m = mass
a = acceleration

Dalton's law of partial pressures
$$PV = V(p_1 + p_2 + p_3 + \cdots + p_n)$$
P = pressure of gas mixture
V = volume
p = pressure of each gas alone

Electrostatic force of attraction
$$F = \dfrac{q_1 q_2}{\varepsilon r^2}$$
r = distance separating q_1 and q_2
ε = dielectric constant

Potential energy
$$E = mgh$$
h = height of mass above surface of Earth

Kinetic energy
$$E = 1/2 mv^2$$
v = velocity of mass

Energy of a charge
$$E = 1/2 qV$$
q = charge
V = electric potential

Perfect gas law
$$PV = nRT$$
P = pressure
V = volume
n = number of moles
R = gas constant
T = absolute temperature

Hooke's law of elasticity
$$F = kT$$
k = spring constant
F = force
T = tension

Energy of a photon
$$E = h\nu$$
h = Planck's constant
ν = frequency

Physical and chemical constants

Avogadro's number	$N_A = 6.022 \times 10^{23}$
Faraday constant	$F = 96{,}487 \ C \cdot mol^{-1}$
Gas constant	$R = 8.314 \ J \cdot K^{-1} \cdot mol$
	$= 1.98 \ cal \cdot K^{-1} \cdot mol$
	$= 0.082 \ L \cdot atm \cdot K^{-1} \cdot mol$
Planck's constant	$h = 6.62 \times 10^{-27} \ ergs \cdot s^{-1}$
	$= 1.58 \times 10^{-34} \ cal \cdot s^{-1}$
Speed of light in a vacuum	$c = 2.997 \times 10^{10} \ cm \cdot s^{-1}$
	$= 186{,}000 \ mi \cdot s^{-1}$

Dimensions of plane and solid figures

Area of a square $= l^2$
Suface area of a cube $= 6l^2$
Volume of a cube $= l^3$
Circumference of a circle $= 2\pi r$
Area of a circle $= \pi r^2$
Surface area of a sphere $= 4\pi r^2$
Volume of a sphere $= 4/3 \pi r^3$
Surface area of a cylinder $= 2\pi rh$
Volume of a cylinder $= \pi r^2 h$

Chemical definitions

1 mol = the mass in grams of a substance equal to its molecular or atomic weight: this mass contains Avogadro's number (N_A) of molecules or atoms

Molar volume = the volume occupied by a mole of gas at standard temperature and pressure (25°C, 1 atm) = 22.414 L

1 molal solution = 1 mol per 1000 g of solvent

1 molar solution = 1 mol of solute in 1 L of solution

1 equivalent = 1 mol of 1 unit charge

1 einstein = 1 mol of photons

REFERENCES CITED

Adams, P. R., S. W. Jones et al. 1986. Slow synaptic transmission in frog sympathetic ganglia. *J. Exp. Biol.* 124:259–285.

Adolph, E. F. 1967. The heart's pacemaker. *Scientific American* 216(3):32–37.

Ahlquist, R.P. 1948. A study of the adrenotropin receptors. *Amer. J. Physiol.* 153:586–600.

Ashley, C. C. 1971. Calcium and the activation of skeletal muscle. *Endeavor* 30:18–25.

Astrup, P., and J. Severinghaus. 1986. *The History of Blood Gases, Acids, and Bases.*

Audesirk, T., and G. Audesirk. 1996. *Biology: Life on Earth.* 4th ed. Upper Saddle River, N.J.: Prentice-Hall, Inc.

Avenet, P., S. C. Kinnamon, and S. D. Roper. 1993. Peripheral transduction mechanisms. In S. A. Simon and S. D. Roper, eds., *Mechanisms of Taste Transduction.* Boca Raton, Fla.: CRC Press.

Baker, J. J. W., and G. E. Allen. 1965. *Matter, Energy, and Life.* Reading, Mass.: Addison-Wesley.

Banko, W. E. 1960. *The Trumpeter Swan.* North American Fauna, No. 63. Washington, D.C.: U.S. Dept. of the Interior, Fish and Wildlife Service.

Bartels, H. 1971. Blood oxygen dissociation curves: mammals. In P. L. Altman and S. W. Dittmer, eds., *Respiration and Circulation.* Bethesda, Md.: Federation of American Societies for Experimental Biology.

Bartholomew, G. A. 1964. *Symposia of the Society for Experimental Biology.* No. 18. New York: Academic Press Inc., pp. 7–29.

Bartholomew, G. A., and R. C. Lasiewski. 1965. Heating and cooling rates, heart rates and simulated diving in the Galapagos marine guana. *Comp. Biochem. Physiol.* 16:573–582.

Baudinette, R. V. 1991. The energetics and cardiorespiratory correlates of mammalian terrestrial locomotion. *J. Exp. Biol.* 160:209–231.

Baylor, D., T. D. Lamb, and K.-W. Yau. 1979. Responses of retinal rods to single photons. *J. Physiol.* 288:613–134.

Beament, J. W. L. 1958. The effect of temperature on the waterproofing mechanism of an insect. *J. Exp. Biol.* 35:494–519.

Bear, M. F., B. W. Connors, and M. A. Paradiso. 1996. *Neuroscience: Exploring the Brain.* Baltimore: Williams and Wilkins.

Beck, W. S. 1971. *Human Design.* New York: Harcourt, Brace, and Jovanovich.

Bell, G. H., J. N. Davidson, and H. Scarborough. 1972. *Textbook of Physiology and Biochemistry.* 8th ed. Edinburgh: Churchill Livingstone.

Bendall, J. R. 1969. *Muscles, Molecules, and Movement.* New York: Elsevier.

Bennett, M.V. L. 1968. Similarities between chemical and electrical mediated transmission. In F. D. Carlson, ed., *Physiological and Biochemical Aspects of Nervous Integration.* Englewood Cliffs, N. J.: Prentice-Hall.

Benzinger, T. H. 1961. The diminution of thermoregulatory sweating during cold reception at the skin. *Proc. Nat. Acad. Sci. USA* 47:1683–1688.

Berg, H. C., and E. M. Purcell. 1977. Physics of chemoreception. *Biophys. J.* 20:193–219.

Bernard, C. 1872. *Physiologie Generale.* Paris: Hachette.

Bernheim, H. A., and M. G. Kluger. 1976. Fever and antipyresis in the lizard *Dipsosaurus dorsalis. Amer. J. Physiol.* 231:198–203.

Berridge, M. 1993. Inositol trisphosphate and calcium signalling. *Nature* 361:315–325.

Berridge, M. J. 1985. The molecular basis of communication within the cell. *Scientific American* 253:124–125.

Berthold, A. A. 1849. Transplantation der hoden. *Arch. Anat. Physiol. Wiss. Med.* 16:42–46.

Biology: An Appreciation of Life. 1972. Del Mar, Calif.: CRM Books.

Block, B., T. Imagawa, K. P. Campbell, and C. Franzini-Armstrong. 1988. Structural evidence for direct interaction between the molecular components of the transverse tubule/sarcoplasmic reticulum junction in skeletal muscle. *J. Cell. Biol.* 107:2587–2600.

Block, B. A. 1994. Thermogenesis in muscle. *Annu. Rev. Physiol.* 56:535–577.

Bortoff, A. 1976. Myogenic control of intestinal motility. *Physiol. Rev.* 56:416–434.

Brand, A. R. 1972. The mechanisms of blood circulation in *Anodonta anatina* L. *(Bivalvia unionidae). J. Exp. Biol.* 56:362–379.

Brenner, B. M., J. L. Troy, and T. M. Daugharty. 1971. The dynamics of glomerular ultrafiltration in the rat. *J. Clin. Invest.* 50:1776–1780.

Bretscher, M. S. 1985. The molecules of the cell membrane. *Scientific American* 86–90.

Brett, J. R. 1971. Role of thermoregulation in salmon physiology and behavior. *Amer. Zool.* 11:99–113.

Brown, K. T. 1974. Physiology of the retina. In V. B. Mountcastle, ed., *Medical Physiology*. 13th ed. St. Louis: Mosby.

Brownell, P., and R. D. Farley. 1979a. Detection of vibrations in sand by tarsal sense organs of the nocturnal scorpion *Paruroctonus mesaensis*. *J. Comp. Physiol*. 131:23–30.

Brownell, P., and R. D. Farley. 1979b. Orientation to vibrations in sand by the nocturnal scorpion *Paruroctonus mesaensis*: mechanism to target location. *J. Comp. Physiol*. 131:31–38.

Bruns, D., and R. Jahn. 1995. Real-time measurement of transmitter release from single synaptic vesicles. *Nature* 377:62–65.

Bülbring, E. 1959. *Lectures on the Scientific Basis of Medicine*. Vol. 7. London: Athlone.

Bülbring, E., and H. Kuriyama. 1963. Effects of changes in ionic environment on the action of acetylcholine and adrenaline on smooth muscle cells of guinea pig. *J. Physiol*. 166:59–74.

Bullock, T. H., and F. P. J. Diecke. 1956. Properties of an infrared receptor. *J. Physiol*. 134:47–87.

Bullock, T. H., and G. A. Horridge. 1965. *Structure and Function in the Nervous Systems of Invertebrates*. New York: W. H. Freeman and Company.

Burggren, W., and K. Johansen.1982. Ventricular hemodynamics in the monitor lizard, *Varanus exanthematicus*: pulmonary and systemic pressure separation. *J. Exp. Biol*. 96:343–354.

Camhi, J. M. 1984. *Neuroethology*. Sunderland, Mass.: Sinauer Associates, Inc.

Cannon, W. 1929. Organization for physiological homeostatics. *Physiol. Rev*. 9:399–431.

Capecchi, M. R. 1994. Targeted gene replacement. *Scientific American* 270:52–59.

Carey, F. G. 1973. Fishes with warm bodies. *Scientific American* 228:36–44.

Carey, F. G., and J. M. Teal. 1966. Heat conservation in tuna fish muscle. *Proc. Nat. Acad. Sci. USA* 56:1464–1469.

Catania, K. C., and J. H. Kaas. 1996. The unusual nose and brain of the star-nosed mole. *BioScience* 46(8):578–586.

Chen, J-N., and M. Fishman. 1996. Genetic dissection of heart development. In W. Burggren and B. Keller, eds., *Development of Cardiovascular Systems: Molecules to Organisms*. New York: Cambridge University Press.

Chess, A., L. Buck, et al. 1992. Molecular biology of smell: expression of the multigene family encoding putative odorant receptors. *Cold Spring Harbor Symp. Quant. Biol*. 57:505–516.

Cheung, W. Y. 1979. Calmodulin plays a pivotal role in cellular regulation. *Science* 207:17–27.

Cole, K. S., and H. J. Curtis. 1939. Electric impedance of the squid giant axon during activity. *J. Gen. Physiol*. 22:640–670.

Comroe, J. H. 1962. *Physiology of Respiration*. Chicago: Year Book Medical Publishers.

Cordina, J., A. Yatani, et al. 1987. The alpha subunit of the GTP binding protein Gk opens atrial potassium channels. *Science* 236:442–445.

Cornwall, I. W. 1956. *Bones for the Archaeologist*. London: Phoenix House.

Curran, P. F. 1965. Ion transport in intestine and its coupling to other transport processes. *Federation Proc*. 24:993–999.

Darnell, J., H. Lodish, and D. Baltimore. 1990. *Molecular Cell Biology*. 2d ed. New York: Scientific American Books.

Davenport, H. W. 1974. *The A.B.C. of Acid-Base Chemistry*. 6th rev. ed. Chicago: University of Chicago Press.

Davenport, H. W. 1977. *Physiology of the Digestive Tract*. Chicago: Year Book Medical Publishers.

Davenport, H. W. 1985. *Physiology of the Digestive Tract*. 5th ed. Chicago: Chicago Yearbook Medical Publishers.

Davis, H. 1968. Mechanisms of the inner ear. *Ann. Otol. Rhinol. Laryngol*. 77:644–655.

Del Castillo, J., and B. Katz. 1954. Quantal components of the endplate potential. *J. Physiol*. 124:560–573.

Denton, E. J. 1961. The buoyancy of fish and cephalopods. *Prog. Biophys*. 11:178–234.

Diamond, J., and K. Hammond. 1992. The matches, achieved by natural selection, between biological capacities and their natural loads. *Experientia* 48:551–557.

Diamond, J. M., and J. McD. Tormey. 1966. Studies on the structural basis of water transport across epithelial membranes. *Federation Proc*. 25:1458–1463.

Douglas, W. W. 1974. Mechanism of release of neurohypophyseal hormones: stimulus-secretion coupling. In R.O. Greep, ed., *Handbook of Physiology*. Section 7. *Endocrinology* (Vol. 4, Part 1, Pituitary Gland). Washington, D.C.: American Physiological Society.

Douglas, W. W., J. Nagasawa, and R. Schulz. 1971. Electron microscopic studies on the mechanism of secretion of posterior pituitary hormones and significance of microvesicles ("synaptic vesicles"): evidence of secretion by exocytosis and formation of microvesicles as a by-product of this process. In H. Heller and K. Lederis, eds., *Subcellular Organization and Function in Endocrine Tissues*. Mem. Soc. Endocrinol. No. 19. New York: Cambridge University Press.

Dudel, J., and S. W. Kuffler. 1961. Presynaptic inhibition at the crayfish neuromuscular junction. *J. Physiol*. 155:543–562.

Eakin, R. 1965. Evolution of photoreceptors. *Cold Spring Harbor Symp. Quant. Biol*. 30:363–370.

Ebashi, S. K. Maruyama, and M. Endo, eds. 1980. *Muscle Contraction: Its Regulatory Mechanisms*. New York: Springer-Verlag; pp. 10–62.

Ebashi, S., M. Endo, and I. Ohtsuki. 1969. Control of muscle contraction. *Quart. Rev. Biophys*. 2:351–384.

Eccles, J. C. 1969. Historical introduction to central cholinergic transmission and its behavioral aspects. *Federation Proc*. 28:90–94.

Eckert, R. O. 1961. Reflex relationships of the abdominal stretch receptors of the crayfish. *J. Cell. Comp. Physiol*. 57:149–162.

Eckert, R. 1972. Bioelectric control of ciliary activity. *Science* 176:473–481.

Edgar, W. M. 1992. Saliva: its secretions, compositions and functions. *Br. Ent. J*. 172:305–312.

Edney, E. B. 1974. Desert arthropods. In G. W. Brown, ed., *Desert Biology*. Vol. 2. New York: Academic.

Edney, E. B., and K. A. Nagy. 1976. Water balance and excretion. In J. Bligh, J. L. Cloudsley-Thompson, and A. G. MacDonald, eds., *Environmental Physiology of Animals*. Oxford: Blackwell Scientific Publications.

Eiduson, S. 1967. The biochemistry of behavior. *Science J*. 3:113–117.

Eyzaguirre, C., and S. W. Kuffler. 1955. Processes of excitation in the dendrites and in the soma of single isolated sensory nerve cells of the lobster and crayfish. *J. Gen. Physiol*. 39:87–119.

Farrell, A. P., S. S. Sobin, D. J. Randall, and S. Crosby. 1980. Intralamellar blood flow patterns in fish gills. *Amer. J. Physiol.* 239:R429–R436.

Fatt, P., and B. Katz. 1951. An analysis of the endplate potential recorded with an intracellular electrode. *J. Physiol.* 115:320–370.

Fatt, P., and B. Katz. 1952. Spontaneous subthreshold activity at motor nerve endings. *J. Physiol.* 117:109–128.

Feigl, E. O. 1974. Physics of the cardiovascular system. In T. C. Ruch and H. D. Patton, eds., *Physiology and Biophysics,* 20th ed., Vol. 2. Philadelphia: Saunders.

Fessenden, R. J., and J. S. Fessenden. 1982. *Organic Chemistry.* 2d ed. Boston: Willard Grant Press.

Firestein, S., G. M. Shepherd, and F. S. Werblin. 1990. Time course of the membrane current underlying sensory transduction in salamander olfactory receptor neurones. *J. Physiol.* 430:135–158.

Flock, A. 1967. Ultrastructure and function in the lateral line organs. In Phyllis H. Cahn, ed., *Lateral Line Detectors.* Bloomington: Indiana University Press.

Florey, E. 1966. *General and Comparative Animal Physiology.* Philadelphia: Saunders.

Frieden, E. H., and H. Lipner. 1971. *Biochemical Endocrinology of the Vertebrates.* Englewood Cliffs, N.J.: Prentice-Hall.

Furshpan, E. J., and D. D. Potter. 1959. Transmission at the giant motor synapses of the crayfish. *J. Physiol.* 145:289–325.

Gesteland, R. C. 1966. The mechanics of smell. *Discovery* 27(2). London: Proprietors, Professional and Industrial Publishing Co.

Gilman, A. G. 1987. G proteins: transducers of receptor-generated signals. *Annu. Rev. Biochem* 56:615–649.

Goldberg, N. D. 1975. Cyclic nucleotides and cell function. In G. Weissman and R. Claiborne, eds., *Cell Membranes: Biochemistry, Cell Biology, and Pathology.* New York: Hospital Practice Publishing Co.

Goldman, D. E. 1943. Potential, impedance, and rectification in membranes. *J. Gen. Physiol.* 27:37–60.

Goldsby, R. A. 1967. *Cells and Energy.* New York: Macmillan.

Goodrich, E. S. 1958. *Studies on the Structure and Development of Vertebrates.* Vol. 2. New York: Dover.

Gordon, A. M., A. F. Huxley, and F. J. Julian. 1966. The variation in isometric tension with sarcomere length in vertebrate muscle fibres. *J. Physiol.* 184:170–192.

Gorski, R. A. 1979. Long-term hormonal modulation of neuronal structure and function. In F. O. Schmitt and F. G. Worden, eds., *The Neurosciences: Fourth Study Program.* Cambridge, Mass.: MIT Press.

Gosline, J., and R. E. Shadwick. 1996. The mechanical properties of fin whale arteries are explained by novel connective tissue. *J. Exp. Biol.* 199:985–995.

Grantham, J. J. 1971. Mode of water transport in mammalian renal collecting tubules. *Fed. Proc.* 30:14–21.

Grell, K. G. 1973. *Protozoology.* New York: Springer-Verlag.

Grigg, G. C. 1970. Water flow through the gills of Port Jackson sharks. *J. Exp. Biol.* 52:565–568.

Hadley, M. E. 1992. *Endocrinology.* 3d ed. Englewood Cliffs, N.J.: Prentice-Hall.

Hadley, N. 1972. Desert species and adaptation. *Amer. Sci.* 60:338–347.

Haggis, G. H., D. Michie, et al. 1964. *Introduction to Molecular Biology.* London: Longmans.

Hagins, W. A. 1972. The visual process: Excitatory mechanisms in the primary receptor cells. *Ann. Rev. Biophys. Bioeng.* 1:131–158.

Hales, J. R. S. 1966. The partition of respiratory ventilation of the panting ox. *J. Physiol.* 188:45–68.

Hall, Z. 1992. *An Introduction to Molecular Neurobiology.* Sunderland, Mass.: Sinauer Associates, Inc.

Ham, A. W. 1957. *Histology.* Philadelphia: Lippincott.

Hanamori, T., I. J. Miller, Jr., and D. V. Smith. 1988. Gustatory responsiveness of fibers in the hamster glossopharyngeal nerve. *J. Neurophysiol.* 60:478–498.

Harris, G. G., and A. Flock. 1967. Spontaneous and evoked activity from *Xenopus laevis* lateral line. In P. H. Cahn, ed., *Lateral Line Detectors.* Bloomington: Indiana University Press.

Hartline, H. K. 1934. Intensity and duration in the excitation of single photoreceptor units. *J Cell. Comp. Physiol.* 5:229–274.

Hartline, H. K., H. G. Wanter, and F. Ratliff. 1956. Inhibition in the eye of *Limulus J. Gen. Physiol.* 39:651–673.

Hayward, J. N., and M. A. Baker. 1969. A comparative study of the role of the cerebral arterial blood in the regulation of brain temperature in five mammals. *Brain Res.* 16:417–440.

Hazel, J. R. 1995. Thermal adaptation in biological membranes: is homeoviscous adaptation the explanation? *Annu. Rev. Physiol.* 57:19–42.

Hebb, D. O. 1949. *The Organization of Behaviour.* New York: Wiley.

Heinrich, B. 1974. Thermoregulation in endothermic insects. *Science* 185:747–756.

Heisler, N., P. Neuman, and G. M. O. Maloiy. 1983. The mechanism of intracardiac shunting in the lizard *Varanus exanthematicus. J. Exp. Biol.* 105:15–31.

Hellam, D. C., and R. J. Podolsky. 1967. Force measurements in skinned muscle fibres. *J. Physiol.* 200:807–819.

Heller, H. C., L. I. Crawshaw, and H. T. Hammel. 1978. The thermostat of vertebrate animals. *Scientific American* 239:102–113.

Hellon, R. F. 1967. Thermal stimulation of hypothalamic neurones in unanaesthetized rabbits. *J. Physiol.* 193:381–395.

Hemmingsen, A. M. 1969. Energy metabolism as related to body size and respiratory surfaces, and its evolution. *Rep. Steno. Mem. Hosp. Nordisk Insulinlaboratorium* 9:1–110.

Herkenham, Miles, et al. 1991. Characterization and localization of cannabinoid receptors in rat brain: A quantitative *in vitro* autoradiographic study. *J. Neuroscience* 11(2):563–583.

Hildebrandt, J., and A. C. Young. 1965. Anatomy and physiology of respiration. In T. C. Ruch and H. D. Patton, eds., *Physiology and Biophysics,* 19th ed. Philadelphia: Saunders.

Hill, A. V. 1938. The heat of shortening and the dynamic constants of muscle. *Proc. Roy. Soc. (London) Ser. B.* 126:136–195.

Hill, A. V. 1964. The efficiency of mechanical power development during muscular shortening and its relation to load. *Proc. Roy. Soc. (Lond.) Ser. B.* 159:319–324.

Hille, B. 1992. *Ionic Channels of Excitable Membranes.* 2d ed. Sunderland, Mass.: Sinauer Associates.

Hirakow, R. 1970. Ultrastructural characteristics of the mammalian and sauropsidan heart. *Amer. J. Cardiol.* 25:195–203.

Hoar, W. S. 1975. *General and Comparative Physiology.* 2d ed. Englewood Cliffs, N.J.: Prentice-Hall.

Hodgkin, A. L. 1937. Evidence for electrical transmission in nerve. *J. Physiol.* 90:183–232.

Hodgkin, A. L., and P. Horowicz. 1960. Potassium contractures in single muscle fibres. *J. Physiol.* 153:386–403.

Hodgkin, A. L., and A. F. Huxley. 1939. Action potentials recorded from inside a nerve fibre. *Nature* 144:710–711.

Hodgkin, A. L., and A. F. Huxley. 1952a. Currents carried by sodium and potassium ions through the membrane of the giant axon of *Loligo*. *J. Physiol.* 116:449–472.

Hodgkin, A. L., and A. F. Huxley. 1952b. A quantitative description of membrane current and its application to conduction and excitation in nerve. *J. Physiol.* 117:500–544.

Hodgkin, A. L., and A. F. Huxley. 1952c. Properties of nerve exons: (I) Movement of sodium and potassium ions during nervous activity. *Cold Spring Harbor Symp. Quant. Biol.* 17:43–52.

Hodgkin, A. L., and B. Katz. 1949. The effect of sodium ions on the electrical activity of the giant axon of the squid. *J. Physiol.* 108:37.

Hodgkin, A. L., A. F. Huxley, and B. Katz. 1952. Measurement of current-voltage relations in the membrane of the giant axon of *Loligo*. *J. Physiol.* 116:424–448.

Hoffman, B. F., and P. F. Cranefield. 1960. *Electrophysiolgy of the Heart.* New York: McGraw-Hill.

Hokin, M. R., and L. E. Hokin. 1953. Enzyme secretion and the incorporation of ^{32}P into phospholipids of pancreas slices. *J. Biol. Chem.* 203:967–977.

Holland, R. A. B., and R. E. Forster. 1966. The effect of size of red cells on the kinetics of their oxygen uptake. *J. Gen. Physiol.* 49:727–742.

Horridge, G. A. 1968. *Interneurons.* New York: W. H. Freeman and Company.

Hoyle, G. 1967. Specificity of muscle. In C. A. G. Wiersma, ed., *Invertebrate Nervous Systems.* Chicago: University of Chicago Press.

Hubbard, R., and A. Kropf. 1967. Molecular isomers in vision. *Scientific American* 216(6):64–76. Offprint 1075.

Hubel, D. H. 1963. The visual cortex of the brain. *Scientific American* 209:54–62.

Hubel, D. H. 1995. *Eye, Brain, and Vision.* New York: Scientific American Library Paperbacks.

Hughes, C. M. 1964. How a fish extracts oxygen from water. *New Scientist* 11:346–348.

Hume, I. D. 1989. Optimal digestive strategies in mammalian herbivores. *Physiol. Zool.* 62(6):1145–1163.

Hutter, O. F., and W. Trautwein. 1956. Vagal and sympathetic effects on the pacemaker fibres in the sinus venosus of the heart. *J. Gen. Physiol.* 39:715–733.

Huxley, A. F., and R. Niedergerke. 1954. Structural changes in muscle during contraction: Interference microscopy of living muscle fibres. *Nature* 173:971–973.

Huxley, A. F., and R. M. Simmons. 1971. Proposed mechanism of force generation in striated muscle. *Nature* 233:533–538.

Huxley, H. E. 1963. Electron microscope studies on the structure of material and synthetic protein filaments from striated muscle. *J. Mol. Biol.* 7:281–308.

Huxley, H. E. 1969. The mechanism of muscular contraction. *Science* 164:1356–1365.

Hyman, L. H. 1940. *The Invertebrates: Protozoa through Ctenophora.* New York: McGraw-Hill.

Imms, A. D. 1949. *Outlines of Entomology.* London: Methuen.

Irvine, B., N. Audsley et al. 1988. Transport properties of locust ileum in vitro: effects of cAMP. *J. Exp. Biol.* 137:361–385.

Irving, L. 1966. Adaptations to cold. *Scientific American* 214:94–101.

Ishimatzu, A., and Y. Itazawa. 1993. Difference in blood oxygen levels in the outflow vessels of the heart of an air-breathing fish, *Channa argus:* Do separate bloodstreams exist in teleostean heart? *J. Comp. Physiol.* 149:435.

Jackson, D. C. 1987. Assigning priorities among interacting physiological systems. In M. E. Feder, A. F. Bennett, W. W. Burggren, and R. B. Huey, eds., *New Directions in Ecological Physiology.* New York: Cambridge University Press.

Jamison, R. L., and R. H. Maffly. 1976. The urinary concentrating mechanism. *N. Engl. J. Med.* 295:1059–1067.

Jan, Y. N., and L. Jan. 1983. Coexistence and corelease of cholinergic and peptidergic transmitters in frog sympathetic ganglia. *Federation Proc.* 42:2929–2933.

Jennings, J. B. 1972. *Feeding, Digestion and Assimilation in Animals.* New York: St. Martin's Press.

Jewell, R. R., and J. C. Ruegg. 1966. Oscillatory contraction of insect fibrillar muscle after glycerol extraction. *Proc. Roy. Soc. (London) Ser. B.* 164:428–459.

Jones, D. R. 1995. Crocodialian cardiac dynamics: a half-hearted attempt. Presented at 4th International Congress of Comparative Physiology and Biochemistry, August, 6–11, 1995, Birmingham, United Kingdom. *Physiol. Zool.* 68(4):9–15.

Jones, D. R., and M. J. Purves. 1970. The effect of carotid body denervation upon the respiratory response to hypoxia and hypercapnia in the duck. *J. Physiol.* 211:295–309.

Jones, D. R., and M. J. Purves. 1970a. The carotid body in the duck and the consequences of its denervation upon the cardiac response to immersion. *J. Physiol.* 211:279–294.

Jones, D. R., P. G. Bushnell, B. K. Evans, and J. Baldwin. 1994. Circulation in the Gippsland giant earthworm *Megascolides australis*. *Physiol. Zool.* 67(4):1383–1401.

Jones, D. R., B. L. Langille, D. J. Randall, and G. Shelton. 1974. Blood flow in dorsal and ventral aortas of the cod *Gadus morhua*. *Amer. J. Physiol.* 226:90–95.

Josephson, R. K. 1985. The mechanical power output of a tettigoniid wing muscle during singing and flight. *J. of Exp. Biol.* 117:357–368.

Kampmeier, O. F. 1969. *Evolution and Comparative Morphology of the Lymphatic System.* Springfield, Ill.: Thomas.

Kandel, E. 1976. *Cellular Basis of Behavior.* New York: W. H. Freeman and Company.

Kandel, E. R., T. Abrams et al. 1983. Classical conditioning and sensitization share aspects of the same molecular cascade in *Aplysia. Cold Spring Harbor Symp. Quant. Biol.* 48:821–830.

Katz, B., and R. Miledi. 1966. Input-output relation of a single synapse. *Nature* 212:1242–1245.

Katz, B., and R. Miledi. 1967. Tetrodotoxin and neuromuscular transmission. *Proc. Roy. Soc. (London) Ser. B.* 167:8–22.

Katz, B., and R. Miledi. 1968. The role of calcium in neuromuscular facilitation. *J. Physiol.* 195:481–492.

Katz, B., and R. Miledi. 1970. Further study of the role of calcium in synaptic transmission. *J. Physiol.* 207:789–801.

Katz, P.S., P. A. Getting, and W. N. Frost. 1994. Dynamic neuromodulation of synaptic strength intrinsic to a central pattern generator circuit. *Nature* 367:729–731.

Kauer, J. S. 1987. Coding in the olfactory system. In T. E. Finger and W. L. Silver, eds., *Neurobiology of Taste and Smell*. New York: Wiley.

Kenagy, G. J., R. D. Stevenson, and D. Masman. 1989. Energy requirements for lactation and postnatal growth in captive golden-mantle ground squirrels. *Physiol. Zool.* 62(2): 470–487.

Kerkut, G. A., and R. C. Thomas. 1964. The effect of anion injection and changes in the external potassium and chloride concentration on the reversal potentials of the IPSP and acetylcholine. *Comp. Physiol. Biochem.* 11:199–213.

Keynes, R. D. 1958. The nerve impulse and the squid. *Scientific American* 199(6):83–90.

Keynes, R. D., and K. J. Aidley. 1981. *Nerve and Muscle*. Cambridge: Cambridge University Press.

Kirschfeld, K. 1971. *Verhandlungen der Gesellschaft Deutscher Naturforscher und Ärtze*. Berlin: Springer-Verlag.

Kleiber, M. 1932. Body size and metabolism. *Hilgardia* 6:315–353.

Kluger, M. J. 1979. *Fever: Its Biology, Evolution, Function*. Princeton, N. J.: Princeton University Press.

Knudsen, E. I. 1981. The hearing of the barn owl. *Scientific American* 245:113–125.

Knudsen, E. I., and M. Konishi. 1978. A neural map of auditory space in the owl. *Science* 200:795–797.

Koefoed-Johnsen, V., and H. H. Ussing. 1958. The nature of frog skin. *Acta Physiol. Scand.* 42:298–308.

Konishi, M. 1993. Listening with two ears. *Scientific American* 268(4):66.

Kooyman, G. L. 1989. Diverse divers: physiology and behaviour. In W. Burggren, D. S. Farner et al., eds., *Zoophysiology*, vol. 23. New York: Springer-Verlag.

Korner, P. I. 1971. Integrative neural cardiovascular control. *Physiol. Revs.* 51(2):312–367.

Kotyk, A., and K. Janáček. 1970. *Cell Membrane Transport*. New York: Plenum.

Krebs, H. A. 1975. The August Krogh principle: "For many problems there is an animal on which it can be most conveniently studied." *J. Exp. Zool.* 194:309–344.

Kuby, J. In press. *Immunology*. 3d ed. New York: W. H. Freeman.

Kuffler, S. W. 1942. Further study on transmission in an isolated nerve-muscle fibre preparation. *J. Neurophysiol.* 6:99–110.

Land, M., and R. Fernald. 1992. The evolution of eyes. *Ann. Rev. Neurosci.*, 15:1–29.

Langille, B. J. 1975. A comparative study of central cardiovascular dynamics in vertebrates. Ph.D. dissertation. University of British Columbia, Vancouver, Canada.

Lehninger, A. L. 1975. *Biochemistry*. 2nd ed. New York: Worth.

Lehninger, A. L., D. L. Nelson, and M. M. Cox. 1993. *Principles of Biochemistry*, 2nd. ed. New York: Worth Publishers.

Lighton, J. R. B. 1994. Discontinuous ventilation in terrestrial insects. *Physiol. Zool.* 67:142–162.

Lindeman, W. 1955. Ueber die Jugendentwicklung beim Luchs (*Lyns l. lynx* Kerr.) und bei der Waldkatze (*Feliss. sylvestris* Schreb). *Behavior* 8:1–45.

Lissman, H. W. 1963. Electric location of fishes. *Scientific American* 208(3):50–59. Offprint 152.

Llinas, R., and C. Nicholson. 1975. Calcium role in depolarization-secretion coupling: an aequorin study in squid giant synapse. *Proc. Nat. Acad. Sci. USA* 72:187–190.

Lodish, H., D. Baltimore, et al. 1995. *Molecular Cell Biology*. 3d ed. New York: Scientific American Books.

Loewi, O. 1921. Uber humoral Ubertragbarkeit der Herznervenwirkung. *Pflugers Arch. Ges. Physiol.* 189:239–242.

Lorenz, K. Z. *Man Meets Dog*. Translated by M. K. Wilson. London: Methuen.

Lorenz, K., and N. Tinbergen. 1938. Taxis und Instinkhandlung in der Eirolbewegung der Graugans. *Z. Tierpsychol.* 2:1–29.

Loumaye, E., J. Thorner, and K.J. Catt. 1982. Yeast mating pheromone activates mammalian gonadotrophs: evolutionary conservation of a reproductive hormone? *Science* 218:1323–1325.

Lowenstein, W. R. 1960. Biological transducers. *Scientific American* 203:98–108.

Lowenstein, W. R. 1971. *Handbook of Sensory Physiology: Principles of Receptor Physiology*. New York: Springer-Verlag.

Lutz, G. J., and L. C. Rome. 1994. Built for jumping: the design of the frog muscular system. *Science* 263:370–372.

Lutz, G. J., and L. C. Rome. 1996. Muscle function during jumping in frogs. I. Sarcomere length change, EMG pattern, and jumping performance. *Am. J. Physiol. (Cell Physiol.)* In press.

Madge, D. S. 1975. *The Mammalian Alimentary System*. London: Arnold.

Marks, W. B. 1965. Visual pigments of single goldfish cones. *J. Physiol.* 178:14–32.

Marshall, P. T., and G. M. Hughes. 1980. *Physiology of Mammals and Other Vertebrates*. 2d ed. Cambridge: Cambridge University Press.

Martini, F., and M. J. Timmons. 1995. *Human Anatomy*. Englewood Cliffs, N.J.: Prentice-Hall.

Matsumoto, A., and S. Ischii, eds. 1992. *Atlas of Endocrine Organs*. Heidelberg: Springer-Verlag.

Mazokhin-Porshnyakov, G. A. 1969. *Insect Vision*. New York: Plenum.

McDonald, D. A. 1960. *Blood Flow in Arteries*. Baltimore: Williams and Wilkins.

McDonald, D. M., and R. A. Mitchell. 1975. The innervation of the glomus cells, ganglion cells, and blood vessels in the rat carotid body: a quantitative ultrastructural analysis. *J. Neurocytol.* 4:177–230.

McGilvery, R. W. 1970. *Biochemistry: A Functional Approach*. Philadelphia: Saunders.

McMahan, U. J., N. C. Spitzer, and K. Peper. 1972. Visual identification of nerve terminals in living isolated skeletal muscle. *Proc. Roy. Soc. (London) Ser. B.* 181:421–430.

McMahon, B. R., W. W. Burggren, A. W. Pinder, and M. G. Wheatly. 1991. Air exposure and physiological compensation in a tropical intertidal chiton, *Chiton stokesii* (Mollusca: Polyplacophora). *Physiol. Zool.* 64(3):728–747.

McMahon, T. A. 1983. *Muscles, Reflexes and Locomotion*. Princeton, N. J.: Princeton University Press.

McMahon, T. A., and Bonner, J. T. 1983. *On Size and Life*. New York: Scientific American Books.

McNaught, A.B., and R. Callander. 1975. *Illustrated Physiology*. New York: Churchill Livingstone.

Menaker, M. 1968. *Proc. 76th. Annu. Convention American Psychologial Assoc.*, pp. 299–300.

Meyrand, P., J. Simmers, and M. Moulins. 1994. Dynamic construction of a neural network from multiple pattern generators in the lobster stomagastric nervous system. *J. Neurosci.* 14:630–644.

Michael, C. R. 1969. Retinal processing of visual images. *Scientific American* 205:104–114.

Mikiten, T.M. 1967. *Electrically Stimulated Release of Vasopressin from Rat Neurohypophyses in Vitro*. Ph.D. dissertation. Yeshiva University, New York.

Miller, W. H., F. Ratliff, and H. K. Hartline. 1961. How cells receive stimuli. *Scientific American* 205:222–238.

Milner, A. 1981. Flamingos, stilts and whales. *Nature* 289:347.

Milsom, W. K. 1992. Control of breathing in hibernating animals. In S. C. Wood, R. E. Weber, A. R. Hargens, and R. W. Millard, eds., *Physiological Adaptations in Vertebrates*. New York: Marcel Dekker, pp. 119–148.

Moffett, D., S. Moffett, and C. L. Schauf. 1993. *Human Physiology—Foundations and Frontiers*. St. Louis: Mosby.

Montagna, W. 1959. *Comparative Anatomy*. New York: Wiley.

Moog, F. 1981. The lining of the small intestine. *Scientific American* 245:154–176.

Morad, M., and R. Orkand. 1971. Excitation-contraction coupling in frog ventricle: Evidence from voltage clamp studies. *J. Physiol.* 219:167–189.

Morris, J. F., and D. V. Pow. 1988. *J. Exp. Biol.* 139:81–103.

Mountcastle, V. B., and R. J. Baldessarini. 1968. Synaptic transmission. In V. B. Mountcastle, ed., *Medical Physiology*. 13th ed. St. Louis: Mosby.

Muller, K. J. 1979. Synapses between neurones in the central nervous system of the leech. *Biol. Rev.* 54:99–134.

Murrary, J. M., and A. Weber. 1974. The cooperative action of muscle proteins. *Scientific American* 230(2):58–71.

Murray, R. G. 1973. The ultrastructure of taste buds. In I. Friedmann, ed. *The Ultrastructure of Sensory Organs*. New York: Elsevier.

Murray, R., and A. Murray. 1970. *Taste and Smell in Vertebrates*. London: Churchill.

Nachtigall, W. 1977. On the significance of Reynolds number and the fluid mechanical phenomena connected to it in swimming physiology and flight biophysics. In W. Nachtigall, ed., *Physiology of Movement—Biomechanics*. Stuttgart: Fischer Verlag.

Nagy, K. A. 1989. Field bioenergetics: accuracy of models and methods. *Physiol. Zool.* 62:237–252.

Nakajima, S., and K. Onodera. 1969. Membrane properties of the stretch receptor neurones of crayfish with particular reference to mechanisms of sensory adaptations. *Amer. J. Physiol.* 200:161–185.

Nathans, J., and D. S. Hogness. 1984. Isolation and nucleotide sequence of the gene encoding human rhodopsin. *Proc. Nat. Acad. Sci. USA* 81:4851–4855.

Nathans, J., D. Thomas, and D. S. Hogness. 1986. Molecular genetics of human color vision: the genes encoding blue, green, and red pigments. *Science* 232:193–202.

Neal, H. V., and H. W. Rand. 1936. *Comparative Anatomy*. Philadelphia: Blakiston.

Neher, E., and B. Sakmann. 1976. Single channel currents recorded from membrane of denervated frog muscle fibres. *Nature* 260:799–802.

Nickel, E., and L. Potter. 1970. Synaptic vesicles in freeze-etched electric tissue of *Torpedo*. *Brain Res.* 23:95–100.

Noback, C. R., and R. J. Demarest. 1972. *The Nervous System: Introduction and Reviews*. New York: McGraw-Hill.

O'Dor, R. K., and D. M. Webber. 1991. Invertebrate athletes: trade-offs between transport efficiency and power density in cephalopod evolution. *J. Exp. Biol.* 160:93–112.

O'Mally, B. W., and W. T. Schrader. 1976. The receptors of steroid hormones. *Scientific American* 234(2):32–43.

Palmer, J. 1973. Tidal rhythms: the clock control of the rhythmic physiology of marine organisms. *Biol. Rev. Cambridge Philos. Soc.* 48:377–418.

Parker, H. W. 1963. *Snakes*. London: Hale.

Patlack, J., and R. Horn. 1982. Effect of *N*-bromoacetamide on single sodium channel currents in excised membrane patches. *J. Gen. Physiol.* 79:333–351.

Peachey, L. D. 1965. Transverse tubules in excitation-contraction coupling. *Federation Proc.* 24:1124–1134.

Pearse, B. 1980. Coated vesicles. *Trends in Biochem. Sci.* 5:131–134.

Penfield, W., and T. Rasmussen. 1950. *The Cerebral Cortex of Man*. New York: Macmillan.

Penry, D. L., and P. A. Jumars. 1986. Chemical reactor analysis and optimal digestion. *BioScience* 36:310–315.

Phillips, G. N., Jr., J. P. Filliers, and C. Cohen. 1986. Tropomyosin cyrstal structure and regulation. *J. Mol. Biol.* 192:111–131.

Phillips, J. E. 1970. Apparent transport of water in insect excretory systems. *Amer. Zool.* 10:416–436.

Phillips, J. G. 1975. *Environmental Physiology*. New York: Wiley.

Pitts, R. F. 1959. *The Physiological Basis of Diuretic Therapy*. Springfield, Ill.: Thomas.

Pitts, R. F. 1968. *Physiology of the Kidney and Body Fluids*. 2d ed. Chicago: Year Book Medical Publishers.

Pitts, R. F. 1974. *Physiology of the Kidney and Body Fluids*. 3d ed. Chicago: Year Book Medical Publishers.

Porter, W. P., and D. M. Gates. 1969. Thermodynamic equilibria of animals with environment. *Ecol. Monogr.* 39:227–244.

Prosser, C. L. 1973. *Comparative Animal Physiology*. Vol. 1. Philadelphia: Saunders.

Rahn, H. 1967. Gas transport from the external environment to the cell. In A. V. S. de Reuck and R. Porter, eds., *Development of the Lung*. London: Churchill.

Randall, D. J. 1968. Functional morphology of the heart in fishes. *Amer. Zool.* 8:179–189.

Randall, D. J. 1970. Gas exchange in fish. In W. S. Hoar and D. J. Randall, eds., *Fish Physiology*. Vol. 4. New York: Academic.

Randall, D. J. 1994. Cardiorespiratory modeling in fishes and the consequences of the evolution of airbreathing. *Cardioscience* 5:167–171.

Randall, D. J., and P. A. Wright. 1989. The interaction between carbon dioxide and ammonia excretion and water pH in fish. *Can. J. Zool.* 67:2936–2942.

Riddiford, L. M., and J. W. Truman. 1978. Biochemistry of insect hormone and insect growth regulators. In Morris Rockstein, ed., *Biochemistry of Insects*. New York: Academic Press.

Romano, L., and H. Passow. 1984. Characterization of anion transport system in trout red blood cell. *Amer. J. Physiol.* 62A:257–271.

Rome, L. C., R. P. Funke, and R. M. Alexander. 1990. The influence of temperature on muscle velocity and sustained performance in swimming carp. *J. Exp. Biol.* 154:163–178.

Rome, L. C., R. P. Funke, R. M. Alexander, et al. 1988. Why animals have different muscle fibre types. *Nature* 355:824–827.

Rome, L. C., and M. J. Kushmerick. 1983. The energetic cost of generating isometric force as a function of temperature in isolated frog muscle. *Amer. J. Physiol.* 244:C100–C109.

Rome, L. C., and A. A. Sosnicki. 1991. Myofilament overlap in swimming carp. II. Sarcomere length changes during swimming. *Amer. J. Physiol. (Cell Physiol.)* 260:C289–C296.

Rome, L. C., D. Swank, and D. Corda. 1993. How fish power swimming. *Science* 261:340–343.

Rome, L. C., D. A. Syme, S. Hollingworth, et al. 1996. The whistle and the rattle: the design of sound producing muscles. *Proc. Nat. Acad. Sci. USA.* In press.

Romer, A. S. 1955. *The Vertebrate Body.* Philadelphia: Saunders.

Romer, A. S. 1962. *The Vertebrate Body.* 3rd ed. Philadelphia: Saunders.

Rosenthal, J. 1969. Post-tetanic potentiation at the neuromuscular junction of the frog. *J. Physiol.* 203:121–133.

Rowell, L. B. 1974. Circulation to skeletal muscle. In T. C. Ruch and H. D. Patton, eds., *Physiology and Biophysics,* 20th ed., Vol. 2. Philadelphia: Saunders.

Rupert, E. W., and R. D. Barnes. 1994. *Invertebrate Zoology.* 6th ed. Philadelphia: Saunders.

Rushmer, R. F. 1965a. The arterial system: arteries and arterioles. In T. C. Ruch and H. D. Patton, eds., *Physiology and Biophysics,* 19th ed. Philadelphia: Saunders.

Rushmer, R. F. 1965b. Control of cardiac output. In T. C. Ruch and H. D. Patton, eds., *Physiology and Biophysics,* 19th ed. Philadelphia: Saunders.

Russell, I. J. 1980. The responses of vertebrate hair cells to mechanical stimulation. In A. Roberts and B. M. Bush, eds., *Neurones Without Impulses.* Cambridge: Cambridge University Press.

Sacca, R., and W. W. Burggren. 1982. Oxygen partitioning between the skin, gills and lungs of the air-breathing reedfish, *Calamoicthys calabaricus. J. Exp. Biol.* 97:179–186.

Sakmann, B. 1992. As described in E. Neher and B. Sakmann. 1992. The patch clamp technique. *Scientific American* 266(3):44–51.

Saudou, F., and R. Hen. 1994. 5-Hydroxytryptamine receptor subtypes in vertebrates and invertebrates. *Neurochem. Int.* 25(6):503–532.

Scheid, P., H. Slama, and J. Piiper. 1972. Mechanisms of unidirectional flow in parabronchi of avian lungs: measurments in duck lung preparations. *Resp. Physiol.* 14:83–95.

Schmidt, R. F. 1971. Möglichkeiten und Grenzen der Hautsinne. *Klin. Wochenschr.* 49:530–540.

Schmidt-Nielsen, B. M., and W. C. Mackay. 1972. Comparative physiology of electrolyte and water regulation, with emphasis on sodium, potassium, chloride, urea, and osmotic pressure. In M. H. Maxwell and C. R. Kleeman, eds., *Clinical Disorders of Fluid and Electrolyte Metabolism.* New York: McGraw-Hill.

Schmidt-Nielsen, K. 1959. Salt Glands. *Scientific American.* 200:109–116.

Schmidt-Nielsen, K. 1960. The salt-secreting gland of marine birds. *Circulation.* 21:955–967.

Schmidt-Nielsen, K. 1964. *Desert Animals: Physiological Problems of Heat and Water.* London: Oxford University Press.

Schmidt-Nielsen, K. 1972. *How Animals Work.* Cambridge: Cambridge University Press.

Schmidt-Nielsen, K. B. 1975. *Animal Physiology, Adaptation and Environment.* New York: Cambridge University Press.

Schmidt-Nielsen, K., W. L. Bretz, and C. R. Taylor. 1970. Panting in dogs: unidirectional air flow over evaporative surfaces. *Science* 169:1102–1104.

Scholander, P. F., W. Flagg, V. Walters, and L. Irving. 1953. Climatic adaptation in arctic and tropical poikilotherms. *Physiol. Zool.* 26:67–92.

Schultz, S. G., and P. F. Curran. 1969. The role of sodium in nonelectrolyte transport across animal cell membranes. *Physiologist* 12:437–452.

Shadwick, R. E. 1992. Circulatory structure and mechanics. In A. A. Biewener, ed., *Biomechanics, Structures and Systems: A Practical Approach.* Oxford, United Kingdom: I. R. L. Press, pp. 233–261.

Shaw, E. A. T. 1974. Transformation of sound pressure level from the free field to the eardrum in the horizontal plane. *J. Acoust. Soc. Am.* 56:1848–1871.

Shelton, G. 1970. The effect of lung ventilation on blood flow to the lungs and body of the amphibian *Xenopus laevis. Resp. Physiol.* 9:183–196.

Shelton, G., and W. Burggren. 1976. Cardiovascular dynamics of the chelonia during apnoea and lung ventilation. *J. Exp. Biol.* 64(2):323–343.

Shephard, G. M. 1994. *Neurobiology,* 3rd edition. New York: Oxford University Press.

Sherrington, C. S. 1906. *The Integrative Activity of the Nervous System.* New Haven, Conn.: Yale University Press.

Sherwood, L. 1993. *Human Physiology, from Cells to Systems.* 2d ed. New York: West Publishing Company.

Sibley, A. P. Strategies of digestion and defecation. In C. R. Townsend and P. Callow, eds., *Physiological Ecology: An Evolutionary Approach to Resource Use.* Sunderland, Mass.: Sinauer Associates, Inc.

Siegelbaum, S. A., J. S. Camardo, and E. R. Kandel. 1982. Serotonin and cyclic AMP close single K^+ channels in *Aplysia* sensory neurones. *Nature* 299:413–417.

Siggaard-Andersen, O. 1963. *The Acid-Base Status of the Blood.* Copenhagen: Munksgaard.

Simmons, J. A., B. M. Fenton, and M. J. O'Farrell. 1979. Ecolocation and pursuit of prey by bats. *Science* 203:16–21.

Smith, D. S. 1965. The flight muscle of insects. *Scientific American* 212(6):76–88.

Smith, E. L., et al. 1983. *Principles of Biochemistry: Mammalian Biochemistry.* 6th ed. New York: McGraw-Hill.

Solomon, A. K. 1962. Pumps in the living cell. *Scientific American.* 207(2):100–108.

Somero, G. N. 1995. Proteins and temperature. *Annu. Rev. Physiol.* 57:43–68.

Sperry, R. W. 1959. The growth of nerve circuits. *Scientific American* 201: 100–108.

Spratt, N. T., Jr. 1971. *Develomental Biology.* Belmont, Calif.: Wadsworth.

Staehelin, L. A. 1974. Structure and function of intercellular junctions. *Int. Rev. Cytol.* 39:191–283.

Starling, E. H. 1908. The chemical control of the body. *Harvey Lectures* 3:115–131.

Steen, J. B. 1963. The physiology of the swimbladder of the eel *Anguilla vulgaris.* I. The solubility of gases and the buffer capacity of the blood. *Acta Physiol. Scand.* 58:124–137.

Steinbrecht, R. A. 1969. Comparative morphology of olfactory receptors. In C. Pfaffman, ed., *Olfaction and Taste,* Vol. 3. New York: Rockefeller University Press.

Stempell, W. 1926. *Zoologie im Grundriss*. Berlin: G. Borntraeger.

Stent, G. S. 1972. Cellular communication. *Scientific American* 227:42–51.

Stevens, C. E. 1988. *Comparative Physiology of the Vertebrate Digestive System*. Cambridge: Cambridge University Press.

Storey, K. B., and J. M. Storey. 1992. Natural freeze tolerance in ectothermic vertebrates. *Annu. Rev. Physiol.* 54:619–637.

Stryer, L. 1988. *Biochemistry*. 3d ed. New York: W. H. Freeman.

Swan, H. 1974. *Thermoregulation and Bioenergetics*. New York: Elsevier.

Taylor, C. R., K. Schmidt-Nielsen, and J. L. Raab. 1970. Scaling of energy costs of running to body size in mammals. *Amer. J. Physiol.* 219:1104–1107.

Tenney, S. M., and J. E. Temmers. 1963. Comparative quantitative morphology of the mammalian lung: diffusing area. *Nature* 197:54–57.

Thomas, D. H., and J. G. Phillips. 1975. Studies in avian and adrenal steroid function, Pts. 4–5. *Gen. Comp. Endocr.* 26:427–450.

Threadgold, L. J. 1967. *Ultra-structure of the Animal Cell*. New York: Academic.

Thurm, U. 1965. An insect mechanoreceptor. *Cold Spring Harbor Symp. Quant. Biol.* 30:75–82.

Tinbergen, N. 1951. *The Study of Instinct*. Oxford: Clarendon.

Toews, D. P., G. Shelton, and D. J. Randall. 1971. Gas tensions in the lungs and major blood vessels of the urodele amphibian, *Amphiuma tridactylum*. *J. Exp. Biol.* 55:47–61.

Tomita, T., A. Kaneko, M. Murakami, and E. L. Pautler. 1967. Spectral response curves of single cones in the carp. *Vision Res.* 7:519–531.

Tootell, R. B., M. S. Silverman, E. Switkes, and R. L. DeValois. 1982. Deoxyglucose analysis of retinotopic organization in primate striate cortex. *Science* 218:902–904.

Tsukada, H., and D. M. Blow. 1985. *J. Mol. Biol.* 184:703.

Tucker, V. A. 1975. The energy cost of moving about. *American Scientist* 63:413–419.

Ullrich, K. J., K. Kramer, and J. W. Boyaln. 1961. Present knowledge of the countercurrent system in the mammalian kidney. *Prog. Cardiovasc. Dis.* 3:395–431.

Unwin, N. 1993. Nicotinic acetylcholine receptor at 9 Å resolution. *J. Mol. Biol.* 229:1101–1124.

Van Vliet, B. N., and N. H. West. 1994. Functional characteristics of arterial chemoreceptors in an amphibian *Bufo marinus*. *Resp. Physiol.* 88:113–127.

Vander, A. J., J. H. Sherman, and D. S. Luciano. 1975. *Human Physiology: The Mechanisms of Body Function*. 2d ed. New York: McGraw-Hill.

Verdugo, P. 1990. Goblet cells secretion and mucogenesis. *Annu. Rev. Physiol.* 52:157–176.

Verdugo, P., M. Aitken, L. Langley, and M. J. Villalon. 1987. Molecular mechanisms of product storage and release in mucin secretion. II. The role of extracellular Ca^{++}. *Biorheology* 24:625–633.

Vogel, S. 1978. Organisms that capture currents. *Scientific American* 239:128–139.

Vollrath, F. 1992. Spider webs and silks. *Scientific American* 266(3):70–76.

von Bekesy, G. 1960. *Experiments in Hearing*. New York: McGraw-Hill.

von Buddenbrock, W. 1956. *The Love of Animals*. London: Muller.

von Euler, U. S., and J. H. Gaddum. 1931. An unidentified depressor substance in certain tissue extracts. *J. Physiol.* 72:74–87.

Wang, T., W. Burggren, and E. Nobrega. 1995. Metabolic, ventilatory and acid-base responses associated with specific dynamic action in the toad, *Bufo marinus*. *Physiol. Zool.* 68(2):192–205.

Wangensteen, O. D. 1972. Gas exchange by a bird's embryo. *Resp. Physiol.* 14:64–74.

Waterman, T. H., and H. R. Fernández. 1970. E-vector and wavelength discrimination by retinular cells of the crayfish *Procamberus Z. Vergl. Physiol.* 68:157–174.

Weibel, E. R. 1973. Morphological basis of alveolar-capillary gas exchange. *Physiol. Rev.* 53:419–495.

Weiderhielm, C. A., and B. U. Weston. 1973. Microvascular lymphatic and tissue pressures in the unanesthetized mammal. *Amer. J. Physiol.* 225:992–996.

Weiderhielm, C. A., J. W. Woodbury, S. Kirk, and R. F. Rushmer, 1964. Pulsatile pressures in the microcirculation of frog's mesentary. *Amer. J. Physiol.* 207:173–176.

Werblin, F. S., and J. E. Dowling. 1969. Organization of the retina of the mudpuppy, *Necturus maculosus:* II. Intracellular recording. *J. Neurophys.* 32:339–355.

West, E. S. 1964. *Textbook of Biophysical Chemistry*. New York: Macmillan.

West, J. B. 1970. *Ventilation/Blood Flow and Gas Exchange*. 2d ed. Oxford: Blackwell Scientific Publications.

West, N. H., and D. R. Jones. 1975. Breathing movements in the frog *Rana pipiens*, I: the mechanical events associated with lung and buccal ventilation. *Can. J. Zool.* 52:332–334.

White, F. N. 1972. Circulation: environmental correlation. In M. S. Gordon, ed., *Animal Physiology: Principles and Adaptation*, 2d ed. New York: Macmillan.

White, J. G., W. Amos, and M. Fordham. 1987. *J. Cell. Biol.* 104:41–48.

Wiedersheim, R. E. 1907. *Comparative Anatomy of Vertebrates*. London: Macmillan.

Wigglesworth, V. B. 1965. *The Principles of Insect Physiology*. 6th ed. London: Methuen.

Wilkie, D. R. 1977. Metabolism and body size. In T. J. Pedley, ed., *Scale Effects in Animal Locomotion*. New York: Academic.

Williams, P. L., ed. 1995. *Gray's Anatomy*. 38th ed. New York: Churchill Livingstone.

Wilson, D. M. 1964. The origin of the flight-motor command in grasshoppers. In R. F. Reiss, ed., *Neural Theory and Modeling: Proceedings of the 1962 Ojai Symposium*. Stanford, Calif.: Stanford University Press.

Wilson, D. M. 1971. Neural operations in arthropod ganglia. In F. O. Schmitt, ed., *The Neurosciences: Second Study Program*. New York: Rockefeller University Press.

Wine, J. J., and F. B. Krasne. 1972. The organization of the escape behavior in the crayfish. *J. Exp. Biol.* 56:1–18.

Wine, J. J., and F. B. Krasne. 1982. The cellular organization of crayfish escape behavior. In D. E. Bliss, H. Atwood, and D. Sandeman, eds., *The Biology of Crustacea*, Vol. IV. *Neural Integration*. New York: Academic Press.

Winlow, W., and E. Kandel. 1976. The morphology of identified neurons in the abdominal ganglion of *Aplysia californica*. *Brain Res.* 112:221–249.

Wood, S. C. 1991. Interactions between hypoxia and hypothermia. *Annu. Rev. Physiol.* 53:71–85.

Wright, E. M. 1993. The intestinal Na^+/glucose cotransporter. *Annu. Rev. Physiol.* 55:575–589.

Wright, P. A. 1995. Nitrogen excretion: three end products, many physiological roles. *J. Exp. Biol.* 198:273–281.

Yau, K.-W. 1976. Receptive fields, geometry and conduction block of sensory neurones in the central nervous system of the leech. *J. Physiol.* 263:513–538.

Yau, K.-W., and K. Nakatani. 1985. Light-suppressible, cyclic GMP-sensitive conductance in the plasma membrane of a truncated rod outer segment. *Nature* 317:252–255.

Young, M. 1971. Changes in human hemoglobins with development. In P. L. Altman and D. W. Dittmer, eds., *Respiration and Cirulation.* Bethesda, Md.: Federation of American Societies for Experimental Biology.

Young, R. W. 1970. Visual cells. *Scientific American* 223:80–91.

Zotterman, Y. 1959. Thermal sensations. In H. W. Magoun, ed., *Handbook of Physiology* (Section 1, Neurophysiology, Vol. I). Baltimore: Williams and Wilkins.

GLOSSARY

a-adrenergic receptors Receptors on cell surfaces that bind norepinephrine and, less effectively, epinephrine; the binding leads to enzymatically mediated responses by the cells.

A band A region of a muscle sarcomere that corresponds to the myosin thick filaments.

abomasum The true digestive stomach of the ruminant digastric stomach.

absolute temperature Temperature measured from absolute zero, the state of no atomic or molecular thermal agitation. The absolute scale is divided into kelvin units (K), with 1°K having the same size as 1 Celsius degree. Thus, 0°K is equal to $-273.15°C$ or $-459.67°F$.

acclimation The persisting change in a specific function due to prolonged exposure to an environmental condition such as high or low temperature.

acclimatization The persisting spectrum of changes due to prolonged exposure to environmental conditions such as high or low temperature.

accommodation The temporary increase in threshold that develops during the course of a subthreshold stimulus.

acetylchofinesterase An enzyme that hydrolyzes ACh and resides on postsynaptic membrane surface.

acetylcholine (ACh) An acetic acid ester of choline ($CH_3—CO—O—CH_2—CH_2—N(CH_3)_3—OH$), important as a synaptic transmitter in most species and in many different types of neurons.

acid Proton donor.

acidosis Excessive body acidity.

acid tide Acidification of the blood following pancreatic secretion.

acini Plural of acinus. *See* acinus.

acinus A small sac or alveolus, sometimes lined with exocrine-secreting cells.

acromegaly Hypersecretion of growth hormone in adulthood, causing enlargement of the skeletal extremities and facial structures.

actin A ubiquitous protein that participates in muscle contraction and other forms of cellular motility. G-actin is the globular monomer that polymerizes to form F-actin, the backbone of the thin filaments of the sarcomere of muscle.

action potential (nerve impulse; spike; AP) Transient all-or-none reversal of a membrane potential produced by regenerative inward current in excitable membranes.

action spectrum The degree of response to different wavelengths of incident light.

activating reaction A reaction that changes an inactive enzyme into an active catalyst.

activation energy The energy required to bring reactant molecules to velocities sufficiently high to break or make chemical bonds.

activation heat Heat produced during excitation and activation of muscle tissue, but independent of shortening.

active hyperemia The increase in blood flow that follows increased activity in a tissue, particularly skeletal muscle.

active site The catalytic region of an enzyme molecule.

active state In muscle, the condition when myosin cross-bridges are attached to actin, causing the muscle fibers to resist a force that would pull them to a greater length.

active transport Energy-requiring translocation of a substance across a membrane, usually against its concentration or electrochemical gradient. *Primary transport:* Transport of a substance directly related to hydrolysis of ATP or other phosphagen. *Secondary transport:* Uphill transport of one substance coupled to and energy derived from the downhill transport of another substance.

active zone Local region, within a presynaptic terminal, at which synaptic vesicles dock and are prepared for release by exocytosis.

activity Capacity of a substance to react with another; the effective concentration of an ionic species in the free state.

activity coefficient A proportionality factor obtained by dividing the effective reactive concentration of an ion (as indicated by its properties in a solution) by its molar concentration.

actomyosin A complex of muscle proteins formed when myosin cross-bridges bind to actin in thin filaments.

acuity Resolving power.

adaptation Evolution through natural selection leading to an organism who's physiology, anatomy and behavior are matched to the demands of its environment.

adaptation (sensory) Decrease in sensitivity during sustained presentation of a stimulus.

adaptive *See* adaptation.

addiction A physiological state of chemical dependence in which neuronal function shifts so that the individual experiences intense—even life-threatening—discomfort unless there are regular doses of the exogenous chemical.

adductor muscle One that brings a limb toward the median plane of the body.

adenine A white, crystalline base 6-amino-purine, $C_5H_5N_5$; purine base constituent of DNA and RNA.

adenohypophysis (anterior pituitary gland; anterior lobe) The glandular anterior lobe of the hypophysis, consisting of the pars tuberalis, pars intermedia, and pars distalis.

adenosine diphosphate (ADP) A nucleotide formed by hydrolysis of ATP, with the release of one high-energy bond.

adenosine triphosphatase *See* ATPase.

adenosine triphosphate (ATP) An energy-rich nucleotide used as a common energy currency by all cells.

adenylate cyclase (adenyl cyclase) A membrane-bound enzyme that catalyzes the conversion of ATP to cAMP.

adipose Fatty.

ADP *See* adenosine diphosphate.

Adrenalin Trade name for epinephrine.

adrenal medulla Central portion of the adrenal gland.

adrenergic Relating to neurons or synapses that release epinephrine, norepinephrine, or other catecholamines.

adrenergic receptors (adrenoreceptors) Receptors on cell surfaces that bind norepinephrine and epinephrine; the binding leads to enzymatically mediated responses by the cells.

adrenocorticotropic hormone (ACTH; adrenocorticotropin; corticotropin) A hormone released by cells in the adenohypophysis that acts mainly on the adrenal cortex, stimulating growth and corticosteroid production and secretion in that organ.

adrenoreceptors *See* adrenergic receptors.

aequorin A protein extracted from the jellyfish *Aequorea*; on combining with Ca^{2+}, it emits a photon of blue-green light.

aerobic Utilizing molecular oxygen.

aerobic metabolic scope The ratio of the maximum sustainable metabolic rate to the BMR (or the SMR).

aerobic metabolism Metabolism utilizing molecular oxygen.

afferent Transporting or conducting toward a central region; centripetal.

afferent fiber (afferent neuron) An axon that relays impulses from a sensory receptor to the central nervous system.

affinity sequence (selectivity sequence) The order of preference with which an electrostatic site will bind different species of counterions.

agiomerular Lacking glomerulae in the kidney.

agonist A substance that can interact with receptor molecules and mimic an endogenous signalling molecule.

aldehydes A large class of substances derived from the primary alcohols by oxidation and containing the —CHO group.

aldosterone A mineralocorticoid secreted by the adrenal cortex; the most important electrolyte-controlling steroid, acting on the renal tubules to increase the reabsorption of sodium.

alimentary canal A hollow, tubular cavity extending through animals open at both ends; for ingestion, digestion and secretion of food materials

alkali earth metals A group of grayish-white, malleable metals easily oxidized in air, comprising Be, Mg, Ca, Sr, Ba, and Ra.

alkaline tide A period of increased body and urinary alkalinity associated with excessive gastric HCl secretion during digestion.

alkaloids A large group of organic nitrogenous bases found in plant tissues, many of which are pharmaceutically active (e.g., codeine, morphine).

alkalosis Excessive body alkalinity.

allantoic membrane One of the membranes within a bird eggshell; important in the respiration of the unhatched chick.

allantoin Waste product of purine metabolism.

allometry Systematic changes in body proportions with increasing species size.

all-or-none Describes a condition in which the magnitude of a cell's response is independent of the strength of a stimulus above some threshold value. If an input signal brings the cell to its threshold, the amplitude of the response is maximal; if the stimulus fails to bring the cell to threshold, there is no response.

allosteric site Area of an enzyme that binds a substance other than the substrate, changing the conformation of the protein so as to alter the catalytic effectiveness of the active site.

α-bungarotoxin neurotoxin in the venom of the krait (a snake) that blocks neuromuscular transmission in vertebrates by binding to nicotinic acetylcholine (ACh) receptors.

alpha helix Helical secondary structure of many proteins in which each NH group is hydrogen bonded to a CO group at a distance equivalent to three amino acid residues; the helix makes a complete turn for each 3.6 residues.

alpha motor neurons Large spinal neurons that innervate extrafusal skeletal muscle fibers of vertebrates.

alveolar ventilation volume The volume of fresh atmospheric air entering the alveoli during each inspiration.

alveoli Small cavities, especially those microscopic cavities that are the functional units of the lung.

amacrine cells Neurons without axons, found in the inner plexiform layer of the vertebrate retina.

ambient Surrounding, prevailing.

amide An organic derivative of ammonia in which a hydrogen atom is replaced by an acyl group.

amine Derivative of ammonia in which at least one hydrogen atom is replaced by an organic group.

amino acids Class of organic compounds containing at least one carboxyl group and one amino group; the alpha-amino acids, $RCH(NH_2)COOH$, make up proteins.

amino group —NH_2.

ammonia NH_3; toxic, water-soluble, alkaline waste product of deamination of amino acids and uric acid.

ammonotelic Pertaining to the excretion of nitrogen in the form of ammonia.

ampere (A) MKS unit of electric current; equal to the current produced through a 1 ohm (Ω) resistance by a potential difference of 1 volt (V); the movement of 1 coulomb (C) of charge per second.

amphipathic Pertaining to molecules bearing groups with different properties, such as hydrophilic or hydrophobic groups.

amphoteric Having opposite characteristics; behaving as either an acid or a base.

amylase Carbohydrases that hydrolyze all but the terminal glycosidic bonds within starch and glycogen, producing disaccharides and oligosaccharides.

anabolism Synthesis by living cells of complex substances from simple substances.

anaerobic Oxygen-free.

anaerobic metabolism Metabolism not utilizing molecular oxygen.

anastomose To interconnect.

anatomical dead space The nonrespiratory conducting pathways in the lung.

androgens Hormones having masculinizing activity.

aneurism Localized dilation of an artery wall.

angiotensin A protein in the blood, converted from angiotensinogen by the action of renin; it first exists as a decapeptide (angiotensin I) that is acted upon by a peptidase, which cleaves it into an octapeptide (angiotensin II), a potent vasopressor and stimulator of aldosterone secretion.

angiotensin II *See* angiotensin.

anion Negatively charged ion; attracted to the anode or positive pole.

anode Positive electrode or pole to which negatively charged ions are attracted.

anoxemia A lack of oxygen in the blood.

anoxia A lack of oxygen.

antagonist remuscle A muscle acting in opposition to the movement of another muscle.

antagonists Agents that inhibit, block, or counteract an effect; for example, antagonists of synaptic transmitters typically block the postsynaptic receptor molecules that bind the neurotransmitter.

antennal gland Crustacean osmoregulatory organ.

antibody An immunoglobulin, a four-chain protein molecule of a specific amino acid sequence; an antibody will interact only with the antigen that brought about its production or one very similar to it.

antidiuretic hormone (ADH, vasopressin) A hormone made in the hypothalamus and liberated from storage in the neurohypophysis; acts on the epithelium of the renal collecting duct by stimulating osmotic reabsorption of water, thereby producing a more concentrated urine; also acts as a vasopressor.

antigen A substance capable of bringing about the production of antibodies and of then reacting with them specifically.

antimycin An antibiotic that is isolated from a *Streptomyces strain*; acts to block electron transport from cytochrome *b* to cytochrome *c* in the electron-transport chain.

antiports Membrane transport proteins that transfer two solutes, each in the opposite direction.

anus The opening of the alimentary canal through which feces are expelled.

aorta The main artery leaving the heart.

aortic body A nodule on the aortic arch containing chemoreceptors that sense the chemical composition of the blood.

apical Pertaining to the apex; opposite the base.

apnea The suspension or absence of breathing.

apoenzyme The protein portion of an enzyme; the apoenzyme and coenzyme form the functioning holoenzyme.

apolysis Release; loosening from.

aporepressor A repressor gene product that, in combination with a corepressor, reduces the activity of particular structural genes.

aquaporin 28kDa protein, tetramers of which form water channels in membranes.

area centralis (fovea) In the mammalian retina, the area with the highest visual resolution due to the small divergence and convergence in the pathway linking photoreceptors to ganglion cells; in primates contains closely packed cone cells.

arginine phosphate A phosphorylated nitrogenous compound found primarily in muscle; acts as a storage form of high-energy phosphate for the rapid phosphorylation of ADP to ATP.

arteriole A tiny branch of an artery; in particular, one nearest a capillary.

arteriosclerosis A class of diseases marked by an increase in thickness and a reduction in elasticity of the arterial walls.

association cortex Areas of the cerebral cortex that neither directly receives sensory information nor directly contributes to motor output; instead neurons of association cortex typically receive input from many sensory modalities and are broadly connected to other areas in the cortex and other brain centers.

asynchronous muscle A type of flight muscle found in the thorax of some insects; contracts at a frequency that bears no one-to-one relation to motor impulses. *See also* fibrillar muscle.

ATP *See* adenosine triphosphate.

ATPase (adenosine triphosphatase) A class of enzymes that catalyze the hydrolysis of ATP.

ATPS Ambient temperature and pressure, saturated with water vapor; referring to gas volume measurements.

atria A chamber that gives entrance to another structure or organ; usually used alone to refer to an atrium of the heart.

atrial natriuretic peptide (ANP) One of a family of peptide hormones, cleaved from a single precursor peptide and produced in the cardiac atria, the physiological effects of which include increased urine output, increased sodium excretion, and a receptor-mediated vasodilation, the net result of which is lowered blood pressure.

atrioventricular node Specialized conduction tissue in the heart, which, along with Purkinje tissue, forms a bridge for electrical conduction of the impulse from atria to ventricles.

auditory cortex Regions of the cerebral cortex that are associated with hearing.

auditory ossicle One of the bones of the middle ear (the malleus, the incus, and the stapes), connecting the tympatic membrane and the oval window.

autocrine A hormonal pathway characterized by the production of a biologically active substance by a cell; the substance then binds to receptors on that same cell to initiate a cellular response.

autoinhibition Self-inhibition.

autonomic nervous system The efferent nerves that control involuntary visceral functions; classically subdivided into the sympathetic and parasympathetic sections.

autoradiography The process of making a photographic record of the internal structures of a tissue by utilizing the radiation emitted from incorporated radioactive material.

autorhythmicity The generation of rhythmic activity without extrinsic control.

autotrophic Pertaining to the ability to synthesize food from inorganic substances by utilizing the energy of the sun or of inorganic compounds.

Avogadro's law Equal volumes of different gases at the same temperature and pressure contain equal numbers of molecules. One mole (mol) of an ideal gas at 0°C and 1 standard atmosphere (atm) occupies 22.414 liters (L). Avogadro's number equals 6.02252×10^{23} molecules/mol.

axon The elongated cylindrical process of a nerve cell along which action potentials are conducted; a nerve fiber.

axonemme Complex of microtubules and associated structures within the flagellar or ciliary shaft.

axon hillock The transitional region between an axon and the nerve cell body.

axon terminal The end of each axon, which typically is the site where signals are passed to another neuron or other cell.

axoplasm The cytoplasm within an axon.

azide Any compound bearing the N_3^- group.

b-adrenergic receptors Epinephrine-binding sites; normally coupled to activation of adenylate cyclase.

baroreceptor A sensory nerve ending that is stimulated by changes in pressure, as those in the walls of blood vessels.

basal body (kinetosome) Microtubular structure from which a cilium or flagellum arises; homologous with centriole.

basal cell A cell in a chemoreceptive organ that regularly gives rise to new chemoreceptor cells during adult life.

basal metabolic rate (BMR) The rate of energy conversion in a homeotherm while it is resting quietly within the thermal neutral zone without food in the intestine.

base Proton acceptor.

basilar membrane The delicate ribbon of tissue bearing the auditory hair cells in the cochlea of the vertebrate ear.

Bell-Magendie rule The dorsal root of the spinal cord contains only sensory axons, whereas the ventral root contains only motor axons.

Bernoulli's effect Fluid pressure drops as fluid velocity increases.

beta (b) adrenergic receptors The class of adrenergic membrane receptors that are blocked by the drug propranolol; their activation is less sensitive to norepinephrine than is that of the alpha receptors.

beta keratin Insoluble, sulfur-rich scleroprotein; constituent of epidermis, horns, hair, feathers, nails, and tooth enamel. *Beta* refers to the protein's secondary structure, which is in pleated sheets.

beta pleated sheet Form of protein secondary structure in which two or more distinct amino acid chains lay side by side, held together by hydrogen bonds.

bile Viscous yellow or greenish alkaline fluid produced by the liver and stored in the gallbladder; containing bile salts, bile pigments, and certain lipids, it is essential for digestion of fats.

bile duct Duct carrying bile fluid from liver to duodenum.

bile fluid Fluid secreted by liver that emulsifies fats and neutralizes acid intestinal contents.

bile pigments Pigments in bile fluid derived from breakdown products of hemoglobin.

bile salt Bile acid such as cholic acid conjugated with glycine or taurine, promoting emulsification and solubilization of intestinal fats.

binocular convergence Positioning of the eyes so that the images formed fall on analogous portions of the two retinas, avoiding double vision.

biogenic amine Any of a number of signalling molecules that are synthesized in the body from single amino acid molecules.

bipolar cell A neuron with two axons emerging from opposite sides of the soma; one class of these neurons is found in the vertebrate retina, where they transmit signals from the photoreceptor cells to the retinal ganglion cells.

birefringence Double refraction; the ability to pass preferentially light that is polarized in one plane.

bleaching Fading of photopigment color upon absorption of light.

blubber A fatty, insulating tissue typically found under the skin in cetaceans.

Bohr effect (Bohr shift) A change in hemoglobin-oxygen affinity due to a change in pH.

Bolus A discrete plug or collection of food material moving through the alimentary canal.

bombykol Sex attractant pheromone of the female silkworm moth (*Bombyx mori*).

book lungs The respiratory surface in spiders.

Bowman's capsule (glomerular capsule) A globular expansion at the beginning of a renal tubule and surrounding the glomerulus.

Boyle's law At a given temperature, the product of pressure and volume of a given mass of gas is constant.

bradycardia A reduction in heart rate from the normal level.

bradykinin A hormone formed from a precursor normally circulating in the blood; a very potent cutaneous vasodilator.

brain The major neuronal center within the body; typically located at the anterior of the body.

brain hormone (prothoracicotropin; activating hormone) A hormone synthesized by the neurosecretory cells of the pars intercerebralis and released by the corpora cardiaca of insects; activates the prothoracic glands to secrete ecdysone.

bronchi Conducting airways in the lung; branches of the tracheae.

bronchioles Small conducting airways in the lung; branches of the bronchi.

brood spot A prolactin-induced bald area on the ventral surface of some brooding birds that receives a rich supply of blood for the incubation of eggs.

brown fat Fat deposits with extensive vascularization, mitochondria and enzyme systems for oxidation. Found in small, specific deposits in a few mammals and used for nonshivering thermogenesis.

Brunner's glands Exocrine glands that are located in the intestinal mucosa and secrete an alkaline mucoid fluid.

brush border A free epithelial cell surface bearing numerous microvilli.

BTPS Body temperature, atmospheric pressure, saturated with water vapor.

buccal Pertaining to the mouth cavity.

buffer A chemical system that stabilizes the concentration of a substance; acid-base systems serve as pH buffers, preventing large changes in hydrogen ion concentration.

bundle of His The conducting tissue within the interventricular septum of the mammalian heart.

bungarotoxin (BuTX) A blocking agent composed of a group of neurotoxins isolated from the venom of members of the snake genus *Bungarus* (the krait) of the cobra family; binds selectively and irreversibly to nicotinic acetylcholine receptors.

Bunsen solubility coefficient The quantity of gas at STPD that will dissolve in a given volume of liquid per unit partial pressure of the gas in the gas phase. This coefficient is used only for gases that do not react chemically with the solvent.

bursicon A hormone secreted by neurosecretory cells of the insect central nervous system; tans and hardens the cuticle of freshly molted insects.

cable properties The passive electrical properties (resistance and capacitance) of a cell; these properties were first worked out for submarine cables.

calcitonin (thyrocalcitonin) A protein hormone secreted by the mammalian parafollicular cells of the thyroid in response to elevated plasma calcium levels.

calcitriol A steroid-like compound produced from vitamin D ingested with some foods and from vitamin D3, or synthesized from cholesterol in the skin. The physiological actions of calcitriol are similar to those of parathyroid hormone.

calcium response A graded depolarization due to a weakly regenerative inward calcium current.

caldesmon A calcium-binding regulatory protein in smooth muscle, which plays a role in the "latch" mechanism of some smooth muscles.

calmodulin A troponinlike calcium-binding regulatory protein found in essentially all tissues.

calorie (cal) The quantity of heat required to raise the temperature of 1 g of water from 14.5° to 15.5°C; most commonly used as kilocalorie (kcal) = 1000 cal.

calorimetry Measurement of heat production in an animal.

calsequestrin A calcium-binding protein that contributes to the regulation of contraction in smooth muscles.

canines Pointed, dagger-like teeth are used for piercing and tearing food.

capacitance The property of storing electric charge by electrostatic means.

capacitive current Current entering and leaving a capacitor.

capacity The ability of a capacitor or other body to store electric charge. The unit of measure is the farad (F), which describes the proportionality between charge stored and potential for a given voltage, $C = q/V =$ coulombs divided by volts.

carbohydrases Enzymes that specifically break down carbohydrates.

carbohydrate Aldehyde or ketone derivative of alcohol; utilized by animal cells primarily for the storage and supply of chemical energy; most important are the sugars and starches.

carbonic anhydrase An enzyme catalyzing the reversible interconversion of carbonic acid to carbon dioxide and water.

carbonyl The organic radical —C═O, which occurs in such compounds as aldehydes, ketones, carboxylic acids, and esters.

carboxyhemoglobin The compound formed when carbon monoxide combines with hemoglobin; carbon monoxide competes successfully with oxygen for combination with hemoglobin, producing tissue anoxia.

carboxylates R—COO—, salts or esters of carboxylic acids.

carboxyl group The radical —COOH, which occurs in the carboxylic acids.

cardiac output The total volume of blood pumped by the heart per unit of time; cardiac output equals heart rate times stroke volume.

carotid body A nodule on the occipital artery just above the carotid sinus, containing chemoreceptors that sense the chemical composition of arterial blood.

carotid rete Countercurrent heat exchanger at base of skull that helps prevent overheating of the brain of certain hoofed animals and carnivores.

carotid sinus A dilation of the internal carotid artery with many baroreceptor (pressure receptor) endings in the wall.

carotid sinus baroreceptors Receptors that sense arterial blood pressure; located in the carotid sinus, a dilatation of the internal carotid artery at its origin.

carrier-mediated transport Transmembrane transport of solutes achieved by membrane embedded carrier (e.g., facilitated diffusion).

carrier molecules Lipid-soluble molecules that act within biological membranes as carriers for certain molecules that have lower mobility in the membrane.

catabolism Disassembly of complex molecules into simpler ones.

catalysis An increase in the rate of a chemical reaction promoted by a substance—the catalyst—not consumed by the reaction.

catalyst A substance that increases the rate of a reaction without being used up in the reaction.

cataract A condition in which proteins of the lens of the eye become cross-linked, causing it to become cloudy and reducing visual acuity.

catecholamines A group of related compounds exerting a sympathomimetic action on nervous tissue; examples are epinephrine, norepinephrine, and dopamine.

catecholandues A group of related compounds exerting a sympathomimetic action on nervous tissue; examples are epinephrine, norepinephrine, and dopamine.

cathode The negative electrode, so called because it is the electrode to which cations are attracted.

cation A positively charged ion; attracted to a negatively charged electrode.

caudal Pertaining to the tail end.

cecum A blind pouch in the alimentary canal.

Cellulase An enzyme that digests cellulose and hemicellulose, produced by gut-borne symbiotic micro-organisms.

central chemoreceptors Sensory structures in the brain monitoring pH and initiating appropriate changes in breathing.

central lacteal Small, blind-ended lymph vessel in center of intestinal villus for the uptake of fats and some vitamins.

central nervous system The collection of neurons and parts of neurons that are contained within the brain and spinal cord of vertebrates, or within the brain, ventral nerve cord, and major ganglia of invertebrates.

central pattern generator A group of neurons that produces and maintains a rhythmic pattern of action, such as breathing, walking, chewing, or swimming.

central pattern-generating network A set of interconnected neurons whose collective activity produces patterned behavior.

central sulcus A deep, almost vertical furrow on the cerebrum, dividing the frontal and parietal lobes.

cephalic Pertaining to the head.

cephalic phase Gastric secretion occurring in response to sight, smell, and/or taste of food, or in response to conditioned reflexes.

cephalization The evolutionary tendency for the neurons of higher organisms to be concentrated in a brain located at the anterior end of the animal.

cerebellum A part of the hindbrain that contributes to the coordination of motor output.

cerebral cortex The thin layer of gray matter that covers the cerebrum of mammals, and probably of birds.

cerebral hemispheres The large paired structures of the cerebrum, connected by the corpus callosum.

cerebral ventricles A series of confluent fluid-filled cavities within the brain of vertebrates; the fluid in the ventricles is cerebrospinal fluid.

cerebrospinal fluid In vertebrates, clear fluid that fills the cavities (ventricles) within the brain and the central canal of the spinal cord; it is a complex filtrate of blood plasma and is modified by brain cells before it returns to the venous system.

cerebrum The largest part and highest center of the mammalian brain; it evolved from the olfactory centers of lower vertebrates.

charge, electric (*q*) Measured in units of coulombs (C). To convert 1 g·equiv weight of a monovalent ion to its elemental form (or vice versa) requires a charge of 96,500 C (1 faraday, *F*). Thus, in loose terms, a coulomb is equivalent to 1/96,500 g·equiv of electrons. The charge on one electron is -1.6×10^{-19} C. If this is multiplied by Avogadro's number, the total charge is 1 *F*, or $-96,487$ C·mol^{-1}.

chelating agent A chemical that binds calcium, or other ions, and removes it from solution.

chemical energy Energy contained in the chemical bonds holding molecules together.

chemical synapse A junction between a neuron and another cell in which the signal from the presynaptic neuron is carried across the synaptic cleft by neurotransmitter molecules.

chemoreceptor A sensory receptor specifically sensitive to certain molecules.

chief cells (zygomatic) Epithelial cells of the gastric epithelium that release pepsin.

chimera An animal with extra copies of a normal gene or a copy of a mutated gene.

chitin A structural polymer of D-glucosamine that serves as the primary constituent of arthropod exoskeletons.

chloride cells Epithelial cells of fish gills that engage in active transport of salts.

chloride shift The movement of chloride ions across the red blood cell membrane to compensate for the movement of bicarbonate ions.

chlorocruorin A green respiratory pigment found in some marine polychaetes; similar to hemoglobin.

choanocytes Flagellated cells lining the body cavity of sponges.

cholecystokinin (CCK, pancreozymin; CCK-PZ) A hormone liberated by the upper intestinal mucosa that induces gallbladder contraction and release of pancreatic enzymes.

cholesterol A natural sterol; precursor to the steroid hormones.

cholinergic Relating to acetylcholine or substances with actions similar to ACh.

choroidplexus Highly vascularized, furrowed projections into the brain ventricles that secrete cerebrospinal fluid.

choroid rete A countercurrent arrangement of arterioles and venules behind the retina in the eyes of teleost fish.

chromaffin cells Epinephrine-secreting cells of the adrenal medulla; named for their high affinity for chromium salt stains.

chromatography A general technique that exploits the fact that different components in a sample will move at different rates through a substrate such as chromatography paper or other porous solid matrices.

chromophore A chemical group that lends a distinct color to a compound containing it.

chronobiology The study of biological rhythms.

chronotropic Pertaining to rate or frequency, especially in reference to the heartbeat.

chylomicrons Protein-coated tiny droplets of triglycerides, phospholipids and cholesterol formed within vesicles of absorptive cells from the digestion produces of fats, monoglycerides, fatty acids, and glycerol.

chyme The mixture of partially digested food and digestive juices found in the stomach and the intestine.

chymotrypsin A proteolytic enzyme that specifically attacks peptide bonds containing the carboxyl groups of tyrosine, phenylalanine, tryptophan, leucine, and methionine.

chymotrypsinogen Inactive precursor of chymotrypsin.

ciliary body A thick region of the anterior vascular tunic of the eye; joins the choroid and the iris.

ciliary muscle A muscle of the ciliary body of the vertebrate eye; influences the shape of the lens in visual accommodation.

ciliary reversal A change in the direction of the power stroke of a cilium, causing it to beat in reverse.

cilium A motile organelle with a "9 + 2" tubular substructure; when present generally in large numbers, a small flagellum.

circadian rhythms Biological rhythms with daily cycles.

circalunar Referring to biorhythms related to lunar cycles.

circannual Referring to biorhythms related to yearly cycles.

circatidal Referring to biorhythms related to tidal cycles.

circular smooth muscle Inner layer of smooth muscle that encircles small intestine.

cis A configuration with similar atoms or groups on the same side of the molecular backbone.

***cis-trans* isomerization** Conversion of a cis isomer into a trans isomer.

citric acid cycle (also Krebs cycle, TCA cycle) A series of eight major reactions following glycolysis, in which acetate residues are degraded to CO_2 and H_2O.

cladogram A form of diagrammatic analysis of taxonomic relationships between organisms that relates animals according to suites of common characters.

clathrin The protein surrounding the cytoplasmic surface of a coated vesicle membrane.

clefts Lateral intercellular spaces between cells in epithelium.

cloaca The terminal area of hindgut of some fishes, amphibians, reptiles, birds and a few mammals; aids in urinary ion and water resorption.

clone A population of genetically identical cells all derived from a single original cell.

coated pits Receptor-lined membrane depressions that eventually form coated vesicles during the process of receptor-mediated endocytosis.

coated vesicle Vesicle with its cytoplasmic surface covered with the protein clathrin, formed in process of receptor-mediated endocytosis.

cochlea A portion of the inner ear, a tapered tube would into a spiral like the shell of a snail, containing hair cell receptors for detecting sound.

cochlear microphonics Electrical signals recorded from the cochlea, having a frequency identical to that of the sound stimulus.

coelenteron A blind-ended tube or cavity in coelenterates that serves as a "batch reactor" site for chemical digestion.

coelom The body cavity of higher metazoans, situated between the gut and the body wall and lined with mesodermal epithelium.

coenzyme An organic molecule that combines with an apoenzyme to form the functioning holoenzyme.

coenzyme A (CoA) A derivative of pantothenic acid to which acetate becomes attached to form acetyl CoA.

cofactor An atom, ion, or molecule that combines with an enzyme to activate it.

coitus Sexual intercourse.

colchicine An antimitotic agent that disrupts microtubules by interfering with the polymerization of tubulin monomers.

collar cells Flagellated cells lining the internal chambers of sponges (Porifera).

collateral processes Branches of an axon that terminate in locations other than the major target location.

collaterals Side branches of a nerve or blood vessel.

collecting duct The portion of the renal tubules in which the final concentration of urine occurs.

colligative properties Characteristics of a solution that depend on the number of molecules in a given volume.

colloid A system in which fine solid particles are suspended in a liquid.

command system A set of neurons that, when stimulated, elicit a set pattern of coordinated movements.

competitive inhibition Reversible inhibition of enzyme activity due to competition between a substrate and an inhibitor for the active site of the enzyme.

compliance The change in length or volume per unit change in the applied force.

compound eye The multifaceted arthropod eye; the functional unit is the ommatidium.

concentration gradient Difference in solute concentration across a membrane, or between two different regions of a solution.

condensation A reaction between two or more organic molecules leading to the formation of a larger molecule and the elimination of a simple molecule, such as water or alcohol.

conditioned reflex Reflexes learned or modified through behavioral repetition.

conductance, electrical (G) A measure of the ease with which a conductor carries an electric current; the unit is the siemen (S), re-ciprocal of the ohm (Ω).

conductance, thermal A quantity describing the ease with which heat flows by conduction under a temperature gradient across a substance or an object.

conductivity The intrinsic property of a substance to conduct electric current; reciprocal of resistivity.

conductor A material that carries electric current.

cone A vertebrate visual receptor cell that has a tapered outer segment in which the lamellar photosensitive membranes remain continuous with the surface membrane.

confocal scanning microscope A microscope using a focused laser beam to rapidly scan different areas of the specimen in a single plane. Light reflected from that plane is assembled by a computer into a composite image of the specimen.

conformer An animal in which internal body fluid condition tends to parallel that of the external environment.

conjugate acid-base pair Two substances related by gain or loss of an H^+ ion (proton).

connective A collection of axons that carry information between neuronal centers, such as ganglia, in many invertebrate nervous systems.

contracture A more or less sustained contraction in response to an abnormal stimulus.

contralateral Pertaining to the opposite side.

conus A chamber invested with cardiac muscle and found in series with and downstream from the ventricle in elasmobranchs.

convection The mass transfer of heat due to mass movement of a gas or liquid.

convergence A pattern in which inputs from many different neurons impinge upon a single neuron.

corepressor A low-molecular-weight molecule that unites with an aporepressor to form a substance that inhibits the synthesis of an enzyme.

cornea The clear surface of the eye through which light passes as it enters the eye.

corneal lens The clear structure at the outside surface of an ommatidium; it admits and focuses light entering the ommatidium.

corpora allata Nonneuronal insect glands existing as paired organs or groups of cells dorsal and posterior to the corpus cardiaca; the corpora allata secrete juvenile hormone (JH).

corpora cardiaca Major insect neurohemal organs existing as paired structures immediately posterior to the brain; they liberate brain hormone.

corpus luteum The yellow ovarian glandular body that arises from a mature follicle that has released its ovum; it secretes progesterone. If the ovum released has been fertilized, the corpus luteum grows and secretes during gestation; if not, it atrophies and disappears.

cortex External or surface layer of an organ.

corticospinal tract A group of axons of neurons whose somata and dendrites are located in the motor cortex and whose axon terminals synapse on motor neurons in the spinal cord.

corticotropin A hormone released by cells in the adenohypophysis that acts mainly on the adrenal cortex, stimulating growth and corticosteroid production and secretion in that organ.

cortisol A steroid hormone secreted by the adrenal cortex.

cotransmitter A second neurotransmitter molecule synthesized in and released from an axon terminal, along with a small transmitter molecule such as acetylcholine or GABA.

cotransport Carrier-mediated transport in which two dissimilar molecules bind to two specific sites on the carrier molecule for transport in the same direction.

coulomb (C) MKS unit of electric charge; equal to the amount of charge transferred in 1 second (s) by 1 ampere (A) of current. *See also* charge, electric.

counter current heat exchanger A specialized parallel arrangement of incoming arteries and outgoing veins forming a heat exchanger that conserves heat in the body core.

countercurrent multiplier A pair of opposed channels containing fluids flowing in opposed directions and having an energetic gradient directed transversely from one of the channels into the other. Since exchange due to the gradient is cumulative with distance, the exchange per unit distance will be multiplied, so to speak, as a function of the total distance over which exchange takes place.

counterion An ion associated with, and of opposite charge to, an ion or an ionized group of a molecule.

countertransport The uphill membrane transport of one substance driven by the downhill diffusion of another substance.

coupled transport Uptake of one substance into a cell that depends on the downhill diffusion exit of another substance into the cell.

covalent bond A bond formed by electron-pair sharing between two atoms.

creatine phosphate (phosphocreatine) A phosphorylated nitrogenous compound found primarily in muscle; acts as a storage form of high-energy phosphate for the rapid phosphorylation of ADP to ATP.

creatinine Nitrogenous waste product of muscle creatine.

cretinism A chronic condition caused by hypothyroidism in childhood; characterized by arrested physical and mental development.

cristae The folds of the inner mitochondrial membrane.

critical fusion frequency The number of light flashes per second at which the light is perceived to be continuous.

critical thermal maximum The temperature above which long-term survival is not possible.

crop A muscular organ in the upper alimentary canal of birds; receives swallowed food items and small stones, and churns the mixture together to break down the food into more digestible particles.

crop milk A nutrient-rich substance fed by regurgitation to pigeon chicks by both parents.

cross-bridges Spirally arranged projections from myosin thick filaments that bind to sites on actin thin filaments during muscle contraction.

crustecdysone A steroid hormone that promotes molting in crabs.

crypt of Lieberkühn A circular depression around the base of each villus in the intestine.

cupula A small upside-down cup or domelike cap; in the lateral-line and the vertebrate organs of equilibrium, the cupula covers hair cells in a gelatinous matrix.

curare (D-tubocurarine) South American arrow poison; blocks synaptic transmission at the motor endplate by competitive inhibition of nicotinic acetylcholine receptors.

current, electric The flow of electric charge. A current of 1 coulomb (C) per second is called an ampere (A). By convention the direction of current flow is the direction that a positive charge moves (i.e., from the anode to the cathode).

cuticle The hard outer coat of insects and crustacea secreted by an epidermal layer, the hypodermis.

cyanide A compound containing cyanogen and one other element; blocks transfer of electrons from the terminal cytochrome a and a_3 to oxygen in the respiratory chain.

cybernetics The science of information communication and control in animals and machines.

cyclic AMP (cAMP) A ubiquitous cyclic nucleotide (adenosine $3'5'$-cyclic monophosphate) produced from ATP by the enzymatic action of adenylate cyclase; important cellular regulatory agent that acts as the second messenger for many hormones and transmitters.

cyclic GMP (cGMP) A cyclic nucleotide (guanosine $3'5'$-cyclic monophosphate) analogous to cAMP but present in cells at a far lower concentration and producing target cell responses that are usually opposite to those of cAMP.

cyclostomes A group of jawless vertebrates, including lampreys and hagfishes.

cytochalasin A drug that disrupts cytoplasmic microfilaments.

cytochromes A group of iron-containing proteins that function in the electron-transport chain in aerobic cells; they accept and pass on electrons.

cytoplasm The semifluid substance within a cell, exclusive of the nucleus but including other organelles.

cytosine Oxyamino-pyrimidine, $C_4H_5N_3O$; base component of nucleic acid.

cytosol The unstructured aqueous phase of the cytoplasm between the structured organelles.

D600 Methoxyverapamil; an organic drug that blocks calcium influx through cell membranes.

D-tubocurarine *See* curare.

Dalton's law The partial pressure of a gas in a mixture is independent of other gases present. The total pressure is the sum of the partial pressures of all gases present.

dark current Steady sodium current that leaks into a vertebrate visual receptor cell at the outer segment. The sodium is actively pumped out of the inner segment, completing the circuit. The dark current is reduced by photoexcitation.

decerebration Experimental elimination of cerebral activity by section of the brain stem or by interruption of the blood supply to the brain.

decussation Crossing over from one side to the other.

defecation The process of expelling feces.

dehydrogenase An enzyme that "loosens" the hydrogen of a substrate in preparation for passage to a hydrogen receptor.

dehydroretinal An aldehyde of dehydroretinol.

dehydroretinol (retinol 2; vitamin A2) The form of vitamin A occurring in the liver and the retina of freshwater fishes, some invertebrates, and amphibians.

delayed outward current (late outward current) Current carried by K^+ through channels that open with a time lag after onset of a depolarization; responsible for repolarization of the action potential.

denaturation Alteration or destruction of the normal nature of a substance by chemical or physical means.

dendrites Fine processes of a neuron, often providing the main receptive area of the cell onto which synaptic contacts are made.

densitometer An instrument that measures the amount of exposure of the film emulsion produced by autoradiography.

deoxyhemoglobin Hemoglobin in which oxygen is not combined to the Fe_{3+} of the heme moiety.

deoxyribonucleic acid *See* DNA.

depolarization The reduction or reversal of the potential difference that exists across the cell membrane at rest.

desmosome A type of cell junction serving primarily to aid the structural bonding of neighboring cells.

diabetes mellitus A metabolic malady in which there is a partial or complete loss of activity in the pancreatic islets; the concomitant insulin insufficiency leads to inadequate uptake of glucose into cells and loss of blood glucose in the urine.

diacylglycerol (DAG) A diglyceride, present as a constituent of cellular phospholipids; when released from these phospholipids by an agonist-activated phospholipase, this molecule serves as the endogenous activator of the calcium- and calmodulin-dependent protein kinase (protein kinase C) and is part of an important intracellular signalling cascade.

dialysis The process by which crystalloids and macromolecules are separated by utilizing the differences in their diffusion rates through a semipermeable membrane.

diaphragm The dome-shaped muscle that separates the thoracic and abdominal cavities and serves as the chief muscle of respiration.

diastole The phase in the heartbeat during which the myocardium is relaxed and the chambers are filling with blood.

dielectric constant A measure of the degree to which a substance is able to store electric charge under an applied voltage; the dielectric constant of a material depends on charge distribution within molecules.

diffusion Dispersion of atoms, molecules, or ions as a result of random thermal motion.

diffusion coefficient Coefficient relating rate of diffusional flux to concentration gradient, path length, and area across which diffusion occurs.

digastric stomach The multichambered stomach of ruminants.

digestive enzymes Enzymes secreted by alimentary canal to aid in chemical digestion.

digestive epithelium Epithelium lining the small intestine.

dimer A molecule made by the joining of two molecules (i.e., monomers) of the same kind.

dinitrophenol (DNP) Any of a group of six isomers, $C_6H_3(OH)(NO_2)_2$, that act as aerobic metabolic inhibitors by virtue of their ability to uncouple oxidation from phosphorylation in mitochondrial electron transport.

dipole A molecule with separate regions of net negative and net positive charge, so that one end acts as a positive pole and the other as a negative pole.

dipole moment The electrostatic force required to align a dipolar molecule parallel to the electrostatic field; the force required increases as the separation of the molecular charges decreases.

disaccharide sugars A double sugar formed when two monosaccharides (single sugars) are joined together by dehydration synthesis.

disinhibition Release of a neuron from inhibitory input when the inhibitory neuron is, itself, synaptically inhibited.

dissociation Separation; resolution by thermal agitation or solvation of a substance into simpler constituents.

dissociation constant $K' = [H][A-]/[HA]$. The empirical measure of the degree of dissociation of a conjugate acid-base pair in solution.

distal More distant from a point of reference in the centrifugal direction; i.e., away from the center.

distal tubule The renal tubules located in the renal cortex leading from (and continuous with) the ascending limb of the loop of Henle to the collecting duct.

distance of closest approach Shortest possible span between the centers of two atoms.

disulfide linkage Bond between sulfide groups that determines protein tertiary structure by linking together portions of polypeptide chains.

diuresis An increase in urine excretion.

diuretic An agent that increases urine secretion.

divalent Carrying an electric charge of two units; a valence of 2.

divergence A pattern in which the axon of a single neuron branches, allowing it to synapse onto more than one synaptic target.

DNA (deoxyribonucleic acid) The class of nucleic acids responsible for hereditary transmission and for the coding of amino acid sequences of proteins.

Donnan equilibrium Electrochemical equilibrium that develops when two solutions are separated by a membrane permeable to only some of the ions of the solutions.

dopanitine (hydroxytyramine) A product of the decarboxylation of dopa, an intermediate in norepinephrine synthesis; a central nervous system transmitter.

dormancy The general term for reduced body activities, including sleep, torpor, hibernation, "winter sleep," and estivation.

dorsal horn The dorsal part of the gray matter in the vertebrate spinal cord; contains the cell bodies of neurons that receive, process, and transmit sensory information.

dorsal root A nerve trunk that enters the spinal cord near the dorsal surface; contains sensory axons only.

dorsal root ganglion (DRG) On the surface of the dorsal root, a collection of the cell bodies of sensory neurons that send processes into the region of the body that is innervated by that spinal segment; each spinal segment contains bilaterally paired DRGs.

drag The resistance to movement of an object through a medium, increasing with the viscosity and density of the medium and the surface area and shape of the object.

ductus arteriosus The vessel connecting the pulmonary artery and the aorta in fetal mammals.

duodenum The initial section of the small intestine, situated between the pylorus of the stomach and the jejunum.

dwarfism An abnormally small size in humans; a result of insufficient growth hormone secretion during childhood and adolescence.

dynamic range The range of energy over which a sensory system is responsive and can encode information about stimulus intensity.

dynein A ciliary protein with magnesium-activated ATPase activity.

dynein arms Projections from tubule A of one microtubule doublet toward tubule B of the next, composed of a protein exhibiting ATPase activity.

dyspnea Labored, difficult breathing.

early inward current Depolarizing current of excitable tissues, carried by Na^+ or Ca^{2+}; responsible for the upstroke of the action potential.

early receptor potential (ERP) An almost instantaneous potential change recorded from the retina in response to a short flash of light that probably corresponds to a movement of charge that occurs as the photopigment undergoes conformational change.

eccentric cell In *Limulus*, the afferent neuron of each ommatidium; it is surrounded by and receives information from photoreceptive retinular cells.

ecdysis Shedding of the outer shell; molting in an arthropod.

ecdysone A steroid hormone secreted by the thoracic gland of arthropods that induces molting.

echolocation The ability of some species to recognize and locate objects in their environment by emitting sounds and receiving the sound that is reflected back to them by the objects.

eclosion hormone Insect hormone that induces the emergence of the adult from the puparium.

ectopic pacemaker A pacemaker situated outside the area where it is normally found.

ectoplasm The gel-state cytoplasm surrounding the endoplasm.

ectotherms Refers to animals whose body temperature is dependent on heat in the environment.

edema Retention of interstitial fluid in organs or tissues.

EDTA Ethylenediaminetetraacetic acid; a calcium- and magnesium-chelating agent.

effector A cell, tissue, or organ that acts to change the condition of an organism (e.g., by contracting muscles or secreting a hormone) in response to neuronal or hormonal signals.

efferent (nerve or neuron) Centrifugal; a neuron that carries information from higher brain centers toward structures in the periphery.

efflux Movement of solute or solvent out of a cell across the cell membrane.

EGTA Ethyleneglycol-bis(β-aminoethylether)-N,N′-tetraacetic acid; a calcium chelating agent.

elastic Capable of being distorted, stretched, or compressed with subsequent spontaneous return to original shape; resilient.

electrical potential Electrostatic force, analogous to water pressure; a potential difference (i.e., voltage) is required for the flow of electrical current across a resistance.

electrical synapse A junction between two cells at which a signal is carried from one cell to the other by the passage of charged particles through gap junctions.

electrocardiogram (ECG) The record of electrical events associated with contractions of the heart; obtained with electrodes placed on other portions of the body.

electrochemical equilibrium The state at which the concentration gradient of an ion across a membrane is precisely balanced by the electric potential across the membrane.

electrochemical gradient The sum of the combined forces of concentration gradient and electrical gradient acting on an ion.

electrochemical potential The electrical potential developed across a membrane due to a chemical concentration gradient of an ion that can diffuse across the membrane.

electrode An electrical circuit element used to make contact with a solution, a tissue, or a cell interior; used either to measure potential or to carry current.

electrogenic Giving rise to an electric current or voltage.

electrolyte A compound that dissociates into ions when dissolved in water.

electromotive force (emf) The potential difference across the terminals of a battery or any other source of electric energy.

electronegativity Affinity for electrons.

electroneutrality rule For a net potential of zero, the positive and negative charges must add up to zero; a solution must contain essentially as many anionic as cationic charges.

electron shells Energy levels of electrons surrounding the nucleus.

electron-transport chain (respiratory chain) A series of enzymes that transfer electrons from substrate molecules to molecular oxygen.

electro-olfactogram An extracellular electrical recording of the summed activity in many olfactory receptors.

electro-osmosis Movement of water through a membrane of fixed charge in response to a potential gradient.

electrophoresis A technique for separating proteins using the net positive or negative charge on their surface amino acids.

electroplax organ An organ in electric fishes such as *Torpedo* that builds up and delivers a powerful electrical charge sufficient to stun or even to kill prey. The electroplax organ is derived from embryonic muscle and uses synapses similar to neuromuscular junctions to build up the charge.

electroreceptors Sensory receptors that detect electrical fields.

electroretinogram (ERG) An extracellular electrical recording, made at the surface of the eye, of activity in many visual receptors and other retinal neurons.

electrotonic potential A graded potential generated locally by currents flowing across the membrane; not actively propagated and not all-or-none.

Embden-Meyerhof pathway *See* glycolysis.

endergonic Characterized by a concomitant absorption of energy.

endocardium The internal lining of the heart chamber.

endocrine A hormonal pathway characterized by the production of a biologically active substance by a ductless gland; the substance then is carried through the bloodstream to initiate a cellular response in a distal target cell or tissue.

endocrine glands Ductless organs or tissues that secrete a hormone into the circulation.

endocytosis Bulk uptake of material into a cell by membrane inpocketing to form an internal vesicle.

endogenous Arising within the body.

endogenous opioids Neurotransmitter or neuromodulator molecules (e.g., endorphins and enkephalins) whose receptors bind the opioid drugs, such as opium and heroin; in some parts of the nervous system, when these transmitters bind to their receptor molecules, the net result is reduction in the perception of pain.

endolymph Aqueous liquid with a high K^+ concentration and a low Na^+ concentration found in the vertebrate organs of equilibrium and in the *scala media* of the cochlea.

endometrium An epithelium that lines the uterus.

endopeptidases Proteolytic enzymes that break up large peptide chains into shorter polypeptide segments.

endoplasm The sol-state cytoplasm that streams within the cell interior.

endorphins Endogenous neuropeptides that exhibit morphinelike actions, found in the central nervous system of vertebrates; different types consist of 16, 17, or 31 amino residues.

endothelium Single cell layer forming the internal lining of blood vessels.

endotherms animals whose body temperature is dependent on heat production by the body.

endplate A traditional name of the vertebrate neuromuscular synapse, where the motor axon forms many fine terminal branches that end over a specialized system of folds in the postsynaptic membrane of the muscle cell.

endplate potential (epp) A postsynaptic potential in the muscle at the neuromuscular junction (or motor endplate).

end-product inhibition (feedback inhibition) Inhibition of a biosynthetic pathway by the end product of the pathway.

energy Capacity to perform work.

energy metabolism The complex collection of biochemical reactions within cells that generate ATP and other high-energy compounds, which serve as the immediate source of energy for all biological events.

enkephalins Endogenous nuropeptides exhibiting morphinelike actions, found in the central nervous system of vertebrates; these peptides consist of five amino acid residues.

enterogastric reflex A reflex that inhibits gastric secretion, triggered when duodenum is stretched by chyme pumped from the stomach.

enterogastrone A hormone that is secreted from the duodenal mucosa in response to fat ingestion and that suppresses a gastric motility and secretion.

enterokinase Intestinal proteolytic enzyme.

enthalpy The heat produced or taken up by a chemical reaction.

entropy Measure of that portion of energy not available for work in a closed system; measure of molecular randomness. Entropy increases with time in all irreversible processes.

enzyme A protein with catalytic properties.

enzyme activity A measure of the catalytic potency of an enzyme: the number of substrate molecules that react per minute per enzyme molecule.

enzyme induction Enzyme production stimulated by the specific substrate (inducer) of that enzyme or by a molecule structurally similar to the substrate.

epicardium The external covering of the heart wall.

epididymis A long, stringlike duct along the dorsal edge of the testis; its function is to store sperm.

epinephrine Generic name for the catecholamine released from the adrenal cortex; also known by the trade name Adrenalin.

equilibrium The state in which a system is in balance as a result of equal action by opposing forces arising from within the system.

equilibrium potential The voltage difference across a membrane at which an ionic species that can diffuse across the membrane is in electrochemical equilibrium; it is dependent on the concentration gradient of the ions and is described by the Nernst equation.

equimolar Having the same molarity.

equivalent pore size Cell membrane pore diameter that accounts for the rate of diffusion of polar substances across the membrane.

eserine (physostigmine) An alkaloid ($C_{15}H_{21}N_3O_2$) of plant origin that blocks the enzyme cholinesterase.

esophagus The region of the alimentary canal that conducts food from the headgut to the digestive areas.

essential amino acids Amino acids that cannot be synthesized by an animal, but are required for synthesis of essential proteins.

esterases Enzymes that hydrolyze esters.

estivation Dormancy in response to high ambient temperatures and/or danger of dehydration.

estradiol-17β The most active natural estrogen.

estrogens A family of female sex steroids responsible for producing estrus and the female secondary sex characteristics; also prepares the reproductive system for fertilization and implantation of the ovum; synthesized primarily in the ovary, although some is made in the adrenal cortex and male testis.

estrous cycle Periodic episodes of "heat," or estrus, marked by sexual receptivity in mature females of most mammalian species.

ethers A class of compounds in which two organic groups are joined by an oxygen atom, R_1—O—R_2.

eupnea Normal breathing.

euryhaline Able to tolerate wide variations in salinity.

exchange diffusion A process by which the movement of one molecule across a membrane enhances the movement of another molecule in the opposite direction; most likely involves a common carrier molecule.

excitability The property of altered membrane conductance (and often membrane potential) in response to stimulation.

excitation-contraction coupling In muscle fibers, the process by which electrical excitation of the surface membrane leads to activation of the contractile process.

excitation-secretion coupling Sequence of molecular events responsible for secretion of stored chemicals from an activated secretory cell.

excitatory In neurophysiology, pertaining to the enhanced probability of producing an action potential.

excitatory postsynaptic potential (epsp) A change in the membrane potential of a postsynaptic cell that increases the probability of an action potential in the cell.

exergonic Characterized by a concomitant release of heat energy.

exocrine Of or relating to organs or structures that secrete substances via a duct.

exocrine gland A gland that secretes a fluid via a duct.

exocytosis Fusion of the vesicle membrane to the surface membrane and subsequent expulsion of the vesicle contents to the cell exterior.

exopeptidases Enzymes that attack only peptide bonds near the end of a peptide chain, providing free amino acids, plus dipeptides and tripeptides.

exothermic *See* exergonic.

expiratory neurons Neurons in the medulla of the brain controlling activity in motor neurons innervating muscles involved in expiration (breathing out).

explants Small pieces of tissue removed from a donor animal kept alive and grown in a flask filled with the appropriate mixture nutrients.

extensor A muscle that extends or straightens a limb or other extremity.

exteroceptors Sense organs that detect stimuli arriving at the surface of the body from a distance.

extracellular digestion Digestion occurring outside of the cell in an alimentary system.

extrafusal fiber Contractile muscle fibers that make up the bulk of skeletal muscle. *See also* intrafusal fibers.

extravasation Forcing fluid out of blood vessels, usually blood, serum, or lymph.

eye An organ of visual reception that includes optical processing of light as well as photoreceptive neurons.

facilitated transport (carrier-mediated transport) Downhill transmembrane diffusion aided by a carrier molecule that enhances the mobility of the diffusing substance in the membrane.

facilitation An increase in the efficacy of a synapse as the result of a preceding activation of that synapse.

factorial scope for locomotion A ratio between the basal metabolic rate and the maximum metabolic rate that can be achieved with intense exercise.

Fahraeus Lindqvist effect The reduction in apparent viscosity of blood as it flows into small arterioles.

farad (F) The unit of electrical capacitance.

faraday Measure of electric charge, $-96,487$ C$\cdot$mol^{-1}.

Faraday's constant (F) The equivalent charge of a mole of electrons, equal to 9.649×10^4 coulombs (C) per mole of electrons.

fast chemical synaptic transmission Synaptic transmission at a chemical synapse that is mediated by neurotransmitters that bind on the postsynaptic membrane to receptor protein complexes, each of which includes an ion channel. Binding of the transmitter to the receptor complex is sufficient to open (or to close) the ion channel.

feces Undigested material and bacteria eliminated from the hindgut.

feedback The return of output to the input part of a system. In *negative feedback,* the sign of the output is inverted before being fed back to the input so as to stabilize the output. In *positive feedback,* the output is unstable because it is returned to the input without a sign inversion, and thus becomes self-reinforcing, or regenerative.

fermentation Enzymatic decomposition; anaerobic transformation of nutrients without net oxidation or electron transfer.

ferritin A large protein molecule opaque to electrons, used as a marker in electron microscopy; normally present in the spleen as a storage protein for iron.

fever Disease-induced increase in body temperature above normal levels.

fibrillar muscle Oscillatory insect flight muscle; also termed asynchronous muscle because contractions are not individually controlled by motor impulses.

fibroblast A connective-tissue cell that can differentiate into chondroblasts, collagenoblasts, and osteoblasts.

Fick diffusion equation An equation defining the rate of solute diffusion through a solvent.

field metabolic rate (FMR) The average rate of energy utilization as the animal goes about its normal activities, which may range from complete inactivity during resting to maximum exertion.

filter feeding *See* suspension feeding.

final common pathway The concept that the sum total of neuronal integrative activity expressed in motor output is channeled through the motor neurons to the muscles.

firing level Potential threshold for the generation of an action potential.

first law of thermodynamics Net energy is conserved in any process.

first-order enzyme kinetics Describes enzymatic reactions, the rates of which are directly proportional to one reactant's concentration (either substrate or product).

fixed action pattern A behavior that is performed in a stereotyped fashion in response to specific stimuli.

flagellum A motile, whiplike organelle similar in organization to a cilium, but longer and generally present on a cell only in small numbers.

flame cells Flagellated cells at the ending of the excretory collecting tubules of flatworms and nemerteans.

flavin adenine dinucleotide (FAD) A coenzyme formed by the condensation of riboflavin phosphate and adenylic acid; it performs an important function in electron transport and is a prosthetic group for some enzymes.

flavoproteins Proteins combined with flavin prosthetic groups that are important as intermediate carriers of electrons between the dehydrogenases and cytochromes in the respiratory chain.

flexer A muscle that flexes or bends an extremity.

fluid mosaic model The accepted model for the cell membrane, in which globular proteins are integrated with the lipid bilayer.

fluorescence The property of emitting light upon molecular excitation by an incident light; the emitted light is always less energetic (has a longer wavelength) than the light producing the excitation.

flux The rate of flow of matter or energy across a unit area.

folds of Kerckring *See* plica circularis.

follicle-stimulating hormone (FSH) An anterior pituitary gonadotropin that stimulates the development of ovarian follicles in the female and testicular spermatogenesis in the male.

follicular phase That part of the estrous and menstrual cycles that is characterized by formation of and secretion by the Graafian follicles.

food vacuole Enzyme-filled vesicle in Protozoa.

foramen An orifice or opening.

foregut The upper region of the alimentary canal involved in food conduction, storage and digestion.

Fourier's law The rate of flow of heat in a conducting body is proportional to its thermal conductance and to the temperature gradient.

fovea (area centralis) In the mammalian retina, the area with the highest visual resolution due to the small divergence and convergence in the pathway linking photoreceptors to ganglion cells; in primates, contains closely packed cone cells.

Frank-Starling mechanism The increase in mechanical work from the ventricle caused by an increase in end-diastolic volume (or venous filling pressure).

free energy The energy available to do work at a given temperature and pressure.

frequency modulated The property of a signal in which information is encoded by varying the frequency at which the strength of the signal changes.

fructose A ketohexose, $C_6H_{12}O_6$, found in honey and many fruits.

fusimotor system The gamma motor neurons and the intrafusal fibers that they innervate.

gain The increase in signal produced by amplification.

gallbladder The organ associated with liver that concentrates and stores bile for eventual discharge into the intestine.

gamma aminobutyric acid (GABA) Inhibitory transmitter identified in crustacean motor synapses and in the vertebrate central nervous system.

gamma efferents The motor axons innervating the intrafusal muscle fibers of spindle organs.

gamma motor neurons Nerve cells of the ventral spinal cord that innervate the intrafusal muscle fibers.

gamma rays Electromagnetic radiation of very short wavelength (10^{-12} cm) and very high energy.

ganglia Collections of neuronal cell bodies.

ganglion An anatomically distinct collection of neuron cell bodies.

ganglion cells A nonspecific term applied to some nerve cell bodies, especially those located in ganglia of invertebrates or outside the vertebrate central nervous system proper.

ganglion cells (retinal) The afferent neurons that carry visual information from the vertebrate retina to higher centers of the brain.

gap junctions Specializations for electrical coupling between cells, where cell membranes are only about 2 nm apart and tubular assemblies of proteins connect the apposed membranes.

gastric Pertaining to the stomach.

gastric cecum The outpouching of insect alimentary canal, lined with enzyme-secreting and phagocytic cells and serving as a stomach.

gastric inhibitory peptide (GIP) A gastrointestinal hormone released into the bloodstream from the duodenal mucosa, inhibiting gastric secretion and motility.

gastric juice Fluid secreted by the cells of the gastric epithelium.

gastric phase Secretion phase of digestion stimulated by presence of food in stomach.

gastrin A protein hormone that is liberated by the gastrin cells of the pyloric gland and induces gastric secretion and motility.

gastrointestinal peptide hormones Hormones that regulate the basic electric rhythm of smooth muscle in the alimentary canal.

Gay-Lussac's law Either the pressure or the volume of a gas is directly proportional to absolute temperature if the other is held constant.

Geiger counter An instrument that detects the presence of ionizing radiation.

gel The stiff, high-viscosity state of cytoplasm.

gene Regions of coded information carried within the subunits of DNA forming chromosomes.

generator potential A change in transmembrane potential within the receptive part of a sensory neuron; its amplitude is graded with stimulus intensity, and the potential is conducted eletrotonically in the neuron. If the potential is sufficiently large at the spike-initiating zone of the axon, action potentials will be generated.

geniculate body A thalamic nucleus that relays incoming sensory (auditory and visual) information to the cortex; named for its kneelike shape in cross section.

gestation Pertaining to pregnancy.

gigantism Excessive growth due to hypersecretion of pituitary growth hormone from birth.

gizzard See crop.

gland An aggregation of specialized cells that secrete or excrete substances, such as the pituitary gland, which produces hormones, and the spleen, which takes part in blood production.

glial cells (neuroglia) Inexcitable supportive cells associated with neurons in nervous tissue.

globin Protein in hemoglobin made of two equal parts each consisting of two polypeptide chains.

glomerular filtrate rate (GFR) The amount of total glomerular filtrate produced per minute by all nephrons of both kidneys; equal to the clearance of a freely filtered and nonreabsorbed substance such as insulin.

glomerulus A coiled mass of capillaries.

glucagon A protein hormone released by the alpha cells of the pancreatic islets; its secretion is induced by low blood sugar or by growth hormone; it stimulates glycogenolysis in the liver.

glucocorticoids Steroids synthesized in the adrenal cortex with wide-ranging metabolic activity; included are cortisone, cortisol, corticosterone, and 11-deoxycorticosterone.

gluconeogenesis Synthesis of carbohydrates from noncarbohydrate sources, such as fatty acids or amino acids.

glucose 6-carbon sugar comprising the cell's primary metabolic fuel; blood sugar.

glutamate A putative excitatory synaptic transmitter in the vertebrate central nervous system and in arthropod neuromuscular junctions.

glutamine Amino acid used, because it is less toxic, to transport ammonia between the liver and kidneys in mammals.

glycocalyx A meshwork of acid mucopolysaccharide and glycoprotein filaments that arise from the membrane covering the microvilli of the intestinal brush border.

glycogen A highly branched D-glucose polymer found in animals.

glycogenesis The synthesis of glycogen.

glycogenolysis The breakdown of glycogen to glucose 6-phosphate.

glycogen synthetase An enzyme that catalyzes the polymerization of glucose to glycogen.

glycolipid A lipid containing carbohydrate groups, in most cases galactose.

glycolysis (Embden-Meyerhof pathway) The metabolic pathway by which hexose and triose sugars are broken down to simpler substances, especially pyruvate or lactate.

glycosidases Carbohydrases that break down disaccharides (sucrose, fructose, maltose, lactose) by hydrolyzing alpha-1,6 and alpha-1,4 glucosidic bonds into their constituent monosaccharides for absorption.

glycosuria (glucosuria) The excretion of excessive amounts of glucose in the urine.

goblet cells See mucous cells.

goiter An abnormal increase in size of the thyroid gland, usually due to a dietary lack of iodine.

Goldman equation The equation describing the equilibrium potential for a system in which more than one species of diffusible ions are separated by a semi-permeable membrane; if only one species can diffuse across the membrane the equation reduces to the Nernst equation.

Golgi tendon organs Tension-sensing nerve endings of the lb afferent fibers found in muscle tendons.

gonadotropic hormones (gonadotrophic hormones; gonadotropins) Hormones that influence the activity of the gonads; in particular, those secreted by the anterior pituitary.

gonadotropins Hormones that act on the gonads.

Graafian follicle A mature ovarian follicle in which fluid is accumulating.

graded response One that increases as a function of the energy applied; a membrane response that is not all-or-none.

Graham's law The rate of diffusion of a gas is proportional to the square root of the density of that gas.

gray matter Tissue of the vertebrate central nervous system consisting of cell bodies, unmyelinated fibers, and glial cells.

growth hormone (GH, somatotropin) A protein hormone that is secreted by the anterior pituitary and stimulates growth; directly influences protein, fat, and carbohydrate metabolism and regulates growth rate.

guanine 2-amino-6-oxypurine ($C_5H_5N_5O$); a white, crystalline base; a breakdown product of nucleic acids.

guano White, pasty waste product of birds and reptiles; high in uric acid content.

guanosine triphosphate (GTP) A high-energy molecule similar to ATP that participates in several energy-requiring processes, such as peptide bond formation.

guanylatecydase An enzyme that converts GTP to cyclic GMP.

gustation The sense of taste; chemoreception of ions and molecules in solution by specialized epithelial sensory receptors.

H zone The light zone in the center of the resting muscle sarcomere, where myosin filaments do not overlap with actin filaments; the region between actin filaments.

habituation The progressive loss of behavioral response probability with repetition of a stimulus.

hair cell A mechanosensory epithelial cell bearing stereocilia and in some cases a kinocilium.

Haldane effect Reduction in total CO_2 content of the blood at constant PCO_2 when hemoglobin is oxygenated.

half-width The length of time during which a transient physiological variable is half of its maximum value or greater.

halide A binary compound of a halogen and another element.

halogens A family of related elements that form similar saltlike compounds with most metals; they are fluorine, chlorine, bromine, and iodine.

headgut The anterior (cranial) region of the alimentary canal providing an external opening for food entry.

heat Energy in the form of molecular or atomic vibration that is transferred by conduction, convection, and radiation down a thermal gradient.

heat capacity Amount of heat required to raise 1 g of substance 1°C.

heater tissues Tissues specialized for heat production, e.g. modified eye muscles in billfishes.

heat of shortening Thermal energy associated with muscle contraction; it is proportional to the distance the muscle has shortened.

heat of vaporization Heat necessary per mass unit of a given liquid to convert all of the liquid to gas at its boiling point.

heavy meromyosin (HMM; H-meromyosin) The "head" and "neck" of the myosin molecule, the part of the myosin molecule that has ATPase activity.

helicotrema The opening that connects the *scala tympani* and the *scala vestibuli* at the cochlear apex.

hematocrit The percentage of total blood volume occupied by red blood cells; in humans, the hematocrit is normally 40 to 50%.

heme $C_{34}H_{33}O_4N_4FeOH$; an iron protoporphyrin portion of many respiratory pigments.

hemerythrin An invertebrate respiratory pigment that is a protein but does not contain heme.

hemimetabolous Refers to insects, like bugs, that show incomplete metamorphosis during their life cycle. *See also* holometabolous.

hemocoel Space between ectoderm and endoderm in many invertebrates containing blood (hemolymph).

hemocyanin An invertebrate respiratory pigment that is a protein, contains copper, and is found in mollusks and crustaceans.

hemoglobin The oxygen-carrying pigment of the erythrocytes, formed by the developing erythrocyte in bone marrow. It is a complex protein composed of four heme groups and four globin polypeptide chains. They are designated α(alpha), β(beta), γ(gamma), and δ(delta) in an adult, and each is composed of several hundred amino acids.

hemolymph The blood of invertebrates with open circulatory systems.

hemopoiesis (hematopoiesis) The processes leading to the production of blood cells.

hemopoietic factor *See* intrinsic factor.

Henderson-Hasselbalch equation $pH = pK + \log([H^+ \text{accep-}$ tor]/$[H^+$ donor]). The formula for calculation of the pH of a buffer solution.

Henry's law The quantity of gas that dissolves in a liquid is nearly proportional to the partial pressure of that gas in the gas phase.

hepatocyte A liver cell.

herbivores Animals that feed solely on plant material.

Hering-Breuer reflex A reflex in which lung inflation activates pulmonary stretch receptors that inhibit further inspiration during that cycle; activity from stretch receptors is carried in the vagus nerve.

hertz (Hz) Cycles per second.

Hess's law The total energy released in the breakdown of a fuel to a given set of end products is always the same, irrespective of the intermediate chemical steps or pathways used.

heterodimers Dimers consisting of two nonidentical subunits.

heterosynaptic facilitation Increased effectiveness of synaptic transmission between two neurons that occurs as a result of activity in a third neuron.

heterosynaptic modulation A change in synaptic efficacy at one synapse that occurs due to activity at another, separate synapse. In *heterosynaptic facilitation* synaptic efficacy increases. In *heterosynaptic depression,* the synaptic efficacy decreases.

heterotherm An animal that derives essentially all of its body heat from the environment.

heterotrophic Depending on energy-yielding carbon compounds derived from the ingestion of other plants or animals.

hexose A six-carbon monosaccharide.

hibernation A period of deep torpor, or winter dormancy, in animals in cold climates, lasting weeks or months.

hindgut The terminal region of the alimentary canal, responsible for storing and eventually eliminating the remnants of digested food.

hindgut fermentation Fermentative digestion occurring in distal portions of the alimentary canal.

histamine The base formed from histidine by decarboxylation; responsible for dilation of blood vessels.

histaminergic Referring to nerves that release histamine.

histone A simple, repeating, basic protein that combines with DNA.

Hodgkin cycle The regenerative, or positive-feedback, loop responsible for the upstroke of the action potential; depolarization causes an increase in the sodium permeability, permitting an increased influx of Na^+, which further depolarizes the membrane.

holometabolous Refers to insects, like flies, that show complete metamorphosis during their life cycle. *See also* hemimetabolous.

homeostasis The condition of relative internal stability maintained by physiological control systems.

homeotherm An animal (mammal, bird) that regulates its own internal temperature within a narrow range, regardless of the ambient temperature, by controlling heat production and heat loss.

homeoviscous adaptation Molecular level adaptations, especially of cell membranes, that help minimize temperature-induced differences in viscosity.

homonymous Pertaining to the same origin.

homosynaptic modulation A change in synaptic efficacy that occurs subsequent to activity at a synapse. In synaptic *facilitation*, efficacy increases. In synaptic *depression*, efficacy decreases.

horizontal cell A nerve cell whose fibers extend horizontally in the outer plexiform layer of the vertebrate retina; interconnects adjacent photoreceptors.

hormone A chemical compound synthesized and secreted by an endocrine tissue into the bloodstream; influences the activity of a target tissue.

horseradish peroxidase A large protein molecule that is opaque in the electron microscope, and is used to trace neuronal pathways in the central nervous system.

hybridomas Hybrid cells that are formed by the fusion of two different cell types, used in the production of monoclonal antibodies and other cellular products.

hydration Combination with water.

hydraulic permeability Refers to the sieve-like properties of the Bowman's capsule in the kidney.

hydride Any compound consisting of an element or a radical combined with hydrogen.

hydrofuge Pertaining to structures with nonwetting surfaces.

hydrogen bond A weak electrostatic attraction between a hydrogen atom bound to a highly electronegative element in a molecule and another highly electronegative atom in the same or a different molecule.

hydrolase transport The mechanism by which monosaccharides are taken up into absorptive cells, using a membrane-bound glycosidase to break down and transport the parent disaccharide across the absorptive cell's membrane.

hydrolysis Fragmentation or splitting of a compound by the addition of water, whereupon the hydroxyl group joins one fragment and the hydrogen atom the other.

hydronium ion (H_3O^+) A hydrogen ion (H^+) combined with a water molecule; H_2O^+.

hydrophilic Having an affinity for water.

hydrophobic Lacking an affinity for water.

hydrostatic pressure Force exerted over an area due to pressure in a fluid.

hydroxyapatite $Ca_{10}(PO_4)_6(OH)_2$, a crystalline material lending hardness and rigidity to the bones of vertebrates and shells of mollusks.

hydroxyl group(radical) The —OH^- group.

hydroxyl ion OH^-.

hypercalcemia Excessive plasma calcium levels.

hypercapnia Increased levels of carbon dioxide.

hyperemia Increased blood flow to a tissue or an organ.

hyperglycemia Excessive blood glucose levels.

hyperosmotic Containing a greater concentration of osmotically active constituents than the solution of reference.

hyperpnea Increased lung ventilation; hyperventilation.

hyperpolarization An increase in potential difference across a membrane, making the cell interior more negative than it is at rest.

hyperthermia A state of abnormally high body temperature.

hypertonic Having a higher tonicity or osmotic pressure than reference solution.

hypertrophy Excessive growth or development of an organ or a tissue.

hyperventilation *See* hyperpnea.

hypoglycemia Low blood glucose levels.

hypoosmotic Containing a lower concentration of osmotically active constituents than the solution of reference.

hypophysis The pituitary gland.

hypopnea Hypoventilation; decreased lung ventilation.

hypothalamic releasing hormones Hormones from the hypothalamus that cause the release of other hormones from the pituitary.

hypothalamus The part of the diencephalon that forms the floor of the median ventricle of the brain; includes the optic chiasma, mammillary bodies, tuber cinereum, and infundibulum; many subregions contribute to the regulation of the autonomic nervous system and of endocrine function.

hypothalmo-hypophyseal relay system Portal veins linking the capillaries of the hypothalamic median eminence with those of the adenohypophysis; these transport hypothalamic neurosecretions directly to the adenohypophysis.

hypothermia A state of abnormally low body temperature.

hypothesis A specific prediction that can then be tested by performing further experiments.

hypothyroidism Reduced thyroid activity.

hypotonic Having a lower tonicity or osmotic pressure than reference solution.

hypoventilation Reduced lung ventilation.

hypoxia Reduced oxygen levels.

hysteresis A nonlinear change in the physical state of a system, such that the state depends in part on the previous history of the system.

I band The region between the A band and Z disk of a resting muscle sarcomere; it appears light when viewed microscopically and contains part of the actin thin filaments that does not overlap with myosin filaments.

ileum Posterior section of the small intestine.

impedance The dynamic resistance to flow met by fluids moving in a pulsatile manner.

impulse-initiating region (spike-initiating zone) The proximal portion of the axon, which has a lower threshold for action potential generation than either the soma or the dendrites.

incus The middle bone of the three bones in the mammalian inner ear; it connects the malleus and the stapes.

influx Movement of solute or solvent into a cell across the cell membrane.

incisors Chisel-like teeth used for gnawing.

inertia The tendency of a mass to resist acceleration.

infradian rhythms Biological rhythms with a periodicity of less than a day in length.

infrared Thermal radiation; electromagnetic radiation of wavelengths greater than 7.7×10^{-5}cm and less than 12×10^{-4}cm; the region between red light and radio waves.

inhibitory In neurophysiology, pertaining to a reduction in probability of generating an action potential.

inhibitory postsynaptic potential (ipsp) A change in the transmembrane potential of a postsynaptic cell that reduces the probability of an action potential in the cell.

initial segment The portion of axon and axon hillock proximal to the first myelinated segment; the spike initiating zone of many neurons.

inner plexiform layer In the vertebrate retina, the layer of connecting processes that lies between the bipolar cells and the ganglion cells.

inner segment The portion of a vertebrate photoreceptor cell that contains the cell organelles and synaptic contacts.

inositol trisphosphate (InsP$_3$) The intracellular second messenger produced by the action of phospholipase C on membrane phosphatidylinositolphosphate in response to stimulation of cell-surface receptors by growth factors, hormones, or neurotransmitters.

inotropic Pertaining to the strength of contraction of the heart.

inspiratory neurons Neurons in the medulla of the brain controlling motor neurons of muscle associated with breathing in.

instars The stages between molts in insect development.

instinct A species-specific set of unlearned behaviors and responses.

insulin A protein hormone synthesized and secreted by the beta cells of the pancreatic islets; controls cellular uptake of carbohydrate and influences lipid and amino acid metabolism.

integral proteins Proteins spanning the cell membrane that form selective filters and active transport devices that nutrients into and cellular products and waste out of the cell.

integration, neuronal Synthesis of an output based on the sum of inputs to a neuron or neuronal network.

intercalated disk The junctional region between two connected cardiac muscle cells.

interferon anti-viral and anti-tumor agent produced by animal cells

interneuron A nerve cell that is entirely contained within the central nervous system; interneurons typically connect two or more other neurons.

internode The space along a myelinated axon that is covered by the myelinating cell (i.e., it is covered by the Schwann cell or the oligodendrocyte).

interoceptive receptors Internal sensory receptors responding to changes inside the body.

interstitial Between cells or tissues.

interstitial cell-stimulating hormone (ICSH) Identical to luteinizing hormone but in the male.

interstitium The tissue space between cells.

intestinal chyme Semifluid mass of partially digested food.

intestinal gastrin Hormone stimulating gastric glands to increase secretion rate.

intestinal juice *See* succus entericus.

intestinal phase Phase of digestion control by gastrin and other hormones.

intracellular digestion Nutrient breakdown occurring within cells.

intracellular milieu The general physio-chemical characteristics within the cell.

intrafusal fibers The muscle fibers within a muscle spindle organ.

intrinsic factor (or hemopoietic factor) A mucoprotein produced by the H$^+$-secreting parietal cells of the stomach; involved in vitamin B$_{12}$ absorption.

inulin An indigestible vegetable starch; used in studies of kidney function because it is freely filtered and not actively transported.

in vitro "In a glass"; in an artificial environment outside the body.

in vivo Within the living organism or tissue.

iodoacetic acid An agent that poisons glycolysis by inhibiting glyceraldehyde phosphate dehydrogenase.

ion An atom bearing a net charge due to loss or gain of electrons.

ion battery The electromotive force capable of driving an ionic current across a membrane; results from unequal concentrations of an ion species in the two compartments separated by the membrane.

ion bonding sites Partially ionized regions of protein and other molecules that electrostatically interact with ions in the surrounding solution.

ion-exchanger site (ion-binding site) An electrostatically charged site that attracts ions of the opposite charge.

ionic bond Electrostatic bond.

ionization The dissociation into ions of a compound in solution.

ipsilateral Relating to the same side.

iris The pigmented circular diaphragm located behind the cornea of the vertebrate eye.

ischemia The absence of blood flow (to an organ or a tissue).

islets of Langerhans Microscopic endocrine structures dispersed throughout the pancreas. They consist of three cell types: the alpha cells, which secrete glucagon; the beta cells, which secrete insulin; and the delta cells, which secrete gastrin.

isoelectric point The pH of a solution at which an amphoteric molecule has a net charge of zero.

isomer A compound having the same chemical formula as another, but with a different arrangement of its atoms.

isometric contraction Contraction during which a muscle does not shorten significantly.

isometry Proportionality of shape regardless of size.

isosmotic Having the same osmotic pressure.

isoteric interaction Chemical interaction involving molecules with the same number of valence electrons in the same configuration, but made up of different types and numbers of atoms.

isotonic Having an equivalent tonicity or osmotic pressure than reference solution.

isotonic contraction Contraction in which the force generated remains constant while the muscle shortens.

isotope Any of two or more forms of an element with the same number of protons (atomic number), but a different number of neutrons (atomic weight).

isovolumic Having the same volume.

isozymes Multiple forms of an enzyme found in the same animal species or even in the same cell.

Jacobs-Stewart cycle The cycling of CO_2 and HCO_3 between intracellular and extracellular compartments, which functions to transfer H$^+$ ions between the cell interior and the extracellular fluid.

jejunum The portion of small intestine between the duodenum and the ileum.

joule (J) SI unit of work equivalent to 0.239 calories (cal).

junctional fold A fold in the cell membrane of a postsynaptic cell that lies under the axon terminals of the presynaptic cell. Junctional folds are typical of skeletal muscle fibers at neuromuscular junctions.

juvenile hormone (JH) A class of insect hormones that are secreted by the corpora allata and that promote retention of juvenile characteristics.

juxtaglomerular apparatus Apparatus comprised of specialized secretory cells situated in the afferent glomerular arterioles; act as receptors that respond to low blood pressure by secreting renin, which converts angiotensinogen to angiotensin, resulting in vasoconstriction and aldosterone secretion.

juxtaglomerular cells Specialized secretory cells situated in the afferent glomerular arterioles; act as receptors that respond to low blood pressure by secreting renin, which converts angiotensinogen to angiotensin, resulting in the stimulation of aldosterone secretion.

juxtapulmonary capillary receptors Sensory receptors found in the lung that, when stimulated, elicit the sensation of breathlessness.

k selection Pattern of energetic investment in reproduction in which small numbers of large offspring are produced, each with a high chance of survival due to extensive parental care.

kelvin (K) *See* absolute temperature.

keratin Structural protein found in skin, feathers, nails, and hoofs.

ketone Any compound having a carbonyl group (CO) attached (by the carbon) to hydrocarbon groups.

ketone bodies Acetone, acetoacetic acid, and β-hydroxybutyric acid; products of fat and pyruvate metabolism formed from acetyl CoA in the liver; oxidized in muscle and by the central nervous system during starvation.

key stimulus (releasing stimulus) The stimulus that is effective in producing a fixed action pattern.

kinematic viscosity Viscosity divided by density; gases of equal kinematic viscosity will become turbulent at equal flow rates in identical airways.

kinetic energy Energy inherent in the motion of a mass.

kininogen Precursor of bradykinin.

kinocilium A true "9 + 2" or "9 + 0" cilium present in sensory hair cells.

Kirchhoff's laws *First law:* The sum of the currents entering a junction in a circuit equals the sum of the currents leaving the junction. *Second law:* The sum of the potential changes encountered in any closed loop in a circuit is equal to zero.

Kleiber's law 0.75 exponent relates metabolic rate to body mass.

knock-outs Animals that lack the ability to express the function originally coded for by the gene.

Krebs cycle *See* tricarboxylic acid (TCA) cycle and citric acid cycle.

labeled line coding A pattern of information processing in the nervous system in which each neuron encodes only one particular type of information (e.g., sour stimuli in the taste system) and all of the axons that carry that type of information project to the same location or locations.

labeled-lines concept The idea that sensory modalities are determined by the stimulus sensitivity of peripheral sense organs and the anatomical specificities of their central connections.

lactation The production of milk by the mammary glands (breasts).

lactogen Hormone that prepares the breasts for milk production.

lagena A structure associated with hair cells in the vertebrate organs of equilibrium.

lamella A thin sheet or leaf.

laminar flow Turbulence-free flow of fluid in a vessel or past a moving object; a gradient of relative velocity exists in which the fluid layers closest to the wall or body have the lowest relative velocity.

Laplace's law The transmural pressure in a thin-walled tube is proportional to the wall tension divided by the inner radius of the tube.

larva The immature, active feeding stage characteristic of many invertebrates.

latent heat of vaporization The amount of energy required to change a liquid to its gaseous form (evaporate) at the same temperature.

latent period The interval between an action potential in a muscle fiber and the initiation of contraction.

lateral geniculate A region of the brain in birds and mammals that processes visual information coming from the retina.

lateral geniculate nuclei The major relay nuclei between the retina and the visual cortex in the mammalian visual system; they are included in the thalamic nuclei.

lateral inhibition Reciprocal suppression of excitation by neighboring neurons in a sensory network; it produces enhanced contrast at boundaries and an increase in dynamic range.

lateral-line system Series of hair cells (*see* neuromast) in canals running the length of the head and body of fishes and many amphibians; these channels have openings to the outside, and the system is sensitive to water movement.

lateral plexus In the compound eye of *Limulus*, the collection of neurons that interconnect eccentric cells of the ommatidia, producing lateral inhibition.

lecithin Any of a group of phospholipids found in animal and plant tissues; composed of choline, phosphoric acid, fatty acids, and glycerol.

length constant (l) The distance along a cell over which a potential change decays in amplitude by $(1 - 1/e)$, or 63%.

lens The major light-focusing structure in the vertebrate eye.

leukocytes White blood cells.

Leydig cells (interstitial cells) Cells of the testes that are stimulated by luteinizing hormone to secrete testosterone.

ligand-gated ion channel An ion channel through the cell membrane that opens when a molecule, or molecules, binds to the extracellular domain of the protein.

light Electromagnetic radiation with wavelengths between those of x-rays and those of heat (infrared radiation).

light meromyosin (LMM) The rodlike fragment of the myosin molecule that constitutes most of the molecule's backbone.

limited heterothermy A survival mechanism used by camels and other normally homeothermic animals, in which body temperature is allowed to rise and fall somewhat as ambient temperature ranges through extremes.

Lineweaver-Burk equation Straight line transformation of the Michaelis-Menton equation.

lipases Enzymes that specifically break down lipids.

lipid Any of the fatty acids, neutral fats, waxes, steroids, and phosphatides; lipids are hydrophobic and feel greasy.

lipid bilayer Continuous double layer of lipid molecules forming the basic structure of the cell membrane.

lipogenesis The formation of fat from nonlipid sources.

lipophilic Having an affinity for lipids.

lipoprotein Protein-lipid complex in the plasma membrane.

local circuit current The current that spreads electrotonically from the excited portion of an axon during conduction of the nerve impulse, flowing longitudinally along the axon, across the membrane, and back to the excited portion.

longitudinal smooth muscle Outer layer of smooth muscle running along the long axis of small intestine.

long-term potentiation An increase in synaptic efficacy that occurs due to sustained synaptic input and that lasts for a relatively long time—even days, weeks, or months.

loop of Henle A U-shaped bend in the portion of a renal tubule that lies in the renal medulla.

lower critical temperature (LCT) ambient temperature below which the BMR becomes insufficient to balance heat loss, resulting in falling body temperature.

lumen The interior of a cavity or duct.

luminosity Brightness; relative quantity of light reflected or emitted.

luteal phase The part of the estrous or menstrual cycle characterized by formation of and secretion by the corpus luteum.

luteinizing hormone (LH) A gonadotropin that is secreted by the adenohypophysis and that acts with follicle-stimulating hormone (FSH) to induce ovulation of the ripe ovum and liberation of estrogen from the ovary; also influences formation of the corpus luteum and stimulates growth in and secretion from the male testicular Leydig cells.

lymph Plasmalike fluid collected from interstitial fluid and returned to the bloodstream via the thoracic duct; contains white, but not red, blood cells.

lymphatic system A collection of blind-ending tubes which drain filtered extracellular fluid from tissues and return it to the blood circulation.

lymph heart A muscular pump found in fish and amphibia causing movement of lymph.

lymph nodes Aggregations of lymphoid tissue in the lymphatic system that produce lymphocytes and filter the lymphatic fluid.

lymphocytes White blood cells, produced in lymphoid tissue, lacking cytoplasmic granules but having a large, round nucleus.

lysolecithin A lecithin without the terminal acid group.

lysosomes Minute electron-opaque organelles that occur in many cell types, contain hydrolytic enzymes, and are normally involved in localized intracellular digestion.

M line In a muscle sarcomere, the darkly staining structure in the middle of the H band.

maculae (Greek for "spots.") Organs of equilibrium in the vertebrate inner ear.

magnetite A magnetic mineral composed of Fe_3O_4 and found in some animals; believed to play a role in geomagnetic orientation.

malleus The outermost bone of the three bones in the mammalian inner ear; it connects the tympanic membrane with the incus.

Malpighian tubules Insect excretory osmoregulatory organs responsible for the active secretion of waste products and the formation of urine.

mass action law The velocity of a chemical reaction is proportional to the active masses of the reactants.

mass-specific metabolic rate The metabolic rate of a unit mass of tissue.

mastication The chewing or grinding of food with the teeth.

mastoid bone The posterior process of the temporal bone, situated behind the ear and in front of the occipital bone.

maximum aerobic velocity (MAV) Locomotor speed at which the maximum rate of aerobic respiration is reached.

mechanism A theory that proposes that life is based purely on the action of physical and chemical laws.

mechanoreceptor A sensory receptor tuned to respond to mechanical distortion or pressure.

median eminence A structure at the base of the hypothalamus that is continuous with the hypophyseal stalk; contains the primary capillary plexus of the hypothalamo-hypophyseal portal system.

medulla oblongata In vertebrates, a cone-shaped neuronal mass that lies between the pons and the spinal cord.

medullary cardiovascular center A group of neurons in the medulla involved in the integration of information used in the control and regulation of circulation.

medullary respiratory centers Groups of neurons in the medulla of the brain controlling the activity in motor neurons associated with breathing.

melanocyte-stimulating hormone (melanophore-stimulating hormone) A peptide hormone released by the adenohypophysis that effects melanin distribution in mammals and creates skin-color changes in fishes, amphibians, and reptiles.

melting point The lowest temperature at which a solid will begin to liquefy.

membrane potential The electric potential measured from within the cell relative to the potential of the extracellular fluid, which is by convention at zero potential; the potential difference between opposite sides of the membrane.

membrane pumps Membrane-based cellular mechanisms that actively transport substances against a gradient.

membrane recycling Recovery and reformation into new secretory vesicles of membrane lost from the cell membrane due to exocytosis.

membrane transport proteins Integral proteins that transport particular classes of molecules across membranes; *see* active transport.

menarche The onset of menstruation during puberty.

menopause The cessation of the menstrual cycle in the mature female human.

menses Shedding of the uterine lining during a menstrual cycle.

menstrual cycle Recurring physiological changes that include menstruation.

menstruation The shedding of the endometrium, an event that usually occurs in the absence of conception throughout the fertile period of the female of certain primate species, including humans.

mesencephalicus lateralis dorsalis (MLD) The nucleus to which auditory information projects in the owl, contributing to the bird's ability to locate objects in its environment based entirely on audition.

messenger molecules Hormones, synaptic transmitters, and other chemicals that regulate biological processes.

messenger RNA (mRNA) A fraction of RNA that is responsible for transmission of the informational base sequence of the DNA to the ribosomes.

metabolic acidosis A decrease in blood pH at constant PCO_2 usually as a result of metabolism or kidney function.

metabolic alkalosis An increase in blood pH at constant PCO_2 usually as a result of metabolism or kidney function.

metabolic intensity The metabolic rate of a unit mass of tissue. *See also* mass-specific metabolic rate.

metabolic pathway A sequence of enzymatic reactions involved in the alteration of one substance into another.

metabolic water Water evolved from cellular oxidation.

metabolism The totality of physical and chemical processes involved in anabolism, catabolism, and cell energetics.

metachronal waves Waves of activity that spread over a population of beating cilia.

metachronism The progression of in-phase activity in a wavelike manner over a population of organelles, such as cilia.

metamorphic climax The last stage of amphibian metamorphosis, in which the adult form is attained.

metamorphosis A change in morphology—in particular, from one stage of development to another, such as juvenile to adult.

metarhodopsin Product of the absorption of light by rhodopsin; decomposes to opsin and trans-retinal.

metarteriole An arterial capillary.

metazoa Multicellular organisms.

methemoglobin Hemoglobin in which the Fe^{3+} of heme has been oxidized to Fe^{2+}.

micelle A microscopic particle made from an aggregation of amphipathic molecules in solution.

Michaelis-Menton equation The rate equation for a single enzyme-catalyzed substrate reaction.

microclimate A small refugium (e.g., burrow, crack in bark) that provides protection from general climatological conditions.

microelectrodes Tiny glass "needles" inserted into tissues or even individual cells for recording physiological data.

microfilaments Actin filaments within the cytoplasmic substance; diameter of less than 10 nm.

micromanipulator A mechanical device that holds and moves microelectrodes incrementally in three different planes.

microtome A device used in microscopy to cut ultrathin sections from small blocks of tissues.

microtubules Cylindrical cytoplasmic structures made of polymerized tubulin and found in many cells, especially motile cells, as constituents of the mitotic spindle, cilia, and flagella.

microvilli Tiny cylindrical projections on a cell surface that greatly increase surface area; frequently found on absorptive epithelia, but also in photoreceptors.

micturition Urination.

midgut Major alimentary canal site for the chemical digestion of protein, fat, and carbohydrates.

mineralocorticoids Steroid hormones that are synthesized and secreted by the adrenal cortex and that influence plasma electrolyte balance—in particular, by sodium and chloride reabsorption in the kidney tubules. *See also* aldosterone.

miniature endplate potentials (mepps) Tiny depolarizations (generally 1 mV or less) of the postsynaptic membrane at a motor endplate; produced by presynaptic release of single packets of transmitter.

miniature postsynaptic potentials (mpps) Potentials produced in a postsynaptic neuron by presynaptic release of single vesicles of transmitter substance.

mitochondria Membrane-enclosed organelle where ATP is produced during aerobic metabolism.

mixed nerve A nerve that contains axons of both sensory and motor neurons.

mobility, electrical A quantity proportional to the migration rate of an ion in an electric field.

mobility, mechanical A quantity proportional to the rate at which a molecule will diffuse in a liquid phase.

modulatory agent One that either increases or decreases the response of a tissue to a physical or chemical signal.

molality The number of moles of solute in a kilogram of a pure solvent.

molarity The number of moles of solute in a liter of solution.

molars Teeth used in a side-by-side grinding motion to break down food.

mole Avogadro's number (6.023×10^{23}) of molecules of an element or a compound; equal to the molecular weight in grams.

molecular phylogeny A system of phylogenetic relations inferred from similarities and differences in the nucleic acid sequences coding for identified proteins.

monoclonal antibody A homogeneous antibody that is produced by a clone of antibody-forming cells and that binds with a single type of antigen.

monocytes White blood cells lacking cytoplasmic granules, but having an indented or horseshoe-shaped nucleus.

monogastric stomach A stomach consisting of a single strong muscular tube or sac.

monomer A compound capable of combining in repeating units to form a dimer, trimer, or polymer.

monopole An object or a particle bearing a single unneutralized electric charge, as, for example, an ion.

monosaccharide sugar An unhydrolyzable carbohydrate, a simple sugar. Such sugars are sweet-tasting, colorless crystalline compounds with the formula $C_n(H_2O)_n$. *See also* saccharide.

monosynaptic Requiring or transmitted through only one synapse; for example, the stretch reflex of vertebrate limbs.

monovalent Having a valence of one.

monozygotic Arising from one ovum or zygote.

motility The ability of the alimentary tract to contract and transport ingested material along its length.

motor cortex The part of the cerebral cortex that controls motor function; situated anterior to the central sulcus, which separates the frontal and parietal lobes.

motor neuron (motoneuron) A nerve cell that innervates muscle fibers.

motor program An endogenous coordinated motor output of central neuronal origin and independent of sensory feedback.

motor unit The unit of motor activity consisting of a motor neuron and the muscle fibers it innervates.

mRNA *See* messenger RNA.

mucin The mucopolysaccharide forming the chief lubricant of mucus.

mucosa Mucous membrane facing a cavity or the exterior of the body.

mucosal *See* mucosa.

mucous cells Mucus-secreting cells of the intestine.

mucus A viscous, protein-containing mixture of mucopolysaccharides secreted from specialized mucous membranes; often plays an important role in filter feeding (invertebrates) or in lubricating or protecting internal or external surfaces.

Mullerian ducts Paired embryonic ducts originating from the peritoneum that connect with the urogenital sinus to develop into the uterus and fallopian tubes.

multineuronal innervation Innervation of a muscle fiber by several motor neurons, as in many invertebrates, especially arthropods.

multiterminal innervation Numerous synapses made by a single motor neuron along the length of a muscle fiber.

multi-unit muscle A smooth muscle in which individual muscle fibers contract only when they receive excitatory input from neurons; contraction of these muscles is neurogenic.

muscarinic Pertaining to muscarine, a toxin derived from mushrooms; refers to acetylcholine receptors that respond to muscarine, but not to nicotine.

muscle fiber A skeletal muscle cell.

muscle spindle (stretch receptor) A length-sensitive receptor organ located between and in parallel with extrafusal muscle fibers; gives rise to the myotatic, or stretch, reflex of vertebrates.

mutagens Compounds that produce mutations in the germ cell line.

mutation A transmissible alteration in genetic material.

myelination Forming a myelin sheath.

myelin sheath A sheath formed by many layers of the membrane of Schwann cells or oligodendrocyte glial cells that are wrapped tightly around segments of axon in vertebrate nerve; serves as electrical insulation in saltatory conduction.

myoblast Embryonic precursor for skeletal muscle fibers.

myocardium Heart muscle.

myofibril A longitudinal unit of muscle fiber made up of sarcomeres and surrounded by sarcoplasmic reticulum.

myogenic Capable of producing an intrinsic cycle of electrical activity.

myogenic pacemaker A pacemaker that is a specialized muscle cell.

myoglobin An iron-containing protoporphyrin-globin complex found in muscle; serves as a reservoir for oxygen and gives some muscles their red or pink color.

myoplasm The cytosol in a muscle cell.

myosin The protein that makes up the thick filaments and cross bridges in muscle fibers; it is also found in many other cell types and is associated with cellular motility.

myotatic reflex (stretch reflex) Reflex contraction of a muscle in response to stretch of the muscle.

myotube A developing muscle fiber.

Na⁺-K⁺ pump *See* sodium-potassium pump.

Naloxone An analog of morphine that acts as an opioid antagonist.

nares Nostrils.

nematocysts Stinging cells of hydras, jellyfish, and anemones.

nephron The morphological and functional unit of the vertebrate kidney; composed of the glomerulus and Bowman's capsule, the proximal and distal tubules, the loop of Henle (birds, mammals), and the collecting duct.

Nernst equation Equation for calculating the electrical potential difference across a membrane that will just balance the concentration gradient of an ion.

nerve *As a noun:* A bundle of axons held together as a unit by connective tissue. *As an adjective:* Neuronal.

nerve net A collection of interconnected neurons that are distributed through the body, rather than concentrated in a central location; these neuronal systems are most typical of lower organisms, such as coelenterates.

nerve-specific energy The term used by Johannes Muller in his hypothesis that the sensory modality of a stimulus is encoded in the projection pattern of sensory neurons and not on particular features of the cellular response in the stimulated neurons.

nervous system The collection of all neurons in an animal's body.

net flux Sum of influx and efflux through a membrane or other material.

neurilemma Connective tissue sheath covering a bundle of axons.

neurites Cell processes extending from the soma of neurons.

neurogenic pacemaker A pacemaker that is a specialized nerve cell.

neuroglia (glia) Inexcitable supporting tissue of the nervous system.

neurohemal organ Organ for storage and discharge into the blood of the products of neurosecretion.

neurohormone A substance that exists within the neurons of the nervous system and exert hormonal effects outside the nervous system.

neurohumor Synaptic transmitters and neurosecretory hormones.

neurohypophysis (pars nervosa) A neuronally derived reservoir for hormones with antidiuretic and oxytocic action; consists of the neural lobe, which makes up its bulk, and the neural stalk, which is connected to and passes neurosecretions from the hypothalamus.

neuromast A collection of hair cells embedded in a cupula in lateral-line mechanoreceptor of the lower vertebrates.

neuromodulation A change in neuronal function caused by chemical messengers *(neuromodulators)* that are released from axon terminals, but that diffuse more widely than do typical neurotransmitters; neuromodulatory effects can be relatively long-lasting.

neuromuscular junction (NMJ) The synapse that connects a motor neuron with a skeletal muscle fiber.

neuron Nerve cell.

neuronal circuit A set of interconnected neurons.

neuronal integration Ongoing summing of all synaptic input onto a postsynaptic cell, which determines whether or not the postsynaptic cell will produce an action potential.

neuronal network (neuronal circuit) A system of interacting nerve cells.

neuronal plasticity Modification of activity in a neuronal circuit based on experience and changes in input.

neuropeptide A peptide molecule identified as a neurotransmitter substance.

neuropeptide Y A 36 amino acid peptide, co-localized with norepinephrine in sympathetic ganglia and adrenergic nerves as well as localized in some nonadrenergic fibers, the physiological effects of which include amelioration of actions of catecholamines on the mammalian heart and potentiation of actions of catecholamines in fish hearts.

neurophysins Proteins associated with neurohypophyseal hormones stored in granules in the neurosecretory terminals; cleaved from the hormones before secretion.

neuropil A dense mass of closely interwoven and synapsing nerve cell processes (axon collaterals and dendrites) and glial cells.

neurosecretory cells Nerve cells that liberate neurohormones.

neurotoxin A substance that interferes with the proper firing of nerve impulses.

neurotransmitter A chemical mediator released by a presynaptic nerve ending that interacts with receptor molecules in the postsynaptic membrane. This process generally induces a permeability increase to an ion or ions and thereby influences the electrical activity of the postsynaptic cell.

nicotinamide adenine dinucleotide (NAD) A coenzyme widely distributed in living organisms, participating in many enzymatic reactions; made up of adenine, nicotinamide, and two molecules each of d-ribose and phosphoric acid.

nicotinic Pertaining to nicotine, an alkaloid derived from tobacco; refers to acetylcholine receptors that respond to nicotine, but not to muscarine.

node of Ranvier Regularly spaced interruption (about every millimeter) of the myelin sheath along an axon.

noncompetitive inhibition Enzyme inhibition due to alteration or destruction of the active site.

nonsaturation kinetics Kinetics occurring when the rate of influx increases in proportion to the concentration of the solute in the extracellular fluid.

nonshivering thermogenesis A thermogenic process in which enzyme systems for fat metabolism are activated, breaking down and oxidizing conventional fats to produce heat.

nonspiking neuron A neuron that receives and transmits information without action potentials; synaptic transmitters are released by these neurons in proportion to their membrane potential, a process called *nonspiking release*.

nonspiking release The release of neurotransmitter from a presynaptic neuron that occurs independent of action potentials; typically graded changes in membrane potential modulate the activation of voltage-gated Ca^{2+} channels, and changes in the intracellular concentration of free Ca^{2+} modulate the release of transmitter.

norepinephrine (noradrenaline) A neurohumor secreted by the peripheral sympathetic nerve terminals, some cells of the central nervous system, and the adrenal medulla.

nucleases Enzymes that hydrolyze nucleic acids and their residues.

nucleic acids Nucleotide polymers of high molecular weight. *See also* DNA; RNA.

nucleosidases Enzymes that hydrolyze nucleic acids and their residues.

nucleotide A product of enzymatic (nuclease) splitting of nucleic acids; made up of a purine or pyrimidine base, a ribose or deoxyribose sugar, and a phosphate group.

nucleus *Of an atom:* The central, positively charged mass surrounded by a cloud of electrons. *Of a cell:* The membrane-bound body within eukaryotic cells that houses the genetic material of the cell. *Of nerve cells:* A related group of neurons in the central nervous system.

nymph A juvenile developmental stage in some arthropods; morphology resembles the adult.

nystatin A rod-shaped, antibiotic molecule that creates channels through membranes that allow the passage of molecules of a diameter less than 0.4 nm.

obligatory osmotic exchange An exchange between an animal and its environment that is determined by physical factors beyond the animal's control.

occipital lobe The most posterior region of the cerebral hemisphere.

occular dominance column A set of neurons arranged vertically through the mammalian visual cortex, all of which receive input from one of the two eyes.

ohm (Ω) MKS unit of electrical resistance, equivalent to the resis-tance of a column of mercury 1 mm^2 in cross-sectional and 106 cm long.

Ohm's law $I = V / R$. The strength of an electric current, I, varies directly as the voltage, V, and inversely as the resistance, R.

olfaction The sense of smell; chemoreception of molecules suspended in air.

oligodendrocytes A class of glial cells with few processes. These cells wrap axons in the central nervous system, forming myelin sheaths.

oligopeptides Polypeptide residues of two or three amino acids.

oligosaccharides Carbohydrates made up of a small number of monosaccharide residues.

omasum The part of the ruminant stomach lying between the rumen and the abomasum.

ommatidium The functional unit of the invertebrate compound eye, consisting of an elongated structure with a lens, a focusing cone, and photoreceptor cells.

oncotic pressure Osmotic pressure plus hydrostatic pressure caused by distribution of ions according to the Donnan equilibrium.

1a afferent fiber An axon with a peripheral sensory ending innervating a muscle spindle organ and responding to stretch of the organ; its central terminals synapse directly onto alpha motor neurons of the homonymous muscle.

1b afferent fiber An axon whose sensory terminals innervate the tendons of skeletal muscle and respond to tension.

1,25-dihydrozycholecalciferol A substance that is converted from vitamin D in the liver and increases Ca^{2+} absorption by the kidney.

onophores Molecules or molecular aggregates that promote the permeation of ions across membranes; these may be carrier molecules or ion-permeable membrane channels.

oocyte A developing ovum.

operator gene A gene that regulates the synthetic activity of closely linked structural genes via its association with a regulator gene.

operon A segment of DNA consisting of an operator gene and its associated structural genes.

opiates Opium-derived narcotic substances.

opioids Substances that exert opiatelike effects; some are synthesized endogenously by neurons within the vertebrate central nervous system.

opsin Protein moiety of visual pigments; it combines with 11-*cis*retinal to become a visual pigment.

optic axis An imaginary straight line passing through the center of curvature of a simple lens.

optic chiasm (optic chiasma) A swelling under the hypothalamus of the vertebrate brain where the two optic nerves meet; depending on the species, some axons cross the midline here and project to the contralateral side of the brain.

optic tectum Region of brain in fish and amphibia involved in processing visual information coming from the retina.

organ of Corti The tissue in the cochlea of the inner ear that contains the hair cells.

organs of equilibrium Regions of the inner ear that sense the position of the body relative to gravity or changes of the body's position with respect to gravity.

ornithine-urea cycle A cyclic succession of reactions that eliminate ammonia and produce urea in the liver of ureotelic organisms.

osmoconformer An organism that exhibits little or no osmoregulation, so that the osmolarity of its body fluids follows changes in the osmolarity of the environment.

osmolarity The effective osmotic pressure.

osmole The standard unit of osmotic pressure.

osmolyte A substance that serves the special purpose of raising the osmotic pressure or lowering the freezing point of a body fluid.

osmometer An instrument for the measurement of the osmotic pressure of a solution.

osmoregulation Maintenance of internal osmolarity with respect to the environment.

osmoregulator An organism that controls its internal osmolarity in the face of changes in environmental osmolarity.

osmosis The movement of pure solvent from a solution of an area of low solvent concentration to an area of high solvent concentration through a semipermeable membrane separating the two solutions.

osmotic flow The solvent flux due to osmotic pressure.

osmotic pressure Pressure that can potentially be created by osmosis between two solutions separated by a semipermeable membrane; the amount of pressure necessary to prevent osmotic flow between the two solutions.

ossicles Little bones. Auditory ossicles are the tiny bones (malleus, incus, stapes) of the middle ear, which transmit sound vibrations from the tympanic membrane to the oval window.

ostia The small, mouth-like openings in the body wall of sponges.

otolith A calcareous particle that lies on hair cells in the organs of equilibrium.

ouabain Cardiac glycoside, a drug capable of blocking some sodium pumps.

outer plexiform layer In the vertebrate retina, the layer of connecting processes that lies between the photoreceptor cells and the bipolar cells.

outer segment The part of a vertebrate photoreceptor that contains the pigmented receptor membranes; it is attached to the inner segment by a thin bridge that has microfilaments arranged as in a cilium.

oval window The connection between the middle ear and the cochlea; it is covered by the base of the stapes.

overshoot The reversal of membrane potential during an action potential; the voltage above zero to the peak of the action potential.

ovulation The release of an ovum from the ovarian follicle.

ovum An egg cell; the reproductive cell (gamete) of the female.

oxidant An electron acceptor in a reaction involving oxidation and reduction.

oxidation Loss of electrons or increase in net positivity of an atom or a molecule. Biological oxidations are usually achieved by removal of a pair of hydrogen atoms from a molecule.

oxidative phosphorylation Respiratory chain phosphorylation; the formation of high-energy phosphate bonds via phosphorylation of ADP to ATP, accompanied by the transport of electrons to oxygen from the substrate.

oxyconformer Animal that allows oxygen consumption to fall as ambient oxygen falls.

oxygen debt The extra oxygen necessary to oxidize the products of anaerobic metabolism that accumulate in the muscle tissues during intense physical activity.

oxygen dissociation curves Curves that describe the relationship between the extent of combination of oxygen with the respiratory pigment and the partial pressure of oxygen in the gas phase.

oxyhemoglobin Hemoglobin with oxygen combined to the Fe atom of the heme group.

oxyntic cells (parietal cells) HCl-secreting cells of the stomach lining.

oxyregulator Animal that maintains oxygen consumption as ambient oxygen falls.

oxytocin An octapeptide hormone secreted by the neurohypophysis; stimulates contractions of the uterus in childbirth and the release of milk from mammary glands.

P-wave That portion of the electrocardiogram associated with depolarization of the atria.

pacemaker An excitable cell or tissue that fires spontaneously and rhythmically.

pacemaker potentials Spontaneous and rhythmical depolarizations produced by pacemaker tissue.

Pacinian corpuscles Pressure receptors found in skin, muscle, joints, and connective tissue of vertebrates; they consist of a nerve ending surrounded by a laminated capsule of connective tissue.

pancreas An organ that produces exocrine secretions, such as digestive enzymes, as well as endocrine secretions, including the hormones insulin and glucagon.

pancreatic duct Duct carrying secretions from the pancreas to the small intestine.

pancreatic juice A secretion of the pancreas containing proteases, lipases, and carbohydrases essential for intestinal digestion.

pancreozymin *See* cholecystokinin.

parabiosis The experimental connection of two individuals to allow mixing of their body fluids.

parabronchi Air-conduction pathways in the bird lung.

paracellular pathways Solvent and solute pathway through epithelium passing between rather than through cells.

paracrine A hormonal pathway characterized by the production of a biologically active substance that passes by diffusion within the extracellular space to a nearby cell where it initiates a response.

parafollicular cells (C cells) Cells in the mammalian thyroid that secrete calcitonin.

parallel processing A pattern of information processing in the nervous system in which multiple pathways simultaneously carry information about a particular input or output; the information carried in multiple channels is synthesized where the pathways converge.

parasympathetic nervous system The craniosacral part of the autonomic nervous system; in general, increased activity of these neurons supports vegetative functions such as digestion.

parathyroid glands Small tissue masses (usually two pairs) close to the thyroid gland that secrete parathormone (parathyroid hormone).

parathyroid hormone (PTH; parathormone) A polypeptide hormone of the parathyroid glands secreted in response to a low plasma calcium level; stimulates calcium release from bone and calcium absorption by the intestines while reducing calcium excretion by the kidneys.

paraventricular nucleus A group of neurosecretory neurons in the supraoptic hypothalamus that send their axons into the neurohypophysis.

parietal cells Cells of the stomach lining that secrete hydrochloric acid. *See also* oxyntic cells.

pars intercerebralis The dorsal part of the insect brain; contains the cell bodies of neurosecretory cells that secrete brain hormones from axon terminals in the corpora cardiaca.

partition coefficient Ratio of the distribution of a substance between two different liquid phases (e.g., oil and water).

parturition The process of giving birth.

parvalbumin Calcium-binding protein found in vertebrate muscle.

patch-clamping A recording technique in which a glass pipette electrode is brought into contact with the outside of a cell membrane and sealed tightly against it. Current is then passed to hold the potential difference across the membrane at a constant value, and the magnitude of the current required is recorded. The method can be used to measure ionic currents through single ion channels or across the membrane of an entire cell.

patch clamp recording A method of investigating epithelial ion current transfer on very localized regions of cells.

patella The bone of the knee-cap.

pentose A five-carbon monosaccharide sugar.

pepsin A proteolytic enzyme secreted by the stomach lining.

pepsinogen Proenzyme of pepsin.

peptide A molecule consisting of a linear array of amino residues. Protein molecules are made of one or more peptides. Short chains are oligopeptides; long chains are polypeptides.

peptide bond The center bond of the —CO—NH— group, created by the condensation of amino acids into peptides.

peptide hormones Hormones that regulate alimentary canal basic electric rhythms.

perfusion The passage of fluid over or through an organ, a tissue, or a cell.

pericardium The connective-tissue sac that encloses the heart.

perilymph The aqueous solution, similar to other body fluids, that is contained within the *scala tympani* and *scala vestibuli* of the cochlea.

peripheral nervous system The set of neurons and parts of neurons that lie outside of the central nervous system.

peripheral resistance units (PRUs) The drop in pressure (in millimeters of mercury, mmHg) along a vascular bed divided by mean flow in milliliters per second.

peristalsis A traveling wave of constriction in tubular tissue produced by contraction of circular muscle.

peritoneum The membrane that lines the abdominal and pelvic cavities.

permeability The ease with which substances can pass through a membrane.

pH scale Negative log scale (base 10) of hydrogen ion concentration of a solution. $pH = -\log[H^+]$.

phagocyte A cell that engulfs other cells, microorganisms, or foreign particulate matter.

phagocytosis The ingestion of particles, cells, or microorganisms by a cell into its cytoplasmic vacuoles.

phase contrast microscopy A microscopic technique using differential light refraction by different components of the specimen to enhance viewed images.

phasic Transient.

pheromone A species-specific substance released into the environment for the purpose of signaling between individuals of the same species.

phlorizin A glycoside that inhibits active transport of glucose.

phonon A quantum of sound energy.

phosphagens High-energy phosphate compounds (e.g., phosphoarginine and phosphocreatine) that serve as phosphate-group donors for rapid rephosphorylation of ADP to ATP.

phosphoarginine A compound that has phosphagen properties similar to those of phosphocreatine and that occurs in the muscles of some invertebrates.

phosphocreatine (creatine phosphate) A phosphorylated nitrogenous compound found primarily in muscle; contains a high-energy phosphate bond, which can be rapidly transferred to ADP, regenerating ATP.

phosphodiesterase An hydrolytic cytoplasmic enzyme that degrades cAMP to AMP.

phosphodiester bonds Bonds that link individual nucleotides in nucleic acids.

phosphodiester group —O—P—O—.

phosphoglycerides Glycerine-based lipids of cell membrane.

phospholipid A phosphorus-containing lipid that hydrolyzes to fatty acids, glycerin, and a nitrogenous compound.

phosphorylase *a* Activated (phosphorylated) form of phosphorylase that catalyzes the cleavage of glycogen to glucose 1-phosphate.

phosphorylase kinase Enzyme that, when phosphorylated by a protein kinase, converts phosphorylase *b* to the more active phosphorylase *a*.

phosphorylation The incorporation of a PO^{3-} group into an organic molecule.

photon A quantum of light energy (the smallest amount of light that can exist at each wavelength).

photopigments Pigment molecules that change their energy state when they absorb one or more photons of light.

photoreceptor A sensory cell that is tuned to receive light energy.

physiological dead space That portion of inhaled air not involved in gas transfer in the lung.

pilomotor Pertaining to the autonomic control of smooth muscle for the erection of body hair.

pinna The outer structure of the mammalian ear, which can be more or less elaborate and which captures and funnels sound into the ear.

pinnate Resembling a feather, with similar parts arranged on opposite sides of the axis.

pinocytosis Fluid intake by cells via surface invaginations that seal off to become vacuoles filled with liquid.

pituitary gland (hypophysis) A complex endocrine organ situated at the base of the brain and connected to the hypothalamus by a stalk. It is of dual origin: the anterior lobe (adenohypophysis) is derived from embryonic buccal epithelium, whereas the posterior lobe is derived from the diencephalon.

pK′ The negative log (base 10) of an ionization constant, $K' \cdot pk' = -\log_{10} K'$.

placebo A physiologically neutral substance that elicits curative or analgesic effects, apparently through psychological means.

placental lactogen A hormone from the placenta that prepares the breasts for milk production.

plane-polarized light Light vibrating in only one plane.

plasma kinins Peptide hormones formed in the blood after injury—for example, bradykinin.

plasmalemma Cell membrane; surface membrane.

plasma membrane Cell membrane; surface membrane.

plasma skimming The separation of plasma from blood within the circulation.

plasticity Compliance to external influence.

plastron The ventral shell of a tortoise or turtle; also a gas film held in place under water by hydrofuge hairs, creating a large air-water interface.

pleura The membranes that line the pleural cavity.

pleural cavity The cavity between the lungs and the wall of the thorax.

plicae circularis The extensive fold of intestinal mucosa bearing villi.

pneumotaxic center A group of neurons in the pons, thought to be involved in the maintenance of rhythmic breathing in mammals.

pneumothorax Collapse of the lung due to a puncture into the pleural cavity of the chest wall or the lung.

poikilotherm An animal whose body temperature tends to fluctuate more or less with the ambient temperature.

Poiseuille's law In laminar flow, the flow is directly proportional to the driving pressure, and resistance is independent of flow.

Poisson distribution A theoretical description of the probability that random, independent events based on a unitary event of a particular size will occur.

polyclonal Derived from different cell lines or clones.

polyestrous The state of having many estrous cycles throughout the year.

polymer A compound composed of a linear sequence of simple molecules or residues.

polypeptide chain A linear arrangement of more than two amino acid residues.

polypnea Increased breathing rate.

polysaccharides Carbohydrases that hydrolyze the glycosidic bonds of long-chain carbohydrates (cellulose, glycogen, and starch).

polysynaptic Referring to transmission through multiple synapses in series.

polytene Having many duplicate chromatin strands.

pons The region of the vertebrate brain that lies just rostral to the medulla oblongata.

pores of Kohn Small holes between adjacent regions of the lung, permitting collateral air flow.

porphyrins A group of cyclic tetrapyrrole derivatives.

porphyropsin A purple photopigment, based on 11-*cis* dehydroretinal, that is present in the retinal rods of some freshwater fishes.

portal vessels Blood vessels that carry blood directly from one capillary bed to another.

positive phototaxis Referring to the movement of an animal toward light.

postsynaptic Located on the receiving side of a synaptic connection.

posttetanic depression Reduced postsynaptic response following prolonged presynaptic stimulation at a high frequency; believed to be due to presynaptic depletion of transmitter.

posttetanic potentiation (PTP) Increased efficacy of synaptic transmission following presynaptic stimulation at a high frequency; often follows posttetanic depression.

potassium activation An increase in the conductance of a membrane to potassium in response to depolarization.

potential The voltage above zero to the peak of the action potential.

potential energy Stored energy that can be released to do work.

premetamorphosis The developmental stage just preceding amphibian metamorphosis, during which iodine binding and hormone synthesis occur in the thyroid gland.

presbyopia The tendency for human eyes to become less able to focus on close objects ("far-sighted") with age; occurs as the lens becomes less compliant.

pressure pulse The difference between the systolic and diastolic pressures.

presynaptic Located on the sending side of a synaptic connection.

presynaptic inhibition Neuronal inhibition resulting from the action of a terminal that ends on the presynaptic terminal of an excitatory synapse, reducing the amount of transmitter released.

primary follicle An immature ovarian follicle.

primary projection cortex A region of cerebral cortex that directly receives sensory signals from lower centers; the first cortical cells to receive sensory information projected to the brain.

primary sensory neurons Neurons that directly receive sensory stimulation.

primary structure The sequence of amino acid residues of a polypeptide chain.

proboscis An elongated, protruding mouth part, typically in sucking insects.

procaine 2-diethylaminoethyl-p-aminobenzoate; a local anesthetic that interferes with some of the ion conductances of excitable membranes.

proenzyme (zymogen) The inactive form of an enzyme before it is activated by removal of a terminal segment of peptide.

progesterone A hormone of the corpus luteum, adrenal cortex, and the placenta that promotes growth of a suitable uterine lining for implantation and development of the fertilized ovum.

prolactin An adenohypophyseal hormone that stimulates milk production and lactation after parturition in mammals.

prometamorphosis The first stage of amphibian metamorphosis, during which there is increased development and activity in the thyroid gland and median eminence.

propeptide A large peptide that contains the amino acid sequences of several smaller peptides, which are released when the large peptide is enzymatically cleaved.

proprioceptors Sensory receptors situated primarily in muscles and tendons that relay information about the position and motion of the body.

prostaglandins A family of natural fatty acids that arise in a variety of tissues and are able to induce contraction in uterine and other smooth muscle, lower blood pressure, and modify the actions of some hormones.

prostate gland A gland located around the neck of the bladder and urethra in males that contributes to the seminal fluid.

prosthetic group An organic compound essential to the function of an enzyme. Prosthetic groups differ from coenzymes in that they are more firmly attached to the enzyme protein.

protagonistic muscles Muscles whose contractions cooperate to produce a movement.

protease Enzymes that break down peptide bonds of proteins and polypeptides.

protein kinase Any enzyme that catalyzes the transfer of a phosphate group from ATP to a protein, creating a phosphoprotein.

proteins Large molecules composed of one or more chains of alpha amino acid residues (i.e., polypeptide chains).

proteolysis The splitting of proteins by hydrolysis of peptide bonds.

proteolytic Protein-hydrolyzing.

prothoracic glands Ecdysone-secreting tissues situated in the anterior thorax of insects.

prothoracicotropic hormone (PTTH) A neurohormone produced by neurosecretory cells in the pars intercerebralis of the brain. PTTH activates the prothoracic gland to synthesize and secrete molting hormones.

proximal tubules Coiled portions of the renal tubules located in the renal cortex, beginning at the glomerulus and leading to (and continuous with) the descending limb of the loop of Henle.

pseudopodium Literally, false foot; a temporary projection of an amoeboid cell for engulfment of food or for locomotion.

pseudopregnancy A false pregnancy.

psychophysics The branch of psychology concerned with relationships between physical stimuli and perception.

pulmonary Pertaining to or affecting the lungs.

pupa A developmental stage of some insect groups; between the larva and the adult.

pupil The opening at the center of the iris through which light passes into the eye.

purinergic Referring to nerve endings that release purines or their derivatives as transmitter substances.

purines A class of nitrogenous heterocyclic compounds, $C_5H_4N_4$, derivatives of which (purine bases) are found in nucleotides; they are colorless and crystalline.

pyloric Pertaining to the caudal portion of the vertebrate stomach where it joins the small intestine.

pyloric sphincter Sphincter guarding the opening of the stomach into the small intestine.

pylorus The distal stomach opening, ringed by a sphincter, that releases the stomach contents into the duodenum.

pyramidal tract A bundle of nerve fibers originating in the motor cortex and descending down the brain stem to the medulla oblongata and to the spinal cord; responsible for mediating control of voluntary muscle movements.

pyrimidine A class of nitrogenous heterocyclic compounds, $C_4H_4N_2$, derivatives of which (pyrimidine bases) are found in nucleotides.

pyrogen Substance that leads to a resetting of a homeotherm's body thermostat to a higher set point, thereby producing fever.

Q_{10} The ratio of the rate of a reaction at a given temperature to its rate at a temperature 10°C lower.

QRS-wave That portion of the electrocardiogram related to depolarization of the ventricle.

quality A property that distinguishes sensory stimuli within a sensory modality; e.g., color is a quality of visual stimuli.

quantal content The number of neurotransmitter molecules in one synaptic vesicle.

quantal release The release of neurotransmitter in discrete packets which correspond to vesicles containing transmitter molecules.

quantal synaptic transmission The concept that neurotransmitter is released in multiples of discrete "packets." It is now apparent that the quantal packets represent individual presynaptic vesicles.

quaternary structure The characteristic ways in which the subunits of a protein containing more than one polypeptide chain are combined.

r selection A pattern of energetic investment in reproduction in which large numbers of very small offspring are produced, each with a low chance of survival due to lack of parental care.

radial finks Extensions from peripheral doublets to the central sheath in cilia and flagella.

radiation, thermal The transfer of heat by electromagnetic radiation without direct contact between objects.

radioimmunoassays (RIAs) An immunological technique for the measurement of minute quantities of antigen or antibody, hormones, certain drugs, and other substances with the use of radioactively labeled reagents.

radioisotope A radioactive isotope.

radula A rasp-like structure in the mouth of many gastropods.

range fractionation The pattern in which receptors within one sensory modality are tuned to receive information within relatively narrow, but not identical, intensity ranges, so the entire dynamic range of the modality is divided among different classes of receptors. For example, in the human eye rods respond to dim light but are saturated in bright light; cones are less sensitive to dim light but remain responsive in bright light.

rate constant (specific reaction rate) The proportionality factor by which the concentration of a reactant in an enzymatic reaction is related to the reaction rate.

reactive hyperemia Higher than normal blood flow that occurs following a brief period of ischemia.

receptive field That area of an organism's body (e.g., on the skin or in the retina) that when stimulated influences the activity of a given neuron is the receptive field of that neuron.

receptor Molecules that are situated on a membrane and that interact specifically with messenger molecules, such as hormones or transmitters.

receptor cell A neuronal cell that is specialized to respond to some particular sensory stimulation.

receptor current A stimulus-induced change in the movement of ions across a receptor cell membrane.

receptor-mediated endocytosis Specialized process of endocytosis that requires solutes being transported to temporarily bind to receptor molecules embedded in the cell membrane prior to transport across the cell membrane.

receptor molecules Molecules that are situated on the outer surface of the cell membrane and that interact specifically with messenger molecules, such as hormones or neurotransmitters.

receptor potential A change in membrane potential elicited in sensory receptor cells by sensory stimulation, which changes the flow of ionic current across the cell membrane.

receptor tyrosine kinases (RTKs) Receptors with intrinsic tyrosine kinase activity, which are known to bind insulin and a number of growth factors. When activated by external signal binding, RTKs transfer the phosphate group from

ATP to the hydroxyl group on a tyrosine residue of selected proteins in the cytosol. RTKs also phosphorylate themselves when activated; this autophosphorylation enhances the activity of the kinase.

reciprocal inhibition Inhibition of the motor neurons innervating one set of muscles during the reflex excitation of their antagonists.

recombinant DNA An engineered DNA molecule resulting from the two different organisms.

recruitment The pattern in which neurons with increasingly high thresholds become active as intensity increases. This pattern can occur in sensory neurons (range fractionation) or in motor neurons.

rectal gland Organ near the rectum of elasmobranchs that excretes a highly concentrated NaCl solution.

redox pair Two compounds, molecules, or atoms involved in mutual reduction and oxidation.

reductant Donor of electrons in a redox reaction.

reduction The addition of electrons to a substance.

reduction potential A measurement of the tendency of a reducer to yield electrons in a redox reaction, expressed in volts.

reflex An action that is generated without the participation of the highest neuronal centers and is thus not voluntary; an involuntary motor response mediated by a neuronal arc in response to sensory input.

reflex arc A neuronal pathway that connects sensory input and motor output; consists of afferent nerve input to a nerve center that produces activity in efferent nerves to an effector organ.

refraction The bending of light rays as they pass from a medium of one density into a medium of another density.

refractive index The refractive power of a medium compared with that of air, designated 1.

refractory period The period of increased membrane threshold immediately following an action potential. *Absolute refractory period:* The initial phase of the refractory period when no AP can be generated. *Relative refractory period:* The later phase of the refractory period when the threshold is elevated, but an AP can be generated with sufficiently intense stimulation.

regenerative Self-reinforcing; utilizing positive feedback; autocatalytic.

regulator Animals that use biochemical, physiological, behavioral, and other mechanisms to maintain internal homeostasis.

regulator gene Genes that code for repressor proteins, which suppress the action of structural genes.

regurgitation Reverse movement of intestinal luminal contents produced by reverse peristalsis.

Reissner's membrane A membrane within the mammalian cochlea.

release-inhibiting hormone (RIH; release-inhibiting factor, RIF) A hypothalamic neurosecretion carried by portal vessels to the adenohypophysis, where it restrains the release of a specific hormone.

releasing hormone (RH, releasing factor) A hypothalamic neurosecretion that stimulates the liberation of a specific hormone from the adenohypophysis.

releasing stimulus (key stimulus) The stimulus that is effective in producing a fixed action pattern.

renal clearance That volume of plasma containing the quantity of a freely filtered substance that appears in the glomerular filtrate per unit time. Total renal clearance is the amount of ultrafiltrate produced by the kidney per unit time.

renin A proteolytic enzyme produced by specialized cells in renal arterioles; converts angiotensinogen to angiotensin.

rennin An endopeptidase enzyme that coagulates milk by promoting the formation of calcium caseinate from the milk protein casein; found especially in the gastric juice of young mammals.

Renshaw cells Small inhibitory interneurons in the ventral horn that are excited by branches of a-motor neuron axons that feed back on the motor neuron pool.

repolarization The return to resting polarity of a cell membrane that has been depolarized.

repression proteins Proteins that can bind to a short region of DNA preceding the structural gene(s), thus preventing transcription.

repressor gene (regulator gene) A gene that produces a substance (repressor) that shuts off the structural-gene activity of an operon by an interaction with its operator gene.

reserpine A botanically derived tranquilizing agent that interferes with the uptake of catecholamine from the cytosol by secretory vesicles; its effect is to deplete the catecholamine content of adrenergic cells.

residual volume The volume of air left in the lungs after maximal expiratory effort.

resistance (R) The property that hinders the flow of current. The unit is the ohm (Ω), defined as the resistance that allows 1 ampere (A) of current to flow when a potential drop of 1 volt (V) exists across the resistance. It is equivalent to the resistance of a column of mercury 1 mm^2 in cross-sectional area and 106.3 cm long. $R = \rho \times$ length $\div$ cross-sectional area.

resistivity (p) The resistance of a conductor 1 cm in length and 1 cm^2 in cross-sectional area.

respiratory acidosis A decrease in blood pH associated with a fall in blood PCO_2 as a result of lung hypoventilation.

respiratory alkalosis An increase in blood pH associated with a rise in blood PCO_2 as a result of lung hyperventilation.

respiratory chain *See* electron-transport chain.

respiratory pigment A substance that combines reversibly with oxygen—for example, hemoglobin.

respiratory quotient (RQ) The ratio of CO_2 production to O_2 consumption; depends on type of food oxidized by the animal.

respirometry Measurement of an animal's respiratory exchange.

resting potential The normal, unstimulated membrane potential of a cell at rest.

rete mirabile An extensive countercurrent arrangement of arterial and venous capillaries.

reticulum A small network.

retina The photosensitive inner surface of the vertebrate eye.

retinal The aldehyde of retinol obtained from the enzymatic oxidative cleavage of carotene; in the 11-*cis* form it unites with opsins in the retina to form the visual pigments.

retinal streak (visual streak) A retinal structure—found in the eyes of some species that inhabit plains—in which photoreceptors are packed into a horizontal streak across the retina, providing high resolution along the visual horizon. This pattern is similar to the fovea of primates, but it has a different shape.

retinol Vitamin A ($C_{20}H_{30}O$), an alcohol of 20 carbons; converted reversibly to retinal by enzymatic dehydrogenation.

retinular cell A photoreceptor cell of the arthropod compound eye.

reversal potential The membrane potential at which no current flows through membrane ion channels, even though the channels are open; it is equal to the equilibrium potential for the ion or ions that are conducted through the open channels.

Reynolds number *(Re)* A unitless number; the tendency of a flowing gas or liquid to become turbulent is proportional to its velocity and density and inversely proportional to its viscosity. Calculated from these parameters, the Reynolds number indicates whether flow will be turbulent or laminar under a particular set of conditions.

rhabdome The aggregate structure consisting of a longitudinal rosette of rhabdomeres located axially in the ommatidium.

rhabdomere The light-absorbing part of a retinular cell that faces the central axis of an ommatidium; the photopigment-bearing surface membrane is expanded into closely packed microvilli, increasing the amount of photosensitive membrane.

rheogenic Producing electric current.

rhodopsin (visual purple) A purplish red, light-sensitive chromoprotein with 11-*cis* retinal as its prosthetic group; found in the rods and cones of the retina; bleaches to "visual yellow" (all-*trans*-retinal) when it absorbs incident light.

ribonucleic acid *See* RNA.

ribose A pentose monosaccharide with the chemical formula $HOCH_2(CHOH)_3CHO$; a constituent of RNA.

ribosome Ribonucleoprotein particles found within the cytoplasm; the sites of intersection of mRNA, tRNA, and the amino acids during the synthesis of polypeptide chains.

rigor mortis The rigidity that develops in dying muscle as ATP becomes depleted and cross bridges remain attached.

Ringer solution Physiological saline solution.

RNA (ribonucleic acid) A nucleic acid made up of adenine, guanine, cytosine, uracil, ribose, and phosphoric acid; responsible for the transcription of DNA and the translation into protein.

rods One class of vertebrate visual receptor cells, the cones being the other; very sensitive to light, based on cellular physiology and on a high degree of convergence onto second-order cells. In most species, there is only one class of rods in the retina, so rods cannot convey information about color.

Root effect (Root shift) A change in blood oxygen capacity as a result of a pH change.

round window A membrane-covered opening, separating the middle ear and the cochlea, through which pressure waves leave after traveling through the cochlea.

rumen The storage and fermentation chamber in the digastric stomach of ruminants.

rumination The chewing of partially digested food brought up by reverse peristalsis from the rumen in ungulate animals and in other ruminants.

saccharide A family of carbohydrates that includes the sugars; they are grouped as to the number of saccharide $(C_nH_{2n}O_{n-1})$ groups comprising them: the mono-, di-, tri-, and polysaccharides.

sacculus One of the vertebrate organs of equilibrium.

safety factor A factor relating input and output in a system, describing how likely transmission through the system will fail; the higher the safety factor, the less likely it is that transmission will fail.

saliva A water-like fluid secreted in the upper alimentary canal (headgut); aids in mechanical and chemical digestion.

salivary glands Glands that secrete saliva into the headgut.

saltatory (conduction) Jumping; discontinuous.

salt glands Osmoregulatory organs of many birds and reptiles that live in desert or marine environments. A hypertonic aqueous exudate is formed by active salt secretion into the small tubules situated above the eyes and is excreted via the nostrils.

"salting out" A decrease in Bunsen solubility coefficient as a result of increased ionic strength of the solvent.

sarcolemma The surface membrane of muscle fibers.

sarcomere The contractile unit of a myofibril, it is bounded by two Z disks.

sarcoplasm Cytosol of a muscle cell.

sarcoplasmic reticulum (SR) A smooth, membrane-limited network surrounding each myofibril. Calcium is stored in the SR and released as free Ca^{2+} during muscle excitation-contraction coupling.

sarcotubular system The sarcoplasmic reticulum plus the transverse tubules.

saturated In reference to fatty acid molecules, indicates that the carbon-carbon bonds are single, with each carbon atom bearing two hydrogens. Without free valence electrons.

scala media The cochlear duct, a membranous labyrinth containing the organ of Corti and the tectorial membrane; it is filled with endolymph.

scala tympani A cochlear chamber connected with the *scala vestibuli* through the helicotrema; it is filled with perilymph.

scala vestibuli A cochlear chamber beginning in the vestibule, connecting with the *scala tympani* through the helicotrema; it is filled with perilymph.

scaling The study of how both anatomical and physiological characteristics change with body mass.

Schwann cell A neuroglial cell outside the central nervous system that wraps its membrane around axons during development to produce the insulating myelin sheath that envelops peripheral axons in the regions between nodes of Ranvier.

scintillation counter An instrument that detects and counts tiny flashes of light produced in scintillation fluid produced by particles emitted from radioisotopes.

SDA (specific dynamic action) The increment in metabolic energy cost that can be ascribed to the digestion and assimilation of food; it is highest for proteins.

secondary structure Refers to the straight or helical configuration of polypeptide chains.

secondary vacuole Vacuole formed when food containing vacuole merges with enzyme-containing lysosomes.

second law of thermodynamics All natural or spontaneous processes are accompanied by an increase in entropy.

second messenger A term applied to cAMP, cGMP, Ca^{2+}, or any other intracellular regulatory agent that is itself under the control of an extracellular first messenger, such as a hormone.

second-order enzyme kinetics Describes enzymatic reactions whose rates are determined by the concentrations of two reactants multiplied together or of one reactant squared.

second-order neuron A neuron that receives input from primary sensory neurons.

secretagogue A substance that stimulates or promotes secretion.

secretin A polypeptide hormone secreted by the duodenal and jejunal mucosa in response to the presence of acid chyme in the intestine; induces pancreatic secretion into the intestine and is chemically identical to enterogastrone.

secretory granules (secretory vesicles) Membrane-bound cytoplasmic granules containing secretory products of a cell.

segmentation Rhythmic contractions of the circular muscle layer of the intestine that mixes the intestinal contents.

selectivity sequence *See* affinity sequence.

self-repair capability Ability of cell organelles and membranes to reseal themselves when chemically or mechanically disturbed.

semicircular canals Three of the vertebrate organs of equilibrium, which sense acceleration of the body with respect to the gravitational field.

seminal vesicles Paired sacs attached to the posterior urinary bladder that have tubes joining the vas deferens in the male.

semipermeable membrane A membrane that allows certain molecules but not others to pass through it.

sensation The perception of a sensory stimulus (as opposed to the reception of stimulus energy in a primary sensory receptor).

sensilla (plural; sensillum, singular) Collections of sensory receptors in the periphery of an organism, usually an invertebrate; sensilla are typically very simple in structure, lacking accessory structures.

sensillum A chitinous, hollow, hairlike projection of the arthropod exoskeleton that serves as an auxiliary structure for sensory neurons.

sensor Mechanical, electrical, or biological device that detects changes in its immediate environment.

sensory adaptation The property of sensory systems to become less sensitive to stimuli during prolonged or repeated stimulation.

sensory fiber An axon that carries sensory information to the central nervous system.

sensory filter network Neuronal circuits that selectively transmit some features of a sensory input and ignore other features.

sensory modality A set of sensory structures that are tuned to receive a particular class of energy; e.g., vision, hearing, and the sense of smell are three sensory modalities.

sensory reception The absorption of stimulus energy by a neuron that produces a receptor potential in the neuron and that may produce action potentials that travel to the central nervous system.

series elastic components (SEC) Elastic structures (such as tendons and other connective tissues) that are arranged in series with the contractile elements in muscle.

serosa Outermost layer of alimentary tract.

serosal Pertaining to the side of an epithelial tissue facing the blood, as opposed to the mucosal side, which faces the exterior or luminal space.

serotonin 5-Hydroxytryptamine, 5-HT; a neurotransmitter, $C_{10}H_{12}N_2O$.

serum The clear component of blood plasma.

servo mechanism A control system that utilizes negative feedback to correct deviations from a selected level, the set point.

set point In a negative-feedback system, the state to which feedback tends to bring the system.

siemen (S) The unit of electrical conductance; reciprocal of the ohm.

sign stimulus The most basic essential pattern of sensory input required to release an instinctual pattern of behavior.

signal-to-noise ratio The relation between a signal and the random background activity that arises as the result of kinetic energy or other irrelevent events; information theory has thoroughly considered this relation and generated rules that describe efficient and effective information transfer.

single-unit muscle A smooth muscle in which individual fibers are coupled through gap junctions, allowing excitation to spread through the muscle independent of neuronal activity; contraction in these muscles is often myogenic, driven by internal pacemaker cells.

sinoatrial node A mass of specialized cardiac tissue that lies at the junction of the superior vena cava with the right atrium; it acts as the pacemaker of the heart in initiating each cardiac contraction.

sinus A cavity or sac; a dilated part of a blood vessel.

sinus node (sinoatrial node, SA node) The junction between the right atrium and the vena cava, the location of the pacemaker.

sinus venosus The membrane chamber attached to the heart that receives venus blood in fish, amphibians, and reptiles, and transmits it to the atrium.

skeletal muscle The striated muscle whose contraction is responsible for moving the bodies of animals. Contraction of these fibers is neurogenic; that is, contraction occurs only when the fibers are excited by synaptic input from motor neurons.

sliding-filament theory The theory that muscle sarcomeres shorten when actin thin filaments are actively pulled toward the middle of myosin thick filaments by the action of myosin crossbridges.

sliding-tubule hypothesis Bending movements of cilia and flagella are produced by active longitudinal sliding of the axonemal microtubules past one another.

slow chemical synaptic transmission Synaptic transmission at a chemical synapse mediated by neurotransmitters that bind on the postsynaptic membrane to receptor molecules that affect intracellular second messenger systems, typically through G proteins. Binding of the transmitter to the receptor complex activates the second messengers, which in turn modify the state of ion channel proteins.

smooth muscle Muscle without sarcomeres and hence without striations. Myofilaments are nonuniformly distributed within small, mononucleated, spindle-shaped cells.

sodium activation An increased conductance of excitable membranes to sodium ions in response to membrane depolarization; believed to result from an opening of sodium gates associated with membrane channels.

sodium hypothesis The upstroke of an action potential is due to an inward movement of Na^+ down its electrochemical gradient as a result of a transient increase in sodium permeability.

sodium inactivation Loss of responsiveness of sodium gates to depolarization; develops with time during a depolarization and persists for a short period after repolarization of the membrane.

sodium-potassium ATPase (sodium pump) Membrane protein responsible for maintaining the asymmetrical concentrations of Na^+ and K^+ ions across the cell membrane; it actively extrudes Na^+ from the cell and takes up K^+ from extracellular fluids at the expense of metabolic energy. In some sodium pumps, there is a $3:2$ exchange of intracellular Na^+ for extracellular K^+.

sol The low-viscosity state of cytoplasm.

solvation The process of dissolving a solute in a solvent; hydration, or clustering, of water molecules around individual ions and polar molecules.

solvent drag Process in which smaller solute molecules are carried passively along with the water as it flows down its osmotic gradient through hydrated channels.

soma The nerve cell body, or perikaryon; in general, the body.

somatic Referring to the body tissues as distinct from the germ cells.

somatic system The part of the nervous system that receives input from the body.

somatosensory cortex The region of the cerebral cortex that receives sensory input from the body surface.

somatostatin Growth-hormone-inhibiting hormone, which inhibits growth-hormone release from the pituitary.

spatial summation Integration by a postsynaptic neuron of simultaneous synaptic currents that arise from the terminals of different presynaptic neurons.

specific dynamic action (SDA) A marked increase in metabolism accompanying the digestion and assimilation of food.

specific resistance (R_m) Resistance per unit area of a membrane in ohms per square centimeter.

spectrophotometer A device that passes a beam of visible or UV light through a fluid-filled vial and measures the emerging wavelengths.

spectrum Specific, charted bands of electromagnetic radiation wavelengths produced by refraction or diffraction.

sphincter A ring-shaped band of muscle fibers capable of constricting an opening or a passageway.

sphingolipid A lipid formed by a fatty acid attached to the nitrogen atom of sphingosine, a long-chain, oily amino alcohol ($C_{18}H_{37}O_2N$). The sphingolipids occur primarily in the membranes of brain and nerve cells.

spike-initiating zone Region of the nerve axon where an action potential is initiated. In many—but not all—neurons, the axon hillock.

spinal canal The fluid-filled cavity that runs longitudinally through the vertebrate spinal cord; it is confluent with the cerebral ventricles.

spinal cord The portion of the vertebrate central nervous system that is encased in the vertebral column, extending from the caudal end of the medulla oblongata to the upper lumbar region; constructed of a core of gray matter and an outer layer of white matter.

spinal root A large bundle of axons that enters or leaves the spinal cord at each spinal segment.

spindle organ A stretch receptor of vertebrate skeletal muscle.

spiracle Surface opening of the tracheal system in insects.

spiral ganglion The ganglion lying near the cochlea that contains the somata of neurons whose axons carry auditory information from the hair cells of the organ of Corti to the auditory centers of the brain.

spirometry Measurement of an animal's O_2 and CO_2 turnover.

standard metabolic rate (SMR) Similar to basal metabolic rate, but utilized for metabolic rate of a heterotherm maintained at a selected body temperature.

standard temperature and pressure (STP; dry, STPD) 25°C, 1 atmosphere (atm).

standing-gradient hypothesis Hypothesis describing process of solute-coupled water transport, involving the active transport of salt across the portions of the epithelial cell membranes facing intercellular clefts.

standing wave A resonating wave with fixed nodes.

stapes (stirrup) The innermost auditory ossicle, which articulates at its apex with the incus and whose base is connected to the oval window.

starch A polysaccharide of plant origin, formula $(C_6H_{10}O_5)_n$.

Starling curves Curves that describe the relationship between heart work and filling pressure.

statocyst Gravity-sensing sensory organ made up of mechanoreceptive hair cells and associated particles called statoliths.

statolith A small, dense, solid granule found in statocysts.

steady state Dynamic equilibrium.

stenohaline Able to tolerate only a narrow range of salinities.

stereocilia Nonmotile filament-filled projections of hair cells; these "hairs" lack the internal structure of motile "9 + 2" cilia.

steric Pertaining to the spatial arrangement of atoms.

steroid hormones Cyclic hydrocarbon derivatives synthesized from cholesterol.

sterols A group of solid, primarily unsaturated polycyclic alcohols.

stimulus A substance, action, or other influence that when applied with sufficient intensity to a tissue causes a response.

stomach The major digestive region of the alimentary canal.

stomatogastric ganglion A ganglion in Crustacea that contains the central pattern generator neurons that control the rhythmic activity of the stomach and associated organs.

STP *See* standard temperature and pressure.

stretch receptor Sensory receptors that respond to stretch, typically associated with lungs or muscle tissue.

stria vascularis Vascular tissue layer over the external wall of the scala media; secretes the endolymph.

striated muscle Characterized by sarcomeres aligned in register. Skeletal and cardiac muscle are striated.

stroke volume The volume of blood pumped by one ventricle during a single heartbeat.

structural gene A gene coding for the sequence of amino acids that make up a polypeptide chain.

strychnine A poisonous alkaloid ($C_{21}H_{22}N_2O_2$) that blocks inhibitory synaptic transmission in the vertebrate central nervous system.

submucosa Second-most inner layer of alimentary canal, underlying inner mucosa.

submucosal plexus Neural plexus that acts to stimulate gut motility and secretion.

substrate A substance that is acted on by an enzyme.

substrate-level phosphorylation Mechanism by which chemical energy of oxidation is stored in the form of ATP.

succus entericus Digestive juice secreted by the glands of Lieberkühn in the small intestine.

sulfhydryl group The radical —SH.

supercooling Cooling of a fluid below its freezing temperature without actual freezing because of the failure of ice crystals fail formation.

supraoptic nucleus A distinct group of neurons in the hypothalamus, just above the optic chiasma; their neurosecretory endings terminate in the neurohypophysis.

surface charge Electric charge at the membrane surface, arising from fixed charged groups associated with the membrane surface.

surface hypothesis Hypothesis of Rubner that metabolic rate of birds and mammals should be proportional to body surface area.

surface tension The elasticity of the surface of a substance (particularly a fluid), which tends to reduce the surface area at each interface.

surfactant A surface-active substance that tends to reduce surface tension, for example, in the lung.

swim bladder A gas-filled bladder used for flotation; found in many teleost fishes.

sympathetic nervous system Thoracolumbar part of the autonomic nervous system; increased activity in sympathetic neurons typically provides metabolic support for vigorous physical activity, so this system has been called "the fight or flight system."

symports Coupled membrane transporters that transfer two solutes in the same direction across the cell membrane.

synapse A specialized area that connects two directly interacting nerve cells. At a synapse, activity in the presynaptic (transmitting) cell influences the activity of the postsynaptic (receiving) cell.

synaptic cleft The space separating the cells at a synapse.

synaptic current The ionic current that flows across a postsynaptic membrane when ion channels open after neurotransmitter molecules bind to membrane receptors; the flow of ions causes the membrane potential to change toward the equilibrium potential of the ion or ions that flow through the synaptic channels.

synaptic delay The time separating the arrival of an impulse at a presynaptic nerve terminal and a change in the membrane potential of the postsynaptic cell.

synaptic efficacy Effectiveness of a presynaptic impulse in producing a postsynaptic potential change.

synaptic facilitation Increase in synaptic efficacy.

synaptic inhibition A change in a postsynaptic cell that reduces the probability of its generating an action potential; produced by a transmitter substance that elicits a postsynaptic current having a reversal potential more negative than the threshold for the action potential.

synaptic noise Irregular changes in the transmembrane potential of a postsynaptic cell, produced by random subthreshold synaptic input.

synaptic transmission Transfer of a signal between two cells, usually between a neuron and another cell, which could be a neuron, a muscle, or some other effector cell such as a gland.

synaptic vesicles Membrane-bound vesicles containing neurotransmitter molecules; located within axon terminals.

syncitium A network of cells that are connected by low-resistance intercellular pathways.

syneresis Contraction of a gelled mixture so that a liquid is squeezed out from molecular interstices.

systemic Pertaining to or affecting the body; for example, systemic circulation.

systole The portion of the cardiac sequence when the heart muscle is contracting; it takes place between the first and second heart sounds as the blood flows through the aorta and pulmonary artery.

T-wave That portion of the electrocardiogram associated with repolarization (and usually relaxation) of the ventricle.

tachycardia An increase in heart rate above the normal level.

target cells Cells that preferentially bind and respond to specific hormones.

taxis Locomotion that is oriented with respect to a stimulus direction or gradient.

tectorial membrane A fine gelatinous sheet lying on the organ of Corti in contact with the cilia of cochlear hair cells.

tectum The highest center for processing visual information in fishes and amphibians.

teleost Bony fish, of the infraclass Teleostei.

temporal Referring to the lateral areas of the head above the zygomatic arch. Also, relating to time; time limited.

temporal lobe A lobe of the cerebral hemisphere, situated in the lower lateral area, at the temples.

temporal summation Summation of postsynaptic membrane potentials that occur close to one another in time, but not simultaneously.

tendon A band of tough fibrous connective tissue that anchors a skeletal muscle to the skeleton, allowing contraction of the muscle to move the body of an animal.

terminal cisternae The closed spaces that make up part of the sarcoplasmic reticulum on both sides of the Z line, making close contact with the T tubules.

tertiary structure Refers to the way a polypeptide chain is folded or bent to produce the overall conformation of the molecule.

testosterone A steroid androgen synthesized by the testicular interstitial cells of the male; responsible for the production and maintenance of male secondary sex characters.

tetanus An uninterrupted muscular contraction caused by high frequency motor impulses. Also the name of a neurotoxin that is retrogradely (toward the cell body) transported in axons and that causes prolonged excitation of muscle fibers, causing tetanic contraction.

tetraethylammonium (TEA) A quarternary ammonium agent, $(C_2H_5)_4N$, that can be used to block some potassium channels in membrane.

tetrodotoxin (TTX) The pufferfish poison, which selectively blocks voltage-gated sodium ion channels in the membranes of excitable cells.

thalamus A major center in the midbrain of birds and mammals that receives and transmits both sensory and motor information.

theca interna The internal vascular layer encasing an ovarian follicle; responsible for the biosynthesis and secretion of estrogen.

theophylline A crystalline alkaloid $(C_7H_8N_4H_2O)$ found in tea; inhibits the enzyme phosphodiesterase, thereby increasing the level of cAMP; also releases Ca^{2+} from calcium-sequestering organelles.

thermal conductance Ability of a material to conduct heat, which is poor in insulating materials such as fur or feathers.

thermal neutral zone (thermoneutral zone) That range of ambient temperatures within which a homeotherm can control its temperature by passive measures and without elevating its metabolic rate to maintain thermal homeostasis.

thermogenesis The production of body heat by metabolic means such as brown-fat metabolism or muscle contraction during shivering.

thermophilic behavior Heat-seeking behavior.

thermophobic behavior Heat-avoiding behavior.

thermoreceptor Sensory nerve ending specifically responsive to temperature changes.

thick filament A myofilament made of myosin.

thin filament A myofilament that contains actin and regulatory proteins.

thoracic cage The chest compartment formed by the ribs and diaphragm containing the lungs and heart.

thoracic duct Duct draining the lymphatic system into the anterior cardural vein.

3-dehydroretinal A derivative of vitamin A that is found in the visual pigments of freshwater fishes and amphibia.

3,5,3-triiodothyronine An iodine-bearing tyrosine derivative synthesized in and secreted by the thyroid gland; raises cellular metabolic rate, as does thyroxine.

threshold potential The potential at which a response (e.g., an action potential or a muscle twitch) is produced.

threshold stimulus The minimum strength of stimulation necessary to produce a detectable response or an all-or-none response.

thymine A pyrimidine base, 5-methyluracil ($C_5H_6N_2O$), a constituent of DNA.

thyroid-stimulating hormone (TSH) An adenohypophyseal hormone that stimulates the secretory activity of the thyroid gland.

thyroxine An iodine-bearing, tyrosine-derived hormone that is synthesized and secreted by the thyroid gland; raises cellular metabolic rate.

tidal volume The volume of air moved in or out of the lungs with each breath.

tight junction An area of membrane fusion between adjoining cells; prevents passage of extracellular material between adjoining cells; prevents passage of extracellular material between the cells.

time constant (t) A measure of the rate of accumulation or decay in an exponential process; the time required for an exponential process to reach 63% completion. In electricity, it is proportional to the product of resistance and capacitance.

tonic Steady; slowly adapting.

tonicity (hyper-, hypo-, iso-) The relative osmotic pressure of a solution under given conditions (e.g., its osmotic effect on a cell relative to the osmotic effect of plasma on the cell).

tonotopic map A pattern of auditory projection to the brain in which neurons are arranged and make synapses based on the frequency of the sound to which they respond.

tonus Sustained resting contraction of muscle, produced by basal neuromotor activity.

torpor A state of inactivity, often with lowered body temperature and reduced metabolism, that some homeotherms enter into so as to conserve energy stores.

trachea The large respiratory passageway that connects the pharynx and bronchioles in the vertebrate lung.

tracheal system Consists of air-filled tubules that carry respiratory gases between the tissues and the exterior in insects.

tracheoles Minute subdivisions of the tracheal system of insects.

tragus The tab that extends from the ventral (anterior) edge of the outer ear and partially covers the opening of the ear.

train of impulses A rapid succession of action potentials propagated down a nerve fiber.

trans A configuration with particular atoms or groups on opposite sides.

transcellular pathways Routes through or around a cell by which substances are actively transported across an epithelium.

transcription The formation of an RNA chain of a complementary base sequence from the informational base sequence of DNA.

transducer A mechanism that translates energy or signals of one form into a different kind of energy or signals.

transducer molecule Intermediate molecule (G protein) within the cell membrane that transmits a hormone-initiated signal from the externally facing hormone receptor to the internally facing enzyme.

transducin The G protein that links the capture of light by rhodopsin molecules with a change in the current flowing across the membrane of photoreceptors.

transduction General term for the modulation of one kind of energy by another kind of energy. Thus, sense organs transduce sensory stimuli into nerve impulses.

transfer RNA (TRNA) A small RNA molecule that is responsible for the transfer of amino acids from their activating enzymes to the ribosomes; there are 20 tRNAs, one for each amino acid.

transgenic animal An animal in which its genetic constitution has been experimentally altered by the addition or substitution of genes from other animals of that species or other species.

translation Utilization of the DNA base sequence for linear organization of amino acid residues on a polypeptide; carried out by mRNA.

transmitter substance A chemical mediator liberated from a presynaptic ending, producing a conductance change or other response in the membrane of the postsynaptic cell.

transmural pressure The difference in pressure across the wall of a structure, for example, a blood vessel.

transphosphorylation The transfer of phosphate groups between organic molecules, bypassing the inorganic phosphate stage.

transverse tubules (T tubules) Branching membrane-limited, intercommunicating tubules that are continuous with the surface membrane and are closely apposed to the terminal cisternae of the sarcoplasmic reticulum.

traveling wave A wave that moves through the propagating medium, as opposed to a standing wave, which remains stationary.

tricarboxylic acid cycle (TCA cycle; Krebs cycle; citric acid cycle) The metabolic cycle responsible for the complete oxidation of the acetyl portion of the acetyl coenzyme A molecule.

trichromacy theory The theory that three kinds of photoreceptor cone cells exist in the human retina, each with a characteristic maximal sensitivity to a different part of the color spectrum.

triglyceride A neutral molecule composed of three fatty acid residues esterified to glycerol; formed in animals from carbohydrates.

triglycerol A neutral molecule composed of three fatty acid residues esterified to glycerol; formed in animals from carbohydrates.

trimer A compound made up of three simpler identical molecules.

trimethylamine oxide A nitrogenous waste product, probably from choline decomposition.

tritium A radioactive isotope of hydrogen with an atomic mass of three (H_3).

Triton X-100 A nonionic detergent used in cell biology to solubilize lipids and certain cell proteins.

trituration Grinding; mastication.

tRNA *See* transfer RNA.

trophic level Individual level within a food chain.

trophic substances Chemical substances believed to be released from neuron terminals and to influence the chemical and functional properties of the postsynaptic cell.

tropomyosin A long protein molecule located in the grooves of actin filaments of muscle; inhibits muscle contraction by blocking the interaction of myosin cross bridges with actin filaments.

troponin A complex of globular calcium-binding proteins associated with actin and tropomyosin in the thin filaments of muscle. When troponin binds Ca^{2+}, it undergoes a conformational change, allowing tropomyosin to reveal myosin-binding sites on the actin filament.

trypsin An enzyme specifically attacking peptide bonds in which the carboxyl group is provided by arginine or lysine.

trypsinogen A proenzyme of trypsin.

tubulin An actinlike, 4 nm globular protein molecule that is the building block of microtubules.

tunica adventitia The fibrous outer layer of arterial blood vessel walls.

tunica intima The inner lining of arterial blood vessel walls.

tunica media The middle layer of arterial blood vessel walls consisting of smooth muscle and elastic tissue.

turbinates Chambers of the nasal passages with olfactory receptors in the surface epithelium.

turbulent flow Flow pattern in which exists sharp gradients and inconsistencies in velocity and direction of flow fluid.

turgor Distension; swollenness.

twitch muscle (fast muscle) The most common striated vertebrate skeletal muscle type, usually pale in color because of its low myoglobin content. It has few mitochondria, and its fibers are constructed of many clearly defined fibrils. These contract rapidly and derive most of their energy from anaerobic metabolism.

2-deoxyribose 5-carbon sugar that forms a major structural component of DNA.

tympanic membrane The eardrum.

tympanum The middle ear cavity; houses the auditory ossicles.

type J receptors Abbreviation for juxtapulmonary capillary receptors found in the lung that, when stimulated, elicit the sensation of breathlessness.

ultradian rhythms Biological rhythms with a periodicity of greater than a day in length.

ultrafiltrate The product of ultrafiltration.

ultrafiltration The process of separating colloidal or molecular particles by filtration, using suction or pressure, by means of a colloidal filter or semipermeable membrane.

ultraviolet light Light of wavelengths between 180 and 390 nm.

uniports Carrier proteins which transport a single solute from one side of the membrane to the other.

unitary currents Electrical currents due to the sudden opening of individual channels in the plasma membrane.

unit membrane The sandwich-like profile of biological membrane seen in electron micrographs and believed to represent the bimolecular leaflet with a hydrophobic center region between hydrophilic surfaces.

unsaturated In reference to fatty acid molecules, indicates that some of the carbon-carbon bonds are double. Having free valence electrons.

upper critical temperature (UCT) Ambient temperature beyond which normal heat loss mechanisms cannot prevent increase in body temperature.

uracil A pyrimidine ($C_4H_4O_2N_2$) constituent of RNA.

urea $(NH_2)_2CO$, the primary nitrogenous waste product in the urine of mammals.

ureotelic Pertaining to the excretion of nitrogen in the form of urea.

ureter A muscular tube passing urine to the bladder from the kidney.

urethra The channel passing urine from the bladder out of the body.

uric acid A crystalline waste product of nitrogen metabolism found in the feces and urine of birds and reptiles; poorly soluble in water.

uricolytic *See* uricotelic.

uricolytic pathway The pathway through which uric acid or urates are cleavaged.

uricotelic Pertaining to the excretion of nitrogen in the form of uric acid.

Ussing chamber A chamber used to suspend tissue such as frog skin and measure its epithelial transport properties.

utriculus One of the vertebrate organs of equilibrium.

vacuole A membrane-limited cavity in the cytoplasm of a cell.

vagus nerve (tenth cranial nerve) A major cranial nerve that sends sensory fibers to the tongue, pharynx, larynx, and car; motor fibers to the esophagus, larynx, and pharynx; and parasympathetic and afferent fibers to the viscera of the thoracic and abdominal regions.

valence The number of missing or extra electrons of an atom or a molecule.

van der Waals forces The close-ranging, relatively weak attraction exhibited between atoms and molecules with hydrophobic properties.

van't Hoff equation An equation used to calculate the Q_{10} of a biological function.

varicosities Swellings along the length of a vessel or fiber.

vasa recta The capillary network that surrounds the loop of Henle in the tubules of the mammalian kidney.

vasa vasorum The tiny arteries and veins that supply nutrients and remove waste products from the tissues in the walls of larger blood vessels.

vas deferens A testicular duct that joins the excretory duct of the seminal vesicle to form the ejaculatory duct.

vasoactive intestinal peptide Peptide hormone regulating intestinal phase of gastric secretion.

vasoconstriction Contraction of circular muscle of arterioles, decreasing their volume and increasing the vascular resistance.

vasodilation A widening of the lumen or interior space of the blood vessels, increasing blood flow.

vasomotor Pertaining to the autonomic control of arteriolar constriction or dilation by contraction or relaxation of circular muscle.

vasopressin *See* antidiuretic hormone.

vasopressor A substance that induces arterial and capillary smooth-muscle contraction.

vector A carrier; an animal transferring an infection from host to host. Also, a mathematical term for a quantity with direction, magnitude, and sign.

venous shunt A direct connection between arterioles and venules, bypassing the capillary network.

ventilation In respiratory physiology, the process of exchange of air between the lungs and the ambient air.

ventral Toward the belly surface.

ventral horn The ventral part of gray matter in the vertebrate spinal cord in which motor neuron cell bodies are situated.

ventral root A nerve trunk leaving the spinal cord near its ventral surface; contains only motor axons.

ventricle A small cavity. Also, a chamber of the vertebrate heart.

ventricular zone The region of the brain that surrounds the cerebral ventricles; in embryonic vertebrates, cells of the ventricular zone remain mitotic and generate the neurons of the brain and spinal cord.

venule A small vessel that connects a capillary bed with a vein.

vestibular apparatus The collection of vertebrate organs of equilibrium in the inner ear.

villi Small, fingerlike projections of the intestinal epithelium.

viscosity A physical property of fluids that determines the ease with which layers of a fluid move past each other.

visible light Light of wavelengths between 390 and 740 nm.

visual cortex The cerebral cortex in the occipital region of the cerebrum; devoted to processing visual information.

visual streak (retinal streak) A retinal structure—found in the eyes of some species that inhabit plains—in which photoreceptors are packed into a horizontal streak across the retina, providing high resolution along the visual horizon. This pattern is similar to the fovea of primates, but it has a different shape.

vital capacity The maximum volume of air that can be inhaled into or exhaled from the lungs.

vitalism The theory that postulates that biological processes cannot adequately be explained by physical and chemical processes and laws.

volt (V) MKS unit of electromotive force; the force required to induce a 1 ampere (A) current to flow through a 1 ohm (Ω) resistance.

voltage (E or V) The electromotive force, or electric potential, expressed in volts. When the work required to move 1 coulomb (C) of charge from one point to a point of higher potential is 1 joule (J), or 1/4.184 calories (cal), the potential difference between these points is said to be 1 volt (V).

voltage clamping An electronic method of imposing a selected membrane potential across a membrane by means of feedback control.

voltage-gated ion channels Protein channels through the cell membrane that allow ions to cross the membrane when they are open. The conductance of these channels depends upon the transmembrane electrical potential difference.

watt (W) A unit of electrical power; the work performed at 1 joule (J) per second.

Weber-Fechner law Sensation increases arithmetically as a stimulus increases geometrically; the least perceptible change in stimulus intensity above any background bears a constant proportion to the intensity of the background.

white matter Tissue of the central nervous system that consists mainly of myelinated nerve fibers.

Wolffian ducts The embryonic ducts that are associated with the primordial kidney and that become the excretory and reproductive ducts in the mate.

work Force exerted upon an object over a distance; force times distance.

X-ray diffraction The method of examining crystalline structure using the pattern of scattered X rays.

Xylocaine The trade name for lidocaine, a local anesthetic related to procaine.

Z disk (Z line, Z band) A narrow zone at either end of a muscle sarcomere, consisting of a latticework into which the actin thin filaments are anchored.

zeitgeber Environmental factors that entrain biological rhythms.

zero-order kinetics Kinetics in which the rate of the reaction is independent of the concentration of any of its reactants. This would occur if the enzyme concentration were the limiting factor.

zonula Zone.

zonula adherens Form of desmosome in epithelial cells that form a belt of cell-to-cell adhesion under tight junctions.

zonula occludens Tight junctions between epithelial cells, usually having a ring-shaped configuration and serving to occlude transepithelial extracellular passages.

zwitterion A molecule carrying both negatively and positively ionized or ionizable sites.

zygomatic cells *See* chief cells.

zygote A fertilized ovum before first cleavage.

zymogen *See* proenzyme.

INDEX

Page numbers in *italics* indicate illustrations.